Representations of Lie Algebras and Partial Differential Equations

Xiaoping Xu

Representations of Lie Algebras and Partial Differential Equations

Springer

Xiaoping Xu
Institute of Mathematics
Academy of Mathematics
and Systems Sciences,
Chinese Academy of Sciences
Beijing
P.R. China

and

School of Mathematics
University of Chinese
Academy of Sciences
Beijing
P.R. China

ISBN 978-981-10-6390-9 ISBN 978-981-10-6391-6 (eBook)
DOI 10.1007/978-981-10-6391-6

Library of Congress Control Number: 2017950037

Mathematics Subject Classification (2010): 17B10, 35C05, 17B20, 33C67, 94B05

Printed on acid-free paper

This Springer imprint is published by Springer Nature
The registered company is Springer Nature Singapore Pte Ltd.
The registered company address is: 152 Beach Road, #21-01/04 Gateway East, Singapore 189721, Singapore

To My Former Ph.D Thesis Advisors:
Profs. James Lepowsky and Robert Lee Wilson

Preface

Symmetry is an important phenomenon in the natural world. Lie algebra is not purely an abstract mathematics but a fundamental tool of studying symmetry. In fact, Norwegian mathematician Sorphus Lie introduced Lie group and Lie algebra in 1874 in order to study the symmetry of differential equations. Lie algebras are the infinitesimal structures (bones) of Lie groups, which are symmetric manifolds. Lie theory has extensive and important applications in many other fields of mathematics, such as geometry, topology, number theory, control theory, integrable systems, operator theory, and stochastic process. Representations of finite-dimensional semisimple Lie algebras play fundamental roles in quantum mechanics. The controllability property of the unitary propagator of an N-level quantum mechanical system subject to a single control field can be described in terms of the structure theory of semisimple Lie algebras. Moreover, Lie algebras were used to explain the degeneracies encountered in genetic codes as the result of a sequence of symmetry breakings that have occurred during its evolution. The structures and representations of simple Lie algebras are connected with solvable quantum many-body system in one-dimension. Our research also showed that the representation theory of Lie algebras is connected with algebraic coding theory.

The existing classical books on finite-dimensional Lie algebras, such as the ones by Jacobson and by Humphreys, purely focus on the algebraic structures of semi simple Lie algebras and their finite-dimensional representations. Explicit irreducible representations of simple Lie algebras had not been addressed extensively. Moreover, the relations of Lie algebras with the other subjects had not been narrated that much. It seems to us that a book on Lie algebras with more extensive view is needed in coupling with modern development of mathematics, sciences, and technology.

This book is mainly an exposition of the author's works and his joint works with his former students on explicit representations of finite-dimensional simple Lie algebras, related partial differential equations, linear orthogonal algebraic codes, combinatorics, and algebraic varieties. Various oscillator generalizations of the classical representation theorem on harmonic polynomials are presented. New functors from the representation category of a simple Lie algebra to that of another simple Lie algebra are given. Partial differential equations play key roles in solving certain representation problems. The weight matrices of the minimal and adjoint representations over the simple Lie algebras of types E and F are proved to generate ternary orthogonal linear codes with large minimal distances. New multivariable hypergeometric functions related to the root systems of simple Lie algebras are introduced in connection with quantum many-body system in one-dimension. Certain equivalent combinatorial properties on representation formulas are found. Irreducibility of representations is proved directly related to algebraic varieties.

This book is self-contained with the minimal prerequisite of calculus and linear algebra. It is our wish that the results this book can also be easily understood by nonalgebraists and applied to the other mathematical fields and physics. It consists of three parts. The first part is mainly the classical structure and finite-dimensional representation theory of finite-dimensional semisimple Lie algebras, written with Humphreys' book "Introduction to Lie Algebras and Representation Theory" as the main reference, where we give more examples and direct constructions of simple Lie algebras of exceptional types, revise some arguments, and prove some new statements. This part serves as the preparation for the main context. The second part is our explicit representation theory of finite-dimensional simple Lie algebras. Many of the irreducible representations in this part are infinite-dimensional, and some of them are even not of highest-weight type. Certain important natural representation problems are solved by means of solving partial differential equations. In particular, some of our irreducible presentations are completely characterized by invariant partial differential equations. New representation functors are constructed from inhomogeneous oscillator representations of simple Lie algebras, which give fractional representations of the corresponding Lie groups, such as the projective representations of special linear Lie groups and the conformal representations of complex orthogonal groups. The third part is an extension of the second part. First we give supersymmetric generalizations of the classical representation theorem on harmonic polynomials. They can also be viewed as certain supersymmetric Howe dualities. Then we present our theory of representation theoretic codes. Finally, we talk about root-related integrable systems and our new multivariable hypergeometric functions, which are natural multivariable analogues of the classical Gauss hypergeometric functions. The corresponding hypergeometric partial differential equations are found.

Part of this book has been taught for many times at the University of Chinese Academy of Sciences. The book can serve as a research reference book for mathematicians and scientists. It can also be treated as a textbook for students after a proper selection of materials.

Beijing, P.R. China Xiaoping Xu
2017

Acknowledgements

The research in this book was partly supported by NSFC Grants 11671381, 11321101 and Hua Loo-Keng Key Mathematical Laboratory, Chinese Academy of Sciences. We thank the reviewers and Dr. Ramond Peng for their helpful comments.

Contents

Notational Conventions

$\mathbb{N}$	$\{0, 1, 2, 3, \ldots\}$, the set of nonnegative integers
$\mathbb{Z}$	The ring of integers
$\mathbb{Q}$	The field of rational numbers
$\mathbb{R}$	The field of real numbers
$\mathbb{C}$	The field of complex numbers
$\mathbb{F}$	A field with characteristic 0, such as, $\mathbb{Q}, \mathbb{R}, \mathbb{C}$
$\overline{i, i+j}$	$\{i, i+1, i+2, \ldots, i+j\}$, an index set. $\delta_{i,j} = 1$ if $i = j$, 0 if $i \neq j$
$\mathscr{A}$	An associative algebra
$\mathscr{G}$	A Lie algebra
Π	The set of positive simple roots
Φ	The set of roots
Ω	Casimir operator
ω	Casimir element
$\mathscr{W}$	The Weyl group
$\mathbb{F}[x_1, x_2, \ldots, x_n]$	The algebra of polynomials in $x_1, x_2, \ldots, x_n$.
$M(\lambda)$	The Verma module of highest weight λ.
$V(\lambda)$	The irreducible module with highest weight λ.
Der $\mathscr{U}$	The derivation Lie algebra of the algebra $\mathscr{U}$.
$\mathbb{W}_n$	The Witt algebra, that is, Der $\mathbb{F}[x_1, x_2, \ldots, x_n]$
σ_i	The *i*th simple reflection.
End V	The algebra of linear transformations on the vector space V.

Introduction

Algebraic study of partial differential equations traces back to Norwegian mathematician Sophus Lie [Lie], who invented the powerful tool of continuous groups (known as Lie groups) in 1874 in order to study symmetry of differential equations. Lie's idea has been carried on mainly by the mathematicians in the former states of Soviet Union, East Europe and some mathematicians in North America. Now it has become an important mathematical field known as "group analysis of differential equations," whose main objective is to find symmetry group of differential equations, related conservation laws, and similarity solutions. One of the classical long-standing problems in the area was to determine all invariant partial differential equation under the natural representations of classical groups (it was also mentioned by Olver [Op]). The problem was eventually settled down theoretically by the author [X5], where we found the complete set of functional generators for the differential invariants of classical groups.

Gel'fand, Dikii, and Dorfman [GDi1, GDi2, GDo1-GDo3] introduced in 1970s a theory of Hamiltonian operators in order to study the integrability of nonlinear evolution partial differential equations. Our first experience with partial differential equation was in the works [X3, X4, X9, X10, X11] on the structure of Hamiltonian operators and their supersymmetric generalizations. In particular, we [X10] proved that Lie conformal algebras in the sense of Kac [Kv2] are equivalent to linear Hamiltonian operators as mathematical structures. Since Borcherds' vertex algebras (cf. [Bo]) are determined by their positive parts (modes in physics terminology) of vertex operators that form Lie conformal algebras (e.g., cf. [Kv2, X6]), our result essentially established an equivalence between vertex algebras and Hamiltonian operators. The result was later generalized by Barakat, De Sole, and Kac [BDK] to vertex Poisson algebras. Moreover, we used the techniques and thoughts from Lie algebras to solve various physical partial differential equations such as the equation of transonic gas flows, the equation of geopotential forecast, the nonlinear Schrödinger equations, the Navier–Stokes equations, and the classical boundary layer equations (cf. [X21] and the references therein). In this book, we present an explicit representation theory of finite-dimensional simple Lie algebras and show that partial differential equations can be used to solve representation problems of Lie algebras.

Abstractly, a Lie algebra $\mathscr{G}$ is a vector space with a bilinear map $[\cdot,\cdot] : \mathscr{G} \times \mathscr{G} \to \mathscr{G}$ such that

$$[u,v] = -[v,u], \quad [u,[v,w]] = [[u,v],\omega] + [v,[u,w]], \quad u,v,w \in \mathscr{G} \tag{1}$$

A Lie algebra $\mathscr{G}$ is called *simple* if it does not contain any nonzero "invariant" proper subspace $\mathcal{I}$; that is,

$$[u,\mathcal{I}] \subset \mathcal{I}, \quad u \in \mathscr{G} \tag{2}$$

Finite-dimensional complex simple Lie algebras were classified by Killing and Cartan in later nineteenth century. For a vector space V, we denote by V^* the space of linear functions on V. It turns out that such a Lie algebra $\mathscr{G}$ must contain a subspace H, called *toral Cartan subalgebra*, such that

$$\mathscr{G} = \bigoplus_{\alpha \in H^*} \mathscr{G}_\alpha, \qquad \mathscr{G}_\alpha = \{u \in \mathscr{G} | [h,u] = \alpha(h)u \quad \text{for } h \in H\}, \tag{3}$$

and $\mathscr{G}_0 = H$, where

$$\Phi = \{\alpha \in H^* \backslash \{0\} | \mathscr{G}_\alpha \neq \{0\}\} \tag{4}$$

is called the *root system of* $\mathscr{G}$. Moreover,

$$\dim \mathscr{G}_\alpha = 1 \quad \text{for} \quad \alpha \in \Phi \tag{5}$$

and there exists a subset Φ_+ of Φ containing a basis $\Pi = \{\alpha_1, \ldots, \alpha_n\}$ of H^* such that

$$\Phi = -\Phi_+ \bigcup \Phi_+, \quad -\Phi_+ \bigcap \Phi_+ = \emptyset \tag{6}$$

and for any $\beta \in \Phi_+$,

$$\beta = \sum_{i=1}^{n} k_i \alpha_i \quad \text{with} \quad 0 \leq k_i \in \mathbb{Z}, \tag{7}$$

where $\mathbb{Z}$ is the ring of integers. Furthermore, there exists $h_i \in [\mathscr{G}_{\alpha_i}, \mathscr{G}_{-\alpha_i}]$ such that

$$\alpha_i(h_i) = 2 \quad \text{for } i = 1, 2, \ldots, n. \tag{8}$$

It turns out that the structure of $\mathscr{G}$ is completely determined by the matrix $(\alpha_i(h_j))_{n \times n}$, which is called the *Cartan matrix* of $\mathscr{G}$. Consequently, there are only nine classes of finite-dimensional simple Lie algebras, which are called of types

$A_n, B_n, C_n, D_n, G_2, F_4, E_6, E_7, E_8$. The first four are infinite series, called "classical Lie algebras." The last five are fixed simple Lie algebras, called "exceptional types." Using the root lattices $\Lambda_r = \sum_{\alpha\in\Pi} \mathbb{Z}\alpha$ of type A and the Coxeter automorphism of Λ_r, we constructed in [X1] two families of self-dual complex lattices that are also real self-dual (unimodular) lattices. These lattices can be used in studying geometry of numbers and conformal field theory.

Since every finite-dimensional complex simple Lie algebra must contain a toral Cartan subalgebra, it is natural to ask whether there exists a complex simple Lie algebra that does not contain any toral Cartan subalgebra. In [X7, X8], we constructed six families of infinite-dimensional complex simple Lie algebras without any toral Cartan subalgebras.

For a vector space M, we denote by End M the space of all linear transformations on M. A *representation* ν of a Lie algebra $\mathscr{G}$ on M is a linear map from $\mathscr{G}$ to *End M* such that

$$\nu([u, v]) = \nu(u)\nu(v) - \nu(v)\nu(u) \quad \text{for} \quad u, v \in \mathscr{G}. \tag{9}$$

The space M is called a *$\mathscr{G}$- module*. In this book, we sometimes use the notions:

$$\nu(\xi) = \xi|_M, \;\; \nu(\xi)(w) = \xi(w) \quad \text{for} \quad \xi \in \mathscr{G}, \; w \in M. \tag{10}$$

A subspace $\mathscr{N}$ of a $\mathscr{G}$-module M is called a *submodule* of M if

$$\xi(w) \in \mathscr{N} \quad \text{for}\, \xi \in \mathscr{G}, \; w \in \mathscr{N}. \tag{11}$$

If M dose not contain any proper nonzero submodule, we say that M is an *irreducible* $\mathscr{G}$-module. Let $\mathscr{G}$ be a Lie algebra with a decomposition (3) satisfying (4), (6), and (7). A $\mathscr{G}$-module M is called a *weight module* if

$$M = \bigoplus_{\mu\in H^*} M_\mu, \;\; M_\mu = \{w \in M \mid h(w) = \mu(h)w\}, \tag{12}$$

where H^* is the space of linear functions on H. The set

$$\Lambda(M) = \{\mu \in H^* | M_\mu \neq \{0\}\} \tag{13}$$

is called the *weight set* of M and the elements in $\Lambda(M)$ are called the *weights* of M. A vector in M_μ with $\mu \in \Lambda(M)$ is called a *weight vector*. A nonzero weight vector ω is called a *singular vector* if

$$\xi(w) = 0 \quad \text{for} \quad \xi \in \bigcup_{\beta\in\Phi_+} \mathscr{G}_\beta. \tag{14}$$

If M is generated by a singular vector v with weight λ, we call M a *highest-weight $\mathscr{G}$-module*, v a *highest-weight vector* and λ the *highest weight* of M.

Finite-dimensional representations of a complex finite-dimensional simple Lie algebra were essentially determined by Cartan in early twentieth century. The approaches were simplified and further developed by Weyl in 1920s. Let $\mathscr{G}$ be a finite-dimensional complex simple Lie algebra and take the settings in (3–8). Set

$$\Lambda^+ = \{\lambda \in H^* | 0 \leq \lambda(h_i) \in \mathbb{Z} \ \text{for}\, i = 1, \ldots, n\} \tag{15}$$

It turns out that any finite-dimensional $\mathscr{G}$-module is a direct sum of its irreducible submodules, and any finite-dimensional irreducible $\mathscr{G}$-module is a highest-weight module with its highest weight in Λ^+. Conversely, for any element $\lambda \in \Lambda^+$, there exists a unique finite-dimensional irreducible $\mathscr{G}$-module with highest weight λ. The above conclusion is now called *Weyl's theorem on completely reducibility*.

Coding theory began with Shannon's famous paper "A mathematical theory of communication" published in 1948 (cf. [Sc]). Naturally, people had speculated the existence of some kind of relations between the representation theory and coding theory. However, it had not been found until our work [X20] published in 2012. For any finite-dimensional $\mathscr{G}$-module M,

$$\mu(h_i) \in \mathbb{Z} \quad \text{for} i \in \{1, \ldots, n\}, \ \mu \in \Lambda(M). \tag{16}$$

Write $\Lambda(M) \setminus \{0\} = \{\mu_1, \mu_2, \ldots, \mu_m\}$. We found in [X20] that the set

$$\{(\mu_1(h_i), \mu_2(h_i), \ldots, \mu_m(h_i)) \,|\, i = 1, \ldots, n\} \tag{17}$$

spans a binary or ternary orthogonal code for certain $\mathscr{G}$ and M. In particular, we showed that when $\mathscr{G}$ is of type F_4, E_6, E_7 and E_8, and M is of minimal dimension or $M = \mathscr{G}$ with left multiplication as the representation, the set (17) spans a ternary orthogonal code with large minimal distance d, which can be used to correct $[\![d/2]\!]$ errors in information technology.

A module of a finite-dimensional simple Lie algebra is called *cuspidal* if it is not induced from its proper "parabolic subalgebras." Infinite-dimensional irreducible weight modules of finite-dimensional simple Lie algebras with finite-dimensional weight subspaces had been intensively studied by the authors in [BBL, BFL, BHL, BL1, BL2, Fs, Fv, Mo]. In particular, Fernando [Fs] proved that such modules must be cuspidal or parabolically induced. Moreover, such cuspidal modules exist only for special linear Lie algebras and symplectic Lie algebras. A similar result was independently obtained by Futorny [Fv]. Mathieu [Mo] proved that these cuspidal modules are irreducible components in the tensor modules of their multiplicity-free modules with finite-dimensional modules. Thus, the structures of irreducible weight modules of finite-dimensional simple Lie algebras with finite-dimensional weight subspaces were essentially determined by Fernando's result in [Fs] and Methieu's result in [Mo]. However, explicit presentations of such modules need further exploration.

An important feature of this book is the connection between representations of simple Lie algebras and partial differential equations. Let $E_{r,s}$ be the square matrix with 1 as its (r,s)-entry and 0 as the others. Denote by $\mathbb{R}$ the field of real numbers. Let $n \geq 3$ be an integer. The compact orthogonal Lie algebra $o(n,\mathbb{R}) = \sum_{1\leq r<s\leq n} \mathbb{R}(E_{r,s} - E_{s,r})$, whose representation on the polynomial algebra $\mathscr{A} = \mathbb{R}[x_1,\ldots,x_n]$ is given by $(E_{r,s} - E_{s,r})|_{\mathscr{A}} = x_r\partial_{x_s} - x_s\partial_{x_r}$. Denote by $\mathscr{A}_k$ the subspace of homogeneous polynomials in $\mathscr{A}$ with degree k. Recall that the *Laplace operator* $\Delta = \partial_{x_1}^2 + \partial_{x_2}^2 + \cdots + \partial_{x_n}^2$ and its corresponding invariant $\eta = x_1^2 + x_2^2 + \cdots + x_n^2$. It is well known that the subspaces

$$\mathscr{H}_k = \{f \in \mathscr{A}_k | \Delta(f) = 0\} \tag{18}$$

of *harmonic polynomials* form irreducible $o(n,\mathbb{R})$-submodules and

$$\mathscr{A} = \bigoplus_{i,k=0}^{\infty} \eta^i \mathscr{H}_k \tag{19}$$

is a direct sum of irreducible submodules. In other words, the irreducible submodules are characterized by the Laplace operator Δ and its dual invariant η gives the complete reducibility of the polynomial algebra $\mathscr{A}$. The above conclusion is called the *classical theorem on harmonic polynomials*.

The *Navier equations*

$$\imath_1\Delta(\vec{u}) + (\imath_1 + \imath_2)(\nabla^t \cdot \nabla)(\vec{u}) = 0 \tag{20}$$

are used to describe the deformation of a homogeneous, isotropic, and linear elastic medium in the absence of body forces, where $\vec{u}$ is an n-dimensional vector-valued function, $\nabla = (\partial_{x_1}, \partial_{x_2}, \ldots, \partial_{x_n})$ is the gradient operator, $\imath_1$ and $\imath_2$ are Lamé constants with $\imath_1 > 0$, $2\imath_1 + \imath_2 > 0$ and $\imath_1 + \imath_2 \neq 0$. In fact, $\nabla^t \cdot \nabla$ is the well-known Hessian operator. Mathematically, the above system is a natural vector generalization of the Laplace equation. In [X16], we found methods of solving linear flag partial differential equations for polynomial solutions. Cao [Cb] used a method of us to prove that the subspaces of homogeneous polynomial-vector solutions are exactly direct sums of three explicitly given irreducible $o(n,\mathbb{R})$-submodules if $n \neq 4$ and of four explicitly given irreducible $o(n,\mathbb{R})$-submodules if $n = 4$. This gave a vector generalization of the classical theorem on harmonic polynomials.

Denote by $\mathscr{B} = \mathbb{C}[x_1,\ldots,x_n,y_1,\ldots,y_n]$ the polynomial algebra in $x_1,\ldots,x_n$ and $y_1,\ldots,y_n$ over the field $\mathbb{C}$ of complex numbers. Fix $n_1, n_2 \in \{1,\ldots,n\}$ with $n_1 \leq n_2$. We redefine

$$\deg x_i = -1, \ \deg x_j = 1, \ \deg y_r = 1, \ \deg y_s = -1 \tag{21}$$

for $i \in \{1,\ldots,n_1\}$, $j \in \{n_1+1,..,n\}$, $r \in \{1,\ldots,n_2\}$ and $s \in \{n_2+1,\ldots,n\}$. Denote by $\mathscr{B}_{\langle \ell_1,\ell_2\rangle}$ the subspace of homogeneous polynomials with degree ℓ_1 in

$\{x_1,\ldots,x_n\}$ and degree ℓ_2 in $\{y_1,\ldots,y_n\}$. We deform the Laplace operator $\sum_{i=1}^n \partial_{x_i}\partial_{y_i}$ to the operator

$$\tilde{\Delta} = -\sum_{i=1}^{n_1} x_i\partial_{y_i} + \sum_{r=n_1+1}^{n_2} \partial_{x_r}\partial_{y_r} - \sum_{s=n_2+1}^{n} y_s\partial_{x_s} \tag{22}$$

and its dual $\sum_{i=1}^n x_i y_i$ to the operator

$$\eta = \sum_{i=1}^{n_1} y_i\partial_{x_i} + \sum_{r=n_1+1}^{n_2} x_r y_r + \sum_{s=n_2+1}^{n} x_s\partial_{y_s}. \tag{23}$$

Define

$$\mathscr{H}_{\langle \ell_1,\ell_2\rangle} = \{f \in \mathscr{B}_{\langle \ell_1,\ell_2\rangle} | \tilde{\Delta}(f) = 0\}. \tag{24}$$

Luo and the author [LX1] constructed a new representation on $\mathscr{B}$ for the simple Lie algebra $\mathscr{G}$ of type A_{n-1} such that the operators in (22) and (23) commute with the elements in $\mathscr{G}|_{\mathscr{B}}$ and used a method in [X16] to prove that $\mathscr{H}_{\langle \ell_1,\ell_2\rangle}$ with $\ell_1,\ell_2 \in \mathbb{Z}$ such that $\ell_1+\ell_2 \leq n_1 - n_2 + 1 - \delta_{n_1,n_2}$ are irreducible $\mathscr{G}$-modules. Moreover, $\mathscr{B}_{\langle \ell_1,\ell_2\rangle} = \oplus_{m=0}^{\infty} \eta^m(\mathscr{H}_{\langle \ell_1-m,\ell_2-m\rangle})$ is a decomposition of irreducible $\mathscr{G}$-submodules. This establishes a $\mathbb{Z}^2$-graded analogue of the classical theorem on harmonic polynomials for the simple Lie algebra of type A_{n-1}.

Set

$$\mathscr{B}_{\langle k\rangle} = \sum_{\ell\in\mathbb{Z}} \mathscr{B}_{\langle \ell,k-\ell\rangle}, \quad \mathscr{H}_{\langle k\rangle} = \sum_{\ell\in\mathbb{Z}} \mathscr{H}_{\langle \ell,k-\ell\rangle} \quad \text{for } k \in \mathbb{Z}. \tag{25}$$

Luo and the author [LX2] extended the above representation of $\mathscr{G}$ on $\mathscr{B}$ to a representation of the simple Lie algebra $\mathscr{G}_1$ of type D_n on $\mathscr{B}$ and proved that $\mathscr{H}_{\langle k\rangle}$ with $n_1 - n_2 + 1 - \delta_{n_1,n_2} \geq k \in \mathbb{Z}$ are irreducible $\mathscr{G}_1$-modules, and $\mathscr{B}_{\langle k\rangle} = \oplus_{i=0}^{\infty} \eta^i(\mathscr{H}_{\langle k-2i\rangle})$ is a decomposition of irreducible $\mathscr{G}_1$-submodules. This gives a $\mathbb{Z}$-graded analogue of the classical theorem on harmonic polynomials for the simple Lie algebra of type D_n.

Let $\mathscr{B}' = \mathbb{C}[x_0,x_1,\ldots,x_n,y_1,\ldots,y_n]$ be the polynomial algebra in $x_0,x_1,\ldots,x_n$ and $y_1,\ldots,y_n$. We define $deg\, x_0 = 1$ and take (21). Write the deformed Laplace operator

$$\tilde{\Delta}' = \partial_{x_0}^2 - 2\sum_{i=1}^{n_1} x_i\partial_{y_i} + 2\sum_{r=n_1+1}^{n_2} \partial_{x_r}\partial_{y_r} - 2\sum_{s=n_2+1}^{n} y_s\partial_{x_s} \tag{26}$$

and its dual operator

$$\eta' = x_0^2 + 2\sum_{i=1}^{n_1} y_i\partial_{x_i} + 2\sum_{r=n_1+1}^{n_2} x_r y_r + 2\sum_{s=n_2+1}^{n} x_s\partial_{y_s}. \tag{27}$$

Set

$$\mathscr{B}'_{\langle k\rangle} = \sum_{i=0}^{\infty} \mathscr{B}_{\langle k-i\rangle} x_0^i, \qquad \mathscr{H}'_{\langle k\rangle} = \{f \in \mathscr{B}'_{\langle k\rangle} | \tilde{\Delta}'(f) = 0\} \tag{28}$$

Luo and the author [LX2] extend the above representation of $\mathscr{G}_1$ on $\mathscr{B}$ to a representation of the simple Lie algebra $\mathscr{G}_2$ of type B_n on $\mathscr{B}'$ and proved that $\mathscr{H}'_{\langle k\rangle}$ with $k \in \mathbb{Z}$ are irreducible $\mathscr{G}_2$-modules, and $\mathscr{B}' = \oplus_{k\in\mathbb{Z}}\oplus_{i=0}^{\infty}\eta'^i(\mathscr{H}'_{\langle k\rangle})$ is a decomposition of irreducible $\mathscr{G}_2$-submodules. This is a $\mathbb{Z}$-graded analogue of the classical theorem on harmonic polynomials for the simple Lie algebra of type B_n.

When $n_1 < n_2$, the bases of the subspaces $\mathscr{H}_{\langle \ell_1,\ell_2\rangle}$, $\mathscr{H}_{\langle k\rangle}$ and $\mathscr{H}'_{\langle k\rangle}$ had been obtained. Using Fourier transformation, we can identify the subspaces $\mathscr{H}_{\langle \ell_1,\ell_2\rangle}$, $\mathscr{H}_{\langle k\rangle}$ and $\mathscr{H}'_{\langle k\rangle}$ with the subspaces of homogeneous solutions for the corresponding usual Laplace equations in certain spaces of generalized functions. In [LX3], Luo and the author extended the above representation of $\mathscr{G}$ on $\mathscr{B}$ to a representation of the simple Lie algebra $\mathscr{G}_3$ of type C_n on $\mathscr{B}$ and proved that if $n_1 < n_2$ or $k \neq 0$, the subspace $\mathscr{B}_{\langle k\rangle}$ with $k \in \mathbb{Z}$ are irreducible $\mathscr{G}_3$-modules, and when $n_1 = n_2$, the subspace $\mathscr{B}_{\langle 0\rangle}$ is a direct sum of two explicitly given irreducible $\mathscr{G}_3$-submodules.

The above irreducible modules except $n_1 = n_2 = n$ in the case of type A_{n-1} are explicit infinite-dimensional weight modules of finite-dimensional weight subspaces. They are not unitary. The irreducible modules for $\mathscr{G}_1, \mathscr{G}_2$ and $\mathscr{G}_3$ are generically neither of highest-weight type nor of unitary type. Bai [B1, B2] proved that some of irreducible representations in [LX1–LX3] have distinguished Gelfand–Kirillov dimensions.

A *supersymmetry* relating mesons and baryons was first proposed by Miyazawa [Mh] in 1966. It was largely ignored because it did not involve space-time. In early 1970s, physicists rediscovered it in the context of quantum field theory, a radically new type of symmetry of space-time and fundamental fields, which establishes a relationship among elementary particles of quantum nature, bosons and fermions, and unifies space-time and internal symmetries of microscopic phenomena (e.g., cf. [GS]). Supersymmetry with a consistent Lie-algebraic-graded structure on which the Gervais–Sakita rediscovery [GS] based directly first arose during 1971 in the context of a early version of string theory (cf. [R1]). It was later widely applied to nuclear physics, critical phenomena, quantum mechanics, and statistical physics. In [LX4], Luo and the author generalized the above results for $\mathscr{G}, \mathscr{G}_1$ and $\mathscr{G}_2$ to those for the corresponding Lie superalgebras.

Dickson [D] (1901) first realized that there exists an E_6-invariant trilinear form on its 27-dimensional basic irreducible module, which corresponds to the unique fundamental cubic polynomial invariant and constant-coefficient differential operator. We proved in [X17] that the space of homogeneous polynomial solutions with degree m for the cubic Dickson invariant differential equation is exactly a direct sum of $[\![m/2+1]\!]$ explicitly determined irreducible E_6-submodules, and the whole polynomial algebra is a free module over the polynomial algebra in the Dickson invariant generated by these solutions. Thus, we obtained a cubic E_6-generalization of the classical theorem on harmonic polynomials.

Next we want to describe our new multivariable hypergeometric functions of type A and their connections with the representations of type-A simple Lie algebras and the Calogero–Sutherland model. For $a \in \mathbb{C}$ and $0<n \in \mathbb{Z}$, we denote

$$(a)_n = a(a+1)\cdots(a+n-1), \qquad (a)_0 = 1 \tag{29}$$

The well-known Gauss hypergeometric function is

$${}_2F_1(\alpha,\beta;\gamma;z) = \sum_{n=0}^{\infty} \frac{(\alpha)_n(\beta)_n}{n!(\gamma)_n} z^n \tag{30}$$

Many well-known elementary functions and orthogonal polynomials are special cases of ${}_2F_1(\alpha,\beta;\gamma;z)$ (e.g., cf. [WG, AAR, X21]). It satisfies the classical hypergeometric equation

$$z(1-z)y'' + [\gamma - (\alpha+\beta+1)z]y' - \alpha\beta y = 0. \tag{31}$$

Denoting $D = z\frac{d}{dz}$, we can rewrite the above equation as

$$(\gamma + D)\frac{d}{dz}(y) = (\alpha + D)(\beta + D)(y) \tag{32}$$

Moreover, it has the differential property:

$$\frac{d}{dz}{}_2F_1(\alpha,\beta;\gamma;z) = \frac{\alpha\beta}{\gamma}{}_2F_1(\alpha+1,\beta+1;\gamma+1;z) \tag{33}$$

Furthermore, the Euler integral representation

$${}_2F_1(\alpha,\beta;\gamma;z) = \frac{\Gamma(\gamma)}{\Gamma(\beta)\Gamma(\gamma-\beta)}\int_0^1 t^{\beta-1}(1-t)^{\gamma-\beta-1}(1-zt)^{-\alpha}dt \tag{34}$$

holds in the z plane cut along the real axis from 1 to ∞.

Let $\mathbb{N}$ be the additive semigroup of nonnegative integers and let

$$\Gamma_A = \sum_{1 \le p < q \le n} \mathbb{N}\epsilon_{q,p} \tag{35}$$

be the additive semigroup of rank $n(n-1)/2$ with $\epsilon_{q,p}$ as base elements. For $\alpha = \sum_{1 \le p<q \le n} \alpha_{q,p}\epsilon_{q,p} \in \Gamma_A$, we denote

$$\alpha_{1*} = \alpha_n^* = 0, \quad \alpha_{k*} = \sum_{r=1}^{k-1} \alpha_{k,r}, \quad \alpha_l^* = \sum_{s=l+1}^{n} \alpha_{s,l} \tag{36}$$

Given $\vartheta \in \mathbb{C} \backslash \{-k | 0 < k \in \mathbb{Z}\}$ and $\tau_r \in \mathbb{C}$ with $r \in \{1, \ldots, n\}$, we defined $(n(n-1)/2)$-variable *hypergeometric function of type A* in [X14] by

$$\mathscr{X}_A(\tau_1, .., \tau_n; \vartheta)\{z_{j,k}\} = \sum_{\beta \in \Gamma_A} \frac{\left[\prod_{s=1}^{n-1} (\tau_s - \beta_{s*})_{\beta_s^*}\right] (\tau_n)_{\beta_{n*}}}{\beta! (\vartheta)_{\beta_{n*}}} z^\beta, \tag{37}$$

where

$$\beta! = \prod_{1 \le k < j \le n} \beta_{j,k}!, \qquad z^\beta = \prod_{1 \le k < j \le n} z_{j,k}^{\beta_{j,k}} \tag{38}$$

The variables $\{z_{j,k}\}$ correspond to the negative root vectors of the simple Lie algebra of type A_{n-1} and the above definition has essentially used the structure of its root system.

Our functions $\mathscr{X}_A(\tau_1, .., \tau_n; \vartheta)\{z_{j,k}\}$ are indeed natural generalizations of the Gauss hypergeometric function ${}_2F_1(\alpha, \beta; \gamma; z)$. Denote

$$\mathcal{D}_{p*} = \sum_{r=1}^{p-1} z_{p,r}\partial_{z_{p,r}}, \quad \mathcal{D}_q^* = \sum_{s=q+1}^{n} z_{s,q}\partial_{z_{s,q}} \text{ for } p \in \{2, \ldots, n\}, \ q \in \{1, \ldots, n-1\} \tag{39}$$

We have the following analogous system of partial differential equations of (32):

$$(\tau_{r_2} - 1 - \mathcal{D}_{r_2*} + \mathcal{D}_{r_2}^*)\partial_{z_{r_2,r_1}}(\mathscr{X}_A) = (\tau_{r_2} - 1 - \mathcal{D}_{r_2*})(\tau_{r_1} - \mathcal{D}_{r_1*} + \mathcal{D}_{r_1}^*)(\mathscr{X}_A) \tag{40}$$

for $1 \le r_1 < r_2 \le n-1$, and

$$(\vartheta + \mathcal{D}_{n*})\partial_{z_{n,r}}(\mathscr{X}_A) = (\tau_n + \mathcal{D}_{n*})(\tau_r - \mathcal{D}_{r*} + \mathcal{D}_r^*)(\mathscr{X}_A) \tag{41}$$

for $r \in \{1, \ldots, n-1\}$. Define

$$\begin{pmatrix} 1 & 0 & 0 & \cdots & 0 \\ P_{[1,2]} & 1 & 0 & \cdots & 0 \\ P_{[1,3]} & P_{[2,3]} & 1 & \ddots & \vdots \\ \vdots & \vdots & \ddots & \ddots & 0 \\ P_{[1,n]} & P_{[2,n]} & \cdots & P_{[n-1,n]} & 1 \end{pmatrix} = \begin{pmatrix} 1 & 0 & 0 & \cdots & 0 \\ z_{2,1} & 1 & 0 & \cdots & 0 \\ z_{3,1} & z_{3,2} & 1 & \ddots & \vdots \\ \vdots & \vdots & \ddots & \ddots & 0 \\ z_{n,1} & z_{n,2} & \cdots & z_{n,n-1} & 1 \end{pmatrix}^{-1} \tag{42}$$

and treat $P_{[i,i]} = 1$. Then $P_{[i,j]}$ are polynomials in $\{z_{r,s} | 1 \leq s < r \leq n\}$. Moreover, for $1 \leq r_1 < r_2 \leq n-1$ and $r \in \{1, \ldots, n-1\}$, we have the differential property:

$$\partial_{z_{r_2,r_1}}(\mathscr{X}_A) = \sum_{s=1}^{r_1} \tau_s P_{[s,r_1]} \mathscr{X}_A[s, r_2], \tag{43}$$

$$\partial_{z_{n,r}}(\mathscr{X}_A) = \frac{\tau_n}{\vartheta} \sum_{s=1}^{r} \tau_s P_{[s,r]} \mathscr{X}_A[s, n], \tag{44}$$

where $\mathscr{X}_A[i,j]$ is obtained from $\mathscr{X}_A$ by changing τ_i to $\tau_i + 1$ and τ_j to $\tau_j - 1$ for $1 \leq i < j \leq n-1$ and $\mathscr{X}_A[k,n]$ is obtained from $\mathscr{X}_A$ by changing τ_i to $\tau_i + 1$, τ_n to $\tau_n + 1$ and ϑ to $\vartheta + 1$ for $k \in \overline{1, n-1}$. The system of (43) and (44) is an analogue of (33). Furthermore, if $Re\ \tau_n > 0$ and $Re\ (\vartheta - \tau_n) > 0$, then we have the Euler integral representation

$$\mathscr{X}_A = \frac{\Gamma(\vartheta)}{\Gamma(\vartheta - \tau_n)\Gamma(\tau_n)} \int_0^1 \left[\prod_{r=1}^{n-1} \left(\sum_{s=r}^{n-1} P_{[r,s]} + t P_{[r,n]} \right)^{-\tau_r} \right] t^{\tau_n - 1} (1-t)^{\vartheta - \tau_n - 1} dt \tag{45}$$

on the region $\left[P_{[r,n]} \Big/ \left(\sum_{s=r}^{n-1} P_{[r,s]} \right) \notin (-\infty, -1) \right]$ for $r \in \{1, \ldots, n-1\}$. Expression (45) is exactly an analogue of (34).

The Calogero–Sutherland model is an exactly solvable quantum many-body system in one-dimension (cf. [Cf, Sb]), whose Hamiltonian is given by

$$H_{CS} = \sum_{\iota=1}^{n} \partial_{x_\iota}^2 + K \sum_{1 \leq p < q \leq n} \frac{1}{\sinh^2(x_p - x_q)}, \tag{46}$$

where K is a constant. Set

$$\xi^A_{r_2,r_1} = \prod_{s=r_1}^{r_2-1} \frac{e^{2x_{r_2}}}{e^{2x_{r_2}} - e^{2x_s}} \quad \text{for } 1 \leq r_1 < r_2 \leq n. \tag{47}$$

Take $(\lambda_1, \ldots, \lambda_n) \in \mathbb{C}^n$ such that

$$\lambda_1 - \lambda_2 = \cdots = \lambda_{n-2} - \lambda_{n-1} = \mu \text{ and } \lambda_{n-1} - \lambda_n = \sigma \notin \mathbb{N}, \tag{48}$$

for some constants μ and σ. Based on Etingof's work [Ep] related to representations of simple Lie algebra of type A_{n-1}, we proved in [X14] that

$$\prod_{r=1}^{n} e^{2(\lambda_r + (n+1)/2 - r)x_r} \mathscr{X}_A(\mu+1, .., \mu+1, -\mu; -\sigma)\{\xi^A_{r_2,r_1}\} \tag{49}$$

is a solution of the Calogero–Sutherland model.

We also introduced in [X14] new multivariable hypergeometric functions related to the simple Lie algebras of types B_n, C_n, D_n, where the functions of type C give rise to the solutions of the corresponding Olshanestsky–Perelomov model (cf. [OP]).

Partial differential equations are also used by us [X12, X17, X18, X19, X22] to explicitly determine singular vectors in certain weight modules of finite-dimensional simple Lie algebras and the structures of the modules (some of the results were abstractly determined by Brion [B] before us). In particular, we obtained combinatorial identities on the dimensions of an infinite family of irreducible modules of F_4, E_6 and E_7, respectively. Differential-operator representations of Lie algebras are called *oscillator representations* in physics terminology (e.g., cf. [FC, FSS, G]). In the oscillator representation of the simple Lie algebra of type F_4 lifted from its basic irreducible module, we showed in [X18] that the number of irreducible submodules contained in the space of homogeneous harmonic polynomials with degree $k \geq 2$ is $\geq [\![k/3]\!] + [\![(k-2)/3]\!] + 2$. Moreover, in the polynomial algebra over the 56-dimensional basic irreducible module of the simple Lie algebra E_7, we found in [X22] two three-parameter families of irreducible submodules in the solution space of the Cartan's fourth-order E_7-invariant partial differential equation.

For any $\lambda \in H^*$, there exists a unique highest-weight $\mathscr{G}$-module $M(\lambda)$ with highest weight λ such that any other highest-weight $\mathscr{G}$-module with highest weight λ must be a quotient module of $M(\lambda)$. The module $M(\lambda)$ is called a *Verma module* (cf. [V1, V2]). Suppose that $\mathscr{G}$ is of type A_{n-1}. The Weyl group of $\mathscr{G}$ in this case is exactly the group S_n of permutations on $\{1, 2, \ldots, n\}$. In [X19], we derived a system of variable-coefficient second-order linear partial differential equations that determine the singular vectors in $M(\lambda)$, and a differential-operator representation of S_n on the related space of truncated power series. We proved that the solution space of the system of partial differential equations is exactly spanned by $\{\sigma(1) | \sigma \in S_n\}$. Moreover, the singular vectors in $M(\lambda)$ are given by those $\sigma(1)$ that are polynomials. Xiao [XW1] found the explicit formula for $\sigma(1)$ when σ is the reflection associated with a positive root and determined the condition for it to be a polynomial. A similar result as that in [X19] for the simple Lie algebra of type C_2 was proved by us [X12]. Xiao [XW2] generalized our result of A_{n-1} in [X19] and that of C_2 in [X12] to all finite-dimensional simple Lie algebras.

The second important feature of this book is the construction of irreducible representations of simple Lie algebras from their natural inhomogeneous oscillator representations, which give fractional representations of the corresponding Lie groups.

Suppose that $n \geq 2$ is an integer. Let $\mathscr{G}$ be the simple Lie algebra of types A_n, B_n, D_n, E_6 and E_7. It is known that it has the following subalgebra decomposition

$$\mathscr{G} = \mathscr{G}_- \oplus \mathscr{G}_0 \oplus \mathscr{G}_+ \tag{50}$$

with

$$[\mathscr{G}_+, \mathscr{G}_-] \subset \mathscr{G}_0, \qquad [\mathscr{G}_0, \mathscr{G}_\pm] \subset \mathscr{G}_\pm, \qquad [\mathscr{G}_\pm, \mathscr{G}_\pm] = \{0\}, \tag{51}$$

where $\mathscr{G}_0$ is a direct sum of a one-dimensional trivial Lie algebra with the simple Lie algebra $\mathscr{G}_0'$ of types A_{n-1}, B_{n-1}, D_{n-1}, D_5 and E_6, respectively. Moreover, $\mathscr{G}_-$ forms the minimal natural irreducible $\mathscr{G}_0'$-module with respect to the left-multiplication representation of $\mathscr{G}$ if $\mathscr{G}$ is not of type E_6, and $\mathscr{G}_-$ gives the "spin representation" of $\mathscr{G}_0'$ when $\mathscr{G}$ is of type E_6. Moreover, $\mathscr{G}_+$ is its dual $\mathscr{G}_{0'}$-module of $\mathscr{G}_-$. We lift the representation of $\mathscr{G}_0'$ on $\mathscr{G}_-$ to an oscillator representation of $\mathscr{G}_0$ on the algebra $\mathscr{A} = \mathbb{C}[x_1, \ldots, x_m]$ of polynomial functions on $\mathscr{G}_-$, where $m = \dim \mathscr{G}_-$.

Denote by $\mathscr{A}_k$ the subspace of homogeneous polynomials with degree k. We extend the oscillator representation of $\mathscr{G}_0$ to an inhomogeneous oscillator representation of $\mathscr{G}$ by fixing $\mathscr{G}_+|_{\mathscr{A}} = \sum_{i=1}^m \mathbb{C}\partial_{x_i}$ and solving $\mathscr{G}_-|_{\mathscr{A}}$ in $\sum_{i=1}^m \mathscr{A}_2\partial_{x_i}$. When $\mathscr{G}$ is of type A_n, the extended representation is exactly the one induced from the corresponding projective transformations (cf. [ZX]). If $\mathscr{G}$ is of types B_n and D_n, the extended representation is exactly the one induced from the corresponding conformal transformations (cf. [XZ]). In the cases of E_6 and E_7, we found the corresponding fractional representations of the corresponding Lie groups (cf. [X23, X26]).

We define a Lie algebra

$$\mathscr{G}_{\mathscr{A}} = \mathscr{G}|_{\mathscr{A}} \oplus \mathscr{A}\mathscr{G}_0 \tag{52}$$

with the Lie bracket

$$[d_1 + f_1 u, d_2 + f_2 v] = (d_1 d_2 - d_2 d_1) + (d_1(f_2)v - d_2(f_1)u + f_1 f_2[u, v]) \tag{53}$$

for $d_1, d_2 \in \mathscr{G}|_{\mathscr{A}}, f_1, f_2 \in \mathscr{A}$ and $u, v \in \mathscr{G}_0$. Using Shen's idea of mixed product (cf. [Sg]), we found a Lie algebra monomorphism $\imath : \mathscr{G} \to \mathscr{G}_{\mathscr{A}}$ such that $\imath(\mathscr{G}) \not\subset \mathscr{G}|_{\mathscr{A}}$. Let M be any $\mathscr{G}_{0'}$-module. We extend it to a $\mathscr{G}_0$-module by letting the central element take a constant map. Define a $\mathscr{G}_{\mathscr{A}}$-module structure on $\widehat{M} = \mathscr{A} \otimes_{\mathbb{C}} M$ by

$$(d+fu)(g\otimes w) = d(g)\otimes w + fg\otimes u(w) \tag{54}$$

for $d \in \mathcal{G}|_{\mathcal{A}}$, $f,g \in \mathcal{A}$, $u \in \mathcal{G}_0$ and $w \in M$. Then $\widehat{M}$ becomes a $\mathcal{G}$-module with the action

$$\xi(\varpi) = \iota(\xi)(\varpi) \qquad \text{for } \xi \in \mathcal{G},\ \varpi \in \widehat{M}. \tag{55}$$

The map $M \mapsto \widehat{M}$ defines a new functor from the category of $\mathcal{G}_{0'}$-modules to the category of $\mathcal{G}$-modules. When M is an irreducible $\mathcal{G}_{0'}$-module, $\mathbb{C}[\mathcal{G}_-](1\otimes M)$ is naturally an irreducible $\mathcal{G}$-module due to the fact $\mathcal{G}_+|_{\mathcal{A}} = \sum_{i=1}^m \mathbb{C}\partial_{x_i}$, where $\mathbb{C}[\mathcal{G}_-]$ is the polynomial algebra generated by $\mathcal{G}_-$. Thus, the less explicit functor $M \mapsto \mathbb{C}[\mathcal{G}_-](1\otimes M)$ gives a polynomial extension from irreducible $\mathcal{G}'_0$-modules to irreducible $\mathcal{G}$-modules. In particular, it can be used to construct Gel'fand–Zetlin-type bases for E_6 from those for D_5 and then construct Gel'fand–Zetlin-type bases for E_7 from those for E_6 (cf. [GZ1, GZ2, Ma]). When M is a finite-dimensional irreducible $\mathcal{G}'_0$-module, Zhao and the author [ZX, XZ] determined the condition for $\widehat{M} = \mathbb{C}[\mathcal{G}_-](1\otimes M)$ if $\mathcal{G}$ is of types A_n, B_n and D_n. The condition for E_6 and E_7 was found by the author in [X23, X26]. In these works, the idea of Kostant's characteristic identities in [Kb] played a key role. Our approaches heavily depend on the explicit decompositions of $\mathcal{A}$ and the tensor module of $\mathcal{G}_-$ with any finite-dimensional $\mathcal{G}'_0$-module into direct sums of irreducible $\mathcal{G}'_0$-submodules. This is a reason why we adopted the case-by-case approach.

The above works also yield a one-parameter (c) family of inhomogeneous first-order differential-representations of $\mathcal{G}$. Letting these operators act on the space of exponential–polynomial functions that depend on a parametric vector $\vec{a} \in \mathbb{C}^m$, we proved that the space forms an irreducible $\mathcal{G}$-module for any $c \in \mathbb{C}$ if $\vec{a}$ is not on an explicitly given algebraic variety when $\mathcal{G}$ is not of type A_n (cf. [X25, X27]). The equivalent combinatorial properties of the representations played key roles. In the case of A_n, we [X24] got new nonweight irreducible modules of the simple Lie algebras of types A_n and C_ℓ if $n = 2\ell$ is an even integer. By partially swapping differential operators and multiplication operators, we obtain more general oscillator representations of $\mathcal{G}$. If $c \notin \mathbb{Z}$, we obtained in [X24] explicit infinite-dimensional weight modules with finite-dimensional weight subspaces for the simple Lie algebras of types A_n and C_ℓ when $n = 2\ell$ is an even integer. When $c \notin \mathbb{Z}/\not\models,$, we [X25] found explicit infinite-dimensional weight modules with finite-dimensional weight subspaces for the simple Lie algebras of types B_n and D_n. These weight modules are not of highest type.

Below we give a chapter-by-chapter introduction. Throughout this book, $\mathbb{F}$ is always a field with characteristic 0 such as the field $\mathbb{Q}$ of rational numbers, the field $\mathbb{R}$ of real numbers, and the field $\mathbb{C}$ of complex numbers. All the vector spaces are assumed over $\mathbb{F}$ unless they are specified.

The first part of this book is the classical theory of finite-dimensional Lie algebras and their representations, which serves as the preparation of our later main contents.

In Chap. 1, we give basic concepts and examples of Lie algebras. Moreover, Engel's theorem on nilpotent Lie algebras and Lie's theorem on solvable Lie algebras are proved. Furthermore, we derive the Jordan–Chevalley decomposition of a linear transformation and use it to show Cartan's criterion on the solvability.

In Chap. 2, we first introduce the Killing form and prove that the semisimplicity of a finite-dimensional Lie algebra over $\mathbb{C}$ is equivalent to the nondegeneracy of its Killing form. Then we use the Killing form to derive the decomposition of a finite-dimensional semisimple Lie algebra over $\mathbb{C}$ into a direct sum of simple ideals. Moreover, it is showed that a derivation of $\mathscr{G}$ must be a left-multiplication operator of $\mathscr{G}$ for such a Lie algebra. Furthermore, we study the completely reducible modules of a Lie algebra and prove the Weyl's theorem of complete reducibility. The equivalence of the complete reducibility of real and complex modules is also given. Cartan's root-space decomposition of a finite-dimensional semisimple Lie algebra over $\mathbb{C}$ is derived. In particular, we prove that such a Lie algebra is generated by two elements. The complete reducibility of finite-dimensional modules of the simple Lie algebra of type A_1 plays an important role in proving the properties of the corresponding root systems.

In Chap. 3, we start with the axiom of root system and give the root systems of special linear algebras, orthogonal Lie algebras, and symplectic Lie algebras. Then we derive some basic properties of root systems; in particular, the existence of the bases of root systems. As finite symmetries of root systems, the Weyl groups are introduced and studied in detail. The classification and explicit constructions of root systems are presented. The automorphism groups of root systems are determined. As a preparation for later representation theory of Lie algebras, the corresponding weight lattices and their saturated subsets are investigated.

In Chap. 4, we show that the structure of a finite-dimensional semisimple Lie algebra over $\mathbb{C}$ is completely determined by its root system. Moreover, we prove that any two Cartan subalgebras of such a Lie algebra $\mathscr{G}$ are conjugated under the group of inner automorphisms of $\mathscr{G}$. In particular, the automorphism group of $\mathscr{G}$ is determined when it is simple. Furthermore, we give explicit constructions of the simple Lie algebras of exceptional types.

By Weyl's Theorem, any finite-dimensional representation of a finite-dimensional semisimple Lie algebra over $\mathbb{C}$ is completely reducible. The main goal in Chap. 5 is to study finite-dimensional irreducible representations of a finite-dimensional semisimple Lie algebra over $\mathbb{C}$. First we introduce the universal enveloping algebra of a Lie algebra and prove the Poincaré–Birkhoff–Witt (PBW) Theorem on its basis. Then we use the universal enveloping algebra of a finite-dimensional semisimple Lie algebra $\mathscr{G}$ to construct the Verma modules of $\mathscr{G}$. Moreover, we prove that any finite-dimensional $\mathscr{G}$-module is the quotient of a Verma module modulo its maximal proper submodule, whose generators are explicitly given. Furthermore, the Weyl's character formula of a finite-dimensional irreducible $\mathscr{G}$-module is derived, and the dimensional formula of the module is determined. Finally, we decompose the tensor module of two finite-dimensional irreducible $\mathscr{G}$-modules into a direct sum of irreducible $\mathscr{G}$-submodules in terms of characters.

Part II is about the explicit representations of finite-dimensional simple Lie algebra over $\mathbb{C}$, which is the main content of this book.

In Chap. 6, we give various explicit representations of the simple Lie algebras of type A. First we present a fundamental lemma of solving flag partial differential equations for polynomial solutions, which was due to our work [X16]. Then we present the canonical bosonic and fermionic oscillator representations over their minimal natural modules and minimal orthogonal modules. Moreover, we determine the structure of the noncanonical oscillator representations obtained from the canonical bosonic oscillator representations over their minimal natural modules and minimal orthogonal modules by partially swapping differential operators and multiplication operators. The case over their minimal natural modules was essentially due to Howe [Hr4]. The results in the case over their minimal orthogonal modules are generalizations of the classical theorem on harmonic polynomials. The results were due to Luo and the author [LX1]. We construct a functor from the category of A_{n-1}-modules to the category of A_n-modules, which is related to n-dimensional projective transformations. This work was due to Zhao and the author [ZX]. Finally, we present multiparameter families of irreducible projective oscillator representations of the algebras given in [X24].

Chapter 7 is devoted to natural explicit representations of the simple Lie algebra of type D. First we present the canonical bosonic and fermionic oscillator representations over their minimal natural modules. Then we determine the structure of the noncanonical oscillator representations obtained from the above bosonic representations by partially swapping differential operators and multiplication operators, which are generalizations of the classical theorem on harmonic polynomials (cf. [LX2]). Furthermore, we speak about a functor from the category of D_n-modules to the category of D_{n+1}-modules, which is related to $2n$-dimensional conformal transformations (cf. [XZ]). In addition, we present multiparameter families of irreducible conformal oscillator representations of the algebras given in [X25], and some of them are related to an explicitly given algebraic variety.

Chapter 8 is about natural explicit representations of the simple Lie algebra of type B. We give the canonical bosonic and fermionic oscillator representations over their minimal natural modules. Moreover, we determine the structure of the noncanonical oscillator representations obtained from the above bosonic representations by partially swapping differential operators and multiplication operators, which are generalizations of the classical theorem on harmonic polynomials (cf. [LX2]). Furthermore, we present a functor from the category of B_n-modules to the category of B_{n+1}-modules, which is related to $(2n+1)$-dimensional conformal transformations (cf. [XZ]). Besides, we present multiparameter families of irreducible conformal oscillator representations of the algebras given in [X25], and some of them are related to an explicitly given algebraic variety.

Chapter 9 determines the structure of the canonical bosonic and fermionic oscillator representations of the simple Lie algebras of type C over their minimal natural modules. Moreover, we study the noncanonical oscillator representations obtained from the above bosonic representations by partially swapping differential operators and multiplication operators, and obtain a two-parameter family of new infinite-dimensional irreducible representations (cf. [LX3]). Finally, we present multiparameter families of irreducible projective oscillator representations of the algebras given in [X24].

In Chap. 10, we determine the structure of the canonical bosonic and fermionic oscillator representations of the simple Lie algebra of type G_2 over its seven-dimensional module. Moreover, we present a one-parameter family of conformal oscillator representations of G_2 derived from those of the simple Lie algebra of type B_3 and determine their irreducibility. The result is newly obtained by the author. Furthermore, we use partial differential equations to find the explicit irreducible decomposition of the space of polynomial functions on 26-dimensional basic irreducible module of the simple Lie algebra of type F_4 (cf. [X18]).

Chapter 11 studies explicit representations of the simple Lie algebra of type E_6. First we prove the cubic E_6-generalization of the classical theorem on harmonic polynomials given [X17]. Then we study the functor from the module category of D_5 to the module category of E_6 developed in [X26]. Finally, we give a family of inhomogeneous oscillator representations of the simple Lie algebra of type E_6 on a space of exponential–polynomial functions and prove that their irreducibility is related to an explicit given algebraic variety. The work was due to an unpublished paper of the author.

Explicit representations of the simple Lie algebra of type E_7 are given in Chap. 12. By solving certain partial differential equations, we find the explicit decomposition of the polynomial algebra over the 56-dimensional basic irreducible module of the simple Lie algebra E_7 into a sum of irreducible submodules (cf. [X22]). Then we study the functor from the module category of E_6 to the module category of E_7 developed in [X23]. Moreover, we construct a family of irreducible inhomogeneous oscillator representations of the simple Lie algebra of type E_7 on a space of exponential–polynomial functions, related to an explicitly given algebraic variety (cf. [X27]).

How to construct explicit irreducible representations of E_8 is still a challenging problem. Part III is an extension of Part II.

In Chap. 13, we first establish two-parameter $\mathbb{Z}^2$-graded supersymmetric oscillator generalizations of the classical theorem on harmonic polynomials for the general linear Lie superalgebra. Then we extend the result to two-parameter $\mathbb{Z}$-graded supersymmetric oscillator generalizations of the classical theorem on harmonic polynomials for the orthosymplectic Lie superalgebras. This chapter is a reformulation of Luo and the author's work [LX4].

Linear codes with large minimal distances are important error correcting codes in information theory. Orthogonal codes have more applications in the other fields of mathematics. In Chap. 14, we study the binary and ternary orthogonal codes generated by the weight matrices of finite-dimensional modules of simple Lie algebras. The Weyl groups of the Lie algebras act on these codes isometrically. It turns out that certain weight matrices of the simple Lie algebras of types A and D generate doubly-even binary orthogonal codes and ternary orthogonal codes with large minimal distances. Moreover, we prove that the weight matrices of F_4, E_6, E_7

and E_8 on their minimal irreducible modules and adjoint modules all generate ternary orthogonal codes with large minimal distances. In determining the minimal distances, we have used the Weyl groups and branch rules of the irreducible representations of the related simple Lie algebras. The above results are taken from [X20].

In Chap. 15, we prove that certain variations of the classical Weyl functions are solutions of the Calogero–Sutherland model and its generalizations—the Olshanestsky–Perelomov model in various cases. New multivariable hypergeometric functions related to the root systems of classical simple Lie algebras are introduced. In particular, those of type A give rise to solutions of the Calogero–Sutherland model based on Etingof's work [Ep] and those of type C yield solutions of the Olshanestsky–Perelomov model of type C based on Etingof and Styrkas' work [ES]. The differential properties and multivariable hypergeometric equations for these multivariable hypergeometric functions are given. The Euler integral representations of the type-A functions are found. These results come from the author's work [X14].

References

[AAR] G. Andrews, R. Askey, R. Roy, *Special Functions* (Cambridge University Press, 1999)
[B] M. Brion, Invariants d'un sous-groupe unipotent maxaimal d'un groupe semi-simple. Ann. Inst. Fourier (Grenoble) **33**(1), 1–27 (1983)
[B1] Z. Bai, Gelfand-Kirillov dimensionas of the Z2-graded oscillator representations of sl(n), Acta Math. Sinica (Engl. Ser.) **31**(6), 921–937 (2015)
[B2] Z. Bai, Gelfand-Kirillov dimensionas of the Z-graded oscillator representations of $o(n,\mathbb{C})$ and $sp(2n,\mathbb{C})$, Commun. Algebra (to appear)
[BBL] G. Benkart, D. Britten, F.W. Lemire, Modules with bounded multiplicities for simple Lie algebras, Math. Z. **225**, 333–353 (1997)
[BDK] A. Barakat, A. De Sole, V.G. Kac, Poisson vertex algebras in the theory of Hamiltonian equations, Jpn. J. Math. **4**(2), 141–252 (2009)
[BFL] D. Britten, V. Futorny, F.W. Lemire, Simple A_2-modules with a finite dimensional weight space, Commun. Algebra **23**, 467–510 (1995)
[BHL] D. Britten, J. Hooper, F.W. Lemire, Simple C_n-modules with multiplicities 1 and applications, Canad. J. Phys. **72**, 326–335 (1994)
[BL1] D. Britten, F.W. Lemire, A classification of simple Lie modules having 1-dimensional weight space. Trans. Amer. Math. Soc. **299**, 683–697 (1987)
[BL2] D. Britten, F.W. Lemire, On modules of bounded multiplicities for symplectic algebras. Trans. Amer. Math. Soc. **351**, 3413–3431 (1999)
[Bo] R. Beerends, E. Opdam, Certain hypergeometric series related to the root system BC. Trans. Amer. Math. Soc. **119**, 581–609 (1993)
[Cb] B. Cao, Solutions of Navier equations and their representation structure. Adv. Appl. Math. **43**, 331–374 (2009)
[Cf] F. Calogero, Solution of the one-dimensional n-body problem with quadratic and /or inversely quadratic pair potentials, J. Math. Phys. **12**, 419–432 (1971)
[D] L. Dickson, A class of groups in an arbitrary realm connected with the configuration of the 27 lines on a cubic surface. J. Math. **33**, 145–123 (1901)
[Ep] P. Etingof, Quantum integrable systems and representations of Lie algebras. J. Math. Phys. **36** (6), 2637–2651 (1995)

[ES] P. Etingof, K. Styrkas, Algebraic integrability of Schrödinger operators and representations of Lie algebras. Compositio Math. **98**(1), 91–112 (1995)

[FC] F.M. Fernández, E.A. Castro, *Algebraic Methods in Quantum Chemistry and Physics* (CRC Press, Inc., 1996)

[Fs] S.L. Fernando, Lie algebra modules with finite-dimensional weight spaces, I. Trans. Amer. Math. Soc. **322**, 757–781 (1990)

[Fv] V. Futorny, The weight representations of semisimple finite-dimensional Lie algebras. Ph.D. Thesis, Kiev University, 1987

[G] H. Georgi, *Lie Algebras in Particle Physics*, 2nd edn. (Perseus Books Group, 1999)

[GDi1] I.M. Gel'fand, L.A. Dikii, Asymptotic behaviour of the resolvent of Sturm-Liouville equations and the algebra of the Korteweg-de Vries equations. Rus. Math. Surv. **30**(5), 77–113 (1975)

[GDi2] I.M. Gel'fand, L.A. Dikii, A Lie algebra structure in a formal variational calculation. Func. Anal. Appl. **10**, 16–22 (1976)

[GDo1] I. M. Gel'fand, I. Ya. Dorfman, Hamiltonian operators and algebraic structures related to them, Funkts. Anal. Prilozhen **13**, 13–30 (1979)

[GDo3] I. M. Gel'fand, I. Ya, Dorfman, Hamiltonian operators and infinite- dimensional Lie algebras, Funkts. Anal. Prilozhen **14**, 23–40 (1981)

[GS] J. Gervais, B. Sakita, Field theory interpretation of supergauges in dual models. Nucl. Phys. **B34**(2), 632–639 (1971)

[Hr4] R. Howe, in *Perspectives on invariant theory: Schur duality, multiplicity-free actions and beyond*. Schur Lect. (1992) (Tel Aviv), 1–182. Israel Mathematics Conference Proceedings, 8, Bar-Ilan University, Ramat Gan (1995)

[Kb] B. Kostant, On the tensor product of a finite and an infinite dimensional representation. J. Fund. Anal. **20**, 257–285 (1975)

[Kv2] V.G. Kac, *Vertex Algebras for Beginners*. University Lecture Series, vol. 10 (AMS, Providendence RI, 1996)

[Lie] S. Lie, Begrüdung einer Invariantentheorie der Berührungstransformationen. Math. Ann. **8**, 215–288 (1874)

[LX1] C. Luo, X. Xu, $\mathbb{Z}_{\not\in}$-graded oscillator generalizations of $sl(n)$, Commun. Algebra **41**, 3147–3173 (2013)

[LX2] C. Luo, X. Xu, Z-Graded oscillator generalizations of the classical theorem on harmonic polynomials. J. Lie Theory **23**, 979–1003 (2013)

[LX3] C. Luo, X. Xu, $\mathbb{Z}$-graded oscillator representations of Symplectic Lie algebras. J. Algebra **403**, 401–425 (2014)

[LX4] C. Luo, X. Xu, Supersymmetric analogues of the classical theorem on harmonic polynomials. J. Algebra Appl. **13**(6),1450011 42pp. (2014)

[Ma] A. Molev, A basis for representations of symplectic Lie algebras. Comm. Math. Phys. **201**, 591–618 (1999)

[Mo] O. Mathieu, Classification of irreducible weight modules. Ann. Inst. Fourier (Grenoble) **50**, 537–592 (2000)

[Mh] H. Miyazawa, Baryon number changing currents. Prog. Theor. Phys. **36**(6), 1266–1276 (1966)

[Op] P.J. Olver, *Equivalence, Invariants and Symmetry* (Cambridge University Press, Cambridge, 1995)

[PHB] V. Pless, W.C. Huffman, R.A. Brualdi, An introduction to algebraic codes, in *Handbook of Coding Theory*, ed. by V. Pless, W. Huffman (Chap. 1) (Elsevier Science B.V., 1998)

[R1] P. Ramond, Dual theory of free fermions. Phys. Review D **3**(10), 2415–2418 (1971)

[Sb] B. Sutherland, Exact results for a quantum many-body problem in one-dimension. Phys. Rev. A **5**, 1372–1376 (1972)

[Sc] C. Shannon, A mathematical theory of communication. Bell System Tech. J. **27**, 379–423, 623–656 (1948)

[Sg] G. Shen, Graded modules of graded Lie algebras of Cartan type (I)—mixed product of modules. Sci. Chin. A **29**, 570–581 (1986)

[V1] D.-N. Verma, Structure of certain induced representations of complex semisimple Lie algebras. Thesis, Yale University, 1966

[V2] D.-N. Verma, Structure of certain induced representations of complex semisimple Lie algebras. Bull. Amer. Math. Soc. **74**, 160–166 (1968)

[WG] Z. Wang, D. Guo, *Special Functions*. (World Scientific, Singapore, 1998)

[X3] X. Xu, Hamiltonian superoperators. J. Phys. A **28**, 1681–1698 (1995)

[X4] X. Xu, Hamiltonian operators and associative algebras with a derivation. Lett. Math.Phys. **33**, 1–6 (1995)

[X5] X. Xu, Differential invariants of classical groups. Duke Math. J. **94**, 543–572 (1998)

[X6] X. Xu, *Introduction to Vertex Operator Superalgebras and their Modules*. (Kluwer Academic Publishers, Dordrecht/Boston/London, 1998)

[X7] X. Xu, Generalizations of the Block algebras. Manuscripta Math. **100**, 489–518 (1999)

[X8] X. Xu, New generalized simple Lie algebras of Cartan type over a field with characteristic 0. J. Algebra **224**, 23–58 (2000)

[X9] X. Xu, Variational calculus of supervariables and related algebraic structures. J. Algebra **223**, 386-437 (2000)

[X10] X. Xu, Equivalence of conformal superalgebras to Hamiltonian superoperators. Algebra Colloq. **9**, 63–92 (2001)

[X11] X. Xu, Poisson and Hamiltonian superpairs over polarized associative algebras. J. Phys. A **34**, 4241–4265 (2001)

[X12] X. Xu, Differential equations for singular vectors of sp(2n). Commun. Algebra **33**, 4177–4196 (2005)

[X14] X. Xu, Path hypergeometric functions. J. Algebra Appl. **6**, 595–653 (2007)

[X16] X.Xu, Flag partial differential equations and representations of Lie algebras. Acta Appl. Math. **102**, 249–280 (2008)

[X17] X. Xu, A cubic E6-generalization of the classical theorem on harmonic polynomials. J. Lie Theory **21**, 145–164 (2011)

[X18] X. Xu, Partial differential equation approach to F4. Front. Math. China **6**, 759–774 (2011)

[X19] X. Xu, Differential-operator representation of S_n and singular vectors in Verma modules. Algebr. Represent. Theor. 15, 211–231 (2012)

[X20] X. Xu, Representations of Lie algebras and coding theory. J. Lie Theory 22(3), 647–682 (2012)

[X21] X. Xu, *Algebraic Approaches to Partial Differential Equations*, (Springer, Heidelberg/New York/Dordrecht/ London, 2013)

[X22] X. Xu, Partial differential equation approach to E7. Acta Math Sin. (Engl. Ser.) **31**, 177–200 (2015)

[X23] X. Xu, A new functor from E6-Mod to E7-Mod. Commun. Algebra **43**(9), 3589–3636 (2015)

[X24] X. Xu, Projective oscillator representations of sl(n + 1) and sp(2m + 2). J. Lie Theory **26**, 96–114 (2016)

[X25] X. Xu, Conformal oscillator representations of orthogonal Lie algebras. Sci. China: Math. (Engl Ser.) **59**(1), 37–48 (2016)

[X26] X. Xu, A new functor from D5-Mod to E6-Mod, Acta Math Sin. (Engl. Ser.) **32**(8), 867–892 (2016)

[X27] X. Xu, Representations of E_7, combinatorics and algebraic varieties. J. Algebra Appl. (to appear)

[XW1] W. Xiao, Differential equations and singular vectors in Verma modules over $sl(n;\mathbb{C})$sl(n; C). Acta Math. Sinica, (Engl. Ser.) **31**(7), 1057–1066 (2015)

[XW2] W. Xiao, Differential-operator representations of Weyl groups and singular vectors in Verma modules. Sci. China: Math. (Engl. Ser.) (to appear)

[XZ] X. Xu, Y. Zhao, Extensions of the conformal representations for orthogonal Lie algebras. J. Algebra **377**, 97–124 (2013)
[ZX] Y. Zhao, X. Xu, Generalized projective representations for sl(n+1). J. Algebra **328**, 132–154 (2011)

Part I
Fundament of Lie Algebras

Chapter 1
Preliminary of Lie Algebras

In this chapter, we give basic concepts of Lie algebras, such as ideal, nilpotency, solvability, simplicity and homomorphism. Moreover, we construct important examples of Lie algebras, such as the Sin algebra, derivation Lie algebras, Lie algebras of Block type, special linear Lie algebras, orthogonal Lie algebras, symplectic Lie algebras and co-adjoint Poisson structures. Engel's theorem on nilpotent Lie algebras and Lie's theorem on solvable Lie algebras are proved. Furthermore, we derive the Jordan-Chevalley decomposition of a linear transformation and use it to show Cartan's criterion on the solvability.

Section 1.1 is about the basic concepts of general algebra, associative algebra, group algebra, Lie algebra and Sin algebra. In Sect. 1.2, we study Lie subalgebras and present Witt algebras, Block algebra, classical Lie algebras and co-adjoint Poisson structures. In Sect. 1.3, we define an ideal of a Lie algebra and simple Lie algebra, and give the basic technique of proving simplicity. Section 1.4 discusses homomorphisms between Lie algebras. Finally in Sect. 1.5, we define nilpotent and solvable Lie algebras, and prove the Engel's theorem, Lie's theorem, Jordan-Chevalley decomposition theorem and Cartan's criterion.

1.1 Basic Definitions and Examples

In this section, we give definitions and examples of algebras.

An *algebra* $\mathscr{A}$ is a vector space with a bilinear map $m(\cdot,\cdot):\mathscr{A}\times\mathscr{A}\rightarrow\mathscr{A}$; that is,

$$m(au+bv,w)=am(u,v)+bm(v,w),\quad m(w,au+bv)=am(w,u)+bm(w,v) \tag{1.1.1}$$

for $a,b\in\mathbb{F}$ and $u,v,w\in\mathscr{A}$. The map $m(\cdot,\cdot)$ is called an *algebraic operation*. The algebra $\mathscr{A}$ is called *associative* if

X. Xu, *Representations of Lie Algebras and Partial Differential Equations*,
DOI 10.1007/978-981-10-6391-6_1

$$m(u, m(v, w)) = m(m(u, v), w) \quad \text{for } u, v, w \in \mathscr{A}. \tag{1.1.2}$$

In this case, we denote $m(u, v) = u \cdot v$ or simply uv. The above equation becomes

$$u \cdot (v \cdot w) = (u \cdot v) \cdot w \quad \text{for } u, v, w \in \mathscr{A}. \tag{1.1.3}$$

Example 1.1.1 Denote by End V the space of all linear transformations (endomorphisms; linear operators) on a vector space V. Then End V becomes an associative algebra with respect to the composition of two linear transformations; that is,

$$(T_1 \cdot T_2)(u) = T_1(T_2(u)) \quad \text{for } T_1, T_2 \in \text{End } V, \ u \in V. \tag{1.1.4}$$

The other familiar examples of associative algebras are the algebra $M_{n\times n}(\mathbb{F})$ of all $n \times n$ matrices with entries in $\mathbb{F}$ and the algebra $\mathbb{F}[x_1, x_2, \ldots, x_n]$ of all the polynomials in variables $x_1, x_2, \ldots, x_n$.

Example 1.1.2 A *group* G is a set with a map $\cdot : G \times G \to G$ and a spacial element 1_G such that

$$(g_1 \cdot g_2) \cdot g_3 = (g_1 \cdot g_2) \cdot g_3, \quad 1_G \cdot g_1 = g_1 \cdot 1_G = g_1 \quad \text{for } g_1, g_2, g_3 \in G, \tag{1.1.5}$$

and for any $g \in G$, there exists an element $g^{-1} \in G$ for which

$$g \cdot g^{-1} = g^{-1} \cdot g = 1_G. \tag{1.1.6}$$

If

$$g_1 \cdot g_2 = g_2 \cdot g_1 \text{ for } g_1, g_2 \in G, \tag{1.1.7}$$

we call $\mathscr{G}$ an *abelian group*.

Given a group G. Denote by $\mathbb{F}[G]$ the vector space with a basis $\{\Im_g \mid g \in G\}$. Define an algebraic operation "$\cdot$" on $\mathbb{F}[G]$ by:

$$\Im_{g_1} \cdot \Im_{g_2} = \Im_{g_1 \cdot g_2} \quad \text{for } g_1, g_2 \in G. \tag{1.1.8}$$

Then $\mathbb{F}[G]$ becomes an associative algebra with respect to (1.1.8), which is called the *group algebra* of G.

A *Lie algebra* $\mathscr{G}$ is a vector space with an algebraic operation $[\cdot, \cdot]$ satisfying the following two axioms:

$$[u, v] = -[v, u] \quad \text{(Skew Symmetry)}, \tag{1.1.9}$$

$$[[u,v],w]+[[v,w],u]+[[w,u],v]=0 \qquad \text{(Jacobi Identity)} \tag{1.1.10}$$

for $u, v, w \in \mathscr{G}$. The algebraic operation $[\cdot,\cdot]$ is called the *Lie bracket* of $\mathscr{G}$. Note

$$[u,u]=-[u,u] \Longrightarrow [u,u]=0 \qquad \text{for } u\in\mathscr{G}. \tag{1.1.11}$$

Example 1.1.3 Let $(\mathscr{A},\cdot)$ be an associative algebra. Define another algebraic operation $[\cdot,\cdot]$ on $\mathscr{A}$:

$$[u,v]=u\cdot v-v\cdot u, \tag{1.1.12}$$

which is called the *commutator* of u and v in $\mathscr{A}$. The skew symmetry is obvious. Moreover,

$$\begin{aligned}&[[u,v],w]=[u\cdot v-v\cdot u,w]=(u\cdot v)\cdot w\\&-(v\cdot u)\cdot w-w\cdot(u\cdot v)+w\cdot(v\cdot u).\end{aligned} \tag{1.1.13}$$

Hence

$$\begin{aligned}&[[u,v],w]+[[v,w],u]+[[w,u],v]=(u\cdot v)\cdot w-(v\cdot u)\cdot w\\&-w\cdot(u\cdot v)+w\cdot(v\cdot u)+(v\cdot w)\cdot u-(w\cdot v)\cdot u-u\cdot(v\cdot w)\\&+u\cdot(w\cdot v)+(w\cdot u)\cdot v-(u\cdot w)\cdot v-v\cdot(w\cdot u)+v\cdot(u\cdot v)=0.\end{aligned} \tag{1.1.14}$$

Thus $(\mathscr{A},[\cdot,\cdot])$ forms a Lie algebra. In particular, we have Lie algebras $(\mathrm{End}\,V,[\cdot,\cdot])$ and $(\mathbb{F}[G],[\cdot,\cdot])$.

Indeed, the above Lie bracket can be generalized as follows. Take any fixed nonzero element $\xi\in\mathscr{A}$ and define an algebraic operation $[\cdot,\cdot]_\xi$ on $\mathscr{A}$ by

$$[u,v]_\xi=u\cdot\xi\cdot v-v\cdot\xi\cdot u \qquad \text{for } u,v\in\mathscr{A}. \tag{1.1.15}$$

It can be verified that $(\mathscr{A},[\cdot,\cdot]_\xi)$ forms a Lie algebra (exercise). This Lie algebra structure is useful in the study of infinite-dimensional simple Lie algebras (cf. [BXZ]) and has also appeared in the other fields of mathematics.

Example 1.1.4 There are so-called "Sin algebras" appeared in geometry and dynamics. We assume $\mathbb{F}=\mathbb{R}$, the field of real numbers. Let $(\Gamma,+)$ be an abelian group. Suppose that $\phi(\cdot,\cdot):\Gamma\times\Gamma\to\mathbb{R}$ is a skew-symmetric $\mathbb{Z}$-bilinear map; that is,

$$\phi(\alpha,\beta)=-\phi(\beta,\alpha), \qquad \phi(\alpha\pm\beta,\gamma)=\phi(\alpha,\gamma)\pm\phi(\beta,\gamma). \tag{1.1.16}$$

On the group algebra $\mathbb{R}[\Gamma]$, we define another algebraic operation $[\cdot,\cdot]$ defined by

$$[\Im_\alpha,\Im_\beta]=\sin\phi(\alpha,\beta)\,\Im_{\alpha+\beta} \qquad \text{for } \alpha,\beta\in\Gamma. \tag{1.1.17}$$

Then $(\mathbb{R}[\Gamma],[\cdot,\cdot])$ forms a Lie algebra (exercise), which is called a *Sin algebra*.

1.2 Lie Subalgebras

In this section, we present derivation Lie algebras, Lie algebras of Block type, special linear Lie algebras, orthogonal Lie algebras, symplectic Lie algebras and co-adjoint Poisson structures.

A subspace $\mathscr{L}$ of a Lie algebra $\mathscr{G}$ is called a *subalgebra* of $\mathscr{G}$ if

$$[u, v] \in \mathscr{L} \qquad \text{for any} \ \ u, v \in \mathscr{L}. \tag{1.2.1}$$

Let $(\mathscr{U}, \circ)$ be any algebra. A linear transformation d on $\mathscr{U}$ is called a *derivation* of $\mathscr{U}$ if

$$d(u \circ v) = d(u) \circ v + u \circ d(v) \qquad \text{for} \ \ u, v \in \mathscr{U}. \tag{1.2.2}$$

Denote

$$\text{Der}\ \mathscr{U} = \text{the set of all derivations on}\ \mathscr{U}. \tag{1.2.3}$$

Then Der $\mathscr{U}$ forms a subspace of End $\mathscr{U}$. Moreover, for $d_1, d_2 \in$ Der $\mathscr{U}$ and $u, v \in \mathscr{U}$, we have

$$\begin{aligned} &d_1 d_2(u \circ v) = d_1(d_2(u) \circ v + u \circ d_2(v)) = d_1 d_2(u) \circ v \\ &+ d_2(u) \circ d_1(v) + d_1(u) \circ d_2(v) + u \circ d_1 d_2(v). \end{aligned} \tag{1.2.4}$$

Thus

$$\begin{aligned} &[d_1, d_2](u \circ v) \\ =\ &(d_1 d_2 - d_2 d_1)(u \circ v) = (d_1 d_2 - d_2 d_1)(u \circ v) = d_1 d_2(u \circ v) - d_2 d_1(u \circ v) \\ =\ &d_1 d_2(u) \circ v + d_2(u) \circ d_1(v) + d_1(u) \circ d_2(v) + u \circ d_1 d_2(v) \\ &-[d_2 d_1(u) \circ v + d_1(u) \circ d_2(v) + d_2(u) \circ d_1(v) + u \circ d_2 d_1(v)] \\ =\ &(d_1 d_2(u) - d_2 d_1(u)) \circ v + u \circ (d_1 d_2(v) - d_2 d_1(v)) \\ =\ &[d_1, d_2](u) \circ v + u \circ [d_1, d_2](v), \end{aligned} \tag{1.2.5}$$

or equivalently, $[d_1, d_2] \in$ Der $\mathscr{U}$. Hence Der $\mathscr{U}$ forms a Lie subalgebra of End $\mathscr{U}$.

Example 1.2.1 Let $\mathscr{A} = \mathbb{F}[x_1, x_2, \ldots, x_n]$ be the polynomial algebra in n variables. It can be verified (exercise) that

$$\mathbb{W}_n = \text{Der}\ \mathscr{A} = \left\{ \sum_{i=1}^{n} f_i \partial_{x_i} \mid f_i \in \mathbb{F}[x_1, x_2, \ldots, x_n] \right\}. \tag{1.2.6}$$

Note that

$$f_1\partial_{x_i}(f_2\partial_{x_j}(g)) = f_1\partial_{x_i}(f_2)\partial_{x_j}(g) + f_1 f_2\partial_{x_i}\partial_{x_j}(g). \tag{1.2.7}$$

Thus

$$\begin{aligned}
&[f_1\partial_{x_i}, f_2\partial_{x_j}](g)\\
&= f_1\partial_{x_i}(f_2\partial_{x_j}(g)) - f_2\partial_{x_j}(f_1\partial_{x_i}(g))\\
&= f_1\partial_{x_i}(f_2)\partial_{x_j}(g) + f_1\partial_{x_i}\partial_{x_j}(g) - [f_2\partial_{x_j}(f_1)\partial_{x_i}(g) + f_2\partial_{x_j}\partial_{x_i}(g)]\\
&= f_1\partial_{x_i}(f_2)\partial_{x_j}(g) - f_2\partial_{x_j}(f_1)\partial_{x_i}(g) = (f_1\partial_{x_i}(f_2)\partial_{x_j} - f_2\partial_{x_j}(f_1)\partial_{x_i})(g).
\end{aligned} \tag{1.2.8}$$

Therefore,

$$[f_1\partial_{x_i}, f_2\partial_{x_j}] = f_1\partial_{x_i}(f_2)\partial_{x_j} - f_2\partial_{x_j}(f_1)\partial_{x_i}. \tag{1.2.9}$$

The Lie algebra $\mathbb{W}_n$ is called a *Witt algebra*.

Example 1.2.2 Let $\mathscr{A}$ be an associative algebra satisfying the commutativity:

$$uv = vu \qquad \text{for } u, v \in \mathscr{A}. \tag{1.2.10}$$

Such an algebra is called a *commutative associative algebra*. Suppose that we have two commuting derivations ∂_1 and ∂_2 on $\mathscr{A}$. Define another algebraic operation $[\cdot, \cdot]$ on $\mathscr{A}$ by:

$$[u, v] = \partial_1(u)\partial_2(v) - \partial_2(u)\partial_1(v) + u\partial_1(v) - \partial_1(u)v \qquad \text{for } u, v \in \mathscr{A}. \tag{1.2.11}$$

Then $(\mathscr{A}, [\cdot, \cdot])$ forms a Lie algebra (exercise), which is called a *Lie algebra of Block type*. When $\mathscr{A}$ is the group algebra of an abelian group, the Lie algebra was discovered by Block [Br] in 1958 and the general case was obtained by the author [X7] in 1999. Chen [Cl] generalized the algebras in [X7] to more general Lie algebras of twisted Hamiltonian type.

The Jacobi identity can be rewritten as

$$[u, [v, w]] = [[u, v], w] + [v, [u, w]], \tag{1.2.12}$$

which shows that the left multiplication operator is a derivation. Let $\mathscr{A}$ be an associative algebra. A linear map $\varphi : \mathscr{A} \to \mathbb{F}$ is called a *trace map* if

$$\varphi(uv) = \varphi(vu) \qquad \text{for } u, v \in \mathscr{A}. \tag{1.2.13}$$

Recall that the trace of an $n \times n$ matrix $B = (b_{i,j})_{n\times n}$ is defined by

$$\text{Tr}\, B = \sum_{i=1} b_{i,i}, \tag{1.2.14}$$

which is a trace map of the algebra $M_{n\times n}(\mathbb{F})$ of all $n\times n$ matrices. For a group algebra $\mathbb{F}[G]$ defined in Example 1.1.2, we have the following trace map:

$$\mathrm{Tr}\,\Im_g=\delta_{g,1_G}\qquad \text{for}\ \ g\in G. \tag{1.2.15}$$

Suppose that $\mathscr{A}$ is an associative algebra with a trace map Tr . Recall that $\mathscr{A}$ forms a Lie algebra with respect to the commutator (1.1.12). Moreover, the set

$$s(\mathscr{A})=\{u\in\mathscr{A}\mid \mathrm{Tr}\,u=0\} \tag{1.2.16}$$

forms a Lie subalgebra, because

$$\mathrm{Tr}\,([u,v])=\mathrm{Tr}\,(uv-vu)=\mathrm{Tr}\,uv-\mathrm{Tr}\,vu=0 \tag{1.2.17}$$

by (1.2.13) for $u,v\in\mathscr{A}$. When we view End V as a Lie algebra with respect to the commutator, we call it a *general linear Lie algebra* and redenote it as $gl(V)$. If $\dim V=n$, End $V\cong M_{n\times n}(\mathbb{F})$. We use $gl(n,\mathbb{F})$ to denote $M_{n\times n}(\mathbb{F})$ as the Lie algebra with respect to the commutator. The Lie subalgebra

$$sl(n,\mathbb{F})=\{B\in gl(n,\mathbb{F})\mid \mathrm{Tr}\,B=0\} \tag{1.2.18}$$

is called a *special linear Lie algebra*.

An *involutive anti-automorphism* σ of an associative algebra $\mathscr{A}$ is a linear transformation on $\mathscr{A}$ such that

$$\sigma^2=\mathrm{Id}_{\mathscr{A}},\quad \sigma(uv)=\sigma(v)\sigma(u). \tag{1.2.19}$$

Set

$$\mathscr{A}_-^\sigma=\{u\in\mathscr{A}\mid \sigma(u)=-u\}, \tag{1.2.20}$$

the set of *skew elements*. Then $\mathscr{A}_-^\sigma$ forms a Lie subalgebra of $\mathscr{A}$ with respect to the commutator. In fact,

$$\sigma([u,v])=\sigma(uv-vu)=\sigma(uv)-\sigma(vu)=\sigma(v)\sigma(u)-\sigma(u)\sigma(v)=-[\sigma(u),\sigma(v)] \tag{1.2.21}$$

for $u,v\in\mathscr{A}$. The transpose of matrix is an involutive anti-automorphism of $M_{n\times n}(\mathbb{F})$. Thus we have the following skew Lie subalgebra

$$o(n,\mathbb{F})=\text{the set of all } n\times n \text{ skew-symmetric matrices with entries in } \mathbb{F}, \tag{1.2.22}$$

which is called an *orthogonal Lie algebra*. In fact, $o(n,\mathbb{R})$ is the Lie algebra of rotation group. To see this, let us show the case $n=2$. Set

$$A = \begin{pmatrix} 0 & 1 \\ -1 & 0 \end{pmatrix}. \tag{1.2.23}$$

Then $o(2,\mathbb{R}) = \mathbb{R}A$ and

$$A^2 = \begin{pmatrix} 0 & 1 \\ -1 & 0 \end{pmatrix}\begin{pmatrix} 0 & 1 \\ -1 & 0 \end{pmatrix} = \begin{pmatrix} -1 & 0 \\ 0 & -1 \end{pmatrix} = -I_2. \tag{1.2.24}$$

Thus

$$\begin{aligned} e^{\theta A} &= \sum_{n=0}^{\infty} \frac{(\theta A)^n}{n!} = \sum_{k=0}^{\infty} \frac{\theta^{2k} A^{2k}}{(2k)!} + \sum_{k=0}^{\infty} \frac{\theta^{2k+1} A^{2k+1}}{(2k+1)!} \\ &= \sum_{k=0}^{\infty} \frac{(-1)^k \theta^{2k}}{(2k)!} I_2 + \sum_{k=0}^{\infty} \frac{(-1)^k \theta^{2k+1}}{(2k+1)!} A \\ &= \cos\theta \; I_2 + \sin\theta \; A = \begin{pmatrix} \cos\theta & \sin\theta \\ -\sin\theta & \cos\theta \end{pmatrix}, \end{aligned} \tag{1.2.25}$$

which is a rotation linear transformation on the plane.

Suppose that $n = 2k$ is even integer. Denote by I_k the $k \times k$ identity matrix. Use up index "t" to denote the transpose of matrix. Define

$$A^{\dagger} = \begin{pmatrix} 0 & I_k \\ -I_k & 0 \end{pmatrix} A^t \begin{pmatrix} 0 & -I_k \\ I_k & 0 \end{pmatrix} \quad \text{for } A \in M_{n\times n}(\mathbb{F}). \tag{1.2.26}$$

It is easy to check that $\dagger$ is an involutive anti-automorphism of $M_{n\times n}(\mathbb{F})$. The corresponding skew Lie subalgebra is:

$$sp(2k,\mathbb{F}) = \left\{ \begin{pmatrix} A & B \\ C & -A^t \end{pmatrix} \mid A, B, C \in M_{k\times k}(\mathbb{F}),\ B^t = B,\ C^t = C \right\}, \tag{1.2.27}$$

which is called a *symplectic Lie algebra*.

On the group algebra $\mathbb{F}[G]$, we have the following involutive anti-automorphism σ:

$$\sigma(\Im_g) = \Im_{g^{-1}} \qquad \text{for } g \in G. \tag{1.2.28}$$

It seems there is not much known on the Lie algebra $\mathbb{F}[G]_-^{\sigma}$.

Suppose that $\{u_1, u_2, \ldots, u_n\}$ is a basis of a Lie algebra $\mathcal{G}$. We write

$$[u_i, u_j] = \sum_{k=1}^{n} c_{i,j}^k u_k \qquad \text{for } i, j \in \{1, 2, \ldots, n\}. \tag{1.2.29}$$

The constants $\{c_{i,j}^k\}$ are called the *structure constants*. Then the skew symmetry (1.1.9) becomes

$$c_{i,j}^k = -c_{j,i}^k \tag{1.2.30}$$

and the Jacobi identity (1.1.10) can be written to:

$$\sum_{r=1}^n (c_{i,j}^r c_{r,k}^s + c_{j,k}^r c_{r,i}^s + c_{k,i}^r c_{r,j}^s) = 0. \tag{1.2.31}$$

We define an algebraic operation $\{\cdot,\cdot\}$ on $\mathbb{F}[x_1, x_2, \ldots, x_n]$ by

$$\{f, g\} = \sum_{i,j,k=1}^n c_{i,j}^k \partial_{x_i}(f)\partial_{x_j}(g)x_k. \tag{1.2.32}$$

Obviously, $\{\cdot,\cdot\}$ is skew-symmetric. Moreover, for $f, g, h \in \mathbb{F}[x_1, x_2, \ldots, x_n]$, we have

$$\begin{aligned}\{\{f,g\},h\} &= \sum_{i,j,k=1}^n c_{i,j}^k \{\partial_{x_i}(f)\partial_{x_j}(g)x_k, h\}\\ &= \sum_{i,j,k,r,s,t=1}^n c_{i,j}^k c_{r,s}^t \partial_{x_r}(\partial_{x_i}(f)\partial_{x_j}(g)x_k)\partial_{x_s}(h)x_t\\ &= \sum_{i,j,r,s,t=1}^n c_{i,j}^r c_{r,s}^t \partial_{x_i}(f)\partial_{x_j}(g)\partial_{x_s}(h)x_t + \sum_{i,j,k,r,s,t=1}^n c_{i,j}^k c_{r,s}^t\\ &\quad \times(\partial_{x_r}\partial_{x_i}(f)\partial_{x_j}(g)\partial_{x_s}(h) + \partial_{x_i}(f)\partial_{x_r}\partial_{x_j}(g)\partial_{x_s}(h))x_k x_t. \end{aligned} \tag{1.2.33}$$

Thus

$$\begin{aligned}&\{\{f,g\},h\} + \{\{g,h\},f\} + \{\{h,f\},g\}\\ &= \sum_{i,j,r,s,t=1}^n c_{i,j}^r c_{r,s}^t \partial_{x_i}(f)\partial_{x_j}(g)\partial_{x_s}(h)x_t + \sum_{i,j,k,r,s,t=1}^n c_{i,j}^k c_{r,s}^t\\ &\quad \times(\partial_{x_r}\partial_{x_i}(f)\partial_{x_j}(g)\partial_{x_s}(h) + \partial_{x_i}(f)\partial_{x_r}\partial_{x_j}(g)\partial_{x_s}(h))x_k x_t\\ &\quad + \sum_{i,j,r,s,t=1}^n c_{i,j}^r c_{r,s}^t \partial_{x_i}(g)\partial_{x_j}(h)\partial_{x_s}(f)x_t + \sum_{i,j,k,r,s,t=1}^n c_{i,j}^k c_{r,s}^t\\ &\quad \times(\partial_{x_r}\partial_{x_i}(g)\partial_{x_j}(h)\partial_{x_s}(f) + \partial_{x_i}(g)\partial_{x_r}\partial_{x_j}(h)\partial_{x_s}(f))x_k x_t\\ &\quad + \sum_{i,j,r,s,t} c_{i,j}^r c_{r,s}^t \partial_{x_i}(h)\partial_{x_j}(f)\partial_{x_s}(g)x_t + \sum_{i,j,k,r,s,t} c_{i,j}^k c_{r,s}^t\\ &\quad \times(\partial_{x_r}\partial_{x_i}(h)\partial_{x_j}(f)\partial_{x_s}(g) + \partial_{x_i}(h)\partial_{x_r}\partial_{x_j}(f)\partial_{x_s}(g))x_k x_t\\ &= \sum_{i,j,r,s,t} (c_{i,j}^r c_{r,s}^t + c_{j,s}^r c_{r,i}^t + c_{s,i}^r c_{r,j}^t)\partial_{x_i}(f)\partial_{x_j}(g)\partial_{x_s}(h)x_t \end{aligned}$$

$$+\sum_{i,j,k,r,s,t} c_{i,j}^{k} c_{r,s}^{t} (\partial_{x_r}\partial_{x_i}(f)\partial_{x_j}(g)\partial_{x_s}(h) + \partial_{x_i}(h)\partial_{x_r}\partial_{x_j}(f)\partial_{x_s}(g))x_k x_t$$
$$+\sum_{i,j,k,r,s,t} c_{i,j}^{k} c_{r,s}^{t} (\partial_{x_i}(f)\partial_{x_r}\partial_{x_j}(g)\partial_{x_s}(h) + \partial_{x_r}\partial_{x_i}(g)\partial_{x_j}(h)\partial_{x_s}(f))x_k x_t$$
$$+\sum_{i,j,k,r,s,t} c_{i,j}^{k} c_{r,s}^{t} (\partial_{x_i}(g)\partial_{x_r}\partial_{x_j}(h)\partial_{x_s}(f) + \partial_{x_r}\partial_{x_i}(h)\partial_{x_j}(f)\partial_{x_s}(g))x_k x_t$$
$$=\sum_{i,j,k,r,s,t} c_{i,j}^{k} (c_{r,s}^{t} + c_{s,r}^{t})\partial_{x_r}\partial_{x_i}(f)\partial_{x_j}(g)\partial_{x_s}(h)x^k x_t$$
$$+\sum_{i,j,k,r,s,t} c_{i,j}^{k} (c_{r,s}^{t} + c_{s,r}^{t})\partial_{x_r}\partial_{x_i}(g)\partial_{x_j}(h)\partial_{x_s}(f)x^k x_t$$
$$+\sum_{i,j,k,r,s,t} c_{i,j}^{k} (c_{r,s}^{t} + c_{s,r}^{t})\partial_{x_r}\partial_{x_i}(h)\partial_{x_j}(f)\partial_{x_s}(g)x^k x_t = 0. \tag{1.2.34}$$

Thus $\{\cdot,\cdot\}$ is a Lie bracket, which is called *co-adjoint Poisson structure* over $\mathscr{G}$. The structure has been used in symplectic geometry and integrable systems.

1.3 Ideals

In this section, we define an ideal of a Lie algebra and simple Lie algebra. Moreover, we present the basic technique of proving simplicity.

If U and V are subspaces of a Lie algebra $\mathscr{G}$, we define

$$[U, V] = \text{Span}\{[u, v] \mid u \in U,\ v \in V\} \subset \mathscr{G}. \tag{1.3.1}$$

A subspace $\mathcal{I}$ of a Lie algebra $\mathscr{G}$ is called an *ideal* if

$$[\mathscr{G}, \mathcal{I}] \subset \mathcal{I}. \tag{1.3.2}$$

In particular, an ideal is a subalgebra. Obvious examples of ideals are $\{0\}$, $\mathscr{G}$ and $[\mathscr{G}, \mathscr{G}]$, which is called *derived algebra* of $\mathscr{G}$. Define the center of $\mathscr{G}$:

$$Z(\mathscr{G}) = \{u \in \mathscr{G} \mid [u, v] = 0 \text{ for any } v \in \mathscr{G}\}. \tag{1.3.3}$$

Easily see that $Z(\mathscr{G})$ is an ideal of $\mathscr{G}$.

Example 1.3.1 Set

$$\mathscr{K}_n = \mathbb{F}\partial_{x_1} + \sum_{i=0}^{n} \mathbb{F}x_1^i \partial_{x_2} \subset \text{Der}\,\mathbb{F}[x_1, x_2]. \tag{1.3.4}$$

Then

$$[\partial_{x_1}, \partial_{x_1}] = 0, \; [\partial_{x_1}, x_1^i \partial_{x_2}] = i x^{i-1} \partial_{x_2}, \quad [x_1^i \partial_{x_2}, x_1^j \partial_{x_2}] = 0. \tag{1.3.5}$$

Thus $\mathscr{K}_n$ forms a Lie algebra. Moreover,

$$\mathcal{I}_k = \sum_{i=0}^{k} \mathbb{F} x_1^i \partial_{x_2}, \tag{1.3.6}$$

is an ideal of $\mathscr{K}_n$ if $k \leq n$. Furthermore,

$$[\mathscr{K}_n, \mathscr{K}_n] = \mathcal{I}_{n-1}, \qquad Z(\mathscr{K}_n) = \mathbb{F}\partial_{x_2}. \tag{1.3.7}$$

A Lie algebra $\mathscr{G}$ is called *simple* if $\dim \mathscr{G} > 1$ and the only ideals of $\mathscr{G}$ are $\{0\}$ and $\mathscr{G}$.

Lemma 1.3.1 *Let T be a linear transformation on a vector space V. Suppose that U is a subspace of V such that $T(U) \subset U$, and $\{u_1, u_2, \ldots, u_k\}$ are eigenvectors of T corresponding to distinct eigenvalues $\{\lambda_1, \lambda_2, \ldots, \lambda_k\}$. If $\sum_{i=1}^{k} u_i \in U$, then each $u_i \in U$ for $i = 1, 2, \ldots, k$.*

Proof Note

$$v_j = T^j \left(\sum_{i=1}^{k} u_i \right) = \sum_{i=1}^{k} \lambda_i^j u_i \in U. \tag{1.3.8}$$

Consider the system

$$\sum_{i=1}^{k} \lambda_i^j u_i = v_j, \qquad j = 0, 1, \ldots, k-1. \tag{1.3.9}$$

Since the coefficient determinant is a Vandermonde determinant

$$\begin{vmatrix} 1 & 1 & \cdots & 1 \\ \lambda_1 & \lambda_2 & \cdots & \lambda_k \\ \vdots & \vdots & \vdots & \vdots \\ \lambda_1^{k-1} & \lambda_2^{k-1} & \cdots & \lambda_k^{k-1} \end{vmatrix} = \prod_{1 \leq i < j \leq k} (\lambda_j - \lambda_i) \neq 0, \tag{1.3.10}$$

we can solve (1.3.9) for $\{u_1, u_2, \ldots, u_k\}$ as linear combinations of $\{v_0, v_1, \ldots, v_{k-1}\} \subset U$. Thus $u_i \in U$. □

Example 1.3.2 Denote by $E_{i,j}$ the square matrix with 1 as its (i, j)-entry and 0 as the others. Note

$$sl(2,\mathbb{F}) = \mathbb{F}E_{1,2} + \mathbb{F}(E_{1,1} - E_{2,2}) + \mathbb{F}E_{2,1} \tag{1.3.11}$$

with the Lie bracket determined by

$$[E_{1,2}, E_{2,1}] = E_{1,1} - E_{2,2}, \quad [E_{1,1} - E_{2,2}, E_{1,2}] = 2E_{1,2}, \quad [E_{1,1} - E_{2,2}, E_{2,1}] = -2E_{2,1}. \tag{1.3.12}$$

Define a linear transformation T on $sl(2,\mathbb{F})$ by

$$T(u) = [E_{1,1} - E_{2,2}, u] \quad \text{for } u \in sl(2,\mathbb{F}). \tag{1.3.13}$$

Then $E_{1,2}$ is an eigenvector of T with eigenvalue 2, $E_{2,1}$ is an eigenvector of T with eigenvalue -2 and $E_{1,1} - E_{2,2}$ is an eigenvector of T with eigenvalue 0. Suppose that $\mathcal{I}$ is a nonzero ideal of $sl(2,\mathbb{F})$. Then $T(\mathcal{I}) \subset \mathcal{I}$. Moreover, there exists a nonzero element

$$u = a_1 E_{1,2} + a_2(E_{1,1} - E_{2,2}) + a_3 E_{2,1} \in \mathcal{I}. \tag{1.3.14}$$

By the above lemma,

$$a_1 E_{1,2}, a_2(E_{1,1} - E_{2,2}), a_3 E_{2,1} \in \mathcal{I}. \tag{1.3.15}$$

Hence one of the elements in $\{E_{1,2}, E_{1,1} - E_{2,2}, E_{2,1}\}$ in $\mathcal{I}$. If one of $E_{1,2}$ and $E_{2,1}$ in $\mathcal{I}$, then the first equation in (1.3.12) implies $E_{1,1} - E_{2,2} \in \mathcal{I}$. Thus we can always assume $E_{1,1} - E_{2,2} \in \mathcal{I}$. The second and third equations in (1.3.12) imply $E_{1,2}, E_{2,1} \in \mathcal{I}$. Therefore, $\mathcal{I} = sl(2,\mathbb{F})$; that is, $sl(2,\mathbb{F})$ is simple.

In order to prove simplicity for more Lie algebras, we need to generalize Lemma 1.3.1.

Lemma 1.3.2 *Let H be a subspace of End V of commuting linear transformations on a vector space V. Denote*

$$H^* = \text{the set of all linear maps (functions) from } H \text{ to } \mathbb{F}, \tag{1.3.16}$$

which forms a vector space with linear operation:

$$(af_1 + bf_2)(h) = af_1(h) + bf_2(h) \quad \text{for } h \in H. \tag{1.3.17}$$

Suppose that U is a subspace of V such that

$$h(U) \subset U \quad \text{for } h \in H, \tag{1.3.18}$$

and $\{u_1, u_2, \ldots, u_k\} \subset V$ are vectors such that there exist distinct elements $\{\alpha_1, \alpha_2, \ldots, \alpha_k\}$ in H^* for which

$$h(u_i) = \alpha_i(h) u_i \quad \text{for } h \in H. \tag{1.3.19}$$

If $\sum_{i=1}^{k} u_i \in U$, *then each* $u_i \in U$ *for* $i = 1, 2, \ldots, k$.

Proof Set

$$\bar{H} = \{h \in H \mid \alpha_1(h) = \alpha_2(h) = \cdots = \alpha_k(h) = 0\}. \tag{1.3.20}$$

Then

$$\dim H/\bar{H} \le k. \tag{1.3.21}$$

Take a subspace H_1 of H such that

$$H = H_1 \oplus \bar{H}, \tag{1.3.22}$$

which implies $\dim H_1 \le k$, and $\{\alpha_1|_{H_1}, \alpha_2|_{H_1}, \ldots, \alpha_k|_{H_1}\}$ are distinct. Replacing H by H_1, we can assume that $\dim H$ is finite.

We prove it by induction on $\dim H$. The lemma is equivalent to Lemma 1.3.1 when $\dim H = 1$. Suppose that the lemma holds for $\dim H \le r$. Assume $\dim H = r + 1$. We take a nonzero vector $T \in H$ and an r-dimensional subspace $\tilde{H}$ of H such that

$$H = \mathbb{F}T + \tilde{H}. \tag{1.3.23}$$

Suppose that $\{\lambda_1, \lambda_2, \ldots, \lambda_s\}$ is the maximal subset of distinct elements in $\{\alpha_1(T), \alpha_2(T), \ldots, \alpha_k(T)\}$. Denote

$$J_i = \{j \in \{1, 2, \ldots, k\} \mid \alpha_j(T) = \lambda_i\} \tag{1.3.24}$$

and set

$$v_i = \sum_{j \in J_i} u_j. \tag{1.3.25}$$

Then

$$\sum_{r=1}^{s} v_r = \sum_{i=1}^{k} u_i \in U, \quad T(v_r) = \lambda_r v_r. \tag{1.3.26}$$

By Lemma 1.3.1, we have

$$v_1, v_2, \ldots, v_s \in U. \tag{1.3.27}$$

Since $\{\alpha_1, \alpha_2, \ldots, \alpha_k\}$ are distinct, $\{\alpha_j|_{\tilde{H}} \mid j \in J_i\}$ are distinct. By induction,

$$u_j \in U \qquad \text{for } j \in J_i. \tag{1.3.28}$$

Thus the lemma holds. □

For two integers m_1, m_2, we use the notation:

$$\overline{m_1, m_2} = \begin{cases} \{m_1, m_1+1, \ldots, m_2\} & \text{if } m_1 \leq m_2 \\ \emptyset & \text{if } m_1 > m_2. \end{cases} \tag{1.3.29}$$

Example 1.3.3 Let $n > 2$. The Lie algebra

$$sl(n, \mathbb{F}) = \sum_{1 \leq i < j \leq n} (\mathbb{F}E_{i,j} + \mathbb{F}E_{j,i}) + \sum_{i=1}^{n-1} \mathbb{F}(E_{i,i} - E_{i+1,i+1}). \tag{1.3.30}$$

Set

$$H = \sum_{i=1}^{n-1} \mathbb{F}(E_{i,i} - E_{i+1,i+1}) = \left\{ \sum_{i=1}^{n} a_i E_{i,i} \mid a_i \in \mathbb{F}, \sum_{i=1}^{n} a_i = 0 \right\}. \tag{1.3.31}$$

For $i \in \overline{1, n}$, we define $\epsilon_i \in H^*$ by

$$\epsilon_i \left(\sum_{j=1}^{n} a_j E_{j,j} \right) = a_i \qquad \text{for } \sum_{j=1}^{n} a_j E_{j,j} \in H. \tag{1.3.32}$$

Since

$$\left[\sum_{r=1}^{n} a_r E_{r,r}, E_{i,j} \right] = (a_i - a_j) E_{i,j} \qquad \text{for } \sum_{r=1}^{n} a_r E_{r,r} \in H, \tag{1.3.33}$$

we have

$$[h, E_{i,j}] = (\epsilon_i - \epsilon_j)(h) E_{i,j} \qquad \text{for } h \in H, \ i, j \in \overline{1, n}. \tag{1.3.34}$$

Let $\mathcal{I}$ be a nonzero ideal of $sl(n, \mathbb{F})$. Since $\{0, \epsilon_i - \epsilon_j \mid i \neq j\}$ are distinct linear functions on H, we have some $E_{i,j} \in \mathcal{I}$ with $i \neq j$ or $0 \neq h \in \mathcal{I} \bigcap H$ by Lemma 1.3.2, (1.3.31), (1.3.32) and (1.3.34). Since the center

$$Z(gl(n, \mathbb{F})) = \mathbb{F} \sum_{i=1}^{n} E_{i,i} \not\subset sl(n, \mathbb{F}), \tag{1.3.35}$$

the case $0 \neq h \in \mathcal{I} \bigcap H$ imply that there exists some $E_{r,s}$ such that

$$0 \neq [h, E_{r,s}] = (\epsilon_r - \epsilon_s)(h) E_{r,s} \in \mathcal{I}. \tag{1.3.36}$$

Thus we can always assume some $E_{i,j} \in \mathcal{I}$ with $i \neq j$.

For any $i, j \neq l \in \overline{1,n}$, we have

$$[E_{l,i}, E_{i,j}] = E_{l,j}, \ [E_{i,j}, E_{j,l}] = E_{i,l} \in \mathcal{I}. \tag{1.3.37}$$

Furthermore, if $i, j, l \neq k \in \overline{1,n}$,

$$[E_{l,j}, E_{j,k}] = E_{l,k} \in \mathcal{I}. \tag{1.3.38}$$

Observe

$$[E_{i,j}, E_{j,i}] = E_{i,i} - E_{j,j} \in \mathcal{I}. \tag{1.3.39}$$

So

$$E_{j,i} = \frac{1}{2}[E_{j,i}, E_{i,i} - E_{j,j}] \in \mathcal{I}. \tag{1.3.40}$$

Similarly as (1.3.37), we have

$$E_{l,i}, E_{j,l} \in \mathcal{I}. \tag{1.3.41}$$

Thus

$$E_{r,s} \in \mathcal{I} \qquad \text{for } \ r, s \in \overline{1,n}, \ r \neq s. \tag{1.3.42}$$

Finally,

$$[E_{i,i+1}, E_{i+1,i}] = E_{i,i} - E_{i+1,i+1} \in \mathcal{I}. \tag{1.3.43}$$

Hence $\mathcal{I} = sl(n, \mathbb{F})$. Therefore, $sl(n, \mathbb{F})$ is simple.

We leave reader to prove that the symplectic algebra $sp(2k, \mathbb{F})$ is simple when $k > 1$.

1.4 Homomorphisms

In this section, we discuss homomorphisms between Lie algebras.

A *homomorphism* ν from a Lie algebra $\mathscr{G}_1$ to a Lie algebra $\mathscr{G}_2$ is a linear map such that

$$\nu([u, v]) = [\nu(u), \nu(v)] \qquad \text{for } \ u, v \in \mathscr{G}_1. \tag{1.4.1}$$

Immediately, we have

$$\nu(\mathscr{G}_1) = \{\nu(u) \mid u \in \mathscr{G}_1\} \text{ forms a subalgebra of } \mathscr{G}_2. \tag{1.4.2}$$

Set

$$\ker \nu = \{u \in \mathscr{G}_1 \mid \nu(u) = 0\}. \tag{1.4.3}$$

If $u \in \ker \nu$, then

$$\nu([u, v]) = [\nu(u), \nu(v)] = [0, \nu(v)] = 0 \qquad \text{for } v \in \mathscr{G}_1. \tag{1.4.4}$$

Thus $[u, v] \in \ker \nu$. So $\ker \nu$ is an ideal of $\mathscr{G}_1$. The map ν is called *injective* if $\ker \nu = \{0\}$ and surjective if $\nu(\mathscr{G}_1) = \mathscr{G}_2$. Moreover, ν is called an *isomorphism* if it is both injective and surjective (i.e. *bijective*). In this case, we say that $\mathscr{G}_1$ is *isomorphic* to $\mathscr{G}_2$ via ν, denoted as

$$\mathscr{G}_1 \overset{\nu}{\cong} \mathscr{G}_2. \tag{1.4.5}$$

Isomorphic Lie algebras are viewed as the same algebra.

Example 1.4.1 Note that

$$[\partial_x, x^2\partial_x] = 2x\partial_x, \quad [2x\partial_x, \partial_x] = -2\partial_x, \quad [2x\partial_x, x^2\partial_x] = 2x^2\partial_x \tag{1.4.6}$$

Thus

$$\mathbb{F}\partial_x + \mathbb{F}x\partial_x + \mathbb{F}x^2\partial_x \tag{1.4.7}$$

forms a Lie algebra isomorphic to $sl(2, \mathbb{F})$ with the isomorphism ν given by:

$$\nu(\partial_x) = -E_{2,1}, \quad \nu(2x\partial_x) = E_{1,1} - E_{2,2}, \quad \nu(x^2\partial_x) = E_{1,2}. \tag{1.4.8}$$

In general, the space

$$\sum_{i,j=1}^{n} \mathbb{F}x_i\partial_{x_j} \tag{1.4.9}$$

forms a Lie algebra isomorphic to $gl(n, \mathbb{F})$ with the isomorphism τ defined by:

$$\tau(x_i\partial_j) = E_{i,j} \qquad \text{for } i, j \in \overline{1, n}. \tag{1.4.10}$$

Example 1.4.2 The symplectic Lie algebra

$$\begin{aligned} sp(2n, \mathbb{F}) = &\sum_{i,j=1}^{n} \mathbb{F}(E_{i,j} - E_{n+j,n+i}) + \sum_{i=1}^{n}(\mathbb{F}E_{i,n+i} + \mathbb{F}E_{n+i,i}) \\ &+ \sum_{1 \leq i < j \leq n} [\mathbb{F}(E_{i,n+j} + E_{j,n+i}) + \mathbb{F}(E_{n+i,j} + E_{n+j,i})]. \end{aligned} \tag{1.4.11}$$

View the elements of $\mathbb{F}[x_1, x_2, \ldots, x_n]$ also as the multiplication operators on $\mathbb{F}[x_1, x_2, \ldots, x_n]$. We have the following Lie subalgebra

$$\sum_{i,j=1}^{n} (\mathbb{F}(x_i\partial_{x_j} + \delta_{i,j}/2) + \mathbb{F}x_ix_j + \mathbb{F}\partial_{x_i}\partial_{x_j}) \tag{1.4.12}$$

of End $\mathbb{F}[x_1, x_2, \ldots, x_n]$, which is isomorphic to $sp(2n, \mathbb{F})$ with the isomorphism ν defined by

$$\nu(x_i\partial_{x_j} + \delta_{i,j}/2) = E_{i,j} - E_{n+j,n+i}, \tag{1.4.13}$$

$$\nu(\partial_{x_i}\partial_{x_j}) = E_{n+i,j} + E_{n+j,i}, \quad \nu(x_ix_j) = -(E_{i,n+j} + E_{j,n+i}) \tag{1.4.14}$$

for $i, j \in \overline{1, n}$.

A homomorphism from a Lie algebra $\mathcal{G}$ to $gl(V)$ is called a *representation* on V. The isomorphisms in the above examples are representations of the Lie algebras on polynomial algebras. For any element u in Lie algebra $\mathcal{G}$, we define $\text{ad}\, u \in gl(\mathcal{G})$ by

$$\text{ad}\, u(v) = [u, v] \qquad \text{for } v \in \mathcal{G}. \tag{1.4.15}$$

Then Jacobi identity implies

$$\text{ad}\, u([v, w]) = [\text{ad}\, u\, (v), w] + [v, \text{ad}\, u\, (w)] \qquad \text{for } v, w \in \mathcal{G}. \tag{1.4.16}$$

Thus ad u is a derivation of $\mathcal{G}$, which is called an *inner derivation* of $\mathcal{G}$. Moreover,

$$\begin{aligned}[\text{ad}\, u, \text{ad}\, v](w) &= (\text{ad}\, u\ \text{ad}\, v - \text{ad}\, v\ \text{ad}\, u)(w) = \text{ad}\, (u)([v, w]) - \text{ad}\, v([u, w]) \\ &= [u, [v, w]] - [v, [u, w]] = [u, [v, w]] + [v, [w, u]] \\ &= -[w, [u, v]] = [[u, v], w] = \text{ad}\, [u, v](w). \end{aligned} \tag{1.4.17}$$

Hence

$$[\text{ad}\, u, \text{ad}\, v] = \text{ad}\, [u, v]. \tag{1.4.18}$$

Thus ad is a representation of $\mathcal{G}$ on $\mathcal{G}$, which is called the *adjoint representation* of $\mathcal{G}$. Furthermore,

$$\ker \text{ad} = Z(\mathcal{G}). \tag{1.4.19}$$

We leave the reader to show that ad $\mathcal{G}$ is an ideal of Der $\mathcal{G}$.

Suppose that $\mathcal{I}$ is an ideal of a Lie algebra $\mathcal{G}$. Define the cosets

$$u + \mathcal{I} = \{u + v \mid v \in \mathcal{I}\} \qquad \text{for } u \in \mathcal{G}. \tag{1.4.20}$$

The quotient space

$$\mathscr{G}/\mathcal{I} = \{u + \mathcal{I} \mid u \in \mathscr{G}\}. \tag{1.4.21}$$

Define an algebraic operation $[\cdot, \cdot]$ on $\mathscr{G}/\mathcal{I}$ by

$$[u + \mathcal{I}, v + \mathcal{I}] = [u, v] + \mathcal{I} \qquad \text{for } u, v \in \mathscr{G}, \tag{1.4.22}$$

which is independent of the choices of representative elements in the cosets. Then $(\mathscr{G}/\mathcal{I}, [\cdot, \cdot])$ forms a Lie algebra. If ν is a homomorphism from a Lie algebra $\mathscr{G}_1$ to a Lie algebra $\mathscr{G}_2$, then

$$\mathscr{G}_1/\ker \nu \cong \nu(\mathscr{G}_1). \tag{1.4.23}$$

Let $\mathscr{L}$ be a subalgebra of a Lie algebra $\mathscr{G}$ and let $\mathcal{I}$ be an ideal of $\mathscr{G}$. Then $\mathscr{L} + \mathcal{I}$ forms a Lie subalgebra and $\mathcal{I}$ is an ideal of $\mathscr{L} + \mathcal{I}$. Moreover,

$$(\mathscr{L} + \mathcal{I})/\mathcal{I} \cong \mathscr{L}/(L \bigcap \mathcal{I}). \tag{1.4.24}$$

Example 1.4.3 In Example 1.3.1,

$$\mathscr{K}_n/\mathcal{I}_k \cong \mathscr{K}_{n-k-1} \tag{1.4.25}$$

through the isomorphism:

$$\partial_{x_1} + \mathcal{I}_k \to \partial_{x_1}, \quad \frac{1}{i!}x_1^i\partial_{x_2} + \mathcal{I}_k \to \frac{1}{(i-k-1)!}x_1^{i-k-1}\partial_{x_2}, \quad i \in \overline{k+1, n}. \tag{1.4.26}$$

An isomorphism from a Lie algebra $\mathscr{G}$ to itself is called an *automorphism*. Suppose that ∂ is a derivation of a Lie algebra $\mathscr{G}$ and e^{∂} is well defined (e.g., ∂ is nilpotent). Note

$$\begin{aligned} e^{\partial}([u, v]) &= \sum_{n=0}^{\infty} \frac{1}{n!}\partial^n([u, v]) = \sum_{n=0}^{\infty} \frac{1}{n!} \sum_{i=0}^{n} \binom{n}{i} [\partial^i(u), \partial^{n-i}(v)] \\ &= \sum_{n=0}^{\infty} \frac{1}{n!} \sum_{i=0}^{n} \frac{n!}{i!(n-i)!}[\partial^i(u), \partial^{n-i}(v)] = \sum_{n=0}^{\infty}\sum_{i=0}^{n} \frac{1}{i!(n-i)!}[\partial^i(u), \partial^{n-i}(v)] \\ &= \left[\sum_{i=0}^{\infty} \frac{1}{i!}\partial^i(u), \sum_{j=0}^{\infty} \frac{1}{j!}\partial^j(v)\right] = [e^{\partial}(u), e^{\partial}(v)]. \end{aligned} \tag{1.4.27}$$

So e^{∂} is an automorphism of $\mathscr{G}$. If $e^{\mathrm{ad}\, u}$ is defined for $u \in \mathscr{G}$, we called $e^{\mathrm{ad}\, u}$ an *inner automorphism*.

Let $\mathscr{A}$ be an associative algebra and let u be an element in $\mathscr{A}$ such that e^u is well defined (e.g., u is nilpotent). Define the left multiplication operator L_u and the right multiplication operator R_u by

$$L_u(v) = uv, \qquad R_u(v) = vu \qquad \text{for } v \in \mathscr{A}. \tag{1.4.28}$$

Thus

$$\text{ad}\, u(v) = uv - vu = (L_u - R_u)(v) \qquad \text{for } v \in \mathscr{A}. \tag{1.4.29}$$

Moreover,

$$L_u R_u = R_u L_u \Longrightarrow e^{\text{ad}\, u} = e^{L_u - R_u} = e^{L_u} e^{-R_u}. \tag{1.4.30}$$

Thus

$$e^{\text{ad}\, u}(v) = e^{L_u} e^{-R_u}(v) = e^u v e^{-u} \tag{1.4.31}$$

(A very useful formula!).

Let $n = 2k$ be an even integer. We define a map $* : M_{n\times n}(\mathbb{F}) \to M_{n\times n}(\mathbb{F})$ by

$$A^* = \begin{pmatrix} 0 & I_k \\ I_k & 0 \end{pmatrix} A^t \begin{pmatrix} 0 & I_k \\ I_k & 0 \end{pmatrix} \qquad \text{for } A \in M_{n\times n}(\mathbb{F}). \tag{1.4.32}$$

Then $*$ is an involutive anti-isomorphism. Thus we have a Lie subalgebra

$$\begin{aligned} o'(2k, \mathbb{F}) &= \{A \in M_{n\times n}(\mathbb{F}) \mid A^* = -A\} \\ &= \left\{ \begin{pmatrix} A & B \\ C & -A^t \end{pmatrix} \mid A, B, C \in M_{k\times k}(\mathbb{F}),\ B^t = -B, C^t = -C \right\}, \end{aligned} \tag{1.4.33}$$

which is simple when $k > 2$, and is called a *Lie algebra of type* D_k for $k > 3$. The algebra $o'(2k, \mathbb{R})$ is the Lie algebra of the Lie group preserving the metric defined by the symmetric matrix

$$\begin{pmatrix} 0 & I_k \\ I_k & 0 \end{pmatrix}. \tag{1.4.34}$$

Suppose $n = 2k + 1$. We define a map $* : M_{n\times n}(\mathbb{F}) \to M_{n\times n}(\mathbb{F})$ by

$$A^* = \begin{pmatrix} 1 & 0 & 0 \\ 0 & 0 & I_k \\ 0 & I_k & 0 \end{pmatrix} A^t \begin{pmatrix} 1 & 0 & 0 \\ 0 & 0 & I_k \\ 0 & I_k & 0 \end{pmatrix} \qquad \text{for } A \in M_{n\times n}(\mathbb{F}). \tag{1.4.35}$$

Then $*$ is an involutive anti-isomorphism. Thus we have a Lie subalgebra

$$o'(2k+1,\mathbb{F}) = \{A \in M_{n\times n}(\mathbb{F}) \mid A^* = -A\} = \{\begin{pmatrix} 0 & -\vec{b}^t & -\vec{a}^t \\ \vec{a} & A & B \\ \vec{b} & C & -A^t \end{pmatrix}$$

$$\mid \vec{a},\vec{b} \in M_{k\times 1}(\mathbb{F});\ A, B, C \in M_{k\times k}(\mathbb{F}),\ B^t = -B, C^t = -C\}, \quad (1.4.36)$$

is simple when $k > 1$, and is called a *Lie algebra of type* B_k for $k > 1$. The algebra $o'(2k+1,\mathbb{R})$ is the Lie algebra of the Lie group preserving the metric defined by the symmetric matrix:

$$\begin{pmatrix} 1 & 0 & 0 \\ 0 & 0 & I_k \\ 0 & I_k & 0 \end{pmatrix}. \quad (1.4.37)$$

Let $\mathscr{G}_1$ and $\mathscr{G}_2$ be Lie algebras. The Lie bracket on

$$\mathscr{G}_1 \oplus \mathscr{G}_2 = \{(u_1, u_2) \mid u_1 \in \mathscr{G}_1,\ u_2 \in \mathscr{G}_2\} \quad (1.4.38)$$

is defined by

$$[(u_1, u_2), (v_1, v_2)] = ([u_1, v_1], [u_2, v_2]) \qquad u_1, v_1 \in \mathscr{G}_1,\ u_2, v_2 \in \mathscr{G}_2. \quad (1.4.39)$$

We leave the reader to prove

$$o(2k,\mathbb{C}) \cong o'(2k,\mathbb{C}), \qquad o(2k+1,\mathbb{C}) \cong o'(2k+1,\mathbb{C}). \quad (1.4.40)$$

Thereby, prove

$$o(3,\mathbb{C}) \cong sl(2,\mathbb{C}),\ \ o(4,\mathbb{C}) \cong sl(2,\mathbb{C}) \oplus sl(2,\mathbb{C}),\ \ o(6,\mathbb{C}) \cong sl(4,\mathbb{C}). \quad (1.4.41)$$

1.5 Nilpotent and Solvable Lie Algebras

In this section, we define nilpotent and solvable Lie algebras. Moreover, we prove the Engel's theorem, Lie's theorem, Jordan-Chevalley decomposition theorem and Cartan's criterion of solvability.

Let $\mathscr{G}$ be a Lie algebra. The *central series* of $\mathscr{G}$ is defined as: $\mathscr{G}^0 = \mathscr{G},\ \mathscr{G}^1 = [\mathscr{G},\mathscr{G}], \mathscr{G}^{i+1} = [\mathscr{G},\mathscr{G}^i]$. Then

$$\mathscr{G}^0 \supset \mathscr{G}^1 \supset \mathscr{G}^2 \supset \cdots \supset \mathscr{G}^i \supset \cdots \quad (1.5.1)$$

forms a descending series of ideals of $\mathcal{G}$, invariant under any automorphism τ; that is,

$$\tau(\mathcal{G}^i) = \mathcal{G}^i. \tag{1.5.2}$$

If $\mathcal{G}^n = \{0\}$ for some positive integer n, then we say that $\mathcal{G}$ is *nilpotent*. For any vector space V, we define a trivial Lie bracket on V by

$$[u, v] = 0 \qquad \text{for } u, v \in V. \tag{1.5.3}$$

We call $(V, [\cdot, \cdot])$ a *trivial Lie algebra* or *abelian Lie algebra*. It is obviously a nilpotent Lie algebra.

Example 1.5.1 The *Heisenberg Lie algebra* $\mathcal{H}_n$ is a vector space with a basis $\{u_i, v_i, \kappa \mid i = 1, 2, \ldots, n\}$ whose Lie bracket is determined by

$$[u_i, v_j] = \delta_{i,j}\kappa, \quad [u_i, u_j] = [v_i, v_j] = [\kappa, u_i] = [\kappa, v_i] = 0. \tag{1.5.4}$$

This algebra is the symmetry of harmonic oscillators in physics. Note

$$\mathcal{H}^1 = \mathbb{F}\kappa, \qquad \mathcal{H}_n^2 = \{0\}. \tag{1.5.5}$$

So it is a nilpotent Lie algebra.

Example 1.5.2 Recall the Lie algebra $\mathcal{K}_n$ defined in Example 1.3.1. We have

$$\mathcal{K}_n^i = \mathcal{I}_{n-i} \tag{1.5.6}$$

(cf. (1.3.7)). In particular,

$$\mathcal{K}^n = \mathbb{F}\partial_{x_2}, \qquad \mathcal{K}^{n+1} = \{0\}. \tag{1.5.7}$$

Thus $\mathcal{K}_n$ is a nilpotent Lie algebra.

Example 1.5.3 The subspace of strict up-triangular matrices

$$\mathcal{N}_n = \left\{ \begin{pmatrix} 0 & a_{1,2} & a_{1,3} & \cdots & a_{1,n} \\ 0 & 0 & a_{2,3} & \ddots & \vdots \\ \vdots & \ddots & \ddots & \ddots & a_{n-2,n} \\ 0 & \cdots & 0 & 0 & a_{n-1,n} \\ 0 & \cdots & 0 & 0 & 0 \end{pmatrix} \mid a_{i,j} \in \mathbb{F} \right\} = \sum_{1 \le i < j \le n} \mathbb{F}E_{i,j} \tag{1.5.8}$$

forms a Lie subalgebra of $gl(n, \mathbb{F})$. Set

$$\mathcal{E}_r = \sum_{j-i>r} \mathbb{F}E_{i,j}. \tag{1.5.9}$$

Then

$$\mathscr{N}_n^r = \mathscr{E}_r. \tag{1.5.10}$$

In particular,

$$\mathscr{N}_n^{n-1} = \mathbb{F}E_{1,n}, \quad \mathscr{N}^n = \{0\}. \tag{1.5.11}$$

Thus $\mathscr{N}_n$ is a nilpotent Lie algebra.

Theorem 1.5.1 *Let $V \neq \{0\}$ be a finite-dimensional space and let $\mathscr{G}$ be a subalgebra of $gl(V)$, consisting of nilpotent linear transformations. There exists nonzero $v \in V$ such that $\mathscr{G}(v) = \{0\}$.*

Proof We prove the conclusion by induction on $\dim \mathscr{G}$. If $\dim \mathscr{G} = 1$, we write $\mathscr{G} = \mathbb{F}T$. Take any eigenvector v of T, we have $\mathscr{G}(v) = \{0\}$.

Assume that the theorem holds for $\dim \mathscr{G} \leq k$. Consider the case $\dim \mathscr{G} = k+1$. Take a maximal proper Lie subalgebra $\mathscr{K}$ of $\mathscr{G}$. For $u \in \mathscr{G}$, we want to prove that $\mathrm{ad}\, u$ is nilpotent on $gl(V)$. Note that

$$(\mathrm{ad} u)^n = (L_u - R_u)^n = \sum_{i=0}^{n} \binom{n}{i} (-1)^i L_u^{n-i} R_u^i = \sum_{i=0}^{n} \binom{n}{i} (-1)^i L_{u^{n-i}} R_{u^i} \tag{1.5.12}$$

by (1.4.28) and (1.4.29). If $u^m = 0$, then $(\mathrm{ad}\, u)^{2m-1} = 0$ by the above equation.

Define the action of $\mathscr{K}$ on the quotient space $\mathscr{G}/\mathscr{K}$ by

$$\mathrm{ad}\, \xi(u + \mathscr{K}) = [\xi, u] + \mathscr{K} \qquad \text{for } \xi \in \mathscr{K},\ u \in \mathscr{G}. \tag{1.5.13}$$

Then

$$\mathrm{ad}\, \mathscr{K}|_{\mathscr{G}/\mathscr{K}} \tag{1.5.14}$$

forms a Lie subalgebra of $gl(\mathscr{G}/\mathscr{K})$. If $\xi^m = 0$, then $(\mathrm{ad}\, \xi)^{2m-1} = 0$ as we have shown in the above. Furthermore,

$$\dim \mathrm{ad}\, \mathscr{K}|_{\mathscr{G}/\mathscr{K}} \leq \dim \mathscr{K} \leq k. \tag{1.5.15}$$

Thus there exists $\xi_0 \in \mathscr{G} \setminus \mathscr{K}$ such that

$$\mathrm{ad}\, \xi(\xi_0 + \mathscr{K}) = \mathscr{K} \qquad \text{for } \xi \in \mathscr{K}, \tag{1.5.16}$$

which implies

$$[\xi, \xi_0] \in \mathscr{K} \qquad \text{for } \xi \in \mathscr{K}. \tag{1.5.17}$$

Hence $\mathbb{F}\xi_0 + \mathscr{K}$ forms a Lie subalgebra of $\mathscr{G}$. Since $\mathscr{K}$ is maximal, we have

$$\mathscr{G} = \mathbb{F}\xi_0 + \mathscr{K}. \tag{1.5.18}$$

Set

$$V_1 = \{u \in V \mid \mathscr{K}(u) = \{0\}\}. \tag{1.5.19}$$

Then $V_1 \neq \{0\}$ by induction. For any $u \in V_1$ and $\xi \in \mathscr{K}$, we have

$$\xi\xi_0(u) = [\xi, \xi_0](u) + \xi_0\xi(u) = 0 \tag{1.5.20}$$

by (1.5.16); that is, $\xi_0(u) \in V_1$. Hence $\xi_0(V_1) \subset V_1$. Take any eigenvector v of $\xi_0|_{V_1}$ and we have $\mathscr{G}(v) = \{0\}$. □

The above theorem and induction show that $\mathscr{G}$ is isomorphic to a Lie algebra of strict up-triangular matrices (cf. (1.5.8)). Hence $\mathscr{G}$ is a nilpotent Lie algebra.

Theorem 1.5.2 (Engel) *If all the elements in a finite-dimensional Lie algebra $\mathscr{G}$ are ad nilpotent, then $\mathscr{G}$ is nilpotent.*

Proof By Theorem 1.5.1, ad $\mathscr{G}$ is nilpotent. Hence there exists a positive integer m such that $(\text{ad}\,\mathscr{G})^m = \text{ad}\,\mathscr{G}^m = \{0\}$. So $\mathscr{G}^m \in Z(\mathscr{G})$. Therefore, $\mathscr{G}^{m+1} = [\mathscr{G}, \mathscr{G}^m] \subset [\mathscr{G}, Z(\mathscr{G})] = \{0\}$. □

Corollary 1.5.3 *Let $\mathscr{G}$ be a finite-dimensional nilpotent Lie algebra and let $\mathcal{I}$ be a nonzero ideal of $\mathscr{G}$. Then $\mathcal{I} \bigcap Z(\mathscr{G}) \neq \{0\}$.*

Proof Note that ad $\mathscr{G}|_{\mathcal{I}}$ is a Lie algebra of nilpotent elements. Thus there exists $0 \neq v \in \mathcal{I}$ such that $[\mathscr{G}, v] = \{0\}$. So $0 \neq v \in \mathcal{I} \bigcap Z(\mathscr{G})$. □

The *derived series* of a Lie algebra $\mathscr{G}$ is defined as: $\mathscr{G}^{(0)} = \mathscr{G}$, $\mathscr{G}^{(1)} = [\mathscr{G}, \mathscr{G}]$, $\mathscr{G}^{(i+1)} = [\mathscr{G}^{(i)}, \mathscr{G}^{(i)}]$. The Lie algebra $\mathscr{G}$ is called *solvable* if $\mathscr{G}^{(n)} = \{0\}$ for some positive integer n. Since $\mathscr{G}^{(i)} \subset \mathscr{G}^i$, any nilpotent Lie algebra is solvable.

Properties:

(1) Let $\mathcal{I}$ be an ideal of $\mathscr{G}$. Then $\mathscr{G}$ is solvable if and only if $\mathcal{I}$ and $\mathscr{G}/\mathcal{I}$ are solvable.

Proof Suppose that $\mathcal{I}$ and $\mathscr{G}/\mathcal{I}$ are solvable. There exist positive integer m and n such that

$$\mathcal{I}^{(m)} = \{0\}, \qquad (\mathscr{G}/\mathcal{I})^{(n)} = \{0\}. \tag{1.5.21}$$

Thus $\mathscr{G}^{(n)} \subset \mathcal{I}$ and $\mathscr{G}^{(m+n)} \subset \mathcal{I}^{(m)} = \{0\}$.

(2) If $\mathcal{I}$ and $\mathcal{J}$ are solvable ideals of $\mathscr{G}$, so is $\mathcal{I} + \mathcal{J}$.

Proof This is because $(\mathcal{I}+\mathcal{J})/\mathcal{J}\cong\mathcal{I}/(\mathcal{I}\bigcap\mathcal{J})$ and $\mathcal{I}\bigcap\mathcal{J}$ are solvable.

We define the *radical* of $\mathscr{G}$ to be the unique maximal solvable ideal, denoted by $\operatorname{Rad}\mathscr{G}$. The Lie algebra $\mathscr{G}$ is called *semisimple* of $\operatorname{Rad}\mathscr{G}=\{0\}$.

Example 1.5.4 The *Diamond Lie algebra* is a vector space $\mathscr{G}_\diamond$ with a basis $\{h,u_+,u_-,\kappa\}$ whose Lie bracket is given by

$$[h,u_\pm]=\pm u_\pm,\quad [u_+,u_-]=\kappa,\quad [\mathscr{G}_\diamond,\kappa]=\{0\}. \tag{1.5.22}$$

Then

$$\mathscr{G}_\diamond^{(1)}=[\mathscr{G}_\diamond,\mathscr{G}_\diamond]=\mathbb{F}u_++\mathbb{F}u_-+\mathbb{F}\kappa, \tag{1.5.23}$$

$$\mathscr{G}_\diamond^{(2)}=\mathbb{F}\kappa,\qquad \mathscr{G}_\diamond^{(3)}=\{0\}. \tag{1.5.24}$$

So $\mathscr{G}_\diamond$ is a solvable Lie algebra. Moreover, we define a symmetric bilinear form on $\mathscr{G}_\diamond$ by

$$\langle u_+,u_-\rangle=\langle h,\kappa\rangle=1, \tag{1.5.25}$$

$$\langle u,v\rangle=0\qquad\text{for the other pairs } u,v\in\{h,u_+,v_-,\kappa\}. \tag{1.5.26}$$

Then $\langle\cdot,\cdot\rangle$ is invariant:

$$\langle[u,v],w\rangle=\langle u,[v,w]\rangle\qquad\text{for}\ \ u,v,w\in\mathscr{G} \tag{1.5.27}$$

(exercise). Thus the diamond Lie algebra $\mathscr{G}_\diamond$ is a solvable Lie algebra with a nondegenerate symmetric invariant bilinear form.

Example 1.5.5 The Lie algebra of up-triangular matrices:

$$\mathscr{B}_n=\left\{\begin{pmatrix} a_{1,1} & a_{1,2} & a_{1,3} & \cdots & a_{1,n}\\ 0 & a_{2,2} & a_{2,3} & \ddots & \vdots\\ \vdots & \ddots & \ddots & \ddots & a_{n-2,n}\\ 0 & \cdots & 0 & a_{n-1,n-1} & a_{n-1,n}\\ 0 & \cdots & 0 & 0 & a_{n,n}\end{pmatrix}\ \Big|\ a_{i,j}\in\mathbb{F}\right\}=\sum_{1\le i\le j\le n}\mathbb{F}E_{i,j} \tag{1.5.28}$$

forms a Lie subalgebra of $gl(n,\mathbb{F})$. Moreover, $[\mathscr{B}_n,\mathscr{B}_n]=\mathscr{N}_n$ is a nilpotent ideal (cf. Example 1.5.3). Thus $\mathscr{B}_n$ is a solvable Lie algebra.

Theorem 1.5.4 (Lie) *Suppose that $\mathbb{F}=\mathbb{C}$ and V is a nonzero finite-dimensional space. Let $\mathscr{G}$ be a solvable Lie subalgebra of $gl(V)$. There exists a common eigenvector for all the elements in $\mathscr{G}$. In particular, $\mathscr{G}$ is isomorphic to a Lie algebra of up-triangular matrices.*

Proof We prove it by induction on $\dim \mathscr{G}$. When $\dim \mathscr{G} = 1$, $\mathscr{G} = \mathbb{F}T$. Take any eigenvector v of T. Then v is a common eigenvector all the elements in $\mathscr{G}$.

Assume that the theorem hold for $\dim \mathscr{G} \leq k$. Consider the case $\dim \mathscr{G} = k+1$. Since $\mathscr{G}$ is solvable, $[\mathscr{G}, \mathscr{G}] = \mathscr{G}^{(1)} \neq \mathscr{G}$. Take any codimension-one subspace $\mathscr{K} \supset \mathscr{G}^{(1)}$ of $\mathscr{G}$. Then $\mathscr{K}$ is a solvable ideal of $\mathscr{G}$. By induction, there exists a common eigenvector v for all the elements in $\mathscr{K}$. Hence there exists a linear function α on $\mathscr{K}$ such that

$$\xi(v) = \alpha(\xi)v \qquad \text{for } \xi \in \mathscr{K}. \tag{1.5.29}$$

Therefore, the subspace

$$V_\alpha = \{v \in V \mid \xi(v) = \alpha(\xi)v \text{ for } \xi \in \mathscr{K}\} \neq \{0\}. \tag{1.5.30}$$

Write $\mathscr{G} = \mathbb{F}\xi_0 + \mathscr{K}$. For any $0 \neq w \in V_\alpha$, we have

$$\xi\xi_0(w) = [\xi, \xi_0](w) + \xi_0\xi(w) = \alpha(\xi)\xi_0(w) + \alpha([\xi, \xi_0])w. \tag{1.5.31}$$

We want to prove $\alpha([\xi, \xi_0]) = 0$. If $\xi_0(w) \in \mathbb{F}w$, we are done. Otherwise, since $\dim V$ is finite, there exists a positive integer n such that $\{w, \xi_0(w), \xi_0^2(w), \ldots, \xi_0^n(w)\}$ are linear independent and

$$\xi_0^{n+1}(w) \in W = \sum_{i=0}^{n} \mathbb{F}\xi_0^i(w). \tag{1.5.32}$$

Thus $\xi_0(W) \in W$. Set

$$W_i = \sum_{r=0}^{i} \mathbb{F}\xi_0^r(w). \tag{1.5.33}$$

For any $\xi \in \mathscr{K}$, we have $\xi(w) = \alpha(\xi)w$. Suppose

$$\xi(\xi_0^i(w)) \equiv \alpha(\xi)\xi_0^i(w) \pmod{W_{i-1}} \tag{1.5.34}$$

for some $i < n$. Now

$$\begin{aligned} \xi(\xi_0^{i+1}(w)) &= \xi\xi_0(\xi_0^i(w)) = [\xi, \xi_0](\xi_0^i(w)) + \xi_0\xi(\xi_0^i(w)) \\ &\equiv \alpha([\xi, \xi_0])\xi_0^i(w) + \alpha(\xi)\xi_0^{i+1}(w) \pmod{\xi_0(W_{i-1}) + W_{i-1}} \\ &\equiv \alpha(\xi)\xi_0^{i+1}(w) \pmod{W_i}. \end{aligned} \tag{1.5.35}$$

By induction, (1.5.34) holds for any $i \leq n$. Thus $\xi(W) \subset W$ and

$$\operatorname{Tr} \xi|_W = (n+1)\alpha(\xi) \qquad \text{for any } \xi \in \mathscr{K}. \tag{1.5.36}$$

Since $[\xi, \xi_0] \in \mathscr{K}$, we have

$$\mathrm{Tr}\,[\xi, \xi_0]|_W = (n+1)\alpha([\xi, \xi_0]) = \mathrm{Tr}\,[\xi|_W, \xi_0|_W] = 0; \tag{1.5.37}$$

that is, $\alpha([\xi, \xi_0]) = 0$. Therefore,

$$\xi(\xi_0 w) = \alpha(\xi)\xi_0 w \tag{1.5.38}$$

for $\xi \in \mathscr{K}$ by (1.5.31). So $\xi_0(V_\alpha) \subset V_\alpha$. Any eigenvector of $\xi_0|_{V_\alpha}$ is a common eigenvector of the elements in $\mathscr{G}$. □

The above theorem shows that $\mathscr{G}$ becomes a Lie subalgebra of up-triangular matrices under a certain basis. Thus $[\mathscr{G}, \mathscr{G}]$ is a Lie algebra of strict up-triangular matrices. Hence $[\mathscr{G}, \mathscr{G}]$ is nilpotent.

Let V be a vector space with a basis $\{v_i \mid i \in \mathbb{Z}\}$. Define $T \in gl(V)$ by

$$T(v_i) = v_{i+1} \qquad \text{for } i \in \mathbb{Z}. \tag{1.5.39}$$

Then T does not have any eigenvector (exercise). So Lie's theorem fail if we allow $\dim V = \infty$.

Corollary 1.5.5 *If $\mathscr{G}$ is a finite-dimensional solvable Lie algebra over $\mathbb{C}$. Then $[\mathscr{G}, \mathscr{G}]$ is nilpotent.*

Proof Note ad $\mathscr{G}$ is a solvable Lie subalgebra of $gl(\mathscr{G})$. Thus

$$\mathrm{ad}\,[\mathscr{G}, \mathscr{G}] = [\mathrm{ad}\,\mathscr{G}, \mathrm{ad}\,\mathscr{G}] \tag{1.5.40}$$

is a nilpotent Lie subalgebra. Since ker ad $= Z(\mathscr{G})$, $[\mathscr{G}, \mathscr{G}]$ is nilpotent. □

An element $T \in \mathrm{End}\, V$ is called *semisimple* if it can be diagonalized under a certain basis of V.

Theorem 1.5.6 (Jordan-Chevalley Decomposition) *Suppose $\mathbb{F} = \mathbb{C}$. Let V be a finite-dimensional vector space. Given $T \in \mathrm{End}\, V$. There exist unique semisimple element T_s and nilpotent element T_n in End V such that $T = T_s + T_n$ and $[T_s, T_n] = 0$. Moreover, T_s and T_n can be written as polynomials of T without constant.*

Proof Suppose that $\{\lambda_1, \lambda_2, \ldots, \lambda_k\}$ are distinct eigenvalues of T. Then the characteristic polynomial is of the form $\prod_{i=1}^k (t - \lambda_i)^{n_i}$. Set

$$V_i = \{u \in V \mid (T - \lambda_i)^{n_i}(u) = 0\}. \tag{1.5.41}$$

We have

$$V = \bigoplus_{i=1}^k V_i. \tag{1.5.42}$$

Suppose that 0 is an eigenvalue of T. Set

$$f_i(t) = \sum_{i \neq j} (t - \lambda_j)^{n_j}. \tag{1.5.43}$$

Then the maximal common divisor of $\{f_1(t), f_2(t), \ldots, f_k(t)\}$ is 1. By long division, there exists polynomials $\{h_1(t), h_2(t), \ldots, h_k(t)\}$ such that

$$\sum_{i=1}^{t} h_i(t) f_i(t) = 1. \tag{1.5.44}$$

Hence

$$h_i(t) f_i(t) \equiv \delta_{i,j} \pmod{(t - \lambda_j)^{n_j}}. \tag{1.5.45}$$

Let

$$P(t) = \sum_{i=1}^{k} \lambda_i h_i(t) f_i(t). \tag{1.5.46}$$

By (1.5.46), we get

$$P(t) \equiv \lambda_i \pmod{(t - \lambda_i)^{n_i}}. \tag{1.5.47}$$

If 0 is not an eigenvalue, then we set

$$f_0(t) = \prod_{j=1}^{k} (t - \lambda_j)^{n_j}, \quad f_i(t) = t \sum_{i \neq j} (t - \lambda_j)^{n_j}, \qquad 0 < i \leq k. \tag{1.5.48}$$

By long division, there exists polynomials $\{h_0, h_1(t), h_2(t), \ldots, h_k(t)\}$ such that

$$\sum_{i=0}^{k} h_i(t) f_i(t) = 1. \tag{1.5.49}$$

Now (1.5.45) holds. We define $P(t)$ by (1.5.46) again. Equation (1.5.48) holds. In any case, we always have $P(0) = 0$.

Set

$$T_s = P(T), \qquad T_n = T - P(T). \tag{1.5.50}$$

Note $P(T)(V_i) \subset V_i$ and

$$T_s|_{Vi} = \lambda_i \mathrm{Id}_{V_i}. \tag{1.5.51}$$

Denote

$$m = \max\{n_1, n_2, \ldots, n_k\}. \tag{1.5.52}$$

Then

$$T_n^m = (T - T_s)^m = 0 \tag{1.5.53}$$

by (1.5.41), (1.5.42) and (1.5.51).

Suppose that $T = A + B$ such that A is a semisimple element in $gl(V)$, B is a nilpotent element in $gl(V)$ and $AB = BA$. Note $AT = A(A + B) = (A + B)A = TA$, which implies $BT = TB$. Thus $T_s A = AT_s$ and $T_n B = BT_n$. Since both T_s and A are diagonalizable and they commute, $T_s - A$ is semsimple. On the other hand, $B - T_n$ is nilpotent. Thus

$$T_s + T_n = A + B \Longrightarrow T_s - A = B - T_n \Longrightarrow T_s - A = B - T_n = 0, \tag{1.5.54}$$

because it is both semisimple and nilpotent. □

Corollary 1.5.7 *If $T = T_s + T_n$ is the Jordan-Chevalley decomposition, so is $ad T = ad\, T_s + ad\, T_n$.*

Proof This is because ad T_s is semisimple, ad T_n is nilpotent and $[\text{ad}\, T_s, \text{ad}\, T_n] = \text{ad}\, [T_s, T_n] = 0$. □

Lemma 1.5.1 *If $(\mathscr{A}, \circ)$ is any finite-dimensional algebra, the semisimple and nilpotent parts of an element in Der $\mathscr{A}$ are also in Der $\mathscr{A}$.*

Proof Let $T \in \text{Der}\, \mathscr{A}$. For any $\lambda \in \mathbb{F}$, we define

$$\mathscr{A}_\lambda = \{u \in \mathscr{A} \mid (T - \lambda)^n(u) = 0 \text{ for some positive integer } n\}. \tag{1.5.55}$$

Then

$$\mathscr{A} = \bigoplus \mathscr{A}_\lambda, \qquad T_s|_{\mathscr{A}_\lambda} = \lambda \text{Id}_{\mathscr{A}_\lambda}. \tag{1.5.56}$$

For $u \in \mathscr{A}_\lambda$ and $v \in \mathscr{A}_\mu$, there exist positive integers m and n such that

$$(T - \lambda)^m(u) = (T - \mu)^n(v) = 0. \tag{1.5.57}$$

Since T is a derivation, we have

$$\begin{aligned}(T - \lambda - \mu)(u \circ v) &= T(u \circ v) - (\lambda + \mu)u \circ v \\ &= T(u) \circ v + u \circ T(v) - (\lambda + \mu)u \circ v = (T - \lambda)(u) \circ v + u \circ (T - \mu)(v).\end{aligned} \tag{1.5.58}$$

The Leibniz rule shows

$$(T-\lambda-\mu)^{m+n}(u\circ v)=\sum_{i=0}^{m+n}\binom{m+n}{i}(T-\lambda)^i(u)\circ(T-\mu)^{m+n-i}(v)=0. \tag{1.5.59}$$

Thus $u\circ v\in\mathscr{A}_{\lambda+\mu}$. Hence

$$T_s(u\circ v)=(\lambda+\mu)u\circ v=(\lambda u)\circ v+u\circ(\mu v)=T_s(u)\circ v+u\circ T_s(v) \tag{1.5.60}$$

by (1.5.56). Hence T_s is a derivation and so is $T_n=T-T_s$. □

In the rest of this section, we assume that $\mathbb{F}=\mathbb{C}$. The trace of linear transformation is trace of its corresponding matrix under a basis.

Lemma 1.5.2 *Let V be a finite-dimensional space and let $\mathscr{A}\subset\mathscr{B}$ be two subspaces of End V. Set $M=\{T\in gl(V)\mid[T,\mathscr{B}]\subset\mathscr{A}\}$. Fix $\xi\in M$. If Tr $\xi T'=0$ for any $T'\in M$, then ξ is nilpotent.*

Proof Let $\xi=\xi_s+\xi_n$ be the Jordan-Chevalley decomposition. Under a certain basis $\{v_1,v_2,\ldots,v_m\}$, $\xi_s=\text{diag}(\lambda_1,\lambda_2,\ldots,\lambda_m)$ and $gl(V)=M_{m\times m}(\mathbb{C})$. Set

$$H=\sum_{i=1}^m\mathbb{Q}\lambda_i, \tag{1.5.61}$$

a subspace of $\mathbb{C}$ over $\mathbb{Q}$. Let $f:H\to\mathbb{Q}$ be any $\mathbb{Q}$-linear map and define

$$T_0=\text{diag}(f(\lambda_1),f(\lambda_2),\ldots,f(\lambda_m)). \tag{1.5.62}$$

We want to prove $T_0\in M$. Note

$$\text{ad}\,\xi_s(E_{i,j})=(\lambda_i-\lambda_j)E_{i,j},\qquad \text{ad}\,T_0(E_{i,j})=(f(\lambda_i)-f(\lambda_j))E_{i,j}. \tag{1.5.63}$$

Suppose that $\{a_1,a_2,\ldots,a_s\}$ is the maximal set of distinct elements in $\{\lambda_i-\lambda_j\mid 1\le i,j\}$. Define

$$\zeta(t)=\sum_{i=1}^s f(a_i)\frac{\prod_{i\neq j\in\{1,2,\ldots,s\}}(t-a_j)}{\prod_{i\neq j\in\{1,2,\ldots,s\}}(a_i-a_j)}. \tag{1.5.64}$$

Then

$$\zeta(a_i)=f(a_i). \tag{1.5.65}$$

Since $0\in\{a_1,a_2,\ldots,a_s\}$, $t|\zeta(t)$. Furthermore,

$$\zeta(\lambda_i-\lambda_j)=f(\lambda_i-\lambda_j)=f(\lambda_i)-f(\lambda_j). \tag{1.5.66}$$

Thus

$$\text{ad } T_0 = \zeta(\text{ad } \xi_s). \tag{1.5.67}$$

By Theorem 1.5.6 and Corollary 1.5.7, ad ξ_s is a polynomial of ad ξ without constant. So is ad T_0. Therefore, $T_0 \in M$.

Now

$$\text{Tr } T_0\xi = \sum_{i=1}^{m} f(\lambda_i)\lambda_i = 0 \tag{1.5.68}$$

by our assumption. Hence

$$0 = f(0) = f\left(\sum_{i=1}^{m} f(\lambda_i)\lambda_i\right) = \sum_{i=1}^{m} f(\lambda_i)f(\lambda_i) = \sum_{i=1}^{m} f(\lambda_i)^2. \tag{1.5.69}$$

Thus all $f(\lambda_i) = 0$. Since f is any linear function, we have $H = \{0\}$ or equivalently, $\xi_s = 0$. □

Theorem 1.5.8 (Cartan's Criterion) *Let V be a finite-dimensional space and let $\mathscr{G}$ be a subalgebra of $gl(V)$. If Tr $uv = 0$ for any $u \in [\mathscr{G}, \mathscr{G}]$ and $v \in \mathscr{G}$, then $\mathscr{G}$ is solvable.*

Proof In the above theorem, we let $\mathscr{A} = [\mathscr{G}, \mathscr{G}]$ and $\mathscr{B} = \mathscr{G}$. Then

$$M = \{T \in gl(V) \mid [T, \mathscr{G}] \subset [\mathscr{G}, \mathscr{G}]\}. \tag{1.5.70}$$

Take any $u, v \in \mathscr{G}$ and $T \in M$. We have

$$\begin{aligned} &\text{Tr } [u, v]T = \text{Tr } (uv - vu)T = \text{Tr } (uvT - vuT) = \text{Tr } uvT - \text{Tr } vuT \\ &= \text{Tr } uvT - \text{Tr } uTv = \text{Tr } (uvT - uTv) = \text{Tr } u(vT - Tv) \\ &= -\text{Tr } u[T, v] = -\text{Tr } [T, v]u = 0. \end{aligned} \tag{1.5.71}$$

By linearity, we have

$$\text{Tr } wT = 0 \qquad \text{for any } w \in [\mathscr{G}, \mathscr{G}]. \tag{1.5.72}$$

According to the above lemma, all the elements in $[\mathscr{G}, \mathscr{G}]$ are nilpotent. So $[\mathscr{G}, \mathscr{G}]$ is a nilpotent Lie algebra. Therefore, $\mathscr{G}$ is solvable. □

Corollary 1.5.9 *If Tr ad ξ ad $u = 0$ for any $\xi \in [\mathscr{G}, \mathscr{G}]$ and $u \in \mathscr{G}$. Then $\mathscr{G}$ is solvable.*

Proof By the above theorem, ad $\mathscr{G}$ is solvable. □

In [X13], we introduced a family of nilpotent Lie algebras associated with a root tree in connection with evolution partial differential equations. L. Luo [L1, L2] generalized our algebras to those associated with more general directed graphs.

References

[BXZ] G. Benkart, X. Xu, K. Zhao, Classical Lie superalgebras over simple associative algebras. Proc. Lond. Math. Soc. **92**(3), 581–600 (2006)

[Br] R. Block, On torsion-free abelian groups and Lie algebras. Proc. Am. Soc. **9**, 613–620 (1958)

[X7] X. Xu, Generalizations of the Block algebras. Manuscri. Math. **100**, 489–518 (1999)

[Cl] L. Chen, Twisted Hamiltonian Lie algebras and their multiplicity-free representations. Acta Math. Sin. (Engl. Ser.)**27**(1), 45–72 (2011)

[X13] X. Xu, Tree diagram Lie algebras of differential operators and evolution partial differential equations. J. Lie Theory **16**, 691–718 (2006)

[L1] L. Luo, Cohomology of oriented tree diagram Lie algebras, Comm. Algebra **37**(3), 965–984 (2009)

[L2] L. Luo, Oriented tree diagram Lie algebras and their abelian ideals. Acta Math. Sin. (Engl. Ser.) **20**(11), 2041–2058 (2010)

Chapter 2
Semisimple Lie Algebras

In this chapter, we first introduce the Killing form and prove that the semisimplicity of a finite-dimensional Lie algebra over $\mathbb{C}$ is equivalent to the nondegeneracy of its Killing form. Then we use the Killing form to derive the decomposition of a finite-dimensional semisimple Lie algebra over $\mathbb{C}$ into a direct sum of simple ideals. Moreover, it is showed that a derivation of such a Lie algebra must be an inner derivation. Furthermore, we study the completely reducible modules of a Lie algebra and prove the Weyl's theorem of complete reducibility. The equivalence of the complete reducibility of real and complex modules is also given. Cartan's root-space decomposition of a finite-dimensional semisimple Lie algebra over $\mathbb{C}$ is derived. In particular, we prove that such a Lie algebra is generated by two elements. The completely reducibility of finite-dimensional modules of $sl(2, \mathbb{C})$ plays an important role in proving the properties of the corresponding root systems.

Section 2.1 gives the relation between the Killing form and the semisimplicity of Lie algebra. In Sect. 2.2, we define a module of a Lie algebra and discuss the complete reducibility. Moreover, the dual module and tensor module are introduced. The lifting from a given representation to an oscillator representation is given. In Sect. 5.3, we prove the equivalence of real and complex complete reducibility. In particular, we find the connection between real and complex finite-dimensional irreducible modules. The Weyl's theorem is proved in Sect. 2.4 by using Schur's Lemma. In Sect. 2.5, we derive the Cartan's root space decomposition. Finally in Sect. 2.6, we use finite-dimensional representations of $sl(2, \mathbb{C})$ to study the properties of roots and root subspaces of a finite-dimensional semisimple Lie algebra, and prove that the algebra is generated by two elements.

2.1 Killing Form

In this section, we assume that Lie algebras are finite dimensional. Moreover, we define the Killing form of a Lie algebra and use it to characterize semisimple Lie algebras.

X. Xu, *Representations of Lie Algebras and Partial Differential Equations*,
DOI 10.1007/978-981-10-6391-6_2

Let $\mathscr{G}$ be a Lie algebra. Define

$$\kappa(u, v) = \text{Tr}\, \text{ad}\, u\, \text{ad}\, v \qquad \text{for } u, v \in \mathscr{G}. \tag{2.1.1}$$

Then $\kappa(\cdot, \cdot)$ is a symmetric bilinear form that is called *Killing form*. Moreover,

$$\begin{aligned}
\kappa([u, v], w) &= \text{Tr}\, \text{ad}\, [u, v]\, \text{ad}\, w = \text{Tr}\, [\text{ad}\, u, \text{ad}\, v]\, \text{ad}\, w \\
&= \text{Tr}\, (\text{ad}\, u\, \text{ad}\, v - \text{ad}\, v\, \text{ad}\, u)\text{ad}\, w = \text{Tr}\, (\text{ad}\, u\, \text{ad}\, v\, \text{ad}\, w - \text{ad}\, v\, \text{ad}\, u\, \text{ad}\, w) \\
&= \text{Tr}\, (\text{ad}\, u\, \text{ad}\, v\, \text{ad}\, w) - \text{Tr}\, (\text{ad}\, v\, \text{ad}\, u\, \text{ad}\, w) \\
&= \text{Tr}\, (\text{ad}\, u\, \text{ad}\, v\, \text{ad}\, w) - \text{Tr}\, (\text{ad}\, u\, \text{ad}\, w\, \text{ad}\, v) \\
&= \text{Tr}\, (\text{ad}\, u\, \text{ad}\, v\, \text{ad}\, w - \text{ad}\, u\, \text{ad}\, w\, \text{ad}\, v) \\
&= \text{Tr}\, \text{ad}\, u(\text{ad}\, v\, \text{ad}\, w - \text{ad}\, w\, \text{ad}\, v) = \text{Tr}\, \text{ad}\, u[\text{ad}\, v, \text{ad}\, w] \\
&= \text{Tr}\, \text{ad}\, u\, \text{ad}\, [v, w] = \kappa(u, [v, w])
\end{aligned} \tag{2.1.2}$$

for $u, v, w \in \mathscr{G}$. Thus κ is *associative* (*invariant*).

Lemma 2.1.1 *Let $\mathcal{I}$ be an ideal of a Lie algebra $\mathscr{G}$. If $\kappa(\cdot, \cdot)$ is the Killing form of $\mathscr{G}$, then $\kappa(|_{\mathcal{I}}, |_{\mathcal{I}})$ is the Killing form of $\mathcal{I}$.*

Proof Assume that $\mathcal{I}$ is a proper nonzero ideal. Take subspace $\mathscr{K}$ of $\mathscr{G}$ such that

$$\mathscr{G} = \mathscr{K} \oplus \mathcal{I}. \tag{2.1.3}$$

Let $\mathscr{S}_1$ be a basis of $\mathscr{K}$ and let $\mathscr{S}_2$ be a basis of $\mathcal{I}$. Then $\mathscr{S} = \mathscr{S}_1 \bigcup \mathscr{S}_2$ forms a basis of $\mathscr{G}$. For each $u \in \mathcal{I}$, the matrix form

$$\text{ad}_{\mathscr{G}} u = \begin{pmatrix} 0 & A_u \\ 0 & B_u \end{pmatrix} \tag{2.1.4}$$

with respect to the basis $\mathscr{S}$ and the matrix form $\text{ad}_{\mathcal{I}} u = B_u$ with respect to $\mathscr{S}_2$. Thus

$$\text{ad}_{\mathscr{G}} u\, \text{ad}_{\mathscr{G}} v = \begin{pmatrix} 0 & A_u \\ 0 & B_u \end{pmatrix} \begin{pmatrix} 0 & A_v \\ 0 & B_v \end{pmatrix} = \begin{pmatrix} 0 & A_u B_v \\ 0 & B_u B_v \end{pmatrix} \qquad \text{for } u, v \in \mathscr{I}. \tag{2.1.5}$$

Hence

$$\kappa(u, v) = \text{Tr}\, \text{ad}_{\mathscr{G}} u\, \text{ad}_{\mathscr{G}} v = \text{Tr}\, B_u B_v = \text{Tr}\, \text{ad}_{\mathcal{I}} u\, \text{ad}_{\mathcal{I}} v \tag{2.1.6}$$

for $u, v \in \mathcal{I}$. □

The *radical* of a symmetric bilinear form $\beta(\cdot, \cdot)$ on a vector space V is defined by

$$\text{Rad}\, \beta = \{u \in V \mid \beta(u, v) = 0 \text{ for any } v \in V\}. \tag{2.1.7}$$

If $\text{Rad}\, \beta = \{0\}$, we call $\beta(\cdot, \cdot)$ *nondegenerate*. Recall the radical of a Lie algebra $\mathscr{G}$ is the unique maximal solvable ideal. The algebra $\mathscr{G}$ is called *semisimple* if $\text{Rad}\, \mathscr{G} = \{0\}$. In the rest of this section, we assume the base field $\mathbb{F} = \mathbb{C}$.

Theorem 2.1.2 *Let $\mathscr{G}$ be a Lie algebra. Then $\mathscr{G}$ is semisimple if and only if the Killing form is nondegenerate.*

Proof Assume that $\mathscr{G}$ is semisimple. For any $u \in \text{Rad}\,\kappa$ and $v, w \in \mathscr{G}$, we have

$$\kappa([u, v], w) = \kappa(u, [v, w]) = 0. \tag{2.1.8}$$

Hence $[u, v] \in \text{Rad}\,\kappa$. So Rad κ is an ideal of $\mathscr{G}$. By Cartan's Criterion, ad Rad κ is solvable, and so is Rad κ because $\ker \text{ad} = Z(\mathscr{G}) = \{0\}$. Hence Rad $\kappa \subset$ Rad $\mathscr{G} = \{0\}$.

Suppose that Rad $\kappa = \{0\}$ and $R = \text{Rad}\,\mathscr{G} \neq \{0\}$. Let n be the minimal positive integer such that $R^{(n)} = \{0\}$. Then $R^{(n-1)}$ is an abelian ideal of $\mathscr{G}$. For any element $u \in R^{(n-1)}$ and $v, w \in \mathscr{G}$, we have

$$(\text{ad}\,u\ \text{ad}\,v)^2(w) = [u, [v, [u, [v, w]]]] \in [R^{(n-1)}, R^{(n-1)}] = \{0\}. \tag{2.1.9}$$

Thus

$$(\text{ad}\,u\ \text{ad}\,v)^2 = 0 \Longrightarrow \text{Tr}\,\text{ad}\,u\ \text{ad}\,v = 0 \sim \kappa(u, v) = 0. \tag{2.1.10}$$

So $R^{(n-1)} \subset \text{Rad}\,\kappa = \{0\}$, which leads to a contradiction. Therefore, Rad $\mathscr{G} = \{0\}$. □

Theorem 2.1.3 *For any semisimple Lie algebra $\mathscr{G}$, there exist ideals $\mathscr{G}_1, \mathscr{G}_2, \ldots, \mathscr{G}_k$ of $\mathscr{G}$, which are simple Lie algebras, such that*

$$\mathscr{G} = \bigoplus_{i=1}^{k} \mathscr{G}_i. \tag{2.1.11}$$

The set $\{\mathscr{G}_1, \mathscr{G}_2, \ldots, \mathscr{G}_k\}$ enumerates all simple ideals of $\mathscr{G}$.

Proof Let $\mathscr{G}_1$ be a minimal nonzero ideal of $\mathscr{G}$. Since $\mathscr{G}_1$ is not solvable by the semisimplicity of $\mathscr{G}$, $\dim \mathscr{G}_1 > 1$. Set

$$\mathscr{G}' = \{u \in \mathscr{G} \mid \kappa(u, v) = 0 \text{ for any } v \in \mathscr{G}_1\}. \tag{2.1.12}$$

For $u \in \mathscr{G}'$, $v \in \mathscr{G}_1$ and $w \in \mathscr{G}$, we have $[w, v] \in \mathscr{G}_1$ and

$$\kappa([u, w], v) = \kappa(u, [w, v]) = 0. \tag{2.1.13}$$

Hence $[u, w] \in \mathscr{G}'$. So $\mathscr{G}'$ is an ideal of $\mathscr{G}$. Since $\kappa(|_{\mathscr{G}_1}, |_{\mathscr{G}_1})$ is also the Killing form of $\mathscr{G}_1$ by Lemma 2.1.1, the radical of the Killing form of $\mathscr{G}_1$ is exactly $\mathscr{G}' \bigcap \mathscr{G}_1$. If $\mathscr{G}' \bigcap \mathscr{G}_1 \neq \{0\}$, then it is an ideal of $\mathscr{G}$ included in $\mathscr{G}_1$. By minimality, $\mathscr{G}' \bigcap \mathscr{G}_1 = \mathscr{G}_1$, which implies that the Killing form of $\mathscr{G}_1$ is zero. By Cartan's Criterion, $\mathscr{G}_1$ is a solvable ideal, which is absurd. Hence $\mathscr{G}' \bigcap \mathscr{G}_1 = \{0\}$. Thus $\kappa(|_{\mathscr{G}_1}, |_{\mathscr{G}_1})$ is nondegenerate. Therefore, we have

$$\mathscr{G} = \mathscr{G}_1 \oplus \mathscr{G}'. \tag{2.1.14}$$

Observe that $[\mathscr{G}_1, \mathscr{G}'] \subset \mathscr{G}_1 \bigcap \mathscr{G}' = \{0\}$. Thus any ideals of $\mathscr{G}_1$ and $\mathscr{G}'$ are ideals of $\mathscr{G}$. In particular, $\mathscr{G}_1$ is a simple Lie algebra and $\mathscr{G}'$ is semisimple if $\mathscr{G}' \neq \{0\}$. Now we work on $\mathscr{G}'$. By induction, there exist simple ideals $\mathscr{G}_2, \ldots, \mathscr{G}_k$ of $\mathscr{G}'$ such that $\mathscr{G}' = \bigoplus_{i=2}^k \mathscr{G}_i$. So we have (2.1.11).

If $\mathscr{L}$ is any simple ideal of $\mathscr{G}$, then $[\mathscr{L}, \mathscr{L}]$ is a nonzero ideal of $\mathscr{L}$. Hence $[\mathscr{L}, \mathscr{L}] = \mathscr{L}$, which implies $\mathscr{L} = [\mathscr{G}, \mathscr{L}]$. Now

$$\mathscr{L} = [\mathscr{G}, \mathscr{L}] = \left[\bigoplus_{i=1}^k \mathscr{G}_i, \mathscr{L}\right]. \tag{2.1.15}$$

So $[\mathscr{G}_i, \mathscr{L}] \neq \{0\}$ for some $i \in \{1, 2, \ldots, k\}$. But $[\mathscr{G}_i, \mathscr{L}]$ is an ideal of $\mathscr{G}$ that is also an ideal of $\mathscr{G}_i$ and $\mathscr{L}$. Thus $\mathscr{L} = [\mathscr{G}_i, \mathscr{L}] = \mathscr{G}_i$. □

Let $\mathcal{I}$ be a nonzero ideal of $\mathscr{G}$. Suppose

$$u = \sum_{i=1}^k u_i \in \mathcal{I}. \tag{2.1.16}$$

If $u_i \neq 0$, then $[u, \mathscr{G}_i] = [u_i, \mathscr{G}_i] \neq 0$, which implies $\{0\} \neq \mathscr{G}_i \bigcap \mathcal{I} = \mathscr{G}_i$. Therefore, $\mathcal{I}$ is a direct sum of some simple ideals.

Theorem 2.1.4 *If $\mathscr{G}$ is a semisimple Lie algebra, then Der $\mathscr{G}$ = ad $\mathscr{G}$.*

Proof We define a symmetric associative bilinear form β on Der $\mathscr{G} \subset$ End G by

$$\beta(d_1, d_2) = \mathrm{Tr}\, d_1 d_2 \qquad \text{for } d_1, d_2 \in \mathrm{Der}\, \mathscr{G}. \tag{2.1.17}$$

Note that

$$(\mathrm{ad}\, \mathscr{G})^{\perp} = \{d \in \mathrm{Der}\, \mathscr{G} \mid \beta(d, \mathrm{ad}\, u) = 0 \text{ for any } u \in \mathscr{G}\} \tag{2.1.18}$$

is an ideal of Der $\mathscr{G}$ because ad $\mathscr{G}$ is an ideal of Der $\mathscr{G}$. Since $\beta(|_{\mathrm{ad}\,\mathscr{G}}, |_{\mathrm{ad}\,\mathscr{G}}) = \kappa(\cdot, \cdot)$, the nondegenerate Killing form of $\mathscr{G}$, we have $\mathrm{ad}\, \mathscr{G} \bigcap (\mathrm{ad}\, \mathscr{G})^{\perp} = \{0\}$. Thus

$$\mathrm{Der}\, \mathscr{G} = \mathrm{ad}\, \mathscr{G} \oplus (\mathrm{ad}\, \mathscr{G})^{\perp}. \tag{2.1.19}$$

For any $d \in (\mathrm{ad}\, \mathscr{G})^{\perp}$ and $u \in \mathscr{G}$, we have $[d, \mathrm{ad}\, u] = \mathrm{ad}\, d(u) \in \mathrm{ad}\, \mathscr{G} \bigcap (\mathrm{ad}\, \mathscr{G})^{\perp} = \{0\}$. Thus $d(u) \in Z(\mathscr{G}) = \{0\}$. So $d = 0$. Hence $(\mathrm{ad}\, \mathscr{G})^{\perp} = \{0\}$. Therefore, $\mathrm{Der}\, \mathscr{G} = \mathrm{ad}\, \mathscr{G}$ by (2.1.19). □

For any element u in a semisimple Lie algebra $\mathcal{G}$, there exist elements $u_s, u_n \in \mathcal{G}$ such that $\text{ad}\, u = \text{ad}\, u_s + \text{ad}\, u_n$ is the Jordan–Chevalley decomposition of $\text{ad}\, u$. Since $Z(\mathcal{G}) = \{0\}$, we have $u = u_s + u_n$, which is called the *abstract Jordan decomposition* of u.

As an exercise, prove that a finite-dimensional Lie algebra $\mathcal{G}$ is solvable if and only if $[\mathcal{G}, \mathcal{G}] \subset \text{Rad}\, \kappa(\cdot, \cdot)$.

2.2 Modules

In this section, we define a module of a Lie algebra and discuss the complete reducibility. Moreover, the dual module and tensor module are introduced. The lifting from a given representation to an oscillator representation is given.

Recall that a representation ν of $\mathcal{G}$ on a vector space V is a Lie algebra homomorphism from $\mathcal{G}$ to $gl(V)$. It is more convenient to work on the elements of $\mathcal{G}$ directly acting on V. We denote

$$\xi(u) = \nu(\xi)(u) \qquad \text{for } \xi \in \mathcal{G},\ u \in V. \tag{2.2.1}$$

The conditions for ν to be a homomorphism becomes

$$\xi_1(au + bv) = a\xi_1(u) + b\xi_1(v), \qquad (a\xi_1 + b\xi_2)(u) = a\xi_1(u) + b\xi_2(u), \tag{2.2.2}$$

$$\xi_1(\xi_2(u)) - \xi_2(\xi_1(u)) = [\xi_1, \xi_2](u) \tag{2.2.3}$$

for $a, b \in \mathbb{F}$, $\xi_1, \xi_2 \in \mathcal{G}$ and $u, v \in V$. We call V a *$\mathcal{G}$-module*. Conversely, a map from $\mathcal{G} \times V \to V$: $(\xi, u) \mapsto \xi(u)$ satisfying (2.2.2) and (2.2.3) gives a representation ν of $\mathcal{G}$ on V defined by (2.2.1). A *submodule* U of a $\mathcal{G}$-module V is a subspace such that $\mathcal{G}(U) \subset U$. A *trivial* $\mathcal{G}$-module V is a vector space with the $\mathcal{G}$-action: $\xi(u) = 0$ for any $\xi \in \mathcal{G}$ and $u \in V$.

A *homomorphism* ϕ from a $\mathcal{G}$-module V to a $\mathcal{G}$-module W is a linear map from V to W such that

$$\phi(\xi(u)) = \xi(\phi(u)) \qquad \text{for } \xi \in \mathcal{G},\ u \in V. \tag{2.2.4}$$

In this case, $\phi(V)$ forms a submodule of W and $\ker \phi$ is a submodule of V. Moreover, $\phi(V) \cong V/\ker \phi$. The map ϕ is called an *isomorphism* if it is bijective. A $\mathcal{G}$-module V is called *irreducible* if it has exactly two submodules $\{0\}$ and V. We do not view $\{0\}$ as an irreducible $\mathcal{G}$-module. A module is called *completely reducible* if it is direct sum of irreducible submodules. A *complement* of a submodule W of V is a submodule W' such that $V = W \oplus W'$.

Lemma 2.2.1 *A module V is completely reducible if and only if any submodule has a complement.*

Proof Suppose that $V \neq \{0\}$ and any submodule of V has a complement. Take any nonzero minimal submodule W_1 of V, which must be irreducible. If $V = W_1$, it is done. Otherwise, it has a nonzero complement W_1'. Suppose that W_2 is any nonzero proper submodule of W_1'. Then $W_1 + W_2$ is a module. By assumption, there exists a complement U of $W_1 + W_2$. This U may not be in W_1'. Define the projection $\mathscr{P} : V \to W_1'$ by

$$\mathscr{P}(u + u') = u' \qquad \text{for } u \in W_1,\ u' \in W_1'. \tag{2.2.5}$$

Then $\mathscr{P}$ is a Lie algebra module homomorphism. In fact, $\mathscr{P}(U)$ is a submodule of W_1'.

If $u \in \mathscr{P}(U) \bigcap W_2$, then there exists $v \in W_1$ such that $v + u \in U$. But $v + u \in W_1 + W_2$. Thus $v + u = 0$, or equivalently, $u = v = 0$. Hence $\mathscr{P}(U) \bigcap W_2 = \{0\}$. Moreover, any element $w \in W_1'$ can be written as $w = w_1 + w_2 + w'$ with $w_1 \in W_1$, $w_2 \in W_2$ and $w' \in U$. Note $w' - \mathscr{P}(w') \in W_1$. We have

$$w - w_2 - \mathscr{P}(w') = w_1 + w' - \mathscr{P}(w') \in W_1 \bigcap W_1' = \{0\}; \tag{2.2.6}$$

that is, $w = w_2 + \mathscr{P}(w')$. Hence

$$W_1' = W_2 \oplus \mathscr{P}(U), \tag{2.2.7}$$

or equivalently, any submodule of W_1' has a complement in W_1'. By induction, $W_1' = \bigoplus_{i=2}^k W_i$ is a direct sum of irreducible submodules $W_2, W_3, \ldots, W_k$ in W_1'. So is $V = \bigoplus_{i=1}^k W_i$.

Assume that $V = \bigoplus_{i=1}^k W_i$ is a direct sum of irreducible submodules $W_1, W_2, \ldots, W_k$ in V and U is any nonzero proper submodule of V. Then there exists some $W_i \not\subset U$. Now

$$\bar{V} = V/W_i = \bigoplus_{i \neq j} (W_j + W_i)/W_i \tag{2.2.8}$$

is completely reducible. By induction, $(U + W_i)/W_i$ has a complement U'/W_i in $\bar{V}$, where U' is a submodule of V containing W_i. If $u \in U \bigcap U'$, then $u + W_i \in [(U + W_i)/W_i] \bigcap (U'/W_i) = W_i$. So $u \in W_i \bigcap U = \{0\}$. That is, $U \bigcap U' = \{0\}$. For any element $v \in V$, there exists $v_1 \in U$ and $v_2 \in U'$ such that

$$v + W_i = (v_1 + W_i) + (v_2 + W_i) = v_1 + v_2 + W_i; \tag{2.2.9}$$

equivalently, there exists $w \in W_i$ such that

$$v = v_1 + v_2 + w. \tag{2.2.10}$$

But $W_i \subset U'$. We have $v_2 + w \in U'$. Thus $v \in U + U'$. Therefore, $V = U \oplus U'$; that is, U' is a complement of U. $\square$

Recall that V^* denotes the space of linear functions on V. Suppose that V is a module of a Lie algebra $\mathscr{G}$. We define an action of $\mathscr{G}$ on V^* by

$$\xi(f)(u) = -f(\xi(u)) \quad \text{for } \xi \in \mathscr{G},\ f \in V^*,\ u \in V. \tag{2.2.11}$$

Note

$$\begin{aligned} & [\xi_1, \xi_2](f)(u) \\ = \;& f(-[\xi_1, \xi_2](u)) = f(-\xi_1(\xi_2(u)) + \xi_2(\xi_1(u))) \\ = \;& -f(\xi_1(\xi_2(u))) + f(\xi_2(\xi_1(u))) = -\xi_1(f)(\xi_2(u)) + \xi_2(f)(\xi_1(u)) \\ = \;& -\xi_2(\xi_1(f))(u) + \xi_1(\xi_2(f))(u) = (\xi_1(\xi_2(f)) - \xi_2(\xi_1(f)))(u) \end{aligned} \tag{2.2.12}$$

for $\xi_1, \xi_2 \in \mathscr{G},\ f \in V^*$ and $u \in V$. Thus

$$[\xi_1, \xi_2](f) = \xi_1(\xi_2(f)) - \xi_2(\xi_1(f)). \tag{2.2.13}$$

Hence V^* forms a $\mathscr{G}$-module, which is called the *dual (contragredient) module* of V. In general, $V \not\cong V^*$ as $\mathscr{G}$-modules. It is in general a difficult problem of determining if they are isomorphic.

A bilinear form $\beta : V \times V \to \mathbb{F}$ on a $\mathscr{G}$-module V is called *$\mathscr{G}$-invariant* if

$$\beta(\xi(u), v) = -\beta(u, \xi(v)) \quad \text{for } \xi \in \mathscr{G},\ u, v \in V. \tag{2.2.14}$$

In fact, we have:

Lemma 2.2.2 *A finite-dimensional module V of a Lie algebra $\mathscr{G}$ is isomorphic to its dual module V^* if and only if it has a nondegenerate invariant bilinear form.*

Proof Suppose that V is isomorphic to V^* via σ. We define a bilinear form β on V by

$$\beta(u, v) = \sigma(u)(v) \quad \text{for } u, v \in V. \tag{2.2.15}$$

Then

$$\beta(\xi(u), v) = \sigma(\xi(u))(v) = \xi(\sigma(u))(v) = -\sigma(u)(\xi(v)) = -\beta(u, \xi(v)). \tag{2.2.16}$$

for $\xi \in \mathscr{G}$ and $u, v \in V$. So β is a nondegenerate invariant bilinear form.

Assume that V has a nondegenerate invariant bilinear form β. Equation (2.2.15) define a linear map $\sigma : V \to V^*$ and (2.2.16) implies that σ is a module homomorphism. The nondegeneracy implies that σ is bijective. □

Next, we want to lift a given representation to an oscillator (differential-operator) representation. Let $\{u_i \mid i \in I\}$ be a basis of a module V of a Lie algebra $\mathscr{G}$. For any $\xi \in \mathscr{G}$, we write

$$\xi(u_i) = \sum_{j \in I} \nu_{i,j}(\xi) u_j. \tag{2.2.17}$$

Then $\nu_{i,j} \in \mathcal{G}^*$. We define an action of $\mathcal{G}$ on the polynomial algebra $\mathbb{F}[x_i \mid i \in I]$ by:

$$\xi(f) = \sum_{i,j\in I} \nu_{i,j}(\xi)x_j\partial_{x_i}(f). \tag{2.2.18}$$

By (2.2.3) and (2.2.17), we have

$$\sum_{j\in I}(\nu_{i,j}(\xi_2)\nu_{j,k}(\xi_1) - \nu_{i,j}(\xi_1)\nu_{j,k}(\xi_2)) = \nu_{i,k}([\xi_1, \xi_2]) \qquad \text{for } \xi_1, \xi_2 \in \mathcal{G}. \tag{2.2.19}$$

On the other hand,

$$\begin{aligned}
&\left[\sum_{i,j\in I} \nu_{i,j}(\xi_1)x_j\partial_{x_i}, \sum_{r,s\in I} \nu_{r,s}(\xi_2)x_s\partial_{x_r}\right] \\
&= \sum_{i,j,r,s\in I} \nu_{i,j}(\xi_1)\nu_{r,s}(\xi_2)x_j\partial_{x_i}(x_s)\partial_{x_r} - \sum_{i,j,r,s\in I} \nu_{i,j}(\xi_1)\nu_{r,s}(\xi_2)x_s\partial_{x_r}(x_j)\partial_{x_i} \\
&= \sum_{i,j,r\in I} \nu_{i,j}(\xi_1)\nu_{r,i}(\xi_2)x_j\partial_{x_r} - \sum_{i,j,s\in I} \nu_{i,j}(\xi_1)\nu_{j,s}(\xi_2)x_s\partial_{x_i} \\
&= \sum_{r,j\in I}\sum_{i\in I}(\nu_{r,i}(\xi_2)\nu_{i,j}(\xi_1) - \nu_{r,i}(\xi_1)\nu_{i,j}(\xi_2))x_j\partial_{x_r} \\
&= \sum_{r,j\in I} \nu_{r,j}([\xi_1, \xi_2])x_j\partial_{x_r}
\end{aligned} \tag{2.2.20}$$

for $\xi_1, \xi_2 \in \mathcal{G}$. Thus $\mathbb{F}[x_i \mid i \in I]$ becomes a $\mathcal{G}$-module. It is important to understand this module structure from the information of the module V. For instance, any one-dimensional trivial submodule gives an invariant of the corresponding Lie group. To find invariants is equivalent to solve the system of partial differential equations:

$$\sum_{i,j\in I} \nu_{i,j}(\xi)x_i\partial_{x_j}(f) = 0, \qquad \xi \in \mathcal{G}. \tag{2.2.21}$$

The above defined module is isomorphic to the "symmetric tensor of V."

Let V and W be modules of a Lie algebra $\mathcal{G}$. We define an action of $\mathcal{G}$ on $V \otimes W$ by

$$\xi(u \otimes v) = \xi(u) \otimes v + u \otimes \xi(v) \qquad \text{for } \xi \in \mathcal{G},\ u \in V,\ v \in W. \tag{2.2.22}$$

Note

$$\begin{aligned}
&(\xi_1\xi_2 - \xi_2\xi_1)(u \otimes v) \\
&= \xi_1\xi_2(u \otimes v) - \xi_2\xi_1(u \otimes v) \\
&= \xi_1(\xi_2(u) \otimes v + u \otimes \xi_2(v)) - \xi_2(\xi_1(u) \otimes v + u \otimes \xi_1(v))
\end{aligned}$$

$$
\begin{aligned}
&= \xi_1(\xi_2(u)) \otimes v + \xi_1(u) \otimes \xi_2(v) + \xi_2(u) \otimes \xi_1(v) + u \otimes \xi_1(\xi_2(v)) \\
&\quad -(\xi_2(\xi_1(u)) \otimes v + \xi_2(u) \otimes \xi_1(v) + \xi_1(u) \otimes \xi_2(v) + u \otimes \xi_2(\xi_1(v)) \\
&= (\xi_1(\xi_2(u)) - \xi_2(\xi_1(u))) \otimes v + u \otimes (\xi_1(\xi_2(v)) - \xi_2(\xi_1(v))) \\
&= [\xi_1, \xi_2](u) \otimes v + u \otimes [\xi_1, \xi_2](v) = [\xi_1, \xi_2](u \otimes v) \qquad (2.2.23)
\end{aligned}
$$

for $\xi_1, \xi_2 \in \mathcal{G}$, $u \in V$ and $v \in W$. Thus $V \otimes W$ becomes a $\mathcal{G}$-module.

Denote by $\mathrm{Hom}_{\mathbb{C}}(V, W)$ the space of all linear maps from V to W. Define $\tau : V^* \otimes W \to \mathrm{Hom}_{\mathbb{C}}(V, W)$ by

$$
\tau\left(\sum_{i=1}^{k} f_i \otimes v_i\right)(u) = \sum_{i=1}^{k} f_i(u)v_i \qquad \text{for } u \in V. \qquad (2.2.24)
$$

If V and W are modules of a Lie algebra $\mathcal{G}$ and $\dim W < \infty$, then τ is a linear isomorphism, $V^* \otimes W$ forms a $\mathcal{G}$-module and τ induces a $\mathcal{G}$-module structure on $\mathrm{Hom}_{\mathbb{C}}(V, W)$. In fact,

$$
\xi(g)(u) = \xi(g(u)) - g(\xi(u)) \qquad \text{for } \xi \in \mathcal{G},\ g \in \mathrm{Hom}_{\mathbb{C}}(V, W),\ u \in V. \quad (2.2.25)
$$

As an exercise, write out the module structure of $\mathrm{Hom}_{\mathbb{C}}(V, W)$ for $sl(2, \mathbb{F})$ if (1) V is the adjoint module and W is the 2-dimensional module of defining $sl(2, \mathbb{F})$; (2) exchange the positions of two modules. Are these two modules irreducible?

2.3 Real and Complex Modules

In this section, we prove the equivalence of real and complex complete reducibility. In particular, we find the connection between real and complex finite-dimensional irreducible modules.

Let V be a module of a Lie algebra $\mathcal{G}$ and let U be a nonzero proper submodule of V. We say that V *splits* at U if U has a complementary submodule in V.

Lemma 2.3.1 *A $\mathcal{G}$-module V splits at a submodule U if and only if there exits a $\mathcal{G}$-module homomorphism $\tau : V \to U$ such that $\tau|_U = Id_U$.*

Proof If V splits at a submodule U, then U has a complement U' such that $V = U \oplus U'$. The projection τ defined by

$$
\tau(u + u') = u \qquad \text{for } u \in U,\ u' \in U', \qquad (2.3.1)
$$

is the required homomorphism. Conversely, if there exits a $\mathcal{G}$-module homomorphism $\tau : V \to U$ such that $\tau|_U = Id_U$, then

$$
v = \tau(v) + (1 - \tau)(v) \qquad \text{for } v \in V, \qquad (2.3.2)
$$

which implies that $(1-\tau)(V)$ is a complement of U because $1-\tau$ is a $\mathscr{G}$-module homomorphism from V to V. □

Let V be a vector space over $\mathbb{R}$ with a basis $\{v_i \mid i \in I\}$. The complexification of V is a vector space over $\mathbb{C}$ with $\{v_i \mid i \in I\}$ as a basis; that is,

$$V_{\mathbb{C}} = \sum_{i\in I} \mathbb{C}v_i. \tag{2.3.3}$$

Moreover, for $u = \sum_{i\in I} a_i v_i \in V$ and $c \in \mathbb{C}$, we define

$$cu = \sum_{i\in I} a_i c v_i. \tag{2.3.4}$$

Suppose that $\mathscr{G}$ is a finite-dimensional real Lie algebra. The Lie bracket on $\mathscr{G}_{\mathbb{C}}$ defined by

$$\left[\sum_{i\in I} a_i u_i, \sum_{j\in I} b_j v_j\right] = \sum_{i,j\in I} a_i b_j [u_i, v_j], \qquad u_i, v_j \in \mathscr{G},\ a_i, b_j \in \mathbb{C}. \tag{2.3.5}$$

In particular, $\mathscr{G}$ is a real subalgebra of $\mathscr{G}_{\mathbb{C}}$. In fact, $\mathscr{G}_{\mathbb{C}} = \mathscr{G} + \sqrt{-1}\mathscr{G}$. For the Killing form $\kappa_{\mathscr{G}}$ of $\mathscr{G}$ and the Killing form $\kappa_{\mathscr{G}_{\mathbb{C}}}$ of $\mathscr{G}_{\mathbb{C}}$, Rad $\kappa_{\mathscr{G}_{\mathbb{C}}} = \mathbb{C}(\text{Rad}\ \kappa_{\mathscr{G}})$ by linear algebra. In particular, Rad $\kappa_{\mathscr{G}}$ is a solvable ideal of $\mathscr{G}$ because Rad $\kappa_{\mathscr{G}_{\mathbb{C}}}$ is. Thus $\mathscr{G}$ is semisimple if and only $\kappa_{\mathscr{G}}$ is nondegenerate. By the proof of Theorem 2.1.3, $\mathscr{G}$ is a direct sum of simple ideals if it is semisimple. Moreover, $\mathscr{G}$ is semisimple if and only if $\mathscr{G}_{\mathbb{C}}$ is.

If V is a finite-dimensional (real) $\mathscr{G}$-module, then $V_{\mathbb{C}}$ becomes a (complex) $\mathscr{G}_{\mathbb{C}}$-module with the action:

$$\left(\sum_i a_i \xi_i\right)\left(\sum_j b_j v_j\right) = \sum_{i,j} a_i b_j \xi_i(v_j), \qquad \text{for } a_i, b_j \in \mathbb{C},\ \xi_i \in \mathscr{G},\ v_j \in V. \tag{2.3.6}$$

We can also view $V_{\mathbb{C}}$ as a real $\mathscr{G}$-module. In fact, $V_{\mathbb{C}} = V + \sqrt{-1}V$ is a decomposition of real $\mathscr{G}$-submodules. As real $\mathscr{G}$-modules, $V \cong \sqrt{-1}V$.

Theorem 2.3.2 *The space V is a completely reducible real $\mathscr{G}$-module if and only if $V_{\mathbb{C}}$ is a completely reducible complex $\mathscr{G}_{\mathbb{C}}$-module.*

Proof Suppose that V is a completely reducible real $\mathscr{G}$-module. Then $V = \bigoplus_{i=1}^k V_i$ is a direct sum of real irreducible $\mathscr{G}$-submodules. Moreover,

$$V_{\mathbb{C}} = \bigoplus_{i=1}^{k} (V_i \oplus \sqrt{-1}V_i) \tag{2.3.7}$$

is a direct sum of $2k$ real irreducible $\mathscr{G}$-submodules. So $V_{\mathbb{C}}$ is a completely reducible real $\mathscr{G}$-module. Let U be a complex $\mathscr{G}_{\mathbb{C}}$-submodule of $V_{\mathbb{C}}$. Note that U is also a real $\mathscr{G}$-submodule of $V_{\mathbb{C}}$. Thus there exists a real $\mathscr{G}$-module homomorphism $\tau : V_{\mathbb{C}} \to U$ such that $\tau|_U = \mathrm{Id}_U$. Define

$$\tilde{\tau}(v) = \frac{1}{2}(\tau(v) - \sqrt{-1}\tau(\sqrt{-1}v)) \qquad \text{for } v \in V. \tag{2.3.8}$$

Note

$$\begin{aligned}\tilde{\tau}(\sqrt{-1}v) &= \frac{1}{2}(\tau(\sqrt{-1}v) + \sqrt{-1}\tau(v)) \\ &= \frac{1}{2}\sqrt{-1}(-\sqrt{-1}\tau(\sqrt{-1}v) + \tau(v)) = \sqrt{-1}\tilde{\tau}(v)\end{aligned} \tag{2.3.9}$$

for $v \in V$. So $\tilde{\tau}$ is a $\mathbb{C}$-linear map from $V_{\mathbb{C}}$ to U. Moreover,

$$\xi(\tilde{\tau}(v)) = \frac{1}{2}(\tau(\xi(v)) - \sqrt{-1}\tau(\sqrt{-1}\xi(v))) = \tilde{\tau}(\xi(v)) \tag{2.3.10}$$

for $v \in V$ and $\xi \in \mathscr{G}$ by (2.3.6). Thus $\tilde{\tau}$ is a $\mathscr{G}_{\mathbb{C}}$-module homomorphism due to $\mathscr{G}_{\mathbb{C}} = \mathbb{C}\mathscr{G}$ by (2.3.4). Furthermore, for $u \in U$, we have $\sqrt{-1}u \in U$ because U is a complex $\mathscr{G}_{\mathbb{C}}$-submodule of $V_{\mathbb{C}}$. Hence $\tau(u) = u$ and $\tau(\sqrt{-1}u) = \sqrt{-1}u$. Thus

$$\tilde{\tau}(u) = \frac{1}{2}(u - \sqrt{-1}(\sqrt{-1}u)) = u; \tag{2.3.11}$$

that is, $\tilde{\tau}|_U = \mathrm{Id}_U$. By the above lemma, $V_{\mathbb{C}}$ is completely reducible.

Conversely, we assume that $V_{\mathbb{C}}$ is a completely reducible complex $\mathscr{G}_{\mathbb{C}}$-module. Let W be a real $\mathscr{G}$-submodule of V. Then $W_{\mathbb{C}}$ forms a $\mathscr{G}_{\mathbb{C}}$-submodule. There exists a $\mathscr{G}_{\mathbb{C}}$-module homomorphism $\bar{\tau} : V_{\mathbb{C}} \to W_{\mathbb{C}}$ such that $\bar{\tau}|_{W_{\mathbb{C}}} = \mathrm{Id}_{W_{\mathbb{C}}}$. So $\bar{\tau}|_V$ is a real $\mathscr{G}$-module homomorphism from V to $W_{\mathbb{C}}$ because $\mathscr{G} \subset \mathscr{G}_{\mathbb{C}}$. But in general, $\bar{\tau}(V) \not\subset W$. Define

$$\nu(w_1 + \sqrt{-1}w_2) = w_1 \qquad w_1, w_2 \in W. \tag{2.3.12}$$

Then $\nu : W_{\mathbb{C}} \to W$ is a real $\mathscr{G}$-module homomorphism. Therefore, $\tau = \nu\bar{\tau}$ is a real $\mathscr{G}$-module homomorphism from V to W and $\tau|_W = \mathrm{Id}_W$. By the above lemma, V is a completely reducible real $\mathscr{G}$-module. □

Let $M = \bigoplus_{i=1}^{k} M_i$ be a completely reducible module of a Lie algebra $\mathscr{G}$ over an arbitrary field, where M_i's are irreducible $\mathscr{G}$-submodules. For $i \in \overline{1,k}$, we define the projection $\mathscr{P}_i : M \to M_i$ by

$$\mathscr{P}_i\left(\sum_{r=1}^{k} u_r\right) = u_i \qquad \text{for } u_r \in M_r. \tag{2.3.13}$$

Then $\mathscr{P}_i$'s are $\mathscr{G}$-module homomorphisms. Let U be any irreducible $\mathscr{G}$-submodule of M. If $\mathscr{P}_i(U) \neq \{0\}$, then $\mathscr{P}_i|_U$ is a $\mathscr{G}$-module isomorphism from U to M_i by the irreducibility of U and M_i; in particular, $\mathscr{P}_i(U) = M_i$. Set

$$I = \{r \in \overline{1,k} \mid \mathscr{P}_r(U) \neq \{0\}\}. \tag{2.3.14}$$

For $r, s \in I$, the map $\mathscr{P}_s(\mathscr{P}_r|_U)^{-1}$ is a $\mathscr{G}$-module isomorphism from M_r to M_s. Re-indexing $\{M_1, M_2, \ldots, M_k\}$ if necessary, we can assume $1 \in I$. Now

$$u = \sum_{r\in I} \mathscr{P}_r(u) = \mathscr{P}_1(u) + \sum_{1\neq r\in I} [\mathscr{P}_r(\mathscr{P}_1|_U)^{-1}](\mathscr{P}_1(u)) \qquad \text{for } u \in U. \tag{2.3.15}$$

Thus

$$U = \left\{ v + \sum_{1\neq r\in I} [\mathscr{P}_r(\mathscr{P}_1|_U)^{-1}](v) \mid v \in M_1 \right\}. \tag{2.3.16}$$

Suppose that $M = \bigoplus_{j=1}^{k'} V_j$ is another decomposition of irreducible $\mathscr{G}$-submodules. Then $\mathscr{P}_i(V_1) \neq \{0\}$ for some $i \in \overline{1,k}$. Re-indexing $\{M_1, M_2, \ldots, M_k\}$ if necessary, we can assume $\mathscr{P}_1(V_1) \neq \{0\}$. So $\mathscr{P}_1|_{V_1}$ is a $\mathscr{G}$-module isomorphism from V_1 to M_1. Moreover,

$$V = V_1 \oplus \bigoplus_{i=2}^{k} M_i. \tag{2.3.17}$$

In particular,

$$\bigoplus_{i=2}^{k} (M_i + V_1)/V_1 = V/V_1 = \bigoplus_{r=2}^{k'} (V_i + V_1)/V_1 \tag{2.3.18}$$

as $\mathscr{G}$-modules. By induction on k, we have $k - 1 = k' - 1$; equivalently, $k = k'$. This shows that although a completely reducible $\mathscr{G}$-module may have more than one decompositions into irreducible submodules, the numbers of irreducible submodules in the decompositions are equal.

Let V be a finite-dimensional real irreducible module of a real Lie algebra $\mathscr{G}$. Then $V_{\mathbb{C}} = \bigoplus_{i=1}^{k} W_i$ is a direct sum of irreducible $\mathscr{G}_{\mathbb{C}}$-submodules. Since W_i are $\mathscr{G}$-modules and $V_{\mathbb{C}} = V \oplus \sqrt{-1}V$ is a direct sum of two irreducible real $\mathscr{G}$-submodules, W_i contain a real $\mathscr{G}$-submodule isomorphic to V. Let M be a finite-dimensional irreducible $\mathscr{G}_{\mathbb{C}}$-module. Take a minimal nonzero real $\mathscr{G}$-submodule U of M. Then U is an irreducible real $\mathscr{G}$-module and so is $\sqrt{-1}U$. If $U \bigcap \sqrt{-1}U \neq \{0\}$, then $U = \sqrt{-1}U$. So U is also a $\mathscr{G}_{\mathbb{C}}$-submodule. Hence $M = U$ is an irreducible real $\mathscr{G}$-module. Assume $U \bigcap \sqrt{-1}U = \{0\}$. Then $U + \sqrt{-1}U$ is a $\mathscr{G}_{\mathbb{C}}$-submodule, and so $M = U + \sqrt{-1}U$. Denote by $\mathscr{P}$ the corresponding projection from M to U. Let U' be any nonzero proper $\mathscr{G}$-submodule such that $\mathscr{P}(U') \neq \{0\}$. Then $\mathscr{P}(U') = U$ by the irreducibility of U. If $\ker \mathscr{P}|_{U'} = \{0\}$, then $U' \overset{\mathscr{P}|_{U'}}{\cong} U$ is an irreducible real $\mathscr{G}$-module.

Suppose $\ker \mathscr{P}|_{U'} \neq \{0\}$. Since $\ker \mathscr{P} = \sqrt{-1}U$, we have $U' \bigcap \sqrt{-1}U \neq \{0\}$. By the irreducibility of $\sqrt{-1}U$, $U' \supset \sqrt{-1}U$. The facts $\mathscr{P}(U') = U$ and $M = U + \sqrt{-1}U$ imply $U' = M$, which contradicts the properness of U'. Note that any nonzero proper $\mathscr{G}$-submodule U' such that $\mathscr{P}(U') = \{0\}$ must be a submodule of $\sqrt{-1}U$. So $U' = \sqrt{-1}U$ by the irreducibility of $\sqrt{-1}U$. In summary, we have:

Theorem 2.3.3 *Let $\mathscr{G}$ be a real Lie algebra. Any finite-dimensional irreducible real $\mathscr{G}$-module must be a real $\mathscr{G}$-submodule of a finite-dimensional irreducible complex $\mathscr{G}_{\mathbb{C}}$-module. A finite-dimensional irreducible complex $\mathscr{G}_{\mathbb{C}}$-module is either an irreducible real $\mathscr{G}$-module or any of its proper $\mathscr{G}$-submodules is an irreducible real $\mathscr{G}$-module.*

Thus we can find all finite-dimensional irreducible real $\mathscr{G}$-modules from finite-dimensional irreducible complex $\mathscr{G}_{\mathbb{C}}$-modules.

2.4 Weyl's Theorem

In this section, we prove the Weyl's theorem of complete reducibility.

Assume the base field $\mathbb{F} = \mathbb{C}$. We will prove that any finite-dimensional module of a finite-dimensional semisimple Lie algebra is completely reducible. A representation is called *irreducible* if the associated module is irreducible. First we have:

Lemma 2.4.1 (Schur) *Let $\mathscr{G}$ be a Lie algebra and let $\phi : \mathscr{G} \to gl(V)$ be a finite-dimensional irreducible representation. If $T \in gl(V)$ commutes with every element in $\phi(\mathscr{G})$, then $T = \lambda \mathrm{Id}_V$ for some $\lambda \in \mathbb{C}$.*

Proof Since we assume $\mathbb{F} = \mathbb{C}$ and V is finite-dimensional, T has an eigenvector u, whose eigenvalue is denoted as λ. Denote by $\mathbb{Z}_+$ the set of positive integers. Set

$$U = \mathrm{Span}\{u, \xi_1(\cdots(\xi_i(u))\cdots) \mid i \in \mathbb{Z}_+,\ \xi_j \in \mathscr{G}\}. \tag{2.4.1}$$

Then U forms a nonzero submodule of V. The irreducibility of V implies $V = U$. Since

$$\begin{aligned} T(\xi_1(\cdots(\xi_i(u))\cdots)) &= \xi_1(T(\xi_2(\cdots(\xi_i(u))\cdots))) \\ &= \xi_1(\cdots(\xi_i(T(u)))\cdots) \\ &= \lambda\xi_1(\cdots(\xi_i(u))\cdots), \end{aligned} \tag{2.4.2}$$

We have $T = \lambda \mathrm{Id}_V$. □

Suppose that $\phi : \mathscr{G} \to gl(V)$ is a finite-dimensional faithful ($\ker \phi = \{0\}$) representation of a semisimple Lie algebra $\mathscr{G}$. Define

$$\beta(\xi, \zeta) = \mathrm{Tr}\, \phi(\xi)\phi(\zeta) \qquad \text{for } \xi, \zeta \in \mathscr{G}. \tag{2.4.3}$$

By Cartan's Criterion, $\phi(\text{Rad}\,\beta)$ is a solvable ideal of $\phi(\mathscr{G})$, which is semisimple. Thus $\text{Rad}\,\beta = \{0\}$, and there exists an orthonormal basis $\{\xi_1, \xi_2, \ldots, \xi_n\}$ of $\mathscr{G}$ with respect to β. Set

$$\Omega_\phi = \sum_{i=1}^{n} \phi(\xi_i)^2 \in \text{End}\, V, \tag{2.4.4}$$

which is called a *Casimir operator*. Write

$$[\xi_i, \xi_j] = \sum_{k=1}^{n} c_{i,j}^k \xi_k. \tag{2.4.5}$$

Then

$$c_{i,j}^k = \beta([\xi_i, \xi_j], \xi_k) = -\beta(\xi_j, [\xi_i, \xi_k]) = -c_{i,k}^j. \tag{2.4.6}$$

Thus

$$\begin{aligned}
[\phi(\xi_i), \Omega_\phi] &= \left[\phi(\xi_i), \sum_{j=1}^{n} \phi_j(u)^2\right] = \sum_{j=1}^{n} [\phi(\xi_i), \phi(\xi_j)^2] \\
&= \sum_{j=1}^{n} ([\phi(\xi_i), \phi(\xi_j)]\phi(\xi_j) + \phi(\xi_j)[\phi(\xi_i), \phi(\xi_j)]) \\
&= \sum_{j=1}^{n} (\phi([\xi_i, \xi_j])\phi(\xi_j) + \phi(\xi_j)\phi([\xi_i, \xi_j])) \\
&= \sum_{j,k=1}^{n} c_{i,j}^k (\phi(\xi_k)\phi(\xi_j) + \phi(\xi_j)\phi(\xi_k)) \\
&= \sum_{j,k=1}^{n} c_{i,j}^k \phi(\xi_k)\phi(\xi_j) + \sum_{j,k=1}^{n} c_{i,j}^k \phi(\xi_j)\phi(\xi_k) \\
&= \sum_{j,k=1}^{n} c_{i,k}^j \phi(\xi_j)\phi(\xi_k) + \sum_{j,k=1}^{n} c_{i,j}^k \phi(\xi_j)\phi(\xi_k) \\
&= \sum_{j,k=1}^{n} (c_{i,j}^k + c_{i,k}^j)\phi(\xi_j)\phi(\xi_k) = 0.
\end{aligned} \tag{2.4.7}$$

So we have

$$\phi(u)\Omega_\phi = \Omega_\phi\phi(u) \qquad \text{for } u \in \mathscr{G}. \tag{2.4.8}$$

If ϕ is irreducible, then $\Omega_\phi = \lambda_\phi \text{Id}_V$ for some $\lambda_\phi \in \mathbb{C}$ by Lemma 2.4.1. In fact,

$$\lambda_\phi \dim V = \text{Tr}\, \Omega_\phi = \sum_{i=1}^{n} \text{Tr}\, \phi(\xi_i)^2 = \sum_{i=1}^{n} \beta(\xi_i, \xi_i) = n. \tag{2.4.9}$$

Hence

$$\lambda_\phi = n/\dim V = \dim \mathcal{G}/\dim V. \tag{2.4.10}$$

Furthermore, we have:

Lemma 2.4.2 *Let $\phi : \mathcal{G} \to gl(V)$ be a finite-dimensional representation of a semisimple Lie algebra $\mathcal{G}$. Then $\phi(\mathcal{G}) \in sl(V)$. In particular, one-dimensional module of $\mathcal{G}$ must be trivial.*

Proof This follows from $\phi(\mathcal{G}) = \phi([\mathcal{G}, \mathcal{G}]) = [\phi(\mathcal{G}), \phi(\mathcal{G})]$ and the fact $\text{Tr}\,[A, B] = 0$ for any $A, B \in gl(V)$. $\square$

Lemma 2.4.3 *Let $\phi : \mathcal{G} \to gl(V)$ be a finite-dimensional representation of a semisimple Lie algebra $\mathcal{G}$. Suppose that V has an irreducible submodule U of codimension one. Then V splits at U.*

Proof If $\phi(\mathcal{G}) = \{0\}$, the lemma holds trivially. Suppose $\phi(\mathcal{G}) \neq \{0\}$. Since $\ker \phi$ is either $\{0\}$ or a direct sum of some of simple ideals of $\mathcal{G}$, $\phi(\mathcal{G}) \cong \mathcal{G}/\ker \phi$ is semisimple. Replacing $\mathcal{G}$ by $\phi(\mathcal{G})$, we may assume that ϕ is faithful. By Lemma 2.4.2, V/U is a trivial module. Thus

$$\xi(V) \subset U \qquad \text{for } \xi \in \mathcal{G}. \tag{2.4.11}$$

If U is a trivial submodule, then V is two-dimensional and $\phi(\mathcal{G})$ is isomorphic to the Lie algebra of strict up-triangular matrices, which is absurd. Thus $\phi(\mathcal{G})|_U \neq \{0\}$ is semisimple.

Expression (2.4.11) implies

$$\beta(\xi, \zeta) = \text{Tr}\,\phi(\xi)\phi(\zeta) = \text{Tr}\,\phi(\xi)|_U\phi(\zeta)|_U \qquad \text{for } \xi, \zeta \in \mathcal{G}. \tag{2.4.12}$$

This shows

$$\Omega_\phi|_U = (\dim \phi(\mathcal{G})|_U/\dim U)\text{Id}_U. \tag{2.4.13}$$

Hence

$$\tau = \frac{\dim U}{\dim \phi(\mathcal{G})|_U}\Omega_\phi \tag{2.4.14}$$

is a module homomorphism from V to U with $\tau|_U = \text{Id}_U$. So V splits at U by Lemma 2.3.1. $\square$

Lemma 2.4.4 *Let $\phi : \mathcal{G} \to gl(V)$ be a finite-dimensional representation of a semisimple Lie algebra $\mathcal{G}$. Suppose that V has a submodule U of codimension one. Then V splits at U.*

Proof We prove it by induction on $\dim V$. It holds for $\dim V = 1$. Suppose that it holds for $\dim V < k$. Consider the case $\dim V = k$. We assume $U \neq \{0\}$. If U is irreducible, this is the above lemma. Suppose that U has a nonzero proper submodule

W. Then U/W is a $\mathscr{G}$-submodule of V/W with codimension one. By induction, there exists a submodule $W' \supset W$ of V such that

$$V/W = W'/W \oplus U/W. \tag{2.4.15}$$

Note $\dim W'/W = 1$. By induction, there exists a one-dimensional submodule U' of W' such that $W' = U' \oplus W$. According to (2.4.15), $U' \not\subset U$. Hence U' is a complement of U in V. □

Theorem 2.4.5 (Weyl) *Any nonzero finite-dimensional module V of a semisimple Lie algebra $\mathscr{G}$ is completely reducible.*

Proof Suppose that U is a nonzero proper submodule of V. Then $\mathrm{Hom}_{\mathbb{C}}(V, U)$ forms a $\mathscr{G}$-module. If $f \in \mathrm{Hom}_{\mathbb{C}}(V, U)$ such that $f|_U = a\mathrm{Id}_U$ with $a \in \mathbb{C}$, then

$$\xi(f)(u) = \xi(f(u)) - f(\xi(u)) = \xi(au) - a\xi(u) = 0 \quad \text{for } u \in U. \tag{2.4.16}$$

Thus

$$W = \{f \in \mathrm{Hom}_{\mathbb{C}}(V, U) \mid f|_U \in \mathbb{C}\mathrm{Id}_U\}, \quad W_1 = \{f \in \mathrm{Hom}_{\mathbb{C}}(V, U) \mid f|_U = 0\} \tag{2.4.17}$$

are submodules. Moreover, $\dim W/W_1 = 1$. By Lemma 2.4.4, W_1 has a complement W_1' in W, which is a one-dimensional trivial module. Hence there exists $\tau \in W_1'$ such that $\tau|_U = \mathrm{Id}_U$ and $\xi(\tau) = 0$ for any $\xi \in \mathscr{G}$; that is,

$$0 = \xi(\tau)(v) = \xi(\tau(v)) - \tau(\xi(v)) \quad \text{for } v \in V. \tag{2.4.18}$$

So τ is a $\mathscr{G}$-homomorphism. By Lemma 2.3.1, V splits at U. Therefore, V is completely reducible by Lemma 2.2.1. □

Theorem 2.4.6 *If $\mathscr{G} \subset gl(V)$ is a semisimple Lie subalgebra and $\dim V < \infty$, then $\mathscr{G}$ includes the semisimple part and nilpotent part of its any element. In particular, the abstract and usual Jordan decompositions coincide.*

Proof By the above theorem, $V = \bigoplus_{i=1}^k V_i$ is a direct sum of irreducible submodules. Set

$$\mathscr{L} = \{T \in gl(V) \mid [T, \mathscr{G}] \subset \mathscr{G};\ T(V_i) \subset V_i,\ \mathrm{Tr}\, T|_{V_i} = 0,\ i = 1, 2, \ldots, k\}. \tag{2.4.19}$$

Since $\mathscr{G}|_{V_i}$ is a homomorphic image of the semisimple Lie algebra $\mathscr{G}$, it is a semisimple Lie algebra. Thus we have $\mathrm{Tr}\, \xi|_{V_i} = 0$ for $\xi \in \mathscr{G}$ and $i = 1, \ldots, k$ by Lemma 2.4.2. Hence $\mathscr{G} \subset \mathscr{L}$. For any $T \in \mathscr{L}$, $\mathrm{ad}_{\mathscr{G}} T \in \mathrm{Der}\, \mathscr{G} = \mathrm{ad}\, \mathscr{G}$ (cf. Theorem 2.1.4). So there exists an element $\xi' \in \mathscr{G}$ such that $\mathrm{ad}_{\mathscr{G}} T = \mathrm{ad}_{\mathscr{G}} \xi'$. Set

$$T_1 = T - \xi'. \tag{2.4.20}$$

Then

$$[T_1, \xi] = 0 \qquad \text{for } \xi \in \mathscr{G}. \tag{2.4.21}$$

In particular, $T_1|_{V_i}$ is a constant map $\lambda \mathrm{Id}_{V_i}$ by Schur's Lemma (Lemma 2.4.1). Now $0 = \mathrm{Tr}\, T_1|_{V_i} = \lambda(\dim V_i)$ by (2.4.19), or equivalently, $\lambda = 0$. Hence $T_1 = 0$, that is, $T = \xi' \in \mathscr{G}$. This shows $\mathscr{L} = \mathscr{G}$.

Given $\xi \in \mathscr{G}$. Let $\xi = \xi_s + \xi_n$ be the usual Jordan–Chevalley decomposition of ξ. Recall that ξ_n is a polynomial of ξ without constant. Hence $\xi_n(V_i) \subset V_i$. Since $\mathrm{ad}_{gl(V)}\xi = \mathrm{ad}_{gl(V)}\xi_s + \mathrm{ad}_{gl(V)}\xi_n$ is also a Jordan–Chevalley decomposition, we have $[\xi_n, \mathscr{G}] \subset \mathscr{G}$. Note that $\xi_n|_{V_i}$ is nilpotent because ξ_n is. So $\mathrm{Tr}\,\xi_n|_{V_i} = 0$; that is, $\xi_n \in \mathscr{L} = \mathscr{G}$. Moreover, $\xi_s = \xi - \xi_n \in \mathscr{G}$. Note that the usual Jordan–Chevalley decomposition $\xi = \xi_s + \xi_n$ implies that $\mathrm{ad}_{\mathscr{G}}\xi_s$ is semisimple, $\mathrm{ad}_{\mathscr{G}}\xi_n$ is nilpotent and $[\mathrm{ad}_{\mathscr{G}}\xi_s, \mathrm{ad}_{\mathscr{G}}\xi_n] = 0$. By the uniqueness of the Jordan–Chevalley decomposition, $\xi = \xi_s + \xi_n$ is also the abstract Jordan–Chevalley decomposition. □

Corollary 2.4.7 *Let $\phi : \mathscr{G} \to gl(V)$ be a finite-dimensional representation of a semisimple Lie algebra $\mathscr{G}$. Given $\xi \in \mathscr{G}$. If $\xi = \xi_s + \xi_n$ is the abstract Jordan–Chevalley decomposition, then $\phi(\xi) = \phi(\xi_s) + \phi(x_n)$ is the usual Jordan–Chevalley decomposition.*

Proof If $\phi(\mathscr{G}) \neq \{0\}$, it is semisimple. Moreover, the usual Jordan–Chevalley decomposition $\phi(\xi) = \phi(\xi)_s + \phi(\xi)_n$ is also the abstract Jordan–Chevalley decomposition of $\phi(\xi)$ by the above theorem. If $\xi = \xi_s + \xi_n$ is the abstract Jordan–Chevalley decomposition, then $\mathrm{ad}\,\xi = \mathrm{ad}\,\xi_s + \mathrm{ad}\,\xi_n$ is the usual Jordan–Chevalley decomposition. Since $\phi(\mathscr{G}) \cong \mathscr{G}/\ker\,\phi$, $\mathrm{ad}\,\phi(\xi_s)$ is semisimple and $\mathrm{ad}\,\phi(\xi_n)$ is nilpotent. Moreover, $[\mathrm{ad}\,\phi(\xi_n), \mathrm{ad}\,\phi(\xi_n)] = \mathrm{ad}\,\phi([\xi_s, \xi_n]) = 0$. Hence $\mathrm{ad}\,\phi(\xi) = \mathrm{ad}\,\phi(\xi_s) + \mathrm{ad}\,\phi(\xi_n)$ is the Jordan–Chevalley decomposition of $\phi(\xi)$ by the uniqueness. Thus $\phi(\xi) = \phi(\xi_s) + \phi(\xi_n)$ is the abstract Jordan–Chevalley decomposition of $\phi(\xi)$. Therefore, $\phi(\xi_s) = \phi(\xi)_s$ and $\phi(\xi_n) = \phi(\xi)_n$. □

Let $\mathscr{G}$ be a simple Lie algebra and let β, γ be two symmetric associative bilinear forms on $\mathscr{G}$. If $\beta \neq 0$, then $\gamma = a\beta$ for some $a \in \mathbb{C}$ (exercise).

We can identify $(\mathscr{G} \otimes \mathscr{G})^*$ with the space of bilinear forms on $\mathscr{G}$. An associative form β is an element in $(\mathscr{G} \otimes \mathscr{G})^*$ such that $\mathscr{G}(\beta) = \{0\}$ (exercise), where we view $\mathscr{G}$ as the adjoint module of $\mathscr{G}$.

2.5 Root Space Decomposition

In this section, we prove the Cartan's root space decomposition of a semisimple Lie algebra.

We always assume that $\mathscr{G}$ is a finite-dimensional semisimple Lie algebra over $\mathbb{C}$. Recall the abstract Jordan decomposition of any element $\xi = \xi_s + \xi_n$ in $\mathscr{G}$, where $\xi_s, \xi_n \in \mathscr{G}$ and $\mathrm{ad}\,\xi = \mathrm{ad}\,\xi_s + \mathrm{ad}\,\xi_n$ is the usual Jordan decomposition. By Engel's Theorem, there exist elements in $\mathscr{G}$ that possess nonzero semisimple parts. Thus $\mathscr{G}$

has nonzero semisimple elements. A Lie subalgebra of $\mathscr{G}$ consisting of semisimple elements is called *toral*.

Lemma 2.5.1 *A toral subalgebra is abelian.*

Proof Let $\mathcal{T}$ be a toral subalgebra of $\mathscr{G}$. Given $\xi \in \mathcal{T}$. The operator $\mathrm{ad}_{\mathcal{T}}\xi$ is diagonalizable. Suppose that ζ is an eigenvector of $\mathrm{ad}_{\mathcal{T}}\xi$ with eigenvalue λ; that is $[\xi, \zeta] = \lambda\zeta$. On the other hand, the fact that $\mathrm{ad}_{\mathcal{T}}\zeta$ is diagonalizable implies that $\mathcal{T}$ has a basis $\{\zeta_1, \zeta_2, \ldots, \zeta_n\}$ consisting eigenvectors of $\mathrm{ad}_{\mathcal{T}}\zeta$ with corresponding eigenvalues $\{\lambda_1, \lambda_2, \ldots, \lambda_n\}$. In particular, we can take $\zeta_1 = \zeta$ with $\lambda_1 = 0$. Write

$$\xi = \sum_{i=1}^{n} a_i\zeta_i, \qquad a_i \in \mathbb{C}. \tag{2.5.1}$$

Then

$$\lambda\zeta = [\xi, \zeta] = \left[\sum_{i=1}^{n} a_i\zeta_i, \zeta\right] = -\sum_{i=2}^{n} \lambda_i a_i \zeta_i. \tag{2.5.2}$$

By linear independence of $\{\zeta, \zeta_2, \ldots, \zeta_n\}$, $\lambda = 0$. Thus all the eigenvalues of $\mathrm{ad}_{\mathcal{T}}\xi$ are 0. Hence $\mathrm{ad}_{\mathcal{T}}u\xi = 0$. □

Let H be a maximal toral subalgebra of $\mathscr{G}$. For instance, we take the algebra of all diagonal elements in $sl(n, \mathbb{C})$. Since $\{\mathrm{ad}\, h \mid h \in H\}$ are commuting diagonalizable operators, by linear algebra,

$$\mathscr{G} = \bigoplus_{\alpha \in H^*} \mathscr{G}_\alpha, \qquad \mathscr{G}_\alpha = \{\xi \in \mathscr{G} \mid [h, \xi] = \alpha(h)\xi \text{ for } h \in H\}. \tag{2.5.3}$$

Lemma 2.5.2 *For* $\alpha, \beta \in H^*$, $[\mathscr{G}_\alpha, \mathscr{G}_\beta] \subset \mathscr{G}_{\alpha+\beta}$. *In particular, for any* $u \in \mathscr{G}_\alpha$ *with* $\alpha \neq 0$, $ad\, u$ *is nilpotent. Moreover,*

$$\kappa(\mathscr{G}_\alpha, \mathscr{G}_\beta) = \{0\} \quad \textit{for} \;\; \alpha, \beta \in H^*, \;\; \alpha \neq -\beta, \tag{2.5.4}$$

which implies that $\kappa(|_{\mathscr{G}_0}, |_{\mathscr{G}_0})$ *is nondegenerate.*

Proof For $h \in H,\ \xi \in \mathscr{G}_\alpha$ and $\zeta \in \mathscr{G}_\beta$, we have:

$$[h, [\xi, \zeta]] = [[h, \xi], \zeta] + [\xi, [h, \zeta]] = (\alpha(h) + \beta(h))[\xi, \zeta] = (\alpha + \beta)(h)[\xi, \zeta]. \tag{2.5.5}$$

Thus $[\mathscr{G}_\alpha, \mathscr{G}_\beta] \subset \mathscr{G}_{\alpha+\beta}$. Suppose that $\alpha, \beta \in H^*$ such that $\alpha \neq -\beta$. Given $h \in H,\ \xi \in \mathscr{G}_\alpha$ and $\zeta \in \mathscr{G}_\beta$, we get

$$\alpha(h)\kappa(\xi, \zeta) = \kappa([h, \xi], \zeta) = -\kappa(\xi, [h, \zeta]) = -\beta(h)\kappa(\xi, \zeta). \tag{2.5.6}$$

Thus

$$(\alpha + \beta)(h)\kappa(\xi, \zeta) = 0 \qquad \text{for any } \ h \in H. \tag{2.5.7}$$

So $\kappa(\xi, \zeta) = 0$; that is, (2.5.4) holds. For $0 \neq \xi \in \mathscr{G}_0$, we have

$$\kappa(\xi, \mathscr{G}_\alpha) = \{0\} \qquad \text{for } \ 0 \neq \alpha \in H^*. \tag{2.5.8}$$

Since κ is nondegenerate, $\kappa(\xi, \mathscr{G}_0) \neq \{0\}$. So $\kappa(|_{\mathscr{G}_0}, |_{\mathscr{G}_0})$ is nondegenerate. □

Lemma 2.5.3 $\mathscr{G}_0 = H$.

Proof For any $\xi \in \mathscr{G}_0$, we have the abstract Jordan decomposition $\xi = \xi_s + \xi_n$. Note $\text{ad}\,\xi(H) = [\xi, H] = \{0\}$. Since $\text{ad}\,\xi_s$ is a polynomial of $\text{ad}\,\xi$ without constant, we have $[\xi_s, H] = \{0\}$. Thus $H + \mathbb{C}\xi_s$ forms a toral subalgebra. But H is a maximal toral subalgebra, we have $\xi_s \in H$, which implies $\xi_n = \xi - \xi_s \in \mathscr{G}_0$ because $H \subset \mathscr{G}_0$. Since $\text{ad}\,\xi_s|_{\mathscr{G}_0} = 0$ and $\text{ad}\,\xi_n$ is nilpotent, $\text{ad}\,\xi|_{\mathscr{G}_0}$ is nilpotent. By Engel's Theorem, $\mathscr{G}_0$ is a nilpotent Lie algebra.

Suppose that $h \in H$ satisfies $\kappa(h, H) = \{0\}$. For any $\xi \in \mathscr{G}_0$, let $\xi = \xi_s + \xi_n$ be the abstract Jordan decomposition. Since $[\text{ad}\,h, \text{ad}\,\xi_n] = \text{ad}\,[h, \xi_n] = 0$, $\text{ad}\,h\,\text{ad}\,\xi_n$ is a nilpotent element. Thus $\kappa(h, \xi_n) = 0$. So $\kappa(h, \xi) = \kappa(h, \xi_s) + \kappa(h, \xi_n) = 0$; that is $h \in \text{Rad}\,\kappa(|_{\mathscr{G}_0}, |_{\mathscr{G}_0}) = \{0\}$. Therefore, $\kappa(|_H, |_H)$ is nondegenerate. For any $h \in H$ and $\xi_1, \xi_2 \in \mathscr{G}_0$,

$$\kappa(h, [\xi_1, \xi_2]) = \kappa([h, \xi_1], \xi_2) = 0. \tag{2.5.9}$$

Hence

$$\kappa(H, [\mathscr{G}_0, \mathscr{G}_0]) = \{0\}. \tag{2.5.10}$$

If $[\mathscr{G}_0, \mathscr{G}_0] \neq \{0\}$, we take $0 \neq \xi \in Z(\mathscr{G}_0) \bigcap [\mathscr{G}_0, \mathscr{G}_0]$. Let $\xi = \xi_s + \xi_n$ be the abstract Jordan decomposition. Then $\xi_s \in H$ and $\kappa(H, \xi_n) = \{0\}$. Hence

$$\kappa(H, \xi_s) = \kappa(H, \xi) = \{0\} \tag{2.5.11}$$

by (2.5.10). The nondegeneracy of $\kappa(|_H, |_H)$ implies $\xi_s = 0$. So ξ is ad-nilpotent. If $[\mathscr{G}_0, \mathscr{G}_0] = \{0\}$ and $H \neq \mathscr{G}_0$, then u_n is a nonzero ad-nilpotent element for any $u \in \mathscr{G}_0 \setminus H$. In summary, if $H \neq \mathscr{G}_0$, we have a nonzero ad-nilpotent element $\xi \in Z(\mathscr{G}_0)$. For any $\xi' \in \mathscr{G}_0$, $\text{ad}\,\xi'\,\text{ad}\,\xi$ is nilpotent. So $\kappa(\xi', \xi) = 0$; that is, $\xi \in \text{Rad}\,\kappa(|_{\mathscr{G}_0}, |_{\mathscr{G}_0}) = \{0\}$, which leads a contradiction. Therefore $H = \mathscr{G}_0$. □

Set

$$\Phi = \{\alpha \in H^* \mid \alpha \neq 0, \ \mathscr{G}_\alpha \neq \{0\}\}. \tag{2.5.12}$$

The elements of Φ are called *roots* of $\mathscr{G}$ and Φ is called the *root system* of $\mathscr{G}$. Since $\kappa(|_H, |_H)$ is nondegenerate, for any $\alpha \in H^*$, there exists a unique $t_\alpha \in H$ such that

$$\alpha(h) = \kappa(t_\alpha, h) \qquad \text{for } \ h \in H. \tag{2.5.13}$$

Moreover, $t_{-\alpha} = -t_\alpha$.

Lemma 2.5.4 *(a) Φ spans H^*.*
(b) If $\alpha \in \Phi$, then $-\alpha \in \Phi$.
(c) Let $\alpha \in \Phi$, $\xi \in \mathscr{G}_\alpha$, $\zeta \in \mathscr{G}_{-\alpha}$. Then $[\xi, \zeta] = \kappa(\xi, \zeta)t_\alpha$. In particular, $[\mathscr{G}_\alpha, \mathscr{G}_{-\alpha}] = \mathbb{C}t_\alpha$.
(d) If $\alpha \in \Phi$ and $0 \neq \xi_\alpha \in \mathscr{G}_\alpha$, there exists $\zeta_\alpha \in \mathscr{G}_{-\alpha}$ such that $\{\xi_\alpha, t_\alpha, \zeta_\alpha\}$ spans a Lie subalgebra isomorphic to $sl(2, \mathbb{C})$.

Proof (a) If Φ does not span H^*, then there exists $0 \neq h \in H$ such that $\alpha(h) = 0$ for any $\alpha \in \Phi$. By (2.5.3), $h \in Z(\mathscr{G})$, which is absurd.
(b) By (2.5.4), $\kappa(\mathscr{G}_\alpha, \mathscr{G}_{-\alpha}) \neq \{0\}$ if $\alpha \in \Phi$.
(c) By Lemmas 2.5.2 and 2.5.3, $[\xi, \zeta] \in H$. For any $h \in H$, we have

$$\kappa(h, [\xi, \zeta]) = \kappa([h, \xi], \zeta) = \alpha(h)\kappa(\xi, \zeta) = \kappa(h, t_\alpha)\kappa(\xi, \zeta), \tag{2.5.14}$$

or equivalently,

$$\kappa(h, [\xi, \zeta] - \kappa(\xi, \zeta)t_\alpha) = \kappa(h, [\xi, \zeta]) - \kappa(h, t_\alpha)\kappa(\xi, \zeta) = 0. \tag{2.5.15}$$

Since $\kappa(|_H, |_H)$ is nondegenerate, we have $[\xi, \zeta] - \kappa(\xi, \zeta)t_\alpha = 0$. $[\mathscr{G}_\alpha, \mathscr{G}_{-\alpha}] = \mathbb{C}t_\alpha$ because $\kappa(\mathscr{G}_\alpha, \mathscr{G}_{-\alpha}) \neq \{0\}$.
(d) By (2.5.3) and the nongeneracy of κ, there exist $\zeta \in \mathscr{G}_{-\alpha}$ such that $\kappa(\xi_\alpha, \zeta) = 1$. So

$$[\xi_\alpha, \zeta] = t_\alpha \tag{2.5.16}$$

by (c). If $\alpha(t_\alpha) = 0$, then

$$[t_\alpha, \xi_\alpha] = \alpha(t_\alpha)\xi_\alpha = 0 = -\alpha(t_\alpha)\zeta = [t_\alpha, \zeta]. \tag{2.5.17}$$

Thus $\mathscr{K} = \mathbb{C}\xi_\alpha + \mathbb{C}t_\alpha + \mathbb{C}\zeta$ forms a solvable Lie subalgebra of $\mathscr{G}$ and $t_\alpha \in [\mathscr{K}, \mathscr{K}]$. By Lie's theorem, $\text{ad}\, t_\alpha$ is nilpotent. But $t_\alpha \in H$ is ad-semisimple. Hence $\text{ad}\, t_\alpha = 0$. Since $Z(\mathscr{G}) = \{0\}$, we have $t_\alpha = 0$. Thereby, $\alpha = 0$, which is absurd. Hence $0 \neq \alpha(t_\alpha) = \kappa(t_\alpha, t_\alpha)$.
Set

$$h_\alpha = \frac{2}{\alpha(t_\alpha)}t_\alpha, \qquad \zeta_\alpha = \frac{2}{\alpha(t_\alpha)}\zeta. \tag{2.5.18}$$

Then

$$[\xi_\alpha, \zeta_\alpha] = h_\alpha, \quad [h_\alpha, \xi_\alpha] = 2\xi_\alpha, \quad [h_\alpha, \zeta_\alpha] = -2\zeta_\alpha. \tag{2.5.19}$$

Hence

$$\mathscr{S}_\alpha = \mathbb{C}\xi_\alpha + \mathbb{C}h_\alpha + \mathbb{C}\zeta_\alpha \tag{2.5.20}$$

forms a Lie subalgebra isomorphic to $sl(2, \mathbb{C})$ via the linear map determined by

$$\xi_\alpha \mapsto E_{1,2}, \quad h_\alpha \mapsto E_{1,1} - E_{2,2}, \quad \zeta_\alpha \mapsto E_{2,1}. \tag{2.5.21}$$

By (2.5.18), $h_{-\alpha} = -h_\alpha$. □

Remark 2.5.5 In [X7, X8], six families of infinite-dimensional simple Lie algebra $\mathscr{G}$ without any toral subalgebra H such that $\mathscr{G}_0 = H$ (cf. (2.5.3)) were constructed.

2.6 Properties of Roots and Root Subspaces

In this section, we use finite-dimensional representations of $sl(2, \mathbb{C})$ to study the properties of roots and root subspaces of finite-dimensional semisimple Lie algebra $\mathscr{G}$. In particular, we prove that such a Lie algebra is generated by two elements.

Again we assume the base field $\mathbb{F} = \mathbb{C}$. For convenience, we also take $\partial_x = d/dx$. For any nonnegative integer n, set

$$V(n) = \sum_{i=0}^{n} \mathbb{C}x^i \subset \mathbb{C}[x]. \tag{2.6.1}$$

Recall

$$sl(2, \mathbb{C}) = \mathbb{C}E_{1,2} + \mathbb{C}(E_{1,1} - E_{2,2}) + \mathbb{C}E_{2,1}. \tag{2.6.2}$$

We define an action of $sl(2, \mathbb{C})$ on $V(n)$ by

$$E_{1,2}|_{V(n)} = x^2\partial_x - nx, \quad E_{2,1}|_{V(n)} = -\partial_x, \quad (E_{1,1} - E_{2,2})|_{V(n)} = 2x\partial_x - n. \tag{2.6.3}$$

It can be verified that $V(n)$ forms an irreducible $sl(2, \mathbb{C})$-module (exercise). For convenience, we denote

$$e = E_{1,2}, \quad f = E_{2,1}, \quad h = E_{1,1} - E_{2,2}. \tag{2.6.4}$$

For any $sl(2, \mathbb{C})$-module W, we denote

$$W_a = \{w \in W \mid h(w) = aw\} \quad \text{for } a \in \mathbb{C}. \tag{2.6.5}$$

For instance,

$$V(n)_{2i-n} = \mathbb{C}x^i, \qquad V(n) = \bigoplus_{i=0}^{n} V(n)_{2i-n}. \tag{2.6.6}$$

Theorem 2.6.1 *Any* $(n+1)$*-dimensional irreducible* $sl(2,\mathbb{C})$*-module is isomorphic to* $V(n)$*. A finite-dimensional* $sl(2,\mathbb{C})$*-module* W *is a direct sum of* $\dim W_0 + \dim W_1$ *irreducible submodules.*

Proof Let V be any $(n+1)$-dimensional irreducible $sl(2,\mathbb{C})$-module. Note that $\mathscr{B} = \mathbb{C}h + \mathbb{C}e$ forms a solvable Lie subalgebra of $sl(2,\mathbb{C})$. By Lie's Theorem, there exists a common eigenvector v of h and e in V; that is,

$$h(v) = \lambda v, \quad e(v) = \mu v, \qquad \lambda, \mu \in \mathbb{C}. \tag{2.6.7}$$

But

$$2\mu v = 2e(v) = [h, e](v) = h(e(v)) - e(h(v)) = \mu\lambda v - \lambda\mu v = 0. \tag{2.6.8}$$

Thus $\mu = 0$; that is, $e(v) = 0$.

Recall that $\mathbb{N}$ is the set of nonnegative integers. Note

$$h(f^i(v)) = (\lambda - 2i) f^i(v) \qquad \text{for } i \in \mathbb{N}. \tag{2.6.9}$$

Moreover,

$$\begin{aligned} e(f^{i+1}(v)) &= e(f((f^i(v)))) = [e, f](f^i(v)) + f(e((f^i(v)))) \\ &= (\lambda - 2i) f^i(v) + f(e((f^i(v)))). \end{aligned} \tag{2.6.10}$$

By induction, we have

$$e(f^{i+1}(v)) = \left((i+1)\lambda - 2\sum_{r=0}^{i} r\right) f^i(v) = (i+1)(\lambda - i) f^i(v). \tag{2.6.11}$$

The above equation shows

$$f^i(v) \neq 0 \qquad \text{if } \lambda \notin \{0, 1, \ldots, i-1\}. \tag{2.6.12}$$

On the other hand, (2.6.9) implies that

$$\{v, f(v), \ldots, f^i(v) \mid \lambda \notin \{0, 1, \ldots, i-1\}\} \tag{2.6.13}$$

is a set of eigenvectors of h with distinct eigenvalues. So it is linearly independent. Since $\dim V = n+1$, $\lambda \in \{0, 1, \ldots, n\}$. By (2.6.11), $e(f^{\lambda+1}(v)) = 0$. Hence

$$\sum_{i=1}^{\infty} \mathbb{C} f^{\lambda+i}(v) \tag{2.6.14}$$

forms a proper $sl(2,\mathbb{C})$ submodule of V. The irreducibility of V implies

$$f^{\lambda+1}(v) = 0. \tag{2.6.15}$$

Now $\sum_{i=0}^{\lambda}\mathbb{C}f^i(v)$ forms a nonzero submodule of V. Thus

$$V = \sum_{i=0}^{\lambda}\mathbb{C}f^i(v) \tag{2.6.16}$$

and $\{v, f(v), \ldots, f^{\lambda}(v)\}$ is a basis by (2.6.9). Therefore, $\lambda = n$.

By (2.6.3) and (2.6.4),

$$f^i(x^n) = (-1)^i n(n-1)\cdots(n-i+1)x^{n-i} \quad \text{for } i \in \{0, 1, \ldots, n\}. \tag{2.6.17}$$

Define a linear map $\sigma : V(n) \to V$ by

$$\sigma(f^i(x^n)) = f^i(v) \quad \text{for } i \in \{0, 1, \ldots, n\}. \tag{2.6.18}$$

The Eqs. (2.6.9), (2.6.11) and (2.6.15) show that the linear map σ is $sl(2,\mathbb{C})$-module isomorphism.

Expressions (2.6.5) and (2.6.6) imply that

$$\dim V_0 = 0, \quad \dim V_1 = 1 \quad \text{if } n \in 2\mathbb{N}+1 \tag{2.6.19}$$

and

$$\dim V_0 = 1, \quad \dim V_1 = 0 \quad \text{if } n \in 2\mathbb{N}. \tag{2.6.20}$$

Now Weyl's Theorem tell us that any finite-dimensional $sl(2,\mathbb{C})$-module W is a direct sum of $\dim W_0 + \dim W_1$ irreducible submodules. $\square$

Let $\mathscr{G}$ be a finite-dimensional semisimple Lie algebra and let H be a maximal toral subalgebra of $\mathscr{G}$. Then we have the root space decomposition:

$$\mathscr{G} = H + \sum_{\alpha\in\Phi}\mathscr{G}_\alpha, \tag{2.6.21}$$

where Φ is the root system of $\mathscr{G}$. Recall Lemma 2.5.4.

Lemma 2.6.2 *(a)* $\dim\mathscr{G}_\alpha = 1$ *and* $\mathbb{C}\alpha\bigcap\Phi = \{\alpha, -\alpha\}$ *for any* $\alpha \in \Phi$. *In particular,* $\mathscr{S}_\alpha = \mathscr{G}_\alpha + \mathbb{C}h_\alpha + \mathscr{G}_{-\alpha}$ *(cf. (2.5.20)).*

(b) If $\alpha, \beta \in \Phi$, *then* $\beta(h_\alpha) \in \mathbb{Z}$ *(which is called* **Cartan integer***) and* $\beta - \beta(h_\alpha)\alpha \in \Phi$.

(c) If $\alpha, \beta, \alpha+\beta \in \Phi$, *then* $[\mathscr{G}_\alpha, \mathscr{G}_\beta] = \mathscr{G}_{\alpha+\beta}$.

(d) Let $\alpha, \beta \in \Phi$, $\beta \neq \pm\alpha$. Let r, q be (respectively) the largest integers such that $\beta - r\alpha$ and $\beta + q\alpha$ are roots. Then all $\beta + i\alpha \in \Phi$ for $-r \leq i \leq q$.

(e) $\mathscr{G}$ is generated by $\{\mathscr{G}_\alpha \mid \alpha \in \Phi\}$.

Proof Denote $\dim H = n$. Let $\alpha \in \Phi$ be any element. Recall the Lie subalgebra $\mathscr{S}_\alpha$ defined in (2.5.20). By the above lemma, all the eigenvalues of $\mathrm{ad}\, h_\alpha$ are integers. If $a\alpha \in \Phi$, then

$$a\alpha(h_\alpha) = 2a \in \mathbb{Z} \Longrightarrow a \in \frac{\mathbb{Z}}{2}. \tag{2.6.22}$$

Note that

$$\mathscr{L} = H + \sum_{m\in\mathbb{Z}} \mathscr{G}_{m\alpha} \tag{2.6.23}$$

forms an $\mathrm{ad}\, \mathscr{S}_\alpha$-submodule of $\mathscr{G}$. Moreover, the eigenvalues of $\mathrm{ad}\, h_\alpha$ in $\mathscr{L}$ are even integers by (2.6.22). Thus $\mathscr{L}$ is direct sum of n irreducible $\mathscr{S}_\alpha$-submodules. Set

$$H' = \{h \in H \mid \alpha(h) = 0\}. \tag{2.6.24}$$

For any $h \in H'$, we have

$$[h, \xi_\alpha] = \alpha(h)\xi_\alpha = 0, \quad [h, \zeta_\alpha] = -\alpha(h)\zeta_\alpha = 0, \quad [h, h_\alpha] = 0 \tag{2.6.25}$$

(cf. Lemma 2.5.4). Since $\mathscr{S}_\alpha = \mathbb{C}\xi_\alpha + \mathbb{C}h_\alpha + \mathbb{C}\zeta_\alpha$, H' is a direct sum of $(n-1)$ one-dimensional irreducible $\mathrm{ad}\, \mathscr{S}_\alpha$-submodules. Thus $\mathscr{S}_\alpha + H'$ is a direct sum of n irreducible $\mathrm{ad}\, \mathscr{S}_\alpha$-submodules. Therefore, the complement of $\mathscr{S}_\alpha + H'$ in $\mathscr{L}$ must be zero; that is, $\mathscr{L} = H' + \mathscr{S}_\alpha$. So $\mathscr{S}_\alpha = \mathscr{G}_\alpha + \mathbb{C}h_\alpha + \mathscr{G}_{-\alpha}$, $\dim \mathscr{G}_\alpha = \dim \mathbb{C}\xi_\alpha = 1$ and $\mathbb{Z}\alpha \bigcap \Phi = \{\alpha, -\alpha\}$. In particular, $2\alpha \notin \Phi$ for any $\alpha \in \Phi$. This shows $\alpha/2 \notin \Phi$; that is, $\mathscr{G}_{\alpha/2} = \{0\}$. Since $\bar{\mathscr{L}} = \sum_{m\in\mathbb{Z}} \mathscr{G}_{m\alpha+\alpha/2}$ forms an $\mathrm{ad}\, \mathscr{S}_\alpha$-submodule, the eigenvalues of $\mathrm{ad}\, h_\alpha$ in $\bar{\mathscr{L}}$ are odd integers and the eigenspace of the eigenvalue 1 is $\mathscr{G}_{\alpha/2} = \{0\}$, we must have $\bar{\mathscr{L}} = \{0\}$ by the above lemma. According to (2.6.22), we get $\mathbb{C}\alpha \bigcap \Phi = \{\alpha, -\alpha\}$ for any $\alpha \in \Phi$. This proves (a).

Let $\alpha, \beta \in \Phi$, $\beta \neq \pm\alpha$. Then

$$\mathscr{K} = \sum_{i\in\mathbb{Z}} \mathscr{G}_{\beta+i\alpha} \tag{2.6.26}$$

forms an $\mathrm{ad}\, \mathscr{S}_\alpha$-submodule of $\mathscr{G}$. Since $\mathscr{G}_\beta$ is an eigenspace of $\mathrm{ad}\, h_\alpha$ with the eigenvalue $\beta(h_\alpha)$, we must have $\beta(h_\alpha) \in \mathbb{Z}$. By (a), $\dim \mathscr{G}_{\beta+i\alpha} \leq 1$. If $\beta + i\alpha \in \Phi$, then $\mathscr{G}_{\beta+i\alpha}$ is an eigenspace of $\mathrm{ad}\, h_\alpha$ with the eigenvalue $\beta(h_\alpha) + 2i$. Thus

$$\dim\{\xi \in \mathscr{K} \mid [h_\alpha, \xi] = 0\} + \dim\{\zeta \in \mathscr{K} \mid [h_\alpha, \zeta] = \zeta\} \leq 1. \tag{2.6.27}$$

By the above lemma, $\mathscr{K}$ is an irreducible $\mathrm{ad}\, \mathscr{S}_\alpha$-module. Let r, q be (respectively) the largest integers such that $\beta - r\alpha$ and $\beta + q\alpha$ are roots. Then all $\beta + i\alpha \in \Phi$ for $-r \leq i \leq q$ by (2.6.6) and the above lemma. This proves (d). If $\alpha + \beta \in \Phi$, then

$[\xi_\alpha, \mathscr{G}_\beta] = \mathscr{G}_{\alpha+\beta}$ by (2.6.11). So (c) holds. Expression (2.6.6) shows

$$-(\beta(h_\alpha) - 2r) = \dim \mathscr{K} - 1 = \beta(h_\alpha) + 2q, \tag{2.6.28}$$

which implies $\beta(h_\alpha) = r - q$. Furthermore,

$$\beta - \beta(h_\alpha)\alpha = \beta - (r-q)\alpha = \beta + q\alpha - r\alpha \in \Phi. \tag{2.6.29}$$

Thus (b) holds. Since Φ spans H^*, $\{t_\alpha \mid \alpha \in \Phi\}$ spans H. The equations in (2.5.18) and (2.5.19) yield (e). □

Proposition 2.6.3 *The semisimple Lie algebra $\mathscr{G}$ is generated by two elements.*

Proof Denote

$$\Psi = \{\alpha - \beta \mid \alpha, \beta \in \Phi,\ \alpha \neq \beta\}. \tag{2.6.30}$$

Take

$$h_0 \in H \setminus \bigcup_{\gamma \in \Psi} \{h \in H \mid \gamma(h) = 0\}. \tag{2.6.31}$$

For each $\alpha \in \Phi$, we pick $0 \neq \xi_\alpha \in \mathscr{G}_\alpha$. Then $\{\xi_\alpha \mid \alpha \in \Phi\}$ are eigenvectors of $\mathrm{ad}\, h_0$ with distinct eigenvalues $\{\alpha(h_0) \mid \alpha \in \Phi\}$. Set

$$\xi = \sum_{\alpha \in \Phi} \xi_\alpha. \tag{2.6.32}$$

Let $\mathscr{L}$ be the Lie subalgebra of $\mathscr{G}$ generated by h_0 and ξ. Then

$$\mathrm{ad}\, h_0(\mathscr{L}) \subset \mathscr{L}, \quad \mathrm{ad}\, h_0(\xi) \in \mathscr{L}. \tag{2.6.33}$$

By Lemma 1.3.1,

$$\xi_\alpha \in \mathscr{L} \qquad \text{for } \alpha \in \Phi. \tag{2.6.34}$$

By Lemma 2.6.2 (e), $\mathscr{L} = \mathscr{G}$. □

Define

$$(\alpha, \beta) = \kappa(t_\alpha, t_\beta) \qquad \text{for } \alpha, \beta \in H^* \tag{2.6.35}$$

(cf. (2.5.13)). Then $(\cdot,\cdot)$ is a nondegenerate symmetric bilinear form on H^*. By the arguments above (2.5.18), $(\alpha, \alpha) = \alpha(t_\alpha) \neq 0$ for $\alpha \in \Phi$. Moreover, we set

$$\langle \beta, \alpha \rangle = \frac{2(\beta, \alpha)}{(\alpha, \alpha)} \qquad \text{for } \alpha \in \Phi,\ \beta \in H^*. \tag{2.6.36}$$

By (2.5.18) and Lemma 2.6.2 (b),

$$\langle \alpha, \beta \rangle \in \mathbb{Z}, \quad \beta - \langle \beta, \alpha \rangle \alpha \in \Phi \qquad \text{for } \alpha, \beta \in \Phi. \tag{2.6.37}$$

Since Φ spans H^*, Φ contains a basis $\{\alpha_1, \dots, \alpha_n\}$ of H^*. For any $\beta \in \Phi$, $\beta = \sum_{i=1}^n c_i\alpha_i$ with $c_i \in \mathbb{C}$. Note

$$\sum_{i=1}^n c_i\langle\alpha_i, \alpha_j\rangle = \langle\beta, \alpha_j\rangle, \qquad j = 1, 2, \dots, n. \tag{2.6.38}$$

View $\{c_1, c_2, \dots, c_n\}$ as unknowns. The coefficients in the above linear systems are integers. The coefficient determinant

$$|(\langle\alpha_i, \alpha_j\rangle)_{n\times n}| = \frac{2^n}{\prod_{r=1}^n(\alpha_r, \alpha_r)}|((\alpha_i, \alpha_j))_{n\times n}| \neq 0 \tag{2.6.39}$$

by the nondegeneracy of $(\cdot, \cdot)$ on H^*. Solving (2.6.38) for $\{c_i\}$ in terms of $\{\langle\alpha_i, \alpha_j\rangle, \langle\beta, \alpha_j\rangle\}$, we obtain

$$c_i \in \mathbb{Q}, \qquad i = 1, 2, \dots, n. \tag{2.6.40}$$

Thus the $\mathbb{Q}$-subspace

$$\Gamma = \sum_{\alpha\in\Phi} \mathbb{Q}\alpha \tag{2.6.41}$$

of H^* is n-dimensional.

By (2.6.21),

$$(\lambda, \mu) = \operatorname{Tr} \operatorname{ad} t_\lambda \operatorname{ad} t_\mu = \sum_{\alpha\in\Phi} \alpha(t_\lambda)\alpha(t_\mu) = \sum_{\alpha\in\Phi}(\alpha, \lambda)(\alpha, \mu) \qquad \text{for } \lambda, \mu \in H^*. \tag{2.6.42}$$

In particular,

$$(\beta, \beta) = \sum_{\alpha\in\Phi}(\alpha, \beta)^2 \qquad \text{for } \beta \in \Phi, \tag{2.6.43}$$

equivalently,

$$\frac{1}{(\beta, \beta)} = \frac{1}{4}\sum_{\alpha\in\Phi}\left(\frac{2(\alpha, \beta)}{(\beta, \beta)}\right)^2 = \frac{1}{4}\sum_{\alpha\in\Phi}\langle\alpha, \beta\rangle^2 \in \mathbb{Q}. \tag{2.6.44}$$

So $(\beta, \beta) \in \mathbb{Q}$. For $\alpha, \beta \in \phi$, we have $\alpha - \langle\alpha, \beta\rangle\beta \in \Phi$ by the above lemma. Hence

$$\langle\alpha, \beta\rangle(\alpha, \beta) = \frac{1}{2}[(\alpha, \alpha) + \langle\alpha, \beta\rangle^2(\beta, \beta) - (\alpha - \langle\alpha, \beta\rangle\beta, \alpha - \langle\alpha, \beta\rangle\beta)] \in \mathbb{Q}. \tag{2.6.45}$$

Therefore,

$$(\alpha, \beta) \in \mathbb{Q} \qquad \text{for } \alpha, \beta \in \Phi. \tag{2.6.46}$$

Thus $(\cdot, \cdot)$ is a positive definite $\mathbb{Q}$-valued symmetric $\mathbb{Q}$-bilinear form on Γ by (2.6.42) and (2.6.46). Extend $(\cdot, \cdot)$ on

$$\mathscr{E} = \Gamma_{\mathbb{R}} = \mathbb{R} \otimes_{\mathbb{Q}} \Gamma \tag{2.6.47}$$

$\mathbb{R}$-bilinearly. Then $\mathscr{E}$ is isomorphic to the Euclidean space $\mathbb{R}^n$. Now we obtain a main theorem.

Theorem 2.6.4 *Let $\mathscr{G}$ be a finite-dimensional semisimple Lie algebra. There exists a maximal toral subalgebra H, whose dimension denoted by n. With respect to H, $\mathscr{G}$ has a root space decomposition (***Cartan decomposition***) (2.6.21), where $\dim \mathscr{G}_\alpha = 1$ and if $\alpha, \beta, \alpha + \beta \in \Phi$, then $[\mathscr{G}_\alpha, \mathscr{G}_\beta] = \mathscr{G}_{\alpha+\beta}$. Moreover, Φ can be identified with a finite subset of the Euclidean space $\mathbb{R}^n$ satisfying: (a) Φ spans $\mathbb{R}^n$ and $0 \notin \Phi$; (b) $\mathbb{R}\alpha \bigcap \Phi = \{\alpha, -\alpha\}$ for $\alpha \in \Phi$; (c) If $\alpha, \beta \in \Phi$, then $2(\beta, \alpha)/(\alpha, \alpha) \in \mathbb{Z}$ and $\beta - [2(\beta, \alpha)/(\alpha, \alpha)]\alpha \in \Phi$.*

As an exercise, find all finite-dimensional real irreducible modules for $o(3, \mathbb{R})$ and $o(4, \mathbb{R})$ defined in (1.2.22).

References

[X7] X. Xu, Generalizations of the Block algebras. Manuscr. Math. **100**, 489–518 (1999)

[X8] X. Xu, New generalized simple Lie algebras of Cartan type over a field with characteristic. J. Algebra **224**, 23–58 (2000)

Chapter 3
Root Systems

As we showed in last chapter, any finite-dimensional semisimple Lie algebra give rise to a root system. Thus it is necessary to study root systems from more general point of view. In this chapter, we start with the axiom of root system and give the root systems of special linear algebras, orthogonal Lie algebras and symplectic Lie algebras. Then we derive some basic properties of root systems; in particular, the existence of the bases of root systems. As finite symmetries of root systems, the Weyl groups are introduced and studied in detail. Classification and explicit construction of root systems are presented. The automorphism groups of roots systems are determined. As a preparation for later representation theory of Lie algebras, the corresponding weight lattices and their saturated subsets are investigated.

In Sect. 3.1, one can find the basic definitions on root systems, examples and properties. In Sect. 3.2, we introduce the Weyl group associated to each root system and use it to study the structure of the root system. We classify all the irreducible root systems in Sect. 3.3 by using graphs in an Euclidean space. Finally in Sect. 3.4, we construct all the irreducible root systems, their weight lattices, automorphism groups and the saturated subsets of the weight lattices.

3.1 Definitions, Examples and Properties

In this section, we present the basic definitions on root systems, examples and properties.

An n-dimensional Euclidean space $\mathscr{E}$ is the vector space $\mathbb{R}^n$ with a positive definite symmetric $\mathbb{R}$-bilinear form $(\cdot,\cdot)$ (called *inner product*). For $\alpha,\beta\in\mathscr{E}$ with $\alpha\neq 0$, we define

$$\langle\beta,\alpha\rangle=\frac{2(\beta,\alpha)}{(\alpha,\alpha)} \tag{3.1.1}$$

X. Xu, *Representations of Lie Algebras and Partial Differential Equations*,
DOI 10.1007/978-981-10-6391-6_3

and

$$P_\alpha = \{\gamma \in \mathscr{E} \mid (\gamma, \alpha) = 0\}. \tag{3.1.2}$$

Note that the number $\langle \beta, \alpha \rangle$ and P_α remain unchanged if we replace $(\cdot, \cdot)$ by $c(\cdot, \cdot)$ for any $0 \neq c \in \mathbb{R}$. Moreover, $(\alpha, \alpha) > 0$ implies

$$\mathscr{E} = \mathbb{R}\alpha + P_\alpha. \tag{3.1.3}$$

Furthermore, we define a linear transformation σ_α by

$$\sigma_\alpha(\beta) = \beta - \langle \beta, \alpha \rangle \alpha \qquad \text{for } \beta \in \mathscr{E}. \tag{3.1.4}$$

Then $\sigma_\alpha|_{P_\alpha} = \mathrm{Id}_{P_\alpha}$ and

$$\sigma_\alpha(\alpha) = \alpha - \langle \alpha, \alpha \rangle \alpha = \alpha - \frac{2(\alpha, \alpha)}{(\alpha, \alpha)} \alpha = -\alpha. \tag{3.1.5}$$

So σ_α is the reflection with respect to the hyperplane P_α. Moreover, $\sigma_{a\alpha} = \sigma_\alpha$ for any $0 \neq a \in \mathbb{R}$. For $\beta, \gamma \in \mathscr{E}$, we write $\beta = \beta_1 + \beta_2$ and $\gamma = \gamma_1 + \gamma_2$ with $\beta_1, \gamma_1 \in \mathbb{R}\alpha$ and $\beta_2, \gamma_2 \in P_\alpha$, and get

$$\langle \sigma_\alpha(\beta), \sigma_\alpha(\gamma) \rangle = \langle -\beta_1 + \beta_2, -\gamma_1 + \gamma_2 \rangle = \langle \beta_1, \gamma_1 \rangle + \langle \beta_2, \gamma_2 \rangle = \langle \beta, \gamma \rangle. \tag{3.1.6}$$

Thus σ_α is an isometry.

A *root system* Φ in $\mathscr{E}$ is a finite set spanning $\mathscr{E}$ such that: (1) $0 \notin \Phi$ and $\mathbb{R}\alpha \bigcap \Phi = \{\alpha, -\alpha\}$ for $\alpha \in \Phi$; (2) $\langle \beta, \alpha \rangle \in \mathbb{Z}$ and $\sigma_\alpha(\Phi) = \Phi$ for $\alpha, \beta \in \Phi$.

Recall that h_α in a semisimple Lie algebra $\mathscr{G} = H + \sum_{\alpha \in \Phi} \mathscr{G}_\alpha$ is uniquely determined by

$$[\mathscr{G}_\alpha, \mathscr{G}_{-\alpha}] = \mathbb{C} h_\alpha, \quad \mathrm{ad}\, h_\alpha|_{\mathscr{G}_\alpha} = 2\mathrm{Id}_{\mathscr{G}_\alpha}. \tag{3.1.7}$$

Moreover,

$$\langle \beta, \alpha \rangle = \beta(h_\alpha) \tag{3.1.8}$$

by (2.5.18), (2.5.19), (2.6.35) and (2.6.36).

Example 3.1.1 The special linear Lie algebra

$$sl(n, \mathbb{C}) = \sum_{1 \leq i < j \leq n} (\mathbb{C}E_{i,j} + \mathbb{C}E_{j,i}) + \sum_{i=1}^{n-1} \mathbb{C}(E_{i,i} - E_{i+1,i+1}) \tag{3.1.9}$$

and recall the notion of indices: $\overline{1, n} = \{1, 2, \ldots, n\}$. Take

$$H = \sum_{i=1}^{n-1} \mathbb{C}(E_{i,i} - E_{i+1,i+1}) = \{\sum_{i=1}^{n} a_i E_{i,i} \mid a_i \in \mathbb{C}, \sum_{j=1}^{n} a_j = 0\}. \tag{3.1.10}$$

For $i \in \overline{1,n}$, we define $\varepsilon_i \in H^*$ by

$$\varepsilon_i(\sum_{j=1}^{n} a_j E_{j,j}) = a_i. \tag{3.1.11}$$

Then the root system of $sl(n, \mathbb{C})$ is

$$\Phi_{A_{n-1}} = \{\varepsilon_i - \varepsilon_j \mid i, j \in \overline{1,n}, \; i \neq j\} \tag{3.1.12}$$

and

$$\mathscr{G}_{\varepsilon_i - \varepsilon_j} = \mathbb{C}E_{i,j}, \quad h_{\varepsilon_i - \varepsilon_j} = E_{i,i} - E_{j,j}, \quad i, j \in \overline{1,n}, \; i \neq j. \tag{3.1.13}$$

Example 3.1.2 The even orthogonal Lie algebra

$$o(2n, \mathbb{C}) = \sum_{i,j=1}^{n} (\mathbb{C}(E_{i,j} - E_{n+j,n+i}) + \sum_{1 \leq i < j \leq n} [\mathbb{C}(E_{i,n+j} - E_{j,n+i}) + \mathbb{C}(E_{n+j,i} - E_{n+i,j})]. \tag{3.1.14}$$

Take

$$H = \sum_{i=1}^{n} \mathbb{C}(E_{i,i} - E_{n+i,n+i}) \tag{3.1.15}$$

and define

$$\varepsilon_i(\sum_{j=1}^{n} a_j (E_{j,j} - E_{n+j,n+j})) = a_i \qquad \text{for } \; i \in \overline{1,n}. \tag{3.1.16}$$

Then the root system of $o(2n, \mathbb{C})$ is

$$\Phi_{D_n} = \{\varepsilon_i - \varepsilon_j, \pm(\varepsilon_i + \varepsilon_j) \mid i, j \in \overline{1,n}, \; i \neq j\}. \tag{3.1.17}$$

Moreover,

$$\mathscr{G}_{\varepsilon_i - \varepsilon_j} = \mathbb{C}(E_{i,j} - E_{n+j,n+i}), \quad \mathscr{G}_{\varepsilon_i + \varepsilon_j} = \mathbb{C}(E_{i,n+j} - E_{j,n+i}), \tag{3.1.18}$$

$$\mathscr{G}_{-\varepsilon_i - \varepsilon_j} = \mathbb{C}(E_{n+i,j} - E_{n+j,i}) \tag{3.1.19}$$

and

$$h_{\varepsilon_i - \varepsilon_j} = E_{i,i} - E_{j,j} - E_{n+i,n+i} + E_{n+j,n+j}, \tag{3.1.20}$$

$$h_{\varepsilon_i+\varepsilon_j} = -h_{-\varepsilon_i-\varepsilon_j} = E_{i,i} + E_{j,j} - E_{n+i,n+i} - E_{n+j,n+j}, \tag{3.1.21}$$

for $i, j \in \overline{1,n},\ i \neq j$.

Example 3.1.3 The symplectic Lie algebra

$$sp(2n,\mathbb{C}) = \sum_{i,j=1}^{n} \mathbb{C}(E_{i,j} - E_{n+j,n+i})$$
$$+ \sum_{1\leq i\leq j\leq n} [\mathbb{C}(E_{i,n+j} + E_{j,n+i}) + \mathbb{C}(E_{n+j,i} + E_{n+i,j})]. \tag{3.1.22}$$

Take H in (3.1.15) and ε_i in (3.1.16). Then the root system of $sp(2n,\mathbb{C})$ is

$$\Phi_{C_n} = \{\varepsilon_i - \varepsilon_j, \pm(\varepsilon_i + \varepsilon_j), \pm 2\varepsilon_i \mid i, j \in \overline{1,n},\ i \neq j\}. \tag{3.1.23}$$

Moreover,

$$\mathscr{G}_{\varepsilon_i-\varepsilon_j} = \mathbb{C}(E_{i,j} - E_{n+j,n+i}),\quad \mathscr{G}_{\varepsilon_i+\varepsilon_j} = \mathbb{C}(E_{i,n+j} + E_{j,n+i}), \tag{3.1.24}$$

$$\mathscr{G}_{-\varepsilon_i-\varepsilon_j} = \mathbb{C}(E_{n+i,j} + E_{n+j,i}),\quad \mathscr{G}_{2\varepsilon_i} = \mathbb{C}E_{i,n+i},\quad \mathscr{G}_{-2\varepsilon_i} = \mathbb{C}E_{n+i,i}, \tag{3.1.25}$$

$$h_{2\varepsilon_i} = -h_{-2\varepsilon_i} = E_{i,i} - E_{n+i,n+i}, \tag{3.1.26}$$

(3.1.20) and (3.1.21) hold, for $i, j \in \overline{1,n},\ i \neq j$.

Example 3.1.4 The odd orthogonal Lie algebra

$$o(2n+1,\mathbb{C}) = o(2n,\mathbb{C}) + \sum_{i=1}^{n} [\mathbb{C}(E_{0,i} - E_{n+i,0}) + \mathbb{C}(E_{0,n+i} - E_{i,0})]. \tag{3.1.27}$$

Take H in (3.1.15) and ε_i in (3.1.16). Then the root system of $o(2n+1,\mathbb{C})$ is

$$\Phi_{B_n} = \{\varepsilon_i - \varepsilon_j, \pm(\varepsilon_i + \varepsilon_j), \pm\varepsilon_i \mid i, j \in \overline{1,n},\ i \neq j\}. \tag{3.1.28}$$

Moreover, (3.1.18)–(3.1.21) hold and

$$\mathscr{G}_{-\varepsilon_i} = \mathbb{C}(E_{0,i} - E_{n+i,0}),\quad \mathscr{G}_{\varepsilon_i} = \mathbb{C}(E_{0,n+i} - E_{i,0}), \tag{3.1.29}$$

$$h_{\varepsilon_i} = -h_{-\varepsilon_i} = 2(E_{i,i} - E_{n+i,n+i}), \tag{3.1.30}$$

for $i \in \overline{1,n}$.

Recall that the symmetric bilinear form $(\cdot,\cdot)$ on H^* for a semisimple Lie algebra $\mathcal{G}$ is induced from its Killing form κ. When $\mathcal{G}$ is simple, any invariant (associative) symmetric bilinear form is a scalar multiple of κ. In a root system Φ, only integers $\langle\beta,\alpha\rangle$ for $\alpha,\beta\in\Phi$ play roles in its structure. But the number $\langle\beta,\alpha\rangle$ and P_α remain unchanged if we replace $(\cdot,\cdot)$ by $c(\cdot,\cdot)$ for any $0\neq c\in\mathbb{R}$. In the above four examples, the bilinear form $\zeta(A,B)=\text{Tr}\,AB$ is a scalar multiple of κ. Replacing κ by ζ in Example 3.1.1, and by $\zeta/2$ in Examples 3.1.2–3.1.4, we have

$$(\varepsilon_i,\varepsilon_j)=\delta_{i,j}\qquad\text{for } i,j\in\overline{1,n}. \tag{3.1.31}$$

For any α,β in the Euclidean space $\mathcal{E}$, we have

$$|(\alpha,\beta)|\leq||\alpha||\,||\beta||,\qquad\text{where } ||\alpha||=\sqrt{(\alpha,\alpha)},\ ||\beta||=\sqrt{(\beta,\beta)}. \tag{3.1.32}$$

We define the angle θ between two nonzero vectors α and β by

$$\cos\theta=\frac{(\alpha,\beta)}{||\alpha||\,||\beta||}. \tag{3.1.33}$$

Then

$$\langle\alpha,\beta\rangle\langle\beta,\alpha\rangle=4\left(\frac{(\alpha,\beta)}{||\alpha||\,||\beta||}\right)^2=4\cos^2\theta. \tag{3.1.34}$$

Moreover,

$$\frac{||\alpha||^2}{||\beta||^2}=\frac{\langle\alpha,\beta\rangle}{\langle\beta,\alpha\rangle}\qquad\text{if } (\alpha,\beta)\neq 0. \tag{3.1.35}$$

Thus we have:

Lemma 3.1.1 *Let α,β be two elements in a root system such that $\alpha\neq\pm\beta$ and $||\alpha||\leq||\beta||$. If $(\alpha,\beta)\neq 0$, then one of the following statements holds:*

Cases	$\langle\alpha,\beta\rangle$	$\langle\beta,\alpha\rangle$	θ	$\|\|\alpha\|\|^2/\|\|\beta\|\|^2$
(1)	1	1	$\pi/3$	1
(2)	-1	-1	$2\pi/3$	1
(3)	1	2	$\pi/4$	1/2
(4)	-1	-2	$3\pi/4$	1/2
(5)	1	3	$\pi/6$	1/3
(6)	-1	-3	$5\pi/6$	1/3

(3.1.36)

Proof By (3.1.34), (3.1.35) and the fact $0<\langle\alpha,\beta\rangle\langle\beta,\alpha\rangle\in\mathbb{Z}$. □

Lemma 3.1.2 *Let α, β be two roots in a root system Φ such that $\alpha \neq \pm\beta$. If $(\alpha, \beta) > 0$, then $\alpha - \beta \in \Phi$. When $(\alpha, \beta) < 0$, $\alpha + \beta \in \Phi$.*

Proof If $(\alpha, \beta) > 0$, then $\langle \alpha, \beta \rangle = 1$ or $\langle \beta, \alpha \rangle = 1$ by (3.1.36). Moreover,

$$\langle \alpha, \beta \rangle = 1 \Longrightarrow \sigma_\beta(\alpha) = \alpha - \langle \alpha, \beta \rangle \beta = \alpha - \beta \in \Phi, \tag{3.1.37}$$

$$\langle \beta, \alpha \rangle = 1 \Longrightarrow \sigma_\alpha(\beta) = \beta - \langle \beta, \alpha \rangle \alpha = \beta - \alpha \in \Phi \Longrightarrow \alpha - \beta = -(\beta - \alpha) \in \Phi. \tag{3.1.38}$$

When $(\alpha, \beta) < 0$, then $\langle \alpha, \beta \rangle = -1$ or $\langle \beta, \alpha \rangle = -1$. Moreover,

$$\langle \alpha, \beta \rangle = -1 \Longrightarrow \sigma_\beta(\alpha) = \alpha - \langle \alpha, \beta \rangle \beta = \alpha + \beta \in \Phi, \tag{3.1.39}$$

$$\langle \beta, \alpha \rangle = -1 \Longrightarrow \sigma_\alpha(\beta) = \beta - \langle \beta, \alpha \rangle \alpha = \beta + \alpha \in \Phi. \quad \square \tag{3.1.40}$$

Lemma 3.1.3 *Let α, β be two roots in a root system Φ such that $\alpha \neq \pm\beta$. Let r and q be the largest integers such that $\beta - r\alpha, \beta + q\alpha \in \Phi$. Then $\beta + i\alpha \in \Phi$ for any $-r \leq i \leq q$, and $\langle \beta, \alpha \rangle = r - q$.*

Proof Suppose that $\beta + i\alpha \notin \Phi$ for some $-r < i < q$. There exists integers s, p with $-r \leq s < i < p \leq q$ such that

$$\beta + s\alpha, \beta + p\alpha \in \Phi \quad \text{and} \quad \beta + (s+1)\alpha, \beta + (p-1)\alpha \notin \Phi. \tag{3.1.41}$$

By the above lemma,

$$(\beta + s\alpha, \alpha) \geq 0 \geq (\beta + p\alpha, \alpha) \Longrightarrow s(\alpha, \alpha) \geq p(\alpha, \alpha) \sim s > p, \tag{3.1.42}$$

which is absurd. Thus $\beta + i\alpha \in \Phi$ for any $-r \leq i \leq q$.

Note that

$$\sigma_\alpha(\beta + i\alpha) = \beta - \langle \beta, \alpha \rangle \alpha - i\alpha \in \Phi \qquad \text{for} \quad -r \leq i \leq q. \tag{3.1.43}$$

Thus σ_α flips the α-string through β:

$$(\beta - r\alpha, \ldots, \beta, \ldots, \beta + q\alpha) \mapsto (\beta - (\langle \beta, \alpha \rangle + q)\alpha, \ldots, \beta, \ldots, \beta + (r - \langle \beta, \alpha \rangle)\alpha). \tag{3.1.44}$$

Hence

$$\langle \beta, \alpha \rangle + q = r \Longrightarrow \langle \beta, \alpha \rangle = r - q. \quad \square \tag{3.1.45}$$

Suppose that Φ is a root system in a Euclidean space $\mathscr{E}$. A subset Π of Φ is called a *base* if Π is a basis of $\mathscr{E}$ and $\beta = \sum_{\alpha \in \Pi} k_\alpha \alpha$ for any $\beta \in \Phi$ such that all $k_\alpha \in \mathbb{N}$ or all $-k_\alpha \in \mathbb{N}$, where $\mathbb{N}$ is the set of all nonnegative integers.

A vector $\gamma \in \mathscr{E} \setminus \bigcup_{\alpha\in\Phi} P_\alpha$ is called *regular*. Since $\dim P_\alpha = \dim \mathscr{E} - 1$, $\mathscr{E} \neq \bigcup_{\alpha\in\Phi} P_\alpha$. So regular vector exists. For a regular vector $\gamma \in \mathscr{E}$, we define

$$\Phi^+(\gamma) = \{\alpha \in \Phi \mid (\gamma, \alpha) > 0\}, \quad \Phi^-(\gamma) = -\Phi^+(\gamma). \tag{3.1.46}$$

Then

$$\Phi = \Phi^+(\gamma) \bigcup \Phi^-(\gamma). \tag{3.1.47}$$

An element $\beta \in \Phi^+(\gamma)$ is called *decomposable* if there exist $\beta_1, \beta_2 \in \Phi^+(\gamma)$ such that $\beta = \beta_1 + \beta_2$. In this case, $(\beta, \gamma) = (\beta_1, \gamma) + (\beta_2, \gamma)$. In particular, (β_1, γ), $(\beta_2, \gamma) < (\beta, \gamma)$. Thus the elements $\alpha \in \Phi^+(\gamma)$ with minimal (α, γ) are indecomposable elements. Set

$$\Pi(\gamma) = \text{the set of all indecomposble elements of } \Phi^+(\gamma). \tag{3.1.48}$$

Since $\{(\alpha, \gamma) \mid \alpha \in \Phi\}$ is a finite set of positive numbers, any element $\beta \in \Phi^+(\gamma)$ can be written as

$$\beta = \sum_{\alpha\in\Pi(\gamma)} k_\alpha \alpha, \qquad k_\alpha \in \mathbb{N}, \tag{3.1.49}$$

by induction on (β, γ).

Theorem 3.1.4 *A subset of Φ is base if and only if it is of the form $\Pi(\gamma)$.*

Proof Let $\alpha, \beta \in \Pi(\gamma)$ such that $\alpha \neq \beta$. If $\alpha - \beta \in \Phi$, then either $\alpha - \beta \in \Phi^+(\gamma)$ or $\beta - \alpha = -(\alpha - \beta) \in \Phi^+(\gamma)$, which is impossible because

$$\alpha = (\alpha - \beta) + \beta, \qquad \beta = (\beta - \alpha) + \alpha. \tag{3.1.50}$$

By Lemma 3.1.2,

$$(\alpha, \beta) \leq 0 \qquad \text{for } \alpha, \beta \in \Pi(\gamma),\ \alpha \neq \beta. \tag{3.1.51}$$

Suppose

$$\sum_{\alpha\in\Pi(\gamma)} a_\alpha \alpha = 0 \qquad \text{with } a_\alpha \in \mathbb{R}. \tag{3.1.52}$$

Write

$$\varepsilon = \sum_{\alpha\in\Pi(\gamma),\ a_\alpha\geq 0} a_\alpha \alpha = -\sum_{\beta\in\Pi(\gamma),\ a_\beta\leq 0} a_\beta \beta. \tag{3.1.53}$$

By (3.1.53), we have

$$(\varepsilon, \varepsilon) = (\sum_{\alpha\in\Pi(\gamma),\, a_\alpha\geq 0} a_\alpha\alpha, -\sum_{\beta\in\Pi(\gamma),\, a_\beta\leq 0} a_\beta\beta)$$
$$= \sum_{\alpha\in\Pi(\gamma),\, a_\alpha\geq 0}\ \sum_{\beta\in\Pi(\gamma),\, a_\beta\leq 0} a_\alpha(-a_\beta)(\alpha, \beta) \leq 0. \quad (3.1.54)$$

Thus $(\varepsilon, \varepsilon) = 0$; that is,

$$\sum_{\alpha\in\Pi(\gamma),\, a_\alpha\geq 0} a_\alpha\alpha = -\sum_{\beta\in\Pi(\gamma),\, a_\beta\leq 0} a_\beta\beta = \varepsilon = 0. \quad (3.1.55)$$

Hence

$$\sum_{\alpha\in\Pi(\gamma),\, a_\alpha\geq 0} a_\alpha(\gamma, \alpha) = -\sum_{\beta\in\Pi(\gamma),\, a_\beta\leq 0} a_\beta(\gamma, \beta) = 0, \quad (3.1.56)$$

or equivalently,

$$a_\alpha = 0 \qquad \text{for } \alpha \in \Pi(\gamma). \quad (3.1.57)$$

Therefore, $\Pi(\gamma)$ is linearly independent. Thus $\Pi(\gamma)$ is a base by (3.1.47) and (3.1.49) because Φ spans $\mathscr{E}$.

Conversely, we assume that Π is a base of Φ. Denote

$$\Pi = \{\alpha_1, \alpha_2, \ldots, \alpha_n\}. \quad (3.1.58)$$

So $\dim \mathscr{E} = n$. Set

$$\mathscr{E}_i = \{\beta \in \mathscr{E} \mid (\beta, \alpha_j) = 0 \text{ for } i \neq j \in \overline{1, n}\}. \quad (3.1.59)$$

Then $\dim \mathscr{E}_i = 1$ and we can take $\gamma_i \in \mathscr{E}_i$ such that $(\gamma_i, \alpha_i) > 0$. Denote

$$\gamma = \sum_{i=1}^{n} \gamma_i. \quad (3.1.60)$$

We have

$$(\gamma, \alpha_i) > 0 \qquad \text{for } i \in \overline{1, n}; \quad (3.1.61)$$

that is, $\Pi \subset \Phi^+(\gamma)$. If some $\alpha_i = \beta_1 + \beta_2$ with $\beta_1, \beta_2 \in \Phi^+(\gamma)$, then

$$\beta_1 = \sum_{j=1}^{n} l_j\alpha_j, \quad \beta_2 = \sum_{j=1}^{n} k_j\alpha_j, \qquad l_j, k_j \in \mathbb{N}, \quad (3.1.62)$$

by (3.1.61) and the definition of base. Note

$$\alpha_i = \sum_{j=1}^{n}(l_j + k_j)\alpha_j \Longrightarrow \sum_{j=1}^{n}(l_i + k_i) = 1 \Longrightarrow \beta_1 = 0 \text{ or } \beta_2 = 0, \quad (3.1.63)$$

which is absurd. Hence all α_i are indecomposable; that is, $\Pi \subset \Pi(\gamma)$. But $|\Pi(\gamma)| = \dim \mathscr{E} = |\Pi|$. Therefore, $\Pi = \Pi(\gamma)$. $\square$

If a base $\Pi = \Pi(\gamma)$, we simply denote $\Phi^+ = \Phi^+(\gamma)$. We call the elements in Φ^+ *positive roots*. Correspondingly, the elements in $\Phi^- = -\Phi^+$ are called *negative roots*. The elements in Π are called *simple roots*.

Lemma 3.1.5 *If α is positive but not simple, then there exists $\beta \in \Pi$ such that $\alpha - \beta \in \Phi^+$.*

Proof Otherwise, we have $(\alpha, \beta) \leq 0$ for any $\beta \in \Pi$. By the arguments in the above theorem, $\{\beta\} \bigcup \Pi$ is linearly independent, which constricts to the fact that Π is base. $\square$

This lemma shows:

Corollary 3.1.6 *Any positive root $\alpha = \sum_{i=1}^{k} \beta_i$ with $\beta_i \in \Pi$ such that $\sum_{i=1}^{s} \beta_i$ are positive roots for $s \in \overline{1, k}$.*

For any α in a root system Φ, we define

$$\check{\alpha} = \frac{2\alpha}{(\alpha, \alpha)}. \tag{3.1.64}$$

Then

$$\check{\Phi} = \{\check{\alpha} \mid \alpha \in \Phi\} \tag{3.1.65}$$

forms a root system (exercise).

If Φ satisfies all the conditions of root system except $\mathbb{R}\alpha \bigcap \Phi = \{\pm\alpha\}$, then the only possible multiples of $\alpha \in \Phi$ in Φ are $\{\pm\alpha/2, \pm\alpha, \pm 2\alpha\}$ and $\{\alpha \in \Phi \mid 2\alpha \notin \Phi\}$ forms a root system (exercise).

Let α, β be two elements in a root system Φ such that $\alpha \neq \pm\beta$. Suppose that the α-string through β is $(\beta - r\alpha, \ldots, \beta + q\alpha)$ and the β-string through α is $(\alpha - r'\beta, \ldots, \alpha + q\beta)$. As an exercise, prove

$$\frac{q(r+1)}{(\beta, \beta)} = \frac{q'(r'+1)}{(\alpha, \alpha)}. \tag{3.1.66}$$

3.2 Weyl Groups

In this section, we introduce the Weyl group associated to each root system and use it to study the structure of the root system.

Let Φ be a root system in a Euclidean space $\mathscr{E}$. Denote by $GL(\mathscr{E})$ the group of all invertible linear transformations on $\mathscr{E}$. The *Weyl group* associated to Φ is

$$\mathscr{W} = \text{the subgroup of } GL(\mathscr{E}) \text{ generated by } \{\sigma_\alpha \mid \alpha \in \Phi\}. \tag{3.2.1}$$

Since $\sigma_\alpha(\Phi) = \Phi$ and Φ spans $\mathscr{E}$, $\mathscr{W}$ is a subgroup of the permutation group on Φ. So $\mathscr{W}$ is a finite group.

Lemma 3.2.1 *Let S be a finite subset spanning $\mathscr{E}$. Suppose $\sigma_\alpha(S) = S$ for some $0 \neq \alpha \in \mathscr{E}$. If an element $\sigma \in GL(\mathscr{E})$ satisfies that $\sigma(\alpha) = -\alpha$, $\sigma(S) = S$ and $\sigma|_P = Id_P$ for some hyperplane P in $\mathscr{E}$, then $\sigma = \sigma_\alpha$ and $P = P_\alpha$.*

Proof Note that

$$\mathscr{E} = P + \mathbb{R}\alpha = P_\alpha + \mathbb{R}\alpha. \tag{3.2.2}$$

So

$$\sigma|_{\mathscr{E}/\mathbb{R}\alpha} = \mathrm{Id}_{\mathscr{E}/\mathbb{R}\alpha}, \quad \sigma_\alpha|_{\mathscr{E}/\mathbb{R}\alpha} = \mathrm{Id}_{\mathscr{E}/\mathbb{R}\alpha}. \tag{3.2.3}$$

Note

$$\sigma\sigma_\alpha(\alpha) = \sigma(-\alpha) = \alpha, \quad \sigma\sigma_\alpha|_{\mathscr{E}/\mathbb{R}\alpha} = \mathrm{Id}_{\mathscr{E}/\mathbb{R}\alpha}. \tag{3.2.4}$$

Take a basis $\mathscr{B}$ of P. Then $\mathscr{B}\bigcup\{\alpha\}$ forms a basis of $\mathscr{E}$, under which

$$\sigma\sigma_\alpha = \begin{pmatrix} I & A \\ 0 & 1 \end{pmatrix}, \tag{3.2.5}$$

where I is an identity matrix. Since $\sigma\sigma_\alpha(S) = S$, it is of finite order; say, k. Then

$$\begin{pmatrix} I & 0 \\ 0 & 1 \end{pmatrix} = (\sigma\sigma_\alpha)^k = \begin{pmatrix} I & A \\ 0 & 1 \end{pmatrix}^k = \begin{pmatrix} I & kA \\ 0 & 1 \end{pmatrix}. \tag{3.2.6}$$

Thus $A = 0$, or equivalently, $\sigma\sigma_\alpha = 1$. Hence $\sigma = \sigma_\alpha^{-1} = \sigma_\alpha$. So $P = (1 + \sigma)(V) = (1 + \sigma_\alpha)(V) = P_\alpha$. $\square$

Lemma 3.2.2 *Let Φ be a root system in a Euclidean space $\mathscr{E}$. If an element $\sigma \in GL(\mathscr{E})$ satisfies $\sigma(\Phi) = \Phi$, then $\sigma\sigma_\alpha\sigma^{-1} = \sigma_{\sigma(\alpha)}$ for any $\alpha \in \Phi$, and $\langle\sigma(\alpha), \sigma(\beta)\rangle = \langle\alpha, \beta\rangle$ for any $\alpha, \beta \in \Phi$ (such σ is called an* **automorphism** *of Φ).*

Proof First we have

$$\sigma\sigma_\alpha\sigma^{-1}(\sigma(\alpha)) = \sigma\sigma_\alpha(\alpha) = \sigma(-\alpha) = -\sigma(\alpha). \tag{3.2.7}$$

Moreover, for any $\gamma \in P_\alpha$, we have

$$\sigma\sigma_\alpha\sigma^{-1}(\sigma(\gamma)) = \sigma\sigma_\alpha(\gamma) = \sigma(\gamma); \tag{3.2.8}$$

that is, $\sigma\sigma_\alpha\sigma^{-1}|_{\sigma(P_\alpha)} = \mathrm{Id}_{\sigma(P_\alpha)}$. Since $\sigma\sigma_\alpha\sigma^{-1}(\Phi) = \Phi$ and $\sigma(P_\alpha)$ is a hyperplane, we have $\sigma\sigma_\alpha\sigma^{-1} = \sigma_{\sigma(\alpha)}$ by the above lemma.

Now for any $\beta \in \Phi$, we have

$$\sigma\sigma_\alpha\sigma^{-1}(\sigma(\beta)) = \sigma(\beta - \langle\beta, \alpha\rangle\alpha) = \sigma(\beta) - \langle\beta, \alpha\rangle\sigma(\alpha), \tag{3.2.9}$$

which is equal to

$$\sigma_{\sigma(\alpha)}(\sigma(\beta)) = \sigma(\beta) - \langle\sigma(\beta), \sigma(\alpha)\rangle\sigma(\alpha). \tag{3.2.10}$$

Thus

$$\langle\sigma(\beta), \sigma(\alpha)\rangle = \langle\beta, \alpha\rangle. \qquad \square \tag{3.2.11}$$

The hyperplanes $\{P_\alpha \mid \alpha \in \Phi\}$ partition $\mathscr{E}$ into finite many regions; the connected components of $\mathscr{E} \setminus \bigcup_{\alpha\in\Phi} P_\alpha$ are called the (open) *Weyl chambers* of $\mathscr{E}$. Each regular vector γ belongs a unique Weyl chamber, denoted by $\mathscr{C}(\gamma)$. If $\Pi = \Pi(\gamma)$ is a base, we call $\mathscr{C}(\gamma) = \mathscr{C}(\Pi)$ the *fundamental Weyl chamber relative to* Π. In fact, we have

$$\begin{aligned}\mathscr{C}(\gamma) &= \{\beta \in \mathscr{E} \mid (\beta, \alpha) > 0 \text{ for } \alpha \in \Phi^+(\gamma)\}\\ &= \{\beta \in \mathscr{E} \mid (\beta, \alpha) > 0 \text{ for } \alpha \in \Pi^+(\gamma)\}.\end{aligned} \tag{3.2.12}$$

Lemma 3.2.3 *For $\sigma \in \mathscr{W}$, $\sigma(\mathscr{C}(\gamma)) = \mathscr{C}(\sigma(\gamma))$ and $\sigma(\Pi(\gamma)) = \Pi(\sigma(\gamma))$.*

Proof Observe that

$$(\sigma(\gamma), \sigma(\alpha)) = (\gamma, \alpha) > 0 \qquad \text{for } \alpha \in \Phi^+(\gamma). \tag{3.2.13}$$

Thus $\sigma(\gamma)$ is regular by (3.1.46) and (3.1.47), and $\sigma(\Phi^+(\gamma)) = \Phi^+(\sigma(\gamma))$. Hence $\sigma(\mathscr{C}(\gamma)) = \mathscr{C}(\sigma(\gamma))$ by (3.2.12). Since σ is linear, $\{\sigma(\alpha) \mid \alpha \in \Pi(\gamma)\}$ are indecomposable elements in $\Phi^+(\sigma(\gamma))$. Thus $\sigma(\Pi(\gamma)) = \Pi(\sigma(\gamma))$. $\square$

Now we fix a base Π of the root system Φ. Then $\Phi^+ = \{\beta \in \Phi \mid \beta = \sum_{\alpha\in\Pi} k_\alpha\alpha,\ k_\alpha \geq 0\}$.

Lemma 3.2.4 *For $\alpha \in \Pi$, $\sigma_\alpha(\Phi^+ \setminus \{\alpha\}) = \Phi^+ \setminus \{\alpha\}$.*

Proof Let $\gamma \in \Phi^+ \setminus \{\alpha\}$. Then $\gamma = \sum_{\beta\in\Pi} k_\beta\beta$ with $k_\beta \in \mathbb{N}$ and $k_{\beta'} > 0$ for some $\alpha \neq \beta' \in \Pi$. Note

$$\sigma_\alpha(\gamma) = \sum_{\alpha\neq\beta} k_\beta\beta + (k_\alpha - \langle\gamma, \alpha\rangle)\alpha, \tag{3.2.14}$$

whose coefficient of β' is still $k_{\beta'} > 0$. Thus $\sigma_\alpha(\gamma) \in \Phi^+ \setminus \{\alpha\}$. $\square$

Corollary 3.2.5 *Set*

$$\rho = \frac{1}{2} \sum_{\alpha \in \Phi^+} \alpha. \tag{3.2.15}$$

Then

$$\sigma_\alpha(\rho) = \rho - \alpha \qquad \textit{for } \alpha \in \Pi, \tag{3.2.16}$$

or equivalently,

$$\langle \rho, \alpha \rangle = 1 \qquad \alpha \in \Pi. \tag{3.2.17}$$

We define partial ordering $\prec$ on $\mathscr{E}$ by

$$\alpha \prec \beta \text{ if } \beta - \alpha = \sum_{\alpha \in \Pi} a_\alpha \alpha \text{ such that } 0 \leq a_\alpha \in \mathbb{R}, \tag{3.2.18}$$

where $\alpha, \beta \in \mathscr{E}$.

Proposition 3.2.6 *Let $\alpha, \beta \in \Phi$ be two positive distinct roots such that $\alpha \prec \beta$. Then there exist $\gamma_1, \ldots, \gamma_s \in \Pi$ such that $\beta - \alpha = \sum_{i=1}^{s} \gamma_i$ and*

$$\alpha + \sum_{i=1}^{r} \gamma_i \in \Phi \qquad \textit{for } r = 1, \ldots, s. \tag{3.2.19}$$

Proof Write $\beta - \alpha = \sum_{\gamma \in \Pi'} k_\gamma \gamma$ with $k_\gamma > 0$ and $\Pi' \subset \Pi$. We prove the proposition by induction on $\sum_{\gamma \in \Pi'} k_\gamma$. If $(\alpha, \gamma_0) < 0$ or $(\beta, \gamma_0) > 0$ for some $\gamma_0 \in \Pi'$, then $\alpha + \gamma_0 \in \Phi$ or $\beta - \gamma_0 \in \Phi$ by Lemma 3.1.2. The conclusion holds for the pair $(\alpha + \gamma_0, \beta)$ or $(\alpha, \beta - \gamma_0)$ by induction, which implies that the conclusion holds for the pair (α, β). Suppose

$$(\alpha, \gamma) \geq 0, \quad (\beta, \gamma) \leq 0 \qquad \text{for } \gamma \in \Pi'. \tag{3.2.20}$$

Then

$$0 \leq (\beta - \alpha, \beta - \alpha) = (\beta - \alpha, \sum_{\gamma \in \Pi'} k_\gamma \gamma) = \sum_{\gamma \in \Pi'} k_\gamma [(\beta, \gamma) - (\alpha, \gamma)] \leq 0. \tag{3.2.21}$$

Thus $(\beta - \alpha, \beta - \alpha) = 0$; that is, $\beta - \alpha = 0$, which contradicts $\alpha \neq \beta$. □

The above proposition had been used in studying the structure of finite-dimensional simple Lie superalgebras. It is also useful in the representations of simple Lie algebras.

Lemma 3.2.7 *Let $\alpha_1, \alpha_2, \ldots, \alpha_k \in \Pi$ (not necessarily distinct). Write $\sigma_i = \sigma_{\alpha_i}$. If $\sigma_1 \cdots \sigma_{k-1}(\alpha_k) \prec 0$, then for some $s \in \overline{1, k-1}$, $\sigma_1 \cdots \sigma_k = \sigma_1 \cdots \sigma_{s-1} \sigma_{s+1} \cdots \sigma_{k-1}$.*

Proof Set

$$\beta_{k-1} = \alpha_k, \quad \beta_i = \sigma_{i+1} \cdots \sigma_{k-1}(\alpha_k) \qquad \text{for } i \in \{0, 1, \ldots, k-2\}. \tag{3.2.22}$$

Then there exists $s \in \overline{1, k-1}$ such that $\beta_s \succ 0$ and $\beta_{s-1} = \sigma_{\alpha_s}(\beta_s) \prec 0$. By the above lemma, $\beta_s = \alpha_s$. Thus

$$\sigma_{s+1} \cdots \sigma_{k-1}\sigma_k(\sigma_{s+1} \cdots \sigma_{k-1})^{-1} = \sigma_s \tag{3.2.23}$$

by Lemma 3.2.2. Hence

$$\sigma_{s+1} \cdots \sigma_{k-1}\sigma_k = \sigma_s\sigma_{s+1} \cdots \sigma_{k-1}, \tag{3.2.24}$$

$$\sigma_s\sigma_{s+1} \cdots \sigma_{k-1}\sigma_k = \sigma_{s+1} \cdots \sigma_{k-1}, \tag{3.2.25}$$

or equivalently, $\sigma_1 \cdots \sigma_k = \sigma_1 \cdots \sigma_{s-1}\sigma_{s+1} \cdots \sigma_{k-1}$. □

Corollary 3.2.8 *If $\sigma = \sigma_1\sigma_2 \cdots \sigma_k$ is a product of minimal number of the reflections of simple roots, then $\sigma(\alpha_k) \prec 0$.*

Proof Otherwise,

$$\sigma(\alpha_k) = \sigma_1\sigma_2 \cdots \sigma_{k-1}(-\alpha_k) = -\sigma_1\sigma_2 \cdots \sigma_{k-1}(\alpha_k) \succ 0. \tag{3.2.26}$$

Thus $\sigma_1\sigma_2 \cdots \sigma_{k-1}(\alpha_k) \prec 0$. By the above lemma, the product $\sigma = \sigma_1\sigma_2 \cdots \sigma_k$ is not minimal. □

Theorem 3.2.9 *Let Π be a base of a root system Φ.*

(a) If $\gamma \in \mathscr{E}$ is regular, there exists $\sigma \in \mathscr{W}$ such that $(\sigma(\gamma), \alpha) > 0$ for $\alpha \in \Pi$ (so $\mathscr{W}$ acts transitively on Weyl chambers).

(b) $\mathscr{W}$ acts transitively on bases.

(c) If $\alpha \in \Phi$, there exists $\sigma \in \mathscr{W}$ such that $\sigma(\alpha) \in \Pi$.

(d) $\mathscr{W}$ is generated by $\{\sigma_\alpha \mid \alpha \in \Pi\}$.

(e) If $\sigma(\Pi) = \Pi$, $\sigma \in \mathscr{W}$, then $\sigma = 1$.

Proof Let $\mathscr{W}'$ be the subgroup of $\mathscr{W}$ generated by $\{\sigma_\alpha \mid \alpha \in \Pi\}$. Since $\mathscr{W}$ is finite, $\mathscr{W}'$ is finite.

(a) Take $\sigma \in \mathscr{W}'$ such that $(\sigma(\gamma), \Pi)$ is maximal in $\{(\tau(\gamma), \Pi) \mid \tau \in \mathscr{W}'\}$. For any $\alpha \in \Pi$, we have

$$(\sigma(\gamma), \Pi) \geq (\sigma_\alpha\sigma(\gamma), \Pi) = (\sigma(\gamma), \sigma_\alpha(\Pi)) = (\sigma(\gamma), \Pi - \alpha) = (\sigma(\gamma), \Pi) - (\sigma(\gamma), \alpha), \tag{3.2.27}$$

which implies $(\sigma(\gamma), \alpha) \geq 0$. Since $\sigma(\gamma)$ is regular, we have

$$(\sigma(\gamma), \alpha) > 0 \qquad \text{for } \alpha \in \Pi. \tag{3.2.28}$$

Thus we have

$$\sigma(\gamma) \in \mathscr{C}(\Pi) \tag{3.2.29}$$

by (3.2.12). Hence $\sigma(\mathscr{C}(\gamma)) = \mathscr{C}(\sigma(\gamma)) = \mathscr{C}(\Pi)$; that is, $\mathscr{W}$ acts transitively on Weyl chambers.

(b) By (a) and (3.2.12), $\sigma(\Phi^+(\gamma)) = \Phi^+$. Since σ is linear, it maps indecomposable elements to indecomposable elements; that is, $\sigma(\Pi(\gamma)) = \Pi(\sigma(\gamma)) = \Pi$.

(c) Since $\sigma_\alpha(\alpha) = -\alpha$, we may assume $\alpha \succ 0$. Suppose $\mathscr{W}'(\alpha) \bigcap \Pi = \emptyset$. By Lemma 3.2.4, $\mathscr{W}'(\alpha)$ is a finite set of positive roots. Choose a minimal element $\gamma \in \mathscr{W}'(\alpha)$ with respect the partial ordering $\prec$ in (3.2.18). Then

$$\sigma_\beta(\gamma) = \gamma - \langle \gamma, \beta \rangle \beta \not\prec \gamma \qquad \text{for } \beta \in \Pi. \tag{3.2.30}$$

Thus

$$\langle \gamma, \beta \rangle \beta \prec 0 \sim (\gamma, \beta) \leq 0 \qquad \text{for } \beta \in \Pi. \tag{3.2.31}$$

According to the proof of the arguments in (3.1.56)–(3.1.62), $\{\gamma, \Pi\}$ is an independent set (the positivity of γ is needed (cf. (3.1.61) and (3.1.62))), which is absurd.

(d) For any $\alpha \in \Phi$, we have $\sigma_\alpha = \sigma_{-\alpha}$. Thus we assume that α is a positive root. By (c), there exists $\sigma \in \mathscr{W}'$ such that $\sigma(\alpha) = \beta \in \Pi$. Thus $\alpha = \sigma^{-1}(\beta)$ and

$$\sigma_\alpha = \sigma_{\sigma^{-1}(\beta)} = \sigma^{-1}\sigma_\beta\sigma \in \mathscr{W}' \tag{3.2.32}$$

by Lemma 3.2.2. Since $\mathscr{W}$ is generated by $\{\sigma_\alpha \mid \alpha \in \Phi\}$, we have $\mathscr{W} = \mathscr{W}'$.

(e) Suppose $\sigma \neq 1$. Then we can write $\sigma = \sigma_1 \cdots \sigma_k$ as a product of minimal number of the reflections of simple roots and $\sigma_k = \sigma_{\alpha_k}$. By Corollary 3.2.8, then $\sigma(\alpha_k) \prec 0$, which contradicts $\sigma(\Pi) = \Pi$. $\square$

For $\sigma \in \mathscr{W}$, if $\sigma = \sigma_1\sigma_2 \cdots \sigma_k$ is a product of minimal number of the reflections of simple roots, we called the product $\sigma_1\sigma_2 \cdots \sigma_k$ *reduced* and call k the *length of* σ, denoted as $\ell(\sigma)$. Moreover, we define

$$n(\sigma) = |\{\alpha \in \Phi^+ \mid \sigma(\alpha) \prec 0\}|. \tag{3.2.33}$$

Lemma 3.2.10 $\ell(\sigma) = n(\sigma)$.

Proof We prove it by induction on $\ell(\sigma)$. The case $\ell(\sigma) = 1$ is Lemma 3.2.4. Suppose the lemma holds for $\ell(\sigma) < k$ for some positive integer k. Consider $\ell(\sigma) = k$. Write $\sigma = \sigma_1\sigma_2 \cdots \sigma_k$ with $\sigma_i = \sigma_{\alpha_i}$ and $\alpha_i \in \Pi$. Let $\sigma' = \sigma_1 \cdots \sigma_{k-1}$. Then $\ell(\sigma') = k - 1$. Set

$$S = \{\alpha \in \Phi^+ \mid \sigma'(\alpha) \prec 0\}. \tag{3.2.34}$$

By assumption, $|S| = k - 1$. According to Lemma 3.2.7, $\sigma'(\alpha_k) \succ 0$. Thus

$$S = \{\beta \in \Phi^+ \setminus \{\alpha_k\} \mid \sigma'(\beta) \prec 0\}. \tag{3.2.35}$$

Since $\sigma_k(\Phi^+ \setminus \{\alpha_k\}) = \Phi^+ \setminus \{\alpha_k\}$ by Lemma 3.2.4, we have

$$\sigma_k(S) = \{\sigma_k(\beta) \in \Phi^+ \setminus \{\alpha_k\} \mid \sigma'(\beta) \prec 0\} = \{\beta' \in \Phi^+ \setminus \{\alpha_k\} \mid \sigma(\beta') \prec 0\}. \tag{3.2.36}$$

Therefore,

$$\sigma_k(S) \bigcup \{\alpha\} = \{\beta \in \Phi^+ \mid \sigma(\beta) \prec 0\}, \tag{3.2.37}$$

which implies $n(\sigma) = k$. □

Remark 3.2.11 According to (3.2.37), if $\sigma = \sigma_1\sigma_2 \cdots \sigma_k$ is a reduced product of the reflections of simple roots with $\sigma_i = \sigma_{\alpha_i}$, then

$$\{\beta \in \Phi^+ \mid \sigma(\beta) \prec 0\} = \{\alpha_k, \sigma_k \cdots \sigma_{i+1}(\alpha_i) \mid i \in \overline{1, k-1}\}. \tag{3.2.38}$$

Lemma 3.2.12 *Let $\lambda, \mu \in \overline{\mathscr{C}(\Pi)}$. If $\sigma(\lambda) = \mu$, then $\lambda = \mu$.*

Proof If $\ell(\sigma) \neq 0$, let $\sigma = \sigma_1\sigma_2 \cdots \sigma_k$ be a reduced product of the reflections of simple roots with $\sigma_i = \sigma_{\alpha_i}$. Then $\sigma(\alpha_k) \prec 0$. Thus

$$0 \geq (\mu, \sigma(\alpha_k)) = (\sigma^{-1}(\mu), \alpha_k) = (\lambda, \alpha_k) \geq 0. \tag{3.2.39}$$

Thus $(\lambda, \alpha_k) = 0$, or equivalently, $\sigma_k(\lambda) = \lambda$. Now $\sigma_1 \cdots \sigma_{k-1}(\lambda) = \mu$. By induction on $\ell(\sigma)$, we have $\lambda = \mu$. □

Lemma 3.2.13 *For any $0 \neq \gamma \in \mathscr{E}$, the orbit $\{\sigma(\gamma) \mid \sigma \in \mathscr{W}\}$ has a unique maximal element β with respect to the partial ordering $\prec$ defined in (3.2.18). Moreover, $\beta \in \overline{\mathscr{C}(\Pi)}$. In particular,*

$$\sigma(\gamma) \prec \gamma \quad \textit{for any} \quad \gamma \in \overline{\mathscr{C}(\Pi)}. \tag{3.2.40}$$

Proof Suppose that $\beta = \sigma_1(\gamma)$ is a maximal element in $\{\sigma(\gamma) \mid \sigma \in \mathscr{W}\}$. For any $\alpha \in \Pi$,

$$\sigma_\alpha(\beta) = \beta - \langle \beta, \alpha \rangle \alpha \not\succ \beta \Longrightarrow \langle \beta, \alpha \rangle \geq 0 \sim (\beta, \alpha) \geq 0. \tag{3.2.41}$$

Thus $\beta \in \overline{\mathscr{C}(\Pi)}$.

Suppose that $\beta' = \sigma_2(\gamma)$ is also a maximal element in $\{\sigma(\gamma) \mid \sigma \in \mathscr{W}\}$. Then we also have $\beta' \in \overline{\mathscr{C}(\Pi)}$. Since $\sigma_1\sigma_2^{-1}(\beta') = \beta$, we have $\beta = \beta'$ by the above lemma. If $\gamma \in \overline{\mathscr{C}(\Pi)}$, we have $\beta = \gamma$ again by the above lemma. □

The above two lemmas shows the $\mathscr{W}$-orbit space

$$\mathscr{E}/\mathscr{W} \cong \overline{\mathscr{C}(\Pi)}. \tag{3.2.42}$$

In other words, there exists a one-to-one correspondence between the functions on $\overline{\mathscr{C}(\Pi)}$ and $\mathscr{W}$-invariant functions on $\mathscr{E}$.

A root system Φ is called *reducible* if it is a union of two nonempty subsets Φ_1 and Φ_2 such that

$$(\alpha, \beta) = 0 \qquad \text{for } \alpha \in \Phi_1, \ \beta \in \Phi_2. \tag{3.2.43}$$

Lemma 3.2.14 *Let Π be a base of a roots system Φ. Then Φ is reducible if and only if Π is a union of two nonempty subsets Π_1 and Π_2 such that*

$$(\alpha, \beta) = 0 \qquad \textit{for } \alpha \in \Pi_1, \ \beta \in \Pi_2. \tag{3.2.44}$$

Proof Suppose that Φ is a union of two nonempty subsets Φ_1 and Φ_2 such that (3.2.43) holds. If $\Pi \subset \Phi_1$, then the elements in Φ_2 are orthogonal to a basis of $\mathscr{E}$. So $\Phi_2 = \{0\}$, which is absurd. Thus $\Pi \not\subset \Phi_i$. Set $\Pi_i = \Pi \bigcap \Phi_i$. Now Π is a union of two nonempty subsets Π_1 and Π_2 such that (3.2.44) holds.

Assume that Π is a union of two nonempty subsets Π_1 and Π_2 such that (3.2.44) holds. Set

$$\Phi_1 = \{\sigma(\alpha) \mid \alpha \in \Pi_1, \ \sigma \in \mathscr{W}\}, \quad \Phi_2 = \{\sigma(\alpha) \mid \alpha \in \Pi_2, \ \sigma \in \mathscr{W}\}. \tag{3.2.45}$$

For $\alpha \in \Pi_1$ and $\beta \in \Pi_2$, we have $\sigma_\alpha(\beta) = \beta - \langle \beta, \alpha \rangle \alpha = \beta$ and

$$\sigma_\alpha \sigma_\beta \sigma_\alpha^{-1} = \sigma_{\sigma_\alpha(\beta)} = \sigma_\beta \sim \sigma_\alpha \sigma_\beta = \sigma_\beta \sigma_\alpha. \tag{3.2.46}$$

Let

$$\mathscr{W}_i = \text{the subgroup of } \mathscr{W} \text{ generated by } \{\sigma_\alpha \mid \alpha \in \Pi_i\}. \tag{3.2.47}$$

Since $\mathscr{W}$ is generated by $\{\sigma_\alpha \mid \alpha \in \Pi\}$ by Theorem 3.2.9 (d), we have

$$\mathscr{W} = \mathscr{W}_1 \mathscr{W}_2 \tag{3.2.48}$$

by (3.2.46). Thus

$$\Phi_i = \{\sigma(\alpha) \mid \alpha \in \Pi_i, \ \sigma \in \mathscr{W}_i\}. \tag{3.2.49}$$

For $\alpha_i \in \Pi_i$ and $\sigma_i \in \mathscr{W}_i$, we have

$$\sigma_2(\alpha_1) = \alpha_1, \qquad \sigma_1(\alpha_2) = \alpha_2. \tag{3.2.50}$$

Hence

$$(\sigma_1(\alpha_1), \sigma_2(\alpha_2)) = (\sigma_2^{-1}(\alpha_1), \sigma_1^{-1}(\alpha_2)) = (\alpha_1, \alpha_2) = 0. \tag{3.2.51}$$

So (3.2.44) holds. By Theorem 3.2.9 (c), $\Phi = \Phi_1 \bigcup \Phi_2$. $\square$

Lemma 3.2.15 *Let Φ be an irreducible root system. Relative to the partial ordering $\prec$, there exists a unique maximal root β. Moreover, $\beta = \sum_{\alpha \in \Pi} k_\alpha \alpha$ with all $k_\alpha > 0$.*

Proof Let $\beta = \sum_{\alpha\in\Pi} k_\alpha \alpha$ be a maximal root. Then $\beta \succ 0$. Set

$$\Pi_1 = \{\alpha \in \Pi \mid k_\alpha > 0\}, \qquad \Pi_2 = \{\alpha \in \Pi \mid k_\alpha = 0\}. \tag{3.2.52}$$

Then $\Pi = \Pi_1 \bigcup \Pi_2$ and $\Pi_1 \neq \emptyset$. If $\Pi_2 \neq \emptyset$, there exist $\alpha_1 \in \Pi_1$ and $\alpha_2 \in \Pi_2$ such that $(\alpha_1, \alpha_2) < 0$ by (3.1.51), the above lemma and the irreducibility of Φ. Again (3.1.51) implies $(\beta, \alpha_2) < 0$. According to Lemma 3.1.2, $\beta \prec \beta + \alpha_2 \in \Phi$, which contradicts the maximality of β. Thus $\Pi_2 = \emptyset$.

Suppose that β' is another maximal root. By Lemma 3.1.2, $(\beta', \alpha) \geq 0$ for $\alpha \in \Pi$. Since β' can not be orthogonal to all the elements of Π, there exists $\alpha_1 \in \Pi$ such that $(\beta', \alpha_1) > 0$. This implies

$$(\beta, \beta') = \sum_{\alpha\in\Pi} k_\alpha(\alpha, \beta') > 0. \tag{3.2.53}$$

Lemma 3.1.2 tells us that $\beta - \beta'$ is a root. Without loss of generality, we can assume $\beta - \beta' \succ 0$. Then

$$\beta = (\beta - \beta') + \beta' \succ \beta', \tag{3.2.54}$$

which contradicts the maximality of β'. □

Lemma 3.2.16 *Let Φ be an irreducible root system. Then $\mathscr{W}$ acts irreducibly on $\mathscr{E}$. In particular, the $\mathscr{W}$-orbit of a root α spans $\mathscr{E}$.*

Proof Let W be a nonzero $\mathscr{W}$-submodule of $\mathscr{E}$ and let $W^\perp$ be its orthogonal complement in $\mathscr{E}$. Suppose $W^\perp \neq \{0\}$. For any $\alpha \in \Phi$, if $\alpha \notin W^\perp$, there exists $\gamma \in W$ such that $(\gamma, \alpha) \neq 0$. Thus

$$\sigma_\alpha(\gamma) = \gamma - \langle\gamma, \alpha\rangle\alpha \in W \Longrightarrow \langle\gamma, \alpha\rangle\alpha \in W \Longrightarrow \alpha \in W. \tag{3.2.55}$$

Symmetrically, if a root $\beta \notin W$, then $\beta \in W^\perp$. Since Φ spans $\mathscr{E}$, we have

$$\Phi_1 = \Phi \bigcap W \neq \emptyset, \qquad \Phi_2 = \Phi \bigcap W^\perp \neq \emptyset \tag{3.2.56}$$

and $\Phi = \Phi_1 \bigcup \Phi_2$, which leads a contradiction to the irreducibility of Φ. □

Lemma 3.2.17 *Let Φ be an irreducible root system. There are only at most two root lengths occurring in Φ, and all roots with the same length are conjugate under $\mathscr{W}$ (in the same $\mathscr{W}$-orbit).*

Proof For any $\alpha, \beta \in \Phi$, $(\alpha, \sigma(\beta)) \neq 0$ for some $\sigma \in \mathscr{W}$ by the above lemma. Replacing β by $\sigma(\beta)$, we may assume $(\alpha, \beta) \neq 0$ and $(\alpha, \alpha) \leq (\beta, \beta)$. By Lemma 3.1.1,

$$\frac{(\beta, \beta)}{(\alpha, \alpha)} \in \{1, 2, 3\}. \tag{3.2.57}$$

If we have $\alpha_1, \alpha_2, \alpha_3 \in \Phi$ with $(\alpha_1, \alpha_1) < (\alpha_2, \alpha_2) < (\alpha_3, \alpha_3)$. By (3.2.57), the pairs (α_1, α_2) and (α_1, α_3) give

$$(\alpha_2, \alpha_2) = 2(\alpha_1, \alpha_1), \qquad (\alpha_3, \alpha_3) = 3(\alpha_1, \alpha_1). \tag{3.2.58}$$

But

$$\frac{(\alpha_3, \alpha_3)}{(\alpha_2, \alpha_2)} = \frac{3}{2}, \tag{3.2.59}$$

which contradicts (3.2.57). Thus there are only at most two root lengths occurring in Φ.

Suppose we have two elements $\alpha, \beta \in \Phi$ with the same length. Replacing β by $\sigma(\beta)$ for some $\sigma \in \mathscr{W}$ if necessary, we may assume $(\alpha, \beta) \neq 0$ and $\alpha \neq \beta$. By Lemma 3.1.1, $\langle \alpha, \beta \rangle = \langle \beta, \alpha \rangle = \pm 1$. Replacing β by $\sigma_\beta(\beta) = -\beta$ if necessarily, we may assume $\langle \alpha, \beta \rangle = \langle \beta, \alpha \rangle = 1$. Now

$$\sigma_\alpha \sigma_\beta \sigma_\alpha(\beta) = \sigma_\alpha \sigma_\beta(\beta - \alpha) = \sigma_\alpha(-\alpha) = \alpha. \qquad \square \tag{3.2.60}$$

Lemma 3.2.18 *Let Φ be an irreducible root system with two distinct lengths. Then the maximal root β is a long root.*

Proof By Lemma 3.1.2, $\beta \in \overline{\mathscr{C}(\Pi)}$. Let $\alpha \in \Phi$ be any root such that $||\alpha|| \neq ||\beta||$. According to Theorem 3.2.9 (a), there exists $\sigma \in \mathscr{W}$ such that $\sigma(\alpha) \in \overline{\mathscr{C}(\Pi)}$. Replacing α by $\sigma(\alpha)$, we can assume $\alpha \in \overline{\mathscr{C}(\Pi)}$. Since $\beta - \alpha \succ 0$, we have

$$(\beta, \beta - \alpha) \geq 0 \Longrightarrow (\beta, \beta) \geq (\beta, \alpha) \tag{3.2.61}$$

and

$$(\beta - \alpha, \alpha) \geq 0 \Longrightarrow (\beta, \alpha) \geq (\alpha, \alpha). \tag{3.2.62}$$

Thus $(\beta, \beta) \geq (\alpha, \alpha)$. $\square$

In the following examples, we let $\mathscr{E}$ be a vector space over $\mathbb{R}$ with a basis $\{\varepsilon_1, \varepsilon_2, \ldots, \varepsilon_n\}$ and an inner product determined by

$$(\varepsilon_i, \varepsilon_j) = \delta_{i,j} \qquad \text{for } i, j \in \overline{1, n}. \tag{3.2.63}$$

Example 3.2.1 Recall the root system $\Phi_{A_{n-1}}$ in (3.1.12) of $sl(n, \mathbb{C})$. It is a root system in the subspace

$$\mathscr{E}' = \{\sum_{i=1}^{n} a_i \varepsilon_i \mid a_i \in \mathbb{R}, \sum_{r=1}^{n} a_r = 0\}. \tag{3.2.64}$$

To understand the Weyl group of $\Phi_{A_{n-1}}$, we consider its action on $\mathscr{E}$. For $\varepsilon_i - \varepsilon_j \in \Phi_{A_{n-1}}$ with $i \neq j$,

$$\sigma_{\varepsilon_i-\varepsilon_j}(\varepsilon_r)=\begin{cases}\varepsilon_r \text{ if } r\neq i,j\\ \varepsilon_j \text{ if } r=i\\ \varepsilon_i \text{ if } r=j\end{cases}\qquad \text{for } r\in\overline{1,n}. \tag{3.2.65}$$

Let S_n be the permutation group on $\{1,2,\ldots,n\}$. Define a representation $\nu: S_n\to GL(\mathscr{E})$ by

$$\nu(\iota)(\varepsilon_r)=\varepsilon_{\iota(r)}\qquad \text{for } \iota\in S_n,\ r\in\overline{1,n}. \tag{3.2.66}$$

Then

$$S_n\overset{\nu}{\cong}\mathscr{W}_{A_{n-1}}, \tag{3.2.67}$$

the Weyl group of $\Phi_{A_{n-1}}$.

Since $\sigma_{\varepsilon_i}=\sigma_{2\varepsilon_i}$, the Weyl groups of Φ_{B_n} in (3.1.28) and Φ_{C_n} in (3.1.27) are the same. Note

$$\sigma_{\varepsilon_i}(\varepsilon_r)=(-1)^{\delta_{i,r}}\varepsilon_r,\quad \sigma_{\varepsilon_i+\varepsilon_j}(\varepsilon_r)=\begin{cases}\varepsilon_r & \text{if } r\neq i,j\\ -\varepsilon_j & \text{if } r=i\\ -\varepsilon_i & \text{if } r=j\end{cases}\qquad \text{for } r\in\overline{1,n} \tag{3.2.68}$$

Let $\mathbb{Z}_2=\mathbb{Z}/2\mathbb{Z}$ be the field of two elements and let Υ be a vector space over $\mathbb{Z}_2$ with a basis $\{\varepsilon_1,\varepsilon_2,\ldots,\varepsilon_n\}$. Define

$$\iota(\varsigma)=\sum_{i=1}^n a_i\varepsilon_{\iota(i)}\qquad \text{for } \iota\in S_n,\ \varsigma=\sum_{i=1}^n a_i\varepsilon_i\in\Upsilon. \tag{3.2.69}$$

Viewing Υ as an additive group, we form a semi-product of S_n with Υ:

$$G=S_n\ltimes\Upsilon, \tag{3.2.70}$$

whose group multiplication is defined by:

$$(\iota_1,\varsigma_1)\cdot(\iota_2,\varsigma_2)=(\iota_1\iota_2,\varsigma_1+\iota_1(\varsigma_2)). \tag{3.2.71}$$

Moreover, we define $\nu: G\to GL(\mathscr{E})$ by

$$\nu[(\iota,\sum_{i=1}^n a_i\varepsilon_i,)](\varepsilon_r)=(-1)^{a_r}\varepsilon_{\iota(r)}\qquad \text{for } r\in\overline{1,n}. \tag{3.2.72}$$

By (3.2.65) and (3.2.68), we have

$$G\overset{\nu}{\cong}\mathscr{W}_{B_n}=\mathscr{W}_{C_n}, \tag{3.2.73}$$

the Weyl groups of Φ_{B_n} and Φ_{C_n}, respectively.

Set

$$\Upsilon' = \{\sum_{i=1}^{n} a_i\varepsilon_i \in \Upsilon \mid \sum_{r=1}^{n} a_i = 0\}, \quad G' = S_n \ltimes \Upsilon'. \tag{3.2.74}$$

Then G' is a subgroup of G. Moreover,

$$G' \overset{\nu}{\cong} \mathscr{W}_{D_n}, \tag{3.2.75}$$

the Weyl groups of Φ_{D_n} in (3.1.17).

The followings are exercises:

(1) If $\sigma \in \mathscr{W}$ can be written as a product of k reflections of simple roots (*simple reflections*), prove $k \equiv \ell(\sigma) \pmod 2$.

(2) Prove that there exists a unique element σ such that $\sigma(\Phi^+) = -\Phi^+ = \Phi^-$. Moreover, any reduced expression for σ must involve all $\{\sigma_\alpha \mid \alpha \in \Pi\}$. What is $\ell(\sigma)$?

(3) Prove that the only reflections in $\mathscr{W}$ are $\{\sigma_\alpha \mid \alpha \in \Phi\}$.

3.3 Classification

In this section, we classify all the irreducible root systems.

Let Φ be an irreducible root system with a base

$$\Pi = \{\alpha_1, \alpha_2, \ldots, \alpha_n\}. \tag{3.3.1}$$

The matrix

$$\mathrm{CM}(\Pi) = (\langle\alpha_i, \alpha_j\rangle)_{n\times n} \tag{3.3.2}$$

is called the *Cartan matrix* of the root system Φ. Since Π is a basis of the corresponding Euclidean space $\mathscr{E}$, each simple reflection σ_{α_i} is determined by the integers

$$\{\langle\alpha_1, \alpha_i\rangle, \langle\alpha_2, \alpha_i\rangle, \ldots, \langle\alpha_n, \alpha_i\rangle\}; \tag{3.3.3}$$

that is, the ith column of CM(Π). Recall that the Weyl group $\mathscr{W}$ is generated by $\{\sigma_{\alpha_1}, \ldots, \sigma_{\alpha_n}\}$ (Theorem 3.2.9 (d)). Thus the root system $\Phi = \mathscr{W}(\Pi)$ is completely determined by CM(Π). Observe that the diagonal entries of CM(Π) are 2. So Φ is determined by

$$\{\langle\alpha_i, \alpha_j\rangle \mid i, j \in \overline{1,n},\ i \neq j\}. \tag{3.3.4}$$

Recall

$$(\alpha_i, \alpha_j) \leq 0, \qquad \langle\alpha_i, \alpha_j\rangle\langle\alpha_j, \alpha_i\rangle = 0, 1, 2, 3 \qquad \text{for } i, j \in \overline{1,n},\ i \neq j \tag{3.3.5}$$

(cf. Lemma 3.1.1 and (3.1.51)). Set

$$\mathfrak{e}_i = \frac{\alpha_i}{||\alpha_i||} \qquad \text{for } i \in \overline{1, n}, \tag{3.3.6}$$

where $||\alpha_i|| = \sqrt{(\alpha_i, \alpha_i)}$. Now $\{\mathfrak{e}_1, \dots, \mathfrak{e}_n\}$ are unit vectors and

$$\langle \mathfrak{e}_i, \mathfrak{e}_j \rangle = 2(\mathfrak{e}_i, \mathfrak{e}_j) = 2\left(\frac{\alpha_i}{||\alpha_i||}, \frac{\alpha_j}{||\alpha_j||}\right) = \frac{2(\alpha_i, \alpha_j)}{||\alpha_i||\,||\alpha_j||}. \tag{3.3.7}$$

Thus

$$\langle \mathfrak{e}_i, \mathfrak{e}_j \rangle \langle \mathfrak{e}_j, \mathfrak{e}_i \rangle = \left(\frac{2(\alpha_i, \alpha_j)}{||\alpha_i||\,||\alpha_j||}\right)^2 = \langle \alpha_i, \alpha_j \rangle \langle \alpha_j, \alpha_i \rangle. \tag{3.3.8}$$

Therefore, (3.3.5) is equivalent to

$$(\mathfrak{e}_i, \mathfrak{e}_j) \le 0, \qquad \langle \mathfrak{e}_i, \mathfrak{e}_j \rangle \langle \mathfrak{e}_j, \mathfrak{e}_i \rangle = 4(\mathfrak{e}_i, \mathfrak{e}_j)^2 = 0, 1, 2, 3 \qquad \text{for } i, j \in \overline{1, n},\ i \neq j \tag{3.3.9}$$

In general, a linearly independent subset of unit vectors $\{\mathfrak{e}_1, \mathfrak{e}_2, \dots, \mathfrak{e}_n\}$ in a Euclidean space is called an *admissible set* if (3.3.9) holds. Note that any subset of an admissible set is still an admissible set. If an admissible set $\mathscr{S}$ can not be written as a union of two orthogonal proper subsets, then we call $\mathscr{S}$ *irreducible*.

Let $\mathscr{S} = \{\mathfrak{e}_1, \mathfrak{e}_2, \dots, \mathfrak{e}_n\}$ be an irreducible admissible set. Define the *Coxeter graph* of $\mathscr{S}$ to be a graph having n vertices corresponding to the vectors in $\mathscr{S}$, in which the ith vertex is jointed to jth vertex ($i \neq j$) by $\langle \mathfrak{e}_i, \mathfrak{e}_j \rangle \langle \mathfrak{e}_j, \mathfrak{e}_i \rangle$ edges. We denote the Coxeter graph of $\mathscr{S}$ by $CG(\mathscr{S})$. Note that

$$0 < (\sum_{i=1}^n \mathfrak{e}_i, \sum_{i=1}^n \mathfrak{e}_i) = \sum_{i=1}^n (\mathfrak{e}_i, \mathfrak{e}_i) + \sum_{1 \le i < j \le n} 2(\mathfrak{e}_i, \mathfrak{e}_j) = n + \sum_{1 \le i < j \le n} 2(\mathfrak{e}_i, \mathfrak{e}_j). \tag{3.3.10}$$

Since

$$2(\mathfrak{e}_i, \mathfrak{e}_j) = 0, -1, -\sqrt{2}, -\sqrt{3} \tag{3.3.11}$$

and the graph is connected by the irreducibility, we have that

$$\text{the number of connected pairs of vertices in } CG(\mathscr{S}) \text{ is } n - 1 = |\mathscr{S}| - 1. \tag{3.3.12}$$

Suppose that there is a cycle in $CG(\mathscr{S})$. Denote by $\mathscr{S}'$ the subset of vectors corresponding to the vertices in the cycle. Then $\mathscr{S}'$ is also an irreducible admissible set. But the number of connected pairs of vertices in $CG(\mathscr{S}')$ is greater than or equal to $|\mathscr{S}'|$, which leads a contradiction to (3.3.12) with $\mathscr{S}$ replaced by $\mathscr{S}'$. Thus

$$CG(\mathscr{S}) \text{ contains no cycles.} \tag{3.3.13}$$

If $(\mathfrak{e}_i, \mathfrak{e}_j) \neq 0$, we say that $\mathfrak{e}_i$ is *connected* to $\mathfrak{e}_j$. For $\mathfrak{e} \in \mathscr{S}$, let $\{\varsigma_1, \ldots, \varsigma_k\}$ are all the vectors connected to $\mathfrak{e}$. By (3.3.13), no pair of ς_i and ς_j is connected; that is,

$$(\varsigma_i, \varsigma_j) = 0 \qquad \text{for } i, j \in \overline{1, k},\ i \neq j. \tag{3.3.14}$$

Set

$$\alpha = \mathfrak{e} - \sum_{i=1}^{k} (\mathfrak{e}, \varsigma_i)\varsigma_i. \tag{3.3.15}$$

Then $\alpha \neq 0$ and

$$(\alpha, \varsigma_i) = 0 \qquad \text{for } i \in \overline{1, k}. \tag{3.3.16}$$

Thus

$$0 < (\alpha, \alpha) = (\alpha, \mathfrak{e} - \sum_{i=1}^{k} (\mathfrak{e}, \varsigma_i)\varsigma_i) = (\alpha, \mathfrak{e}) = 1 - \sum_{i=1}^{k} (\mathfrak{e}, \varsigma_i)^2; \tag{3.3.17}$$

that is,

$$\sum_{i=1}^{k} (\mathfrak{e}, \varsigma_i)^2 < 1 \sim \sum_{i=1}^{k} 4(\mathfrak{e}, \varsigma_i)^2 < 4. \tag{3.3.18}$$

By (3.3.9),

$$\text{no more than three edges can oringinate at a vertex in } CG(\mathscr{S}), \tag{3.3.19}$$

which implies that if $CG(\mathscr{G})$ contains a triple edges, then it must be 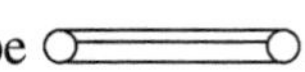.
Moreover, $CG(\mathscr{S})$ contains no following subgraphs:

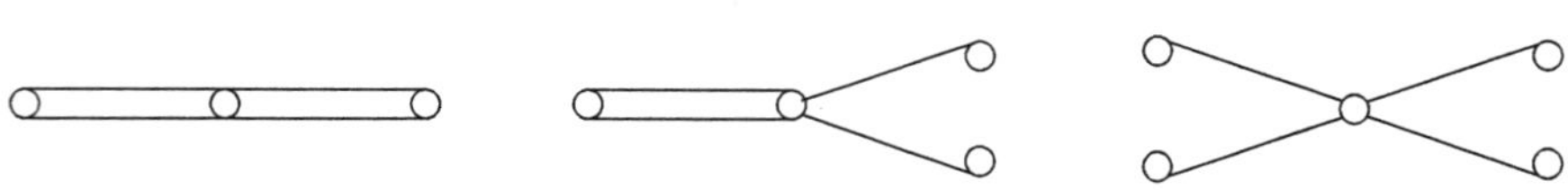

Suppose that $CG(\mathscr{S})$ contains a simple chain: 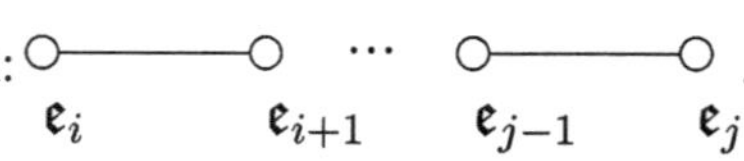

.

Set

$$\mathfrak{e} = \sum_{r=i}^{j} \mathfrak{e}_r. \tag{3.3.20}$$

Then

$$(\mathfrak{e}, \mathfrak{e}) = j - i + 1 + \sum_{r=i}^{j-1} 2(\mathfrak{e}_r, \mathfrak{e}_{r+1}) = j - i + 1 - (j - i) = 1, \tag{3.3.21}$$

$$(\mathfrak{e}_r, \mathfrak{e}) = (\mathfrak{e}_r, \mathfrak{e}_i) \text{ or } (\mathfrak{e}_r, \mathfrak{e}_j) \qquad \text{for } r < i \text{ or } r > j. \tag{3.3.22}$$

Thus $\{\mathfrak{e}\} \bigcup \mathscr{S} \setminus \{\mathfrak{e}_i, \ldots, \mathfrak{e}_j\}$ is an irreducible admissible set. By the arguments in the above paragraph, $CG(\mathscr{S})$ contains no subgraphs of the forms:

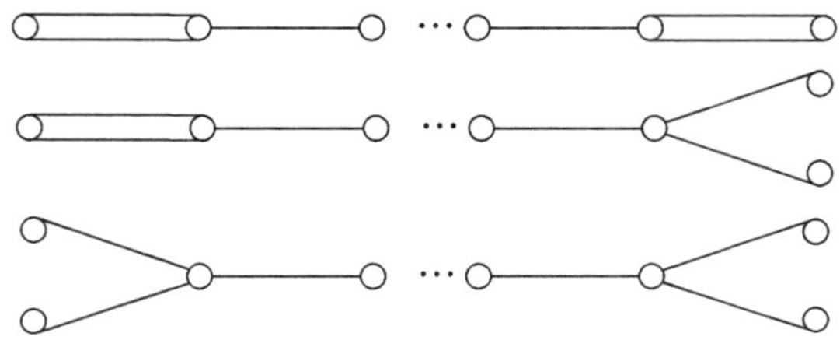

The remaining possible cases of $CG(\mathscr{G})$ are:

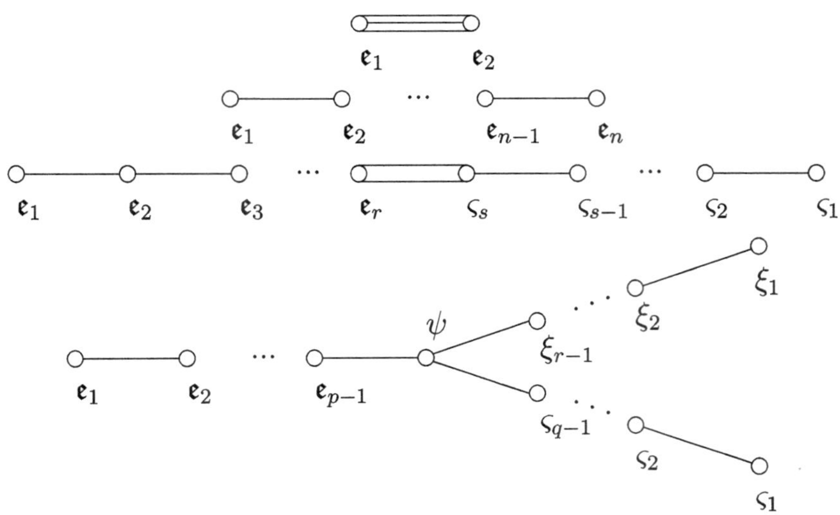

In the third case, we let

$$\mathfrak{e} = \sum_{i=1}^{r} i\mathfrak{e}_i, \qquad \varsigma = \sum_{i=1}^{s} i\varsigma_s. \tag{3.3.23}$$

Moreover,

$$2(\mathfrak{e}_i, \mathfrak{e}_{i+1}) = -1 = 2(\varsigma_i, \varsigma_{i+1}), \qquad 4(\mathfrak{e}_r, \varsigma_s)^2 = 2. \tag{3.3.24}$$

Thus

$$(\mathfrak{e}, \mathfrak{e}) = \sum_{i=1}^{r} i^2 - \sum_{s=1}^{r-1}(i+1)i = r^2 - \sum_{i=1}^{r-1} i = \frac{r(r+1)}{2}, \tag{3.3.25}$$

$$(\varsigma, \varsigma) = \frac{s(s+1)}{2}, \quad (\mathfrak{e}, \varsigma)^2 = r^2 s^2 (\mathfrak{e}_r, \varsigma_s)^2 = \frac{r^2 s^2}{2}. \tag{3.3.26}$$

By the Schwartz inequality, we have:

$$(\mathfrak{e}, \varsigma)^2 < (\mathfrak{e}, \mathfrak{e})(\varsigma, \varsigma) \sim \frac{r^2 s^2}{2} < \frac{rs(r+1)(s+1)}{4} \tag{3.3.27}$$

$$\sim 2rs < (r+1)(s+1) \sim (r-1)(s-1) < 2 \sim r = 1 \text{ or } s = 1 \text{ or } r = s = 2. \tag{3.3.28}$$

Suppose that we have the fourth case with $p \geq q \geq r > 1$. Let

$$\mathfrak{e} = \sum_{i=1}^{p-1} i\mathfrak{e}_i, \quad \varsigma = \sum_{i=1}^{q-1} i\varsigma_i, \quad \xi = \sum_{i=1}^{r-1} i\xi_i. \tag{3.3.29}$$

By (3.3.25), we have

$$(\mathfrak{e}, \mathfrak{e}) = \frac{p(p-1)}{2}, \quad (\varsigma, \varsigma) = \frac{q(r-1)}{2}, \quad (\xi, \xi) = \frac{r(r-1)}{2}. \tag{3.3.30}$$

Moreover, $\mathfrak{e}$, ς and ξ are mutually orthogonal. Set

$$\beta = \psi - \frac{(\psi, \mathfrak{e})}{(\mathfrak{e}, \mathfrak{e})}\mathfrak{e} - \frac{(\psi, \varsigma)}{(\varsigma, \varsigma)}\varsigma - \frac{(\psi, \xi)}{(\xi, \xi)}\xi. \tag{3.3.31}$$

Then $0 \neq \beta$ is orthogonal to $\mathfrak{e}$, ς and ξ. Furthermore,

$$0 < (\beta, \beta) = (\beta, \psi) = 1 - \frac{(\psi, \mathfrak{e})^2}{(\mathfrak{e}, \mathfrak{e})} - \frac{(\psi, \varsigma)^2}{(\varsigma, \varsigma)} - \frac{(\psi, \xi)^2}{(\xi, \xi)}. \tag{3.3.32}$$

On the other hand,

$$4(\psi, \mathfrak{e})^2 = 4(p-1)^2(\psi, \mathfrak{e}_{p-1})^2 = (p-1)^2, \tag{3.3.33}$$

$$4(\psi, \varsigma)^2 = (q-1)^2, \quad 4(\psi, \xi)^2 = (r-1)^2. \tag{3.3.34}$$

Thus

$$\frac{(p-1)^2}{(\mathfrak{e}, \mathfrak{e})} + \frac{(q-1)^2}{(\varsigma, \varsigma)} + \frac{(r-1)^2}{(\xi, \xi)} < 4 \sim \frac{p-1}{p} + \frac{q-1}{q} + \frac{r-1}{r} < 2 \tag{3.3.35}$$

$$\Longrightarrow \frac{1}{p} + \frac{1}{q} + \frac{1}{r} > 1. \tag{3.3.36}$$

In particular,

$$\frac{3}{r} > 1 \sim r < 3 \Longrightarrow r = 2. \tag{3.3.37}$$

Now (3.3.35) becomes

$$\frac{1}{p} + \frac{1}{q} > \frac{1}{2} \Longrightarrow \frac{2}{q} > \frac{1}{2} \sim q < 4 \sim q = 2, 3. \tag{3.3.38}$$

If $q = 3$, then (3.3.36) becomes

$$\frac{1}{p} > \frac{1}{6} \sim p < 6 \sim p = 2, 3, 4, 5. \tag{3.3.39}$$

In summary, the fourth case yields:

$$(p, q, r) = (p, 2, 2), \ (3, 3, 2), \ (4, 3, 2), \ (5, 3, 2). \tag{3.3.40}$$

This gives all possible $CG(\mathscr{S})$. Whenever a double edge or a triple edge occurs in $CG(\Pi)$, we add an arrow pointing to the shorter of two roots. The resulting figure is called the *Dynkin diagram* of Φ.

Theorem 3.3.1 *Let Φ be an irreducible root system. The Dynkin diagram of Φ must be one of the followings:*

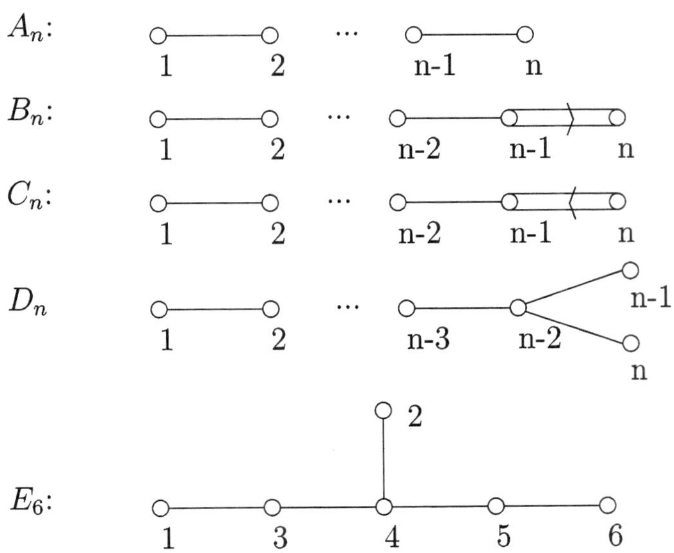

E_7:

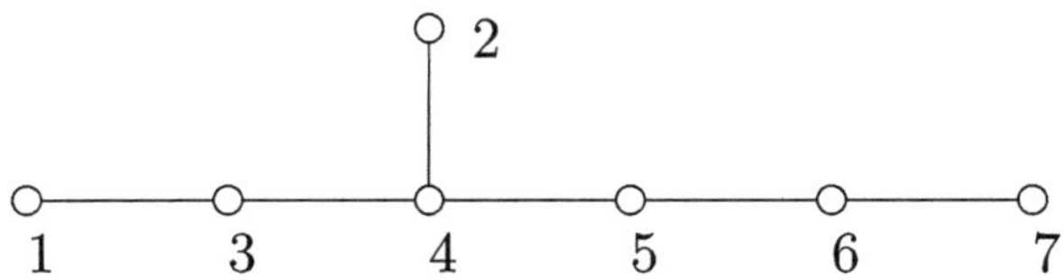

E_8:

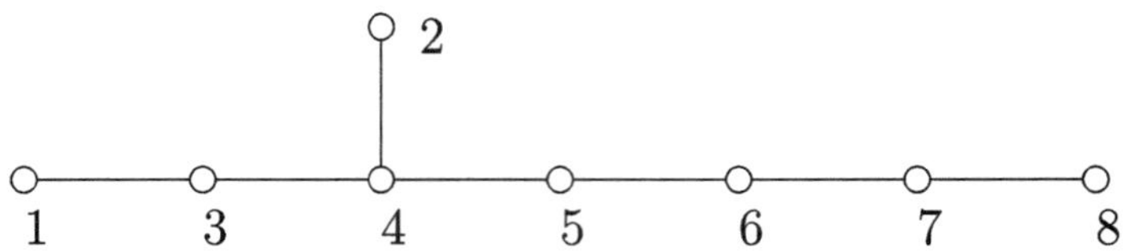

F_4:

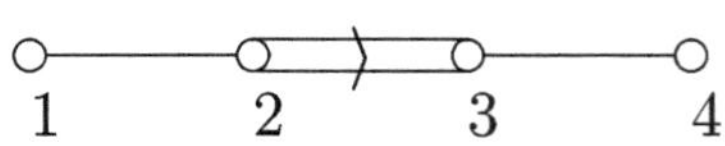

G_2:

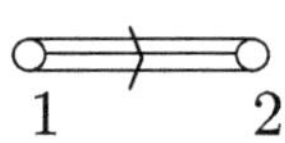

As an exercise, calculate the determinants of the Cartan matrices of the above root systems, which are as follows:

$$A_n : n+1;\ B_n : 2;\ C_n : 2;\ D_n : 4;\ E_6 : 3;\ E_7 : 2;\ E_8,\ F_4,\ G_2 : 1. \tag{3.3.41}$$

3.4 Automorphisms, Constructions and Weights

In this section, we construct all the irreducible root systems and their associated weight lattices. Moreover, their automorphism groups are determined and the saturated subsets of weight lattices are investigated.

Let Φ be a root system in a Euclidean space $\mathscr{E}$ and let $\Pi = \{\alpha_1, \alpha_2, \ldots, \alpha_n\}$ be a base of Φ. The additive subgroup

$$\Lambda_r = \sum_{i=1}^{n} \mathbb{Z}\alpha_i \subset \mathscr{E} \tag{3.4.1}$$

is called the *root lattice* of Φ. Moreover,

$$\Lambda = \{\gamma \in \mathscr{E} \mid \langle \gamma, \alpha \rangle \in \mathbb{Z} \text{ for } \alpha \in \Phi\} = \{\gamma \in \mathscr{E} \mid \langle \gamma, \alpha \rangle \in \mathbb{Z} \text{ for } \alpha \in \Pi\} \tag{3.4.2}$$

is called the *weight lattice* of Φ, whose elements are called *weights*. Obviously, $\Lambda \supset \Lambda_r$. We call Λ/Λ_r the *fundamental group* of Φ.

Since the inner product of $\mathscr{E}$ is nondegenerate, there exist $\lambda_1, \lambda_2, \ldots, \lambda_n \in \mathscr{E}$ such that

$$\langle \lambda_i, \alpha_j \rangle = \delta_{i,j} \qquad \text{for } i, j \in \overline{1, n}. \tag{3.4.3}$$

We call $\{\lambda_1, \ldots, \lambda_n\}$ *fundamental dominant weights relative to* Π. Moreover,

$$\Lambda = \sum_{i=1}^{n} \mathbb{Z}\lambda_i. \tag{3.4.4}$$

Write

$$\alpha_i = \sum_{r=1}^{n} m_{i,r}\lambda_r \qquad \text{for } i \in \overline{1,n}. \tag{3.4.5}$$

Then

$$\langle \alpha_i, \alpha_j \rangle = \langle \sum_{r=1}^{n} m_{i,r}\lambda_r, \alpha_j \rangle = m_{i,j}. \tag{3.4.6}$$

Thus

$$\alpha_i = \sum_{j=1}^{n} \langle \alpha_i, \alpha_j \rangle \lambda_j \qquad \text{for } i \in \overline{1,n}. \tag{3.4.7}$$

By the fundamental theorem of finitely generated abelian groups,

$$|\Lambda/\Lambda_r| = |(\langle \alpha_i, \alpha_j \rangle)_{n\times n}| \tag{3.4.8}$$

is finite. Moreover,

$$\lambda_i \in \frac{1}{|(\langle \alpha_i, \alpha_j \rangle)_{n\times n}|}\Lambda_r \qquad \text{for } i \in \overline{1,n}. \tag{3.4.9}$$

By (3.2.15) and (3.2.17),

$$\rho = \frac{1}{2}\sum_{\alpha\in\Phi^+} \alpha = \sum_{i=1}^{n} \lambda_i. \tag{3.4.10}$$

Example 3.4.1 Let $\mathscr{E}$ be the Euclidean space with a basis $\{\varepsilon_1, \varepsilon_2, \ldots, \varepsilon_n\}$ such that $(\varepsilon_i, \varepsilon_j) = \delta_{i,j}$. The root system of A_{n-1} and a base are as follows:

$$\Phi_{A_{n-1}} = \{\varepsilon_i - \varepsilon_j \mid i, j \in \overline{1,n},\ i \neq j\}, \quad \Pi_{A_{n-1}} = \{\varepsilon_i - \varepsilon_{i+1} \mid i \in \overline{1,n-1}\}. \tag{3.4.11}$$

Set

$$Q = \sum_{i=1}^{n} \mathbb{Z}\varepsilon_i. \tag{3.4.12}$$

The root lattice of $\Phi_{A_{n-1}}$:

$$\Lambda_r(A_{n-1}) = \{\sum_{i=1}^{n} a_i\varepsilon_i \in Q \mid \sum_{i=1}^{n} a_i = 0\} \tag{3.4.13}$$

and

$$\lambda_i = \frac{1}{n}(\sum_{r=1}^{i}(n-i)\varepsilon_r - \sum_{s=i+1}^{n} i\varepsilon_s) \quad \text{for } i \in \overline{1, n-1}. \tag{3.4.14}$$

The group $\Lambda(\Phi_{A_{n-1}})/\Lambda_r(\Phi_{A_{n-1}}) \cong \mathbb{Z}_n = \mathbb{Z}/n\mathbb{Z}$, the cyclic group of order n(exercise).

The root system of B_n is

$$\Phi_{B_n} = \{\varepsilon_i - \varepsilon_j, \pm\varepsilon_i, \pm(\varepsilon_i + \varepsilon_j) \mid i, j \in \overline{1, n},\ i \neq j\} \tag{3.4.15}$$

with a base

$$\Pi_{B_n} = \{\varepsilon_1 - \varepsilon_2, \ldots, \varepsilon_{n-1} - \varepsilon_n, \varepsilon_n\}. \tag{3.4.16}$$

The root lattice

$$\Lambda_r(B_n) = Q \tag{3.4.17}$$

and

$$\lambda_i = \sum_{r=1}^{i} \varepsilon_r, \quad \lambda_n = \frac{1}{2}\sum_{s=1}^{n} \varepsilon_s, \quad \text{for } i \in \overline{1, n-1}. \tag{3.4.18}$$

Thus the weight lattice:

$$\Lambda(B_n) = Q \bigcup (\lambda_n + Q). \tag{3.4.19}$$

The fundamental group: $\Lambda(B_n)/\Lambda_r(B_n) \cong \mathbb{Z}_2$.

The root system of C_n is

$$\Phi_{C_n} = \{\varepsilon_i - \varepsilon_j, \pm 2\varepsilon_i, \pm(\varepsilon_i + \varepsilon_j) \mid i, j \in \overline{1, n},\ i \neq j\} \tag{3.4.20}$$

with a base

$$\Pi_{C_n} = \{\varepsilon_1 - \varepsilon_2, \ldots, \varepsilon_{n-1} - \varepsilon_n, 2\varepsilon_n\}. \tag{3.4.21}$$

The root lattice

$$\Lambda_r(C_n) = \{\sum_{i=1}^{n} a_i\varepsilon_i \in Q \mid \sum_{i=1}^{n} a_i \in 2\mathbb{Z}\} \tag{3.4.22}$$

and

$$\lambda_i = \sum_{r=1}^{i} \varepsilon_i, \quad \lambda_n = \sum_{s=1}^{n} \varepsilon_s, \quad \text{for } i \in \overline{1, n-1}. \tag{3.4.23}$$

Thus the weight lattice:

$$\Lambda(C_n) = Q. \tag{3.4.24}$$

The fundamental group: $\Lambda(C_n)/\Lambda_r(C_n) \cong \mathbb{Z}_2$.

The root system of D_n is

$$\Phi_{D_n} = \{\varepsilon_i - \varepsilon_j, \pm(\varepsilon_i + \varepsilon_j) \mid i, j \in \overline{1,n},\ i \neq j\} \tag{3.4.25}$$

with a base

$$\Pi_{D_n} = \{\varepsilon_1 - \varepsilon_2, \ldots, \varepsilon_{n-1} - \varepsilon_n, \varepsilon_{n-1} + \varepsilon_n\}. \tag{3.4.26}$$

The root lattice and weight lattice:

$$\Lambda_r(D_n) = \Lambda_r(C_n). \tag{3.4.27}$$

$$\lambda_i = \sum_{r=1}^{i} \varepsilon_i \quad \lambda_{n-1} = \frac{1}{2}(\sum_{r=1}^{n-1} \varepsilon_i - \varepsilon_n), \quad \lambda_n = \frac{1}{2}\sum_{s=1}^{n} \varepsilon_s, \qquad \text{for } i \in \overline{1, n-2}. \tag{3.4.28}$$

Thus the weight lattice

$$\Lambda(D_n) = \Lambda(B_n) = Q \bigcup (\lambda_n + Q). \tag{3.4.29}$$

The fundamental group:

$$\Lambda(D_n)/\Lambda_r(D_n) \cong \begin{cases} \mathbb{Z}_4 & \text{if } n \in 2\mathbb{Z}+1 \\ \mathbb{Z}_2 \times \mathbb{Z}_2 & \text{if } n \in 2\mathbb{Z} \end{cases} \tag{3.4.30}$$

Example 3.4.2 The root and weight lattices of E_8:

$$\Lambda_r(E_8) = \Lambda(D_8) = \Lambda(E_8), \tag{3.4.31}$$

which is a self-dual integral even lattice. The root system

$$\Phi_{E_8} = \left\{\frac{1}{2}\sum_{i=1}^{8}(-1)^{\iota_i}\varepsilon_i \mid \iota_i \in \mathbb{Z}_2,\ \sum_{r=1}^{8}\iota_r = 0 \text{ in } \mathbb{Z}_2\right\} \bigcup \Phi_{D_8} \tag{3.4.32}$$

with a base

$$\Pi_{E_8} = \left\{\varepsilon_1 - \varepsilon_2, \frac{1}{2}(\sum_{s=4}^{8}\varepsilon_s - \sum_{r=1}^{3}\varepsilon_r), \varepsilon_{i+1} - \varepsilon_{i+2} \mid i \in \overline{1,6}\right\}. \tag{3.4.33}$$

Moreover,

$$\lambda_1 = \frac{1}{2}(3\varepsilon_1 + \sum_{i=2}^{8} \varepsilon_i), \quad \lambda_2 = \sum_{i=1}^{8} \varepsilon_i, \quad \lambda_3 = 2(\varepsilon_1 + \varepsilon_2) + \sum_{i=3}^{8} \varepsilon_i, \tag{3.4.34}$$

$$\lambda_4 = \frac{1}{2}(5(\varepsilon_1 + \varepsilon_2 + \varepsilon_3) + 3\sum_{i=4}^{8} \varepsilon_i), \quad \lambda_5 = \sum_{i=1}^{4}(2\varepsilon_i + \varepsilon_{4+i}), \tag{3.4.35}$$

$$\lambda_6 = \frac{1}{2}(3\sum_{i=1}^{5} \varepsilon_i + \varepsilon_6 + \varepsilon_7 + \varepsilon_8), \quad \lambda_7 = \sum_{i=1}^{6} \varepsilon_i, \quad \lambda_8 = \frac{1}{2}(\sum_{i=1}^{7} \varepsilon_i - \varepsilon_8). \tag{3.4.36}$$

Observe

$$\Pi_{E_7} = \left\{ \varepsilon_1 - \varepsilon_2, \frac{1}{2}(\sum_{r=1}^{3} \varepsilon_r - \sum_{s=4}^{8} \varepsilon_s), \varepsilon_{i+1} - \varepsilon_{i+2} \mid i \in \overline{1,5} \right\} \tag{3.4.37}$$

and

$$\Pi_{E_6} = \left\{ \varepsilon_1 - \varepsilon_2, \frac{1}{2}(\sum_{r=1}^{3} \varepsilon_r - \sum_{s=4}^{8} \varepsilon_s), \varepsilon_{i+1} - \varepsilon_{i+2} \mid i \in \overline{1,4} \right\}. \tag{3.4.38}$$

Moreover, the fundamental groups:

$$\Lambda(E_7)/\Lambda_r(E_7) \cong \mathbb{Z}_2, \qquad \Lambda(E_6)/\Lambda_r(E_6) \cong \mathbb{Z}_3 \tag{3.4.39}$$

(exercise).

Example 3.4.3 The root and weight lattices of F_4:

$$\Lambda_r(F_4) = \Lambda(F_4) = \Lambda(D_4). \tag{3.4.40}$$

Moreover,

$$\Phi_{F_4} = \Phi_{B_4} \bigcup \left\{ \frac{1}{2} \sum_{i=1}^{4} (-1)^{\iota_i} \varepsilon_i \mid \iota_i \in \mathbb{Z}_2 \right\} \tag{3.4.41}$$

with a base

$$\Pi_{F_4} = \left\{ \varepsilon_2 - \varepsilon_3, \varepsilon_3 - \varepsilon_4, \varepsilon_4, \frac{1}{2}(\varepsilon_1 - \varepsilon_2 - \varepsilon_3 - \varepsilon_4) \right\}. \tag{3.4.42}$$

Moreover,

$$\lambda_1 = \varepsilon_1 + \varepsilon_2, \quad \lambda_2 = 2\varepsilon_1 + \varepsilon_2 + \varepsilon_3, \quad \lambda_3 = \frac{1}{2}(3\varepsilon_1 + \varepsilon_2 + \varepsilon_3 + \varepsilon_4), \quad \lambda_4 = \varepsilon_1. \tag{3.4.43}$$

Example 3.4.4 The root and weight lattices of G_2:

$$\Lambda_r(G_2) = \Lambda(G_2) = \Lambda_r(A_2). \tag{3.4.44}$$

Moreover,

$$\Phi_{G_2} = \{\pm(2\varepsilon_i - \varepsilon_j - \varepsilon_k), \varepsilon_i - \varepsilon_j \mid \{i, j, k\} = \{1, 2, 3\}\} \tag{3.4.45}$$

with a base

$$\Pi_{G_2} = \{\varepsilon_2 + \varepsilon_3 - 2\varepsilon_1, \varepsilon_1 - \varepsilon_2\}. \tag{3.4.46}$$

Moreover,

$$\lambda_1 = 2\varepsilon_3 - \varepsilon_1 - \varepsilon_2, \qquad \lambda_2 = \varepsilon_3 - \varepsilon_2. \tag{3.4.47}$$

Let Φ be a root system in a Euclidean space $\mathscr{E}$ and let Π be a base of Φ. Recall that Λ is the set of weights of Φ. Set

$$\Lambda^+ = \Lambda \bigcap \overline{\mathscr{C}(\Pi)} \tag{3.4.48}$$

and call the elements in Λ^+ *dominant weights*. Moreover, we call the elements of $\Lambda \bigcap \mathscr{C}(\Pi)$ *strong dominant weights*. We define another partial ordering $\lhd$ on Λ by

$$\lambda \lhd \mu \text{ if } \mu - \lambda = \sum_{\alpha \in \Pi} a_\alpha \alpha \text{ such that } 0 \leq a_\alpha \in \mathbb{Z}, \tag{3.4.49}$$

where $\lambda, \mu \in \Lambda$. By (3.1.4) and (3.4.2), two partial orderings coincide on $\mathscr{W}(\lambda)$ for any $\lambda \in \Lambda$. According to Lemma 3.2.12, any weight is conjugated to a dominant weight under the Weyl group $\mathscr{W}$. If λ is a dominant weight, then $\sigma(\lambda) \lhd \lambda$ for $\sigma \in \mathscr{W}$.

If λ is a strong dominant weight and $\sigma(\lambda) = \lambda$ with $\sigma \in \mathscr{W}$, then $\sigma = 1$ by Theorem 3.2.8 (e).

Lemma 3.4.1 *For* $\lambda \in \Lambda^+$, $|\{\mu \in \Lambda^+ \mid \mu \lhd \lambda\}| < \infty$.

Proof Note

$$0 \leq (\lambda, \lambda - \mu) = (\lambda, \lambda) - (\lambda, \mu) \tag{3.4.50}$$

and

$$0 \leq (\lambda - \mu, \mu) = (\lambda, \mu) - (\mu, \mu). \tag{3.4.51}$$

Adding two inequalities, we obtain

$$0 \leq (\lambda, \lambda) - (\mu, \mu) \sim (\mu, \mu) \leq (\lambda, \lambda). \tag{3.4.52}$$

Since the intersection of Λ with the sphere of radius $\|\lambda\|$ is a finite set, so is its subset $\{\mu \in \Lambda^+ \mid \mu \lhd \lambda\}$. $\square$

A subset $\mathbb{S}$ of Λ is called *saturated* if $\mu - i\alpha \in \mathbb{S}$ for $\mu \in \mathbb{S},\ \alpha \in \Phi$ and i between 0 and $\langle \mu, \alpha \rangle$. By definition $\mathscr{W}(\mathbb{S}) = \mathbb{S}$. If $\mathbb{S}$ has a unique maximal weight λ, we call λ the *highest weight* of $\mathbb{S}$. Lemma 3.2.12 says that $\lambda \in \Lambda^+$. By Lemmas 3.2.12 and 3.4.1, we have:

Lemma 3.4.2 *A saturated set of weights with a highest weight must be finite.*

Lemma 3.4.3 *Let $\mathbb{S}$ be a saturated subset with a highest weight λ. If $\lambda \rhd \mu \in \Lambda^+$, then $\mu \in \mathbb{S}$.*

Proof Assume $\mu \neq \lambda$. Suppose $\lambda' = \mu + \sum_{\alpha\in\Pi} k_\alpha \alpha \in \mathbb{S}$ with $0 \leq k_\alpha \in \mathbb{Z}$ and $\sum_{\alpha\in\Pi} k_\alpha > 0$. For instance λ is such a λ'. Since

$$0 < (\sum_{\alpha\in\Pi} k_\alpha \alpha, \sum_{\alpha\in\Pi} k_\alpha \alpha) = \sum_{\beta\in\Pi} k_\beta (\sum_{\alpha\in\Pi} k_\alpha \alpha, \beta), \tag{3.4.53}$$

there exists $\gamma \in \Pi$ such that

$$k_\gamma (\sum_{\alpha\in\Pi} k_\alpha \alpha, \gamma) > 0 \sim k_\gamma,\ (\sum_{\alpha\in\Pi} k_\alpha \alpha, \gamma) > 0. \tag{3.4.54}$$

So

$$\langle \lambda', \gamma \rangle = \langle \mu, \gamma \rangle + \langle \sum_{\alpha\in\Pi} k_\alpha \alpha, \gamma \rangle > 0. \tag{3.4.55}$$

By definition, $\lambda' - \gamma \in \mathbb{S}$. An induction on $\sum_{\alpha\in\Pi} k_\alpha$ shows $\mu \in \mathbb{S}$. $\square$

Lemma 3.4.4 *The subset $\mathbb{S}$ is a saturated subset of weights with a highest weight λ if and only if*

$$\lambda \in \Lambda^+, \quad \mathbb{S} = \mathscr{W}(\{\mu \in \Lambda^+ \mid \mu \lhd \lambda\}). \tag{3.4.56}$$

Proof Set $\mathbb{S}_\lambda = \mathscr{W}(\{\mu \in \Lambda^+ \mid \mu \lhd \lambda\})$. If $\mathbb{S}$ is a saturated subset of weights with a highest weight λ, then (3.4.56) holds by Lemmas 3.2.12 and 3.4.3.

Suppose $\lambda \in \Lambda^+$. Then $\mathscr{W}(\mathbb{S}_\lambda) = \mathbb{S}_\lambda$ and λ is the highest weight of $\mathbb{S}_\lambda$. For any $\lambda \rhd \mu \in \Lambda^+$, $\sigma \in \mathscr{W}$, $\alpha \in \Phi$ and i between 0 and $\langle \sigma(\mu), \alpha \rangle$, there exists $\tau \in \mathscr{W}$ such that $\tau(\sigma(\mu) - i\alpha) \in \Lambda^+$. Note $\tau\sigma(\mu) \lhd \mu \lhd \lambda$ by Lemma 3.2.12. First we assume $\tau(\alpha) \in \Phi^+$. If $\langle \sigma(\mu), \alpha \rangle \geq 0$, then $i \geq 0$. So

$$\tau(\sigma(\mu) - i\alpha) = \tau\sigma(\mu) - i\tau(\alpha) \lhd \tau\sigma(\mu) \lhd \lambda. \tag{3.4.57}$$

When $\langle \sigma(\mu), \alpha \rangle < 0$, we have $i < 0$ and

$$\tau\sigma(\mu) - i\tau(\alpha) \lhd \tau\sigma(\mu) - \langle \sigma(\mu), \alpha \rangle \tau(\alpha) = \tau\sigma_\alpha\sigma(\mu) \lhd \mu \lhd \lambda \tag{3.4.58}$$

by Lemma 3.2.12. Hence $\tau(\sigma(\mu) - i\alpha) \in \mathbb{S}_\lambda$. Thus $\sigma(\mu) - i\alpha \in \mathbb{S}_\lambda$.

Assume $\tau(\alpha) \in \Phi^-$. Then

$$\sigma(\mu) - i\alpha = \sigma(\mu) - (-i)(-\alpha) \tag{3.4.59}$$

and $-i$ is between 0 and $\langle \sigma(\mu), -\alpha \rangle$. Replacing α by $-\alpha$ in the above arguments, we have

$$\sigma(\mu) - i\alpha = \sigma(\mu) - (-i)(-\alpha) \in \mathbb{S}_\lambda. \tag{3.4.60}$$

Therefore, $\mathbb{S}_\lambda$ is saturated. □

The automorphism group of Φ:

$$\text{Aut}\ \Phi = \{\sigma \in GL(\mathscr{E}) \mid \sigma(\Phi) = \Phi\}. \tag{3.4.61}$$

By Lemma 3.2.2, $\mathscr{W}$ is a normal subgroup of Aut Φ and

$$\langle \sigma(\alpha), \sigma(\beta) \rangle = \langle \alpha, \beta \rangle \qquad \text{for}\ \ \sigma \in \text{Aut}\ \Phi,\ \ \alpha, \beta \in \Phi. \tag{3.4.62}$$

Since Φ spans $\mathscr{E}$, the above equality holds for any $\alpha \in \mathscr{E}$ and $\beta \in \Phi$. So σ maps regular elements to regular elements. Thus $\sigma(\Pi)$ is a base of Φ for any $\sigma \in$ Aut Φ. By Theorem 3.2.8 (b), there exists $\sigma_1 \in \mathscr{W}$ such that $\sigma_1\sigma(\Pi) = \Pi$. Hence Aut $\Phi/\mathscr{W}$ is completely determined by

$$\{\sigma \in GL(\mathscr{E}) \mid \sigma(\Pi) = \Pi,\ \langle \sigma(\alpha), \sigma(\beta) \rangle = \langle \alpha, \beta \rangle \ \text{for}\ \ \alpha, \beta \in \Pi\} \tag{3.4.63}$$

because $\Phi = \mathscr{W}(\Pi)$. By Theorem 3.3.1, we have:

Theorem 3.4.5 *Let Φ be an irreducible root system. Then Aut $\Phi/\mathscr{W}$ is completely determined as follows:*

$$Aut\ \Phi_{A_n}/\mathscr{W} \cong Aut\ \Phi_{E_6}/\mathscr{W} \cong \mathbb{Z}_2, \tag{3.4.64}$$

$$Aut\ \Phi_{D_4}/\mathscr{W} \cong S_3,\ \ Aut\ \Phi_{D_n}/\mathscr{W} \cong \mathbb{Z}_2 \qquad for\ \ n > 4, \tag{3.4.65}$$

and Aut $\Phi/\mathscr{W} = \{1\}$ for the other types of Φ.

Exercise Prove the Weyl group of G_2 is isomorphic the dihedral group $\mathscr{D}_6$.

Chapter 4
Isomorphisms, Conjugacy and Exceptional Types

Finite-dimensional complex simple Lie algebras were classified by Killing and Cartan in later nineteen century. The arguments were simplified by Weyl in 1920s. In this chapter, we show that the structure of a finite-dimensional semisimple Lie algebra over $\mathbb{C}$ is completely determined by its root system. Moreover, we prove that any two Cartan subalgebras of such a Lie algebra $\mathscr{G}$ are conjugated under the group of inner automorphisms of $\mathscr{G}$, where maximal solvable subalgebras play important roles. In particular, we prove that the intersection of all the maximal solvable subalgebras of any complex Lie algebra is exactly its maximal solvable ideal. The automorphism group of a simple Lie algebra is determined. Furthermore, we give explicit constructions of the simple Lie algebras of exceptional types.

In Sect. 4.1, we prove that semisimple Lie algebras with isomorphic root systems are isomorphic. In Sect. 4.2, we give characterizations of Cartan subalgebras. Moreover, we prove a correspondence between Cartan subalgebras under homomorphisms. In Sect. 4.3, we prove that all the Cartan subalgebras of a finite-dimensional complex semisimple Lie algebra are conjugated under the group of its inner automorphisms. The relation between maximal solvable subalgebras and the maximal solvable ideal of a complex Lie algebra is determined. Finally in Sect. 4.4, the finite-dimensional complex simple Lie algebras of types G_2, F_4, E_6, E_7 and E_8 are constructed.

4.1 Isomorphisms

In this section, we prove that two semisimple Lie algebras are isomorphic if their root systems are.

Let $\mathscr{G}$ be a finite-dimensional semisimple Lie algebra over $\mathbb{C}$ and let H be a maximal toral subalgebra of $\mathscr{G}$. Then we have the Contan root space decomposition:

X. Xu, *Representations of Lie Algebras and Partial Differential Equations*,
DOI 10.1007/978-981-10-6391-6_4

$$\mathscr{G} = H + \sum_{\alpha\in\Phi} \mathscr{G}_\alpha, \tag{4.1.1}$$

where $\Phi \subset H^*$ is a root system of $\mathscr{G}$.

Lemma 4.1.1 *If $\mathscr{G}$ is simple, then Φ is irreducible.*

Proof Suppose $\Phi = \Phi_1 \bigcup \Phi_2$, $\Phi_1 \neq \emptyset \neq \Phi_2$ and $\Phi_1 \perp \Phi_2$. For any $\alpha_1 \in \Phi_1$ and $\alpha_2 \in \Phi_2$, $\alpha_1 + \alpha_2$ is not orthogonal to Φ_1 and Φ_2. So $\alpha_1 + \alpha_2 \notin \Phi$. Hence $[\mathscr{G}_{\alpha_1}, \mathscr{G}_{\alpha_2}] = \{0\}$. Let $\mathscr{G}_i$ be the Lie algebra generated by $\{\mathscr{G}_\alpha \mid \alpha \in \Phi_i\}$. Then $\mathscr{G} = \mathscr{G}_1 + \mathscr{G}_2$ and $[\mathscr{G}_1, \mathscr{G}_2] = \{0\}$, which contradicts the simplicity of $\mathscr{G}$. □

Corollary 4.1.2 *In general, if $\mathscr{G} = \mathscr{G}_1 \oplus \mathscr{G}_2 \oplus \cdots \oplus \mathscr{G}_k$ is the decomposition of $\mathscr{G}$ into simple ideals, then $H_i = H \bigcap \mathscr{G}_i$ is a maximal toral subalgebra of $\mathscr{G}_i$ whose corresponding root system Φ_i are irreducible components of Φ; that is, $\Phi = \bigcup_{i=1}^k \Phi_i$.*

Lemma 4.1.3 *If $\mathscr{G}$ is simple, then it is generated by $\{\mathscr{G}_\alpha, \mathscr{G}_{-\alpha} \mid \alpha \in \Pi\}$.*

Proof According to Corollary 3.1.6, any positive root $\alpha = \sum_{i=1}^k \beta_i$ with $\beta_i \in \Pi$ such that $\sum_{i=1}^s \beta_i$ are positive roots for $s \in \overline{1, k}$. By Lemma 2.6.2 (c),

$$\mathscr{G}_\alpha = [\mathscr{G}_{\beta_k}, [\mathscr{G}_{\beta_{k-1}}, ..., [\mathscr{G}_{\beta_2}, \mathscr{G}_{\beta_1}]...]]. \tag{4.1.2}$$

Symmetrically,

$$\mathscr{G}_{-\alpha} = [\mathscr{G}_{-\beta_k}, [\mathscr{G}_{-\beta_{k-1}}, ..., [\mathscr{G}_{-\beta_2}, \mathscr{G}_{-\beta_1}]...]]. \tag{4.1.3}$$

Since $\mathscr{G}$ is generated by $\{\mathscr{G}_\alpha \mid \alpha \in \Phi\}$ by Lemma 2.6.2 (e), it is generated by $\{\mathscr{G}_\alpha, \mathscr{G}_{-\alpha} \mid \alpha \in \Pi\}$. □

Theorem 4.1.4 *Suppose that $\mathscr{G}$ is simple. Let $\mathscr{G}'$ be another finite-dimensional simple Lie algebra and H' is a maximal toral subalgebra $\mathscr{G}'$ with corresponding root system Φ'. If $\Phi \cong \Phi'$, then $\mathscr{G} \cong \mathscr{G}'$.*

Proof Let $\mathscr{E}$ be the Euclidean space associated with Φ and let $\mathscr{E}'$ be the Euclidean space associated with Φ'. Suppose that $\tau : \alpha \mapsto \alpha'$ is an isomorphism from Φ to Φ'; that is, τ is a linear map from $\mathscr{E}$ to $\mathscr{E}'$ such that $\tau(\Phi) = \Phi'$ and

$$\langle\alpha', \beta'\rangle = \langle\alpha, \beta\rangle \qquad \text{for } \alpha, \beta \in \Phi. \tag{4.1.4}$$

Note that $\Pi' = \tau(\Pi)$ is a base of Φ'.

Let $\mathscr{L} = \mathscr{G} \oplus \mathscr{G}'$ be the direct sum of two Lie algebras. So $\mathscr{L}$ is semisimple, and $\mathscr{G}$, $\mathscr{G}'$ are the only simple ideals. Take

$$\{\xi_\alpha \in \mathscr{G}_\alpha,\ \zeta_\alpha \in \mathscr{G}_{-\alpha}, \xi'_{\alpha'} \in \mathscr{G}'_{\alpha'}, \zeta'_{\alpha'} \in \mathscr{G}'_{-\alpha'} \mid \alpha \in \Pi\} \tag{4.1.5}$$

such that

$$[\xi_\alpha, \zeta_\alpha] = h_\alpha, \quad [h_\alpha, \xi_\alpha] = 2\xi_\alpha, \quad [h_\alpha, \zeta_\alpha] = -2\zeta_\alpha \tag{4.1.6}$$

and

$$[\xi'_{\alpha'}, \zeta'_{\alpha'}] = h'_{\alpha'}, \quad [h'_{\alpha'}, \xi'_{\alpha'}] = 2\xi_{\alpha'}, \quad [h'_{\alpha'}, \zeta'_{\alpha'}] = -2\zeta'_{\alpha'} \tag{4.1.7}$$

for $\alpha \in \Pi$. Let

$$\mathscr{K} = \text{the Lie subalgebra generated by } \{\xi_\alpha + \xi'_{\alpha'}, \zeta_\alpha + \zeta'_{\alpha'} \mid \alpha \in \Pi\}. \tag{4.1.8}$$

Define $\mathscr{P}$, $\mathscr{P}' \in \text{End}\,\mathscr{L}$ by

$$\mathscr{P}(u + u') = u, \quad \mathscr{P}'(u + u') = u' \qquad \text{for } u \in \mathscr{G},\ u' \in \mathscr{G}'. \tag{4.1.9}$$

Then

$$\mathscr{P}(\mathscr{K}) = \mathscr{G}, \quad \mathscr{P}'(\mathscr{K}) = \mathscr{G}' \tag{4.1.10}$$

by Lemma 4.1.3.

Let β be the maximal root in Φ relative to Π. Then β' is the maximal root in Φ' relative to Π'. Take any $0 \neq \zeta \in \mathscr{G}_\beta$ and $0 \neq \zeta' \in \mathscr{G}'_{\beta'}$. Set

$$M = \text{span}\{\zeta + \zeta', \text{ad}\,(\zeta_{\alpha_1} + \zeta'_{\alpha'_1}) \cdots \text{ad}\,(\zeta_{\alpha_k} + \zeta'_{\alpha'_k})(\zeta + \zeta') \mid \alpha_i \in \Pi,\ k \in \mathbb{Z}_+\}. \tag{4.1.11}$$

Note

$$[\xi_\alpha + \xi'_{\alpha'}, \zeta_\gamma + \zeta'_{\gamma'}] = \delta_{\alpha,\gamma}(h_\alpha + h'_{\alpha'}) \qquad \text{for } \alpha, \gamma \in \Pi \tag{4.1.12}$$

by (4.1.6) and (4.1.7), because $\alpha - \gamma$ is not a root. Moreover,

$$[h_\alpha + h'_{\alpha'}, \zeta_\gamma + \zeta'_{\gamma'}] = \gamma(h_\alpha)\zeta_\gamma + \gamma'(h'_{\alpha'})\zeta'_{\gamma'} = \langle\gamma, \alpha\rangle\zeta_\gamma + \langle\gamma', \alpha'\rangle\zeta'_{\gamma'} = \langle\gamma, \alpha\rangle(\zeta_\gamma + \zeta'_{\gamma'}) \tag{4.1.13}$$

and

$$[h_\alpha + h'_{\alpha'}, \zeta + \zeta'] = \langle\beta, \alpha\rangle(\zeta + \zeta') \tag{4.1.14}$$

by (4.1.4). Furthermore,

$$[\xi_\alpha + \xi'_{\alpha'}, \zeta + \zeta'] = 0 \qquad \text{for } \alpha \in \Pi, \tag{4.1.15}$$

due to $\beta + \alpha \notin \Phi$ and $\beta' + \alpha' \notin \Phi'$ by the maximality of β. Equations (4.1.12)–(4.1.15) imply that

$$[\mathscr{K}, M] \subset M. \tag{4.1.16}$$

Since

$$M \bigcap (\mathscr{G}_\beta + \mathscr{G}'_{\beta'}) = \mathbb{F}(\zeta + \zeta'), \tag{4.1.17}$$

we have

$$M \neq \mathscr{L}, \mathscr{G}, \mathscr{G}'. \tag{4.1.18}$$

Thus M is not an ideal of $\mathscr{L}$. This means $\mathscr{K} \neq \mathscr{L}$. If $\mathscr{K} \bigcap \mathscr{G} \neq \{0\}$, then

$$[\mathscr{K}, \mathscr{K} \bigcap \mathscr{G}] = [\mathscr{G}, \mathscr{K} \bigcap \mathscr{G}] \subset \mathscr{K} \bigcap \mathscr{G} \tag{4.1.19}$$

by (4.1.10). So $\mathscr{K} \bigcap \mathscr{G}$ is a nonzero ideal of $\mathscr{G}$. By the simplicity of $\mathscr{G}$, $\mathscr{G} = \mathscr{K} \bigcap \mathscr{G}$, which implies $\mathscr{L} = \mathscr{K}$ by (4.1.10). This contradicts $\mathscr{K} \neq \mathscr{L}$. Thus $\mathscr{K} \bigcap \mathscr{G} = \{0\}$. Symmetrically, $\mathscr{K} \bigcap \mathscr{G}' = \{0\}$. These facts show that the map

$$u \mapsto \Im(u) \text{ determined by } u + \Im(u) \in \mathscr{K}, \qquad u \in \mathscr{G}, \tag{4.1.20}$$

is a Lie isomorphism from $\mathscr{G}$ to $\mathscr{G}'$. □

Note

$$\Im(\xi_\alpha) = \xi'_{\alpha'}, \quad \Im(\zeta_\alpha) = \zeta'_{\alpha'}, \quad \Im(h_\alpha) = h'_{\alpha'} \tag{4.1.21}$$

by (4.1.6)–(4.1.8). In particular, any automorphism σ of Φ gives rise to an automorphism $\Im_\sigma$ (not unique) of $\mathscr{G}$ such that

$$\Im_\sigma(H) = H, \quad \Im_\sigma(\mathscr{G}_\alpha) = \mathscr{G}_{\sigma(\alpha)} \qquad \text{for } \alpha \in \Phi. \tag{4.1.22}$$

Note that $-\mathrm{Id}_{\mathscr{E}} \in \mathrm{Aut}\,\Phi$. We have a unique automorphism ϑ of $\mathscr{G}$ such that

$$\vartheta(\xi_\alpha) = -\zeta_\alpha, \quad \vartheta(\zeta_\alpha) = -\xi_\alpha, \quad \vartheta(h_\alpha) = -h_\alpha \qquad \text{for } \alpha \in \Pi. \tag{4.1.23}$$

The automorphism ϑ is called a *Cartan involution*. For any $\alpha \in \Pi$, ad ξ_α and ad ζ_α are nilpotent by Lemma 2.5.2. The linear transformation

$$\tau_\alpha = e^{\mathrm{ad}\,\xi_\alpha} e^{-\mathrm{ad}\,\zeta_\alpha} e^{\mathrm{ad}\,\xi_\alpha} \tag{4.1.24}$$

is an automorphism of $\mathscr{G}$ by (1.4.27) satisfying

$$\tau_\alpha(\mathscr{G}_\beta) = \mathscr{G}_{\sigma_\alpha(\beta)} \qquad \text{for } \beta \in \Phi \tag{4.1.25}$$

and $\tau_\alpha = \vartheta$ when $\mathscr{G} = sl(2, \mathbb{C})$ (exercise).

4.2 Cartan Subalgebras

In this section, we give characterizations of Cartan subalgebras. Moreover, we prove a correspondence between Cartan subalgebras under homomorphisms.

Let $\mathscr{G}$ be a finite-dimensional Lie algebra over $\mathbb{C}$. For any $\xi \in \mathscr{G}$, we have

$$\mathscr{G} = \bigoplus_{a \in \mathbb{C}} \mathscr{G}_\alpha(\text{ad}\,\xi), \tag{4.2.1}$$

where

$$\mathscr{G}_a(\text{ad}\,\xi) = \{u \in \mathscr{G} \mid (\text{ad}\,\xi - a)^m(u) = 0 \ \text{ for some } m \in \mathbb{N}\}. \tag{4.2.2}$$

Moreover,

$$[\mathscr{G}_a(\text{ad}\,\xi), \mathscr{G}_b(\text{ad}\,\xi)] = \mathscr{G}_{a+b}(\text{ad}\,\xi) \qquad \text{for } a, b \in \mathbb{C} \tag{4.2.3}$$

by (1.5.60). In particular, $\mathscr{G}_0(\text{ad}\,\xi)$ forms a Lie subalgebra of $\mathscr{G}$, which is called an *Engel subalgebra.*

For a subalgebra $\mathscr{K}$ of $\mathscr{G}$, we define the *normalizer* of $\mathscr{K}$ in $\mathscr{G}$ by

$$N_{\mathscr{G}}(\mathscr{K}) = \{\xi \in \mathscr{G} \mid [\xi, \mathscr{K}] \subset \mathscr{K}\}. \tag{4.2.4}$$

Then $N_{\mathscr{G}}(\mathscr{K})$ is the unique maximal subalgebra of $\mathscr{G}$ that contains $\mathscr{K}$ as an ideal.

Lemma 4.2.1 *If $\mathscr{K}$ is a subalgebra of $\mathscr{G}$ containing an Engel subalgebra. Then $N_{\mathscr{G}}(\mathscr{K}) = \mathscr{K}$.*

Proof Suppose $\mathscr{K} \supset \mathscr{G}_0(\text{ad}\,\xi)$ for some $\xi \in \mathscr{G}$. Obviously, $\xi \in \mathscr{G}_0(\text{ad}\,\xi) \subset \mathscr{K}$. If $N_{\mathscr{G}}(\mathscr{K})/\mathscr{K} \neq \{0\}$, then ad$\xi$ acts on it as an invertible linear transformation because $\mathscr{G}_0(\text{ad}\,\xi) \subset \mathscr{K}$. This contradicts $[\xi, N_{\mathscr{G}}(\mathscr{K})] \subset \mathscr{K}$. $\square$

We call a subalgebra $\mathscr{K}$ of $\mathscr{G}$ *self-normalizing* if $N_{\mathscr{G}}(\mathscr{K}) = \mathscr{K}$.

Lemma 4.2.2 *Let $\mathscr{K}$ be a subalgebra of $\mathscr{G}$. Take $\upsilon \in \mathscr{K}$ such that $\mathscr{G}_0(ad\,\upsilon)$ is a minimal element in $\{\mathscr{G}_0(ad\,\xi) \mid \xi \in \mathscr{K}\}$ with respect to inclusion. If $\mathscr{K} \subset \mathscr{G}_0(ad\,\upsilon)$, then $\mathscr{G}_0(ad\,\upsilon) \subset \mathscr{G}_0(ad\,\xi)$ for any $\xi \in \mathscr{K}$.*

Proof Let $0 \neq \xi \in \mathscr{K}$ be a fixed element. Suppose

$$\dim \mathscr{G} = n, \qquad \dim \mathscr{G}_0(\text{ad}\, v) = m. \tag{4.2.5}$$

For any $c \in \mathbb{C}$, we have

$$\text{ad}\,(v + c\xi)(\mathscr{G}_0(\text{ad}\, v)) \subset \mathscr{G}_0(\text{ad}\, v). \tag{4.2.6}$$

Let

$$f(t, c) = t^m + f_1(c)t^{m-1} + \cdots + f_m(c) \tag{4.2.7}$$

be the characteristic polynomial of $\text{ad}\,(v + c\xi)|_{\mathscr{G}_0(\text{ad}\, v)}$ and let

$$g(t, c) = t^{n-m} + g_1(c)t^{n-m-1} + \cdots + g_{n-m}(c) \tag{4.2.8}$$

be the characteristic polynomial of $\text{ad}\,(v + c\xi)|_{\mathscr{G}/\mathscr{G}_0(\text{ad}\, v)}$. Then the characteristic polynomial of $\text{ad}\,(v + c\xi)$ on $\mathscr{G}$ is $f(t, c)g(t, c)$. Moreover, as polynomials of c,

$$\deg f_i(c) \leq i, \qquad \deg g_j(c) \leq j. \tag{4.2.9}$$

Since $\text{ad}\, v|_{\mathscr{G}/\mathscr{G}_0(\text{ad}\, v)}$ is invertible, we have $g_{n-m}(0) \neq 0$. So $g_{n-m}(c)$ is a nonzero polynomial. There exists distinct $c_1, c_2, \ldots, c_{n+1} \in \mathbb{C}$ such that

$$g_{n-m}(c_i) \neq 0 \qquad \text{for } i \in \overline{1, n+1}. \tag{4.2.10}$$

Thus

$$\mathscr{G}_0(\text{ad}\,(v + c_i\xi)) \subset \mathscr{G}_0(\text{ad}\, v). \tag{4.2.11}$$

By the minimality of $\mathscr{G}_0(\text{ad}\, v)$, we have

$$\mathscr{G}_0(\text{ad}\,(v + c_i\xi)) = \mathscr{G}_0(\text{ad}\, v) \qquad \text{for } i \in \overline{1, n+1}, \tag{4.2.12}$$

or equivalently,

$$f(t, c_i) = t^m \qquad \text{for } i \in \overline{1, n+1}. \tag{4.2.13}$$

By (4.2.7),

$$f_j(c_i) = 0 \qquad \text{for } i \in \overline{1, n+1},\ j \in \overline{1, m}. \tag{4.2.14}$$

Then (4.2.9) and (4.2.14) imply that

$$f_1(c) \equiv \cdots \equiv f_m(c) \equiv 0 \text{ as polynomials of } c. \tag{4.2.15}$$

So

$$\mathscr{G}_0(\text{ad } \upsilon) \subset \mathscr{G}_0(\text{ad } (\upsilon + c\xi)) \quad \text{for any } \xi \in \mathscr{K},\ c \in \mathbb{C}. \tag{4.2.16}$$

But $\upsilon + \mathscr{K} = \mathscr{K}$ because $\upsilon \in \mathscr{K}$. Therefore, $\mathscr{G}_0(\text{ad } \upsilon) \subset \mathscr{G}_0(\text{ad } \eta)$ for any $\eta \in \mathscr{K}$. □

A *Cartan subalgebra* (abbreviated CSA) of a Lie algebra $\mathscr{G}$ is a self-normalizing nilpotent subalgebra.

Theorem 4.2.3 *A subalgebra H of $\mathscr{G}$ is a CSA if and only if it is a minimal Engel subalgebra of $\mathscr{G}$.*

Proof Suppose that $H = \mathscr{G}_0(\text{ad } \upsilon)$ is a minimal Engel subalgebra. It is self-normalizing by Lemma 4.2.1. Taking $\mathscr{K} = H$ in the above lemma, we have

$$H \subset \mathscr{G}_0(\text{ad } \xi) \quad \text{for any } \xi \in H, \tag{4.2.17}$$

or equivalently, $\text{ad}\xi|_H$ is nilpotent for any $\xi \in H$. By Engel's Theorem, H is nilpotent.

Conversely, we assume that H is a CSA. Let $\mathscr{G}_0(\text{ad } \upsilon)$ be a minimal element in $\{\mathscr{G}_0(\text{ad } \xi) \mid \xi \in H\}$. Since H is nilpotent, we have

$$H \subset \mathscr{G}_0(\text{ad } \upsilon). \tag{4.2.18}$$

By the above lemma with $\mathscr{K} = H$, $\text{ad } \xi|_{\mathscr{G}_0(\text{ad } \upsilon)}$ is nilpotent and so is $\text{ad } \xi|_{\mathscr{G}_0(\text{ad } \upsilon)/H}$ for any $\xi \in H \subset \mathscr{G}_0(\text{ad } \upsilon)$. If $H \neq \mathscr{G}_0(\text{ad } \upsilon)$, then there exists $u \in \mathscr{G}_0(\text{ad } \upsilon) \setminus H$ such that

$$[H, u + H] \subset H \tag{4.2.19}$$

by Theorem 1.5.1. So $u \in N_{\mathscr{G}}(H) \setminus H$, which contradicts that H is self-normalizing. Thus $H = \mathscr{G}_0(\text{ad } \upsilon)$.

If $\mathscr{G}_0(\text{ad } u) \subset H$ for some $u \in \mathscr{G}$, then $u \in H$. Since H is nilpotent, we have $H \subset \mathscr{G}_0(\text{ad } u)$. Thus $H = \mathscr{G}_0(\text{ad } \upsilon)$ is a minimal Engel subalgebra. □

Corollary 4.2.4 *Let $\mathscr{G}$ be a finite-dimensional semisimple Lie algebra over $\mathbb{C}$. Then the CSA's of $\mathscr{G}$ are precisely the maximal toral subalgebras of $\mathscr{G}$.*

Proof Suppose that H is a maximal toral subalgebra of $\mathscr{G}$. Then we have the Cartan decomposition

$$\mathscr{G} = H + \sum_{\alpha \in \Phi} \mathscr{G}_\alpha, \tag{4.2.20}$$

which implies that H is self-normalizing. So H is a CSA.

Conversely, we assume that H is a CSA of $\mathscr{G}$. By the above theorem, $H = \mathscr{G}_0(\text{ad}\, \upsilon)$ is a minimal Engel subalgebra. Let $\upsilon = \upsilon_s + \upsilon_n$ be the abstract Jordan decomposition. Then

$$C_{\mathscr{G}}(\upsilon_s) = \{u \in \mathscr{G} \mid [\upsilon_s, u] = 0\} = \mathscr{G}_0(\text{ad}\, \upsilon) = H. \tag{4.2.21}$$

On the other hand, there exists a maximal toral subalgebra H' containing υ_s. But $H' \subset C_{\mathscr{G}}(\upsilon_s) = H$ is a CSA and it is also a minimal Engel suablgebra of $\mathscr{G}$ by the above theorem. By the minimality of $\mathscr{G}_0(\text{ad}\, \upsilon) = H$, we have $H = H'$. □

In the above, we have $\upsilon = \upsilon_s$, which is called a *regular semisimple* element.

Lemma 4.2.5 *Let $\phi : \mathscr{G} \to \mathscr{G}'$ be an epimorphism of Lie algebras. If H is a CSA of $\mathscr{G}$, then $\phi(H)$ is a CSA of $\mathscr{G}'$. Conversely, if H' is a CSA of $\mathscr{G}'$, then any CSA of $\phi^{-1}(H')$ is a CSA of $\mathscr{G}$.*

Proof Set $\ker \phi = \mathscr{K}$. Then $\mathscr{G}' \cong \mathscr{G}/\mathscr{K}$. Obviously, $\phi(H)$ is a nilpotent subalgebra of $\mathscr{G}'$ because H is nilpotent. Let $\xi' \in N_{\mathscr{G}'}(\phi(H))$. Take an element $\xi \in \phi^{-1}(\xi')$. Then

$$\phi([\xi, H + \mathscr{K}]) = [\xi', \phi(H)] \subset \phi(H). \tag{4.2.22}$$

Thus

$$\xi \in N_{\mathscr{G}}(H + \mathscr{K}). \tag{4.2.23}$$

Since H is an Engel subalgebra by Theorem 4.2.3, we have

$$N_{\mathscr{G}}(H + \mathscr{K}) = H + \mathscr{K} \tag{4.2.24}$$

by Lemma 4.2.1. So $\xi' = \phi(\xi) \in \phi(H + \mathscr{K}) = \phi(H)$; that is, $\phi(H)$ is self-normalizing. Thus $\phi(H)$ is a CSA of $\mathscr{G}'$.

Suppose that H' is a CSA of $\mathscr{G}'$ and H is a CSA of $\phi^{-1}(H')$. By the above argument, $\phi(H)$ is a CSA of H'. So $\phi(H)$ is an Engel subalgebra of H'. Since H' is nilpotent, we have $\phi(H) = H'$. If $\xi \in \mathscr{G}$ such that $[\xi, H] \subset H$, then $[\phi(\xi), H'] \subset H'$. Thus $\phi(\xi) \in H'$ because H' is self-normalizing. Hence $\xi \in N_{\phi^{-1}(H')}(H) = H$; that is, $H = N_{\mathscr{G}}(H)$. Therefore, H is a CSA of $\mathscr{G}$ because it is nilpotent as a CSA of $\phi^{-1}(H')$. □

Exercises:

1. Let $\mathscr{G}$ be a finite-dimensional semisimple Lie algebra over $\mathbb{C}$. Prove that a nonzero semisimple element $v \in \mathscr{G}$ is regular if and only if v lies in exactly one CSA of $\mathscr{G}$.
2. Prove that a CSA is a maximal nilpotent subalgebra, but a maximal nilpotent subalgebra may not be a CSA.

4.3 Conjugacy Theorems

In this section, we always assume that $\mathscr{G}$ is a finite-dimensional Lie algebra over $\mathbb{C}$. We want to prove that all CSA's of $\mathscr{G}$ are conjugated under the group $\text{Int}\,\mathscr{G}$ of inner automorphisms of $\mathscr{G}$.

An element $u \in \mathscr{G}$ is called *strongly nilpotent* if $u \in \mathscr{G}_a(\text{ad}\,\xi)$ for some $\xi \in \mathscr{G}$ and $0 \neq a \in \mathbb{C}$. Set

$$\mathscr{N}(\mathscr{G}) = \text{the set of all strongly nilpotent elements in } \mathscr{G}. \tag{4.3.1}$$

Then $\mathscr{N}(\mathscr{G})$ is invariant under $\text{Aut}\,\mathscr{G}$. Denote

$$\mathscr{E}(\mathscr{G}) = \text{the subgroup of Aut}\,\mathscr{G} \text{ generated by } \{e^{\text{ad}\,u} \mid u \in \mathscr{N}(\mathscr{G})\}. \tag{4.3.2}$$

For any $\tau \in \text{Aut}\,\mathscr{G}$ and $u \in \mathscr{N}(\mathscr{G})$, we have $\tau(u) \in \mathscr{N}(\mathscr{G})$ and

$$\tau e^{\text{ad}\,u}\tau^{-1} = e^{\text{ad}\,\tau(u)} \in \mathscr{E}(\mathscr{G}). \tag{4.3.3}$$

Thus $\mathscr{E}(\mathscr{G})$ is a normal subgroup of $\text{Aut}\,\mathscr{G}$.

If $\mathscr{K}$ is a subalgebra of $\mathscr{G}$, then

$$\mathscr{N}(\mathscr{K}) \subset \mathscr{N}(\mathscr{G}). \tag{4.3.4}$$

Set

$$\mathscr{E}(\mathscr{G};\mathscr{K}) = \text{the subgroup of Aut}\,\mathscr{G} \text{ generated by } \{e^{\text{ad}_{\mathscr{G}}\,u} \mid u \in \mathscr{N}(\mathscr{K})\}, \tag{4.3.5}$$

which is a subgroup of $\mathscr{E}(\mathscr{G})$.

Let $\phi : \mathscr{G} \to \mathscr{G}'$ be a Lie algebra epimorphism. For any $\xi \in \mathscr{G}$, we have $\phi(\mathscr{G}_a(\text{ad}\,\xi)) \subset \mathscr{G}'_a(\text{ad}\,\phi(\xi))$. Since $\mathscr{G} = \sum_{b\in\mathbb{C}} \mathscr{G}_b(\text{ad}\,\xi)$ and ϕ is surjective, we have

$$\phi(\mathscr{G}_a(\text{ad}\,\xi)) = \mathscr{G}'_a(\text{ad}\,\phi(\xi)) \quad \text{for } a \in \mathbb{C}. \tag{4.3.6}$$

Thus

$$\phi(\mathscr{N}(\mathscr{G})) = \mathscr{N}(\mathscr{G}'). \tag{4.3.7}$$

For any $u \in \mathscr{N}(\mathscr{G})$,

$$e^{\text{ad}\,\phi(u)}(\phi(v)) = \phi(e^{\text{ad}\,u}(v)). \tag{4.3.8}$$

Hence for any $\tau \in \mathscr{E}(\mathscr{G})$, there exists a unique $\tau' \in \mathscr{E}(\mathscr{G}')$ such that

$$\tau'(\phi(v)) = \phi(\tau(v)) \qquad \text{for} \;\; v \in \mathscr{G}. \tag{4.3.9}$$

Thus we have:

Lemma 4.3.1 *The map $\tau \mapsto \tau'$ is a group epimorphism from $\mathscr{E}(\mathscr{G})$ to $\mathscr{E}(\mathscr{G}')$.*

Theorem 4.3.2 *If $\mathscr{G}$ is solvable, then any two CSA's are conjugated under $\mathscr{E}(\mathscr{G})$.*

Proof Let H and H' be two CSA's of $\mathscr{G}$. We prove by induction on $\dim \mathscr{G}$. If $\mathscr{G}$ is nilpotent, then $H = H' = \mathscr{G}$ (This includes $\dim \mathscr{G} = 1$. So the theorem holds). Assume that the theorem holds for $\dim \mathscr{G} < k$ with $1 < k \in \mathbb{Z}$. Consider $\dim \mathscr{G} = k$ and $\mathscr{G}$ is not nilpotent. By Lie's theorem, there exists a common eigenvector $v \in \mathscr{G}$ for ad $\mathscr{G}$. So $\mathbb{C}v$ is an ideal of $\mathscr{G}$. By Lemma 4.2.5, $H + \mathbb{C}v$ and $H' + \mathbb{C}v$ are two CSA's of $\mathscr{G}/\mathbb{C}v$. According to the above lemma, there exists $\tau \in \mathscr{E}(\mathscr{G})$ such that

$$\tau(H') + \mathbb{C}v = H + \mathbb{C}v. \tag{4.3.10}$$

Now $\tau(H')$ and H are two CSA's of the Lie subalgebra $\mathscr{K} = H+\mathbb{C}v$. If $\mathscr{K} \neq \mathscr{G}$, then $\dim \mathscr{K} < k$. By assumption, there exists $\sigma \in \mathscr{E}(\mathscr{G}; \mathscr{K})$ such that $\sigma\tau(H') = H$. We are done.

Suppose $\mathscr{K} = \mathscr{G}$. So $H+\mathbb{C}v = \mathscr{G} = H'+\mathbb{C}v$. By Theorem 4.2.3, $H = \mathscr{G}_0(\text{ad}\xi)$, which implies $[\xi, v] = av$ with $0 \neq a \in \mathbb{C}$. Thus $v \in \mathscr{N}(\mathscr{G})$. Write $\xi = \zeta + bv$ with $\zeta \in H'$. Then

$$\zeta = \xi - bv = e^{\text{ad}\,a^{-1}bv}(\xi). \tag{4.3.11}$$

Hence

$$e^{\text{ad}\,a^{-1}bv}(H) = \mathscr{G}_0(\text{ad}\,\zeta) \supset H'. \tag{4.3.12}$$

Since $e^{\text{ad}\,a^{-1}bv}(H)$ and H' are both CSA's, they are minimal Engel subalgebras. Thus $e^{\text{ad}\,a^{-1}bv}(H) = H'$. $\square$

A maximal solvable subalgebra of $\mathscr{G}$ is called a *Borel subalgebra.*

Lemma 4.3.3 *A Borel subalgebra $\mathscr{B}$ is self-normalizing.*

Proof For any $u \in N_{\mathscr{G}}(\mathscr{B})$, $\mathscr{B} + \mathbb{C}u$ is also a solvable subalgebra. By maximality, $u \in \mathscr{B}$. $\square$

Recall that Rad $\mathscr{G}$ is the unique maximal solvable ideal of $\mathscr{G}$. For any solvable subalgebra $\mathscr{K}$, $\mathscr{K} + \text{Rad}\,\mathscr{G}$ is still a solvable subalgebra. Thus any Borel subalgebra contains Rad $\mathscr{G}$. By Lemma 4.3.1, we have:

Lemma 4.3.4 *If all the Borel subalgebras of $\mathscr{G}/Rad\ \mathscr{G}$ are conjugated under $\mathscr{E}(\mathscr{G}/Rad\ \mathscr{G})$, then all the Borel subalgebras of $\mathscr{G}$ are conjugated under $\mathscr{E}(\mathscr{G})$.*

Note that $\mathscr{G}/\text{Rad}\,\mathscr{G}$ is semisimple.

Theorem 4.3.5 *All the Borel subalgebras of $\mathscr{G}$ are conjugated under $\mathscr{E}(\mathscr{G})$.*

Proof Set

$$m_0(\mathscr{G}) = \max\{\dim \mathscr{K} \mid \mathscr{K} \text{ is a Borel subalgebra of } \mathscr{G}\}. \tag{4.3.13}$$

Let $\mathscr{B}$ and $\mathscr{B}'$ be two Borel subalgebras of $\mathscr{G}$. Define

$$m(\mathscr{B}, \mathscr{B}') = m_0(\mathscr{G}) - \dim \mathscr{B} \bigcap \mathscr{B}'. \tag{4.3.14}$$

We prove the theorem by induction on $\dim \mathscr{G}$ and on $m(\mathscr{B}, \mathscr{B}')$. The theorem holds if $\dim \mathscr{G} = 1$ or $m(\mathscr{B}, \mathscr{B}') = 0$, which implies $\mathscr{B} = \mathscr{B}'$. Suppose that the theorem holds for $\dim \mathscr{G} < k_1$ or $m(\mathscr{B}, \mathscr{B}) < k_2$. Now we assume $\dim \mathscr{G} = k_1$ and $\mathscr{B} \neq \mathscr{B}'$ with $m(\mathscr{B}, \mathscr{B}') = k_2$. If Rad $\mathscr{G} \neq \{0\}$, then the theorem holds by Lemma 4.3.4 because any two Borel subalgebras of $\mathscr{G}/\text{Rad}\,\mathscr{G}$ are conjugated under $\mathscr{E}(\mathscr{G}/\text{Rad}\,\mathscr{G})$ due to $\dim \mathscr{G}/\text{Rad}\,\mathscr{G} < k$.

Now we assume that $\mathscr{G}$ is semsimple and H is a maximal toral subalgebra of $\mathscr{G}$ with root system Φ. Let Π be a base of Φ. With respect to Π, we have positive roots and negative roots. Set

$$\mathscr{B}(\Pi) = H + \sum_{\alpha \in \Phi^+} \mathscr{G}_\alpha, \quad \mathscr{N}(\Pi) = \sum_{\alpha \in \Phi^+} \mathscr{G}_\alpha. \tag{4.3.15}$$

Suppose $\Pi = \{\alpha_1, \alpha_2, ..., \alpha_n\}$. For any $\beta = \sum_{i=1}^n b_i \alpha_i \in \Phi$, we have

$$\text{ht}\,\beta = \sum_{i=1}^n b_i. \tag{4.3.16}$$

Let

$$\ell = \max\{\text{ht}\,\beta \mid \beta \in \Phi\}. \tag{4.3.17}$$

Set

$$\mathscr{N}_i(\Pi) = \sum_{\beta\in\Phi^+;\,\mathrm{ht}\,\beta\geq i} \mathscr{G}_\beta. \tag{4.3.18}$$

Then we have

$$[\mathscr{B}(\Pi), \mathscr{B}(\Pi)] = \mathscr{N}(\Pi), \quad (\mathscr{N}(\Pi))^i \subset \mathscr{N}_i(\Pi), \tag{4.3.19}$$

where $(\mathscr{N}(\Pi))^i = [\mathscr{N}(\Pi), (\mathscr{N}(\Pi))^{i-1}]$ is the central series. In particular, $(\mathscr{N}(\Pi))^{\ell+1} = \{0\}$. So $\mathscr{N}(\Pi)$ is nilpotent and $\mathscr{B}(\Pi)$ is solvable. If $\mathscr{K}$ is a subalgebra strictly containing $\mathscr{B}(\Pi)$, then

$$\mathscr{K} \supset \mathscr{G}_\alpha + \mathbb{C}h_\alpha + \mathscr{G}_{-\alpha} \cong sl(2,\mathbb{C}) \qquad \text{for some } \alpha \in \Phi. \tag{4.3.20}$$

Thus $\mathscr{K}$ is not solvable. Hence $\mathscr{B}(\Pi)$ is a Borel subalgebra, which is called *standard* relative to H. To prove that any two Borel subalgebras are conjugated under $\mathscr{E}(\mathscr{G})$, it is enough to prove that any Borel subalgebra $\mathscr{B}'$ is conjugated to $\mathscr{B}(\Pi)$ under $\mathscr{E}(\mathscr{G})$. So we may assume $\mathscr{B} = \mathscr{B}(\Pi)$ and simply denote $\mathscr{N} = \mathscr{N}(\Pi)$. Note that

$$\mathscr{N} = \text{the set of all } \mathrm{ad}_{\mathscr{G}}\text{-nilpotent elements in } \mathscr{B}. \tag{4.3.21}$$

If u is an element of a Borel subalgebra $\overline{\mathscr{B}}$ and $u = u_s + u_n$ is the abstract Jordan decomposition, then $[u_s, \overline{\mathscr{B}}] \subset \overline{\mathscr{B}}$ because ad u_s is a polynomial of ad u without constant. Hence $\mathbb{C}u_s + \overline{\mathscr{B}}$ is a solvable subalgebra. By maximality of $\bar{\mathscr{B}}$, $u_s \in \overline{\mathscr{B}}$ and also $u_n = u - u_s \in \overline{\mathscr{B}}$. Thus $\mathscr{B} \bigcap \mathscr{B}'$ contains the semsimple and nilpotent parts of its elements. Moreover,

$$N' = \mathscr{N} \bigcap \mathscr{B}' = \text{the set of all } \mathrm{ad}_{\mathscr{G}}\text{-nilpotent elements in } \mathscr{B} \bigcap \mathscr{B}' \tag{4.3.22}$$

by (4.3.21). According to (4.3.19),

$$N' \supset [\mathscr{B} \bigcap \mathscr{B}', \mathscr{B} \bigcap \mathscr{B}']. \tag{4.3.23}$$

So N' is an ideal of $\mathscr{B} \bigcap \mathscr{B}'$.

Case 1. $N' \neq \{0\}$.

Set

$$\mathscr{K} = N_{\mathscr{G}}(N'). \tag{4.3.24}$$

Then $\mathscr{B} \bigcap \mathscr{B}' \subset \mathscr{K}$ because N' is an ideal of $\mathscr{B} \bigcap \mathscr{B}'$. Note that $\mathscr{B} \neq \mathscr{B}'$ implies $\mathscr{B}/\mathscr{B} \bigcap \mathscr{B}' \neq \{0\}$ and $\mathscr{B}'/\mathscr{B} \bigcap \mathscr{B}' \neq \{0\}$. Since ad ${}_{\mathscr{G}}N'$ consists of nilpotent elements, there exist $u \in \mathscr{B} \setminus \mathscr{B} \bigcap \mathscr{B}'$ and $u' \in \mathscr{B}' \setminus \mathscr{B} \bigcap \mathscr{B}'$ such that

$$[u, N'], [u', N'] \subset \mathscr{B} \bigcap \mathscr{B}'. \tag{4.3.25}$$

By Lie's Theorem, the elements $\mathrm{ad}_{\mathscr{G}}[\mathscr{B}', \mathscr{B}']$ are nilpotent. So

$$[u, N'], [u', N'] \subset N' \tag{4.3.26}$$

by (4.3.22). Thus

$$\mathscr{K} \bigcap \mathscr{B} \setminus \mathscr{B} \bigcap \mathscr{B}' \neq \emptyset, \quad \mathscr{K} \bigcap \mathscr{B}' \setminus \mathscr{B} \bigcap \mathscr{B}' \neq \emptyset. \tag{4.3.27}$$

Let $\mathscr{C} \supset \mathscr{K} \bigcap \mathscr{B}$ and $\mathscr{C}' \supset \mathscr{K} \bigcap \mathscr{B}'$ be Borel subalgebras of $\mathscr{K}$. Since $\mathscr{G}$ is semisimple, N' can not be an ideal of $\mathscr{G}$ because it is nilpotent. So $\dim \mathscr{K} < k$. By assumption, there exists $\nu \in \mathscr{E}(\mathscr{G}; \mathscr{K})$ such that $\nu(\mathscr{C}) = \mathscr{C}'$.

Take Borel subalgebras $\mathscr{C}_1 \supset \mathscr{C}$ and $\mathscr{C}_2 \supset \mathscr{C}'$ of $\mathscr{G}$. Then

$$\mathscr{C}_1 \bigcap \mathscr{B} \supset \mathscr{K} \bigcap \mathscr{B}, \ \nu(\mathscr{C}_1) \bigcap \mathscr{C}_2 \supset \mathscr{C}' \supset \mathscr{K} \bigcap \mathscr{B}', \ \mathscr{C}_2 \bigcap \mathscr{B}' \supset \mathscr{K} \bigcap \mathscr{B}'. \tag{4.3.28}$$

Thus

$$m(\mathscr{B}, \mathscr{C}_1), \ m(\nu(\mathscr{C}_1), \mathscr{C}_2), \ m(\mathscr{C}_2, \mathscr{B}') < m_1. \tag{4.3.29}$$

By assumption, there exist $\sigma_1, \sigma_2, \sigma_3 \in \mathscr{E}(\mathscr{G})$ such that

$$\sigma_1(\mathscr{B}) = \mathscr{C}_1, \quad \sigma_2\nu(\mathscr{C}_1) = \mathscr{C}_2, \quad \sigma_3(\mathscr{C}_2) = \mathscr{B}'. \tag{4.3.30}$$

Therefore,

$$\sigma_3\sigma_2\nu\sigma_1(\mathscr{B}) = \mathscr{B}'. \tag{4.3.31}$$

Case 2. $N' = \{0\}$ and $\mathscr{B} \bigcap \mathscr{B}' \neq \{0\}$.

In this case, $T = \mathscr{B} \bigcap \mathscr{B}'$ is a subalgebra consisting of semisimple elements. So T is abelian by Lemma 2.4.1. Let $\mathscr{C}$ be a CSA of

$$C_{\mathscr{B}}(T) = \{u \in \mathscr{B} \mid [u, T] = \{0\}\}. \tag{4.3.32}$$

Since $\mathscr{C}$ is self-normalizing and $[T, \mathscr{C}] = \{0\}$, we have $T \subset \mathscr{C}$. For any $v \in N_{\mathscr{B}}(\mathscr{C})$ and $h \in T$, we have $[h, v] \in \mathscr{C}$ and $(\mathrm{ad}\, h)^2(v) = 0$. The semisimplicity of h implies $[h, v] = 0$. Hence $v \in C_{\mathscr{B}}(T)$. So $v \in N_{C_{\mathscr{B}}(T)}(\mathscr{C}) = \mathscr{C}$. Thus $\mathscr{C}$ is a self-normalizing nilpotent subalgebra of $\mathscr{B}$; that is, a CSA of $\mathscr{B}$. On the other hand, H is a CSA of $\mathscr{B}$. By Theorem 4.3.2, there exists $\tau \in \mathscr{E}(\mathscr{G}; \mathscr{B})$ such that $\tau(\mathscr{C}) = H$. Note $\tau(\mathscr{B}) = \mathscr{B}$. So

$$\tau(\mathscr{B}')\bigcap\mathscr{B}=\tau(\mathscr{B}')\bigcap\tau(\mathscr{B})=\tau(\mathscr{B}'\bigcap\mathscr{B})=\tau(T)\subset\tau(\mathscr{C})=H. \quad (4.3.33)$$

Replacing $\mathscr{B}'$ by $\tau(\mathscr{B}')$, we may assume $T\subset H$.

Suppose that $T\neq H$. Note $H\subset C_{\mathscr{G}}(T)=\{u\in\mathscr{G}\mid[u,T]=\{0\}\}$. If $\mathscr{B}'\subset C_{\mathscr{G}}(T)$, we take a Borel subalgebra $\overline{\mathscr{B}}\supset H$ of $C_{\mathscr{G}}(T)$. Since $Z(\mathscr{G})=\{0\}$, we have $\dim C_{\mathscr{G}}(T)<k$. By assumption, there exists $\sigma\in\mathscr{E}(\mathscr{G};C_{\mathscr{G}}(T))$ such that $\sigma(\mathscr{B}')=\overline{\mathscr{B}}$. Replacing $\mathscr{B}'$ by $\sigma(\mathscr{B}')$, we may assume $T=H$. When $\mathscr{B}'\not\subset C_{\mathscr{G}}(T)$, there exist $u\in\mathscr{B}'$ and $h\in T$ such that $[h,u]=u$. Set

$$\mathscr{C}=H+\sum_{\alpha\in\Phi;\ 0<\alpha(h)\in\mathbb{Z}}\mathscr{G}_\alpha. \quad (4.3.34)$$

Then $\mathscr{C}$ is a solvable subalgebra and $u\in\mathscr{C}$. Take a Borel subalgebra $\mathscr{C}_1\supset\mathscr{C}$ of $\mathscr{G}$. We have $\dim\mathscr{C}_1\bigcap\mathscr{B}'>\dim T=\dim\mathscr{B}\bigcap\mathscr{B}'$. So $m(\mathscr{B}',\mathscr{C}_1)<m_1$. By assumption, there exists $\varrho\in\mathscr{E}(\mathscr{G})$ such that $\varrho(\mathscr{B}')=\mathscr{C}_1$. Replacing $\mathscr{B}'$ by $\varrho(\mathscr{B}')$, we may again assume $T=H$.

Now we only need to deal with the case $T=H$. Since $\mathscr{B}'\neq\mathscr{B}$, we have $\mathscr{G}_{-\alpha}\subset\mathscr{B}'$ for some $\alpha\in\Phi^+$. Take $\xi\in\mathscr{G}_\alpha$ and $\zeta\in\mathscr{G}_{-\alpha}$ such that

$$[\xi,\zeta]=h_\alpha,\quad [h_\alpha,\xi]=2\xi. \quad (4.3.35)$$

Denote

$$\tau_\alpha=e^{\mathrm{ad}\,\xi}e^{-\mathrm{ad}\,\zeta}e^{\mathrm{ad}\,\xi}\in\mathscr{E}(\mathscr{G}). \quad (4.3.36)$$

Then

$$\tau_\alpha(\mathscr{G}_{-\alpha})=\mathscr{G}_\alpha,\quad \tau_\alpha(H)=H. \quad (4.3.37)$$

Now $m(\mathscr{B},\tau_\alpha(\mathscr{B}'))<m_1$. By assumption, there exists $\sigma\in\mathscr{E}(\mathscr{G})$ such that $\sigma\tau_\alpha(\mathscr{B}')=\mathscr{B}$.

Case 3. $\mathscr{B}\bigcap\mathscr{B}'=\{0\}$.

Let T be a maximal toral subalgebra of $\mathscr{B}'$. If $T=\{0\}$, then $\mathscr{B}'$ consists of ad-nilpotent elements because it contains the semisimple parts of all its elements. By Engel's Theorem and Lemma 4.3.3, $\mathscr{B}'$ is a CSA. But Corollary 4.2.4 says that $\mathscr{B}'$ is a toral subalgebra. Thus $\mathscr{B}'=\{0\}$, which is absurd. Thus $T\neq\{0\}$. Take $\bar{H}\supset T$ to be a maximal toral subalgebra of $\mathscr{G}$. Take any standard Borel subalgebra $\overline{\mathscr{B}}$ of $\mathscr{G}$ relative to $\bar{H}$. Then $\mathscr{B}'\bigcap\overline{\mathbf{B}}\neq\{0\}$. By Case 2, $\mathscr{B}'$ is conjugated to $\overline{\mathscr{B}}$. So

$$\dim\mathscr{B}'=\dim\overline{\mathscr{B}}=\frac{1}{2}(\dim\mathscr{G}+\dim\bar{H})>\frac{1}{2}\dim\mathscr{G}. \quad (4.3.38)$$

On the other hand,

$$\dim \mathscr{B} = \frac{1}{2}(\dim \mathscr{G} + \dim H) > \frac{1}{2}\dim \mathscr{G}. \tag{4.3.39}$$

Thus

$$\dim \mathscr{G} \geq \dim \mathscr{B} + \dim \mathscr{B}' > \frac{1}{2}\dim \mathscr{G} + \frac{1}{2}\dim \mathscr{G} = \dim \mathscr{G}, \tag{4.3.40}$$

which is absurd. Therefore, this case does not exist. □

Corollary 4.3.6 *Any Borel subalgebra $\mathscr{B}$ of a semisimple Lie algebra $\mathscr{G}$ is standard.*

Proof Take a maximal toral subalgebra H of $\mathscr{G}$ and let Π be a base of the corresponding root system Φ. Then there exists a $\sigma \in \mathscr{E}(\mathscr{G})$ such that $\sigma(\mathscr{B}(\Pi)) = \mathscr{B}$ by the above theorem (cf. (4.3.15)). Note that $\sigma(H) \subset \mathscr{B}$ is also a maximal toral subalgebra of $\mathscr{G}$. Replacing H by $\sigma(H)$, we may assume $\mathscr{B} \supset H$. Again there exists $\sigma_1 \in \mathscr{E}(\mathscr{G})$ such that $\sigma_1(\mathscr{B}(\Pi)) = \mathscr{B}$. Now H and $\sigma_1(H)$ are CSA's of $\mathscr{B}$. By Lemma 4.3.2, there exists $\tau \in \mathscr{E}(\mathscr{G}; \mathscr{B})$ such that $\tau\sigma_1(H) = H$.

Since $\tau(\mathscr{B}) = \mathscr{B}$, we have $\tau\sigma_1(\mathscr{B}(\Pi)) = \mathscr{B}$. Denote $\nu = \tau\sigma_1$. So we have $\nu(H) = H$ and $\nu(\mathscr{B}(\Pi)) = \mathscr{B}$. Define the action of ν on H^* by

$$\nu(\alpha)(h) = \alpha(\nu^{-1}(h)) \qquad \text{for } \alpha \in H^*. \tag{4.3.41}$$

For any $\alpha \in \Phi$ and $u \in \mathscr{G}_\alpha$, we have

$$[h, \nu(u)] = \nu([\nu^{-1}(h), u]) = \alpha(\nu^{-1}(h))\nu(u) \qquad \text{for } h \in H. \tag{4.3.42}$$

Thus

$$\nu(\mathscr{G}_\alpha) = \mathscr{G}_{\nu(\alpha)} \qquad \text{for } \alpha \in \Phi. \tag{4.3.43}$$

Hence $\nu(\Phi) = \Phi$. Since ν is linear, $\nu(\Pi)$ is also a base of Φ. Thus $\mathscr{B} = \mathscr{B}(\nu(\Pi))$ is a standard Borel subalgebra. □

Corollary 4.3.7 *Any two CSA's of $\mathscr{G}$ are conjugated under $\mathscr{E}(\mathscr{G})$.*

Proof Let H and H' be two CSA's of $\mathscr{G}$. Take Borel subalgebras $\mathscr{B} \supset H$ and $\mathscr{B}' \supset H'$. By Theorem 4.3.5, there exists $\tau \in \mathscr{E}(\mathscr{G})$ such that $\tau(\mathscr{B}') = \mathscr{B}$. Now H and $\tau(H')$ are two CSA's in the solvable Lie algebra $\mathscr{B}$. According to Lemma 4.3.2, there exists $\sigma \in \mathscr{E}(\mathscr{G}; \mathscr{B})$ such that $\sigma\tau(H') = H$. □.

Let H and H' be two maximal toral subalgebras (which are CSA's by Corollary 4.2.4) of a finite-dimensional semisimple Lie algebra $\mathscr{G}$. Their associated root systems are denoted by Φ and Φ', respectively. For $\alpha \in \Phi$ and $\alpha' \in \Phi'$, we denote

$$\mathscr{G}_\alpha = \{u \in \mathscr{G} \mid [h, u] = \alpha(h)u \text{ for } h \in H\}, \tag{4.3.44}$$

$$\mathscr{G}'_{\alpha'} = \{u \in \mathscr{G} \mid [h', u] = \alpha'(h')u \text{ for } h' \in H'\} \tag{4.3.45}$$

and take $h_\alpha \in [\mathscr{G}_\alpha, \mathscr{G}_{-\alpha}]$ and $h'_{\alpha'} \in [\mathscr{G}'_{\alpha'}, \mathscr{G}'_{-\alpha'}]$ such that

$$\text{ad } h_\alpha|_{\mathscr{G}_\alpha} = 2\text{Id}_{\mathscr{G}_\alpha}, \quad \text{ad } h'_{\alpha'}|_{\mathscr{G}'_{\alpha'}} = 2\text{Id}_{\mathscr{G}'_{\alpha'}}. \tag{4.3.46}$$

Then

$$\beta(h_\alpha) = \langle \beta, \alpha \rangle, \quad \beta'(h'_{\alpha'}) = \langle \beta', \alpha' \rangle \text{ for } \alpha, \beta \in \Phi, \; \alpha', \beta' \in \Phi'. \tag{4.3.47}$$

By the above corollary, there exists $\nu \in \mathscr{E}(\mathscr{G})$ such that $\nu(H) = H'$. We define $\nu : H^* \to (H')^*$ by

$$\nu(\alpha)(h') = \alpha(\nu^{-1}(h')) \qquad \text{for } \alpha \in H^*, \; h' \in H'. \tag{4.3.48}$$

For any $\alpha \in \Phi$, $v \in \mathscr{G}_\alpha$ and $h' \in H'$, we have

$$\begin{aligned}[h', \nu(v)] &= [\nu(\nu^{-1}(h')), \nu(v)] = \nu([\nu^{-1}(h'), v]) \\ &= \alpha(\nu^{-1}(h'))\nu(v) = \nu(\alpha)(h')\nu(v).\end{aligned} \tag{4.3.49}$$

Thus

$$\nu(\mathscr{G}_\alpha) = \mathscr{G}'_{\nu(\alpha)} \qquad \text{for } \alpha \in \Phi. \tag{4.3.50}$$

In particular, $\nu(\Phi) = \Phi'$. Since ν is an automorphism,

$$\nu(h_\alpha) = h'_{\nu(\alpha)} \qquad \text{for } \alpha \in \Phi. \tag{4.3.51}$$

Now for $\alpha, \beta \in \Phi$ and $0 \neq u \in \mathscr{G}_\beta$, we have

$$\langle \beta, \alpha \rangle \nu(u) = \nu([h_\alpha, u]) = [\nu(h_\alpha), \nu(u)] = \nu(\beta)(h'_{\nu(\alpha)})\nu(u) = \langle \nu(\beta), \nu(\alpha) \rangle \nu(u). \tag{4.3.52}$$

Therefore,

$$\langle \nu(\beta), \nu(\alpha) \rangle = \langle \beta, \alpha \rangle \qquad \text{for } \alpha, \beta \in \Phi; \tag{4.3.53}$$

that is, $\nu : \Phi \to \Phi'$ is an isomorphism of two root systems. This shows that the root system of $\mathscr{G}$ is independent of the choice of its maximal toral subalgebra.

The dimension of a CSA of a finite-dimensional semisimple Lie algebra $\mathscr{G}$ is called the *rank* of $\mathscr{G}$. For instance, $sl(n, \mathbb{C})$ for $n > 1$ is a finite-dimensional simple Lie algebra of rank $n - 1$. While the ranks of $o(2n, \mathbb{C})$, $o(2n + 1, \mathbb{C})$ and $sp(2n, \mathbb{C})$ are n. The Lie algebra $sl(2, \mathbb{C})$ is the unique rank-1 finite-dimensional simple Lie algebra. There are three rank-2 finite-dimensional simple Lie algebras: $sl(3,\mathbb{C})$, $o(5,\mathbb{C}) \cong sp(4,\mathbb{C})$ and the Lie algebra of type G_2.

Finally, we want to study the structure of Aut $\mathscr{G}$ when $\mathscr{G}$ is a finite-dimensional simple Lie algebra over $\mathbb{C}$. Note that $sl(2,\mathbb{C})$ has the following representation on the polynomial algebra $\mathscr{A} = \mathbb{C}[x_1, x_2]$:

$$E_{1,2}|_{\mathscr{A}} = x_1\partial_{x_2}, \quad E_{2,1}|_{\mathscr{A}} = x_2\partial_{x_1}, \quad (E_{1,1} - E_{2,2})|_{\mathscr{A}} = x_1\partial_{x_1} - x_2\partial_{x_2}. \tag{4.3.54}$$

For $0 \neq a \in \mathbb{C}$, we define the operator

$$\sigma(a) = e^{aE_{1,2}}e^{-a^{-1}E_{2,1}}e^{aE_{1,2}}e^{E_{1,2}}e^{-E_{2,1}}e^{E_{1,2}} \tag{4.3.55}$$

on $\mathscr{A}$. For any $f(x_1, x_2) \in \mathscr{A}$,

$$\begin{aligned} & e^{aE_{1,2}}e^{-a^{-1}E_{2,1}}e^{aE_{1,2}}(f(x_1, x_2)) \\ &= e^{ax_1\partial_{x_2}}e^{-a^{-1}x_2\partial_{x_1}}e^{ax_1\partial_{x_2}}(f(x_1, x_2)) = e^{ax_1\partial_{x_2}}e^{-a^{-1}x_2\partial_{x_1}}(f(x_1, x_2 + ax_1)) \\ &= e^{ax_1\partial_{x_2}}(f(x_1 - a^{-1}x_2, ax_1)) = f(-a^{-1}x_2, ax_1) \end{aligned} \tag{4.3.56}$$

by Taylor's expansion. Thus

$$\sigma(a)(f(x_1, x_2)) = f(-a^{-1}x_1, -ax_2). \tag{4.3.57}$$

Hence

$$(E_{1,1} - E_{2,2})(x_1^{\ell_1}x_2^{\ell_2}) = (\ell_1 - \ell_2)x_1^{\ell_1}x_2^{\ell_2}, \quad \sigma(a)(x_1^{\ell_1}x_2^{\ell_2}) = (-a)^{\ell_2-\ell_1}x_1^{\ell_1}x_2^{\ell_2} \tag{4.3.58}$$

for $\ell_1, \ell_2 \in \mathbb{N}$. It is easy to verify that the subspace $\mathscr{A}_m$ of homogeneous polynomials with degree m in $\mathscr{A}$ forms an $(m+1)$-dimensional irreducible $sl(2,\mathbb{C})$-submodule for $m \in \mathbb{N}$. By (4.3.58) and Weyl's theorem of complete reducibility (cf. Theorem 2.4.5), we have:

Lemma 4.3.8 *Let u be an element of a finite-dimensional $sl(2,\mathbb{C})$-module such that $(E_{1,1} - E_{2,2})(u) = ku$ with $k \in \mathbb{Z}$. Then $\sigma(a)(u) = (-a)^{-k}u$.*

Take a maximal toral subalgebra H of $\mathscr{G}$ and let Π be a base of the corresponding root system Φ. For any $\tau \in \mathrm{Aut}\mathscr{G}$, $\tau(\mathscr{B}(\Pi))$ is a Borel subalgebra of $\mathscr{G}$. By Theorem 4.3.5, there exists $\sigma \in \mathscr{E}(\mathscr{G})$ such that $\sigma\tau(\mathscr{B}(\Pi)) = \mathscr{B}(\Pi)$. Since $\sigma\tau(H)$ and H are two CSA's in $\mathscr{B}$. By Lemma 4.3.2, there exists $\tau_1 \in \mathscr{E}(\mathscr{G}; \mathscr{B}(\Pi))$ such that $\tau_1\sigma\tau(H) = H$. Denote $\nu = \tau_1\sigma\tau$. By the proof of Corollary 4.3.6, (4.3.43) holds. According to Theorem 3.2.9 (e), (4.1.24) and (4.1.25), there exists $\sigma_1 \in \mathscr{W}$ and $\Im_{\sigma_1} \in \mathscr{E}(\mathscr{G})$ such that $\sigma_1\nu(\Pi) = \Pi$ and

$$\Im_{\sigma_1}\nu(\mathscr{G}_\alpha) = \mathscr{G}_{\sigma_1\nu(\alpha)} \qquad \text{for } \alpha \in \Phi. \tag{4.3.59}$$

This shows that

$$\tau \in \Im_\iota \mathscr{E}(\mathscr{G}), \tag{4.3.60}$$

where ι is an automorphism of Φ such that $\iota(\Pi) = \Pi$ and $\Im_\iota \in \text{Aut}\,\mathscr{G}$ such that $\Im_\iota(H) = H$ and $\Im_\iota(\mathscr{G}_\alpha) = \mathscr{G}_{\iota(\alpha)}$ for $\alpha \in \Phi$.

If ϑ is another automorphism of $\mathscr{G}$ such that $\vartheta(H) = H$ and $\vartheta(\mathscr{G}_\alpha) = \mathscr{G}_{\iota(\alpha)}$ for $\alpha \in \Phi$, then

$$\vartheta^{-1}\Im_\iota(h) = h, \quad \vartheta^{-1}\Im_\iota(\mathscr{G}_\alpha) = \mathscr{G}_\alpha \qquad \text{for } h \in H, \ \alpha \in \Phi. \tag{4.3.61}$$

Suppose $\Pi = \{\alpha_1, \alpha_2, ..., \alpha_n\}$. We get

$$\vartheta^{-1}\Im_\iota|_{\mathscr{G}_{\alpha_i}} = b_i \text{Id}_{\mathscr{G}_{\alpha_i}}, \quad \vartheta^{-1}\Im_\iota|_{\mathscr{G}_{\alpha_i}} = b_i^{-1}\text{Id}_{\mathscr{G}_{\alpha_i}}, \qquad 0 \neq b_i \in \mathbb{C}, \tag{4.3.62}$$

for $i \in \overline{1,n}$. By Lemma 4.1.3, $\vartheta^{-1}\Im_\iota$ is completely determined by (4.3.62). Take $\xi_1 \in \mathscr{G}_{\alpha_i}$ and $\zeta_i \in \mathscr{G}_{-\alpha_i}$ such that $[\xi_i, \zeta_i] = h_{\alpha_i}$ for $i \in \overline{1,n}$. Moreover, we define

$$\sigma_i(a_i) = e^{a_i \text{ad}\,\xi_i} e^{-a^{-1}\text{ad}\,\zeta_i} e^{a_i \text{ad}\,\xi_i} e^{\text{ad}\,\xi_i} e^{-\text{ad}\,\zeta_i} e^{\text{ad}\,\xi_i} \in \mathscr{E}(\mathscr{G}), \quad a_i \in \mathbb{C}. \tag{4.3.63}$$

By Lemma 4.3.8, $\sigma_i(a_i)|_H = \text{Id}_H$ and

$$\sigma_i(a_i)|_{\mathscr{G}_{\alpha_j}} = (-a_i)^{-\langle \alpha_j, \alpha_i \rangle}\text{Id}_{\mathscr{G}_{\alpha_j}} \qquad \text{for } i, j \in \overline{1,n}. \tag{4.3.64}$$

Since the Cartan matrix $(\langle \alpha_h, \alpha_i \rangle)_{n\times n}$ is nondegenerate, there exist $a_1, a_2, ..., a_n \in \mathbb{C}$ such that

$$\prod_{i=1}^{n}(-a_i)^{-\langle \alpha_j, \alpha_i \rangle} = e^{-\sum_{i=1}^{n}\langle \alpha_j, \alpha_i \rangle \ln(-a_i)} = e^{\ln b_j} = b_j \qquad \text{for } j \in \overline{1,n}. \tag{4.3.65}$$

In this case,

$$\vartheta^{-1}\Im_\iota = \sigma_1(a_1)\sigma_2(a_2)\cdots\sigma_n(a_n) \in \mathscr{E}(\mathscr{G}). \tag{4.3.66}$$

For each automorphism ι of Φ induced by an automorphism of the Dynkin diagram of $\mathscr{G}$, we fix an automorphism $\Im_\iota$ of $\mathscr{G}$ such that such that $\Im_\iota(H) = H$ and $\Im_\iota(\mathscr{G}_\alpha) = \mathscr{G}_{\iota(\alpha)}$ for $\alpha \in \Phi$. Then

$$\tau \in \Im_\iota \mathscr{E}(\mathscr{G}) \tag{4.3.67}$$

for some automorphism ι of Φ induced by an automorphism of the Dynkin diagram of $\mathscr{G}$. Therefore,

$$\text{Aut}\,\mathscr{G}/\mathscr{E}(\mathscr{G}) \cong \text{the automorphism group of the Dynkin diagram of } \mathscr{G}. \tag{4.3.68}$$

In particular,

$$\text{Aut}\,\mathscr{G} = \mathscr{E}(\mathscr{G}) \ \text{ if } \mathscr{G} \text{ is of types: } B_n, C_n, E_8, E_7, F_4, G_2. \tag{4.3.69}$$

Let $\mathbb{B}(\mathscr{G})$ be the set of Borel subalgebras of a Lie algebra $\mathscr{G}$.

Lemma 4.3.9 *We have $\bigcap_{\mathscr{B}\in\mathbb{B}(\mathscr{G})} \mathscr{B} = Rad\,\mathscr{G}$.*

Proof Since any Borel subalgebra contains Rad $\mathscr{G}$, the conclusion is equivalent to $\bigcap_{\overline{\mathscr{B}}\in\mathbb{B}(\mathscr{G}/\text{Rad}\,\mathscr{G})} \overline{\mathscr{B}} = \{0\}$. So we only need to consider the case when $\mathscr{G}$ is semisimple. Let H be a maximal toral subalgebra $\mathscr{G}$ with root system Φ. Take a base Π of Φ. We have the following two Borel subalgebras:

$$\mathscr{B}_1 = H + \sum_{\alpha\in\Phi^+} \mathscr{G}_\alpha, \quad \mathscr{B}_2 = H + \sum_{\alpha\in\Phi^-} \mathscr{G}_\alpha. \tag{4.3.70}$$

Note

$$\mathscr{B}_1 \bigcap \mathscr{B}_2 = H. \tag{4.3.71}$$

Let H be the set of maximal toral subalgebras of $\mathscr{G}$. By (4.3.71),

$$\bigcap_{\mathscr{B}\in\mathbb{B}(\mathscr{G})} \mathscr{B} \subset \bigcap_{H'\in\mathbb{H}} H'. \tag{4.3.72}$$

For any $\alpha \in \Phi$, take $0 \neq \xi_\alpha \in \mathscr{G}_\alpha$. We have

$$e^{\text{ad}\,\xi_\alpha}(h) = h - \alpha(h)\xi_\alpha \qquad \text{for } \ h \in H. \tag{4.3.73}$$

Thus

$$e^{\text{ad}\,\xi_\alpha}(H) \bigcap H = \{h \in H \mid \alpha(h) = 0\}. \tag{4.3.74}$$

Since $\{e^{\text{ad}\,\xi_\alpha}(H) \mid \alpha \in \Phi\}$ are maximal toral subalgebras of $\mathscr{G}$, we have

$$\bigcap_{\mathscr{B}\in\mathbb{B}(\mathscr{G})} \mathscr{B} \subset H \bigcap \bigcap_{\alpha\in\Phi} e^{\text{ad}\,\xi_\alpha}(H) = \{h \in H \mid \alpha(h) = 0 \text{ for any } \alpha \in \Phi\} = \{0\} \tag{4.3.75}$$

by (4.3.72). □

Exercises:

1. Prove that a regular semisimple element of a semisimple Lie algebra lies in only finitely many Borel subalgebras.

2. Prove that the intersection of two Borel subalgebras in a semisimple Lie algebra contains a CSA (Hint: if $\mathscr{B}$ and $\mathscr{B}'$ are Borel subalgebras and $\mathscr{N} = [\mathscr{B}, \mathscr{B}]$, $\mathscr{N}' = [\mathscr{B}', \mathscr{B}']$, then $\mathscr{B}^{\perp} = \mathscr{N}$, $\mathscr{N}^{\perp} = \mathscr{B}$ and $\mathscr{B}'^{\perp} = \mathscr{N}'$, $\mathscr{N}'^{\perp} = \mathscr{B}'$. Moreover, $\mathscr{B} \bigcap \mathscr{N}' \subset \mathscr{N}$ and $\mathscr{B}' \bigcap \mathscr{N} \subset \mathscr{N}'$).

4.4 Simple Lie Algebra of Exceptional Types

In this section, we construct finite-dimensional complex simple Lie algebra of exceptional types.

We use $\mathscr{G}^X$ to denote the simple Lie algebra of type X. It is well known that the simple Lie algebra of type G_2 is exactly the derivation Lie algebra of a special 8-dimensional algebra with an identity element, which is called *octonion algebra*. Since any derivation annihilates the identity element (exercise), the octonion algebra gives a 7-dimensional representation of $\mathscr{G}^{G_2}$, which is presented as follows. Denote by $\mathbb{Z}_3 = \mathbb{Z}/3\mathbb{Z}$ and identify $i + 3\mathbb{Z}$ with i for $i \in \{1, 2, 3\}$. Write

$$o(7,\mathbb{C}) = \sum_{i,j\in\mathbb{Z}_3} [\mathbb{C}(E_{i,j} - E_{j',i'}) + \mathbb{C}(E_{i,j'} - E_{j,i'}) + \mathbb{C}(E_{i',j} - E_{j',i})]$$
$$+ \sum_{i\in\mathbb{Z}_3} [\mathbb{C}(E_{0,i} - E_{i',0}) + \mathbb{C}(E_{0,i'} - E_{i,0})]. \tag{4.4.1}$$

Then the Lie subalgebra

$$\mathscr{K} = \sum_{i,j\in\mathbb{Z}_3; i\neq j} \mathbb{C}(E_{i,j} - E_{j',i'}) + \sum_{r=1,2} \mathbb{C}(E_{r,r} - E_{r+1,r+1} - E_{r',r'} + E_{(r+1)',(r+1)'}) \tag{4.4.2}$$

is isomorphic to $sl(3,\mathbb{C})$. Denote

$$C_r = E_{r,(r+2)'} - E_{r+2,r'} + \sqrt{2}(E_{0,r+1} - E_{(r+1)',0}), \tag{4.4.3}$$
$$C_r' = E_{r',r+2} - E_{(r+2)',r} + \sqrt{2}(E_{0,(r+1)'} - E_{r+1,0}) \tag{4.4.4}$$

for $r \in \mathbb{Z}_3$. Then

$$[C_r, C_{r+1}] = 2C_{r+2}', \qquad [C_r', C_{r+1}'] = 2C_{r+2}, \tag{4.4.5}$$

$$[C_r, C_r'] = E_{r',r'} - E_{r,r} + E_{(r+2)',(r+2)'} - E_{r+2,r+2} + 2(E_{r+1,r+1} - E_{(r+1)',(r+1)'}) \tag{4.4.6}$$

$$[C_r, C_{r+1}'] = 3(E_{r+2,r+1} - E_{(r+1)',(r+2)'}), \;\; [C_{r+1}, C_r'] = 3(E_{r+1,r+2} - E_{(r+2)',(r+1)'}) \tag{4.4.7}$$

for $r \in \mathbb{Z}_3$. Moreover, $\sum_{r=1}^{3} \mathbb{C}C'_r$ and $\sum_{r=1}^{3} \mathbb{C}C_r$ form (ad $\mathscr{K}$)-submodules, which are isomorphic to the three dimensional natural $sl(3, \mathbb{C})$-module and its dual, respectively. Thus

$$\mathscr{G}^{G_2} = \mathscr{K} + \sum_{r=1}^{3} (\mathbb{C}C_r + \mathbb{C}C'_r) \tag{4.4.8}$$

forms a Lie subalgebra of $o(7, \mathbb{C})$, which is a simple Lie algebra of type G_2.

Note that if $|\beta(h_\alpha)| < |\alpha(h_\beta)|$, then $|\langle\beta, \alpha\rangle| < |\langle\alpha, \beta\rangle|$, which implies $(\alpha, \alpha) > (\beta, \beta)$.

Thus we set

$$h_1 = E_{1,1} - E_{2,2} - E_{1',1'} + E_{2',2'}, \; h_2 = -2E_{1,1} + E_{2,2} + E_{3,3} + 2E_{1',1'} - E_{2',2'} - E_{3',3'}, \tag{4.4.9}$$

Take the Cartan subalgebra

$$H = \mathbb{C}h_1 + \mathbb{C}h_2. \tag{4.4.10}$$

Then

$$\mathscr{G}^{G_2}_{\alpha_1} = \mathbb{C}(E_{1,2} - E_{2',1'}), \;\; \mathscr{G}^{G_2}_{\alpha_2} = \mathbb{C}C_3, \;\; \mathscr{G}^{G_2}_{\alpha_1+\alpha_2} = \mathbb{C}C_1, \tag{4.4.11}$$

$$\mathscr{G}^{G_2}_{\alpha_1+2\alpha_2} = \mathbb{C}C'_2, \;\; \mathscr{G}^{G_2}_{\alpha_1+3\alpha_2} = \mathbb{C}(E_{3,1} - E_{1',3'}), \;\; \mathscr{G}^{G_2}_{2\alpha_1+3\alpha_2} = \mathbb{C}(E_{3,2} - E_{2',3'}), \tag{4.4.12}$$

$$\mathscr{G}^{G_2}_{-\alpha_1} = \mathbb{C}(E_{2,1} - E_{1',2'}), \;\; \mathscr{G}^{G_2}_{-\alpha_2} = \mathbb{C}C'_3, \;\; \mathscr{G}^{G_2}_{-\alpha_1-\alpha_2} = \mathbb{C}C'_1, \tag{4.4.13}$$

$$\mathscr{G}^{G_2}_{-\alpha_1-2\alpha_1} = \mathbb{C}C_2, \;\; \mathscr{G}^{G_2}_{-\alpha_1-3\alpha_2} = \mathbb{C}(E_{1,3} - E_{3',1'}), \;\; \mathscr{G}^{G_2}_{-2\alpha_1-3\alpha_2} = \mathbb{C}(E_{2,3} - E_{3',2'}). \tag{4.4.14}$$

Let Λ_r be the root lattices of types $X = E_6, \; E_7$ and E_8 (cf. (3.4.33), (3.4.37) and (3.4.38)). Recall that $(\cdot, \cdot)$ is a symmetric $\mathbb{Z}$-bilinear form on Λ_r such that $(\alpha, \alpha) = 2$ for $\alpha \in \Phi_X$. In fact,

$$0 < \left(\sum_{i=1}^{n} k_i\alpha_i, \sum_{i=1}^{n} k_i\alpha_i\right) = 2\left(\sum_{p=1}^{n} k_p^2 + \sum_{i<j} k_i k_j (\alpha_i, \alpha_j)\right) \tag{4.4.15}$$

for $0 \neq \sum_{i=1}^{n} k_i\alpha_i \in \Lambda_r$. So (α, α) is a positive even integer for any $0 \neq \alpha \in \Lambda_r$. For $\alpha, \beta \in \Phi$,

$$\alpha + \beta \in \Phi \Rightarrow 2 = (\alpha + \beta, \alpha + \beta) = (\alpha, \alpha) + (\beta, \beta) + 2(\alpha, \beta) = 2 + 2 + 2(\alpha, \beta), \tag{4.4.16}$$

So $(\alpha, \beta) = -1$. By Lemma 3.1.2,

$$\alpha + \beta \in \Phi \Leftrightarrow (\alpha, \beta) = -1 \qquad \text{for } \alpha, \beta \in \Phi. \tag{4.4.17}$$

Denote $\Pi = \{\alpha_1, \alpha_2, ..., \alpha_n\}$ according to the Dynkin diagrams of E_6, E_7 and E_8 in Theorem 3.3.1. Define $F(\cdot, \cdot) : \Lambda_r \times \Lambda_r \to \{\pm 1\}$ by

$$F\left(\sum_{i=1}^{n} k_i\alpha_i, \sum_{j=1}^{n} l_j\alpha_j\right) = (-1)^{\sum_{i=1}^{n} k_i l_i + \sum_{i>j} k_i l_j(\alpha_i, \alpha_j)}, \qquad k_i, l_j \in \mathbb{Z}. \tag{4.4.18}$$

Then for $\alpha, \beta, \gamma \in \Lambda_r$,

$$F(\alpha + \beta, \gamma) = F(\alpha, \gamma)F(\beta, \gamma), \quad F(\alpha, \beta + \gamma) = F(\alpha, \beta)F(\alpha, \gamma), \tag{4.4.19}$$

$$F(\alpha, \beta)F(\beta, \alpha)^{-1} = (-1)^{(\alpha, \beta)}, \quad F(\alpha, \alpha) = (-1)^{(\alpha, \alpha)/2}. \tag{4.4.20}$$

In particular,

$$F(\alpha, \beta) = -F(\beta, \alpha) \qquad \text{if } \alpha, \beta, \alpha + \beta \in \Phi \tag{4.4.21}$$

by (4.4.17).

Denote

$$H = \sum_{i=1}^{n} \mathbb{C}\alpha_i. \tag{4.4.22}$$

Define

$$\mathscr{G}^X = H \oplus \bigoplus_{\alpha \in \Phi} \mathbb{C}E_\alpha, \tag{4.4.23}$$

an algebraic operation $[\cdot, \cdot]$ and a symmetric bilinear form $(\cdot, \cdot)$ on $\mathscr{G}$ by:

$$[H, H] = 0, \;\; [h, E_\alpha] = -[E_\alpha, h] = (h, \alpha)E_\alpha, \;\; [E_\alpha, E_{-\alpha}] = -\alpha, \tag{4.4.24}$$

$$[E_\alpha, E_\beta] = \begin{cases} 0 & \text{if } \alpha + \beta \notin \Phi, \\ F(\alpha, \beta)E_{\alpha+\beta} & \text{if } \alpha + \beta \in \Phi. \end{cases} \tag{4.4.25}$$

Theorem 4.4.1 *The pair $(\mathscr{G}^X, [\cdot, \cdot])$ forms a simple Lie algebra of type X.*

Proof The skew symmetry of $[\cdot, \cdot]$ follows from (4.4.21). Recall the Jacobi identity:

$$[[x, y], z] + [[y, z], x] + [[z, x], y] = 0. \tag{4.4.26}$$

It is trivial if $x, y, z \in H$. For $x, y \in H$ and $z = E_\alpha$ with $\alpha \in \Phi$, (4.4.26) becomes

$$0 + (y, \alpha)(-(x, \alpha))E_\alpha + (-(x, \alpha))(-(y, \alpha))E_\alpha = 0, \tag{4.4.27}$$

which obviously holds. When $x \in H,\ y = E_\alpha$ and $z = E_\beta$ with $\alpha, \beta \in \Phi$, (4.4.26) gives

$$(x, \alpha)[E_\alpha, E_\beta] + (-(x, \alpha+\beta))[E_\alpha, E_\beta] + (-(\beta, x))[E_\beta, E_\alpha] = 0, \tag{4.4.28}$$

which holds because of the skew symmetry. Observe that (4.4.26) holds whenever two of $\{x, y, z\}$ are equal.

Consider $x = E_\alpha,\ y = E_{-\alpha}$ and $z = E_\beta$ with $\alpha, \beta \in \Phi$ and $\beta \neq \alpha, -\alpha$. If $\alpha + \beta \in \Phi$, then $(\alpha, \beta) = -1$, and $(-\alpha, \beta) = 1$. So $\beta - \alpha \notin \Phi$ by (4.4.17). Now (4.4.26) yields

$$(-\alpha, \beta)E_\beta + 0 + F(\beta, \alpha)F(\beta+\alpha, -\alpha)E_\beta = 0, \tag{4.4.29}$$

or equivalently, $1 + F(\beta, \alpha)F(\beta+\alpha, -\alpha) = 0$. This holds because

$$\begin{aligned} F(\beta, \alpha)F(\beta+\alpha, -\alpha) &= F(\beta, \alpha)F(\beta, -\alpha)F(\alpha, -\alpha) \\ &= F(\beta, 0)F(\alpha, \alpha)^{-1} = -1 \end{aligned} \tag{4.4.30}$$

by (4.4.20) and (4.4.21). Assume $\alpha + \beta, \beta - \alpha \notin \Phi$, or equivalently,

$$(\beta \pm \alpha, \beta \pm \alpha) = 4 \pm 2(\beta, \alpha) > 2 \Rightarrow \pm(\beta, \alpha) > -1 \sim -1 < (\beta, \alpha) < 1. \tag{4.4.31}$$

Thus we have $(\beta, \alpha) = 0$ because $(\beta, \alpha) \in \mathbb{Z}$. So all three terms in (4.4.26) are 0.

Suppose that $x = E_\alpha,\ y = E_\beta,\ z = E_\gamma$ with $\alpha, \beta, \gamma \in \Phi$ such that $\alpha \neq \pm\beta, \pm\gamma$ and $\beta \neq \pm\gamma$, and also at least one of the terms in (4.4.26) is nonzero. We may assume $\alpha + \beta \in \Phi$ and $\alpha + \beta + \gamma = 0$ or $\alpha + \beta + \gamma \in \Phi$. First we have $(\alpha, \beta) = -1$ by (4.4.17). Observe that

$$F(\beta, -\alpha) = F(\beta, \alpha)^{-1} = F(-\beta, \alpha) = -F(\alpha, \beta) \tag{4.4.32}$$

and

$$F(\beta, -\beta) = F(-\alpha, \alpha) = -1 \tag{4.4.33}$$

by (4.4.20). If $\alpha + \beta + \gamma = 0$, then $\gamma = -\alpha - \beta$ and (4.4.26) gives

$$F(\alpha, \beta)(-(\alpha+\beta)) + F(\beta, -(\alpha+\beta))\alpha + F(-(\alpha+\beta), \alpha)\beta = 0, \tag{4.4.34}$$

or equivalently,

$$F(\alpha, \beta) = F(\beta, -(\alpha + \beta)) = F(-(\alpha + \beta), \alpha), \tag{4.4.35}$$

which holds by (4.4.18)–(4.4.20). Assume $\alpha + \beta + \gamma \in \Phi$. So

$$(\alpha + \beta, \gamma) = (\alpha, \gamma) + (\beta, \gamma) = -1. \tag{4.4.36}$$

Since $[(\alpha, \gamma)]^2 < (\alpha, \alpha)(\gamma, \gamma) = 4$, we have $(\alpha, \gamma) = 0, \pm 1$. Similarly, $(\beta, \gamma) = 0, \pm 1$. Thus (4.4.36) forces $(\alpha, \gamma) = -1,\ (\beta, \gamma) = 0$ or $(\alpha, \gamma) = 0,\ (\beta, \gamma) = -1$. We may assume $(\alpha, \gamma) = -1$ and $(\beta, \gamma) = 0$. Then (4.4.26) becomes

$$F(\alpha, \beta)F(\alpha + \beta, \gamma)E_{\alpha+\beta+\gamma} + 0 + F(\gamma, \alpha)F(\alpha + \gamma, \beta)E_{\alpha+\beta+\gamma} = 0, \tag{4.4.37}$$

or equivalently,

$$\begin{aligned} & F(\alpha, \beta)F(\alpha + \beta, \gamma) + F(\gamma, \alpha)F(\alpha + \gamma, \beta) = 0 \\ \sim\ & F(\alpha + \beta, \gamma) + F(\gamma, \alpha)F(\gamma, \beta) = 0 \\ \sim\ & F(\alpha + \beta, \gamma) + F(\gamma, \alpha + \beta) = 0, \end{aligned} \tag{4.4.38}$$

which is implied by (4.4.21). This completes the proof of the Jacobi identity (4.4.26). The simplicity follows from the fact that Φ is an irreducible root system. □

Now to understand the simple Lie algebras of types $E_6,\ E_7$ and E_8 explicitly, we need to write all the positive roots as a linear combinations of simple positive roots. In terms of the order in the Dynkin diagram of E_8 in Theorem 3.3.1, we have:

$$\alpha_1 = \varepsilon_1 - \varepsilon_2, \qquad \alpha_2 = \frac{1}{2}\left(\sum_{s=4}^{8} \varepsilon_s - \sum_{r=1}^{3} \varepsilon_r\right), \tag{4.4.39}$$

$$\alpha_{i+2} = \varepsilon_{i+1} - \varepsilon_{i+2} \qquad \text{for } i \in \overline{1, 6}, \tag{4.4.40}$$

where $\{\varepsilon_i \mid i \in \overline{1, 8}\}$ is the standard basis in the Euclidean space $\mathbb{R}^8$. Recall the root system

$$\Phi_{E_8} = \left\{\frac{1}{2}\sum_{i=1}^{8}(-1)^{\iota_i}\varepsilon_i \mid \iota_i \in \mathbb{Z}_2,\ \sum_{r=1}^{8} \iota_r = 0 \text{ in } \mathbb{Z}_2\right\} \bigcup \Phi_{D_8}. \tag{4.4.41}$$

Let $\Phi_{E_8}^+$ be the set of positive roots. For convenience, we also denote

$$\alpha_{(k_1,\ldots,k_r)} = \sum_{s=1}^{r} k_s\alpha_s,\ k_r \neq 0 \tag{4.4.42}$$

and

$$E_{(k_1,\ldots,k_r)} = E_{\alpha_{(k_1,\ldots,k_r)}} \quad \text{if } \alpha_{(k_1,\ldots,k_r)} \in \Phi_{E_8}^+. \tag{4.4.43}$$

Moreover, we write

$$(aE_\beta + bE_\gamma)' = aE'_\beta + bE'_\gamma = aE_{-\beta} + bE_{-\gamma} \quad \text{for } a,b \in \mathbb{F},\ \beta,\gamma \in \Phi_{E_8}^+. \tag{4.4.44}$$

Take the convention $\sum_{r=i}^{j} \alpha_r = 0$ if $j < i$. We calculate

$$\varepsilon_1 - \varepsilon_j = \alpha_1 + \sum_{r=3}^{j} \alpha_r \quad \text{for } j \in \overline{2,8}, \tag{4.4.45}$$

$$\varepsilon_i - \varepsilon_j = \sum_{r=i+1}^{j} \alpha_r \quad \text{for } 2 \leq i < j \leq 8, \tag{4.4.46}$$

$$\varepsilon_1 + \varepsilon_j = \alpha_{(2,2,3,4,3,2,1)} + \sum_{s=j+1}^{8} \alpha_s \quad \text{for } j \in \overline{2,8}, \tag{4.4.47}$$

$$\varepsilon_7 + \varepsilon_8 = \alpha_{(1,2,2,3,2,1)}, \quad \frac{1}{2}\left(\sum_{r=2}^{8} \varepsilon_r - \varepsilon_1\right) = \alpha_{(1,3,3,5,4,3,2,1)} \tag{4.4.48}$$

$$\varepsilon_i + \varepsilon_j = \alpha_{(1,2,2,3,2,1)} + \sum_{r=i+1}^{7} \alpha_r + \sum_{s=j+1}^{8} \alpha_s \quad \text{for } 2 \leq i < j \leq 8, \tag{4.4.49}$$

$$\frac{1}{2}\left(\sum_{r\neq 1,j,k} \varepsilon_r - \varepsilon_1 - \varepsilon_j - \varepsilon_k\right) = \sum_{s=2}^{j} \alpha_s + \sum_{t=4}^{k} \alpha_t \quad \text{for } 2 \leq j < k \leq 8, \tag{4.4.50}$$

$$\frac{1}{2}\left(\sum_{r\neq i} \varepsilon_r - \varepsilon_i\right) = \alpha_{(2,2,3,5,4,3,2,1)} + \sum_{r=2}^{i} \alpha_r \quad \text{for } i \in \overline{2,8}, \tag{4.4.51}$$

$$\frac{1}{2}\left(\sum_{r\neq i,j,k}\varepsilon_r-\varepsilon_i-\varepsilon_j-\varepsilon_k\right)=\sum_{\iota=1}^{i}\alpha_\iota+\sum_{s=3}^{j}\alpha_s+\sum_{t=4}^{k}\alpha_t \tag{4.4.52}$$

for $2\leq i<j<k\leq 8$.

Proposition 4.4.2 *The positive roots of type E_8 in terms of linear combinations of Π_{E_8} are given in (4.4.39), (4.4.40) and (4.4.45)–(4.4.52). Moreover, $|\Phi^+_{E_8}|=120$. In particular,*

$$\dim \mathscr{G}^{E_8}=248. \tag{4.4.53}$$

According to Example 3.4.2 and the above proposition, the positive roots of type E_7 in terms of linear combinations of Π_{E_7} are:

$$\alpha_{(2,2,3,4,3,2,1)},\quad \left\{\alpha_{(1,2,2,3,2,1)}+\sum_{r=i+1}^{7}\alpha_r\mid i\in\overline{2,7}\right\}, \tag{4.4.54}$$

$$\left\{\alpha_1+\sum_{r=3}^{j}\alpha_r\mid j\in\overline{2,7}\right\},\quad \left\{\sum_{r=i+1}^{j}\alpha_r\mid 2\leq i<j\leq 7\right\}, \tag{4.4.55}$$

$$\left\{\sum_{s=2}^{j}\alpha_s+\sum_{t=4}^{k}\alpha_t\mid 2\leq j<k\leq 7\right\}, \tag{4.4.56}$$

$$\left\{\sum_{\iota=1}^{i}\alpha_\iota+\sum_{s=3}^{j}\alpha_s+\sum_{t=4}^{k}\alpha_t\mid 2\leq i<j<k\leq 7\right\}. \tag{4.4.57}$$

Thus

$$|\Phi^+_{E_7}|=63,\qquad \dim \mathscr{G}^{E_7}=133. \tag{4.4.58}$$

Moreover, the positive roots of type E_6 in terms of linear combinations of Π_{E_6} are:

$$\left\{\alpha_1+\sum_{r=3}^{j}\alpha_r\mid j\in\overline{2,6}\right\}\bigcup\left\{\sum_{r=i+1}^{j}\alpha_r\mid 2\leq i<j\leq 6\right\}, \tag{4.4.59}$$

$$\alpha_{(1,2,2,3,2,1)},\quad \left\{\sum_{s=2}^{j}\alpha_s+\sum_{t=4}^{k}\alpha_t\mid 2\leq j<k\leq 6\right\} \tag{4.4.60}$$

and

$$\left\{\sum_{\iota=1}^{i}\alpha_\iota+\sum_{s=3}^{j}\alpha_s+\sum_{t=4}^{k}\alpha_t \mid 2\leq i<j<k\leq 6\right\}. \tag{4.4.61}$$

Hence

$$|\Phi_{E_6}^+|=36, \qquad \dim\mathscr{G}^{E_6}=78. \tag{4.4.62}$$

Finally we give a construction of the simple Lie algebra of type F_4. From the Dynkin diagram of E_6 in Theorem 3.3.1, we have the following automorphism of Φ_{E_6}:

$$\sigma\left(\sum_{i=1}^{6}k_i\alpha_i\right)=k_6\alpha_1+k_2\alpha_2+k_5\alpha_3+k_4\alpha_4+k_3\alpha_5+k_1\alpha_6 \tag{4.4.63}$$

for $\sum_{i=1}^{6}k_i\alpha_i\in\Phi_{E_6}$. In the construction of the Lie algebra $\mathscr{G}$ in (4.4.22)–(4.4.25), we take $\Lambda_r=\Lambda_r(E_6)$ and replace the F in (4.4.18) by

$$F\left(\sum_{i=1}^{6}k_i\alpha_i,\sum_{j=1}^{6}l_j\alpha_j\right)=(-1)^{\sum_{i=1}^{6}k_il_i+k_1l_3+k_4l_2+k_3l_4+k_5l_4+k_6l_5}, \qquad k_i,l_j\in\mathbb{Z}. \tag{4.4.64}$$

Then (4.4.19)–(4.4.21) holds and

$$F\left(\sigma(\alpha),\sigma(\beta)\right)=F(\alpha,\beta) \qquad \text{for }\ \alpha,\beta\in\Lambda_r(E_6). \tag{4.4.65}$$

Thus we have the following automorphism $\hat{\sigma}$ of the Lie algebra $\mathscr{G}$:

$$\hat{\sigma}\left(\sum_{i=1}^{6}b_i\alpha_i\right)=\sum_{i=1}^{6}b_i\sigma(\alpha_i), \qquad b_i\in\mathbb{C}, \tag{4.4.66}$$

$$\hat{\sigma}(E_\alpha)=E_{\sigma(\alpha)} \qquad \text{for }\ \alpha\in\Phi_{E_6}. \tag{4.4.67}$$

Using the notation in (4.4.43), we write

$$F_{(1)}=E_{\alpha_2},\quad F_{(0,1)}=E_{\alpha_4},\quad F_{(0,0,1)}=E_{\alpha_3}+E_{\alpha_5},\quad F_{(0,0,0,1)}=E_{\alpha_1}+E_{\alpha_6}, \tag{4.4.68}$$

$$F_{(1,1)} = E_{(0,1,0,1)}, \quad F_{(0,1,1)} = E_{(0,0,1,1)} + E_{(0,0,0,1,1)}, \quad F_{(0,0,1,1)} = E_{(1,0,1)} + E_{(0,0,0,0,1,1)}, \tag{4.4.69}$$

$$F_{(0,0,1,1)} = E_{(1,0,1)} + E_{(0,0,0,0,1,1)}, \quad F_{(1,1,1)} = E_{(0,1,1,1)} + E_{(0,1,0,1,1)}, \tag{4.4.70}$$

$$F_{(0,1,1,1)} = E_{(1,0,1,1)} + E_{(0,0,0,1,1,1)}, \quad F_{(0,1,2)} = E_{(0,0,1,1,1)}, \quad F_{(1,1,2)} = E_{(0,1,1,1,1)}, \tag{4.4.71}$$

$$F_{(0,1,2,1)} = E_{(1,0,1,1,1)} + E_{(0,0,1,1,1,1)}, \quad F_{(1,1,1,1)} = E_{(1,1,1,1)} + E_{(0,1,0,1,1,1)}, \tag{4.4.72}$$

$$F_{(1,2,2)} = E_{(0,1,1,2,1)}, \quad F_{(1,1,2,1)} = E_{(1,1,1,1,1)} + E_{(0,1,1,1,1,1)}, \quad F_{(0,1,2,2)} = E_{(1,0,1,1,1,1)}, \tag{4.4.73}$$

$$F_{(1,2,2,1)} = E_{(1,1,1,2,1)} + E_{(0,1,1,2,1,1)}, \quad F_{(1,1,2,2)} = E_{(1,1,1,1,1,1)}, \quad F_{(1,2,2,2)} = E_{(1,1,1,2,1,1)}, \tag{4.4.74}$$

$$F_{(1,2,3,1)} = E_{(1,1,2,2,1)} + E_{(0,1,1,2,2,1)}, \quad F_{(1,2,3,2)} = E_{(1,1,2,2,1,1)} + E_{(1,1,1,2,2,1)}, \tag{4.4.75}$$

$$F_{(1,2,4,2)} = E_{(1,1,2,2,2,1)}, \quad F_{(1,3,4,2)} = E_{(1,1,2,3,2,1)}, \quad F_{(2,3,4,2)} = E_{(1,2,2,3,2,1)}. \tag{4.4.76}$$

Moreover, we set

$$h_1 = \alpha_2, \qquad h_2 = \alpha_4, \qquad h_3 = \alpha_3 + \alpha_5, \qquad h_4 = \alpha_1 + \alpha_6. \tag{4.4.77}$$

The Dynkin diagram of F_4 is
F_4:

F_4:

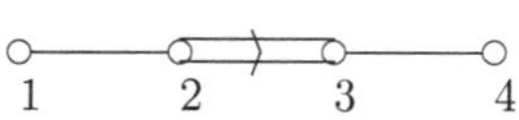

In order to make notation distinguishable, we add a bar on the roots in the root system Φ_{F_4} of type F_4. In particular, we let $\{\bar{\alpha}_1, \bar{\alpha}_2, \bar{\alpha}_3, \bar{\alpha}_4\}$ be the simple positive roots corresponding to the above Dynkin diagram of F_4, where $\bar{\alpha}_1, \bar{\alpha}_2$ are long roots and $\bar{\alpha}_3, \bar{\alpha}_4$ are short roots. Let $\Phi_{F_4}^+$ be the set of positive roots of F_4. Denote

$$S_{F_4} = \left\{ (k_1, ..., k_r) \mid \bar{\alpha}_{(k_1,...,k_r)} = \sum_{i=1}^{r} k_i \bar{\alpha}_i \in \Phi_{F_4}^{+}, \ k_r \neq 0 \right\}. \tag{4.4.78}$$

Moreover, we denote

$$F'_{(k_1,...,k_r)} = (F_{(k_1,...,k_r)})' \quad \text{for } (k_1, ..., k_r) \in S_{F_4} \tag{4.4.79}$$

(cf. (4.4.44)). Then the simple Lie algebra of type F_4 is

$$\mathscr{G}^{F_4} = \{u \in \mathscr{G}^{E_6} \mid \hat{\sigma}(u) = u\} = \sum_{\varpi \in S_{F_4}} (\mathbb{F}F_{\varpi} + \mathbb{F}F'_{\varpi}) + \sum_{i=1}^{4} \mathbb{F}h_i. \tag{4.4.80}$$

In fact, F_{ϖ} is a root vector of root $\bar{\alpha}_{\varpi}$, F'_{ϖ} is a root vector of root $-\bar{\alpha}_{\varpi}$, and $H_{F_4} = \sum_{i=1}^{4} \mathbb{F}h_i$ is the corresponding Cartan subalgebra of $\mathscr{G}^{F_4}$.

Chapter 5
Highest-Weight Representation Theory

Finite-dimensional representations of a complex finite-dimensional simple Lie algebra were essentially determined by Cartan in early twenty century. The approaches were simplified and further developed by Weyl in 1920s. In this chapter, we always assume that the base field $\mathbb{F} = \mathbb{C}$. By Weyl's Theorem, any finite-dimensional representation of a finite-dimensional semisimple Lie algebra $\mathscr{G}$ is completely reducible. The main goal in this chapter is to study finite-dimensional irreducible representations of a finite-dimensional semisimple Lie algebra. First we introduce the universal enveloping algebra of a Lie algebra and prove the Poincaré-Birkhoff-Witt (PBW) Theorem on its basis. Then we use the universal enveloping algebra of $\mathscr{G}$ to construct the Verma module of $\mathscr{G}$, which is the maximal module of a given highest weight. Moreover, we prove that any finite-dimensional irreducible $\mathscr{G}$-module is the quotient of a Verma module modulo its maximal proper submodule, whose generators are explicitly given. Furthermore, the Weyl's character formula of a finite-dimensional irreducible $\mathscr{G}$-module is derived and the dimensional formula of the module is determined. Finally, we decompose the tensor module of two finite-dimensional irreducible $\mathscr{G}$-modules into a direct sum of irreducible $\mathscr{G}$-submodules in terms of characters.

In Sect. 5.1, we use tensor algebra to construct the universal enveloping algebra of a Lie algebra and prove the PBW theorem by defining a representation of the algebra on the corresponding polynomial algebra. In Sect. 5.2, we use the universal enveloping algebra of $\mathscr{G}$ to construct the Verma module of $\mathscr{G}$. Moreover, we prove that any finite-dimensional irreducible $\mathscr{G}$-module is the quotient of a Verma module modulo its maximal proper submodule. In Sect. 5.3, we introduce the notion of formal character of a weight module of $\mathscr{G}$ and prove that the character of a finite-dimensional module is invariant under the Weyl group of $\mathscr{G}$. In Sect. 5.4, we prove Weyl's character formula and use it to prove the tensor formula of two finite-dimensional $\mathscr{G}$-modules. Finally in Sect. 5.5, we find the dimensional formula of a finite-dimensional irreducible module of $\mathscr{G}$.

X. Xu, *Representations of Lie Algebras and Partial Differential Equations*,
DOI 10.1007/978-981-10-6391-6_5

5.1 Universal Enveloping Algebras

In this section, we construct the universal enveloping algebra of a Lie algebra and prove the Poincaré-Birkhoff-Witt Theorem on its basis.

Let V be a vector space. The *free associative algebra generated by* V is an associative algebra $\mathscr{A}(V)$ and a linear injective map $\iota : V \to \mathscr{A}(V)$ such that if $\varphi : V \to \mathscr{B}$ is linear map from V to an associative algebra $\mathscr{B}$, then there exists a unique algebra homomorphism $\phi : \mathscr{A}(V) \to \mathscr{B}$ for which $\phi(1_{\mathscr{A}(V)}) = 1_{\mathscr{B}}$ and $\phi\iota = \varphi$. If S is a basis of V, we also say that $\mathscr{A}(V)$ is a free associative algebra generated by S.

The algebra $\mathscr{A}(V)$ is indeed the *tensor algebra* over V as follows. Suppose that $S = \{u_i \mid i \in I\}$ is a basis of V. We set $\mathcal{I}(V)$ to be the vector space with

$$\{1, u_{l_1} \otimes u_{l_2} \otimes \cdots \otimes u_{l_s} \mid 0 < s \in \mathbb{Z},\ l_1, ..., l_s \in I\} \text{ as a basis} \tag{5.1.1}$$

and define an associative algebraic operation $\cdot$ on $\mathcal{I}(V)$ by $1 \cdot 1 = 1$,

$$1 \cdot (u_{l_1} \otimes \cdots \otimes u_{l_r}) = (u_{l_1} \otimes \cdots \otimes u_{l_r}) \cdot 1 = u_{l_1} \otimes \cdots \otimes u_{l_r} \tag{5.1.2}$$

and

$$(u_{l_1} \otimes \cdots \otimes u_{l_r}) \cdot (u_{k_1} \otimes \cdots \otimes u_{k_s}) = u_{l_1} \otimes \cdots \otimes u_{l_r} \otimes u_{k_1} \otimes \cdots \otimes u_{k_s}. \tag{5.1.3}$$

Given

$$v_i = \sum_{j \in I} a_{i,j} u_j \in V \qquad \text{for } i \in \overline{1, r}, \tag{5.1.4}$$

we define

$$v_1 \otimes v_2 \otimes \cdots \otimes v_r = \sum_{j_1,...,j_r \in I} a_{1,j_1} a_{2,j_2} \ldots a_{r,j_r} u_{j_1} \otimes u_{j_1} \otimes \cdots \otimes u_{j_r}. \tag{5.1.5}$$

As associative algebras, $\mathscr{A}(V) \cong \mathcal{I}(V)$. The map $\iota : V \to \mathcal{I}$ is the inclusion map.

Denote by

$$\mathscr{J}_0(V) = \text{the ideal of } \mathcal{I}(V) \text{ generated by } \{u \otimes v - v \otimes u \mid u, v \in V\}. \tag{5.1.6}$$

The *symmetric tensor algebra* over V (or *free commutative associative algebra generated by* V) is defined as the quotient algebra

$$\mathscr{S}(V) = \mathcal{I}(V)/\mathscr{J}_0(V). \tag{5.1.7}$$

In fact, we have the following algebra isomorphism: $1_{\mathscr{S}(V)} \mapsto 1$ and

$$u_{j_1} \otimes u_{j_2} \cdots \otimes u_{j_r} + \mathscr{J}_0(V) \mapsto x_{j_1} x_{j_2} \ldots x_{j_r} \tag{5.1.8}$$

from $\mathscr{S}(V)$ to the polynomial algebra $\mathbb{C}[x_i \mid i \in I]$.

Suppose that $\mathscr{G}$ is a Lie algebra. We have the tensor algebra $\mathcal{I}(\mathscr{G})$. Set

$$\mathscr{J}_1(\mathscr{G}) = \text{the ideal of } \mathcal{I}(\mathscr{G}) \text{ generated by } \{u \otimes v - v \otimes u - [u, v] \mid u, v \in \mathscr{G}\}. \tag{5.1.9}$$

Define

$$U(\mathscr{G}) = \mathcal{I}(\mathscr{G})/\mathscr{J}_1(\mathscr{G}). \tag{5.1.10}$$

Suppose that $\{u_i \mid i \in I\}$ is a basis of $\mathscr{G}$ and I has a total ordering $\preceq$. If $I = \{1, 2, ..., n\}$ is finite, we can take a total ordering just usual $\leq$. Recall that $\mathbb{Z}_+$ is the set of positive integers. Denote $\bar{u} = u + \mathscr{J}_1(\mathscr{G})$ for $u \in \mathscr{G}$.

Theorem 5.1.1 (Poincaré-Birkhoff-Witt (PBW) Theorem). *The set*

$$\{1, \bar{u}_{i_1}^{m_1} \bar{u}_{i_2}^{m_2} \ldots \bar{u}_{i_r}^{m_r} \mid r, m_1, ..., m_r \in \mathbb{Z}_+,\ i_1, ..., i_r \in I;\ i_1 \prec i_2 \prec \ldots \prec i_r\} \tag{5.1.11}$$

forms a basis of $U(\mathscr{G})$.

Proof Let $\mathscr{A} = \mathbb{F}[x_i \mid i \in I]$, the polynomial algebra in $\{x_i \mid i \in I\}$. For $r \in \mathbb{Z}_+$, we denote

$$I_r = \{\vec{i} = (i_1, i_2, ..., i_r) \in I^r \mid i_1 \preceq i_2 \preceq \ldots \preceq i_r\}. \tag{5.1.12}$$

Write

$$x^{\vec{i}} = x_{i_1} x_{i_2} \ldots x_{i_r} \qquad \text{for } \vec{i} \in I_r. \tag{5.1.13}$$

Let

$$\mathscr{A}_{(0)} = \mathbb{F},\quad \mathscr{A}_{(k)} = \mathscr{A}_{(0)} + \text{Span}\{x^{\vec{i}} \mid \vec{i} \in I_r;\ r \leq k\}. \tag{5.1.14}$$

Then $\mathscr{A} = \bigcup_{k=0}^{\infty} \mathscr{A}_{(k)}$. Define an action of $\mathscr{G}$ on $\mathscr{A}$ by induction on $\mathscr{A}_{(r)}$ as follows. On $\mathscr{A}_{(0)}$, we define

$$u_i(1) = x_i \qquad \text{for } i \in I. \tag{5.1.15}$$

Assume that we have defined the action of $\mathscr{G}$ on $\mathscr{A}_{(r-1)}$. For $\ell \in I$ and $\vec{i} \in I_r$, we define

$$u_\ell(x^{\vec{i}}) = x_\ell x^{\vec{i}} + \sum_{i_s \prec \ell} u_{i_1}(\ldots(u_{i_{s-1}}([u_\ell, u_{i_s}](x_{i_{s+1}} \ldots x_{i_r})))\ldots). \tag{5.1.16}$$

By induction, we have defined an action $\mathscr{G}$ on $\mathscr{A}$. Note that (5.1.16) implies

$$u_\ell(x^{\vec{i}}) = u_{i_1}(u_\ell(x_{i_2} \ldots x_{i_r})) + [u_\ell, u_{i_1}](x_{i_2} \ldots x_{i_r}) \text{ if } i_1 \prec \ell. \tag{5.1.17}$$

Next we want to prove that $\mathscr{A}$ forms a $\mathscr{G}$-module with respect to the above action. Again we prove

$$u(v(f)) - v(u(f)) = [u, v](f) \qquad \text{for } f \in \mathscr{A}_{(r)} \tag{5.1.18}$$

by induction on r. Suppose $\ell, j \in I$ with $\ell \prec j$. Then

$$u_\ell(u_j)(1) - u_j(u_\ell)(1) = u_\ell(x_j) - u_j(x_\ell) = x_\ell x_j - (x_j x_\ell + [u_j, u_\ell](1)) = [u_\ell, u_j](1). \tag{5.1.19}$$

Thus (5.1.18) holds for $r = 0$. Assume that (5.1.18) holds for $r = k$. Consider the case $r = k + 1$. Let $\vec{i} \in I_r$ with $r = k + 1$.

If $\ell \preceq i_1$, then

$$u_j(u_\ell x^{\vec{i}}) = u_j(x_\ell x^{\vec{i}}) = [u_j, u_\ell](x^{\vec{i}}) + u_\ell(u_j(x^{\vec{i}})) \tag{5.1.20}$$

by (5.1.17). Thus

$$u_\ell(u_j)(x^{\vec{i}}) - u_j(u_\ell)(x^{\vec{i}}) = [u_\ell, u_j](x^{\vec{i}}). \tag{5.1.21}$$

Suppose $i_1 \prec \ell$. We have

$$\begin{aligned} u_\ell(u_j x^{\vec{i}}) &= u_\ell\Big[x_j x_{i_1} x_{i_2} \ldots x_{i_r} + [u_j, u_{i_1}](x_{i_2} \ldots x_{i_r}) \\ &\quad + u_{i_1}\Big(\sum_{2 \le s; i_s \prec \ell} u_{i_2}(\ldots(u_{i_{s-1}}([u_j, u_{i_s}](x_{i_{s+1}} \ldots x_{i_r})))\ldots)\Big)\Big] \\ &= u_{i_1}(u_\ell(x_j x_{i_2} \ldots x_{i_r})) + [u_\ell, u_{i_1}](x_j x_{i_2} \ldots x_{i_r}) + u_\ell([u_j, u_{i_1}](x_{i_2} \ldots x_{i_r})) \\ &\quad + [u_\ell, u_{i_1}]\Big(\sum_{2 \le s; i_s \prec \ell} u_{i_2}(\ldots(u_{i_{s-1}}([u_j, u_{i_s}](x_{i_{s+1}} \ldots x_{i_r})))\ldots)\Big) \\ &\quad + u_{i_1}\Big(u_\ell\Big(\sum_{2 \le s; i_s \prec \ell} u_{i_2}(\ldots(u_{i_{s-1}}([u_j, u_{i_s}](x_{i_{s+1}} \ldots x_{i_r})))\ldots)\Big)\Big) \\ &= u_{i_1}(u_\ell(u_j(x_{i_2} \ldots x_{i_r}))) + [u_\ell, u_{i_1}](u_j(x_{i_2} \ldots x_{i_r})) \\ &\quad + u_\ell([u_j, u_{i_1}](x_{i_2} \ldots x_{i_r})) \end{aligned} \tag{5.1.22}$$

by (5.1.16), (5.1.17) and the inductional assumption. Similarly, we get

$$\begin{aligned} u_j(u_\ell x^{\vec{i}}) &= u_j\Big[x_\ell x_{i_1} x_{i_2} \ldots x_{i_r} + [u_\ell, u_{i_1}](x_{i_2} \ldots x_{i_r}) \\ &\quad + u_{i_1}\Big(\sum_{2 \le s; i_s \prec \ell} u_{i_2}(\ldots(u_{i_{s-1}}([u_\ell, u_{i_s}](x_{i_{s+1}} \ldots x_{i_r})))\ldots)\Big)\Big] \\ &= u_{i_1}(u_j(x_\ell x_{i_2} \ldots x_{i_r})) + [u_j, u_{i_1}](x_\ell x_{i_2} \ldots x_{i_r}) + u_j([u_\ell, u_{i_1}](x_{i_2} \ldots x_{i_r})) \\ &\quad + [u_j, u_{i_1}]\Big(\sum_{2 \le s; i_s \prec \ell} u_{i_2}(\ldots(u_{i_{s-1}}([u_\ell, u_{i_s}](x_{i_{s+1}} \ldots x_{i_r})))\ldots)\Big) \end{aligned}$$

$$
\begin{aligned}
&+u_{i_1}(u_j\Big(\sum_{2\leq s;i_s\prec\ell} u_{i_2}(\ldots(u_{i_{s-1}}([u_\ell,u_{i_s}](x_{i_{s+1}}\ldots x_{i_r})))\ldots))\Big)\\
&= u_{i_1}(u_j(u_\ell(x_{i_2}\ldots x_{i_r}))) + [u_j,u_{i_1}](u_\ell(x_{i_2}\ldots x_{i_r})) + u_j([u_\ell,u_{i_1}](x_{i_2}\ldots x_{i_r}))\\
&= u_{i_1}(u_j(u_\ell(x_{i_2}\ldots x_{i_r}))) + [u_j,u_{i_1}](u_\ell(x_{i_2}\ldots x_{i_r}))\\
&\quad +[u_j,[u_\ell,u_{i_1}]](x_{i_2}\ldots x_{i_r}) + [u_\ell,u_{i_1}](u_j(x_{i_2}\ldots x_{i_r}))\\
&= u_{i_1}(u_j(u_\ell(x_{i_2}\ldots x_{i_r}))) + [u_j,u_{i_1}](u_\ell(x_{i_2}\ldots x_{i_r})) + [[u_j,u_\ell],u_{i_1}](x_{i_2}\ldots x_{i_r})\\
&\quad +[u_\ell,[u_j,u_{i_1}]](x_{i_2}\ldots x_{i_r}) + [u_\ell,u_{i_1}](u_j(x_{i_2}\ldots x_{i_r}))\\
&= u_{i_1}(u_j(u_\ell(x_{i_2}\ldots x_{i_r}))) + [u_j,u_\ell](u_{i_1}(x_{i_2}\ldots x_{i_r})) - u_{i_1}([u_j,u_\ell](x_{i_2}\ldots x_{i_r}))\\
&\quad +u_\ell([u_j,u_{i_1}](x_{i_2}\ldots x_{i_r}) + [u_\ell,u_{i_1}](u_j(x_{i_2}\ldots x_{i_r}))\\
&= u_{i_1}(u_\ell(u_j(x_{i_2}\ldots x_{i_r}))) + [u_j,u_\ell](x^{\vec{i}}) + u_\ell([u_j,u_{i_1}](x_{i_2}\ldots x_{i_r})\\
&\quad +[u_\ell,u_{i_1}](u_j(x_{i_2}\ldots x_{i_r})),
\end{aligned}
\tag{5.1.23}
$$

where we have used Jacobi identity for the fifth equality. Thus (5.1.21) holds again. By induction on r, (5.1.18) holds or equivalently $\mathscr{A}$ forms a $\mathcal{G}$-module.

We define $\mathscr{U}_0 = \mathbb{C}1$, $\mathscr{U}_1 = \mathscr{G}$ and

$$
\mathscr{U}_i = \text{Span}\,\{1, \bar{u}_{i_1}\bar{u}_{i_2}\ldots\bar{u}_{i_r} \mid i_s \in I,\ i \geq r \in \mathbb{Z}\}. \tag{5.1.24}
$$

Then we have

$$
U(\mathscr{G}) = \bigcup_{i=0}^{\infty} \mathscr{U}_i. \tag{5.1.25}
$$

Note

$$
\mathscr{U}_{k+1} = \text{Span}\,\{\bar{u}_{i_1}\ldots\bar{u}_{i_{k+1}} \mid i_s \in I\} + \mathscr{U}_k. \tag{5.1.26}
$$

Suppose $i_{s+1} \prec i_s$ for some $s \in \overline{1,k}$. Then

$$
\begin{aligned}
&\bar{u}_{i_1}\ldots\bar{u}_{i_{s-1}}\bar{u}_{i_s}\bar{u}_{i_{s+1}}\bar{u}_{i_{s+2}}\ldots\bar{u}_{_{k+1}}\\
&= \bar{u}_{i_1}\ldots\bar{u}_{i_{s-1}}\bar{u}_{i_{s+1}}\bar{u}_{i_s}\bar{u}_{i_{s+2}}\ldots\bar{u}_{_{k+1}} + \bar{u}_{i_1}\ldots\bar{u}_{i_{s-1}}(\bar{u}_{i_s}\bar{u}_{i_{s+1}} - \bar{u}_{i_{s+1}}\bar{u}_{i_s})\bar{u}_{i_{s+2}}\ldots\bar{u}_{_{k+1}}\\
&= \bar{u}_{i_1}\ldots\bar{u}_{i_{s-1}}\bar{u}_{i_{s+1}}\bar{u}_{i_s}\bar{u}_{i_{s+2}}\ldots\bar{u}_{_{k+1}} + \bar{u}_{i_1}\ldots\bar{u}_{i_{s-1}}[\bar{u}_{i_s},\bar{u}_{i_{s+1}}]\bar{u}_{i_{s+2}}\ldots\bar{u}_{_{k+1}}\\
&\equiv \bar{u}_{i_1}\ldots\bar{u}_{i_{s-1}}\bar{u}_{i_{s+1}}\bar{u}_{i_s}\bar{u}_{i_{s+2}}\ldots\bar{u}_{_{k+1}} \pmod{\mathscr{U}_k}.
\end{aligned}
\tag{5.1.27}
$$

This shows that we can change the orders of adjacent factors in the product $\bar{u}_{i_1}\ldots\bar{u}_{i_k}$ modulo $\mathscr{U}_k$. After finite steps of exchanging the positions of adjacent factors, we get

$$
\bar{u}_{i_1}\ldots\bar{u}_{i_{k+1}} = \bar{u}_{j_1}\ldots\bar{u}_{j_{k+1}} \pmod{\mathscr{U}_k} \text{ with } \vec{j} \in I_{k+1} \tag{5.1.28}
$$

(cf. (5.1.12)). This shows

$$
U(\mathscr{G}) = \text{span}\{1, \bar{u}_{i_1}\ldots\bar{u}_{i_r} \mid r \in \mathbb{Z}_+; \vec{i} \in I_r\}. \tag{5.1.29}
$$

On the other hand, the map $\varphi : u \mapsto u|_{\mathscr{A}}$ is a Lie homomorphism from $\mathscr{G}$ to End $\mathscr{A}$, whose Lie bracket is the commutator of linear transformations. By the universality of the tensor algebra $\mathcal{I}(\mathscr{G})$, φ can be extended to an associative algebra homomorphism $\widehat{\varphi} : \mathcal{I}(\mathscr{G}) \to \mathrm{End}\,\mathscr{A}$. By (5.1.9) and (5.1.18), $\mathscr{J}_1(\mathscr{G}) \subset \ker_{\widehat{\varphi}}$. Hence $\widehat{\varphi}$ induces an associative algebra homomorphism $\overline{\varphi} : U(\mathscr{G}) \to \mathrm{End}\,\mathscr{A}$. Since

$$\overline{\varphi}(\bar{u}_{i_1} \dots \bar{u}_{i_r})(1) = \bar{u}_{i_1}(\dots(\bar{u}_{i_r}(1))\dots) = x^{\vec{i}} \qquad \text{for } \vec{i} \in I_k \tag{5.1.30}$$

and $\{1, x^{\vec{i}} \mid r \in \mathbb{Z}_+; \vec{i} \in I_r\}$ is a basis of $\mathscr{A}$, the set $\{1, \bar{u}_{i_1} \dots \bar{u}_{i_r} \mid r \in \mathbb{Z}_+; \vec{i} \in I_r\}$ must be linearly independent. Hence it is a basis of $U(\mathscr{G})$; that is, theorem holds. □

In particular, the above theorem implies the map $u \mapsto \bar{u}$ is a Lie algebra monomorphism from $\mathscr{G}$ to $U(\mathscr{G})$, whose Lie bracket is the commutator in $U(\mathscr{G})$. From now on, we make the following identification

$$u \longleftrightarrow \bar{u} = u + \mathscr{J}_1(\mathscr{G}) \qquad \text{for } u \in \mathscr{G}. \tag{5.1.31}$$

Let $\mathscr{A}$ be an associative algebra. Suppose that $\varphi : \mathscr{G} \to \mathscr{A}$ is a Lie algebra homomorphism, where the Lie bracket of $\mathscr{A}$ is the commutator. There exists a unique associative algebra homomorphism $\varphi_1 : \mathcal{I}(\mathscr{G}) \to \mathscr{A}$ such that $\varphi_1(1) = 1$ and $\varphi_1|_{\mathscr{G}} = \varphi$. Moreover,

$$\varphi_1(u \otimes v - v \otimes u - [u, v]) = \varphi(u)\varphi(v) - \varphi(v)\varphi(u) - \varphi([u, v]) = 0 \tag{5.1.32}$$

for $u, v \in \mathscr{G}$ because φ is a Lie algebra homomorphism. Hence $\mathscr{J}_1(\mathscr{G}) \subset \ker \varphi_1$. So we have a unique associative algebra homomorphism $\phi : U(\mathscr{G}) \to \mathscr{A}$ defined by

$$\phi(w + \mathscr{J}_1(\mathscr{G})) = \varphi_1(w) \qquad \text{for } w \in \mathcal{I}(\mathscr{G}) \tag{5.1.33}$$

with $\phi|_{\mathscr{G}} = \varphi$.

Suppose that $\overline{\mathscr{U}}$ is an associative algebra with an injective linear map $\iota : \mathscr{G} \to \overline{\mathscr{U}}$ such that given a Lie algebra homomorphism φ' from $\mathscr{G}$ to an associative $\mathscr{A}$, there exists a unique associative algebra homomorphism $\phi' : \overline{\mathscr{U}} \to \mathscr{A}$ for which $\phi'\iota = \varphi'$. Such an algebra $\overline{\mathscr{U}}$ is called an *universal enveloping algebra* of $\mathscr{G}$. By the above paragraph, there exists an associative algebra homomorphism $\tau : U(\mathscr{G}) \to \overline{\mathscr{U}}$ such that $\tau|_{\mathscr{G}} = \iota$. Since the inclusion map from $\mathscr{G}$ to $U(\mathscr{G})$ is a Lie algebra monomorphism, there exists an associative algebra homomorphism $\tau' : \overline{\mathscr{U}} \to U(\mathscr{G})$ such that $\tau'\iota = \mathrm{Id}_{\mathscr{G}}$.

Now $\tau\tau' : \overline{\mathscr{U}} \to \overline{\mathscr{U}}$ is an associative algebra endomorphism such that $\tau\tau'\iota = \iota$. Since $\mathrm{Id}_{\overline{\mathscr{U}}}\iota = \iota$, we have $\tau\tau' = \mathrm{Id}_{\overline{\mathscr{U}}}$ by the uniqueness from the universality of $\overline{\mathscr{U}}$. On the other hand, $\tau'\tau : U(\mathscr{G}) \to U(\mathscr{G})$ an associative algebra endomorphism such that $\tau'\tau|_{\mathscr{G}} = \tau'\iota = \mathrm{Id}_{\mathscr{G}}$. Since $U(\mathscr{G})$ is generated by $\mathscr{G}$, we have $\tau'\tau = \mathrm{Id}_{U(\mathscr{G})}$. Thus $\overline{\mathscr{U}} \cong U(\mathscr{G})$; that is, $U(\mathscr{G})$ is the unique universal enveloping algebra of $\mathscr{G}$.

5.2 Highest-Weight Modules

In this section, we use the universal enveloping algebra of a finite-dimensional semi-simple Lie algebra $\mathscr{G}$ to construct the Verma module of $\mathscr{G}$. Moreover, we prove that any finite-dimensional irreducible $\mathscr{G}$-module is the quotient of a Verma module modulo its maximal proper submodule.

Let $\mathscr{G}$ be a Lie algebra (may not be finite-dimensional) with a toral Cartan subalgebra H and two subalgebras $\mathscr{G}_\pm$ such that

$$\mathscr{G} = \mathscr{G}_- \oplus H \oplus \mathscr{G}_+, \qquad [H, \mathscr{G}_\pm] \subset \mathscr{G}_\pm. \tag{5.2.1}$$

By PBW Theorem,

$$U(\mathscr{G}) = U(\mathscr{G}_-)U(H)U(\mathscr{G}_+). \tag{5.2.2}$$

The universality of $U(\mathscr{G})$ implies that a vector space forms a module of a Lie algebra $\mathscr{G}$ if and only if it forms a module of the associative algebra $U(\mathscr{G})$. Suppose that V is a $\mathscr{G}$-module. For any $\lambda \in H^*$, we define

$$V_\lambda = \{v \in V \mid h(v) = \lambda(h)v \text{ for } h \in H\}. \tag{5.2.3}$$

Any nonzero vector in the *weight subspace* V_λ is called a *weight vector* with *weight* λ. A weight vector v is called *singular* if

$$\mathscr{G}_+(v) = \{0\}. \tag{5.2.4}$$

If v is a singular vector, then

$$U(\mathscr{G})(v) = U(\mathscr{G}_-)(v) \tag{5.2.5}$$

is a submodule by (5.2.2). A $\mathscr{G}$-module generated by a singular vector v of weight λ is called a *highest-weight module*. In this case, v is called a *highest-weight vector* with highest weight λ. In particular, an irreducible module containing a singular vector is a highest weight module. The module V is called a *weight module* if

$$V = \sum_{\lambda \in H^*} V_\lambda. \tag{5.2.6}$$

A highest-weight module is naturally a weight module. One of fundamental problems in the representation theory of Lie algebra is to investigate the structure of weight modules. For instance, an interesting but difficult problem is to find all the singular vectors in a weight module.

Observe that

$$\mathscr{B} = H + \mathscr{G}_+ \tag{5.2.7}$$

is a Lie subalgebra of $\mathscr{G}$. For $\lambda \in H^*$, we define a one-dimensional $\mathscr{B}$-module $\mathbb{C}v_\lambda$ by

$$\mathscr{G}_+(v_\lambda) = \{0\}, \quad h(v_\lambda) = \lambda(h)v_\lambda \qquad \text{for } h \in H. \tag{5.2.8}$$

Since $\mathscr{G} = \mathscr{G}_- \oplus \mathscr{B}$, we have $U(\mathscr{G}) = U(\mathscr{G}_-)U(\mathscr{B})$ by PBW Theorem. Now we form an *induced* $U(\mathscr{G})$*-module*

$$M(\lambda) = U(\mathscr{G}) \otimes_{U(\mathscr{B})} \mathbb{C}v_\lambda \cong U(\mathscr{G}_-) \otimes_{\mathbb{C}} \mathbb{C}v_\lambda \text{ (as vector spaces)}, \tag{5.2.9}$$

which is also a module of the Lie algebra $\mathscr{G}$. In fact,

$$\xi_1(u \otimes v_\lambda) = \xi_1 u \otimes v_\lambda, \quad \xi_2(u \otimes v_\lambda) = [\xi_2, u] \otimes v_\lambda + u \otimes \xi_2(v_\lambda) \tag{5.2.10}$$

for $\xi_1 \in \mathscr{G}_-$, $\xi_2 \in \mathscr{B}$ and $u \in U(\mathscr{G}_-)$. Identify $1 \otimes v_\lambda$ with v_λ. The module $M(\lambda)$ is called a *Verma module*. Obviously, $M(\lambda)$ is a highest weight module with v_λ as a highest-weight vector and λ as the highest weight.

Note that

$$U(\mathscr{G}_-) = \bigoplus_{\mu \in H^*} U(\mathscr{G}_-)_\mu, \qquad U(\mathscr{G}_-)_\mu = \{u \in U(\mathscr{G}_-) \mid [h, u] = \mu(h)u \text{ for } h \in H\}. \tag{5.2.11}$$

Then

$$M(\lambda)_{\lambda+\mu} = \{w \in M(\lambda) \mid h(w) = (\lambda + \mu)(h)w \text{ for } h \in H\} = U(\mathscr{G}_-)_\mu v_\lambda. \tag{5.2.12}$$

Assume

$$U(\mathscr{G}_-)_0 = \mathbb{C}1. \tag{5.2.13}$$

Then

$$M(\lambda)_\lambda = \mathbb{C}v_\lambda. \tag{5.2.14}$$

Since any proper submodule U of $M(\lambda)$ must be a weight module, we have

$$U \subset \sum_{0 \neq \mu \in H^*} M(\lambda)_{\lambda+\mu}. \tag{5.2.15}$$

Thus the sum of all proper submodules of $M(\lambda)$ is the unique maximal proper submodule $N(\lambda)$ of $M(\lambda)$. The quotient module

$$V(\lambda) = M(\lambda)/N(\lambda) \tag{5.2.16}$$

is an irreducible highest-weight $\mathscr{G}$-module. In general, it is difficult to determine $N(\lambda)$ explicitly.

Let $\mathscr{G}$ be a finite-dimensional semisimple Lie algebra and let H be a CSA of $\mathscr{G}$ with root system Φ. Fix a base $\Pi = \{\alpha_1, \alpha_2, ..., \alpha_n\}$ of Φ, and so we have positive and negative roots. Set

$$\mathscr{G}_{\pm} = \sum_{\alpha \in \Phi^{\pm}} \mathscr{G}_{\alpha}. \tag{5.2.17}$$

Then $\mathscr{G}_{\pm}$ are nilpotent subalgebras of $\mathscr{G}$ and (5.2.1) holds. Moreover, $\mathscr{B} = \mathscr{B}(\Pi)$ (cf. (5.2.7)) is the standard Borel subalgebra.

Let V be any finite-dimensional irreducible $\mathscr{G}$-module. By Lie's Theorem, there exists a common eigenvector v of $\mathscr{B}$. Since $\mathscr{G}_+ = [\mathscr{B}, \mathscr{B}]$, we have $\mathscr{G}_+(v) = \{0\}$; that is, v is a singular vector. Hence

$$V = U(\mathscr{G})(v) = U(\mathscr{G}_-)(v). \tag{5.2.18}$$

Let λ be the weight of v. Define

$$\tau(w \otimes v_\lambda) = wv \qquad \text{for } \; w \in U(\mathscr{G}_-). \tag{5.2.19}$$

The map τ is a Lie algebra module epimorphism from $M(\lambda)$ to V by (5.2.9). By (5.2.17), (5.2.13) holds. Now $\ker \tau \subset N(\lambda)$. Furthermore,

$$M(\lambda)/\ker \tau \cong V \tag{5.2.20}$$

is irreducible, which implies that $\ker \tau$ is a maximal proper submodule of $M(\lambda)$. Thus $N(\lambda) = \ker \tau$, or equivalently,

$$V \cong V(\lambda). \tag{5.2.21}$$

Take $0 \neq \xi_i \in \mathscr{G}_{\alpha_i}$, $\zeta_i \in \mathscr{G}_{-\alpha_i}$ such that

$$[\xi_i, \zeta_i] = h_{\alpha_i} = h_i, \;\; [h_i, \xi_i] = 2\xi_i \qquad \text{for } \; i \in \overline{1, n}. \tag{5.2.22}$$

Then

$$\mathscr{S}_i = \mathbb{C}\xi_i + \mathbb{C}h_i + \mathbb{C}\zeta_i \tag{5.2.23}$$

forms a Lie subalgebra isomorphic to $sl(2, \mathbb{C})$. By the proof of Theorem 2.6.1, v generates a finite-dimensional irreducible $\mathscr{S}_i$-module. Thus

$$0 \leq \lambda(h_i) = \langle \lambda, \alpha_i \rangle \in \mathbb{Z} \qquad \text{for } \; i \in \overline{1, n}; \tag{5.2.24}$$

that is, λ is a dominant integral weight. Therefore, we obtain:

Lemma 5.2.1 *Any finite-dimensional irreducible $\mathscr{G}$-module is of the form $V(\lambda)$ for some dominant integral weight λ.*

Now given a dominant weight λ, we want to determine $N(\lambda)$ and prove that $V(\lambda)$ is a finite-dimensional irreducible $\mathscr{G}$-module.

Lemma 5.2.2 *The elements*

$$v_{i,\lambda} = \zeta_i^{\langle\lambda,\alpha_i\rangle+1} v_\lambda \qquad \textit{for } i \in \overline{1,n} \tag{5.2.25}$$

are singular vectors in $M(\lambda)$.

Proof Since $\mathscr{G}_+$ is generated by $\{\xi_1, ..., \xi_n\}$, we only need to prove

$$\xi_j(v_{i,\lambda}) = 0 \qquad \text{for } j \in \overline{1,n}, \tag{5.2.26}$$

which is obvious if $i \neq j$ because $\xi_j\zeta_i = \zeta_i\xi_j$ in $U(\mathscr{G})$. Moreover,

$$\begin{aligned}\xi_i(v_{i,\lambda}) &= \sum_{s=0}^{\langle\lambda,\alpha_i\rangle} \zeta_i^{\langle\lambda,\alpha_i\rangle-s}[\xi_i, \zeta_i]\zeta_i^s v_\lambda = \sum_{s=0}^{\langle\lambda,\alpha_i\rangle} \zeta_i^{\langle\lambda,\alpha_i\rangle-s} h_i \zeta_i^s v_\lambda \\ &= \sum_{s=0}^{\langle\lambda,\alpha_i\rangle} (\langle\lambda, \alpha_i\rangle - 2s)\zeta_i^{\langle\lambda,\alpha_i\rangle-s}\zeta_i^s v_\lambda = 0. \qquad \square\end{aligned} \tag{5.2.27}$$

Set

$$N' = \sum_{i=1}^{n} U(\mathscr{G})v_{i,\lambda} = \sum_{i=1}^{n} U(\mathscr{G}_-)v_{i,\lambda}. \tag{5.2.28}$$

Then N' forms a $\mathscr{G}$-submodule of $M(\lambda)$. Form a quotient $\mathscr{G}$-module:

$$V' = M(\lambda)/N'. \tag{5.2.29}$$

We can identify v_λ with its image in V'. Given $\alpha, \beta \in \Phi$. Assume that q is the largest nonnegative integer such that $\beta + q\alpha \in \Phi$. By Lemmas 3.1.1 and 3.1.2,

$$0 \leq \langle\beta + q\alpha, \alpha\rangle \leq 3. \tag{5.2.30}$$

Moreover, by Lemma 3.1.3,

$$\beta + q\alpha - (\langle\beta + q\alpha, \alpha\rangle + 1)\alpha \notin \Phi \Longrightarrow \beta + (q - 4)\alpha \notin \Phi \Longrightarrow q \leq 3. \tag{5.2.31}$$

Thus

$$(\text{ad}\, u_\alpha)^4(u_\beta) = 0 \qquad \text{for } u_\alpha \in \mathscr{G}_\alpha,\ u_\beta \in \mathscr{G}_\beta. \tag{5.2.32}$$

Hence

$$\xi_i^{4r}(\zeta_{\beta_1}\zeta_{\beta_2}\ldots\zeta_{\beta_r}v_\lambda) = \sum_{m_1+\cdots+m_{r+1}=4r} \frac{(4r)!}{m_1!m_2!\cdots m_{r+1}!}$$
$$(\text{ad }\xi_i)^{m_1}(\zeta_{\beta_1})(\text{ad }\xi_i)^{m_2}(\zeta_{\beta_2})\ldots(\text{ad }\xi_i)^{m_r}(\zeta_{\beta_r})\xi_i^{m_{r+1}}v_\lambda = 0 \tag{5.2.33}$$

and

$$\zeta_i^{4r+\langle\lambda,\alpha_i\rangle}(\zeta_{\beta_1}\zeta_{\beta_2}\ldots\zeta_{\beta_r}v_\lambda) = \sum_{m_1+\cdots+m_{r+1}=4r+\langle\lambda,\alpha_i\rangle} \frac{(4r+\langle\lambda,\alpha_i\rangle)!}{m_1!m_2!\cdots m_{r+1}!}$$
$$(\text{ad }\zeta_i)^{m_1}(\zeta_{\beta_1})(\text{ad }\zeta_i)^{m_2}(\zeta_{\beta_2})\ldots(\text{ad }\zeta_i)^{m_r}(\zeta_{\beta_r})\zeta_i^{m_{r+1}}v_\lambda = 0 \tag{5.2.34}$$

for $i \in \overline{1,n}$ and $\zeta_{\beta_j} \in \mathcal{G}_{\beta_i}$ with $\beta_j \in \Phi^-$. Therefore, $\{\xi_i, \zeta_i \mid i \in \overline{1,n}\}$ are locally nilpotent on V'. So

$$\hat{\tau}_i = e^{\xi_i}e^{-\zeta_i}e^{\xi_i} \qquad \text{for } i \in \overline{1,n} \tag{5.2.35}$$

are linear automorphisms of V'.

Recall

$$V'_\mu = \{w \in V' \mid h(w) = \mu(h)w \text{ for } h \in H\} \qquad \text{for } \mu \in H^*. \tag{5.2.36}$$

Set

$$\mathbb{S}(\lambda) = \{\mu \in H^* \mid V'_\mu \neq 0\}, \tag{5.2.37}$$

the set of weights of V'. Then

$$\mu \lhd \lambda \qquad \text{for } \mu \in \mathbb{S}(\lambda) \tag{5.2.38}$$

because $V' = U(\mathcal{G}_-)v_\lambda$. Moreover, as operators on V',

$$\hat{\tau}_i h = \hat{\tau}_i h \hat{\tau}_i^{-1}\hat{\tau}_i = e^{\xi_i}e^{-\zeta_i}e^{\xi_i}he^{-\xi_i}e^{\zeta_i}e^{-\xi_i}\hat{\tau}_i = [e^{\text{ad }\xi_i}e^{-\text{ad }\zeta_i}e^{\text{ad }\xi_i}(h)]\hat{\tau}_i = \tau_{\alpha_i}(h)\hat{\tau}_i \tag{5.2.39}$$

by (1.4.31) (cf. (4.1.24) and (4.1.25)). So

$$\hat{\tau}_i(V'_\mu) = V'_{\sigma_{\alpha_i}(\mu)} \qquad \text{for } \mu \in H^* \tag{5.2.40}$$

by (5.2.39). For each $\sigma \in \mathcal{W}$, we pick an expression $\sigma = \sigma_{\alpha_{i_1}}\sigma_{\alpha_{i_2}}\ldots\sigma_{\alpha_{i_r}}$ and define

$$\hat{\tau}_\sigma = \hat{\tau}_{i_1}\hat{\tau}_{i_2}\ldots\hat{\tau}_{i_r}. \tag{5.2.41}$$

By (5.2.40), we get

$$\hat{\tau}_\sigma(V'_\mu) = V'_{\sigma(\mu)} \qquad \text{for } \mu \in \mathbb{S}(\lambda). \tag{5.2.42}$$

So $\sigma(\mathbb{S}(\lambda)) = \mathbb{S}(\lambda)$.

Recall that Λ^+ denotes the set of dominant integral weights. Set

$$\mathbb{S}^+(\lambda) = \mathbb{S}(\lambda) \bigcap \Lambda^+. \tag{5.2.43}$$

According to Lemma 3.4.1 and (5.2.38), $\mathbb{S}^+(\lambda)$ is a finite set. Moreover, any element in $\mathbb{S}(\lambda)$ is conjugated to an element in $\mathbb{S}^+(\lambda)$ under the Weyl group $\mathscr{W}$ by Lemma 3.2.13. Lemma 2.4.4 and (5.2.38) imply that $\mathbb{S}(\lambda)$ is a saturated set of weights. Thus

$$|\mathbb{S}(\lambda)| \leq |\mathscr{W}||\mathbb{S}^+(\lambda)| < \infty. \tag{5.2.44}$$

Given $0 \lhd \gamma = \sum_{i=1}^n k_i\alpha_i$, we define $\mathscr{P}(\gamma)$ to be the number of distinct sets of nonnegative integers $\{l_\alpha \mid \alpha \in \Phi^+\}$ such that $\sum_{\alpha\in\Phi^+} l_\alpha\alpha = \gamma$. Then

$$\dim M(\lambda)_\mu = \mathscr{P}(\lambda - \mu) \qquad \text{for } \lambda \succ \mu \in H^* \tag{5.2.45}$$

(exercise). Hence

$$\dim V'_\mu \leq \dim M(\lambda)_\mu = \mathscr{P}(\lambda - \mu) \qquad \text{for } \mu \in \Pi(\lambda). \tag{5.2.46}$$

Expressions (5.2.44) and (5.2.46) imply that V' is finite-dimensional. By Weyl's Theorem, V' is a direct sum of irreducible submodules, which are weight modules. Since

$$\dim V'_\lambda = \dim \mathbb{C}v_\lambda = 1, \tag{5.2.47}$$

v_λ must be in one of irreducible summands, say U. But

$$V' = U(\mathscr{G})v_\lambda \subset U. \tag{5.2.48}$$

Thus $V' = U$ is a finite-dimensional irreducible $\mathscr{G}$-module. In summary, we obtain:

Theorem 5.2.3 *The set $\{V(\lambda) \mid \lambda \in \Lambda^+\}$ are all finite-dimensional irreducible $\mathscr{G}$-modules. Moreover,*

$$N(\lambda) = \sum_{i=1}^n U(\mathscr{G}_-)v_{i,\lambda}. \tag{5.2.49}$$

The set $\mathbb{S}(\lambda)$ is a saturated set of weights.

Suppose that

$$\mathscr{G} = \mathscr{G}_1 \oplus \mathscr{G}_2 \oplus \cdots \oplus \mathscr{G}_s \tag{5.2.50}$$

is a direct sum of simple ideals. Take a CSA H_i of $\mathscr{G}_i$ with root system Φ_i for $i \in \overline{1,s}$. Then

$$H = H_1 \oplus H_2 \oplus \cdots \oplus H_s \tag{5.2.51}$$

is a CSA of $\mathscr{G}$ with root system

$$\Phi = \bigcup_{i=1}^{s} \Phi_i. \tag{5.2.52}$$

Pick a base Π_i for Φ_i for $i \in \overline{1,s}$. The set

$$\Pi = \bigcup_{i=1}^{s} \Pi_i \tag{5.2.53}$$

is a base for Φ. Moreover,

$$\Phi^{\pm} = \bigcup_{i=1}^{s} \Phi_i^{\pm}. \tag{5.2.54}$$

For $\lambda_i \in H_i^*$, denote

$$\lambda = \bigoplus_{i=1}^{s} \lambda_i \in H^*. \tag{5.2.55}$$

The highest-weight vector v_λ for $\mathscr{G}$ is equivalent to

$$v_\lambda = v_{\lambda_1} \otimes v_{\lambda_2} \otimes \cdots \otimes v_{\lambda_s}, \tag{5.2.56}$$

where v_{λ_i} are the highest weight vectors of $\mathscr{G}_i$. By PBW Theorem,

$$U(\mathscr{G}) \cong U(\mathscr{G}_1) \otimes_{\mathbb{C}} U(\mathscr{G}_2) \otimes_{\mathbb{C}} \cdots \otimes_{\mathbb{C}} U(\mathscr{G}_s). \tag{5.2.57}$$

In particular, the Verma module:

$$M(\lambda) \cong M(\lambda_1) \otimes_{\mathbb{C}} M(\lambda_2) \otimes_{\mathbb{C}} \cdots \otimes_{\mathbb{C}} M(\lambda_s) \tag{5.2.58}$$

with the action:

$$\left(\sum_{i=1}^{s} u_i\right)(w_1 \otimes w_2 \otimes \cdots \otimes w_s) = \sum_{i=1}^{s} w_1 \otimes \cdots \otimes w_{i-1} \otimes u_i(w_i) \otimes w_{i+1} \otimes \cdots \otimes w_s \tag{5.2.59}$$

for

$$u_i \in \mathscr{G}_i, \quad w_i \in M(\lambda_i) \qquad \text{for } i \in \overline{1,s}. \tag{5.2.60}$$

Moreover, it can be proved that the maximal proper submodule:

$$N(\lambda) \cong \sum_{i=1}^{k} M(\lambda_1) \otimes_{\mathbb{C}} \cdots M(\lambda_{i-1}) \otimes_{\mathbb{C}} N(\lambda_i) \otimes_{\mathbb{C}} M(\lambda_{i+1}) \otimes_{\mathbb{C}} \cdots \otimes_{\mathbb{C}} M(\lambda_s). \tag{5.2.61}$$

Thus the irreducible highest weight module:

$$V(\lambda) \cong V(\lambda_1) \otimes_{\mathbb{C}} V(\lambda_2) \otimes_{\mathbb{C}} \cdots \otimes_{\mathbb{C}} V(\lambda_s), \tag{5.2.62}$$

where $V(\lambda_i)$ is the irreducible highest weight $\mathscr{G}_i$-module with the highest weight λ_i.

Remark 5.2.4 Following Lepowsky [Lj], the Verma module can be generalized as follows. Let

$$\mathscr{G} = \mathscr{G}_- \oplus \mathscr{G}_0 \oplus \mathscr{G}_+ \tag{5.2.63}$$

be a Lie algebra that is a direct sum of its subalgebras $\mathscr{G}_\pm$ and $\mathscr{G}_0$ such that

$$[\mathscr{G}_-, \mathscr{G}_+] \subset \mathscr{G}_0, \qquad [\mathscr{G}_0, \mathscr{G}_\pm] \subset \mathscr{G}_\pm. \tag{5.2.64}$$

Denote

$$\mathscr{B} = \mathscr{G}_0 + \mathscr{G}_+. \tag{5.2.65}$$

For any $\mathscr{G}_0$-module V, we extend it to a $\mathscr{B}$-module by letting

$$\mathscr{G}_+(V) = \{0\}. \tag{5.2.66}$$

Then we have the induced $\mathscr{G}$-module

$$M(V) = U(\mathscr{G}) \otimes_{U(\mathscr{B})} V \cong U(\mathscr{G}_-) \otimes_{\mathbb{C}} V \ \text{ (as vector spaces)}, \tag{5.2.67}$$

which is called a *generalized Verma module*.

5.3 Formal Characters

In this section, we introduce the notion of formal character of a weight module of a finite-dimensional semisimple Lie algebra.

Let $\mathscr{G}$ be a finite-dimensional semisimple Lie algebra and let H be a CSA of $\mathscr{G}$ with root system Φ. Denote by $\mathscr{F}(H^*)$ the set of $\mathbb{C}$-valued functions on H^*, which becomes a vector space with the linear operation:

$$(af + bg)(\lambda) = af(\lambda) + bg(\lambda) \qquad \text{for } f, g \in \mathscr{F}(H^*),\ a, b \in \mathbb{C}. \tag{5.3.1}$$

Set

$$\text{supp}\, f = \{\lambda \in H^* \mid f(\lambda) \neq 0\} \qquad \text{for } f \in \mathscr{F}(H^*). \tag{5.3.2}$$

Given $\mu_1, ..., \mu_r \in H^*$, we denote

$$S(\mu_1, ..., \mu_r) = \bigcup_{i=1}^{r} \Big(\mu_i - \sum_{s=1}^{n} \mathbb{N}\alpha_s\Big). \tag{5.3.3}$$

Define

$$\mathscr{X}(H^*) = \{f \in \mathscr{F}(H^*) \mid \text{supp}\, f \subset S(\mu_1, ..., \mu_r) \text{ for some } \mu_1, ..., \mu_r \in H^*\}. \tag{5.3.4}$$

It can be verified that $\mathscr{X}(H^*)$ is a subspace of $\mathscr{F}(H^*)$ (exercise). Moreover, we define an algebraic operation $*$ (*convolution*) on $\mathscr{X}(H^*)$ by:

$$(f * g)(\lambda) = \sum_{\mu_1,\mu_2 \in H^*;\ \mu_1+\mu_2=\lambda} f(\mu_1)g(\mu_2) \quad \text{for } f, g \in \mathscr{X}(H^*),\ \lambda \in H^*. \tag{5.3.5}$$

The above equation makes sense due to (5.3.4). The convolution $*$ is commutative and associative. Given $\lambda \in H^*$, we define $\varepsilon_\lambda \in \mathscr{X}(H^*)$ by

$$\varepsilon_\lambda(\mu) = \begin{cases} 1 \text{ if } \mu = \lambda \\ 0 \text{ if } \mu \neq \lambda. \end{cases} \tag{5.3.6}$$

Then

$$\varepsilon_\lambda * \varepsilon_\mu = \varepsilon_{\lambda+\mu} \quad \text{for } \lambda, \mu \in H^*. \tag{5.3.7}$$

In fact, ε_0 is the identity element of $\mathscr{X}(H^*)$.

Define an action of the Weyl group $\mathscr{W}$ on $\mathscr{X}(H^*)$ by

$$\sigma(f)(\lambda) = f(\sigma^{-1}(\lambda)) \quad \text{for } f \in \mathscr{X}(H^*),\ \sigma \in \mathscr{W},\ \lambda \in H^*. \tag{5.3.8}$$

In particular,

$$\sigma(\varepsilon_\lambda)(\mu) = \varepsilon_\lambda(\sigma^{-1}(\mu)) = \begin{cases} 1 \text{ if } \sigma^{-1}(\mu) = \lambda \\ 0 \text{ if } \sigma^{-1}(\mu) \neq \lambda \end{cases}$$
$$= \begin{cases} 1 \text{ if } \mu = \sigma(\lambda) \\ 0 \text{ if } \mu \neq \sigma(\lambda) \end{cases} = \varepsilon_{\sigma(\lambda)}(\mu). \tag{5.3.9}$$

So $\sigma(\varepsilon_\lambda) = \varepsilon_{\sigma(\lambda)}$.

Let $V = \bigoplus_{\mu \in H^*} V_\mu$ be a weight $\mathscr{G}$-module such that $\dim V_\mu < \infty$ for $\mu \in H^*$ (V may not be finite-dimensional). Define the *formal character* of V by

$$\text{ch}_V = \sum_{\mu \in H^*} m_{_V}(\mu)\varepsilon_\mu, \qquad m_{_V}(\mu) = \dim V_\mu. \tag{5.3.10}$$

For any $\lambda \in H^*$, we have

$$\text{ch}_{M(\lambda)} \in \mathscr{X}(H^*), \tag{5.3.11}$$

where $M(\lambda)$ is the Verma module (cf. (5.2.9)). Furthermore, we define the *Kostant function*:

$$p = \mathrm{ch}_{M(0)} = \prod_{\alpha\in\Phi^+}\left(\sum_{i=0}^{\infty}\varepsilon_{-i\alpha}\right) \tag{5.3.12}$$

and the *Weyl function*:

$$q = \prod_{\alpha\in\Phi^+}\left(\varepsilon_{\alpha/2}-\varepsilon_{-\alpha/2}\right) = \varepsilon_\rho * \prod_{\alpha\in\Phi^+}\left(\varepsilon_0-\varepsilon_{-\alpha}\right). \tag{5.3.13}$$

Note that

$$\mathrm{ch}_{M(\lambda)} = \varepsilon_\lambda * p. \tag{5.3.14}$$

Moreover,

$$\left(\varepsilon_0-\varepsilon_{-\alpha}\right) * \sum_{i=0}^{\infty}\varepsilon_{-i\alpha} = \sum_{i=0}^{\infty}\varepsilon_{-i\alpha} - \sum_{i=0}^{\infty}\varepsilon_{-(i+1)\alpha} = \varepsilon_0. \tag{5.3.15}$$

Thus

$$q * p = \varepsilon_\rho * \prod_{\alpha\in\Phi^+}\left[\left(\varepsilon_0-\varepsilon_{-\alpha}\right) * \left(\sum_{i=0}^{\infty}\varepsilon_{-i\alpha}\right)\right] = \varepsilon_\rho. \tag{5.3.16}$$

By (5.3.14) and (5.3.16), we obtain:

Lemma 5.3.1 *For any $\lambda \in H^*$,*

$$q * \mathrm{ch}_{M(\lambda)} = \varepsilon_{\lambda+\rho}. \tag{5.3.17}$$

Recall that the weight lattice of $\mathscr{G}$ is defined by:

$$\Lambda = \{\lambda \in H^* \mid \langle\lambda,\alpha\rangle \in \mathbb{Z} \text{ for } \alpha \in \Phi\}. \tag{5.3.18}$$

Take a base Π of Φ. The set of dominant integral weights

$$\Lambda^+ = \{\lambda \in \Lambda \mid 0 \le \langle\lambda,\alpha\rangle \text{ for } \alpha \in \Pi\}. \tag{5.3.19}$$

If both U and V are finite-dimensional $\mathscr{G}$-modules, we have

$$(U\otimes V)_\lambda = \sum_{\mu_1,\mu_2\in\Lambda;\ \mu_1+\mu_2=\lambda} U_{\mu_1}\otimes V_{\mu_2} \qquad \text{for } \lambda\in\Lambda \tag{5.3.20}$$

by (2.2.22). Thus we have

$$\mathrm{ch}_{U\otimes V} = \mathrm{ch}_U \cdot \mathrm{ch}_V. \tag{5.3.21}$$

For any $\lambda \in \Lambda^+$, the space $V(\lambda)$ is the finite-dimensional irreducible module with highest weight λ. By (5.2.41) and (5.2.42), we have

$$m_{V(\lambda)}(\sigma(\mu)) = m_{V(\lambda)}(\mu) \qquad \text{for } \sigma \in \mathscr{W},\ \mu \in \Lambda, \tag{5.3.22}$$

which is equivalent to:

$$\sigma(\mathrm{ch}_{V(\lambda)}) = \mathrm{ch}_{V(\lambda)} \qquad \text{for } \sigma \in \mathscr{W}. \tag{5.3.23}$$

So $\mathrm{ch}_{V(\lambda)}$ is $\mathscr{W}$-invariant.

Remark 5.3.2 We can treat ε_μ as the following function on H:

$$\varepsilon_\mu(h) = e^{\mu(h)} \qquad \text{for } h \in H. \tag{5.3.24}$$

If V is a finite-dimensional $\mathscr{G}$-module, then

$$\mathrm{ch}_V(h) = \mathrm{tr}\, e^h \qquad \text{for } h \in H. \tag{5.3.25}$$

5.4 Weyl's Character Formula

In this section, we continue to study the characters of finite-dimensional irreducible modules of a finite-dimensional semisimple Lie algebra $\mathscr{G}$. In particular, we prove Weyl's character formula and use it to prove the tensor formula of two finite-dimensional $\mathscr{G}$-modules.

Pick an orthonormal basis $\{h_i \in \overline{1,n}\}$ of the Cartan subalgebra H of $\mathscr{G}$ with respect to its Killing form κ. For any $\alpha \in \Phi^+$, we take $\xi_\alpha \in \mathscr{G}_\alpha$ and $\zeta_\alpha \in \mathscr{G}_{-\alpha}$ such that $\kappa(\xi_\alpha, \zeta_\alpha) = 1$. Then

$$\omega = \sum_{i=1}^{n} h_i^2 + \sum_{\alpha\in\Phi^+} \left(\xi_\alpha\zeta_\alpha + \zeta_\alpha\xi_\alpha\right) \tag{5.4.1}$$

is a quadratic central element of the universal enveloping algebra $U(\mathscr{G})$; that is,

$$u\omega = \omega u \qquad \text{for } u \in U(\mathscr{G}) \tag{5.4.2}$$

(exercise). We call ω a *Casimir element*. Indeed, for any representation ϕ of $\mathscr{G}$, $\phi(\omega) = \omega_\phi$ the Casimir operator defined in (2.4.4).

Let V be a highest-weight $\mathscr{G}$-module with a highest-weight vector v of weight λ. According to Lemma 2.5.4, $\left[\xi_\alpha, \zeta_\alpha\right] = t_\alpha$ for $\alpha \in \Phi^+$. Thus

$$\omega(v) = \left[\sum_{i=1}^{n} h_i^2 + \sum_{\alpha\in\Phi^+} \xi_\alpha\zeta_\alpha\right](v) = \left[\sum_{i=1}^{n} \lambda(h_i)^2 + \sum_{\alpha\in\Phi^+} \lambda(t_\alpha)\right]v = [(\lambda, \lambda) + 2(\rho, \lambda)]v \tag{5.4.3}$$

(cf. (3.2.15)). Since $V = U(\mathscr{G})(v)$, (5.4.2) yields

$$\omega|_V = [(\lambda, \lambda) + 2(\rho, \lambda)]\mathrm{Id}_V. \tag{5.4.4}$$

Suppose that u is a singular vector of V with weight μ. Then we have

$$\omega(u) = [(\mu, \mu) + 2(\rho, \mu)]u. \tag{5.4.5}$$

Hence

$$(\mu, \mu) + 2(\rho, \mu) = (\lambda, \lambda) + 2(\rho, \lambda) \sim (\rho + \mu, \rho + \mu) = (\rho + \lambda, \rho + \lambda). \tag{5.4.6}$$

Recall the base $\Pi = \{\alpha_1, \alpha_2, ..., \alpha_n\}$. Given $\lambda \in H^*$, we denote

$$B_\lambda = \left\{\mu \in \lambda - \sum_{i=1}^{n} \mathbb{N}\alpha_i \mid \mu \neq \lambda,\ (\rho + \mu, \rho + \mu) = (\rho + \lambda, \rho + \lambda)\right\}, \tag{5.4.7}$$

where $\mathbb{N}$ is the additive semigroup of nonnegative integers. The nondegeneracy of the bilinear form $(\cdot, \cdot)$ (cf. Theorem 2.1.2 and (2.6.35)) implies that B_λ is a finite set. Moreover,

$$B_{\lambda'} \subset B_\lambda \qquad \text{for } \lambda' \in B_\lambda. \tag{5.4.8}$$

Let V be a highest weight $\mathscr{G}$-module with the highest weight λ. We define $m_{_V} = 0$ if $B_\lambda = \emptyset$ and otherwise,

$$m_{_V} = \sum_{\mu\in B_\lambda} \dim V_\mu. \tag{5.4.9}$$

Lemma 5.4.1 *A highest weight $\mathscr{G}$-module V with highest weight λ has a sequence of submodules:*

$$V_0 = \{0\} \subset V_1 \subset V_2 \subset \ldots \subset V_r = V \tag{5.4.10}$$

such that $V_r/V_{r-1} = V(\lambda)$ and

$$V_i/V_{i-1} \cong V(\mu_i) \quad \textit{with} \quad \mu_i \in B_\lambda \qquad \textit{for} \quad i \in \overline{1, r-1}. \tag{5.4.11}$$

Proof Suppose that V has a proper nonzero submodule U, which is also a weight module. Let $0 \neq v \in U$ be a weight vector with weight $\mu = \lambda - \sum_{i=1}^{n} m_i\alpha_i$, where $\Pi = \{\alpha_1, \alpha_2, ..., \alpha_n\}$. Denote

$$m = \sum_{i=1}^{n} m_i. \tag{5.4.12}$$

Take $0 \neq \xi_i \in \mathcal{G}_{\alpha_i}$ for $i \in \overline{1,n}$. Since $V_\lambda \not\subset U$,

$$\xi_{i_1}\xi_{i_2}\ldots\xi_{i_m}(v) = 0 \qquad \text{for } \ i_r \in \overline{1,n}. \tag{5.4.13}$$

If v is not a singular vector, then there exist $j_1, ..., j_s \in \overline{1,n}$ such that $\xi_{j_1}\xi_{j_2}\ldots\xi_{j_s}(v) \in U$ is a singular vector by (5.4.13). So U has a singular vector u with weight $\mu \in B_\lambda$.

If $m_{_V} = 0$, then V has a unique singular vector up to a scalar. Thus $V = V(\lambda)$ is irreducible and the lemma holds. Suppose that it holds for V with $m_{_V} < k$. Assume that $m_{_V} = k$ and V is not irreducible. Take a singular vector u of V with weight $\mu \in B_\lambda$. Set

$$W = U(\mathcal{G}_-)u. \tag{5.4.14}$$

Now both V/W and W are highest weight modules with $m_{_{V/W}}, m_{_W} < k$. Moreover, $\mu \in B_\lambda$. By assumption, W has a sequence of submodules

$$W_0 = \{0\} \subset W_1 \subset \cdots \subset W_{r_1} = W \tag{5.4.15}$$

such that $W_i/W_{i-1} \cong V(\mu_i)$ with $\mu_i \in B_\mu$ for $i \in \overline{1, r_1 - 1}$ and $W_{r_1}/W_{r_1-1} \cong V(\mu)$. Furthermore, V contains a sequence of submodules

$$V_0' = W \subset V_1' \subset \cdots \subset V_{r_2}' = V \tag{5.4.16}$$

such that

$$V_{r_2}'/V_{r_2-1}' \cong (V_{r_2}'/W)/(V_{r_2-1}'/W) \cong V(\lambda) \tag{5.4.17}$$

and

$$V_j'/V_{j-1}' \cong (V_j'/W)/(V_{j-1}'/W) \cong V(\mu_j') \tag{5.4.18}$$

with $\mu_j' \in B_\lambda$ for $j \in \overline{1, r_2}$. Now the sequence:

$$W_0 = \{0\} \subset W_1 \subset \cdots \subset W_{r_1} = W \subset V_1' \subset \cdots \subset V_{r_2}' = V \tag{5.4.19}$$

is the required. □

By the above lemma,

$$\mathrm{ch}_V = \sum_{i=1}^{r} \mathrm{ch}_{V_i/V_{i-1}} = \mathrm{ch}_{V(\lambda)} + \sum_{i=1}^{r-1} \mathrm{ch}_{V(\mu_i)} \tag{5.4.20}$$

with $\mu_i \in B_\lambda$. In particular,

$$\mathrm{ch}_{M(\lambda)} = \mathrm{ch}_{V(\lambda)} + \sum_{\mu \in B_\lambda} c(\lambda, \mu)\mathrm{ch}_{V(\mu)}, \qquad c(\lambda, \mu) \in \mathbb{N}, \tag{5.4.21}$$

for any $\lambda \in H^*$.

Fix $\lambda \in H^*$. If $\mu \in B_\lambda$ is minimal with respect to "$\lhd$" in (3.4.49), then $B_\mu = \emptyset$. So $M(\mu) = V(\mu)$, which implies $\mathrm{ch}_{M(\mu)} = \mathrm{ch}_{V(\mu)}$. If μ is not minimal, we have

$$\mathrm{ch}_{V(\mu)} = \mathrm{ch}_{M(\mu)} - \sum_{\mu' \in B_\mu} c(\mu, \mu')\mathrm{ch}_{V(\mu')}. \tag{5.4.22}$$

By induction,

$$\mathrm{ch}_{V(\mu)} = \mathrm{ch}_{M(\mu)} + \sum_{\mu' \in B_\mu} d(\mu, \mu')\mathrm{ch}_{M(\mu')}, \qquad d(\mu, \mu') \in \mathbb{Z}. \tag{5.4.23}$$

In particular,

$$\mathrm{ch}_{V(\lambda)} = \mathrm{ch}_{M(\lambda)} + \sum_{\mu \in B_\lambda} a(\mu)\mathrm{ch}_{M(\mu)}, \qquad a(\mu) \in \mathbb{Z}. \tag{5.4.24}$$

Thus

$$q * \mathrm{ch}_{V(\lambda)} = \varepsilon_{\lambda+\rho} + \sum_{\mu \in B_\lambda} a(\mu)\varepsilon_{\mu+\rho} \tag{5.4.25}$$

by (5.3.17).

Note that Lemma 3.2.10 implies

$$\sigma(q) = (-1)^{\ell(\sigma)}q \qquad \text{for } \sigma \in \mathscr{W}. \tag{5.4.26}$$

In fact,

$$\mathrm{sn}(\sigma) = \text{the determinant of } \sigma = (-1)^{\ell(\sigma)}. \tag{5.4.27}$$

So

$$\sigma(q) = \mathrm{sn}(\sigma)q \qquad \text{for } \sigma \in \mathscr{W}. \tag{5.4.28}$$

Observe that $\mathscr{W}$ is an automorphism subgroup of the algebra $(\mathscr{X}(H^*), *)$. Suppose $\lambda \in \Lambda^+$. Then $\mathrm{ch}_{V(\lambda)}$ is $\mathscr{W}$-invariant. Applying $\sigma \in \mathscr{W}$ to (5.4.25), we obtain

$$\begin{aligned}
&\operatorname{sn}(\sigma)\left(\varepsilon_{\lambda+\rho}+\sum_{\mu\in B_\lambda}a(\mu)\varepsilon_{\mu+\rho}\right)\\
&=\sigma(q)*\sigma(\mathrm{ch}_{V(\lambda)})=\sigma(q*\mathrm{ch}_{V(\lambda)})\\
&=\varepsilon_{\sigma(\lambda+\rho)}+\sum_{\mu\in B_\lambda}a(\mu)\varepsilon_{\sigma(\mu+\rho)}. \qquad (5.4.29)
\end{aligned}$$

On the other hand, if $\mu\in B_\lambda$ and $\rho+\mu\in\Lambda^+$, then

$$(\rho+\lambda,\rho+\lambda)-(\rho+\mu,\rho+\mu)=(\rho,\lambda-\mu)+(\lambda,\lambda-\mu)+(\rho+\mu,\lambda-\mu)>0, \quad (5.4.30)$$

which contradicts (5.4.7). Thus

$$\{\rho+\mu\mid\mu\in B_\lambda\}\bigcap\Lambda^+=\emptyset. \qquad (5.4.31)$$

For any $\mu'\in B_\lambda$, there exists $\sigma\in\mathscr{W}$ such that $\sigma(\rho+\mu')\in\Lambda^+$ by Lemma 3.2.13. Then (5.4.29) and (5.4.31) imply

$$\sigma(\rho+\mu')=\rho+\lambda \ \text{ and } \ a(\mu')=\operatorname{sn}(\sigma). \qquad (5.4.32)$$

Since $\lambda+\rho$ is strongly dominant (regular), $\{\sigma(\lambda+\rho)\mid\sigma\in\mathscr{W}\}$ are distinct by Theorem 3.2.9 (e). Observe that $\operatorname{sn}(\sigma)=\operatorname{sn}(\sigma^{-1})$. Therefore, we have:

Theorem 5.4.2 (Weyl) *For* $\lambda\in\Lambda^+$,

$$q*\mathrm{ch}_{V(\lambda)}=\sum_{\sigma\in\mathscr{W}}\operatorname{sn}(\sigma)\varepsilon_{\sigma(\lambda+\rho)}. \qquad (5.4.33)$$

Taking $\lambda=0$ *in the above equation, we get the Weyl denominator identity:*

$$\prod_{\alpha\in\Phi^+}\left(\varepsilon_{\alpha/2}-\varepsilon_{-\alpha/2}\right)=\sum_{\sigma\in\mathscr{W}}sn(\sigma)\varepsilon_{\sigma(\rho)}. \qquad (5.4.34)$$

Example 5.4.1 Let $n>1$ be a integer and let $\mathscr{E}$ be the Euclidean space with a basis $\{\varepsilon_1,\varepsilon_2,...,\varepsilon_n\}$ such that $(\varepsilon_i,\varepsilon_j)=\delta_{i,j}$. The root system of $sl(n,\mathbb{C})$ and a base are as follows:

$$\Phi=\{\varepsilon_i-\varepsilon_j\mid i,j\in\overline{1,n},\ i\neq j\},\quad \Pi=\{\varepsilon_i-\varepsilon_{i+1}\mid i\in\overline{1,n-1}\}. \qquad (5.4.35)$$

Denote

$$\Lambda_1^+=\left\{\sum_{i=1}^{n-1}m_i\varepsilon_i\mid m_i\in\mathbb{N},\ m_1\geq m_2\geq\ldots\geq m_{n-1}\right\}. \qquad (5.4.36)$$

Define

$$|\mu| = \sum_{i=1}^{n-1} m_i \qquad \text{for } \mu = \sum_{i=1}^{n-1} m_i \varepsilon_i \in \Lambda_1^+ \tag{5.4.37}$$

and

$$\mathbf{1} = \sum_{i=1}^{n} \varepsilon_i. \tag{5.4.38}$$

Then

$$\Lambda^+ = \left\{ \mu - \frac{|\mu|}{n}\mathbf{1} \mid \mu \in \Lambda_1^+ \right\} \tag{5.4.39}$$

(exercise, cf. (3.4.11)–(3.4.14)). In particular,

$$\rho = \frac{1}{2} \sum_{1 \le i \le j \le n} \left(\varepsilon_i - \varepsilon_j\right) = \frac{1}{2} \sum_{i=1}^{n} \left(n + 1 - 2i\right)\varepsilon_i = \sum_{i=1}^{n-1} \left(n - i\right)\varepsilon_i - \frac{n-1}{2}\mathbf{1}. \tag{5.4.40}$$

As we showed in Example 3.2.1, the Weyl group of $sl(n, \mathbb{C})$ isomorphic to the permutation group S_n on $\{1, 2, ..., n\}$. Note

$$\sigma(\varepsilon_i) = \varepsilon_{\sigma(i)} \qquad \text{for } i \in \overline{1, n},\ \sigma \in S_n. \tag{5.4.41}$$

Denote

$$x_i = \varepsilon_{\varepsilon_i} \qquad \text{for } i \in \overline{1, n}. \tag{5.4.42}$$

Moreover, we write

$$x^{\mu} = x_1^{a_1} x_2^{a_2} \dots x_n^{a_n} \qquad \text{for } \mu = \sum_{i=1}^{n} a_i \varepsilon_i \in H^*. \tag{5.4.43}$$

So we can treat

$$\varepsilon_{\mu} = x^{\mu}. \tag{5.4.44}$$

Thus

$$\sigma(x_1^{a_1} x_2^{a_2} \dots x_n^{a_n}) = x_{\sigma(1)}^{a_1} x_{\sigma(2)}^{a_2} \cdots x_{\sigma(n)}^{a_n} \qquad \text{for } \sigma \in S_n,\ a_i \in \mathbb{C}. \tag{5.4.45}$$

In particular,

$$\sigma(x^{a\mathbf{1}}) = \sigma(x_1^a x_2^a \dots x_n^a) = x^{a\mathbf{1}} \qquad \text{for } \sigma \in S_n. \tag{5.4.46}$$

Now

$$q = \prod_{1\le i<j\le n}\left(x_i^{1/2}x_j^{-1/2} - x_i^{-1/2}x_j^{1/2}\right) = \left[\prod_{1\le i<j\le n}\left(x_ix_j\right)\right]^{-1/2}\left[\prod_{1\le i<j\le n}\left(x_i - x_j\right)\right]$$
$$= x^{-(n-1)\mathbf{1}/2}\prod_{1\le i<j\le n}\left(x_i - x_j\right) \qquad (5.4.47)$$

and

$$\sum_{\sigma\in\mathscr{W}} \mathrm{sn}(\sigma)\varepsilon_{\sigma(\rho)} = \sum_{\sigma\in S_n} \mathrm{sn}(\sigma)\sigma(x^{-(n-1)\mathbf{1}/2}x_1^{n-1}x_2^{n-2}\dots x_{n-1})$$
$$= x^{-(n-1)\mathbf{1}/2}\sum_{\sigma\in S_n}\mathrm{sn}(\sigma)x_{\sigma(1)}^{n-1}x_{\sigma(2)}^{n-2}\dots x_{\sigma(n-1)}$$
$$= x^{-(n-1)\mathbf{1}/2}\begin{vmatrix} x_1^{n-1} & x_2^{n-1} & \dots & x_n^{n-1} \\ x_1^{n-2} & x_2^{n-2} & \dots & x_n^{n-2} \\ \vdots & \vdots & \vdots & \vdots \\ 1 & 1 & \dots & 1 \end{vmatrix} \qquad (5.4.48)$$

by (5.4.40). So the Weyl denominator identity (5.4.34) is equivalent to the identity of Vandermonde determinant:

$$\prod_{1\le i<j\le n}\left(x_i - x_j\right) = \begin{vmatrix} x_1^{n-1} & x_2^{n-1} & \dots & x_n^{n-1} \\ x_1^{n-2} & x_2^{n-2} & \dots & x_n^{n-2} \\ \vdots & \vdots & \vdots & \vdots \\ 1 & 1 & \dots & 1 \end{vmatrix}. \qquad (5.4.49)$$

For $\mu = \sum_{i=1}^{n-1} m_i\varepsilon_i \in \Lambda_1^+$, the Schur's symmetric polynomial

$$s(\mu) = \frac{\sum_{\sigma\in S_n}\mathrm{sn}(\sigma)x_{\sigma(1)}^{m_1+n-1}x_{\sigma(2)}^{m_2+n-2}\dots x_{\sigma(n-1)}^{m_{n-1}}}{\prod_{1\le i<j\le n}(x_i - x_j)}$$
$$= \frac{1}{\prod_{1\le i<j\le n}(x_i - x_j)}\begin{vmatrix} x_1^{m_1+n-1} & x_2^{m_1+n-1} & \dots & x_n^{m_1+n-1} \\ x_1^{m_2+n-2} & x_2^{m_2+n-2} & \dots & x_n^{m_2+n-2} \\ \vdots & \vdots & \vdots & \vdots \\ x_1^{m_{n-1}} & x_2^{m_{n-1}} & \dots & x_n^{m_{n-1}} \end{vmatrix}. \qquad (5.4.50)$$

Let $\lambda = \mu - |\mu|\mathbf{1}/n \in \Lambda^+$ with $\mu \in \Lambda_1^+$. By (5.4.33), (5.4.40), (5.4.46), (5.4.47) and (5.4.50),

$$\mathrm{ch}_{V(\lambda)} = x^{-|\mu|\mathbf{1}/n}s(\mu). \qquad (5.4.51)$$

Let $\lambda', \lambda'' \in \Lambda^+$. Since $V(\lambda') \otimes V(\lambda'')$ is completely reducible,

$$\mathrm{ch}_{V(\lambda')} * \mathrm{ch}_{V(\lambda'')} = \sum_{\lambda \in \Lambda^+} n(\lambda)\mathrm{ch}_{V(\lambda)}, \qquad n(\lambda) \in \mathbb{N}. \tag{5.4.52}$$

Our goal is to find a formula for $n(\lambda)$. Multiplying q to both sides of the above equation, we get

$$\mathrm{ch}_{V(\lambda')} * \Big(\sum_{\sigma \in \mathscr{W}} \mathrm{sn}(\sigma)\varepsilon_{\sigma(\lambda'+\rho)}\Big) = \sum_{\lambda \in \Lambda^+} \sum_{\sigma \in \mathscr{W}} n(\lambda)\mathrm{sn}(\sigma)\varepsilon_{\sigma(\lambda+\rho)} \tag{5.4.53}$$

by (5.4.33).

For any $\mu \in \Lambda$, we define

$$\mathscr{W}(\mu) \bigcap \Lambda^+ = \{\bar{\mu}\} \tag{5.4.54}$$

by Lemma 3.2.12 and

$$\mathscr{W}_\mu = \{\sigma \in \mathscr{W} \mid \sigma(\mu) = \mu\}. \tag{5.4.55}$$

If $\mathscr{W}_\mu \neq \{1\}$, then μ is not regular by Theorem 3.2.9 (e). Thus there exists $\alpha \in \Phi$ such that $(\mu, \alpha) = 0$. Hence $\sigma_\alpha \in \mathscr{W}_\mu$. Note $\mathrm{sn}(\sigma_\alpha) = -1$, $\mathscr{W}_\mu = \mathscr{W}_\mu\sigma_\alpha$ and so

$$\sum_{\tau \in \mathscr{W}_\mu} \mathrm{sn}(\tau) = \sum_{\tau \in \mathscr{W}_\mu} \mathrm{sn}(\tau\sigma_\alpha) = \sum_{\tau \in \mathscr{W}_\mu} \mathrm{sn}(\tau)\mathrm{sn}(\sigma_\alpha) = -\sum_{\tau \in \mathscr{W}_\mu} \mathrm{sn}(\tau). \tag{5.4.56}$$

Thus

$$\sum_{\tau \in \mathscr{W}_\mu} \mathrm{sn}(\tau) = 0. \tag{5.4.57}$$

Write

$$\mathscr{W} = \bigcup_{i=1}^{r} \tau_i \mathscr{W}_\mu, \qquad \tau_i\tau_j^{-1} \notin \mathscr{W}_\mu \ \text{ if } i \neq j. \tag{5.4.58}$$

Then

$$\sum_{\sigma \in \mathscr{W}} \mathrm{sn}(\sigma)\varepsilon_{\sigma(\mu)} = \sum_{i=1}^{r} \sum_{\tau \in \mathscr{W}_\mu} \mathrm{sn}(\tau_i\tau)\varepsilon_{\tau_i(\mu)} = \sum_{i=1}^{r} \Big(\sum_{\tau \in \mathscr{W}_\mu} \mathrm{sn}(\tau)\Big)\mathrm{sn}(\tau_i)\varepsilon_{\tau_i(\mu)} = 0. \tag{5.4.59}$$

Define

$$\psi(\mu) = \begin{cases} 0 & \text{if } \mathscr{W}_\mu \neq \{1\} \\ \mathrm{sn}(\sigma) & \text{if } \mathscr{W}_\mu = \{1\} \text{ and } \sigma(\mu) = \bar{\mu} \end{cases} \qquad \text{for } \mu \in \Lambda. \tag{5.4.60}$$

Since

$$m_{V(\lambda')}(\sigma(\mu)) = m_{V(\lambda')}(\mu) \qquad \text{for } \sigma \in \mathscr{W},\ \mu \in \mathbb{S}(\lambda'), \tag{5.4.61}$$

we have

$$\mathrm{ch}_{V(\lambda')} = \sum_{\mu\in\mathbb{S}(\lambda')} m_{V(\lambda')}(\mu)\varepsilon_{\sigma(\mu)} \qquad \text{for } \sigma \in \mathscr{W}. \tag{5.4.62}$$

So

$$\mathrm{ch}_{V(\lambda')} * \left(\sum_{\sigma\in\mathscr{W}} \mathrm{sn}(\sigma)\varepsilon_{\sigma(\lambda''+\rho)}\right) = \sum_{\mu\in\mathbb{S}(\lambda')}\sum_{\sigma\in\mathscr{W}} m_{V(\lambda')}(\mu)\mathrm{sn}(\sigma)\varepsilon_{\sigma(\lambda''+\rho+\mu)}. \tag{5.4.63}$$

Note that $\mathscr{W}_{\lambda'+\rho+\mu} = \{1\}$ if and only if $\overline{\lambda''+\rho+\mu} \in \Lambda^+ + \rho$. Comparing the coefficients of $\varepsilon_{\lambda+\rho}$ in (5.4.53) and using (5.4.59), we get:

Theorem 5.4.3 *For* $\lambda', \lambda'' \in \Lambda^+$,

$$ch_{V(\lambda')} * ch_{V(\lambda'')} = \sum_{\mu\in\mathbb{S}(\lambda'),\ \overline{\lambda''+\rho+\mu}\in\Lambda^++\rho} m_{V(\lambda')}(\mu)\psi(\lambda''+\rho+\mu)ch_{V(\overline{\lambda''+\rho+\mu}-\rho)}. \tag{5.4.64}$$

Exercise Calculate (5.4.64) for $sl(n,\mathbb{C})$ with $\lambda' = \lambda_1$ and $\lambda'' = \lambda_2$.

5.5 Dimensional Formula

In this section, we find the dimensional formula of a finite-dimensional irreducible module of the finite-dimensional semisimple Lie algebra $\mathscr{G}$.

For $\alpha \in \Phi^+$, we define $\partial_\alpha \in \mathrm{Der}\,\mathscr{X}(H^*)$ by:

$$\partial_\alpha(\varepsilon_\lambda) = (\lambda, \alpha)\varepsilon_\lambda \qquad \text{for } \lambda \in H^*, \tag{5.5.1}$$

where $(\alpha, \lambda) = \kappa(t_\alpha, t_\lambda)$. Set

$$\partial = \prod_{\alpha\in\Phi^+} \partial_\alpha. \tag{5.5.2}$$

Note that

$$\begin{aligned}\partial\Big(\sum_{\sigma\in\mathscr{W}}\mathrm{sn}(\sigma)\varepsilon_{\sigma(\lambda+\rho)}\Big) &= \sum_{\sigma\in\mathscr{W}}\mathrm{sn}(\sigma)\prod_{\alpha\in\Phi^+}\Big(\sigma(\lambda+\rho),\alpha\Big)\varepsilon_{\sigma(\lambda+\rho)}\\ &= \sum_{\sigma\in\mathscr{W}}\mathrm{sn}(\sigma)\prod_{\alpha\in\Phi^+}\Big(\lambda+\rho,\sigma^{-1}(\alpha)\Big)\varepsilon_{\sigma(\lambda+\rho)}\\ &= \sum_{\sigma\in\mathscr{W}}\prod_{\alpha\in\Phi^+}\Big(\lambda+\rho,\alpha\Big)\varepsilon_{\sigma(\lambda+\rho)}\\ &= \prod_{\alpha\in\Phi^+}\Big(\lambda+\rho,\alpha\Big)\sum_{\sigma\in\mathscr{W}}\varepsilon_{\sigma(\lambda+\rho)} \end{aligned} \tag{5.5.3}$$

by Lemma 3.2.10.

If $f=\sum_{\mu\in H^*}a_\mu\varepsilon_\mu\in\mathscr{X}(H^*)$ has only finite number of $a_\mu\neq 0$, we define

$$\jmath(f)=\sum_{\mu\in H^*}a_\mu. \tag{5.5.4}$$

If $g\in\mathscr{X}(H^*)$ has the same property as f, then

$$\jmath(f*g)=\jmath(f)\jmath(g). \tag{5.5.5}$$

In particular,

$$\jmath\Big[\partial\Big(\sum_{\sigma\in\mathscr{W}}\mathrm{sn}(\sigma)\varepsilon_{\sigma(\lambda+\rho)}\Big)\Big]=|\mathscr{W}|\prod_{\alpha\in\Phi^+}\Big(\lambda+\rho,\alpha\Big) \tag{5.5.6}$$

by (5.5.3). Taking $\lambda=0$, we have

$$\jmath(\partial(q))=|\mathscr{W}|\prod_{\alpha\in\Phi^+}(\rho,\alpha) \tag{5.5.7}$$

by (5.4.33).

On the other hand, $q=\prod_{\beta\in\Phi^+}(\varepsilon_{\beta/2}-\varepsilon_{-\beta/2})$. If S is a proper subset of Φ^+, then $(\prod_{\alpha\in S}\partial_\alpha)(q)$ is a sum of the terms with a factor $(\varepsilon_{\beta/2}-\varepsilon_{-\beta/2})$ for some $\beta\in\Phi$. Since $\jmath(\varepsilon_{\beta/2}-\varepsilon_{-\beta/2})=0$, we have

$$\jmath\Big(\Big(\prod_{\alpha\in S}\partial_\alpha\Big)(q)\Big)=0 \tag{5.5.8}$$

by (5.5.5). This shows

$$\jmath(\partial(q*\mathrm{ch}_{V(\lambda)}))=\jmath(\partial(q))\jmath(\mathrm{ch}_{V(\lambda)})=|\mathscr{W}|\Big[\prod_{\alpha\in\Phi^+}(\rho,\alpha)\Big]\dim V(\lambda). \tag{5.5.9}$$

Expressions (5.4.33), (5.5.6) and (5.5.9) yield:

Theorem 5.5.1 *For* $\lambda \in \Lambda^+$,

$$dim\ V(\lambda) = \frac{\prod_{\alpha \in \Phi^+}(\lambda + \rho, \alpha)}{\prod_{\alpha \in \Phi^+}(\rho, \alpha)} = \frac{\prod_{\alpha \in \Phi^+}\langle \lambda + \rho, \alpha \rangle}{\prod_{\alpha \in \Phi^+}\langle \rho, \alpha \rangle}. \tag{5.5.10}$$

Example 5.5.1 As in Example 3.4.1, the simple roots of $sl(n, \mathbb{C})$ are:

$$\alpha_i = \varepsilon_i - \varepsilon_{i+1} \qquad \text{for } i \in \overline{1, n-1}. \tag{5.5.11}$$

The positive roots:

$$\varepsilon_i - \varepsilon_j = \sum_{r=i}^{j-1} \alpha_r \qquad \text{for } 1 \leq i < j \leq n. \tag{5.5.12}$$

For $\lambda = \sum_{i=1}^{n-1} m_i \lambda_i \in \Lambda^+$, we have

$$\langle \lambda + \rho, \varepsilon_i - \varepsilon_j \rangle = \sum_{r=i}^{j-1} m_r + j - i. \tag{5.5.13}$$

Thus

$$\dim V(\lambda) = \prod_{1 \leq i < j \leq n} \frac{\sum_{r=i}^{j-1} m_r + j - i}{j - i}. \tag{5.5.14}$$

Reference

[Lj] J. Lepowsky, Generalized Verma modules, the Cartan-Helgason theorem, and the Harish-Chandra homomorphism. J. Algebra **49**(2), 470–495 (1977)

Part II
Explicit Representations

Chapter 6
Representations of Special Linear Algebras

Harmonic polynomials are important objects in analysis, differential geometry and physics. A fundamental theorem in classical harmonic analysis says that the spaces of homogeneous harmonic polynomials (solutions of Laplace equation) are irreducible modules of the corresponding orthogonal Lie group (algebra) and the whole polynomial algebra is a free module over the invariant polynomials generated by harmonic polynomials. Bases of these irreducible modules can be obtained easily (e.g., cf. [X16]). The algebraic beauty of the above theorem is that Laplace equation characterizes the irreducible submodules of the polynomial algebra and the corresponding quadratic invariant gives a decomposition of the polynomial algebra into a direct sum of irreducible submodules. This actually forms an $(sl(2, \mathbb{F}), o(n, \mathbb{F}))$ Howe duality.

Lie algebras (Lie groups) serve as the symmetries in quantum physics (e.g., cf. [FC, LF, G, L]). Their various representations provide distinct concrete practical physical models. Many important physical and biological phenomena have been interpreted as the consequences of symmetry breaking (e.g., cf. [HH, LF]). Harmonic oscillators are basic objects in quantum mechanics (e.g., cf. [FC, G]). Oscillator representations of finite-dimensional simple Lie algebras are the most fundamental ones in quantum physics. In [LX1], Luo and the author established certain two-parameter oscillator variations of the classical theorem on harmonic polynomials, associated with noncanonical oscillator representations of special linear Lie algebras, where the Laplace equation deforms to a variable-coefficient invariant partial differential equation and the quadratic invariant changes to an invariant differential operator.

It is well-known that a transformation of mapping straight lines to lines must be a projective transformation. The group of projective transformations is the fundamental symmetry of n-dimensional projective geometry. Physically, the group with $n = 4$ and $\mathbb{F} = \mathbb{R}$ consists of all the transformations of keeping free particles including light signals moving with constant velocities along straight lines (e.g., cf. [GHW1, GHW2]). Based on the embeddings of the poincaré group and De Sitter group into the projective group with $n = 4$ and $\mathbb{F} = \mathbb{R}$, Guo, Huang and Wu [GHW1, GHW2] proposed three kinds of special relativity.

X. Xu, *Representations of Lie Algebras and Partial Differential Equations*,
DOI 10.1007/978-981-10-6391-6_6

It is well known that n-dimensional projective group gives rise to an inhomogeneous first-order differential operator representation of $sl(n+1, \mathbb{F})$. Using Shen's technique of mixed product (cf. [Sg]) (also known as Larsson functor (cf. [La])), Zhao and the author [ZX] generalized the above representation of $sl(n+1, \mathbb{F})$ to an inhomogeneous representation on the tensor space of any finite-dimensional irreducible $gl(n, \mathbb{F})$-module with the polynomial space. It turned out that a hidden central transformation is involved. More importantly, we found a condition on the constant value taken by the central transformation such that the generalized representation is irreducible. In our approach, Pieri's formulas and the idea of Kostant's characteristic identities played key roles (cf. [Kb]).

In this chapter, we give various explicit representations of special linear Lie algebras. In Sect. 6.1, we present a fundamental lemma of solving flag partial differential equations for polynomial solutions, which was due to our work [X16]. In Sect. 6.2, we talk about the canonical bosonic oscillator representations and fermionic oscillator representations of special linear Lie algebras over their minimal natural modules and minimal orthogonal modules. Weyl's Theorem of complete reducibility in Sect. 2.4 plays an important role. In Sect. 6.3, we give a general set-up of noncanonical oscillator representations of special linear Lie algebras and Howe's result on the representations obtained from the canonical bosonic oscillator representations over minimal modules by partially swapping differential operators and multiplication operators (cf. [Hr4]). Sections 6.4, 6.5 and 6.6 are devoted to determining the structure of the noncanonical oscillator representations obtained from the canonical bosonic oscillator representations over minimal orthogonal modules by partially swapping differential operators and multiplication operators, which are generalizations of the classical theorem on harmonic polynomials. The results in Sects. 6.4, 6.5 and 6.6 were due to Luo and the author [LX1].

In Sect. 6.7, we construct a functor from the category of A_{n-1}-modules to the category of A_n-modules, which is related to n-dimensional projective transformations. Its irreducibility is proved by Theorem 5.4.3 of decomposing the tensor modules. This work was due to Zhao and the author [ZX]. As a consequence of Sect. 6.7, we obtain a one-parameter family of inhomogeneous first-order differential operator representations of special linear Lie algebras. By partially swapping differential operators and multiplication operators, we obtain in Sect. 6.8 more general differential operator representations. Letting these differential operators act on the corresponding polynomial algebra and the space of exponential-polynomial functions, we construct multi-parameter families of explicit infinite-dimensional irreducible representations. These results are taken from our work [X24].

6.1 Fundamental Lemma on Polynomial Solutions

In this section, we present a fundamental lemma of solving for polynomial solutions of linear partial differential equations. In terms of representations, it will be useful in obtaining a basis for irreducible oscillator representations.

Lemma 6.1.1 *Let $\mathscr{B}$ be a commutative associative algebra and let $\mathscr{A}$ be a free $\mathscr{B}$-module generated by a filtrated subspace $V = \bigcup_{r=0}^{\infty} V_r$ (i.e., $V_r \subset V_{r+1}$). Let T_1 be a linear operator on $\mathscr{B} \oplus \mathscr{A}$ with a right inverse T_1^- such that*

$$T_1(\mathscr{B}),\ T_1^-(\mathscr{B}) \subset \mathscr{B}, \qquad T_1(\zeta_1\zeta_2) = T_1(\zeta_1)\zeta_2, \qquad T_1^-(\zeta_1\zeta_2) = T_1^-(\zeta_1)\zeta_2 \quad (6.1.1)$$

for $\zeta_1 \in \mathscr{B}$, $\zeta_2 \in V$, and let T_2 be a linear operator on $\mathscr{B} \oplus \mathscr{A}$ such that $T_2(V_0) = T_2(\mathscr{B}) = \{0\}$,

$$T_2(V_{r+1}) \subset \mathscr{B}V_r, \quad T_2(f\zeta) = fT_2(\zeta) \quad \textit{for } r \in \mathbb{N},\ f \in \mathscr{B},\ \zeta \in \mathscr{A}. \quad (6.1.2)$$

Then we have

$$\begin{aligned}&\{f \in \mathscr{A} \mid (T_1 + T_2)(f) = 0\}\\ &= Span\{\sum_{i=0}^{\infty}(-T_1^- T_2)^i(hg) \mid g \in V,\ h \in \mathscr{B};\ T_1(h) = 0\}, \qquad (6.1.3)\end{aligned}$$

where the summation is finite due to (6.1.2). Moreover, the operator $\sum_{i=0}^{\infty} (-T_1^- T_2)^i T_1^-$ is a right inverse of $T_1 + T_2$ on $\mathscr{B} \oplus \mathscr{A}$.

Proof For $h \in \mathscr{B}$ such that $T_1(h) = 0$ and $g \in V$, we have

$$\begin{aligned}&(T_1 + T_2)(\sum_{i=0}^{\infty}(-T_1^- T_2)^i(hg))\\ &= T_1(hg) - \sum_{i=1}^{\infty} T_1[T_1^- T_2(-T_1^- T_2)^{i-1}(hg)] + \sum_{i=0}^{\infty} T_2[(-T_1^-)^i(hg)]\\ &= T_1(h)g - \sum_{i=1}^{\infty}(T_1T_1^-)T_2(-T_1^- T_2)^{i-1}(hg) + \sum_{i=0}^{\infty} T_2(-T_1^- T_2)^i(hg)\\ &= -\sum_{i=1}^{\infty} T_2(-T_1^- T_2)^{i-1}(hg) + \sum_{i=0}^{\infty} T_2(-T_1^- T_2)^i(hg) = 0 \qquad (6.1.4)\end{aligned}$$

by (6.1.1). Set $V_{-1} = \{0\}$. For $k \in \mathbb{N}$, we take $\{\psi_i \mid i \in I_k\} \subset V_k$ such that

$$\{\psi_i + V_{k-1} \mid i \in I_k\} \text{ forms a basis of } V_k/V_{k-1}, \qquad (6.1.5)$$

where I_k is an index set. Let

$$\mathscr{A}^{(m)} = \mathscr{B}V_m = \sum_{s=0}^{m}\sum_{i \in I_s} \mathscr{B}\psi_i. \qquad (6.1.6)$$

Obviously,

$$T_1(\mathscr{A}^{(m)}),\ T_1^-(\mathscr{A}^{(m)}),\ T_2(\mathscr{A}^{(m+1)}) \subset \mathscr{A}^{(m)} \qquad \text{for } m \in \mathbb{N} \tag{6.1.7}$$

by (6.1.1) and (6.1.2), and

$$\mathscr{A} = \bigcup_{m=0}^{\infty} \mathscr{A}^{(m)}. \tag{6.1.8}$$

Suppose $\phi \in \mathscr{A}^{(m)}$ such that $(T_1 + T_2)(\phi) = 0$. If $m = 0$, then

$$\phi = \sum_{i \in I_0} h_i \psi_i, \qquad h_i \in \mathscr{B}. \tag{6.1.9}$$

Now

$$0 = (T_1 + T_2)(\phi) = \sum_{i \in I_0} T_1(h_i)\psi_i + \sum_{i \in I_0} h_i T_2(\psi_i) = \sum_{i \in I_0} T_1(h_i)\psi_i. \tag{6.1.10}$$

Since $T_1(h_i) \in \mathscr{B}$ by (6.1.1) and $\mathscr{A}$ is a free $\mathscr{B}$-module generated by V, we have $T_1(h_i) = 0$ for $i \in I_0$. Denote by $\mathscr{S}$ the right hand side of the equation (6.1.3). Then

$$\phi = \sum_{i \in I_0} \sum_{m=0}^{\infty} (-T_1^- T_2)^m (h_i \psi_i) \in \mathscr{S}. \tag{6.1.11}$$

Suppose $m > 0$. We write

$$\phi = \sum_{i \in I_m} h_i \psi_i + \phi', \qquad h_i \in \mathscr{B},\ \phi' \in \mathscr{A}^{(m-1)}. \tag{6.1.12}$$

Then

$$0 = (T_1 + T_2)(\phi) = \sum_{i \in I_m} T_1(h_i)\psi_i + T_1(\phi') + T_2(\phi). \tag{6.1.13}$$

Since $T_1(\phi') + T_2(\phi) \in \mathscr{A}^{(m-1)}$, we have $T_1(h_i) = 0$ for $i \in I_m$. Now

$$\phi - \sum_{i \in I_m} \sum_{k=0}^{\infty} (-T_1^- T_2)^k (h_i \psi_i) = \phi' - \sum_{i \in I_m} \sum_{k=1}^{\infty} (-T_1^- T_2)^k (h_i \psi_i) \in \mathscr{A}^{(m-1)} \tag{6.1.14}$$

and (6.1.4) implies

$$(T_1 + T_2)(\phi - \sum_{i \in I_m} \sum_{k=0}^{\infty} (-T_1^- T_2)^k (h_i \psi_i)) = 0. \tag{6.1.15}$$

By induction on m,

$$\phi - \sum_{i\in I_m}\sum_{k=0}^{\infty}(-T_1^- T_2)^k(h_i\psi_i) \in \mathscr{S}. \tag{6.1.16}$$

Therefore, $\phi \in \mathscr{S}$.

For any $f \in \mathscr{B} \oplus \mathscr{A}$, we have:

$$\begin{aligned}&(T_1+T_2)(\sum_{i=0}^{\infty}(-T_1^- T_2)^i T_1^-)(f)\\ &= f - \sum_{i=1}^{\infty} T_2(-T_1^- T_2)^{i-1}T_1^-(f) + \sum_{i=0}^{\infty} T_2(-T_1^- T_2)^i T_1^-(f) = f.\end{aligned} \tag{6.1.17}$$

Thus the operator $\sum_{i=0}^{\infty}(-T_1^- T_2)^i T_1^-$ is a right inverse of $T_1 + T_2$. □

We remark that the above operator T_1 and T_2 may not commute.

Define

$$x^\alpha = x_1^{\alpha_1}x_2^{\alpha_2}\cdots x_n^{\alpha_n} \quad \text{for} \quad \alpha = (\alpha_1, \ldots, \alpha_n) \in \mathbb{N}^n. \tag{6.1.18}$$

Moreover, we denote

$$\epsilon_i = (0, \ldots, 0, \overset{i}{1}, 0, \ldots, 0) \in \mathbb{N}^n. \tag{6.1.19}$$

For each $i \in \overline{1, n}$, we define the linear operator $\int_{(x_i)}$ on $\mathscr{A}$ by:

$$\int_{(x_i)}(x^\alpha) = \frac{x^{\alpha+\epsilon_i}}{\alpha_i + 1} \quad \text{for} \quad \alpha \in \mathbb{N}^n. \tag{6.1.20}$$

Furthermore, we let

$$\int_{(x_i)}^{(0)} = 1, \qquad \int_{(x_i)}^{(m)} = \overbrace{\int_{(x_i)}\cdots\int_{(x_i)}}^{m} \qquad \text{for} \quad 0 < m \in \mathbb{Z} \tag{6.1.21}$$

and denote

$$\partial^\alpha = \partial_{x_1}^{\alpha_1}\partial_{x_2}^{\alpha_2}\cdots\partial_{x_n}^{\alpha_n}, \qquad \int^{(\alpha)} = \int_{(x_1)}^{(\alpha_1)}\int_{(x_2)}^{(\alpha_2)}\cdots\int_{(x_n)}^{(\alpha_n)} \qquad \text{for} \quad \alpha \in \mathbb{N}^n. \tag{6.1.22}$$

Obviously, $\int^{(\alpha)}$ is a right inverse of ∂^α for $\alpha \in \mathbb{N}^n$. We remark that $\int^{(\alpha)}\partial^\alpha \neq 1$ if $\alpha \neq 0$ due to $\partial^\alpha(1) = 0$.

Consider the *wave equation in Riemannian space* with a nontrivial conformal group (cf. [I1]):

$$u_{tt} - u_{x_1x_1} - \sum_{i,j=2}^{n} g_{i,j}(x_1 - t)u_{x_ix_j} = 0, \tag{6.1.23}$$

where we assume that $g_{i,j}(z)$ are one-variable polynomials. Change variables:

$$z_0 = x_1 + t, \qquad z_1 = x_1 - t. \tag{6.1.24}$$

Then

$$\partial_t^2 = (\partial_{z_0} - \partial_{z_1})^2, \qquad \partial_{x_1}^2 = (\partial_{z_0} + \partial_{z_1})^2. \tag{6.1.25}$$

So the Eq. (6.1.23) changes to:

$$2u_{z_0z_1} + \sum_{i,j=2}^{n} g_{i,j}(z_1)u_{x_ix_j} = 0. \tag{6.1.26}$$

Denote

$$T_1 = 2\partial_{z_0}\partial_{z_1}, \qquad T_2 = \sum_{i,j=2}^{n} g_{i,j}(z_1)\partial_{x_i}\partial_{x_j}. \tag{6.1.27}$$

Take $T_1^- = \frac{1}{2}\int_{(z_0)}\int_{(z_1)}$, and

$$\mathscr{B} = \mathbb{F}[z_0, z_1], \qquad V = \mathbb{F}[x_2, \ldots, x_n], \qquad V_r = \{f \in V \mid \deg f \leq r\}. \tag{6.1.28}$$

Then the conditions in Lemma 6.1.1 hold. Thus we have:

Theorem 6.1.2 *The space of all polynomial solutions for the Eq. (6.1.23) is:*

$$\begin{aligned} Span &\left\{\sum_{m=0}^{\infty}(-2)^{-m}\left(\sum_{i,j=2}^{n}\int_{(z_0)}\int_{(z_1)} g_{i,j}(z_1)\partial_{x_i}\partial_{x_j}\right)^m (f_0g_0 + f_1g_1)\right. \\ &\left.\mid f_0 \in \mathbb{F}[z_0],\ f_1 \in \mathbb{F}[z_1],\ g_0, g_1 \in \mathbb{F}[x_2, \ldots, x_n]\right\} \end{aligned} \tag{6.1.29}$$

with z_0, z_1 defined in (6.1.24).

Let $m_1, m_2, \ldots, m_n$ be positive integers. According to Lemma 6.1.1, the set

$$\begin{aligned} &\left\{\sum_{k_2,\ldots,k_n=0}^{\infty}(-1)^{k_2+\cdots+k_n}\binom{k_2+\cdots+k_k}{k_2,\ldots,k_n}\int_{(x_1)}^{((k_2+\cdots+k_n)m_1)}(x_1^{\ell_1})\right. \\ &\left.\times\partial_{x_2}^{k_2m_2}(x_2^{\ell_2})\cdots\partial_{x_n}^{k_nm_n}(x_n^{\ell_n}) \mid \ell_1 \in \overline{0, m_1-1},\ \ell_2, \ldots, \ell_n \in \mathbb{N}\right\} \end{aligned} \tag{6.1.30}$$

forms a basis of the space of polynomial solutions for the equation

$$(\partial_{x_1}^{m_1} + \partial_{x_2}^{m_2} + \cdots + \partial_{x_n}^{m_n})(u) = 0 \tag{6.1.31}$$

in $\mathscr{A}$.

Next we give an example of applying Lemma 6.1.1 iteratively. Let

$$f_i \in \mathbb{F}[x_1, \ldots, x_i] \qquad \text{for } i \in \overline{1, n-1}. \tag{6.1.32}$$

Consider the equation:

$$(\partial_{x_1}^{m_1} + f_1\partial_{x_2}^{m_2} + \cdots + f_{n-1}\partial_{x_n}^{m_n})(u) = 0 \tag{6.1.33}$$

Denote

$$d_1 = \partial_{x_1}^{m_1}, \;\; d_r = \partial_{x_1}^{m_1} + f_1\partial_{x_2}^{m_2} + \cdots + f_{r-1}\partial_{x_r}^{m_r} \qquad \text{for } r \in \overline{2, n}. \tag{6.1.34}$$

We will successively apply Lemma 6.1.1 with $T_1 = d_r$, $T_2 = f_r\partial_{x_{r+1}}^{m_{r+1}}$ and $\mathscr{B} = \mathbb{F}[x_1, \ldots, x_r]$, $V = \mathbb{F}[x_{r+1}]$,

$$V_k = \sum_{s=0}^{k} \mathbb{F}x_{r+1}^s. \tag{6.1.35}$$

Take a right inverse $d_1^- = \int_{(x_1)}^{(m_1)}$. Suppose that we have found a right inverse d_s^- of d_s for some $s \in \overline{1, n-1}$ such that

$$x_i d_s^- = d_s^- x_i, \;\; \partial_{x_i} d_s^- = d_s^- \partial_{x_i} \qquad \text{for } i \in \overline{s+1, n}. \tag{6.1.36}$$

Lemma 6.1.1 enables us to take

$$d_{s+1}^- = \sum_{i=0}^{\infty} (-d_s^- f_s)^i d_s^- \partial_{x_{s+1}}^{im_{s+1}} \tag{6.1.37}$$

as a right inverse of d_{s+1}. Obviously,

$$x_i d_{s+1}^- = d_{s+1}^- x_i, \;\; \partial_{x_i} d_{s+1}^- = d_{s+1}^- \partial_{x_i} \qquad \text{for } i \in \overline{s+2, n} \tag{6.1.38}$$

according to (6.1.34). By induction, we have found a right inverse d_s^- of d_s such that (6.1.36) holds for each $s \in \overline{1, n}$.

We set

$$\mathscr{S}_r = \{g \in \mathbb{F}[x_1, \ldots, x_r] \mid d_r(g) = 0\} \qquad \text{for } r \in \overline{1, n}. \tag{6.1.39}$$

By (6.1.34),

$$\mathscr{S}_1 = \sum_{i=0}^{m_1-1} \mathbb{F}x_1^i. \tag{6.1.40}$$

Suppose that we have found $\mathscr{S}_r$ for some $r \in \overline{1, n-1}$. Given $h \in \mathscr{S}_r$ and $\ell \in \mathbb{N}$, we define

$$\sigma_{r+1,\ell}(h) = \sum_{i=0}^{\infty} (-d_r^- f_r)^i (h) \partial_{x_{r+1}}^{im_{r+1}} (x_{r+1}^\ell), \tag{6.1.41}$$

which is actually a finite summation. Lemma 6.1.1 says

$$\mathscr{S}_{r+1} = \sum_{\ell=0}^{\infty} \sigma_{r+1,\ell}(\mathscr{S}_r). \tag{6.1.42}$$

By induction, we obtain:

Theorem 6.1.3 *The set*

$$\{\sigma_{n,\ell_n}\sigma_{n-1,\ell_{n-1}} \cdots \sigma_{2,\ell_2}(x_1^{\ell_1}) \mid \ell_1 \in \overline{0, m_1 - 1},\ \ell_2, \ldots, \ell_n \in \mathbb{N}\} \tag{6.1.43}$$

forms a basis of the polynomial solution space $\mathscr{S}_n$ of the partial differential equation (6.1.33).

6.2 Canonical Oscillator Representations

In this section, we present the canonical oscillator representations of special linear Lie algebras.

Recall that we denote by $E_{i,j}$ the square matrix with 1 as its (i, j)-entry and 0 as the others. The special linear Lie algebra

$$sl(n, \mathbb{F}) = \sum_{1 \le i < j \le n} (\mathbb{F}E_{i,j} + \mathbb{F}E_{j,i}) + \sum_{r=1}^{n-1} \mathbb{F}(E_{r,r} - E_{r+1,r+1}) \tag{6.2.1}$$

with the Lie bracket:

$$[A, B] = AB - BA \qquad \text{for } A, B \in sl(n, \mathbb{F}). \tag{6.2.2}$$

Set

$$h_i = E_{i,i} - E_{i+1,i+1}, \qquad i \in \overline{1, n-1}. \tag{6.2.3}$$

The subspace

$$H = \sum_{i=1}^{n-1} \mathbb{F}h_i \tag{6.2.4}$$

forms a Cartan subalgebra of $sl(n, \mathbb{F})$. We choose

$$\{E_{i,j} \mid 1 \le i < j \le n\} \text{ as positive root vectors.} \tag{6.2.5}$$

In particular, we have

$$\{E_{i,i+1} \mid i = 1, 2, \ldots, n-1\} \text{ as positive simple root vectors.} \tag{6.2.6}$$

Accordingly,

$$\{E_{i,j} \mid 1 \le j < i \le n\} \text{ are negative root vectors} \tag{6.2.7}$$

and we have

$$\{E_{i+1,i} \mid i = 1, 2, \ldots, n-1\} \text{ as negative simple root vectors.} \tag{6.2.8}$$

In particular,

$$sl(n, \mathbb{F})_+ = \sum_{1 \le i < j \le n} \mathbb{F}E_{i,j} \text{ and } sl(n, \mathbb{F})_- = \sum_{1 \le i < j \le n} \mathbb{F}E_{j,i} \tag{6.2.9}$$

are the nilpotent subalgebra of positive root vectors and the nilpotent subalgebra of negative root vectors, respectively. Recall that an $sl(n, \mathbb{F})$-module V is called a weight module if for any $h \in H$, $h|_V$ is diagonalizable. A singular vector of V is a weight vector annihilated by the elements in $sl(n, \mathbb{F})_+$. The fundamental weights $\lambda_i \in H^*$ are

$$\lambda_i(h_r) = \delta_{i,r}. \tag{6.2.10}$$

For $\alpha = (\alpha_1, \alpha_2, \ldots, \alpha_n) \in \mathbb{N}^n$, we define

$$|\alpha| = \sum_{i=1}^{n} \alpha_n. \tag{6.2.11}$$

Denote $\mathscr{A} = \mathbb{F}[x_1, x_2, \ldots, x_n]$. Define the *basic bosonic oscillator representation* of $sl(n, \mathbb{F})$ on $\mathscr{A}$ by:

$$E_{i,j}|_{\mathscr{A}} = x_i \partial_j \qquad \text{for } i, j \in \overline{1, n}. \tag{6.2.12}$$

For any nonnegative integer m, we set

$$\mathscr{A}_m = \text{Span}\,\{x^\alpha \mid \alpha \in \mathbb{N}^n,\ |\alpha| = m\}. \tag{6.2.13}$$

Note that a polynomial $f \in \mathscr{A}$ is a solution of the system

$$0 = E_{i,i+1}(f) = x_i \partial_{x_{i+1}}(f) \quad \text{for } i \in \overline{1, n-1} \tag{6.2.14}$$

if and only if f is a one-variable polynomial in x_1. For $m \in \mathbb{N}$, $\mathscr{A}_m$ is a finite-dimensional $sl(n, \mathbb{F})$-module and it has a unique singular vector x_1^m up to scalar multiple. Since $\mathscr{A}_m$ is weight module and any submodule must contain a singular vector, we obtain:

Proposition 6.2.1 *The subspace $\mathscr{A}_m$ is an irreducible $sl(n, \mathbb{F})$-module with highest weight $m\lambda_1$.*

Let Ψ be the exterior algebra generated by $\{\theta_1, \theta_2, \ldots, \theta_n\}$; that is, an associative algebra with the defining relations:

$$\theta_i \theta_j = -\theta_j \theta_i. \tag{6.2.15}$$

Set

$$\Psi_r = \text{Span}\,\{\theta_{i_1}\theta_{i_2}\cdots\theta_{i_r} \mid i_1, \ldots, i_r \in \overline{1, n}\} \tag{6.2.16}$$

For each $i \in \overline{1, n}$, we define a linear transformation ∂_{θ_i} by:

$$\partial_{\theta_i}(\theta_j) = \delta_{i,j} \tag{6.2.17}$$

and

$$\partial_{\theta_i}(\theta_{j_1}\cdots\theta_{j_r}) = \sum_{s=1}^{r}(-1)^{s-1}\theta_{j_1}\cdots\theta_{j_{s-1}}\partial_{\theta_i}(\theta_{j_s})\theta_{j_{s+1}}\cdots\theta_{j_r}. \tag{6.2.18}$$

Now $sl(n, \mathbb{F})$ has the *basic fermionic oscillator representation* on Ψ defined by:

$$E_{i,j}|_{\Psi} = \theta_i \partial_{\theta_j}. \tag{6.2.19}$$

Observe that $\theta_1\theta_2\cdots\theta_r$ is a singular vector of Ψ_r and

$$\theta_{i_1}\theta_{i_2}\cdots\theta_{i_r} = E_{i_1,1}E_{i_2,2}\cdots E_{i_r,r}(\theta_1\theta_2\cdots\theta_r) \tag{6.2.20}$$

for any $1 \le i_1 < i_2 < \cdots < i_r \le n$. So Ψ_r is an $sl(n, \mathbb{F})$-module generated by $\theta_1\theta_2\cdots\theta_r$.

Proposition 6.2.2 *The subspace Ψ_r forms an irreducible $sl(n, \mathbb{F})$-module with highest weight λ_r.*

Let

$$\mathscr{Q} = \mathbb{F}(x_1, \ldots, x_n, y_1, \ldots, y_n), \tag{6.2.21}$$

the space of rational functions in $x_1, \ldots, x_n, y_1, \ldots, y_n$. Define the *bosonic orthogonal oscillator representation* of $sl(n, \mathbb{F})$ on $\mathscr{Q}$ via

$$E_{i,j}|_{\mathcal{Q}} = x_i\partial_{x_j} - y_j\partial_{y_i} \qquad \text{for } i, j \in \overline{1,n}. \tag{6.2.22}$$

Set

$$\eta = \sum_{i=1}^{n} x_i y_i. \tag{6.2.23}$$

Then

$$\xi(\eta) = 0 \qquad \text{for } \xi \in sl(n, \mathbb{F}). \tag{6.2.24}$$

Lemma 6.2.3 *Any singular function in $\mathcal{Q}$ is a rational function in x_1, y_n, η.*

Proof Let $f \in \mathcal{Q}$ be a singular function. We can write

$$f = g(x_1, \ldots, x_{n-1}, \eta, y_1, \ldots, y_n) \tag{6.2.25}$$

as a rational function in $x_1, \ldots, x_{n-1}, \eta, y_1, \ldots, y_n$. By (6.2.23) and (6.2.24), we have:

$$E_{i,n}(f) = (x_i\partial_{x_n} - y_n\partial_{y_i})(g) = y_n g_{y_i} = 0 \qquad \text{for } i \in \overline{1, n-1}, \tag{6.2.26}$$

or equivalently,

$$g_{y_i} = 0 \qquad \text{for } i \in \overline{1, n-1}. \tag{6.2.27}$$

Thus (6.2.22), (6.2.24) and (6.2.27) imply

$$E_{1,i}(g) = (x_1\partial_{x_i} - y_i\partial_{y_1})(g) = x_1 g_{x_i} = 0 \qquad \text{for } i \in \overline{2, n-1}; \tag{6.2.28}$$

that is,

$$g_{x_i} = 0 \qquad \text{for } i \in \overline{2, n-1}. \tag{6.2.29}$$

Therefore, g is independent of $x_2, \ldots, x_{n-1}$ and $y_1, \ldots, y_{n-1}$. □

Denote $\mathscr{B} = \mathbb{F}[x_1, \ldots, x_n, y_1, \ldots, y_n] \subset \mathcal{Q}$. Set

$$\mathscr{B}_{\ell_1,\ell_2} = \sum_{\alpha,\beta\in\mathbb{N}^n;\ |\alpha|=\ell_1,\ |\beta|=\ell_2} \mathbb{F}x^\alpha y^\beta \qquad \text{for } \ell_1, \ell_2 \in \mathbb{N}. \tag{6.2.30}$$

Then $\mathscr{B}_{\ell_1,\ell_2}$ is a finite-dimensional $sl(n, \mathbb{F})$-submodule by (6.2.22) and $\mathscr{B} = \bigoplus_{\ell_1,\ell_2=0}^{\infty} \mathscr{B}_{\ell_1,\ell_2}$. The function $x_1^{\ell_1} y_n^{\ell_2}$ is a singular function of weight $\ell_1\lambda_1 + \ell_2\lambda_{n-1}$. According to the above lemma, any singular polynomial in $\mathscr{B}_{\ell_1,\ell_2}$ must be of the form $ax_1^{\ell_1-i} y_n^{\ell_2-i}\eta^i$ for some $0 \neq a \in \mathbb{F}$ and $i \in \mathbb{N}$. Define

$$V_{\ell_1,\ell_2} = \text{the submodule generated by } x_1^{\ell_1} y_n^{\ell_2}. \tag{6.2.31}$$

Since $\mathscr{B}_{\ell_1,\ell_2}$ is a weight module, Weyl's Theorem 2.3.6 of complete reducibility implies that $\mathscr{B}_{\ell_1,\ell_2}$ is a direct sum of its irreducible submodules, which are generated by its singular polynomials. So V_{ℓ_1,ℓ_2} is an irreducible highest weight module with the highest weight $\ell_1\lambda_1 + \ell_2\lambda_{n-1}$ and

$$\mathscr{B}_{\ell_1,\ell_2} = V_{\ell_1,\ell_2} \oplus \eta\mathscr{B}_{\ell_1-1,\ell_2-1} \tag{6.2.32}$$

as a direct sum of two $sl(n,\mathbb{F})$-submodules, where we treat $V_{i,j} = \{0\}$ if $\{i, j\} \not\subset \mathbb{N}$.

Denote

$$\Delta = \sum_{i=1}^{n} \partial_{x_i}\partial_{y_i}. \tag{6.2.33}$$

It can be verified that

$$\xi\Delta = \Delta\xi \qquad \text{for } \xi \in sl(n,\mathbb{F}), \tag{6.2.34}$$

as operators on $\mathcal{Q}$. Set

$$\mathscr{H}_{\ell_1,\ell_2} = \{f \in \mathscr{B}_{\ell_1,\ell_2} \mid \Delta(f) = 0\} \tag{6.2.35}$$

Since $\Delta(x_1^{\ell_1} y_n^{\ell_2}) = 0$, we have

$$V_{\ell_1,\ell_2} \subset \mathscr{H}_{\ell_1,\ell_2} \tag{6.2.36}$$

by (6.2.31) and (6.2.34). On the other hand,

$$\mathscr{B}_{\ell_1,\ell_2} = \bigoplus_{i=0}^{\infty} \eta^i V_{\ell_1-i,\ell_2-i} \tag{6.2.37}$$

by (6.2.32) and induction. Note

$$\Delta\eta = n + \eta\Delta + \sum_{i=1}^{n}(x_i\partial_{x_i} + y_i\partial_{y_i}) \tag{6.2.38}$$

as operators on $\mathcal{Q}$. Thus

$$\Delta(\eta^i g) = \sum_{r=1}^{i}(n + \ell_1 + \ell_2 - 2r)(\eta^{i-1}g) = i(n + \ell_1 + \ell_2 - i + 1)\eta^{i-1}g \tag{6.2.39}$$

for $i \in \mathbb{N}+1,\ g \in V_{\ell_1-i,\ell_2-i}$. Hence

$$\mathscr{H}_{\ell_1,\ell_2} \bigcap \eta\mathscr{B}_{\ell_1-1,\ell_2-1} = \{0\}. \tag{6.2.40}$$

Therefore,

$$V_{\ell_1,\ell_2} = \mathscr{H}_{\ell_1,\ell_2} \tag{6.2.41}$$

by (6.2.32) and (6.2.36). Now Lemma 6.1.1 gives:

Theorem 6.2.4 *The subspace $\mathscr{H}_{\ell_1,\ell_2}$ is an irreducible $sl(n,\mathbb{F})$-module with highest weight $\ell_1\lambda_1 + \ell_2\lambda_{n-1}$. Moreover, it has a basis*

$$\left\{\sum_{i=0}^{\infty}(-1)^i \frac{(x_1y_1)^i(\sum_{i=2}^{n}\partial_{x_i}\partial_{y_i})^i(x^\alpha y^\beta)}{\prod_{r=1}^{i}(\alpha_1+i)(\beta_1+i)} \mid \alpha,\beta\in\mathbb{N}^n; \alpha_1\beta_1=0, |\alpha|=\ell_1, |\beta|=\ell_2\right\}. \tag{6.2.42}$$

Furthermore, $\mathscr{B} = \bigoplus_{\ell_1,\ell_2,m=0}^{\infty}\eta^m\mathscr{H}_{\ell_1,\ell_2}$ is a direct sum of irreducible $sl(n,\mathbb{F})$-submodules.

The above result was due to the author [X16]. According to (6.2.38), the space $\mathbb{F}\Delta + \mathbb{F}[\Delta,\eta] + \mathbb{F}\eta$ forms a Lie subalgebra of linear operators on $\mathscr{B}$, which is isomorphic to $sl(2,\mathbb{F})$. The above result can be interpreted as an $(sl(2), sl(n))$-Howe duality (cf. [Hr1, Hr2, Hr3, Hr4]).

Consider the exterior algebra $\check{\mathscr{A}}$ generated by $\{\theta_1,\ldots,\theta_n,\vartheta_1,\ldots,\vartheta_n\}$ (cf. (6.2.15)). We define the *fermionic orthogonal oscillator representation* of $sl(n,\mathbb{F})$ on $\check{\mathscr{A}}$ by

$$E_{i,j}|_{\check{\mathscr{A}}} = \theta_i\partial_{\theta_j} - \vartheta_j\partial_{\vartheta_i} \qquad \text{for } i,j\in\overline{1,n}. \tag{6.2.43}$$

Denote

$$\Theta_1 = \sum_{i=1}^{n}\mathbb{F}\theta_i, \qquad \Theta_2 = \sum_{i=1}^{n}\mathbb{F}\vartheta_i. \tag{6.2.44}$$

For $\ell_1,\ell_2\in\overline{1,n}$, we define

$$\check{\mathscr{A}}_{\ell_1,\ell_2} = \Theta_1^{\ell_1}\Theta_2^{\ell_2}. \tag{6.2.45}$$

Then $\check{\mathscr{A}}_{\ell_1,\ell_2}$ is a finite-dimensional $\check{\mathscr{G}}$-module and

$$\check{\mathscr{A}} = \bigoplus_{\ell_1,\ell_2=0}^{n}\check{\mathscr{A}}_{\ell_1,\ell_2}. \tag{6.2.46}$$

Moreover, we define an ordering:

$$\theta_1 \prec \theta_2 \prec \cdots \prec \theta_n \prec \vartheta_n \prec \vartheta_{n-1} \prec \cdots \prec \vartheta_1. \tag{6.2.47}$$

On the basis

$$\{\theta_{i_1}\cdots\theta_{i_r}\vartheta_{j_1}\cdots\vartheta_{j_s} \mid r,s\in\overline{0,n};$$

$$1 \leq i_1 < i_2 < \cdots < i_r \leq n; n \geq j_1 > j_2 > \cdots j_s \geq 1\} \tag{6.2.48}$$

of $\check{\mathscr{A}}$, we define the partial ordering "$\prec$" lexically.

Write

$$\check{\Delta} = \sum_{r=1}^{n} \partial_{\theta_r} \partial_{\vartheta_r}, \qquad \check{\eta} = \sum_{r=1}^{n} \theta_r \vartheta_r. \tag{6.2.49}$$

For $r \in \overline{1,n}$, we denote

$$\vec{\theta}_r = \theta_1 \cdots \theta_r, \qquad \vec{\vartheta}_r = \vartheta_n \cdots \vartheta_r. \tag{6.2.50}$$

For convenience, we let

$$\vec{\theta}_0 = 1 = \vec{\vartheta}_{n+1}. \tag{6.2.51}$$

It can easily proved that a minimal term of any singular vector in $\check{\mathscr{A}}_{\ell_1,\ell_2}$ is of the form $\vec{\theta}_r \vec{\vartheta}_s$ for some $r \in \overline{0,n}$ and $s \in \overline{r+1,n+1}$ or

$$\vec{\theta}_r \theta_{r+1} \cdots \theta_{s_1} \vec{\vartheta}_{s_2} \vartheta_{s_1} \cdots \vartheta_{r+1} \tag{6.2.52}$$

for some $0 \leq r < s_1 < s_2 \leq n+1$. By comparing minimal terms, we can prove that

$$\{\check{\eta}^\ell \vec{\theta}_r \vec{\vartheta}_s \mid 0 \leq r < s \leq n+1; \ell \in \overline{0, s-r-1}; r + \ell = \ell_1; \ell + n - s + 1 = \ell_2\} \tag{6.2.53}$$

is the set of all $sl(n,\mathbb{F})$-singular vectors in $\check{\mathscr{A}}_{\ell_1,\ell_2}$. Let $V_{r,s}$ be the finite-dimensional irreducible $sl(n,\mathbb{F})$-submodule generated by $\vec{\theta}_r \vec{\vartheta}_s \in \check{\mathscr{A}}_{r,n+1-s}$. Since $\check{\mathscr{A}}$ is a weight module, Theorem 2.4.5 of Weyl's complete reducibility gives

$$\check{\mathscr{A}} = \bigoplus_{0 \leq r < s \leq n+1} \bigoplus_{\ell=0}^{s-r-1} \check{\eta}^\ell V_{r,s} \tag{6.2.54}$$

is a direct sum of irreducible $sl(n,\mathbb{F})$-submodules.

Define

$$\check{\mathscr{H}} = \{f \in \check{\mathscr{A}} \mid \check{\Delta}(f) = 0\}. \tag{6.2.55}$$

Note

$$E_{r,s} \check{\Delta} = \check{\Delta} E_{r,s}, \quad E_{r,s} \check{\eta} = \check{\eta} E_{r,s} \ \text{ on } \ \check{\mathscr{A}} \tag{6.2.56}$$

for $r, s \in \overline{1,n}$. Moreover,

$$\check{\Delta} \check{\eta} = \check{\eta} \check{\Delta} - n + \sum_{r=1}^{n} (\theta_r \partial_{\theta_r} + \vartheta_r \partial_{\vartheta_r}) \tag{6.2.57}$$

by (6.2.49). Furthermore,

$$\check{\Delta}(\vec{\theta}_r\vec{\vartheta}_s) = 0 \quad \text{if } r < s. \tag{6.2.58}$$

Hence

$$V_{r,s} \subset \check{\mathscr{H}} \quad \text{for } 0 \leq r < s \leq n+1 \tag{6.2.59}$$

by (6.2.56). Suppose $0 \leq r+1 < s \leq n+1$ and $\ell \in \overline{1, s-r-1}$. For any $f \in V_{r,s}$, we have

$$\check{\Delta}(\check{\eta}^\ell f) = \left(\sum_{p=0}^{\ell-1}(2p+r+1-s)\right)\check{\eta}^{\ell-1} f = \ell(\ell+r-s)\check{\eta}^{\ell-1} f. \tag{6.2.60}$$

Therefore,

$$\check{\mathscr{H}} = \bigoplus_{0 \leq r < s \leq n+1} V_{r,s}. \tag{6.2.61}$$

In particular,

$$\check{\mathscr{H}}_{r,n+1-s} = \{f \in \check{\mathscr{A}}_{r,n+1-s} \mid \check{\Delta}(f) = 0\} = \check{\mathscr{A}}_{r,n+1-s} \bigcap \check{\mathscr{H}} = V_{r,s} \tag{6.2.62}$$

for $0 \leq r < s \leq n+1$ and

$$\check{\mathscr{A}}_{\ell_1,\ell_2} \bigcap \check{\mathscr{H}} = \{0\} \quad \text{if } \ell_1 + \ell_2 \geq n+1. \tag{6.2.63}$$

Treat $\lambda_0 = \lambda_n = 0$. We have:

Theorem 6.2.5 *For $0 \leq r < s \leq n+1$, $\check{\mathscr{H}}_{r,n+1-s}$ is a finite-dimensional irreducible $sl(n, \mathbb{F})$-module with the highest-weight vector $\vec{\theta}_r\vec{\vartheta}_s$ of weight $\lambda_r + \lambda_{s-1}$. Moreover,*

$$\check{\mathscr{A}} = \bigoplus_{0 \leq r < s \leq n+1} \bigoplus_{\ell=0}^{s-r-1} \check{\eta}^\ell \check{\mathscr{H}}_{r,n+1-s}. \tag{6.2.64}$$

The above theorem was due to Luo and the author's work [LX4]. According to (6.2.57), the space $\mathbb{F}\check{\Delta} + \mathbb{F}[\check{\Delta}, \check{\eta}] + \mathbb{F}\check{\eta}$ forms a Lie subalgebra of linear operators on $\check{\mathscr{A}}$, which is isomorphic to $sl(2, \mathbb{F})$. The above result can also be interpreted as an $(sl(2), sl(n))$-Howe duality (cf. [Hr1, Hr2, Hr3, Hr4]).

6.3 Noncanonical Representations I: General

In this section, we study the noncanonical oscillator representation of deformed from the canonical ones in (6.2.12) and (6.2.22).

Fix $1 \leq n_1 < n$. Note the symmetry

$$[\partial_{x_r}, x_r] = 1 = [-x_r, \partial_{x_r}]. \tag{6.3.1}$$

Changing operators $\partial_{x_r} \mapsto -x_r$ and $x_r \mapsto \partial_{x_r}$ in (6.2.12) for $r \in \overline{1, n_1}$, we obtain the following *basic noncanonical oscillator representation* of $sl(n, \mathbb{F})$ (also $gl(n, \mathbb{F})$) on $\mathscr{A} = \mathbb{F}[x_1, x_2, \ldots, x_n]$:

$$E_{i,j}|_{\mathscr{A}} = \begin{cases} -x_j\partial_{x_i} - \delta_{i,j} & \text{if } i, j \in \overline{1, n_1}; \\ \partial_{x_i}\partial_{x_j} & \text{if } i \in \overline{1, n_1},\ j \in \overline{n_1 + 1, n}; \\ -x_i x_j & \text{if } i \in \overline{n_1 + 1, n},\ j \in \overline{1, n_1}; \\ x_i\partial_{x_j} & \text{if } i, j \in \overline{n_1 + 1, n}. \end{cases} \tag{6.3.2}$$

For any $k \in \mathbb{Z}$, we denote

$$\mathscr{A}_{\langle k \rangle} = \text{Span}\,\{x^\alpha \mid \alpha \in \mathbb{N}^n;\ \sum_{r=n_1+1}^{n} \alpha_r - \sum_{i=1}^{n_1} \alpha_i = k\}. \tag{6.3.3}$$

Then

$$\mathscr{A} = \bigoplus_{k \in \mathbb{Z}} \mathscr{A}_{\langle k \rangle}. \tag{6.3.4}$$

Set

$$\tilde{D} = \sum_{r=n_1+1}^{n} x_r\partial_{x_r} - \sum_{i=1}^{n_1} x_i\partial_{x_i}. \tag{6.3.5}$$

We have

$$\mathscr{A}_{\langle k \rangle} = \{f \in \mathscr{A} \mid \tilde{D}(f) = kf\}. \tag{6.3.6}$$

Moreover, we have

$$\tilde{D}E_{i,j} = E_{j,i}\tilde{D} \text{ on } \mathscr{A} \qquad \text{for } i, j \in \overline{1, n}. \tag{6.3.7}$$

Thus $\mathscr{A}_{\langle k \rangle}$ forms a $\mathscr{G}$-module for any subalgebra $\mathscr{G}$ of $gl(n, \mathbb{F})$.

For $\alpha \in \mathbb{N}^n$, we denote

$$\alpha! = \prod_{i=1}^{n} \alpha_i!. \tag{6.3.8}$$

Define a symmetric bilinear form $(\cdot|\cdot)$ on $\mathscr{A}$ by

$$(x^\alpha | x^\beta) = \delta_{\alpha,\beta}(-1)^{\sum_{i=1}^{n_1} \alpha_i}\alpha! \qquad \text{for } \alpha, \beta \in \mathbb{N}^n. \tag{6.3.9}$$

Then we have:

Lemma 6.3.1 *For any $A \in gl(n, \mathbb{F})$ and $f, g \in \mathscr{A}$, we have*

$$(A(f)|g) = (f|A^t(g)), \tag{6.3.10}$$

where A^t denote the transpose of the matrix A.

Proof Let $\alpha, \beta \in \mathbb{N}^n$. For $i, j \in \overline{1, n_1}$,

$$(E_{i,j}(x^\alpha)|x^\beta) = -\alpha_i(x^{\alpha+\epsilon_j-\epsilon_i}|x^\beta) - \delta_{i,j}(x^\alpha|x^\beta) \tag{6.3.11}$$

and

$$(x^\alpha|E_{j,i}(x^\beta)) = -\beta_j(x^\alpha|x^{\beta+\epsilon_i-\epsilon_j}) - \delta_{i,j}(x^\alpha|x^\beta) \tag{6.3.12}$$

by (6.3.2). Note

$$\begin{aligned}\alpha_i(x^{\alpha+\epsilon_j-\epsilon_i}|x^\beta) &= \delta_{\alpha+\epsilon_j-\epsilon_i,\beta}(-1)^{\sum_{r=1}^{n_1}\alpha_r}(\alpha_j+1)\alpha! \\ &= \beta_j\delta_{\alpha,\beta+\epsilon_i-\epsilon_j}(-1)^{\sum_{r=1}^{n_1}\alpha_r}\alpha! = \beta_j(x^\alpha|x^{\beta+\epsilon_i-\epsilon_j})\end{aligned} \tag{6.3.13}$$

by (6.3.9). Hence

$$(E_{i,j}(x^\alpha)|x^\beta) = (x^\alpha|E_{j,i}(x^\beta)). \tag{6.3.14}$$

If $i, j \in \overline{n_1+1, n}$, then (6.3.13) holds and so does (6.3.14).

Consider $i \in \overline{1, n_1}$ and $j \in \overline{n_1+1, n}$.

$$(E_{i,j}(x^\alpha)|x^\beta) = \alpha_i\alpha_j(x^{\alpha-\epsilon_i-\epsilon_j}|x^\beta) = -\delta_{\alpha-\epsilon_i-\epsilon_j,\beta}(-1)^{\sum_{r=1}^{n_1}\alpha_r}\alpha! \tag{6.3.15}$$

and

$$(x^\alpha|E_{j,i}(x^\beta)) = -(x^\alpha|x^{\beta+\epsilon_i+\epsilon_j}) = -\delta_{\alpha,\beta+\epsilon_i+\epsilon_j}(-1)^{\sum_{r=1}^{n_1}\alpha_r}\alpha! \tag{6.3.16}$$

by (6.3.2) and (6.3.9). So (6.3.14) holds. Therefore, (6.3.10) holds by the symmetry of the form. □

For a subspace V of $\mathscr{A}$, the *radical* of V with respect to the symmetric bilinear form $(\cdot|\cdot)$ is:

$$\mathscr{R}_V = \{v \in V \mid (v|u) = 0 \text{ for any } u \in V\}. \tag{6.3.17}$$

If $\mathscr{R}_V = \{0\}$, we call V *nondegenerate*.

Let $\mathscr{G}$ be simple Lie subalgebra of $gl(n, \mathbb{F})$ such that the transpose

$$A^t \in \mathscr{G} \qquad \text{if } A \in \mathscr{G}. \tag{6.3.18}$$

Suppose that H is a Cartan subalgebra of $\mathscr{G}$ and $\Pi = \{\alpha_1, \alpha_2, \ldots, \alpha_\ell\}$ is a base of the corresponding root system Φ. Take $0 \neq \xi_i \in \mathscr{G}_{\alpha_i}$ and assume $\zeta_i = \xi_i^t \in \mathscr{G}_{-\alpha_i}$ for $i \in \overline{1, \ell}$. Set

$$\mathscr{G}_\pm = \sum_{\alpha \in \Phi_\pm} \mathscr{G}_\alpha. \tag{6.3.19}$$

Then $\mathscr{G}_+$ is a subalgebra generated by $\{\xi_i \mid i \in \overline{1, \ell}\}$, $\mathscr{G}_-$ is a subalgebra generated by $\{\zeta_i \mid i \in \overline{1, \ell}\}$ and

$$[\xi_i, \zeta_j] = 0, \quad [\xi_i, \zeta_i] \in H \qquad \text{for } i, j \in \overline{1, \ell}, \ i \neq j. \tag{6.3.20}$$

Assume that $\mathscr{A}$ forms a weight $\mathscr{G}$-module with respect to H. A linear transformation (operator) T on $\mathscr{A}$ is called *locally nilpotent* if for any $f \in \mathscr{A}$, there exists a positive integer k such that $T^k(f) = 0$. Moreover, an element $g \in \mathscr{A}$ is called *nilpotent with respect to* $\mathscr{G}_+$ if there exist a positive integer m such that

$$u_1 \cdots u_m(g) = 0 \qquad \text{for any } u_1, \ldots, u_m \in \mathscr{G}_+. \tag{6.3.21}$$

A subspace V of $\mathscr{A}$ is called *nilpotent with respect to* $\mathscr{G}_+$ if all its elements are nilpotent with respect to $\mathscr{G}_+$. If the elements of $\mathscr{G}_+|_{\mathscr{A}}$ are locally nilpotent and

$$\mathscr{G}_+(\mathscr{A}_i) \subset \sum_{r=0}^{i} \mathscr{A}_r \qquad \text{for any } i \in \mathbb{N}, \tag{6.3.22}$$

then any element of $\mathscr{A}$ is nilpotent with respect to $\mathscr{G}_+$ by Theorem 1.5.1.

Lemma 6.3.2 *If a submodule M of $\mathscr{A}$ is nilpotent with respect to $\mathscr{G}_+$, M contains a unique singular vector v (up to a scaler) and $(v|v) \neq 0$, then M is an irreducible summand of $\mathscr{A}$.*

Proof First $V = U(\mathcal{G})(v)$ is an irreducible submodule by the uniqueness of singular vector. For any $g \in M$, there exists $m \in \mathbb{N} + 1$ such that (6.3.21) holds. Set

$$V_1 = \text{Span}\,\{v, \zeta_{i_1} \cdots \zeta_{i_s}(v) \mid m > s \in \mathbb{N} + 1; \ i_1, \ldots, i_s \in \overline{1, \ell}\} \tag{6.3.23}$$

and

$$V_2 = \text{Span}\,\{\zeta_{i_1} \cdots \zeta_{i_s}(v) \mid m \leq s \in \mathbb{N} + 1; \ i_1, \ldots, i_s \in \overline{1, \ell}\}. \tag{6.3.24}$$

Then

$$V = U(\mathscr{G}_-)(v) = V_1 \oplus V_2 \tag{6.3.25}$$

because $\mathscr{G}_-$ is generated by $\{\zeta_i \mid i \in \overline{1, \ell}\}$. Note that $\mathscr{R}_V$ is a submodule of V by Lemma 6.3.1 and the assumption (6.3.18). If $\mathscr{R}_V \neq \{0\}$, then it contains a singular vector under the nilpotent assumption, which must be v by the uniqueness. This contradicts the assumption $(v|v) \neq 0$. Therefore, $\mathscr{R}_V = \{0\}$; that is V is nondegenerate.

According to Lemma 6.3.1

$$(\zeta_{i_1} \cdots \zeta_{i_s}(v)|\zeta_{j_1} \cdots \zeta_{j_r}(v)) = (\xi_{j_r} \cdots \xi_{j_1}\zeta_{i_1} \cdots \zeta_{i_s}(v)|v) = 0 \text{ if } s < r, \tag{6.3.26}$$

by (6.3.20) and

$$(\zeta_{j_1} \cdots \zeta_{j_r}(v)|g) = (v|\xi_{j_r} \cdots \xi_{j_1}(g)) = 0 \text{ when } r \geq m \tag{6.3.27}$$

by the assumption of (6.3.21). Hence

$$(V_1|V_2) = \{0\}, \quad (V_2|g) = \{0\}. \tag{6.3.28}$$

Thus V_1 and V_2 are nondegenerate. Since dim V_1 is finite, there exists $g' \in V_1$ such that $g - g' \in V_1^{\perp}$ by linear algebra. Moreover, $g - g' \in V_2^{\perp}$ by (6.3.28). So $g - g' \in V^{\perp}$. Thus we have

$$M = V \oplus M \bigcap V^{\perp}. \tag{6.3.29}$$

By Lemma 6.3.1 and the assumption (6.3.18), $V^{\perp}$ is a submodule of $\mathscr{A}$. So $M \bigcap V^{\perp}$ is a submodule of M, which contains v if it is nonzero. The assumption $(v|v) \neq 0$ leads $M \bigcap V^{\perp} = \{0\}$. Therefore, $M = V$ is irreducible. □

Let $f \in \mathscr{A}$ be a singular vector of $sl(n, \mathbb{F})$. Then

$$E_{i,n_1}(f) = -x_{n_1}\partial_{x_i}(f) = 0 \qquad \text{for } i \in \overline{1, n_1 - 1} \tag{6.3.30}$$

and

$$E_{n_1+1,j}(f) = x_{n_1+1}\partial_{x_j}(f) = 0 \qquad \text{for } j \in \overline{n_1 + 2, n} \tag{6.3.31}$$

by (6.3.2). Thus $f = g(x_{n_1}, x_{n_1+1})$ only depends on x_{n_1} and x_{n_1+1}. Moreover,

$$0 = E_{n_1,n_1+1}(f) = \partial_{x_{n_1}}\partial_{x_{n_1+1}}(g). \tag{6.3.32}$$

This shows that g is a one-variable function in x_{n_1} or x_{n_1+1}. Since f is a weight vector, $f = x_{n_1}^{m_1}$ or $x_{n_1+1}^{m_2}$ up to a scalar multiple for $m_1 \in \mathbb{N}$ and $0 \neq m_2 \in \mathbb{N}$. Note $x_{n_1}^{m_1} \in \mathscr{A}_{\langle -m_1 \rangle}$ and $x_{n_1+1}^{m_2} \in \mathscr{A}_{\langle m_2 \rangle}$. So $\mathscr{A}_{\langle k \rangle}$ has a unique singular vector for any $k \in \mathbb{Z}$. By the above lemma, $\mathscr{A}_{\langle k \rangle}$ is an irreducible $sl(n, \mathbb{F})$-submodule.

For any graded subspace V of $\mathscr{A}$ (i.e. $V = \bigoplus_{k=0}^{\infty} V \bigcap \mathscr{A}_k$), we define it q-*dimension* by

$$\dim_q V = \sum_{k=0}^{\infty} (\dim V \bigcap \mathscr{A}_k) q^k. \tag{6.3.33}$$

Recall that λ_i is the ith fundamental weight of $sl(n+1,\mathbb{F})$. For convenience, we treat $\lambda_0 = 0$.

Theorem 6.3.3 *Let $m_1, m_2 \in \mathbb{N}$ with $m_1 > 0$, $\mathscr{A}_{\langle -m_1\rangle}$ is an irreducible highest-weight $sl(n,\mathbb{F})$-submodule with highest weight $m_1\lambda_{n_1-1} - (m_1+1)\lambda_{n_1}$ and $\mathscr{A}_{\langle m_2\rangle}$ is an irreducible highest-weight $sl(n,\mathbb{F})$-submodule with highest weight $-(m_2+1)\lambda_{n_1} + m_2\lambda_{n_1+1}$. Moreover,*

$$\sum_{m\in\mathbb{Z}} z^m ch_{\mathscr{A}_{\langle m\rangle}} = \frac{\varepsilon_{-\lambda_{n_1}}}{(1-z^{-1}\varepsilon_{-\lambda_1})(1-z\varepsilon_{-\lambda_{n-1}})} \times \frac{1}{[\prod_{i=2}^{n_1}(1-z^{-1}\varepsilon_{\lambda_{i-1}-\lambda_i})][\prod_{r=n_1+1}^{n-1}(1-z\varepsilon_{\lambda_r-\lambda_{r-1}})]}, \tag{6.3.34}$$

$$\sum_{m\in\mathbb{Z}} (\dim_q \mathscr{A}_{\langle m\rangle}) z^m = \frac{1}{(1-z^{-1}q)^{n_1}(1-zq)^{n-n_1}}. \tag{6.3.35}$$

The above result was essentially due to Howe [Hr4], where he used the symmetric tensor over several copies of the natural module and its dual to construct the representation. Since the structure constants are integers, the result works for any field with characteristic 0.

Fix $n_1, n_2 \in \overline{1,n}$ such that $n_1 \le n_2$. Recall that $\mathcal{Q}$ is the space of rational functions in $x_1,\dots,x_n,y_1,\dots,y_n$. Changing operators $\partial_{x_r} \mapsto -x_r,\ x_r \mapsto \partial_{x_r}$ for $r \in \overline{1,n_1}$ and $\partial_{y_s} \mapsto -y_s,\ y_s \mapsto \partial_{y_s}$ for $s \in \overline{n_2+1,n}$ in the representation (6.2.22) of $sl(n,\mathbb{F})$, we get a *noncanonical orthogonal oscillator representation* of $sl(n,\mathbb{F})$ on $\mathcal{Q}$ determined by

$$E_{i,j}|_{\mathcal{Q}} = E^x_{i,j} - E^y_{j,i} \qquad \text{for } i,j \in \overline{1,n} \tag{6.3.36}$$

with

$$E^x_{i,j} = \begin{cases} -x_j\partial_{x_i} - \delta_{i,j} & \text{if } i,j \in \overline{1,n_1};\\ \partial_{x_i}\partial_{x_j} & \text{if } i \in \overline{1,n_1},\ j \in \overline{n_1+1,n};\\ -x_ix_j & \text{if } i \in \overline{n_1+1,n},\ j \in \overline{1,n_1};\\ x_i\partial_{x_j} & \text{if } i,j \in \overline{n_1+1,n} \end{cases} \tag{6.3.37}$$

and

$$E^y_{i,j} = \begin{cases} y_i\partial_{y_j} & \text{if } i,j \in \overline{1,n_2};\\ -y_iy_j & \text{if } i \in \overline{1,n_2},\ j \in \overline{n_2+1,n};\\ \partial_{y_i}\partial_{y_j} & \text{if } i \in \overline{n_2+1,n},\ j \in \overline{1,n_2};\\ -y_j\partial_{y_i} - \delta_{i,j} & \text{if } i,j \in \overline{n_2+1,n}. \end{cases} \tag{6.3.38}$$

Recall $\tilde{D}$ in (6.3.5) and define

$$\tilde{D}' = \sum_{i=1}^{n_2} y_i\partial_{y_i} - \sum_{r=n_2+1}^{n} y_r\partial_{y_r}. \tag{6.3.39}$$

We set

$$\tilde{\Delta}=\sum_{i=1}^{n_1}x_i\partial_{y_i}-\sum_{r=n_1+1}^{n_2}\partial_{x_r}\partial_{y_r}+\sum_{s=n_2+1}^{n}y_s\partial_{x_s} \tag{6.3.40}$$

and

$$\eta=\sum_{i=1}^{n_1}y_i\partial_{x_i}+\sum_{r=n_1+1}^{n_2}x_ry_r+\sum_{s=n_2+1}^{n}x_s\partial_{y_s}. \tag{6.3.41}$$

Then

$$TE_{i,j}|_{\mathcal{Q}}=E_{i,j}|_{\mathcal{Q}}T\qquad\text{for }\; T=\tilde{D},\tilde{D}',\tilde{\Delta},\eta;\; i,j\in\overline{1,n}. \tag{6.3.42}$$

Moreover,

$$[\tilde{D},\tilde{\Delta}]=[\tilde{D}',\tilde{\Delta}]=-\tilde{\Delta},\qquad [\tilde{D},\eta]=[\tilde{D}',\eta]=\eta. \tag{6.3.43}$$

Furthermore, we have

$$E_{i,r}|_{\mathcal{Q}}=-x_r\partial_{x_i}-y_r\partial_{y_i}\qquad\text{for }\; 1\le i<r\le n_1, \tag{6.3.44}$$

$$E_{i,n_1+s}|_{\mathcal{Q}}=\partial_{x_i}\partial_{x_{n_1+s}}-y_{n_1+s}\partial_{y_i}\qquad\text{for }\; i\in\overline{1,n_1},\; s\in\overline{1,n_2-n_1}, \tag{6.3.45}$$

$$E_{r,s}|_{\mathcal{Q}}=x_r\partial_{x_s}-y_s\partial_{y_r}\qquad\text{for }\; n_1<r<s\le n_2, \tag{6.3.46}$$

$$E_{n_2,n_2+1}|_{\mathcal{Q}}=x_{n_2}\partial_{x_{n_2+1}}-\partial_{y_{n_2}}\partial_{y_{n_2+1}}, \tag{6.3.47}$$

$$E_{i,r}|_{\mathcal{Q}}=x_i\partial_{x_r}+y_i\partial_{y_r}\qquad\text{for }\; n_2+1\le i<r\le n. \tag{6.3.48}$$

Thus $\mathscr{B}=\mathbb{F}[x_1,\dots,x_n,y_1,\dots,y_n]$ is nilpotent with respect to $sl(n,\mathbb{F})_+$ in (6.2.9).

Denote

$$\zeta_1=x_{n_1-1}y_{n_1}-x_{n_1}y_{n_1-1},\quad \zeta=\sum_{r=n_1+1}^{n_2}x_ry_r,\quad \zeta_2=x_{n_2+1}y_{n_2+2}-x_{n_2+2}y_{n_2+1}. \tag{6.3.49}$$

Then

$$E_{i,j}(\zeta_1)=0\qquad\text{for }\; 1\le i<j\le n_1, \tag{6.3.50}$$

$$E_{p,q}(\zeta)=0\qquad\text{for }\; n_1+1\le p<q\le n_2, \tag{6.3.51}$$

$$E_{r,s}(\zeta_2)=0\qquad\text{for }\; n_2+1\le r<s\le n. \tag{6.3.52}$$

We will process according to three cases.

6.4 Noncanonical Representations II: $n_1 + 1 < n_2$

In this section, we study the representation of $sl(n, \mathbb{F})$ given in (6.3.36)–(6.3.38) with $n_1 + 1 < n_2$.

Assume $n_1 + 1 < n_2 < n$. Suppose that $f \in \mathcal{Q}$ is a singular vector. By Lemma 6.2.3, f can be written as a rational function in

$$\{x_i, y_r, \zeta_1, \zeta, \zeta_2 \mid n_2 + 2 \neq i \in \overline{1, n_1 + 1} \bigcup \overline{n_2 + 1, n},\ n_1 - 1 \neq r \in \overline{1, n_1} \bigcup \overline{n_2, n}\}. \tag{6.4.1}$$

Note that (6.3.44) and (6.3.50) give

$$E_{n_1-1,n_1}(f) = -(x_{n_1}\partial_{x_{n_1-1}} + y_{n_1}\partial_{y_{n_1-1}})(f) = -x_{n_1} f_{x_{n_1-1}} = 0. \tag{6.4.2}$$

Moreover, (6.3.48) and (6.3.52) yield

$$E_{n_2+1,n_2+2}(f) = (x_{n_2+1}\partial_{x_{n_2+2}} + y_{n_2+1}\partial_{y_{n_2+2}})(f) = y_{n_2+1} f_{y_{n_2+2}} = 0. \tag{6.4.3}$$

So f is independent of x_{n_1-1} and y_{n_2+2} in terms of (6.4.1). For $i \in \overline{1, n_1 - 2}$, we have

$$\begin{aligned} E_{i,n_1-1}(f) &= -(x_{n_1-1}\partial_{x_i} + y_{n_1-1}\partial_{y_i})(f) \\ &= -x_{n_1-1} f_{x_i} - y_{n_1-1} f_{y_i} \\ &= -x_{n_1-1}(f_{x_i} + x_{n_1}^{-1} y_{n_1} f_{y_i}) + x_{n_1}^{-1}\zeta_1 f_{y_i} = 0 \end{aligned} \tag{6.4.4}$$

by (6.3.44), (6.4.49) and (6.4.50). Since both $f_{x_i} + x_{n_1}^{-1} y_{n_1} f_{y_i}$ and $x_{n_1}^{-1}\zeta_1 f_{y_i}$ are independent of x_{n_1-1} in terms of (6.4.1), we have $f_{y_i} = 0$, and so $f_{y_i} = 0$. Thus f is independent of $\{x_i, y_i \mid i \in \overline{1, n_1 - 1}\}$ in terms of (6.4.1). Similarly, we can prove that f is independent of $\{x_i, y_i \mid i \in \overline{n_2 + 1, n}\}$ in terms of (6.4.1). Therefore, f only depends on

$$\{x_{n_1}, x_{n_1+1}, x_{n_2+1}, y_{n_1}, y_{n_2}, y_{n_2+1}, \zeta_1, \zeta, \zeta_2\}. \tag{6.4.5}$$

According to (6.3.45) and (6.3.49), $E_{n_1,n_1+1}|_{\mathcal{Q}} = \partial_{x_{n_1}}\partial_{x_{n_1+1}} - y_{n_1+1}\partial_{y_{n_1}}$ and

$$\begin{aligned} &E_{n_1,n_1+1}(f) = f_{x_{n_1}x_{n_1+1}} - y_{n_1-1} f_{\zeta_1 x_{n_1+1}} \\ &+ y_{n_1+1}(f_{x_{n_1}\zeta} - y_{n_1-1} f_{\zeta_1\zeta} - f_{y_{n_1}} - x_{n_1-1} f_{\zeta_1}) = 0. \end{aligned} \tag{6.4.6}$$

Applying $E_{n_1+1,n_2}|_{\mathcal{Q}} = x_{n_1+1}\partial_{x_{n_2}} - y_{n_2}\partial_{y_{n_1+1}}$ in (6.3.45) to the above equation, we get

$$f_{x_{n_1}\zeta} - y_{n_1-1} f_{\zeta_1\zeta} - f_{y_{n_1}} - x_{n_1-1} f_{\zeta_1} = 0 \tag{6.4.7}$$

by (6.3.52). According to (6.3.49),

$$x_{n_1-1} = y_{n_1}^{-1}\zeta_1 + x_{n_1} y_{n_1}^{-1} y_{n_1-1}. \tag{6.4.8}$$

Substituting it into (6.4.7), we get

$$f_{x_{n_1}\zeta} - y_{n_1-1}(f_{\zeta_1\zeta} + y_{n_1}^{-1}x_{n_1}f_{\zeta_1}) - f_{y_{n_1}} - y_{n_1}^{-1}\zeta_1 f_{\zeta_1} = 0. \tag{6.4.9}$$

Since f is independent of y_{n_1-1} in terms of (6.4.5), we have

$$f_{\zeta_1\zeta} + y_{n_1}^{-1}x_{n_1}f_{\zeta_1} = 0. \tag{6.4.10}$$

Thus

$$f_{\zeta_1} = e^{-y_{n_1}^{-1}x_{n_1}\zeta}g \tag{6.4.11}$$

for some rational function g in the variables of (6.4.5) except ζ, i.e., $g_\zeta = 0$. But f is a rational function in the variables of (6.4.5) and so is f_{ζ_1}. Hence (6.4.11) forces $f_{\zeta_1} = 0$; that is, f is independent of ζ_1 in terms of (6.4.5). Similarly, we can prove that f is independent of ζ_2 in terms of (6.4.5). Now f only depends on

$$\{x_{n_1}, x_{n_1+1}, x_{n_2+1}, y_{n_1}, y_{n_2}, y_{n_2+1}, \zeta\}. \tag{6.4.12}$$

Next we only consider $f \in \mathscr{B} = \mathbb{F}[x_1, \ldots, x_n, y_1, \ldots, y_n]$. Since $\zeta = \sum_{i=n_1+1}^{n_2} x_i y_i$, f must be a polynomial in the variables (6.4.12). Now (6.4.6) and (6.4.7) are equivalent to

$$f_{x_{n_1}x_{n_1+1}} = 0, \qquad f_{x_{n_1}\zeta} - f_{y_{n_1}} = 0. \tag{6.4.13}$$

Similarly, we can prove

$$f_{y_{n_2}y_{n_2+1}} = 0, \qquad f_{y_{n_2+1}\zeta} - f_{x_{n_2+1}} = 0. \tag{6.4.14}$$

Set

$$\phi(m_1, m_2) = \sum_{i=0}^{\infty} \frac{y_{n_1}^i (\partial_{x_{n_1}}\partial_\zeta)^i (x_{n_1}^{m_1}\zeta^{m_2})}{i!} \qquad for \;\; m_1, m_2 \in \mathbb{N}. \tag{6.4.15}$$

By Lemma 6.1.1 with $T_1 = \partial_{y_{n_1}}$ $T_1^- = \int_{(y_{n_1})}$ (cf. (6.1.20)) and $T_2 = -\partial_{x_{n_1}}\partial_\zeta$, the polynomial solution space of the second equation in (6.4.13) is

$$\left[\sum_{m_1=0}^{\infty}\sum_{m_2=0}^{\infty} \mathbb{F}\phi(m_1, m_2)\right]\left[\mathbb{F}[x_{n_1+1}, x_{n_2+1}, y_{n_2}, y_{n_2+1}]\right]. \tag{6.4.16}$$

The first equation in (6.4.13) says that f can not contain the monomials with $x_{n_1}x_{n_1+1}$ as a factor. Moreover, $\phi(0, m_2) = \zeta^{m_2}$. Thus the polynomial solution space of (6.4.13) is

$$\left[\mathbb{F}[x_{n_1+1}, \zeta] + \sum_{m_1=1}^{\infty}\sum_{m_2=0}^{\infty} \mathbb{F}\phi(m_1, m_2)][\mathbb{F}[x_{n_2+1}, y_{n_2}, y_{n_2+1}]\right]. \tag{6.4.17}$$

Denote

$$\psi(m_1, m_2) = \sum_{i=0}^{\infty} \frac{x_{n_2+1}^i (\partial_{y_{n_2+1}} \partial_\zeta)^i (y_{n_2+1}^{m_1} \zeta^{m_2})}{i!} \qquad \text{for } m_1, m_2 \in \mathbb{N}, \tag{6.4.18}$$

$$\begin{aligned}\phi(m_1, m_2, m_3) &= \sum_{r=0}^{\infty} \frac{x_{n_2+1}^r (\partial_{y_{n_2+1}} \partial_\zeta)^r (\phi(m_1, m_2) y_{n_2+1}^{m_3})}{r!} \\ &= \sum_{i,r=0}^{\infty} \frac{y_{n_i}^i x_{n_2+1}^r \partial_{x_{n_1}}^i \partial_{y_{n_2+1}}^r \partial_\zeta^{i+r} (x_{n_1}^{m_1} \zeta^{m_2} y_{n_2+1}^{m_3})}{i!r!}.\end{aligned} \tag{6.4.19}$$

By Lemma 6.1.1 with $T_1 = \partial_{x_{n_2+1}}$, $T_1^- = \int_{(x_{n_2+1})}$ (cf. (6.1.20)) and $T_2 = -\partial_{y_{n_2+1}} \partial_\zeta$, the polynomial solution space of the system (6.4.13) and the second equation in (6.4.14) is

$$\left[\sum_{m_1,m_2=0}^{\infty} \mathbb{F}[x_{n_1+1}] \psi(m_1, m_2) + \sum_{m_1=1}^{\infty} \sum_{m_2,m_3=0}^{\infty} \mathbb{F}\phi(m_1, m_2, m_3)][\mathbb{F}[y_{n_2}] \right]. \tag{6.4.20}$$

The first equation in (6.4.14) says that f can not contain the monomials with $y_{n_2} y_{n_2+1}$ as a factor. Moreover, $\psi(0, m_2) = \zeta^{m_2}$ and $\phi(m_1, m_2, 0) = \phi(m_1, m_2)$. Therefore the polynomial solution space of the system (6.4.13) and (6.4.14) is

$$\begin{aligned}&\mathbb{F}[x_{n_1+1}, y_{n_2}, \zeta] + \sum_{m_1,m_3=1}^{\infty} \sum_{m_2=0}^{\infty} \mathbb{F}\phi(m_1, m_2, m_3) \\ &+ \sum_{m_1=1}^{\infty} \sum_{m_2=0}^{\infty} (\mathbb{F}[y_{n_2}] \phi(m_1, m_2) + \mathbb{F}[x_{n_1+1}] \psi(m_1, m_2)).\end{aligned} \tag{6.4.21}$$

According to (6.3.41),

$$x_{n_1+1}^{m_1} y_{n_2}^{m_2} \zeta^{m_3} = \eta^{m_3} (x_{n_1+1}^{m_1} y_{n_2}^{m_2}), \tag{6.4.22}$$

$$\eta^{m_2} (x_{n_1}^{m_1} y_{n_2}^{m_3}) = (\zeta + y_{n_1} \partial_{x_{n_1}})^{m_2} (x_{n_1}^{m_1} y_{n_2}^{m_3}) = \phi(m_1, m_2) y_{n_2}^{m_3}, \tag{6.4.23}$$

$$\eta^{m_2} (y_{n_2+1}^{m_1} x_{n_1+1}^{m_3}) = (\zeta + x_{n_2+1} \partial_{y_{n_2+1}})^{m_2} (y_{n_2+1}^{m_1} x_{n_1+1}^{m_3}) = \psi(m_1, m_2) x_{n_1+1}^{m_3}, \tag{6.4.24}$$

$$\eta^{m_2} (x_{n_1}^{m_1} y_{n_2+1}^{m_3}) = (\zeta + y_{n_1} \partial_{x_{n_1}} + x_{n_2+1} \partial_{y_{n_2+1}})^{m_2} (x_{n_1}^{m_1} y_{n_2+1}^{m_3}) = \phi(m_1, m_2, m_3). \tag{6.4.25}$$

It can be verified that $\{\eta^{m_1} (x_i^{m_2} y_j^{m_3}) \mid m_1, m_2, m_3 \in \mathbb{N}; i = n_1, n_1 + 1; j = n_2, n_2 + 1\}$ are singular vectors. By (6.4.19)–(6.4.25), we have:

Lemma 6.4.1 *The nonzero vectors in*

$$\{\mathbb{F}[\eta](x_i^{m_1} y_j^{m_2}) \mid m_1, m_2 \in \mathbb{N}; i = n_1, n_1 + 1; j = n_2, n_2 + 1\} \tag{6.4.26}$$

are all the singular vectors of $sl(n, \mathbb{F})$ *in* $\mathscr{B} = \mathbb{F}[x_1, \ldots, x_{n_1}, y_1, \ldots, y_{n_2}]$.

Similarly, when $n_2 = n$ *and* $n_1 \leq n - 2$, *the nonzero vectors in*

$$\{\mathbb{F}[\eta](x_i^{m_1} y_n^{m_2}) \mid m_1, m_2 \in \mathbb{N}; i = n_1, n_1 + 1\} \tag{6.4.27}$$

are all the singular vectors of $sl(n, \mathbb{F})$ *in* $\mathscr{B}$.

Denote

$$\mathscr{H} = \{f \in \mathscr{B} \mid \tilde{\Delta}(f) = 0\}. \tag{6.4.28}$$

By (6.3.42), $\mathscr{H}$ forms an $sl(n, \mathbb{F})$-submodule. Set

$$\mathscr{B}_{\langle \ell_1, \ell_2 \rangle} = \text{Span}\{x^\alpha y^\beta \mid \alpha, \beta \in \mathbb{N}^n; \sum_{r=n_1+1}^{n} \alpha_r - \sum_{i=1}^{n_1} \alpha_i = \ell_1; \sum_{i=1}^{n_2} \beta_i - \sum_{r=n_2+1}^{n} \beta_r = \ell_2\} \tag{6.4.29}$$

for $\ell_1, \ell_2 \in \mathbb{Z}$. Then

$$\mathscr{B}_{\langle \ell_1, \ell_2 \rangle} = \{f \in \mathscr{B} \mid \tilde{D}(f) = \ell_1 f; \tilde{D}'(f) = \ell_2 f\}. \tag{6.4.30}$$

By (6.3.42), $\mathscr{B}_{\langle \ell_1, \ell_2 \rangle}$ forms an $sl(n, \mathbb{F})$-submodule, and so does

$$\mathscr{H}_{\langle \ell_1, \ell_2 \rangle} = \mathscr{B}_{\langle \ell_1, \ell_2 \rangle} \bigcap \mathscr{H}. \tag{6.4.31}$$

Next (6.3.40) and (6.3.41) imply

$$[\tilde{\Delta}, \eta] = n_1 - n_2 - \tilde{D} - \tilde{D}', \quad \tilde{\Delta}(x_i^{m_1} y_j^{m_2}) = 0 \tag{6.4.32}$$

for $m_1, m_2 \in \mathbb{N}$, $i = n_1, n_1 + 1$ and $j = n_2, n_2 + 1$. Thus

$$x_{n_1+1}^{m_1} y_{n_2}^{m_2} \in \mathscr{H}_{\langle m_1, m_2 \rangle}, \qquad x_{n_1+1}^{m_1} y_{n_2+1}^{m_2} \in \mathscr{H}_{\langle m_1, -m_2 \rangle}, \tag{6.4.33}$$

$$x_{n_1}^{m_1} y_{n_2}^{m_2} \in \mathscr{H}_{\langle -m_1, m_2 \rangle}, \qquad x_{n_1}^{m_1} y_{n_2+1}^{m_2} \in \mathscr{H}_{\langle -m_1, -m_2 \rangle}. \tag{6.4.34}$$

For any $g \in \mathscr{H}_{\langle \ell_1, \ell_2 \rangle}$ and $0 < m \in \mathbb{N}$, we have

$$\eta^m(g) \in \mathscr{B}_{\ell_1+m, \ell_2+m} \tag{6.4.35}$$

and

$$\tilde{\Delta}(\eta^m(g)) = m(-\ell_1 - \ell_2 + n_1 - n_2 - m + 1)\eta^{m-1}(g). \tag{6.4.36}$$

Thus

$$\tilde{\Delta}(\eta^m(g)) = 0 \text{ if and only if } \ell_1 + \ell_2 \le n_1 - n_2 \text{ and } m = n_1 - n_2 - \ell_1 - \ell_2 + 1. \tag{6.4.37}$$

If so,

$$\eta^m(g) \in \mathscr{H}_{n_1-n_2-\ell_2+1, n_1-n_2-\ell_1+1}. \tag{6.4.38}$$

Note

$$(n_1 - n_2 - \ell_2 + 1) + (n_1 - n_2 - \ell_1 + 1) = n_1 - n_2 + 2 + (n_1 - n_2 - \ell_1 - \ell_2) \ge n_1 - n_2 + 2. \tag{6.4.39}$$

Let $f_{\langle \ell_1, \ell_2 \rangle} \in \mathscr{H}_{\langle \ell_1, \ell_2 \rangle}$ be a singular vector in (6.4.33) and (6.4.34). Then the singular vectors in $\mathscr{H}$ are nonzero weight vectors in

$$\text{Span}\{f_{\langle \ell_1, \ell_2 \rangle}, \eta^{n_1-n_2+1-r_1-r_2}(f_{\langle r_1, r_2 \rangle}) \mid \ell_1, \ell_2, r_1, r_2 \in \mathbb{Z}; r_1 + r_2 \le n_1 - n_2\} \tag{6.4.40}$$

by Lemma 6.4.1, where

$$\eta^{n_1-n_2+1-r_1-r_2}(f_{\langle r_1, r_2 \rangle}) \in \mathscr{H}_{\langle n_1-n_2+1-r_2, n_1-n_2+1-r_1 \rangle}. \tag{6.4.41}$$

Thus we get:

Lemma 6.4.2 *When $n_1 + 1 < n_2 < n$, we have*

$$\mathscr{H}_{\langle \ell_1, \ell_2 \rangle} \text{ has a unique singular vector if } \ell_1 + \ell_2 \le n_1 - n_2 + 1 \tag{6.4.42}$$

and

$$\mathscr{H}_{\langle \ell_1, \ell_2 \rangle} \text{ has exactly two singular vectors if } \ell_1 + \ell_2 > n_1 - n_2 + 1. \tag{6.4.43}$$

In the case $n_1 + 1 < n_2 = n$, $\mathscr{B}_{\langle \ell_1, \ell_2 \rangle} = 0$ if $\ell_2 < 0$, and for $\ell_1 \in Z$ and $\ell_2 \in \mathbb{N}$,

$$\mathscr{H}_{\langle \ell_1, \ell_2 \rangle} \text{ has a unique singular vector if } \ell_1 \ge n_1 - n + 2 \text{ or } \ell_1 + \ell_2 \le n_1 - n + 1, \tag{6.4.44}$$

$$\mathscr{H}_{\langle \ell_1, \ell_2+1 \rangle} \text{ has exactly two singular vector if and } n_1 - n + 1 - \ell_2 \le \ell_1 \le n_1 - n + 1. \tag{6.4.45}$$

According to the bilinear form $(\cdot|\cdot)$ in Sect. 6.3 with n replaced by $2n$, $\overline{1, n_1}$ replaced by $\overline{1, n_1} \bigcup \overline{n + n_2 + 1, 2n}$ and x_{n+i} replaced by y_i for $i \in \overline{1, n}$, we have

$$(x^\alpha y^\beta | x^{\alpha'} y^{\beta'}) = \delta_{\alpha,\alpha'}\delta_{\beta,\beta'}(-1)^{\sum_{i=1}^{n_1}\alpha_i+\sum_{r=n_2+1}^{n}\beta_r}\alpha!\beta! \qquad \text{for } \alpha,\beta,\alpha',\beta' \in \mathbb{N}^n. \tag{6.4.46}$$

It can be verified that

$$(\tilde{\Delta}(x^\alpha y^\beta)|x^{\alpha'} y^{\beta'}) = -(x^\alpha y^\beta|\eta(x^{\alpha'} y^{\beta'})). \tag{6.4.47}$$

We remind the concerned Lie algebra $\mathcal{G} = sl(n,\mathbb{F})$ in this section. The Cartan subalgebra H is given in (6.2.4). Suppose that $f_{\langle \ell_1,\ell_2\rangle} \in \mathcal{H}_{\langle \ell_1,\ell_2\rangle}$ is a singular vector in (6.4.33) and (6.4.34). Then

$$(f_{\langle \ell_1,\ell_2\rangle}|f_{\langle \ell_1,\ell_2\rangle}) \neq 0 \tag{6.4.48}$$

and

$$(f_{\langle \ell_1,\ell_2\rangle}|f_{\langle \ell_1',\ell_2'\rangle}) = 0 \qquad \text{if } (\ell_1,\ell_2) \neq (\ell_1',\ell_2'). \tag{6.4.49}$$

Recall that $\mathcal{G}_+ = \sum_{1\le i<j\le n}\mathbb{F}E_{i,j}$ is the subalgebra spanned by the positive root vectors and $\mathcal{G}_- = \sum_{1\le i<j\le n}\mathbb{F}E_{j,i}$ is the subalgebra spanned by the negative root vectors. Moreover,

$$\mathcal{G}_-^t = \mathcal{G}_+. \tag{6.4.50}$$

According to (6.3.46)–(6.3.50), $\mathcal{B}$ is nilpotent with respect to $\mathcal{G}_+$. Thus all $\mathcal{H}_{\langle \ell_1,\ell_2\rangle}$ with $\ell_1 + \ell_2 \le n_1 - n_2 + 1$ are irreducible $sl(n,\mathbb{F})$-submodules by Lemmas 6.3.2 and 6.4.2, and so are $\eta^m(\mathcal{H}_{\langle \ell_1,\ell_2\rangle})$ for any $m \in \mathbb{N}$ by (6.3.44).

We extend the transpose to an algebraic anti-isomorphism on $U(\mathcal{G})$ by $1^t = 1$ and

$$(A_1A_2\cdots A_r)^t = A_r^t\cdots A_2^tA_1^t \qquad \text{for } A_i \in \mathcal{G}. \tag{6.4.51}$$

By the irreducibility,

$$\mathcal{H}_{\langle \ell_1,\ell_2\rangle} = U(\mathcal{G}_-)(f_{\langle \ell_1,\ell_2\rangle}) \qquad \text{if } \ell_1+\ell_2 \le n_1 - n_2 + 1. \tag{6.4.52}$$

Let $\ell_1, \ell_2, \ell_1', \ell_2' \in \mathbb{Z}$ such that $\ell_1+\ell_2, \ell_1'+\ell_2' \le n_1 - n_2 + 1$ and $(\ell_1,\ell_2) \neq (\ell_1',\ell_2')$. Then

$$(w(f_{\langle \ell_1,\ell_2\rangle})|f_{\langle \ell_1',\ell_2'\rangle}) = (f_{\langle \ell_1,\ell_2\rangle}|w^t(f_{\langle \ell_1',\ell_2'\rangle})) = 0 \qquad \text{for } w \in U(\mathcal{G}_-)\mathcal{G}- \tag{6.4.53}$$

by Lemma 6.3.1. Since $f_{\langle \ell_1,\ell_2\rangle}$ is a weight vector, we have

$$U(H)(f_{\langle \ell_1,\ell_2\rangle}) \subset \mathbb{F}f_{\langle \ell_1,\ell_2\rangle}. \tag{6.4.54}$$

Thus for any $w_1, w_2 \in U(\mathcal{G}_-)$,

$$(w_1(f_{\langle \ell_1,\ell_2\rangle})|w_2(f_{\langle \ell_1',\ell_2'\rangle})) = (w_2^tw_1(f_{\langle \ell_1,\ell_2\rangle})|f_{\langle \ell_1',\ell_2'\rangle}) = c(f_{\langle \ell_1,\ell_2\rangle})|f_{\langle \ell_1',\ell_2'\rangle}) \tag{6.4.55}$$

for some $c \in \mathbb{F}$ by (6.4.53) and (6.4.54). Hence (6.4.52) implies

$$(\mathscr{H}_{\langle \ell_1,\ell_2\rangle}|\mathscr{H}_{\langle \ell_1',\ell_2'\rangle}) = \{0\}. \tag{6.4.56}$$

For $f \in \mathscr{H}_{\langle \ell_1,\ell_2\rangle}$, $g \in \mathscr{B}$ and $m, m' \in \mathbb{N}$ such that $m \leq m'$, we find

$$\begin{aligned}(\eta^m(f)|\eta^{m'}(g)) &= (-1)^{m'}(\tilde{\Delta}^{m_1}\eta^m(f)|g)\\ &= (-1)^{m'}\delta_{m',m}m![\prod_{r=0}^{m-1}(-\ell_1-\ell_2+n_1-n_2-r)](f|g)\end{aligned} \tag{6.4.57}$$

by (6.4.36) and (6.4.47). In particular, the singular vectors $\eta^{n_1-n_2+1-r_1-r_2}(f_{\langle r_1,r_2\rangle})$ for $r_1, r_2 \in \mathbb{Z}$ with $r_1+r_2 \leq n_1-n_2$ are isotropic polynomials. Moreover, for $m, m' \in \mathbb{N}$ and $\ell_1, \ell_2, \ell_1', \ell_2' \in \mathbb{Z}$ such that $\ell_1+\ell_2, \ell_1'+\ell_2' \leq n_1-n_2+1$,

$$(\eta^m(\mathscr{H}_{\langle \ell_1,\ell_2\rangle})|\eta^{m'}(\mathscr{H}_{\langle \ell_1',\ell_2'\rangle})) = \{0\} \quad \text{if } (m,\ell_1,\ell_1) \neq (m',\ell_1',\ell_2'). \tag{6.4.58}$$

On the other hand, for $0 < m \in \mathbb{N}$,

$$\begin{aligned}&(\eta^m(f_{\langle \ell_1-m,\ell_2-m\rangle})|\eta^m(f_{\langle \ell_1-m,\ell_2-m\rangle}))\\ &= (-1)^m m![\prod_{r=0}^{m-1}(2m-\ell_1-\ell_2+n_1-n_2-r)]\\ &\quad \times (f_{\langle \ell_1-m,\ell_2-m\rangle}|f_{\langle \ell_1-m,\ell_2-m\rangle}) \neq 0\end{aligned} \tag{6.4.59}$$

by (6.4.57). Since the radical of $(\cdot|\cdot)$ on $\eta^m(\mathscr{H}_{\langle \ell_1-m,\ell_2-m\rangle})$ is a proper submodule by Lemma 6.3.1, the irreducibility of $\eta^m(\mathscr{H}_{\langle \ell_1-m,\ell_2-m\rangle})$ implies that

$$(\cdot|\cdot) \text{ is nondegenerate restricted to } \eta^m(\mathscr{H}_{\langle \ell_1-m,\ell_2-m\rangle}). \tag{6.4.60}$$

Fix $\ell_1, \ell_2 \in \mathbb{Z}$ with $\ell_1+\ell_2 \leq n_1-n_2+1$. Set

$$\hat{\mathscr{B}}_{\langle \ell_1,\ell_2\rangle} = \sum_{m=0}^{\infty} \eta^m(\mathscr{H}_{\langle \ell_1-m,\ell_2-m\rangle}). \tag{6.4.61}$$

By (6.4.58) and (6.4.60), the above sum is a direct sum and $(\cdot|\cdot)$ is nondegenerate restricted to $\hat{\mathscr{B}}_{\langle \ell_1,\ell_2\rangle}$. For any $g \in \mathscr{B}_{\langle \ell_1,\ell_2\rangle}$, there exists an $m_0 \in \mathbb{N}$ such that

$$\tilde{\Delta}^{m_0+1}(g) = 0 \tag{6.4.62}$$

by (6.3.40). So

$$(g|\eta^m(\mathscr{H}_{\langle \ell_1-m,\ell_2-m\rangle})) = (-1)^m(\tilde{\Delta}^m(g)|\mathscr{H}_{\langle \ell_1-m,\ell_2-m\rangle}) = \{0\} \quad \text{for } m_0 < m \in \mathbb{N}. \tag{6.4.63}$$

By the arguments in (6.3.23)–(6.3.29), there exists

$$g' \in \sum_{m=0}^{m_0} \eta^m(\mathscr{H}_{\langle \ell_1 - m, \ell_2 - m\rangle}) \text{ such that } \left(g - g' | \sum_{m=0}^{m_0} \eta^m(\mathscr{H}_{\langle \ell_1 - m, \ell_2 - m\rangle})\right) = \{0\}. \tag{6.4.64}$$

On the other hand,

$$(g - g' | \sum_{k=m_0+1}^{\infty} \eta^k(\mathscr{H}_{\langle \ell_1 - k, \ell_2 - k\rangle}) = \{0\} \tag{6.4.65}$$

by (6.4.58) and (6.4.63). Thus $g - g' \in \hat{\mathscr{B}}^{\perp}_{\langle \ell_1, \ell_2\rangle}$. Therefore

$$\mathscr{B}_{\langle \ell_1, \ell_2\rangle} = \hat{\mathscr{B}}_{\langle \ell_1, \ell_2\rangle} \oplus (\hat{\mathscr{B}}^{\perp}_{\langle \ell_1, \ell_2\rangle} \bigcap \mathscr{B}_{\langle \ell_1, \ell_2\rangle}). \tag{6.4.66}$$

If $\hat{\mathscr{B}}^{\perp}_{\langle \ell_1, \ell_2\rangle} \bigcap \mathscr{B}_{\langle \ell_1, \ell_2\rangle} \neq \{0\}$, then it contains a singular vector, which must be of the form $\eta^{m_1}(f_{\langle \ell_1 - m_1, \ell_2 - m_1\rangle})$ for some $m_1 \in \mathbb{N}$ by Lemma 6.4.1. This contradicts (6.4.59). Thus $\hat{\mathscr{B}}^{\perp}_{\langle \ell_1, \ell_2\rangle} \bigcap \mathscr{B}_{\langle \ell_1, \ell_2\rangle} = \{0\}$, or equivalently

$$\mathscr{B}_{\langle \ell_1, \ell_2\rangle} = \bigoplus_{m=0}^{\infty} \eta^m(\mathscr{H}_{\langle \ell_1 - m, \ell_2 - m\rangle}) \tag{6.4.67}$$

is completely reducible. Applying (6.4.67) to $\mathscr{B}_{\langle \ell_1 - 1, \ell_2 - 1\rangle}$, we have

$$\mathscr{B}_{\ell_1, \ell_2} = \mathscr{H}_{\langle \ell_1, \ell_2\rangle} \oplus \eta(\mathscr{B}_{\langle \ell_1 - 1, \ell_2 - 1\rangle}) \quad \text{if } \ell_1 + \ell_2 \leq n_1 - n_2 + 1. \tag{6.4.68}$$

Assume $n_1 + 1 < n_2 = n$. For $\ell_1 \in \mathbb{Z}$ and $\ell_2 \in \mathbb{N}$ such that $\ell_1 \geq n_1 - n + 2$ or $\ell_1 + \ell_2 \leq n_1 - n + 1$, all $\mathscr{H}_{\ell_1, \ell_2}$ are irreducible submodules of $\mathscr{B}_{\ell_1, \ell_2}$ by Lemma 6.3.2, (6.4.44) and (6.4.45). In summary, we have:

Theorem 6.4.3 *Suppose $n_1 + 1 < n_2$. For $\ell_1, \ell_2 \in \mathbb{Z}$ with $\ell_1 + \ell_2 \leq n_1 - n_2 + 1$, $\mathscr{H}_{\langle \ell_1, \ell_2\rangle}$ is an infinite-dimensional irreducible highest-weight $sl(n, \mathbb{F})$-module and*

$$\mathscr{B}_{\langle \ell_1, \ell_2\rangle} = \bigoplus_{m=0}^{\infty} \eta^m(\mathscr{H}_{\langle \ell_1 - m, \ell_2 - m\rangle}) \tag{6.4.69}$$

is an orthogonal decomposition of irreducible submodules. In particular, $\mathscr{B}_{\langle \ell_1, \ell_2\rangle} = \mathscr{H}_{\langle \ell_1, \ell_2\rangle} \oplus \eta(\mathscr{B}_{\langle \ell_1 - 1, \ell_2 - 1\rangle})$. The symmetric bilinear form restricted to $\eta^m(\mathscr{H}_{\langle \ell_1 - m, \ell_2 - m\rangle})$ is nondegenerate. If $n_2 < n$, all $\mathscr{H}_{\langle \ell_1, \ell_2\rangle}$ for $\ell_1, \ell_2 \in \mathbb{Z}$ with $\ell_1 + \ell_2 > n_1 - n_2 + 1$ have exactly two singular vectors.

Assume $n_2 = n$. Then $\mathscr{B}_{\langle \ell, 0\rangle} = \mathscr{H}_{\langle \ell, 0\rangle}$ with $\ell \in \mathbb{Z}$ are infinite-dimensional irreducible highest-weight $sl(n, \mathbb{F})$-modules. All $\mathscr{H}_{\langle \ell_1, \ell_2\rangle}$ for $\ell_1 \in \mathbb{Z}$ and $\ell_2 \in \mathbb{N}$ such that $\ell_1 \geq n_1 - n + 2$ are also infinite-dimensional irreducible highest-weight $sl(n, \mathbb{F})$-modules. Moreover, for $\ell_2 \in 1 + \mathbb{N}$ and $n_1 - n_2 + 1 + \ell_2 \leq \ell_1 \in \mathbb{Z}$, the orthogonal *decompositions in (6.4.69) hold. Furthermore,* $\mathscr{H}_{\langle \ell_1, \ell_2+1\rangle}$ *for* $\ell_1 \in \mathbb{Z}$ *and* $\ell_2 \in \mathbb{N}$ *such that* $n_1 - n + 1 - \ell_2 \leq \ell_1 \leq n_1 - n + 1$ *have exactly two singular vectors.*

Indeed, we have more detailed information. Suppose $n_1 + 1 < n_2 < n$. For $m_1, m_2 \in \mathbb{N}$ with $m_1 + m_2 \geq n_2 - n_1 - 1$, $\mathscr{H}_{\langle -m_1, -m_2\rangle}$ has a highest-weight vector $x_{n_1}^{m_1} y_{n_2+1}^{m_2}$ of weight $m_1\lambda_{n_1-1} - (m_1+1)\lambda_{n_1} - (m_2+1)\lambda_{n_2} + m_2(1-\delta_{n_2,n-1})\lambda_{n_2+1}$. When $m_1, m_2 \in \mathbb{N}$ with $m_2 - m_1 \geq n_2 - n_1 - 1$, $\mathscr{H}_{\langle m_1, -m_2\rangle}$ has a highest-weight vector $x_{n_1+1}^{m_1} y_{n_2+1}^{m_2}$ of weight $-(m_1+1)\lambda_{n_1} + m_1\lambda_{n_1+1} - (m_2+1)\lambda_{n_2} + m_2(1-\delta_{n_2,n-1})\lambda_{n_2+1}$. If $m_1, m_2 \in \mathbb{N}$ with $m_1 - m_2 \geq n_2 - n_1 - 1$, $\mathscr{H}_{\langle -m_1, m_2\rangle}$ is has a highest-weight vector $x_{n_1}^{m_1} y_{n_2}^{m_2}$ of weight $m_1\lambda_{n_1-1} - (m_1+1)\lambda_{n_1} + m_2\lambda_{n_2-1} - (m_2+1)\lambda_{n_2}$.

Assume $n_1 + 1 < n_2 = n$. When $m_1, m_2 \in \mathbb{N}$, $\mathscr{H}_{\langle m_1, m_2\rangle}$ has a highest-weight vector $x_{n_1+1}^{m_1} y_n^{m_2}$ of weight $-(m_1+1)\lambda_{n_1} + m_1\lambda_{n_1+1} + m_2\lambda_{n-1}$. If $m_1, m_2 \in \mathbb{N}$ with $m_1 \leq n - n_1 - 2$ or $m_2 - m_1 \leq n_1 - n + 1$, $\mathscr{H}_{\langle -m_1, m_2\rangle}$ has a highest-weight vector $x_{n_1}^{m_1} y_n^{m_2}$ of weight $m_1\lambda_{n_1-1} - (m_1+1)\lambda_{n_1} + m_2\lambda_{n-1}$.

By Lemma 6.1.1, $\mathscr{H}_{\langle \ell_1, \ell_2\rangle}$ has a basis

$$\left\{ \sum_{i=0}^{\infty} \frac{(x_{n_1+1}y_{n_1+1})^i (\tilde{\Delta} + \partial_{x_{n_1+1}}\partial_{y_{n_1+1}})^i (x^\alpha y^\beta)}{\prod_{r=1}^{i}(\alpha_{n_1+1}+r)(\beta_{n_1+1}+r)} \mid \alpha, \beta \in \mathbb{N}^n; \right.$$
$$\left. \alpha_{n_1+1}\beta_{n_1+1} = 0; \sum_{i=1}^{n_1} \alpha_i + \ell_1 = \sum_{r=n_1+1}^{n} \alpha_r; \sum_{i=1}^{n_2} \beta_i = \sum_{r=n_2+1}^{n} \beta_r + \ell_2 \right\}. \tag{6.4.70}$$

6.5 Noncanonical Representations III: $n_1 + 1 = n_2$

In this section, we study the representation of $sl(n, \mathbb{F})$ given in (6.3.36)–(6.3.38) with $n_1 + 1 = n_2$.

First we consider the subcase $n_2 < n$. Suppose that $f \in \mathcal{Q}$ is a singular vector. Note $\zeta = x_{n_1+1}y_{n_1+1}$ and $y_{n_2} = y_{n_1+1}$ in this case. According to the arguments in (6.4.1)–(6.4.5), f is a rational function in

$$\{x_{n_1}, x_{n_1+1}, x_{n_1+2}, y_{n_1}, y_{n_1+2}, \zeta, \zeta_1, \zeta_2\}. \tag{6.5.1}$$

By (6.4.6),

$$f_{x_{n_1}x_{n_1+1}} - y_{n_1-1}f_{\zeta_1 x_{n_1+1}} + y_{n_1+1}(f_{x_{n_1}\zeta} - y_{n_1-1}f_{\zeta_1\zeta} - f_{y_{n_1}} - x_{n_1-1}f_{\zeta_1}) = 0. \tag{6.5.2}$$

Substituting

$$x_{n_1-1} = y_{n_1}^{-1}\zeta_1 + x_{n_1}y_{n_1}^{-1}y_{n_1-1}, \quad y_{n_1+1} = x_{n_1+1}^{-1}\zeta \tag{6.5.3}$$

into (6.5.2), we get

$$\begin{aligned} &f_{x_{n_1}x_{n_1+1}}+x_{n_1+1}^{-1}\zeta(f_{x_{n_1}\zeta}-f_{y_{n_1}}-y_{n_1}^{-1}\zeta_1 f_{\zeta_1})\\ &-y_{n_1-1}[f_{\zeta_1 x_{n_1+1}}+x_{n_1+1}^{-1}\zeta(f_{\zeta_1\zeta}+x_{n_1}y_{n_1}^{-1}f_{\zeta_1})]=0. \end{aligned} \tag{6.5.4}$$

Since f is independent of y_{n_1-1} in terms of (6.5.1), we have

$$f_{\zeta_1 x_{n_1+1}}+x_{n_1+1}^{-1}\zeta(f_{\zeta_1\zeta}+x_{n_1}y_{n_1}^{-1}f_{\zeta_1})]=0; \tag{6.5.5}$$

equivalently

$$x_{n_1+1}f_{\zeta_1 x_{n_1+1}}+\zeta(f_{\zeta_1\zeta}+x_{n_1}y_{n_1}^{-1}f_{\zeta_1})=0. \tag{6.5.6}$$

Change variable

$$f_{\zeta_1}=e^{-x_{n_1}y_{n_1}^{-1}\zeta}g. \tag{6.5.7}$$

Then

$$x_{n_1+1}f_{\zeta_1 x_{n_1+1}}=x_{n_1+1}e^{-x_{n_1}y_{n_1}^{-1}\zeta}g_{x_{n_1+1}},\quad f_{\zeta_1\zeta}+x_{n_1}y_{n_1}^{-1}f_{\zeta_1}=e^{-x_{n_1}y_{n_1}^{-1}\zeta}g_\zeta. \tag{6.5.8}$$

Thus (6.5.6) is equivalent to

$$x_{n_1+1}g_{x_{n_1+1}}+\zeta g_\zeta=0, \tag{6.5.9}$$

which says that the total degree of g with respect to x_{n_1+1} and ζ is zero. In particular, x_{n_1+1}/ζ is a solution. Changing variables $(x_{n_1+1},\zeta)\mapsto(x_{n_1+1}/\zeta,\zeta)$, we can prove

$$g=h(x_{n_1+1}/\zeta) \tag{6.5.10}$$

for some one-variable function $h(z)$ (it can also be derived by the method of characteristic line in partial differential equation). Since

$$f_{\zeta_1}=e^{-x_{n_1}y_{n_1}^{-1}\zeta}h(x_{n_1+1}/\zeta) \tag{6.5.11}$$

is rational in x_{n_1+1}, $h(z)$ must be a rational function in z. Now f_{ζ_1} and $h(x_{n_1+1}/\zeta)$ are rational in ζ. Equation (6.5.11) forces $h(z)=0$ and so $f_{\zeta_1}=0$. Thus f is independent of ζ_1 in terms of (6.5.1). Symmetrically, f is independent of ζ_2 in terms of (6.5.1). Hence f a rational function in

$$\{x_{n_1},x_{n_1+1},x_{n_1+2},y_{n_1},y_{n_1+2},\zeta\}. \tag{6.5.12}$$

The fact $\zeta=x_{n_1+1}y_{n_1+1}$ allows us to rewrite f as a rational function in

$$\{x_{n_1},x_{n_1+1},x_{n_1+2},y_{n_1},y_{n_1+1},y_{n_1+2}\}. \tag{6.5.13}$$

Now f is a singular vector if and only if it is a weight vector satisfying the following system of differential equations

$$E_{n_1,n_1+1}(f) = (\partial_{x_{n_1}}\partial_{x_{n_1+1}} - y_{n_1+1}\partial_{y_{n_1}})(f) = 0, \tag{6.5.14}$$

$$E_{n_1+1,n_1+2}(f) = (x_{n_1+1}\partial_{x_{n_1+2}} - \partial_{y_{n_1+1}}\partial_{y_{n_1+2}})(f) = 0. \tag{6.5.15}$$

Since the differential operators in the above two equations do not commute, we have to solve them according to the proof of Theorem 1.5.1. Note

$$E_{n_1,n_1+2}|_{\mathcal{Q}} = [E_{n_1,n_1+1}|_{\mathcal{Q}}, E_{n_1+1,n_1+2}|_{\mathcal{Q}}] = \partial_{x_{n_1}}\partial_{x_{n_1+2}} - \partial_{y_{n_1}}\partial_{y_{n_1+2}}. \tag{6.5.16}$$

So

$$(\partial_{x_{n_1}}\partial_{x_{n_1+2}} - \partial_{y_{n_1}}\partial_{y_{n_1+2}})(f) = 0. \tag{6.5.17}$$

For our purpose of representation, we only consider f is a polynomial in $\{x_i, y_i \mid i = n_1, n_1+1, n_1+2\}$. Now

$$\tilde{\Delta} = \sum_{i=1}^{n_1} x_i\partial_{y_i} - \partial_{x_{n_1+1}}\partial_{y_{n_1+1}} + \sum_{s=n_1+2}^{n} y_s\partial_{x_s} \tag{6.5.18}$$

and

$$\eta = \sum_{i=1}^{n_1} y_i\partial_{x_i} + x_{n_1+1}y_{n_1+1} + \sum_{s=n_1+2}^{n} x_s\partial_{y_s}. \tag{6.5.19}$$

Set

$$\begin{aligned}\phi_{m_1,m_2,m_3} &= \left[\prod_{s=1}^{m_2}(m_1+s)\right]\sum_{i=0}^{\infty}\frac{x_{n_1}^{m_1+i}x_{n_1+2}^{i}(\partial_{y_{n_1}}\partial_{y_{n_1+2}})^i(y_{n_1}^{m_2}y_{n_1+2}^{m_3})}{i!\prod_{r=1}^{i}(m_1+r)}\\ &= \sum_{i=0}^{\infty}\binom{m_2}{i}\partial_{x_{n_1}}^{m_2-i}(x_{n_1}^{m_1+m_2})x_{n_1+2}^{i}y_{n_1}^{m_2-i}\partial_{y_{n_1+2}}^{i}(y_{n_1+2}^{m_3})\\ &= (y_{n_1}\partial_{x_{n_1}} + x_{n_1+2}\partial_{y_{n_1+2}})^{m_2}(x_{n_1}^{m_1+m_2}y_{n_1+2}^{m_3})\end{aligned} \tag{6.5.20}$$

and

$$\begin{aligned}\psi_{m_1,m_2,m_3} &= \frac{(m_1+m_2)!\prod_{s=1}^{m_1}(m_3+s)}{m_1!}\sum_{i=0}^{\infty}\frac{x_{n_1}^{i}x_{n_1+2}^{m_1+i}(\partial_{y_{n_1}}\partial_{y_{n_1+2}})^i(y_{n_1}^{m_2}y_{n_1+2}^{m_3})}{i!\prod_{r=1}^{i}(m_1+r)}\\ &= \sum_{i=0}^{m_2}\binom{m_2}{i}\frac{(m_1+m_2)!x_{n_1}^{i}x_{n_1+2}^{m_1+i}y_{n_1}^{m_2-i}\partial_{y_{n_1+2}}^{m_1+i}(y_{n_1+2}^{m_1+m_3})}{(m_1+i)!}\end{aligned}$$

$$
\begin{aligned}
&= \sum_{i=0}^{m_2} \binom{m_2}{m_2 - i} \frac{(m_1+m_2)! x_{n_1}^i y_{n_1}^{m_2-i} (x_{n_1+2}\partial_{y_{n_1+2}})^{m_1+i}(y_{n_1+2}^{m_1+m_3})}{(m_1+i)!} \\
&= \sum_{i=0}^{m_2} \frac{(m_1+m_2)!(y_{n_1}\partial_{x_{n_1}})^{m_2-i}(x_{n_1}^{m_2})(x_{n_1+2}\partial_{y_{n_1+2}})^{m_1+i}(y_{n_1+2}^{m_1+m_3})}{(m_2-i)!(m_1+i)!} \\
&= \sum_{r=0}^{\infty} \frac{(m_1+m_2)!(y_{n_1}\partial_{x_{n_1}})^{r}(x_{n_1}^{m_2})(x_{n_1+2}\partial_{y_{n_1+2}})^{m_1+m_2-r}(y_{n_1+2}^{m_1+m_3})}{r!(m_1+m_2-r)!} \\
&= (y_{n_1}\partial_{x_{n_1}} + x_{n_1+2}\partial_{y_{n_1+2}})^{m_1+m_2}(x_{n_1}^{m_2} y_{n_1+2}^{m_1+m_3}).
\end{aligned} \tag{6.5.21}
$$

By Lemma 6.1.1 with $T_1 = \partial_{x_{n_1}}\partial_{x_{n_1+2}}$, $T_1^- = \int_{(x_{n_1+1})}\int_{(x_{n_1+2})}$ (cf. (6.1.20)) and $T_2 = \partial_{y_{n_1}}\partial_{y_{n_1+2}}$, the polynomial solution space of (6.5.17) is

$$\mathscr{S} = \text{Span}\{\phi_{m_1,m_2,m_3} x_{n_1+1}^{m_4} y_{n_1+1}^{m_5}, \psi_{m_1+1,m_2,m_3} x_{n_1+1}^{m_4} y_{n_1+1}^{m_5} \mid m_i \in \mathbb{N}\}. \tag{6.5.22}$$

On the other hand,

$$\partial_{x_{n_1}}(\phi_{0,m_2,m_3}) = m_2 \psi_{1,m_2-1,m_3-1}, \tag{6.5.23}$$

$$\partial_{x_{n_1}}(\phi_{m_1,m_2,m_3}) = (m_1+m_2)\phi_{m_1-1,m_2,m_3} \qquad \text{if } m_1 > 0, \tag{6.5.24}$$

$$\partial_{x_{n_1}}(\psi_{m_1,m_2,m_3}) = m_2 \psi_{m_1+1,m_2-1,m_3-1}, \tag{6.5.25}$$

$$\partial_{y_{n_1}}(\phi_{m_1,m_2,m_3}) = m_2(m_1+m_2)\phi_{m_1,m_2-1,m_3}, \tag{6.5.26}$$

$$\partial_{y_{n_1}}(\psi_{m_1,m_2,m_3}) = m_2(m_1+m_2)\psi_{m_1,m_2-1,m_3}. \tag{6.5.27}$$

In Lemma 6.1.1, $\mathscr{A}$ is a free $\mathscr{B}$-module, and so the condition (6.1.1) yields

$$\{hg \mid g \in V,\ h \in \mathscr{B};\ T_1(h) = 0\} = \{f \in \mathscr{A} \mid T_1(f) = 0\}. \tag{6.5.28}$$

Thus

$$\{f \in \mathscr{A} \mid (T_1 + T_2)(f) = 0\} = (\sum_{i=0}^{\infty}(-T_1^- T_2)^i)[\{g \in \mathscr{A} \mid T_1(g) = 0\}]. \tag{6.5.29}$$

Next we want to solve (6.5.14) in the space $\mathscr{S}$ in (6.5.22). Let $T_1 = \partial_{x_{n_1}}\partial_{x_{n_1+1}}$ and $T_2 = -y_{n_1+1}\partial_{y_{n_1}}$. First we want to find the solution space

$$\{g \in \mathscr{S} \mid T_1(g) = 0\} = \{g \in \mathscr{S} \mid \partial_{x_{n_1}}\partial_{x_{n_1+1}}(g) = 0\}. \tag{6.5.30}$$

Note that

$$\partial_{x_{n_1}}(\phi_{m_1,m_2,m_3}) = 0 \Longleftrightarrow m_1 = m_2 m_3 = 0 \tag{6.5.31}$$

and

$$\partial_{x_{n_1}}(\psi_{m_1,m_2,m_3}) = 0 \Longleftrightarrow m_2 m_3 = 0. \tag{6.5.32}$$

Moreover,

$$\phi_{0,m_2,0} = m_2! y_{n_1}^{m_2}, \quad \phi_{0,0,m_3} = y_{n_1+2}^{m_3}, \tag{6.5.33}$$

$$\psi_{m_1,m_2,0} = x_{n_1+2}^{m_1} y_{n_1}^{m_2}, \quad \psi_{m_1,0,m_3} = x_{n_1+2}^{m_1} y_{n_+2}^{m_3}. \tag{6.5.34}$$

Since

$$\partial_{x_{n_1+1}}(\phi_{m_1,m_2,m_3}) = \partial_{x_{n_1+1}}(\phi_{m_1,m_2,m_3}) = 0 \tag{6.5.35}$$

by (6.5.20) and (6.5.21), we have

$$\{g \in \mathscr{S} \mid T_1(g) = 0\} = \text{span}\{\phi_{m_1,m_2,m_3} y_{n_1+1}^{m_4}, \psi_{m_1,m_2,m_3} y_{n_1+1}^{m_4}, y_{n_1}^{m_1} x_{n_1+1}^{m_2} y_{n_1+1}^{m_3} x_{n_1+2}^{m_4}, x_{n_1+1}^{m_1} y_{n_1+1}^{m_2} x_{n_1+2}^{m_3} y_{n_1}^{m_4} \mid m_1, m_2, m_3, m_4 \in \mathbb{N}\}. \tag{6.5.36}$$

By (6.5.23)–(6.5.25), we take T_1^- as follows:

$$T_1^-(\phi_{m_1,m_2,m_3} x_{n_1+1}^{m_4} y_{n_1+1}^{m_5}) = \frac{1}{(m_1+m_2+1)(m_4+1)} \phi_{m_1+1,m_2,m_3} x_{n_1+1}^{m_4+1} y_{n_1+1}^{m_5}, \tag{6.5.37}$$

$$T_1^-(\psi_{1,m_2,m_3} x_{n_1+1}^{m_4} y_{n_1+1}^{m_5}) = \frac{1}{(m_2+1)(m_4+1)} \phi_{0,m_2+1,m_3+1} x_{n_1+1}^{m_4+1} y_{n_1+1}^{m_5}, \tag{6.5.38}$$

$$T_1^-(\psi_{m_1+2,m_2,m_3} x_{n_1+1}^{m_4} y_{n_1+1}^{m_5}) = \frac{1}{(m_2+1)(m_4+1)} \psi_{m_1+1,m_2+1,m_3+1} x_{n_1+1}^{m_4+1} y_{n_1+1}^{m_5} \tag{6.5.39}$$

for $m_i \in \mathbb{N}$ with $i \in \overline{1,5}$. In particular,

$$T_1^- \partial_{y_{n_1}}(\phi_{m_1,m_2,m_3} x_{n_1+1}^{m_4} y_{n_1+1}^{m_5}) = \frac{m_2}{m_4+1} \phi_{m_1+1,m_2-1,m_3} x_{n_1+1}^{m_4+1} y_{n_1+1}^{m_5}, \tag{6.5.40}$$

$$T_1^- \partial_{y_{n_1}}(\psi_{1,m_2,m_3} x_{n_1+1}^{m_4} y_{n_1+1}^{m_5}) = \frac{m_2+1}{m_4+1} \phi_{0,m_2,m_3+1} x_{n_1+1}^{m_4+1} y_{n_1+1}^{m_5}, \tag{6.5.41}$$

$$T_1^- \partial_{y_{n_1}}(\psi_{m_1+2,m_2,m_3} x_{n_1+1}^{m_4} y_{n_1+1}^{m_5}) = \frac{m_1+m_2+2}{m_4+1} \psi_{m_1+1,m_2,m_3+1} x_{n_1+1}^{m_4+1} y_{n_1+1}^{m_5} \tag{6.5.42}$$

for $m_i \in \mathbb{N}$ with $i \in \overline{1,5}$.

By Lemma 6.1.1, applying the operator $\sum_{i=0}^{\infty}(T_1^- y_{n_1+1}\partial_{y_{n_1}})^i$ to (6.5.36) will lead to the solution space of (6.5.14) in $\mathscr{S}$. We calculate

$$\begin{aligned}&\sum_{i=0}^{\infty}(T_1^- y_{n_1+1}\partial_{y_{n_1}})^i(\phi_{m_1,m_2,m_3})\\&=\sum_{i=0}^{m_2}\binom{m_2}{i}\phi_{m_1+i,m_2-i,m_3}(x_{n_1+1}y_{n_1+1})^i\\&=\sum_{i=0}^{m_2}\binom{m_2}{i}(x_{n_1+1}y_{n_1+1})^i(y_{n_1}\partial_{x_{n_1}}+x_{n_1+2}\partial_{y_{n_1+2}})^{m_2-i}(x_{n_1}^{m_1+m_2}y_{n_1+2}^{m_3})\\&=\eta^{m_2}(x_{n_1}^{m_1+m_2}y_{n_1+2}^{m_3}),\end{aligned}\tag{6.5.43}$$

$$\begin{aligned}&\sum_{i=0}^{\infty}(T_1^- y_{n_1+1}\partial_{y_{n_1}})^i(\psi_{m_1,m_2,m_3})\\&=\sum_{i=0}^{m_1-1}\binom{m_1+m_2}{i}\psi_{m_1-i,m_2,m_3+i}(x_{n_1+1}y_{n_1+1})^i\\&\quad+\sum_{r=0}^{m_2}\binom{m_1+m_2}{m_1+r}\phi_{r,m_2-r,m_1+m_3}(x_{n_1+1}y_{n_1+1})^{m_1+r}\\&=\sum_{i=0}^{m_1-1}\binom{m_1+m_2}{i}(x_{n_1+1}y_{n_1+1})^i(y_{n_1}\partial_{x_{n_1}}+x_{n_1+2}\partial_{y_{n_1+2}})^{m_1+m_2-i}(x_{n_1}^{m_2}y_{n_1+2}^{m_1+m_3})\\&\quad+\sum_{r=0}^{m_2}\binom{m_1+m_2}{m_1+r}(x_{n_1+1}y_{n_1+1})^{m_1+r}(y_{n_1}\partial_{x_{n_1}}+x_{n_1+2}\partial_{y_{n_1+2}})^{m_2-r}(x_{n_1}^{m_2}y_{n_1+2}^{m_1+m_3})\\&=\eta^{m_1+m_2}(x_{n_1}^{m_2}y_{n_1+2}^{m_1+m_3})\end{aligned}\tag{6.5.44}$$

by (6.5.19). Moreover,

$$\begin{aligned}&\left[\prod_{j=1}^{m_1}(m_2+j)\right]\sum_{r=0}^{\infty}(T_1^- y_{n_1+1}\partial_{y_{n_1}})^r(y_{n_1}^{m_1}x_{n_1+1}^{m_2}y_{n_1+1}^{m_3})\\&=\left[\prod_{j=1}^{m_1}(m_2+j)\right]\sum_{r=0}^{m_1}\binom{m_1}{r}\frac{y_{n_1}^{m_1-r}x_{n_1}^r x_{n_1+1}^{m_2+r}y_{n_1+1}^{m_3+r}}{\prod_{s=1}^{r}(m_2+s)}\\&=\sum_{r=0}^{m_1}\binom{m_1+m_2}{m_1-r}(y_{n_1}\partial_{x_{n_1}})^{m_1-r}(x_{n_1}^{m_1})x_{n_1+1}^{m_2+r}y_{n_1+1}^{m_3+r}\\&=\sum_{s=0}^{m_1+m_2}\binom{m_1+m_2}{s}(y_{n_1}\partial_{x_{n_1}})^s(x_{n_1}^{m_1})x_{n_1+1}^{m_1+m_2-r}y_{n_1+1}^{m_1+m_3-r}\\&=\eta^{m_1+m_2}(x_{n_1}^{m_1}y_{n_1+1}^{m_3-m_2}),\end{aligned}\tag{6.5.45}$$

Note that $\eta^{m_2}(x_{n_1}^{m_1}y_{n_1+1}^{m_3}y_{n_1+2}^{m_4})$ is a solution of the system of (6.5.14) and (6.5.17) for any $m_1, m_2, m_3, m_4 \in \mathbb{N}$. Therefore, the polynomial solution space of the system of (6.5.14) and (6.5.17) is

$$\begin{aligned}\mathscr{S}_1 = \text{span}\{&\eta^{m_2}(x_{n_1}^{m_1}y_{n_1+1}^{m_3}y_{n_1+2}^{m_4}), \eta^{m_1+m_2}(x_{n_1}^{m_1}y_{n_1+1}^{m_3-m_2}x_{n_1+2}^{m_4}),\\ &x_{n_1+1}^{m_1}y_{n_1+1}^{m_2}x_{n_1+2}^{m_3}y_{n_1+2}^{m_4} \mid m_1, m_2, m_3, m_4 \in \mathbb{N}\}.\end{aligned} \tag{6.5.46}$$

Recall $E_{n_1+1,n_1+2}|_{\mathscr{B}} = x_{n_1+1}\partial_{x_{n_1+2}} - \partial_{y_{n_1+1}}\partial_{y_{n_1+2}}$. So the solution space of the Eq. (6.5.15) in $\mathscr{S}_1$ is the span of

$$\text{the solutions of (6.5.15) in span}\{\eta^{m_2}(x_{n_1}^{m_1}y_{n_1+1}^{m_3}y_{n_1+2}^{m_4}) \mid m_i \in \mathbb{N}; m_1 > 0\}, \tag{6.5.47}$$

$$\begin{aligned}\text{the solutions of (6.5.15) in span}\{&\eta^{m_1+m_2}(x_{n_1}^{m_1}y_{n_1+1}^{m_3-m_2}x_{n_1+2}^{m_4})\\ &\mid m_i \in \mathbb{N}; m_1 > 0; m_3 - m_2 < 0 \text{ or } m_4 > 0\}\end{aligned} \tag{6.5.48}$$

and

$$\text{the solutions of (6.5.15) in span}\{x_{n_1+1}^{m_1}y_{n_1+1}^{m_2}x_{n_1+2}^{m_3}y_{n_1+2}^{m_4} \mid m_i \in \mathbb{N}\}. \tag{6.5.49}$$

Moreover,

$$E_{n_1+1,n_1+2}[\eta^{m_2}(x_{n_1}^{m_1}y_{n_1+1}^{m_3}y_{n_1+2}^{m_4})] = \eta^{m_2}(E_{n_1+1,n_1+2}(x_{n_1}^{m_1}y_{n_1+1}^{m_3}y_{n_1+2}^{m_4})) = 0 \tag{6.5.50}$$

if and only if $m_3m_4 = 0$, and

$$\begin{aligned}&E_{n_1+1,n_1+2}[\eta^{m_1+m_2}(x_{n_1}^{m_1}y_{n_1+1}^{m_3-m_2}x_{n_1+2}^{m_4})]\\ &= \eta^{m_1+m_2}(E_{n_1+1,n_1+2}(x_{n_1}^{m_1}y_{n_1+1}^{m_3-m_2}x_{n_1+2}^{m_4})) = 0\end{aligned} \tag{6.5.51}$$

if and only if $m_4 = 0$.

Let $T_1 = \partial_{y_{n_1+1}}\partial_{y_{n_1+2}}$, $T_1^- = \int_{(y_{n_1+1})}\int_{(y_{n_1+2})}$ and $T_2 = -x_{n_1+1}\partial_{x_{n_1+2}}$ in Lemma 6.1.1. Then (6.5.49) is determined by

$$\begin{aligned}&\left[\prod_{j=1}^{m_3}(m_2+j)\right]\sum_{r=0}^{\infty}(T_1^- x_{n_1+1}\partial_{x_{n_1+2}})^r(x_{n_1+1}^{m_1}y_{n_1+1}^{m_2}x_{n_1+2}^{m_3})\\ &= \left[\prod_{j=1}^{m_3}(m_2+j)\right]\sum_{r=0}^{m_3}\binom{m_3}{r}\frac{(x_{n_1+1}^{m_1+r}y_{n_1+1}^{m_2+r}y_{n_1+2}^{r}x_{n_1+2}^{m_3-r})}{\prod_{s=1}^{r}(m_2+s)}\\ &= \sum_{r=0}^{m_3}\binom{m_2+m_3}{m_3-r}x_{n_1+1}^{m_1+r}y_{n_1+1}^{m_2+r}x_{n_1+2}^{m_3-r}\partial_{y_{n_2+2}}^{m_3-r}(y_{n_1+2}^{m_3})\end{aligned}$$

$$= \sum_{k=0}^{m_2+m_3} \binom{m_2+m_3}{k} x_{n_1+1}^{m_1+m_2-k} y_{n_1+1}^{m_2+m_3-k} x_{n_1+2}^{k} \partial_{y_{n_2+2}}^{k}(y_{n_1+2}^{m_3})$$
$$= \eta^{m_2+m_3}(x_{n_1+1}^{m_1-m_3} y_{n_1+2}^{m_3}), \tag{6.5.52}$$

$$\left[\prod_{j=1}^{m_3}(m_2+j)\right] \sum_{r=0}^{\infty}(T_1^- x_{n_1+1}\partial_{x_{n_1+2}})^r (x_{n_1+1}^{m_1} y_{n_2+1}^{m_2} x_{n_1+2}^{m_3})$$
$$= \left[\prod_{j=1}^{m_3}(m_2+j)\right] \sum_{r=0}^{m_3} \binom{m_3}{r} \frac{(x_{n_1+1}^{m_1+r} y_{n_1+1}^{r} y_{n_1+2}^{m_2+r} x_{n_1+2}^{m_3-r})}{\prod_{s=1}^{r}(m_2+s)}$$
$$= \sum_{r=0}^{m_3} \binom{m_3}{m_r} x_{n_1+1}^{m_1+r} y_{n_1+1}^{m_2+r} x_{n_1+2}^{m_3-r} \partial_{y_{n_2+2}}^{m_3-r}(y_{n_1+2}^{m_2+m_3})$$
$$= \eta^{m_3}(x_{n_1+1}^{m_1} y_{n_1+2}^{m_2+m_3}). \tag{6.5.53}$$

Therefore, the singular vectors in $\mathscr{B}$ are nonzero vectors in

$$\text{Span}\{\eta^{m_2}(x_i^{m_1} y_j^{m_3}), \eta^{m_1+m_2}(x_{n_1}^{m_2} y_{n_1+1}^{m_3-m_1}), \eta^{m_1+m_2}(y_{n_1+2}^{m_2} x_{n_1+1}^{m_3-m_1})$$
$$\mid m_r \in \mathbb{N}; (i,j) = (n_1, n_1+1), (n_1, n_1+2), (n_1+1, n_1+2)\}. \tag{6.5.54}$$

According to (6.4.36),

$$\tilde{\Delta}[\eta^{m_1+m_2}(x_{n_1}^{m_2} y_{n_1+1}^{m_3-m_1})] = -(m_1+m_2)m_3\eta^{m_1+m_2-1}(x_{n_1}^{m_2} y_{n_1+1}^{m_3-m_1}). \tag{6.5.55}$$

Thus we find a singular

$$\eta^{m_1+m_2}(x_{n_1}^{m_2} y_{n_1+1}^{-m_1}) \in \mathscr{H}_{\langle m_1, m_2\rangle} \tag{6.5.56}$$

of new type if $m_1, m_2 \geq 1$. Symmetrically, $\eta^{m_1+m_2}(y_{n_1+2}^{m_2} x_{n_1+1}^{-m_1}) \in \mathscr{H}_{\langle m_2, m_1\rangle}$ is a singular vector. Recall the singular vectors

$$f_{\langle -m_1,-m_2\rangle} = x_{n_1}^{m_1} y_{n_1+2}^{m_2} \in \mathscr{H}_{\langle -m_1,-m_2\rangle}, \qquad f_{\langle -m_1,m_2\rangle} = x_{n_1}^{m_1} y_{n_1+1}^{m_2} \in \mathscr{H}_{\langle -m_1,m_2\rangle}, \tag{6.5.57}$$
$$f_{\langle m_1,-m_2\rangle} = x_{n_1+1}^{m_1} y_{n_1+2}^{m_2} \in \mathscr{H}_{\langle m_1,-m_2\rangle}. \tag{6.5.58}$$

Moreover, we have the singular vectors

$$\eta^{-\ell_1-\ell_2}(f_{\langle \ell_1,\ell_2\rangle}) \in \mathscr{H}_{\langle -\ell_2,-\ell_1\rangle} \qquad \text{for } \ell_1,\ell_2 \in \mathbb{Z} \text{ with } \ell_1+\ell_2 \leq -1 \tag{6.5.59}$$

by (6.4.35)–(6.4.39). Therefore, any singular vector in $\mathscr{H}$ (cf. (6.4.28)) is a nonzero weight vector in

$$\text{Span}\{f_{\langle \ell_1,\ell_2\rangle}, \eta^{-\ell_1'-\ell_2'}(f_{\langle \ell_1',\ell_2'\rangle}), \eta^{m_1+m_2}(x_{n_1}^{m_2}y_{n_1+1}^{-m_1}), \eta^{m_1+m_2}(y_{n_1+2}^{m_2}x_{n_1+1}^{-m_1})$$
$$\mid \ell_1,\ell_2,\ell_1',\ell_2' \in \mathbb{Z},\ m_1,m_2 \in \mathbb{N}+1; \ell_1 \leq 0 \text{ or } \ell_2 \leq 0; \ell_1'+\ell_2' \leq -1\}. \tag{6.5.60}$$

Assume $n_2 = n$. We similarly find that the singular vectors in $\mathscr{B}$ are nonzero vectors in

$$\text{Span}\{\eta^{m_2}(x_{n-1}^{m_1}y_n^{m_3}), x_n^{m_1}y_n^{m_2}, \eta^{m_1+m_2}(x_{n-1}^{m_2}y_n^{m_3-m_1}) \mid m_i \in \mathbb{N}\}. \tag{6.5.61}$$

In particular, any singular vector in $\mathscr{H}$ (cf. (6.4.28)) is a nonzero weight vector in

$$\text{Span}\{x_{n-1}^{m_1}y_n^{m_2}, x_n^{m_1}, \eta^{m_1+1}(x_{n-1}^{m_1+m_2+1}y_n^{m_2}),$$
$$\eta^{m_1+m_2+2}(x_{n-1}^{m_2+1}y_n^{-m_1-1}) \mid m_1,m_2 \in \mathbb{N}\}. \tag{6.5.62}$$

By the arguments of (6.4.46)–(6.4.68), we have:

Theorem 6.5.1 *Suppose $n_1 + 1 = n_2$. For $\ell_1, \ell_2 \in \mathbb{Z}$ with $\ell_1 + \ell_2 \leq 0$ or $n_2 = n$ and $0 \leq \ell_2 \leq \ell_1$, $\mathscr{H}_{\langle \ell_1,\ell_2\rangle}$ is an infinite-dimensional irreducible highest-weight $sl(n,\mathbb{F})$-module and*

$$\mathscr{B}_{\langle \ell_1,\ell_2\rangle} = \bigoplus_{m=0}^{\infty} \eta^m(\mathscr{H}_{\langle \ell_1-m,\ell_2-m\rangle}) \tag{6.5.63}$$

is an orthogonal decomposition of irreducible submodules. In particular, $\mathscr{B}_{\langle \ell_1,\ell_2\rangle} = \mathscr{H}_{\langle \ell_1,\ell_2\rangle} \oplus \eta(\mathscr{B}_{\langle \ell_1-1,\ell_2-1\rangle})$. The symmetric bilinear form restricted to $\eta^m(\mathscr{H}_{\langle \ell_1-m,\ell_2-m\rangle})$ is nondegenerate.

Assume $n_2 < n$. For $m_1, m_2 \in \mathbb{N}+1$, $\mathscr{H}_{\langle m_1,m_2\rangle}$ has exactly three singular vectors. All the submodules $\mathscr{H}_{\langle \ell_1,\ell_2\rangle}$ for $\ell_1, \ell_2 \in \mathbb{Z}$ such $\ell_1+\ell_2 > 0$ and $\ell_1\ell_2 \leq 0$ have exactly two singular vectors. Consider $n_2 = n$. For $m_1, m_2 \in \mathbb{N}$ with $m_1 < m_2$, $\mathscr{H}_{\langle m_1,m_2\rangle}$ and $\mathscr{H}_{-m_2-m_1-1,m_1+1}$ are also infinite-dimensional irreducible highest-weight $sl(n,\mathbb{F})$-modules. All submodules $\mathscr{H}_{\langle -m_1,m_1+m_2+1\rangle}$ with $m_1, m_2 \in \mathbb{N}$ have exactly two singular vectors.

Indeed, we have more detailed information. Suppose $n_2 < n$. For $m_1, m_2 \in \mathbb{N}$, $\mathscr{H}_{\langle -m_1,-m_2\rangle}$ has a highest-weight vector $x_{n_1}^{m_1}y_{n_1+2}^{m_2}$ of weight $m_1\lambda_{n_1-1} - (m_1+1)\lambda_{n_1} - (m_2+1)\lambda_{n_1+1} + m_2\lambda_{n_1+2}$. When $m_1, m_2 \in \mathbb{N}$ with $m_2 - m_1 \geq 0$, $\mathscr{H}_{\langle m_1,-m_2\rangle}$ has a highest-weight vector $x_{n_1+1}^{m_1}y_{n_1+2}^{m_2}$ of weight $-(m_1+1)\lambda_{n_1} + (m_1-m_2-1)\lambda_{n_1+1} + m_2\lambda_{n_1+2}$. If $m_1, m_2 \in \mathbb{N}$ with $m_1 - m_2 \geq 0$, $\mathscr{H}_{\langle -m_1,m_2\rangle}$ is has a highest-weight vector $x_{n_1}^{m_1}y_{n_1+1}^{m_2}$ of weight $m_1\lambda_{n_1-1} + (m_2-m_1-1)\lambda_{n_1} - (m_2+1)\lambda_{n_1+1}$.

Assume $n_2 = n$. For $m_1, m_2 \in \mathbb{N}$ with $m_2 \leq m_1$, $\mathscr{H}_{\langle -m_1,m_2\rangle}$ has a highest-weight vector $x_{n-1}^{m_1}y_n^{m_2}$ of weight $m_1\lambda_{n-2} + (m_2-m_1-1)\lambda_{n-1}$. Moreover, $\mathscr{H}_{\langle m,0\rangle}$ has a highest-weight vector x_{n-1}^{m} of weight $m\lambda_{n-2} - (m+1)\lambda_{n-1}$ for $m \in \mathbb{Z}$. For $m_1, m_2 \in \mathbb{N}+1$, $\mathscr{H}_{\langle m_1,m_2\rangle}$ has a highest-weight vector $\eta^{m_1+m_2}(x_{n-1}^{m_2}y_n^{-m_1})$ of weight $m_2\lambda_{n-2} + (m_1-m_2-1)\lambda_{n-1}$. Again $\mathscr{H}_{\langle \ell_1,\ell_2\rangle}$ has a basis of the form (6.4.70).

6.6 Noncanonical Representations VI: $n_1 = n_2$

In this section, we continue the discussion from last section.

Recall $n \geq 2$. In this case,

$$\tilde{\Delta} = \sum_{i=1}^{n_1} x_i \partial_{y_i} + \sum_{s=n_1+1}^{n} y_s \partial_{x_s} \tag{6.6.1}$$

and

$$\eta = \sum_{i=1}^{n_1} y_i \partial_{x_i} + \sum_{s=n_1+1}^{n} x_s \partial_{y_s}. \tag{6.6.2}$$

First we consider the subcase $1 < n_1 < n-1$. Suppose that $f \in \mathcal{Q}$ is a singular vector. According to the arguments in (6.4.1)–(6.4.4), f is a rational function in

$$\{x_{n_1}, x_{n_1+1}, y_{n_1}, y_{n_1+1}, \zeta_1, \zeta_2\} \tag{6.6.3}$$

(cf. (6.4.5)). Note

$$E_{n_1,n_1+1}|_{\mathcal{Q}} = \partial_{x_{n_1}} \partial_{x_{n_1+1}} - \partial_{y_{n_1}} \partial_{y_{n_1+1}}. \tag{6.6.4}$$

Now $E_{n_1,n_1+1}(f) = 0$ implies

$$(\partial_{x_{n_1}} \partial_{x_{n_1+1}} - \partial_{y_{n_1}} \partial_{y_{n_1+1}})(f) = 0, \tag{6.6.5}$$

or equivalently,

$$\begin{aligned}(x_{n_1-1}x_{n_1+2} - y_{n_1-1}y_{n_1+2}) f_{\zeta_1\zeta_2} - y_{n_1-1} f_{\zeta_1 x_{n_1+1}} - x_{n_1-1} f_{\zeta_1 y_{n_1+1}} \\ + y_{n_1+2} f_{\zeta_2 x_{n_1}} + x_{n_1+2} f_{\zeta_2 y_{n_1}} + f_{x_{n_1} x_{n_1+1}} - f_{y_{n_1} y_{n_1+1}} = 0.\end{aligned} \tag{6.6.6}$$

According (6.3.49),

$$y_{n_1-1} = x_{n_1}^{-1} y_{n_1} x_{n_1-1} - x_{n_1}^{-1}\zeta_1, \qquad y_{n_1+2} = x_{n_1+1}^{-1}\zeta_2 + x_{n_1+1}^{-1} y_{n_1+1} x_{n_1+2}. \tag{6.6.7}$$

Substituting (6.6.7) into (6.6.6), the coefficient of $x_{n_1-1}x_{n_1+2}$ implies $f_{\zeta_1\zeta_2} = 0$. Thus

$$f = g + h \quad \text{with } g_{\zeta_2} = h_{\zeta_1} = 0. \tag{6.6.8}$$

Now (6.6.6) becomes

$$\begin{aligned}x_{n_1}^{-1}\zeta_1 g_{\zeta_1 x_{n_1+1}} - (x_{n_1}^{-1} y_{n_1} g_{\zeta_1 x_{n_1+1}} + g_{\zeta_1 y_{n_1+1}}) x_{n_1-1} + (x_{n_1+1}^{-1} y_{n_1+1} h_{\zeta_2 x_{n_1}} + h_{\zeta_2 y_{n_1}}) x_{n_1+2} \\ + x_{n_1+1}^{-1}\zeta_2 h_{\zeta_2 x_{n_1}} + g_{x_{n_1} x_{n_1+1}} - g_{y_{n_1} y_{n_1+1}} + h_{x_{n_1} x_{n_1+1}} - h_{y_{n_1} y_{n_1+1}} = 0,\end{aligned} \tag{6.6.9}$$

which implies

$$x_{n_1}^{-1} y_{n_1} g_{\zeta_1 x_{n_1+1}} + g_{\zeta_1 y_{n_1+1}} = 0, \qquad x_{n_1+1}^{-1} y_{n_1+1} h_{\zeta_2 x_{n_1}} + h_{\zeta_2 y_{n_1}} = 0. \tag{6.6.10}$$

For the representation purpose, we assume that g is polynomial in ζ_1 with $g|_{\zeta_1=0} = 0$ and h is polynomial in ζ_2. The first assumption and the first equation in (6.6.10) yield

$$y_{n_1} g_{x_{n_1+1}} + x_{n_1}^{-1} g_{y_{n_1+1}} = 0. \tag{6.6.11}$$

Note

$$\zeta_3 = x_{n_1} y_{n_1+1} - x_{n_1+1} y_{n_1} \tag{6.6.12}$$

is a solution of (6.6.11). Write g as a function in $\{x_{n_1}, x_{n_1+1}, \ y_{n_1}, \zeta_1, \zeta_3\}$. Then (6.6.11) gives $g_{x_{n_1+1}}$=0. So

$$g \ \text{ is a function in } \ x_{n_1}, y_{n_1}, \zeta_1, \zeta_3. \tag{6.6.13}$$

Moreover, (6.6.9) says

$$-x_{n_1}^{-1} y_{n_1} \zeta_1 g_{\zeta_1 \zeta_3} - y_{n_1} g_{x_{n_1} \zeta_3} - x_{n_1} g_{y_{n_1} \zeta_3} = 0. \tag{6.6.14}$$

Again we can assume that $g = \hat{g} + \tilde{g}$ is polynomial in $x_{n_1}, y_{n_1}, \zeta_1, \zeta_3$ with $\hat{g}|_{\zeta_1=0} = \hat{g}|_{\zeta_3=0} = 0$ and $\tilde{g}_{\zeta_3} = 0$. Then (6.6.14) is equivalent to

$$y_{n_1} \zeta_1 \hat{g}_{\zeta_1} + x_{n_1} y_{n_1} \hat{g}_{x_{n_1}} + x_{n_1}^2 \hat{g}_{y_{n_1}} = 0. \tag{6.6.15}$$

Observe that ζ_1/x_{n_1} and $x_{n_1}^2 - y_{n_1}^2$ are solutions of (6.6.15). Write $\hat{g}$ as a function in $\{x_{n_1}, \zeta_1/x_{n_1}, x_{n_1}^2 - y_{n_1}^2, \zeta_3\}$. Then (6.6.15) yields $\hat{g}_{x_{n_1}} = 0$. This shows that $\hat{g}$ is a function in $\{\zeta_1/x_{n_1}, x_{n_1}^2 - y_{n_1}^2, \zeta_3\}$. If $\hat{g} \in \mathscr{B}$, then $\hat{g}$ is independent of ζ_1. The assumption $\hat{g}|_{\zeta_1=0} = 0$ implies $\hat{g} = 0$. So the polynomial solution of g must be a polynomial in $x_{n_1}, y_{n_2}, \zeta_1$ with $g_{\zeta_1} \neq 0$. Similarly, if $h_{\zeta_2} \neq 0$ and $h|_{\zeta_2=0} = 0$, the polynomial solution of h must be a polynomial in $x_{n+1}, y_{n+1}, \zeta_2$. Assume $h_{\zeta_2} = 0$. Then

$$h_{x_{n_1} x_{n_1+1}} - h_{y_{n_1} y_{n_1+1}} = 0. \tag{6.6.16}$$

By (6.6.2) and (6.5.17), (6.5.20), (6.5.21) with $n_1 + 2$ replaced by $n_1 + 1$, the polynomial solution of h must be in

$$\text{Span}\{\eta^{m_3}(x_{n_1}^{m_1} y_{n_1+1}^{m_2}) \mid m_1, m_2, m_3 \in \mathbb{N}\}. \tag{6.6.17}$$

Therefore, we have:

Lemma 6.6.1 *Any singular vector in $\mathscr{B}$ must be a nonzero weight vector in*

$$Span\{x_{n_1}^{m_1} y_{n_1}^{m_2} \zeta_1^{m_3+1}, x_{n_1+1}^{m_1} y_{n_1+1}^{m_2} \zeta_2^{m_3+1}, \eta^{m_3}(x_{n_1}^{m_1} y_{n_1+1}^{m_2}) \mid m_i \in \mathbb{N}\}. \tag{6.6.18}$$

Note

$$x_{n_1}^{m_1} y_{n_1}^{m_2} \zeta_1^{m_3+1} \in \mathscr{B}_{\langle -m_1-m_3-1, m_2+m_3+1\rangle}, \tag{6.6.19}$$

$$x_{n_1+1}^{m_1} y_{n_1+1}^{m_2} \zeta_2^{m_3+1} \in \mathscr{B}_{\langle m_1+m_3+1, -m_2-m_3-1\rangle}. \tag{6.6.20}$$

Moreover,

$$\tilde{\Delta}(x_{n_1}^{m_1} y_{n_1}^{m_2} \zeta_1^{m_3+1}) = m_2 x_{n_1}^{m_1+1} y_{n_1}^{m_2-1} \zeta_1^{m_3+1} = 0 \Longleftrightarrow m_2 = 0 \tag{6.6.21}$$

and

$$\tilde{\Delta}(x_{n_1+1}^{m_1} y_{n_1+1}^{m_2} \zeta_2^{m_3+1}) = m_1 x_{n_1+1}^{m_1-1} y_{n_1+1}^{m_2+1} \zeta_2^{m_3+1} = 0 \Longleftrightarrow m_1 = 0 \tag{6.6.22}$$

by (6.3.49) and (6.6.1). Furthermore,

$$x_{n_1}^{m_1} y_{n_1}^{m_2} \zeta_1^{m_3+1} = \frac{\eta^{m_2}(x_{n_1}^{m_1+m_2} \zeta_1^{m_3+1})}{\prod_{r=1}^{m_2}(m_1+r)}, \quad x_{n_1+}^{m_1} y_{n_1+1}^{m_2} \zeta_2^{m_3+1} = \frac{\eta^{m_1}(y_{n_1+1}^{m_1+m_2} \zeta_2^{m_3+1})}{\prod_{r=1}^{m_1}(m_2+r)} \tag{6.6.23}$$

by (6.6.2). Indeed,

$$\eta^{m_1+1}(x_{n_1}^{m_1} \zeta_1^{m_2}) = \eta^{m_1+1}(y_{n_1+1}^{m_1} \zeta_2^{m_2}) = 0 \qquad \text{for } m_1, m_2 \in \mathbb{N}. \tag{6.6.24}$$

Since $x_{n_1}^{m_1} y_{n_1+1}^{m_2} \in \mathscr{H}_{\langle -m_1, -m_2\rangle}$, (6.4.36) says that $\eta^m(x_{n_1}^{m_1} y_{n_1+1}^{m_2})$ with $m > 0$ is a singular vector only if $m = m_1 + m_2 + 1$. But $\eta^{m_1+m_2+1}(x_{n_1}^{m_1} y_{n_1+1}^{m_2}) = 0$ by (6.6.2). Thus we have:

Lemma 6.6.2 *Any singular vector in $\mathscr{H}$ (cf. (6.6.28)) is a nonzero weight vector in*

$$Span\{x_{n_1}^{m_1} \zeta_1^{m_2+1}, y_{n_1+1}^{m_1} \zeta_2^{m_2+1}, x_{n_1}^{m_1} y_{n_1+1}^{m_2} \mid m_1, m_2 \in \mathbb{N}\}. \tag{6.6.25}$$

Since $\mathscr{B}$ is nilpotent with respect to $\mathscr{G}_+$, any nonzero submodule of $\mathscr{B}$ has a singular vector. The above fact implies

$$\mathscr{H}_{\langle \ell_1, \ell_2\rangle} = \{0\} \qquad \text{for } \ell_1, \ell_2 \in \mathbb{Z} \text{ such that } \ell_1 + \ell_2 > 0. \tag{6.6.26}$$

Observe that

$$\sum_{i=0}^{m_1}(-1)^{m_1-i} i!(m_2+i)! = \sum_{i+0}^{m_1-1}[(m_1-1)!(m_1+m_2)! + (-1)^i i!(m_2+i)!] > 0 \tag{6.6.27}$$

for $m_1, m_2 \in \mathbb{N}$. Thus

$$(x_{n_1}^{m_1} \zeta_1^{m_2} | x_{n_1}^{m_1} \zeta_1^{m_2})$$

$$= \left(\sum_{i=0}^{m_2}\binom{m_2}{i}(-1)^i x_{n_1-1}^{m_2-i} x_{n_1}^{m_1+i} y_{n_1-1}^{i} y_{n_1}^{m_2-i} \,|\, \sum_{i=0}^{m_2}\binom{m_2}{i}(-1)^i x_{n_1-1}^{m_2-i} x_{n_1}^{m_1+i} y_{n_1-1}^{i} y_{n_1}^{m_2-i}\right)$$

$$= (-1)^{m_1} m_2! \sum_{i=0}^{m_2}(-1)^{m_2-i}(m_1+i)!(m_2-i))! \neq 0 \tag{6.6.28}$$

by (6.6.27). Similarly, $(y_{n_1+1}^{m_1}\zeta_2^{m_2}|y_{n_1}^{m_1}\zeta_2^{m_2}) \neq 0$.

Next we assume $n_1 = n_2 = 1$ and $n \geq 3$. By the arguments in the above, a singular vector in $\mathscr{B}$ must be a nonzero weight vector in

$$\text{Span}\{x_{n_1+1}^{m_1} y_{n_1+1}^{m_2} \zeta_2^{m_3+1}, \eta^{m_3}(x_{n_1}^{m_1} y_{n_1+1}^{m_2}) \mid m_i \in \mathbb{N}\}. \tag{6.6.29}$$

Thus any singular vector in $\mathscr{H}$ (cf. (6.4.28)) is a nonzero weight vector in

$$\text{Span}\{y_{n_1+1}^{m_1} \zeta_2^{m_2+1}, x_{n_1}^{m_1} y_{n_1+1}^{m_2} \mid m_1, m_2 \in \mathbb{N}\}. \tag{6.6.30}$$

The above fact implies

$$\mathscr{H}_{\langle \ell_1,\ell_2\rangle} = \{0\} \qquad \text{for } \ell_1,\ell_2 \in \mathbb{Z} \text{ such that } \ell_1+\ell_2 > 0 \text{ or } \ell_2 > 0. \tag{6.6.31}$$

Consider the subcase $n_1 = n_2 = n-1$ and $n \geq 3$. A singular vector in $\mathscr{B}$ must be a nonzero weight vector in

$$\text{Span}\{x_{n_1}^{m_1} y_{n_1}^{m_2} \zeta_1^{m_3+1}, \eta^{m_3}(x_{n_1}^{m_1} y_{n_1+1}^{m_2}) \mid m_i \in \mathbb{N}\}. \tag{6.6.32}$$

Thus any singular vector in $\mathscr{H}$ (cf. (6.4.28)) is a nonzero weight vector in

$$\text{Span}\{x_{n_1}^{m_1} \zeta_1^{m_2+1}, x_{n_1}^{m_1} y_{n_1+1}^{m_2} \mid m_1, m_2 \in \mathbb{N}\}. \tag{6.6.33}$$

The above fact implies

$$\mathscr{H}_{\langle \ell_1,\ell_2\rangle} = \{0\} \qquad \text{for } \ell_1,\ell_2 \in \mathbb{Z} \text{ such that } \ell_1+\ell_2 > 0 \text{ or } \ell_1 > 0. \tag{6.6.34}$$

Suppose $n_1 = n_2 = 1$ and $n = 2$. A singular vector in $\mathscr{B}$ must be a nonzero weight vector in

$$\text{Span}\{\eta^{m_3}(x_1^{m_1} y_2^{m_2}) \mid m_i \in \mathbb{N}\}. \tag{6.6.35}$$

Thus any singular vector in $\mathscr{H}$ (cf. (6.4.28)) is a nonzero weight vector in

$$\text{Span}\{x_1^{m_1} y_2^{m_2} \mid m_1, m_2 \in \mathbb{N}\}. \tag{6.6.36}$$

The above fact implies

$$\mathscr{H}_{\langle \ell_1,\ell_2\rangle} = \{0\} \qquad \text{for } \ell_1,\ell_2 \in \mathbb{Z} \text{ such that } \ell_1 > 0 \text{ or } \ell_2 > 0. \tag{6.6.37}$$

Finally, we assume $n_1 = n_2 = n$. A singular vector in $\mathscr{B}$ must be a nonzero weight vector in

$$\mathrm{Span}\{x_{n_1}^{m_1} y_{n_1}^{m_2} \zeta_1^{m_3} \mid m_i \in \mathbb{N}\}. \tag{6.6.38}$$

Thus any singular vector in $\mathscr{H}$ (cf. (6.4.29)) is a nonzero weight vector in

$$\mathrm{Span}\{x_{n_1}^{m_1} \zeta_1^{m_2} \mid m_1, m_2 \in \mathbb{N}\}. \tag{6.6.39}$$

The above fact implies

$$\mathscr{H}_{\langle \ell_1, \ell_2 \rangle} = \{0\} \qquad \text{for } \ell_1, \ell_2 \in \mathbb{Z} \text{ such that } \ell_1 + \ell_2 > 0. \tag{6.6.40}$$

By the arguments of (6.4.46)–(6.4.68), we obtain:

Theorem 6.6.3 *Suppose $n_1 = n_2$. Let $\ell_1, \ell_2 \in \mathbb{Z}$ such that $\ell_2 \geq 0$ when $n_1 = n$, or $\ell_1 + \ell_2 \leq 0$ and: (a) $\ell_2 \leq 0$ if $n_1 = 1$ and $n \geq 3$; (b) $\ell_1 \leq 0$ if $n_1 = n - 1$ and $n \geq 3$; (c) $\ell_1, \ell_2 \leq 0$ when $n_1 = 1$ and $n = 2$. Then $\mathscr{H}_{\langle \ell_1, \ell_2 \rangle}$ is an irreducible highest-weight $sl(n, \mathbb{F})$-module and*

$$\mathscr{B}_{\langle \ell_1, \ell_2 \rangle} = \bigoplus_{m=0}^{\infty} \eta^m (\mathscr{H}_{\langle \ell_1 - m, \ell_2 - m \rangle}) \tag{6.6.41}$$

is an orthogonal decomposition of irreducible submodules. The symmetric bilinear form restricted to $\eta^m(\mathscr{H}_{\langle \ell_1 - m, \ell_2 - m \rangle})$ is nondenerate. In particular, $\mathscr{B}_{\langle \ell_1, \ell_2 \rangle} = \mathscr{H}_{\langle \ell_1, \ell_2 \rangle} \oplus \eta(\mathscr{B}_{\langle \ell_1 - 1, \ell_2 - 1 \rangle})$. If the conditions fails, $\mathscr{H}_{\langle \ell_1, \ell_2 \rangle} = \{0\}$.

Suppose $n_1 < n - 1$. Let $m_1, m_2 \in \mathbb{N}$. The subspace $\mathscr{H}_{\langle -m_1, -m_2 \rangle}$ has a highest-weight vector $x_{n_1}^{m_1} y_{n_1+1}^{m_2}$ of weight $m_1(1 - \delta_{1,n_1})\lambda_{n_1-1} - (m_1 + m_2 + 2)\lambda_{n_1} + m_2 \lambda_{n_1+1}$. If $n_1 \geq 2$, the subspace $\mathscr{H}_{\langle -m_1 - m_2 - 1, m_2 + 1 \rangle}$ has a highest-weight vector $x_{n_1}^{m_1} \zeta_1^{m_2+1}$ of weight $(m_2 + 1)\lambda_{n_1-2} - m_1 \lambda_{n_1-1} - (m_1 + m_2 + 3)\lambda_{n_1}$. The subspace $\mathscr{H}_{\langle m_1 + 1, -m_2 - m_1 - 1 \rangle}$ has a highest-weight vector $y_{n_1+1}^{m_2} \zeta_2^{m_1+1}$ of weight $-(m_1 + m_2 + 3)\lambda_{n_1} + m_2 \lambda_{n_1+1} - (m_1 + 1)(1 - \delta_{n_1, n-2})\lambda_{n_1+2}$.

Consider $n_1 = n - 1$. The subspace $\mathscr{H}_{\langle -m_1, -m_2 \rangle}$ has a highest-weight vector $x_{n_1}^{m_1} y_{n_1+1}^{m_2}$ of weight $m_1(1 - \delta_{n,2})\lambda_{n-2} - (m_1 + m_2 + 2)\lambda_{n-1}$. If $n \geq 3$, the subspace $\mathscr{H}_{\langle -m_1 - m_2 - 1, m_2 + 1 \rangle}$ has a highest-weight vector $x_{n_1}^{m_1} \zeta_1^{m_2+1}$ of weight $(m_2 + 1)(1 - \delta_{n,3})\lambda_{n-3} - m_1 \lambda_{n-2} - (m_1 + m_2 + 3)\lambda_{n-1}$.

Assume $n_1 = n$. The subspace $\mathscr{H}_{\langle -m_1 - m_2, m_2 \rangle}$ has a highest-weight vector $x_n^{m_1} \zeta_1^{m_2}$ of weight $m_2(1 - \delta_{n,2})\lambda_{n-2} + m_1 \lambda_{n-1}$.

Now we want to find an explicit expression for $\mathscr{H}_{\langle \ell_1, \ell_2 \rangle}$ when it is irreducible. First we assume $n_1 < n$. Note

$$-E_{n_1+r,i}|_{\mathscr{B}} = x_i x_{n_1+r} - y_i y_{n_r} \qquad \text{for } i \in \overline{1, n_1},\ r \in \overline{1, n - n_1} \tag{6.6.42}$$

by (6.3.36)–(6.3.38). Thus

$$(x_i x_{n_1+r} - y_i y_{n_r})(\mathscr{H}_{\langle \ell_1, \ell_2 \rangle}) \subset \mathscr{H}_{\langle \ell_1, \ell_2 \rangle}. \tag{6.6.43}$$

For $m_1, m_2 \in \mathbb{N}$, we set

$$\mathscr{H}'_{\langle -m_1,-m_2\rangle} = \text{Span}\left\{\left[\prod_{r=1}^{n_1} x_r^{l_r}\right]\left[\prod_{s=1}^{n-n_1} y_{n_1+s}^{k_s}\right]\left[\prod_{r=1}^{n_1}\prod_{s=1}^{n-n_1}(x_r x_{n_1+s} - y_r y_{n_1+s})^{l_{r,s}}\right]\right.$$
$$\left. \mid l_r, k_s, l_{r,s} \in \mathbb{N}; \sum_{r=1}^{n_1} l_r = m_1; \sum_{s=1}^{n-n_1} k_s = m_2\right\}. \qquad (6.6.44)$$

By (6.6.1), $\mathscr{H}'_{\langle m_1,m_2\rangle} \subset \mathscr{H}_{\langle -m_1,-m_2\rangle}$. Moreover,

$$E_{i,r}|_{\mathscr{B}} = -x_r\partial_{x_i} - y_r\partial_{y_i} - \delta_{i,r} \qquad \text{for } i, r \in \overline{1, n_1}, \qquad (6.6.45)$$

$$E_{n_1+j,n_1+s}|_{\mathscr{B}} = x_{n_1+j}\partial_{x_{n_1+s}} + y_{n_1+j}\partial_{y_{n_1+s}} + \delta_{j,s} \qquad \text{for } j, s \in \overline{1, n-n_1} \quad (6.6.46)$$

and

$$E_{i,n_1+s}|_{\mathscr{B}} = \partial_{x_i}\partial_{x_{n_1+s}} - \partial_{y_i}\partial_{y_{n_1+s}} \qquad \text{for } i \in \overline{1, n_1},\ s \in \overline{1, n-n_1} \qquad (6.6.47)$$

by (6.3.36)–(6.3.38). It is obvious that

$$E_{p,p'}(\mathscr{H}'_{\langle -m_1,-m_2\rangle}), E_{n_1+q,n_1+q'}(\mathscr{H}'_{\langle -m_1,-m_2\rangle}), E_{n_1+q,p'}(\mathscr{H}'_{\langle -m_1,-m_2\rangle}) \subset \mathscr{H}'_{\langle -m_1,-m_2\rangle} \qquad (6.6.48)$$

for $p, p' \in \overline{1, n_1}$ and $q, q' \in \overline{1, n-n_1}$ by (6.6.42), (6.6.45) and (6.6.46). Furthermore,

$$E_{p,n_1+q}\left\{\left[\prod_{r=1}^{n_1} x_r^{l_r}\right]\left[\prod_{s=1}^{n-n_1} y_{n_1+s}^{k_s}\right]\left[\prod_{r=1}^{n_1}\prod_{s=1}^{n-n_1}(x_r x_{n_1+s} - y_r y_{n_1+s})^{l_{r,s}}\right]\right\}$$
$$= (\partial_{x_p}\partial_{x_{n_1+q}} - \partial_{y_p}\partial_{y_{n_1+q}})\left\{\left[\prod_{r=1}^{n_1} x_r^{l_r}\right]\left[\prod_{s=1}^{n-n_1} y_{n_1+s}^{k_s}\right]\left[\prod_{r=1}^{n_1}\prod_{s=1}^{n-n_1}(x_r x_{n_1+s} - y_r y_{n_1+s})^{l_{r,s}}\right]\right\}$$
$$= \left(l_p\sum_{\iota=1}^{n_1}\frac{l_{\iota,q}x_\iota}{x_p(x_\iota x_{n_1+q} - y_\iota y_{n_1+q})} + k_q\sum_{\iota_1=1}^{n-n_1}\frac{l_{p,\iota_1}y_{n_1+\iota_1}}{y_{n_1+q}(x_p x_{n_1+\iota_1} - y_p y_{n_1+\iota_1})}\right.$$
$$\left. + \frac{l_{p,q}}{x_p x_{n_1+q} - y_p y_{n_1+q}} + \sum_{\iota_1=1}^{n_1}\sum_{\iota_2=1}^{n-n_1}\frac{l_{\iota_1,q}l_{p,\iota_2}(x_{\iota_1}x_{n_1+\iota_2} - y_{\iota_1}y_{n_1+\iota_2})}{(x_{\iota_1}x_{n_1+q} - y_{\iota_1}y_{n_1+q})(x_p x_{n_1+\iota_2} - y_p y_{n_1+\iota_2})}\right)$$
$$\times\left[\prod_{r=1}^{n_1} x_r^{l_r}\right]\left[\prod_{s=1}^{n-n_1} y_{n_1+s}^{k_s}\right]\left[\prod_{r=1}^{n_1}\prod_{s=1}^{n-n_1}(x_r x_{n_1+s} - y_r y_{n_1+s})^{l_{r,s}}\right] \in \mathscr{H}'_{\langle -m_1,-m_2\rangle}. \qquad (6.6.49)$$

Thus $\mathscr{H}'_{\langle -m_1,-m_2\rangle}$ is a nonzero submodule of $\mathscr{H}_{\langle -m_1,-m_2\rangle}$, which is irreducible. So we have $\mathscr{H}'_{\langle -m_1,-m_2\rangle} = \mathscr{H}_{\langle -m_1,-m_2\rangle}$.

Next we assume $n_1 > 1$. We let

$$\begin{aligned}
&\mathscr{H}'_{\langle -m_1-m_2,m_2\rangle}\\
&= \text{Span}\left\{\left[\prod_{r=1}^{n_1} x_r^{l_r}\right]\left[\prod_{1\le p<q\le n_1}(x_p y_q - x_q y_p)^{k_{p,q}}\right]\left[\prod_{r=1}^{n_1}\prod_{s=1}^{n-n_1}(x_r x_{n_1+s} - y_r y_{n_1+s})^{l_{r,s}}\right]\right.\\
&\qquad\left.\mid l_r, k_{p,q}, l_{r,s}\in\mathbb{N};\ \sum_{r=1}^{n_1} l_r = m_1;\ \sum_{1\le p<q\le n_1} k_{p,q} = m_2\right\}. \qquad (6.6.50)
\end{aligned}$$

For convenience, we treat $k_{p,q} = -k_{q,p}$. To prove $\mathscr{H}'_{\langle -m_1-m_2,m_2\rangle} = \mathscr{H}_{\langle -m_1-m_2,m_2\rangle}$, it is enough to calculate

$$\begin{aligned}
&\partial_{x_i}\left[\prod_{1\le p<q\le n_1}(x_p y_q - x_q y_p)^{k_{p,q}}\right]\partial_{x_{n_1+j}}\left[\prod_{r=1}^{n_1}\prod_{s=1}^{n-n_1}(x_r x_{n_1+s} - y_r y_{n_1+s})^{l_{r,s}}\right]\\
&-\partial_{y_i}\left[\prod_{1\le p<q\le n_1}(x_p y_q - x_q y_p)^{k_{p,q}}\right]\partial_{y_{n_1+j}}\left[\prod_{r=1}^{n_1}\prod_{s=1}^{n-n_1}(x_r x_{n_1+s} - y_r y_{n_1+s})^{l_{r,s}}\right]\\
&=\left(\sum_{i\neq \iota_1\in\overline{1,n_1}}\sum_{\iota_2=1}^{n_1}\frac{k_{i,i_1} l_{\iota_2,j}(x_{\iota_1} y_{\iota_2} - x_{\iota_2} y_{\iota_1})}{(x_i y_{\iota_1} - x_{\iota_1} y_i)(x_{\iota_2} x_{n_1+j} - y_{\iota_2} y_{n_1+j})}\right)\\
&\quad\times\left[\prod_{1\le p<q\le n_1}(x_p y_q - x_q y_p)^{k_{p,q}}\right]\left[\prod_{r=1}^{n_1}\prod_{s=1}^{n-n_1}(x_r x_{n_1+s} - y_r y_{n_1+s})^{l_{r,s}}\right] \qquad (6.6.51)
\end{aligned}$$

for $i \in \overline{1,n_1}$ and $j \in \overline{1,n_1}$ by (6.6.45)–(6.6.47) and (6.6.49).

Symmetrically, similar expression can obtained for $\mathscr{H}_{\langle m_2,-m_1-m_2\rangle}$. In summary, we have:

Theorem 6.6.4 *Let $m_1, m_2 \in \mathbb{N}$. If $n_1 = n_2 < n$, we have*

$$\begin{aligned}
\mathscr{H}_{\langle -m_1,-m_2\rangle} &= \mathit{Span}\left\{\left[\prod_{r=1}^{n_1} x_r^{l_r}\right]\left[\prod_{s=1}^{n-n_1} y_{n_1+s}^{k_s}\right]\left[\prod_{r=1}^{n_1}\prod_{s=n_1+1}^{n}(x_r x_s - y_r y_s)^{l_{r,s}}\right]\right.\\
&\qquad\left.\mid l_r, k_s, l_{r,s}\in\mathbb{N};\ \sum_{r=1}^{n_1} l_r = m_1;\ \sum_{s=1}^{n-n_1} k_s = m_2\right\}. \qquad (6.6.52)
\end{aligned}$$

When $2 \le n_1 = n_2 < n$,

$$\begin{aligned}
&\mathscr{H}_{\langle -m_1-m_2,m_2\rangle}\\
&= \mathit{Span}\left\{\left[\prod_{r=1}^{n_1} x_r^{l_r}\right]\left[\prod_{1\le p<q\le n_1}(x_p y_q - x_q y_p)^{k_{p,q}}\right]\left[\prod_{r=1}^{n_1}\prod_{s=n_1+1}^{n}(x_r x_s - y_r y_s)^{l_{r,s}}\right]\right.
\end{aligned}$$

$$\left. \mid l_r, k_{p,q}, l_{r,s} \in \mathbb{N}; \sum_{r=1}^{n_1} l_r = m_1; \sum_{1\le p<q\le n_1} k_{p,q} = m_2 \right\}. \tag{6.6.53}$$

In the case $n_1 = n_2 < n-1$,

$$\mathscr{H}_{\langle m_2,-m_1-m_2\rangle} = Span\left\{\left[\prod_{n_1+1\le p<q\le n}(x_py_q - x_qy_p)^{k_{p,q}}\right]\left[\prod_{r=1}^{n_1}\prod_{s=n_1+1}^{n}(x_rx_s - y_ry_s)^{l_{r,s}}\right]\right.$$
$$\left.\times\left[\prod_{r=1}^{n-n_1} y_{n_1+r}^{l_r}\right] \mid l_r, k_{p,q}, l_{r,s} \in \mathbb{N}; \sum_{r=1}^{n_1} l_r = m_1; \sum_{n_1+1\le p<q\le n} k_{p,q} = m_2 \right\}. \tag{6.6.54}$$

If $n_1 = n_2 = n$,

$$\mathscr{H}_{\langle -m_1-m_2,m_2\rangle} = Span\left\{\left[\prod_{r=1}^{n} x_r^{l_r}\right]\left[\prod_{1\le p<q\le n}(x_py_q - x_qy_p)^{k_{p,q}}\right]\right.$$
$$\left. \mid l_r, k_{p,q} \in \mathbb{N}; \sum_{r=1}^{n} l_r = m_1; \sum_{1\le p<q\le n} k_{p,q} = m_2 \right\}. \tag{6.6.55}$$

6.7 Extensions of the Projective Representations

In this section, we construct a functor from the category of A_{n-1}-modules to the category of A_n-modules, which is related to n-dimensional projective transformations.

Denote by $GL(n+1,\mathbb{F})$ the group of $(n+1)\times(n+1)$ invertible matrices. A *projective transformation* on $\mathbb{F}^n$ for a field $\mathbb{F}$ is given by

$$u \mapsto \frac{Au+\vec{b}}{\vec{c}^{\,t}u+d} \quad \text{for } u \in \mathbb{F}^n, \tag{6.7.1}$$

where all the vector in $\mathbb{F}^n$ are in column form and

$$\begin{pmatrix} A & \vec{b} \\ \vec{c}^{\,t} & d \end{pmatrix} \in GL(n+1,\mathbb{F}). \tag{6.7.2}$$

It is well-known that a transformation of mapping straight lines to lines must be a projective transformation. The group of projective transformations is the fundamental group of n-dimensional projective geometry. Physically, the group with $n = 4$ and $\mathbb{F} = \mathbb{R}$ consists of all the transformations of keeping free particles including light signals moving with constant velocities along straight lines (e.g., cf. [GWZ1-2]). Based on the embeddings of the poincaré group and De Sitter group into the projective group with $n = 4$ and $\mathbb{F} = \mathbb{R}$, Guo, Wu and Zhou [GWZ1-2] proposed three kinds of special relativity.

Now we consider the Lie algebra

$$sl(n+1,\mathbb{F}) = \sum_{1\leq i<j\leq n+1} (\mathbb{F}E_{i,j} + \mathbb{F}E_{j,i}) + \sum_{r=1}^{n} \mathbb{F}(E_{r,r} - E_{r+1,r+1}). \tag{6.7.3}$$

Recall that

$$D = \sum_{r=1}^{n} x_r \partial_{x_r} \tag{6.7.4}$$

is the degree operator on the polynomial algebra $\mathscr{A} = \mathbb{F}[x_1, \ldots, x_n]$. Note

$$[\partial_{x_j}, x_i D] = \begin{cases} x_i \partial_{x_j}, & i \neq j, \\ D + x_i \partial_{x_i}, & i = j, \end{cases} \tag{6.7.5}$$

$$[x_i \partial_{x_j}, x_k D] = \delta_{j,k} x_i D, \quad [x_i \partial_{x_j}, \partial_{x_k}] = -\delta_{i,k} \partial_{x_j} \tag{6.7.6}$$

for $i, j, k \in \overline{1, n}$. Thus we have the following inhomogeneous representation of $sl(n+1,\mathbb{F})$ on $\mathscr{A}$ determined by

$$E_{i,j}|_{\mathscr{A}} = x_i \partial_{x_j}, \qquad E_{i,n+1}|_{\mathscr{A}} = x_i D, \tag{6.7.7}$$

$$E_{n+1,i}|_{\mathscr{A}} = -\partial_{x_i}, \quad (E_{i,i} - E_{n+1,n+1})|_{\mathscr{A}} = D + x_i \partial_{x_i} \tag{6.7.8}$$

for $i, j \in \overline{1, n}$ with $i \neq j$. The above representation is exactly the one of $sl(n+1,\mathbb{F})$ induced by the projective transformations in (6.7.1).

Recall the classical Witt algebra $\mathbb{W}_n = \sum_{i=1}^{n} \mathscr{A}\partial_{x_i}$ with the following Lie bracket:

$$\left[\sum_{i=1}^{n} f_i \partial_{x_i}, \sum_{j=1}^{n} g_j \partial_{x_j}\right] = \sum_{i,j=1}^{n} (f_j \partial_{x_j}(g_i) - g_j \partial_{x_j}(f_i))\partial_{x_i} \tag{6.7.9}$$

(cf. Example 1.2.1). Acting on the entries of the elements of the Lie algebra $gl(n,\mathscr{A})$, $\mathbb{W}_n$ becomes a Lie subalgebra of the Lie algebra of derivations of $gl(n,\mathscr{A}) = \sum_{i,j=1}^{n} \mathscr{A} E_{i,j}$. In particular,

$$\widehat{\mathbb{W}}_n = \mathbb{W}_n \oplus gl(n,\mathscr{A}) \tag{6.7.10}$$

becomes a Lie algebra with the Lie bracket

$$[d_1 \oplus A_1, d_2 \oplus A_2] = [d_1, d_2] \oplus [d_1(A_2) - d_2(A_1) + [A_1, A_2]] \tag{6.7.11}$$

for $d_1, d_2 \in \mathbb{W}_n$ and $A_1, A_2 \in gl(n, \mathscr{A})$.

Lemma 6.7.1 (Shen) *The map* $\Im$ *given by*

$$\Im\left(\sum_{i=1}^{n} f_i \partial_{x_i}\right) = \sum_{i=1}^{n} f_i \partial_{x_i} \oplus \sum_{i,j=1}^{n} \partial_{x_i}(f_j) E_{i,j} \tag{6.7.12}$$

is a Lie algebra monomorphism from $\mathbb{W}_n$ *to* $\widehat{\mathbb{W}}_n$.

Proof Note that

$$\begin{aligned} &\Im\left(\left[\sum_{i=1}^{n} f_i \partial_{x_i}, \sum_{j=1}^{n} g_j \partial_{x_j}\right]\right) \\ &= \Im\left(\sum_{i,k=1}^{n} (f_k \partial_{x_k}(g_i) - g_k \partial_{x_k}(f_i)) \partial_{x_i}\right) \\ &= \sum_{i,k=1}^{n} (f_k \partial_{x_k}(g_i) - g_k \partial_{x_k}(f_i)) \partial_{x_i}) \oplus \sum_{i,j,k=1}^{n} \partial_{x_i}[f_k \partial_{x_k}(g_j) - g_k \partial_{x_k}(f_j)] E_{i,j} \end{aligned} \tag{6.7.13}$$

and

$$\begin{aligned} &\left[\Im\left(\sum_{i=1}^{n} f_i \partial_{x_i}\right), \Im\left(\sum_{j=1}^{n} g_j \partial_{x_j}\right)\right] \\ &= \left[\sum_{i=1}^{n} f_i \partial_{x_i} \oplus \sum_{i,r=1}^{n} \partial_{x_i}(f_r) E_{i,r}, \sum_{j=1}^{n} g_j \partial_{x_j} \oplus \sum_{j,s=1}^{n} \partial_{x_j}(g_s) E_{j,s}\right] \\ &= \sum_{i,k=1}^{n} (f_k \partial_{x_k}(g_i) - g_k \partial_{x_k}(f_i)) \partial_{x_i}) \oplus \left[\sum_{i,j,s=1}^{n} f_i \partial_{x_i} \partial_{x_j}(g_s) E_{j,s} - \sum_{i,j,r} \sum_{i,r=1}^{n} g_j \partial_{x_j} \partial_{x_i}(f_r) E_{i,r}\right. \\ &\quad \left. + \sum_{i,r,s=1}^{n} \partial_{x_i}(f_r) \partial_{x_r}(g_s) E_{i,s} - \sum_{j,r,s=1}^{n} \partial_{x_j}(g_s) \partial_{x_s}(f_r) E_{j,r}\right]. \end{aligned} \tag{6.7.14}$$

Moreover,

$$\begin{aligned} &\sum_{i,j,s=1}^{n} f_i \partial_{x_i} \partial_{x_j}(g_s) E_{j,s} - \sum_{i,j,r} \sum_{i,r=1}^{n} g_j \partial_{x_j} \partial_{x_i}(f_r) E_{i,r} \\ &+ \sum_{i,r,s=1}^{n} \partial_{x_i}(f_r) \partial_{x_r}(g_s) E_{i,s} - \sum_{j,r,s=1}^{n} \partial_{x_j}(g_s) \partial_{x_s}(f_r) E_{j,r} \end{aligned}$$

$$= \sum_{i,j,k=1}^{n} [f_k\partial_{x_k}\partial_{x_i}(g_j) - g_k\partial_{x_k}\partial_{x_i}(f_j) + \partial_{x_i}(f_k)\partial_{x_k}(g_j) - \partial_{x_i}(g_k)\partial_{x_k}(f_j)]E_{i,j}$$

$$= \sum_{i,j,k=1}^{n} \partial_{x_i}[f_k\partial_{x_k}(g_j) - g_k\partial_{x_k}(f_j)]E_{i,j} \tag{6.7.15}$$

Thus

$$\Im\left(\left[\sum_{i=1}^{n} f_i\partial_{x_i}, \sum_{j=1}^{n} g_j\partial_{x_j}\right]\right) = \left[\Im\left(\sum_{i=1}^{n} f_i\partial_{x_i}\right), \Im\left(\sum_{j=1}^{n} g_j\partial_{x_j}\right)\right]. \quad \square \tag{6.7.16}$$

The above lemma was an important discovery of Shen [Sg]. Now we have the Lie algebra monomorphism $\nu : sl(n+1,\mathbb{F}) \to \widehat{\mathbb{W}}_n$ given by

$$\nu(\xi) = \Im(\xi|_{\mathscr{A}}) \qquad \text{for } \xi \in sl(n+1,\mathbb{F}). \tag{6.7.17}$$

In particular,

$$\nu(E_{i,n+1}) = \Im(x_i D) = x_i D \oplus \left(\sum_{r=1}^{n}(x_i E_{r,r} + x_r E_{i,r})\right) \qquad \text{for } i \in \overline{1,n} \tag{6.7.18}$$

and

$$\nu(E_{i,j}) = x_i\partial_{x_j} \oplus E_{i,j}, \;\; \nu(E_{i,i} - E_{j,j}) = (x_i\partial_{x_i} - x_j\partial_{x_j}) \oplus (E_{i,i} - E_{j,j}) \tag{6.7.19}$$

for $i, j \in \overline{1,n}$ with $i \neq j$.

Let M be a $gl(n,\mathbb{F})$-module. Set

$$\widehat{M} = \mathscr{A} \otimes_{\mathbb{F}} M. \tag{6.7.20}$$

According to (6.7.11), $\widehat{M}$ becomes a $\widehat{\mathbb{W}}_n$-module with the action

$$(d \oplus fA)(g \otimes v) = d(g) \otimes v + fg \otimes A(v) \tag{6.7.21}$$

for $d \in \mathbb{W}_n$, $f, g \in \mathscr{A}$, $A \in gl(n,\mathbb{F})$ and $v \in M$. Thus we have the following $sl(n+1,\mathbb{F})$-module structure on $\widehat{M}$:

$$\xi(g \otimes v) = \nu(\xi)(g \otimes v) \qquad \text{for } \xi \in sl(n+1,\mathbb{F}),\; g \in \mathscr{A},\; v \in M. \tag{6.7.22}$$

For convenience, we denote

$$\mathscr{G}_0 = sl(n, \mathbb{F}) + \mathbb{F}(E_{n,n} - E_{n+1,n+1}),\; \mathscr{G}_+ = \sum_{i=1}^{n} \mathbb{F}E_{n+1,i},\; \mathscr{G}_- = \sum_{i=1}^{n} \mathbb{F}E_{i,n+1}. \tag{6.7.23}$$

Then $\mathscr{G}_0$ is a Lie subalgebra and $\mathscr{G}_\pm$ are abelian Lie subalgebras of $sl(n+1, \mathbb{F})$. Moreover,

$$\nu(E_{n+1,i}) = -\partial_{x_i} \qquad \text{for } i \in \overline{1, n}. \tag{6.7.24}$$

Thus

$$\mathscr{G}_+(1 \otimes M) = \{0\}. \tag{6.7.25}$$

Furthermore,

$$\mathscr{G}_0 = sl(n, \mathbb{F}) + \mathbb{F}(\sum_{r=1}^{n} E_{r,r} - nE_{n+1,n+1}) \tag{6.7.26}$$

and

$$\nu\left(\sum_{r=1}^{n} E_{r,r} - nE_{n+1,n+1}\right) = (n+1)\left(D \oplus \sum_{r=1}^{n} E_{r,r}\right) \in \widehat{\mathbb{W}}_n. \tag{6.7.27}$$

So

$$U(\mathscr{G}_0)(1 \otimes M) = 1 \otimes U(sl(n, \mathbb{F}))(M) = 1 \otimes M. \tag{6.7.28}$$

Therefore, the $sl(n+1, \mathbb{F})$-submodule

$$U(\mathscr{G})(1 \otimes M) = U(\mathscr{G}_-)(1 \otimes M). \tag{6.7.29}$$

By (6.7.24), we have:

Proposition 6.7.2 *The map $M \mapsto U(\mathscr{G}_-)(1 \otimes M)$ gives rise to a functor from the category of $gl(n, \mathbb{F})$-modules to the category of $sl(n+1, \mathbb{F})$-modules. In particular, it maps irreducible $gl(n, \mathbb{F})$-modules to irreducible $sl(n+1, \mathbb{F})$-modules.*

From another point view, $U(\mathscr{G}_-)(1 \otimes M)$ is a polynomial extension from $gl(n, \mathbb{F})$-module M to an $sl(n+1, \mathbb{F})$-module. Next we want to study when $\widehat{M} = U(\mathscr{G}_-)(1 \otimes M)$. Set

$$E_i = E_{i,i} - \frac{1}{1+n} \sum_{r=1}^{n+1} E_{r,r} \qquad \text{for } i \in \overline{1, n}. \tag{6.7.30}$$

Then we have a Lie algebra isomorphism $\varsigma : gl(n, \mathbb{F}) \to \mathscr{G}_0$ determined by

$$\varsigma(E_{r,s}) = E_{r,s},\; \varsigma(E_{i,i}) = E_i \qquad \text{for } i, r, s \in \overline{1, n},\; r \neq s. \tag{6.7.31}$$

Moreover,

$$E_i = E_{i,i} - E_{n+1,n+1} - \frac{1}{1+n}\left(\sum_{r=1}^{n} E_{r,r} - nE_{n+1,n+1}\right) \qquad \text{for } i \in \overline{1,n}. \tag{6.7.32}$$

According to (6.7.8),

$$E_i|_{\mathscr{A}} = x_i\partial_{x_i} \qquad \text{for } i \in \overline{1,n}. \tag{6.7.33}$$

Thus

$$\nu(E_i) = x_i\partial_{x_i} \oplus E_{i,i} \qquad \text{for } i \in \overline{1,n}. \tag{6.7.34}$$

We calculate

$$[E_i, E_{j,n+1}] = \delta_{i,j}E_{j,n+1} \qquad \text{for } i, j \in \overline{1,n}. \tag{6.7.35}$$

In the rest of this section, we use ς as the identification of $gl(n,\mathbb{F})$ with $\mathscr{G}_0$. In particular,

$$E_{i,j}|_{\mathscr{A}} = x_i\partial_{x_j} \qquad \text{for } i, j \in \overline{1,n} \tag{6.7.36}$$

by (6.7.7) and (6.7.33). So $\widehat{M}$ is a usual tensor module of $gl(n,\mathbb{F})$.

Denote

$$E^{\alpha} = E_{1,n+1}^{\alpha_1}E_{2,n+1}^{\alpha_2}\cdots E_{n,n+1}^{\alpha_n} \qquad \text{for } \alpha = (\alpha_1,\ldots,\alpha_n) \in \mathbb{N}^n. \tag{6.7.37}$$

We write

$$\widehat{M}_k = \mathscr{A}_k \otimes_{\mathbb{F}} M, \;\; (U(\mathscr{G}_-)(1\otimes M))_k = \widehat{M}_k \bigcap (U(\mathscr{G}_-)(1\otimes M)) \qquad \text{for } k \in \mathbb{N}. \tag{6.7.38}$$

According to (6.7.18),

$$(U(\mathscr{G}_-)(1\otimes M))_k = \sum_{\alpha\in\mathbb{N}^n,\ |\alpha|=k} E^{\alpha}(1\otimes M), \tag{6.7.39}$$

where $|\alpha| = \sum_{r=1}^{n}\alpha_r$. Define a linear map $\varphi : \widehat{M} \to U(\mathscr{G}_-)(1\otimes M)$ by

$$\varphi(x^{\alpha}\otimes v) = E^{\alpha}(1\otimes v) \qquad \text{for } \alpha \in \mathbb{N}^n,\ v \in M. \tag{6.7.40}$$

Then (6.7.19) and (6.7.35) implies that φ is $gl(n,\mathbb{F})$-module homomorphism.

Note that

$$\omega = \sum_{i,j=1}^{n} E_{i,j}E_{j,i} \in U(gl(n,\mathbb{F})) \tag{6.7.41}$$

is a Casimir element of $gl(n, \mathbb{F})$. Set

$$\tilde{\omega} = \frac{1}{2}(\omega|_{\widehat{M}} - \omega|_{\mathscr{A}} \otimes \mathrm{Id}_M - \mathrm{Id}_{\mathscr{A}} \otimes \omega|_M) = \sum_{i,j=1}^{n} E_{i,j}|_{\mathscr{A}} \otimes E_{j,i}|_M. \tag{6.7.42}$$

Note that $\sum_{r=1}^{n} E_{r,r}$ is the central element of $gl(n, \mathbb{F})$.

Lemma 6.7.3 *If* $(\sum_{r=1}^{n} E_{r,r})|_M = c\, Id_M$ *for* $c \in \mathbb{F}$, *then we have* $\varphi|_{\widehat{M}_1} = (\tilde{\omega} + c)|_{\widehat{M}_1}$.

Proof For $s \in \overline{1,n}$ and $v \in M$,

$$\begin{aligned}\varphi(x_s \otimes v) &= \nu(E_{s,n+1})(1 \otimes v) = \left[P_s \oplus \left(\sum_{r=1}^{n}(x_s E_{r,r} + x_r E_{s,r}\right)\right](1 \otimes v) \\ &= c x_s \otimes v + \sum_{r=1}^{n} x_r \otimes E_{s,r}(v).\end{aligned} \tag{6.7.43}$$

Moreover,

$$\tilde{\omega}(x_s \otimes v) = \sum_{i,j=1}^{n} x_i \partial_{x_j}(x_s) \otimes E_{j,i}(v) = \sum_{i=1}^{n} x_i \otimes E_{s,i}(v). \qquad \square \tag{6.7.44}$$

Fix the Cartan subalgebra $H = \sum_{i=1}^{n} \mathbb{F} E_{i,i}$ of $gl(n, \mathbb{F})$ and define $\varepsilon_i \in H^*$ by

$$\varepsilon_i(E_{j,j}) = \delta_{i,j} \qquad \text{for } for \;\; i, j \in \overline{1,n}. \tag{6.7.45}$$

For any $\mu = \sum_{i=1}^{n} \mu_i \varepsilon_i$, $\mu' = \sum_{i=1}^{n} \mu_i' \varepsilon_i$, we define

$$(\mu, \mu') = \sum_{i=1}^{n} \mu_i \mu_i', \quad |\mu| = \sum_{i=1}^{n} \mu_i. \tag{6.7.46}$$

Set

$$\Lambda^+ = \{\mu = \sum_{i=1}^{n} \mu_i \varepsilon_i \in H^* \mid \mu_i - \mu_{i+1} \in \mathbb{N} \text{ for } i \in \overline{1, n-1}\}. \tag{6.7.47}$$

Let $V(\mu)$ be a highest-weight irreducible $gl(n, \mathbb{F})$-module with μ as the highest weight. According to Sect. 5.2, any finite-dimensional $gl(n, \mathbb{F})$-module is a highest-weight irreducible module $V(\mu)$ for some $\mu \in \Lambda^+$ when $\mathbb{F}$ is algebraically closed. Moreover,

$$(\sum_{r=1}^{n} E_{r,r})|_{V(\mu)} = |\mu| \, \mathrm{Id}_{V(\mu)} \tag{6.7.48}$$

by its action on the highest-weight vector. By (3.2.15),

$$\rho = \frac{1}{2}\sum_{r=1}^{n}(n-2r+1)\varepsilon_r. \tag{6.7.49}$$

Consider the action of ω on the highest-weight vector, we have:

$$\omega|_{V(\mu)} = (\mu, \mu + 2\rho) \, \mathrm{Id}_{V(\mu)} \tag{6.7.50}$$

As a $gl(n, \mathbb{F})$-module,

$$\mathscr{A}_1 \cong V(\varepsilon_1), \tag{6.7.51}$$

whose weights are

$$\{\varepsilon_1, \varepsilon_2, \ldots, \varepsilon_n\}. \tag{6.7.52}$$

Fix $\mu \in \Lambda^+$ and take

$$M = V(\mu). \tag{6.7.53}$$

Moreover, we define

$$\mathcal{I}(\mu) = \{i_1, i_2, \ldots, i_{s+1}\} \subset \overline{1,n} \;\text{ such that }\; 1 = i_1 < i_2 < \cdots < i_{s+1} = n+1, \tag{6.7.54}$$

$$\mu_{i_r} = \mu_{i_{r+1}-1} \quad \text{for } r \in \overline{1,s} \text{ and } \mu_{i_r} > \mu_{i_{r+1}} \text{ if } r < s. \tag{6.7.55}$$

According to Theorem 5.4.3, we have:

Lemma 6.7.4 (Pieri's formula) *As $gl(n, \mathbb{F})$-modules,*

$$\widehat{V(\mu)}_1 = \mathscr{A}_1 \otimes_{\mathbb{F}} V(\mu) \cong \bigoplus_{r=1}^{s} V(\mu + \varepsilon_{i_r}), \tag{6.7.56}$$

where $\mathcal{I}(\mu) = \{i_1, i_2, \ldots, i_{s+1}\}$.

By (6.7.50), the eigenvalues of ω on $\widehat{V(\mu)}_1$ are:

$$\{(\mu + \varepsilon_{i_1}, \mu + \varepsilon_{i_1} + 2\rho), \ldots, (\mu + \varepsilon_{i_s}, \mu + \varepsilon_{i_s} + 2\rho)\}. \tag{6.7.57}$$

On the other hand, $\omega|_{\mathscr{A}_1} \otimes \mathrm{Id}_{V(\mu)}$ has the only eigenvalue $(\varepsilon_1, \varepsilon_1 + 2\rho) = n$ and $\mathrm{Id}_{\mathscr{A}_1} \otimes \omega|_{V(\mu)}$ has the only eigenvalue $(\mu, \mu + 2\rho)$. Calculate

$$(\mu + \varepsilon_{i_r}, \mu + \varepsilon_{i_r} + 2\rho) - (\mu, \mu + 2\rho) - n = 2(\mu_{i_r} + 1 - i_r). \tag{6.7.58}$$

Thus we get:

Lemma 6.7.5 *The eigenvalues of* $\tilde{\omega}$ *on* $\widehat{V(\mu)}_1$ *are*

$$\{\mu_{i_r} + 1 - i_r \mid r \in \overline{1, s}\}. \tag{6.7.59}$$

For $f \in \mathscr{A}$, we define the action

$$f(g \otimes v) = fg \otimes v \qquad \text{for } g \in \mathscr{A},\ v \in M. \tag{6.7.60}$$

Now we have:

Theorem 6.7.6 *The* $sl(n+1, \mathbb{F})$*-module* $\widehat{V(\mu)}$ *is irreducible if and only if* $|\mu| + \mu_{i_s} - i_s \notin -\mathbb{N} - 1$.

Proof Note

$$\varphi[\widehat{V(\mu)}_k] = (U(\mathscr{G}_-)(1 \otimes V(\mu)))_k \qquad \text{for } k \in \mathbb{N}. \tag{6.7.61}$$

By Proposition 6.7.2, $\widehat{V(\mu)}$ is irreducible if and only if φ is injective. Observe that

$$\widehat{V(\mu)}_0 = (U(\mathscr{G}_-)(1 \otimes V(\mu))_0 = 1 \otimes V(\mu) \tag{6.7.62}$$

and

$$\varphi(1 \otimes v) = 1 \otimes v \qquad \text{for } v \in V(\mu). \tag{6.7.63}$$

So $\varphi|_{\widehat{V(\mu)}_0}$ is injective.

Suppose that $\varphi|_{\widehat{V(\mu)}_k}$ is injective for some $k \in \mathbb{N}$. So

$$\widehat{V(\mu)}_k = (U(\mathscr{G}_-)(1 \otimes V(\mu)))_k. \tag{6.7.64}$$

Thus $\varphi|_{\widehat{V(\mu)}_{k+1}}$ is injective if and only if

$$\sum_{i=1}^{n} E_{i,n+1}[\widehat{V(\mu)}_k] = \widehat{V(\mu)}_{k+1}. \tag{6.7.65}$$

For $f \in \mathscr{A}_k$ and $v \in V(\mu)$, we have

$$\begin{aligned} E_{i,n+1}(f \otimes v) &= (x_i D \oplus (\sum_{r=1}^{n}(x_i E_{r,r} + x_r E_{i,r})))(f \otimes v) \\ &= kx_i f \otimes v + f \sum_{r=1}^{n}(x_i E_{r,r} + x_r E_{i,r})(1 \otimes v) \\ &= f[(k + \varphi)(x_i \otimes v)] \end{aligned} \tag{6.7.66}$$

by (6.7.18). Thus (6.7.64) holds if and only if

$$(k+\varphi)|_{\widehat{V(\mu)}_1} \quad \text{is injective.} \tag{6.7.67}$$

According to (6.7.48) and Lemmas 6.7.3, 6.7.5, the eigenvalues of $(k+\varphi)|_{\widehat{V(\mu)}_1}$ are

$$\{k+|\mu|+\mu_{i_r}+1-i_r \mid r\in\overline{1,s}\}. \tag{6.7.68}$$

Therefore, φ is injective if and only if

$$0\notin\{k+|\mu|+\mu_{i_r}+1-i_r \mid r\in\overline{1,s},\ k\in\mathbb{N}\}. \tag{6.7.69}$$

Observe that

$$\mu_{i_r}-\mu_{i_s}+i_s-i_r\in\mathbb{N} \tag{6.7.70}$$

by (6.7.47) and (6.7.54). Suppose $|\mu|+\mu_{i_s}-i_s\notin-\mathbb{N}-1$. Then

$$|\mu|+\mu_{i_r}-i_r=|\mu|+\mu_{i_s}-i_s+\mu_{i_r}-\mu_s+i_s-i_r\notin-\mathbb{N}-1 \tag{6.7.71}$$

for $r\in\overline{1,s}$ by (6.7.70). This shows that

$$(6.7.69)\ \text{ holds for any } k\in\mathbb{N} \text{ if and only if } \ |\mu|+\mu_{i_s}-i_s\notin-\mathbb{N}-1. \quad \square \tag{6.7.72}$$

Let λ_i be the ith fundamental weight of $sl(n+1,\mathbb{F})$ with respect to the order $n+1,1,2,\ldots,n$. Since $\nu(E_{1,1}-E_{n+1,n+1})=(D+x_1\partial_{1,1})\oplus(\sum_{r=1}^n E_{r,r}+E_{1,1})$ by (6.7.8), (6.7.12) and (6.7.17), the irreducible $sl(n+1,\mathbb{F})$-module $U(\mathscr{G}_-)(V(\mu))$ is a highest-weight module with highest weight $-(|\mu|+\mu_1)\lambda_1+\sum_{r=1}^{s-1}(\mu_{i_r}-\mu_{i_{r+1}})\lambda_{i_r+1}$, and so is $\widehat{V(\mu)}$ when $|\mu|+\mu_{i_s}-i_s\notin-\mathbb{N}-1$. In particular, $U(\mathscr{G}_-)(V(\mu))$ is a finite-dimensional irreducible $sl(n+1,\mathbb{F})$-module if and only if $|\mu|+\mu_1\in-\mathbb{N}$.

Let $\mathscr{B}=\mathbb{F}[y_1,\ldots,y_n,z_1,\ldots,z_n]$. Given $c\in\mathbb{F}$, we define an action of $gl(n,\mathbb{F})$ on $\mathscr{B}$ by

$$E_{i,j}|_{\mathscr{B}}=c\delta_{i,j}+y_i\partial_{y_j}-z_j\partial_{z_i} \qquad \text{for } i,j\in\overline{1,n}. \tag{6.7.73}$$

For $\ell_1,\ell_2\in\mathbb{N}$, we set

$$\mathscr{B}_{\ell_1,\ell_2}=\sum_{\alpha,\beta\in\mathbb{N}^n;\ |\alpha|=\ell_1,\ |\beta|=\ell_2}\mathbb{F}y^\alpha z^\beta. \tag{6.7.74}$$

Moreover, we define

$$\mathscr{H}_{\ell_1,\ell_2}=\{f\in\mathscr{B}_{\ell_1,\ell_2}\mid(\sum_{i=1}^n\partial_{y_i}\partial_{z_i})(f)=0\}. \tag{6.7.75}$$

According to Theorem 6.2.4, $\mathscr{H}_{\ell_1,\ell_2}$ forms a finite-dimensional irreducible $gl(n,\mathbb{F})$-module and $y_1^{\ell_1}z_n^{\ell_2}$ is a highest-weight vector with the weight $\ell_1\varepsilon_1-\ell_2\varepsilon_n+c\sum_{i=1}^n\varepsilon_i$. By the above Theorem, we get:

Corollary 6.7.7 *Suppose $\ell_1>0$. The $sl(n+1,\mathbb{F})$-module $\widehat{\mathscr{H}_{\ell_1,\ell_2}}$ is irreducible if and only if $c\notin-(\mathbb{N}+\ell_1-1)/(n+1)$ when $\ell_2=0$ and $c\notin(2\ell_2+n-\ell_1-1-\mathbb{N})/(n+1)$ when $\ell_2\neq 0$.*

Let Ψ be the exterior algebra generated by $\{\theta_1,\theta_2,\ldots,\theta_n\}$ (cf. (6.2.15)) and take the settings (6.2.16)–(6.2.18). Given $c\in\mathbb{F}$, we define an action of $gl(n,\mathbb{F})$ on Ψ by

$$E_{i,j}|_\Psi=c\delta_{i,j}+\theta_i\partial_{\theta_j}. \tag{6.7.76}$$

For any $k\in\overline{1,n-1}$, Ψ_k is a finite-dimensional irreducible $gl(n,\mathbb{F})$-module and $\theta_1\theta_2\cdots\theta_r$ is a highest-weight vector with the weight $\sum_{r=1}^k\varepsilon_r+c\sum_{i=1}^n\varepsilon_i$. By Theorem 6.7.6, we have:

Corollary 6.7.8 *The $sl(n+1,\mathbb{F})$-module $\widehat{\Psi_k}$ is irreducible if and only if $c\notin-\mathbb{N}/(n+1)$.*

Suppose that $M=\mathbb{F}v$ is a one-dimensional $gl(n,\mathbb{F})$-module. Then

$$E_{i,j}|_M=c\delta_{i,j}\,\mathrm{Id}_M\qquad\text{for } i,j\in\overline{1,n}. \tag{6.7.77}$$

We identify $\mathscr{A}$ with $\widehat{M}$ via

$$f\leftrightarrow f\otimes v\qquad\text{for } f\in\mathscr{A}. \tag{6.7.78}$$

Denote

$$\kappa=(n+1)c. \tag{6.7.79}$$

Recall that $D=\sum_{s=1}^n x_s\partial_{x_s}$ is the degree operator on $\mathscr{A}=\mathbb{F}[x_1,\ldots,x_n]$. Then we have the following one-parameter generalization π_κ of the *projective oscillator representation* of $sl(n+1,\mathbb{F})$ in (6.7.7) and (6.7.8):

$$\pi_\kappa(E_{i,j})=x_i\partial_{x_j},\qquad \pi_\kappa(E_{i,n+1})=x_i(D+\kappa), \tag{6.7.80}$$

$$\pi_\kappa(E_{n+1,i})=-\partial_{x_i},\quad \pi_\kappa(E_{i,i}-E_{n+1,n+1})=D+\kappa+x_i\partial_{x_i} \tag{6.7.81}$$

for $i,j\in\overline{1,n}$ with $i\neq j$. According to Theorem 6.7.6, we have:

Corollary 6.7.9 *The representation π_κ of $sl(n+1,\mathbb{F})$ on $\mathscr{A}$ is irreducible if and only if $\kappa\notin-\mathbb{N}$.*

For $\ell\in\mathbb{N}$, we set

$$\mathscr{A}_{(\ell)}=\sum_{i=0}^{\ell}\mathscr{A}_i. \tag{6.7.82}$$

Reordering $\overline{1, n+1}$ as $\{n+1, 1, 2, \ldots, n\}$, we get directly by (6.7.80) and (6.7.81):

Corollary 6.7.10 *If $\kappa = -\ell$ for some $\ell \in \mathbb{N}$, then $\mathscr{A}_{(\ell)}$ is a finite-dimensional irreducible $sl(n+1, \mathbb{F})$-module with highest weight $\ell\lambda_1$ and $\mathscr{A}/\mathscr{A}_{(\ell)}$ is an infinite-dimensional highest-weight irreducible $sl(n+1, \mathbb{F})$-module with highest weight $-(\ell+2)\lambda_1 + (\ell+1)\lambda_2$.*

6.8 Projective Oscillator Representations

In this section, we study the oscillator generalizations of the representation π_κ of $sl(n+1, \mathbb{F})$ in (6.7.80) and (6.7.81).

Note the symmetry

$$[\partial_{x_r}, x_r] = 1 = [-x_r, \partial_{x_r}]. \tag{6.8.1}$$

Fix $n_1 \in \overline{0, n-1}$. We define $\pi_{\kappa,0} = \pi_\kappa$. Changing operators $\partial_{x_r} \mapsto -x_r$ and $x_r \mapsto \partial_{x_r}$ for $r \in \overline{1, n_1}$ in (6.7.80) and (6.7.81) when $n_1 > 0$, we get another differential-operator representation π_{κ,n_1} of $sl(n+1, \mathbb{F})$. We call π_{κ,n_1} *noncanonical projective oscillator representations* in terms of physics terminology. For $\vec{a} = (a_1, a_2, \ldots, a_n)^t \in \mathbb{F}^n$, we denote $\vec{a} \cdot \vec{x} = \sum_{i=1}^n a_i x_i$. Recall $\mathscr{A} = \mathbb{F}[x_1, x_2, \ldots, x_n]$. Set

$$\mathscr{A}_{\vec{a}} = \{f e^{\vec{a}\cdot\vec{x}} \mid f \in \mathscr{A}\}. \tag{6.8.2}$$

Denote by $\pi_{\kappa,n_1}^{\vec{a}}$ the representation π_{κ,n_1} of $sl(n+1, \mathbb{F})$ on $\mathscr{A}_{\vec{a}}$. Corollary 6.7.9 says that the representation $\pi_{\kappa,0}^{\vec{0}}$ of $sl(n+1, \mathbb{F})$ is irreducible if and only if $\kappa \notin -\mathbb{N}$. The following is the main theorem in this section.

Theorem 6.8.1 *The representation $\pi_{\kappa,n_1}^{\vec{0}}$ is irreducible for any $\kappa \in \mathbb{F} \setminus \mathbb{Z}$, and the underlying module $\mathscr{A}$ is an infinite-dimensional weight $sl(n+1, \mathbb{F})$-module with finite-dimensional weight subspaces. If $a_i \neq 0$ for some $i \in \overline{n_1+1, n}$, then the representation $\pi_{\kappa,n_1}^{\vec{a}}$ of $sl(n+1, \mathbb{F})$ is always irreducible for any $\kappa \in \mathbb{F}$. When $n_1 > 1$, $\vec{a} \neq 0$ and $a_i = 0$ for any $i \in \overline{n_1+1, n}$, the representation $\pi_{\kappa,n_1}^{\vec{a}}$ of $sl(n+1, \mathbb{F})$ is irreducible for $\kappa \in \mathbb{F} \setminus \mathbb{Z}$.*

We will prove Theorem 6.8.1 case by case.

Case 1. The representation $\pi_{\kappa,n_1}^{\vec{0}}$ with $0 < n_1 < n$.

As in Sect. 6.3, we set

$$\tilde{D} = \sum_{r=n_1+1}^{n} x_r \partial_{x_r} - \sum_{i=1}^{n_1} x_i \partial_{x_i}. \tag{6.8.3}$$

Then the representation $\pi^{\vec{0}}_{\kappa,n_1}$ of $sl(n+1,\mathbb{F})$ is the representation π_{κ,n_1} on $\mathscr{A}$ with

$$\pi_{\kappa,n_1}(E_{i,j}) = \begin{cases} -x_j\partial_{x_i} - \delta_{i,j} & \text{if } i,j \in \overline{1,n_1}; \\ \partial_{x_i}\partial_{x_j} & \text{if } i \in \overline{1,n_1},\ j \in \overline{n_1+1,n}; \\ -x_i x_j & \text{if } i \in \overline{n_1+1,n},\ j \in \overline{1,n_1}; \\ x_i\partial_{x_j} & \text{if } i,j \in \overline{n_1+1,n}, \end{cases} \tag{6.8.4}$$

$$\pi_{\kappa,n_1}(E_{i,n+1}) = \begin{cases} (\tilde{D}+\kappa-n_1-1)\partial_{x_i} & \text{if } i \leq n_1, \\ x_i(\tilde{D}+\kappa-n_1) & \text{if } i > n_1, \end{cases} \tag{6.8.5}$$

$$\pi_{\kappa,n_1}(E_{n+1,i}) = \begin{cases} x_i & \text{if } i \leq n_1, \\ -\partial_{x_i} & \text{if } i > n_1, \end{cases} \tag{6.8.6}$$

$$\pi_{\kappa,n_1}(E_{n,n} - E_{n+1,n+1}) = \tilde{D} - n_1 + \kappa + x_n\partial_{x_n} \tag{6.8.7}$$

Recall

$$\mathscr{A}_{\langle k\rangle} = \text{Span}\,\{x^\alpha = \prod_{i=1}^{n} x_i^{\alpha_i} \mid \alpha = (\alpha_1,\ldots,\alpha_n) \in \mathbb{N}^n; \sum_{i=1}^{n_1}\alpha_i - \sum_{r=n_1+1}^{n}\alpha_r = k\}. \tag{6.8.8}$$

Then $\mathscr{A} = \bigoplus_{k\in\mathbb{Z}} \mathscr{A}_{\langle k\rangle}$ and

$$\mathscr{A}_{\langle k\rangle} = \{f \in \mathscr{A} \mid \tilde{D}(f) = kf\}. \tag{6.8.9}$$

Note that

$$\mathscr{G}_0 = \sum_{1\leq i<j\leq n} (\mathbb{F}E_{i,j} + \mathbb{F}E_{j,i}) + \sum_{r=1}^{n-1} \mathbb{F}(E_{r,r} - E_{r+1,r+1}) \tag{6.8.10}$$

is a Lie subalgebra of $sl(n+1,\mathbb{F})$ isomorphic to $sl(n,\mathbb{F})$.

Lemma 6.8.2 *The representation $\pi^{\vec{0}}_{\kappa,n_1}$ of $sl(n+1,\mathbb{F})$ is irreducible for any $\kappa \in \mathbb{F}\setminus\mathbb{Z}$.*

Proof Let k be any integer. For any $0 \neq f \in \mathscr{A}_{\langle k\rangle}$, we have

$$0 \neq E_{n+1,1}(f) = x_1 f \in \mathscr{A}_{\langle k-1\rangle} \tag{6.8.11}$$

by (6.8.6), and

$$0 \neq E_{n,n+1}(f) = (k+\kappa-n_1)x_n f \in \mathscr{A}_{\langle k+1\rangle} \tag{6.8.12}$$

by (6.8.5). Let M be a nonzero $sl(n+1,\mathbb{F})$-submodule of $\mathscr{A}$. If $k_1, k_2 \in \mathbb{Z}$ with $k_1 \neq k_2$, then the highest weights of $\mathscr{A}_{\langle k_1\rangle}$ and $\mathscr{A}_{\langle k_2\rangle}$ are different as $\mathscr{G}_0$-modules by Theorem 6.3.3. So $\mathscr{A}_{\langle k_0\rangle} \subset M$ for some $k_0 \in \mathbb{Z}$. Moreover, (6.8.11) and (6.8.12) imply $\mathscr{A}_{\langle k\rangle} \subset M$ for any $k \in \mathbb{Z}$. Hence $\mathscr{M} = \mathscr{A}$. $\square$

Expressions (6.8.4)–(6.8.7) imply the above representation is not of highest-weight type. Moreover, $\mathscr{A}$ is a weight $sl(n+1,\mathbb{F})$-module with finite-dimensional weight subspaces.

Case 2. The representation $\pi^{\vec{a}}_{\kappa,0}$ with $\vec{0} \neq \vec{a} \in \mathbb{F}^n$.

In this case,

$$E_{n+1,i}(fe^{\vec{a}\cdot\vec{x}}) = -(\partial_{x_i} + a_i)(f)e^{\vec{a}\cdot\vec{x}} \qquad \text{for } i \in \overline{1,n},\ f \in \mathscr{A}. \tag{6.8.13}$$

Thus

$$(E_{n+1,i} + a_i)(fe^{\vec{a}\cdot\vec{x}}) = -\partial_{x_i}(f)e^{\vec{a}\cdot\vec{x}} \qquad \text{for } i \in \overline{1,n},\ f \in \mathscr{A}. \tag{6.8.14}$$

The second result in this section.

Lemma 6.8.3 *The representation $\pi^{\vec{a}}_{\kappa,0}$ with $\vec{0} \neq \vec{a} \in \mathbb{F}^n$ is an irreducible representation of $sl(n+1,\mathbb{F})$ for any $\kappa \in \mathbb{F}$.*

Proof Let $\mathscr{A}_k$ be the subspace of homogeneous polynomials with degree k. Set

$$\mathscr{A}_{\vec{a},k} = \mathscr{A}_k e^{\vec{a}\cdot\vec{x}} \qquad \text{for } k \in \mathbb{N}. \tag{6.8.15}$$

Without loss of generality, we assume $a_1 \neq 0$. Let M be a nonzero $sl(n+1,\mathbb{F})$-submodule of $\mathscr{A}_{\vec{a}}$. Take any $0 \neq fe^{\vec{a}\cdot\vec{x}} \in M$ with $f \in \mathscr{A}$. By (6.8.14),

$$\partial_{x_i}(f)e^{\vec{a}\cdot\vec{x}} \in M \qquad \text{for } i \in \overline{1,n}. \tag{6.8.16}$$

By induction, we have $e^{\vec{a}\cdot\vec{x}} \in M$; that is, $\mathscr{A}_{\vec{a},0} \subset M$.

Suppose $\mathscr{A}_{\vec{a},\ell} \subset M$ for some $\ell \in \mathbb{N}$. For any $ge^{\vec{a}\cdot\vec{x}} \in \mathscr{A}_{\vec{a},\ell}$,

$$E_{i,1}(ge^{\vec{a}\cdot\vec{x}}) = x_i(\partial_{x_1} + a_1)(g)e^{\vec{a}\cdot\vec{x}} = a_1x_ige^{\vec{a}\cdot\vec{x}} + x_i\partial_{x_1}(g)e^{\vec{a}\cdot\vec{x}} \in M \qquad \text{for } i \in \overline{2,n} \tag{6.8.17}$$

by (6.7.80). Since $x_i\partial_{x_1}(g)e^{\vec{a}\cdot\vec{x}} \in \mathscr{A}_{\vec{a},\ell} \subset M$, we have

$$x_ige^{\vec{a}\cdot\vec{x}} \in M \qquad \text{for } i \in \overline{2,n}. \tag{6.8.18}$$

On the other hand,

$$(E_{1,1} - E_{2,2})(ge^{\vec{a}\cdot\vec{x}}) = a_1x_1ge^{\vec{a}\cdot\vec{x}} + (x_1\partial_{x_1} - x_2\partial_{x_2} - a_2x_2)(g)e^{\vec{a}\cdot\vec{x}} \in M \tag{6.8.19}$$

by (6.7.81). Our assumption says that $(x_1\partial_{x_1} - x_2\partial_{x_2})(g)e^{\vec{a}\cdot\vec{x}} \in \mathscr{A}_{\vec{a},\ell} \subset M$. According to (6.8.18), $-a_2x_2(g)e^{\vec{a}\cdot\vec{x}} \in M$. Therefore,

$$x_1ge^{\vec{a}\cdot\vec{x}} \in M. \tag{6.8.20}$$

Expressions (6.8.19) and (6.8.20) imply $\mathscr{A}_{\vec{a},\ell+1} \subset M$. By induction, $\mathscr{A}_{\vec{a},\ell} \subset M$ for any $\ell \in \mathbb{N}$. So $\mathscr{A}_{\vec{a}} = M$. Hence $\mathscr{A}_{\vec{a}}$ is an irreducible $sl(n+1,\mathbb{F})$-module. □

Case 3. The representation $\pi^{\vec{a}}_{\kappa,n_1}$ with $n_1 > 0$, and $a_i \neq 0$ for some $i \in \overline{n_1+1,n}$.
The following is the third result in this section.

Lemma 6.8.4 *Under the above assumption, the representation $\pi^{\vec{a}}_{\kappa,n_1}$ is an irreducible representation of $sl(n+1,\mathbb{F})$ for any $\kappa \in \mathbb{F}$.*

Proof Let M be a nonzero $sl(n+1,\mathbb{F})$-submodule of $\mathscr{A}_{\vec{a}}$. By (6.8.6) and (6.8.13)–(6.8.16), there exists $0 \neq fe^{\vec{a}\cdot\vec{x}} \in M$ with $f \in \mathbb{F}[x_1,\ldots,x_{n_1}]$. By symmetry, we can assume $a_n \neq 0$. According to (6.8.4),

$$E_{i,n}(fe^{\vec{a}\cdot\vec{x}}) = (\partial_{x_i}+a_i)(\partial_{x_n}+a_n)(f)e^{\vec{a}\cdot\vec{x}} = a_i a_n fe^{\vec{a}\cdot\vec{x}} + a_n\partial_{x_i}(f)e^{\vec{a}\cdot\vec{x}} \quad \text{for } i \in \overline{1,n_1}. \tag{6.8.21}$$

Thus

$$(a_n^{-1}E_{i,n} - a_i)(fe^{\vec{a}\cdot\vec{x}}) = \partial_{x_i}(f)e^{\vec{a}\cdot\vec{x}} \in M \quad \text{for } i \in \overline{1,n_1}. \tag{6.8.22}$$

By induction on the degree of f, we get $e^{\vec{a}\cdot\vec{x}} \in M$; that is, $\mathscr{A}_{\vec{a},0} \subset M$.
The arguments in (6.8.17)–(6.8.20) yield

$$\mathbb{F}[x_{n_1+1},\ldots,x_n]e^{\vec{a}\cdot\vec{x}} \subset M. \tag{6.8.23}$$

According to (6.8.6),

$$E^{\ell_1}_{n+1,1}\cdots E^{\ell_{n_1}}_{n+1,n_1}(\mathbb{F}[x_{n_1+1},\ldots,x_n]e^{\vec{a}\cdot\vec{x}}) = x_1^{\ell_1}\cdots x_{n_1}^{\ell_{n_1}}(\mathbb{F}[x_{n_1+1},\ldots,x_n]e^{\vec{a}\cdot\vec{x}}) \subset M \tag{6.8.24}$$

for $\ell_i \in \mathbb{N}$ with $i \in \overline{1,n_1}$. Thus $\mathscr{A}_{\vec{a}} = M$. So $\mathscr{A}_{\vec{a}}$ is an irreducible $sl(n+1,\mathbb{F})$-module. □

Case 4. $n_1 > 1$, $\vec{a} \neq \vec{0}$ and $a_i = 0$ for any $i \in \overline{n_1+1,n}$.
The following is the fourth result in this section.

Lemma 6.8.5 *Under the above assumption, the representation $\pi^{\vec{a}}_{\kappa,n_1}$ is an irreducible representation of $sl(n+1,\mathbb{F})$ for any $\kappa \in \mathbb{F}\setminus\mathbb{Z}$.*

Proof Applying the transformation

$$\vec{x} \mapsto T\vec{x},\quad A \mapsto TAT^{-1},\quad E_{n,n} - E_{n+1,n+1} \mapsto E_{n,n} - E_{n+1,n+1}. \tag{6.8.25}$$

$$(E_{1,n+1},\ldots,E_{n,n+1}) \mapsto (E_{1,n+1},\ldots,E_{n,n+1})T^{-1}, \tag{6.8.26}$$

$$(E_{n+1,1},\ldots,E_{n+1,n}) \mapsto (E_{n+1,n},\ldots,E_{n+1,n})T^{-1} \tag{6.8.27}$$

with $A \in sl(n,\mathbb{F})$ for some $n \times n$ orthogonal matrix T, we can assume $a_1 \neq 0$ and $a_i = 0$ for $i \in \overline{2,n}$.

Let M be a nonzero $sl(n+1,\mathbb{F})$-submodule of $\mathscr{A}_{\vec{a}}$. Take any $0 \neq fe^{\vec{a}\cdot\vec{x}} \in M$. Note that

$$E_{1,2}(fe^{\vec{a}\cdot\vec{x}}) = -x_2(\partial_{x_1}+a_1)(f)e^{\vec{a}\cdot\vec{x}} \in M \tag{6.8.28}$$

by (6.8.4) and

$$E_{n+1,2}(fe^{\vec{a}\cdot\vec{x}}) = x_2fe^{\vec{a}\cdot\vec{x}} \in M \tag{6.8.29}$$

by (6.8.6). Thus

$$x_2\partial_{x_1}(f)e^{\vec{a}\cdot\vec{x}} \in M. \tag{6.8.30}$$

Repeatedly applying (6.8.29) if necessary, we can assume $f \in \mathbb{F}[x_2,\dots,x_n]$. We apply the arguments in the proof of Lemma 6.8.2 to the Lie subalgebra

$$\mathscr{L} = \sum_{2\leq i<j\leq n+1}(\mathbb{F}E_{i,j}+\mathbb{F}E_{j,i}) + \sum_{r=2}^{n}\mathbb{F}(E_{r,r}-E_{r+1,r+1}) \tag{6.8.31}$$

and obtain

$$\mathbb{F}[x_2,\dots,x_n]e^{\vec{a}\cdot\vec{x}} \subset M \tag{6.8.32}$$

when $\kappa \notin \mathbb{Z}$. According to (6.8.6),

$$E_{n+1,1}^{\ell}(\mathbb{F}[x_2,\dots,x_n]e^{\vec{a}\cdot\vec{x}}) = x_1^{\ell}(\mathbb{F}[x_2,\dots,x_n]e^{\vec{a}\cdot\vec{x}}) \subset M \tag{6.8.33}$$

for any $\ell \in \mathbb{N}$. Therefore $M = \mathscr{A}_{\vec{a}}$. So $\mathscr{A}_{\vec{a}}$ is an irreducible $sl(n+1,\mathbb{F})$-module. □

With respect to the representation $\pi_{\kappa,n_1}^{\vec{a}}$ with $\vec{a} \neq \vec{0}$, $\mathscr{A}_{\vec{a}}$ is not a weight $sl(n+1,\mathbb{F})$-module. Now Theorem 6.8.1 follows from Lemmas 6.8.2–6.8.5.

References

[X16] X. Xu, Flag partial differential equations and representations of Lie algebras. Acta Appl. Math. **102**, 249–280 (2008)

[FC] F.M. Fernández, E.A. Castro, *Algebraic Methods in Quantum Chemistry and Physics* (CRC Press, Inc., Boca Raton, 1996)

[LF] W. Ludwig, C. Falter, *Symmetries in Physics*, 2nd edn. (Springer, Berlin, 1996)

[G] H. Georgi, *Lie Algebras in Particle Physics*, 2nd edn. (Perseus Books Group, 1999)

[HH] J.E.M. Homos, Y.M.M. Homos, Algebraic model for the evolution of the generic code. Phys. Rev. Lett. **71**, 4401–4404 (1991)

[LX1] C. Luo, X. Xu, $\mathbb{Z}^2$-Graded oscillator generalizations of $sl(n)$. Commun. Algebra **41**, 3147–3173 (2013)

[Sg] G. Shen, Graded modules of graded Lie algebras of Cartan type (I)–mixed product of modules. Sci. China A **29**, 570–581 (1986)

[La] T. Larsson, Conformal fields: a class of representations of Vect(N)[J]. Int. J. Modern Phys. A **7**(26), 6493–6508 (1992)

[ZX] Y. Zhao, X. Xu, Generalized projective representations for sl(n+1). J. Algebra **328**, 132–154 (2011)

[Kb] B. Kostant, On the tensor product of a finite and an infinite dimensional representation. J. Fund. Anal. **20**, 257–285 (1975)

[Hr1] R. Howe, Dual pairs in physics: harmonic oscillators, photons, electrons, and singletons, in *Applications of GroupTheory in Physics and Mathematical Physics (Chicago, 1982)*. Lectures in Appl. Mathe. vol. 21 (Amer. Math. Soc., Providence, RI, 1985), pp. 179–207

[Hr2] R. Howe, Remarks on classical invariant theory. Trans. Am. Math. Soc. **313**, 539–570 (1989)

[Hr3] R. Howe, Transcending classical invariant theory. J. Am. Math. Soc. **2**, 535–552 (1989)

[Hr4] R. Howe, Perspectives on invariant theory: Schur duality, multiplicity-free actions and beyond, *The Schurlectures* (1992) (*Tel Aviv*), 1–182, *Israel Math. Conf. Proc.*, 8, *Bar-Ilan Univ., Ramat Gan,* (1995)

[X24] X. Xu, Projective oscillator representations of $sl(n+1)$ and $sp(2m+2)$. J. Lie Theory **26**, 96–114 (2016)

[I1] N.H. Ibragimov, *Transformation Groups Applied to Mathematical Physics* (Nauka, 1983)

[LX4] C. Luo, X. Xu, Supersymmetric analogues of the classical theorem on harmonic polynomials. J. Algebra Appl. **13**(6), 1450011, 42pp (2014)

[GHW1] H. Guo, C. Huang, H. Wu, Yang's model as triply special relativity and the Snyder's model–de Sitter special relativity duality. Phys. Lett. B **663**, 270–274 (2008)

[GHW2] H. Guo, C. Huang, H. Wu, The principle of relativity and the special relativity triple. Phys. Lett. B **670**, 437–441 (2009)

[L] F.S. Levin, *An Introduction to Quantum Theory* (Cambridge University Press, 2002)

Chapter 7
Representations of Even Orthogonal Lie Algebras

A *quantum field* is an operator-valued function on a certain Hilbert space, which is often a direct sum of infinite-dimensional irreducible modules of a certain Lie algebra (group). The Lie algebra of two-dimensional conformal group is exactly the *Virasoro algebra*, which is infinite-dimensional. The minimal models of two-dimensional conformal field theory were constructed from direct sums of certain infinite-dimensional irreducible modules of the Virasoro algebra, where a distinguished module called, the *vacuum module*, gives rise to a *vertex operator algebra*. When $n \geq 3$, n-dimension conformal groups (depending on the metric) are finite-dimensional. Higher-dimensional conformal field theory is not so well understood partly because we are lack of enough knowledge on the infinite-dimensional irreducible modules of orthogonal Lie algebras that carry the information of conformal transformations. This motivated Zhao and the author [sg] to study explicit infinite-dimensional irreducible representations of the orthogonal Lie algebra $o(n+2, \mathbb{C})$ by using the inhomogeneous differential-operator representations arising from conformal transformations with respect to the metric defining $o(n, \mathbb{C})$ and Shen's technique of mixed product (cf. [Ra]) (also known as Larsson functor (cf. [La])).

It turned out that a hidden central transformation is involved. More importantly, we found a condition on the constant value taken by the central transformation such that the generalized representation is irreducible. In our approach, Pieri's formulas, invariant operators and the idea of Kostant's characteristic identities played key roles (cf. [Kb]). The result could be useful in understanding higher-dimensional conformal field theory with the constant value taken by the central transformation as the central charge. Our representations virtually provide natural extensions of the conformal transformations on a Riemannian manifold to its vector bundles. The modules are also "modules of tensor densities" in the terminology of Rudakov [LX2].

In this chapter, we focus on the natural explicit representations of even orthogonal Lie algebras. In Sect. 7.1, we study the canonical bosonic and fermionic oscillator representations over their minimal natural modules. The spin representations are also presented. Weyl's Theorem of complete reducibility in Sect. 2.4 plays an important role. In Sect. 7.2, we determine the structure of the noncanonical oscillator representations obtained from the above bosonic representations by partially swapping

X. Xu, *Representations of Lie Algebras and Partial Differential Equations*,
DOI 10.1007/978-981-10-6391-6_7

differential operators and multiplication operators, which are generalizations of the classical theorem on harmonic polynomials. The noncanonical oscillator representations of $sl(n,\mathbb{F})$ in Sects. 6.3–6.6 are used. The results were due to Luo and the author [XZ2].

In Sect. 7.3, we construct a new functor from the category of $o(2n,\mathbb{C})$-modules to the category of $o(2n+2,\mathbb{C})$-modules, which is an extension of the conformal representation of $o(2n+2,\mathbb{C})$. Moreover, we find the condition for the functor to map a finite-dimensional irreducible $o(2n,\mathbb{C})$-module to an infinite-dimensional irreducible $o(2n+2,\mathbb{C})$-module by Theorem 5.4.3 of decomposing the tensor module. It is the second sequential work of constructing functors among the module categories of different simple Lie algebras after Sect. 6.7. This work was due to Zhao and the author [sg].

The work in Sect. 7.3 gives rise to a one-parameter (c) family of inhomogeneous first-order differential operator representations of $o(2n+2,\mathbb{C})$. Letting these operators act on the space of exponential-polynomial functions that depend on a parametric vector $\vec{a}\in\mathbb{C}^{2n}$, we prove in Sect. 7.4 that the space forms an irreducible $o(2n+2,\mathbb{C})$-module for any $c\in\mathbb{C}$ if $\vec{a}$ is not on a certain hypersurface. By partially swapping differential operators and multiplication operators, we obtain more general differential operator representations of $o(2n+2,\mathbb{C})$ on the polynomial algebra $\mathscr{B}$ in $2n$ variables. We prove that $\mathscr{B}$ forms an infinite-dimensional irreducible weight $o(2n+2,\mathbb{C})$-module with finite-dimensional weight subspaces if $c\notin\mathbb{Z}/2$. These results are taken from [X25].

7.1 Canonical Oscillator Representations

In this section, we present the canonical bosonic and fermionic oscillator representations of even orthogonal Lie algebras over their minimal natural modules.

Let $n>1$ be an integer. Recall that the split even orthogonal Lie algebra

$$\begin{aligned} o(2n,\mathbb{F}) = & \sum_{1\le p<q\le n}[\mathbb{F}(E_{p,n+q}-E_{q,n+p})+\mathbb{F}(E_{n+p,q}-E_{n+q,p})] \\ & +\sum_{i,j=1}^{n}\mathbb{F}(E_{i,j}-E_{n+j,n+i}). \end{aligned} \tag{7.1.1}$$

We take the subspace

$$H=\sum_{i=1}^{n}\mathbb{F}(E_{i,i}-E_{n+i,n+i}) \tag{7.1.2}$$

as a Cartan subalgebra and define $\{\varepsilon_i\mid i\in\overline{1,n}\}\subset H^*$ by

$$\varepsilon_i(E_{j,j}-E_{n+j,n+j})=\delta_{i,j}. \tag{7.1.3}$$

The inner product $(\cdot,\cdot)$ on the $\mathbb{Q}$-subspace

$$L_{\mathbb{Q}} = \sum_{i=1}^{n} \mathbb{Q}\varepsilon_i \tag{7.1.4}$$

is given by

$$(\varepsilon_i, \varepsilon_j) = \delta_{i,j} \quad \text{for } i, j \in \overline{1,n}. \tag{7.1.5}$$

Then the root system of $o(2n, \mathbb{F})$ is

$$\Phi_{D_n} = \{\pm\varepsilon_i \pm \varepsilon_j \mid 1 \le i < j \le n\}. \tag{7.1.6}$$

We take the set of positive roots

$$\Phi_{D_n}^{+} = \{\varepsilon_i \pm \varepsilon_j \mid 1 \le i < j \le n\}. \tag{7.1.7}$$

In particular,

$$\Pi_{D_n} = \{\varepsilon_1 - \varepsilon_2, \ldots, \varepsilon_{n-1} - \varepsilon_n, \varepsilon_{n-1} + \varepsilon_n\} \text{ is the set of positive simple roots.} \tag{7.1.8}$$

Recall the set of dominate integral weights

$$\Lambda^{+} = \{\mu \in L_{\mathbb{Q}} \mid (\varepsilon_{n-1} + \varepsilon_n, \mu), (\varepsilon_i - \varepsilon_{i+1}, \mu) \in \mathbb{N} \text{ for } i \in \overline{1, n-1}\}. \tag{7.1.9}$$

According to (7.1.5),

$$\Lambda^{+} = \{\mu = \sum_{i=1}^{n} \mu_i \varepsilon_i \mid \mu_i \in \mathbb{Z}/2;\ \mu_i - \mu_{i+1}, \mu_{n-1} + \mu_n \in \mathbb{N}\}. \tag{7.1.10}$$

Note that if $\mu \in \Lambda^{+}$, then $\mu_{n-1} \ge |\mu_n|$. For $\lambda \in \Lambda^{+}$, we denote by $V(\lambda)$ the finite-dimensional irreducible $o(2n, \mathbb{F})$-module with highest weight λ.

Let $\mathscr{B} = \mathbb{F}[x_1, \ldots, x_n, y_1, \ldots, y_n]$. Recall (6.2.22)–(6.2.42). The *canonical bosonic oscillator representation* of $o(2n, \mathbb{F})$ is given by

$$(E_{i,j} - E_{n+j,n+i})|_{\mathscr{B}} = x_i\partial_{x_j} - y_j\partial_{y_i}, \quad (E_{n+i,j} - E_{n+j,i})|_{\mathscr{B}} = y_i\partial_{x_j} - y_j\partial_{x_i}, \tag{7.1.11}$$

$$(E_{i,n+j} - E_{j,n+i})|_{\mathscr{B}} = x_i\partial_{y_j} - x_j\partial_{y_i} \tag{7.1.12}$$

for $i, j \in \overline{1,n}$. The positive root vectors of $o(2n, \mathbb{F})$ are

$$\{E_{i,j} - E_{n+j,n+i}, E_{i,n+j} - E_{j,n+i} \mid 1 \le i < j \le n\}. \tag{7.1.13}$$

It can be verified that η in (6.2.23) is an $o(2n,\mathbb{F})$-invariant and Δ in (6.2.33) is an $o(2n,\mathbb{F})$-invariant differential operator, or equivalently,

$$\xi\eta=\eta\xi,\ \xi\Delta=\Delta\xi\quad\text{on }\mathscr{B}\ \text{ for }\ \xi\in o(2n,\mathbb{F}). \tag{7.1.14}$$

Recall the notion $\mathscr{B}_{\ell_1,\ell_2}$ in (6.2.30) and set

$$\mathscr{B}_k=\sum_{\ell=0}^{k}\mathscr{B}_{\ell,k-\ell},\quad \mathscr{H}_k=\{f\in\mathscr{B}_k\mid\Delta(f)=0\}\qquad\text{for }\ k\in\mathbb{N}. \tag{7.1.15}$$

Theorem 7.1.1 *For $k\in\mathbb{N}$, the subspace $\mathscr{H}_k$ forms a finite-dimensional irreducible $o(2n,\mathbb{F})$-submodule and x_1^k is a highest-weight vector with the weight $k\varepsilon_1$. Moreover, it has a basis*

$$\left\{\sum_{i=0}^{\infty}(-1)^i\frac{(x_1y_1)^i(\sum_{i=2}^n\partial_{x_i}\partial_{y_i})^i(x^\alpha y^\beta)}{\prod_{r=1}^i(\alpha_1+i)(\beta_1+i)}\mid\alpha,\beta\in\mathbb{N}^n;\alpha_1\beta_1=0,|\alpha|+|\beta|=k\right\}. \tag{7.1.16}$$

Furthermore,

$$\mathscr{B}=\bigoplus_{k_1,k_2=0}^{\infty}\eta^{k_1}\mathscr{H}_{k_2} \tag{7.1.17}$$

is a direct sum of irreducible $o(2n,\mathbb{F})$-submodules.

Proof According to (6.2.35),

$$\mathscr{H}_k=\sum_{\ell=0}^{k}\mathscr{H}_{\ell,k-\ell}. \tag{7.1.18}$$

Thus (7.1.16) follows from (6.2.42) and (7.1.17) follows from (6.2.37) and (6.2.41). Moreover, the second equation in (7.1.14) implies that $\mathscr{H}_k$ forms an $o(2n,\mathbb{F})$-submodule. Note that $o(2n,\mathbb{F})$ contains the Lie subalgebra

$$\sum_{i,j=1}^{n}\mathbb{F}(E_{i,j}-E_{n+j,n+i})\cong gl(n,\mathbb{F}) \tag{7.1.19}$$

and the representation of $o(2n,\mathbb{F})$ in (7.1.11) and (7.1.12) is essentially an extension of that for $sl(n,\mathbb{F})$ given in (6.2.22). Thus the only possible singular vectors for $o(2n,\mathbb{F})$ in $\mathscr{H}_k$ are:

$$\{x_1^\ell y_n^{k-\ell}\mid\ell\in\overline{0,k}\} \tag{7.1.20}$$

by Lemma 6.2.3, (6.2.39) and (7.1.13). Note

$$(E_{1,2n}-E_{n,n+1})(x_1^\ell y_n^{k-\ell}) = (x_1\partial_{y_n}-x_n\partial_{y_1})(x_1^\ell y_n^{k-\ell}) = (k-\ell)x_1^\ell y_n^{k-\ell-1}. \quad (7.1.21)$$

Thus $\mathscr{H}_k$ has a unique (up to a scalar multiple) singular vector x_1^k of weight $k\varepsilon_1$. By Weyl's Theorem 2.3.6 of complete reducibility if $\mathbb{F} = \mathbb{C}$ or Lemma 6.3.2 with $n_1 = 0$, it is irreducible. The first equation in (7.1.14) shows that all $\eta^{k_1}\mathscr{H}_k$ are irreducible $o(2n, \mathbb{F})$-submodules. □

Consider the exterior algebra $\check{\mathscr{A}}$ generated by $\{\theta_1, \dots, \theta_n, \vartheta_1, \dots, \vartheta_n\}$ (cf. (6.2.15)) and take the settings in (6.2.44)–(6.2.51). We have the *basic fermionic oscillator representation* $o(2n, \mathbb{F})$ on $\check{\mathscr{A}}$ by

$$(E_{i,j} - E_{n+j,n+i})|_{\check{\mathscr{A}}} = \theta_i\partial_{\theta_j} - \vartheta_j\partial_{\vartheta_i}, \quad (E_{n+i,j} - E_{n+j,i})|_{\check{\mathscr{A}}} = \vartheta_i\partial_{\theta_j} - \vartheta_j\partial_{\theta_i}, \quad (7.1.22)$$

$$(E_{i,n+j} - E_{j,n+i})|_{\check{\mathscr{A}}} = \theta_i\partial_{\vartheta_j} - \theta_j\partial_{\vartheta_i} \quad (7.1.23)$$

for $i, j \in \overline{1,n}$. By (7.1.19), (7.1.22) and (7.1.23), the above representation of $o(2n, \mathbb{F})$ is essentially an extension of that for $sl(n, \mathbb{F})$ given in (6.2.43). Set

$$\check{\mathscr{A}}_k = \sum_{\ell=0}^{\min\{k,n\}} \check{\mathscr{A}}_{\ell,k-\ell} \quad \text{for } k \in \overline{0,2n} \quad (7.1.24)$$

(cf. (6.2.45)). According to (6.2.53), the $o(2n, \mathbb{F})$-singular vectors of $\check{\mathscr{A}}_k$ are in the set

$$\{\check{\eta}^\ell\vec{\theta}_r\vec{\vartheta}_s \mid 0 \le r < s \le n+1; \ell \in \overline{0, s-r-1}; r+2\ell+n-s+1 = k\} \quad (7.1.25)$$

(cf. (6.2.50) and (6.2.51)).

For $0 < i < j \le n$,

$$[E_{i,n+j} - E_{j,n+i}, \check{\eta}] = \theta_j\theta_i - \theta_i\theta_j = 2\theta_j\theta_i \quad (7.1.26)$$

by (6.2.49) and (7.1.23). If $0 < r+1 < s \le n$ and $\ell \in \overline{0, s-r-1}$, then

$$(E_{r+1,n+s} - E_{s,n+r+1})(\check{\eta}^\ell\vec{\theta}_r\vec{\vartheta}_s) = 2\ell\theta_s\theta_{r+1}\check{\eta}^{\ell-1}\vec{\theta}_r\vec{\vartheta}_s + (-1)^{n-s}\check{\eta}^\ell\vec{\theta}_{r+1}\vec{\vartheta}_{s+1} \neq 0 \quad (7.1.27)$$

When $1 \le r+1 = s \le n-1$,

$$(E_{n,n+s} - E_{s,2n})(\vec{\theta}_{s-1}\vec{\vartheta}_s) = \vec{\theta}_{s-1}\vec{\vartheta}_{s+1}\theta_n - \vec{\theta}_s\vartheta_{n-1}\cdots\vartheta_s \neq 0. \quad (7.1.28)$$

Note that $\eta^\ell\vec{\theta}_r = 0$ if $\ell > n-r$ by (6.2.49). For $r \in \overline{1,n}$ and $\ell \in \overline{0, n-r}$, (7.1.23) yields

$$(E_{n-1,2n} - E_{n,2n-1})(\check{\eta}^{\ell}\vec{\theta}_r) = 2\ell\theta_n\theta_{n-1}\eta^{\ell-1}\vec{\theta}_r = 0 \Leftrightarrow \ell = 0 \text{ or } \ell \geq n-r. \tag{7.1.29}$$

Furthermore,

$$(E_{n-1,2n} - E_{n,2n-1})(\vec{\theta}_{n-1}\vartheta_n) = \vec{\theta}_{n-1}\theta_{n-1} = 0. \tag{7.1.30}$$

Since $\{E_{i,i+1} - E_{n+i+1,n+i}, E_{n-1,2n} - E_{n,2n-1} \mid i \in \overline{1, n-1}\}$ is the set of simple positive root vectors, the singular vectors of $o(2n, \mathbb{F})$ in $\check{\mathscr{A}}$ are

$$\{\vec{\theta}_n, \vec{\theta}_{n-1}\vartheta_n, \vec{\theta}_r, \check{\eta}^{n-r}\vec{\theta}_r \mid r \in \overline{0, n-1}\}. \tag{7.1.31}$$

By Weyl's Theorem 2.3.6 of complete reducibility if $\mathbb{F} = \mathbb{C}$ or an analogue of Lemma 6.3.2, we have:

Theorem 7.1.2 *For $r \in \overline{1, n-1}$, $\check{\mathscr{A}}_r$ and $\check{\mathscr{A}}_{2n-r}$ are finite-dimensional irreducible $o(2n, \mathbb{F})$-submodules with highest weight $\sum_{i=1}^{r} \varepsilon_i$. Moreover, $\check{\mathscr{A}}_n$ is the direct sum of a finite-dimensional irreducible $o(2n, \mathbb{F})$-submodule with highest weight $\sum_{i=1}^{n} \varepsilon_i$ and a finite-dimensional irreducible $o(2n, \mathbb{F})$-submodule with highest weight $\sum_{i=1}^{n-1} \varepsilon_i - \varepsilon_n$. Furthermore, $\check{\mathscr{A}}_{2n}$ and $\check{\mathscr{A}}_0$ are isomorphic to the one-dimensional trivial $o(2n, \mathbb{F})$-module.*

Let Ψ be the exterior algebra generated by $\{\theta_1, \theta_2, \ldots, \theta_n\}$ (cf. (6.2.15)). Take the settings as those in (6.2.16)–(6.2.18). Observe that

$$[\theta_i\theta_j, \partial_{\theta_j}\partial_{\theta_i}] = \theta_i\partial_{\theta_i} + \theta_j\partial_{\theta_j} - 1 \tag{7.1.32}$$

and

$$[\theta_i\theta_j, \partial_{\theta_j}\partial_{\theta_l}] = \theta_i\partial_{\theta_l} \tag{7.1.33}$$

for distinct $i, j, l \in \overline{1, n}$. Thus we have the following representation of $o(2n, \mathbb{F})$:

$$(E_{i,j} - E_{n+j,n+i})|_{\Psi} = \theta_i\partial_{\theta_j} - \frac{\delta_{i,j}}{2} \quad \text{for } i, j \in \overline{1, n}, \tag{7.1.34}$$

$$(E_{r,n+s} - E_{s,n+r})|_{\Psi} = \theta_r\theta_s, \qquad (E_{n+r,s} - E_{n+s,r})|_{\Psi} = \partial_{\theta_r}\partial_{\theta_s} \tag{7.1.35}$$

for $1 \leq r < s \leq n$. The above representation is called the *spin representation* of $o(2n, \mathbb{F})$.

Recall the notion Ψ_r defined in (6.2.16) and we set

$$\Psi_{(0)} = \sum_{i=0}^{[\![n/2]\!]} \Psi_{2i}, \qquad \Psi_{(1)} = \sum_{s=0}^{[\![(n-1)/2]\!]} \Psi_{2s+1}. \tag{7.1.36}$$

Then

$$\Psi = \Psi_{(0)} \oplus \Psi_{(1)}. \tag{7.1.37}$$

Moreover, we have:

Theorem 7.1.3 *The subspaces $\Psi_{(0)}$ and $\Psi_{(1)}$ form 2^{n-1}-dimensional irreducible $o(2n, \mathbb{F})$-submodules. If n is even, then $\theta_1 \cdots \theta_n$ is a highest-weight vector of $\Psi_{(0)}$ with the weight $(\sum_{i=1}^{n} \varepsilon_i)/2$ and $\theta_1 \cdots \theta_{n-1}$ is a highest-weight vector of $\Psi_{(1)}$ with the weight $(\sum_{i=1}^{n-1} \varepsilon_i - \varepsilon_n)/2$. When n is odd, then $\theta_1 \cdots \theta_n$ is a highest-weight vector of $\Psi_{(1)}$ with weight $(\sum_{i=1}^{n} \varepsilon_i)/2$ and $\theta_1 \cdots \theta_{n-1}$ is a highest-weight vector of $\Psi_{(0)}$ with weight $(\sum_{i=1}^{n-1} \varepsilon_i - \varepsilon_n)/2$.*

7.2 Noncanonical Oscillator Representations

In this section, we determine the structure of the noncanonical oscillator representations of even orthogonal Lie algebras obtained from the above bosonic representations by partially swapping differential operators and multiplication operators, which are generalizations of the classical theorem on harmonic polynomials.

Let $\mathscr{B} = \mathbb{F}[x_1, \ldots, x_n, y_1, \ldots, y_n]$. Recall the canonical oscillator representation of $o(2n, \mathbb{F})$ is given in (7.1.11) and (7.1.12). Fix $n_1, n_2 \in \overline{1, n}$ with $n_1 \leq n_2$. Swapping operators $\partial_{x_r} \mapsto -x_r$, $x_r \mapsto \partial_{x_r}$ for $r \in \overline{1, n_1}$ and $\partial_{y_s} \mapsto -y_s$, $y_s \mapsto \partial_{y_s}$ for $s \in \overline{n_2 + 1, n}$ in the canonical oscillator representation (7.1.11) and (7.1.12), we get a *noncanonical oscillator representation* of $o(2n, \mathbb{F})$ on $\mathscr{B}$ determined by

$$(E_{i,j} - E_{n+j,n+i})|_{\mathscr{B}} = E^x_{i,j} - E^y_{j,i} \tag{7.2.1}$$

with

$$E^x_{i,j} = \begin{cases} -x_j \partial_{x_i} - \delta_{i,j} & \text{if } i, j \in \overline{1, n_1}; \\ \partial_{x_i} \partial_{x_j} & \text{if } i \in \overline{1, n_1},\ j \in \overline{n_1 + 1, n}; \\ -x_i x_j & \text{if } i \in \overline{n_1 + 1, n},\ j \in \overline{1, n_1}; \\ x_i \partial_{x_j} & \text{if } i, j \in \overline{n_1 + 1, n}; \end{cases} \tag{7.2.2}$$

and

$$E^y_{i,j} = \begin{cases} y_i \partial_{y_j} & \text{if } i, j \in \overline{1, n_2}; \\ -y_i y_j & \text{if } i \in \overline{1, n_2},\ j \in \overline{n_2 + 1, n}; \\ \partial_{y_i} \partial_{y_j} & \text{if } i \in \overline{n_2 + 1, n},\ j \in \overline{1, n_2}; \\ -y_j \partial_{y_i} - \delta_{i,j} & \text{if } i, j \in \overline{n_2 + 1, n}; \end{cases} \tag{7.2.3}$$

and

$$E_{i,n+j}|_{\mathscr{B}} = \begin{cases} \partial_{x_i}\partial_{y_j} & \text{if } i \in \overline{1,n_1},\ j \in \overline{1,n_2}; \\ -y_j\partial_{x_i} & \text{if } i \in \overline{1,n_1},\ j \in \overline{n_2+1,n}; \\ x_i\partial_{y_j} & \text{if } i \in \overline{n_1+1,n},\ j \in \overline{1,n_2}; \\ -x_i y_j & \text{if } i \in \overline{n_1+1,n},\ j \in \overline{n_2+1,n}; \end{cases} \tag{7.2.4}$$

and

$$E_{n+i,j}|_{\mathscr{B}} = \begin{cases} -x_j y_i & \text{if } j \in \overline{1,n_1},\ i \in \overline{1,n_2}; \\ -x_j\partial_{y_i} & \text{if } j \in \overline{1,n_1},\ i \in \overline{n_2+1,n}; \\ y_i\partial_{x_j} & \text{if } j \in \overline{n_1+1,n},\ i \in \overline{1,n_2}; \\ \partial_{x_j}\partial_{y_i} & \text{if } j \in \overline{n_1+1,n},\ i \in \overline{n_2+1,n}. \end{cases} \tag{7.2.5}$$

Recall

$$\tilde{\Delta} = \sum_{i=1}^{n_1} x_i\partial_{y_i} - \sum_{r=n_1+1}^{n_2} \partial_{x_r}\partial_{y_r} + \sum_{s=n_2+1}^{n} y_s\partial_{x_s} \tag{7.2.6}$$

and its dual

$$\eta = \sum_{i=1}^{n_1} y_i\partial_{x_i} + \sum_{r=n_1+1}^{n_2} x_r y_r + \sum_{s=n_2+1}^{n} x_s\partial_{y_s}. \tag{7.2.7}$$

Moreover,

$$[\tilde{\Delta}, \eta] = n_1 - n_2 - \widehat{D}, \quad \widehat{D} = -\sum_{i=1}^{n_1} x_i\partial_{x_i} + \sum_{r=n_1+1}^{n} x_r\partial_{x_r} + \sum_{j=1}^{n_2} y_j\partial_{y_j} - \sum_{s=n_2+1}^{n} y_s\partial_{y_s} \tag{7.2.8}$$

and as operators on $\mathscr{B}$,

$$\xi\tilde{\Delta} = \tilde{\Delta}\xi, \quad \xi\eta = \eta\xi, \quad \xi\widehat{D} = \widehat{D}\xi \qquad \text{for } \xi \in o(2n, \mathbb{F}). \tag{7.2.9}$$

Set

$$\mathscr{B}_{\langle k\rangle} = \text{Span}\{x^\alpha y^\beta \mid \alpha, \beta \in \mathbb{N}^n;\ \sum_{r=n_1+1}^{n} \alpha_r - \sum_{i=1}^{n_1} \alpha_i + \sum_{i=1}^{n_2} \beta_i - \sum_{r=n_2+1}^{n} \beta_r = k\} \tag{7.2.10}$$

for $k \in \mathbb{Z}$. Then

$$\mathscr{B}_{\langle k\rangle} = \{u \in \mathscr{B} \mid \widehat{D}(u) = ku\} \tag{7.2.11}$$

forms an $o(2n, \mathbb{F})$-submodule by the third equation (7.2.9). Define

$$\mathscr{H}_{\langle k\rangle} = \{f \in \mathscr{B}_{\langle k\rangle} \mid \tilde{\Delta}(f) = 0\}. \tag{7.2.12}$$

According to the first equation (7.2.9), $\mathscr{H}_{\langle k\rangle}$ forms an $o(2n, \mathbb{F})$-submodule. The following is our main theorem:

Theorem 7.2.1 *For any $n_1 - n_2 + 1 - \delta_{n_1,n_2} \geq k \in \mathbb{Z}$, $\mathscr{H}_{\langle k\rangle}$ is an infinite-dimensional irreducible $o(2n, \mathbb{F})$-weight submodule with finite-dimensional weight subspaces and $\mathscr{B}_{\langle k\rangle} = \bigoplus_{i=0}^{\infty} \eta^i(\mathscr{H}_{\langle k-2i\rangle})$ is a decomposition of irreducible submodules. In particular, $\mathscr{B}_{\langle k\rangle} = \mathscr{H}_{\langle k\rangle} \oplus \eta(\mathscr{B}_{\langle k-2\rangle})$. The module $\mathscr{H}_{\langle k\rangle}$ under the assumption is of highest-weight type only if $n_2 = n$, and $x_{n_1}^{-k}$ is a highest-weight vector with weight $k\varepsilon_{n_1} - \sum_{i=1}^{n_1} \varepsilon_i$ (cf. (7.1.3)).*

Proof We will prove it case by case. According to (7.1.19), the representation of $o(2n, \mathbb{F})$ given in (7.2.1)–(7.2.5) is an extension of that for $sl(n, \mathbb{F})$ given in (6.3.36)–(6.3.38) on $\mathscr{B}$. Moreover,

$$\mathscr{B}_{\langle k\rangle} = \bigoplus_{\ell\in\mathbb{Z}} \mathscr{B}_{\langle \ell,k-\ell\rangle}, \qquad \mathscr{H}_{\langle k\rangle} = \bigoplus_{\ell\in\mathbb{Z}} \mathscr{H}_{\langle \ell,k-\ell\rangle} \tag{7.2.13}$$

by (6.4.29), (6.4.31), (7.2.10) and (7.2.12).

Case 1. $n_1 + 1 < n_2$ *and* $n_1 - n_2 + 1 \geq k \in \mathbb{Z}$.

According to (6.4.40) and (6.4.41), the $sl(n, \mathbb{F})$-singular vectors in $\mathscr{H}_{\langle k\rangle}$ are: for $m_1, m_2 \in \mathbb{N}$,

$$x_{n_1}^{m_1} y_{n_2+1}^{m_2} \qquad \text{with } -(m_1 + m_2) = k, \tag{7.2.14}$$

$$x_{n_1+1}^{m_1} y_{n_2+1}^{m_2} \qquad \text{with } m_1 - m_2 = k, \tag{7.2.15}$$

$$x_{n_1}^{m_1} y_{n_2}^{m_2} \qquad \text{with } -m_1 + m_2 = k. \tag{7.2.16}$$

Note

$$(E_{n+n_2+1,n_1} - E_{n+n_1,n_2+1})|_{\mathscr{B}} = -x_{n_1}\partial_{y_{n_2+1}} - y_{n_1}\partial_{x_{n_2+1}} \tag{7.2.17}$$

by (7.2.5). So

$$(E_{n+n_2+1,n_1} - E_{n+n_1,n_2+1})^{m_2}(x_{n_1}^{m_1} y_{n_2+1}^{m_2}) = (-1)^{m_2} m_2! x_{n_1}^{-k} \tag{7.2.18}$$

for the vectors in (7.2.14). Moreover,

$$(E_{n+n_2+1,n_1+1} - E_{n+n_1+1,n_2+1})|_{\mathscr{B}} = \partial_{x_{n_1+1}}\partial_{y_{n_2+1}} - y_{n_1+1}\partial_{x_{n_2+1}} \tag{7.2.19}$$

again by (7.2.5), which implies

$$(E_{n+n_2+1,n_1+1}-E_{n+n_1+1,n_2+1})^{m_1}(x_{n_1+1}^{m_1}y_{n_2+1}^{m_2})=m_1!\left[\prod_{r=0}^{m_1-1}(m_2-r)\right]y_{n_2+1}^{-k} \tag{7.2.20}$$

for the vectors in (7.2.15). Furthermore,

$$(E_{n_1,n+n_2}-E_{n_2,n+n_1})|_{\mathscr{B}}=\partial_{x_{n_1}}\partial_{y_{n_2}}-x_{n_2}\partial_{y_{n_1}} \tag{7.2.21}$$

by (7.2.4), which implies

$$(E_{n_1,n+n_2}-E_{n_2,n+n_1})^{m_2}(x_{n_1}^{m_1}y_{n_2}^{m_2})=m_2!\left[\prod_{r=0}^{m_2-1}(m_1-r)\right]x_{n_1}^{-k} \tag{7.2.22}$$

for the vectors in (7.2.16).

On the other hand,

$$(E_{n_1,n+n_2+1}-E_{n_2+1,n+n_1})|_{\mathscr{B}}=-y_{n_2+1}\partial_{x_{n_1}}-x_{n_2+1}\partial_{y_{n_1}} \tag{7.2.23}$$

by (7.2.4), which implies

$$(E_{n_1,n+n_2+1}-E_{n_2+1,n+n_1})^{m_2}(x_{n_1}^{-k})=(-1)^{m_2}\left[\prod_{r=0}^{m_2-1}(-k-r)\right]x_{n_1}^{m_1}y_{n_2+1}^{m_2} \tag{7.2.24}$$

for the vectors in (7.2.14). Moreover,

$$(E_{n_1+1,n+n_2+1}-E_{n_2+1,n+n_1+1})|_{\mathscr{B}}=-x_{n_1+1}y_{n_2+1}-x_{n_2+1}\partial_{y_{n_1+1}} \tag{7.2.25}$$

by (7.2.4), which implies

$$(E_{n_1+1,n+n_2+1}-E_{n_2+1,n+n_1+1})^{m_1}(y_{n_2+1}^{-k})=(-1)^{m_1}x_{n_1+1}^{m_1}y_{n_2+1}^{m_2} \tag{7.2.26}$$

for the vectors in (7.2.15). Furthermore,

$$(E_{n+n_2,n_1}-E_{n+n_1,n_2})|_{\mathscr{B}}=-x_{n_1}y_{n_2}-y_{n_1}\partial_{x_{n_2}} \tag{7.2.27}$$

by (7.2.5), which implies

$$(E_{n+n_2,n_1}-E_{n+n_1,n_2})^{m_2}(x_{n_1}^{-k})=(-1)^{m_2}x_{n_1}^{m_1}y_{n_2}^{m_2} \tag{7.2.28}$$

for the vectors in (7.2.16). Thus for any two vectors in (7.2.14)–(7.2.16), there exists an element in the universal enveloping algebra $U(o(2n,\mathbb{F}))$ which carries one to another. On the other hand, the vectors in (7.2.14)–(7.2.16) have distinct weights.

Thus any nonzero $o(2n, \mathbb{F})$-submodule of $\mathscr{H}_{\langle k\rangle}$ must contain one of the vectors in (7.2.14)–(7.2.16). Hence all the vectors in (7.2.14)–(7.2.16) are in the submodule by (7.2.17)–(7.2.28). Therefore, the submodule must contains all $\mathscr{H}_{\langle \ell,k-\ell\rangle}$ for $\ell \in \mathbb{Z}$ by Theorem 6.4.3, or equivalently, it is equal to $\mathscr{H}_{\langle k\rangle}$ due to (7.2.13). So $\mathscr{H}_{\langle k\rangle}$ is irreducible. By (7.2.25) and (7.2.27), $\mathscr{H}_{\langle k\rangle}$ is not of highest-weight type. For any $m \in \mathbb{N}$, $n_1 - n_2 + 1 \geq k - m$ and so $\mathscr{H}_{\langle k-m\rangle}$ is an irreducible $o(2n, \mathbb{F})$-submodule.

Since $\tilde{\Delta}$ is locally nilpotent, for any $0 \neq u \in \mathscr{B}_{\langle k\rangle}$, there exists an element $\kappa(u) \in \mathbb{N}$ such that

$$\tilde{\Delta}^{\kappa(u)}(u) \neq 0 \ \text{ and } \ \tilde{\Delta}^{\kappa(u)+1}(u) = 0. \tag{7.2.29}$$

Set

$$V = \sum_{i=0}^{\infty} \eta^i(\mathscr{H}_{\langle k-2i\rangle}). \tag{7.2.30}$$

Given $0 \neq u \in \mathscr{B}_{\langle k\rangle}$, $\kappa(u) = 0$ implies $u \in \mathscr{H}_{\langle k\rangle} \subset V$. Suppose that $u \in V$ whenever $\kappa(u) < r$ for some positive integer r. Assume $\kappa(u) = r$. First

$$v = \tilde{\Delta}^r(u) \in \mathscr{H}_{\langle k-2r\rangle} \subset V. \tag{7.2.31}$$

Note

$$\tilde{\Delta}^r[\eta^r(v)] = r!\left[\prod_{i=1}^{r}(n_1 - n_2 - k + r + i)\right] v \tag{7.2.32}$$

by (7.2.8) and (7.2.11). Thus we have either

$$u = \frac{1}{r![\prod_{i=1}^{r}(n_1 - n_2 - k + r + i)]}\eta^r(v) \in V \tag{7.2.33}$$

or

$$\kappa\left(u - \frac{1}{r![\prod_{i=1}^{r}(n_1 - n_2 - k + r + i)]}\eta^r(v)\right) < r. \tag{7.2.34}$$

By induction,

$$u - \frac{1}{r![\prod_{i=1}^{r}(n_1 - n_2 - k + r + i)]}\eta^r(v) \in V, \tag{7.2.35}$$

which implies $u \in V$. Therefore, we have $V = \mathscr{B}_{\langle k\rangle}$. Since the weight of any $sl(n, \mathbb{F})$-singular vector in $\eta^i(\mathscr{H}_{\langle k-2i\rangle})$ is different from the weight of any $sl(n, \mathbb{F})$-singular vector in $\eta^j(\mathscr{H}_{\langle k-2j\rangle})$ when $i \neq j$, the sums in Theorem 7.2.1 are direct sums.

Case 2. $n_1 + 1 = n_2$ *and* $0 \geq k \in \mathbb{Z}$.

Assume $n_1 + 1 = n_2 < n$. By (6.5.60), (7.2.8) and (7.2.11), the $sl(n,\mathbb{F})$-singular vectors in $\mathscr{H}_{\langle k\rangle}$ are those in (7.2.14)–(7.2.16). So Theorem 7.2.1 holds by the arguments in Case 1. Suppose $n_1 < n_2 = n$. According to (6.5.62), (7.2.8) and (7.2.11), the $sl(n,\mathbb{F})$-singular vectors in $\mathscr{H}_{\langle k\rangle}$ are those in (7.2.16). Expressions (7.2.22) and (7.2.28)–(7.2.35) imply the conclusions in the Theorem 7.2.1.

Case 3. $n_1 = n_2$ *and* $0 \geq k \in \mathbb{Z}$.

Recall

$$\zeta_1 = x_{n_1-1}y_{n_1} - x_{n_1}y_{n_1-1}, \quad \zeta_2 = x_{n_1+1}y_{n_1+2} - x_{n_1+2}y_{n_1+1}. \tag{7.2.36}$$

Suppose $n_1 = n_2 < n - 1$. Lemma 6.6.2 tells us that the $sl(n,\mathbb{F})$-singular vectors in $\mathscr{H}_{\langle k\rangle}$ are those in (7.2.14) and

$$x_{n_1}^{-k}\zeta_1^{m+1} \quad \text{for } m \in \mathbb{N}, \tag{7.2.37}$$

$$y_{n_1+1}^{-k}\zeta_2^{m+1} \quad \text{for } m \in \mathbb{N}. \tag{7.2.38}$$

Again all the singular vectors have distinct weights. If $\mathscr{N}$ is a nonzero submodule of $\mathscr{H}_{\langle k\rangle}$, then $\mathscr{N}$ must contain one of the above $sl(n,\mathbb{F})$-singular vectors. If $\mathscr{N}$ contains a singular vector in (7.2.14), then $x_{n_1}^{-k} \in N$ by (7.2.18). Suppose $x_{n_1}^{-k}\zeta_1^{m+1} \in N$ for some $m \in \mathbb{N}$. Note

$$(E_{n_1-1,n+n_1} - E_{n_1,n+n_1-1})|_{\mathscr{B}} = \partial_{x_{n_1-1}}\partial_{y_{n_1}} - \partial_{x_{n_1}}\partial_{y_{n_1-1}} \tag{7.2.39}$$

by (7.2.4). Thus

$$\begin{aligned}
&(E_{n_1-1,n+n_1} - E_{n_1,n+n_1-1})^{m+1}(x_{n_1}^{-k}\zeta_1^{m+1})\\
&= \left[\sum_{r=0}^{m+1}(-1)^r\binom{m+1}{r}(\partial_{x_{n_1-1}}\partial_{y_{n_1}})^{m+1-r}(\partial_{x_{n_1}}\partial_{y_{n_1-1}})^r\right]\\
&\quad\times\left[\sum_{s=0}^{m+1}(-1)^s\binom{m+1}{s}(x_{n_1-1}y_{n_1})^{m+1-s}x_{n_1}^{-k+s}y_{n_1-1}^s\right]\\
&= \left(\sum_{r=0}^{m+1}\binom{m+1}{r}^2[(m+1-r!)]^2 r![\prod_{i=1}^{r}(-k+i)]\right)x_{n_1}^{-k}\\
&= [(m+1)!]^2\left(\sum_{r=0}^{m+1}\binom{-k+r}{r}\right)x_{n_1}^{-k} \in N.
\end{aligned} \tag{7.2.40}$$

So we have $x_{n_1}^{-k} \in \mathscr{N}$ again. Symmetrically, $y_{n_1+1}^{-k} \in \mathscr{N}$ if $y_{n_1+1}^{-k}\zeta_2^{m+1} \in \mathscr{N}$ for some $m \in \mathbb{N}$. Observe

$$(E_{n+n_1,n_1+1} - E_{n+n_1+1,n_1})|_{\mathscr{B}} = y_{n_1}\partial_{x_{n_1+1}} + x_{n_1}\partial_{y_{n_1+1}} \tag{7.2.41}$$

by (7.2.5). Thus

$$(E_{n+n_1,n_1+1} - E_{n+n_1+1,n_1})^{-k}(y_{n_1+1}^{-k}) = (-k)!x_{n_1}^{-k} \in \mathscr{N}. \tag{7.2.42}$$

Hence we always have $x_{n_1}^{-k} \in \mathscr{N}$.

According to (7.2.24), $\mathscr{N}$ contains all the singular vectors in (7.2.14). Observe that

$$(E_{n+n_1-1,n_1} - E_{n+n_1,n_1-1})|_{\mathscr{B}} = \zeta_1, \quad (E_{n_1+2,n+n_1+1} - E_{n_1+1,n+n_1+2})|_{\mathscr{B}} = \zeta_2 \tag{7.2.43}$$

as multiplication operators on $\mathscr{B}$ by (7.2.4) and (7.2.5). Thus

$$(E_{n+n_1-1,n_1} - E_{n+n_1,n_1-1})^{m+1}(x_{n_1}^{-k}) = x_{n_1}^{-k}\zeta_1^{m+1}, \tag{7.2.44}$$

$$(E_{n_1+2,n+n_1+1} - E_{n_1+1,n+n_1+2})^{m+1}(x_{n_1}^{-k}) = x_{n_1}^{-k}\zeta_2^{m+1} \in \mathscr{N}. \tag{7.2.45}$$

Note that

$$(E_{n_1+1,n+n_1} - E_{n_1,n+n_1+1})|_{\mathscr{B}} = x_{n_1+1}\partial_{y_{n_1}} + y_{n_1+1}\partial_{x_{n_1}} \tag{7.2.46}$$

by (7.2.4). So

$$\frac{1}{(-k)!}(E_{n_1+1,n+n_1} - E_{n_1,n+n_1+1})^{-k}(x_{n_1}^{-k}\zeta_2^{m+1}) = y_{n_1+1}^{-k}\zeta_2^{m+1} \in \mathscr{N}. \tag{7.2.47}$$

Thus $\mathscr{N}$ contains all the $sl(n,\mathbb{F})$-singular vectors in $\mathscr{H}_{\langle k\rangle}$, which implies that it contains all $\mathscr{H}_{\langle \ell,k-\ell\rangle} \subset \mathscr{H}_{\langle k\rangle}$ by Theorem 6.6.3. So $\mathscr{N} = \mathscr{H}_{\langle k\rangle}$ by (7.2.13); that is, $\mathscr{H}_{\langle k\rangle}$ is an irreducible $o(2n,\mathbb{F})$-module. By (7.2.29)–(7.2.35), the direct sums in Theorem 7.2.1 holds. Observe that Theorem 7.2.1 under the subcase $n_1 = n-1, n_2 = n$ is implied by (6.6.30), (6.6.33), (6.6.36), (6.6.39) and the related partial arguments in the above. This completes the proof of Theorem 7.2.1. □

Suppose $n_1 < n_2$. Taking $T_1 = -\partial_{x_{n_1+1}}\partial_{y_{n_1+1}}$, $T_1^- = -\int_{(x_{n_1+1})}\int_{(y_{n_1+1})}$ and $T_2 = \tilde{\delta} + \partial_{x_{n_1}+1}\partial_{y_{n_1}+1}$ in Lemma 6.1.1, $\mathscr{H}_{\langle k\rangle}$ has a basis

$$\left\{\sum_{i=0}^{\infty}\frac{(x_{n_1+1}y_{n_1+1})^i(\tilde{\Delta}+\partial_{x_{n_1}+1}\partial_{y_{n_1}+1})^i(x^\alpha y^\beta)}{\prod_{r=1}^{i}(\alpha_{n_1+1}+r)(\beta_{n_1+1}+r)} \mid \alpha,\beta\in\mathbb{N}^n;\right.$$
$$\left.\alpha_{n_1+1}\beta_{n_1+1} = 0; -\sum_{i=1}^{n_1}\alpha_i + \sum_{r=n_1+1}^{n}\alpha_r + \sum_{i=1}^{n_2}\beta_i - \sum_{r=n_2+1}^{n}\beta_r = k\right\} \tag{7.2.48}$$

Finally, we want to find an expression for $\mathscr{H}_{\langle k\rangle}$ for $0 \geq k \in \mathbb{Z}$ when $n_1 = n_2$. First we assume $n_1 = n_2 = 1$ and $n = 2$. According to (6.6.36), (6.6.52) and (7.2.13),

$$\mathscr{H}_{\langle -k\rangle} = \text{Span}\{[x_1^r y_2^s(x_1x_2 - y_1y_2)^l \mid r, s, l \in \mathbb{N}; r + s = k\}. \tag{7.2.49}$$

Next we consider the subcase $n_1 = n_2 = 1$ and $n \geq 3$. According to (6.6.30), (6.6.52), (6.6.54) and (7.2.13)

$$\begin{aligned}&\mathscr{H}_{\langle -k\rangle}\\ &= \text{Span}\left\{\left[\prod_{r=2}^n y_r^{\hat{l}_r}\right]\left[\prod_{2\leq p<q\leq n}(x_py_q - x_qy_p)^{\hat{k}_{p,q}}\right]\left[\prod_{s=2}^n(x_1x_s - y_1y_s)^{\hat{l}_s}\right], x_1^l\left[\prod_{s=2}^n y_s^{k_s}\right]\right.\\ &\left.\times\left[\prod_{s=2}^n(x_1x_s - y_1y_s)^{l_s}\right] \mid l, k_s, l_s, \hat{l}, \hat{k}_{p,q}, \hat{l}_s \in \mathbb{N}; l + \sum_{s=2}^n k_s = \sum_{r=2}^n \hat{l}_r = k\right\}.\end{aligned} \tag{7.2.50}$$

Assume $1 < n_1 = n_2 < n - 1$. By (6.6.25), (6.6.52)–(6.6.54) and (7.2.13), we have

$$\begin{aligned}&\mathscr{H}_{\langle -k\rangle}\\ &= \text{Span}\left\{\left[\prod_{r=1}^{n_1} x_r^{l'_r}\right]\left[\prod_{1\leq p<q\leq n_1}(x_py_q - x_qy_p)^{k'_{p,q}}\right]\left[\prod_{r=1}^{n_1}\prod_{s=n_1+1}^{n}(x_rx_s - y_ry_s)^{l'_{r,s}}\right],\right.\\ &\left[\prod_{r=1}^{n-n_1} y_{n_1+r}^{\hat{l}_r}\right]\left[\prod_{n_1+1\leq p<q\leq n}(x_py_q - x_qy_p)^{\hat{k}_{p,q}}\right]\left[\prod_{r=1}^{n_1}\prod_{s=n_1+1}^{n}(x_rx_s - y_ry_s)^{\hat{l}_{r,s}}\right],\\ &\left[\prod_{r=1}^{n_1} x_r^{l_r}\right]\left[\prod_{s=1}^{n-n_1} y_{n_1+s}^{k_s}\right]\left[\prod_{r=1}^{n_1}\prod_{s=1}^{n-n_1}(x_rx_{n_1+s} - y_ry_{n_1+s})^{l_{r,s}}\right] \mid l_r, k_s, l_{r,s}, l'_r, k'_{p,q},\\ &\left. l'_{r,s}, \hat{l}_r, \hat{k}_{p,q}, \hat{l}_{r,s} \in \mathbb{N}; \sum_{r=1}^{n_1} l_r + \sum_{s=1}^{n-n_1} k_s = \sum_{r=1}^{n_1} l'_r = \sum_{r=1}^{n-n_1} \hat{l}_r = k\right\}.\end{aligned} \tag{7.2.51}$$

Consider the subcase $n_1 = n_2 = n - 1$ and $n \geq 3$. By (6.6.33), (6.6.52), (6.6.53) and (7.2.13), we obtain

$$\begin{aligned}&\mathscr{H}_{\langle -k\rangle}\\ &= \text{Span}\left\{\left[\prod_{r=1}^{n-1} x_r^{l'_r}\right]\left[\prod_{1\leq p<q\leq n-1}(x_py_q - x_qy_p)^{k'_{p,q}}\right]\left[\prod_{r=1}^{n-1}(x_rx_n - y_ry_n)^{\bar{l}'_r}\right], \left[\prod_{r=1}^{n-1} x_r^{l_r}\right] y_n^{\hat{k}}\right.\\ &\left.\times\left[\prod_{r=1}^{n-1}(x_rx_n - y_ry_n)^{\bar{l}_r}\right] \mid l_r, \hat{k}, \bar{l}_r, l'_r, k'_{p,q}, \bar{l}'_r \in \mathbb{N}; \sum_{r=1}^{n-1} l_r + \hat{k} = \sum_{r=1}^{n-1} l'_r = k\right\}.\end{aligned} \tag{7.2.52}$$

At last, we assume $n_1 = n_2 = n$. By (6.6.39), (6.6.53) and (7.2.13),

$$\mathscr{H}_{\langle -k\rangle} = \text{Span}\left\{\left[\prod_{r=1}^{n} x_r^{l_r}\right]\left[\prod_{1\le p<q\le n} (x_p y_q - x_q y_p)^{k_{p,q}}\right] \mid l_r, k_{p,q} \in \mathbb{N}; \sum_{r=1}^{n} l_r = k\right\}. \tag{7.2.53}$$

7.3 Extensions of the Conformal Representation

In this section, we construct a new functor from the category $o(2n, \mathbb{C})$-**mod** to the category $o(2n+2, \mathbb{C})$-**mod**, which is an extension of the conformal representation of $o(2n+2, \mathbb{C})$.

The n-dimensional conformal group with respect to Euclidean metric $(\cdot, \cdot)$ is generated by the translations, rotations, dilations and *special conformal transformations*

$$\vec{x} \mapsto \frac{\vec{x} - (\vec{x}, \vec{x})\vec{b}}{(\vec{b}, \vec{b})(\vec{x}, \vec{x}) - 2(\vec{b}, \vec{x}) + 1}. \tag{7.3.1}$$

The above transformations yield an inhomogeneous representation of the real Lie algebra $o(2, n)$. We assume the base field $\mathbb{F} = \mathbb{C}$.

For $\lambda \in \Lambda^+$, we denote by $V(\lambda)$ the finite-dimensional irreducible $o(2n, \mathbb{C})$-module with highest weight λ. The $2n$-dimensional natural module of $o(2n, \mathbb{C})$ is $V(\varepsilon_1)$ with the weights $\{\pm\varepsilon_i \mid i \in \overline{1, n}\}$. Recall

$$\rho = \frac{1}{2} \sum_{\nu\in\Phi_{D_n}^+} \nu. \tag{7.3.2}$$

Then

$$(\rho, \nu) = 1 \qquad \text{for } \nu \in \Pi_{D_n} \tag{7.3.3}$$

by (3.2.17). Expression (7.1.7) yields

$$\rho = \sum_{i=1}^{n-1} (n-i)\varepsilon_i. \tag{7.3.4}$$

The following result can be derived from Theorem 5.4.3:

Lemma 7.3.1 (Pieri's formula) *Given* $\mu \in \Lambda^+$ *with* $\mathscr{S}(\mu) = \{i_1, i_2, \ldots, i_{s+1}\}$ *(cf. (6.7.54) and (6.7.55)),*

$$V(\varepsilon_1)\otimes_{\mathbb{C}} V(\mu)\cong\bigoplus_{r=1}^{s}(V(\mu+\varepsilon_{i_r})\oplus V(\mu-\varepsilon_{i_{r+1}-1})) \tag{7.3.5}$$

if $\mu_{n-1}+\mu_n>0$ *and*

$$V(\varepsilon_1)\otimes_{\mathbb{C}} V(\mu)\cong\bigoplus_{r=1}^{s-2+\delta_{\mu_n,0}} V(\mu-\varepsilon_{i_{r+1}-1})\oplus\bigoplus_{r=1}^{s} V(\mu+\varepsilon_{i_r}) \tag{7.3.6}$$

when $\mu_{n-1}+\mu_n=0$.

Denote by $U(\mathscr{G})$ the universal enveloping algebra of a Lie algebra $\mathscr{G}$. The algebra $U(\mathscr{G})$ can be imbedded into the tensor algebra $U(\mathscr{G})\otimes U(\mathscr{G})$ by the associative algebra homomorphism $\mathfrak{d}: U(\mathscr{G})\to U(\mathscr{G})\otimes_{\mathbb{C}} U(\mathscr{G})$ determined by

$$\mathfrak{d}(u)=u\otimes 1+1\otimes u\qquad\text{for } u\in\mathscr{G}. \tag{7.3.7}$$

Note that the Casimir element of $o(2n,\mathbb{C})$ is

$$\omega=\sum_{1\le i<j\le n}[(E_{i,n+j}-E_{j,n+i})(E_{n+j,i}-E_{n+i,j})+(E_{n+j,i}-E_{n+i,j})(E_{i,n+j}-E_{j,n+i})]$$
$$+\sum_{i,j=1}^{n}(E_{i,j}-E_{n+j,n+i})(E_{j,i}-E_{n+i,n+j})\in U(o(2n,\mathbb{C})). \tag{7.3.8}$$

Set

$$\tilde{\omega}=\frac{1}{2}(\mathfrak{d}(\omega)-\omega\otimes 1-1\otimes\omega)\in U(o(2n,\mathbb{C}))\otimes_{\mathbb{C}} U(o(2n,\mathbb{C})). \tag{7.3.9}$$

By (7.3.8),

$$\begin{aligned}\tilde{\omega}=\sum_{1\le i<j\le n}&[(E_{i,n+j}-E_{j,n+i})\otimes(E_{n+j,i}-E_{n+i,j})\\&+(E_{n+j,i}-E_{n+i,j})\otimes(E_{i,n+j}-E_{j,n+i})]\\&+\sum_{i,j=1}^{n}(E_{i,j}-E_{n+j,n+i})\otimes(E_{j,i}-E_{n+i,n+j}).\end{aligned} \tag{7.3.10}$$

For any $\mu\in\Lambda^+$, we have

$$\omega|_{V(\mu)}=(\mu+2\rho,\mu)\mathrm{Id}_{V(\mu)}. \tag{7.3.11}$$

Denote

$$\ell_i^+(\mu)=\dim V(\mu+\varepsilon_i)\quad\text{ if }\ \mu+\varepsilon_i\in\Lambda^+ \tag{7.3.12}$$

and

$$\ell_i^-(\mu) = \dim V(\mu - \varepsilon_i) \quad \text{if } \mu - \varepsilon_i \in \Lambda^+. \tag{7.3.13}$$

Observe that

$$(\mu + \varepsilon_i + 2\rho, \mu + \varepsilon_i) - (\mu + 2\rho, \mu) - (\varepsilon_1 + 2\rho, \varepsilon_1) = 2(\mu_i + 1 - i) \tag{7.3.14}$$

and

$$(\mu - \varepsilon_i + 2\rho, \mu - \varepsilon_i) - (\mu + 2\rho, \mu) - (\varepsilon_1 + 2\rho, \varepsilon_1) = 2(1 + i - 2n - \mu_i) \tag{7.3.15}$$

for $\mu = \sum_{r=1}^n \mu_r \varepsilon_r$ by (7.3.10). Moreover, the algebra $U(o(2n, \mathbb{C})) \otimes_{\mathbb{C}} U(o(2n, \mathbb{C}))$ acts on $V(\varepsilon_1) \otimes_{\mathbb{C}} V(\mu)$ by

$$(\xi_1 \otimes \xi_2)(v \otimes u) = \xi_1(v) \otimes \xi_2(u) \quad \text{for } \xi_1, \xi_2 \in U(o(2n, \mathbb{F})),\ v \in V(\varepsilon_1),\ u \in V(\mu). \tag{7.3.16}$$

By Lemma 7.3.1, (7.3.9) and (7.3.12)–(7.3.15), we obtain:

Lemma 7.3.2 *Let* $\mu = \sum_{i=1}^n \mu_i \varepsilon_i \in \Lambda^+$ *with* $\mathscr{S}(\mu) = \{i_1, i_2, \ldots, i_{s+1}\}$ *(cf. (6.7.54) and (6.7.55)). If* $\mu_{n-1} + \mu_n > 0$*, the characteristic polynomial of* $\tilde{\omega}|_{V(\varepsilon_1) \otimes_{\mathbb{C}} V(\mu)}$ *is*

$$\prod_{r=1}^{s} [(t - \mu_{i_r} + i_r - 1)^{\ell_{i_r}^+(\mu)} (t + \mu_{i_r} + 2n - i_{r+1})^{\ell_{i_{r+1}-1}^-(\mu)}]. \tag{7.3.17}$$

When $\mu_{n-1} + \mu_n = 0$*, the characteristic polynomial of* $\tilde{\omega}|_{V(\varepsilon_1) \otimes_{\mathbb{C}} V(\mu)}$ *is*

$$[\prod_{r=1}^{s} [(t - \mu_{i_r} + i_r - 1)^{\ell_{i_r}^+(\mu)}]][\prod_{\iota=1}^{s-2+\delta_{\mu_n,0}} (t + \mu_{i_\iota} + 2n - i_{\iota+1})^{\ell_{i_{\iota+1}-1}^-(\mu)}]. \tag{7.3.18}$$

We remark that the above lemma is equivalent to an explicit version of Kostant's characteristic identity (cf. [Kb]).

Denote

$$A_{i,j} = E_{i,j} - E_{n+1+j,n+1+i}, \quad B_{i,j} = E_{i,n+1+j} - E_{j,n+1+i}, \quad C_{i,j} = E_{n+1+i,j} - E_{n+1+j,i} \tag{7.3.19}$$

for $i, j \in \overline{1, n+1}$. Then the split orthogonal Lie algebra

$$o(2n+2, \mathbb{C}) = \sum_{i,j=1}^{n+1} \mathbb{C}A_{i,j} + \sum_{1 \le i < j \le n+1} (\mathbb{C}B_{i,j} + \mathbb{C}C_{j,i}) \tag{7.3.20}$$

Recall

$$\mathscr{A} = \mathbb{C}[x_1, x_2, \ldots, x_{2n}]. \tag{7.3.21}$$

Set

$$\eta = \sum_{i=1}^{n} x_i x_{n+i}, \qquad D = \sum_{r=1}^{2n} x_r \partial_{x_r}. \tag{7.3.22}$$

Replace n by $2n$ and take the product $(\vec{x}, \vec{y}) = (1/2)\sum_{i=1}^{n}(x_i y_{n+i} + x_{n+i} y_i)$ in (7.3.1). Using the derivatives of one-parameter conformal transformations just as one derives the Lie algebra of a Lie group, we get the following *conformal representation* of $o(2n+2, \mathbb{C})$ determined by

$$A_{i,j}|_{\mathscr{A}} = x_i\partial_{x_j} - x_{n+j}\partial_{x_{n+i}}, \qquad B_{i,j}|_{\mathscr{A}} = x_i\partial_{x_{n+j}} - x_j\partial_{x_{n+i}}, \tag{7.3.23}$$

$$C_{i,j}|_{\mathscr{A}} = x_{n+i}\partial_{x_j} - x_j\partial_{x_{n+i}}, \qquad A_{n+1,n+1}|_{\mathscr{A}} = -D, \tag{7.3.24}$$

$$A_{n+1,i}|_{\mathscr{A}} = \partial_{x_i}, \qquad B_{i,n+1}|_{\mathscr{A}} = -\partial_{x_{n+i}}, \tag{7.3.25}$$

$$A_{i,n+1}|_{\mathscr{A}} = -x_i D + \eta\partial_{x_{n+i}}, \quad C_{n+1,i}|_{\mathscr{A}} = x_{n+i}D - \eta\partial_{x_i} \tag{7.3.26}$$

for $i, j \in \overline{1, n}$.

Note that

$$\mathscr{L} = \sum_{i,j=1}^{n} \mathbb{C}A_{i,j} + \sum_{1\leq i<j\leq n} (\mathbb{C}B_{i,j} + \mathbb{C}C_{j,i}) \tag{7.3.27}$$

forms a Lie subalgebra of $o(2n+2, \mathbb{C})$ that is isomorphic to $o(2n, \mathbb{C})$. For convenience, we make the identification of $o(2n, \mathbb{C})$ with $\mathscr{L}$ as follows:

$$E_{i,j} - E_{n+j,n+i} \leftrightarrow A_{i,j}, \quad E_{i,n+j} - E_{j,n+i} \leftrightarrow B_{i,j}, \quad E_{n+i,j} - E_{n+j,i} \leftrightarrow C_{i,j} \tag{7.3.28}$$

for $i, j \in \overline{1, n}$. Recall the Witt algebra $\mathbb{W}_{2n} = \sum_{i=1}^{2n} \mathscr{A}\partial_{x_i}$, and Shen's monomorphism $\Im : \mathbb{W}_{2n} \to \widehat{\mathbb{W}}_{2n} = \mathbb{W}_{2n} \oplus gl(2n, \mathscr{A})$ (cf. (6.7.12)) given by

$$\Im\left(\sum_{i=1}^{2n} f_i\partial_{x_i}\right) = \sum_{i=1}^{2n} f_i\partial_{x_i} \oplus \sum_{i,j=1}^{n} \partial_{x_i}(f_j)E_{i,j}. \tag{7.3.29}$$

Now we have the Lie algebra monomorphism $\nu : o(2n+2, \mathbb{C}) \to \widehat{\mathbb{W}}_{2n}$ given by

$$\nu(\xi) = \Im(\xi|_{\mathscr{A}}) \qquad \text{for } \xi \in o(2n+2, \mathbb{C}). \tag{7.3.30}$$

Denote

$$I_{2n} = \sum_{r=1}^{2n} E_{r,r}. \tag{7.3.31}$$

According to the identification (7.3.28), we have

$$\nu(A_{i,j}) = (x_i\partial_{x_j} - x_{n+j}\partial_{x_{n+i}}) \oplus A_{i,j}, \quad \nu(B_{i,j}) = (x_i\partial_{x_{n+j}} - x_j\partial_{x_{n+i}}) \oplus B_{i,j}, \tag{7.3.32}$$

$$\nu(C_{i,j}) = (x_{n+i}\partial_{x_j} - x_{n+j}\partial_{x_i}) \oplus C_{i,j}, \quad \nu(A_{n+1,n+1}) = -(D \oplus I_{2n}), \tag{7.3.33}$$

$$\nu(A_{n+1,i}) = \partial_{x_i}, \qquad \nu(B_{i,n+1}) = -\partial_{x_{n+i}}, \tag{7.3.34}$$

$$\nu(A_{i,n+1}) = (-x_i D + \eta\partial_{x_{n+i}}) \oplus \left[\sum_{p=1}^{n}(x_{n+p}B_{p,i} - x_p A_{i,p}) - x_i I_{2n}\right], \tag{7.3.35}$$

$$\nu(C_{n+1,i}) = (x_{n+i} D - \eta\partial_{x_i}) \oplus \left[\sum_{p=1}^{n}(x_p C_{i,p} - x_{n+p} A_{p,i}) + x_{n+i} I_{2n}\right] \tag{7.3.36}$$

for $i, j \in \overline{1, n}$.

Observe that

$$\widehat{\mathbb{W}}_{2n}^{o} = \widehat{\mathbb{W}}_{2n} \oplus [o(2n, \mathscr{A}) + \mathscr{A} I_{2n}] \tag{7.3.37}$$

form a Lie subalgebra of $\widehat{\mathbb{W}}_{2n}$ and

$$\nu(o(2n+2, \mathbb{C})) \subset \widehat{\mathbb{W}}_{2n}^{o}. \tag{7.3.38}$$

Let M be an $o(2n, \mathbb{C})$-module and let $c \in \mathbb{C}$ be a fixed constant. Then

$$\widehat{M} = \mathscr{A} \otimes_{\mathbb{C}} M \tag{7.3.39}$$

becomes a $\widehat{\mathbb{W}}_{2n}^{o}$-module with the action:

$$(d + f_1 A + f_2 I_{2n})(g \otimes v) = (d(g) + c f_2 g) \otimes v + f_1 g \otimes A(v) \tag{7.3.40}$$

for $f_1, f_2, g \in \mathscr{A}$, $A \in o(2n, \mathbb{C})$ and $v \in M$. Moreover, we make $\widehat{M}$ an $o(2n+2, \mathbb{C})$-module with the action:

$$\xi(\varpi) = \nu(\xi)(\varpi) \qquad \text{for } \xi \in o(2n+2, \mathbb{C}),\ \varpi \in \widehat{M}. \tag{7.3.41}$$

Denote

$$\mathscr{G} = o(2n+2,\mathbb{C}), \qquad \mathscr{G}_0 = o(2n,\mathbb{C}) + \mathbb{C}A_{n+1,n+1} \tag{7.3.42}$$

$$\mathscr{G}_+ = \sum_{i=1}^{n}(\mathbb{C}A_{n+1,i} + \mathbb{C}B_{i,n+1}), \qquad \mathscr{G}_- = \sum_{i=1}^{n}(\mathbb{C}A_{i,n+1} + \mathbb{C}C_{n+1,i}) \tag{7.3.43}$$

(cf. (7.3.28)). Then $\mathscr{G}_\pm$ are abelian Lie subalgebras of $\mathscr{G}$ and $\mathscr{G}_0$ is a Lie subalgebra of $\mathscr{G}$. Moreover,

$$\mathscr{G} = \mathscr{G}_- + \mathscr{G}_0 + \mathscr{G}_+, \qquad [\mathscr{G}_0, \mathscr{G}_\pm] \subset \mathscr{G}_\pm. \tag{7.3.44}$$

By (7.3.32)–(7.3.34),

$$\mathscr{G}_+(1\otimes M) = \{0\}, \quad U(\mathscr{G}_0)(1\otimes M) = 1\otimes M. \tag{7.3.45}$$

Thus

$$U(\mathscr{G})(1\otimes M) = U(\mathscr{G}_-)(1\otimes M). \tag{7.3.46}$$

Using (7.3.34), we can prove:

Proposition 7.3.3 *The map $M \mapsto U(\mathscr{G}_-)(1\otimes M)$ gives rise to a functor from the category of $o(2n,\mathbb{C})$-modules to the category of $o(2n+2,\mathbb{C})$-modules. In particular, it maps irreducible $o(2n,\mathbb{C})$-modules to irreducible $o(2n+2,\mathbb{C})$-modules.*

From another point view, $U(\mathscr{G}_-)(1\otimes M)$ is a polynomial extension from $o(2n,\mathbb{C})$-module M to an $o(2n+2,\mathbb{C})$-module. Next we want to study when $\widehat{M} = U(\mathscr{G}_-)(1\otimes M)$. Write

$$x^\alpha = \prod_{i=1}^{2n} x_i^{\alpha_i}, \quad E^\alpha = \prod_{r=1}^{n}[(-A_{r,n+1})^{\alpha_r} C_{n+1,r}^{\alpha_{n+r}}] \qquad \text{for } \alpha = (\alpha_1,\alpha_2,\ldots,\alpha_{2n}) \in \mathbb{N}^{2n}. \tag{7.3.47}$$

For $k \in \mathbb{N}$, we set

$$\mathscr{A}_k = \text{Span}\{x^\alpha \mid \alpha \in \mathbb{N}^{2n},\ |\alpha| = k\}, \quad \widehat{M}_k = \mathscr{A}_k \otimes_{\mathbb{C}} M \tag{7.3.48}$$

(recall $|\alpha| = \sum_{i=1}^{2n}\alpha_i$) and

$$\begin{aligned}(U(\mathscr{G}_-)(1\otimes M))_k &= U(\mathscr{G}_-)(1\otimes M)\bigcap \widehat{M}_k \\ &= \text{Span}\{E^\alpha(1\otimes M) \mid \alpha \in \mathbb{N}^{2n},\ |\alpha| = k\}\end{aligned} \tag{7.3.49}$$

by (7.3.35) and (7.3.36). Moreover,

$$(U(\mathscr{G}_-)(1\otimes M))_0 = \widehat{M}_0 = 1\otimes M. \tag{7.3.50}$$

Furthermore,

$$\widehat{M} = \bigoplus_{k=0}^{\infty} \widehat{M}_k, \qquad U(\mathscr{G}_-)(1 \otimes M) = \bigoplus_{k=0}^{\infty} (U(\mathscr{G}_-)(1 \otimes M))_k. \tag{7.3.51}$$

According to (7.3.26), we define a $\mathscr{G}_0$-module homomorphism $\varphi : \widehat{M} \to U(\mathscr{G}_-)(1 \otimes M)$ by

$$\varphi(x^\alpha \otimes v) = E^\alpha(1 \otimes v) \qquad \text{for } \alpha \in \mathbb{N}^{2n}, \ v \in M. \tag{7.3.52}$$

Then we have

$$\varphi(\widehat{M}_k) = (U(\mathscr{G}_-)(1 \otimes M))_k \qquad \text{for } k \in \mathbb{N}. \tag{7.3.53}$$

Under the identification (7.3.28), we have:

Lemma 7.3.4 *We have* $\varphi|_{\widehat{M}_1} = (c + \tilde{\omega})|_{\widehat{M}_1}$ *(cf. (7.3.8)–(7.3.10)).*

Proof Let $i \in \overline{1, n}$ and $v \in M$. Expressions (7.3.35) and (7.3.36) give

$$\varphi(x_i \otimes v) = -A_{i,n+1}(1 \otimes v) = \sum_{p=1}^{n} (x_p \otimes A_{i,p}(v) + x_{n+p} \otimes B_{i,p}(v)) + c x_i \otimes v, \tag{7.3.54}$$

$$\varphi(x_{n+i} \otimes v) = C_{n+1,i}(1 \otimes v) = \sum_{p=1}^{n} (x_p \otimes C_{i,p}(v) - x_{n+p} \otimes A_{p,i}(v)) + c x_{n+i} \otimes v. \tag{7.3.55}$$

On the other hand, (7.3.10), (7.3.28), (7.3.32) and (7.3.33) yield

$$\begin{aligned} \tilde{\omega}(x_i \otimes v) = & \sum_{1 \le p < q \le n} [(B_{p,q} \otimes C_{q,p})(x_i \otimes v) + (C_{q,p} \otimes B_{p,q})(x_i \otimes v)] \\ & + \sum_{r,s=1}^{n} (A_{r,s} \otimes A_{s,r})(x_i \otimes v) \\ = & \sum_{q=1}^{n} x_{n+q} \otimes B_{i,q}(v) + \sum_{r=1}^{n} x_r \otimes A_{i,r}(v), \end{aligned} \tag{7.3.56}$$

$$\begin{aligned} \tilde{\omega}(x_{n+i} \otimes v) = & \sum_{1 \le p < q \le n} [(B_{p,q} \otimes C_{q,p})(x_{n+i} \otimes v) + (C_{q,p} \otimes B_{p,q})(x_{n+i} \otimes v)] \\ & + \sum_{r,s=1}^{n} (A_{r,s} \otimes A_{s,r})(x_{n+i} \otimes v) \end{aligned}$$

$$= \sum_{p=1}^{n} x_p \otimes C_{i,p}(v) - \sum_{s=1}^{n} x_{n+s} \otimes A_{s,i}(v). \tag{7.3.57}$$

Comparing the above four expressions, we get the conclusion in the lemma. □

For $f \in \mathscr{A}$, we define the action

$$f(g \otimes v) = fg \otimes v \qquad \text{for } g \in \mathscr{A},\ v \in M. \tag{7.3.58}$$

Then we have the $o(2n, \mathbb{C})$-invariant operator

$$T = \sum_{i=1}^{n} [\nu(C_{n+1,i})x_i - \nu(A_{i,n+1})x_{n+i}] \tag{7.3.59}$$

on $\widehat{M}$.

Lemma 7.3.5 *We have* $T|_{\widehat{M}_k} = (2c + 2 - 2n + k)\eta|_{\widehat{M}_k}$.

Proof Let f be a homogeneous polynomial with degree k and let $v \in M$. According to (7.3.22), we find

$$\sum_{i=1}^{n} [(x_i D - \eta\partial_{x_{n+i}})(x_{n+i} f) + (x_{n+i} D - \eta\partial_{x_i})(x_i f)]$$
$$= 2(k+1)\eta f - 2n\eta f - \eta D(f) = (k + 2 - 2n)\eta f. \tag{7.3.60}$$

Moreover, the skew-symmetry of $C_{i,p}$ and $B_{p,i}$ implies

$$\sum_{i=1}^{n} \left[\sum_{p=1}^{n} (x_p C_{i,p} - x_{n+p} A_{p,i}) + x_{n+i} I_{2n} \right] x_i = - \sum_{i,p=1}^{n} x_i x_{n+p} A_{p,i} + \eta I_{2n}, \tag{7.3.61}$$

$$-\sum_{i=1}^{n} \left[\sum_{p=1}^{n} (x_{n+p} B_{p,i} - x_p A_{i,p}) - x_i I_{2n} \right] x_{n+i} = \sum_{i,p=1}^{n} x_{n+i} x_p A_{i,p} + \eta I_{2n}. \tag{7.3.62}$$

Now the lemma follows from (7.3.35) and (7.3.36). □

For $0 \neq \mu = \sum_{i=1}^{n} \mu_i \varepsilon_i \in \Lambda^+$ with $\mathcal{I}(\mu) = \{i_1, i_2, \ldots, i_{s+1}\}$ (cf. (6.7.54) and (6.7.55)), we define

$$\Theta(\mu) = \begin{cases} \mu_1 + n - 1 - \mathbb{N} & \text{if } \mu_{n-1} = -\mu_n > 0 \text{ and } s = 2, \\ \mu_1 + 2n - i_2 - \mathbb{N} & \text{otherwise.} \end{cases} \tag{7.3.63}$$

Theorem 7.3.6 *For $0 \neq \mu \in \Lambda^+$, the $o(2n+2, \mathbb{C})$-module $\widehat{V(\mu)}$ defined by (7.3.32)–(7.3.41) is irreducible if $c \in \mathbb{C} \setminus \{n - 1 - \mathbb{N}/2, \Theta(\mu)\}$.*

Proof By Proposition 7.3.3, it is enough to prove that the homomorphism φ defined in (7.3.47) and (7.3.52) satisfies $\varphi(\widehat{V(\mu)}) = \widehat{V(\mu)}$. According to (7.3.53), we only need to prove

$$\varphi(\widehat{V(\mu)}_k) = \widehat{V(\mu)}_k \tag{7.3.64}$$

for any $k \in \mathbb{N}$. We will prove it by induction on k.

When $k = 0$, (7.3.64) holds by (7.3.50) and (7.3.52). Consider $k = 1$. Write $\mu = \sum_{i=1}^n \mu_i \varepsilon_i \in \Lambda^+$ with $\mathscr{S}(\mu) = \{i_1, i_2, \ldots, i_{s+1}\}$. According to Lemmas 7.3.2 and 7.3.4 with $M = V(\mu)$, the eigenvalues of $\varphi|_{\widehat{V(\mu)}_1}$ are

$$c + \mu_{i_r} - i_r + 1,\ c - \mu_{i_r} - 2n + i_{r+1} \ \text{ for } \ r \in \overline{1,s} \ \text{ if } \ \mu_{n-1} + \mu_n > 0 \tag{7.3.65}$$

and

$$c + \mu_{i_r} - i_r + 1,\ c - \mu_{i_\iota} - 2n + i_{\iota+1} \ \text{ for } \ r \in \overline{1,s},\ \iota \in \overline{1, s-2+\delta_{\mu_n,0}} \tag{7.3.66}$$

when $\mu_{n-1} + \mu_n = 0$. Recall that $\mu_i \in \mathbb{Z}/2$ for $i \in \overline{1,n}$,

$$\mu_\iota - \mu_{\iota+1} \in \mathbb{N} \ \text{ for } \ \iota \in \overline{1, n-1}, \qquad \mu_{n-1} \geq |\mu_n|. \tag{7.3.67}$$

So $c \notin \Theta(\mu)$ implies that the eigenvalues of $\varphi|_{\widehat{V(\mu)}_1}$ are nonzero. Thus (7.3.64) holds for $k = 1$.

Suppose that (7.3.64) holds for $k \leq \ell$ with $\ell \geq 1$. Consider $k = \ell + 1$. Note that

$$\begin{aligned} \varphi(\widehat{V(\mu)}_{\ell+1}) &= \sum_{i=1}^{2n} \varphi(x_i \widehat{V(\mu)}_\ell) = \sum_{i=1}^{n} [A_{i,n+1}[\varphi(\widehat{V(\mu)}_\ell)] + C_{n+1,i}[\varphi(\widehat{V(\mu)}_\ell)]] \\ &= \sum_{i=1}^{n} [A_{i,n+1}(\widehat{V(\mu)}_\ell) + C_{n+1,i}(\widehat{V(\mu)}_\ell)] \end{aligned} \tag{7.3.68}$$

by the inductional assumption. To prove (7.3.64) with $k = \ell + 1$ is equivalent to prove

$$\sum_{i=1}^{n} [A_{i,n+1}(\widehat{V(\mu)}_\ell) + C_{n+1,i}(\widehat{V(\mu)}_\ell)] = \widehat{V(\mu)}_{\ell+1}. \tag{7.3.69}$$

For any $u \in \widehat{V(\mu)}_{\ell-1}$, Lemma 7.3.5 says that

$$\sum_{i=1}^{n} [C_{n+1,i}(x_i u) - A_{i,n+1}(x_{n+i} u)] = (2c + 2 - 2n + \ell)\eta u. \tag{7.3.70}$$

Since $c \notin n-1-\mathbb{N}/2$, $2c+2-2n+\ell \neq 0$ and (7.3.70) gives

$$\eta u \in \sum_{i=1}^{n}[A_{i,n+1}(\widehat{V(\mu)}_{\ell}) + C_{n+1,i}(\widehat{V(\mu)}_{\ell})] \qquad \text{for } u \in \widehat{V(\mu)}_{\ell-1}. \tag{7.3.71}$$

Let $g \otimes v \in \widehat{V(\mu)}_{\langle \ell \rangle}$. According to (7.3.35), (7.3.36) and Lemma 7.3.4,

$$\begin{aligned} -A_{i,n+1}(g\otimes v) &= x_i D(g)\otimes v - \eta\partial_{x_{n+i}}(g)\otimes v + x_i g\otimes I_{2n}(v) \\ &\quad + \sum_{p=1}^{n}(x_{n+p}g\otimes B_{i,n+p}(v) + x_p g\otimes A_{i,p}(v)) \\ &= \ell x_i g\otimes v - \eta\partial_{x_{n+i}}(g)\otimes v + g[x_i\otimes I_{2n}(v) \\ &\quad + \sum_{p=1}^{n}(x_{n+p}\otimes B_{i,n+p}(v) + x_p\otimes A_{i,p}(v))] \\ &= \ell x_i g\otimes v - \eta\partial_{x_{n+i}}(g)\otimes v - g[A_{i,n+1}(1\otimes v)] \\ &= \ell x_i g\otimes v - \eta\partial_{x_{n+i}}(g)\otimes v + g\varphi(x_i\otimes v) \\ &= -\eta\partial_{x_{n+i}}(g)\otimes v + g[(\ell+c+\tilde{\omega})(x_i\otimes v)] \end{aligned} \tag{7.3.72}$$

and

$$\begin{aligned} C_{n+i,i}(g\otimes v) &= x_{n+i} D(g)\otimes v - \eta\partial_{x_i}(g)\otimes v + x_{n+i} g\otimes I_{2n}(v) \\ &\quad + \sum_{p=1}^{n}(x_p g\otimes C_{i,p}(v) - x_{n+p} g\otimes A_{p,i}(v)) \\ &= \ell x_{n+i} g\otimes v - \eta\partial_{x_i}(g)\otimes v + g[x_{n+i}\otimes I_{2n}(v) \\ &\quad + \sum_{p=1}^{n}(x_p\otimes C_{i,p}(v) - x_{n+p}\otimes A_{p,i}(v))] \\ &= \ell x_{n+i} g\otimes v - \eta\partial_{x_i}(g)\otimes v + g[C_{n+1,i}(1\otimes v)] \\ &= \ell x_{n+i} g\otimes v - \eta\partial_{x_i}(g)\otimes v + g\varphi(x_{n+i}\otimes v) \\ &= -\eta\partial_{x_i}(g)\otimes v + g[(\ell+c+\tilde{\omega})(x_{n+i}\otimes v)] \end{aligned} \tag{7.3.73}$$

for $i \in \overline{1,n}$. Since (7.3.71) says that

$$\eta\partial_{x_r}(g)\otimes v \in \sum_{i=1}^{n}[A_{i,n+1}(\widehat{V(\mu)}_{\ell}) + C_{n+1,i}(\widehat{V(\mu)}_{\ell})]) \tag{7.3.74}$$

for $r \in \overline{1, 2n}$, Expressions (7.3.72) and (7.3.73) show

$$g[(\ell + c + \tilde{\omega})(x_r \otimes v)] \in \sum_{i=1}^{n}[A_{i,n+1}(\widehat{V(\mu)}_\ell) + C_{n+1,i}(\widehat{V(\mu)}_\ell)] \tag{7.3.75}$$

for $r \in \overline{1, 2n}$ and $g \in \mathscr{A}_\ell$.

According to Lemmas 7.3.2, the eigenvalues of $(\ell + c + \tilde{\omega})|_{\widehat{V(\mu)}_{\langle 1\rangle}}$ are among

$$\ell + c + \mu_{i_r} - i_r + 1,\ \ell + c - \mu_{i_r} - 2n + i_{r+1} \quad \text{for} \quad r \in \overline{1, s}. \tag{7.3.76}$$

Again

$$-\ell - \mu_{i_r} + i_r - 1,\ -\ell + \mu_{i_r} + 2n - i_{r+1} \in \mu_1 + 2n - i_2 - \mathbb{N} \quad \text{for} \quad i \in \overline{1, s}. \tag{7.3.77}$$

If $c \notin \mu_1 + 2n - i_2 - \mathbb{N}$, then all the eigenvalues of $(\ell + c + \tilde{\omega})|_{\widehat{V(\mu)}_1}$ are nonzero. In the case $\mu_n = -\mu_{n-1} > 0$ and $s = 2$, $\mu_1 = \mu_{n-1} = -\mu_n$ and the eigenvalues $(\ell + c + \tilde{\omega})|_{\widehat{V(\mu)}_1}$ are $c + \mu_1 + \ell$ and $c + \mu_n - n + 1 + \ell = c - \mu_1 - n + 1 + \ell$, which are not equal to 0 because of $c \notin \Theta(\mu) = \mu_1 + n - 1 - \mathbb{N}$. Hence

$$(\ell + c + \tilde{\omega})(\widehat{V(\mu)}_1) = \widehat{V(\mu)}_1. \tag{7.3.78}$$

By (7.3.75) and (7.3.78),

$$g(\widehat{V(\mu)}_{\langle 1\rangle}) \subset \sum_{i=1}^{n}[A_{i,n+1}(\widehat{V(\mu)}_\ell) + C_{n+1,i}(\widehat{V(\mu)}_\ell)] \quad \text{for} \quad g \in \mathscr{A}_\ell, \tag{7.3.79}$$

or equivalently, (7.3.64) holds for $k = \ell + 1$ by (7.3.68). By induction, (7.3.64) holds for any $k \in \mathbb{N}$. □

Let λ_i be the ith fundamental weight of $o(2n+2, \mathbb{C})$ with respect to the following simple positive roots

$$\{\varepsilon_{n+1} - \varepsilon_n, \varepsilon_n - \varepsilon_{n-1}, \ldots, \varepsilon_2 - \varepsilon_1, \varepsilon_2 + \varepsilon_1\}. \tag{7.3.80}$$

By the Weyl group's actions on $o(2n, \mathbb{C})$ and $V(\mu)$, $\widehat{V(\mu)}$ is a highest-weight $o(2n + 2, \mathbb{C})$-module with highest weight $-(c + \mu_1)\lambda_1 + \sum_{r=1}^{s-1}(\mu_{i_r} - \mu_{i_{r+1}})\lambda_{i_{r+1}} + (\mu_{n-1} + \mu_n)\lambda_{n+1}$.

Up to this stage, we do not known if the condition in Theorem 7.3.6 is necessary for the $o(2n + 2, \mathbb{C})$-module $\widehat{V(\mu)}$ to be irreducible if $\mu \neq 0$. We will deal with the case $\mu = 0$ in next section.

7.4 Conformal Oscillator Representations

In this section, we study the $o(2n+2,\mathbb{C})$-module $\widehat{V(0)}$ and its oscillator generalizations.

In order to use the results in Sects. 7.1 and 7.2, we redenote

$$y_i = x_{n+i} \qquad \text{for } i \in \overline{1,n} \tag{7.4.1}$$

and use $\mathscr{B} = \mathbb{C}[x_1,\ldots,x_n,y_1,\ldots,y_n]$ to replace $\mathscr{A}$ in last section. Fix $c \in \mathbb{C}$ and identify $\widehat{V(0)} = \mathscr{B} \otimes v_0$ with $\mathscr{B}$ by

$$f \otimes v_0 \leftrightarrow f \qquad \text{for } f \in \mathscr{B}, \tag{7.4.2}$$

where $V(0) = \mathbb{C}v_0$. Then we have the following one-parameter generalization π_c of the conformal representation of $o(2n+2,\mathbb{C})$:

$$\pi_c(A_{i,j}) = x_i\partial_{x_j} - y_j\partial_{y_i},\ \pi_c(B_{i,j}) = x_i\partial_{y_j} - x_j\partial_{y_i},\ \pi_c(C_{i,j}) = y_i\partial_{x_j} - y_j\partial_{x_i}, \tag{7.4.3}$$

$$\pi_c(A_{n+1,i}) = \partial_{x_i},\quad \pi_c(A_{n+1,n+1}) = -D - c,\quad \pi_c(B_{i,n+1}) = -\partial_{y_i}, \tag{7.4.4}$$

$$\pi_c(A_{i,n+1}) = \eta\partial_{y_i} - x_i(D+c),\quad \pi_c(C_{n+1,i}) = y_i(D+c) - \eta\partial_{x_i} \tag{7.4.5}$$

for $i,j \in \overline{1,n}$. Recall that $\mathscr{B}_r$ denotes the subspace of homogeneous polynomials in $\mathscr{B}$ with degree r. For any $k \in \mathbb{N}$, we write

$$\mathscr{B}_{[k]} = \mathscr{B}_0 + \sum_{i=1}^{k}(\mathscr{B}_i + \eta^i\mathscr{B}_{k-i}). \tag{7.4.6}$$

The following is an improved result of that in Zhao and the author's work [sg].

Theorem 7.4.1 *The representation π_c of $o(2n+2,\mathbb{C})$ on $\mathscr{B}$ is irreducible with highest weight $-c\lambda_1$ if $c \notin -\mathbb{N}$. When $c = -k$ with $k \in \mathbb{N}$, $\mathscr{B}_{[k]}$ is an irreducible $o(2n+2,\mathbb{C})$-submodule with highest weight $k\lambda_1$ and $\mathscr{B}/\mathscr{B}_{[k]}$ is an irreducible highest-weight $o(2n+2,\mathbb{C})$-module with highest weight $-(k+2)\lambda_1 + (k+1)\lambda_2$ with respect to (7.3.80).*

Proof Recall

$$\Delta = \sum_{i=1}^{n}\partial_{x_i}\partial_{y_i},\quad D = \sum_{r=1}^{n}x_r\partial_{x_r} + \sum_{s=1}^{n}y_s\partial_{y_s},\quad \eta = \sum_{i=1}^{n}x_iy_i \tag{7.4.7}$$

in terms of (7.4.1) and the identification (7.3.28). Moreover,

$$\mathscr{H}_r = \{f \in \mathscr{B}_r \mid \Delta(f) = 0\} \qquad \text{for } r \in \mathbb{N}. \tag{7.4.8}$$

According to Theorem 7.1.1, $\mathscr{H}_r$ is an irreducible $o(2n, \mathbb{C})$-submodule with the highest-weight vector x_1^r of weight $r\varepsilon_1$ for any $r \in \mathbb{N}$, and

$$\mathscr{B} = \bigoplus_{m,r=0}^{\infty} \eta^m \mathscr{H}_r \tag{7.4.9}$$

is a direct sum of irreducible $o(2n, \mathbb{C})$-submodules. On the other hand, $U(\mathscr{G}_-)(1_{\mathscr{B}})$ forms an irreducible $o(2n+2, \mathbb{C})$-submodule of $\mathscr{B}$ due to the first and third equations in (7.4.4) (cf. (7.3.43) for $\mathscr{G}_-$).

Next we assume $c \notin -\mathbb{N}$. Let M be a nonzero $o(2n+2, \mathbb{C})$-submodule of $\mathscr{B}$. Repeatedly applying the first and third equations in (7.4.4) if necessary, we have $1_{\mathscr{B}} \in M$. Note

$$-A_{i,n+1}(1_{\mathscr{B}}) = cx_i, \quad C_{n+1,i}(1_{\mathscr{B}}) = cy_i \qquad \text{for } i \in \overline{1,n}. \tag{7.4.10}$$

Thus $\mathscr{B}_1 \subset M$. Suppose that

$$\mathscr{B}_r \subset M \qquad \text{for } r < \ell \tag{7.4.11}$$

with $2 \leq \ell \in \mathbb{N}$. According to (7.4.8),

$$\mathscr{B}_\ell = \bigoplus_{m=0}^{[\![\ell/2]\!]} \eta^m \mathscr{H}_{\ell-2m}. \tag{7.4.12}$$

Moreover,

$$[\Delta, \eta] = n + D. \tag{7.4.13}$$

Set

$$\mathscr{B}_{\ell,r} = \bigoplus_{m=0}^{r} \eta^m \mathscr{H}_{\ell-2m} \tag{7.4.14}$$

for $r \in \overline{0, [\![\ell/2]\!]}$. Then

$$\mathscr{B}_{\ell,r} = \{w \in \mathscr{B}_\ell \mid \Delta^{r+1}(w) = 0\} \tag{7.4.15}$$

and

$$\Delta^r(\mathscr{B}_{\ell,r}) = \mathscr{H}_{\ell-2r} \tag{7.4.16}$$

by (7.4.14).

Since

$$-A_{1,n+1}(x_1^{\ell-1}) = (c+\ell-1)x_1^{\ell} \in M \tag{7.4.17}$$

and $c \notin -\mathbb{N}-1$, we have

$$x_1^{\ell} \in M \tag{7.4.18}$$

by (7.4.5). Hence

$$\mathscr{H}_\ell \subset M \tag{7.4.19}$$

because $\mathscr{H}_\ell$ is an irreducible $o(2n,\mathbb{C})$-submodule generated by x_1^ℓ. Recall $n \geq 2$ by our assumption. For $r \in \overline{1, [\![\ell/2]\!]}$,

$$-A_{2,n+1}(x_1^{\ell-r-1}y_1^r) = (c+\ell-1)x_1^{\ell-r-1}x_2y_1^r \in M \tag{7.4.20}$$

by (7.4.5). So $x_1^{\ell-r-1}x_2y_1^r \in M$. Moreover,

$$\Delta^r(x_1^{\ell-r-1}x_2y_1^r) = r!\left[\prod_{s=1}^{r}(\ell-r-s)\right]x_1^{\ell-2r-1}x_2 \in \mathscr{H}_{\ell-2r}. \tag{7.4.21}$$

Observe that (7.4.12) is a direct sum of irreducible $o(2n,\mathbb{C})$-submodules with distinct highest weights. So

$$\mathscr{B}_\ell \bigcap M = \bigoplus_{m=0}^{[\![\ell/2]\!]}(\eta^m\mathscr{H}_{\ell-2m})\bigcap M. \tag{7.4.22}$$

By (7.4.17)–(7.4.23), $(\eta^r\mathscr{H}_{\ell-2r})\bigcap M$ is a nonzero $o(2n,\mathbb{C})$-submodule. Since $\eta^r\mathscr{H}_{\ell-2r}$ is an irreducible $o(2n,\mathbb{C})$-module, we have $\eta^r\mathscr{H}_{\ell-2r} = (\eta^r\mathscr{H}_{\ell-2r})\bigcap M$. Therefore, $\mathscr{B}_\ell \subset M$. By induction, $\mathscr{B}_r \subset M$ for any $r \in \mathbb{N}$, or equivalently, $M = \mathscr{B}$. This proves the theorem for $c \notin -\mathbb{N}$; that is, $\mathscr{B}$ is an irreducible highest-weight $o(2n+2,\mathbb{C})$-module with $1_{\mathscr{B}}$ as a highest-weight vector of weight $-c\lambda_1$ with respect to (7.3.80).

Suppose that $c = -k$ with $k \in \mathbb{N}$, then $\sum_{r=0}^{k}\mathscr{B}_r \subset U(\mathscr{G}_-)(1_{\mathscr{B}})$ (cf. (7.3.43)). According to (7.4.5), $\mathscr{G}_-(\mathscr{B}_k) = \eta\mathscr{B}_{k-1} \subset U(\mathscr{G}_-)(1_{\mathscr{B}})$. Assume $\eta^r\mathscr{B}_{k-r} \subset U(\mathscr{G}_-)(1_{\mathscr{B}})$. Note $\eta^r\mathscr{B}_{k-r} \in \mathscr{B}_{r+k}$. So

$$A_{i,n+1}(\eta^r f) = \eta^{r+1}\partial_{y_i}(f),\quad C_{n+1,i}(\eta^r f) = -\eta^{r+1}\partial_{x_i}(f)\qquad \text{for } f \in \mathscr{B}_{k-r} \tag{7.4.23}$$

by (7.4.5). Thus $\eta^{r+1}\mathscr{B}_{k-r-1} \subset U(\mathscr{G}_-)(1_{\mathscr{B}})$. By induction, $\mathscr{B}_{[k]} \subset U(\mathscr{G}_-)(1_{\mathscr{B}})$.

Recall that we identify $o(2n,\mathbb{C})$ with $\mathscr{G}_0$. First $\mathscr{B}_{[k]}$ is an $o(2n,\mathbb{C})$-submodule by (7.4.6) and (7.4.9). According to (7.4.4),

$$A_{n+1,i}(\eta^r f) = \eta^{r-1}(ry_i f + \eta\partial_{x_i}(f)),\ \ B_{i,n+1}(\eta^r f) = -\eta^{r-1}(rx_i f + \eta\partial_{y_i}(f)) \tag{7.4.24}$$

for $f \in \mathscr{B}$. Hence $\mathscr{G}_+(\eta^r\mathscr{B}_{k-r}) \subset \eta^{r-1}\mathscr{B}_{k-r+1}$. By (7.4.23), $\mathscr{G}_-((\eta^r\mathscr{B}_{k-r}) \subset \eta^{r+1}\mathscr{B}_{k-r-1}$ if $r < k$ and $\mathscr{G}_-(\eta^k) = \{0\}$. Therefore, $\mathscr{B}_{[k]} = U(\mathscr{G}_-)(1_{\mathscr{B}})$ is finite-dimensional irreducible $o(2n+2,\mathbb{C})$-module with $1_{\mathscr{B}}$ as a highest-weight vector of weight $k\lambda_1$ with respect to (7.3.80).

Let M' be an $o(2n+2,\mathbb{C})$-submodule containing $\mathscr{B}_{[k]}$ as a proper submodule. Note that $\mathscr{B}_{k+1} = \mathscr{H}_k \oplus \mathscr{B}_{k+1} \bigcap \mathscr{B}_{[k]}$. Applying the first and third equations in (7.4.4), we get $M' \bigcap \mathscr{H}_{k+1} \neq \{0\}$. Since $\mathscr{H}_{k+1}$ is an irreducible $o(2n,\mathbb{C})$-module, we have $\mathscr{H}_{k+1} = M' \bigcap \mathscr{H}_{k+1}$. So $\mathscr{B}_{k+1} \subset M'$. Using the arguments in (7.4.17)–(7.4.24) with $\ell \geq k+2$, we can prove $\mathscr{B}_\ell \subset M'$ for any $k+2 \leq \ell \in \mathbb{N}$. Hence $M' = \mathscr{B}$. Thus $\mathscr{B}/\mathscr{B}_{[k]}$ is an irreducible highest-weight $o(2n+2,\mathbb{C})$-module with $x_n^{k+1} + \mathscr{B}_{[k]}$ as a highest-weight vector of weight $-(k+2)\lambda_1 + (k+1)\lambda_2$ with respect to (7.3.80). □

The rest of this section is taken from our work [X25]. For $\vec{a} = (a_1, a_2, \ldots, a_n)^t$, $\vec{b} = (b_1, b_2, \ldots, b_n)^t \in \mathbb{C}^n$, we put

$$\vec{a}\cdot\vec{x} = \sum_{i=1}^n a_i x_i, \qquad \vec{b}\cdot\vec{y} = \sum_{i=1}^n b_i y_i. \tag{7.4.25}$$

Set

$$\mathscr{B}_{\vec{a},\vec{b}} = \{f e^{\vec{a}\cdot\vec{x}+\vec{b}\cdot\vec{y}} \mid f \in \mathscr{B}\}. \tag{7.4.26}$$

Denote by $\pi_{c,\vec{a},\vec{b}}$ the representation π_c of $o(2n+2,\mathbb{C})$ on $\mathscr{B}_{\vec{a},\vec{b}}$.

Theorem 7.4.2 *The representation $\pi_{c,\vec{a},\vec{b}}$ of $o(2n+2,\mathbb{C})$ is irreducible for any $c \in \mathbb{C}$ if $\sum_{i=1}^n a_i b_i \neq 0$.*

Proof By symmetry, we may assume $a_1 \neq 0$. Let M be a nonzero $o(2n+2,\mathbb{C})$-submodule of $\mathscr{B}_{\vec{a},\vec{b}}$. Take any $0 \neq f e^{\vec{a}\cdot\vec{x}+\vec{b}\cdot\vec{y}} \in M$ with $f \in \mathscr{B}$. Set

$$\mathscr{B}_{\vec{a},\vec{b},k} = \mathscr{B}_k e^{\vec{a}\cdot\vec{x}+\vec{b}\cdot\vec{y}} \qquad \text{for } k \in \mathbb{N}. \tag{7.4.27}$$

According to (7.4.4),

$$(A_{n+1,i} - a_i)(f e^{\vec{a}\cdot\vec{x}+\vec{b}\cdot\vec{y}}) = \partial_{x_i}(f) e^{\vec{a}\cdot\vec{x}+\vec{b}\cdot\vec{y}},\ \ -(B_{i,n+1} + b_i)(f e^{\vec{a}\cdot\vec{x}+\vec{b}\cdot\vec{y}}) = \partial_{y_i}(f) e^{\vec{a}\cdot\vec{x}+\vec{b}\cdot\vec{y}} \tag{7.4.28}$$

for $i \in \overline{1,n}$. Repeatedly applying (7.4.28), we obtain $e^{\vec{a}\cdot\vec{x}+\vec{b}\cdot\vec{y}} \in M$. Equivalently, $\mathscr{B}_{\vec{a},\vec{b},0} \subset M$.

Suppose $\mathscr{B}_{\vec{a},\vec{b},\ell} \subset M$ for some $\ell \in \mathbb{N}$. Take any $g e^{\vec{a}\cdot\vec{x}+\vec{b}\cdot\vec{y}} \in \mathscr{B}_{\vec{a},\vec{b},\ell}$. Since

$$(x_i\partial_{x_1} - y_1\partial_{y_i})(g) e^{\vec{a}\cdot\vec{x}+\vec{b}\cdot\vec{y}},\ (y_i\partial_{x_1} - y_1\partial_{x_i})(g) e^{\vec{a}\cdot\vec{x}+\vec{b}\cdot\vec{y}} \in \mathscr{B}_{\vec{a},\vec{b},\ell} \subset M, \tag{7.4.29}$$

we have

$$A_{i,1}(ge^{\vec{a}\cdot\vec{x}+\vec{b}\cdot\vec{y}}) \equiv (a_1x_i - b_iy_1)ge^{\vec{a}\cdot\vec{x}+\vec{b}\cdot\vec{y}} \equiv 0 \pmod{M} \tag{7.4.30}$$

and

$$C_{i,1}(ge^{\vec{a}\cdot\vec{x}+\vec{b}\cdot\vec{y}}) \equiv (a_1y_i - a_iy_1)ge^{\vec{a}\cdot\vec{x}+\vec{b}\cdot\vec{y}} \equiv 0 \pmod{M} \tag{7.4.31}$$

for $i \in \overline{1,n}$ by (7.4.3). On the other hand,

$$(D+c)(g)e^{\vec{a}\cdot\vec{x}+\vec{b}\cdot\vec{y}} \in \mathscr{B}_{\vec{a},\vec{b},\ell} \subset M, \tag{7.4.32}$$

and so the second equation in (7.4.4) implies

$$-A_{n+1,n+1}(ge^{\vec{a}\cdot\vec{x}+\vec{b}\cdot\vec{y}}) \equiv \left[\sum_{i=1}^{n}(a_ix_i + b_iy_i)\right]ge^{\vec{a}\cdot\vec{x}+\vec{b}\cdot\vec{y}} \equiv 0 \pmod{M} \tag{7.4.33}$$

Substituting (7.4.30) and (7.4.31) into (7.4.33), we get

$$\left(\sum_{i=1}^{n}a_ib_i\right)y_1ge^{\vec{a}\cdot\vec{x}+\vec{b}\cdot\vec{y}} \equiv 0 \pmod{M}. \tag{7.4.34}$$

Equivalently, $y_1ge^{\vec{a}\cdot\vec{x}+\vec{b}\cdot\vec{y}} \in M$. Substituting it to (7.4.30) and (7.4.31), we obtain

$$x_ige^{\vec{a}\cdot\vec{x}+\vec{b}\cdot\vec{y}},\ y_ige^{\vec{a}\cdot\vec{x}+\vec{b}\cdot\vec{y}} \in M \tag{7.4.35}$$

for $i \in \overline{1,n}$. Therefore, $\mathscr{B}_{\vec{a},\vec{b},\ell+1} \subset M$. By induction, $\mathscr{B}_{\vec{a},\vec{b},\ell} \subset M$ for any $\ell \in \mathbb{N}$. So $\mathscr{B}_{\vec{a},\vec{b}} = M$. Hence $\mathscr{B}_{\vec{a},\vec{b}}$ is an irreducible $o(2n+2,\mathbb{C})$-module. □

In the above theorem, $\mathscr{B}_{\vec{a},\vec{b}}$ is not a weight $o(2n+2,\mathbb{C})$-module.

Fix $n_1, n_2 \in \overline{1,n}$ with $n_1 \leq n_2$. We take (7.2.6), (7.2.8) and write

$$\tilde{\eta} = \sum_{i=1}^{n_1} y_i\partial_{x_i} + \sum_{r=n_1+1}^{n_2} x_ry_r + \sum_{s=n_2+1}^{n} x_s\partial_{y_s}. \tag{7.4.36}$$

Moreover, we denote

$$\tilde{c} = c + n_2 - n_1 - n. \tag{7.4.37}$$

Changing operators $\partial_{x_r} \mapsto -x_r,\ x_r \mapsto \partial_{x_r}$ for $r \in \overline{1,n_1}$ and $\partial_{y_s} \mapsto -y_s,\ y_s \mapsto \partial_{y_s}$ for $s \in \overline{n_2+1,n}$ in the above representation of $o(2n+2,\mathbb{C})$ (7.4.3)–(7.4.5), we have the following representation $\pi_c^{n_1,n_2}$ of the Lie algebra $o(2n+2,\mathbb{C})$ determined by

$$\pi_c^{n_1,n_2}(A_{i,j}) = E_{i,j}^x - E_{j,i}^y \tag{7.4.38}$$

with

$$E_{i,j}^x = \begin{cases} -x_j\partial_{x_i} - \delta_{i,j} & \text{if } i,\, j \in \overline{1, n_1}; \\ \partial_{x_i}\partial_{x_j} & \text{if } i \in \overline{1, n_1},\ j \in \overline{n_1+1, n}; \\ -x_i x_j & \text{if } i \in \overline{n_1+1, n},\ j \in \overline{1, n_1}; \\ x_i\partial_{x_j} & \text{if } i,\, j \in \overline{n_1+1, n}; \end{cases} \tag{7.4.39}$$

and

$$E_{i,j}^y = \begin{cases} y_i\partial_{y_j} & \text{if } i,\, j \in \overline{1, n_2}; \\ -y_i y_j & \text{if } i \in \overline{1, n_2},\ j \in \overline{n_2+1, n}; \\ \partial_{y_i}\partial_{y_j} & \text{if } i \in \overline{n_2+1, n},\ j \in \overline{1, n_2}; \\ -y_j\partial_{y_i} - \delta_{i,j} & \text{if } i,\, j \in \overline{n_2+1, n}; \end{cases} \tag{7.4.40}$$

and

$$\pi_c^{n_1,n_2}(E_{i,n+1+j}) = \begin{cases} \partial_{x_i}\partial_{y_j} & \text{if } i \in \overline{1, n_1},\ j \in \overline{1, n_2}; \\ -y_j\partial_{x_i} & \text{if } i \in \overline{1, n_1},\ j \in \overline{n_2+1, n}; \\ x_i\partial_{y_j} & \text{if } i \in \overline{n_1+1, n},\ j \in \overline{1, n_2}; \\ -x_i y_j & \text{if } i \in \overline{n_1+1, n},\ j \in \overline{n_2+1, n}; \end{cases} \tag{7.4.41}$$

$$\pi_c^{n_1,n_2}(E_{n+1+i,j}) = \begin{cases} -x_j y_i & \text{if } j \in \overline{1, n_1},\ i \in \overline{1, n_2}; \\ -x_j\partial_{y_i} & \text{if } j \in \overline{1, n_1},\ i \in \overline{n_2+1, n}; \\ y_i\partial_{x_j} & \text{if } j \in \overline{n_1+1, n},\ i \in \overline{1, n_2}; \\ \partial_{x_j}\partial_{y_i} & \text{if } j \in \overline{n_1+1, n},\ i \in \overline{n_2+1, n}; \end{cases} \tag{7.4.42}$$

$$\pi_c^{n_1,n_2}(A_{n+1,n+1}) = -\widehat{D} - \tilde{c}; \tag{7.4.43}$$

$$\pi_c^{n_1,n_2}(A_{n+1,i}) = \begin{cases} -x_i & \text{if } i \in \overline{1, n_1}; \\ \partial_{x_i} & \text{if } i \in \overline{n_1+1, n}; \end{cases} \tag{7.4.44}$$

$$\pi_c^{n_1,n_2}(B_{i,n+1}) = \begin{cases} -\partial_{y_i} & \text{if } \in \overline{1, n_2}; \\ y_i & \text{if } i \in \overline{n_2+1, n}; \end{cases} \tag{7.4.45}$$

$$\pi_c^{n_1,n_2}(A_{i,n+1}) = \begin{cases} \tilde{\eta}\partial_{y_i} - (\widehat{D} + \tilde{c} - 1)\partial_{x_i} & \text{if } i \in \overline{1, n_1}; \\ \tilde{\eta}\partial_{y_i} - x_i(\widehat{D} + \tilde{c}) & \text{if } i \in \overline{n_1+1, n_2}; \\ -\tilde{\eta} y_i - x_i(\widehat{D} + \tilde{c}) & \text{if } i \in \overline{n_2+1, n}; \end{cases} \tag{7.4.46}$$

$$\pi_c^{n_1,n_2}(C_{n+1,i}) = \begin{cases} \tilde{\eta}x_i + y_i(\widehat{D}+\tilde{c}) & \text{if } i \in \overline{1,n_1}; \\ -\tilde{\eta}\partial_{x_i} + y_i(\widehat{D}+\tilde{c}) & \text{if } i \in \overline{n_1+1,n_2}; \\ -\tilde{\eta}\partial_{x_i} + (\widehat{D}+\tilde{c}-1)\partial_{y_i} & \text{if } i \in \overline{n_2+1,n} \end{cases} \tag{7.4.47}$$

for $i, j \in \overline{1,n}$.

Recall that as operators on $\mathscr{B}$, (7.2.9) holds with η replaced by $\tilde{\eta}$. We take (7.2.10) and (7.2.12). Moreover, we have (7.2.11).

Theorem 7.4.3 *The representation $\pi_c^{n_1,n_2}$ of $o(2n+2,\mathbb{C})$ on $\mathscr{B}$ is irreducible if $c \notin \mathbb{Z}/2$.*

Proof Let M be a nonzero $o(2n+2,\mathbb{C})$-submodule of $\mathscr{B}$. By (7.2.11) and (7.4.43),

$$M = \bigoplus_{k\in\mathbb{Z}} \mathscr{B}_{\langle k\rangle} \bigcap M. \tag{7.4.48}$$

Thus $\mathscr{B}_{\langle k\rangle} \bigcap M \neq \{0\}$ for some $k \in \mathbb{Z}$. If $k > n_1 - n_2 + 1 - \delta_{n_1,n_2}$, then

$$\{0\} \neq (-x_1)^{k-(n_1-n_2+1-\delta_{n_1,n_2})}(\mathscr{B}_{\langle k\rangle} \bigcap M) = A_{n+1,1}^{k-(n_1-n_2+1-\delta_{n_1,n_2})}(\mathscr{A}_{\langle k\rangle} \bigcap M) \tag{7.4.49}$$

by (7.4.44), which implies $\mathscr{B}_{\langle n_1-n_2+1-\delta_{n_1,n_2}\rangle} \bigcap M \neq \{0\}$. Thus we can assume $k \leq n_1 - n_2 + 1 - \delta_{n_1,n_2}$. Observe that the Lie subalgebra

$$\mathscr{L} = \sum_{i,j=1}^{n} \mathbb{C}A_{i,j} \cong gl(n,\mathbb{C}). \tag{7.4.50}$$

By Theorem 7.2.1, $\mathscr{B}_{\langle k\rangle} = \bigoplus_{i=0}^{\infty} \tilde{\eta}^i(\mathscr{H}_{\langle k-2i\rangle})$ is a decomposition of irreducible $o(2n,\mathbb{C})$-submodules. Moreover, the weight sets of $\mathscr{L}$-singular vectors in $\tilde{\eta}^i(\mathscr{H}_{\langle k-2i\rangle})$ are distinct by Sects. 6.4–6.6. Hence

$$\tilde{\eta}^i(\mathscr{H}_{\langle k-2i\rangle}) \subset M \quad \text{for some } i \in \mathbb{N}. \tag{7.4.51}$$

Observe that

$$x_1^{-k+2i} \in \mathscr{H}_{\langle k-2i\rangle}. \tag{7.4.52}$$

By (7.4.35) and (7.4.44),

$$i!(-1)^i\left(\prod_{r=1}^{i}(-k+i+r)\right)x_1^{-k+i} = B_{1,n+1}^i(\eta_{n_1,n_2}^i(x_1^{-k+2i})) \in M. \tag{7.4.53}$$

Thus

$$\mathscr{H}_{\langle k-i\rangle} \subset M. \tag{7.4.54}$$

So we can just assume

$$\mathscr{H}_{\langle k\rangle} \subset M. \tag{7.4.55}$$

According to (7.4.43),

$$x_1^{-k+s} = (-1)^s A_{n+1,1}^s(x_1^{-k}) \in M \qquad \text{for } s \in \mathbb{N}. \tag{7.4.56}$$

So Theorem 7.2.1 gives

$$\mathscr{H}_{\langle k-s\rangle} \subset M \qquad \text{for } s \in \mathbb{N}. \tag{7.4.57}$$

For any $r \in k - \mathbb{N}$, we suppose $\tilde{\eta}^s(x_1^{-r+s}), \tilde{\eta}^s(x_1^{-r+s+1}) \in M$ for some $s \in \mathbb{N}$. Applying (7.4.47) to it, we get

$$C_{n+1,1}[\tilde{\eta}^s(x_1^{-r+s})] = \tilde{\eta}^{s+1}(x_1^{-r+s+1}) + (r+\tilde{c})\tilde{\eta}^s(y_1 x_1^{-r+s}) \in M. \tag{7.4.58}$$

By (7.4.36) and (7.4.47),

$$C_{n+1,i}[\tilde{\eta}^s(x_1^{-r+s+1})] = (r-1+\tilde{c})\tilde{\eta}^s(y_i x_1^{-r+s+1}) \in M \tag{7.4.59}$$

for $i \in \overline{n_1+1, n_2}$. According to (7.4.36) and (7.4.46),

$$A_{i,n+1}[\tilde{\eta}^s(y_i x_1^{-r+s+1})] = \tilde{\eta}^{s+1}(x_1^{-r+s+1}) - (r+\tilde{c})\tilde{\eta}^s(x_i y_i x_1^{-r+s+1}) \in M \tag{7.4.60}$$

for $i \in \overline{n_1+1, n_2}$. Again (7.4.36), and $(-r+s+1) \times (7.4.58) - \sum_{i=n_1+1}^{n_2}(7.4.60)$ lead to

$$(1+s+\tilde{c}-n_2+n_1)\tilde{\eta}^{s+1}(x_1^{-r+s+1}) \in M \Rightarrow \tilde{\eta}^{s+1}(x_1^{-r+s+1}) \in M \;\; \text{for } r \in k-\mathbb{N}. \tag{7.4.61}$$

Replacing r by $r-1$ in the above expression, we get $\tilde{\eta}^{s+1}(x_1^{-r+s+2}) \in M$. By induction,

$$\tilde{\eta}^{\ell}(x_1^{-r+\ell}) \in M \qquad \text{for } r \in k-\mathbb{N},\ \ell \in \mathbb{N}. \tag{7.4.62}$$

Since $\tilde{\eta}^{\ell}(\mathscr{H}_{\langle r-\ell\rangle}) \ni \tilde{\eta}^{\ell}(x_1^{-r+\ell})$ is an irreducible $o(2n, \mathbb{C})$-module by Theorem 7.2.1, we have

$$\tilde{\eta}^{\ell}(\mathscr{H}_{\langle r-\ell\rangle}) \subset M \qquad \text{for } \ell \in \mathbb{N}. \tag{7.4.63}$$

Taking $r = m - \ell$ with $m \in k - \mathbb{N}$, we get

$$\tilde{\eta}^{\ell}(\mathscr{H}_{\langle m-2\ell\rangle}) \subset M \qquad \text{for } \ell \in \mathbb{N}. \tag{7.4.64}$$

According to Theorem 7.2.1,

$$\mathscr{B}_{\langle m\rangle} = \bigoplus_{\ell=0}^{\infty} \tilde{\eta}^{\ell}(\mathscr{H}_{\langle m-2\ell\rangle}) \subset M \qquad \text{for } m \in k - \mathbb{N}. \tag{7.4.65}$$

Expression (7.4.46) gives

$$\pi_c^{n_1,n_2}(A_{i,n+1})y_i = \begin{cases} \tilde{\eta}(y_i\partial_{y_i}+1) - y_i\partial_{x_i}(\widehat{D}+\tilde{c}+1) \text{ if } i \in \overline{1,n_1}, \\ \tilde{\eta}(y_i\partial_{y_i}+1) - x_i y_i(\widehat{D}+\tilde{c}+1) \text{ if } i \in \overline{n_1+1,n_2}, \end{cases} \tag{7.4.66}$$

$$\pi_c^{n_1,n_2}(A_{j,n+1})\partial_{y_j} = -\tilde{\eta}y_j\partial_{y_j} - x_j\partial_{y_j}(\widehat{D}+\tilde{c}+1) \qquad \text{for } j \in \overline{n_2+1,n}. \tag{7.4.67}$$

Moreover, (7.4.47) yields

$$\pi_c^{n_1,n_2}(C_{n+1,r})\partial_{x_r} = \tilde{\eta}x_r\partial_{x_i} + y_r\partial_{x_r}(\widehat{D}+\tilde{c}+1) \qquad \text{for } r \in \overline{1,n_1}, \tag{7.4.68}$$

$$\begin{aligned} &\pi_c^{n_1,n_2}(C_{n+1,s})x_s \\ &= \begin{cases} -\tilde{\eta}(x_s\partial_{x_s}+1) + x_s y_s(\widehat{D}+\tilde{c}+1) \text{ if } s \in \overline{n_1+1,n_2}, \\ -\tilde{\eta}(x_s\partial_{x_s}+1) + x_s\partial_{y_s}(\widehat{D}+\tilde{c}+1) \text{ if } s \in \overline{n_2+1,n}. \end{cases} \end{aligned} \tag{7.4.69}$$

Thus

$$\begin{aligned} &\sum_{i=1}^{n_2} \pi_c^{n_1,n_2}(A_{i,n+1})y_i + \sum_{j=n_2+1}^{n} \pi_c^{n_1,n_2}(A_{j,n+1})\partial_{y_j} \\ &- \sum_{r=1}^{n_1} \pi_c^{n_1,n_2}(C_{n+1,r})\partial_{x_r} - \sum_{s=n_1+1}^{n} \pi_c^{n_1,n_2}(C_{n+1,s})x_s \\ &= \tilde{\eta}(-\widehat{D} + n_2 + n - n_1 - 2(\tilde{c}+1)) \end{aligned} \tag{7.4.70}$$

as operators on $\mathscr{B}$. Suppose that $\mathscr{B}_{\langle \ell - s\rangle} \subset M$ for some $k \leq \ell \in \mathbb{Z}$ and any $s \in \mathbb{N}$. For any $f \in \mathscr{B}_{\langle \ell-1\rangle}$, we apply the above equation to it and get

$$(1 - \ell + n_2 + n - n_1 - 2(\tilde{c}+1))\tilde{\eta}(f) \in M. \tag{7.4.71}$$

Since $c \notin \mathbb{Z}/2$, we have

$$\tilde{\eta}(f) \in M. \tag{7.4.72}$$

Now for any $g \in \mathscr{B}_{\langle \ell \rangle}$, we have $\partial_{y_1}(g) \in \mathscr{B}_{\langle \ell-1 \rangle}$. By (7.4.46),

$$A_{1,n+1}(g) = \tilde{\eta}(\partial_{y_1}(g)) - (\ell + \tilde{c})\partial_{x_1}(g) \in M. \tag{7.4.73}$$

Moreover, (7.4.72) and (7.4.73) yield

$$\partial_{x_1}(g) \in M \qquad \text{for } g \in \mathscr{B}_{\langle \ell \rangle}. \tag{7.4.74}$$

Since

$$\partial_{x_1}(\mathscr{B}_{\langle \ell \rangle}) = \mathscr{B}_{\langle \ell+1 \rangle}, \tag{7.4.75}$$

we obtain

$$\mathscr{B}_{\langle \ell+1 \rangle} \subset M. \tag{7.4.76}$$

By induction on ℓ, we find

$$\mathscr{B}_{\langle \ell \rangle} \subset M \qquad \text{for } \ell \in \mathbb{Z}, \tag{7.4.77}$$

or equivalently, $\mathscr{B} = \bigoplus_{\ell \in \mathbb{Z}} \mathscr{B}_{\langle \ell \rangle} = M$. Thus $\mathscr{B}$ is an irreducible $o(2n+2, \mathbb{C})$-module. □

Remark 7.4.4 The above irreducible representation depends on the three parameters $c \in \mathbb{F}$ and $n_1, n_2 \in \overline{1, n}$. It is not highest-weight type because of the mixture of multiplication operators and differential operators in (7.4.39)–(7.4.42) and (7.4.44)–(7.4.47). Since $\mathscr{B}$ is not completely reducible as an $\mathscr{L}$-module by Sects. 6.4–6.6 when $n \geq 2$ and $n_1 < n$, $\mathscr{B}$ is not a unitary $o(2n+2, \mathbb{C})$-module. Expression (7.4.43) shows that $\mathscr{B}$ is a weight $o(2n+2, \mathbb{C})$-module with finite-dimensional weight subspaces.

References

[Kb] T. Larsson, Conformal fields: a class of representations of Vect(N)[J]. Int. J. Mod. Phys. A **7**(26), 6493–6508 (1992)

[La] B. Kostant, On the tensor product of a finite and an infinite dimensional representation. J. Fund. Anal. **20**, 257–285 (1975)

[LX2] C. Luo, X. Xu, Z-Graded oscillator generalizations of the classical theorem on harmonic polynomials. J. Lie Theory **23**, 979–1003 (2013)

[Ra] A.N. Rudakov, Irreducible representations of infinite-dimensional Lie algebras of Cartan type. Izv. Akad. Nauk SSSR Ser. Mat. **38**, 835–866 (1974)

[sg] G. Shen, Graded modules of graded Lie algebras of Cartan type (I)-mixed product of modules. Sci. China A **29**, 570–581 (1986)

[XZ2] X. Xu, Y. Zhao, Extensions of the conformal representations for orthogonal Lie algebras. J. Algebra **377**, 97–124 (2013)
[X25] X. Xu, Conformal oscillator representations of orthogonal Lie algebras. Sci. China: Math. (Engl. Ser.) **59**(1), 37–48 (2016)

Chapter 8
Representations of Odd Orthogonal Lie Algebras

In this chapter, we focus on the natural explicit representations of odd orthogonal Lie algebras. It is an analogue of Chap. 7. In Sect. 8.1, we study the canonical bosonic and fermionic oscillator representations over their minimal natural modules. The spin representations are also presented. Weyl's Theorem of complete reducibility in Sect. 2.4 plays an important role. In Sect. 8.2, we determine the structure of the noncanonical oscillator representations obtained from the above bosonic representations by partially swapping differential operators and multiplication operators, which are generalizations of the classical theorem on harmonic polynomials. The fundamental lemma on polynomial solutions in Sect. 6.1 and the noncanonical oscillator representations of $sl(n, \mathbb{F})$ in Sects. 6.3–6.6 are used. The results were due to Luo and the author [LX2]. In Sect. 8.3, we construct a new functor from the category of $o(2n+1, \mathbb{C})$-modules to the category of $o(2n+3, \mathbb{C})$-modules, which is an extension of the conformal representation of $o(2n + 3, \mathbb{C})$. Moreover, we find the condition for the functor to map a finite-dimensional irreducible $o(2n + 1, \mathbb{C})$-module to an infinite-dimensional irreducible $o(2n + 3, \mathbb{C})$-module by Theorem 5.4.3 of decomposing the tensor module. It is the third sequential work of constructing functors among the module categories of different simple Lie algebras after Sects. 6.7 and 7.3. This work was due to Zhao and the author [XZ].

The work in Sect. 8.3 gives rise to a one-parameter (c) family of inhomogeneous first-order differential-operator representations of $o(2n + 3, \mathbb{C})$. Letting these operators act on the space of exponential-polynomial functions that depend on a parametric vector $\vec{a} \in \mathbb{C}^{2n+1}$, we prove in Sect. 8.4 that the space forms an irreducible $o(2n+3, \mathbb{C})$-module for any $c \in \mathbb{C}$ if $\vec{a}$ is not on a certain hypersurface. By partially swapping differential operators and multiplication operators, we obtain more general differential-operator representations of $o(2n + 3, \mathbb{C})$ on the polynomial algebra $\mathscr{B}'$ in $2n + 1$ variables. We prove that $\mathscr{B}'$ forms an infinite-dimensional irreducible weight $o(2n + 3, \mathbb{C})$-module with finite-dimensional weight subspaces if $c \notin \mathbb{Z}/2$. These results are taken from our work [X25]. This is the second sequential work of connecting the irreducibility and algebraic varieties after Sect. 7.4.

X. Xu, *Representations of Lie Algebras and Partial Differential Equations*,
DOI 10.1007/978-981-10-6391-6_8

8.1 Canonical Oscillator Representations

In this section, we study the canonical bosonic and fermionic oscillator representations of odd orthogonal Lie algebras over their minimal natural modules and the spin representations.

Let $n \geq 1$ be an integer. The odd split orthogonal Lie algebra

$$o(2n+1,\mathbb{F}) = o(2n,\mathbb{F}) + \sum_{i=1}^{n}[\mathbb{F}(E_{0,i} - E_{n+i,0}) + \mathbb{F}(E_{0,n+i} - E_{i,0})] \quad (8.1.1)$$

(cf. (7.1.1)). We take (7.1.2) as a Cartan subalgebra and use the settings in (7.1.3)–(7.1.5). Then the root system of $o(2n+1,\mathbb{F})$ is

$$\Phi_{B_n} = \{\pm\varepsilon_i \pm \varepsilon_j, \pm\varepsilon_r \mid 1 \leq i < j \leq n; r \in \overline{1,n}\}. \quad (8.1.2)$$

We take the set of positive roots

$$\Phi_{B_n}^{+} = \{\varepsilon_i \pm \varepsilon_j, \varepsilon_r \mid 1 \leq i < j \leq n, r \in \overline{1,n}\}. \quad (8.1.3)$$

In particular,

$$\Pi_{B_n} = \{\varepsilon_1 - \varepsilon_2, \ldots, \varepsilon_{n-1} - \varepsilon_n, \varepsilon_n\} \text{ is the set of positive simple roots.} \quad (8.1.4)$$

Recall the set of dominant integral weights

$$\Lambda^{+} = \{\mu \in L_{\mathbb{Q}} \mid 2(\varepsilon_n, \mu), (\varepsilon_i - \varepsilon_{i+1}, \mu) \in \mathbb{N} \text{ for } i \in \overline{1,n-1}\} \quad (8.1.5)$$

(cf. (7.1.4)). According to (7.1.5),

$$\Lambda^{+} = \{\mu = \sum_{i=1}^{n} \mu_i\varepsilon_i \mid \mu_i \in \mathbb{N}/2; \mu_i - \mu_{i+1} \in \mathbb{N} \text{ for } i \in \overline{1,n-1}\}. \quad (8.1.6)$$

Let $\mathscr{B}' = \mathbb{F}[x_0, x_1, \ldots, x_n, y_1, \ldots, y_n]$. Recall that the canonical oscillator representation of $o(2n+1,\mathbb{F})$ is given by

$$(E_{i,j}-E_{n+j,n+i})|_{\mathscr{B}'} = x_i\partial_{x_j} - y_j\partial_{y_i}, \quad (E_{n+i,j}-E_{n+j,i})|_{\mathscr{B}'} = y_i\partial_{x_j} - y_j\partial_{x_i}, \quad (8.1.7)$$

$$(E_{i,n+j} - E_{j,n+i})|_{\mathscr{B}'} = x_i\partial_{y_j} - x_j\partial_{y_i}, \quad (E_{0,i} - E_{n+i,0})|_{\mathscr{B}'} = x_0\partial_{x_i} - y_i\partial_{x_0}, \quad (8.1.8)$$

$$(E_{0,n+i} - E_{i,0})|_{\mathscr{B}'} = x_0\partial_{y_i} - x_i\partial_{x_0} \quad (8.1.9)$$

for $i, j \in \overline{1,n}$. The positive root vectors of $o(2n+1,\mathbb{F})$ are

$$\{E_{i,j} - E_{n+j,n+i}, E_{i,n+j} - E_{j,n+i}, E_{0,n+r} - E_{r,0} \mid 1 \le i < j \le n; r \in \overline{1,n}\}. \tag{8.1.10}$$

Denote

$$\Delta' = \partial_{x_0}^2 + 2\sum_{i=1}^n \partial_{x_i}\partial_{y_i}, \quad \eta' = x_0^2 + 2\sum_{i=1}^n x_i y_i. \tag{8.1.11}$$

It can be verified that η' is an $o(2n+1, \mathbb{F})$-invariant and Δ' is an $o(2n+1, \mathbb{F})$-invariant differential operator, or equivalently,

$$\xi\eta' = \eta'\xi, \; \xi\Delta' = \Delta'\xi' \quad \text{on } \mathscr{B}' \text{ for } \xi' \in o(2n+1, \mathbb{F}). \tag{8.1.12}$$

Let $\mathscr{B}'_k$ be the subspace of homogeneous polynomials in $\mathscr{B}'$ with degree k. Set

$$\mathscr{H}'_k = \{f \in \mathscr{B}'_k \mid \Delta'(f) = 0\} \qquad \text{for } k \in \mathbb{N}. \tag{8.1.13}$$

Theorem 8.1.1 *For $k \in \mathbb{N}$, the subspace $\mathscr{H}'_k$ forms a finite-dimensional irreducible $o(2n+1, \mathbb{F})$-submodule and x_1^k is a highest-weight vector with the weight $k\varepsilon_1$. Moreover, it has a basis*

$$\left\{\sum_{i=0}^\infty (-2)^i \frac{x_0^{2i+\iota}(\sum_{i=1}^n \partial_{x_i}\partial_{y_i})^i (x^\alpha y^\beta)}{(2i+\iota)!} \mid \iota = 0, 1; \alpha, \beta \in \mathbb{N}^n; |\alpha| + |\beta| + \iota = k\right\}. \tag{8.1.14}$$

Furthermore,

$$\mathscr{B}' = \bigoplus_{k_1,k_2=0}^\infty \eta'^{k_1} \mathscr{H}'_{k_2} \tag{8.1.15}$$

is a direct sum of irreducible $o(2n+1, \mathbb{F})$-submodules.

Proof Take $\eta = \sum_{i=1}^n x_i y_i$. According to Theorem 7.1.1, the $o(2n+1, \mathbb{F})$-singular vectors in $\mathscr{B}'$ are in

$$\sum_{i,j,k=0}^\infty \mathbb{F} x_0^i x_1^j \eta^k. \tag{8.1.16}$$

Since

$$\eta = \frac{1}{2}(\eta' - x_0^2), \tag{8.1.17}$$

the $o(2n+1, \mathbb{F})$-singular vectors in $\mathscr{B}'$ are in

$$\sum_{i,j,k=0}^\infty \mathbb{F} x_0^i x_1^j \eta'^k. \tag{8.1.18}$$

Suppose that

$$g=\sum_{i,j,k} a_{i,j,k} x_0^i x_1^j \eta'^k \quad \text{is a singular vector.} \tag{8.1.19}$$

Then

$$(E_{0,n+1}-E_{1,0})(g)=-x_1\sum_{i,j,k} i a_{i,j,k} x_0^{i-1} x_1^j \eta'^k=0 \tag{8.1.20}$$

by the first equation in (8.1.12). So g is independent of x_0. Thus the $o(2n+1,\mathbb{F})$-singular vectors in $\mathscr{B}'_k$ are the set

$$\{\eta'^{\ell} x_1^{k-2\ell} \mid \ell\in\mathbb{N}\}. \tag{8.1.21}$$

Note

$$[\Delta',\eta']=2(1+2n)+4D',\quad D'=x_0\partial_{x_0}+\sum_{i=1}^n(x_i\partial_{x_i}+y_i\partial_{y_i}). \tag{8.1.22}$$

Hence

$$\Delta'[\eta'^{\ell} x_1^{k-2\ell}]=2\ell[1+2(n+k-1)]\eta'^{\ell-1} x_1^{k-2\ell}. \tag{8.1.23}$$

Thus $\mathscr{H}'_k$ has a unique $o(2n+1,\mathbb{F})$-singular vector x_1^k. According to the second equation in (8.1.12), $\mathscr{H}'_k$ is a finite-dimensional $o(2n+1,\mathbb{F})$-module. By Weyl's Theorem 2.3.6 of complete reducibility if $\mathbb{F}=\mathbb{C}$ or more generally by Lemma 6.3.2 with $n_1=0$, it is irreducible. The Eq. (8.1.15) follows from the first equation in (8.1.12) and the fact that $o(2n+1,\mathbb{F})$-singular vector are polynomials in x_1 and η'. Expression (8.1.14) follows from Lemma 6.1.1 with $T_1=\partial_{x_0}^2$, $T_1^-=\int_{(x_0)}^2$ and $T_2=2\sum_{i=1}^n\partial_{x_i}\partial_{y_i}$. □

Consider the exterior algebra $\check{\mathscr{A}}'$ generated by $\{\theta_0,\theta_1,\ldots,\theta_n,\vartheta_1,\ldots,\vartheta_n\}$ (cf. (6.2.15)) and take the settings in (6.2.44)–(6.2.51). Define the *basic fermionic representation* of $o(2n+1,\mathbb{F})$ on $\check{\mathscr{A}}'$ by

$$(E_{i,j}-E_{n+j,n+i})|_{\check{\mathscr{A}}'}=\theta_i\partial_{\theta_j}-\vartheta_j\partial_{\vartheta_i},\quad (E_{n+i,j}-E_{n+j,i})|_{\check{\mathscr{A}}'}=\vartheta_i\partial_{\theta_j}-\vartheta_j\partial_{\theta_i}, \tag{8.1.24}$$

$$(E_{i,n+j}-E_{j,n+i})|_{\check{\mathscr{A}}'}=\theta_i\partial_{\vartheta_j}-\theta_j\partial_{\vartheta_i},\quad (E_{0,i}-E_{n+i,0})|_{\check{\mathscr{A}}'}=\theta_0\partial_{\theta_i}-\vartheta_i\partial_{\theta_0}, \tag{8.1.25}$$

$$(E_{0,n+i}-E_{i,0})|_{\check{\mathscr{A}}'}=\theta_0\partial_{\vartheta_i}-\theta_i\partial_{\theta_0} \tag{8.1.26}$$

for $i,j\in\overline{1,n}$. By (8.1.24) and (8.1.25), the above representation of $o(2n+1,\mathbb{F})$ is essentially an extension of the fermionic oscillator representation for $o(2n,\mathbb{F})$ given in (7.1.22) and (7.1.23). Recall the notion $\check{\mathscr{A}}_k$ in (7.1.24). We set

$$\check{\mathscr{A}}_0' = \{1\}, \quad \check{\mathscr{A}}_{k+1}' = \theta_0 \check{\mathscr{A}}_k + \check{\mathscr{A}}_{k+1} \qquad \text{for } k \in \overline{0, 2k}. \tag{8.1.27}$$

According to (7.1.29), the $o(2n+1, \mathbb{F})$-singular vectors of $\check{\mathscr{A}}_k'$ are in the set

$$\{\theta_0 \vec{\theta}_{k-1}, \vec{\theta}_k\} \tag{8.1.28}$$

if $k < n$. Applying (8.1.26), we get that $\vec{\theta}_k$ is the unique $o(2n+1, \mathbb{F})$-singular vector in $\check{\mathscr{A}}_k'$. By (7.1.29) again, the $o(2n+1, \mathbb{F})$-singular vectors of $\check{\mathscr{A}}_n'$ are in the set

$$\{\vec{\theta}_n, \vec{\theta}_{n-1}\vartheta_n, \theta_0\vec{\theta}_{n-1}\}. \tag{8.1.29}$$

Since

$$(E_{0,2n} - E_{n,0})(\vec{\theta}_{n-1}\vartheta_n) = \vec{\theta}_{n-1}\theta_0, \quad (E_{0,2n} - E_{n,0})(\theta_0\vec{\theta}_{n-1}) = (-1)^n\vec{\theta}_n \tag{8.1.30}$$

by (8.1.26), $\vec{\theta}_n$ is the unique $o(2n+1, \mathbb{F})$-singular vector in $\check{\mathscr{A}}_n'$.

Next (7.1.29) implies that the $o(2n+1, \mathbb{F})$-singular vectors of $\check{\mathscr{A}}_{n+1}'$ are in the set

$$\{\theta_0\vec{\theta}_n, \theta_0\vec{\theta}_{n-1}\vartheta_n, \check{\eta}\vec{\theta}_{n-1}\}. \tag{8.1.31}$$

Since

$$(E_{0,2n} - E_{n,0})(\theta_0\vec{\theta}_{n-1}\vartheta_n) = (-1)^n\vec{\theta}_n\vartheta_n, \quad (E_{0,2n} - E_{n,0})(\check{\eta}\vec{\theta}_{n-1}) = (-1)^n\theta_0\vec{\theta}_n \tag{8.1.32}$$

by (6.2.49) and (8.1.26), $\theta_0\vec{\theta}_n$ is the unique $o(2n+1, \mathbb{F})$-singular vector in $\check{\mathscr{A}}_{n+1}'$. For $k \in \overline{1, n-1}$, the $o(2n+1, \mathbb{F})$-singular vectors of $\check{\mathscr{A}}_{2n+1-k}'$ are in the set

$$\{\theta_0\check{\eta}^{n-k}\vec{\theta}_k, \check{\eta}^{n+1-k}\vec{\theta}_{k-1}\} \tag{8.1.33}$$

by (7.1.29). Note

$$\check{\eta}^{n-r}\vec{\theta}_r = (n-r)!\theta_{r+1}\vartheta_{r+1}\cdots\theta_n\vartheta_n\vec{\theta}_r. \tag{8.1.34}$$

So $\theta_0\check{\eta}^{n-k}\vec{\theta}_k$ is an $o(2n+1, \mathbb{F})$-singular vector by (8.1.26). Moreover,

$$(E_{0,n+k} - E_{k,0})(\check{\eta}^{n+1-k}\vec{\theta}_{k-1}) = (n+1-k)\check{\eta}^{n-k}\vec{\theta}_k\theta_0 \tag{8.1.35}$$

by (6.2.49) and (8.1.26). Thus $\theta_0\check{\eta}^{n-k}\vec{\theta}_k$ is the unique $o(2n+1, \mathbb{F})$-singular vector in $\check{\mathscr{A}}_{2n+1-k}'$. Furthermore, $\check{\mathscr{A}}_{2n+1}'$ is the one-dimensional trivial module. By Weyl's Theorem 2.3.6 of complete reducibility or more generally by an analogue of Lemma 6.3.2, we have:

Theorem 8.1.2 *For $r \in \overline{1, n}$, $\check{\mathscr{A}}_r'$ and $\check{\mathscr{A}}_{2n+1-r}'$ are finite-dimensional irreducible $o(2n+1, \mathbb{F})$-submodules with the highest weight $\sum_{i=1}^r \varepsilon_i$. Moreover, $\check{\mathscr{A}}_{2n+1}'$ and $\check{\mathscr{A}}_0'$ are isomorphic to the one-dimensional trivial $o(2n+1, \mathbb{F})$-module.*

Let Ψ' be the associative algebra generated by $\{\theta_0, \theta_1, \dots, \theta_n\}$ with the defining relations:

$$\theta_0^2 = 1, \quad \theta_0\theta_i = -\theta_i\theta_0, \quad \theta_i\theta_j = -\theta_j\theta_i \qquad \text{for } i, j \in \overline{1, n}. \tag{8.1.36}$$

The algebra Ψ' is an extension of the exterior algebra Ψ (cf. (6.2.15)). Take the settings as those in (6.2.16)–(6.2.18), where $i \in \overline{1, n}$ and $j, j_s \in \overline{0, n}$. Note

$$[\theta_0\partial_{\theta_i}, \theta_0\theta_i] = 2\theta_i\partial_{\theta_i} - 1 \qquad \text{for } i \in \overline{1, n}. \tag{8.1.37}$$

Thus we have the following representation of $o(2n + 1, \mathbb{F})$ on Ψ':

$$(E_{i,j} - E_{n+j,n+i})|_{\Psi'} = \theta_i\partial_{\theta_j} - \frac{\delta_{i,j}}{2} \qquad \text{for } i, j \in \overline{1, n}. \tag{8.1.38}$$

$$(E_{r,n+s}, E_{s,n+r})|_{\Psi'} = \theta_r\theta_s, \qquad (E_{n+r,s}, E_{n+s,r})|_{\Psi'} = \partial_{\theta_r}\partial_{\theta_s} \tag{8.1.39}$$

for $1 \leq r < s \leq n$, and

$$(E_{0,i} - E_{n+i,0})|_{\Psi'} = \frac{1}{\sqrt{2}}\theta_0\partial_{\theta_i}, \quad (E_{0,n+i} - E_{i,0})|_{\Psi'} = \frac{1}{\sqrt{2}}\theta_0\theta_i \ \text{ for } i \in \overline{1, n}. \tag{8.1.40}$$

The above representation is called the *spin representation* of $o(2n + 1, \mathbb{F})$, which is an extension of the spin representation of $o(2n, \mathbb{F})$ given (7.1.32) and (7.1.33).

Set

$$\Psi'_{\pm} = \Psi(1 \pm \theta_0). \tag{8.1.41}$$

Then $\Psi'_{\pm}$ are $o(2n + 1, \mathbb{F})$-submodules. Indeed, we have:

Theorem 8.1.3 *The subspaces $\Psi'_{\pm}$ are 2^n-dimensional irreducible $o(2n + 1, \mathbb{F})$-submodules with the highest-weight vector $\theta_1\theta_2 \cdots \theta_n(1 \pm \theta_0)$ of weight $(\sum_{i=1}^{n} \varepsilon_i)/2$.*

8.2 Noncanonical Oscillator Representations

In this section, we study the noncanonical oscillator representations of $o(2n + 1, \mathbb{F})$ obtained from the bosonic oscillator representation in (8.1.7)–(8.1.9) by partially swapping differential operators and multiplication operators.

Recall that $\mathscr{B}' = \mathbb{F}[x_0, x_1, \dots, x_n, y_1, \dots, y_n]$ and the canonical oscillator representations of $o(2n+1, \mathbb{F})$ given in (8.1.7)–(8.1.9). Fix $n_1, n_2 \in \overline{1, n}$ with $n_1 \leq n_2$. Changing operators $\partial_{x_r} \mapsto -x_r$, $x_r \mapsto \partial_{x_r}$ for $r \in \overline{1, n_1}$ and $\partial_{y_s} \mapsto -y_s$, $y_s \mapsto \partial_{y_s}$ for $s \in \overline{n_2 + 1, n}$ in the canonical oscillator representation, we get a *noncanonical oscillator representation* of $o(2n + 1, \mathbb{F})$ on $\mathscr{B}'$ determined by

$$(E_{i,j} - E_{n+j,n+i})|_{\mathscr{B}'} = E^x_{i,j} - E^y_{j,i} \tag{8.2.1}$$

with

$$E^x_{i,j} = \begin{cases} -x_j\partial_{x_i} - \delta_{i,j} & \text{if } i, j \in \overline{1, n_1}; \\ \partial_{x_i}\partial_{x_j} & \text{if } i \in \overline{1, n_1},\ j \in \overline{n_1+1, n}; \\ -x_i x_j & \text{if } i \in \overline{n_1+1, n},\ j \in \overline{1, n_1}; \\ x_i\partial_{x_j} & \text{if } i, j \in \overline{n_1+1, n}; \end{cases} \tag{8.2.2}$$

and

$$E^y_{i,j} = \begin{cases} y_i\partial_{y_j} & \text{if } i, j \in \overline{1, n_2}; \\ -y_i y_j & \text{if } i \in \overline{1, n_2},\ j \in \overline{n_2+1, n}; \\ \partial_{y_i}\partial_{y_j} & \text{if } i \in \overline{n_2+1, n},\ j \in \overline{1, n_2}; \\ -y_j\partial_{y_i} - \delta_{i,j} & \text{if } i, j \in \overline{n_2+1, n}; \end{cases} \tag{8.2.3}$$

and

$$E_{i,n+j}|_{\mathscr{B}'} = \begin{cases} \partial_{x_i}\partial_{y_j} & \text{if } i \in \overline{1, n_1},\ j \in \overline{1, n_2}; \\ -y_j\partial_{x_i} & \text{if } i \in \overline{1, n_1},\ j \in \overline{n_2+1, n}; \\ x_i\partial_{y_j} & \text{if } i \in \overline{n_1+1, n},\ j \in \overline{1, n_2}; \\ -x_i y_j & \text{if } i \in \overline{n_1+1, n},\ j \in \overline{n_2+1, n}; \end{cases} \tag{8.2.4}$$

$$E_{n+i,j}|_{\mathscr{B}'} = \begin{cases} -x_j y_i & \text{if } j \in \overline{1, n_1},\ i \in \overline{1, n_2}; \\ -x_j\partial_{y_i} & \text{if } j \in \overline{1, n_1},\ i \in \overline{n_2+1, n}; \\ y_i\partial_{x_j} & \text{if } j \in \overline{n_1+1, n},\ i \in \overline{1, n_2}; \\ \partial_{x_j}\partial_{y_i} & \text{if } j \in \overline{n_1+1, n},\ i \in \overline{n_2+1, n}. \end{cases} \tag{8.2.5}$$

$$E_{0,i}|_{\mathscr{B}'} = \begin{cases} -x_0 x_i & \text{if } i \in \overline{1, n_1}; \\ x_0\partial_{x_i} & \text{if } i \in \overline{n_1+1, n}; \\ x_0\partial_{y_{i-n}} & \text{if } i \in \overline{n+1, n+n_2}; \\ -x_0 y_{i-n} & \text{if } i \in \overline{n+n_2+1, 2n}; \end{cases} \tag{8.2.6}$$

and

$$E_{i,0}|_{\mathscr{B}'} = \begin{cases} \partial_{x_0}\partial_{x_i} & \text{if } i \in \overline{1, n_1}; \\ x_i\partial_{x_0} & \text{if } i \in \overline{n_1+1, n}; \\ y_{i-n}\partial_{x_0} & \text{if } i \in \overline{n+1, n+n_2}; \\ \partial_{x_0}\partial_{y_{i-n}} & \text{if } i \in \overline{n+n_2+1, 2n}. \end{cases} \tag{8.2.7}$$

Correspondingly we obtain a variation of the Laplace operator:

$$\tilde{\Delta}' = \partial_{x_0}^2 - 2\sum_{i=1}^{n_1} x_i\partial_{y_i} + 2\sum_{r=n_1+1}^{n_2} \partial_{x_r}\partial_{y_r} - 2\sum_{s=n_2+1}^{n} y_s\partial_{x_s} \tag{8.2.8}$$

and its dual operator

$$\eta' = x_0^2 + 2\sum_{i=1}^{n_1} y_i\partial_{x_i} + 2\sum_{r=n_1+1}^{n_2} x_r y_r + 2\sum_{s=n_2+1}^{n} x_s\partial_{y_s}. \tag{8.2.9}$$

Set

$$\mathscr{B}'_{\langle k\rangle}=\sum_{i=0}^{\infty}\mathscr{B}_{\langle k-i\rangle}x_0^i,\qquad \mathscr{H}'_{\langle k\rangle}=\{f\in\mathscr{B}'_{\langle k\rangle}\mid\tilde{\Delta}'(f)=0\}\tag{8.2.10}$$

(cf. (7.2.10)). Denote

$$\tilde{D}'=x_0\partial_{x_0}-\sum_{i=1}^{n_1}x_i\partial_{x_i}+\sum_{r=n_1+1}^{n}x_r\partial_{x_r}+\sum_{j=1}^{n_2}y_j\partial_{y_j}-\sum_{s=n_2+1}^{n}y_s\partial_{y_s}.\tag{8.2.11}$$

Then

$$\mathscr{B}'_{\langle k\rangle}=\{f\in\mathscr{B}'\mid\tilde{D}'(f)=kf\}.\tag{8.2.12}$$

A straightforward verification shows

$$\tilde{\Delta}'\xi=\xi\tilde{\Delta}',\ \xi\eta'=\eta'\xi,\ \xi\tilde{D}'=\tilde{D}'\xi\text{ on }\mathscr{B}'\ \text{ for }\ \xi\in o(2n+1,\mathbb{F}).\tag{8.2.13}$$

Thus all $\mathscr{B}'_{\langle k\rangle}$ and $\mathscr{H}'_{\langle k\rangle}$ with $k\in\mathbb{Z}$ are $o(2n+1,\mathbb{F})$-submodules and $\mathscr{B}'=\bigoplus_{k\in\mathbb{Z}}\mathscr{B}'_{\langle k\rangle}$ forms a $\mathbb{Z}$-graded algebra.

We take $\tilde{\Delta}$ in (7.2.6) and η in (7.2.7). Then

$$\tilde{\Delta}'=\partial_{x_0}^2+2\tilde{\Delta},\qquad \eta'=x_0^2+2\eta.\tag{8.2.14}$$

For $\iota=0,1$, we define

$$\widehat{T}_\iota=\sum_{i=0}^{\infty}\frac{(-2)^i x_0^{2i+\iota}}{(2i+\iota)!}\tilde{\Delta}^i.\tag{8.2.15}$$

By Lemma 6.1.1 with $T_1=\partial_{x_0}^2,\ T_1^-=\int_{(x_0)}^{(2)}$ and $T_2=2\tilde{\Delta}$, we obtain

$$\mathscr{H}'_{\langle k\rangle}=\widehat{T}_0(\mathscr{B}_{\langle k\rangle})\oplus\widehat{T}_1(\mathscr{B}_{\langle k-1\rangle}).\tag{8.2.16}$$

Lemma 8.2.1 *Suppose $n_1<n_2$. For $k\in\mathbb{N}$, $\mathscr{H}'_{\langle k\rangle}$ is an $o(2n+1,\mathbb{F})$-submodule generated by $x_{n_1+1}^k$ and $\mathscr{H}'_{\langle -k\rangle}$ is an $o(2n+1,\mathbb{F})$-submodule generated by $x_{n_1}^k$.*

Proof Let V be an $o(2n+1,\mathbb{F})$-submodule generated by $x_{n_1+1}^k$. Since $x_{n_1+1}^k\in\mathscr{H}'_{\langle k\rangle}$ that is an $o(2n+1,\mathbb{F})$-submodule, we have $V\subset\mathscr{H}'_{\langle k\rangle}$.

Observe

$$(E_{n_1+1,n+j}-E_{j,n+n_1+1})|_{\mathscr{B}'}=-x_{n_1+1}y_j-x_j\partial_{y_{n_1+1}}\qquad\text{for }\ j\in\overline{n_2+1,n}\tag{8.2.17}$$

by (8.2.4). Repeatedly applying (8.2.17) to $x_{n_1+1}^k$ with various $j\in\overline{n_2+1,n}$, we obtain

$$x_{n_1+1}^{k+\ell}\prod_{s=n_2+1}^{n} y_s^{\beta_s} \in V \qquad \text{for } \ell, \beta_{n_2+1}, \ldots, \beta_n \in \mathbb{N}, \ \sum_{s=n_2+1}^{n} \beta_s = \ell. \tag{8.2.18}$$

Note

$$(E_{n+i,n+n_1+1} - E_{n_1+1,i})|_{\mathscr{B}'} = y_i \partial_{y_{n_1+1}} + x_i x_{n_1+1} \qquad \text{for } i \in \overline{1, n_1} \tag{8.2.19}$$

by (8.2.1)–(8.2.3). Repeatedly applying (8.2.19) to (8.2.18) with various $i \in \overline{1, n_1}$, we have

$$\left[\prod_{i=1}^{n_1+1} x_i^{\alpha_i}\right]\left[\prod_{s=n_2+1}^{n} y_s^{\beta_s}\right] \in V \tag{8.2.20}$$

for $\alpha_1, \ldots, \alpha_{n_1+1}, \beta_{n_2+1}, \ldots, \beta_n \in \mathbb{N}$ such that $k + \sum_{r=1}^{n_1} \alpha_r + \sum_{s=n_2+1}^{n} \beta_s = \alpha_{n_1+1}$.

Denote

$$I = \{0, \overline{n_1+1, n_2}, \overline{n+n_1+1, n+n_2}\}. \tag{8.2.21}$$

Observe the Lie subalgebra

$$\mathscr{G} = o(2n+1, \mathbb{F}) \bigcap \left(\sum_{i,j \in I} \mathbb{F}E_{i,j}\right) \cong o(2(n_2 - n_1) + 1, \mathbb{F}). \tag{8.2.22}$$

Set

$$\bar{\mathscr{H}}_{\langle \ell \rangle} = \mathscr{H}'_{\langle \ell \rangle} \bigcap \mathbb{F}[x_0, x_{n_1+1}, \ldots, x_{n_2}, y_{n_1+1}, \ldots, y_{n_2}] \qquad \text{for } \ell \in \mathbb{N}. \tag{8.2.23}$$

Then $\bar{\mathscr{H}}_{\langle \ell \rangle}$ forms an irreducible $\mathscr{G}$-module by Theorem 7.1.1. Repeatedly applying $\mathscr{G}|_{\mathscr{B}'}$ to (8.2.20), we get

$$\widehat{T}_\iota(x^\alpha y^\beta) \in V \tag{8.2.24}$$

for $\alpha, \beta \in \mathbb{N}^n$ such that $\alpha_i = 0$ if $i > n_2$, $\beta_j = 0$ if $j \leq n_1$, and

$$\iota + \sum_{r=n_1+1}^{n_2} \alpha_r - \sum_{i=1}^{n_1} \alpha_i + \sum_{s=n_1+1}^{n_2} \beta_s - \sum_{j=n_2+1}^{n} \beta_j = k. \tag{8.2.25}$$

Repeatedly applying (8.2.19) to (8.2.24) satisfying the above conditions, we get that (8.2.24) holds for $\alpha, \beta \in \mathbb{N}^n$ such that $\alpha_i = 0$ if $i > n_2$, and

$$\iota + \sum_{r=n_1+1}^{n_2} \alpha_r - \sum_{i=1}^{n_1} \alpha_i + \sum_{s=1}^{n_2} \beta_s - \sum_{j=n_2+1}^{n} \beta_j = k. \tag{8.2.26}$$

According to (8.2.1)–(8.2.3),

$$(E_{i,n_1+1} - E_{n+n_1+1,n+i})|_{\mathscr{B}'} = x_i\partial_{x_{n_1+1}} + y_{n_1+1}y_i \qquad \text{for } i \in \overline{n_2+1,n}. \tag{8.2.27}$$

Repeatedly applying (8.2.27) to (8.2.24) satisfying (8.2.26), we get (8.2.25) for any $\alpha, \beta \in \mathbb{N}^n$ such that

$$\iota + \sum_{r=n_1+1}^{n} \alpha_r - \sum_{i=1}^{n_1} \alpha_i + \sum_{s=1}^{n_2} \beta_s - \sum_{j=n_2+1}^{n} \beta_j = k. \tag{8.2.28}$$

Thus $V = \mathscr{H}'_{\langle k \rangle}$.

Let U be an $o(2n+1,\mathbb{F})$-submodule generated by $x_{n_1}^k$. Since $x_{n_1}^k \in \mathscr{H}'_{\langle -k \rangle}$ that is an $o(2n+1,\mathbb{F})$-submodule, we have $U \subset \mathscr{H}'_{\langle -k \rangle}$. Repeatedly applying (8.2.17) and (8.2.19) to $x_{n_1}^k$ with various $i \in \overline{1,n_1}$ and $j \in \overline{n_2+1,n}$, we have

$$x_{n_1}^k \left[\prod_{i=1}^{n_1+1} x_i^{\alpha_i}\right]\left[\prod_{s=n_2+1}^{n} y_s^{\beta_s}\right] \in U \text{ for } \alpha_1,\ldots,\alpha_{n_1+1},\beta_{n_2+1},\ldots,\beta_n \in \mathbb{N} \tag{8.2.29}$$

such that $\sum_{r=1}^{n_1} \alpha_r + \sum_{s=n_2+1}^{n} \beta_s = \alpha_{n_1+1}$. Note

$$(E_{n_1,i} - E_{n+i,n+n_1})|_{\mathscr{B}'} = -x_i\partial_{x_{n_1}} - y_i\partial_{y_{n_1}} \qquad \text{for } i \in \overline{n_1-1} \tag{8.2.30}$$

by (8.2.1)–(8.2.3). Repeatedly applying (8.2.31) to (8.2.30), we have

$$\left[\prod_{i=1}^{n_1+1} x_i^{\alpha_i}\right]\left[\prod_{s=n_2+1}^{n} y_s^{\beta_s}\right] \in U \text{ for } \alpha_1,\ldots,\alpha_{n_1+1},\beta_{n_2+1},\ldots,\beta_n \in \mathbb{N} \tag{8.2.31}$$

such that $\sum_{r=1}^{n_1} \alpha_r + \sum_{s=n_2+1}^{n} \beta_s = \alpha_{n_1+1} + k$.

Repeatedly applying $\mathscr{G}|_{\mathscr{B}'}$ to (8.2.31), we get

$$\widehat{T}_\iota(x^\alpha y^\beta) \in U \tag{8.2.32}$$

for $\alpha, \beta \in \mathbb{N}^n$ such that $\alpha_i = 0$ if $i > n_2$, $\beta_j = 0$ if $j \leq n_1$, and

$$\iota + \sum_{r=n_1+1}^{n_2} \alpha_r - \sum_{i=1}^{n_1} \alpha_i + \sum_{s=n_1+1}^{n_2} \beta_s - \sum_{j=n_2+1}^{n} \beta_j = -k. \tag{8.2.33}$$

Repeatedly applying (8.2.20) to (8.2.33) satisfying the above conditions, we get that (8.2.32) holds for $\alpha, \beta \in \mathbb{N}^n$ such that $\alpha_i = 0$ if $i > n_2$, and

$$\iota + \sum_{r=n_1+1}^{n_2} \alpha_r - \sum_{i=1}^{n_1} \alpha_i + \sum_{s=1}^{n_2} \beta_s - \sum_{j=n_2+1}^{n} \beta_j = -k. \tag{8.2.34}$$

Repeatedly applying (8.2.27) to (8.2.32) satisfying (8.2.34), we get (8.2.32) for any $\alpha, \beta \in \mathbb{N}^n$ such that

$$\iota + \sum_{r=n_1+1}^{n} \alpha_r - \sum_{i=1}^{n_1} \alpha_i + \sum_{s=1}^{n_2} \beta_s - \sum_{j=n_2+1}^{n} \beta_j = -k. \tag{8.2.35}$$

Thus $U = \mathscr{H}'_{\langle -k\rangle}$. $\square$

Lemma 8.2.2 *Suppose $n_1 = n_2$. For $k \in \mathbb{N}$, $\mathscr{H}'_{\langle -k\rangle}$ is an $o(2n+1, \mathbb{F})$-submodule generated by $x_{n_1}^k$ and $\mathscr{H}'_{\langle k\rangle}$ is an $o(2n+1, \mathbb{F})$-submodule generated by $\widehat{T}_1(y_{n_1}^{k-1})$ when $k > 0$.*

Proof In this case,

$$\tilde{\Delta} = -\sum_{i=1}^{n_1} x_i \partial_{y_i} - \sum_{s=n_1+1}^{n} y_s \partial_{x_s} \tag{8.2.36}$$

Note

$$(E_{n+n_1,0} - E_{0,n_1})|_{\mathscr{B}'} = y_{n_1}\partial_{x_0} + x_0 x_{n_1} \tag{8.2.37}$$

by (8.2.6) and (8.2.7). Thus for $u \in \mathscr{B}$,

$$\begin{aligned}
&(E_{n+n_1,0} - E_{0,n_1})[\widehat{T}_0(u)] \\
&= (y_{n_1}\partial_{x_0} + x_0 x_{n_1})\left(\sum_{i=0}^{\infty} \frac{(-2)^i x_0^{2i}}{(2i)!} \tilde{\Delta}^i(u)\right) \\
&= \sum_{i=1}^{\infty} \frac{(-2)^i x_0^{2i-1}}{(2i-1)!} y_{n_1} \tilde{\Delta}^i(u) + \sum_{i=0}^{\infty} \frac{(-2)^i x_0^{2i+1}}{(2i)!} \tilde{\Delta}^i(x_{n_1}u) \\
&= \sum_{i=1}^{\infty} \frac{(-2)^i i x_0^{2i-1}}{(2i-1)!} \tilde{\Delta}^{i-1}(x_{n_1}u) + \sum_{i=1}^{\infty} \frac{(-2)^i x_0^{2i-1}}{(2i-1)!} \tilde{\Delta}^{i-1}[\tilde{\Delta}(y_{n_1}u)] \\
&\quad + \sum_{i=0}^{\infty} \frac{(-2)^i x_0^{2i+1}}{(2i)!} \tilde{\Delta}^i(x_{n_1}u) \\
&= -\sum_{i=0}^{\infty} \frac{(-2)^i 2(i+1) x_0^{2i+1}}{(2i+1)!} \tilde{\Delta}^i(x_{n_1}u) - 2\sum_{i=0}^{\infty} \frac{(-2)^i x_0^{2i+1}}{(2i+1)!} \tilde{\Delta}^i[\tilde{\Delta}(y_{n_1}u)] \\
&\quad + \sum_{i=0}^{\infty} \frac{(-2)^i x_0^{2i+1}}{(2i)!} \tilde{\Delta}^i(x_{n_1}u) = -\widehat{T}_1(x_{n_1}u) - 2\widehat{T}_1[\tilde{\Delta}(y_{n_1}u)]
\end{aligned} \tag{8.2.38}$$

and

$$(E_{n+n_1,0} - E_{0,n_1})[\widehat{T}_1(u)]$$

$$= (y_{n_1}\partial_{x_0} + x_0 x_{n_1})\left(\sum_{i=0}^{\infty}\frac{(-2)^i x_0^{2i+1}}{(2i+1)!}\tilde{\Delta}^i(u)\right)$$
$$= \sum_{i=0}^{\infty}\frac{(-2)^i x_0^{2i}}{(2i)!} y_{n_1}\tilde{\Delta}^i(u) + \sum_{i=0}^{\infty}\frac{(-2)^i x_0^{2i+2}}{(2i+1)!}\tilde{\Delta}^i(x_{n_1}u)$$
$$= -\sum_{i=1}^{\infty}\frac{(-2)^{i-1} x_0^{2i}}{(2i-1)!}\tilde{\Delta}^{i-1}(x_{n_1}u) + \sum_{i=0}^{\infty}\frac{(-2)^i x_0^{2i}}{(2i)!}\tilde{\Delta}^i(y_{n_1}u)$$
$$+ \sum_{i=0}^{\infty}\frac{(-2)^i x_0^{2i+2}}{(2i+1)!}\tilde{\Delta}^i(x_{n_1}u)$$
$$= \sum_{i=0}^{\infty}\frac{(-2)^i x_0^{2i}}{(2i)!}\tilde{\Delta}^i(y_{n_1}u) = \widehat{T}_0(y_{n_1}u). \tag{8.2.39}$$

Symmetrically, we have,

$$(E_{n_1+1,0} - E_{0,n+n_1+1})|_{\mathscr{B}'} = x_{n_1+1}\partial_{x_0} + x_0 y_{n_1+1} \tag{8.2.40}$$

by (8.2.6) and (8.2.7). So

$$(E_{n_1+1,0} - E_{0,n+n_1+1})[\widehat{T}_0(u)] = -\widehat{T}_1(y_{n_1+1}u) - 2\widehat{T}_1[\tilde{\Delta}(x_{n_1+1}u)] \tag{8.2.41}$$

and

$$(E_{n_1+1,0} - E_{0,n+n_1+1})[\widehat{T}_1(u)] = \widehat{T}_0(x_{n_1+1}u). \tag{8.2.42}$$

Let V be an $o(2n+1,\mathbb{F})$-submodule generated by $x_{n_1}^k$. Since $x_{n_1}^k \in \mathscr{H}'_{\langle -k\rangle}$ that is an $o(2n+1,\mathbb{F})$-submodule, we have $V \subset \mathscr{H}'_{\langle -k\rangle}$. Note

$$(E_{n_1+1,n+n_1} - E_{n_1,n+n_1+1})|_{\mathscr{B}'} = x_{n_1+1}\partial_{y_{n_1}} + y_{n_1+1}\partial_{x_{n_1}} \tag{8.2.43}$$

by (8.2.4). Repeatedly applying (8.2.43) to $x_{n_1}^k$, we have

$$x_{n_1}^{k-r} y_{n_1+1}^r = \widehat{T}_0(x_{n_1}^{k-r} y_{n_1+1}^r) \in V \qquad \text{for } r \in \overline{0,k}. \tag{8.2.44}$$

According to (8.2.38),

$$(E_{n+n_1,0} - E_{0,n_1})[\widehat{T}_0(x_{n_1}^{k-r} y_{n_1+1}^r)]$$
$$= -\widehat{T}_1(x_{n_1}^{k-r+1} y_{n_1+1}^r)) - 2\widehat{T}_1[\tilde{\Delta}(x_{n_1}^{k-r} y_{n_1} y_{n_1+1}^r)] = \widehat{T}_1(x_{n_1}^{k-r+1} y_{n_1+1}^r) \in V. \tag{8.2.45}$$

Similarly, (8.2.41) shows

$$(E_{n_1+1,0} - E_{0,n+n_1+1})[\widehat{T}_0(x_{n_1}^{k-r} y_{n_1+1}^r)] = \widehat{T}_1(x_{n_1}^{k-r} y_{n_1+1}^{r+1}) \in V. \tag{8.2.46}$$

Suppose that

$$\widehat{T}_1(x_{n_1}^{m_1} y_{n_1}^{m_2} x_{n_1+1}^{m_3} y_{n_1+1}^{m_4}) \in V \qquad \text{for } m_i \in \mathbb{N},\ m_1+m_4-m_2-m_3 = k+1 \quad (8.2.47)$$

and $\sum_{i=1}^{4} m_i = \ell$. Then (8.2.39) implies

$$(E_{n+n_1,0}-E_{0,n_1})[\widehat{T}_1(x_{n_1}^{m_1} y_{n_1}^{m_2} x_{n_1+1}^{m_3} y_{n_1+1}^{m_4})] = \widehat{T}_0(x_{n_1}^{m_1} y_{n_1}^{m_2+1} x_{n_1+1}^{m_3} y_{n_1+1}^{m_4}) \in V \quad (8.2.48)$$

and (8.2.42) yields

$$(E_{n_1+1,0} - E_{0,n+n_1+1})[\widehat{T}_1(x_{n_1}^{m_1} y_{n_1}^{m_2} x_{n_1+1}^{m_3} y_{n_1+1}^{m_4})] = \widehat{T}_0(x_{n_1}^{m_1} y_{n_1}^{m_2} x_{n_1+1}^{m_3+1} y_{n_1+1}^{m_4}) \in V. \quad (8.2.49)$$

For $i_1, i_2 \in \mathbb{N}$ such that $i_1 + i_2 = 1$, (8.2.36), (8.2.37), (8.2.48) and (8.2.49) give

$$\begin{aligned}
&(E_{n+n_1,0} - E_{0,n_1})[\widehat{T}_0(x_{n_1}^{m_1} y_{n_1}^{m_2+i_1} x_{n_1+1}^{m_3+i_2} y_{n_1+1}^{m_4})] \\
&= -\widehat{T}_1(x_{n_1}^{m_1+1} y_{n_1}^{m_2+i_1} x_{n_1+1}^{m_3+i_2} y_{n_1+1}^{m_4}) - 2\widehat{T}_1[\tilde{\Delta}(x_{n_1}^{m_1} y_{n_1}^{m_2+i_1+1} x_{n_1+1}^{m_3+i_2} y_{n_1+1}^{m_4})] \\
&= (2(m_2 + i_1) + 1)\widehat{T}_1(x_{n_1}^{m_1+1} y_{n_1}^{m_2+i_1} x_{n_1+1}^{m_3+i_2} y_{n_1+1}^{m_4}) \\
&\quad +2(m_3 + i_2)\widehat{T}_1(x_{n_1}^{m_1} y_{n_1}^{m_2+i_1+1} x_{n_1+1}^{m_3+i_2-1} y_{n_1+1}^{m_4+1}) \in V. \qquad (8.2.50)
\end{aligned}$$

On the other hand,

$$(E_{n_1+1,n_1} - E_{n+n_1,n+n_1+1})|_{\mathscr{B}'} = -x_{n_1} x_{n_1+1} + y_{n_1} y_{n_1+1} \quad (8.2.51)$$

by (8.2.1)–(8.2.3). Thus

$$\begin{aligned}
&(E_{n_1+1,n_1} - E_{n+n_1,n+n_1+1})[\widehat{T}_1(x_{n_1}^{m_1} y_{n_1}^{m_2+i_1} x_{n_1+1}^{m_3+i_2-1} y_{n_1+1}^{m_4})] \\
&= -\widehat{T}_1(x_{n_1}^{m_1+1} y_{n_1}^{m_2+i_1} x_{n_1+1}^{m_3+i_2} y_{n_1+1}^{m_4}) + \widehat{T}_1(x_{n_1}^{m_1} y_{n_1}^{m_2+i_1+1} x_{n_1+1}^{m_3+i_1-1} y_{n_1+1}^{m_4+1}) \in V. \quad (8.2.52)
\end{aligned}$$

Solving the system (8.2.50) and (8.2.52) for $\widehat{T}_1(x_{n_1}^{m_1+1} y_{n_1}^{m_2+i_1} x_{n_1+1}^{m_3+i_2} y_{n_1+1}^{m_4})$, we get

$$\widehat{T}_1(x_{n_1}^{m_1+1} y_{n_1}^{m_2+i_1} x_{n_1+1}^{m_3+i_2} y_{n_1+1}^{m_4}) \in V. \quad (8.2.53)$$

Symmetrically, (8.2.36), (8.2.41), (8.2.48) and (8.2.49) give

$$\begin{aligned}
&(E_{n_1+1,0} - E_{0,n+n_1+1})[\widehat{T}_0(x_{n_1}^{m_1} y_{n_1}^{m_2+i_1} x_{n_1+1}^{m_3+i_2} y_{n_1+1}^{m_4})] \\
&= -\widehat{T}_1(x_{n_1}^{m_1} y_{n_1}^{m_2+i_1} x_{n_1+1}^{m_3+i_2} y_{n_1+1}^{m_4+1}) - 2\widehat{T}_1[\mathscr{D}(x_{n_1}^{m_1} y_{n_1}^{m_2+i_1} x_{n_1+1}^{m_3+i_2+1} y_{n_1+1}^{m_4})] \\
&= (2(m_3 + i_2) + 1)\widehat{T}_1(x_{n_1}^{m_1} y_{n_1}^{m_2+i_1} x_{n_1+1}^{m_3+i_2} y_{n_1+1}^{m_4+1}) \\
&\quad +2(m_2 + i_1)\widehat{T}_1(x_{n_1}^{m_1+1} y_{n_1}^{m_2+i_1-1} x_{n_1+1}^{m_3+i_2+1} y_{n_1+1}^{m_4}) \in V \qquad (8.2.54)
\end{aligned}$$

and (8.2.51) yields

$$(E_{n_1+1,n_1} - E_{n+n_1,n+n_1+1})[\widehat{T}_1(x_{n_1}^{m_1} y_{n_1}^{m_2+i_1-1} x_{n_1+1}^{m_3+i_2} y_{n_1+1}^{m_4})]$$
$$= -\widehat{T}_1(x_{n_1}^{m_1+1} y_{n_1}^{m_2+i_1-1} x_{n_1+1}^{m_3+i_2+1} y_{n_1+1}^{m_4}) + \widehat{T}_1(x_{n_1}^{m_1} y_{n_1}^{m_2+i_1} x_{n_1+1}^{m_3+i_1} y_{n_1+1}^{m_4+1}) \in V. \quad (8.2.55)$$

Solving the system (8.2.54) and (8.2.55) for $\widehat{T}_1(x_{n_1}^{m_1} y_{n_1}^{m_2+i_1} x_{n_1+1}^{m_3+i_2} y_{n_1+1}^{m_4+1})$, we find

$$\widehat{T}_1(x_{n_1}^{m_1} y_{n_1}^{m_2+i_1} x_{n_1+1}^{m_3+i_2} y_{n_1+1}^{m_4+1}) \in V. \quad (8.2.56)$$

By induction on $\sum_{i=1}^4 m_i$, we conclude

$$\widehat{T}_\iota(x_{n_1}^{m_1} y_{n_1}^{m_2} x_{n_1+1}^{m_3} y_{n_1+1}^{m_4}) \in V \qquad \text{for } m_i \in \mathbb{N},\ m_1+m_4-m_2-m_3 = k+\delta_{\iota,1}. \quad (8.2.57)$$

Set

$$\mathscr{A}' = \mathbb{F}[x_1, \dots, x_{n_1}, y_1, \dots, y_{n_1}], \quad \mathscr{A}^* = \mathbb{F}[x_{n_1+1}, \dots, x_n, y_{n_1+1}, \dots, y_n], \quad (8.2.58)$$

$$\mathscr{K}_1 = \sum_{i,j=1}^{n_1} [\mathbb{F}(E_{i,j} - E_{n+j,n+i}) + \mathbb{F}(E_{i,n+j} - E_{j,n+i}) + \mathbb{F}(E_{n+i,j} - E_{n+j,i})] \quad (8.2.59)$$

and

$$\mathscr{K}_2 = \sum_{i,j=n_1+1}^{n} [\mathbb{F}(E_{i,j} - E_{n+j,n+i}) + \mathbb{F}(E_{i,n+j} - E_{j,n+i}) + \mathbb{F}(E_{n+i,j} - E_{n+j,i})]. \quad (8.2.60)$$

Then $\mathscr{K}_1 \cong gl(n_1, \mathbb{F})$ and $\mathscr{K}_2 \cong gl(n - n_1, \mathbb{F})$ are Lie subalgebras of $o(2n+1, \mathbb{F})$ and $\mathscr{B} = \mathscr{A}'\mathscr{A}^*$. Since

$$\mathscr{A}'_{\langle \ell_1, \ell_2 \rangle} = \mathscr{A}' \bigcap \mathscr{B}_{\langle \ell_1, \ell_2 \rangle}, \quad \mathscr{A}^*_{\langle \ell_1, \ell_2 \rangle} = \mathscr{A}^* \bigcap \mathscr{B}_{\langle \ell_1, \ell_2 \rangle} \quad (8.2.61)$$

are of finite-dimensional for any $\ell_1, \ell_2 \in \mathbb{Z}$ (cf. (6.4.29)) and

$$\mathscr{A}' = \bigoplus_{\ell_1, \ell_2 \in \mathbb{Z}} \mathscr{A}'_{\langle \ell_1, \ell_2 \rangle}, \quad \mathscr{A}^* = \bigoplus_{\ell_1, \ell_2 \in \mathbb{Z}} \mathscr{A}^*_{\langle \ell_1, \ell_2 \rangle}, \quad (8.2.62)$$

$\mathscr{A}'$ is a $\mathscr{K}_1$-module generated by its $\mathscr{K}_1$-singular vectors and $\mathscr{A}^*$ is a $\mathscr{K}_2$-module generated by its $\mathscr{K}_2$-singular vectors. According to (6.6.18), (6.6.29) and (6.6.32)

$$\{x_{n_1}^{m_1} y_{n_1}^{m_2} \zeta_1^{\delta_{1,n_1} m_3} \mid m_i \in \mathbb{N}\} \quad (8.2.63)$$

is the set of homogeneous $\mathscr{K}_1$-singular vectors in $\mathscr{A}'$ and

$$\{x_{n_1+1}^{m_1} y_{n_1+1}^{m_2} \zeta_2^{\delta_{1,n-n_1} m_3} \mid m_i \in \mathbb{N}\} \quad (8.2.64)$$

is the set of homogeneous $\mathscr{K}_2$ singular vectors in $\mathscr{A}^*$. Therefore, $\mathscr{A}$ is a $(\mathscr{K}_1+\mathscr{K}_2)$-module generated by

$$\{x_{n_1}^{m_1}y_{n_1}^{m_2}\zeta_1^{\delta_{1,n_1}m_3}x_{n_1+1}^{m_4}y_{n_1+1}^{m_5}\zeta_2^{\delta_{1,n-n_1}m_6}\mid m_i\in\mathbb{N}\}. \tag{8.2.65}$$

Observe

$$(E_{n+n_1-1,n_1}-E_{n+n_1,n_1-1})|_{\mathscr{B}'}=\zeta_1,\quad (E_{n_1+2,n+n_1+1}-E_{n_1+1,n+n_1+2})|_{\mathscr{B}'}=\zeta_2 \tag{8.2.66}$$

as multiplication operators on $\mathscr{B}'$ by (8.2.4) and (8.2.5). Repeatedly applying (8.2.66) to (8.2.57), we obtain

$$\widehat{T}_\iota(x_{n_1}^{m_1}y_{n_1}^{m_2}\zeta_1^{\delta_{1,n_1}m_5}x_{n_1+1}^{m_3}y_{n_1+1}^{m_4}\zeta_2^{\delta_{1,n-n_1}m_6})\in V \tag{8.2.67}$$

for $m_i\in\mathbb{N}$ such that $m_1+m_4-m_2-m_3=k+\delta_{\iota,1}$. Applying $U(\mathscr{K}_1+\mathscr{K}_2)$ to (8.2.67), we have

$$\widehat{T}_0(\mathscr{A}_{\langle -k\rangle}),\ \widehat{T}_1(\mathscr{A}_{\langle -k-1\rangle})\subset V. \tag{8.2.68}$$

According to (8.2.17), $\mathscr{H}'_{\langle -k\rangle}\subset V$. Thus $\mathscr{H}'_{\langle -k\rangle}=V$.

Let U be an $o(2n+1,\mathbb{F})$-submodule generated by $\widehat{T}_0(y_{n_1}^k)$ with $k>0$. Since $T_1(y_{n_1}^{k-1})\in\mathscr{H}'_{\langle k\rangle}$ that is an $o(2n+1,\mathbb{F})$-submodule, we have $U\subset\mathscr{H}'_{\langle k\rangle}$. Repeatedly applying (8.2.43) to $\widehat{T}_1(y_{n_1}^{k-1})$, we have

$$\widehat{T}_1(x_{n_1+1}^r y_{n_1}^{k-r-1})\in U\qquad \text{for } r\in\overline{0,k-1}. \tag{8.2.69}$$

By the same arguments as (8.2.46)–(8.2.68) with V replaced by U and k replaced by $-k$, we prove $\mathscr{H}'_{\langle k\rangle}=U$. □

Theorem 8.2.3 *For any $k\in\mathbb{Z}$, $\mathscr{H}'_{\langle k\rangle}$ is an infinite-dimensional irreducible $o(2n+1,\mathbb{F})$-weight submodule with finite-dimensional weight subspaces and is of highest-weight type only if $n_2=n$. Moreover, $\mathscr{B}'=\bigoplus_{i,k=0}^{\infty}\eta'^i(\mathscr{H}'_{\langle k\rangle})$ is a decomposition of irreducible submodules.*

Proof Note that

$$\mathscr{K}=\sum_{i,j=1}^{n}\mathbb{F}(E_{i,j}-E_{n+j,n+i}) \tag{8.2.70}$$

forms a Lie subalgebra of $o(2n+1,\mathbb{F})$, which is isomorphic to $gl(n,\mathbb{F})$. Our representation (8.2.1)–(8.2.7) of $o(2n+1,\mathbb{F})$ is an extension the representation (6.3.36)–(6.3.38) of $sl(n,\mathbb{F})$ with $\mathcal{Q}$ replaced by $\mathscr{B}$. First we prove that $\mathscr{H}'_{\langle m\rangle}$ is an irreducible $o(2n+1,\mathbb{F})$-submodule for any $m\in\mathbb{Z}$. We divide it into two cases.

Case 1. $n_1+1\le n_2$.

Consider $\mathscr{H}'_{\langle k\rangle}$ with $k\in\mathbb{N}$. Let U be any nonzero $o(2n+1,\mathbb{F})$-submodule of $\mathscr{H}'_{\langle k\rangle}$. Since U is a $\mathscr{K}$-module, it must contain a $\mathscr{K}$-singular vector. According to

Lemma 6.4.1, (6.5.54) and (6.5.61), U must contain some vectors in

$$\{\mathbb{F}[\eta](x_i^{m_1}y_j^{m_2}) \mid m_1, m_2 \in \mathbb{N}; i = n_1, n_1+1; j = n_2, n_2+1\} \tag{8.2.71}$$

if $n_2 < n$, or

$$\{\mathbb{F}[\eta](x_i^{m_1}y_n^{m_2}) \mid m_1, m_2 \in \mathbb{N}; i = n_1, n_1+1\} \tag{8.2.72}$$

when $n_2 = n$.

Note

$$(E_{n+1,n+n_2} - E_{n_2,1})|_{\mathscr{B}'} = y_1\partial_{y_{n_2}} + x_1x_{n_2} \tag{8.2.73}$$

by (8.2.1)–(8.2.3). Moreover,

$$(E_{n_1+1,n+1} - E_{1,n+n_1+1})|_{\mathscr{B}'} = x_{n_1+1}\partial_{y_1} - x_1\partial_{y_{n_1+1}} \tag{8.2.74}$$

by (8.2.4). If $\widehat{T}_\iota(\eta^\ell(x_j^{k_1}y_{n_2}^{k_2})) \in U$ with $j = n_1, n_1+1$, we have

$$\begin{aligned}&(k_2!)^{-2}(E_{n_1+1,n+1} - E_{1,n+n_1+1})^{k_2}(E_{n+1,n+n_2} - E_{n_2,1})^{k_2}[\widehat{T}_\iota(\eta^\ell(x_j^{k_1}y_{n_2}^{k_2}))]\\ &= \widehat{T}_\iota(\eta^\ell(x_j^{k_1}x_{n_1+1}^{k_2})) \in U.\end{aligned} \tag{8.2.75}$$

We use (8.2.72) because n_2 may be equal to n_1+1. Consider the case $\widehat{T}_\iota(\eta^\ell(x_j^{k_1}y_{n_2+1}^{k_2})) \in U$ with $j = n_1, n_1+1$. Observe

$$(E_{n+n_1,n_2+1} - E_{n+n_2+1,n_1})_{\mathscr{B}'} = y_{n_1}\partial_{x_{n_2+1}} + x_{n_1}\partial_{y_{n_2+1}} \tag{8.2.76}$$

by (8.2.5). Hence

$$\frac{1}{k_2!}(E_{n+n_1,n_2+1} - E_{n+n_2+1,n_1})^{k_2}[\widehat{T}_\iota(\eta^\ell(x_j^{k_1}y_{n_2+1}^{k_2}))] = \widehat{T}_\iota(\eta^\ell(x_j^{k_1}x_{n_1}^{k_2})) \in U. \tag{8.2.77}$$

Therefore, we always have

$$\widehat{T}_\iota(\eta^\ell(x_{n_1}^{k_1}x_{n_1+1}^{k_2})) \in U. \tag{8.2.78}$$

Note

$$(E_{n_1+1,0} - E_{0,n+n_1+1})|_{\mathscr{B}'} = x_{n_1+1}\partial_{x_0} - x_0\partial_{y_{n_1+1}} \tag{8.2.79}$$

by (8.2.6) and (8.2.7). Thus (7.2.7) and (7.2.8) give

$$\begin{aligned}&(E_{n_1+1,0} - E_{0,n+n_1+1})[\widehat{T}_0(\eta^\ell(x_{n_1}^{k_1}x_{n_1+1}^{k_2}))]\\ &= (x_{n_1+1}\partial_{x_0} - x_0\partial_{y_{n_1+1}})\left(\sum_{i=0}^{\infty}\frac{(-2)^i x_0^{2i}}{(2i)!}\tilde{\Delta}^i(\eta^\ell(x_{n_1}^{k_1}x_{n_1+1}^{k_2}))\right)\end{aligned}$$

$$= \sum_{i=1}^{\infty}\frac{(-2)^i x_0^{2i-1}}{(2i-1)!}x_{n_1+1}\tilde{\Delta}^i(\eta^\ell(x_{n_1}^{k_1}x_{n_1+1}^{k_2})) - \sum_{i=0}^{\infty}\frac{(-2)^i x_0^{2i+1}}{(2i)!}\tilde{\Delta}^i(\partial_{y_{n_1+1}}(\eta^\ell(x_{n_1}^{k_1}x_{n_1+1}^{k_2})))$$

$$= \sum_{i=1}^{\infty}\frac{(-2)^i x_0^{2i-1}}{(2i-1)!}\tilde{\Delta}^i(\eta^\ell(x_{n_1}^{k_1}x_{n_1+1}^{k_2+1})) - \sum_{i=1}^{\infty}\frac{(-2)^i i x_0^{2i-1}}{(2i-1)!}\tilde{\Delta}^{i-1}(\partial_{y_{n_1+1}}\eta^\ell(x_{n_1}^{k_1}x_{n_1+1}^{k_2}))$$

$$-\ell\sum_{i=0}^{\infty}\frac{(-2)^i x_0^{2i+1}}{(2i)!}\tilde{\Delta}^i(\eta^{\ell-1}(x_{n_1}^{k_1}x_{n_1+1}^{k_2+1})))$$

$$= \ell(n_2 - n_1 + \ell + k_2 - k_1)\sum_{i=1}^{\infty}\frac{(-2)^i x_0^{2i-1}}{(2i-1)!}\tilde{\Delta}^{i-1}(\eta^{\ell-1}(x_{n_1}^{k_1}x_{n_1+1}^{k_2+1}))$$

$$+\ell\widehat{T}_1((\eta^{\ell-1}(x_{n_1}^{k_1}x_{n_1+1}^{k_2+1})))$$

$$= \ell[1 - 2(n_2 - n_1 + \ell + k_2 - k_1)]\widehat{T}_1((\eta^{\ell-1}(x_{n_1}^{k_1}x_{n_1+1}^{k_2+1}))) \tag{8.2.80}$$

and

$$(E_{n_1+1,0} - E_{0,n+n_1+1})[\widehat{T}_1(\eta^\ell(x_{n_1}^{k_1}x_{n_1+1}^{k_2}))]$$

$$= (x_{n_1+1}\partial_{x_0} - x_0\partial_{y_{n_1+1}})\left(\sum_{i=0}^{\infty}\frac{(-2)^i x_0^{2i+1}}{(2i+1)!}\tilde{\Delta}^i(\eta^\ell(x_{n_1}^{k_1}x_{n_1+1}^{k_2}))\right)$$

$$= \sum_{i=0}^{\infty}\frac{(-2)^i x_0^{2i}}{(2i)!}x_{n_1+1}\tilde{\Delta}^i(\eta^\ell(x_{n_1}^{k_1}x_{n_1+1}^{k_2})) - \sum_{i=0}^{\infty}\frac{(-2)^i x_0^{2i+2}}{(2i+1)!}\tilde{\Delta}^i(\partial_{y_{n_1+1}}(\eta^\ell(x_{n_1}^{k_1}x_{n_1+1}^{k_2})))$$

$$= \sum_{i=0}^{\infty}\frac{(-2)^i x_0^{2i}}{(2i)!}\tilde{\Delta}^i(\eta^\ell(x_{n_1}^{k_1}x_{n_1+1}^{k_2+1})) + \sum_{i=1}^{\infty}\frac{(-2)^{i-1} x_0^{2i}}{(2i-1)!}\tilde{\Delta}^{i-1}(\partial_{y_{n_1+1}}(\eta^\ell(x_{n_1}^{k_1}x_{n_1+1}^{k_2})))$$

$$-\ell\sum_{i=0}^{\infty}\frac{(-2)^i x_0^{2i+2}}{(2i+1)!}\tilde{\Delta}^i(\eta^{\ell-1}(x_{n_1}^{k_1}x_{n_1+1}^{k_2+1}))) = \widehat{T}_0(\eta^\ell(x_{n_1}^{k_1}x_{n_1+1}^{k_2+1})). \tag{8.2.81}$$

By (8.2.78), (8.2.80) and (8.2.81), we get

$$\widehat{T}_0(x_{n_1}^{k_1}x_{n_1+1}^{k_2}) \in U \qquad \text{for some } k_1, k_2 \in \mathbb{N} \text{ such that } k_2 - k_1 = k. \tag{8.2.82}$$

Observe

$$(E_{n_1,n_1+1} - E_{n+n_1+1,n+n_1})_{\mathscr{B}'} = \partial_{x_{n_1}}\partial_{x_{n_1+1}} - y_{n_1+1}\partial_{y_{n_1}} \tag{8.2.83}$$

by (8.2.1)–(8.2.3). Repeatedly applying (8.2.83) to (8.2.82), we obtain $x_{n_1+1}^k = T_0(x_{n_1+1}^k) \in U$. Thanks to Lemma 8.2.1, $U = \mathscr{H}'_{\langle k\rangle}$. So $\mathscr{H}'_{\langle k\rangle}$ is an irreducible $o(2n+1,\mathbb{F})$-submodule.

Let V be any nonzero $o(2n+1,\mathbb{F})$-submodule of $\mathscr{H}'_{\langle -k\rangle}$ with $k \in \mathbb{N}+1$. By the above arguments, $x_{n_1}^k \in V$. Thanks to Lemma 8.2.1, $V = \mathscr{H}'_{\langle -k\rangle}$. So $\mathscr{H}'_{\langle -k\rangle}$ is an irreducible $o(2n+1,\mathbb{F})$-submodule.

Case 2. $n_1 = n_2$.
In this case

$$\eta = \sum_{i=1}^{n_1} y_i \partial_{x_i} + \sum_{s=n_1+1}^{n} x_s \partial_{y_s}. \tag{8.2.84}$$

Thus

$$x_{n_1}^{m_1} y_{n_1}^{m_2} \zeta_1^{m_3+1} = \frac{1}{\prod_{i=1}^{m_2}(m_1+i)} \eta^{m_2}(x_{n_1}^{m_1+m_2}\zeta_1^{m_3+1}) \tag{8.2.85}$$

and

$$x_{n_1+1}^{m_1} y_{n_1+1}^{m_2} \zeta_2^{m_3+1} = \frac{1}{\prod_{i=1}^{m_1}(m_2+i)} \eta^{m_1}(y_{n_1+1}^{m_1+m_2}\zeta_2^{m_3+1}). \tag{8.2.86}$$

First we consider $\mathscr{H}'_{\langle k\rangle}$ with $k \in \mathbb{N}+1$. Let U be any nonzero $o(2n+1,\mathbb{F})$-submodule of $\mathscr{H}'_{\langle k\rangle}$. Then U must contain a $\mathscr{K}$-singular vector. Moreover, (8.2.76) becomes

$$(E_{n+n_1,n_1+1} - E_{n+n_1+1,n_1})_{\mathscr{B}'} = y_{n_1}\partial_{x_{n_1+1}} + x_{n_1}\partial_{y_{n_1+1}}. \tag{8.2.87}$$

If $\widehat{T}_\iota(\eta^\ell(x_{n_1}^{k_1} y_{n_1+1}^{k_2})) \in U$, then

$$\frac{1}{k_2!}(E_{n+n_1,n_1+1} - E_{n+n_1+1,n_1})^{k_2}[\widehat{T}_\iota(\eta^\ell(x_{n_1}^{k_1} y_{n_1+1}^{k_2}))] = \widehat{T}_\iota(\eta^\ell(x_{n_1}^{k_1+k_2})) \in U. \tag{8.2.88}$$

When $\widehat{T}_\iota(\eta^\ell(x_{n_1}^{k_1}\zeta_1^{k_2})) \in U$, we have $\widehat{T}_\iota(\eta^\ell(x_{n_1}^{k_1})) \in U$ by (7.2.40). Symmetrically, we have $\widehat{T}_\iota(\eta^\ell(y_{n_1+1}^{k_1})) \in U$ if $\widehat{T}_\iota(\eta^\ell(y_{n_1+1}^{k_1}\zeta_2^{k_2}) \in U$, and (8.2.88) implies $0 \neq \widehat{T}_\iota(\eta^\ell(x_{n_1}^{k_1})) \in U$. By (6.6.18), (6.6.29) and (6.6.32), we always have some $\widehat{T}_\iota(\eta^\ell(x_{n_1}^{k_1})) \in U$. According to (8.2.84),

$$\eta^m(x_{n_1}^{m_1}) = 0 \qquad \text{for } m, m_1 \in \mathbb{N} \text{ such that } m > m_1. \tag{8.2.89}$$

Thus $\ell \leq k_1$.

Note that

$$(E_{n_1,0} - E_{0,n+n_1})|_{\mathscr{B}'} = \partial_{x_0}\partial_{x_{n_1}} - x_0\partial_{y_{n_1}} \tag{8.2.90}$$

by (8.2.6) and (8.2.7). Moreover, (8.2.37), (8.2.84) and (8.2.90) imply that

$$\begin{aligned}
&(E_{n_1,0} - E_{0,n+n_1})[\widehat{T}_0(\eta^\ell(x_{n_1}^{k_1}))] \\
&= (\partial_{x_0}\partial_{x_{n_1}} - x_0\partial_{y_{n_1}})\left(\sum_{i=0}^{\infty}\frac{(-2)^i x_0^{2i}}{(2i)!}\tilde{\Delta}^i(\eta^\ell(x_{n_1}^{k_1}))\right) \\
&= -\sum_{i=1}^{\infty}\frac{(-2)^i i x_0^{2i-1}}{(2i-1)!}\tilde{\Delta}^{i-1}(\partial_{y_{n_1}}(\eta^\ell(x_{n_1}^{k_1}))) + k_1\sum_{i=1}^{\infty}\frac{(-2)^i x_0^{2i-1}}{(2i-1)!}\tilde{\Delta}^i(\eta^\ell(x_{n_1}^{k_1-1})))
\end{aligned}$$

$$-\sum_{i=0}^{\infty}\frac{(-2)^i x_0^{2i+1}}{(2i)!}\tilde{\Delta}^i(\partial_{y_{n_1}}(\eta^\ell(x_{n_1}^{k_1})))$$
$$= \ell k_1\sum_{i=0}^{\infty}\frac{(-2)^i x_0^{2i+1}}{(2i+1)!}\tilde{\Delta}^i(\eta^{\ell-1}(x_{n_1}^{k_1-1})) - 2\ell k_1(\ell-k_1)\sum_{i=0}^{\infty}\frac{(-2)^i x_0^{2i+1}}{(2i+1)!}\tilde{\Delta}^i(\eta^{\ell-1}(x_{n_1}^{k_1-1})))$$
$$= \ell k_1[1-2(\ell-k_1)]\widehat{T}_1(\eta^{\ell-1}(x_{n_1}^{k_1-1})), \tag{8.2.91}$$

$$(E_{n_1,0}-E_{0,n+n_1})[\widehat{T}_1(\eta^\ell(x_{n_1}^{k_1}))]$$
$$= (\partial_{x_0}\partial_{x_{n_1}} - x_0\partial_{y_{n_1}})\left(\sum_{i=0}^{\infty}\frac{(-2)^i x_0^{2i+1}}{(2i+1)!}\tilde{\Delta}^i(\eta^\ell(x_{n_1}^{k_1}))\right)$$
$$= \sum_{i=1}^{\infty}\frac{(-2)^{i-1}x_0^{2i}}{(2i-1)!}\tilde{\Delta}^{i-1}(\partial_{y_{n_1}}(\eta^\ell(x_{n_1}^{k_1}))) + \sum_{i=0}^{\infty}\frac{(-2)^i x_0^{2i}}{(2i)!}\tilde{\Delta}^i(\partial_{x_{n_1}}(\eta^\ell(x_{n_1}^{k_1})))$$
$$-\sum_{i=0}^{\infty}\frac{(-2)^i x_0^{2i+2}}{(2i+1)!}\tilde{\Delta}^i(\partial_{y_{n_1}}(\eta^\ell(x_{n_1}^{k_1}))) = k_1\widehat{T}_0(\eta^\ell(x_{n_1}^{k_1-1})). \tag{8.2.92}$$

Therefore, we have

$$\frac{1}{(k-1)!}\widehat{T}_1(\eta^{k-1}(x_{n_1}^{k-1})) = \widehat{T}_1(y_{n_1}^{k-1}) \in U. \tag{8.2.93}$$

By Lemma 8.2.2, $U = \mathscr{H}'_{\langle k\rangle}$. Thus $\mathscr{H}'_{\langle k\rangle}$ is an irreducible $o(2n+1,\mathbb{F})$-submodule.

Let V be any nonzero $o(2n+1,\mathbb{F})$-submodule of $\mathscr{H}'_{\langle -k\rangle}$ with $k\in\mathbb{N}$. By the above arguments (8.2.84)–(8.2.92), $x_{n_1}^k \in V$. Thanks to Lemma 8.2.2, $V = \mathscr{H}'_{\langle -k\rangle}$. So $\mathscr{H}'_{\langle -k\rangle}$ is an irreducible $o(2n+1,\mathbb{F})$-submodule.

Fix $k\in\mathbb{Z}$. Since $\tilde{\Delta}'$ in (8.2.8) is locally nilpotent, for any $0\neq u\in\mathscr{B}'_{\langle k\rangle}$, there exists an element $\kappa(u)\in\mathbb{N}$ such that

$$(\tilde{\Delta}')^{\kappa(u)}(u)\neq 0 \;\text{ and }\; (\tilde{\Delta}')^{\kappa(u)+1}(u)=0. \tag{8.2.94}$$

Set

$$W = \sum_{i=0}^{\infty}\eta'^i(\mathscr{H}'_{\langle k-2i\rangle}). \tag{8.2.95}$$

Given $0\neq u\in\mathscr{B}'_{\langle k\rangle}$, $\kappa(u)=1$ implies $u\in\mathscr{H}'_{\langle k\rangle}\subset W$. Suppose that $u\in W$ whenever $\kappa(u)<r$ for some positive integer r. Assume $\kappa(u)=r$. First

$$v = (\tilde{\Delta}')^r(u)\in\mathscr{H}'_{\langle k-2r\rangle}\subset W. \tag{8.2.96}$$

Note that $[\partial_{x_0}^2, x_0^2] = 2+4x_0\partial_{x_0}$ and so

$$(\tilde{\Delta}')^r[\eta'^r(v)] = 2^r r! \left[\prod_{i=1}^{r}[1+2(n_1-n_2-k+r+i)]\right] v \tag{8.2.97}$$

by (7.2.8) and (8.2.14).

Thus we have either

$$u = \frac{1}{2^r r! \prod_{i=1}^{r}[1+2(n_1-n_2-k+r+i)]}\eta'^r(v) \in W \tag{8.2.98}$$

or

$$\kappa\left(u - \frac{1}{2^r r! \prod_{i=1}^{r}[1+2(n_1-n_2-k+r+i)]}\eta'^r(v)\right) < r. \tag{8.2.99}$$

By induction,

$$u - \frac{1}{2^r r! \prod_{i=1}^{r}[1+2(n_1-n_2-k+r+i)]}\eta'^r(v) \in W \tag{8.2.100}$$

which implies $u \in W$. Therefore, we have $W = \mathscr{B}'_{\langle k\rangle}$. Since the weight of any $\mathscr{K}$-singular vector in $\eta'^i(\mathscr{H}'_{\langle k-2i\rangle})$ is different from the weight of any $\mathscr{K}$-singular vector in $\eta'^j(\mathscr{H}'_{\langle k-2j\rangle})$ when $i \neq j$, the sum $\mathscr{B}'_{\langle k\rangle} = \sum_{i=0}^{\infty}\eta'^i(\mathscr{H}'_{\langle k-2i\rangle})$ is a direct sum. Thus $\mathscr{B}' = \bigoplus_{k\in\mathbb{Z}}\mathscr{B}'_{\langle k\rangle} = \bigoplus_{i,r=0}^{\infty}\eta'^i(\mathscr{H}'_{\langle r\rangle})$. This completes the proof of Theorem 8.2.3. □

Suppose $n_2 = n$ and $k \in \mathbb{N}$. Then $x_{n_1}^k$ is a highest-weight vector in $\mathscr{H}'_{\langle -k\rangle}$ with weight $-k\varepsilon_{n_1} - \sum_{i=1}^{n_1}\varepsilon_i$ (cf. (7.1.3)). If $n_1 < n$, $x_{n_1+1}^k$ is a highest-weight vector in $\mathscr{H}'_{\langle k\rangle}$ with weight $k\varepsilon_{n_1+1} - \sum_{i=1}^{n_1}\varepsilon_i$. When $n_1 = n$, $\widehat{T}_1(y_n^k)$ is a highest-weight vector in $\mathscr{H}'_{\langle k+1\rangle}$ with weight $-k\varepsilon_n - \sum_{i=1}^{n}\varepsilon_i$.

8.3 Extensions of the Conformal Representation

This section is to study extensions of the conformal representation of the odd orthogonal Lie algebra $o(2n+3,\mathbb{C})$ via $o(2n+1,\mathbb{C})$-modules.

For $\lambda \in \Lambda^+$, we denote by $V(\lambda)$ the finite-dimensional irreducible $o(2n+1,\mathbb{C})$-module with highest weight λ. The $(2n+1)$-dimensional natural module of $o(2n+1,\mathbb{C})$ is $V(\varepsilon_1)$ with weights $\{0, \pm\varepsilon_i \mid i \in \overline{1,n}\}$. The following result can be derived from Theorem 5.4.3:

Lemma 8.3.1 (Pieri's formula) *Given $\mu \in \Lambda^+$ with $\mathscr{S}(\mu) = \{i_1, i_2, \ldots, i_{s+1}\}$ (cf. (6.7.55) and (6.7.56)),*

$$V(\varepsilon_1)\otimes_{\mathbb{C}} V(\mu)\cong(1-\delta_{\mu_n,0})V(\mu)\oplus\bigoplus_{\iota=1}^{s-\delta_{\mu_n,0}-\delta_{\mu_n,1/2}} V(\mu-\varepsilon_{i_{\iota+1}-1}))\oplus\bigoplus_{r=1}^{s} V(\mu+\varepsilon_{i_r}). \tag{8.3.1}$$

Note that the Casimir element of $o(2n+1,\mathbb{C})$ is

$$\begin{aligned}\omega=&\sum_{1\le i<j\le n}[(E_{i,n+j}-E_{j,n+i})(E_{n+j,i}-E_{n+i,j})+(E_{n+j,i}-E_{n+i,j})(E_{i,n+j}-E_{j,n+i})]\\&+\sum_{i=1}^{n}[(E_{0,i}-E_{n+i,0})(E_{i,0}-E_{0,n+i})+(E_{i,0}-E_{0,n+i})(E_{0,i}-E_{n+i,0})]\\&+\sum_{i,j=1}^{n}(E_{i,j}-E_{n+j,n+i})(E_{j,i}-E_{n+i,n+j})\in U(o(2n+1,\mathbb{C})).\end{aligned} \tag{8.3.2}$$

Recall the operator $\mathfrak{d}$ defined in (7.3.7). Set

$$\tilde{\omega}=\frac{1}{2}(\mathfrak{d}(\omega)-\omega\otimes 1-1\otimes\omega)\in U(o(2n+1,\mathbb{C}))\otimes_{\mathbb{C}} U(o(2n+1,\mathbb{C})). \tag{8.3.3}$$

By (8.3.2),

$$\begin{aligned}\tilde{\omega}=&\sum_{1\le i<j\le n}[(E_{i,n+j}-E_{j,n+i})\otimes(E_{n+j,i}-E_{n+i,j})+(E_{n+j,i}-E_{n+i,j})\otimes(E_{i,n+j}-E_{j,n+i})]\\&+\sum_{i=1}^{n}[(E_{0,i}-E_{n+i,0})\otimes(E_{i,0}-E_{0,n+i})+(E_{i,0}-E_{0,n+i})\otimes(E_{0,i}-E_{n+i,0})]\\&+\sum_{i,j=1}^{n}(E_{i,j}-E_{n+j,n+i})\otimes(E_{j,i}-E_{n+i,n+j}).\end{aligned} \tag{8.3.4}$$

Moreover, (7.3.11) also holds. Take the settings in (7.3.12) and (7.3.13). Denote

$$\ell(\mu)=\dim V(\mu). \tag{8.3.5}$$

Recall

$$\rho=\frac{1}{2}\sum_{\nu\in\Phi_{B_n}^{+}}\nu. \tag{8.3.6}$$

Then

$$\frac{2(\rho,\nu)}{(\nu,\nu)}=1\qquad\text{for }\ \nu\in\Pi_{B_n} \tag{8.3.7}$$

by (3.2.17). Expression (8.1.3) yields

$$\rho=\sum_{i=1}^{n-1}(n-i+1/2)\varepsilon_i. \tag{8.3.8}$$

Observe that

$$(\mu+2\rho,\mu)-(\mu+2\rho,\mu)-(\varepsilon_1+2\rho,\varepsilon_1)=-2n, \tag{8.3.9}$$

$$(\mu+\varepsilon_i+2\rho,\mu+\varepsilon_i)-(\mu+2\rho,\mu)-(\varepsilon_1+2\rho,\varepsilon_1)=2(\mu_i+1-i) \tag{8.3.10}$$

and

$$(\mu-\varepsilon_i+2\rho,\mu-\varepsilon_i)-(\mu+2\rho,\mu)-(\varepsilon_1+2\rho,\varepsilon_1)=2(i-2n-\mu_i) \tag{8.3.11}$$

for $\mu=\sum_{r=1}^n \mu_r\varepsilon_r$ by (8.3.8). Moreover, the algebra $U(o(2n+1,\mathbb{C}))\otimes_{\mathbb{C}} U(o(2n+1,\mathbb{C}))$ acts on $V(\varepsilon_1)\otimes_{\mathbb{C}} V(\mu)$ by

$$(\xi_1\otimes\xi_2)(v\otimes u)=\xi_1(v)\otimes\xi_2(u)\ \text{for}\ \xi_1,\xi_2\in U(o(2n+1,\mathbb{C})),\ v\in V(\varepsilon_1),\ u\in V(\mu). \tag{8.3.12}$$

Note that (7.3.11) also holds for $o(2n+1,\mathbb{C})$. By Lemma 8.3.1, (7.3.11), (8.3.3) and (8.3.9)–(8.3.11), we get:

Lemma 8.3.2 *Let $\mu=\sum_{i=1}^n \mu_i\varepsilon_i\in\Lambda^+$ with $\mathscr{S}(\mu)=\{i_1,i_2,\dots,i_{s+1}\}$ (cf. (6.7.55) and (6.7.56)). The characteristic polynomial of $\tilde{\omega}|_{V(\varepsilon_1)\otimes_{\mathbb{C}}V(\mu)}$ is*

$$(t+n)^{\ell(\mu)(1-\delta_{\mu_n,0})}\left[\prod_{\iota=1}^{s-\delta_{\mu_n,0}-\delta_{\mu_n,1/2}}(t+\mu_{i_\iota}+2n-i_{\iota+1})^{\ell^-_{i_{\iota+1}-1}(\mu)}\right]\times\left[\prod_{r=1}^{s-1}(t-\mu_{i_r}+i_r-1)^{\ell^+_{i_r}(\mu)}\right]. \tag{8.3.13}$$

We remark that the above lemma is also equivalent to a special detailed version of Kostant's characteristic identity.

Set

$$\mathscr{A}=\mathbb{C}[x_0,x_1,x_2,\dots,x_{2n}]. \tag{8.3.14}$$

Then $\mathscr{A}$ forms an $o(2n+1,\mathbb{C})$-module with the action determined via

$$E_{i,j}|_{\mathscr{A}}=x_i\partial_{x_j}\qquad\text{for}\ \ i,j\in\overline{0,2n}. \tag{8.3.15}$$

The corresponding Laplace operator, its dual invariant and the degree operator are

$$\Delta=\partial_{x_0}^2+2\sum_{r=1}^n\partial_{x_r}\partial_{x_{n+r}},\quad \eta=\frac{1}{2}x_0^2+\sum_{i=1}^n x_ix_{n+i},\quad D=\sum_{r=0}^{2n}x_r\partial_{x_r}, \tag{8.3.16}$$

respectively. Denote

$$A_{i,j} = E_{i,j} - E_{n+1+j,n+1+i}, \quad B_{i,j} = E_{i,n+1+j} - E_{j,n+1+i}, \quad C_{i,j} = E_{n+1+i,j} - E_{n+1+j,i}, \tag{8.3.17}$$

$$K_i = E_{0,i} - E_{n+1+i,0}, \qquad K_{n+1+i} = E_{0,n+1+i} - E_{i,0} \tag{8.3.18}$$

for $i, j \in \overline{1, n+1}$. Then the split odd orthogonal Lie algebra

$$o(2n+3, \mathbb{C}) = \sum_{i,j=1}^{n+1} \mathbb{C}A_{i,j} + \sum_{1 \le i < j \le n+1} (\mathbb{C}B_{i,j} + \mathbb{C}C_{j,i}) + \sum_{i=1}^{n+1} (\mathbb{C}K_i + \mathbb{C}K_{n+1+i}). \tag{8.3.19}$$

Replace n by $2n+1$ and take the product $(\vec{x}, \vec{y}) = (1/2)(x_0 y_0 + \sum_{i=1}^{n} (x_i y_{n+i} + x_{n+i} y_i)$ in (7.3.1). Using the derivatives of one-parameter conformal transformations just as one derives the Lie algebra of a Lie group, we get the following conformal representation of $o(2n+3, \mathbb{C})$ determined by

$$A_{i,j}|_{\mathscr{A}} = x_i \partial_{x_j} - x_{n+j} \partial_{x_{n+i}}, \qquad B_{i,j}|_{\mathscr{A}} = x_i \partial_{x_{n+j}} - x_j \partial_{x_{n+i}}, \tag{8.3.20}$$

$$C_{i,j}|_{\mathscr{A}} = x_{n+i} \partial_{x_j} - x_j \partial_{x_{n+i}}, \qquad A_{n+1,n+1}|_{\mathscr{A}} = -D, \tag{8.3.21}$$

$$A_{n+1,i}|_{\mathscr{A}} = \partial_{x_i}, \qquad B_{i,n+1}|_{\mathscr{A}} = -\partial_{x_{n+i}}, \tag{8.3.22}$$

$$A_{i,n+1}|_{\mathscr{A}} = -x_i D + \eta \partial_{x_{n+i}}, \quad C_{n+1,i}|_{\mathscr{A}} = x_{n+i} D - \eta \partial_{x_i}, \tag{8.3.23}$$

$$K_i|_{\mathscr{A}} = x_0 \partial_{x_i} - x_{n+i} \partial_{x_0}, \qquad K_{n+1+i}|_{\mathscr{A}} = x_0 \partial_{x_{n+i}} - x_i \partial_{x_0}, \tag{8.3.24}$$

$$K_{2n+2}|_{\mathscr{A}} = -\partial_{x_0}, \qquad K_{n+1}|_{\mathscr{A}} = x_0 D - \eta \partial_{x_0} \tag{8.3.25}$$

for $i, j \in \overline{1, n}$.

Note that

$$\mathscr{L} = \sum_{i,j=1}^{n} \mathbb{C}A_{i,j} + \sum_{1 \le i < j \le n} (\mathbb{C}B_{i,j} + \mathbb{C}C_{j,i} + \sum_{i=1}^{n} (\mathbb{C}K_i + \mathbb{C}K_{n+1+i})) \tag{8.3.26}$$

forms a Lie subalgebra of $o(2n+3, \mathbb{C})$ that is isomorphic to $o(2n+1, \mathbb{C})$. For convenience, we make the identification of $o(2n+1, \mathbb{C})$ with $\mathscr{L}$ as follows:

$$E_{i,j} - E_{n+j,n+i} \leftrightarrow A_{i,j}, \quad E_{i,n+j} - E_{j,n+i} \leftrightarrow B_{i,j}, \quad E_{n+i,j} - E_{n+j,i} \leftrightarrow C_{i,j}, \tag{8.3.27}$$

$$E_{0,i} - E_{n+i,0} \leftrightarrow K_i, \qquad E_{0,n+i} - E_{i,0} \leftrightarrow K_{n+1+i} \tag{8.3.28}$$

for $i, j \in \overline{1, n}$. Recall the Witt algebra $\mathbb{W}_{2n+1} = \sum_{i=0}^{2n} \mathscr{A} \partial_{x_i}$, and Shen's monomorphism $\Im : \mathbb{W}_{2n+1} \to \widehat{\mathbb{W}}_{2n+1} = \mathbb{W}_{2n+1} \oplus gl(2n+1, \mathscr{A})$ (cf. (6.7.13)) given by

$$\Im\left(\sum_{i=0}^{2n} f_i\partial_{x_i}\right)=\sum_{i=0}^{2n} f_i\partial_{x_i}\oplus\sum_{i,j=0}^{n}\partial_{x_i}(f_j)E_{i,j}. \tag{8.3.29}$$

Now we have the Lie algebra monomorphism $\nu : o(2n+3,\mathbb{C})\to\widehat{\mathbb{W}}_{2n+1}$ given by

$$\nu(\xi)=\Im(\xi|_{\mathscr{A}})\qquad\text{for }\ \xi\in o(2n+3,\mathbb{C}). \tag{8.3.30}$$

Denote

$$I_{2n+1}=\sum_{r=0}^{2n}E_{r,r}. \tag{8.3.31}$$

According to the identification (8.3.27) and (8.3.28), we have

$$\nu(A_{i,j})=(x_i\partial_{x_j}-x_{n+j}\partial_{x_{n+i}})\oplus A_{i,j},\quad \nu(B_{i,j})=(x_i\partial_{x_{n+j}}-x_j\partial_{x_{n+i}})\oplus B_{i,j}, \tag{8.3.32}$$

$$\nu(C_{i,j})=(x_{n+i}\partial_{x_j}-x_{n+j}\partial_{x_i})\oplus C_{i,j},\quad \nu(A_{n+1,n+1})=-(D\oplus I_{2n+1}), \tag{8.3.33}$$

$$\nu(A_{n+1,i})=\partial_{x_i},\qquad \nu(B_{i,n+1})=-\partial_{x_{n+i}},\qquad \nu(K_{2n+2})=-\partial_{x_0}, \tag{8.3.34}$$

$$\nu(A_{i,n+1})=(-x_iD+\eta\partial_{x_{n+i}})\oplus\left[\sum_{p=1}^{n}(x_{n+p}B_{p,i}-x_pA_{i,p})-x_iI_{2n+1}+x_0K_{n+1+i}\right], \tag{8.3.35}$$

$$\nu(C_{n+1,i})=(x_{n+i}D-\eta\partial_{x_i})\oplus\left[\sum_{p=1}^{n}(x_pC_{i,p}-x_{n+p}A_{p,i})+x_{n+i}I_{2n+1}-x_0K_i\right], \tag{8.3.36}$$

$$\nu(K_i)=(x_0\partial_{x_i}-x_{n+i}\partial_{x_0})\oplus K_i,\qquad \nu(K_{n+1+i})=(x_0\partial_{x_{n+i}}-x_i\partial_{x_0})\oplus K_{n+i+i}, \tag{8.3.37}$$

$$\nu(K_{n+1})=(x_0D-\eta\partial_{x_0})\oplus\left[\sum_{s=1}^{n}(x_sK_s+x_{n+s}K_{n+1+s})+x_0I_{2n+1}\right] \tag{8.3.38}$$

for $i, j\in\overline{1,n}$.

Observe that

$$\widehat{\mathbb{W}}^o_{2n+1}=\widehat{\mathbb{W}}_{2n+1}\oplus[o(2n+1,\mathscr{A})+\mathscr{A}I_{2n+1}] \tag{8.3.39}$$

form a Lie subalgebra of $\widehat{\mathbb{W}}_{2n+1}$ and

$$\nu(o(2n+3,\mathbb{C})) \subset \widehat{\mathbb{W}}^o_{2n+1}. \tag{8.3.40}$$

Let M be an $o(2n+1,\mathbb{C})$-module and let $c \in \mathbb{C}$ be a fixed constant. Then

$$\widehat{M} = \mathscr{A} \otimes_{\mathbb{C}} M \tag{8.3.41}$$

becomes a $\widehat{\mathbb{W}}^o_{2n+1}$-module with the action:

$$(d + f_1 A + f_2 I_{2n})(g \otimes v) = (d(g) + c f_2 g) \otimes v + f_1 g \otimes A(v) \tag{8.3.42}$$

for $f_1, f_2, g \in \mathscr{A}$, $A \in o(2n+1,\mathbb{C})$ and $v \in M$. Moreover, we make $\widehat{M}$ an $o(2n+3,\mathbb{C})$-module with the action:

$$\xi(\varpi) = \nu(\xi)(\varpi) \qquad \text{for } \xi \in o(2n+3,\mathbb{C}),\ \varpi \in \widehat{M}. \tag{8.3.43}$$

Denote

$$\mathscr{G} = o(2n+3,\mathbb{C}), \qquad \mathscr{G}_0 = o(2n+1,\mathbb{C}) + \mathbb{C}A_{n+1,n+1} \tag{8.3.44}$$

$$\mathscr{G}_+ = \mathbb{C}K_{2n+2} + \sum_{i=1}^{n}(\mathbb{C}A_{n+1,i} + \mathbb{C}B_{i,n+1}),\ \ \mathscr{G}_- = \mathbb{C}K_{n+1}\sum_{i=1}^{n}(\mathbb{C}A_{i,n+1} + \mathbb{C}C_{n+1,i}) \tag{8.3.45}$$

(cf. (8.3.27) and (8.3.28)). Then $\mathscr{G}_\pm$ are abelian Lie subalgebras of $\mathscr{G}$ and $\mathscr{G}_0$ is a Lie subalgebra of $\mathscr{G}$. Moreover,

$$\mathscr{G} = \mathscr{G}_- + \mathscr{G}_0 + \mathscr{G}_+, \qquad [\mathscr{G}_0, \mathscr{G}_\pm] \subset \mathscr{G}_\pm. \tag{8.3.46}$$

By (8.3.32)–(8.3.38),

$$\mathscr{G}_+(1 \otimes M) = \{0\},\ \ U(\mathscr{G}_0)(1 \otimes M) = 1 \otimes M. \tag{8.3.47}$$

Thus

$$U(\mathscr{G})(1 \otimes M) = U(\mathscr{G}_-)(1 \otimes M). \tag{8.3.48}$$

Using (8.3.34), we can prove:

Proposition 8.3.3 *The map $M \mapsto U(\mathscr{G}_-)(1 \otimes M)$ gives rise to a functor from the category of $o(2n+1,\mathbb{C})$-modules to the category of $o(2n+3,\mathbb{C})$-modules. In particular, it maps irreducible $o(2n+1,\mathbb{C})$-modules to irreducible $o(2n+3,\mathbb{C})$-modules.*

From another point view, $U(\mathscr{G}_-)(1\otimes M)$ is a polynomial extension from $o(2n+1,\mathbb{C})$-module M to an $o(2n+3,\mathbb{C})$-module. Next we want to study when $\widehat{M}=U(\mathscr{G}_-)(1\otimes M)$. Write

$$x^\alpha=\prod_{i=0}^{2n}x_i^{\alpha_i},\quad E^\alpha=K_{n+1}^{\alpha_0}\prod_{r=1}^{n}[(-A_{r,n+1})^{\alpha_r}C_{n+1,r}^{\alpha_{n+r}}] \tag{8.3.49}$$

for $\alpha=(\alpha_0,\alpha_1,\ldots,\alpha_{2n})\in\mathbb{N}^{2n+1}$. For $k\in\mathbb{N}$, we set

$$\mathscr{A}_k=\text{Span}\{x^\alpha\mid\alpha\in\mathbb{N}^{2n+1},\ |\alpha|=k\},\quad \widehat{M}_k=\mathscr{A}_k\otimes_{\mathbb{C}}M \tag{8.3.50}$$

(recall $|\alpha|=\sum_{i=0}^{2n}\alpha_i$) and

$$\begin{aligned}(U(\mathscr{G}_-)(1\otimes M))_k&=U(\mathscr{G}_-)(1\otimes M)\bigcap\widehat{M}_k\\&=\text{Span}\{E^\alpha(1\otimes M)\mid\alpha\in\mathbb{N}^{2n+1},\ |\alpha|=k\}\end{aligned} \tag{8.3.51}$$

by (8.3.36)–(8.3.38). Moreover,

$$(U(\mathscr{G}_-)(1\otimes M))_0=\widehat{M}_0=1\otimes M. \tag{8.3.52}$$

Furthermore,

$$\widehat{M}=\bigoplus_{k=0}^{\infty}\widehat{M}_k,\qquad U(\mathscr{G}_-)(1\otimes M)=\bigoplus_{k=0}^{\infty}(U(\mathscr{G}_-)(1\otimes M))_k. \tag{8.3.53}$$

According to (8.3.23) and (8.3.25), we define a $\mathscr{G}_0$-module homomorphism $\varphi:\widehat{M}\to U(\mathscr{G}_-)(1\otimes M)$ by

$$\varphi(x^\alpha\otimes v)=E^\alpha(1\otimes v)\qquad\text{for }\ \alpha\in\mathbb{N}^{2n+1},\ v\in M. \tag{8.3.54}$$

Then we have

$$\varphi(\widehat{M}_k)=(U(\mathscr{G}_-)(1\otimes M))_k\qquad\text{for }\ k\in\mathbb{N}. \tag{8.3.55}$$

Under the identification (8.3.27) and (8.3.28), we have:

Lemma 8.3.4 *We have* $\varphi|_{\widehat{M}_1}=(c+\tilde{\omega})|_{\widehat{M}_1}$ *(cf. (8.3.2)–(8.3.4)).*

Proof Recall the identification (8.3.27) and (8.3.28). Moreover, $\widehat{M}_1=\mathscr{A}_1\otimes_{\mathbb{C}}M$. Let $i\in\overline{1,n}$ and $v\in M$. Expressions (8.3.38), (8.3.49) and (8.3.54) give

$$\varphi(x_0\otimes v)=\sum_{s=1}^{n}[x_s\otimes K_s(v)+x_{n+s}\otimes K_{n+1+s}(v)]+cx_0\otimes v. \tag{8.3.56}$$

Moreover, (8.3.35), (8.3.49) and (8.3.54) imply

$$\varphi(x_i \otimes v) = \sum_{p=1}^{n} x_{n+p} \otimes B_{i,p}(v) - x_0 \otimes K_{n+1+i}(v) + \sum_{q=1}^{n} x_q \otimes A_{i,q}(v) + cx_i \otimes v. \tag{8.3.57}$$

for $i \in \overline{1,n}$. Furthermore, (8.3.36), (8.3.49) and (8.3.54) yield

$$\varphi(x_{n+i} \otimes v) = -\sum_{p=1}^{n} x_{n+p} \otimes A_{p,i}(v) - x_0 \otimes K_i(v) + \sum_{q=1}^{n} x_q \otimes C_{i,q}(v) + cx_{n+i} \otimes v \tag{8.3.58}$$

for $i \in \overline{1,n}$.

On the other hand, (8.3.4), (8.3.15), (8.3.27) and (8.3.28) yield

$$\tilde{\omega}(x_0 \otimes v) = \sum_{i=1}^{n} [x_{n+i} \otimes K_{n+1+i}(v) + x_i \otimes K_i(v)], \tag{8.3.59}$$

$$\tilde{\omega}(x_i \otimes v) = \sum_{p=1}^{n} x_{n+p} \otimes B_{i,p}(v) - x_0 \otimes K_{n+1+i}(v) + \sum_{r=1}^{n} x_r \otimes A_{i,r}(v), \tag{8.3.60}$$

$$\tilde{\omega}(x_{n+i} \otimes v) = \sum_{p=1}^{n} x_p \otimes C_{i,p}(v) - x_0 \otimes K_i(v) - \sum_{s=1}^{n} x_{n+s} \otimes A_{s,i}(v). \tag{8.3.61}$$

Comparing the above six expressions, we get the conclusion in the lemma. □

For $f \in \mathscr{A}$, we define the action

$$f(g \otimes v) = fg \otimes v \qquad \text{for } g \in \mathscr{A},\ v \in M. \tag{8.3.62}$$

Then we have the $o(2n+1, \mathbb{C})$-invariant operator

$$T = \nu(K_{n+1})x_0 + \sum_{i=1}^{n} [\nu(C_{n+1,i})x_i - \nu(A_{i,n+1})x_{n+i}] \tag{8.3.63}$$

on $\widehat{M}$.

Lemma 8.3.5 *We have* $T|_{\widehat{M}_k} = (2c - 2n + k + 2)\eta$.

Proof Let $f \in \mathscr{A}_k$ and $v \in M$. According to (8.3.38),

$$\nu(K_{n+1})x_0(f\otimes v)=x_0\sum_{s=1}^{n}[x_sf\otimes K_s(v)+x_{n+s}f\otimes K_{n+1+s}(v)]$$
$$+[(k+1+c)x_0^2-\eta]f\otimes v-\eta x_0\partial_{x_0}(f)\otimes v. \tag{8.3.64}$$

Moreover, (8.3.35) gives

$$-\nu(A_{i,n+1})x_{n+i}(f\otimes v)$$
$$=x_{n+i}\sum_{p=1}^{n}[x_p\otimes A_{i,p}(v)-x_{n+p}\otimes B_{p,i}(v)]-x_0x_{n+i}f\otimes K_{n+1+i}(v)$$
$$+[(k+1+c)x_ix_{n+i}-\eta]f\otimes v-\eta x_{n+i}\partial_{x_{n+i}}(f)\otimes v \tag{8.3.65}$$

and (8.3.36) yields

$$\nu(C_{n+1,i})x_i(f\otimes v)$$
$$=x_i\sum_{p=1}^{n}[x_p\otimes C_{i,p}(v)-x_{n+p}\otimes A_{p,i}(v)]-x_0x_if\otimes K_i(v)$$
$$+[(k+1+c)x_ix_{n+i}-\eta]f\otimes v-\eta x_i\partial_{x_i}(f)\otimes v \tag{8.3.66}$$

for $i\in\overline{1,n}$. Then the lemma follows from (8.3.63)–(8.3.66) and the skew-symmetry $B_{i,j}=-B_{j,i}$ and $C_{i,j}=-C_{j,i}$. □

For $0\neq\mu=\sum_{i=1}^{n}\mu_i\varepsilon_i\in\Lambda^+$ with $\mathcal{I}(\mu)=\{i_1,i_2,\ldots,i_{s+1}\}$ (cf. (6.7.54) and (6.7.55)), we define

$$\Theta(\mu)=\begin{cases}\emptyset & \text{if } \mu=(\sum_{i=1}^{n}\varepsilon_i)/2,\\ \mu_1+2n-i_2-\mathbb{N} & \text{otherwise.}\end{cases} \tag{8.3.67}$$

Theorem 8.3.6 *For* $0\neq\mu\in\Lambda^+$*, the* $o(2n+3,\mathbb{C})$*-module* $\widehat{V(\mu)}$ *defined by (8.3.32)–(8.3.43) is irreducible if* $c\in\mathbb{C}\setminus\{n-\mathbb{N}/2,\Theta(\mu)\}$

Proof By Lemma 8.3.3, it is enough to prove that the homomorphism φ defined in (8.3.49) and (8.3.54) satisfies $\varphi(\widehat{V(\mu)})=\widehat{V(\mu)}$. According to (8.3.55), we only need to prove

$$\varphi(\widehat{V(\mu)}_k)=\widehat{V(\mu)}_k \tag{8.3.68}$$

for any $k\in\mathbb{N}$. We will prove it by induction on k.

When $k=0$, (8.3.68) holds by the definition (8.3.47). Consider $k=1$. Write $\mu=\sum_{i=1}^{n}\mu_i\varepsilon_i\in\Lambda^+$ with $\mathscr{S}(\mu)=\{i_1,i_2,\ldots,i_{s+1}\}$. According to Lemmas 8.3.2 and 8.3.4 with $M=V(\mu)$, the eigenvalues of $\varphi|_{\widehat{V(\mu)}_{(1)}}$ are among

$$\{c-n,\ c+\mu_{i_r}-i_r+1, c-\mu_{i_r}-2n+i_{r+1}\mid \text{for } i\in\overline{1,s}\}. \tag{8.3.69}$$

Recall that $\mu_r \in \mathbb{N}/2$ for $r \in \overline{1,n}$,

$$\mu_{\iota+1} - \mu_\iota \in \mathbb{N} \quad \text{for} \quad \iota \in \overline{1, n-1} \tag{8.3.70}$$

such that (6.7.55) holds. So

$$-\mu_{i_r} + i_r - 1,\ \mu_{i_r} + 2n - i_{r+1} \in \mu_1 + 2n - i_2 - \mathbb{N} \quad \text{for} \quad i \in \overline{1,s}. \tag{8.3.71}$$

If $c \notin \mu_1 + 2n - n_1 - \mathbb{N}$ and $c \neq n$, then all the eigenvalues of $\varphi|_{\widehat{V(\mu)}_{\langle 1\rangle}}$ are nonzero. In the case $\mu = (\sum_{i=1}^n \varepsilon_i)/2$, the eigenvalues $\varphi|_{\widehat{V(\mu)}_{\langle 1\rangle}}$ are $c - n$ and $c + 1/2$, which are not equal to 0 because of $c \notin n - \mathbb{N}/2$. Thus (8.3.68) holds for $k = 1$.

Suppose that (8.3.68) holds for $k \leq \ell$ with $\ell \geq 1$. Consider $k = \ell + 1$. Note that

$$\begin{aligned}
\varphi(\widehat{V(\mu)}_{\ell+1}) &= \sum_{i=0}^{2n} \varphi(x_i \widehat{V(\mu)}_\ell) \\
&= K_{n+1}[\varphi(\widehat{V(\mu)}_\ell)] + \sum_{i=1}^{n} [A_{i,n+1}[\varphi(\widehat{V(\mu)}_\ell)] + C_{n+1,i}[\varphi(\widehat{V(\mu)}_\ell)]] \\
&= K_{n+1}(\widehat{V(\mu)}_\ell) + \sum_{i=1}^{n} [A_{i,n+1}(\widehat{V(\mu)}_\ell) + C_{n+1,i}(\widehat{V(\mu)}_\ell)]
\end{aligned} \tag{8.3.72}$$

by the inductional assumption. To prove (8.3.68) with $k = \ell + 1$ is equivalent to prove

$$K_{n+1}(\widehat{V(\mu)}_\ell) + \sum_{i=1}^{n} [A_{i,n+1}(\widehat{V(\mu)}_\ell) + C_{n+1,i}(\widehat{V(\mu)}_\ell)] = \widehat{V(\mu)}_{\ell+1}. \tag{8.3.73}$$

For any $u \in \widehat{V(\mu)}_{\ell-1}$, Lemma 8.3.5 says that

$$K_{n+1}(x_0 u) + \sum_{i=1}^{n} [C_{n+1,i}(x_i u) - A_{i,n+1}(x_{n+i} u)] = (2c + 2 - 2n + \ell)\eta u. \tag{8.3.74}$$

Since $c \notin n - \mathbb{N}/2$, $2c + 2 - 2n + \ell \neq 0$ and (8.3.73) gives

$$\eta u \in K_{n+1}(\widehat{V(\mu)}_\ell) + \sum_{i=1}^{n} [A_{i,n+1}(\widehat{V(\mu)}_\ell) + C_{n+1,i}(\widehat{V(\mu)}_\ell)] \quad \text{for} \quad u \in \widehat{V(\mu)}_{\ell-1}. \tag{8.3.75}$$

Let $g \otimes v \in \widehat{V(\mu)}_{\langle \ell \rangle}$. According to (8.3.35), (8.3.36), (8.3.38) and Lemma 8.3.4,

$$\begin{aligned}
-A_{i,n+1}(g\otimes v) &= x_i D(g)\otimes v - \eta\partial_{x_{n+i}}(g)\otimes v + x_i g\otimes I_{2n+1}(v) - x_0 g\otimes K_{n+1+i}(v)\\
&\quad + \sum_{p=1}^{n}(x_{n+p}g\otimes B_{i,n+p}(v) + x_p g\otimes A_{i,p}(v))\\
&= \ell x_i g\otimes v - \eta\partial_{x_{n+i}}(g)\otimes v + g[x_i\otimes I_{2n+1}(v) - x_0 g\otimes K_{n+1+i}(v)\\
&\quad + \sum_{p=1}^{n}(x_{n+p}\otimes B_{i,n+p}(v) + x_p\otimes A_{i,p}(v))]\\
&= \ell x_i g\otimes v - \eta\partial_{x_{n+i}}(g)\otimes v - g[A_{i,n+1}(1\otimes v)]\\
&= \ell x_i g\otimes v - \eta\partial_{x_{n+i}}(g)\otimes v + g\varphi(x_i\otimes v)\\
&= -\eta\partial_{x_{n+i}}(g)\otimes v + g[(\ell + c + \tilde{\omega})(x_i\otimes v)],
\end{aligned} \tag{8.3.76}$$

$$\begin{aligned}
C_{n+i,i}(g\otimes v) &= x_{n+i}D(g)\otimes v - \eta\partial_{x_i}(g)\otimes v + x_{n+i}g\otimes I_{2n+1}(v) - x_0 g\otimes K_i(v)\\
&\quad + \sum_{p=1}^{n}(x_p g\otimes C_{i,p}(v) - x_{n+p}g\otimes A_{p,i}(v))\\
&= \ell x_{n+i}g\otimes v - \eta\partial_{x_i}(g)\otimes v + g[x_{n+i}\otimes I_{2n+1}(v) - x_0 g\otimes K_i(v)\\
&\quad + \sum_{p=1}^{n}(x_p\otimes C_{i,p}(v) - x_{n+p}\otimes A_{p,i}(v))]\\
&= \ell x_{n+i}g\otimes v - \eta\partial_{x_i}(g)\otimes v + g[C_{n+1,i}(1\otimes v)]\\
&= \ell x_{n+i}g\otimes v - \eta\partial_{x_i}(g)\otimes v + g\varphi(x_{n+i}\otimes v)\\
&= -\eta\partial_{x_i}(g)\otimes v + g[(\ell + c + \tilde{\omega})(x_{n+i}\otimes v)]
\end{aligned} \tag{8.3.77}$$

for $i \in \overline{1,n}$, and

$$\begin{aligned}
K_{n+1}(g\otimes v) &= x_0 D(g)\otimes v - \eta\partial_{x_0}(g)\otimes v + x_0 g\otimes I_{2n+1}(v)\\
&\quad + \sum_{s=1}^{n}(x_s g\otimes K_s(v) + x_{n+s}g\otimes K_{n+1+s}(v))\\
&= \ell x_0 g\otimes v - \eta\partial_{x_0}(g)\otimes v + g[x_0\otimes I_{2n+1}(v)\\
&\quad + \sum_{s=1}^{n}(x_s\otimes K_s(v) + x_{n+s}\otimes K_{n+1+s}(v))]\\
&= \ell x_0 g\otimes v - \eta\partial_{x_0}(g)\otimes v + g[K_{n+1}(1\otimes v)]\\
&= \ell x_0 g\otimes v - \eta\partial_{x_0}(g)\otimes v + g\varphi(x_0\otimes v)\\
&= -\eta\partial_{x_0}(g)\otimes v + g[(\ell + c + \tilde{\omega})(x_0\otimes v)].
\end{aligned} \tag{8.3.78}$$

Since (8.3.75) says that

$$\eta\partial_{x_r}(g)\otimes v \in K_{n+1}(\widehat{V(\mu)}_\ell) + \sum_{i=1}^{n}[A_{i,n+1}(\widehat{V(\mu)}_\ell) + C_{n+1,i}(\widehat{V(\mu)}_\ell)]) \tag{8.3.79}$$

for $r \in \overline{0, 2n}$, Expressions (8.3.76)–(8.3.78) show

$$g[(\ell + c + \tilde{\omega})(x_i \otimes v)] \in K_{n+1}(\widehat{V(\mu)}_\ell) + \sum_{i=1}^{n}[A_{i,n+1}(\widehat{V(\mu)}_\ell) + C_{n+1,i}(\widehat{V(\mu)}_\ell)] \tag{8.3.80}$$

for $i \in \overline{0, 2n}$ and $g \in \mathscr{A}_\ell$.

According to Lemma 8.3.2, the eigenvalues of $(\ell + c + \tilde{\omega})|_{\widehat{V(\mu)}_{\langle 1\rangle}}$ are among

$$\ell + c + \mu_{i_r} - i_r + 1,\ \ \ell + c - \mu_{i_r} - 2n + i_{r+1} \quad \text{for} \quad r \in \overline{1, s}. \tag{8.3.81}$$

Again

$$-\ell - \mu_{i_r} + i_r - 1,\ \ -\ell + \mu_{i_r} + 2n - i_{r+1} \in \mu_1 + 2n - i_2 - \mathbb{N} \quad \text{for} \quad i \in \overline{1, s}. \tag{8.3.82}$$

If $c \notin \mu_1 + 2n - i_2 - \mathbb{N}$, then all the eigenvalues of $(\ell + c + \tilde{\omega})|_{\widehat{V(\mu)}_1}$ are nonzero. Hence

$$(\ell + c + \tilde{\omega})(\widehat{V(\mu)}_1) = \widehat{V(\mu)}_1. \tag{8.3.83}$$

By (8.3.80) and (8.3.83),

$$g(\widehat{V(\mu)}_{\langle 1\rangle}) \subset \sum_{i=1}^{n}[A_{i,n+1}(\widehat{V(\mu)}_\ell) + C_{n+1,i}(\widehat{V(\mu)}_\ell)] \quad \text{for} \quad g \in \mathscr{A}_\ell, \tag{8.3.84}$$

or equivalently, (8.3.68) holds for $k = \ell + 1$ by (8.3.84). By induction, (8.3.68) holds for any $k \in \mathbb{N}$. □

Let λ_i be the ith fundamental weight of $o(2n+3, \mathbb{C})$ with respect to the following simple positive roots

$$\{\varepsilon_{n+1} - \varepsilon_n, \varepsilon_n - \varepsilon_{n-1}, \ldots, \varepsilon_2 - \varepsilon_1, \varepsilon_1\}. \tag{8.3.85}$$

By the Weyl group's actions on $o(2n + 1, \mathbb{C})$ and $V(\mu)$, $\widehat{V(\mu)}$ is a highest-weight $o(2n + 3, \mathbb{C})$-module with highest weight $-(c + \mu_1)\lambda_1 + \sum_{r=1}^{s-1}(\mu_{i_r} - \mu_{i_{r+1}})\lambda_{i_{r+1}} + 2\mu_n\lambda_{n+1}$.

Up to this stage, we do not known if the condition in Theorem 8.3.6 is necessary for the $o(2n + 3, \mathbb{C})$-module $\widehat{V(\mu)}$ to be irreducible if $\mu \neq 0$. We will deal with the case $\mu = 0$ in next section.

8.4 Conformal Oscillator Representations

In this section, we study the $o(2n + 3, \mathbb{C})$-module $\widehat{V(0)}$ and its oscillator generalizations.

In order to use the results in Sects. 8.1 and 8.2, we redenote

$$y_i = x_{n+i} \qquad \text{for } i \in \overline{1,n} \tag{8.4.1}$$

and use $\mathscr{B}' = \mathbb{C}[x_0, x_1, \ldots, x_n, y_1, \ldots, y_n]$ to replace $\mathscr{A}$ in last section. Recall

$$\Delta' = \partial_{x_0}^2 + 2\sum_{i=1}^{n} \partial_{x_i}\partial_{y_i}, \quad \eta' = x_0^2/2 + \sum_{i=1}^{n} x_i y_i, \tag{8.4.2}$$

$$D' = \sum_{r=0}^{n} x_r \partial_{x_r} + \sum_{s=1}^{n} y_s \partial_{y_s}. \tag{8.4.3}$$

Fix $c \in \mathbb{C}$ and identify $\widehat{V(0)} = \mathscr{B}' \otimes v_0$ with $\mathscr{B}'$ by

$$f \otimes v_0 \leftrightarrow f \qquad \text{for } f \in \mathscr{B}', \tag{8.4.4}$$

where $V(0) = \mathbb{C}v_0$. Then we have the following one-parameter generalization π_c of the conformal representation of $o(2n+3, \mathbb{C})$:

$$\pi_c(A_{i,j}) = x_i\partial_{x_j} - y_j\partial_{y_i},\ \pi_c(B_{i,j}) = x_i\partial_{y_j} - x_j\partial_{y_i},\ \pi_c(C_{i,j}) = y_i\partial_{x_j} - y_j\partial_{x_i}, \tag{8.4.5}$$

$$\pi_c(A_{n+1,i}) = \partial_{x_i}, \quad \pi_c(B_{i,n+1}) = -\partial_{y_i}, \quad \pi_c(K_{2n+2}) = -\partial_{x_0}, \tag{8.4.6}$$

$$\pi_c(A_{i,n+1}) = \eta'\partial_{y_i} - x_i(D' + c), \quad \pi_c(C_{n+1,i}) = y_i(D' + c) - \eta'\partial_{x_i} \tag{8.4.7}$$

$$\pi_c(K_i) = x_0\partial_{x_i} - x_{n+i}\partial_{x_0}, \qquad \pi_c(K_{n+1+i}) = x_0\partial_{x_{n+i}} - x_i\partial_{x_0}, \tag{8.4.8}$$

$$\pi_c(A_{n+1,n+1}) = -D' - c, \qquad \pi_c(K_{n+1}) = x_0(D' + c) - \eta'\partial_{x_0} \tag{8.4.9}$$

for $i, j \in \overline{1,n}$. Recall that $\mathscr{B}'_r$ denotes the subspace of homogeneous polynomials in $\mathscr{B}'$ with degree r. For any $k \in \mathbb{N}$, we write

$$\mathscr{B}'_{[k]} = \mathscr{B}'_0 + \sum_{i=1}^{k} (\mathscr{B}'_i + \eta'^i \mathscr{B}'_{k-i}). \tag{8.4.10}$$

The following is an improved result of that in Zhao and the author's work [XZ].

Theorem 8.4.1 *The representation π_c of $o(2n+3,\mathbb{C})$ on $\mathscr{B}'$ is irreducible with highest weight $-c\lambda_1$ if $c \notin -\mathbb{N}$. When $c=-k$ with $k \in \mathbb{N}$, $\mathscr{B}'_{[k]}$ is an irreducible $o(2n+2,\mathbb{C})$-submodule with highest weight $k\lambda_1$ and $\mathscr{B}'/\mathscr{B}'_{[k]}$ is an irreducible highest-weight $o(2n+3,\mathbb{C})$-module with highest weight $-(k+2)\lambda_1+(k+1)\lambda_2$ with respect to (8.3.85)*

Proof Recall the identification (8.3.27) and (8.3.28). Moreover, $\mathscr{B}'_k$ denotes the subspace of homogeneous polynomials in $\mathscr{B}'$ with degree k and

$$\mathscr{H}'_k = \{f \in \mathscr{B}'_k \mid \Delta'(f)=0\} \quad \text{for } k \in \mathbb{N}. \tag{8.4.11}$$

According to Theorem 8.1.1, $\mathscr{H}'_k$ is an irreducible $o(2n+1,\mathbb{C})$-submodule with the highest-weight vector x_1^k of weight $k\varepsilon_1$ for any $k \in \mathbb{N}$, and

$$\mathscr{B}' = \bigoplus_{m,k=0}^{\infty} \eta'^m \mathscr{H}'_k \tag{8.4.12}$$

is a direct sum of irreducible $o(2n+1,\mathbb{C})$-submodules. On the other hand, $U(\mathscr{G}_-)(1_{\mathscr{B}'})$ forms an irreducible $o(2n+3,\mathbb{C})$-submodule of $\mathscr{B}'$ due to (8.4.6).

Assume $c \notin -\mathbb{N}$. Let M be a nonzero $o(2n+3,\mathbb{C})$-submodule of $\mathscr{B}'$. Repeatedly applying (8.4.6) if necessary, we have $1_{\mathscr{B}'} \in M$. Note

$$K_{n+1}(1_{\mathscr{B}'}) = cx_0, \quad -A_{i,n+1}(1_{\mathscr{B}'}) = cx_i, \quad C_{n+1,i}(1_{\mathscr{B}'}) = cy_i \qquad \text{for } i \in \overline{1,n}. \tag{8.4.13}$$

Thus $\mathscr{B}'_1 \subset M$. Suppose that

$$\mathscr{B}'_k \subset M \qquad \text{for } k < \ell \tag{8.4.14}$$

with $2 \leq \ell \in \mathbb{N}$. According to (8.4.12),

$$\mathscr{B}'_\ell = \bigoplus_{m=0}^{[\![\ell/2]\!]} \eta'^m \mathscr{H}'_{\ell-2m}. \tag{8.4.15}$$

Moreover,

$$[\Delta', \eta'] = 1 + 2n + 2D'. \tag{8.4.16}$$

Set

$$\mathscr{B}'_{\ell,r} = \bigoplus_{m=0}^{r} \eta'^m \mathscr{H}'_{\ell-2m} \tag{8.4.17}$$

for $r \in \overline{0, [\![\ell/2]\!]}$. Then

$$\mathscr{B}'_{\ell,r} = \{w \in \mathscr{B}'_\ell \mid \Delta'^{r+1}(w) = 0\} \tag{8.4.18}$$

and

$$\Delta''^r(\mathscr{B}'_{\ell,r}) = \mathscr{H}'_{\ell-2r} \tag{8.4.19}$$

by (8.4.17).

Since

$$-A_{1,n+1}(x_1^{\ell-1}) = (c+\ell-1)x_1^\ell \in M \tag{8.4.20}$$

and $c \notin -\mathbb{N}-1$, we have

$$x_1^\ell \in M \tag{8.4.21}$$

by (8.4.7). Hence

$$\mathscr{H}'_\ell \subset M \tag{8.4.22}$$

because $\mathscr{H}_\ell'$ is an irreducible $o(2n+1,\mathbb{C})$-submodule generated by x_1^ℓ. Furthermore,

$$K_{n+1}(x_1^{\ell-r-1}y_1^r) = (c+\ell-1)x_1^{\ell-r-1}x_0y_1^r \in M. \tag{8.4.23}$$

So $x_1^{\ell-r-1}x_0y_1^r \in M$. Moreover,

$$\Delta''^r(x_1^{\ell-r-1}x_0y_1^r) = 2^r r!\left[\prod_{s=1}^r(\ell-r-s)\right]x_1^{\ell-2r-1}x_0 \in \mathscr{H}'_{\ell-2r}. \tag{8.4.24}$$

Observe that (8.4.15) is a direct sum of irreducible $o(2n+1,\mathbb{C})$-submodules with distinct highest weights. So

$$\mathscr{B}'_\ell \bigcap M = \bigoplus_{m=0}^{[\![\ell/2]\!]}(\eta'^m\mathscr{H}'_{\ell-2m})\bigcap M. \tag{8.4.25}$$

By (8.4.18)–(8.4.24), $(\eta'^r\mathscr{H}'_{\ell-2r})\bigcap M$ is a nonzero $o(2n+1,\mathbb{C})$-submodule. Since $\eta'^r H'_{\ell-2r}$ is an irreducible $o(2n+1,\mathbb{C})$-module, we have $\eta'^r\mathscr{H}'_{\ell-2r} = (\eta'^r\mathscr{H}'_{\ell-2r}) \bigcap M$. Therefore, $\mathscr{B}'_\ell \subset M$. By induction, $\mathscr{B}'_k \subset M$ for any $k\in\mathbb{N}$, or equivalently, $M = \mathscr{B}'$. This proves the theorem for $c\notin -\mathbb{N}$; that is, $\mathscr{B}$ is an irreducible highest-weight $o(2n+2,\mathbb{C})$-module with $1_{\mathscr{B}}$ as a highest-weight vector of weight $-c\lambda_1$ with respect to (8.3.85).

Suppose that $c=-k$ with $k\in\mathbb{N}$, then $\sum_{r=0}^k\mathscr{B}'_r \subset U(\mathscr{G}_-)(1_{\mathscr{B}'})$ (cf. (8.3.45)). According to (8.4.7) and (8.4.9), $\mathscr{G}_-(\mathscr{B}'_k) = \eta'\mathscr{B}'_{k-1} \subset U(\mathscr{G}_-)(1_{\mathscr{B}'})$. Assume $\eta'^r\mathscr{B}'_{k-r} \subset U(\mathscr{G}_-)(1_{\mathscr{B}'})$. Note $\eta'^r\mathscr{B}'_{k-r} \in \mathscr{B}'_{r+k}$. So $K_{n+1}(\eta'^r f) = -\eta'^{r+1}\partial_{x_0}(f)$ and

$$A_{i,n+1}(\eta'^r f) = \eta'^{r+1}\partial_{y_i}(f),\;\; C_{n+1,i}(\eta'^r f) = -\eta'^{r+1}\partial_{x_i}(f) \tag{8.4.26}$$

for $f\in\mathscr{B}'_{k-r}$ by (8.4.7) and (8.4.9). Thus $\eta'^{r+1}\mathscr{B}'_{k-r-1}\subset U(\mathscr{G}_-)(1_{\mathscr{B}'})$. By induction, $\mathscr{B}'_{[k]}\subset U(\mathscr{G}_-)(1_{\mathscr{B}'})$.

Recall that we identify $o(2n+1,\mathbb{C})$ with $\mathcal{G}_0$ (cf. (8.3.44)). First $\mathcal{B}'_{[k]}$ is an $o(2n+1,\mathbb{C})$-submodule by (8.4.10) and (8.4.12). According to (8.4.6), $K_{2n+2}(\eta'^r f) = -\eta'^{r-1}(rx_0 f + \eta'\partial_{x_0}(f))$ and

$$A_{n+1,i}(\eta'^r f) = \eta'^{r-1}(ry_i f + \eta'\partial_{x_i}(f)),\quad B_{i,n+1}(\eta'^r f) = -\eta'^{r-1}(rx_i f + \eta'\partial_{y_i}(f)) \tag{8.4.27}$$

for $f \in \mathcal{B}'$. Hence $\mathcal{G}_+(\eta'^r\mathcal{B}'_{k-r}) \subset \eta'^{r-1}\mathcal{B}'_{k-r+1}$. By (8.4.26) and its above equation, $\mathcal{G}_-((\eta'^r\mathcal{B}'_{k-r}) \subset \eta'^{r+1}\mathcal{B}'_{k-r-1}$ if $r < k$ and $\mathcal{G}_-(\eta'^k) = \{0\}$. Therefore, $\mathcal{B}'_{[k]} = U(\mathcal{G}_-)(1_{\mathcal{B}'})$ is finite-dimensional irreducible $o(2n+3,\mathbb{C})$-module with $1_{\mathcal{B}'}$ as a highest-weight vector of weight $k\lambda_1$ with respect to (8.3.85).

Let M' be an $o(2n+3,\mathbb{C})$-submodule containing $\mathcal{B}'_{[k]}$ as a proper submodule. Note that $\mathcal{B}'_{k+1} = \mathcal{H}'_k \oplus (\mathcal{B}'_{k+1} \bigcap \mathcal{B}'_{[k]})$. Applying the first and third equations in (8.4.6), we get $M' \bigcap \mathcal{H}'_{k+1} \neq \{0\}$. Since $\mathcal{H}'_{k+1}$ is an irreducible $o(2n+1,\mathbb{C})$-module, we have $\mathcal{H}'_{k+1} = M' \bigcap \mathcal{H}_{k+1}$. So $\mathcal{B}'_{k+1} \subset M'$. Using the arguments in (8.4.14)–(8.4.25) with $\ell \geq k+2$, we can prove $\mathcal{B}'_\ell \subset M'$ for any $k+2 \leq \ell \in \mathbb{N}$. Hence $M' = \mathcal{B}'$. Thus $\mathcal{B}'/\mathcal{B}'_{[k]}$ is an irreducible highest-weight $o(2n+3,\mathbb{C})$-module with $x_n^{k+1} + \mathcal{B}'_{[k]}$ as a highest-weight vector of weight $-(k+2)\lambda_1 + (k+1)\lambda_2$ with respect to (8.3.85). □

For $\vec{a} = (a_0, a_1, \ldots, a_n)^t \in \mathbb{C}^{n+1}$ and $\vec{b} = (b_1, b_2, \ldots, b_n)^t \in \mathbb{C}^n$, we put

$$\vec{a}\cdot\vec{x} = \sum_{i=0}^n a_i x_i, \qquad \vec{b}\cdot\vec{y} = \sum_{i=1}^n b_i y_i. \tag{8.4.28}$$

Set

$$\mathcal{B}'_{\vec{a},\vec{b}} = \{fe^{\vec{a}\cdot\vec{x}+\vec{b}\cdot\vec{y}} \mid f \in \mathcal{B}'\}. \tag{8.4.29}$$

Denote by $\pi_{c,\vec{a},\vec{b}}$ the representation π_c of $o(2n+3,\mathbb{C})$ on $\mathcal{B}'_{\vec{a},\vec{b}}$.

Theorem 8.4.2 *If $a_0^2 + 2\sum_{i=1}^n a_i b_i \neq 0$, then representation $\pi_{c,\vec{a},\vec{b}}$ of $o(2n+3,\mathbb{C})$ is irreducible for any $c \in \mathbb{C}$*

Proof Set

$$\mathcal{B}'_{\vec{a},\vec{b},k} = \mathcal{B}'_k e^{\vec{a}\cdot\vec{x}+\vec{b}\cdot\vec{y}} \qquad \text{for } k \in \mathbb{N}. \tag{8.4.30}$$

Let M be a nonzero $o(2n+3,\mathbb{C})$-submodule of $\mathcal{B}'_{\vec{a},\vec{b}}$. Take any $0 \neq fe^{\vec{a}\cdot\vec{x}+\vec{b}\cdot\vec{y}} \in M$ with $f \in \mathcal{B}$. According to (8.4.6),

$$(A_{n+1,i} - a_i)(fe^{\vec{a}\cdot\vec{x}+\vec{b}\cdot\vec{y}}) = \partial_{x_i}(f)e^{\vec{a}\cdot\vec{x}+\vec{b}\cdot\vec{y}},\quad -(B_{i,n+1} + b_i)(fe^{\vec{a}\cdot\vec{x}+\vec{b}\cdot\vec{y}}) = \partial_{y_i}(f)e^{\vec{a}\cdot\vec{x}+\vec{b}\cdot\vec{y}} \tag{8.4.31}$$

for $i \in \overline{1,n}$, and

$$-(K_{2n+2} + a_0)(fe^{\vec{a}\cdot\vec{x}+\vec{b}\cdot\vec{y}}) = \partial_{x_0}(f)e^{\vec{a}\cdot\vec{x}+\vec{b}\cdot\vec{y}}. \tag{8.4.32}$$

Repeatedly applying (8.4.31) and (8.4.32), we obtain $e^{\vec{a}\cdot\vec{x}+\vec{b}\cdot\vec{y}} \in M$. Equivalently, $\mathscr{B}'_{\vec{a},\vec{b},0} \subset M$. Suppose $\mathscr{B}'_{\vec{a},\vec{b},\ell} \subset M$ for some $\ell \in \mathbb{N}$. Let $ge^{\vec{a}\cdot\vec{x}+\vec{b}\cdot\vec{y}}$ be any element in $\mathscr{B}'_{\vec{a},\vec{b},\ell}$.

Case 1. $a_i \neq 0$ *or* $b_i \neq 0$ *for some* $i \in \overline{1,n}$.

By symmetry, we may assume $a_1 \neq 0$. Since

$$(x_i\partial_{x_1} - y_1\partial_{y_i})(g)e^{\vec{a}\cdot\vec{x}+\vec{b}\cdot\vec{y}},\ (y_i\partial_{x_1} - y_1\partial_{x_i})(g)e^{\vec{a}\cdot\vec{x}+\vec{b}\cdot\vec{y}} \in \mathscr{B}'_{\vec{a},\vec{b},\ell} \subset M \tag{8.4.33}$$

by inductional assumption, Expression (8.4.5) yields

$$A_{i,1}(ge^{\vec{a}\cdot\vec{x}+\vec{b}\cdot\vec{y}}) \equiv (a_1x_i - b_iy_1)ge^{\vec{a}\cdot\vec{x}+\vec{b}\cdot\vec{y}} \equiv 0 \pmod M \tag{8.4.34}$$

and

$$C_{i,1}(ge^{\vec{a}\cdot\vec{x}+\vec{b}\cdot\vec{y}}) \equiv (a_1y_i - a_iy_1)ge^{\vec{a}\cdot\vec{x}+\vec{b}\cdot\vec{y}} \equiv 0 \pmod M. \tag{8.4.35}$$

Moreover, the first equation in (8.4.8) gives

$$K_1(ge^{\vec{a}\cdot\vec{x}+\vec{b}\cdot\vec{y}}) \equiv (a_1x_0 - a_0y_1)ge^{\vec{a}\cdot\vec{x}+\vec{b}\cdot\vec{y}} \equiv 0 \pmod M \tag{8.4.36}$$

because

$$(x_0\partial_{x_1} - y_1\partial_{x_0})(g)e^{\vec{a}\cdot\vec{x}+\vec{b}\cdot\vec{y}} \in \mathscr{B}'_{\vec{a},\vec{b},\ell} \subset M. \tag{8.4.37}$$

On the other hand,

$$(D' + c)(g)e^{\vec{a}\cdot\vec{x}+\vec{b}\cdot\vec{y}} = (\ell + c)ge^{\vec{a}\cdot\vec{x}+\vec{b}\cdot\vec{y}} \in \mathscr{B}'_{\vec{a},\vec{b},\ell} \subset M \tag{8.4.38}$$

and the first equation in (8.4.9) implies

$$-A_{n+1,n+1}(ge^{\vec{a}\cdot\vec{x}+\vec{b}\cdot\vec{y}}) \equiv [a_0x_0 + \sum_{i=1}^{n}(a_ix_i + b_iy_i)]ge^{\vec{a}\cdot\vec{x}+\vec{b}\cdot\vec{y}} \equiv 0 \pmod M. \tag{8.4.39}$$

Substituting (8.4.34)–(8.4.36) into (8.4.39), we get

$$(a_0^2 + 2\sum_{i=1}^{n} a_ib_i)y_1ge^{\vec{a}\cdot\vec{x}+\vec{b}\cdot\vec{y}} \equiv 0 \pmod M. \tag{8.4.40}$$

Equivalently, $y_1ge^{\vec{a}\cdot\vec{x}+\vec{b}\cdot\vec{y}} \in M$. Substituting it to (8.4.34)–(8.4.36), we obtain

$$x_0ge^{\vec{a}\cdot\vec{x}+\vec{b}\cdot\vec{y}},\ x_ige^{\vec{a}\cdot\vec{x}+\vec{b}\cdot\vec{y}},\ y_ige^{\vec{a}\cdot\vec{x}+\vec{b}\cdot\vec{y}} \in M \tag{8.4.41}$$

for $i \in \overline{1,n}$. Therefore, $\mathscr{B}'_{\vec{a},\vec{b},\ell+1} \subset M$. By induction, $\mathscr{B}'_{\vec{a},\vec{b},\ell} \subset M$ for any $\ell \in \mathbb{N}$. So $\mathscr{B}'_{\vec{a},\vec{b}} = M$. Hence $\mathscr{B}'_{\vec{a},\vec{b}}$ is an irreducible $o(2n+3,\mathbb{C})$-module.

Case 2. $a_0 \neq 0$ *and* $a_i = b_i = 0$ for $i \in \overline{1,n}$.
Under the above assumption,

$$K_i(ge^{\vec{a}\cdot\vec{x}+\vec{b}\cdot\vec{y}}) = (x_0\partial_{x_i} - y_i\partial_{x_0} - a_0 y_i)(g)e^{\vec{a}\cdot\vec{x}+\vec{b}\cdot\vec{y}} \in M \tag{8.4.42}$$

and

$$K_{n+1+i}(ge^{\vec{a}\cdot\vec{x}+\vec{b}\cdot\vec{y}}) = (x_0\partial_{y_i} - x_i\partial_{x_0} - a_0 x_i)(g)e^{\vec{a}\cdot\vec{x}+\vec{b}\cdot\vec{y}} \in M \tag{8.4.43}$$

for $i \in \overline{1,n}$. Note

$$(x_0\partial_{x_i} - y_i\partial_{x_0})(g)e^{\vec{a}\cdot\vec{x}+\vec{b}\cdot\vec{y}}, (x_0\partial_{y_i} - x_i\partial_{x_0})(g)e^{\vec{a}\cdot\vec{x}+\vec{b}\cdot\vec{y}} \in \mathscr{B}'_{\vec{a},\vec{b},\ell} \subset M \tag{8.4.44}$$

by the inductional assumption. Thus (8.4.42) and (8.4.43) imply

$$y_i ge^{\vec{a}\cdot\vec{x}+\vec{b}\cdot\vec{y}}, x_i ge^{\vec{a}\cdot\vec{x}+\vec{b}\cdot\vec{y}} \in M \qquad \text{for } i \in \overline{1,n}. \tag{8.4.45}$$

Now (8.4.39) yields $x_0 ge^{\vec{a}\cdot\vec{x}+\vec{b}\cdot\vec{y}} \in M$. So $\mathscr{B}'_{\vec{a},\vec{b},\ell+1} \subset M$. By induction, $\mathscr{B}' = M$; that is, $\mathscr{B}'$ is irreducible. $\square$

Fix $n_1, n_2 \in \overline{1,n}$ with $n_1 \leq n_2$. Set

$$\tilde{D}' = x_0\partial_{x_0} - \sum_{i=1}^{n_1} x_i\partial_{x_i} + \sum_{r=n_1+1}^{n} x_r\partial_{x_r} + \sum_{j=1}^{n_2} y_j\partial_{y_j} - \sum_{s=n_2+1}^{n} y_s\partial_{y_s}, \tag{8.4.46}$$

$$\tilde{\Delta}' = \partial_{x_0}^2 - 2\sum_{i=1}^{n_1} x_i\partial_{y_i} + 2\sum_{r=n_1+1}^{n_2} \partial_{x_r}\partial_{y_r} - 2\sum_{s=n_2+1}^{n} y_s\partial_{x_s} \tag{8.4.47}$$

and

$$\tilde{\eta}' = \frac{x_0^2}{2} + \sum_{i=1}^{n_1} y_i\partial_{x_i} + \sum_{r=n_1+1}^{n_2} x_r y_r + \sum_{s=n_2+1}^{n} x_s\partial_{y_s}. \tag{8.4.48}$$

Then the representation $\pi_c^{n_1,n_2}$ of $o(2n+3,\mathbb{C})$ is determined as follows: $\pi_c|_{o(2n+2,\mathbb{C})}$ is given by (7.4.38)–(7.4.47) with $\widehat{D}$ replaced by $\tilde{D}'$ and $\tilde{\eta}$ by $\tilde{\eta}'$, and

$$\pi_c^{n_1,n_2}(K_i) = \begin{cases} -x_0x_i - y_i\partial_{x_0} & \text{if } i \in \overline{1,n_1}, \\ x_0\partial_{x_i} - y_i\partial_{x_0} & \text{if } i \in \overline{n_1+1,n_2}, \\ x_0\partial_{x_i} - \partial_{x_0}\partial_{y_i} & \text{if } i \in \overline{n_2+1,n}, \end{cases} \tag{8.4.49}$$

$$\pi_c^{n_1,n_2}(K_{n+1+i}) = \begin{cases} x_0\partial_{y_i} - \partial_{x_0}\partial_{x_i} & \text{if } i \in \overline{1,n_1}, \\ x_0\partial_{y_i} - x_i\partial_{x_0} & \text{if } i \in \overline{n_1+1,n_2}, \\ -x_0y_i - x_i\partial_{x_0} & \text{if } i \in \overline{n_2+1,n}, \end{cases} \tag{8.4.50}$$

$$\pi_c^{n_1,n_2}(K_{n+1}) = x_0(\tilde{D}' + \tilde{c}) - \tilde{\eta}'\partial_{x_0}, \qquad \pi_c^{n_1,n_2}(K_{2n+2}) = -\partial_{x_0}, \tag{8.4.51}$$

where

$$\tilde{c} = c + n_2 - n_1 - n. \tag{8.4.52}$$

Recall (8.2.10) and (8.2.12). We have the following third result in this section:

Theorem 8.4.3 *The representation $\pi_c^{n_1,n_2}$ of $o(2n+3,\mathbb{C})$ on $\mathscr{A}$ is irreducible if $c \notin \mathbb{Z}/2$*

Proof Let M be a nonzero $o(2n+3,\mathbb{C})$-submodule of $\mathscr{B}$. Note

$$\pi_c^{n_1,n_2}(A_{n+1,n+1}) = -\tilde{D}' - \tilde{c}. \tag{8.4.53}$$

According to (8.2.12),

$$M = \bigoplus_{k\in\mathbb{Z}} \mathscr{B}_{\langle k\rangle} \bigcap M. \tag{8.4.54}$$

Thus $\mathscr{B}_{\langle k\rangle} \bigcap M \neq \{0\}$ for some $k \in \mathbb{Z}$. On the other hand, $\mathscr{B}'_{\langle k\rangle} = \bigoplus_{i=0}^{\infty} (\tilde{\eta}')^i(\mathscr{H}'_{\langle k-2i\rangle})$ is a decomposition of irreducible $o(2n+1,\mathbb{C})$-submodules by Theorem 8.2.3 and (8.3.26). Moreover, the weights of the $sl(n,\mathbb{C})$-singular vectors in $(\tilde{\eta}')^i(\mathscr{H}'_{\langle k-2i\rangle})$ are distinct by Sects. 6.4, 6.5 and 6.6. Hence

$$(\tilde{\eta}')^i(\mathscr{H}'_{\langle k-2i\rangle}) \subset M \quad \text{for some } \; i \in \mathbb{N}. \tag{8.4.55}$$

Theorem 8.2.3 and the arguments in (7.4.52)–(7.4.57) show that we can assume

$$\mathscr{H}'_{\langle k-s\rangle} \subset M \qquad \text{for } \; s \in \mathbb{N}. \tag{8.4.56}$$

Suppose $(\tilde{\eta}')^s(x_1^{-r+s}) \in M$ for any $r \in k - \mathbb{N}$ and some $s \in \mathbb{N}$. Then,

$$K_{n+1}((\tilde{\eta}')^s(x_1^{-r+s+1})) = (r - 1 + \tilde{c})(\tilde{\eta}')^s(x_0x_1^{-r+s+1}) \in M \tag{8.4.57}$$

by (8.4.48) and the first equation in (8.4.51), which implies $\eta^s_{n_1,n_2}(x_0x_1^{-r+s+1}) \in M$. Moreover,

$$K_{n+1}((\tilde{\eta}')^s(x_0x_1^{-r+s+1})) = -(\tilde{\eta}')^{s+1}(x_1^{-r+s+1}) + (r+\tilde{c})(\tilde{\eta}')^s(x_0^2x_1^{-r+s+1}). \tag{8.4.58}$$

Now (7.4.59) and (7.4.61) with $\tilde{\eta}$ replaced by $\tilde{\eta}'$, and (8.4.58) lead to

$$(1/2 + s + \tilde{c} - n_2 + n_1)(\tilde{\eta}')^{s+1}(x_1^{-r+s+1}) \in M \Rightarrow \eta^{s+1}_{n_1,n_2}(x_1^{-r+s+1}) \in M. \tag{8.4.59}$$

By induction,

$$(\tilde{\eta}')^{\ell}(x_1^{-r+\ell}) \in M \qquad \text{for } \; \ell \in \mathbb{N},\; r \in k - \mathbb{N}. \tag{8.4.60}$$

According to Theorem 8.2.1,

$$\mathscr{B}'_{\langle m\rangle} = \bigoplus_{\ell=0}^{\infty}(\tilde{\eta}')^{\ell}(\mathscr{H}'_{\langle m-2\ell\rangle}) \subset M \qquad \text{for } m \in k - \mathbb{N}. \tag{8.4.61}$$

Observe that

$$\pi_c^{n_1,n_2}(K_{n+1})x_0 = x_0^2(D_{n_1,n_2} + \tilde{c} + 1) - \eta_{n_1,n_2}(x_0\partial_{x_0} + 1) \tag{8.4.62}$$

by (8.4.51). Then (8.4.62) and (7.4.66)–(7.4.69) with $\tilde{\eta}$ replaced by $\tilde{\eta}'$ and $\widehat{D}$ replaced by $\tilde{D}'$ yield

$$\begin{aligned}&-\pi_c^{n_1,n_2}(K_{n+1})x_0 + \sum_{i=1}^{n_2}\pi_c^{n_1,n_2}(A_{i,n+1})y_i + \sum_{j=n_2+1}^{n}\pi_c^{n_1,n_2}(A_{j,n+1})\partial_{y_j}\\ &-\sum_{r=1}^{n_1}\pi_c^{n_1,n_2}(C_{n+1,r})\partial_{x_r} - \sum_{s=n_1+1}^{n}\pi_c^{n_1,n_2}(C_{n+1,s})x_s\\ &= \tilde{\eta}'(1 - \tilde{D}' + n_2 + n - n_1 - 2(\tilde{c}+1))\end{aligned} \tag{8.4.63}$$

as operators on $\mathscr{B}'$. The arguments in (7.4.71)–(7.4.77) show $M = \mathscr{B}'$; that is, $\mathscr{B}'$ is an irreducible $o(2n+3, \mathbb{C})$-module. □

Remark 8.4.4 The above irreducible representation depends on the three parameters $c \in \mathbb{C}$ and $n_1, n_2 \in \overline{1,n}$. It is not highest-weight type because of the mixture of multiplication operators and differential operators in (7.4.39)–(7.4.42), (7.4.44)–(7.4.47), (8.4.49) and (8.4.50). Since $\mathscr{B}'$ is not completely reducible as a $sl(n, \mathbb{C})$-module by Sects. 6.4–6.6 when $n \geq 2$ and $n_1 < n$, $\mathscr{B}'$ is not a unitary $o(2n+3, \mathbb{C})$-module. Expression (8.4.53) shows that $\mathscr{B}'$ is a weight $o(2n+3, \mathbb{C})$-module with finite-dimensional weight subspaces.

References

[LX2] C. Luo, X. Xu, Z-Graded oscillator generalizations of the classical theorem on harmonic polynomials. J. Lie Theory **23**, 979–1003 (2013)

[XZ] X. Xu, Y. Zhao, Extensions of the conformal representations for orthogonal Lie algebras. J. Algebra **377**, 97–124 (2013)

[X25] X. Xu, Conformal oscillator representations of orthogonal Lie algebras. Sci. China: Math. (Engl. Ser.) **59**(1), 37-48 (2016)

Chapter 9
Representations of Symplectic Lie Algebras

In this chapter, we study the natural explicit representations of symplectic Lie algebras. First in Sect. 9.1, we determine the structure of the canonical bosonic and fermionic oscillator representations of the Lie algebras over their minimal natural modules. Moreover, the Weil's full quadratic operator oscillator representation are presented. Weyl's Theorem of complete reducibility in Sect. 2.4 plays an important role.

In Sect. 9.2, we study the noncanonical oscillator representations obtained from the above bosonic representations by partially swapping differential operators and multiplication operators, and obtain a two-parameter family of infinite-dimensional irreducible representations, where the noncanonical oscillator representations of $sl(n, \mathbb{F})$ in Sects. 6.3–6.6 are used. The results were due to Luo and the author [LX3], and can be viewed as generalizations of the Weil representation. In Sect. 9.3, we determine the structures of the projective oscillator representations in Sect. 6.8 restricted on symplectic Lie algebras, which were due to our work [X24].

9.1 Canonical Oscillator Representations

In this section, we present the canonical bosonic and fermionic oscillator representations of symplectic Lie algebras over their minimal natural modules.

Let $n > 1$ be an integer. The symplectic Lie algebra

$$sp(2n, \mathbb{F}) = \sum_{i,j=1}^{n} \mathbb{F}(E_{i,j} - E_{n+j,n+i}) + \sum_{i=1}^{n}(\mathbb{F}E_{i,n+i} + \mathbb{F}E_{n+i,i})$$
$$+ \sum_{1\leq i<j\leq n} [\mathbb{F}(E_{i,n+j} + E_{n+j,i}) + \mathbb{F}(E_{n+i,j} + E_{n+j,i})]. \quad (9.1.1)$$

X. Xu, *Representations of Lie Algebras and Partial Differential Equations*,
DOI 10.1007/978-981-10-6391-6_9

We again take

$$H=\sum_{i=1}^{n}\mathbb{F}(E_{i,i}-E_{n+i,n+i}) \tag{9.1.2}$$

as a Cartan subalgebra of $sp(2n,\mathbb{F})$ and the settings in (7.1.3)–(7.1.5). Then the root system of $sp(2n,\mathbb{F})$ is

$$\Phi_{C_n}=\{\pm\varepsilon_i\pm\varepsilon_j,\pm2\varepsilon_r\mid 1\leq i<j\leq n;r\in\overline{1,n}\}. \tag{9.1.3}$$

We choose the set of positive roots

$$\Phi_{C_n}^{+}=\{\varepsilon_i\pm\varepsilon_j,2\varepsilon_r\mid 1\leq i<j\leq n;r\in\overline{1,n}\}. \tag{9.1.4}$$

In particular,

$$\Pi_{C_n}=\{\varepsilon_1-\varepsilon_2,...,\varepsilon_{n-1}-\varepsilon_n,2\varepsilon_n\}\ \text{ is the set of positive simple roots.} \tag{9.1.5}$$

Recall the set of dominant integral weights

$$\Lambda^{+}=\{\mu\in L_{\mathbb{Q}}\mid(\varepsilon_n,\mu),(\varepsilon_i-\varepsilon_{i+1},\mu)\in\mathbb{N}\text{ for }i\in\overline{1,n-1}. \tag{9.1.6}$$

According to (7.1.5),

$$\Lambda^{+}=\left\{\mu=\sum_{i=1}^{n}\mu_i\varepsilon_i\mid\mu_i\in\mathbb{N};\mu_i-\mu_{i+1}\in\mathbb{N}\text{ for }i\in\overline{1,n-1}\right\}. \tag{9.1.7}$$

For $\lambda\in\Lambda^{+}$, we denote by $V(\lambda)$ the finite-dimensional irreducible $sp(2n,\mathbb{F})$-module with highest weight λ.

Let $\mathscr{B}=\mathbb{F}[x_1,...,x_n,y_1,...,y_n]$. Recall (6.2.22)–(6.2.42). The *canonical bosonic oscillator representation of* $sp(2n,\mathbb{F})$ is given by

$$(E_{i,j}-E_{n+j,n+i})|_{\mathscr{B}}=x_i\partial_{x_j}-y_j\partial_{y_i},\ (E_{n+i,j}+E_{n+j,i})|_{\mathscr{B}}=y_i\partial_{x_j}+y_j\partial_{x_i}, \tag{9.1.8}$$

$$(E_{i,n+j}+E_{j,n+i})|_{\mathscr{B}}=x_i\partial_{y_j}+x_j\partial_{y_i} \tag{9.1.9}$$

for $i,j\in\overline{1,n}$.

The positive root vectors of $sp(2n,\mathbb{F})$ are

$$\{E_{i,j}-E_{n+j,n+i},E_{i,n+j}+E_{j,n+i},E_{r,n+r}\mid 1\leq i<j\leq n;r\in\overline{1,n}\}. \tag{9.1.10}$$

Recall the notion $\mathscr{B}_{\ell_1,\ell_2}$ in (6.2.30) and $\mathscr{B}_k$ in (7.1.15).

Theorem 9.1.1 *For $k \in \mathbb{N}$, the subspace $\mathscr{B}_k$ forms a finite-dimensional irreducible $sp(2n, \mathbb{F})$-submodule and x_1^k is a highest-weight vector with the weight $k\varepsilon_1$.*

Proof Note that $sp(2n, \mathbb{F})$ contains the Lie subalgebra

$$\sum_{i,j=1}^{n} \mathbb{F}(E_{i,j} - E_{n+j,n+i}) \cong gl(n, \mathbb{F}) \tag{9.1.11}$$

and the representation of $sp(2n, \mathbb{F})$ in (9.1.8) and (9.1.9) is essentially an extension of that for $sl(n, \mathbb{F})$ given in (6.2.22). Thus the only possible highest-weight vectors for $sp(2n, \mathbb{F})$ in $\mathscr{B}_k$ are:

$$\{\eta^i x_1^\ell y_n^{k-\ell-2i} \mid i \in \overline{1, [\![k/2]\!]};\ \ell \in \overline{0, k-2i}\} \tag{9.1.12}$$

by Lemma 6.2.3 and (9.1.10), where $\eta = \sum_{i=1}^{n} x_i y_i$. Note

$$\begin{aligned} 0 &= E_{n,2n}(\eta^i x_1^\ell y_n^{k-\ell-2i}) = x_n \partial_{y_n}(\eta^i x_1^\ell y_n^{k-\ell-2i}) \\ &= i x_n^2 \eta^{i-1} x_1^\ell y_n^{k-\ell-2i} + (k-\ell-2i) x_n \eta^i x_1^\ell y_n^{k-\ell-2i-1}, \end{aligned} \tag{9.1.13}$$

which forces $i = 0$ and $\ell = k$. Thus $\mathscr{B}_k$ has a unique (up to a scalar multiple) singular vector x_1^k of weight $k\varepsilon_1$. By Weyl's Theorem 2.3.6 of complete reducibility if $\mathbb{F} = \mathbb{C}$ or more generally by Lemma 6.3.2 with $n_1 = 0$, it is irreducible. □

Consider the exterior algebra $\check{\mathscr{A}}$ generated by $\{\theta_1, ..., \theta_n, \vartheta_1, ..., \vartheta_n\}$ (cf. 6.2.15) and take the settings in (6.2.44)–(6.2.51). Define the *basic fermionic oscillator representation* of $sp(2n, \mathbb{F})$ on $\check{\mathscr{A}}$ by

$$(E_{i,j} - E_{n+j,n+i})|_{\check{\mathscr{A}}} = \theta_i \partial_{\theta_j} - \vartheta_j \partial_{\vartheta_i}, \quad (E_{n+i,j} + E_{n+j,i})|_{\check{\mathscr{A}}} = \vartheta_i \partial_{\theta_j} + \vartheta_j \partial_{\theta_i}, \tag{9.1.14}$$

$$(E_{i,n+j} + E_{j,n+i})|_{\check{\mathscr{A}}} = \theta_i \partial_{\vartheta_j} + \theta_j \partial_{\vartheta_i} \tag{9.1.15}$$

for $i, j \in \overline{1, n}$. By (9.1.11), (9.1.14) and (9.1.15), the above representation of $sp(2n, \mathbb{F})$ is essentially an extension of that for $sl(n, \mathbb{F})$ given in (6.2.43). Recall

$$\check{\Delta} = \sum_{r=1}^{n} \partial_{\theta_r} \partial_{\vartheta_r}, \qquad \check{\eta} = \sum_{r=1}^{n} \theta_r \vartheta_r. \tag{9.1.16}$$

It can be verified that

$$\xi \check{\Delta} = \check{\Delta} \xi, \qquad \xi \check{\eta} = \check{\eta} \xi \qquad \text{for } \xi \in sp(2n, \mathbb{F}) \tag{9.1.17}$$

as operators on $\check{\mathscr{A}}$.

Recall the notation $\check{\mathscr{A}}_k$ in (7.1.24). We denote

$$\check{\mathscr{H}}_k = \{u \in \check{\mathscr{A}}_k \mid \check{\Delta}(u) = 0\}. \tag{9.1.18}$$

Then $\check{\mathscr{H}}_k$ forms an $sp(2n, \mathbb{F})$-submodule. According to (6.2.62),

$$\check{\mathscr{H}}_k = \sum_{\ell=0}^{\min\{k,n\}} \check{\mathscr{H}}_{\ell,k-\ell} \qquad \text{for } k \in \overline{0, 2n}. \tag{9.1.19}$$

Expressions (6.2.53) and (6.2.60) tell us that the $sp(2n, \mathbb{F})$-singular vectors of $\check{\mathscr{H}}_k$ are in the set

$$\{\vec{\theta}_r \vec{\vartheta}_s \mid 0 \le r < s \le n+1; r+n-s+1 = k\} \tag{9.1.20}$$

(cf. (6.2.50) and (6.2.51)). If $s < n+1$, then

$$E_{s,n+s}(\vec{\theta}_r \vec{\vartheta}_s) = \theta_s \partial_{\vartheta_s}(\vec{\theta}_r \vec{\vartheta}_s) = \vec{\theta}_r \vec{\vartheta}_{s+1} \theta_s \ne 0. \tag{9.1.21}$$

Thus $\check{\mathscr{H}}_k$ has an unique (up to a scalar multiple) singular vector $\vec{\theta}_k$ of weight $\sum_{i=1}^k \varepsilon_i$. By (6.2.64), (9.1.21) and Weyl's Theorem 2.3.6 of completely reducibility if $\mathbb{F} = \mathbb{C}$ or more generally by Lemma 6.3.2 with $n_1 = 0$, we have:

Theorem 9.1.2 *For $k \in \overline{0, n}$, $\check{\mathscr{H}}_k$ is a finite-dimensional irreducible $sp(2n, \mathbb{F})$-module with the highest-weight vector $\vec{\theta}_k$ of weight $\sum_{i=1}^k \varepsilon_i$. Moreover,*

$$\check{\mathscr{A}} = \bigoplus_{k=0}^{n} \bigoplus_{\ell=0}^{[\![n-k/2]\!]} \check{\eta}^\ell \check{\mathscr{H}}_k \tag{9.1.22}$$

is a direct sum of irreducible $sp(2n, \mathbb{F})$-submodules.

We let $\mathscr{A} = \mathbb{F}[x_1, ..., x_n]$. The *Weil representation* of $sp(2n, \mathbb{F})$ on $\mathscr{A}$ by

$$(E_{i,j} - E_{n+j,n+i})|_{\mathscr{A}} = x_i \partial_{x_j} + \frac{1}{2}\delta_{i,j} \tag{9.1.23}$$

$$(E_{i,n+j} + E_{j,i+n})|_{\mathscr{A}} = x_i x_j, \quad (E_{n+i,j} + E_{n+j,i})|_{\mathscr{A}} = -\partial_{x_i}\partial_{x_j} \tag{9.1.24}$$

for $i, j \in \overline{1, n}$. Denote

$$\mathscr{A}_{(0)} = \text{Span}\{x^\alpha \mid \alpha \in \mathbb{N}^n, \ |\alpha| \text{ is even}\}, \tag{9.1.25}$$

$$\mathscr{A}_{(1)} = \text{Span}\{x^\alpha \mid \alpha \in \mathbb{N}^n, \ |\alpha| \text{ is odd}\}. \tag{9.1.26}$$

Theorem 9.1.3 *The subspace $\mathscr{A}_{(0)}$ is an irreducible $sp(2n,\mathbb{F})$-module and the identity element $1_{\mathscr{A}}$ is a highest-weight vector with weight $(1/2)\sum_{i=1}^{n}\varepsilon_i$. Moreover, $\mathscr{A}_{(1)}$ is also an irreducible $sp(2n,\mathbb{F})$-module and x_1 element is a highest-weight vector with weight $\varepsilon_1+(1/2)\sum_{i=1}^{n}\varepsilon_i$.*

9.2 Noncanonical Oscillator Representations

In this section, we study the noncanonical oscillator representations of symplectic Lie algebras obtained from the canonical bosonic oscillator representations of sympletic Lie algebras over their minimal natural modules by partially swapping differential operators and multiplication operators.

9.2.1 Main Theorem

Denote $\mathscr{B}=\mathbb{F}[x_1,...,x_n,y_1,...,y_n]$. Fix $n_1,n_2\in\overline{1,n}$ with $n_1\leq n_2$. Swapping operators $\partial_{x_r}\mapsto -x_r,\ x_r\mapsto\partial_{x_r}$ for $r\in\overline{1,n_1}$ and $\partial_{y_s}\mapsto -y_s,\ y_s\mapsto\partial_{y_s}$ for $s\in\overline{n_2+1,n}$ in the canonical bonsonic oscillator representation (9.1.8) and (9.1.9), we have the *nonocanoical oscillator representation* of the Lie algebra $sp(2n,\mathbb{F})$ on $\mathscr{B}$ determined by

$$(E_{i,j}-E_{n+j,n+i})|_{\mathscr{B}}=E_{i,j}^x-E_{j,i}^y \tag{9.2.1}$$

with

$$E_{i,j}^x=\begin{cases}-x_j\partial_{x_i}-\delta_{i,j} & \text{if } i,j\in\overline{1,n_1};\\ \partial_{x_i}\partial_{x_j} & \text{if } i\in\overline{1,n_1},\ j\in\overline{n_1+1,n};\\ -x_ix_j & \text{if } i\in\overline{n_1+1,n},\ j\in\overline{1,n_1};\\ x_i\partial_{x_j} & \text{if } i,j\in\overline{n_1+1,n};\end{cases} \tag{9.2.2}$$

and

$$E_{i,j}^y=\begin{cases}y_i\partial_{y_j} & \text{if } i,j\in\overline{1,n_2};\\ -y_iy_j & \text{if } i\in\overline{1,n_2},\ j\in\overline{n_2+1,n};\\ \partial_{y_i}\partial_{y_j} & \text{if } i\in\overline{n_2+1,n},\ j\in\overline{1,n_2};\\ -y_j\partial_{y_i}-\delta_{i,j} & \text{if } i,j\in\overline{n_2+1,n};\end{cases} \tag{9.2.3}$$

$$E_{i,n+j}|_{\mathscr{B}}=\begin{cases}\partial_{x_i}\partial_{y_j} & \text{if } i\in\overline{1,n_1},\ j\in\overline{1,n_2},\\ -y_j\partial_{x_i} & \text{if } i\in\overline{1,n_1},\ j\in\overline{n_2+1,n},\\ x_i\partial_{y_j} & \text{if } i\in\overline{n_1+1,n},\ j\in\overline{1,n_2},\\ -x_iy_j & \text{if } i\in\overline{n_1+1,n},\ j\in\overline{n_2+1,n};\end{cases} \tag{9.2.4}$$

$$E_{n+i,j}|_{\mathscr{B}} = \begin{cases} -x_j y_i & \text{if } j \in \overline{1,n_1},\ i \in \overline{1,n_2}, \\ -x_j \partial_{y_i} & \text{if } j \in \overline{1,n_1},\ i \in \overline{n_2+1,n}, \\ y_i \partial_{x_j} & \text{if } j \in \overline{n_1+1,n},\ i \in \overline{1,n_2}, \\ \partial_{x_j}\partial_{y_i} & \text{if } j \in \overline{n_1+1,n},\ i \in \overline{n_2+1,n}. \end{cases} \tag{9.2.5}$$

As Sect. 7.2, we set

$$\mathscr{B}_{\langle k\rangle} = \text{Span}\left\{x^\alpha y^\beta \mid \alpha,\beta \in \mathbb{N}^n;\ \sum_{r=n_1+1}^{n} \alpha_r - \sum_{i=1}^{n_1} \alpha_i + \sum_{i=1}^{n_2} \beta_i - \sum_{r=n_2+1}^{n} \beta_r = k\right\} \tag{9.2.6}$$

for $k \in \mathbb{Z}$. Then $\mathscr{B} = \bigoplus_{k\in\mathbb{Z}} \mathscr{B}_{\langle k\rangle}$ is a $\mathbb{Z}$-graded space. Let

$$\widehat{D} = \sum_{r=n_1+1}^{n} x_r\partial_{x_r} - \sum_{i=1}^{n_1} x_i\partial_{x_i} + \sum_{i=1}^{n_2} y_i\partial_{y_i} - \sum_{r=n_2+1}^{n} y_r\partial_{y_r}. \tag{9.2.7}$$

It is straightforward to verify

$$\mathscr{B}_{\langle k\rangle} = \{f \in \mathscr{B} \mid \widehat{D}(f) = kf\} \tag{9.2.8}$$

and

$$[E^x_{i,j}, \widehat{D}] = [E^y_{i,j}, \widehat{D}] = [E_{i,n+j}|_{\mathscr{B}}, \widehat{D}] = [E_{n+i,j}|_{\mathscr{B}}, \widehat{D}] = 0 \tag{9.2.9}$$

for $i, j \in \overline{1,n}$. Thus $\mathscr{B}_{\langle k\rangle}$ forms an $sp(2n,\mathbb{F})$-submodule for any $k \in \mathbb{Z}$.

The bilinear form $(\cdot|\cdot)$ on $\mathscr{B}$ is defined by (6.4.46). The main theorem in this section is:

Theorem 9.2.1 *Let $k \in \mathbb{Z}$. If $n_1 < n_2$ or $k \neq 0$, the subspace $\mathscr{B}_{\langle k\rangle}$ is an irreducible $sp(2n,\mathbb{F})$-module. Moreover, it is a highest-weight module only if $n_2 = n$, in which case for $m \in \mathbb{N}$, $x_{n_1}^m$ is a highest-weight vector of $\mathscr{B}_{\langle -m\rangle}$ with weight $m\varepsilon_{n_1} - \sum_{i=1}^{n_1}\varepsilon_i$, $x_{n_1+1}^{m+1}$ is a highest-weight vector of $\mathscr{B}_{\langle m+1\rangle}$ with weight $(m+1)\varepsilon_{n_1+1} - \sum_{i=1}^{n_1}\varepsilon_i$ if $n_1 < n$ and y_n^{m+1} is a highest-weight vector of $\mathscr{B}_{\langle m+1\rangle}$ with weight $-(m+1)\varepsilon_n - \sum_{i=1}^{n_1}\varepsilon_i$ when $n_1 = n$.*

When $n_1 = n_2$, the subspace $\mathscr{B}_{\langle 0\rangle}$ is a direct sum of two irreducible $sp(2n,\mathbb{F})$-submodules. If $n_1 = n_2 = n$, they are highest-weight modules with a highest-weight vector 1 of weight $-\sum_{i=1}^{n}\varepsilon_i$ and with a highest-weight vector $x_{n-1}y_n - x_ny_{n-1}$ of weight $-\varepsilon_{n-1} - \varepsilon_n - \sum_{i=1}^{n}\varepsilon_i$, respectively.

When $n_1 - n_2 + 1 - \delta_{n_1,n_2} < k \in \mathbb{Z}$, $\mathscr{B}_{\langle k\rangle}$ is not completely reducible with respect to its Lie subalgebra in (9.1.11) by Sects. 6.4–6.6. Thus the above representation (9.2.1)–(9.2.5) is not unitary. Observe that

$$(E_{i,i}-E_{n+i,n+i})|_{\mathscr{B}}=\begin{cases}-x_i\partial_{x_i}-y_i\partial_{y_i}-1 & \text{if } 1\leq i\leq n_1,\\ x_i\partial_{x_i}-y_i\partial_{y_i} & \text{if } n_1<i\leq n_2,\\ x_i\partial_{x_i}+y_i\partial_{y_i}+1 & \text{if } n_2<i\leq n\end{cases}\tag{9.2.10}$$

by (9.2.1)–(9.2.3). So only finite number of monomials $x_i^{\ell_1}y_i^{\ell_2}$ appear in a weight subspace when $1\leq i\leq n_1$ or $n_2<i\leq n$. If $n_1<i\leq n_2$, the degree of $x_i^{\ell_1}y_i^{\ell_2}$ is $\ell_1+\ell_2$. This shows that all $\mathscr{B}_{\langle k\rangle}$ with $k\in\mathbb{Z}$ are infinite-dimensional weight $sp(2n,\mathbb{F})$-modules with finite-dimensional weight subspaces.

Due to the mixture of differential operators with multiplications operators, the modules $\mathscr{B}_{\langle k\rangle}$ with $k\in\mathbb{Z}$ are cuspidal modules without highest-weight vectors if $n_2<n$.

Note

$$(E_{i,n+j}+E_{j,n+i})|_{\mathscr{B}}=\partial_{x_i}\partial_{y_j}+\partial_{x_j}\partial_{y_i},\quad (E_{i,n+r}+E_{r,n+i})|_{\mathscr{B}}=\partial_{x_i}\partial_{y_r}+x_r\partial_{y_i},\tag{9.2.11}$$

$$(E_{r,n+s}+E_{s,n+r})|_{\mathscr{B}}=x_r\partial_{y_s}+x_s\partial_{y_r}\tag{9.2.12}$$

for $i,j\in\overline{1,n_1}$ and $r,s\in\overline{n_1+1,n_2}$ by (9.2.4). Moreover,

$$(E_{i,j}-E_{n+j,n+i})|_{\mathscr{B}}=-x_j\partial_{x_i}-y_j\partial_{y_i}-\delta_{i,j}\tag{9.2.13}$$

and

$$(E_{i,r}-E_{n+r,n+i})|_{\mathscr{B}}=\partial_{x_i}\partial_{x_r}-y_r\partial_{y_i}\tag{9.2.14}$$

for $i,j\in\overline{1,n_1}$ and $r\in\overline{n_1+1,n_2}$ by (9.2.1)–(9.2.3). Recall

$$\eta=\sum_{i=1}^{n_1}y_i\partial_{x_i}+\sum_{r=n_1+1}^{n_2}x_ry_r+\sum_{s=n_2+1}^{n}x_s\partial_{y_s}.\tag{9.2.15}$$

9.2.2 Proof of the Theorem When $n_2 = n$

In this subsection, we assume $n_2=n$.

Under the assumption, any element of $\mathscr{B}$ is nilpotent with respect to $sp(2n,\mathbb{F})_+$ by (6.3.21) and (9.2.11)–(9.2.14). Thus any $sp(2n,\mathbb{F})$-submodule of $\mathscr{B}$ must contain an $sp(2n,\mathbb{F})$-singular vector.

First we assume $n_1+1<n$. According to Lemma 6.4.1, the nonzero weight vectors in

$$\text{Span}\{\eta^{m_3}(x_i^{m_1}y_n^{m_2})\mid m_r\in\mathbb{N};\ i=n_1,n_1+1\}\tag{9.2.16}$$

are all the $sl(n,\mathbb{F})$-singular vectors in $\mathscr{B}$. The singular vectors of $sp(2n,\mathbb{F})$ in $\mathscr{B}$ must be among them by (9.1.11). Moreover, the subalgebra $sp(2n,\mathbb{F})_+$ is generated by $sl(n,\mathbb{F})_+$ and $E_{n,2n}$. According to (9.2.12), $E_{n,2n}|_{\mathscr{B}} = x_n\partial_{y_n}$. Hence

$$E_{n,2n}(\eta^{m_3}(x_i^{m_1}y_n^{m_2})) = x_n[m_3x_n\eta^{m_3-1}(x_i^{m_1}y_n^{m_2}) + m_2\eta^{m_3}(x_i^{m_1}y_n^{m_2-1})] \quad (9.2.17)$$

for $i = n_1, n_1+1$ by (9.2.15). Considering weights, we conclude that the vectors $\{x_{n_1}^m, x_{n_1+1}^{m+1} \mid m \in \mathbb{N}\}$ are all the singular vectors of $sp(2n,\mathbb{F})$ in $\mathscr{B}$. Furthermore,

$$x_{n_1}^m \in \mathscr{B}_{\langle -m\rangle} \text{ and } x_{n_1+1}^{m+1} \in \mathscr{B}_{\langle m+1\rangle} \quad \text{for } m \in \mathbb{N}. \quad (9.2.18)$$

Thus each $\mathscr{B}_{\langle k\rangle}$ has a unique non-isotropic singular vector for $k \in \mathbb{Z}$. By Lemma 6.3.2, all $\mathscr{B}_{\langle k\rangle}$ with $k \in \mathbb{Z}$ are irreducible highest-weight $sp(2n,\mathbb{F})$-submodules.

Consider the case $n_1 + 1 = n$. According to (6.5.61), the nonzero weight vectors in

$$\text{Span}\{\eta^{m_2}(x_{n-1}^{m_1}y_n^{m_3}), x_n^{m_1}y_n^{m_2}, \eta^{m_1+m_2}(x_{n-1}^{m_2}y_n^{m_3-m_1}) \mid m_i \in \mathbb{N}\} \quad (9.2.19)$$

are all the $sl(n,\mathbb{F})$-singular vectors in $\mathscr{B}$. By (9.2.17) and considering weights, we again conclude that the vectors $\{x_{n-1}^m, x_n^{m+1} \mid m \in \mathbb{N}\}$ are all the singular vectors of $sp(2n,\mathbb{F})$ in $\mathscr{B}$. Again all $\mathscr{B}_{\langle k\rangle}$ with $k \in \mathbb{Z}$ are irreducible highest-weight $sp(2n,\mathbb{F})$-submodules.

Suppose $n_1 = n$. By (9.2.11), we have $E_{n,2n}|_{\mathscr{B}} = \partial_{x_n}\partial_{y_n}$ in this case. According to (6.6.38), the nonzero weight vectors in

$$\text{Span}\{x_n^{m_1}y_n^{m_2}\zeta_1^{m_3} \mid m_i \in \mathbb{N}\} \quad (9.2.20)$$

are all the $sl(n,\mathbb{F})$-singular vectors in $\mathscr{B}$, where $\zeta_1 = x_{n-1}y_n - x_ny_{n-1}$ in this case. Moreover,

$$\begin{aligned}
&E_{n,2n}(x_n^{m_1}y_n^{m_2}\zeta_1^{m_3}) \\
&= m_1m_2x_n^{m_1-1}y_n^{m_2-1}\zeta_1^{m_3} + m_1m_3x_{n-1}x_n^{m_1-1}y_n^{m_2}\zeta_1^{m_3-1} \\
&\quad -m_2m_3y_{n-1}x_n^{m_1}y_n^{m_2-1}\zeta_1^{m_3-1} - m_3(m_3-1)x_{n-1}y_{n-1}x_n^{m_1}y_n^{m_2}\zeta_1^{m_3-2} \\
&= m_1(m_2+m_3)x_n^{m_1-1}y_n^{m_2-1}\zeta_1^{m_3} + m_3(m_1-m_2-m_3+1)y_{n-1}x_n^{m_1}y_n^{m_2-1}\zeta_1^{m_3-1} \\
&\quad -m_3(m_3-1)y_{n-1}^2x_n^{m_1+1}y_n^{m_2-1}\zeta_1^{m_3-2}. \qquad (9.2.21)
\end{aligned}$$

Considering weights, we again conclude that the vectors $\{x_n^m, y_n^{m+1}, \zeta_1 \mid m \in \mathbb{N}\}$ are all the singular vectors of $sp(2n,\mathbb{F})$ in $\mathscr{B}$. Moreover,

$$x_n^m \in \mathscr{B}_{\langle -m\rangle}, \quad \zeta_1 \in \mathscr{B}_{\langle 0\rangle} \text{ and } y_n^{m+1} \in \mathscr{B}_{\langle m+1\rangle} \quad \text{for } m \in \mathbb{N}. \quad (9.2.22)$$

Thus each $\mathscr{B}_{\langle k\rangle}$ with $k \neq 0$ has a unique non-isotropic singular vector for $k \in \mathbb{Z}$. By Lemma 6.3.2, all $\mathscr{B}_{\langle k\rangle}$ with $0 \neq k \in \mathbb{Z}$ are irreducible highest-weight $sp(2n,\mathbb{F})$-submodules.

Set

$$\mathscr{B}_{\langle 0,1\rangle} = \text{Span}\left\{\left[\prod_{1\le r\le s\le n}(x_r y_s + x_s y_r)^{l_{r,s}}\right] \mid l_{r,s}\in\mathbb{N}\right\} \tag{9.2.23}$$

and

$$\mathscr{B}_{\langle 0,2\rangle} = \text{Span}\left\{\left[\prod_{1\le r\le s\le n}(x_r y_s + x_s y_r)^{l_{r,s}}\right](x_p y_q - x_q y_p) \mid l_{r,s}\in\mathbb{N};\ 1\le p<q\le n\right\}. \tag{9.2.24}$$

Let

$$\mathscr{G}' = \sum_{1\le r\le s\le n}\mathbb{F}(E_{n+s,r}+E_{n+r,s}) \tag{9.2.25}$$

and

$$\widehat{\mathscr{G}} = \sum_{i,j=1}^{n}\mathbb{F}(E_{i,j}-E_{n+j,n+i}) + \sum_{1\le r\le s\le n}\mathbb{F}(E_{r,n+s}+E_{s,n+r}). \tag{9.2.26}$$

Then $\mathscr{G}'$ and $\widehat{\mathscr{G}}$ are Lie subalgebras of $sp(2n,\mathbb{F})$ and $sp(2n,\mathbb{F}) = \mathscr{G}'\oplus\widehat{\mathscr{G}}$. By PBW Theorem

$$U(sp(2n,\mathbb{F})) = U(\mathscr{G}')U(\widehat{\mathscr{G}}). \tag{9.2.27}$$

Note

$$(E_{n+s,r}+E_{n+r,s})|_{\mathscr{B}} = -(x_r y_s + x_s y_r) \qquad \text{for } r,s\in\overline{1,n} \tag{9.2.28}$$

by (9.2.5). According to (9.2.11), (9.2.13) and (9.2.28),

$$\mathscr{B}_{\langle 0,1\rangle} = U(\mathscr{G}')(1) = U(sp(2n,\mathbb{F}))(1) \tag{9.2.29}$$

and

$$\mathscr{B}_{\langle 0,2\rangle} = \sum_{1\le p<q\le n} U(\mathscr{G}')(x_p y_q - x_q y_p) = U(sp(2n,\mathbb{F}))(\zeta_1) \tag{9.2.30}$$

are $sp(2n,\mathbb{F})$-submodules.

It is obvious, $1\notin\mathscr{B}_{\langle 0,2\rangle}$. On the other hand, $(\mathscr{B}_{\langle 0,1\rangle}|x_{n-1}y_n - x_n y_{n-1}) = \{0\}$. Hence $x_{n-1}y_n - x_n y_{n-1}\notin\mathscr{B}_{\langle 0,1\rangle}$. Thus $\mathscr{B}_{\langle 0,1\rangle}$ and $\mathscr{B}_{\langle 0,0\rangle}$ have a unique non-isotropic singular vector. By Lemma 6.3.2, they are irreducible. Note that

$$\widehat{\mathscr{B}}_{\langle 0\rangle} = \mathscr{B}_{\langle 0,1\rangle} + \mathscr{B}_{\langle 0,2\rangle} \tag{9.2.31}$$

is a direct sum of $sp(2n,\mathbb{F})$-submodules and the restriction of the symmetric bilinear form $(\cdot|\cdot)$ on $\widehat{\mathscr{B}}_{\langle 0\rangle}$ is nondegenerate because $\mathscr{B}_{\langle 0,1\rangle}$ and $\mathscr{B}_{\langle 0,2\rangle}$ are irreducible $sp(2n,\mathbb{F})$-submodules whose singular vectors are non-isotropic. Thus

$$\mathscr{B}_{\langle 0\rangle} = \widehat{\mathscr{B}}_{\langle 0\rangle} \oplus \left((\widehat{\mathscr{B}}_{\langle 0\rangle})^{\perp} \bigcap \mathscr{B}_{\langle 0\rangle}\right) \tag{9.2.32}$$

and $(\mathscr{B}'_{\langle 0\rangle})^{\perp} \bigcap \mathscr{B}_{\langle 0\rangle}$ is an $sp(2n, \mathbb{F})$-submodule. Thus $(\widehat{\mathscr{B}}_{\langle 0\rangle})^{\perp} \bigcap \mathscr{B}_{\langle 0\rangle}$ contains an $sp(2n, \mathbb{F})$-singular vector. Since 1 and $x_{n-1}y_n - x_n y_{n-1}$ are the only singular vectors in $\mathscr{B}_{\langle 0\rangle}$, it contains 1 or $x_{n-1}y_n - x_n y_{n-1}$. This contradicts the fact that 1 and $x_{n-1}y_n - x_n y_{n-1}$ are non-isotropic. Thus $(\widehat{\mathscr{B}}_{\langle 0\rangle})^{\perp} \bigcap \mathscr{B}_{\langle 0\rangle} = \{0\}$. This proves the theorem when $n_2 = n$.

9.2.3 Proof of the Theorem When $n_1 < n_2 < n$

In this subsection, we assume $n_1 < n_2 < n$.

Case 1. $n_1 + 1 < n_2$

Set

$$\begin{aligned}\mathscr{G}_1 = &\sum_{i,j=1}^{n_1+1} \mathbb{F}(E_{i,j} - E_{n+j,n+i}) + \sum_{i=1}^{n_1+1} (\mathbb{F}E_{i,n+i} + \mathbb{F}E_{n+i,i}) \\ &+ \sum_{1\leq i<j\leq n_1+1} [\mathbb{F}(E_{i,n+j} + E_{n+j,i}) + \mathbb{F}(E_{n+i,j} + E_{n+j,i})]\end{aligned} \tag{9.2.33}$$

and

$$\begin{aligned}\mathscr{G}_2 = &\sum_{i,j=n_1+2}^{n} \mathbb{F}(E_{i,j} - E_{n+j,n+i}) + \sum_{i=n_1+2}^{n} (\mathbb{F}E_{i,n+i} + \mathbb{F}E_{n+i,i}) \\ &+ \sum_{n_1+2\leq i<j\leq n} [\mathbb{F}(E_{i,n+j} + E_{n+j,i}) + \mathbb{F}(E_{n+i,j} + E_{n+j,i})].\end{aligned} \tag{9.2.34}$$

Then $\mathscr{G}_1 = sp(2n_1 + 2, \mathbb{F})$ and $\mathscr{G}_2 \cong sp(2(n - n_1) - 2, \mathbb{F})$ are Lie subalgebras of $sp(2n, \mathbb{F})$. Denote

$$M^1 = \mathbb{F}[x_1, ..., x_{n_1+1}, y_1, ..., y_{n_1+1}], \qquad M^2 = \mathbb{F}[x_{n_1+2}, ..., x_n, y_{n_1+2}, ..., y_n]. \tag{9.2.35}$$

Observe that M^1 is exactly the $\mathscr{G}_1$-module as $\mathscr{B}$ in Sect. 9.2.2 with $n \to n_1 + 1$ and M^2 is exactly the $\mathscr{G}_1$-module as $\mathscr{B}$ in Sect. 9.2.2 with $n \to n - n_1 - 1$. Write

$$M^i_{\langle k\rangle} = M^i \bigcap \mathscr{B}_{\langle k\rangle}. \tag{9.2.36}$$

By Sect. 9.2.2, $\{M^i_{\langle k\rangle} \mid k \in \mathbb{Z}\}$ are irreducible $\mathscr{G}_i$-submodules.

According to Lemma 6.4.1, the nonzero weight vectors in

$$\text{Span}\{\eta^{m_3}(x_i^{m_1} y_j^{m_2}) \mid m_r \in \mathbb{N}; i = n_1, n_1 + 1; j = n_2, n_2 + 1\} \tag{9.2.37}$$

are all the $sl(n,\mathbb{F})$-singular vectors in $\mathscr{B}$. Fix $k \in \mathbb{N}$. Then the $sl(n,\mathbb{F})$-singular vectors in $\mathscr{B}_{\langle -k\rangle}$ are

$$\{\eta^{m_3}(x_{n_1}^{k+m_2+2m_3}y_{n_2}^{m_2}),\ \eta^{m_3}(x_{n_1+1}^{m_1}y_{n_2+1}^{k+m_1+2m_3}),\\ \eta^{m_3}(x_{n_1}^{m_4}y_{n_2+1}^{m_5}) \mid m_i \in \mathbb{N};\ m_4+m_5-2m_3=k\}. \tag{9.2.38}$$

Let M be a nonzero $sp(2n,\mathbb{F})$-submodule of $\mathscr{B}_{\langle -k\rangle}$. Then M contains an $sl(n,\mathbb{F})$-singular vector by (9.1.11). Suppose some $\eta^{m_3}(x_{n_1}^{k+m_2+2m_3}y_{n_2}^{m_2}) \in M$. We have $E_{n_1,n+n_1}|_{\mathscr{B}} = \partial_{x_{n_1}}\partial_{y_{n_1}}$ and

$$E_{n_1,n+n_1}^{m_3}[\eta^{m_3}(x_{n_1}^{k+m_2+2m_3}y_{n_2}^{m_2})] = m_3!\left[\prod_{r=1}^{2m_3}(k+m_2+r)\right]x_{n_1}^{k+m_2}y_{n_2}^{m_2} \in M \tag{9.2.39}$$

by (9.2.11) and (9.2.15). Moreover, $(E_{n_1,n+n_2}+E_{n_2,n+n_1})|_{\mathscr{B}} = \partial_{x_{n_1}}\partial_{y_{n_2}} + x_{n_2}\partial_{y_{n_1}}$ and

$$(E_{n_1,n+n_2}+E_{n_2,n+n_1})^{m_2}(x_{n_1}^{k+m_2}y_{n_2}^{m_2}) = m_2!\left[\prod_{r=1}^{m_2}(k+r)\right]x_{n_1}^{k} \in M \tag{9.2.40}$$

by (9.2.11). Thus

$$x_{n_1}^{k} \in M. \tag{9.2.41}$$

Assume some $\eta^{m_3}(x_{n_1+1}^{m_1}y_{n_2+1}^{k+m_1+2m_3}) \in M$. According to (9.2.5),

$$(E_{n+i,j}+E_{n+j,i})|_{\mathscr{B}} = \partial_{x_i}\partial_{y_j} + \partial_{x_j}\partial_{y_i} \qquad \text{for } i \in \overline{n_2+1,n}. \tag{9.2.42}$$

So

$$E_{n+n_2+1,n_2+1}^{m_3}[\eta^{m_3}(x_{n_1+1}^{m_1}y_{n_2+1}^{k+m_1+2m_3})] = m_3!\left[\prod_{r=1}^{2m_3}(k+m_1+r)\right]x_{n_1+1}^{m_1}y_{n_2+1}^{k+m_1} \in M. \tag{9.2.43}$$

Moreover,

$$(E_{n+n_2+1,n_1+1}+E_{n+n_1+1,n_2+1})|_{\mathscr{B}} = \partial_{x_{n_1+1}}\partial_{y_{n_2+1}} + y_{n_1+1}\partial_{x_{n_2+1}} \tag{9.2.44}$$

by (9.2.5). Hence

$$(E_{n+n_2+1,n_1+1}+E_{n+n_1+1,n_2+1})^{m_1}(x_{n_1+1}^{m_1}y_{n_2+1}^{k+m_1}) = m_1!\left[\prod_{r=1}^{m_1}(k+r)\right]y_{n_2+1}^{k} \in M. \tag{9.2.45}$$

Furthermore,

$$(E_{n+n_2+1,n_1}+E_{n+n_1,n_2+1})|_{\mathscr{B}}=-x_{n_1}\partial_{y_{n_2+1}}+y_{n_1}\partial_{x_{n_2+1}} \tag{9.2.46}$$

by (9.2.5). Thus

$$(E_{n+n_2+1,n_1}+E_{n+n_1,n_2+1})^k(y_{n_2+1}^k)=(-1)^k k!x_{n_1}^k\in M. \tag{9.2.47}$$

Thus (9.2.41) holds again.

Consider $\eta^{m_3}(x_{n_1}^{m_4}y_{n_2+1}^{m_5})$ for some $m_3, m_3, m_4\in\mathbb{N}$ such that $m_4+m_5-2m_3=k$. Note that $E_{n_1+1,n+n_1+1}|_{\mathscr{B}}=x_{n_1+1}\partial_{y_{n_1+1}}$ by (9.2.12) and

$$E_{n_1+1,n+n_1+1}^{m_3}[\eta^{m_3}(x_{n_1}^{m_4}y_{n_2+1}^{m_5})]=m_3!x_{n_1+1}^{2m_3}x_{n_1}^{m_4}y_{n_2+1}^{m_5}\in M. \tag{9.2.48}$$

There exist $r_1, r_2\in\mathbb{N}$ such that $r_1+r_2=2m_3$ and $r_1\le m_4,\ r_2\le m_5$. Moreover,

$$(E_{n_1,n_1+1}-E_{n+n_1+1,n+n_1})|_{\mathscr{B}}=\partial_{x_{n_1}}\partial_{x_{n_1+1}}-y_{n_1+1}\partial_{y_{n_1}} \tag{9.2.49}$$

by (9.2.1)–(9.2.3). So (9.2.44) and (9.2.49) yield

$$\begin{aligned}&(E_{n_1,n_1+1}-E_{n+n_1+1,n+n_1})^{r_1}(E_{n+n_2+1,n_1+1}+E_{n+n_1+1,n_2+1})^{r_2}(x_{n_1+1}^{2m_3}x_{n_1}^{m_4}y_{n_2+1}^{m_5})\\&=(2m_3)!\left[\prod_{s_1=0}^{r_1-1}(m_4-s_1)\right]\left[\prod_{s_2=0}^{r_2-1}(m_5-s_2)\right]x_{n_1}^{m_4-r_1}y_{n_2+1}^{m_5-r_2}\in M.\end{aligned} \tag{9.2.50}$$

Furthermore, (9.2.5) yields

$$\begin{aligned}&(E_{n+n_2+1,n_1}+E_{n+n_1,n_2+1})^{m_5-r_2}(x_{n_1}^{m_4-r_1}y_{n_2+1}^{m_5-r_2})\\&=(-1)^{m_5-r_2}(m_5-r_2)!x_{n_1}^k\in M.\end{aligned} \tag{9.2.51}$$

Thus we always have $x_{n_1}^k\in M$.

Note that $M_{\langle -k\rangle}^1\ni x_{n_1}^k$ is an irreducible $\mathscr{G}_1$-module (cf. (9.2.33) and (9.2.35)) by Sect. 9.2.2. So

$$M_{\langle -k\rangle}^1\subset M. \tag{9.2.52}$$

Observe that

$$(E_{n_1,n+n_2+1}+E_{n_2+1,n+n_1})|_{\mathscr{B}}=-y_{n_2+1}\partial_{x_{n_1}}+x_{n_2+1}\partial_{y_{n_1}} \tag{9.2.53}$$

by (9.2.4). Moreover,

$$(E_{n+n_2,n_1+1}+E_{n+n_1+1,n_2})|_{\mathscr{B}}=y_{n_2}\partial_{x_{n_1+1}}+y_{n_1+1}\partial_{x_{n_2}} \tag{9.2.54}$$

by (9.2.5).

For any $r\in\mathbb{N}$, we have $x_{n_1}^{k+r}x_{n_1+1}^r\in M_{\langle -k\rangle}^1\subset M$. Take any $s_1\in\overline{0,k+r}$ and $r_1\in\overline{0,r}$. Using (9.2.53) and (9.2.54), we get

$$(E_{n+n_2,n_1+1}+E_{n+n_1+1,n_2})^{r_1}(E_{n_1,n+n_2+1}+E_{n_2+1,n+n_1})^s(x_{n_1}^{k+r}x_{n_1+1}^r)$$
$$=(-1)^s\left[\prod_{i_1=0}^{r_1}(r-i_1)\right]\left[\prod_{i_2=0}^{s-1}(k+r-i_2)\right]x_{n_1}^{k+r-s}x_{n_1+1}^{r-r_1}y_{n_2}^{r_1}y_{n_2+1}^s\in M. \tag{9.2.55}$$

Thus

$$x_{n_1}^{r_1}x_{n_1+1}^{r_2}y_{n_2}^{s_1}y_{n_2+1}^{s_2}\in M \text{ whenever } r_2-r_1+s_1-s_2=-k. \tag{9.2.56}$$

Applying $U(\mathscr{G}_1)U(\mathscr{G}_2)$ to (9.2.56), we get

$$M_{\langle k_1\rangle}^1 M_{\langle k_2\rangle}^2\subset M \text{ whenever } k_1+k_2=-k. \tag{9.2.57}$$

Therefore, $\mathscr{B}_{\langle -k\rangle}=\sum_{k_1\in\mathbb{Z}}M_{\langle k_1\rangle}^1M_{\langle -k-k_1\rangle}^2=M$; that is, $\mathscr{B}_{\langle -k\rangle}$ is an irreducible $sp(2n,\mathbb{F})$-submodule.

Fix $0<k\in\mathbb{N}$. Then the singular vectors of $\mathscr{K}$ in $\mathscr{B}_{\langle k\rangle}$ are

$$\{\eta^{m_2}(x_{n_1+1}^{k+m_1-2m_2}y_{n_2+1}^{m_1}),\ \eta^{m_2}(x_{n_1}^{m_1}y_{n_2}^{k+m_1-2m_2}),\ \eta^{m_3}(x_{n_1+1}^{m_4}y_{n_2}^{m_5})$$
$$\mid m_i\in\mathbb{N};\ 2m_2\le k+m_1;\ m_4+m_5+2m_3=k\} \tag{9.2.58}$$

by (9.2.39). Let M be a nonzero $sp(2n,\mathbb{F})$-submodule of $\mathscr{B}_{\langle k\rangle}$. Then M contains an $sl(n,\mathbb{F})$-singular vector by (9.1.11). Suppose some $\eta^{m_2}(x_{n_1+1}^{k+m_1-2m_2}y_{n_2+1}^{m_1})\in M$ with $2m_2\le k+m_1$. We have $E_{n_1+1,n+n_1+1}|_{\mathscr{B}}=x_{n_1+1}\partial_{y_{n_1+1}}$ and

$$E_{n_1+1,n+n_1+1}^{m_2}[\eta^{m_2}(x_{n_1+1}^{k+m_1-2m_2}y_{n_2}^{m_1})]=m_2!x_{n_1+1}^{k+m_1}y_{n_2+1}^{m_1}\in M \tag{9.2.59}$$

by (9.2.12) and (9.2.15). Moreover, (9.2.5) gives

$$(E_{n+n_2+1,n_1+1}+E_{n+n_1+1,n_2+1})^{m_1}(x_{n_1+1}^{k+m_1}y_{n_2+1}^{m_1})=m_1!\left[\prod_{r=1}^{m_1}(k+r)\right]x_{n_1+1}^k\in M. \tag{9.2.60}$$

Thus

$$x_{n_1+1}^k\in M. \tag{9.2.61}$$

Assume some $\eta^{m_2}(x_{n_1}^{m_1}y_{n_2}^{k+m_1-2m_2})\in M$ with $2m_2\le k+m_1$. Observe $E_{n+n_2,n_2}=y_{n_2}\partial_{x_{n_2}}$ by (9.2.5). So

$$E_{n+n_2,n_2}^{m_2}[\eta^{m_2}(x_{n_1}^{m_1}y_{n_2}^{k+m_1-2m_2})]=m_2!x_{n_1}^{m_1}y_{n_2}^{k+m_1}\in M. \tag{9.2.62}$$

Moreover, (9.2.11) gives that $(E_{n_1,n+n_2}+E_{n_2,n+n_1})|_{\mathscr{B}}=\partial_{x_{n_1}}\partial_{y_{n_2}}+x_{n_2}\partial_{y_{n_1}}$ and

$$(E_{n_1,n+n_2}+E_{n_2,n+n_1})^{m_1}(x_{n_1}^{m_1}y_{n_2}^{k+m_1})=m_1!\left[\prod_{r=1}^{m_1}(k+r)\right]y_{n_2}^k\in M. \tag{9.2.63}$$

Furthermore, (9.2.12) yields that $(E_{n_1+1,n+n_2}+E_{n_2,n+n_1+1})|_{\mathscr{B}} = x_{n_1+1}\partial_{y_{n_2}} + x_{n_2}\partial_{y_{n_1+1}}$ and

$$(E_{n_1+1,n+n_2}+E_{n_2,n+n_1+1})^k(y_{n_2}^k) = k!x_{n_1+1}^k \in M. \tag{9.2.64}$$

Thus (9.2.61) holds again.

Consider $\eta^{m_3}(x_{n_1+1}^{m_4}y_{n_2}^{m_5})$ for some $m_3, m_3, m_4 \in \mathbb{N}$ such that $m_4+m_5+2m_3 = k$. Note $E_{n_1+1,n+n_1+1} = x_{n_1+1}\partial_{y_{n_1+1}}$ by (9.2.12). So

$$E_{n_1+1,n+n_1+1}^{m_3}[\eta^{m_3}(x_{n_1+1}^{m_4}y_{n_2}^{m_5})] = m_3!x_{n_1+1}^{m_4+2m_3}y_{n_2}^{m_5} \in M. \tag{9.2.65}$$

According to (9.2.12),

$$(E_{n_1+1,n+n_2}+E_{n_2,n+n_1+1})^{m_5}(x_{n_1+1}^{m_4+2m_3}y_{n_2}^{m_5}) = m_5!x_{n_1+1}^k \in M. \tag{9.2.66}$$

Therefore, we always have $x_{n_1+1}^k \in M$.

Observe that $M_{\langle k\rangle}^1 \ni x_{n_1+1}^k$ is an irreducible $\mathscr{G}_1$-module (cf. (9.2.34) and (9.2.35)) by Sect. 9.2.2. So

$$M_{\langle k\rangle}^1 \subset M. \tag{9.2.67}$$

For any $r \in \mathbb{N}$, we have $x_{n_1}^r x_{n_1+1}^{k+r} \in M_{\langle k\rangle}^1 \subset M$. Take any $s_1 \in \overline{0,r}$ and $r_1 \in \overline{0,k+r}$. Using (9.2.53) and (9.2.54), we get

$$\begin{aligned}&(E_{n+n_2,n_1+1}+E_{n+n_1+1,n_2})^{r_1}(E_{n_1,n+n_2+1}+E_{n_2+1,n+n_1})^s(x_{n_1}^r x_{n_1+1}^{k+r})\\ &= (-1)^s\left[\prod_{i_1=0}^{r_1}(k+r-i_1)\right]\left[\prod_{i_2=0}^{s-1}(r-i_2)\right]x_{n_1}^{r-s}x_{n_1+1}^{k+r-r_1}y_{n_2}^{r_1}y_{n_2+1}^s \in M. \end{aligned}\tag{9.2.68}$$

Thus

$$x_{n_1}^{r_1}x_{n_1+1}^{r_2}y_{n_2}^{s_1}y_{n_2+1}^{s_2} \in M \quad \text{whenever} \quad r_2-r_1+s_1-s_2 = k. \tag{9.2.69}$$

Applying $U(\mathscr{G}_1)U(\mathscr{G}_2)$ to (9.2.69), we get

$$M_{\langle k_1\rangle}^1 M_{\langle k_2\rangle}^2 \subset M \quad \text{whenever} \quad k_1+k_2 = k. \tag{9.2.70}$$

Therefore, $\mathscr{B}_{\langle k\rangle} = \sum_{k_1\in\mathbb{Z}} M_{\langle k_1\rangle}^1 M_{\langle k-k_1\rangle}^2 = M$; that is, $\mathscr{B}_{\langle k\rangle}$ is an irreducible $sp(2n,\mathbb{F})$-submodule.

Case 2. $n_2 = n_1+1$.

According to (6.5.54), the nonzero weight vectors in

$$\begin{aligned}\text{Span}\{&\eta^{m_2}(x_i^{m_1}y_j^{m_3}), x_{n_1+1}^{m_1}y_{n_1+1}^{m_2}, \eta^{m_1+m_2}(x_{n_1}^{m_2}y_{n_1+1}^{m_3-m_1}), \eta^{m_1+m_2}(y_{n_1+2}^{m_2}x_{n_1+1}^{m_3-m_1})\\ &\mid m_r \in \mathbb{N}; (i,j) = (n_1,n_1+1), (n_1,n_1+2), (n_1+1,n_1+2)\}. \end{aligned}\tag{9.2.71}$$

are all the singular vectors of $sl(n,\mathbb{F})$ in $\mathscr{B}$. Fix $k\in\mathbb{N}$. Then the $sl(n,\mathbb{F})$-singular vectors in $\mathscr{B}_{\langle -k\rangle}$ are those in (9.2.38). Set

$$\mathscr{G}_3=\sum_{i,j=n_1+1}^{n}\mathbb{F}(E_{i,j}-E_{n+j,n+i})+\sum_{i=n_1+1}^{n}(\mathbb{F}E_{i,n+i}+\mathbb{F}E_{n+i,i})+\sum_{n_1+1\le i<j\le n}[\mathbb{F}(E_{i,n+j}+E_{n+j,i})+\mathbb{F}(E_{n+i,j}+E_{n+j,i})] \qquad (9.2.72)$$

and

$$M^3=\mathbb{F}[x_{n_1+1},...,x_n,y_{n_1+1},...,y_n],\quad M^3_{\langle\ell\rangle}=M^3\bigcap\mathscr{B}_{\langle\ell\rangle}\ \text{ for }\ \ell\in\mathbb{Z}. \qquad (9.2.73)$$

According to the arguments in Case 1, (9.2.56) holds. Applying $U(\mathscr{G}_3)$ to (9.2.56), we have

$$x_{n_1}^{r_1}M^3_{\langle r_2+s_1+s_2\rangle}\subset M \qquad (9.2.74)$$

by Sect. 9.2.2. In particular,

$$x_{n_1}^{r_1}x_{n_1+1}^{r_2}M^2_{\langle s_1+s_2\rangle}\subset M. \qquad (9.2.75)$$

Applying $U(\mathscr{G}_1)$ to (9.2.74), we get (9.2.57). Thus $\mathscr{B}_{\langle -k\rangle}$ is an irreducible $sp(2n,\mathbb{F})$-submodule.

Let $0<k\in\mathbb{N}$. Then the $\mathscr{K}$-singular vectors in $\mathscr{B}_{\langle k\rangle}$ are

$$\{\eta^{m_2}(x_{n_1+1}^{k+m_1-2m_2}y_{n_1+2}^{m_1}),\eta^{m_2}(x_{n_1}^{m_1}y_{n_1+1}^{k+m_1-2m_2}),\eta^{m_5+m_6}(x_{n_1}^{m_6}y_{n_1+1}^{m_7-m_5}),\eta^{m_5+m_6}(y_{n_1+2}^{m_6}x_{n_1+1}^{m_7-m_5}),\\ x_{n_1+1}^{m_3}y_{n_1+1}^{m_4}\mid m_i\in\mathbb{N};\,2m_2\le k+m_1;\,m_3+m_4=k=m_5+m_6+m_7\} \qquad (9.2.76)$$

by (9.2.71). Let M be a nonzero $sp(2n,\mathbb{F})$-submodule of $\mathscr{B}_{\langle k\rangle}$. As an $sl(n,\mathbb{F})$-module, M contains an $sl(n,\mathbb{F})$-singular vector. If $x_{n_1+1}^{m_3}y_{n_1+1}^{m_4}\in M$ with $m_3+m_4=k$, then $E_{n_1+1,n+n_1+1}|_{\mathscr{B}}=x_{n_1+1}\partial_{y_{n_1+1}}$ and

$$E^{m_4}_{n_1+1,n+n_1+1}(x_{n_1+1}^{m_3}y_{n_1+1}^{m_4})=m_4!x_{n_1+1}^{k}\in M\Longrightarrow x_{n_1+1}^{k}\in M \qquad (9.2.77)$$

by (9.2.12). Suppose some $\eta^{m_5+m_6}(x_{n_1}^{m_6}y_{n_1+1}^{m_7-m_5})\in M$ with $m_5+m_6+m_7=k$. According to (9.2.5), $E_{n+n_1+1,n_1+1}=y_{n_1+1}\partial_{x_{n_1+1}}$. So

$$E^{m_5+m_6}_{n+n_1+1,n_1+1}[\eta^{m_5+m_6}(x_{n_1}^{m_6}y_{n_1+1}^{m_7-m_5})]=(m_5+m_6)!x_{n_1}^{m_6}y_{n_1+1}^{k+m_6}\in M. \qquad (9.2.78)$$

Moreover, (9.2.4) yields that $(E_{n_1,n+n_1+1}+E_{n_1+1,n+n_1})|_{\mathscr{B}}=\partial_{x_{n_1}}\partial_{y_{n_1+1}}+x_{n_1+1}\partial_{y_{n_1}}$ and

$$(E_{n_1,n+n_1+1}+E_{n_1+1,n+n_1})^{m_6}(x_{n_1}^{m_6}y_{n_1+1}^{k+m_6})=m_6!\left[\prod_{r=1}^{m_6}(k+r)\right]y_{n_1+1}^k\in M. \quad (9.2.79)$$

Assume some $\eta^{m_5+m_6}(y_{n_1+2}^{m_6}x_{n_1+1}^{m_7-m_5})\in M$ with $m_5+m_6+m_7=k$. By (9.2.12) and (9.2.15),

$$E_{n_1+1,n+n_1+1}^{m_5+m_6}[\eta^{m_5+m_6}(y_{n_1+2}^{m_6}x_{n_1+1}^{m_7-m_5})]=(m_5+m_6)!y_{n_1+2}^{m_6}x_{n_1+1}^{k+m_6}\in M. \quad (9.2.80)$$

Observe

$$(E_{n+n_1+2,n_1+1}+E_{n+n_1+1,n_1+2})|_{\mathscr{B}}=\partial_{x_{n_1+1}}\partial_{y_{n_1+2}}+y_{n_1+1}\partial_{x_{n_1+2}} \quad (9.2.81)$$

by (9.2.5). Hence

$$(E_{n+n_1+2,n_1+1}+E_{n+n_1+1,n_1+2})^{m_6}(y_{n_1+2}^{m_6}x_{n_1+1}^{k+m_6})=m_6!\left[\prod_{r=1}^{m_6}(k+r)\right]x_{n_1+1}^k\in M. \quad (9.2.82)$$

Expressions (9.2.59)–(9.2.66), (9.2.77), (9.2.79) and (9.2.82) show that we always have $x_{n_1+1}^k\in M$. According to the arguments in Case 1, (7.2.69) holds. Then (9.2.75) holds for $r_2-r_1+s_1-s_2=k$. Applying $U(\mathscr{G}_1)$ to it, we get (9.2.70). Thus $\mathscr{B}_{\langle k\rangle}$ is an irreducible $sp(2n,\mathbb{F})$-module.

This completes the proof of the theorem when $n_1<n_2<n$.

9.2.4 Proof of the Theorem When $n_1 = n_2 < n$

In this subsection, we assume $n_1=n_2<n$.

Under the assumption,

$$\eta=\sum_{i=1}^{n_1}y_i\partial_{x_i}+\sum_{s=n_2+1}^{n}x_s\partial_{y_s}. \quad (9.2.83)$$

First we consider the subcase $1<n_1<n-1$. Lemma 6.6.1 says that the nonzero weight vectors in

$$\text{Span}\{x_{n_1}^{m_1}y_{n_1}^{m_2}\zeta_1^{m_3+1},x_{n_1+1}^{m_1}y_{n_1+1}^{m_2}\zeta_2^{m_3+1},\eta^{m_3}(x_{n_1}^{m_1}y_{n_1+1}^{m_2})\mid m_i\in\mathbb{N}\} \quad (9.2.84)$$

are all the $sl(n,\mathbb{F})$-singular vectors in $\mathscr{B}$, where

$$\zeta_1=x_{n_1-1}y_{n_1}-x_{n_1}y_{n_1-1},\quad \zeta_2=x_{n_1+1}y_{n_1+2}-x_{n_1+2}y_{n_1+1}. \quad (9.2.85)$$

Fix $k \in \mathbb{N}+1$. Then the $sl(n,\mathbb{F})$-singular vectors in $\mathscr{B}_{\langle -k\rangle}$ are

$$\{x_{n_1}^{k+m_1}y_{n_1}^{m_1}\zeta_1^{m_2+1}, x_{n_1+1}^{m_1}y_{n_1+1}^{k+m_1}\zeta_2^{m_2+1}, \eta^{m_3}(x_{n_1}^{m_4}y_{n_1+1}^{m_5}) \mid m_i \in \mathbb{N}; m_4+m_5-2m_3=k\}. \tag{9.2.86}$$

Let M be a nonzero $sp(2n,\mathbb{F})$-submodule of $\mathscr{B}_{\langle -k\rangle}$. As a $\mathscr{K}$-module, M contains an $sl(n,\mathbb{F})$-singular vector. Suppose some $x_{n_1}^{k+m_1}y_{n_1}^{m_1}\zeta_1^{m_2+1} \in M$. Note $E_{n_1,n+n_1}|_{\mathscr{B}} = \partial_{x_{n_1}}\partial_{y_{n_1}}$ by (9.2.11), and so

$$\begin{aligned}&E_{n_1,n+n_1}(x_{n_1}^{k+m_1}y_{n_1}^{m_1}\zeta_1^{m_2})\\ &= (k+m_1)m_1x_{n_1}^{k+m_1-1}y_{n_1}^{m_1-1}\zeta_1^{m_2} - m_2(m_2-1)x_{n_1}^{k+m_1}y_{n_1}^{m_1}x_{n_1-1}y_{n_1-1}\zeta_1^{m_2-2}\\ &\quad +(k+m_1)m_2x_{n_1}^{k+m_1-1}y_{n_1}^{m_1}x_{n_1-1}\zeta_1^{m_2-1} - m_1m_2x_{n_1}^{k+m_1}y_{n_1}^{m_1-1}y_{n_1-1}\zeta_1^{m_2-1}.\end{aligned} \tag{9.2.87}$$

Moreover,

$$(E_{n_1-1,n_1} - E_{n+n_1,n+n_1-1})|_{\mathscr{B}} = -(x_{n_1}\partial_{x_{n_1-1}} + y_{n_1}\partial_{y_{n_1-1}}) \tag{9.2.88}$$

by (9.2.1)–(9.2.3). Thus

$$\begin{aligned}&(E_{n_1-1,n_1} - E_{n+n_1,n+n_1-1})^2E_{n_1,n+n_1}(x_{n_1}^{k+m_1}y_{n_1}^{m_1}\zeta_1^{m_2})\\ &= -2m_2(m_2-1)x_{n_1}^{k+m_1+1}y_{n_1}^{m_1+1}\zeta_1^{m_2-2} \in M.\end{aligned} \tag{9.2.89}$$

Hence

$$x_{n_1}^{k+m_1+1}y_{n_1}^{m_1+1}\zeta_1^{m_2-2} \in M \qquad \text{if } m_2 > 1. \tag{9.2.90}$$

Furthermore,

$$(E_{n_1-1,n_1} - E_{n+n_1,n+n_1-1})E_{n_1,n+n_1}(x_{n_1}^{k+m_1}y_{n_1}^{m_1}\zeta_1) = -kx_{n_1}^{k+m_1}y_{n_1}^{m_1} \in M. \tag{9.2.91}$$

So we always have $x_{n_1}^{k+m}y_{n_1}^{m} \in M$ for some $m \in \mathbb{N}$ by induction on m_2.

Observe

$$E_{n_1,n+n_1}(x_{n_1}^{k+m}y_{n_1}^{m}) = \partial_{x_{n_1}}\partial_{y_{n_1}}(x_{n_1}^{k+m}y_{n_1}^{m}) = m!\left[\prod_{r=1}^{m}(k+r)\right]x_{n_1}^{k} \tag{9.2.92}$$

by (9.2.11). Thus

$$x_{n_1}^{k} \in M. \tag{9.2.93}$$

Symmetrically, if some $x_{n_1+1}^{m_1}y_{n_1+1}^{k+m_1}\zeta_2^{m_2+1} \in M$, we have $y_{n_1+1}^{k} \in M$. But

$$(E_{n+n_1+1,n_1} + E_{n+n_1,n_1+1})|_{\mathscr{B}} = -x_{n_1}\partial_{y_{n_1+1}} + y_{n_1}\partial_{x_{n_1+1}} \tag{9.2.94}$$

by (9.2.5), which gives

$$(E_{n+n_1+1,n_1}+E_{n+n_1,n_1+1})^k(y_{n_1+1}^k)=(-1)^k k!x_{n_1}^k\in M. \tag{9.2.95}$$

Thus (9.2.93) holds again.

Assume that some $\eta^{m_3}(x_{n_1}^{m_4}y_{n_1+1}^{m_5})\in M$ with $m_4+m_5-2m_3=k$. Note there exists $r_1,r_2\in\mathbb{N}$ such that $r_1+r_2=m_3$ and $2r_1\leq m_4,\ \ 2r_2\leq m_5$. Moreover, $E_{n_1,n+n_1}|_{\mathscr{B}}=\partial_{x_{n_1}}\partial_{y_{n_1}}$ by (9.2.11) and $E_{n+n_1+1,n_1+1}|_{\mathscr{B}}=\partial_{x_{n_1+1}}\partial_{y_{n_1+1}}$ by (9.2.5). Thus

$$\begin{aligned}&E_{n_1,n+n_1}^{r_1}E_{n+n_1+1,n_1+1}^{r_2}[\eta^{m_3}(x_{n_1}^{m_4}y_{n_1+1}^{m_5})]\\&=m_3!\left[\prod_{s_1=0}^{2r_1-1}(m_4-s_1)\right]\left[\prod_{s_2=0}^{2r_2-1}(m_5-s_2)\right]x_{n_1}^{m_4-2r_1}y_{n_1+1}^{m_5-2r_2}\in M.\end{aligned} \tag{9.2.96}$$

Furthermore, (9.2.11) gives $(E_{n+n_1+1,n_1}+E_{n+n_1,n_1+1})|_{\mathscr{B}}=-x_{n_1}\partial_{y_{n_1+1}}+y_{n_1}\partial_{x_{n_1+1}}$, and so

$$\begin{aligned}&(E_{n+n_1+1,n_1}+E_{n+n_1,n_1+1})^{m_5-2r_2}(x_{n_1}^{m_4-2r_1}y_{n_1+1}^{m_5-2r_2})\\&=(-1)^{m_5-2r_2}(m_5-2r_2)!x_{n_1}^k\in M.\end{aligned} \tag{9.2.97}$$

Thus we always have $x_{n_1}^k\in M$.

Now

$$(E_{n_1,n+n_1+1}+E_{n_1+1,n+n_1})|_{\mathscr{B}}=-y_{n_1+1}\partial_{x_{n_1}}+x_{n_1+1}\partial_{y_{n_1}} \tag{9.2.98}$$

by (9.2.11). For any $r\in\mathbb{N}+1$,

$$\frac{(-1)^r}{\prod_{s=0}^{r-1}(k-r)}(E_{n_1,n+n_1+1}+E_{n_1+1,n+n_1})^r(x_{n_1}^k)=x_{n_1}^{k-r}y_{n_1+1}^r\in M. \tag{9.2.99}$$

Set

$$\begin{aligned}\mathscr{G}_4=&\sum_{i,j=1}^{n_1}\mathbb{F}(E_{i,j}-E_{n+j,n+i})+\sum_{i=1}^{n_1}(\mathbb{F}E_{i,n+i}+\mathbb{F}E_{n+i,i})\\&+\sum_{1\leq i<j\leq n_1}[\mathbb{F}(E_{i,n+j}+E_{n+j,i})+\mathbb{F}(E_{n+i,j}+E_{n+j,i})]\end{aligned} \tag{9.2.100}$$

and

$$M^4=\mathbb{F}[x_1,...,x_{n_1},y_1,...,y_{n_1}],\ \ M_{\langle\ell\rangle}^4=M^4\bigcap\mathscr{B}_{\langle\ell\rangle}\ \text{ for }\ \ell\in\mathbb{Z}. \tag{9.2.101}$$

If $k\geq 2$ and $r\in\overline{1,k-1}$, then

$$M^4_{\langle -k+r\rangle}M^3_{\langle -r\rangle} = U(\mathscr{G}_3)U(\mathscr{G}_4)(x_{n_1}^{k-r}y_{n_1+1}^r) \subset M \tag{9.2.102}$$

(cf. (9.2.72) and (9.2.73)) because $M^4_{\langle -k+r\rangle}$ is an irreducible $\mathscr{G}_4$-module and $M^3_{\langle -r\rangle}$ is an irreducible $\mathscr{G}_3$-module by Sect. 9.2.2. Moreover,

$$M^4_{\langle -k\rangle} = U(\mathscr{G}_4)(x_{n_1}^k),\ \ M^3_{\langle -k\rangle} = U(\mathscr{G}_3)(y_{n_1+1}^k) \subset M. \tag{9.2.103}$$

Furthermore,

$$M^4_{\langle -k\rangle}M^3_{\langle 0\rangle} = U(\mathscr{G}_3)U(\mathscr{G}_4)(x_{n_1}^k) \subset M \ \text{ if } \ n_1 = n-1 \tag{9.2.104}$$

and

$$M^4_{\langle 0\rangle}M^3_{\langle -k\rangle} = U(\mathscr{G}_3)U(\mathscr{G}_4)(y_{n_1+1}^k) \subset M \ \text{ if } \ n_1 = 1. \tag{9.2.105}$$

Note

$$(E_{r,i} - E_{n+i,n+r})|_{\mathscr{B}} = y_iy_r - x_ix_r \qquad \text{for } i \in \overline{1,n_1},\ r \in \overline{n_1+1,n} \tag{9.2.106}$$

by (9.2.1)–(9.2.3). In particular, if $k > 1$ or $n_1 = 1$, we have

$$(E_{n_1+1,n_1} - E_{n+n_1,n+n_1+1})(x_{n_1}^k) = y_{n_1}x_{n_1}^k y_{n_1+1} - x_{n_1}^{k+1}x_{n_1+1} \in M. \tag{9.2.107}$$

Since

$$y_{n_1}x_{n_1}^k y_{n_1+1} \in M^4_{\langle -k+1\rangle}M^3_{\langle -1\rangle} \subset M, \tag{9.2.108}$$

we get

$$x_{n_1}^{k+1}x_{n_1+1} \in M. \tag{9.2.109}$$

Suppose $k = 1$ and $n_1 > 1$. By (9.2.103),

$$\zeta_1 x_{n_1} = (x_{n_1-1}y_{n_1} - x_{n_1}y_{n_1-1})x_{n_1} \in M. \tag{9.2.110}$$

Observe

$$(E_{n_1+1,n+n_1-1} + E_{n_1-1,n+n_1+1})|_{\mathscr{B}} = x_{n_1+1}\partial_{y_{n_1-1}} - y_{n_1+1}\partial_{x_{n_1-1}} \tag{9.2.111}$$

by (9.2.4). So

$$-(E_{n_1+1,n+n_1-1} + E_{n_1-1,n+n_1+1})(\zeta_1 x_{n_1}) = x_{n_1}^2 x_{n_1+1} - x_{n_1}y_{n_1}y_{n_1+1} \in M. \tag{9.2.112}$$

On the other hand, (9.2.5) gives

$$(E_{n+i,j} + E_{n+j,i})|_{\mathscr{B}} = -(x_iy_j + x_jy_i) \qquad \text{for } i, j \in \overline{1,n_1}, \tag{9.2.113}$$

which implies

$$-E_{n+n_1,n_1}(y_{n_1+1}) = x_{n_1}y_{n_1}y_{n_1+1} \in M. \tag{9.2.114}$$

By (9.2.112), we have $x_{n_1}^2 x_{n_1+1} \in M$. So (9.2.109) always holds.

By Sect. 9.2.2,

$$M^4_{\langle -k-1\rangle}M^3_{\langle 1\rangle} = U(\mathscr{G}_3)U(\mathscr{G}_4)(x_{n_1}^{k+1}x_{n_1+1}) \subset M. \tag{9.2.115}$$

Suppose

$$M^4_{\langle -k-i\rangle}M^3_{\langle i\rangle} \subset M \tag{9.2.116}$$

for $1 \leq i \leq m$. Then

$$\begin{aligned}(E_{n_1+1,n_1} &- E_{n+n_1,n+n_1+1})(x_{n_1}^{k+m}x_{n_1+1}^m)\\ &= y_{n_1}x_{n_1}^{k+m}x_{n_1+1}^m y_{n_1+1} - x_{n_1}^{k+m+1}x_{n_1+1}^{m+1} \in M\end{aligned} \tag{9.2.117}$$

by (9.2.106). If $m > 1$, we have

$$y_{n_1}x_{n_1}^{k+m}x_{n_1+1}^m y_{n_1+1} \in M^4_{\langle -k-(m-1)\rangle}M^3_{\langle m-1\rangle} \subset M. \tag{9.2.118}$$

Note

$$(E_{r,n+s} + E_{s,n+r})|_{\mathscr{B}} = -(x_r y_s + x_s y_r) \qquad \text{for } r, s \in \overline{n_1+1, n} \tag{9.2.119}$$

by (9.2.4). If $m = 1$, we have

$$y_{n_1}x_{n_1}^{k+1}x_{n_1+1}y_{n_1+1} = -E_{n_1+1,n+n_1+1}(y_{n_1}x_{n_1}^{k+1}) \subset E_{n_1+1,n+n_1+1}(M^1_{\langle -k\rangle}) \subset M. \tag{9.2.120}$$

Then (9.2.117), (9.2.118) and (9.2.120) give

$$x_{n_1}^{k+m+1}x_{n_1+1}^{m+1} \in M. \tag{9.2.121}$$

Furthermore,

$$M^4_{\langle -k-m-1\rangle}M^3_{\langle m+1\rangle} = U(\mathscr{G}_3)U(\mathscr{G}_4)(x_{n_1}^{k+m+1}x_{n_1+m+1}) \subset M. \tag{9.2.122}$$

Thus (9.2.116) holds for any $i \in \mathbb{N}+1$. Symmetrically, we have

$$M^4_{\langle i\rangle}M^3_{\langle -k-i\rangle} \subset M \qquad \text{for } i \in \mathbb{N}+1. \tag{9.2.123}$$

Suppose $n_1 < n-1$. Then $x_{n_1}^{k+1}x_{n_1+1}\zeta_2 \in M$ by (9.2.116). Moreover,

$$(k+1)y_{n_1}x_{n_1}^k y_{n_1+1}\zeta_2 = -(k+1)E_{n+n_1,n_1}(x_{n_1}^{k-1}y_{n_1+1}\zeta_2) \in M \tag{9.2.124}$$

by (9.2.5) and (9.2.102). According to (9.2.1)–(9.2.3),

$$(E_{n_1,n_1+1} - E_{n+n_1+1,n+n_1})|_{\mathscr{B}} = \partial_{x_{n_1}}\partial_{x_{n_1+1}} - \partial_{y_{n_1}}\partial_{y_{n_1+1}}. \tag{9.2.125}$$

Thus

$$\begin{aligned}&(E_{n_1,n_1+1} - E_{n+n_1+1,n+n_1})[(x_{n_1}^{k+1}x_{n_1+1} - (k+1)y_{n_1}x_{n_1}^k y_{n_1+1})\zeta_2]\\ &= 3(k+1)x_{n_1}^k\zeta_2 \in M\end{aligned} \tag{9.2.126}$$

by (9.2.85). Hence

$$M^4_{\langle -k\rangle}M^3_{\langle 0\rangle} = U(\mathscr{G}_3)U(\mathscr{G}_4)(x_{n_1}^k) + U(\mathscr{G}_3)U(\mathscr{G}_4)(x_{n_1}^k\zeta_2) \subset M \tag{9.2.127}$$

by (9.2.93) and (9.2.126). Symmetrically,

$$M^4_{\langle 0\rangle}M^3_{\langle -k\rangle} \subset M. \tag{9.2.128}$$

By (9.2.102)–(9.2.105), (9.2.116), (9.2.122), (9.2.123), (9.2.127) and (9.2.128),

$$M^4_{\langle -k-r\rangle}M^3_{\langle r\rangle} \subset M \qquad \text{for } r \in \mathbb{Z}. \tag{9.2.129}$$

Therefore,

$$\mathscr{B}_{\langle -k\rangle} = \bigoplus_{r\in\mathbb{Z}} M^1_{\langle -k-r\rangle}M^2_{\langle r\rangle} \subset M. \tag{9.2.130}$$

We get $M = \mathscr{B}_{\langle -k\rangle}$; that is, $\mathscr{B}_{\langle -k\rangle}$ is an irreducible $sp(2n,\mathbb{F})$-module. We can similarly prove that $\mathscr{B}_{\langle k\rangle}$ is an irreducible $sp(2n,\mathbb{F})$-module.

Finally, we study $\mathscr{B}_{\langle 0\rangle}$. We first consider the generic case $1 < n_1 < n-1$. Set

$$\begin{aligned}\mathscr{B}_{\langle 0,1\rangle} = \text{Span}\Bigg\{&\Bigg[\prod_{1\leq r\leq s\leq n_1 \text{ or } n_1+1\leq r\leq s\leq n}(x_r y_s + x_s y_r)^{l_{r,s}}\Bigg]\\ &\times\Bigg[\prod_{p=1}^{n_1}\prod_{q=n_1+1}^{n}(x_p x_q - y_p y_q)^{k_{p,q}}\Bigg] \mid l_{r,s}, k_{p,q} \in \mathbb{N}\Bigg\}\end{aligned} \tag{9.2.131}$$

and

$$\begin{aligned}\mathscr{B}_{\langle 0,2\rangle} = &\sum_{1\leq r<s\leq n_1 \text{ or } n_1+1\leq r<s\leq n}\mathscr{B}_{\langle 0,1\rangle}(x_r y_s - x_s y_r)\\ &+\sum_{p=1}^{n_1}\sum_{q=n_1+1}^{n}\mathscr{B}_{\langle 0,1\rangle}(x_p x_q + y_p y_q).\end{aligned} \tag{9.2.132}$$

We want to prove that $\mathscr{B}_{\langle 0,1\rangle}$ and $\mathscr{B}_{\langle 0,2\rangle}$ form $sp(2n,\mathbb{F})$-submodules.

Let

$$\mathscr{G}^* = \sum_{1\le r\le s\le n_1} \mathbb{F}(E_{n+s,r} + E_{n+r,s}) + \sum_{n_1+1\le p\le q\le n} \mathbb{F}(E_{p,n+q} + E_{q,n+p}) + \sum_{r=1}^{n_1}\sum_{p=n_1+1}^{n} \mathbb{F}(E_{p,r} - E_{n+r,n+p}) \tag{9.2.133}$$

and

$$\tilde{\mathscr{G}} = \sum_{i,j=1}^{n_1} \mathbb{F}(E_{i,j} - E_{n+j,n+i}) + \sum_{r,s=n_1+1}^{n} \mathbb{F}(E_{r,s} - E_{n+s,n+r}) + \sum_{1\le r\le s\le n_1} \mathbb{F}(E_{r,n+s} + E_{s,n+r}) + \sum_{n_1+1\le p\le q\le n} \mathbb{F}(E_{n+q,p} + E_{n+p,q}) + \sum_{r=1}^{n_1}\sum_{p=n_1+1}^{n} [\mathbb{F}(E_{r,p} - E_{n+p,n+r}) + \mathbb{F}(E_{r,n+p} + E_{p,n+r}) + \mathbb{F}(E_{n+r,p} + E_{n+p,r})]. \tag{9.2.134}$$

Then $\mathscr{G}^*$ and $\tilde{\mathscr{G}}$ are Lie subalgebras of $sp(2n,\mathbb{F})$ and $sp(2n,\mathbb{F}) = \mathscr{G}^* \oplus \tilde{\mathscr{G}}$. By PBW Theorem (Theorem 5.1.1), $U(sp(2n,\mathbb{F})) = U(\mathscr{G}^*)U(\tilde{\mathscr{G}})$.

Expressions (9.2.106), (9.2.113) and (9.2.119) show

$$U(\mathscr{G}^*)|_{\mathscr{B}} = \mathscr{B}_{\langle 0,1\rangle} \text{ as multiplication operators on } \mathscr{B}. \tag{9.2.135}$$

Moreover,

$$(E_{r,s} - E_{n+s,n+r})|_{\mathscr{B}} = x_r\partial_{x_s} + y_r\partial_{y_s} + \delta_{r,s}, \tag{9.2.136}$$

$$(E_{n+r,s} + E_{n+s,r})|_{\mathscr{B}} = \partial_{x_r}\partial_{y_s} + \partial_{x_s}\partial_{y_r}, \tag{9.2.137}$$

$$(E_{n+r,i} + E_{n+i,r})|_{\mathscr{B}} = -x_i\partial_{y_r} + y_i\partial_{x_r}, \tag{9.2.138}$$

$$(E_{i,n+r} + E_{r,n+i})|_{\mathscr{B}} = -y_r\partial_{x_i} + x_r\partial_{y_i}, \tag{9.2.139}$$

$$(E_{i,r} - E_{n+r,n+i})|_{\mathscr{B}} = \partial_{x_i}\partial_{x_r} - \partial_{y_i}\partial_{y_r} \tag{9.2.140}$$

for $i \in \overline{1,n_1}$ and $r,s \in \overline{n_1+1,n}$. According to (9.2.11), (9.2.14), (9.2.134) and (9.2.136)–(9.2.140), $U(\tilde{\mathscr{G}})(1) = \mathbb{F}$. Thus

$$\mathscr{B}_{\langle 0,1\rangle} = U(\mathscr{G}^*)(1) = U(sp(2n,\mathbb{F}))(1) \tag{9.2.141}$$

forms an $sp(2n,\mathbb{F})$-submodule.

Let

$$W = \sum_{1\leq r<s\leq n_1 \text{ or } n_1+1\leq r<s\leq n} \mathbb{F}(x_r y_s - x_s y_r) + \sum_{p=1}^{n_1}\sum_{q=n_1+1}^{n} \mathbb{F}(x_p x_q + y_p y_q). \tag{9.2.142}$$

By (9.2.11), (9.2.14), (9.2.134) and (9.2.136)–(9.2.140), we can verify that W forms an irreducible $\tilde{\mathscr{G}}$-submodule. Hence

$$\mathscr{B}_{\langle 0,2\rangle} = U(\mathscr{G}^*)(W) = U(sp(2n,\mathbb{F}))(W) \tag{9.2.143}$$

forms an $sp(2n,\mathbb{F})$-submodule. Moreover,

$$\mathscr{B}_{\langle 0,1\rangle} \bigcap W = \{0\}. \tag{9.2.144}$$

Next we want to prove that $\mathscr{B}_{\langle 0,1\rangle}$ and $\mathscr{B}_{\langle 0,2\rangle}$ are irreducible $sp(2n,\mathbb{F})$-submodules. According to (9.2.84), the $sl(n,\mathbb{F})$-singular vectors in $\mathscr{B}_{\langle 0\rangle}$ are

$$\{x_{n_1}^{m_1} y_{n_1}^{m_1}\zeta_1^{m_2+1}, x_{n_1+1}^{m_1} y_{n_1+1}^{m_1}\zeta_2^{m_2+1}, \eta^{m_3}(x_{n_1}^{m_4} y_{n_1+1}^{m_5}) \mid m_i \in \mathbb{N};\ m_4+m_5 = 2m_3\}. \tag{9.2.145}$$

Let M be a nonzero submodule of $\mathscr{B}_{\langle 0,1\rangle}$. Then M contains an $sl(n,\mathbb{F})$-singular vector. Suppose some $x_{n_1}^{m_1} y_{n_1}^{m_1}\zeta_1^{m_2} \in M$. By (9.2.87)–(9.2.90), we can assume $m_2 = 0, 1$. If $m_2 = 0$, (9.2.92) yields $1 \in M$. Then $M = \mathscr{B}_{\langle 0,1\rangle}$ by (9.2.141). Suppose $m_2 = 1$. We have $E_{n_1,n+n_1}|_{\mathscr{B}} = \partial_{x_{n_1}}\partial_{y_{n_1}}$ by (9.2.11), and

$$E_{n_1,n+n_1}[x_{n_1}^{m_1} y_{n_1}^{m_1}\zeta_1] = m_1(m_1+m_2)x_{n_1}^{m_1-1} y_{n_1}^{m_1-1}\zeta_1 \tag{9.2.146}$$

by (9.2.87). By induction on m_1, we have $\zeta_1 \in M \subset \mathscr{B}_{\langle 0,1\rangle}$, which contradicts (9.2.144). Similarly, if some $x_{n_1+1}^{m_1} y_{n_1+1}^{m_1}\zeta_2^{m_2+1} \in M$, we have $M = \mathscr{B}_{\langle 0,1\rangle}$. Assume some $\eta^{m_3}(x_{n_1}^{m_4} y_{n_1+1}^{m_5})$ in M with $m_4 + m_5 = 2m_3$. Note that m_4 and m_5 are both even or odd. If $m_4 = 2r_1$ and $m_5 = 2r_2$ are even, then (9.2.96) gives $1 \in M$, or equivalently $M = \mathscr{B}_{\langle 0,1\rangle}$. Suppose that $m_4 = 2r_1 + 1$ and $m_5 = 2r_2 + 1$ are odd. As (9.2.96), we find

$$\eta(x_{n_1} y_{n_1+1}) = x_{n_1}x_{n_1+1} + y_{n_1}y_{n_1+1} \in M \subset \mathscr{B}_{\langle 0,1\rangle} \tag{9.2.147}$$

by (9.2.83), which contradicts (9.2.144) again. Thus we always have $M = \mathscr{B}_{\langle 0,1\rangle}$; that is, $\mathscr{B}_{\langle 0,1\rangle}$ is irreducible. Similarly, we can prove that $\mathscr{B}_{\langle 0,2\rangle}$ is irreducible.

If $n_1 = 1$ and $n = 2$, we let

$$\mathscr{B}_{\langle 0,1\rangle} = \text{Span}\{\left[\prod_{i=1}^{n}(x_i y_i)^{m_i}\right](x_1x_2 - y_1y_2)^{\ell} \mid \ell, m_i \in \mathbb{N}\} \tag{9.2.148}$$

and $\mathscr{B}_{\langle 0,2\rangle} = \mathscr{B}_{\langle 0,1\rangle}(x_1x_2 + y_1y_2)$. When $n_1 = 1$ and $n > 2$, we set

$$\mathscr{B}_{\langle 0,1\rangle} = \text{Span}\left\{\left[(x_1y_1)^l \prod_{2\le r\le s\le n}(x_ry_s+x_sy_r)^{l_{r,s}}\right]\left[\prod_{q=2}^{n}(x_1x_q-y_1y_q)^{k_q}\right] \mid l, l_{r,s}, k_q \in \mathbb{N}\right\} \tag{9.2.149}$$

and

$$\mathscr{B}_{\langle 0,2\rangle} = \sum_{2\le r<s\le n}\mathscr{B}_{\langle 0,1\rangle}(x_ry_s-x_sy_r) + \sum_{q=2}^{n}\mathscr{B}_{\langle 0,1\rangle}(x_1x_q+y_1y_q). \tag{9.2.150}$$

In the case $1 < n_1 = n-1$, we put

$$\mathscr{B}_{\langle 0,1\rangle} = \text{Span}\left\{(x_ny_n)^l\left[\prod_{1\le r\le s\le n-1}(x_ry_s+x_sy_r)^{l_{r,s}}\right]\times\left[\prod_{p=1}^{n-1}(x_px_n-y_py_n)^{k_p}\right] \mid l, l_{r,s}, k_p \in \mathbb{N}\right\} \tag{9.2.151}$$

and

$$\mathscr{B}_{\langle 0,2\rangle} = \sum_{1\le r<s\le n_1}\mathscr{B}_{\langle 0,1\rangle}(x_ry_s-x_sy_r) + \sum_{p=1}^{n_1}\mathscr{B}_{\langle 0,1\rangle}(x_px_n+y_py_n). \tag{9.2.152}$$

By the arguments similar to part of those in the above proof of the irreducibility of $\mathscr{B}_{\langle 0,1\rangle}$ when $1 < n_1 < n-1$, we can show that $\mathscr{B}_{\langle 0,1\rangle}$ and $\mathscr{B}_{\langle 0,2\rangle}$ are irreducible.

Recall the symmetric bilinear form $(\cdot|\cdot)$ in (6.4.46). Now 1 is a non-isotropic element in $\mathscr{B}_{\langle 0,1\rangle}$ and $x_{n_1}x_{n_1+1}+y_{n_1}y_{n_1+1}$ a non-isotropic element in $\mathscr{B}_{\langle 0,2\rangle}$. By Lemma 6.3.2, $\mathscr{B}_{\langle 0,r\rangle}\bigcap(\mathscr{B}_{\langle 0,r\rangle})^{\perp}$ is a proper $sp(2n,\mathbb{F})$-submodule of $\mathscr{B}_{\langle 0,r\rangle}$, the irreducibility of $\mathscr{B}_{\langle 0,r\rangle}$ implies $\mathscr{B}_{\langle 0,r\rangle}\bigcap(\mathscr{B}_{\langle 0,r\rangle})^{\perp}=\{0\}$; that is, the bilinear form $(\cdot|\cdot)$ is nondegenerate on $\mathscr{B}_{\langle 0,1\rangle}$ and $\mathscr{B}_{\langle 0,2\rangle}$.

Since $(1|\mathscr{B}_{\langle 0,2\rangle})=\{0\}$, $\mathscr{B}_{\langle 0,1\rangle}$ is orthogonal to $\mathscr{B}_{\langle 0,2\rangle}$. Thus the symmetric bilinear form $(\cdot|\cdot)$ on $\mathscr{B}_{\langle 0,1\rangle}+\mathscr{B}_{\langle 0,2\rangle}$ is nondegenerate. Recall that $\mathscr{B}_k$ is the subspace of homogeneous polynomials in $\mathscr{B}$ with degree k and $\mathscr{B}=\bigoplus_{k=0}^{\infty}\mathscr{B}_k$. According to (6.4.46),

$$(\mathscr{B}_{k_1}|\mathscr{B}_{k_2})=\{0\} \qquad \text{if } k_1\neq k_2. \tag{9.2.153}$$

Moreover,

$$\mathscr{B}_{\langle 0\rangle}=\bigoplus_{k=0}^{\infty}\mathscr{B}_k\bigcap\mathscr{B}_{\langle 0\rangle}, \tag{9.2.154}$$

$$\mathscr{B}_{\langle 0,r\rangle} = \bigoplus_{k=0}^{\infty} \mathscr{B}_k \bigcap \mathscr{B}_{\langle 0,r\rangle}. \tag{9.2.155}$$

The above three expressions show

$$\mathscr{B}_{\langle 0\rangle} = (\mathscr{B}_{\langle 0,1\rangle} + \mathscr{B}_{\langle 0,2\rangle}) \oplus (\mathscr{B}_{\langle 0,1\rangle} + \mathscr{B}_{\langle 0,2\rangle})^{\perp} \bigcap \mathscr{B}_{\langle 0\rangle}. \tag{9.2.156}$$

If $(\mathscr{B}_{\langle 0,1\rangle} + \mathscr{B}_{\langle 0,2\rangle})^{\perp} \bigcap \mathscr{B}_{\langle 0\rangle} \neq \{0\}$, then it contains an $sl(n,\mathbb{F})$-singular vector. Our above arguments in proving the irreducibility of $\mathscr{B}_{\langle 0,1\rangle}$ show that it contains either $\mathscr{B}_{\langle 0,1\rangle}$ or $\mathscr{B}_{\langle 0,2\rangle}$, which is absurd. Therefore, $\mathscr{B}_{\langle 0\rangle} = \mathscr{B}_{\langle 0,1\rangle} \oplus \mathscr{B}_{\langle 0,2\rangle}$ is an orthogonal decomposition of irreducible $sp(2n,\mathbb{F})$-submodules.

This completes the proof of the theorem when $n_1 = n_2 < n$. Therefore, the theorem holds by Sects. 9.2.2–9.2.4.

9.3 Projective Oscillator Representations

Let us go back to Sect. 6.8 and assume $n = 2m+1 > 1$ is an odd integer. In this section, we study the restrictions of the projective oscillator representations on $sp(2m+2,\mathbb{F})$.

Set

$$A_{i,j} = E_{i,j} - E_{m+1+j,m+1+i},\quad B^s_{i,j} = E_{i,m+1+j} + E_{j,m+1+i},\quad C^s_{i,j} = E_{m+1+i,j} + E_{m+1+j,i} \tag{9.3.1}$$

for $i, j \in \overline{1, m+1}$. Then the symplectic Lie algebra

$$sp(2m+2,\mathbb{F}) = \sum_{i,j=1}^{m+1} \mathbb{F}A_{i,j} + \sum_{1\leq p\leq q\leq m+1} (\mathbb{F}B^s_{p,q} + \mathbb{F}C^s_{q,p}). \tag{9.3.2}$$

For convenience, we rednote

$$x_0 = x_{m+1},\quad y_i = x_{m+1+i} \qquad \text{for } i \in \overline{1,m}. \tag{9.3.3}$$

In particular,

$$D = \sum_{i=0}^{m} x_i \partial_{x_i} + \sum_{r=1}^{m} y_r \partial_{y_r}. \tag{9.3.4}$$

According to (6.7.80) and (6.7.81), we have the representation π_κ of $sp(2m+2,\mathbb{F})$:

$$\pi_\kappa(A_{i,j}) = x_i\partial_{x_j} - y_j\partial_{y_i},\quad \pi_\kappa(B^s_{i,j}) = x_i\partial_{y_j} + x_j\partial_{y_i},\quad \pi_\kappa(C^s_{i,j}) = y_i\partial_{x_j} + y_j\partial_{x_i}, \tag{9.3.5}$$

$$\pi_\kappa(A_{i,m+1}) = x_i\partial_{x_0} + \partial_{y_i},\quad \pi_\kappa(C^s_{m+1,i}) = y_i\partial_{x_0} - \partial_{x_i},\quad \pi_\kappa(C^s_{m+1,m+1}) = -2\partial_{x_0}, \tag{9.3.6}$$

$$\pi_\kappa(A_{m+1,i}) = x_0\partial_{x_i} - y_i(D+\kappa),\quad \pi_\kappa(B^s_{i,m+1}) = x_0\partial_{y_i} + x_i(D+\kappa), \tag{9.3.7}$$

$$\pi_\kappa(B^s_{m+1,m+1}) = 2x_0(D+\kappa),\quad \pi_\kappa(A_{m+1,m+1}) = D + x_0\partial_{x_0} + \kappa \tag{9.3.8}$$

for $i, j \in \overline{1, m}$.

Denote

$$\mathscr{K} = \sum_{i,j=1}^{m} \mathbb{F}A_{i,j} + \sum_{1\leq p\leq q\leq m} (\mathbb{F}B^s_{p,q} + \mathbb{F}C^s_{q,p}), \tag{9.3.9}$$

which is a Lie subalgebra of $sp(2m+2,\mathbb{F})$ isomorphic to $sp(2m,\mathbb{F})$. Again we take

$$\mathscr{B} = \mathbb{F}[x_1, ..., x_m, y_1, ..., y_m],\quad \mathscr{B}' = \mathbb{F}[x_0, x_1, ..., x_m, y_1, ..., y_m]. \tag{9.3.10}$$

Let $\mathscr{B}_k$ be the subspace of homogeneous polynomials in $\mathscr{B}$ with degree k and let $\mathscr{B}'_k$ be the subspace of homogeneous polynomials in $\mathscr{B}'$ with degree k. Set

$$\mathscr{B}'_{(\ell)} = \sum_{i=0}^{\ell} \mathscr{B}'_i \qquad \text{for } \ell \in \mathbb{N}. \tag{9.3.11}$$

Take the Cartan subalgebra $H = \sum_{i=1}^{m+1} \mathbb{F}A_{i,i}$ of $sp(2m+2,\mathbb{F})$. Define $\{\varepsilon_1, ..., \varepsilon_{m+1}\} \subset H^*$ by:

$$\varepsilon_j(A_{i,i}) = \delta_{i,j} \qquad \text{for } i, j \in \overline{1, m+1}. \tag{9.3.12}$$

First we have:

Theorem 9.3.1 . *If $\kappa \notin -\mathbb{N}$, the representation π_κ of $sp(2m+2,\mathbb{F})$ on $\mathscr{B}'$ is a highest-weight irreducible representation with highest-weight $-\kappa\lambda_1$. When $-\kappa = \ell \in \mathbb{N}$, $\mathscr{B}'_{(\ell)}$ is a finite-dimensional irreducible $sp(2m+2,\mathbb{F})$-module with highest weight $\ell\lambda_1$ and $\mathscr{B}'/\mathscr{B}'_{(\ell)}$ is an irreducible highest weight $sp(2m+2,\mathbb{F})$-module with highest weight $-(\ell+2)\lambda_1 + (\ell+1)\lambda_2$, where λ_i is the ith fundamental weight of $sp(2m+2,\mathbb{F})$.*

Proof Observe that

$$\mathscr{B}'_k = \sum_{s=0}^{k} x_0^s \mathcal{B}_{k-s} \qquad \text{for } k \in \mathbb{N}. \tag{9.3.13}$$

Let M be a nonzero $sp(2m+2,\mathbb{F})$-submodule of $\mathscr{B}'$. Take any $0 \neq f \in M$. Repeatedly applying (9.3.6) to f, we obtain $1 \in M$. Note

$$(B^s_{m+1,m+1})^k(1) = 2^k\left[\prod_{r=0}^{k-1}(r+\kappa)\right]x_0^k \in M \tag{9.3.14}$$

by (9.3.8) and

$$(A_{1,m+1})^r(x_0^k) = \left[\prod_{i=0}^{r-1}(k-i)x_0^{k-r}\right]x_1^r \quad \text{for } r \in \overline{1,k} \tag{9.3.15}$$

by the first equation in (9.3.5). Suppose $c \notin -\mathbb{N}$. Then (9.3.14) yields

$$x_0^k \in M \quad \text{for } k \in \mathbb{N}. \tag{9.3.16}$$

Moreover, (9.3.15) with $k = r + p$ gives

$$x_0^r x_1^p \in V \quad \text{for } r, p \in \mathbb{N}. \tag{9.3.17}$$

Furthermore,

$$U(\mathcal{K})(x_0^r x_1^p) = x_0^r \mathscr{B}_p \subset M \tag{9.3.18}$$

by Theorem 9.1.1. Thus

$$\mathscr{B}' = \sum_{r,p=0}^{\infty} x_0^r \mathscr{B}_p \subset M; \tag{9.3.19}$$

that is, $M = \mathscr{B}'$. So $\mathscr{B}'$ is an irreducible $sp(2m+2,\mathbb{F})$-module and 1 is its highest-weight vector with weight $-c\lambda_1$ with respect to the following simple positive roots

$$\{\varepsilon_{n+1} - \varepsilon_n, \varepsilon_n - \varepsilon_{n-1}, \ldots, \varepsilon_2 - \varepsilon_1, 2\varepsilon_1\}. \tag{9.3.20}$$

Next we assume $c = -\ell$ with $\ell \in \mathbb{N}$. Since

$$B^s_{m+1,m+1}|_{\mathscr{B}'_\ell} = 0, \quad A_{m+1,i}|_{\mathscr{B}'_\ell} = x_0\partial_{x_i}, \quad B_{m+1,i}|_{\mathscr{B}'_\ell} = x_0\partial_{y_i} \tag{9.3.21}$$

by (9.3.7) and (9.3.8), $\mathscr{B}'_{(\ell)}$ is a finite-dimensional $sp(2m+2,\mathbb{F})$-module. Let M be a nonzero $sp(2m+2,\mathbb{F})$-submodule of $\mathscr{B}'_{(\ell)}$. By (9.3.14),

$$x_0^k \in M \quad \text{for } k \in \overline{0,\ell}. \tag{9.3.22}$$

Moreover, (9.3.15) with $k = r + s$ gives

$$x_0^r x_1^s \in M \quad \text{for } r, s \in \overline{0,\ell} \text{ such that } r + s \le \ell. \tag{9.3.23}$$

Thus

$$\mathscr{B}'_{(\ell)} = \sum_{r=0}^{\ell}\sum_{s=0}^{\ell-r} x_0^r \mathcal{B}_s \subset M \tag{9.3.24}$$

by Lemma 3.1; that is, $M = \mathscr{B}'_{(\ell)}$. So $\mathscr{B}'_{(\ell)}$ is an irreducible $sp(2n+2,\mathbb{F})$-module and 1 is again its highest-weight vector.

Consider the quotient $sp(2n+2,\mathbb{F})$-module $\mathscr{B}'/\mathscr{B}'_{(\ell)}$. Let $W \supset \mathscr{B}'_{(\ell)}$ be an $sp(2n+2,\mathbb{F})$-submodule of $\mathscr{B}'$ such that $W \neq \mathscr{B}'_{(\ell)}$. Take any $f \in W \setminus \mathscr{B}'_{(\ell)}$. Repeatedly applying (9.3.6) to f if necessary, we can assume $f \in \mathcal{B}_{\ell+1}$. Since $\mathcal{B}_{\ell+1}$ is an irreducible $\mathscr{K}$-module by Theorem 9.1.1, we have

$$\mathcal{B}_{\ell+1} \subset W. \tag{9.3.25}$$

In particular, $x_1^{\ell+1} \in W$. According to (9.3.7),

$$(B^s_{1,m+1})^r(x_1^{\ell+1}) = r!x_1^{\ell+1+r} \in W \qquad \text{for } 0 < r \in \mathbb{Z}. \tag{9.3.26}$$

Since $\mathcal{B}_{\ell+1+r} \ni x_1^{\ell+1+r}$ is an irreducible $\mathscr{K}$-module by Theorem 9.1.1, we have

$$\mathcal{B}_{\ell+1+r} \subset W. \tag{9.3.27}$$

Suppose that

$$x_0^r\mathcal{B}_p \subset W \qquad \text{for } r \in \overline{0,k} \text{ and } p \in \mathbb{N} \text{ such that } r+p \geq \ell+1. \tag{9.3.28}$$

Fix such r and p. Observe $x_0^r x_1^{p-1} y_1 \in x_0^r\mathcal{B}_p \subset W$. Using the second equation in (9.3.7), we get

$$B^s_{1,m+1}(x_0^r x_1^{p-1}y_1) = (r+p-\ell)x_0^r x_1^p y_1 + x_0^{r+1}x_1^{p-1} \in W. \tag{9.3.29}$$

By the assumption (9.3.28), $(r+p-\ell)x_0^r x_1^p y_1 \in x_0^r\mathcal{B}_{p+1} \subset W$. So

$$x_0^{r+1}x_1^{p-1} \in W \bigcap x_0^{r+1}\mathcal{B}_{p-1}. \tag{9.3.30}$$

Since $x_0^{r+1}\mathcal{B}_{p-1}$ is an irreducible $\mathscr{K}$-module by Theorem 9.1.1, we get

$$x_0^{r+1}\mathcal{B}_{p-1} \subset W. \tag{9.3.31}$$

By induction on r, we prove

$$x_0^r\mathcal{B}_p \subset W \qquad \text{for } r, p \in \mathbb{N} \text{ such that } r+p \geq \ell+1. \tag{9.3.32}$$

According to (9.3.13),

$$\sum_{k=\ell+1}^{\infty} \mathscr{B}'_k \subset W. \tag{9.3.33}$$

Since $W \supset \mathscr{B}'_{(\ell)}$, we have $W = \mathscr{B}'$. So $\mathscr{B}'/\mathscr{B}'_{(\ell)}$ is an irreducible $sp(2m+2,\mathbb{F})$-module. Moreover, $x_n^{\ell+1}$ is a highest-weight vector of weight $-(\ell+2)\lambda_1+(\ell+1)\lambda_2$ with respect to (9.3.20). □

For $\vec{a}=(a_1,...,a_m)^t$, $\vec{b}=(b_1,b_2,...,b_m)^t \in \mathbb{F}^m$, we put

$$\vec{a}\cdot\vec{x}=\sum_{i=1}^{m} a_i x_i, \qquad \vec{b}\cdot\vec{y}=\sum_{i=1}^{m} b_i y_i. \tag{9.3.34}$$

Set

$$\mathscr{B}'_{\vec{a},\vec{b}} = \{f e^{\vec{a}\cdot\vec{x}+\vec{b}\cdot\vec{y}} \mid f \in \mathscr{B}'\}. \tag{9.3.35}$$

Denote by $\pi_{\kappa,\vec{a},\vec{b}}$ the representation π_κ of $sp(2m+m,\mathbb{F})$ on $\mathscr{B}'_{\vec{a},\vec{b}}$. Our second result in this section is:

Theorem 9.3.2 *The representation $\pi_{\kappa,\vec{a},\vec{b}}$ with $(\vec{a},\vec{b}) \neq (\vec{0},\vec{0})$ is an irreducible representation of $sp(2m+2,\mathbb{F})$ for any $\kappa \in \mathbb{F}$.*

Proof By symmetry, we may assume $a_1 \neq 0$. Set

$$\mathscr{B}'_{\vec{a},\vec{b},k} = \mathscr{B}'_k e^{\vec{a}\cdot\vec{x}+\vec{b}\cdot\vec{y}} \qquad \text{for } k \in \mathbb{N}. \tag{9.3.36}$$

Let M be a nonzero $sp(2m+2,\mathbb{F})$-submodule of $\mathscr{B}'_{\vec{a},\vec{b}}$. Take any $0 \neq f e^{\vec{a}\cdot\vec{x}} \in M$ with $f \in \mathscr{B}'$. By the assumption $a_0=0$ and (9.3.6),

$$C^s_{m+1,m+1}(f e^{\vec{a}\cdot\vec{x}}) = -2\partial_{x_0}(f) e^{\vec{a}\cdot\vec{x}} \in M, \tag{9.3.37}$$

$$(A_{i,m+1}-b_i)(f e^{\vec{a}\cdot\vec{x}}) = [\partial_{y_i}(f)+x_i\partial_{x_0}(f)] e^{\vec{a}\cdot\vec{x}} \in M, \tag{9.3.38}$$

$$(C^s_{m+1,i}+a_i)(f e^{\vec{a}\cdot\vec{x}}) = [-\partial_{x_i}(f)+y_i\partial_{x_0}(f)] e^{\vec{a}\cdot\vec{x}} \in M \tag{9.3.39}$$

for $i \in \overline{1,m}$. Repeatedly applying (9.3.37)–(9.3.39), we obtain $e^{\vec{a}\cdot\vec{x}} \in M$. Equivalently, $\mathscr{B}'_{\vec{a},\vec{b},0} \subset M$.

Suppose $\mathscr{B}'_{\vec{a},\vec{b},\ell} \subset M$ for some $\ell \in \mathbb{N}$. For any $g e^{\vec{a}\cdot\vec{x}} \in \mathscr{B}'_{\vec{a},\vec{b},\ell}$,

$$A_{i,1}(g e^{\vec{a}\cdot\vec{x}}) = [a_1 x_i - b_i y_1 + x_i\partial_{x_1} - y_1\partial_{x_i}](g) e^{\vec{a}\cdot\vec{x}} \in M \tag{9.3.40}$$

and
$$C_{i,1}^s(ge^{\vec{a}\cdot\vec{x}}) = [a_1y_i + a_iy_1 + y_i\partial_{x_1} + y_1\partial_{x_i}](g)e^{\vec{a}\cdot\vec{x}} \in M \tag{9.3.41}$$

by (9.3.5), where $i \in \overline{1,m}$. Since

$$(x_i\partial_{x_1} - y_1\partial_{x_i})(g)e^{\vec{a}\cdot\vec{x}},\ (y_i\partial_{x_1} + y_1\partial_{x_i})(g)e^{\vec{a}\cdot\vec{x}} \in \mathscr{B}'_{\vec{a},\vec{b},\ell} \subset M, \tag{9.3.42}$$

we have
$$(a_1x_i - b_iy_1)ge^{\vec{a}\cdot\vec{x}},\ (a_1y_i + a_iy_1)ge^{\vec{a}\cdot\vec{x}} \in M \tag{9.3.43}$$

for $i \in \overline{1,m}$. The above second equation with $i = 1$ gives

$$2a_1y_1ge^{\vec{a}\cdot\vec{x}} \in M \Rightarrow y_1ge^{\vec{a}\cdot\vec{x}} \in M. \tag{9.3.44}$$

Thus (9.3.43) yields

$$x_ige^{\vec{a}\cdot\vec{x}},\ y_ige^{\vec{a}\cdot\vec{x}} \in M \qquad \text{for } i \in \overline{1,m}. \tag{9.3.45}$$

According to the first equation in (9.3.7),

$$A_{m+1,1}(ge^{\vec{a}\cdot\vec{x}}) = [a_1x_0 - \sum_{i=1}^m (a_ix_i + b_iy_i)y_1 + x_0\partial_{x_1} - y_1(D+\kappa)](g)e^{\vec{a}\cdot\vec{x}} \in M. \tag{9.3.46}$$

Replacing $ge^{\vec{a}\cdot\vec{x}} \in \mathscr{B}'_{\vec{a},\vec{b},\ell}$ by $ge^{\vec{a}\cdot\vec{x}} \in \sum_{i=1}^m (x_i\mathscr{B}'_{\vec{a},\vec{b},\ell} + y_i\mathscr{B}'_{\vec{a},\vec{b},\ell})$ in (9.3.40)–(9.3.45), we obtain
$$x_iy_1ge^{\vec{a}\cdot\vec{x}},\ y_iy_1ge^{\vec{a}\cdot\vec{x}} \in M \qquad \text{for } i \in \overline{1,m}. \tag{9.3.47}$$

Since $D(g) = \ell g$ and $x_0\partial_{x_1}(g)e^{\vec{a}\cdot\vec{x}} \in \mathscr{B}'_{\vec{a},\vec{b},\ell} \subset M$, we have

$$[-\sum_{i=1}^m (a_ix_i + b_iy_i)y_1 + x_0\partial_{x_1} - y_1(D+\kappa)](g)e^{\vec{a}\cdot\vec{x}} \in M. \tag{9.3.48}$$

Hence (9.3.46) yields $x_0ge^{\vec{a}\cdot\vec{x}} \in M$. Therefore, $\mathscr{B}'_{\vec{a},\vec{b},\ell+1} \subset M$. By induction, $\mathscr{B}'_{\vec{a},\vec{b},\ell} \subset M$ for any $\ell \in \mathbb{N}$. So $\mathscr{B}'_{\vec{a},\vec{b}} = M$. Hence $\mathscr{B}'_{\vec{a},\vec{b}}$ is an irreducible $sp(2m+2,\mathbb{F})$-module. □

Fix $m_1, m_2 \in \overline{1,m}$ with $m_1 \le m_2$. Set

$$\tilde{D}' = x_0\partial_{x_0} + \sum_{r=m_1+1}^m x_r\partial_{x_r} - \sum_{i=1}^{m_1} x_i\partial_{x_i} + \sum_{i=1}^{m_2} y_i\partial_{y_i} - \sum_{r=m_2+1}^m y_r\partial_{y_r} \tag{9.3.49}$$

and

$$\tilde{\kappa} = \kappa + m_2 - m_1 - m. \tag{9.3.50}$$

Swapping operators $\partial_{x_r} \mapsto -x_r$, $x_r \mapsto \partial_{x_r}$ for $r \in \overline{1, m_1}$ and $\partial_{y_p} \mapsto -y_p$, $y_p \mapsto \partial_{y_p}$ for $s \in \overline{n_2 + 1, n}$ in the projective oscillator representation (9.3.5)–(9.3.8), we get the following *noncanonical projective oscillator representation* $\pi_\kappa^{m_1,m_2}$ of the Lie algebra $sp(2m+2, \mathbb{F})$ determined by

$$\pi_\kappa^{m_1,m_2}(A_{i,j}) = E_{i,j}^x - E_{j,i}^y \tag{9.3.51}$$

with

$$E_{i,j}^x = \begin{cases} -x_j\partial_{x_i} - \delta_{i,j} & \text{if } i, j \in \overline{1, m_1}; \\ \partial_{x_i}\partial_{x_j} & \text{if } i \in \overline{1, m_1},\ j \in \overline{m_1 + 1, m}; \\ -x_i x_j & \text{if } i \in \overline{m_1 + 1, m},\ j \in \overline{1, m_1}; \\ x_i\partial_{x_j} & \text{if } i, j \in \overline{m_1 + 1, m} \end{cases} \tag{9.3.52}$$

and

$$E_{i,j}^y = \begin{cases} y_i\partial_{y_j} & \text{if } i, j \in \overline{1, m_2}; \\ -y_i y_j & \text{if } i \in \overline{1, m_2},\ j \in \overline{m_2 + 1, m}; \\ \partial_{y_i}\partial_{y_j} & \text{if } i \in \overline{m_2 + 1, m},\ j \in \overline{1, m_2}; \\ -y_j\partial_{y_i} - \delta_{i,j} & \text{if } i, j \in \overline{m_2 + 1, m}; \end{cases} \tag{9.3.53}$$

$$\pi_\kappa^{m_1,m_2}(E_{i,m+1+j}) = \begin{cases} \partial_{x_i}\partial_{y_j} & \text{if } i \in \overline{1, m_1},\ j \in \overline{1, m_2}; \\ -y_j\partial_{x_i} & \text{if } i \in \overline{1, m_1},\ j \in \overline{m_2 + 1, m}; \\ x_i\partial_{y_j} & \text{if } i \in \overline{m_1 + 1, m},\ j \in \overline{1, m_2}; \\ -x_i y_j & \text{if } i \in \overline{m_1 + 1, m},\ j \in \overline{m_2 + 1, m}; \end{cases} \tag{9.3.54}$$

$$\pi_\kappa^{m_1,m_2}(E_{m+1+i,j}) = \begin{cases} -x_j y_i & \text{if } j \in \overline{1, m_1},\ i \in \overline{1, m_2}; \\ -x_j\partial_{y_i} & \text{if } j \in \overline{1, m_1},\ i \in \overline{m_2 + 1, m}; \\ y_i\partial_{x_j} & \text{if } j \in \overline{m_1 + 1, m},\ i \in \overline{1, m_2}; \\ \partial_{x_j}\partial_{y_i} & \text{if } j \in \overline{m_1 + 1, m},\ i \in \overline{m_2 + 1, m}; \end{cases} \tag{9.3.55}$$

$$\pi_\kappa^{m_1,m_2}(C_{m+1,m+1}^s) = -2\partial_{x_0}, \quad \pi_\kappa^{m_1,m_2}(B_{m+1,m+1}^s) = 2x_0(\tilde{D} + \tilde{\kappa}); \tag{9.3.56}$$

$$\pi_\kappa^{m_1,m_2}(A_{i,m+1}) = \begin{cases} \partial_{x_0}\partial_{x_i} + \partial_{y_i} & \text{if } i \in \overline{1, m_1}; \\ x_i\partial_{x_0} + \partial_{y_i} & \text{if } i \in \overline{m_1 + 1, m_2}; \\ x_i\partial_{x_0} - y_i & \text{if } i \in \overline{m_2 + 1, m}, \end{cases} \tag{9.3.57}$$

$$\pi_\kappa^{m_1,m_2}(C_{m+1,i}^s) = \begin{cases} y_i\partial_{x_0} + x_i & \text{if } i \in \overline{1, m_1}; \\ y_i\partial_{x_0} - \partial_{x_i} & \text{if } i \in \overline{m_1 + 1, m_2}, \\ \partial_{x_0}\partial_{y_i} - \partial_{x_i} & \text{if } i \in \overline{m_2 + 1, m}; \end{cases} \tag{9.3.58}$$

$$\pi_\kappa^{m_1,m_2}(A_{m+1,i}) = \begin{cases} -x_0x_i - y_i(\tilde{D}' + \tilde{\kappa}) & \text{if } i \in \overline{1, m_1}; \\ x_0\partial_{x_i} - y_i(\tilde{D}' + \tilde{\kappa}) & \text{if } i \in \overline{m_1+1, m_2}; \\ x_0\partial_{x_i} - (\tilde{D}' + \tilde{c} - 1)\partial_{y_i} & \text{if } i \in \overline{m_2+1, m}; \end{cases} \tag{9.3.59}$$

$$\pi_\kappa^{m_1,m_2}(B^s_{i,m+1}) = \begin{cases} x_0\partial_{y_i} + (\tilde{D}' + \tilde{c} - 1)\partial_{x_i} & \text{if } i \in \overline{1, m_1}; \\ x_0\partial_{y_i} + x_i(\tilde{D}' + \tilde{\kappa}) & \text{if } i \in \overline{m_1+1, m_2}; \\ -x_0y_i + x_i(\tilde{D}' + \tilde{\kappa}) & \text{if } i \in \overline{m_2+1, m}; \end{cases} \tag{9.3.60}$$

$$\pi_\kappa^{m_1,m_2}(A_{m+1,m+1}) = \tilde{D}' + x_0\partial_{x_0} + \tilde{\kappa} \tag{9.3.61}$$

for $i, j \in \overline{1, m}$.

For $k \in \mathbb{Z}$, we write

$$\mathscr{B}_{\langle k\rangle} = \text{Span}\left\{x^\alpha y^\beta \mid \alpha, \beta \in \mathbb{N}^m;\ \sum_{r=m_1+1}^{m} \alpha_r - \sum_{i=1}^{m_1}\alpha_i + \sum_{i=1}^{m_2}\beta_i - \sum_{r=m_2+1}^{n}\beta_r = k\right\} \tag{9.3.62}$$

and

$$\mathscr{B}'_{\langle k\rangle} = \sum_{i=0}^{\infty} \mathscr{B}_{\langle k-i\rangle} x_0^i. \tag{9.3.63}$$

Then

$$\mathscr{B}'_{\langle k\rangle} = \{u \in \mathscr{B} \mid \tilde{D}'(u) = ku\}. \tag{9.3.64}$$

Recall the Lie subalgebra $\mathscr{K}$ in (9.3.9). Suppose $m > 1$. Theorem 9.2.1 says that if $m_1 < m_2$ or $k \neq 0$, the subspace $\mathscr{B}_{\langle k\rangle}$ is an irreducible $\mathscr{K}$-module; when $m_1 = m_2$, the subspace $\mathscr{B}_{\langle 0\rangle}$ is a direct sum of two irreducible $\mathscr{K}$-submodules.

Assume $m = m_1 = m_2 = 1$, $\mathscr{B} = \mathbb{F}[x_1, y_1]$ and

$$\pi_\kappa^{m_1,m_2}(\mathscr{K}) = \mathbb{F}(x_1\partial_{x_1} + y_1\partial_{y_1} + 1) + \mathbb{F}x_1y_1 + \mathbb{F}\partial_{x_1}\partial_{y_1}. \tag{9.3.65}$$

So all $\mathscr{B}_{\langle k\rangle}$ with $k \in \mathbb{Z}$ are irreducible $\mathscr{K}$-submodules. The following is the third result in this section.

Theorem 9.3.3 *The representation $\pi_\kappa^{m_1,m_2}$ of $sp(2m+2, \mathbb{F})$ on $\mathscr{B}'$ is irreducible if $\kappa \notin \mathbb{Z}$.*

Proof Let M be any nonzero $sp(2m+2, \mathbb{F})$-submodule of $\mathscr{B}'$. Repeatedly applying the first equation in (9.3.56) to M, we get

$$M \bigcap \mathscr{B} \neq \{0\}. \tag{9.3.66}$$

According to (9.3.61),

$$\mathscr{B}_{\langle k\rangle} = \{f \in \mathscr{B} \mid A_{m+1,m+1}(f) = (k+\tilde{\kappa})f\}. \tag{9.3.67}$$

Hence

$$M = \bigoplus_{k\in\mathbb{Z}} M \bigcap \mathscr{B}_{\langle k\rangle}. \tag{9.3.68}$$

If $M \bigcap \mathscr{B}_{\langle 0\rangle} \neq \{0\}$, then (9.3.58) gives

$$C^s_{m+1,1}(M \bigcap \mathscr{B}_{\langle 0\rangle}) = x_1(M \bigcap \mathscr{B}_{\langle 0\rangle}) \subset M \bigcap \mathscr{B}_{\langle -1\rangle}. \tag{9.3.69}$$

Thus we always have $M \bigcap \mathscr{B}_{\langle k\rangle} \neq \{0\}$ for some $0 \neq k \in \mathbb{Z}$. Note that $\mathscr{B}_{\langle k\rangle}$ is an irreducible $\mathscr{K}$-module. So

$$\mathscr{B}_{\langle k\rangle} \subset M. \tag{9.3.70}$$

Next (9.3.57) yields

$$\mathscr{B}_{\langle k-r\rangle} = (\partial_{y_1})^r(\mathscr{B}_{\langle k\rangle}) = A^r_{1,m+1}(\mathscr{B}_{\langle k\rangle}) \subset M \qquad \text{for } r \in \mathbb{N}. \tag{9.3.71}$$

On the other hand, if $\mathscr{B}_{\langle \ell\rangle} \subset \mathscr{M}$, then the assumption $\kappa \notin \mathbb{Z}$ and the second equation in (9.3.56) give

$$x_0^r\mathscr{B}_{\langle \ell\rangle} = (C^s_{m+1,m+1})^r(\mathscr{B}_{\langle \ell\rangle}) \subset M \qquad \text{for } r \in \mathbb{N}. \tag{9.3.72}$$

Suppose that for some $p \in \mathbb{Z}$,

$$x_0^r\mathscr{B}_{\langle p\rangle}, x_0^r\mathscr{B}_{\langle p-1\rangle} \subset M \qquad \text{for } r \in \mathbb{N}. \tag{9.3.73}$$

For any $\ell \in \mathbb{N}$,

$$x_0^\ell\mathscr{B}_{\langle p+1\rangle} = (\tilde{D}' + \tilde{c} - 1)\partial_{x_1}(x_0^\ell\mathscr{B}_{\langle p\rangle}) = [B^s_{1,m+1} - x_0\partial_{y_1}](x_0^\ell\mathscr{B}_{\langle p\rangle}) \tag{9.3.74}$$

by (9.3.60). Note

$$x_0\partial_{y_1}(x_0^\ell\mathscr{B}_{\langle p\rangle}) = x_0^{\ell+1}\mathscr{B}_{\langle p-1\rangle} \subset M. \tag{9.3.75}$$

Thus (9.3.74) leads to

$$x_0^\ell\mathscr{B}_{\langle p+1\rangle} \subset M. \tag{9.3.76}$$

By (9.3.71)–(9.3.76) and induction on p, we prove

$$x_0^r\mathscr{B}_{\langle k\rangle} \subset M \qquad \text{for } x_0 \in \mathbb{N},\ k \in \mathbb{Z}. \tag{9.3.77}$$

So $M = \mathscr{B}'$. Therefore, $\mathscr{B}'$ is an irreducible $sp(2m+2,\mathbb{F})$-module. $\square$

Remark 9.3.4 The above irreducible representation depends on the three parameters $\kappa \in \mathbb{F}$ and $m_1, m_2 \in \overline{1,n}$. It is not highest-weight type because of the mixture of multiplication operators and differential operators in (9.3.54), (9.3.55) and (9.3.57)–(9.3.60). Since $\mathscr{B}$ is not completely reducible as a module of the Lie subalgebra $\sum_{i,j=1}^{m} \mathbb{F}A_{i,j}$ by Sects. 6.4–6.6 when $m \geq 2$ and $m_1 < m$, $\mathscr{B}'$ is not a unitary $sp(2m+2,\mathbb{F})$-module. The constraints on the degrees of the monomials with a fixed weight via the operators $\pi_\kappa^{m_1,m_2}(A_{i,i})$ in (3.51)–(3.53) with $i \in \overline{1,m}$ and $\pi_\kappa^{m_1,m_2}(A_{m+1,m+1})$ in (3.61) show that the weight subspaces are finite-dimensional. Thus $\mathscr{B}'$ is a weight $sp(2m+2,\mathbb{F})$-module with finite-dimensional weight subspaces.

Next we assume that

$$b_{j_0} \neq 0 \ \text{ for some } \ j_0 \in \overline{1,m_1} \tag{9.3.78}$$

and

$$a_{i_0} \neq 0 \ \text{ for some } \ i_0 \in \overline{m_2+1,m} \ \text{ if } \ m_2 < m. \tag{9.3.79}$$

Under the assumption, we have the following fourth result in this section:

Theorem 9.3.5 *The representation $\pi_\kappa^{m_1,m_2}$ of $sp(2m+2,\mathbb{F})$ on $\mathscr{B}'_{\vec{a},\vec{b}}$ is irreducible for any $\kappa \in \mathbb{F}$.*

Proof Let M be a nonzero $sp(2m+2,\mathbb{F})$-submodule of $\mathscr{B}'_{\vec{a},\vec{b}}$. Take any $0 \neq fe^{\vec{a}\cdot\vec{x}} \in M$ with $f \in \mathscr{B}'$. Repeatedly applying the first equation in (9.3.56) to $fe^{\vec{a}\cdot\vec{x}}$ if necessary, we may assume $f \in \mathscr{B} = \mathbb{F}[x_1,...,x_m,y_1,...,y_m]$. Then (9.3.58) yields

$$(C_{m+1,i}^s + a_i)(fe^{\vec{a}\cdot\vec{x}}) = -\partial_{x_i}(f)e^{\vec{a}\cdot\vec{x}} \in M \ \text{ for } \ i \in \overline{m_1+1,m}. \tag{9.3.80}$$

Moreover, (9.3.57) yields

$$(A_{i,m+1} - b_j)(fe^{\vec{a}\cdot\vec{x}}) = \partial_{y_j}(f)e^{\vec{a}\cdot\vec{x}} \in M \ \text{ for } \ j \in \overline{1,m_2}. \tag{9.3.81}$$

Repeatedly applying (9.3.80) and (9.3.81) if necessary, we can assume

$$f \in \mathbb{F}[x_1,...,x_{m_1},y_{m_2+1},...,y_m]. \tag{9.3.82}$$

According to (9.3.54),

$$(B_{i,j_0}^s - a_{j_0}b_i - a_ib_{j_0})(fe^{\vec{a}\cdot\vec{x}}) = (b_{j_0}\partial_{x_i} + b_i\partial_{x_{j_0}})(f)e^{\vec{a}\cdot\vec{x}} \in M \tag{9.3.83}$$

for $i \in \overline{1,m_1}$. Taking $i = j_0$ in (9.3.83), we get $\partial_{x_{j_0}}(f)e^{\vec{a}\cdot\vec{x}} \in M$. Substituting it to (9.3.83) for general i, we obtain

$$\partial_{x_i}(f)e^{\vec{a}\cdot\vec{x}} \in M \qquad \text{for } \ i \in \overline{1,m_1}. \tag{9.3.84}$$

Moreover, (9.3.55) yields

$$(C^s_{j,i_0} - a_j b_{i_0} - a_{i_0} b_j)(f e^{\vec{a}\cdot\vec{x}}) = (a_{i_0}\partial_{y_j} + a_j\partial_{y_{i_0}})(f)e^{\vec{a}\cdot\vec{x}} \in M \tag{9.3.85}$$

for $j \in \overline{m_2+1, m}$. Letting $j = i_0$ in (9.3.85), we find $\partial_{y_{i_0}}(f)e^{\vec{a}\cdot\vec{x}} \in M$. Substituting it to (9.3.85) for general j, we get

$$\partial_{y_j}(f)e^{\vec{a}\cdot\vec{x}} \in M \qquad \text{for } j \in \overline{m_2+1, m}. \tag{9.3.86}$$

Repeatedly applying (9.3.84) and (9.3.86) if necessary, we obtain $e^{\vec{a}\cdot\vec{x}} \in M$. Equivalently, $\mathscr{B}'_{\vec{a},\vec{a},0} \subset M$ (cf. (9.3.36)).

Suppose that for some $\ell \in \mathbb{N}$, $\mathscr{B}'_{\vec{a},\vec{a},k} \subset M$ whenever $\ell \geq k \in \mathbb{N}$. For any $ge^{\vec{a}\cdot\vec{x}} \in \mathscr{B}'_{\vec{a},\vec{b},\ell}$, (9.3.58) implies

$$(C^s_{m+1,i} - y_i\partial_{x_0})(ge^{\vec{a}\cdot\vec{x}}) = x_i g e^{\vec{a}\cdot\vec{x}} \in M \qquad \text{for } i \in \overline{1, m_1} \tag{9.3.87}$$

and (9.3.57) leads to

$$(-A_{i,m+1} + x_i\partial_{x_0})(ge^{\vec{a}\cdot\vec{x}}) = y_j g e^{\vec{a}\cdot\vec{x}} \in M \qquad \text{for } j \in \overline{m_2+1, m}. \tag{9.3.88}$$

Moreover, (9.3.54) gives

$$B^s_{i,m+1+j_0}(ge^{\vec{a}\cdot\vec{x}}) = [b_{j_0}x_i + x_i\partial_{y_{j_0}} + (\partial_{x_{j_0}} + a_{j_0})(\partial_{y_i} + b_i)](g)e^{\vec{a}\cdot\vec{x}} \in M \tag{9.3.89}$$

if $i \in \overline{m_1+1, m_2}$, and

$$B^s_{i,j_0}(ge^{\vec{a}\cdot\vec{x}}) = [b_{j_0}x_i - a_{j_0}y_i + x_i\partial_{y_{j_0}} - y_i\partial_{x_{j_0}}](g)e^{\vec{a}\cdot\vec{x}} \in M \tag{9.3.90}$$

if $i \in \overline{m_2+1, m}$. Note that the inductional assumption imply

$$[x_i\partial_{y_{j_0}} + (\partial_{x_{j_0}} + a_{j_0})(\partial_{y_i} + b_i)](g)e^{\vec{a}\cdot\vec{x}} \in M \tag{9.3.91}$$

if $i \in \overline{m_1+1, m_2}$, and

$$[-a_{j_0}y_i + x_i\partial_{y_{j_0}} - y_i\partial_{x_{j_0}}](g)e^{\vec{a}\cdot\vec{x}} \in M \tag{9.3.92}$$

by (9.3.90) if $i \in \overline{m_2+1, m}$. Thus

$$x_i g e^{\vec{a}\cdot\vec{x}} \in M \qquad \text{for } i \in \overline{m_1+1, m}. \tag{9.3.93}$$

On the other hand, (9.3.55) yields

$$C^s_{j,i_0}(ge^{\vec{a}\cdot\vec{x}}) = (a_{i_0}y_j - b_{i_0}x_j + y_j\partial_{x_{i_0}} - x_j\partial_{y_{i_0}})(g)e^{\vec{a}\cdot\vec{x}} \in M \tag{9.3.94}$$

if $j \in \overline{1, m_1}$, and

$$C^s_{j,i_0}(ge^{\vec{a}\cdot\vec{x}}) = [a_{i_0}y_j + y_j\partial_{x_{i_0}} + (\partial_{x_j} + a_j)(\partial_{y_{i_0}} + b_{i_0})](g)e^{\vec{a}\cdot\vec{x}} \in M \qquad (9.3.95)$$

if $j \in \overline{m_1 + 1, m_2}$. Observe that the inductional assumption imply

$$(-b_{i_0}x_j + y_j\partial_{x_{i_0}} - x_j\partial_{y_{i_0}})(g)e^{\vec{a}\cdot\vec{x}} \in M \qquad (9.3.96)$$

by (9.3.87) if $j \in \overline{1, m_1}$, and

$$[y_j\partial_{x_{i_0}} + (\partial_{x_j} + a_j)(\partial_{y_{i_0}} + b_{i_0})](g)e^{\vec{a}\cdot\vec{x}} \in M \qquad (9.3.97)$$

if $j \in \overline{m_1 + 1, m_2}$. Hence

$$y_jge^{\vec{a}\cdot\vec{x}} \in M \qquad \text{for } j \in \overline{1, m_2}. \qquad (9.3.98)$$

Moreover, (9.3.60) yields

$$\begin{aligned} &B^s_{j_0,m+1}(ge^{\vec{a}\cdot\vec{x}}) \\ &= \Bigg[b_{j_0}x_0 + x_0\partial_{y_{j_0}} + \Bigg(\tilde{D}' - \sum_{i=1}^{m_1} a_ix_i + \sum_{j=m_1+1}^{m} a_jx_j + \sum_{r=1}^{m_2} b_ry_r \\ &\quad - \sum_{s=m_2+1}^{m} b_sy_s + \tilde{\kappa} + 1\Bigg)(a_{j_0} + \partial_{x_{j_0}})\Bigg](g)e^{\vec{a}\cdot\vec{x}} \in M. \end{aligned} \qquad (9.3.99)$$

Note that

$$x_0\partial_{y_{j_0}}(g)e^{\vec{a}\cdot\vec{x}} \in \mathscr{B}'_{\vec{a},\vec{b},\ell} \subset M; \quad (\tilde{D} + \tilde{\kappa} + 1)(\partial_{x_{j_0}}(g))e^{\vec{a}\cdot\vec{x}} \in \mathscr{B}'_{\vec{a},\ell-1} \subset M. \qquad (9.3.100)$$

Now (9.3.87), (9.3.88), (9.3.93) and (9.3.98)–(9.3.100) imply $x_0ge^{\vec{a}\cdot\vec{x}} \in M$. Therefore, $\mathscr{B}'_{\vec{a},\vec{b},\ell+1} \subset M$. By induction, $\mathscr{B}'_{\vec{a},\vec{b},\ell} \subset M$ for any $\ell \in \mathbb{N}$. So $\mathscr{B}'_{\vec{a},\vec{b}} = M$. Hence $\mathscr{B}'_{\vec{a},\vec{b}}$ is an irreducible $sp(2m + 2, \mathbb{F})$-module. □

References

[LX3] C. Luo, X. Xu, $\mathbb{Z}$-graded oscillator representations of symplectic Lie algebras. J. Algebr. **403**, 401–425 (2014)

[X24] X. Xu, Projective oscillator representations of $sl(n + 1)$ and $sp(2m + 2)$. J. Lie Theory **26**, 96–114 (2016)

Chapter 10
Representations of G_2 and F_4

The simple Lie group (Lie algebra) of type G_2 is a popular object both in geometry and physics. In fact, the history of G_2-geometry is almost as old as that of the simple Lie group (Lie algebra) of type G_2 itself (cf. [IA] for a nice exposition). The existence of metrics of G_2 holonomy was proved by Bryant [RLB]. Moreover, Bryant and Salamon [BS] gave a construction of metrics of G_2 holonomy on the total space of the bundle of anti-self-dual 2-forms over a 4-dimensional self-dual Einstein manifold. Hitchin [NH] introduced a diffeomorphism-invariant functional on the space of differential 3-forms on a closed 11-dimensional manifold and found that a generic critical point defines a Riemannian manifold with G_2 holonomy. There are also the other developments in G_2-geometry such as Laplacian flows of G_2-structures. Algebraically, simple Lie algebra of type G_2 is exactly the derivation Lie algebra of the octonion algebra (e.g. cf. [A]).

In physics, the group chain $SO(7) \supset G_2 \supset SO(3)$ was introduced by Racah in 1951 for the study of atomic electrons. Physicists discovered supergravity in 1976 (cf. FZV,[DZ]). In 1978, Cremmer, Julia and Scherk [CJS] introduced supergravity theory in 11-dimension, which was later viewed as the low energy limit of M-theory. Witten [EW] laid down the criterion for spacetime supersymmetry in Kaluza-Klein theory. These works lead the manifolds with G_2 holonomy to play fundamental roles in M-theory (e.g., cf. [AW, HM]). Besides, physicists used G_2 to prove the equivalence of the casual description and noncasual description in photon emissions. Moreover, they also described the behavior of meson by means of G_2.

Simple Lie algebra of type F_4 is exactly the derivation Lie algebra of the 27-dimensional exceptional simple Jordan algebra (e.g. cf. [A]). Its geometric construction was found by Figueroa-O'Farrill [FO]. Moreover, Coquereaux and Isasi [CI] described quantum symmetries associated with the F_4 Dynkin diagram.

The goal of this chapter is to give a basic picture of G_2 and F_4 symmetries. First we determine the structure of the canonical bosonic and fermionic oscillator representations of the simple Lie algebra of type G_2 over its 7-dimensional module. Moreover, we present a one-parameter family of conformal oscillator representations of G_2

X. Xu, *Representations of Lie Algebras and Partial Differential Equations*,
DOI 10.1007/978-981-10-6391-6_10

derived from those of $o(7,\mathbb{C})$ in Sect. 8.4 and determine their irreducibility. This is a new result obtained by the author, which might lead to a 5-dimensional new G_2 geometry. Furthermore, we use partial differential equations to find the explicit irreducible decomposition of the space of polynomial functions on 26-dimensional basic irreducible module of the simple Lie algebra of type F_4 (cf. [X18]). In Sects. 10.1 and 10.4, Weyl's Theorem of complete reducibility in Sect. 2.4 plays a key role.

10.1 Basic Oscillator Representations of G_2

In this section, we study the structure of the canonical bosonic and fermionic oscillator representations of the simple Lie algebra of type G_2 over its 7-dimensional module.

Denote by $\mathbb{Z}_3=\mathbb{Z}/3\mathbb{Z}$ and identify $i+3\mathbb{Z}$ with i for $i\in\{1,2,3\}$. Denote

$$A_{i,j}=E_{i,j}-E_{j',i'}\qquad \text{for}\;\; i,\,j\in\mathbb{Z}_3. \tag{10.1.1}$$

Recall the simple Lie algebra

$$\mathscr{G}^{G_2}=\mathbb{F}h_1+\mathbb{F}h_2+\sum_{i,j\in\mathbb{Z}_3,\;i\neq j}\mathbb{F}A_{i,j}+\sum_{r\in\mathbb{Z}_3}(\mathbb{F}C_r+\mathbb{F}C_r') \tag{10.1.2}$$

constructed in (4.4.1)–(4.4.10). Let $\mathcal{Q}$ be the space of rational function in $x_0,x_1,x_2,x_3,x_{1'},x_{2'},x_{3'}$. The corresponding *bosonic oscillator representation* is given by

$$h_1|_{\mathcal{Q}}=x_1\partial_{x_1}-x_2\partial_{x_2}-x_{1'}\partial_{x_{1'}}+x_{2'}\partial_{x_{2'}}, \tag{10.1.3}$$

$$h_2|_{\mathcal{Q}}=-2x_1\partial_{x_1}+x_2\partial_{x_2}+x_3\partial_{x_3}+2x_{1'}\partial_{x_{1'}}-x_{2'}\partial_{x_{2'}}-x_{3'}\partial_{x_{3'}}, \tag{10.1.4}$$

$$A_{i,j}|_{\mathcal{Q}}=x_i\partial_{x_j}-x_{j'}\partial_{x_{i'}}\qquad\text{for}\;\; i,\,j\in\mathbb{Z}_3,\;\; i\neq j, \tag{10.1.5}$$

$$C_r|_{\mathcal{Q}}=x_r\partial_{x_{(r+2)'}}-x_{r+2}\partial_{x_{r'}}+\sqrt{2}(x_0\partial_{x_{r+1}}-x_{(r+1)'}\partial_{x_0}), \tag{10.1.6}$$

$$C_r'|_{\mathcal{Q}}=x_{r'}\partial_{x_{r+2}}-x_{(r+2)'}\partial_{x_r}+\sqrt{2}(x_0\partial_{x_{(r+1)'}}-x_{r+1}\partial_{x_0}) \tag{10.1.7}$$

for $r\in\mathbb{Z}_3$. Moreover, the vectors

$$C_1,\;C_2',\;C_3,\;A_{1,2},\;A_{3,1},\;A_{3,2} \tag{10.1.8}$$

are positive root vectors.

Define

$$\eta=x_0^2+2x_1x_{1'}+2x_2x_{2'}+2x_3x_{3'}. \tag{10.1.9}$$

It can be verified that

$$\xi(\eta) = 0 \qquad \text{for } \xi \in \mathcal{G}^{G_2}. \tag{10.1.10}$$

Lemma 10.1.1 *Any singular function in $\mathcal{Q}$ with respect to $\mathcal{G}^{G_2}$ must be a rational function in x_3 and η.*

Proof Let f be a $\mathcal{G}^{G_2}$-singular function in $\mathcal{Q}$. By Lemma 6.2.3 and

$$A_{3,1}(f) = 0, \qquad A_{1,2}(f) = 0, \tag{10.1.11}$$

we have

$$f = \varphi(x_0, x_3, x_{2'}, \eta) \tag{10.1.12}$$

as a rational function in $x_0, x_3, x_{2'}, \eta$. Note

$$C_1(f) = [x_1\partial_{x_{3'}} - x_3\partial_{x_{1'}} + \sqrt{2}(x_0\partial_{x_2} - x_{2'}\partial_{x_0})](f) = -\sqrt{2}x_{2'}\partial_{x_0}(\varphi) = 0. \tag{10.1.13}$$

So φ is independent of x_0. Moreover,

$$C_3(f) = [x_3\partial_{x_{2'}} - x_2\partial_{x_{3'}} + \sqrt{2}(x_0\partial_{x_1} - x_{1'}\partial_{x_0})] = x_3\partial_{x_{2'}}(\varphi) = 0. \tag{10.1.14}$$

Thus φ is independent of $x_{2'}$. Equivalently, f is a function in x_3 and η. □

Recall that

$$\Delta' = \partial_{x_0}^2 + 2\partial_{x_1}\partial_{x_{1'}} + 2\partial_{x_2}\partial_{x_{2'}} + 2\partial_{x_3}\partial_{x_{3'}} \tag{10.1.15}$$

is an $o(7, \mathbb{F})$-invariant differential operator (cf. (8.1.11)) and $\mathcal{G}^{G_2} \subset o(7, \mathbb{F})$. So

$$\xi\mathcal{D}' = \mathcal{D}'\xi \qquad \text{for } \xi \in \mathcal{G}^{G_2}. \tag{10.1.16}$$

Let $\mathcal{B}' = \mathbb{F}[x_0, x_1, x_2, x_3, x_{1'}, x_{2'}, x_{3'}] \subset \mathcal{Q}$ be the polynomial algebra in $x_0, x_1, x_2, x_3, x_{1'}, x_{2'}, x_{3'}$. The above lemma implies that any $\mathcal{G}^{G_2}$-singular vector in $\mathcal{B}'$ is a polynomial in x_3 and η. Denote by $\mathcal{B}'_k$ the subspace of homogeneous polynomials in $\mathcal{B}'$ with degree k. Set

$$\mathcal{H}'_k = \{f \in \mathcal{B}'_k \mid \mathcal{D}'(f) = 0\}. \tag{10.1.17}$$

By Theorem 8.1.1, we have the following theorem.

Theorem 10.1.2 *For any $k \in \mathbb{N}$, the subspace $\mathcal{H}'_k$ forms an irreducible $\mathcal{G}^{G_2}$-module with highest-weight vector x_3^k of weight $k\lambda_2$. Moreover,*

$$\mathcal{B}' = \bigoplus_{i,k=0}^{\infty} \eta^i \mathcal{H}'_k \tag{10.1.18}$$

is a direct sum of irreducible $\mathcal{G}^{G_2}$-submodules.

Consider the exterior algebra $\check{\mathscr{A}}'$ generated by $\{\theta_0,\theta_1,\theta_2,\theta_3,\theta_{1'},\theta_{2'},\theta_{3'}\}$ (cf. (6.2.15)–(6.2.18)). The *basic fermionic oscillator representation* of $\mathcal{G}^{G_2}$ on $\check{\mathscr{A}}'$ by

$$h_1|_{\check{\mathscr{A}}'}=\theta_1\partial_{\theta_1}-\theta_2\partial_{\theta_2}-\theta_{1'}\partial_{\theta_{1'}}+\theta_{2'}\partial_{\theta_{2'}},\tag{10.1.19}$$

$$h_2|_{\check{\mathscr{A}}'}=-2\theta_1\partial_{\theta_1}+\theta_2\partial_{\theta_2}+\theta_3\partial_{\theta_3}+2\theta_{1'}\partial_{\theta_{1'}}-\theta_{2'}\partial_{\theta_{2'}}-\theta_{3'}\partial_{\theta_{3'}},\tag{10.1.20}$$

$$A_{i,j}|_{\check{\mathscr{A}}'}=\theta_i\partial_{\theta_j}-\theta_{j'}\partial_{\theta_{i'}}\qquad\text{for } i,j\in\mathbb{Z}_3,\ i\neq j,\tag{10.1.21}$$

$$C_r|_{\check{\mathscr{A}}'}=\theta_r\partial_{\theta_{(r+2)'}}-\theta_{r+2}\partial_{\theta_{r'}}+\sqrt{2}(\theta_0\partial_{\theta_{r+1}}-\theta_{(r+1)'}\partial_{\theta_0}),\tag{10.1.22}$$

$$C'_r|_{\check{\mathscr{A}}'}=\theta_{r'}\partial_{\theta_{r+2}}-\theta_{(r+2)'}\partial_{\theta_r}+\sqrt{2}(\theta_0\partial_{\theta_{(r+1)'}}-\theta_{r+1}\partial_{\theta_0})\tag{10.1.23}$$

for $r\in\mathbb{Z}_3$.

Note that

$$\Theta=\sum_{i=0}^{6}\mathbb{F}\theta_i\tag{10.1.24}$$

is the 7-dimensional basic irreducible $\mathcal{G}^{G_2}$-module. Define

$$\check{\mathscr{A}}'_k=\Theta^k\qquad\text{for } k\in\overline{0,7}.\tag{10.1.25}$$

Then $\check{\mathscr{A}}'_k$ forms a $\mathcal{G}^{G_2}$-module. By the Dynkin diagram of G_2, any finite-dimensional $\mathcal{G}^{G_2}$-module is isomorphic to its dual module which consists of its linear functions. In particular,

$$\check{\mathscr{A}}'_k\cong\check{\mathscr{A}}'_{7-k}\ \text{ as } \mathcal{G}_2\text{-modules.}\tag{10.1.26}$$

We only need to study the structure $\mathcal{G}^{G_2}$-modules $\check{\mathscr{A}}'_2$ and $\check{\mathscr{A}}'_3$.

Recall that

$$\check{\eta}=\theta_1\theta_{1'}+\theta_2\theta_{2'}+\theta_3\theta_{3'}\tag{10.1.27}$$

is a $\mathcal{G}_0$-invariant. According to (6.2.53) with $n=3$ and $\vartheta_i=\theta_{i'}$ for $i\in\overline{1,3}$, the $\mathcal{G}^{G_2}$-singular vectors in $\check{\mathscr{A}}'_2$ are in the set

$$\{a\theta_0\theta_3+b\theta_{1'}\theta_{2'},a\theta_0\theta_{2'}+b\theta_1\theta_3,\theta_3\theta_{2'},\check{\eta}\mid(0,0)\neq(a,b)\in\mathbb{F}\}.\tag{10.1.28}$$

Moreover,

$$C_3(a\theta_0\theta_3+b\theta_{1'}\theta_{2'})=-\sqrt{2}a\theta_{1'}\theta_3+b\theta_3\theta_{1'},\tag{10.1.29}$$

$$C_3(a\theta_0\theta_{2'}+b\theta_1\theta_3)=a\theta_0\theta_3+\sqrt{2}b\theta_0\theta_3,\tag{10.1.30}$$

and

$$C_3(\check{\eta})=\sqrt{2}\theta_0\theta_{1'}+2\theta_1\theta_3\tag{10.1.31}$$

by (10.1.22). Thus $\theta_3\theta_{2'}$ is the only $\mathscr{G}^{G_2}$-singular vector in $\check{\mathscr{A}}_2'$, which has weight λ_1.

Again by (6.2.53), the $\mathscr{G}^{G_2}$-singular vectors in $\check{\mathscr{A}}_3'$ are in the set

$$\{a\theta_0\check{\eta}+b\theta_1\theta_2\theta_3+c\theta_{1'}\theta_{2'}\theta_{3'},\, a\theta_0\theta_1\theta_3+b\theta_{2'}\check{\eta},\, a\theta_0\theta_{1'}\theta_{2'}+b\theta_1\check{\eta},\\ \theta_0\theta_3\theta_{2'},\, \theta_1\theta_3\theta_{2'},\, \theta_3\theta_{1'}\theta_{2'} \mid a,b,c\in\mathbb{F}\}. \tag{10.1.32}$$

Calculate

$$C_3(a\theta_0\check{\eta}+b\theta_1\theta_2\theta_3+c\theta_{1'}\theta_{2'}\theta_{3'})\\ =-\sqrt{2}a\theta_{1'}\check{\eta}+(2a+\sqrt{2}b)\theta_0\theta_2\theta_3+c\theta_{1'}(\theta_2\theta_{2'}+\theta_3\theta_{3'}), \tag{10.1.33}$$

$$C_3(a\theta_0\theta_1\theta_3+b\theta_{2'}\check{\eta})=-a\sqrt{2}\theta_1\theta_2\theta_{1'}+b(\theta_3\check{\eta}+\sqrt{2}\theta_0\theta_{1'}\theta_{2'}+2\theta_2\theta_3\theta_{2'}), \tag{10.1.34}$$

$$C_3(a\theta_0\theta_{1'}\theta_{2'}+b\theta_3\check{\eta})=-a\theta_0\theta_3\theta_{1'}-b\sqrt{2}\theta_0\theta_3\theta_{1'}, \tag{10.1.35}$$

$$C_3(\theta_0\theta_3\theta_{2'})=\sqrt{2}\theta_3\theta_{1'}\theta_{2'},\;\; C_3(\theta_1\theta_3\theta_{2'})=\sqrt{2}\theta_0\theta_3\theta_{2'},\;\; C_3(\theta_3\theta_{1'}\theta_{2'})=0 \tag{10.1.36}$$

by (10.1.22). Thus we have three $\mathscr{G}^{G_2}$-singular vectors $\theta_0\check{\eta}+\sqrt{2}(-\theta_1\theta_2\theta_3+\theta_{1'}\theta_{2'}\theta_{3'})$ of weight 0, $\sqrt{2}\theta_0\theta_{1'}\theta_{2'}-\theta_3\check{\eta}$ of weight λ_2 and $\theta_3\theta_{1'}\theta_{2'}$ of weight $2\lambda_2$ by (10.1.19) and (10.1.20). By Weyl's Theorem 2.3.6 of completely reducibility, we have the following theorem:

Theorem 10.1.3 *The subspace $\check{\mathscr{A}}_2'$ is an irreducible $\mathscr{G}^{G_2}$-module of highest weight λ_1. The subspace $\check{\mathscr{A}}_3'$ is a direct sum of a one-dimensional trivial $\mathscr{G}^{G_2}$-module, an irreducible $\mathscr{G}^{G_2}$-module of highest weight λ_2 and an irreducible $\mathscr{G}^{G_2}$-module of highest weight $2\lambda_2$.*

10.2 Conformal Oscillator Representations of G_2

In this section, we study the restricted representation of $\mathscr{G}^{G_2}$ from the conformal oscillator representation of $o(7,\mathbb{C})$ in (8.4.5)–(8.4.9) with $n=2$ and $4=1', 5=2', 6=3'$.

Recall

$$\mathscr{B}'=\mathbb{C}[x_0,x_1,x_2,y_1,y_2] \tag{10.2.1}$$

and

$$D'=x_0\partial_{x_0}+x_1\partial_{x_1}+x_2\partial_{x_2}+y_1\partial_{y_1}+y_2\partial_{y_2},\quad \eta'=\frac{1}{2}x_0^2+x_1y_1+x_2y_2. \tag{10.2.2}$$

For any $c \in \mathbb{C}$, (8.4.5)–(8.4.9) lead to the *conformal oscillator representation* π_c of $\mathcal{G}^{G_2}$ on $\mathscr{B}'$:

$$\pi_c(A_{1,2}) = x_1\partial_{x_2} - y_2\partial_{y_1}, \quad \pi_c(A_{2,1}) = x_2\partial_{x_1} - y_1\partial_{y_2}, \quad \pi_c(A_{3,1}) = \partial_{x_1} \tag{10.2.3}$$

$$\pi_c(A_{3,2}) = \partial_{x_2}, \quad \pi_c(h_1) = x_1\partial_{x_1} - x_2\partial_{x_2} - y_1\partial_{y_1} + y_2\partial_{y_2}, \tag{10.2.4}$$

$$\pi_c(A_{1,3}) = \eta'\partial_{y_1} - x_1(D'+c), \quad \pi_c(A_{2,3}) = \eta'\partial_{y_2} - x_2(D'+c), \tag{10.2.5}$$

$$\pi_c(h_2) = -2x_1\partial_{x_1} + x_2\partial_{x_2} + 2y_1\partial_{y_1} - y_2\partial_{y_2} - D' - c, \tag{10.2.6}$$

$$\pi_c(C_1) = -\partial_{y_1} + \sqrt{2}(x_0\partial_{x_2} - y_2\partial_{x_0}), \tag{10.2.7}$$

$$\pi_c(C_2) = x_2\partial_{y_1} - x_1\partial_{y_2} + \sqrt{2}(x_0(D'+c) - \eta'\partial_{x_0}), \tag{10.2.8}$$

$$\pi_c(C_3) = \partial_{y_2} + \sqrt{2}(x_0\partial_{x_1} - y_1\partial_{x_0}), \tag{10.2.9}$$

$$\pi_c(C_1') = -y_1(D'+c) + \eta'\partial_{x_1} + \sqrt{2}(x_0\partial_{y_2} - x_2\partial_{x_0}), \tag{10.2.10}$$

$$\pi_c(C_2') = y_2\partial_{x_1} - y_1\partial_{x_2} - \sqrt{2}\partial_{x_0}, \tag{10.2.11}$$

$$\pi_c(C_3') = y_2(D'+c) - \eta'\partial_{x_2} + \sqrt{2}(x_0\partial_{y_1} - x_1\partial_{x_0}). \tag{10.2.12}$$

Recall that $\mathbb{N}$ is the set of nonnegative integers.

Theorem 10.2.1 *The representation π_c of $\mathcal{G}^{G_2}$ on $\mathscr{B}'$ is irreducible if and only if $c \notin -\mathbb{N}$.*

Proof Set

$$\mathcal{G}_0 = \mathbb{C}h_1 + \mathbb{C}h_2 + \mathbb{C}A_{1,2} + \mathbb{C}A_{2,1}, \tag{10.2.13}$$

$$\mathcal{G}_- = \mathbb{C}A_{1,3} + \mathbb{C}A_{2,3} + \mathbb{C}C_1' + \mathbb{C}C_3' + \mathbb{C}C_2, \tag{10.2.14}$$

$$\mathcal{G}_+ = \mathbb{C}A_{3,1} + \mathbb{C}A_{3,2} + \mathbb{C}C_1 + \mathbb{C}C_3 + \mathbb{C}C_2'. \tag{10.2.15}$$

Then $\mathcal{G}_0$ and $\mathcal{G}_\pm$ are Lie subalgebras of $\mathcal{G}^{G_2}$ by (4.4.5)–(4.4.7). Moreover,

$$[\mathcal{G}_0, \mathcal{G}_\pm] \subset \mathcal{G}_\pm, \quad \mathcal{G}^{G_2} = \mathcal{G}_- + \mathcal{G}_0 + \mathcal{G}_+. \tag{10.2.16}$$

In particular, the universal enveloping algebra

$$U(\mathcal{G}^{G_2}) = U(\mathcal{G}_-)U(\mathcal{G}_0)U(\mathcal{G}_+). \tag{10.2.17}$$

Note that

$$V = U(\mathcal{G}^{G_2})(1_{\mathcal{B}'}) = U(\mathcal{G}_-)U(\mathcal{G}_0)U(\mathcal{G}_+)(1_{\mathcal{B}'}) = U(\mathcal{G}_-)(1_{\mathcal{B}'}) \tag{10.2.18}$$

is a $\mathcal{G}^{G_2}$-submodule. Assume $c = -\ell$ with $0 < \ell \in \mathbb{Z}$. Let $\mathcal{B}'_{(k)}$ be the subspace of polynomials in $\mathcal{B}'$ with degree $\leq k$. By (10.2.5), (10.2.8), (10.2.10), (10.2.12) and (10.2.14), we have

$$V \bigcap \mathcal{B}'_{(\ell+1)} \subset \eta' \mathcal{B}'_{(\ell-1)} + \mathcal{B}'_{(\ell)}. \tag{10.2.19}$$

Thus $V \neq \mathcal{B}'$. Hence $\mathcal{B}'$ is again not an irreducible $\mathcal{G}^{G_2}$-module.

Next we assume $c \notin -\mathbb{N}$. Let M be a nonzero $\mathcal{G}^{G_2}$-submodule of $\mathcal{B}'$. Repeatedly applying $\pi_c(A_{3,1}) = \partial_{x_1}$ in (10.2.3) and $\pi_c(A_{3,2}) = \partial_{x_2}$ in (10.2.4) to M, we can assume

$$M_1 = M \bigcap \mathbb{C}[x_0, y_1, y_2] \neq \{0\}. \tag{10.2.20}$$

According to (10.2.11), we have

$$\pi_c(C_2')|_{M_1} = (y_2\partial_{x_1} - y_1\partial_{x_2} - \sqrt{2}\partial_{x_0})|_{M_1} = -\sqrt{2}\partial_{x_0}|_{M_1}. \tag{10.2.21}$$

Repeatedly applying (10.2.21) to M_1, we can assume

$$M_2 = M \bigcap \mathbb{C}[y_1, y_2] \neq \{0\}. \tag{10.2.22}$$

Moreover, (10.2.7) gives

$$\pi_c(C_1)|_{M_2} = (-\partial_{y_1} + \sqrt{2}(x_0\partial_{x_2} - y_2\partial_{x_0}))|_{M_2} = -\partial_{y_1}|_{M_2} \tag{10.2.23}$$

and (10.2.9) yields

$$\pi_c(C_3)|_{M_2} = (\partial_{y_2} + \sqrt{2}(x_0\partial_{x_1} - y_1\partial_{x_0}))|_{M_2} = \partial_{y_2}|_{M_2}. \tag{10.2.24}$$

Repeatedly applying (10.2.23) and (10.2.24) to M_2, we obtain $1_{\mathcal{B}'} \in M$.

Recall that $\mathcal{B}'_k$ the subspace of homogeneous polynomials in $\mathcal{B}'$ with degree k. First we have

$$A_{1,3}(1_{\mathcal{B}'}) = -cx_1,\ A_{2,3}(1_{\mathcal{B}'}) = -cx_2,\ C_2(1_{\mathcal{B}'}) = \sqrt{2}cx_0 \in M \tag{10.2.25}$$

by (10.2.5) and (10.2.8). Moreover,

$$C_1'(1_{\mathcal{B}'}) = -cy_1,\ C_3'(1_{\mathcal{B}'}) = cy_2 \in M. \tag{10.2.26}$$

Since $c \neq 0$, $\mathcal{B}'_1 \subset M$.

Observe that

$$\begin{aligned}\mathcal{L} = \mathcal{G}_0 &+ \mathbb{C}(E_{0,1} - E_{1',0}) + \mathbb{C}(E_{0,2} - E_{2',0}) + \mathbb{C}(E_{0,1'} - E_{1,0}) \\ &+ \mathbb{C}(E_{0,2'} - E_{2,0}) + \mathbb{C}(E_{2,1'} - E_{1,2'}) + \mathbb{C}(E_{1',2} - E_{2',1})\end{aligned} \tag{10.2.27}$$

is a Lie subalgebra of $o(7,\mathbb{C})$ which is isomorphic to $o(5,\mathbb{C})$. We extend the action of $\mathscr{G}_0$ on $\mathbb{A}$ via $E_{i,j}|_{\mathscr{A}} = x_i\partial_{x_j}$, where $x_{r'} = y_r$. Then $\mathscr{B}'$ becomes a $\mathscr{L}$-module with the corresponding representation is the basic oscillator representation of $o(5,\mathbb{C})$. Recall the Laplace operator

$$\Delta' = \partial_{x_0}^2 + 2\partial_{x_1}\partial_{y_1} + 2\partial_{x_2}\partial_{y_2}. \tag{10.2.28}$$

and

$$\mathscr{H}'_k = \{f \in \mathscr{B}'_k \mid \Delta'(f) = 0\}. \tag{10.2.29}$$

Take

$$A_{2,1},\ E_{0,1'} - E_{1,0},\ E_{0,2'} - E_{2,0},\ E_{2,1'} - E_{1,2'} \tag{10.2.30}$$

as the positive root vectors of $\mathscr{L}$. According to Theorem 8.1.1, each $\mathscr{H}'_k$ is an irreducible $\mathscr{L}$-submodule with highest-weight vector x_2^k and

$$\mathscr{B}' = \sum_{r,k=0} \eta'^r \mathscr{H}'_k \tag{10.2.31}$$

is a direct sum of irreducible $\mathscr{L}$-submodules.

Up to now, we have proved $\mathscr{B}'_{(1)} = \mathscr{B}'_1 + \mathbb{C}1_{\mathscr{B}'} \subset M$. Assume that we have $\mathscr{B}'_{(k)} \subset M$ for some $1 \le k \in \mathbb{Z}$. For any $k/2 \ge \ell \in \mathbb{N}$,

$$A_{2,3}(\eta^\ell x_2^{k-2\ell}) = (\ell - k - c)\eta^\ell x_2^{k+1-2\ell} \in M \Longrightarrow \eta^\ell x_2^{k+1-2\ell} \in M \tag{10.2.32}$$

by (10.2.18). Assume that $k = 2s+1$ is an odd integer. The above expression yields $\eta'^{2s} x_2^2 \in M$. Similarly

$$A_{1,3}(\eta'^{2s} x_1) = -(1+c)\eta'^{2s} x_1^2 \in M \Longrightarrow \eta'^{2s} x_1^2 \in M, \tag{10.2.33}$$

$$C_2(\eta'^{2s} x_0) = \sqrt{2}(1+c)\eta'^{2s} x_0^2 \in M \Longrightarrow \eta'^{2s} x_0^2 \in M, \tag{10.2.34}$$

$$C'_1(\eta'^{2s} y_1) = -(1+c)\eta'^{2s} y_1^2 \in M \Longrightarrow \eta'^{2s} y_1^2 \in M, \tag{10.2.35}$$

$$C'_3(\eta'^{2s} y_2) = (1+c)\eta'^{2s} y_2^2 \in M \Longrightarrow \eta'^{2s} y_2^2 \in M. \tag{10.2.36}$$

Thus

$$\eta'^{s+1} = \frac{1}{2}\eta'^{2s} x_0^2 + \eta'^{2s} x_1^2 + \eta'^{2s} x_2^2 + \eta'^{2s} y_1^2 + \eta'^{2s} y_2^2 \in M. \tag{10.2.37}$$

Therefore, we have proved that in any case,

$$\eta'^\ell x_2^{k+1-2\ell} \in M \qquad \text{for any } 0 \le \ell \le [\![\frac{k+1}{2}]\!]. \tag{10.2.38}$$

Let

$$\mathscr{L}_- = \mathbb{C}A_{1,2} + \mathbb{C}(E_{0,1} - E_{1',0}) + \mathbb{C}(E_{0,2} - E_{2',0}) + \mathbb{C}(E_{1',2} - E_{2',1}) \quad (10.2.39)$$

be the Lie subalgebra of $\mathscr{L}$ spanned by negative root vectors. According to highest-weight representation theory,

$$\eta'^{\ell}\mathscr{H}_{k+1-2\ell} = U(\mathscr{L}_-)(\eta'^{\ell}x_2^{k+1-2\ell}). \quad (10.2.40)$$

For any $f \in \mathscr{B}'_{k+1}$, we have

$$C_1(f) \equiv \sqrt{2}(E_{0,2} - E_{2',0})(f) \pmod{\mathscr{B}'_{(k)}}, \quad (10.2.41)$$

$$C_3(f) \equiv \sqrt{2}(E_{0,1} - E_{1',0})(f) \pmod{\mathscr{B}'_{(k)}} \quad (10.2.42)$$

and

$$C'_2(f) \equiv (E_{2',1} - E_{1',2})(f) \pmod{\mathscr{B}'_{(k)}} \quad (10.2.43)$$

by (10.2.7), (10.2.9) and (10.2.11). Since $\mathscr{B}'_{(k)} \subset M$ by inductional assumption, we get

$$A_{1,2}(f),\ (E_{0,1} - E_{1',0})(f),\ (E_{0,2} - E_{2',0})(f),\ (E_{2',1} - E_{1',2})(f) \in M \text{ if } f \in M. \quad (10.2.44)$$

Now (10.2.39), (10.2.40) and (10.2.44) yield

$$\eta'^{\ell}\mathscr{H}_{k+1-2\ell} \subset M \qquad \text{for any } 0 \leq \ell \leq [\![\frac{k+1}{2}]\!]. \quad (10.2.45)$$

According to (10.2.31),

$$\mathscr{B}'_{k+1} = \sum_{0\leq\ell\leq[\![\frac{k+1}{2}]\!]} \eta'^{\ell}\mathscr{H}'_{k+1-2\ell} \subset M. \quad (10.2.46)$$

Therefore,

$$\mathscr{B}'_{(k+1)} = \mathscr{B}'_{k+1} + \mathscr{B}'_{(k)} \subset M. \quad (10.2.47)$$

By induction,

$$\mathscr{B}'_{(k)} \subset M \qquad \text{for any } k \in \mathbb{N}, \quad (10.2.48)$$

which implies $M = \mathscr{B}'$. This proves the theorem. □

According to ((10.1.18), (10.2.3), (10.2.4), (10.2.6), (10.2.7), (10.2.9) and (10.2.11), $\mathscr{B}'$ is a highest-weight irreducible $\mathscr{G}^{G_2}$-module with highest-weight vector $1_{\mathscr{B}'}$ of weight $-c\lambda_2$.

10.3 Basic Oscillator Representation of F_4

In this section, we present the basic oscillator representation of F_4 over its 26-dimensional irreducible module.

Let us go back to the construction of the simple Lie algebra $\mathscr{G}^{F_4}$ of type F_4 via the root lattice construction of the simple Lie algebra $\mathscr{G}^{E_6}$ of type E_6 in (4.4.63)–(4.4.80).

Recall the Dynkin diagram of E_6:

E_6: 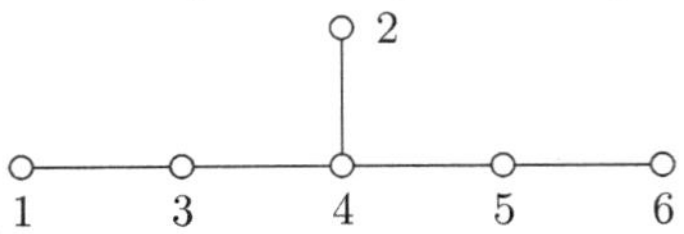

Let $\Pi_{E_6} = \{\alpha_1, ..., \alpha_6\}$ be the set of positive simple roots corresponding to the above diagram and let Φ_{E_6} be the root system. The simple Lie algebra $\mathscr{G}^{E_6}$ of type E_6 is

$$\mathscr{G}^{E_6} = H \oplus \bigoplus_{\alpha \in \Phi_{E_6}} \mathbb{C}E_\alpha, \quad H = \sum_{i=1}^{6} \mathbb{F}\alpha, \tag{10.3.1}$$

with the Lie bracket given in (4.4.24) and (4.4.25) and $F(\cdot, \cdot)$ given in (4.4.64). In terms of (4.4.68)–(4.4.80), the simple Lie algebra of type F_4 is a Lie subalgebra of $\mathscr{G}^{E_6}$:

$$\mathscr{G}^{F_4} = \sum_{\varpi \in S_{F_4}} (\mathbb{F}F_\varpi + \mathbb{F}F'_\varpi) + \sum_{i=1}^{4} \mathbb{F}h_i \tag{10.3.2}$$

(cf. (4.4.78), (4.4.79)), whose Dynkin diagram is

F_4: 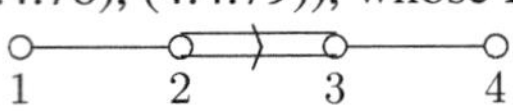

Recall the notions in (4.4.42)–(4.4.44). Set

$$\xi_1 = E_{(1,1,2,2,1,1)} - E_{(1,1,1,2,2,1)}, \qquad \xi_2 = E_{(1,1,2,2,1)} - E_{(0,1,1,2,2,1)}, \tag{10.3.3}$$

$$\xi_3 = E_{(1,1,1,2,1)} - E_{(0,1,1,2,1,1)}, \qquad \xi_4 = E_{(1,1,1,1,1)} - E_{(0,1,1,1,1,1)}, \tag{10.3.4}$$

$$\xi_5 = E_{(1,1,1,1)} - E_{(0,1,0,1,1,1)}, \qquad \xi_6 = E_{(1,0,1,1,1)} - E_{(0,0,1,1,1,1)}, \tag{10.3.5}$$

$$\xi_7 = E_{(0,1,1,1)} - E_{(0,1,0,1,1)}, \ \xi_8 = E_{(1,0,1,1)} - E_{(0,0,0,1,1,1)}, \ \xi_9 = E_{(0,0,1,1)} - E_{(0,0,0,1,1)}, \tag{10.3.6}$$

$$\xi_{10} = E_{(1,0,1)} - E_{(0,0,0,0,1,1)}, \ \ \xi_{11} = E_{\alpha_3} - E_{\alpha_5}, \ \ \xi_{12} = E_{\alpha_1} - E_{\alpha_6}, \tag{10.3.7}$$

$$\xi_{13} = \alpha_1 - \alpha_6, \qquad \xi_{14} = \alpha_3 - \alpha_5, \qquad \xi_{15} = E_{-\alpha_1} - E_{-\alpha_6}, \tag{10.3.8}$$

$$\xi_{16} = E_{-\alpha_3} - E_{-\alpha_5}, \ \ \xi_{17} = E'_{(1,0,1)} - E'_{(0,0,0,0,1,1)}, \ \ \xi_{18} = E'_{(0,0,1,1)} - E'_{(0,0,0,1,1)}, \tag{10.3.9}$$

$$\xi_{19} = E'_{(1,0,1,1)} - E'_{(0,0,0,1,1,1)}, \quad \xi_{20} = E'_{(0,1,1,1)} - E'_{(0,1,0,1,1)}, \tag{10.3.10}$$

$$\xi_{21} = E'_{(1,0,1,1,1)} - E'_{(0,0,1,1,1,1)}, \quad \xi_{22} = E_{(1,1,1,1)} - E_{(0,1,0,1,1,1)}, \tag{10.3.11}$$

$$\xi_{23} = E'_{(1,1,1,1,1)} - E'_{(0,1,1,1,1,1)}, \quad \xi_{24} = E'_{(1,1,1,2,1)} - E'_{(0,1,1,2,1,1)}, \tag{10.3.12}$$

$$\xi_{25} = E'_{(1,1,2,2,1)} - E'_{(0,1,1,2,2,1)}, \quad \xi_{26} = E'_{(1,1,2,2,1,1)} - E'_{(1,1,1,2,2,1)}. \tag{10.3.13}$$

Recall the automorphism $\hat{\sigma}$ of $\mathscr{G}^{E_6}$ defined in (4.4.63), (4.4.66) and (4.4.67). Then the subspace

$$V = \sum_{i=1}^{26} \mathbb{F}\xi_i = \{v \in \mathscr{G}^{E_6} \mid \hat{\sigma}(v) = -v\} \tag{10.3.14}$$

forms an irreducible $\mathscr{G}^{F_4}$-module with respect to the adjoint representation of $\mathscr{G}^{E_6}$, ξ_1 is a highest-weight vector of weight λ_4 and

$$\mathscr{G}^{E_6} = \mathscr{G}^{F_4} \oplus V. \tag{10.3.15}$$

Write

$$[u, \xi_i] = \sum_{j=1}^{26} \varphi_{i,j}(u)\xi_j \qquad \text{for } u \in \mathscr{G}^{F_4}. \tag{10.3.16}$$

Set

$$\mathscr{A} = \mathbb{F}[x_1, x_2, ..., x_{26}] \tag{10.3.17}$$

and define the *basic oscillator representation of* $\mathscr{G}^{F_4}$ on $\mathscr{A}$ by

$$u(g) = \sum_{i,j=1}^{26} \varphi_{i,j}(u)x_j\partial_{x_i}(g) \qquad \text{for } u \in \mathscr{G}^{F_4},\ g \in \mathscr{A} \tag{10.3.18}$$

(cf. (2.2.17)–(2.2.20)). Then $\mathscr{A}$ forms a $\mathscr{G}^{F_4}$-module isomorphic to the symmetric tensor $S(V)$ over V. According to (4.4.68)–(4.4.77), we have

$$F_{(1)}|_{\mathscr{A}} = x_4\partial_{x_6} + x_5\partial_{x_8} + x_7\partial_{x_9} - x_{18}\partial_{x_{20}} - x_{19}\partial_{x_{22}} - x_{21}\partial_{x_{23}}, \tag{10.3.19}$$

$$F_{(0,1)}|_{\mathscr{A}} = x_3\partial_{x_4} + x_8\partial_{x_{10}} + x_9\partial_{x_{11}} - x_{16}\partial_{x_{18}} - x_{17}\partial_{x_{19}} - x_{23}\partial_{x_{24}}, \tag{10.3.20}$$

$$\begin{aligned}F_{(0,0,1)}|_{\mathscr{A}} = &-x_2\partial_{x_3} - x_4\partial_{x_5} - x_6\partial_{x_8} + x_{10}\partial_{x_{12}} + x_{11}(\partial_{x_{13}} - 2\partial_{x_{14}}) \\ &-x_{14}\partial_{x_{16}} - x_{15}\partial_{x_{17}} + x_{19}\partial_{x_{21}} + x_{22}\partial_{x_{23}} + x_{24}\partial_{x_{25}},\end{aligned} \tag{10.3.21}$$

$$F_{(0,0,0,1)}|_{\mathscr{A}} = -x_1\partial_{x_2} - x_5\partial_{x_7} - x_8\partial_{x_9} - x_{10}\partial_{x_{11}} + x_{12}(\partial_{x_{14}} - 2\partial_{x_{13}}) -x_{13}\partial_{x_{15}} + x_{16}\partial_{x_{17}} + x_{18}\partial_{x_{19}} + x_{20}\partial_{x_{22}} + x_{25}\partial_{x_{26}}, \quad (10.3.22)$$

$$F_{(1,1)}|_{\mathscr{A}} = -x_3\partial_{x_6} + x_5\partial_{x_{10}} + x_7\partial_{x_{11}} - x_{16}\partial_{x_{20}} - x_{17}\partial_{x_{22}} + x_{21}\partial_{x_{24}}, \quad (10.3.23)$$

$$F_{(0,1,1)}|_{\mathscr{A}} = x_2\partial_{x_4} + x_3\partial_{x_5} + x_6\partial 10 + x_8\partial_{x_{12}} + x_9(\partial_{x_{13}} - 2\partial_{x_{14}}) -x_{14}\partial_{x_{18}} - x_{15}\partial_{x_{19}} - x_{17}\partial_{x_{21}} - x_{22}\partial_{x_{24}} - x_{23}\partial_{x_{25}}, \quad (10.3.24)$$

$$F_{(0,0,1,1)}|_{\mathscr{A}} = -x_1\partial_{x_3} + x_4\partial_{x_7} + x_6\partial_{x_9} - x_{10}(\partial_{x_{13}} + \partial_{x_{14}}) - x_{11}\partial_{x_{15}} +x_{12}\partial_{x_{16}} - (x_{13} + x_{14})\partial_{x_{17}} - x_{18}\partial_{x_{21}} - x_{20}\partial_{x_{23}} + x_{24}\partial_{x_{26}}, \quad (10.3.25)$$

$$F_{(1,1,1)}|_{\mathscr{A}} = -x_2\partial_{x_6} + x_3\partial_{x_8} + x_4\partial_{x_{10}} + x_5\partial_{x_{12}} + x_7(\partial_{x_{13}} - 2\partial_{x_{14}}) - x_{14}\partial_{x_{20}} -x_{15}\partial_{x_{22}} - x_{17}\partial_{x_{23}} - x_{19}\partial_{x_{24}} + x_{21}\partial_{x_{25}}, \quad (10.3.26)$$

$$F_{(0,1,1,1)}|_{\mathscr{A}} = x_1\partial_{x_4} + x_3\partial_{x_7} - x_6\partial_{x_{11}} - x_8(\partial_{x_{13}} + \partial_{x_{14}}) - x_9\partial_{x_{15}} + x_{12}\partial_{x_{18}} -(x_{13} + x_{14})\partial_{x_{19}} + x_{16}\partial_{x_{21}} - x_{20}\partial_{x_{24}} - x_{23}\partial_{x_{26}}, \quad (10.3.27)$$

$$F_{(0,1,2)}|_{\mathscr{A}} = -x_2\partial_{x_5} + x_6\partial_{x_{12}} + x_9\partial_{x_{16}} - x_{11}\partial_{x_{18}} - x_{15}\partial_{x_{21}} + x_{22}\partial_{x_{25}}, \quad (10.3.28)$$

$$F_{(1,1,2)}|_{\mathscr{A}} = x_2\partial_{x_8} + x_4\partial_{x_{12}} + x_7\partial_{x_{16}} - x_{11}\partial_{x_{20}} - x_{15}\partial_{x_{23}} - x_{19}\partial_{x_{25}}, \quad (10.3.29)$$

$$F_{(0,1,2,1)}|_{\mathscr{A}} = -x_1\partial_{x_5} + x_2\partial_{x_7} + x_6(\partial_{x_{14}} - 2\partial_{x_{13}}) + x_8\partial_{x_{16}} + x_9\partial_{x_{17}} -x_{10}\partial_{x_{18}} - x_{11}\partial_{x_{19}} - x_{13}\partial_{x_{21}} - x_{20}\partial_{x_{25}} + x_{22}\partial_{x_{26}}, \quad (10.3.30)$$

$$F_{(1,1,1,1)}|_{\mathscr{A}} = -x_1\partial_{x_6} - x_3\partial_{x_9} - x_4\partial_{x_{11}} - x_5(\partial_{x_{13}} + \partial_{x_{14}}) - x_7\partial_{x_{15}} + x_{12}\partial_{x_{20}} -(x_{13} + x_{14})\partial_{x_{22}} + x_{16}\partial_{x_{23}} + x_{18}\partial_{x_{24}} + x_{21}\partial_{x_{26}}, \quad (10.3.31)$$

$$F_{(1,2,2)}|_{\mathscr{A}} = -x_2\partial_{x_{10}} + x_3\partial_{x_{12}} + x_7\partial_{x_{18}} - x_9\partial_{x_{20}} - x_{15}\partial_{x_{24}} + x_{17}\partial_{x_{25}}, \quad (10.3.32)$$

$$F_{(1,1,2,1)}|_{\mathscr{A}} = x_1\partial_{x_8} - x_2\partial_{x_9} + x_4(\partial_{x_{14}} - 2\partial_{x_{13}}) + x_5\partial_{x_{16}} + x_7\partial_{x_{17}} - x_{10}\partial_{x_{20}} -x_{11}\partial_{x_{22}} - x_{13}\partial_{x_{23}} + x_{18}\partial_{x_{25}} - x_{19}\partial_{x_{26}}, \quad (10.3.33)$$

$$F_{(0,1,2,2)}|_{\mathscr{A}} = x_1\partial_{x_7} + x_6\partial_{x_{15}} + x_8\partial_{x_{17}} - x_{10}\partial_{x_{19}} - x_{12}\partial_{x_{21}} - x_{20}\partial_{x_{26}}, \quad (10.3.34)$$

$$F_{(1,2,2,1)}|_{\mathscr{A}} = -x_1\partial_{x_{10}} + x_2\partial_{x_{11}} + x_3(\partial_{x_{14}} - 2\partial_{x_{13}}) + x_5\partial_{x_{18}} + x_7\partial_{x_{19}} - x_8\partial_{x_{20}} -x_9\partial_{x_{22}} - x_{13}\partial_{x_{24}} - x_{16}\partial_{x_{25}} + x_{17}\partial_{x_{26}}, \quad (10.3.35)$$

$$F_{(1,1,2,2)}|_{\mathscr{A}} = -x_1\partial_{x_9} + x_4\partial_{x_{15}} + x_5\partial_{x_{17}} - x_{10}\partial_{x_{22}} - x_{12}\partial_{x_{23}} + x_{18}\partial_{x_{26}}, \quad (10.3.36)$$

$$F_{(1,2,2,2)}|_{\mathscr{A}} = x_1\partial_{x_{11}} + x_3\partial_{x_{15}} + x_5\partial_{x_{19}} - x_8\partial_{x_{22}} - x_{12}\partial_{x_{24}} - x_{16}\partial_{x_{26}}, \quad (10.3.37)$$

$$E_{(1,2,3,1)}|_{\mathscr{A}} = -x_1\partial_{x_{12}} - x_2(\partial_{x_{14}} + \partial_{x_{13}}) - x_3\partial_{x_{16}} + x_4\partial_{x_{18}} - x_6\partial_{x_{20}} + x_7\partial_{x_{21}} -x_9\partial_{x_{23}} + x_{11}\partial_{x_{24}} - (x_{13} + x_{14})\partial_{x_{25}} + x_{15}\partial_{x_{26}}, \quad (10.3.38)$$

$$F_{(1,2,3,2)}|_{\mathscr{A}} = x_1(\partial_{x_{13}} - 2\partial_{x_{14}}) + x_2\partial_{x_{15}} - x_3\partial_{x_{17}} + x_4\partial_{x_{19}} + x_5\partial_{x_{21}} - x_6\partial_{x_{22}} - x_8\partial_{x_{23}} + x_{10}\partial_{x_{24}} - x_{12}\partial_{x_{25}} - x_{14}\partial_{x_{26}}, \quad (10.3.39)$$

$$F_{(1,2,4,2)}|_{\mathscr{A}} = x_1\partial_{x_{16}} - x_2\partial_{x_{17}} + x_4\partial_{x_{21}} - x_6\partial_{x_{23}} + x_{10}\partial_{x_{25}} - x_{11}\partial_{x_{26}}, \quad (10.3.40)$$

$$F_{(1,3,4,2)}|_{\mathscr{A}} = -x_1\partial_{x_{18}} + x_2\partial_{x_{19}} - x_3\partial_{x_{21}} + x_6\partial_{x_{24}} - x_8\partial_{x_{25}} + x_9\partial_{x_{26}}, \quad (10.3.41)$$

$$F_{(2,3,4,2)}|_{\mathscr{A}} = x_1\partial_{x_{20}} - x_2\partial_{x_{22}} + x_3\partial_{x_{23}} - x_4\partial_{x_{24}} + x_5\partial_{x_{25}} - x_7\partial_{x_{26}}, \quad (10.3.42)$$

$$h_1|_{\mathscr{A}} = x_4\partial_{x_4} + x_5\partial_{x_5} - x_6\partial_{x_6} + x_7\partial_{x_7} - x_8\partial_{x_8} - x_9\partial_{x_9} + x_{18}\partial_{x_{18}} + x_{19}\partial_{x_{19}} - x_{20}\partial_{x_{20}} + x_{21}\partial_{x_{21}} - x_{22}\partial_{x_{22}} - x_{23}\partial_{x_{23}}, \quad (10.3.43)$$

$$h_2|_{\mathscr{A}} = x_3\partial_{x_3} - x_4\partial_{x_4} + x_8\partial_{x_8} + x_9\partial_{x_9} - x_{10}\partial_{x_{10}} - x_{11}\partial_{x_{11}} + x_{16}\partial_{x_{16}} + x_{17}\partial_{x_{17}} - x_{18}\partial_{x_{18}} - x_{19}\partial_{x_{19}} + x_{23}\partial_{x_{23}} - x_{24}\partial_{x_{24}}, \quad (10.3.44)$$

$$h_3|_{\mathscr{A}} = x_2\partial_{x_2} - x_3\partial_{x_3} + x_4\partial_{x_4} - x_5\partial_{x_5} + x_6\partial_{x_6} - x_8\partial_{x_8} + x_{10}\partial_{x_{10}} + 2x_{11}\partial_{x_{11}} - x_{12}\partial_{x_{12}} + x_{15}\partial_{x_{15}} - 2x_{16}\partial_{x_{16}} - x_{17}\partial_{x_{17}} + x_{19}\partial_{x_{19}} - x_{21}\partial_{x_{21}} + x_{22}\partial_{x_{22}} - x_{23}\partial_{x_{23}} + x_{24}\partial_{x_{24}} - x_{25}\partial_{x_{25}}, \quad (10.3.45)$$

$$h_4|_{\mathscr{A}} = x_1\partial_{x_1} - x_2\partial_{x_2} + x_5\partial_{x_5} - x_7\partial_{x_7} + x_8\partial_{x_8} - x_9\partial_{x_9} + x_{10}\partial_{x_{10}} - x_{11}\partial_{x_{11}} + 2x_{12}\partial_{x_{12}} - 2x_{15}\partial_{x_{15}} + x_{16}\partial_{x_{16}} - x_{17}\partial_{x_{17}} + x_{18}\partial_{x_{18}} - x_{19}\partial_{x_{19}} + x_{20}\partial_{x_{20}} - x_{22}\partial_{x_{22}} + x_{25}\partial_{x_{25}} - x_{26}\partial_{x_{26}}. \quad (10.3.46)$$

Denote

$$\bar{r} = 27 - r \qquad \text{for } r \in \overline{1, 26}. \quad (10.3.47)$$

Define an algebraic isomorphism τ on $\mathscr{A}[\partial_{x_1}, ..., \partial_{x_{26}}]$ by

$$\tau(x_r) = x_{\bar{r}}, \ \tau(\partial_{x_r}) = \partial_{x_{\bar{r}}} \qquad \text{for } r \in \overline{1, 26} \setminus \{13, 14\} \quad (10.3.48)$$

and

$$\tau(x_{13}) = -x_{13}, \ \tau(x_{14}) = -x_{14}, \ \tau(\partial_{x_{13}}) = -\partial_{x_{13}}, \ \tau(\partial_{x_{14}}) = -\partial_{x_{14}}. \quad (10.3.49)$$

Then

$$F'_{\varpi}|_{\mathscr{A}} = \tau(F_{\varpi}|_{\mathscr{A}}) \qquad \text{for } \varpi \in S_{F_4}. \quad (10.3.50)$$

Thus we have given the full explicit formulas for the basic oscillator representation of F_4.

10.4 Decomposition of the Representation of F_4

In this section, we decompose the polynomial algebra $\mathscr{A}$ into a direct sum of irreducible $\mathscr{G}^{F_4}$-submodules.

Suppose that

$$\eta_1 = 3\sum_{r=1}^{12} x_r x_{\bar{r}} + a x_{13}^2 + b x_{13} x_{14} + c x_{14}^2 \tag{10.4.1}$$

is an F_4-invariant (cf. (10.3.47)), where a, b, c are constants to be determined. Then $F_{(1)}(\eta_1) = F_{(0,1)}(\eta_1) = 0$ naturally hold. Moreover,

$$0 = F_{(0,0,1)}(\eta_1) = 2(a-b)x_{11}x_{13} + [b - 4c - 3)]x_{11}x_{14}, \tag{10.4.2}$$

which gives

$$a = b = 4c + 3. \tag{10.4.3}$$

Similarly, the constraint $E_{(0,0,0,1)}(\eta_1) = 0$ yield $b = c = 4a + 3$. So we have the quadratic invariant

$$\eta_1 = 3\sum_{r=1}^{12} x_r x_{\bar{r}} - x_{13}^2 - x_{13}x_{14} - x_{14}^2. \tag{10.4.4}$$

This invariant also gives a symmetric $\mathscr{G}^{F_4}$-invariant bilinear form on V.

By (10.3.43)–(10.3.46), we try to find a quadratic singular vector of the form

$$\zeta_1 = x_1(a_1 x_{13} + a_2 x_{14}) + a_3 x_2 x_{12} + a_4 x_3 x_{10} + a_5 x_4 x_8 + a_6 x_5 x_6, \tag{10.4.5}$$

where a_r are constants. Observe

$$0 = F_{(1)}(\zeta_1) = (a_5 + a_6)x_4 x_5 \Longrightarrow a_6 = -a_5. \tag{10.4.6}$$

Moreover,

$$0 = F_{(0,1)}(\zeta_1) = (a_4 + a_5)x_3 x_8 \Longrightarrow a_5 = -a_4. \tag{10.4.7}$$

Note

$$0 = F_{(0,0,1)}(\zeta_1) = (a_1 - 2a_2)x_1 x_{11} + (a_3 - a_4)x_2 x_{10} \Longrightarrow a_1 = 2a_2, \quad a_3 = a_4. \tag{10.4.8}$$

Furthermore,

$$0 = F_{(0,0,0,1)}(\zeta_1) = (a_2 - 2a_1 - a_3)x_1 x_{12} \Longrightarrow 2a_1 = a_2 - a_3. \tag{10.4.9}$$

Hence we have the singular vector

$$\zeta_1 = x_1(2x_{13} + x_{14}) - 3x_2x_{12} - 3x_3x_{10} + 3x_4x_8 - 3x_5x_6 \tag{10.4.10}$$

of weight λ_4. So it generates an irreducible module that is isomorphic to the basic module V. Note

$$F'_{(1)}|_{\mathscr{A}} = -x_6\partial_{x_4} - x_8\partial_{x_5} - x_9\partial_{x_7} + x_{20}\partial_{x_{18}} + x_{22}\partial_{x_{19}} + x_{23}\partial_{x_{21}}, \tag{10.4.11}$$

$$F'_{(0,1)}|_{\mathscr{A}} = -x_4\partial_{x_3} - x_{10}\partial_{x_8} - x_{11}\partial_{x_9} + x_{18}\partial_{x_{16}} + x_{19}\partial_{x_{17}} + x_{24}\partial_{x_{23}}, \tag{10.4.12}$$

$$\begin{aligned} F'_{(0,0,1)}|_{\mathscr{A}} = {} & x_3\partial_{x_2} + x_5\partial_{x_4} + x_8\partial_{x_6} - x_{12}\partial_{x_{10}} + x_{16}(2\partial_{x_{14}} - \partial_{x_{13}}) \\ & + x_{14}\partial_{x_{11}} + x_{17}\partial_{x_{15}} - x_{21}\partial_{x_{19}} - x_{23}\partial_{x_{22}} - x_{25}\partial_{x_{24}}, \end{aligned} \tag{10.4.13}$$

$$\begin{aligned} F'_{(0,0,0,1)}|_{\mathscr{A}} = {} & x_2\partial_{x_1} + x_7\partial_{x_5} + x_9\partial_{x_8} + x_{11}\partial_{x_{10}} + x_{15}(2\partial_{x_{13}} - \partial_{x_{14}}) \\ & + x_{13}\partial_{x_{12}} - x_{17}\partial_{x_{16}} - x_{19}\partial_{x_{18}} - x_{22}\partial_{x_{20}} - x_{26}\partial_{x_{25}} \end{aligned} \tag{10.4.14}$$

by (10.3.48)–(10.3.50). To get a basis of the module generated by ζ_1 compatible to $\{x_i \mid i \in \overline{1,26}\}$, we set

$$\zeta_2 = F'_{(0,0,0,1)}(\zeta_1) = x_2(-x_{13} + x_{14}) + 3x_1x_{15} - 3x_3x_{11} + 3x_4x_9 - 3x_6x_7, \tag{10.4.15}$$

$$\zeta_3 = F'_{(0,0,1)}(\zeta_2) = -x_3(x_{13} + 2x_{14}) + 3x_1x_{17} + 3x_2x_{16} + 3x_5x_9 - 3x_7x_8, \tag{10.4.16}$$

$$\zeta_4 = -F'_{(0,1)}(\zeta_3) = -x_4(x_{13} + 2x_{14}) - 3x_1x_{19} - 3x_2x_{18} + 3x_5x_{11} - 3x_7x_{10}, \tag{10.4.17}$$

$$\zeta_5 = F'_{(0,0,1)}(\zeta_4) = x_5(-x_{13} + x_{14}) + 3x_1x_{21} - 3x_3x_{18} - 3x_4x_{16} + 3x_7x_{12}, \tag{10.4.18}$$

$$\zeta_6 = -F'_{(1)}(\zeta_4) = -x_6(x_{13} + 2x_{14}) + 3x_1x_{22} + 3x_2x_{20} + 3x_8x_{11} - 3x_9x_{10}, \tag{10.4.19}$$

$$\zeta_7 = F'_{(0,0,0,1)}(\zeta_5) = x_7(2x_{13} + x_{14}) + 3x_2x_{21} + 3x_3x_{19} + 3x_4x_{17} - 3x_5x_{15}, \tag{10.4.20}$$

$$\zeta_8 = -F'_{(1)}(\zeta_5) = x_8(-x_{13} + x_{14}) - 3x_1x_{23} + 3x_3x_{20} - 3x_6x_{16} + 3x_9x_{12}, \tag{10.4.21}$$

$$\zeta_9 = -F'_{(1)}(\zeta_7) = x_9(2x_{13} + x_{14}) - 3x_2x_{23} - 3x_3x_{22} + 3x_6x_{17} - 3x_8x_{15}, \tag{10.4.22}$$

$$\zeta_{10} = -F'_{(0,1)}(\zeta_8) = x_{10}(-x_{13} + x_{14}) + 3x_1x_{24} + 3x_4x_{20} + 3x_6x_{18} + 3x_{11}x_{12}, \tag{10.4.23}$$

$$\zeta_{11} = -F'_{(0,1)}(\zeta_9) = x_{11}(2x_{13} + x_{14}) + 3x_2x_{24} - 3x_4x_{22} - 3x_6x_{19} - 3x_{10}x_{15}, \tag{10.4.24}$$

$$\zeta_{12} = -F'_{(0,0,1)}(\zeta_{10}) = -x_{12}(x_{13} + 2x_{14}) + 3x_1x_{25} - 3x_5x_{20} - 3x_8x_{18} - 3x_{10}x_{16}, \tag{10.4.25}$$

$$\begin{aligned} \zeta_{13} = F'_{(0,0,0,1)}(\zeta_{12}) = {} & -x_{13}(x_{13} + 2x_{14}) - 3x_1x_{26} + 3x_2x_{25} + 3x_5x_{22} - 3x_7x_{20} \\ & + 3x_8x_{19} - 3x_9x_{18} + 3x_{10}x_{17} - 3x_{11}x_{16}, \end{aligned} \tag{10.4.26}$$

$$\zeta_{14} = F'_{(0,0,1)}(\zeta_{11}) = x_{14}(2x_{13} + x_{14}) - 3x_2x_{25} + 3x_3x_{24} + 3x_4x_{23} - 3x_5x_{22} + 3x_6x_{21} - 3x_8x_{19} - 3x_{10}x_{17} + 3x_{12}x_{15}, \quad (10.4.27)$$

$$\zeta_r = \tau(\zeta_{\bar{r}}) \quad \text{for } r \in \overline{15, 27}, \quad (10.4.28)$$

where τ is an algebra automorphism determined by (10.3.48) and (10.3.49). The above construction shows that the map $x_r \mapsto \zeta_r$ determine a module isomorphism from V to the module generated by ζ_1. In particular,

$$u(x_i) = ax_j \Leftrightarrow u(\zeta_i) = a\zeta_j, \qquad a \in \mathbb{F}, \ u \in \mathscr{G}^{F_4}. \quad (10.4.29)$$

First

$$\vartheta = (x_1\zeta_2 - x_2\zeta_1)/3 = x_1(-x_2x_{13} + x_1x_{15} - x_3x_{11} + x_4x_9 - x_6x_7) + x_2(x_2x_{12} + x_3x_{10} - x_4x_8 + x_5x_6) \quad (10.4.30)$$

is a singular vector of weight λ_3. Recall that the invariant η_1 in (10.4.4) define an invariant bilinear form on V. Thus we have the following cubic invariant

$$\eta_2 = 3\sum_{r=1}^{12}(\zeta_r x_{\bar{r}} + x_r\zeta_{\bar{r}}) - 2x_{13}\zeta_{13} - x_{13}\zeta_{14} - x_{14}\zeta_{13} - 2x_{14}\zeta_{14}. \quad (10.4.31)$$

According to (10.4.10) and (10.4.15)–(10.4.28), we find

$$\begin{aligned}\eta_2 = {} & 9(1+\tau)[(x_2x_{12} + x_3x_{10} - x_4x_8 + x_5x_6)x_{26} + (x_3x_{11} - x_4x_9 + x_6x_7)x_{25} \\ & + (x_7x_8 - x_5x_9)x_{24} + x_{10}(x_4x_{23} + x_9x_{21}) - x_{11}(x_5x_{23} + x_8x_{21} + x_{12}x_{17}) \\ & - x_{12}(x_7x_{22} + x_9x_{19})] + 2x_{13}^3 + 3x_{13}^2x_{14} - 3x_{13}x_{14}^2 - 2x_{14}^3 + 3x_1(2x_{13} + x_{14})x_{26} \\ & + 3x_2(x_{13} + 2x_{14})x_{25} - 3x_{13}[x_3x_{24} + x_4x_{23} + x_5x_{22} + x_6x_{21} - 2x_7x_{20} + x_8x_{19} \\ & - 2x_9x_{18} + x_{10}x_{17} - 2x_{11}x_{16} + x_{12}x_{15}] - 3x_{14}[2x_3x_{24} + 2x_4x_{23} - x_5x_{22} \\ & + 2x_6x_{21} - x_7x_{20} - x_8x_{19} - x_9x_{18} - x_{10}x_{17} - x_{11}x_{16} + 2x_{12}x_{15}],\end{aligned} \quad (10.4.32)$$

where τ is an algebra automorphism defined in (10.3.47)–(10.3.49).

Theorem 10.3.1 *Any polynomial f in $\mathscr{A}$ satisfying the system of partial differential equations*

$$F_{\varpi}(f) = 0 \quad \textit{for } \varpi \in S^+_{F_4} \quad (10.4.33)$$

must be a polynomial in $x_1, \zeta_1, \vartheta, \eta_1, \eta_2$. In particular, the elements

$$\{x_1^{m_1}\zeta_1^{m_2}\vartheta^{m_3}\eta_1^{m_4}\eta_2^{m_5} \mid m_1, m_2, m_3, m_4, m_5 \in \mathbb{N}\} \quad (10.4.34)$$

are linearly independent singular vectors and any singular vector is a linear combination of those in (10.4.34) with the same weight. The weight of $x_1^{m_1}\zeta_1^{m_2}\vartheta^{m_3}\eta_1^{m_4}\eta_2^{m_5}$ *is* $m_3\lambda_3+(m_1+m_2)\lambda_4$.

Proof First we note

$$x_1x_{14}=\zeta_1-2x_1x_{13}+3x_2x_{12}+3x_3x_{10}-3x_4x_8+3x_5x_6, \tag{10.4.35}$$

$$3x_1x_{15}=\zeta_2+x_2(x_{13}-x_{14})+3x_3x_{11}-3x_4x_9+3x_6x_7, \tag{10.4.36}$$

$$3x_1x_{17}=\zeta_3+x_3(x_{13}+2x_{14})-3x_2x_{16}-3x_5x_9+3x_7x_8, \tag{10.4.37}$$

$$3x_1x_{19}=3x_5x_{11}-\zeta_4-x_4(x_{13}+2x_{14})-3x_2x_{18}-3x_7x_{10}, \tag{10.4.38}$$

$$3x_1x_{21}=\zeta_5+x_5(x_{13}-x_{14})+3x_3x_{18}+3x_4x_{16}-3x_7x_{12}, \tag{10.4.39}$$

$$3x_1x_{22}=\zeta_6+x_6(x_{13}+2x_{14})-3x_2x_{20}-3x_8x_{11}+3x_9x_{10}, \tag{10.4.40}$$

$$3x_1x_{23}=\zeta_8+x_8(x_{13}-x_{14})-3x_3x_{20}+3x_6x_{16}-3x_9x_{12}, \tag{10.4.41}$$

$$3x_1x_{24}=\zeta_{10}+x_{10}(x_{13}-x_{14})-3x_4x_{20}-3x_6x_{18}-3x_{11}x_{12}, \tag{10.4.42}$$

$$3x_2x_{25}+3x_1x_{26}=\eta_1-3\sum_{r=3}^{12}x_rx_{\bar r}+x_{13}^2+x_{13}x_{14}+x_{14}^2, \tag{10.4.43}$$

$$\begin{aligned}&3[3(x_1x_{15}+x_3x_{11}-x_4x_9+x_6x_7)+x_2(x_{13}+2x_{14})]x_{25}\\&+3[3(x_2x_{12}+x_3x_{10}-x_4x_8+x_5x_6)+x_1(2x_{13}+x_{14})]x_{26}\\=\;&\eta_2-9x_1(x_{17}x_{24}-x_{19}x_{23}+x_{21}x_{22})-9x_2(x_{16}x_{24}-x_{18}x_{23}+x_{20}x_{21})\\&-3x_{13}^2x_{14}-9(1+\tau)[(x_7x_8-x_5x_9)x_{24}+x_{10}(x_4x_{23}+x_9x_{21})-2x_{13}^3\\&-x_{11}(x_5x_{23}+x_8x_{21}+x_{12}x_{17})-x_{12}(x_7x_{22}+x_9x_{19})]+3x_{13}x_{14}^2+2x_{14}^3\\&+3x_{13}[x_3x_{24}+x_4x_{23}+x_5x_{22}+x_6x_{21}-2x_7x_{20}+x_8x_{19}-2x_9x_{18}\\&+x_{10}x_{17}-2x_{11}x_{16}+x_{12}x_{15}]-3x_{14}[2x_3x_{24}+2x_4x_{23}-x_5x_{22}\\&+2x_6x_{21}-x_7x_{20}-x_8x_{19}-x_9x_{18}-x_{10}x_{17}-x_{11}x_{16}+2x_{12}x_{15}]\end{aligned} \tag{10.4.44}$$

by (10.4.4), (10.4.10), (10.4.15)–(10.4.19), (10.4.21), (10.4.23) and (10.4.32). Thus $\{x_r \mid 16,18,20\neq r\in\overline{14,26}\}$ are rational functions in

$$\{x_r,\zeta_s,\eta_1,\eta_2 \mid r\in\{\overline{1,13},16,18,20\};\,7,9\neq s\in\overline{1,10}\}. \tag{10.4.45}$$

Suppose that $f\in\mathscr{A}$ is a solution of (10.4.33). Write f as a rational function f_1 in the variables of (10.4.45). In the following calculations, we will always use (10.4.29). By (10.3.42),

$$0 = F_{(2,3,4,2)}(f_1) = x_1\partial_{x_{20}}(f_1). \tag{10.4.46}$$

So f_1 is independent of x_{20}. Moreover, (10.3.41) gives

$$0 = F_{(1,3,4,2)}(f_1) = -x_1\partial_{x_{18}}(f_1). \tag{10.4.47}$$

Hence f_1 is independent of x_{18}. Furthermore, (10.3.40) yields

$$0 = F_{(1,2,4,2)}(f_1) = x_1\partial_{x_{16}}(f_1). \tag{10.4.48}$$

Thus f_1 is independent of x_{16}. Successively applying (10.3.39), (10.3.38), (10.3.37), (10.3.36) and (10.3.34) to f_1, we obtain that f_1 is independent of $x_{13}, x_{12}, x_{11}, x_9$ and x_7. Therefore, f_1 is a rational function in

$$\{x_r, \zeta_s, \eta_1, \eta_2 \mid 7, 9 \neq r \in \overline{1,10}; 7, 9 \neq s \in \overline{1,10}\}. \tag{10.4.49}$$

By (10.3.25), (10.3.27), (10.3.30), (10.3.31), (10.3.33) and (10.3.35),

$$0 = F_{(0,0,1,1)}(f_1) = -x_1\partial_{x_3}(f_1) - \zeta_1\partial_{\zeta_3}(f_1), \tag{10.4.50}$$

$$0 = F_{(0,1,1,1)}(f_1) = x_1\partial_{x_4}(f_1) + \zeta_1\partial_{\zeta_4}(f_1), \tag{10.4.51}$$

$$0 = F_{(0,1,2,1)}(f_1) = -x_1\partial_{x_5}(f_1) - \zeta_1\partial_{\zeta_5}(f_1), \tag{10.4.52}$$

$$0 = F_{(1,1,1,1)}(f_1) = -x_1\partial_{x_6}(f_1) - \zeta_1\partial_{\zeta_6}(f_1), \tag{10.4.53}$$

$$0 = F_{(1,1,2,1)}(f_1) = x_1\partial_{x_8}(f_1) + \zeta_1\partial_{\zeta_8}(f_1), \tag{10.4.54}$$

$$0 = F_{(1,2,2,1)}(f_1) = -x_1\partial_{x_{10}}(f_1) - \zeta_1\partial_{\zeta_{10}}(f_1). \tag{10.4.55}$$

Set

$$\eta_r = x_1\zeta_r - x_r\zeta_1, \quad r \in \overline{3,6}; \quad \eta_7 = x_1\zeta_8 - x_8\zeta_1, \quad \eta_8 = x_1\zeta_{10} - x_{10}\zeta_1. \tag{10.4.56}$$

By the characteristic method of solving linear partial differential equations (e.g., cf. [X21]), we get that f_1 can be written as a rational function f_2 in

$$\{x_r, \zeta_s, \eta_q \mid r, s = 1, 2; q \in \overline{1,8}\}. \tag{10.4.57}$$

Next applying (10.3.28), (10.3.29) and (10.3.32) to f_2, we get

$$0 = F_{(0,1,2)}(f_2) = -(x_1\zeta_2 - \zeta_1 x_2)\partial_{\eta_5}(f_2) = -3\vartheta\partial_{\eta_5}(f_2), \tag{10.4.58}$$

$$0 = F_{(1,1,2)}(f_2) = 3\vartheta\partial_{\eta_7}(f_2), \qquad 0 = F_{(1,2,2)}(f_2) = -3\vartheta\partial_{\eta_8}(f_2) = 0 \tag{10.4.59}$$

(cf. (10.4.30)). Thus f_2 is independent of η_5, η_7 and η_8. Furthermore, we apply (10.3.21), (10.3.24) and (10.3.26) to f_2 and obtain

$$0 = F_{(0,0,1)}(f_2) = -3\vartheta\partial_{\eta_3}(f_2), \qquad 0 = F_{(0,1,1)}(f_2) = 3\vartheta\partial_{\eta_4}(f_2), \tag{10.4.60}$$

$$0 = F_{(1,1,1)}(f_2) = -3\vartheta\partial_{\eta_6}(f_2). \tag{10.4.61}$$

Therefore, f_2 is a rational function in $x_1, x_2, \zeta_1, \zeta_2, \eta_1, \eta_2$. By (10.3.22),

$$0 = F_{(0,0,0,1)}(f_2) = -x_1\partial_{x_2}(f_2) - \zeta_1\partial_{\zeta_2}(f_2). \tag{10.4.62}$$

Again the characteristic method (e.g., cf. [X21]) tells us that f_2 can be written as a rational function f_3 in $x_1, \zeta_1, \vartheta, \eta_1, \eta_2$. Since $f_2 = f$ is a polynomial in $\{x_r \mid r \in \overline{1, 26}\}$, Expressions (10.4.30), (10.4.35), (10.4.43) and (10.4.44) imply that f_2 must be a polynomial in $x_1, \zeta_1, \vartheta, \eta_1, \eta_2$. The other statements follow directly. □

Calculating the weights of the singular vectors in the above theorem, we have:

Corollary 10.3.2 *The space of polynomial $\mathscr{G}^{F_4}$-invariants over its basic module is a subalgebra of $\mathscr{A}$ generated by η_1 and η_2.*

Let $L(m_1, m_2, m_3, m_4, m_5)$ be the $\mathscr{G}^{F_4}$-submodule generated by $x_1^{m_1}\zeta_1^{m_2}\vartheta^{m_3}\eta_1^{m_4}\eta_2^{m_5}$. Note that (10.3.2) is a Cartan root space decomposition $\mathscr{G}^{F_4}$ over $\mathbb{F}$. Let $\mathscr{A}_k$ be the subspace of polynomials in $\mathscr{A}$ with degree k. Then $\mathscr{A}_k$ is a finite-dimensional weight $\mathscr{G}^{F_4}$-module by (10.3.43)–(10.3.46). Thus $L(m_1, m_2, m_3, m_4, m_5)$ is a finite-dimensional irreducible $\mathscr{G}^{F_4}$-submodule with the highest weight $m_3\lambda_3 + (m_1 + m_2)\lambda_4$. By Weyl's Theorem 2.3.6 of complete reducibility if $\mathbb{F} = \mathbb{C}$ or more generally by Lemma 6.3.2 with $n_1 = 0$,

$$\mathscr{A} = \bigoplus_{k=0}^{\infty} \mathscr{A}_k = \bigoplus_{m_1,m_2,m_3,m_4,m_5=0}^{\infty} L(m_1, m_2, m_3, m_4, m_5). \tag{10.4.63}$$

Denote by $d(k, l)$ the dimension of the highest weight irreducible module with the weight $k\lambda_3 + l\lambda_4$. The above equation implies the following combinatorial identity:

$$\frac{1}{(1-t)^{26}} = \frac{1}{(1-t^2)(1-t^3)} \sum_{k_1,k_2,k_3=0}^{\infty} d(k_1, k_2 + k_3)t^{3k_1+2k_2+k_3}. \tag{10.4.64}$$

Multiplying $(1-t)^2$ to the above equation, we obtain a new combinatorial identity about twenty-four:

$$\frac{1}{(1-t)^{24}} = \frac{1}{(1+t)(1+t+t^2)} \sum_{k_1,k_2,k_3=0}^{\infty} d(k_1, k_2 + k_3)t^{3k_1+2k_2+k_3}. \tag{10.4.65}$$

Equivalently, we have:

Corollary 10.3.3 *The dimensions $d(k,l)$ of the irreducible module with the weights $k\lambda_3 + l\lambda_4$ are linearly correlated by the following identity:*

$$(1+t)(1+t+t^2) = (1-t)^{24} \sum_{k_1,k_2,k_3=0}^{\infty} d(k_1, k_2+k_3)t^{3k_1+2k_2+k_3}. \qquad (10.4.66)$$

Recall the quadratic invariants η_1 in (10.4.4). Dually we have $\mathcal{G}^{F_4}$-invariant Laplace operator

$$\Delta_{F_4} = 3\sum_{r=1}^{12} \partial_{x_r}\partial x_{\bar{r}} - \partial^2_{x_{13}} - \partial_{x_{13}}\partial_{x_{14}} - \partial^2_{x_{14}}. \qquad (10.4.67)$$

Now the subspace of complex homogeneous harmonic polynomials with degree k is

$$\mathcal{H}_k^{F_4} = \{f \in \mathcal{A}_k \mid \Delta_{F_4}(f) = 0\}. \qquad (10.4.68)$$

Then $\mathcal{H}_1^{F_4} = V$. Assume $k \geq 2$. Suppose that k_1, k_2, m_1, m_2 are nonnegative integers such that

$$k_1 + 3k_2 = m_1 + 3m_2 + 2 = k. \qquad (10.4.69)$$

If $\Delta_{F_4}(x_1^{k_1}\vartheta^{k_2}) \neq 0$, then it is a singular vector of degree $k-2$ with weight $k_2\lambda_3 + k_1\lambda_4$. By Theorem 10.3.1, $\mathcal{A}_{k-2}$ does not contain a singular vector of such weight. This leads a contradiction. Thus $\Delta_{F_4}(x_1^{k_1}\vartheta^{k_2}) = 0$. By the same reason, $\Delta_{F_4}(x_1^{m_1}\zeta_1\vartheta^{m_2}) = 0$. Hence the irreducible submodules

$$L(k_1, 0, k_2, 0, 0),\ \ L(m_1, 1, m_2, 0, 0, 0) \subset \mathcal{H}_k^{F_4}. \qquad (10.4.70)$$

This gives the following corollary:

Corollary 10.3.4 *The number of irreducible submodules contained in the subspace $\mathcal{H}_k^{F_4}$ of complex homogeneous harmonic polynomials with degree $k \geq 2$ is $\geq [\![k/3]\!] + [\![(k-2)/3]\!] + 2$.*

References

[A] J. Adams, *Lectures on Exceptional Lie Groups* (The University of Chicago Press Ltd., London, 1996)

[IA] I. Agricola, Old and new on the exceptional group G_2. Not. AMS **55**, 922 (2008)

[AW] M. Atiyah, E. Witten, M-Theory dynamics on a manifold of G_2 holonomy. Adv. Theor. Math. Phys. **6**, 1–106 (2002)

[RLB] R.L. Bryant, Metrics with exceptional holonomy. Ann. Math. **126**, 525–567 (1987)

[BS] R.L. Bryant, S.S. Salamon, On the construction of some complete metrics with exceptional holonomy. Duke Math. J **58**, 829–850 (1989)

[CI] R. Coquereaux, E. Isasi, On quantum symmetries of the non-ADE graph F_4. Adv. Theor. Math. Phys. **8**, 955–985 (2004)

[CJS] E. Cremmer, B. Julia and J. Scherk, Supergravity theory in eleven-dimension. Phys. Lett. B **76**, 409 (1978) [[CJS]]

[DZ] S. Doser, B. Zumino, Consistent supergravity. Phys. Lett. B **62**, 335 (1976)

[FO] J. Figueroa-O'Farrill, A geometric construction of the exceptional Lie algebras of F_4 and E_8. Commun. Math. Phys. **283**, 663–674 (2008)

[NH] N. Hitchin, The geometry of three forms in six and seven dimensions. Diff. Geom. **55**(3), 547–576 (2000)

[EW] E. Witten, Search for a realistic Kalazu-Klein theory. Nucl. Phys. B **186**, 412–428 (1981)

[X18] X. Xu, Partial differential equation approach to F_4. Front. Math. China **6**, 759–774 (2011)

[X21] X. Xu, *Algebraic Approaches to Partial Differential Equations* (Springer, Heidelberg, 2013)

[HM] J. Halverson, D. R. Morrison, The landscape of M-theory compactifications on seven manifolds, J. High Ener. Phys. **4**, 047 (2015)

Chapter 11
Representations of E_6

The E_6 Lie algebra and group are popular mathematical objects with broad applications. Dickson [D] (1901) first realized that there exists an E_6-invariant trilinear form on its 27-dimensional basic irreducible module. The 78-dimensional simple Lie algebra of type E_6 can be realized by all the derivations and multiplication operators with trace zero on the 27-dimensional exceptional simple Jordan algebra (e.g., cf. [A]). Brylinski and Kostant [BK] obtained a generalized Capelli identity on the minimal representation of E_6. Ginzburg [Gd] proved that the twisted partial L-function on the 27-dimensional representation of $GE_6(\mathbb{C})$ is entire except the points 0 and 1. Iltyakov [I] showed that the field of invariant rational functions of E_6 on the direct sum of finite copies of the basic module and its dual is purely transcendental. Suzuki and Wakui [SW] studied the Turaev-Viro-Ocneanu invariant of 3-manifilds derived from the E_6-subfactor.

Okamoto and Marshak [OM] constructed a grand unification preson model with E_6 metacolor. The E_6 Lie algebra was used in [HH] to explain the degeneracies encountered in the genetic code as the result of a sequence of symmetry breakings that have occurred during its evolution. Wang [Wa] identified Geoner's model with twisted LG model and E_6 singlets. Morrison, Pieruschka and Wybourne [MPW] constructed the E_6 interacting boson model. Berglund, Candelas et al. [BCDH] studied instanton contributions to the masses and couplings of E_6 singles. Haba and Matsuoka [HaMt] found large lepton flavor mixing in the E_6-type unification models. Ghezelbash, Shafiekhani and Abolbasani [GSA] derived explicitly a set of Picard–Fuchs equations of $N = 2$ supersymmetric E_6 Yang–Mills theory. Howl and King [HK] proposed a minimal E_6 supersymmetric standard model which allows Planck scale unification, provides a solution to the μ problem and predicts a new Z'. Das and Laperashvili [DL] studied Preon model related to family replicated E_6 unification.

This chapter presents the basic knowledge of E_6 symmetries. Firstly in this chapter, we prove that the space of homogeneous polynomial solutions with degree m in 27 variables for the Dickson invariant differential equation is exactly a direct sum of $[\![m/2]\!] + 1$ explicitly determined irreducible E_6-submodules and the whole polynomial algebra is a free module over the polynomial algebra in the Dickson invariant generated by these solutions. Thus we obtain a cubic E_6-generalization

X. Xu, *Representations of Lie Algebras and Partial Differential Equations*,
DOI 10.1007/978-981-10-6391-6_11

of the classical theorem on harmonic polynomials. Weyl's Theorem of complete reducibility in Sect. 2.4 plays an important role. The result was due to us [X17].

Secondly we construct a representation of the simple Lie algebra of type E_6 on the polynomial algebra in 16 variables, which gives a fractional representation of the corresponding Lie group on 16-dimensional space. Using this representation and Shen's idea of mixed product (cf. [Sg]), we construct a functor from the category of D_5-modules to the category of E_6-modules. A condition for the functor to map a finite-dimensional irreducible D_5-module to an infinite-dimensional irreducible E_6-module is obtained by Theorem 5.4.3 of decomposing the tensor module. Our results yield explicit constructions of certain infinite-dimensional irreducible weight E_6-modules with finite-dimensional weight subspaces. In our approach, the idea of Kostant's characteristic identities plays a key role. It is the fourth sequential work of constructing functors among the module categories of different simple Lie algebras after Sects. 6.7, 7.3 and 8.3. This part is taken from [X26].

In the above work, we found a one-parameter ($\mathfrak{c}$) family of inhomogeneous first-order differential operator representations of the simple Lie algebra of type E_6 in 16 variables. Letting these operators act on the space of exponential-polynomial functions that depend on a parametric vector $\vec{a} \in \mathbb{F}^{16} \setminus \{\vec{0}\}$, we prove that the space forms an irreducible E_6-module for any constant $\mathfrak{c}$ if $\vec{a}$ is not on an explicitly given projective algebraic variety. Certain equivalent combinatorial properties of the spin oscillator representation of D_5 play key roles in our proof. The result was due to an unpublished work of us. It is the third sequential work of connecting the irreducibility and algebraic varieties after Sects. 7.4 and 8.4.

In Sect. 11.1, we construct the basic oscillator representation of E_6 via the lattice construction of the simple Lie algebra E_7 included in (4.4.15)–(4.4.25). Section 11.2 is a proof of the cubic E_6-generalization of the classical theorem on harmonic polynomials. As a preparation for the later exponential-polynomial representation of E_6, we determine in Sect. 11.3 the structure the spin oscillator representation of D_5. In Sect. 11.4, we give an inhomogeneous first-order differential-operator representation of E_6. A functor from the category of D_5-modules to the category of E_6-modules is constructed in Sect. 11.5. Moreover, a condition for the functor to map a finite-dimensional irreducible D_5-module to an infinite-dimensional irreducible E_6-module is obtained in Sect. 11.6. Finally in Sect. 11.7, we prove that the irreducibility of certain modules of E_6 is related to an algebraic variety.

11.1 Basic Oscillator Representation

In this section, we present the bosonic oscillator tepresentation of E_6 over its 27-dimensional irreducible module.

First we go back to the construction of the simple Lie algebra $\mathscr{G}^X$ in (4.4.15)–(4.4.25) with $X = E_6,\ E_7$ and E_8. Recall the Dynkin diagram of E_7:

E_7:

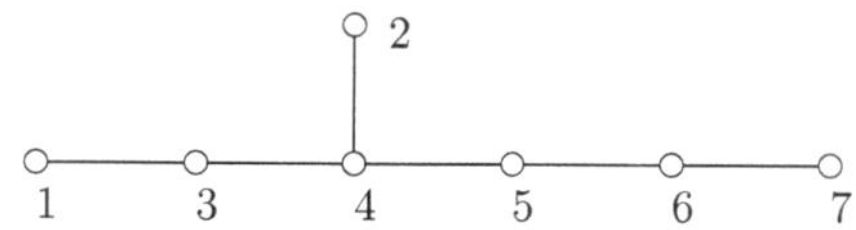

Let $\{\alpha_i \mid i \in \overline{1,7}\}$ be the simple positive roots corresponding to the vertices in the diagram, and let Φ_{E_7} be the root system of E_7. The simple Lie algebra of type E_7 is

$$\mathscr{G}^{E_7} = H \oplus \bigoplus_{\alpha\in\Phi_{E_7}} \mathbb{F}E_\alpha, \qquad H = H_{E_7} = \sum_{i=1}^{7} \mathbb{F}\alpha_i, \tag{11.1.1}$$

with the Lie bracket given in (4.4.24) and (4.4.25). Note that the Dynkin diagram of E_6 is a sub-diagram of that of E_7. Set

$$H_{E_6} = \sum_{i=1}^{6} \mathbb{F}\alpha_i, \qquad \Phi_{E_6} = \Phi_{E_7} \bigcap H_{E_6}. \tag{11.1.2}$$

We take the simple Lie algebra $\mathscr{G}^{E_6}$ of type E_6 as the Lie subalgebra

$$\mathscr{G}^{E_6} = H_{E_6} \oplus \bigoplus_{\alpha\in\Phi_{E_6}} \mathbb{F}E_\alpha \tag{11.1.3}$$

of $\mathscr{G}^{E_7}$.

Recall the notion in (4.4.42)–(4.4.44). Denote

$$\mathfrak{b}_1 = E'_{(0,0,0,0,0,0,1)}, \quad \mathfrak{b}_2 = E'_{(0,0,0,0,0,1,1)}, \quad \mathfrak{b}_3 = E'_{(0,0,0,0,1,1,1)}, \tag{11.1.4}$$

$$\mathfrak{b}_4 = E'_{(0,0,0,1,1,1,1)},, \quad \mathfrak{b}_5 = E'_{(0,0,1,1,1,1,1)}, \quad \mathfrak{b}_6 = E'_{(0,1,0,1,1,1,1)}, \tag{11.1.5}$$

$$\mathfrak{b}_7 = E'_{(0,1,1,1,1,1,1)}, \quad \mathfrak{b}_8 = E'_{(1,0,1,1,1,1,1)}, \quad \mathfrak{b}_9 = E'_{(0,1,1,2,1,1,1)}, \tag{11.1.6}$$

$$\mathfrak{b}_{10} = E'_{(1,1,1,1,1,1,1)}, \quad \mathfrak{b}_{11} = E'_{(0,1,1,2,2,1,1)}, \quad \mathfrak{b}_{12} = E'_{(1,1,1,2,1,1,1)}, \tag{11.1.7}$$

$$\mathfrak{b}_{13} = E'_{(1,1,1,2,2,1,1)}, \quad \mathfrak{b}_{14} = E'_{(0,1,1,2,2,2,1)}, \quad \mathfrak{b}_{15} = E'_{(1,1,2,2,1,1,1)}, \tag{11.1.8}$$

$$\mathfrak{b}_{16} = E'_{(1,1,2,2,2,1,1)}, \quad \mathfrak{b}_{17} = E'_{(1,1,1,2,2,2,1)}, \quad \mathfrak{b}_{18} = E'_{(1,1,2,3,2,1,1)}, \tag{11.1.9}$$

$$\mathfrak{b}_{19} = E'_{(1,1,2,2,2,2,1)}, \quad \mathfrak{b}_{20} = E'_{(1,2,2,3,2,1,1)}, \quad \mathfrak{b}_{21} = E'_{(1,1,2,3,2,2,1)}, \tag{11.1.10}$$

$$\mathfrak{b}_{22} = E'_{(1,1,2,3,3,2,1)}, \quad \mathfrak{b}_{23} = E'_{(1,2,2,3,2,2,1)}, \quad \mathfrak{b}_{24} = E'_{(1,2,2,3,3,2,1)}, \tag{11.1.11}$$

$$\mathfrak{b}_{25} = E'_{(1,2,2,4,3,2,1)}, \quad \mathfrak{b}_{26} = E'_{(1,2,3,4,3,2,1)}, \quad \mathfrak{b}_{27} = E'_{(2,2,3,4,3,2,1)}, \tag{11.1.12}$$

Then the subspace

$$V = \sum_{i=1}^{27} \mathbb{F}\mathfrak{b}_i \tag{11.1.13}$$

forms an irreducible $\mathscr{G}^{E_6}$-module with respect to the adjoint representation of $\mathscr{G}^{E_7}$, $\mathfrak{b}_1$ is a highest-weight vector of weight λ_6 and

$$\sum_{\alpha\in\Phi^-_{E_7}} \mathbb{F}E_\alpha = \sum_{\beta\in\Phi^-_{E_6}} \mathbb{F}E_\beta \oplus V. \tag{11.1.14}$$

Write

$$[u, \mathfrak{b}_i] = \sum_{j=1}^{27} \varphi_{i,j}(u)\mathfrak{b}_j \qquad \text{for } u \in \mathscr{G}^{F_4}. \tag{11.1.15}$$

Set

$$\mathscr{A} = \mathbb{F}[x_1, x_2, \ldots, x_{27}] \tag{11.1.16}$$

and define the *basic oscillator representation* $\mathfrak{r}$ *of* $\mathscr{G}^{E_6}$ on $\mathscr{A}$ by

$$\mathfrak{r}(u) = \sum_{i,j=1}^{27} \varphi_{i,j}(u)x_j\partial_{x_i} \tag{11.1.17}$$

(cf. (2.2.17)–(2.2.20)). Then $\mathscr{A}$ forms a $\mathscr{G}^{E_6}$-module isomorphic to the symmetric tensor $S(V)$ over V. More explicitly, we have the following representation formulas for the positive root vectors in $\mathscr{G}^{E_6}$ by (4.4.59)–(4.4.61):

$$\mathfrak{r}(E_{\alpha_1}) = x_5\partial_{x_8} + x_7\partial_{x_{10}} + x_9\partial_{x_{12}} + x_{11}\partial_{x_{13}} + x_{14}\partial_{x_{17}} + x_{26}\partial_{x_{27}}, \tag{11.1.18}$$

$$\mathfrak{r}(E_{\alpha_2}) = x_4\partial_{x_6} + x_5\partial_{x_7} + x_8\partial_{x_{10}} - x_{18}\partial_{x_{20}} - x_{21}\partial_{x_{23}} - x_{22}\partial_{x_{24}}, \tag{11.1.19}$$

$$\mathfrak{r}(E_{\alpha_3}) = x_4\partial_{x_5} + x_6\partial_{x_7} + x_{12}\partial_{x_{15}} + x_{13}\partial_{x_{16}} + x_{17}\partial_{x_{19}} - x_{25}\partial_{x_{26}}, \tag{11.1.20}$$

$$\mathfrak{r}(E_{\alpha_4}) = x_3\partial_{x_4} - x_7\partial_{x_9} - x_{10}\partial_{x_{12}} - x_{16}\partial_{x_{18}} - x_{19}\partial_{x_{21}} - x_{24}\partial_{x_{25}}, \tag{11.1.21}$$

$$\mathfrak{r}(E_{\alpha_5}) = x_2\partial_{x_3} - x_9\partial_{x_{11}} - x_{12}\partial_{x_{13}} - x_{15}\partial_{x_{16}} - x_{21}\partial_{x_{22}} - x_{23}\partial_{x_{24}}, \tag{11.1.22}$$

$$\mathfrak{r}(E_{\alpha_6}) = x_1\partial_{x_2} - x_{11}\partial_{x_{14}} - x_{13}\partial_{x_{17}} - x_{16}\partial_{x_{19}} - x_{18}\partial_{x_{21}} - x_{20}\partial_{x_{23}}, \tag{11.1.23}$$

$$\mathfrak{r}(E_{(1,0,1)}) = x_4\partial_{x_8} + x_6\partial_{x_{10}} - x_9\partial_{x_{15}} - x_{11}\partial_{x_{16}} - x_{14}\partial_{x_{19}} - x_{25}\partial_{x_{27}}, \tag{11.1.24}$$

$$\mathfrak{r}(E_{(0,1,0,1)}) = x_3\partial_{x_6} + x_5\partial_{x_9} + x_8\partial_{x_{12}} + x_{16}\partial_{x_{20}} + x_{19}\partial_{x_{23}} - x_{22}\partial_{x_{25}}, \tag{11.1.25}$$

$$r(E_{(0,0,1,1)}) = x_3\partial_{x_5} + x_6\partial_{x_9} - x_{10}\partial_{x_{15}} + x_{13}\partial_{x_{18}} + x_{17}\partial_{x_{21}} + x_{24}\partial_{x_{26}}, \tag{11.1.26}$$

$$\mathfrak{r}(E_{(0,0,0,1,1)}) = x_2\partial_{x_4} - x_7\partial_{x_{11}} - x_{10}\partial_{x_{13}} + x_{15}\partial_{x_{18}} - x_{19}\partial_{x_{22}} + x_{23}\partial_{x_{25}}, \tag{11.1.27}$$

$$\mathfrak{r}(E_{(0,0,0,0,1,1)}) = x_1\partial_{x_3} - x_9\partial_{x_{14}} - x_{12}\partial_{x_{17}} - x_{15}\partial_{x_{19}} + x_{18}\partial_{x_{22}} + x_{20}\partial_{x_{24}}, \tag{11.1.28}$$

$$\mathfrak{r}(E_{(1,0,1,1)}) = x_3\partial_{x_8} + x_6\partial_{x_{12}} + x_7\partial_{x_{15}} - x_{11}\partial_{x_{18}} - x_{14}\partial_{x_{21}} + x_{24}\partial_{x_{27}}, \tag{11.1.29}$$

$$\mathfrak{r}(E_{(0,1,1,1)}) = x_3\partial_{x_7} - x_4\partial_{x_9} + x_8\partial_{x_{15}} - x_{13}\partial_{x_{20}} - x_{17}\partial_{x_{23}} + x_{22}\partial_{x_{26}}, \tag{11.1.30}$$

$$\mathfrak{r}(E_{(0,1,0,1,1)}) = x_2\partial_{x_6} + x_5\partial_{x_{11}} + x_8\partial_{x_{13}} - x_{15}\partial_{x_{20}} + x_{19}\partial_{x_{x_{24}}} + x_{21}\partial_{x_{25}}, \tag{11.1.31}$$

$$\mathfrak{r}(E_{(0,0,1,1,1)}) = x_2\partial_{x_5} + x_6\partial_{x_{11}} - x_{10}\partial_{x_{16}} - x_{12}\partial_{x_{18}} + x_{17}\partial_{x_{22}} - x_{23}\partial_{x_{26}}, \tag{11.1.32}$$

$$\mathfrak{r}(E_{(0,0,0,1,1,1)}) = x_1\partial_{x_4} - x_7\partial_{x_{14}} - x_{10}\partial_{x_{17}} + x_{15}\partial_{x_{21}} + x_{16}\partial_{x_{22}} - x_{20}\partial_{x_{25}}, \tag{11.1.33}$$

$$\mathfrak{r}(E_{(1,1,1,1)}) = x_3\partial_{x_{10}} - x_4\partial_{x_{12}} - x_5\partial_{x_{15}} + x_{11}\partial_{x_{20}} + x_{14}\partial_{x_{23}} + x_{22}\partial_{x_{27}}, \tag{11.1.34}$$

$$\mathfrak{r}(E_{(1,0,1,1,1)}) = x_2\partial_{x_8} + x_6\partial_{x_{13}} + x_7\partial_{x_{16}} + x_9\partial_{x_{18}} - x_{14}\partial_{x_{22}} - x_{23}\partial_{x_{27}}, \tag{11.1.35}$$

$$\mathfrak{r}(E_{(0,1,1,1,1)}) = x_2\partial_{x_7} - x_4\partial_{x_{11}} + x_8\partial_{x_{16}} + x_{12}\partial_{x_{20}} - x_{17}\partial_{x_{24}} - x_{21}\partial_{x_{26}}, \tag{11.1.36}$$

$$\mathfrak{r}(E_{(0,1,0,1,1,1)}) = x_1\partial_{x_6} + x_5\partial_{x_{14}} + x_8\partial_{x_{17}} - x_{15}\partial_{x_{23}} - x_{16}\partial_{x_{24}} - x_{18}\partial_{x_{25}}, \tag{11.1.37}$$

$$\mathfrak{r}(E_{(0,0,1,1,1,1)}) = x_1\partial_{x_5} + x_6\partial_{x_{14}} - x_{10}\partial_{x_{19}} - x_{12}\partial_{x_{21}} - x_{13}\partial_{x_{22}} + x_{20}\partial_{x_{26}}, \tag{11.1.38}$$

$$\mathfrak{r}(E_{(1,1,1,1,1}) = x_2\partial_{x_{10}} - x_4\partial_{x_{13}} - x_5\partial_{x_{16}} - x_9\partial_{x_{20}} + x_{14}\partial_{x_{24}} - x_{21}\partial_{x_{27}}, \tag{11.1.39}$$

$$\mathfrak{r}(E_{(1,0,1,1,1,1)}) = x_1\partial_{x_8} + x_6\partial_{x_{17}} + x_7\partial_{x_{19}} + x_9\partial_{x_{21}} + x_{11}\partial_{x_{22}} + x_{20}\partial_{x_{27}}, \tag{11.1.40}$$

$$\mathfrak{r}(E_{(0,1,1,2,1)}) = x_2\partial_{x_9} - x_3\partial_{x_{11}} + x_8\partial_{x_{18}} - x_{10}\partial_{x_{20}} - x_{17}\partial_{x_{25}} + x_{19}\partial_{x_{26}}, \quad (11.1.41)$$

$$\mathfrak{r}(E_{(0,1,1,1,1,1)}|_{\mathscr{A}} = x_1\partial_{x_7} - x_4\partial_{x_{14}} + x_8\partial_{x_{19}} + x_{12}\partial_{x_{23}} + x_{13}\partial_{x_{24}} + x_{18}\partial_{x_{26}}, \quad (11.1.42)$$

$$\mathfrak{r}(E_{(1,1,1,2,1)}) = x_2\partial_{x_{12}} - x_3\partial_{x_{13}} - x_5\partial_{x_{18}} + x_7\partial_{x_{20}} + x_{14}\partial_{x_{25}} + x_{19}\partial_{x_{27}}, \quad (11.1.43)$$

$$\mathfrak{r}(E_{(1,1,1,1,1,1)}) = x_1\partial_{x_{10}} - x_4\partial_{x_{17}} - x_5\partial_{x_{19}} - x_9\partial_{x_{23}} - x_{11}\partial_{x_{24}} + x_{18}\partial_{x_{27}}, \quad (11.1.44)$$

$$\mathfrak{r}(E_{(0,1,1,2,1,1)}) = x_1\partial_{x_9} - x_3\partial_{x_{14}} + x_8\partial_{x_{21}} - x_{10}\partial_{x_{23}} + x_{13}\partial_{x_{25}} - x_{16}\partial_{x_{26}}, \quad (11.1.45)$$

$$\mathfrak{r}(E_{(1,1,2,2,1)}) = x_2\partial_{x_{15}} - x_3\partial_{x_{16}} + x_4\partial_{x_{18}} - x_6\partial_{x_{20}} - x_{14}\partial_{x_{26}} - x_{17}\partial_{x_{27}}, \quad (11.1.46)$$

$$\mathfrak{r}(E_{(1,1,1,2,1,1)}) = x_1\partial_{x_{12}} - x_3\partial_{x_{17}} - x_5\partial_{x_{21}} + x_7\partial_{x_{23}} - x_{11}\partial_{x_{25}} - x_{16}\partial_{x_{27}}, \quad (11.1.47)$$

$$\mathfrak{r}(E_{(0,1,1,2,2,1)}) = x_1\partial_{x_{11}} - x_2\partial_{x_{14}} + x_8\partial_{x_{22}} - x_{10}\partial_{x_{24}} - x_{12}\partial_{x_{25}} + x_{15}\partial_{x_{26}}, \quad (11.1.48)$$

$$\mathfrak{r}(E_{(1,1,2,2,1,1)}) = x_1\partial_{x_{15}} - x_3\partial_{x_{19}} + x_4\partial_{x_{21}} - x_6\partial_{x_{23}} + x_{11}\partial_{x_{26}} + x_{13}\partial_{x_{27}}, \quad (11.1.49)$$

$$\mathfrak{r}(E_{(1,1,1,2,2,1)}) = x_1\partial_{x_{13}} - x_2\partial_{x_{17}} - x_5\partial_{x_{22}} + x_7\partial_{x_{24}} + x_9\partial_{x_{25}} + x_{15}\partial_{x_{27}}, \quad (11.1.50)$$

$$\mathfrak{r}(E_{(1,1,2,2,2,1)}) = x_1\partial_{x_{16}} - x_2\partial_{x_{19}} + x_4\partial_{x_{22}} - x_6\partial_{x_{24}} - x_9\partial_{x_{26}} - x_{12}\partial_{x_{27}}, \quad (11.1.51)$$

$$\mathfrak{r}(E_{(1,1,2,3,2,1)}) = x_1\partial_{x_{18}} - x_2\partial_{x_{21}} + x_3\partial_{x_{22}} - x_6\partial_{x_{25}} + x_7\partial_{x_{26}} + x_{10}\partial_{x_{27}}, \quad (11.1.52)$$

$$\mathfrak{r}(E_{(1,2,2,3,2,1)}) = x_1\partial_{x_{20}} - x_2\partial_{x_{23}} + x_3\partial_{x_{24}} - x_4\partial_{x_{25}} + x_5\partial_{x_{26}} + x_8\partial_{x_{27}}. \quad (11.1.53)$$

Recall that we also view α_i as the elements of $\mathscr{G}^{E_6}$ (cf. (11.1.1)). Then

$$\mathfrak{r}(\alpha_r) = \sum_{i=1}^{27} a_{i,r} x_i \partial_{x_i} \qquad \text{for } r \in \overline{1,6} \quad (11.1.54)$$

with $a_{i,r}$ given in the following table 11.1:

We define a symmetric linear operation τ on the space $\sum_{i,j=1}^{27} \mathbb{F} x_i \partial_{x_j}$ by

$$\tau(x_i \partial_{x_j}) = x_j \partial_{x_i}. \quad (11.1.55)$$

Then

$$\mathfrak{r}(E_{-\alpha}) = -\tau(\mathfrak{r}(E_\alpha)) \qquad \text{for } \alpha \in \Phi_{E_6}^+ \quad (11.1.56)$$

by the second equations in (4.4.19), (4.4.20) and (4.4.25).

11.2 Decomposition of the Oscillator Representation

In this section, we decompose $\mathscr{A}$ into a direct sum of irreducible $\mathscr{G}^{E_6}$-submodules and prove that $\mathscr{A}$ is a free module over the algebra of $\mathscr{G}^{E_6}$-invariants generated by the solutions of Dickson's cubic invariant partial differential equation.

Recall that a singular vector of $\mathscr{G}^{E_6}$ is a nonzero weight vector annihilated by positive root vectors. According to Table 11.1 and (11.1.18)–(11.1.23), we find a singular vector

$$\zeta_1 = x_1x_{14} + x_2x_{11} + x_3x_9 + x_4x_7 - x_5x_6 \quad (11.2.1)$$

of weight λ_1, which generates a 27-dimensional irreducible $\mathscr{G}^{E_6}$-module

Table 11.1 Weights of the Basic Representation

i	$a_{i,1}$	$a_{i,2}$	$a_{i,3}$	$a_{i,4}$	$a_{i,5}$	$a_{i,6}$	i	$a_{i,1}$	$a_{i,2}$	$a_{i,3}$	$a_{i,4}$	$a_{i,5}$	$a_{i,6}$
1	0	0	0	0	0	1	2	0	0	0	0	1	−1
3	0	0	0	1	−1	0	4	0	1	1	−1	0	0
5	1	1	−1	0	0	0	6	0	−1	1	0	0	0
7	1	−1	−1	1	0	0	8	−1	1	0	0	0	0
9	1	0	0	−1	1	0	10	−1	−1	0	1	0	0
11	1	0	0	0	−1	1	12	−1	0	1	−1	1	0
13	−1	0	1	0	−1	1	14	1	0	0	0	0	−1
15	0	0	−1	0	1	0	16	0	0	−1	1	−1	1
17	−1	0	1	0	0	−1	18	0	1	0	−1	0	1
19	0	0	−1	1	0	−1	20	0	−1	0	0	0	1
21	0	1	0	−1	1	−1	22	0	1	0	0	−1	0
23	0	−1	0	0	1	−1	24	0	−1	0	1	−1	0
25	0	0	1	−1	0	0	26	1	0	−1	0	0	0
27	−1	0	0	0	0	0							

$$U = \sum_{i=1}^{27} \mathbb{F}\zeta_i \tag{11.2.2}$$

with

$$\zeta_2 = x_1x_{17} + x_2x_{13} + x_3x_{12} + x_4x_{10} - x_6x_8, \tag{11.2.3}$$
$$\zeta_3 = x_1x_{19} + x_2x_{16} + x_3x_{15} + x_5x_{10} - x_7x_8, \tag{11.2.4}$$
$$\zeta_4 = -x_1x_{21} - x_2x_{18} + x_4x_{15} - x_5x_{12} + x_8x_9, \tag{11.2.5}$$
$$\zeta_5 = x_1x_{22} - x_3x_{18} - x_4x_{16} + x_5x_{13} - x_8x_{11}, \tag{11.2.6}$$
$$\zeta_6 = x_1x_{23} + x_2x_{20} + x_6x_{15} - x_7x_{12} + x_9x_{10}, \tag{11.2.7}$$
$$\zeta_7 = -x_1x_{24} + x_3x_{20} - x_6x_{16} + x_7x_{13} - x_{10}x_{11}, \tag{11.2.8}$$
$$\zeta_8 = x_2x_{22} + x_3x_{21} + x_4x_{19} - x_5x_{17} + x_8x_{14}, \tag{11.2.9}$$
$$\zeta_9 = x_1x_{25} + x_4x_{20} + x_6x_{18} - x_9x_{13} + x_{11}x_{12}, \tag{11.2.10}$$
$$\zeta_{10} = -x_2x_{24} - x_3x_{23} + x_6x_{19} - x_7x_{17} + x_{10}x_{14}, \tag{11.2.11}$$
$$\zeta_{11} = x_1x_{26} - x_5x_{20} - x_7x_{18} + x_9x_{16} - x_{11}x_{15}, \tag{11.2.12}$$
$$\zeta_{12} = x_2x_{25} - x_4x_{23} - x_6x_{21} + x_9x_{17} - x_{12}x_{14}, \tag{11.2.13}$$
$$\zeta_{13} = x_3x_{25} + x_4x_{24} + x_6x_{22} - x_{11}x_{17} + x_{13}x_{14}, \tag{11.2.14}$$
$$\zeta_{14} = x_1x_{27} - x_8x_{20} - x_{10}x_{18} + x_{12}x_{16} - x_{13}x_{15}, \tag{11.2.15}$$
$$\zeta_{15} = x_2x_{26} + x_5x_{23} + x_7x_{21} - x_9x_{19} + x_{14}x_{15}, \tag{11.2.16}$$
$$\zeta_{16} = x_3x_{26} - x_5x_{24} - x_7x_{22} + x_{11}x_{19} - x_{14}x_{16}, \tag{11.2.17}$$
$$\zeta_{17} = -x_2x_{27} - x_8x_{23} - x_{10}x_{21} + x_{12}x_{19} - x_{15}x_{17}, \tag{11.2.18}$$

$$\zeta_{18} = x_4x_{26} + x_5x_{25} + x_9x_{22} - x_{11}x_{21} + x_{14}x_{18}, \quad (11.2.19)$$
$$\zeta_{19} = -x_3x_{27} + x_8x_{24} + x_{10}x_{22} - x_{13}x_{19} + x_{16}x_{17}, \quad (11.2.20)$$
$$\zeta_{20} = -x_6x_{26} - x_7x_{25} + x_9x_{24} - x_{11}x_{23} + x_{14}x_{20}, \quad (11.2.21)$$
$$\zeta_{21} = -x_4x_{27} - x_8x_{25} - x_{12}x_{22} + x_{13}x_{21} - x_{17}x_{18}, \quad (11.2.22)$$
$$\zeta_{22} = x_5x_{27} - x_8x_{26} + x_{15}x_{22} - x_{16}x_{21} + x_{18}x_{19}, \quad (11.2.23)$$
$$\zeta_{23} = x_6x_{27} + x_{10}x_{25} - x_{12}x_{24} + x_{13}x_{23} - x_{17}x_{20}, \quad (11.2.24)$$
$$\zeta_{24} = -x_7x_{27} + x_{10}x_{26} + x_{15}x_{24} - x_{16}x_{23} + x_{19}x_{20}, \quad (11.2.25)$$
$$\zeta_{25} = -x_9x_{27} + x_{12}x_{26} + x_{15}x_{25} - x_{18}x_{23} + x_{20}x_{21}, \quad (11.2.26)$$
$$\zeta_{26} = -x_{11}x_{27} + x_{13}x_{26} + x_{16}x_{25} - x_{18}x_{24} + x_{20}x_{22}, \quad (11.2.27)$$
$$\zeta_{27} = -x_{14}x_{27} + x_{17}x_{26} + x_{19}x_{25} - x_{21}x_{24} + x_{22}x_{23}. \quad (11.2.28)$$

By (11.1.18)–(11.1.23) and (11.2.1)–(11.2.28), we calculate

$$E_{\alpha_1}|_U = \zeta_1\partial_{\zeta_2} + \zeta_{11}\partial_{\zeta_{14}} - \zeta_{15}\partial_{\zeta_{17}} - \zeta_{16}\partial_{\zeta_{19}} - \zeta_{18}\partial_{\zeta_{21}} - \zeta_{20}\partial_{\zeta_{23}}, \quad (11.2.29)$$
$$E_{\alpha_2}|_U = \zeta_4\partial_{\zeta_6} + \zeta_5\partial_{\zeta_7} + \zeta_8\partial_{\zeta_{10}} - \zeta_{18}\partial_{\zeta_{20}} - \zeta_{21}\partial_{\zeta_{23}} - \zeta_{22}\partial_{\zeta_{24}}, \quad (11.2.30)$$
$$E_{\alpha_3}|_U = \zeta_2\partial_{\zeta_3} - \zeta_9\partial_{\zeta_{11}} - \zeta_{12}\partial_{\zeta_{15}} - \zeta_{13}\partial_{\zeta_{16}} - \zeta_{21}\partial_{\zeta_{22}} - \zeta_{23}\partial_{\zeta_{24}}, \quad (11.2.31)$$
$$E_{\alpha_4}|_U = \zeta_3\partial_{\zeta_4} + \zeta_7\partial_{\zeta_9} + \zeta_{10}\partial_{\zeta_{12}} + \zeta_{16}\partial_{\zeta_{18}} + \zeta_{19}\partial_{\zeta_{21}} - \zeta_{24}\partial_{\zeta_{25}}, \quad (11.2.32)$$
$$E_{\alpha_5}|_U = \zeta_4\partial_{\zeta_5} + \zeta_6\partial_{\zeta_7} + \zeta_{12}\partial_{\zeta_{13}} + \zeta_{15}\partial_{\zeta_{16}} + \zeta_{17}\partial_{\zeta_{19}} - \zeta_{25}\partial_{\zeta_{26}}, \quad (11.2.33)$$
$$E_{\alpha_6}|_U = \zeta_5\partial_{\zeta_8} + \zeta_7\partial_{\zeta_{10}} + \zeta_9\partial_{\zeta_{12}} + \zeta_{11}\partial_{\zeta_{15}} - \zeta_{14}\partial_{\zeta_{17}} - \zeta_{26}\partial_{\zeta_{27}}. \quad (11.2.34)$$

Define an anti-homomorphism † from the associative algebra

$$\mathbb{A} = \sum_{i_1,\dots,i_{27}=0}^{\infty} \mathscr{A}\partial_{x_1}^{i_1}\cdots\partial_{x_{27}}^{i_{27}} \quad (11.2.35)$$

of differential operators to the associative algebra

$$\mathbb{A}' = \sum_{\ell_1,\dots,\ell_{27},i_1,\dots,i_{27}=0}^{\infty} \mathbb{F}\zeta_1^{\ell_1}\cdots\zeta_{27}^{\ell_{27}}\partial_{\zeta_1}^{i_1}\cdots\partial_{\zeta_{27}}^{i_{27}} \quad (11.2.36)$$

determined by

$$(x_1)^\dagger = \partial_{\zeta_{27}}, \quad (\partial_{x_1})^\dagger = \zeta_{27}, \quad (x_{27})^\dagger = -\partial_{\zeta_1}, \quad (\partial_{x_{27}})^\dagger = -\zeta_1, \quad (11.2.37)$$
$$(x_{14})^\dagger = -\partial_{\zeta_{14}}, \quad (\partial_{x_{14}})^\dagger = -\zeta_{14}, \quad (x_i)^\dagger = \partial_{\zeta_{28-i}}, \quad (\partial_{x_i})^\dagger = \zeta_{28-i} \quad (11.2.38)$$

for $14 \neq i \in \overline{2,26}$. Then the map $d \mapsto -d^\dagger$ gives a Lie algebra monomorphism from $\mathfrak{r}(\mathscr{G}^{E_6}) \to \mathbb{A}'$. By (11.1.18)–(11.1.23), (11.1.54)–(11.1.56), Table 11.1 and (11.2.29)–(11.2.34), we have

Table 11.2 Weights of the Dual Representation

i	$b_{i,1}$	$b_{i,2}$	$b_{i,3}$	$b_{i,4}$	$b_{i,5}$	$b_{i,6}$	i	$b_{i,1}$	$b_{i,2}$	$b_{i,3}$	$b_{i,4}$	$b_{i,5}$	$b_{i,6}$
1	1	0	0	0	0	0	2	−1	0	1	0	0	0
3	0	0	−1	1	0	0	4	0	1	0	−1	1	0
5	0	1	0	0	−1	1	6	0	−1	0	0	1	0
7	0	−1	0	1	−1	1	8	0	1	0	0	0	−1
9	0	0	1	−1	0	1	10	0	−1	0	1	0	−1
11	1	0	−1	0	0	1	12	0	0	1	−1	1	−1
13	0	0	1	0	−1	0	14	−1	0	0	0	0	1
15	1	0	−1	0	1	−1	16	1	0	−1	1	−1	0
17	−1	0	0	0	1	−1	18	1	1	0	−1	0	0
19	−1	0	0	1	−1	0	20	1	−1	0	0	0	0
21	−1	1	1	−1	0	0	22	0	1	−1	0	0	0
23	−1	−1	1	0	0	0	24	0	−1	−1	1	0	0
25	0	0	0	−1	1	0	26	0	0	0	0	−1	1
27	0	0	0	0	0	−1							

$$u|_U = -(\mathfrak{r}(u))^{\dagger} \qquad \text{for } u \in \mathcal{G}^{E_6}. \tag{11.2.39}$$

In particular,

$$\alpha_r|_U = \sum_{i=1}^{27} b_{i,r}\zeta_i\partial_{\zeta_i} \qquad \text{for } r \in \overline{1,6} \tag{11.2.40}$$

with $b_{i,r}$ given in the following table 11.2:

According to Tables 11.1, 11.2 and (11.1.18)–(11.1.23), we find the following invariant

$$\sum_{14\neq i\in\overline{1,26}} x_i\zeta_{28-i} - x_{14}\zeta_{14} - x_{27}\zeta_1 = -3\eta \tag{11.2.41}$$

with

$$\begin{aligned}\eta = {} & x_1(x_{17}x_{26} + x_{19}x_{25} - x_{21}x_{24} + x_{22}x_{23}) - x_{14}(x_1x_{27} - x_8x_{20} - x_{10}x_{18} + x_{12}x_{16} \\ & - x_{13}x_{15}) + x_2(x_{13}x_{26} + x_{16}x_{25} - x_{18}x_{24} + x_{20}x_{22}) - x_{11}(x_2x_{27} + x_8x_{23} + x_{10}x_{21} \\ & - x_{12}x_{19} + x_{15}x_{17}) + x_3(x_{12}x_{26} + x_{15}x_{25} - x_{18}x_{23} + x_{20}x_{21}) - x_9(x_3x_{27} - x_8x_{24} \\ & - x_{10}x_{22} + x_{13}x_{19} - x_{16}x_{17}) + x_4(x_{10}x_{26} + x_{15}x_{24} - x_{16}x_{23} + x_{19}x_{20}) - x_7(x_4x_{27} \\ & + x_8x_{25} + x_{12}x_{22} - x_{13}x_{21} + x_{17}x_{18}) + x_5(x_{10}x_{25} - x_{12}x_{24} + x_{13}x_{23} - x_{17}x_{20}) \\ & + x_6(x_5x_{27} - x_8x_{26} + x_{15}x_{22} - x_{16}x_{21} + x_{18}x_{19}),\end{aligned} \tag{11.2.42}$$

which is the Dickson's cubic invariant (cf. [D]).

Lemma 11.2.1 *Any homogeneous singular vector in* $\mathcal{A}$ *is a monomial in* x_1, ζ_1 *and* η.

Proof Note that

$$x_1x_{14} = \zeta_1 - x_2x_{11} - x_3x_9 - x_4x_7 + x_5x_6 \tag{11.2.43}$$
$$x_1x_{17} = \zeta_2 - x_2x_{13} - x_3x_{12} - x_4x_{10} + x_6x_8, \tag{11.2.44}$$
$$x_1x_{19} = \zeta_3 - x_2x_{16} - x_3x_{15} - x_5x_{10} + x_7x_8, \tag{11.2.45}$$
$$x_1x_{21} = -\zeta_4 - x_2x_{18} + x_4x_{15} - x_5x_{12} + x_8x_9, \tag{11.2.46}$$
$$x_1x_{22} = \zeta_5 + x_3x_{18} + x_4x_{16} - x_5x_{13} + x_8x_{11}, \tag{11.2.47}$$
$$x_1x_{23} = \zeta_6 - x_2x_{20} - x_6x_{13} + x_7x_{12} - x_9x_{10}, \tag{11.2.48}$$
$$x_1x_{24} = -\zeta_7 + x_3x_{20} - x_6x_{16} + x_7x_{13} - x_{10}x_{11}, \tag{11.2.49}$$
$$x_1x_{25} = \zeta_9 - x_4x_{20} - x_6x_{18} + x_9x_{13} - x_{11}x_{12}, \tag{11.2.50}$$
$$x_1x_{26} = \zeta_{11} + x_5x_{20} + x_7x_{18} - x_9x_{16} + x_{11}x_{15} \tag{11.2.51}$$

by (11.2.1), (11.2.3)–(11.2.7), (11.2.9), (11.2.11) and (11.2.15). Moreover, (11.2.42) can be written as

$$\begin{aligned}&(x_5x_6 - x_1x_{14} - x_2x_{11} - x_3x_9 - x_4x_7)x_{27}\\ &= \eta - x_1(x_{17}x_{26} + x_{19}x_{25} - x_{21}x_{24} + x_{22}x_{23}) - x_{14}(x_8x_{20} + x_{10}x_{18} - x_{12}x_{16}\\ &\quad - x_{13}x_{15}) - x_2(x_{13}x_{26} + x_{16}x_{25} - x_{18}x_{24} + x_{20}x_{22}) + x_{11}(x_8x_{23} + x_{10}x_{21}\\ &\quad - x_{12}x_{19} + x_{15}x_{17}) - x_3(x_{12}x_{26} + x_{15}x_{25} - x_{18}x_{23} + x_{20}x_{21}) - x_9(x_8x_{24}\\ &\quad - x_{10}x_{22} - x_{13}x_{19} + x_{16}x_{17}) - x_4(x_{10}x_{26} + x_{15}x_{24} - x_{16}x_{23} + x_{19}x_{20})\\ &\quad + x_7(x_8x_{25} + x_{12}x_{22} - x_{13}x_{21} + x_{17}x_{18}) - x_5(x_{10}x_{25} - x_{12}x_{24} + x_{13}x_{23}\\ &\quad - x_{17}x_{20}) + x_6(x_8x_{26} - x_{15}x_{22} + x_{16}x_{21} - x_{18}x_{19}).\end{aligned} \tag{11.2.52}$$

Let f be any homogenous singular vector in $\mathscr{A}$. According to the above equations, f can be written as a rational function f_1 in

$$\{x_i, \zeta_r, \eta \mid i \in \{\overline{1,13}, 15, 16, 18, 20\};\ r \in \{\overline{1,7}, 9, 11\}\}. \tag{11.2.53}$$

By (11.1.47)–(11.1.53) and (11.2.37)–(11.2.39), we have

$$0 = E_{(1,1,1,2,1,1)}(f_1) = x_1\partial_{x_{12}}(f_1), \tag{11.2.54}$$
$$0 = E_{(0,1,1,2,2,1)}(f_1) = x_1\partial_{x_{11}}(f_1), \tag{11.2.55}$$
$$0 = E_{(1,1,2,2,1,1)}(f_1) = x_1\partial_{x_{15}}(f_1), \tag{11.2.56}$$
$$0 = E_{(1,1,1,2,2,1)}(f_1) = x_1\partial_{x_{13}}(f_1), \tag{11.2.57}$$
$$0 = E_{(1,1,2,2,2,1)}(f_1) = x_1\partial_{x_{16}}(f_1), \tag{11.2.58}$$
$$0 = E_{(1,1,2,3,2,1)}(f_1) = x_1\partial_{x_{18}}(f_1), \tag{11.2.59}$$
$$0 = E_{(1,2,2,3,2,1)}(f) = x_1\partial_{x_{20}}(f). \tag{11.2.60}$$

So f_1 is independent of $x_{11}, x_{12}, x_{13}, x_{15}, x_{16}, x_{18}, x_{20}$; that is, f_1 is a rational function in

$$\{x_i, \zeta_r, \eta \mid i \in \overline{1,10};\ r \in \{\overline{1,7}, 9, 11\}\}. \tag{11.2.61}$$

Next (11.1.40), (11.1.42)–(11.1.46) and (11.2.37)–(11.2.39) imply that

$$0 = E_{(1,0,1,1,1,1)}(f_1) = x_1\partial_{x_8}(f_1), \tag{11.2.62}$$
$$0 = E_{(0,1,1,1,1,1)}(f_1) = x_1\partial_{x_7}(f_1), \tag{11.2.63}$$
$$0 = E_{(1,1,1,2,1)}(f_1) = \zeta_1\partial_{\zeta_9}, \tag{11.2.64}$$
$$0 = E_{(1,1,1,1,1,1)}(f_1) = x_1\partial_{x_{10}}(f_1), \tag{11.2.65}$$
$$0 = E_{(0,1,1,2,1,1)}(f_1) = x_1\partial_{x_9}(f_1), \tag{11.2.66}$$
$$0 = E_{(1,1,2,2,1)}(f_1) = -\zeta_1\partial_{\zeta_{11}}(f_1). \tag{11.2.67}$$

Hence f_1 is independent of $x_7, x_8, x_9, x_{10}, \zeta_9$ and ζ_{11}; that is, f_1 is a rational function in

$$\{x_i, \zeta_r, \eta \mid i \in \overline{1,6},\ r \in \overline{1,7}\}. \tag{11.2.68}$$

Expressions (11.1.18), (11.1.23), (11.1.24), (11.1.28), (11.1.29), (11.1.33)–(11.1.35), (11.1.37)–(11.1.39) and (11.2.37)–(11.2.39) yield that

$$0 = E_{\alpha_1}(f_1) = \zeta_1\partial_{\zeta_2}(f_1) \tag{11.2.69}$$
$$0 = E_{\alpha_6}(f_1) = x_1\partial_{x_2}(f_1), \tag{11.2.70}$$
$$0 = E_{(1,0,1)}(f_1) = -\zeta_1\partial_{\zeta_3}(f_1), \tag{11.2.71}$$
$$0 = E_{(0,0,0,0,1,1)}(f_1) = x_1\partial_{x_3}(f_1), \tag{11.2.72}$$
$$0 = E_{(1,0,1,1)}(f_1) = \zeta_1\partial_{\zeta_4}(f_1), \tag{11.2.73}$$
$$0 = E_{(0,0,0,1,1,1)}(f_1) = x_1\partial_{x_4}(f_1), \tag{11.2.74}$$
$$0 = E_{(1,1,1,1)}(f_1) = \zeta_1\partial_{\zeta_6}(f_1), \tag{11.2.75}$$
$$0 = E_{(1,0,1,1,1)}(f_1) = -\zeta_1\partial_{\zeta_5}(f_1), \tag{11.2.76}$$
$$0 = E_{(0,1,0,1,1,1)}(f_1) = x_1\partial_{x_6}(f_1), \tag{11.2.77}$$
$$0 = E_{(0,0,1,1,1,1)}(f_1) = x_1\partial_{x_5}(f_1), \tag{11.2.78}$$
$$0 = E_{(1,1,1,1,1}(f_1) = -\zeta_1\partial_{\zeta_7}(f_1). \tag{11.2.79}$$

Thus f_1 is independent of $\{x_i, \zeta_r \mid i \in \overline{2,6},\ r \in \overline{2,7}\}$; that is, $f = f_1$ is a rational function in x_1, ζ_1 and η. By (11.2.1) and (11.2.42), it must be a polynomial in x_1, ζ_1 and η. Recall that the weights of x_1, ζ_1 and η are λ_6, λ_1 and 0, respectively. The homogeneity of f implies that it must be a monomial in x_1, ζ_1 and η. □

Let $L(m_1, m_2, m_3)$ be the $\mathscr{G}^{E_6}$-submodule generated by $x_1^{m_1}\zeta_1^{m_2}\eta^{m_3}$. Note that (11.1.3) is a Cartan root space decomposition of $\mathscr{G}^{E_6}$ over $\mathbb{F}$. Moreover, (11.1.54) implies that $\mathscr{A}$ is a direct sum of weigh subspaces of $\mathscr{G}^{E_6}$ and the subspaces of homogeneous polynomials are finite-dimensional $\mathscr{G}^{E_6}$-submodules. Thus $L(m_1, m_2, m_3)$

is a finite-dimensional irreducible submodule of highest weight $m_1\lambda_1 + m_2\lambda_6$. By Weyl's Theorem 2.3.6 of complete reducibility if $\mathbb{F} = \mathbb{C}$ or more generally by Lemma 6.3.2 with $n_1 = 0$, we have

$$\mathscr{A} = \sum_{m_1,m_2,m_2=0}^{\infty} L(m_1, m_2, m_3). \tag{11.2.80}$$

Recall we denote by $V(\lambda)$ the finite-dimensional irreducible module of highest weight λ. The above equation implies

$$\frac{1}{(1-q)^{27}} = \frac{1}{1-q^3} \sum_{m_1,m_2=0}^{\infty} (\dim V(m_1\lambda_6 + m_2\lambda_1))q^{m_1+2m_2}. \tag{11.2.81}$$

Equivalently, we have:

Corollary 11.2.2 *The following dimensional property of irreducible $\mathscr{G}^{E_6}$-modules holds:*

$$(1-q)^{26} \sum_{m_1,m_2=0}^{\infty} (\dim V(m_1\lambda_1 + m_2\lambda_6))q^{m_1+2m_2} = 1 + q + q^2. \tag{11.2.82}$$

Set

$$W = \sum_{i=1}^{27} \mathbb{F}\partial_{x_i}. \tag{11.2.83}$$

Then W is isomorphic to the module of linear functions on V via $\partial_{x_i}(x_j) = \delta_{i,j}$. Indeed, the linear map determined by $\partial_{x_i} \mapsto (\partial_{x_i})^\dagger$ (cf. (11.2.37), (11.2.38)) is a $\mathscr{G}^{E_6}$-module isomorphism from W to U by (11.2.41). We define a linear map $\mathfrak{I}$: $\mathscr{A} \to \mathbb{R}[\partial_{x_1}, \ldots, \partial_{x_{27}}]$ by

$$\mathfrak{I}(x_1^{\alpha_1} x_2^{\alpha_2} \cdots x_{27}^{\alpha_{27}}) = \partial_{x_1}^{\alpha_1} \partial_{x_2}^{\alpha_2} \cdots \partial_{x_{27}}^{\alpha_{27}}. \tag{11.2.84}$$

Set

$$\mathscr{D} = \mathfrak{I}(\eta), \qquad \mathscr{D}_1 = \sum_{i=1}^{27} x_i\partial_{x_i}, \qquad \mathscr{D}_2 = \sum_{i=1}^{27} \zeta_i\mathfrak{I}(\zeta_i). \tag{11.2.85}$$

Then $\mathscr{D}$, $\mathscr{D}_1$ and $\mathscr{D}_2$ are invariant differential operators; that is, as operators on $\mathscr{A}$,

$$d\xi = \xi d \qquad \text{for } d = \mathscr{D}, \mathscr{D}_1, \mathscr{D}_2;\ \xi \in \mathscr{G}^{E_6}. \tag{11.2.86}$$

Note that Lemma 11.2.1 implies

$$V^2 = L(2, 0, 0) + L(0, 1, 0). \tag{11.2.87}$$

Symmetrically,

$$W^2 = L'(0,2,0) + L'(1,0,0), \tag{11.2.88}$$

where $L'(0,2,0)$ is a module generated by the highest-weight vector ∂_{27}^2 with weight $2\lambda_1$ and $L'(1,0,0)$ is a module generated by the highest-weight vector $\Im(\zeta_{27})$ with weight λ_6. Thus the subspace of invariants (the trivial submodule) in V^2W^2 is two-dimensional. The trivial submodule of $L(0,1,0)L'(1,0,0)$ is $\mathbb{F}\mathscr{D}_2$. In $L(2,0,0)L'(0,2,0)$, there exists an invariant $\mathscr{D}_3$ with a term $x_1^2\partial_{x_1}^2$. So any invariant in V^2W^2 must be in $\mathbb{F}\mathscr{D}_2 + \mathbb{F}\mathscr{D}_3$. In particular, the invariant differential operator

$$[\mathscr{D},\eta] = \mathscr{D}\eta - \eta\mathscr{D} = b_0 + b_1\mathscr{D}_1 + b_2\mathscr{D}_2 + b_3\mathscr{D}_3 \tag{11.2.89}$$

for some $b_s \in \mathbb{F}$. According to (11.2.42), η does not contain x_1^2. So $b_3 = 0$. Moreover, (11.2.42) also implies $b_0 = 45$.

According to (11.2.52), the coefficient of $x_{27}\partial_{x_{27}}$ in $[\mathscr{D},\eta]$ must be 5, which implies $b_1 = 5$. Observe that there exists a unique monomial in η containing x_1x_{14}, which is $x_1x_{14}x_{27}$. Thus the coefficient of $x_1x_{14}\partial_{x_1}\partial_{x_{14}}$ in $[\mathscr{D},\eta]$ must be 1; that is, $b_2 = 1$. So we have:

Lemma 11.2.3 *As operators on* $\mathscr{A}$,

$$[\mathscr{D},\eta] = 45 + 5\mathscr{D}_1 + \mathscr{D}_2. \tag{11.2.90}$$

Let m_1 and m_2 be nonnegative integers. If $\mathscr{D}(x_1^{m_1}\zeta_1^{m_2}) \neq 0$, then it is also a singular of degree $m_1 + 2m_2 - 3$ with the same weight $m_1\lambda_6 + m_2\lambda_1$. But Lemma 11.2.1 implies that any singular vector with weight $m_1\lambda_6 + m_2\lambda_1$ must has degree $\geq m_1 + 2m_2$. This leads a contradiction. Thus

$$\mathscr{D}(x_1^{m_1}\zeta_1^{m_2}) = 0 \qquad \text{for } m_1, m_2 \in \mathbb{N}. \tag{11.2.91}$$

Moreover, (11.2.76) implies

$$\mathscr{D}(L(m_1,m_2,0)) = \{0\} \qquad \text{for } m_1, m_2 \in \mathbb{N}. \tag{11.2.92}$$

Since $\mathscr{D}_2(x_1^{m_1}\zeta_1^{m_2})$ is also a singular vector of degree $m_1 + 2m_2$ with the same weight $m_1\lambda_6 + m_2\lambda_1$, we have

$$\mathscr{D}_2(x_1^{m_1}\zeta_1^{m_2}) = c x_1^{m_1}\zeta_1^{m_2} \tag{11.2.93}$$

for some $c \in \mathbb{F}$. Let

$$x_i = 0 \qquad \text{for } 1, 14 \neq i \in \overline{1,27} \tag{11.2.94}$$

in (11.2.83) and we get

$$\begin{aligned} c x_1^{m_1+m_2} x_{14}^{m_2} = & \lim_{x_i \to 0;\ 8,10 \neq i \in \overline{2,11}} x_1x_{14}(\partial_{x_1}\partial_{x_{14}} + \partial_{x_2}\partial_{x_{11}} + \partial_{x_3}\partial_{x_9} + \partial_{x_4}\partial_{x_7} - \partial_{x_5}\partial_{x_6})[x_1^{m_1} \\ & \times (x_1x_{14} + x_2x_{11} + x_3x_9 + x_4x_7 - x_5x_6)^{m_2}] \end{aligned}$$

$$= m_2(m_1 + m_2 + 4)x_1^{m_1+m_2}x_{14}^{m_2} \tag{11.2.95}$$

by (11.2.1)–(11.2.29); that is, $c = m_2(m_1 + m_2 + 4)$. We get:

Lemma 11.2.4 *For* $m_1, m_2 \in \mathbb{N}$,

$$\mathscr{D}_2(x_1^{m_1}\zeta_1^{m_2}) = m_2(m_1 + m_2 + 4)x_1^{m_1}\zeta_1^{m_2}. \tag{11.2.96}$$

According to Lemma 11.2.1,

$$V^4 = L(4,0,0) + L(2,1,0) + L(1,0,1). \tag{11.2.97}$$

Moreover, $L(1,0,1) = \eta V$. Thus the invariants in V^4W are $\mathbb{F}\eta\mathscr{D}_1$. Hence

$$[\mathscr{D}_2, \eta] = c_1\eta + c_2\eta\mathscr{D}_1 \qquad \text{for some } \ c_1, c_2 \in \mathbb{F}. \tag{11.2.98}$$

Letting the above equation act on 1, we have

$$\mathscr{D}_2(\eta) = c_1\eta. \tag{11.2.99}$$

By (11.2.1)–(11.2.28) and (11.2.42),

$$\begin{aligned} c_1x_1x_{14}x_{27} &= -\lim_{x_i\to 0;\ 14\neq i\in\overline{2,16}}\eta = \lim_{x_i\to 0;\ 14\neq i\in\overline{2,16}}\mathscr{D}_2(\eta) \\ &= (x_1x_{14}\partial_{x_1}\partial_{x_{14}} + x_{14}x_{27}\partial_{x_{14}}\partial_{x_{27}} + x_1x_{27}\partial_{x_1}\partial_{x_{27}})(x_1x_{14}x_{27}) = 3x_1x_{14}x_{27}, \end{aligned} \tag{11.2.100}$$

So $c_1 = 3$. Letting (11.2.98) act on x_1, we have:

$$\mathscr{D}_2(\eta x_1) = (3 + c_2)\eta x_1. \tag{11.2.101}$$

As (11.2.100),

$$\begin{aligned} (3 + c_2)x_1^2x_{14}x_{27} &= -\lim_{x_i\to 0;\ 14\neq i\in\overline{2,16}}(3 + c_2)\eta x_1 = -\lim_{x_i\to 0;\ 14\neq i\in\overline{2,16}}\mathscr{D}_2(\eta x_1) \\ &= (x_1x_{14}\partial_{x_1}\partial_{x_{14}} + x_{14}x_{27}\partial_{x_{14}}\partial_{x_{27}} + x_1x_{27}\partial_{x_1}\partial_{x_{27}})(x_1^2x_{14}x_{27}) = 4x_1^2x_{14}x_{27}, \end{aligned} \tag{11.2.102}$$

Hence $c_2 = 1$. We get:

Lemma 11.2.5 *As operators on* $\mathscr{A}$,

$$[\mathscr{D}_2, \eta] = \eta(3 + \mathscr{D}_1). \tag{11.2.103}$$

For $m, m_1, m_2 \in \mathbb{N}$ with $m > 0$, we have

$$\mathscr{D}(\eta^m x_1^{m_1} \zeta_1^{m_2}) = [m(45 + 5m_1 + m_2(m_1 + m_2 + 4)) + \sum_{s=1}^{m-1} s(3 + 3(s+1)/2 + m_1 + 2m_2)]\eta^{m-1} x_1^{m_1} \zeta_1^{m_2} \neq 0 \tag{11.2.104}$$

by Lemmas 11.2.3–11.2.5. According to (11.2.80) and (11.2.104), we have:

Lemma 11.2.6 *For any* $0 \neq f \in \mathscr{A}$,

$$\mathscr{D}(\eta f) \neq 0. \tag{11.2.105}$$

The above lemma implies that

$$\{f \in \mathscr{A} \mid \mathscr{D}(f)\} = \sum_{m_1, m_2 = 0}^{\infty} L(m_1, m_2, 0). \tag{11.2.106}$$

Recall that $\mathscr{A}_m$ is the subspace of homogeneous polynomials of degree m in $\mathscr{A}$. Denote

$$\Phi_m = \{f \in \mathscr{A}_m \mid (f) = 0\}. \tag{11.2.107}$$

In summary, we have the following version of the main theorem.

Theorem 11.2.7 *The set* $\{x_1^{m_1} \zeta_1^{m_2} \eta^{m_3} \mid n_1, m_2, m_3 \in \mathbb{N}\}$ *is the set of all singular vectors in* $\mathscr{A}$ *up to a scalar multiple. In particular,* η *is the unique fundamental invariant (up to constant) and the identity*

$$(1 - q)^{26} \sum_{m_1, m_2 = 0}^{\infty} (dim\ V(m_1 \lambda_1 + m_2 \lambda_6)) q^{m_1 + 2m_2} = 1 + q + q^2 \tag{11.2.108}$$

holds. Furthermore,

$$\mathscr{A}_k = \Phi_k \oplus \eta \mathscr{A}_{k-3} \quad for\ \ k \in \mathbb{N} \tag{11.2.109}$$

and

$$\Phi_m = \sum_{i=0}^{[\![m/2]\!]} L(m - 2i, i, 0) \quad for\ \ m \in \mathbb{N}, \tag{11.2.110}$$

where we treat $\mathscr{A}_r = \{0\}$ *if* $r < 0$.

As a consequence, the space of homogeneous polynomial solutions with degree m for the Dickson's invariant cubic partial differential equation is exactly a direct sum of $[\![m/2]\!] + 1$ irreducible E_6-submodules.

11.3 Spin Oscillator Representation of D_5

In this section, we give another construction of the spin oscillator representation of D_5 which will be used later.

Let we go back to the construction of the simple simple Lie algebra $\mathscr{G}^X$ in (4.4.15)–(4.4.25) with $X = E_6$ and $F(\cdot, \cdot)$ given in (4.4.64).

Recall the Dynkin diagram of E_6:

E_6:

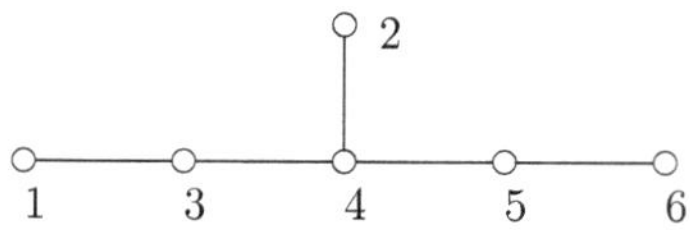

Let $\{\alpha_i \mid i \in \overline{1,6}\}$ be the simple positive roots corresponding to the vertices in the diagram, and let Φ_{E_6} be the root system of E_6. The simple Lie algebra of type E_6 is

$$\mathscr{G}^{E_6} = H \oplus \bigoplus_{\alpha \in \Phi_{E_6}} \mathbb{F}E_\alpha, \qquad H = H_{E_6} = \sum_{i=1}^{6} \mathbb{F}\alpha_i, \tag{11.3.1}$$

with the Lie bracket given in (4.4.24) and (4.4.25). Moreover, we define a bilinear form $(\cdot|\cdot)$ on $\mathscr{G}^{E_6}$ by

$$h_1|h_2) = (h_1, h_2), \ \ (h|E_\alpha) = 0, \ \ (E_\alpha|E_\beta) = -\delta_{\alpha+\beta,0} \tag{11.3.2}$$

for $h_1, h_2 \in H$ and $\alpha, \beta \in \Phi_{E_6}$. It can be verified that $(\cdot|\cdot)$ is a $\mathscr{G}^{E_6}$-invariant form; that is,

$$([u, v]|w) = (u|[v, w]) \qquad \text{for } u, v \in \mathscr{G}^{E_6}. \tag{11.3.3}$$

11.3.1 Construction

Let $E_{r,s}$ be the 10×10 matrix with 1 as its (r, s)-entry and 0 as the others. Denote

$$A_{i,j} = E_{i,j} - E_{5+j,5+i}, \ \ B_{i,j} = E_{i,5+j} - E_{j,5+i}, \ \ C_{i,j} = E_{5+i,j} - E_{5+j,i} \tag{11.3.4}$$

for $i, j \in \overline{1,5}$. Then the orthogonal Lie algebra

$$o(10, \mathbb{F}) = \sum_{i,j=1}^{5} (\mathbb{F}A_{i,j} + \mathbb{F}B_{i,j} + \mathbb{F}C_{i,j}). \tag{11.3.5}$$

Write

$$Q^{D_5} = \sum_{i=1}^{5} \mathbb{Z}\alpha_i, \qquad \Phi_{D_5} = \Phi_{E_6} \bigcap Q^{D_5}. \tag{11.3.6}$$

Then

$$\mathcal{G}^{D_5} = \sum_{i=1}^{5} \mathbb{F}\alpha_i + \sum_{\beta\in\Phi_{D_5}} \mathbb{F}E_\beta \tag{11.3.7}$$

forms a Lie subalgebra of $\mathcal{G}^{E_6}$, which is isomorphic to the orthogonal Lie algebra $o(10,\mathbb{F})$.

Denote by $\Phi_{E_6}^+$ the set of positive roots of E_6 and by $\Phi_{D_5}^+$ the set of positive roots of D_5. Recall the notion

$$\alpha_{(r_1,\ldots,r_s)} = \sum_{i=1}^{s} r_i\alpha_i \in \Phi_{E_6}^+ \quad \text{with} \quad r_s \neq 0. \tag{11.3.8}$$

We find the elements of $\Phi_{D_5}^+$: α_r $(r \in \overline{1,5})$ and

$$\alpha_{(1,0,1)},\ \alpha_{(0,1,0,1)},\ \alpha_{(0,0,1,1)},\ \alpha_{(0,0,0,1,1)},\ \alpha_{(1,0,1,1)},\ \alpha_{(0,1,1,1)},\ \alpha_{(0,1,0,1,1)},\ \alpha_{(0,0,1,1,1)}, \tag{11.3.9}$$

$$\alpha_{(1,1,1,1)},\ \alpha_{(1,0,1,1,1)},\ \alpha_{(0,1,1,1,1)},\ \alpha_{(1,1,1,1,1)},\ \alpha_{(0,1,1,2,1)},\ \alpha_{(1,1,1,2,1)},\ \alpha_{(1,1,2,2,1)}. \tag{11.3.10}$$

Moreover, the elements in $\Phi_{E_6}^+ \setminus \Phi_{D_5}^+$ are:

$$\alpha_6,\ \alpha_{(0,0,0,0,1,1)},\ \alpha_{(0,0,0,1,1,1)},\ \alpha_{(0,0,1,1,1,1)},\ \alpha_{(0,1,0,1,1,1)},\ \alpha_{(0,1,1,1,1,1)}, \tag{11.3.11}$$

$$\alpha_{(1,0,1,1,1,1)},\ \alpha_{(1,1,1,1,1,1)},\ \alpha_{(0,1,1,2,1,1)},\ \alpha_{(1,1,1,2,1,1)},\ \alpha_{(0,1,1,2,2,1)}, \tag{11.3.12}$$

$$\alpha_{(1,1,2,2,1,1)},\ \alpha_{(1,1,1,2,2,1)},\ \alpha_{(1,1,2,2,2,1)},\ \alpha_{(1,1,2,3,2,1)},\ \alpha_{(1,2,2,3,2,1)}. \tag{11.3.13}$$

Recall the notions

$$E_{(r_1,\ldots,r_s)} = E_{\alpha_{(r_1,\ldots,r_s)}},\quad (E_\alpha)' = E_{-\alpha}. \tag{11.3.14}$$

Set

$$\xi_1 = E_{\alpha_6},\ \xi_2 = E_{(0,0,0,0,1,1)},\ \xi_3 = E_{(0,0,0,1,1,1)},\ \xi_4 = E_{(0,0,1,1,1,1)}, \tag{11.3.15}$$

$$\xi_5 = E_{(0,1,0,1,1,1)},\ \xi_6 = E_{(0,1,1,1,1,1)},\ \xi_7 = E_{(1,0,1,1,1,1)},\ \xi_8 = E_{(1,1,1,1,1,1)}, \tag{11.3.16}$$

$$\xi_9 = E_{(0,1,1,2,1,1)},\ \xi_{10} = E_{(1,1,1,2,1,1)},\ \xi_{11} = E_{(0,1,1,2,2,1)},\ \xi_{12} = E_{(1,1,2,2,1,1)}, \tag{11.3.17}$$

$$\xi_{13} = E_{(1,1,1,2,2,1)},\ \xi_{14} = E_{(1,1,2,2,2,1)},\ \xi_{15} = E_{(1,1,2,3,2,1)},\ \xi_{16} = E_{(1,2,2,3,2,1)}. \tag{11.3.18}$$

$$\eta_i = (\xi_i)' \qquad \text{for } i \in \overline{1, 16}. \tag{11.3.19}$$

Write

$$\mathcal{G}_+ = \sum_{i=1}^{16} \mathbb{F}\xi_i, \qquad \mathcal{G}_- = \sum_{i=1}^{16} \mathbb{F}\eta_i, \qquad \mathcal{G}_0 = \mathcal{G}^{D_5} + \mathbb{F}\alpha_6. \tag{11.3.20}$$

It is straightforward to verify that $\mathcal{G}_\pm$ are ableian Lie subalgebras of $\mathcal{G}^{E_6}$, $\mathcal{G}_0$ is a reductive Lie subalgebra of $\mathcal{G}^{E_6}$ and

$$\mathcal{G}^{E_6} = \mathcal{G}_- \oplus \mathcal{G}_0 \oplus \mathcal{G}_+. \tag{11.3.21}$$

Moreover, $\mathcal{G}_\pm$ form irreducible $\mathcal{G}_0$-submodules with respect to the adjoint representation of $\mathcal{G}^{E_6}$. Furthermore,

$$\xi_i|\eta_j) = -\delta_{i,j} \qquad \text{for } i, j \in \overline{1, 16} \tag{11.3.22}$$

by (11.3.2). Expression (11.3.3) shows that $\mathcal{G}_+$ is isomorphic to the dual $\mathcal{G}_0$-module of $\mathcal{G}_-$.

Set

$$\mathcal{B} = \mathbb{F}[x_1, x_2, \ldots, x_{16}], \tag{11.3.23}$$

the polynomial algebra in $x_1, x_2, \ldots, x_{16}$. Write

$$[u, \eta_i] = \sum_{j=1}^{16} \varphi_{i,j}(u)\eta_j \qquad \text{for } i \in \overline{1, 16},\ u \in \mathcal{G}_0, \tag{11.3.24}$$

where $\varphi_{i,j}(u) \in \mathbb{F}$. Define an action of $\mathcal{G}_0$ on $\mathcal{B}$ by

$$u(f) = \sum_{i,j=1}^{16} \varphi_{i,j}(u)x_j\partial_{x_i}(f) \qquad \text{for } u \in \mathcal{G}_0,\ f \in \mathcal{B}. \tag{11.3.25}$$

Then $\mathcal{B}$ forms a $\mathcal{G}_0$-module and the subspace

$$V = \sum_{i=1}^{16} \mathbb{F}x_i \tag{11.3.26}$$

forms a $\mathcal{G}_0$-submodule isomorphic to $\mathcal{G}_-$, where the isomorphism is determined by $x_i \mapsto \eta_i$ for $i \in \overline{1, 16}$.

Recall that $\mathbb{N}$ is the set of nonnegative integers and

$$x^{\alpha} = \prod_{i=1}^{16} x_i^{\alpha_i}, \quad \partial^{\alpha} = \prod_{i=1}^{16} \partial_{x_i}^{\alpha_i} \qquad \text{for } \alpha = (\alpha_1, \alpha_2, \ldots, \alpha_{16}) \in \mathbb{N}^{16}. \tag{11.3.27}$$

Let

$$\mathbb{B} = \sum_{\alpha \in \mathbb{N}^{16}} \mathscr{B}\partial^{\alpha} \tag{11.3.28}$$

be the algebra of differential operators on $\mathscr{B}$. Then the linear transformation τ determined by

$$\tau(x^{\beta}\partial^{\gamma}) = x^{\gamma}\partial^{\beta} \qquad \text{for } \beta, \gamma \in \mathbb{N}^{16} \tag{11.3.29}$$

is an involutive anti-automorphism of $\mathbb{B}$.

According to (4.4.24) and (4.4.25), we have the Lie algebra isomorphism ν : $o(10, \mathbb{F}) \to \mathscr{G}^{D_5}$ determined by the generators:

$$\nu(A_{1,2}) = E_{\alpha_1}, \; \nu(A_{2,3}) = E_{\alpha_3}, \; \nu(A_{3,4}) = E_{\alpha_4}, \; \nu(A_{4,5}) = E_{\alpha_5}, \; \nu(B_{4,5}) = E_{\alpha_2}, \tag{11.3.30}$$

$$\nu(A_{2,1}) = -E_{-\alpha_1}, \; \nu(A_{3,2}) = -E_{-\alpha_3}, \;\; \nu(A_{4,3}) = -E_{-\alpha_4}, \;\; \nu(A_{5,4}) = -E_{-\alpha_5}, \tag{11.3.31}$$

$$\nu(C_{5,4}) = -E_{-\alpha_2}, \;\; \nu(A_{1,1}) = \alpha_1 + \alpha_3 + \alpha_4 + (\alpha_2 + \alpha_5)/2, \tag{11.3.32}$$

$$\nu(A_{2,2}) = \alpha_3 + \alpha_4 + (\alpha_2 + \alpha_5)/2, \;\; \nu(A_{3,3}) = \alpha_4 + \frac{1}{2}(\alpha_2 + \alpha_5), \tag{11.3.33}$$

$$\nu(A_{4,4}) = (\alpha_2 + \alpha_5)/2, \qquad \nu(A_{5,5}) = (\alpha_2 - \alpha_5)/2. \tag{11.3.34}$$

Then $\mathscr{B}$ becomes an $o(10, \mathbb{F})$-module with respect to the action

$$A(f) = \nu(A)(f) \qquad \text{for } A \in o(10, \mathbb{F}), \; f \in \mathscr{B}. \tag{11.3.35}$$

Thanks to (4.4.24), (4.4.25), (11.3.14)–(11.3.19), (11.3.24) and (11.3.25), we have the following differential-operator representation π of $o(10, \mathbb{F})$ on $\mathscr{B}$:

$$\pi(A_{1,2}) = x_4\partial_{x_7} + x_6\partial_{x_8} + x_9\partial_{x_{10}} + x_{11}\partial_{x_{13}}, \tag{11.3.36}$$

$$\pi(A_{2,3}) = x_3\partial_{x_4} + x_5\partial_{x_6} + x_{10}\partial_{x_{12}} + x_{13}\partial_{x_{14}}, \tag{11.3.37}$$

$$\pi(A_{3,4}) = -x_2\partial_{x_3} - x_6\partial_{x_9} - x_8\partial_{x_{10}} + x_{14}\partial_{x_{15}}, \tag{11.3.38}$$

$$\pi(A_{4,5}) = -x_1\partial_{x_2} + x_9\partial_{x_{11}} + x_{10}\partial_{x_{13}} + x_{12}\partial_{x_{14}}, \tag{11.3.39}$$

$$\pi(B_{4,5}) = -x_3\partial_{x_5} - x_4\partial_{x_6} - x_7\partial_{x_8} + x_{15}\partial_{x_{16}}, \tag{11.3.40}$$

$$\pi(A_{1,3}) = -x_3\partial_{x_7} - x_5\partial_{x_8} + x_9\partial_{x_{12}} + x_{11}\partial_{x_{14}}, \quad (11.3.41)$$

$$\pi(A_{2,4}) = x_2\partial_{x_4} - x_5\partial_{x_9} + x_8\partial_{x_{12}} + x_{13}\partial_{x_{15}}, \quad (11.3.42)$$

$$\pi(A_{3,5}) = -x_1\partial_{x_3} - x_6\partial_{x_{11}} - x_8\partial_{x_{13}} - x_{12}\partial_{x_{15}}, \quad (11.3.43)$$

$$\pi(B_{3,5}) = x_2\partial_{x_5} - x_4\partial_{x_9} - x_7\partial_{x_{10}} + x_{14}\partial_{x_{16}}, \quad (11.3.44)$$

$$\pi(A_{1,4}) = -x_2\partial_{x_7} + x_5\partial_{x_{10}} + x_6\partial_{x_{12}} + x_{11}\partial_{x_{15}}, \quad (11.3.45)$$

$$\pi(A_{2,5}) = x_1\partial_{x_4} - x_5\partial_{x_{11}} + x_8\partial_{x_{14}} - x_{10}\partial_{x_{15}}, \quad (11.3.46)$$

$$\pi(B_{2,5}) = -x_2\partial_{x_6} - x_3\partial_{x_9} + x_7\partial_{x_{12}} + x_{13}\partial_{x_{16}}, \quad (11.3.47)$$

$$\pi(B_{3,4}) = -x_1\partial_{x_5} + x_4\partial_{x_{11}} + x_7\partial_{x_{13}} + x_{12}\partial_{x_{16}}, \quad (11.3.48)$$

$$\pi(A_{1,5}) = -x_1\partial_{x_7} + x_5\partial_{x_{13}} + x_6\partial_{x_{14}} - x_9\partial_{x_{15}}, \quad (11.3.49)$$

$$\pi(B_{1,5}) = x_2\partial_{x_8} + x_3\partial_{x_{10}} + x_4\partial_{x_{12}} + x_{11}\partial_{x_{16}}, \quad (11.3.50)$$

$$\pi(B_{2,4}) = x_1\partial_{x_6} + x_3\partial_{x_{11}} - x_7\partial_{x_{14}} + x_{10}\partial_{x_{16}}, \quad (11.3.51)$$

$$\pi(B_{1,4}) = -x_1\partial_{x_8} - x_4\partial_{x_{14}} - x_3\partial_{x_{13}} + x_9\partial_{x_{16}}, \quad (11.3.52)$$

$$\pi(B_{2,3}) = x_1\partial_{x_9} - x_2\partial_{x_{11}} + x_7\partial_{x_{15}} - x_8\partial_{x_{16}}, \quad (11.3.53)$$

$$\pi(B_{1,3}) = -x_1\partial_{x_{10}} + x_2\partial_{x_{13}} + x_4\partial_{x_{15}} - x_6\partial_{x_{16}}, \quad (11.3.54)$$

$$\pi(B_{1,2}) = x_1\partial_{x_{12}} - x_2\partial_{x_{14}} + x_3\partial_{x_{15}} - x_5\partial_{x_{16}}, \quad (11.3.55)$$

$$\pi(A_{j,i}) = \tau(\pi(A_{i,j})), \quad \pi(C_{j,i}) = \tau(\pi(B_{i,j})) \quad \text{for } 1 \le i < j \le 5, \quad (11.3.56)$$

$$\pi(A_{r,r}) = \sum_{i=1}^{16}(1/2 + a_{r,i})x_i\partial_{x_i}, \qquad r \in \overline{1,5}, \quad (11.3.57)$$

where $a_{r,i}$ are given in Table 11.3.

Note that (11.3.36)–(11.3.55) are the representation formulas of all the positive root vectors. In particular, x_1 is a highest-weight vector of V with weight λ_4, the forth fundamental weight of $o(10, \mathbb{F})$, and V gives a spin representation of $o(10, \mathbb{C})$.

11.3.2 Decomposition

Recall that the representation of $\mathscr{G}^{D_5}$ on $\mathscr{A}$ is given by (11.3.25) and the representation of $o(10, \mathbb{C})$ is given in (11.3.35). We calculate

$$\alpha_r|_{\mathscr{A}} = \sum_{i=1}^{16} b_{r,i}x_i\partial_{x_i} \qquad \text{for } r \in \overline{1,5}, \quad (11.3.58)$$

where $b_{r,i}$ are given in Table 11.4:

Recall that a singular vector of $o(10, \mathbb{F})$ is a nonzero weight vector annihilated by positive root vectors $\{A_{i,j}, B_{i,j} \mid 1 \le i < j \le 5\}$. Note that the weight of a singular vector in $\mathscr{B}$ must be dominate integral. Using the above table, we find the singular vector

$$\zeta_1 = x_1x_{11} + x_2x_9 - x_3x_6 + x_4x_5 \tag{11.3.59}$$

of weight λ_1, the first fundamental weight of $o(10,\mathbb{F})$. Thus ζ_1 generates the 10-dimensional natural $o(10,\mathbb{F})$-module U. According to (11.3.36)–(11.3.40) and (11.3.56),

$$\pi(A_{2,1}) = x_7\partial_{x_4} + x_8\partial_{x_6} + x_{10}\partial_{x_9} + x_{13}\partial_{x_{11}}, \tag{11.3.60}$$
$$\pi(A_{3,2}) = x_4\partial_{x_3} + x_6\partial_{x_5} + x_{12}\partial_{x_{10}} + x_{14}\partial_{x_{13}}, \tag{11.3.61}$$
$$\pi(A_{4,3}) = -x_3\partial_{x_2} - x_9\partial_{x_6} - x_{10}\partial_{x_8} + x_{15}\partial_{x_{14}}, \tag{11.3.62}$$
$$\pi(A_{5,4}) = -x_2\partial_{x_1} + x_{11}\partial_{x_9} + x_{13}\partial_{x_{10}} + x_{14}\partial_{x_{12}}, \tag{11.3.63}$$
$$\pi(C_{5,4}) = -x_5\partial_{x_3} - x_6\partial_{x_4} - x_8\partial_{x_7} + x_{16}\partial_{x_{15}}. \tag{11.3.64}$$

We take

$$\zeta_2 = A_{2,1}(\zeta_1) = x_1x_{13} + x_2x_{10} - x_3x_8 + x_5x_7, \tag{11.3.65}$$
$$\zeta_3 = A_{3,2}(\zeta_2) = x_1x_{14} + x_2x_{12} - x_4x_8 + x_6x_7, \tag{11.3.66}$$
$$\zeta_4 = A_{4,3}(\zeta_3) = x_1x_{15} - x_3x_{12} + x_4x_{10} - x_7x_9, \tag{11.3.67}$$
$$\zeta_5 = A_{5,4}(\zeta_4) = -x_2x_{15} - x_3x_{14} + x_4x_{13} - x_7x_{11}, \tag{11.3.68}$$
$$\zeta_{10} = C_{5,4}(\zeta_4) = x_1x_{16} + x_5x_{12} - x_6x_{10} + x_8x_9, \tag{11.3.69}$$
$$\zeta_9 = -C_{5,4}(\zeta_5) = x_2x_{16} - x_5x_{14} + x_6x_{13} - x_8x_{11}, \tag{11.3.70}$$
$$\zeta_8 = -A_{4,3}(\zeta_9) = x_3x_{16} + x_5x_{15} + x_9x_{13} - x_{10}x_{11}, \tag{11.3.71}$$
$$\zeta_7 = -A_{3,2}(\zeta_8) = -x_4x_{16} - x_6x_{15} - x_9x_{14} + x_{11}x_{12}, \tag{11.3.72}$$
$$\zeta_6 = -A_{2,1}(\zeta_7) = x_7x_{16} + x_8x_{15} + x_{10}x_{14} - x_{12}x_{13}. \tag{11.3.73}$$

Then $U = \sum_{i=1}^{10} \mathbb{F}\zeta_i$ forms an $o(10,\mathbb{F})$-module isomorphic to the 10-dimensional natural $o(10,\mathbb{F})$-module with $\{\zeta_1,\ldots,\zeta_{10}\}$ as the standard basis.

Theorem 11.3.1 *Any $o(10,\mathbb{F})$-singular vector in $\mathscr{B}$ is a polynomial in x_1 and ζ_1.*

Proof Note

$$x_{11} = x_1^{-1}(\zeta_1 - x_2x_9 + x_3x_6 - x_4x_5), \tag{11.3.74}$$
$$x_{13} = x_1^{-1}(\zeta_2 - x_2x_{10} + x_3x_8 - x_5x_7), \tag{11.3.75}$$
$$x_{14} = x_1^{-1}(\zeta_3 - x_2x_{12} + x_4x_8 - x_6x_7), \tag{11.3.76}$$
$$x_{15} = x_1^{-1}(\zeta_4 + x_3x_{12} - x_4x_{10} + x_7x_9), \tag{11.3.77}$$
$$x_{16} = x_1^{-1}(\zeta_{10} - x_5x_{12} + x_6x_{10} - x_8x_9). \tag{11.3.78}$$

Let f be a singular vector in $\mathscr{B}$. Substituting (11.3.74)–(11.3.78) into it, we can write

$$f = g(x_i, \zeta_1, \zeta_2, \zeta_3, \zeta_4, \zeta_{10} \mid 11, 13, 14, 15, 16 \neq i \in \overline{1,16}). \tag{11.3.79}$$

By (11.3.39), (11.3.43), (11.3.46), (11.3.48), (11.3.49), (11.3.51)–(11.3.55), and the weights of (11.3.59), (11.3.65)–(11.3.67) and (11.3.69),

$$A_{4,5}(f) = -x_1\partial_{x_2}(g) = 0 \Longrightarrow g_{x_2} = 0, \tag{11.3.80}$$
$$A_{3,5}(f) = -x_1\partial_{x_3}(g) = 0 \Longrightarrow g_{x_3} = 0, \tag{11.3.81}$$
$$A_{2,5}(f) = x_1\partial_{x_4}(g) = 0 \Longrightarrow g_{x_4} = 0, \tag{11.3.82}$$
$$B_{3,4}(f) = -x_1\partial_{x_5}(g) = 0 \Longrightarrow g_{x_5} = 0, \tag{11.3.83}$$
$$A_{1,5}(f) = -x_1\partial_{x_7}(g) = 0 \Longrightarrow g_{x_7} = 0, \tag{11.3.84}$$
$$B_{2,4}(f) = x_1\partial_{x_6}(g) = 0 \Longrightarrow g_{x_6} = 0, \tag{11.3.85}$$
$$B_{1,4}(f) = -x_1\partial_{x_8}(g) = 0 \Longrightarrow g_{x_8} = 0, \tag{11.3.86}$$
$$B_{2,3}(f) = x_1\partial_{x_9}(g) = 0 \Longrightarrow g_{x_9} = 0, \tag{11.3.87}$$
$$B_{1,3}(f) = -x_1\partial_{x_{10}}(g) = 0 \Longrightarrow g_{x_{10}} = 0, \tag{11.3.88}$$
$$B_{1,2}(f) = x_1\partial_{x_{12}}(g) = 0 \Longrightarrow g_{x_{12}} = 0. \tag{11.3.89}$$

Thus f is a function in x_1, ζ_1, ζ_2, ζ_3, ζ_4 and ζ_{10}.

According to (11.3.36)–(11.3.38), (11.3.40), (11.3.59), (11.3.65)–(11.3.67) and (11.3.69),

$$A_{1,2}(f) = \zeta_1\partial_{\zeta_2}(g) = 0 \Longrightarrow g_{\zeta_2} = 0, \tag{11.3.90}$$
$$A_{2,3}(f) = \zeta_2\partial_{\zeta_3}(g) = 0 \Longrightarrow g_{\zeta_3} = 0, \tag{11.3.91}$$
$$A_{3,4}(f) = \zeta_3\partial_{\zeta_4}(g) = 0 \Longrightarrow g_{\zeta_4} = 0, \tag{11.3.92}$$
$$B_{4,5}(f) = \zeta_4\partial_{\zeta_{10}}(g) = 0 \Longrightarrow g_{\zeta_{10}} = 0. \tag{11.3.93}$$

Hence f is a function in x_1 and ζ_1. Thanks to (11.3.79), it must be a polynomial in x_1 and ζ_1. □

Let V_{m_1,m_2} be the $o(10,\mathbb{F})$-submodule generated by $x_1^{m_1}\zeta^{m_2}$. Note that (11.3.7) is a Cartan root space decomposition of $o(10,\mathbb{F})$. Since $\mathscr{B}$ is a weight $o(10,\mathbb{F})$-module, Weyl's Theorem 2.3.6 of completely reducibility yields that

$$\mathscr{B} = \bigoplus_{m_1,m_2=0}^{\infty} V_{m_1,m_2} \tag{11.3.94}$$

is a decomposition of irreducible $o(10,\mathbb{F})$-submodules. Denote by $V(\lambda)$ the highest-weight irreducible $o(10,\mathbb{F})$-module with the highest weight λ. The above equation leads to the following combinatorial identity:

$$\sum_{m_1,m_2=0}^{\infty} (\dim V(m_2\lambda_1 + m_1\lambda_4))q^{m_1+2m_2} = \frac{1}{(1-q)^{16}}, \tag{11.3.95}$$

which was proved in (11.3.74)–(11.3.93) by partial differential equations.

Table 11.3 Weights of the Spin Representation I

i	1	2	3	4	5	6	7	8	9	10	11	12	13	14	15	16
$a_{1,i}$	0	0	0	0	0	0	−1	−1	0	−1	0	−1	−1	−1	−1	−1
$a_{2,i}$	0	0	0	−1	0	−1	0	0	−1	0	−1	−1	0	−1	−1	−1
$a_{3,i}$	0	0	−1	0	−1	0	0	0	−1	−1	−1	0	−1	0	−1	−1
$a_{4,i}$	0	−1	0	0	−1	−1	0	−1	0	0	−1	0	−1	−1	0	−1
$a_{5,i}$	−1	0	0	0	−1	−1	0	−1	−1	−1	0	−1	0	0	0	−1

11.3.3 Symmetry and Equivalent Combinatorics

Next we want to study the symmetry of $\overline{1,16}$ with respect to the representation in (11.3.36)–(11.3.57). Set

$$\mathfrak{W} = \{\pi(A_{i,j}), \pi(A_{j,i}), \pi(B_{i,j}), \pi(C_{j,i}) \mid 1 \le i < j \le 5\}, \tag{11.3.96}$$

which is the set of the representations of root vectors in $o(10, \mathbb{F})$. Write

$$\mathfrak{Z} = \{\zeta_i \mid i \in \overline{1,10}\} \tag{11.3.97}$$

(cf. (11.3.59) and (11.3.65)–(11.3.73)). First we have:

Lemma 11.3.2 *The set $\overline{1,16}$ is symmetric with respect to the sets $\mathfrak{W}$ and $\mathfrak{Z}$.*

Proof Take $n = 5$ in the spin representation of $o(2n, \mathbb{F})$ in (7.1.32)–(7.1.37) and Theorem 7.1.3. By (11.3.57) and Table 11.3, we have an $o(10, \mathbb{F})$-module isomorphism $\sigma : U \to \Psi_{(0)}$ such that $\sigma(x_1) = \theta_1\theta_2\theta_3\theta_4$. According to (7.1.34), (7.1.35), (11.3.36)–(11.3.40) and (11.3.56), the following sets are symmetric with respect to the representation π of $o(10, \mathbb{F})$:

$$S_1 = \{x_1, x_2, x_3, x_4, x_7\},\;\; S_2 = \{x_5, x_6, x_8, x_9, x_{10}, x_{11}, x_{12}, x_{13}, x_{14}, x_{15}\} \tag{11.3.98}$$

and $S_3 = \{x_{16}\}$.

We define an anti-automorphism $\dagger$ of the associative algebra

$$\mathbb{B} = \sum_{i_1,\ldots,i_{16}=0}^{\infty} \mathscr{B}\partial_{x_1}^{i_1} \cdots \partial_{x_{16}}^{16} \tag{11.3.99}$$

of differential operators by

$$(x_6)^\dagger = -\partial_{x_{10}},\;\; (\partial_{x_6})^\dagger = -x_{10},\;\; (x_{10})^\dagger = -\partial_{x_6},\;\; (\partial_{x_{10}})^\dagger = -x_6, \tag{11.3.100}$$

$$(x_i)^\dagger = -\partial_{x_{17-i}},\;\; (\partial_{x_i})^\dagger = -x_{17-i} \;\text{ for } i = 4, 13, \tag{11.3.101}$$

$$(x_7)^\dagger = \partial_{x_{11}}, \quad (\partial_{x_7})^\dagger = x_{11}, \quad (x_{11})^\dagger = \partial_{x_7}, \quad (\partial_{x_{11}})^\dagger = x_7, \quad (11.3.102)$$

$$(x_r)^\dagger = \partial_{x_{17-r}}, \quad (\partial_{x_r})^\dagger = x_{17-r} \quad \text{for } r = 1, 2, 3, 5, 8, 9, 12, 14, 15, 16. \quad (11.3.103)$$

It turns out that the map $d \mapsto -d^\dagger$ forms a Lie algebra automorphism of $\pi(o(10, \mathbb{F}))$ by (11.3.36)–(11.3.40), (11.3.56), (11.3.57) and Table 11.3, which corresponds to an automorphism of $o(10, \mathbb{F})$ inducing the automorphism of its Dynkin diagram. Moreover,

$$-d^\dagger \in \mathfrak{W} \qquad \text{for any } d \in \mathfrak{W}. \qquad (11.3.104)$$

Define

$$\mathfrak{s}(d) = -\tau(d^\dagger) \qquad \text{for } d \in \mathbb{B}. \qquad (11.3.105)$$

Then $\mathfrak{s}$ is an associative algebra automorphism of $\mathbb{B}$ with order 2. In fact,

$$\mathfrak{s}(x_6) = -x_{10}, \quad \mathfrak{s}(x_4) = -x_{13}, \quad \mathfrak{s}(x_7) = x_{11}, \quad \mathfrak{s}(x_r)^\dagger = x_{17-r} \qquad (11.3.106)$$

for $r = 1, 2, 3, 5, 8$. Moreover,

$$\mathfrak{s}(\zeta_i) = \zeta_{5+i}, \quad \mathfrak{s}(\zeta_r) = \zeta_r \qquad \text{for } i \in \overline{1, 4}, \ r = 5, 10. \qquad (11.3.107)$$

This shows that $\{1, 16\}$ and $\{2, 15\}$ are symmetric with respect to the sets $\mathfrak{W}$ and $\mathfrak{Z}$. Thus the lemma follows from the conclusion of the first paragraph. □

For any $i \in \overline{1, 16}$, we define

$$\Upsilon_i = \{(r, s) \mid \zeta_r \text{ contains } x_i x_s\}. \qquad (11.3.108)$$

We have the following equivalent combinatorial property:

Lemma 11.3.3 *Each polynomial ζ_r contains exactly eight x_i's. Moreover,*

$$|\Upsilon_i| = 5 \qquad \textit{for } i \in \overline{16}. \qquad (11.3.109)$$

Proof According to (11.3.59) and (11.3.65)–(11.3.73), we calculate

$$\Upsilon_1 = \{(1, 11), (2, 13), (3, 14), (4, 15), (10, 16)\}, \qquad (11.3.110)$$

Then (11.3.108) follows from Lemma 11.3.2. The first statement is obtained by checking (11.3.59) and (11.3.65)–(11.3.73) one-by-one. □

If we represent $\{x_i \mid i \in \overline{1, 16}\}$ by 16 vertices and represent $\{\zeta_r \mid r \in \overline{1, 10}\}$ by 10 lines, then we obtain a graph of 16 vertices and 10 lines such that each line contains 8 vertices and each vertex is exactly on 5 lines.

For $i \in \overline{1, 16}$, we denote

$$I_i = \{r \in \overline{1, 16} \mid \text{some } d \in \mathfrak{W} \text{ contains } x_r \partial_{x_i}\} \qquad (11.3.111)$$

and

$$\mathfrak{W}_i = \{d \in \mathfrak{W} \mid d \text{ does not contain } x_i \text{ and } \partial_{x_i}\}. \tag{11.3.112}$$

Write

$$J_i = \{s \in \overline{1, 16} \mid (r, s) \in \Upsilon_i \text{ for some } r \in \overline{1, 16}\}. \tag{11.3.113}$$

The following equivalent combinatorial properties will be crucial to our main result in Sect. 11.7.

Lemma 11.3.4 *For any $i \in \overline{1, 16}$, we have*

$$|I_i| = 10, \quad |\mathfrak{W}_i| = 20, \quad |J_i| = 5. \tag{11.3.114}$$

In fact,

$$I_i \bigcup J_i = \overline{1, 16} \setminus \{i\}. \tag{11.3.115}$$

Moreover, every element in $\mathfrak{W}_i$ contains exactly one x_s with $s \in J_i$, and for any $r \in J_i$, x_r is contained in exactly four elements in $\mathfrak{W}_i$.

Proof By Lemma 11.3.2, we only need to prove it for $i = 1$. Note that the elements in $\mathfrak{W}$ containing ∂_{x_1} are

$$\pi(A_{5,4}), \quad \pi(A_{5,3}), \quad \pi(A_{5,2}), \quad \pi(A_{5,1}), \quad \pi(C_{4,3}), \tag{11.3.116}$$

$$\pi(C_{4,2}), \quad \pi(C_{4,1}), \quad \pi(C_{3,2}), \quad \pi(C_{3,1}), \quad \pi(C_{2,1}) \tag{11.3.117}$$

by (11.3.36)–(11.3.56). Thus $I_1 = \overline{2, 10} \bigcup \{12\}$. On the other hand,

$$J_1 = \{11, 13, 14, 15, 16\} \tag{11.3.118}$$

by (11.3.110). Since $\tau(\mathfrak{W}) = \mathfrak{W}$ (cf. (11.3.29)), $|\mathfrak{W}_1| = 20$. So (11.3.114) and (11.3.115) holds for $i = 1$.

The elements in $\mathfrak{W}_1$ containing x_{11} are

$$\pi(A_{1,2}), \pi(A_{1,3}), \pi(A_{1,4}), \pi(B_{1,5}). \tag{11.3.119}$$

The elements in $\mathfrak{W}_1$ containing x_{13} are

$$\pi(A_{2,1}), \pi(A_{2,3}), \pi(A_{2,4}), \pi(B_{2,5}). \tag{11.3.120}$$

The elements in $\mathfrak{W}_1$ containing x_{14} are

$$\pi(A_{3,1}), \pi(A_{3,2}), \pi(A_{3,4}), \pi(B_{3,5}). \tag{11.3.121}$$

The elements in $\mathfrak{W}_1$ containing x_{15} are

$$\pi(A_{4,1}),\ \pi(A_{4,2}),\ \pi(A_{4,3}),\ \pi(B_{4,5}). \tag{11.3.122}$$

The elements in $\mathfrak{W}_1$ containing x_{16} are

$$\pi(B_{5,1}),\ \pi(B_{5,2}),\ \pi(B_{5,3}),\ \pi(B_{5,4}). \tag{11.3.123}$$

Therefore, every element in $\mathfrak{W}_1$ contains exactly one x_s with $s \in J_1$, and for any $r \in J_1$, x_r is contained in exactly four elements in $\mathfrak{W}_1$. □

11.4 Realization of E_6 in 16-Dimensional Space

In this section, we want to find a differential-operator representation of $\mathcal{G}^{E_6}$, or equivalently, a fractional representation on 16-dimensional space of the Lie group of type E_6.

Recall the Lie subalgebra $\mathcal{G}_0 = \mathcal{G}^{D_5} \oplus \mathbb{F}\alpha_6$ of $\mathcal{G}^{E_6}$. We can view the linear map ν determined in (11.3.30)–(11.3.34) as an identification of $o(10, \mathbb{F})$ with $\mathcal{G}^{D_5}$. We want to find the extension of the representation π of $o(10, \mathbb{F})$ given in (11.3.36)–(11.3.57) to $\mathcal{G}^{E_6}$. According to (11.3.24) and (11.3.25), we calculate

$$\pi(\alpha_6) = -2x_1\partial_{x_1} - \sum_{i=2}^{10} x_i\partial_{x_i} - x_{12}\partial_{x_{12}}. \tag{11.4.1}$$

Write

$$\widehat{\alpha} = 2\alpha_1 + 4\alpha_3 + 6\alpha_4 + 3\alpha_2 + 5\alpha_5 + 4\alpha_6. \tag{11.4.2}$$

Then

$$(\widehat{\alpha}, \alpha_r) = 0 \qquad \text{for } r \in \overline{1,5} \tag{11.4.3}$$

by the Dynkin diagram of E_6, where $(\cdot,\cdot)$ is a symmetric $\mathbb{Z}$-bilinear form on the root lattice of E_6 such that $(\alpha, \alpha) = 2$ for $\alpha \in \Phi_{E_6}$.

Thanks to (4.4.24),

$$[\widehat{\alpha}, \mathcal{G}^{D_5}] = 0. \tag{11.4.4}$$

So $\widehat{\alpha}$ is a central element of $\mathcal{G}_0$. By Schur's Lemma, $\widehat{\alpha}|_V = c\sum_{i=1}^{16} x_i\partial_{x_i}$. According to the coefficients of $x_1\partial_{x_1}$ in (11.3.58) with the data in Table 11.4 and (11.4.2), we have

$$\pi(\widehat{\alpha}) = -3D, \quad \text{where } D = \sum_{i=1}^{16} x_i\partial_{x_i} \tag{11.4.5}$$

is the degree operator on $\mathcal{B}$.

Table 11.4 Weights of the Spin Representation II

i	1	2	3	4	5	6	7	8	9	10	11	12	13	14	15	16
$b_{1,i}$	0	0	0	1	0	1	−1	−1	1	−1	1	0	−1	0	0	0
$b_{2,i}$	0	0	1	1	−1	−1	1	−1	0	0	0	0	0	0	1	−1
$b_{3,i}$	0	0	1	−1	1	−1	0	0	0	1	0	−1	1	−1	0	0
$b_{4,i}$	0	1	−1	0	0	1	0	1	−1	−1	0	0	0	1	−1	0
$b_{5,i}$	1	−1	0	0	0	0	0	0	1	1	−1	1	−1	−1	0	0

Recall that the Lie bracket in the algebra $\mathbb{B}$ (cf. (11.3.28)) is given by the commutator

$$[d_1, d_2] = d_1 d_2 - d_2 d_1. \tag{11.4.6}$$

Set

$$\mathscr{T} = \sum_{i=1}^{16} \mathbb{F}\partial_{x_i}. \tag{11.4.7}$$

Then $\mathscr{T}$ forms an $o(10,\mathbb{F})$-module with respect to the action

$$B(\partial) = [\pi(B), \partial] \quad \text{for } B \in o(10,\mathbb{F}),\ \partial \in \mathscr{T}. \tag{11.4.8}$$

On the other hand, $\mathscr{G}_\pm$ (cf. (11.3.14)–(11.3.20)) form $o(10,\mathbb{F})$-modules with respect to the action

$$B(u) = [\nu(B), u] \quad \text{for } B \in o(10,\mathbb{F}),\ u \in \mathscr{G}_\pm \tag{11.4.9}$$

(cf. (11.3.30)–(11.3.34)). According to (11.3.24) and (11.3.25), the linear map determined by $\eta_i \mapsto x_i$ for $i \in \overline{1,16}$ gives an $o(10,\mathbb{F})$-module isomorphism from $\mathscr{G}_-$ to V. Moreover, (11.3.3) and (11.3.22) imply that the linear map determined by $\xi_i \mapsto \partial_{x_i}$ for $i \in \overline{1,16}$ gives an $o(10,\mathbb{F})$-module isomorphism from $\mathscr{G}_+$ to $\mathscr{T}$. Hence we define the representation π of $\mathscr{G}_+$ on $\mathscr{B}$ by

$$\pi(\xi_i) = \partial_{x_i} \quad \text{for } i \in \overline{1,16}. \tag{11.4.10}$$

Recall the Witt Lie subalgebra of $\mathbb{B}$:

$$\mathbb{W}_{16} = \sum_{i=1}^{16} \mathscr{B}\partial_{x_i}. \tag{11.4.11}$$

Now we want to find the differential operators $P_1, P_2, \ldots, P_{16} \in \mathbb{W}_{16}$ such that the following representation π matches the structure of $\mathscr{G}^{E_6}$:

$$\pi(\eta_i) = P_i \quad \text{for } i \in \overline{1,16}. \tag{11.4.12}$$

Imposing

$$[\partial_{x_1}, P_1] = \pi([E_{\alpha_6}, E_{-\alpha_6}]) = -\pi(\alpha_6) = 2x_1\partial_{x_1} + \sum_{i=2}^{10} x_i\partial_{x_i} + x_{12}\partial_{x_{12}}, \quad (11.4.13)$$

we take

$$P_1 = x_1\left(\sum_{i=1}^{10} x_i\partial_{x_i} + x_{12}\partial_{x_{12}}\right) + P_1', \quad (11.4.14)$$

where P_1' is a differential operator such that $[\partial_{x_1}, P_1'] = 0$. Moreover,

$$\left[\partial_{x_r}, x_1\left(\sum_{i=1}^{10} x_i\partial_{x_i} + x_{12}\partial_{x_{12}}\right)\right] = x_1\partial_{x_r} \qquad \text{for } r \in \{\overline{2,10}, 12\}. \quad (11.4.15)$$

Wanting $[\partial_{x_r}, P_1] \in \pi(o(10, \mathbb{F}))$ (cf. (11.3.36)–(11.3.57)), we take

$$\begin{aligned} P_1 &= x_1\left(\sum_{i=1}^{10} x_i\partial_{x_i} + x_{12}\partial_{x_{12}}\right) - (x_2x_9 - x_3x_6 + x_4x_5)\partial_{x_{11}} \\ &\quad -(x_2x_{10} - x_3x_8 + x_5x_7)\partial_{x_{13}} - (x_2x_{12} - x_4x_8 + x_6x_7)\partial_{x_{14}} \\ &\quad +(x_3x_{12} - x_4x_{10} + x_7x_9)\partial_{x_{15}} - (x_5x_{12} - x_6x_{10} + x_8x_9)\partial_{x_{16}} \\ &= x_1D - \zeta_1\partial_{x_{11}} - \zeta_2\partial_{x_{13}} - \zeta_3\partial_{x_{14}} - \zeta_4\partial_{x_{15}} - \zeta_{10}\partial_{x_{16}} \end{aligned} \quad (11.4.16)$$

by (11.3.39), (11.3.43), (11.3.46), (11.3.48), (11.3.49), (11.3.51)–(11.3.55), (11.3.59), (11.3.65)–(11.3.67), (11.3.69) and (11.4.5). Then

$$[\partial_{x_s}, P_1] = \pi(\nu^{-1}([\xi_s, \eta_1])) \qquad \text{for } s \in \overline{1,16} \quad (11.4.17)$$

due to (11.3.30)–(11.3.35).

Since $[E_{-\alpha_5}, \eta_1] = \eta_2$ by (4.4.25), we take

$$\begin{aligned} P_2 &= [\pi(\nu^{-1}(E_{-\alpha_5})), \pi(\eta_1)] = -[\pi(A_{5,4}), P_1] \\ &= x_2D - \zeta_1\partial_{x_9} - \zeta_2\partial_{x_{10}} - \zeta_3\partial_{x_{12}} + \zeta_5\partial_{x_{15}} - \zeta_9\partial_{x_{16}} \end{aligned} \quad (11.4.18)$$

by (11.3.31) and (11.3.63). Similarly, we take

$$P_3 = x_3D + \zeta_1\partial_{x_6} + \zeta_2\partial_{x_8} + \zeta_4\partial_{x_{12}} + \zeta_5\partial_{x_{14}} - \zeta_8\partial_{x_{16}}, \quad (11.4.19)$$

$$P_4 = x_4D - \zeta_1\partial_{x_5} + \zeta_3\partial_{x_8} - \zeta_4\partial_{x_{10}} - \zeta_5\partial_{x_{13}} + \zeta_7\partial_{x_{16}}, \quad (11.4.20)$$

$$P_5 = x_5D - \zeta_1\partial_{x_4} - \zeta_2\partial_{x_7} - \zeta_{10}\partial_{x_{12}} + \zeta_9\partial_{x_{14}} - \zeta_8\partial_{x_{15}}, \quad (11.4.21)$$

$$P_6 = x_6D + \zeta_1\partial_{x_3} - \zeta_3\partial_{x_7} + \zeta_{10}\partial_{x_{10}} - \zeta_9\partial_{x_{13}} + \zeta_7\partial_{x_{15}}, \quad (11.4.22)$$

$$P_7 = x_7D - \zeta_2\partial_{x_5} - \zeta_3\partial_{x_6} + \zeta_4\partial_{x_9} + \zeta_5\partial_{x_{11}} - \zeta_6\partial_{x_{16}}, \quad (11.4.23)$$

$$P_8 = x_8 D + \zeta_2 \partial_{x_3} + \zeta_3 \partial_{x_4} - \zeta_{10} \partial_{x_9} + \zeta_9 \partial_{x_{11}} - \zeta_6 \partial_{x_{15}}, \tag{11.4.24}$$
$$P_9 = x_9 D - \zeta_1 \partial_{x_2} + \zeta_4 \partial_{x_7} - \zeta_{10} \partial_{x_8} - \zeta_8 \partial_{x_{13}} + \zeta_7 \partial_{x_{14}}, \tag{11.4.25}$$
$$P_{10} = x_{10} D - \zeta_2 \partial_{x_2} - \zeta_4 \partial_{x_4} + \zeta_{10} \partial_{x_6} + \zeta_8 \partial_{x_{11}} - \zeta_6 \partial_{x_{14}}, \tag{11.4.26}$$
$$P_{11} = x_{11} D - \zeta_1 \partial_{x_1} + \zeta_5 \partial_{x_7} + \zeta_9 \partial_{x_8} + \zeta_8 \partial_{x_{10}} - \zeta_7 \partial_{x_{12}}, \tag{11.4.27}$$
$$P_{12} = x_{12} D - \zeta_3 \partial_{x_2} + \zeta_4 \partial_{x_3} - \zeta_{10} \partial_{x_5} - \zeta_7 \partial_{x_{11}} + \zeta_6 \partial_{x_{13}}, \tag{11.4.28}$$
$$P_{13} = x_{13} D - \zeta_2 \partial_{x_1} - \zeta_5 \partial_{x_4} - \zeta_9 \partial_{x_6} - \zeta_8 \partial_{x_9} + \zeta_6 \partial_{x_{12}}, \tag{11.4.29}$$
$$P_{14} = x_{14} D - \zeta_3 \partial_{x_1} + \zeta_5 \partial_{x_3} + \zeta_9 \partial_{x_5} + \zeta_7 \partial_{x_9} - \zeta_6 \partial_{x_{10}}, \tag{11.4.30}$$
$$P_{15} = x_{15} D - \zeta_4 \partial_{x_1} + \zeta_5 \partial_{x_2} - \zeta_8 \partial_{x_5} + \zeta_7 \partial_{x_6} - \zeta_6 \partial_{x_8}, \tag{11.4.31}$$
$$P_{16} = x_{16} D - \zeta_{10} \partial_{x_1} - \zeta_9 \partial_{x_2} - \zeta_8 \partial_{x_3} + \zeta_7 \partial_{x_4} - \zeta_6 \partial_{x_7}. \tag{11.4.32}$$

Set

$$\mathscr{P} = \sum_{i=1}^{16} \mathbb{F} P_i, \qquad \mathfrak{C}_0 = \pi(o(10, \mathbb{F})) + \mathbb{F} D \tag{11.4.33}$$

(cf. (11.3.36)–(11.3.57) and (11.4.5)) and

$$\mathfrak{C} = \mathscr{P} + \mathfrak{C}_0 + \mathscr{T} \tag{11.4.34}$$

(cf. (11.4.7)). Then we have:

Theorem 11.4.1 *The space $\mathfrak{C}$ forms a Lie subalgebra of the Witt algebra $\mathbb{W}_{16}$ (cf.* (11.4.11)*). Moreover, the linear map ϑ determined by*

$$\vartheta(\xi_i) = \partial_{x_i}, \quad \vartheta(\eta_i) = P_i, \quad \vartheta(u) = \pi(\nu^{-1}(u)) \quad \text{for } i \in \overline{1, 16}, \; u \in \mathscr{G}^{D_5} \tag{11.4.35}$$

(cf. (11.3.30)–(11.3.34)*) and*

$$\vartheta(\alpha_6) = -2x_1 \partial_{x_1} - \sum_{i=2}^{10} x_i \partial_{x_i} - x_{12} \partial_{x_{12}} \tag{11.4.36}$$

(cf. (11.4.1)*) gives a Lie algebra isomorphism from $\mathscr{G}^{E_6}$ to $\mathfrak{C}$, which is a first-order inhomogeneous differential-operator representation of $\mathscr{G}^{E_6}$.*

Proof Since $\mathscr{T} \cong \mathscr{G}_+$ as $\mathscr{G}^{D_5}$-modules, we have

$$\mathscr{G}_0 + \mathscr{G}_+ \overset{\vartheta}{\cong} \mathfrak{C}_0 + \mathscr{T} \tag{11.4.37}$$

as Lie algebras. Recall that $U(\mathscr{G})$ stands for the universal enveloping algebra of a Lie algebra $\mathscr{G}$ (cf. Sect. 5.1). Note that

$$\mathscr{B}_- = \mathscr{G}_0 + \mathscr{G}_-, \qquad \mathscr{B}_+ = \mathscr{G}_0 + \mathscr{G}_+ \tag{11.4.38}$$

are also Lie subalgebras of $\mathscr{G}^{E_6}$ and

$$\mathscr{G}^{E_6} = \mathscr{B}_- \oplus \mathscr{G}_+ = \mathscr{G}_- \oplus \mathscr{B}_+. \tag{11.4.39}$$

We define a one-dimensional $\mathscr{B}_-$-module $\mathbb{F}u_0$ by

$$w(u_0) = 0 \quad \text{for } w \in \mathscr{G}_- + \mathscr{G}^{D_5}, \quad \widehat{\alpha}(u_0) = 48u_0 \tag{11.4.40}$$

(cf. (11.4.2)). Let

$$M = U(\mathscr{G}^{E_6}) \otimes_{\mathscr{B}_-} \mathbb{F}u_0 \cong U(\mathscr{G}_+) \otimes_{\mathbb{F}} \mathbb{F}u_0 \tag{11.4.41}$$

be the induced $\mathscr{G}^{E_6}$-module.

Recall that $\mathbb{N}$ is the set of nonnegative integers. Let

$$\widehat{B} = \mathbb{F}[\partial_{x_1}, \partial_{x_2}, \ldots, \partial_{x_{16}}]. \tag{11.4.42}$$

We define an action of the associative algebra $\mathbb{B}$ (cf. (11.3.28)) on $\widehat{B}$ by

$$\partial_{x_i}\left(\prod_{j=1}^{16} \partial_{x_j}^{\beta_j}\right) = \partial_{x_i}^{\beta_i+1} \prod_{i\neq j\in\overline{1,16}} \prod_{j=1}^{16} \partial_{x_j}^{\beta_j} \tag{11.4.43}$$

and

$$x_i\left(\prod_{j=1}^{16} \partial_{x_j}^{\beta_j}\right) = -\beta_i \partial_{x_i}^{\beta_i-1} \prod_{i\neq j\in\overline{1,16}} \prod_{j=1}^{16} \partial_{x_j}^{\beta_j} \tag{11.4.44}$$

for $i \in \overline{1,16}$. Since

$$[-x_i, \partial_{x_j}] = [\partial_{x_i}, x_j] = \delta_{i,j} \qquad \text{for } i, j \in \overline{1,16}, \tag{11.4.45}$$

the above action gives an associative algebra representation of $\mathbb{B}$. Thus it also gives a Lie algebra representation of $\mathbb{B}$ (cf. (11.4.6)). It is straightforward to verify that

$$[d|_{\widehat{B}}, \partial|_{\widehat{B}}] = [d, \partial]|_{\widehat{B}} \qquad \text{for } d \in \mathfrak{C}_0, \ \partial \in \mathscr{T}. \tag{11.4.46}$$

Define linear map $\mathfrak{F} : M \to \widehat{B}$ by

$$\mathfrak{F}\left(\prod_{i=1}^{16} \xi_i^{\ell_i} \otimes u_0\right) = \prod_{i=1}^{16} \partial_{x_i}^{\ell_i} \qquad (\ell_1, \ldots, \ell_{16}) \in \mathbb{N}^{16}. \tag{11.4.47}$$

According to (11.3.36)–(11.3.57), (11.4.43) and (11.4.44),

$$D(1) = -16, \quad d(1) = 0 \qquad \text{for } d \in \pi(o(10, \mathbb{F})). \tag{11.4.48}$$

Moreover, (11.4.40), (11.4.41), (11.4.43), (11.4.44) and (11.4.48) imply

$$\mathfrak{F}(\xi(v)) = \vartheta(\xi)[\mathfrak{F}(v)] \qquad \text{for } \xi \in \mathscr{G}_0, \ v \in M. \tag{11.4.49}$$

Now (11.4.41) and (11.4.43) yield

$$\mathfrak{F}(w(u)) = \vartheta(w)[\mathfrak{F}(u)] \qquad \text{for } w \in \mathscr{B}_+, \ u \in M. \tag{11.4.50}$$

Thus we have

$$\varsigma w|_\Psi \varsigma^{-1} = \vartheta(w)|_{\mathscr{A}'} \qquad \text{for } w \in \mathscr{B}_+. \tag{11.4.51}$$

On the other hand, the linear map

$$\psi(v) = \mathfrak{F}(v|_M)\mathfrak{F}^{-1} \qquad \text{for } v \in \mathscr{G}^{E_6} \tag{11.4.52}$$

is a Lie algebra monomorphism from $\mathscr{G}^{E_6}$ to $\mathbb{B}|_{\widehat{B}}$. According to (11.4.17) and (11.4.45),

$$\psi(\eta_1) = P_1|_{\widehat{B}}. \tag{11.4.53}$$

By the constructions of $P_2, \ldots, P_{16}$ in (11.4.18)–(11.4.32), we have

$$\psi(\eta_i) = P_i|_{\widehat{B}} \qquad \text{for } i \in \overline{2, 16}. \tag{11.4.54}$$

Therefore, we have

$$\psi(v) = \vartheta(v)|_{\widehat{B}} \qquad \text{for } v \in \mathscr{G}^{E_6}. \tag{11.4.55}$$

In particular, $\mathfrak{C}|_{\widehat{B}} = \vartheta(\mathscr{G}^{E_6})|_{\widehat{B}} = \psi(\mathscr{G}^{E_6})$ forms a Lie algebra. Since the linear map $d \mapsto d|_{\widehat{B}}$ for $d \in \mathfrak{C}$ is injective, we have that $\mathfrak{C}$ forms a Lie subalgebra of $\mathbb{B}$ and ϑ is a Lie algebra isomorphism. □

By the above theorem, a Lie group of type E_6 is generated by the linear transformations $\{e^{b\pi(u)} \mid b \in \mathbb{R}, \ u \in o(10, \mathbb{R})\}$ associated with (11.3.36)–(11.3.57), the real translations and dilations in $\mathbb{R}^{16}$ with x_r as the rth coordinate function, and the fractional transformations $\{e^{bP_s} : (x_1, x_2, \ldots, x_{16}) \mapsto (e^{bP_s}(x_1), e^{bP_s}(x_2), \ldots, e^{bP_s}(x_{16})) \mid b \in \mathbb{R}, \ s \in \overline{1, 16}\}$, where $e^{bP_s}(x_i)$ are of the forms as the following forms of the case $s = 1$:

$$e^{bP_1}(x_i) = \frac{x_i}{1 - bx_1}, \quad i \in \{\overline{1, 10}, 12\}, \tag{11.4.56}$$

$$e^{bP_1}(x_{11}) = x_{11} - \frac{b(x_2x_9 - x_3x_6 + x_4x_5)}{1 - bx_1}, \tag{11.4.57}$$

$$e^{bP_1}(x_{13}) = x_{13} - \frac{b(x_2x_{10} - x_3x_8 + x_5x_7)}{1 - bx_1}, \tag{11.4.58}$$

$$e^{bP_1}(x_{14}) = x_{14} - \frac{b(x_2x_{12} - x_4x_8 + x_6x_7)}{1 - bx_1}, \tag{11.4.59}$$

$$e^{bP_1}(x_{15}) = x_{15} + \frac{b(x_3x_{12} - x_4x_{10} + x_7x_9)}{1 - bx_1}, \tag{11.4.60}$$

$$e^{bP_1}(x_{16}) = x_{16} - \frac{b(x_5x_{12} - x_6x_{10} + x_8x_9)}{1 - bx_1}. \tag{11.4.61}$$

This gives a fractional representation of the Lie group E_6.

11.5 Functor from D_5-Mod to E_6-Mod

In this section, we construct a functor from the category of D_5-modules to the category of E_6-modules.

Note that

$$o(10, \mathscr{B}) = \sum_{i,j=1}^{n} (\mathscr{B}A_{i,j} + \mathscr{B}B_{i,j} + \mathscr{B}C_{i,j}) \tag{11.5.1}$$

(cf. (11.3.4)) forms a Lie subalgebra of the matrix algebra $gl(10, \mathscr{B})$ over $\mathscr{B}$ with respect to the commutator, i.e.

$$[fB_1, gB_2] = fg[B_1, B_2] \quad \text{for } f, g \in \mathscr{B},\ B_1, B_2 \in gl(10, \mathbb{F}). \tag{11.5.2}$$

Moreover, we define the Lie algebra

$$\mathscr{K} = o(10, \mathscr{B}) \oplus \mathscr{B}\kappa \tag{11.5.3}$$

with the Lie bracket:

$$[\xi_1 + f\kappa, \xi_2 + g\kappa] = [\xi_1, \xi_2] \quad \text{for } \xi_1, \xi_2 \in o(10, \mathscr{B}),\ f, g \in \mathscr{B}. \tag{11.5.4}$$

Similarly, $gl(16, \mathscr{B})$ becomes a Lie algebra with the Lie bracket as that in (11.5.2). Recall the Witt algebra $\mathbb{W}_{16} = \sum_{i=1}^{16} \mathscr{B}\partial_{x_i}$, and Lemma 6.7.1 says that there exists a monomorphism $\Im$ from the Lie algebra $\mathbb{W}_{16}$ to the Lie algebra of semi-product $\mathbb{W}_{16} + gl(16, \mathscr{B})$ defined by

$$\Im\left(\sum_{i=1}^{16} f_i\partial_{x_i}\right) = \sum_{i=1}^{16} f_i\partial_{x_i} + \Im_1\left(\sum_{i=1}^{16} f_i\partial_{x_i}\right), \quad \Im_1\left(\sum_{i=1}^{16} f_i\partial_{x_i}\right) = \sum_{i,j=1}^{16} \partial_{x_i}(f_j)E_{i,j}. \tag{11.5.5}$$

According to our construction of P_1–P_{16} in (11.4.12)–(11.4.32),

$$\Im_1(P_i) = \sum_{r=1}^{16} x_r \Im_1(\pi([\xi_r, \eta_i])) \qquad \text{for } i \in \overline{1,16} \tag{11.5.6}$$

(cf. (11.3.36)–(11.3.57) and (11.4.4)). On the other hand,

$$\widehat{\mathscr{K}} = \mathbb{W}_{16} \oplus \mathscr{K} \tag{11.5.7}$$

becomes a Lie algebra with the Lie bracket

$$\begin{aligned}&[d_1 + f_1 B_1 + f_2 \kappa, d_2 + g_1 B_2 + g_2 \kappa]\\ &= [d_1, d_2] + f_1 g_1 [B_1, B_2] + d_1(g_2) B_2 - d_2(g_1) B_1 + (d_1(g_2) - d_2(g_1))\kappa\end{aligned} \tag{11.5.8}$$

for $f_1, f_2, g_1, g_2 \in \mathscr{B}$, $B_1, B_2 \in o(10, \mathbb{F})$ and $d_1, d_2 \in \mathbb{W}_{16}$. Note

$$\mathscr{G}_0 = \mathscr{G}^{D_5} \oplus \mathbb{F}\widehat{\alpha} \tag{11.5.9}$$

by (11.3.20) and (11.4.2). So there exists a Lie algebra monomorphism $\mathfrak{k} : \mathscr{G}_0 \to \mathscr{K}$ determined by

$$\mathfrak{k}(\widehat{\alpha}) = 2\kappa, \quad \mathfrak{k}(u) = \nu^{-1}(u) \qquad \text{for } u \in \mathscr{G}^{D_5} \tag{11.5.10}$$

(cf. (11.3.30)–(11.3.34)). Since $\Im$ is a Lie algebra monomorphism, our construction of P_1–P_{16} in (11.4.12)–(11.4.32) and (11.5.6) show that we have a Lie algebra monomorphism $\iota : \mathscr{G}^{E_6} \to \widehat{\mathscr{K}}$ given by

$$\iota(u) = \pi(u) + \mathfrak{k}(u) \qquad \text{for } u \in \mathscr{G}_0, \tag{11.5.11}$$

$$\iota(\xi_i) = \partial_{x_i}, \quad \iota(\eta_i) = P_i + \sum_{r=1}^{16} x_r \mathfrak{k}([\xi_r, \eta_i]) \qquad \text{for } i \in \overline{1,16}. \tag{11.5.12}$$

In order to calculate $\{\iota(\eta_1), \ldots, \iota(\eta_{16})\}$ explicitly, we need the more formulas of ν on the positive root vectors of $o(10, \mathbb{F})$ extended from (11.3.30)–(11.3.34). We calculate

$$\nu(A_{3,5}) = E_{(0,0,0,1,1)}, \quad \nu(A_{2,5}) = -E_{(0,0,1,1,1)}, \quad \nu(B_{1,4}) = E_{(1,1,1,1,1)}, \tag{11.5.13}$$

$$\nu(B_{3,4}) = E_{(0,1,0,1,1)}, \quad \nu(B_{2,4} = -E_{(0,1,1,1,1)}, \quad \nu(A_{1,5}) = E_{(1,0,1,1,1)}, \tag{11.5.14}$$

$$\nu(B_{2,3}) = -E_{(0,1,1,2,1)}, \quad \nu(B_{1,3}) = E_{(1,1,1,2,1)}, \quad \nu(B_{1,2}) = -E_{(1,1,2,2,1)}. \tag{11.5.15}$$

Using these formulas, we find

$$\iota(\eta_1) = P_1 + x_1\left(\sum_{i=1}^{4} A_{i,i} - A_{5,5} - \kappa\right)/2 - x_2A_{4,5} - x_3A_{3,5} + x_4A_{2,5} - x_5B_{3,4}$$
$$+ x_6B_{2,4} - x_7A_{1,5} - x_8B_{1,4} + x_9B_{2,3} - x_{10}B_{1,3} + x_{12}B_{1,2}, \qquad (11.5.16)$$

$$\iota(\eta_2) = P_2 + x_2\left(\sum_{i\neq 4}(A_{i,i} - A_{4,4} - \kappa\right)/2 - x_1A_{5,4} - x_3A_{3,4} + x_4A_{2,4} + x_5B_{3,5}$$
$$- x_6B_{2,5} - x_7A_{1,4} + x_8B_{1,5} - x_{11}B_{2,3} + x_{13}B_{1,3} - x_{14}B_{1,2}, \qquad (11.5.17)$$

$$\iota(\eta_3) = P_3 + x_3\left(\sum_{i\neq 3} A_{i,i} - A_{3,3} - \kappa\right)/2 - x_1A_{5,3} - x_2A_{4,3} + x_4A_{2,3} - x_5B_{4,5}$$
$$- x_7A_{1,3} - x_9B_{2,5} + x_{10}B_{1,5} + x_{11}B_{2,4} - x_{13}B_{1,4} + x_{15}B_{1,2}, \qquad (11.5.18)$$

$$\iota(\eta_4) = P_4 + x_4\left(\sum_{i\neq 2} A_{i,i} - A_{2,2} - \kappa\right)/2 + x_1A_{5,2} + x_2A_{4,2} + x_3A_{3,2} - x_6B_{4,5}$$
$$+ x_7A_{1,2} - x_9B_{3,5} + x_{11}B_{3,4} + x_{12}B_{1,5} - x_{14}B_{1,4} + x_{15}B_{1,3}, \qquad (11.5.19)$$

$$\iota(\eta_5) = P_5 + x_5\left(A_{1,1} + A_{2,2} - \sum_{i=3}^{5} A_{i,i} - \kappa\right)/2 - x_1C_{4,3} + x_2C_{5,3} - x_3C_{5,4} + x_6A_{2,3}$$
$$- x_8A_{1,3} - x_9A_{2,4} + x_{10}A_{1,4} - x_{11}A_{2,5} + x_{13}A_{1,5} - x_{16}B_{1,2}, \qquad (11.5.20)$$

$$\iota(\eta_6) = P_6 + x_6\left(A_{1,1} + A_{3,3} - \sum_{i=2,4,5} A_{i,i} - \kappa\right)/2 + x_1C_{4,2} - x_2C_{5,2} - x_4C_{5,4} + x_5A_{3,2}$$
$$+ x_8A_{1,2} - x_9A_{3,4} - x_{11}A_{3,5} + x_{12}A_{1,4} + x_{14}A_{1,5} - x_{16}B_{1,3}, \qquad (11.5.21)$$

$$\iota(\eta_7) = P_7 + x_7\left(\sum_{i2}^{5} A_{i,i} - A_{1,1} - \kappa\right)/2 - x_1A_{5,1} - x_2A_{4,1} - x_3A_{3,1} + x_4A_{2,1}$$
$$- x_8B_{4,5} - x_{10}B_{3,5} + x_{12}B_{2,5} + x_{13}B_{3,4} - x_{14}B_{2,4} + x_{15}B_{2,3}, \qquad (11.5.22)$$

$$\iota(\eta_8) = P_8 + x_8\left(A_{2,3} + A_{3,3} - \sum_{i=1,4,5} A_{i,i} - \kappa\right)/2 - x_1C_{4,1} + x_2C_{5,1} - x_5A_{3,1} + x_6A_{2,1}$$
$$- x_7C_{5,4} - x_{10}A_{3,4} + x_{12}A_{2,4} - x_{13}A_{3,5} + x_{14}A_{2,5} - x_{16}B_{2,3}, \qquad (11.5.23)$$

$$\iota(\eta_9) = P_9 + x_9\left(A_{1,1}+A_{4,4}-\sum_{i=2,3,5}A_{i,i}-\kappa\right)/2 + x_1C_{3,2} - x_3C_{5,2} - x_4C_{5,3} - x_5A_{4,2} \\ -x_6A_{4,3} + x_{10}A_{1,2} + x_{11}A_{4,5} + x_{12}A_{1,3} - x_{15}A_{1,5} + x_{16}B_{1,4}, \quad (11.5.24)$$

$$\iota(\eta_{10}) = P_{10} + x_{10}\left(A_{2,2}+A_{4,4}-\sum_{i=1,3,5}A_{i,i}-\kappa\right)/2 - x_1C_{3,1} + x_3C_{5,1} + x_5A_{4,1} - x_7C_{5,3} \\ -x_8A_{4,3} + x_9A_{2,1} + x_{12}A_{2,3} + x_{13}A_{4,5} - x_{15}A_{2,5} + x_{16}B_{2,4}, \quad (11.5.25)$$

$$\iota(\eta_{11}) = P_{11} + x_{11}\left(A_{1,1}+A_{5,5}-\sum_{i=2,3,4}A_{i,i}-\kappa\right)/2 - x_2C_{3,2} + x_3C_{4,2} + x_4C_{4,3} - x_5A_{5,2} \\ -x_6A_{5,3} + x_9A_{5,4} + x_{13}A_{1,2} + x_{14}A_{1,3} + x_{15}A_{1,4} + x_{16}B_{1,5}, \quad (11.5.26)$$

$$\iota(\eta_{12}) = P_{12} + x_{12}\left(A_{3,3}+A_{4,4}-\sum_{i=1,2,5}A_{i,i}-\kappa\right)/2 + x_1C_{2,1} + x_4C_{5,1} + x_6A_{4,1} + x_7C_{5,2} \\ +x_8A_{4,2} + x_9A_{3,1} + x_{10}A_{3,2} + x_{14}A_{4,5} - x_{15}A_{3,5} + x_{16}B_{3,4}, \quad (11.5.27)$$

$$\iota(\eta_{13}) = P_{13} + x_{13}\left(A_{2,2}+A_{5,5}-\sum_{i=1,3,4}A_{i,i}-\kappa\right)/2 + x_2C_{3,1} - x_3C_{4,1} + x_5A_{5,1} + x_7C_{4,3} \\ -x_8A_{5,3} + x_{10}A_{5,4} + x_{11}A_{2,1} + x_{14}A_{2,3} + x_{15}A_{2,4} + x_{16}B_{2,5}, \quad (11.5.28)$$

$$\iota(\eta_{14}) = P_{14} + x_{14}\left(A_{2,2}+A_{5,5}-\sum_{i=1,2,4}A_{i,i}-\kappa\right)/2 - x_2C_{2,1} - x_4C_{4,1} + x_6A_{5,1} - x_7C_{4,2} \\ +x_8A_{5,2} + x_{11}A_{3,1} + x_{12}A_{5,4} + x_{13}A_{3,2} + x_{15}A_{3,4} + x_{16}B_{3,5}, \quad (11.5.29)$$

$$\iota(\eta_{15}) = P_{15} + x_{15}\left(A_{2,2}+A_{5,5}-\sum_{i=1,2,3}A_{i,i}-\kappa\right)/2 + x_3C_{2,1} + x_4C_{3,1} + x_7C_{3,2} - x_9A_{5,1} \\ -x_{10}A_{5,2} + x_{11}A_{4,1} - x_{12}A_{5,3} + x_{13}A_{4,2} + x_{14}A_{4,3} + x_{16}B_{4,5}, \quad (11.5.30)$$

$$\iota(\eta_{16}) = P_{16} - x_{16}\left(\sum_{i=1}^{5}A_{i,i}+\kappa\right)/2 - x_5C_{2,1} - x_6C_{3,1} - x_8C_{3,2} + x_9C_{4,1} \\ +x_{10}C_{4,2} + x_{11}C_{5,1} + x_{12}C_{4,3} + x_{13}C_{5,2} + x_{14}C_{5,3} + x_{15}C_{5,4}. \quad (11.5.31)$$

Recall $\mathscr{B} = \mathbb{F}[x_1, x_2, \ldots, x_{16}]$. Let M be an $o(10, \mathbb{F})$-module and set

$$\widehat{M} = \mathscr{B} \otimes_{\mathbb{F}} M. \quad (11.5.32)$$

We identify

$$f \otimes v = fv \qquad \text{for } f \in \mathscr{B},\ v \in M. \tag{11.5.33}$$

Recall the Lie algebra $\widehat{\mathscr{K}}$ defined via (11.5.1)–(11.5.8). Fix $c \in \mathbb{F}$. Then $\widehat{M}$ becomes a $\widehat{\mathscr{K}}$-module with the action defined by

$$d(fv) = d(f)v, \quad \kappa(fv) = cfv, \quad (gB)(fv) = fgB(v) \tag{11.5.34}$$

for $d \in \mathscr{W}_{16},\ f, g \in \mathscr{B},\ v \in M$ and $B \in o(10, \mathbb{F})$.

Since the linear map $\iota : \mathscr{G}^{E_6} \to \widehat{\mathscr{K}}$ defined in (11.5.10)–(11.5.12) is a Lie algebra monomorphism, $\widehat{M}$ becomes a $\mathscr{G}^{E_6}$-module with the action defined by

$$\xi(w) = \iota(\xi)(w) \qquad \text{for } \xi \in \mathscr{G}^{E_6},\ w \in \widehat{M}. \tag{11.5.35}$$

In fact, we have:

Theorem 11.5.1 *The map $M \mapsto \widehat{M}$ gives a functor from $o(10, \mathbb{F})$-**Mod** to $\mathscr{G}^{E_6}$-**Mod**, where the morphisms are $o(10, \mathbb{F})$-module homomorphisms and $\mathscr{G}^{E_6}$-module homomorphisms, respectively.*

We remark that the module $\widehat{M}$ is not a generalized Verma module in general because it may not be equal to $U(\mathscr{G})(M) = U(\mathscr{G}_-)(M)$.

Proposition 11.5.2 *If M is an irreducible $o(10, \mathbb{F})$-module, then $U(\mathscr{G}_-)(M)$ is an irreducible $\mathscr{G}^{E_6}$-module.*

Proof Note that for any $i \in \overline{1, 16},\ f \in \mathscr{B}$ and $v \in M$, (11.5.12), (11.5.34) and (11.5.35) imply

$$\xi_i(fv) = \partial_{x_i}(f)v. \tag{11.5.36}$$

Let W be any nonzero $\mathscr{G}^{E_6}$-submodule. The above expression shows that $W \bigcap M \neq \{0\}$. According to (11.5.34), $W \bigcap M$ is an $o(10, \mathbb{F})$-submodule. By the irreducibility of M, $M \subset W$. Thus $U(\mathscr{G}_-)(M) \subset W$. So $U(\mathscr{G}_-)(M) = W$ is irreducible. □

By the above proposition, the map $M \mapsto U(\mathscr{G}_-)(M)$ is a polynomial extension from irreducible $o(10, \mathbb{F})$-modules to irreducible $\mathscr{G}^{E_6}$-modules. We can use it to derive Gel'fand-Zetlin bases for E_6 from those for $o(10, \mathbb{F})$.

11.6 Irreducibility of the Functor

In this section, we want to determine the irreducibility of the $\mathscr{G}^{E_6}$-module $\widehat{M}$.

Note that $\widehat{M}$ can be viewed as an $o(10, \mathbb{F})$-module with the representation $\iota(\nu(B))|_{\widehat{M}}$ (cf. (11.3.30)–(11.3.34)). Indeed, (11.5.11) and (11.5.35) show

$$\nu(B(fv) = B(f)v + fB(v) \qquad \text{for } B \in o(10, \mathbb{F}),\ f \in \mathscr{A},\ v \in M \tag{11.6.1}$$

(cf. (11.3.36)–(11.3.57)). So $\widehat{M} = \mathscr{B} \otimes_{\mathbb{F}} M$ is a tensor module of $o(10, \mathbb{F})$. Write

$$\eta^{\alpha} = \prod_{i=1}^{16} \eta_i^{\alpha_i}, \quad |\alpha| = \sum_{i=1}^{16} \alpha_i \ \text{ for } \ \alpha = (\alpha_1, \alpha_2, \dots, \alpha_{16}) \in \mathbb{N}^{16} \tag{11.6.2}$$

(cf. (11.3.14)–(11.3.19)). Recall the Lie subalgebras $\mathscr{G}_{\pm}$ and $\mathscr{G}_0$ of $\mathscr{G}^{E_6}$ defined in (11.3.20). For $k \in \mathbb{N}$, we set

$$\mathscr{B}_k = \text{Span}\{x^{\alpha} \mid \alpha \in \mathbb{N}^{16}; |\alpha| = k\}, \quad \widehat{M}_k = \mathscr{B}_k M \tag{11.6.3}$$

(cf. (11.3.27), (11.5.34)) and

$$(U(\mathscr{G}_-)(M))_k = \text{Span}\{\eta^{\alpha}(M) \mid \alpha \in \mathbb{N}^{16}, \ |\alpha| = k\}. \tag{11.6.4}$$

Moreover,

$$(U(\mathscr{G}_-)(M))_0 = \widehat{M}_0 = M. \tag{11.6.5}$$

Furthermore,

$$\widehat{M} = \bigoplus_{k=0}^{\infty} \widehat{M}_k, \qquad U(\mathscr{G}_-)(M) = \bigoplus_{k=0}^{\infty} (U(\mathscr{G}_-)(M))_k. \tag{11.6.6}$$

Next we define a linear transformation φ on $\widehat{M}$ determined by

$$\varphi(x^{\alpha} v) = \eta^{\alpha}(v) \qquad \text{for } \ \alpha \in \mathbb{N}^{16}, \ v \in M. \tag{11.6.7}$$

Note that $\mathscr{B}_1 = \sum_{i=1}^{16} \mathbb{F} x_i$ forms the 16-dimensional $\mathscr{G}_0$-module (equivalently, the $o(10, \mathbb{F})$ spin module). According to (4.4.24) and (4.4.25), $\mathscr{G}_-$ forms a $\mathscr{G}_0$-module with respect to the adjoint representation, and the linear map from $\mathscr{B}_1$ to $\mathscr{G}_0$ determined by $x_i \mapsto \eta_i$ for $i \in \overline{1, 16}$ gives a $\mathscr{G}_0$-module isomorphism. Thus φ can also be viewed as a $\mathscr{G}_0$-module homomorphism from $\widehat{M}$ to $U(\mathscr{G}_0)(M)$. Moreover,

$$\varphi(\widehat{M}_k) = (U(\mathscr{G}_-)(M))_k \qquad \text{for } \ k \in \mathbb{N}. \tag{11.6.8}$$

Note that the Casimir element of $o(10, \mathbb{F})$ is

$$\omega = \sum_{1 \le i < j \le 5} (B_{i,j} C_{j,i} + C_{j,i} B_{i,j}) + \sum_{r,s=1}^{5} A_{r,s} A_{s,r} \in U(o(10, \mathbb{F})) \tag{11.6.9}$$

(cf. (11.3.4)). The algebra $U(o(10, \mathbb{F}))$ can be imbedded into the tensor algebra $U(o(10, \mathbb{F})) \otimes U(o(10, \mathbb{F}))$ by the associative algebra homomorphism $\mathfrak{d} : U(o(10, \mathbb{F})) \to U(o(10, \mathbb{F})) \otimes_{\mathbb{F}} U(o(10, \mathbb{F}))$ determined by

$$\mathfrak{d}(u) = u \otimes 1 + 1 \otimes u \qquad \text{for } u \in o(10, \mathbb{F}). \tag{11.6.10}$$

Set

$$\tilde{\omega} = \frac{1}{2}(\mathfrak{d}(\omega) - \omega \otimes 1 - 1 \otimes \omega) \in U(o(10, \mathbb{F})) \otimes_{\mathbb{F}} U(o(2n, \mathbb{F})). \tag{11.6.11}$$

By (11.6.9),

$$\tilde{\omega} = \sum_{1 \leq i < j \leq 5} (B_{i,j} \otimes C_{j,i} + C_{j,i} \otimes B_{i,j}) + \sum_{r,s=1}^{5} A_{r,s} \otimes A_{s,r}. \tag{11.6.12}$$

Furthermore, $\tilde{\omega}$ acts on $\widehat{M}$ as an $o(10, \mathbb{F})$-module homomorphism via

$$(B_1 \otimes B_2)(fv) = B_1(f)B_2(v) \qquad \text{for } B_1, B_2 \in o(10, \mathbb{F}),\ f \in \mathscr{B},\ v \in M. \tag{11.6.13}$$

Lemma 11.6.1 *We have* $\varphi|_{\widehat{M}_1} = (\tilde{\omega} - c/2)|_{\widehat{M}_1}$.

Proof For any $v \in M$, (11.3.36)–(11.3.57), (11.5.16), (11.6.12) and (11.6.13) give

$$\begin{aligned}\tilde{\omega}(x_1 v) &= \Big[-x_2 A_{4,5} - x_3 A_{3,5} + x_4 A_{2,5} - x_5 B_{3,4} - x_7 A_{1,5} + x_6 B_{2,4} \\ &\qquad -x_8 B_{1,4} + x_9 B_{2,3} - x_{10} B_{1,3} + x_{12} B_{1,7} + x_1 \Big(\sum_{i=1}^{4} A_{i,i} - A_{5,5}\Big)\Big] v \\ &= \eta_1(v) + (c/2) x_1 v = (\varphi + c/2)(x_1 v),\end{aligned} \tag{11.6.14}$$

or equivalently, $\varphi(x_1 v) = (\tilde{\omega} - c/2)(x_1 v)$. According to (11.3.63), $-A_{5,4}(x_1) = x_2$. So

$$\begin{aligned}&\varphi(x_2 v) - \varphi(x_1 A_{5,4}(v)) \\ &= -A_{5,4}(\varphi(x_1 v) = -A_{5,4}[(\tilde{\omega} - c/2)(x_1 v)] = (c/2 - \tilde{\omega}) A_{5,4}(x_1 v) \\ &= (\tilde{\omega} - c/2)(x_2 v - x_1 A_{5,4}(v)) = (\tilde{\omega} - c/2)(x_2 v) - (\tilde{\omega} - c/2)(x_1 A_{5,4}(v)) \\ &= (\tilde{\omega} - c/2)(x_2 v) - \varphi(x_1 A_{5,4}(v)),\end{aligned} \tag{11.6.15}$$

or equivalently, $\varphi(x_2 v) = (\tilde{\omega} - c/2)(x_2 v)$.

Observe that

$$-A_{4,3}(x_2) = x_3,\quad A_{3,2}(x_3) = x_4,\quad -C_{5,4}(x_3) = x_5,\quad -A_{3,1}(x_3) = x_7, \tag{11.6.16}$$

$$C_{5,1}(x_2) = x_8,\quad -C_{5,2}(x_2) = x_6,\quad -C_{3,1}(x_1) = x_{10},\quad -C_{3,2}(x_2) = x_{11}, \tag{11.6.17}$$

$$A_{3,2}(x_{10}) = x_{12},\quad C_{3,2}(x_1) = x_9,\quad C_{3,1}(x_2) = x_{13}, \tag{11.6.18}$$

$$A_{3,2}(x_{13}) = x_{14},\quad A_{4,3}(x_{14}) = x_{15},\quad C_{10,4}(x_{15}) = x_{16} \tag{11.6.19}$$

by (11.3.60)–(11.3.64). Using the argument similar to (11.6.14) and induction, we can prove

$$\varphi(x_i v) = (\tilde{\omega} - c/2)(x_i v) \qquad \text{for } i \in \overline{1, 16}, \tag{11.6.20}$$

or equivalently, the lemma holds. □

We take (7.1.2)–(7.1.10) and (7.3.2)–(7.2.4) with $n = 5$. For any $\mu \in \Lambda^+$, we denote by $V(\mu)$ the finite-dimensional irreducible $o(10, \mathbb{F})$-module with the highest weight μ and have

$$\omega|_{V(\mu)} = (\mu + 2\rho, \mu)\mathrm{Id}_{V(\mu)} \tag{11.6.21}$$

by (11.6.9).

Let $\mathbb{Z}_2 = \mathbb{Z}/2\mathbb{Z} = \{0, 1\}$. According to (11.3.57) and Table 11.3, the weight set of the $o(10, \mathbb{F})$-module $\mathscr{B}_1$ is

$$\Lambda(\mathscr{B}_1) = \left\{ (1/2) \sum_{i=1}^{5} (-1)^{k_i} \varepsilon_i \mid k_i \in \mathbb{Z}_2, \ \sum_{i=1}^{5} k_i = 1 \right\}. \tag{11.6.22}$$

Fixing $\lambda \in \Lambda^+$, we define

$$\mathfrak{H}(\lambda) = \{\lambda + \mu \mid \mu \in \Lambda(\mathscr{A}_1), \ \lambda + \mu \in \Lambda^+\}. \tag{11.6.23}$$

Lemma 11.6.2 *We have:*

$$\mathscr{B}_1 \otimes V(\lambda) \cong \bigoplus_{\lambda' \in \mathfrak{H}(\lambda)} V(\lambda'). \tag{11.6.24}$$

Proof Note that all the weight subspaces of $\mathscr{B}_1$ are one-dimensional. Thus all the irreducible components of $\mathscr{A}_1 \otimes V(\lambda)$ are of multiplicity one. Since

$$\rho + \lambda + \mu \in \Lambda^+ \qquad \text{for } \mu \in \Lambda(\mathscr{A}_1), \tag{11.6.25}$$

Theorem 5.4.3 says that $V(\lambda')$ is a component of $\mathscr{B}_1 \otimes V(\lambda)$ if and only if $\lambda' \in \mathfrak{H}(\lambda)$. □

Recall

$$\text{the highest weight of } \mathscr{B}_1 = \frac{1}{2}(\varepsilon_1 + \varepsilon_2 + \varepsilon_3 + \varepsilon_4 - \varepsilon_5) = \lambda_4, \tag{11.6.26}$$

the forth fundamental weight of $o(10, \mathbb{F})$, by (11.3.57) and Table 11.3. Thus the eigenvalues of $\tilde{\omega}|_{\widehat{V(\lambda)}_1}$ are

$$\{[(\lambda' + 2\rho, \lambda') - (\lambda + 2\rho, \lambda) - (\lambda_4 + 2\rho, \lambda_4)]/2 \mid \lambda' \in \mathfrak{H}(\lambda)\} \tag{11.6.27}$$

by (11.6.11) and (11.6.13). We remark that the above fact is equivalent to special detailed version of Kostant's characteristic identity (cf. [Kb]). Define

$$\ell_\omega(\lambda) = \min\{[(\lambda' + 2\rho, \lambda') - (\lambda + 2\rho, \lambda) - (\lambda_4 + 2\rho, \lambda_4)]/2 \mid \lambda' \in \mathfrak{H}(\lambda)\}, \tag{11.6.28}$$

which will be used to determine the irreducibility of $\widehat{V(\lambda)}$. If $\lambda' = \lambda + \lambda_4 - \alpha \in \mathfrak{H}(\lambda)$ with $\alpha \in \Phi^+_{D_5}$, then

$$(\lambda' + 2\rho, \lambda') - (\lambda + 2\rho, \lambda) - (\lambda_4 + 2\rho, \lambda_4) = 2[(\lambda, \lambda_4) + 1 - (\rho + \lambda + \lambda_4, \alpha)]. \tag{11.6.29}$$

Recall the differential operators $P_1, \ldots, P_{16}$ given in (11.4.16)–(11.4.32). We also view the elements of $\mathscr{B}$ as the multiplication operators on $\mathscr{B}$. Recall ζ_1 in (11.3.59). It turns out that we need the following lemma in order to determine the irreducibility of $\widehat{V(\lambda)}$.

Lemma 11.6.3 *As operators on* $\mathscr{B}$:

$$P_{11}x_1 + P_1x_{11} + P_9x_2 + P_2x_9 - P_6x_3 - P_3x_6 + P_5x_4 + P_4x_5 = \zeta_1(D - 6). \tag{11.6.30}$$

Proof According to (11.4.16), (11.4.18)–(11.4.22), (11.4.25) and (11.4.27), we find that

$$\begin{aligned} &P_{11}x_1 + P_1x_{11} + P_9x_2 + P_2x_9 - P_6x_3 - P_3x_6 + P_5x_4 + P_4x_5 \\ &= -6\zeta_1 + x_1P_{11} + x_{11}P_1 + x_2P_9 + x_9P_2 - x_3P_6 - x_6P_3 + x_4P_5 + x_5P_4 \end{aligned} \tag{11.6.31}$$

and

$$\begin{aligned} &x_1P_{11} + x_{11}P_1 + x_2P_9 + x_9P_2 - x_3P_6 - x_6P_3 + x_4P_5 + x_5P_4 \\ &= x_1(x_{11}D - \zeta_1\partial_{x_1} + \zeta_5\partial_{x_7} + \zeta_9\partial_{x_8} + \zeta_8\partial_{x_{10}} - \zeta_7\partial_{x_{12}}) \\ &\quad + x_{11}(x_1D - \zeta_1\partial_{x_{11}} - \zeta_2\partial_{x_{13}} - \zeta_3\partial_{x_{14}} - \zeta_4\partial_{x_{15}} - \zeta_{10}\partial_{x_{16}}) \\ &\quad + x_2(x_9D - \zeta_1\partial_{x_2} + \zeta_4\partial_{x_7} - \zeta_{10}\partial_{x_8} - \zeta_8\partial_{x_{13}} + \zeta_7\partial_{x_{14}}) \\ &\quad + x_9(x_2D - \zeta_1\partial_{x_9} - \zeta_2\partial_{x_{10}} - \zeta_3\partial_{x_{12}} + \zeta_5\partial_{x_{15}} - \zeta_9\partial_{x_{16}}) \\ &\quad - x_3(x_6D + \zeta_1\partial_{x_3} - \zeta_3\partial_{x_7} + \zeta_{10}\partial_{x_{10}} - \zeta_9\partial_{x_{13}} + \zeta_7\partial_{x_{15}}) \\ &\quad - x_6(x_3D + \zeta_1\partial_{x_6} + \zeta_2\partial_{x_8} + \zeta_4\partial_{x_{12}} + \zeta_5\partial_{x_{14}} - \zeta_8\partial_{x_{16}}) \\ &\quad + x_4(x_5D - \zeta_1\partial_{x_4} - \zeta_2\partial_{x_7} - \zeta_{10}\partial_{x_{12}} - \zeta_9\partial_{x_{14}} - \zeta_8\partial_{x_{15}}) \\ &\quad + x_5(x_4D - \zeta_1\partial_{x_5} + \zeta_3\partial_{x_8} - \zeta_4\partial_{x_{10}} - \zeta_5\partial_{x_{13}} + \zeta_7\partial_{x_{16}}) \\ &= 2\zeta_1D - \zeta_1\sum_{i=1,2,3,4,5,6,9,11} x_i\partial_{x_i} + (x_1\zeta_5 + x_2\zeta_4 + x_3\zeta_3 - x_4\zeta_2)\partial_{x_7} \\ &\quad + (x_1\zeta_9 - x_2\zeta_{10} - x_6\zeta_2 + x_5\zeta_3)\partial_{x_8} + (x_1\zeta_8 - x_9\zeta_2 - x_3\zeta_{10} - x_5\zeta_4)\partial_{x_{10}} \end{aligned}$$

$$-(x_1\zeta_7+x_9\zeta_3+x_6\zeta_4+x_4\zeta_{10})\partial_{x_{12}}-(x_{11}\zeta_2+x_2\zeta_8-x_3\zeta_9+x_5\zeta_5)\partial_{x_{13}}$$
$$-(x_{11}\zeta_3-x_2\zeta_7+x_6\zeta_5-x_4\zeta_9)\partial_{x_{14}}-(x_{11}\zeta_4-x_9\zeta_5+x_3\zeta_7+x_4\zeta_8)\partial_{x_{15}}$$
$$-(x_{11}\zeta_{10}+x_9\zeta_9-x_6\zeta_8-x_5\zeta_7)\partial_{x_{16}}=\zeta_1 D. \tag{11.6.32}$$

□

We define the multiplication

$$f(gv)=(fg)v \qquad \text{for } f,g\in\mathscr{B},\ v\in M. \tag{11.6.33}$$

Then (11.5.16)–(11.5.21), (11.5.24) and (11.5.26) gives

$$\sum_{r=1}^{16} x_r\varrho([\xi_r, x_1\eta_{11}+x_{11}\eta_1+x_2\eta_9+x_9\eta_2-x_3\eta_6-x_6\eta_3+x_4\eta_5+x_5\eta_4])$$
$$=\sum_{i=1}^{5}\zeta_i A_{1,i}+\sum_{r=2}^{5}\zeta_{5+r}B_{1,r}-c\zeta_1 \tag{11.6.34}$$

as operators on $\widehat{M}$ (cf. (11.5.33)), where ζ_i are defined in (11.3.59) and (11.3.65)–(11.3.83). By Lemma 11.6.3, (11.5.5), (11.5.6) and (11.6.32),

$$T_1=\iota(\eta_{11})x_1+\iota(\eta_1)x_{11}+\iota(\eta_9)x_2+\iota(\eta_2)x_9-\iota(\eta_6)x_3-\iota(\eta_3)x_6+\iota(\eta_5)x_4+\iota(\eta_4)x_5$$
$$=\zeta_1(D-c-6)+\sum_{i=1}^{5}\zeta_i A_{1,i}+\sum_{r=2}^{5}\zeta_{5+r}B_{1,r} \tag{11.6.35}$$

as operators on $\widehat{M}$. We define an $o(10,\mathbb{F})$-module structure on the space End $\widehat{M}$ of linear transformations on $\widehat{M}$ by

$$B(T)=[\iota(\nu(B)),T]=\iota(\nu(B))T-T\iota(\nu(B)) \qquad \text{for } B\in o(10,\mathbb{F}),\ T\in \text{End}\,\widehat{M} \tag{11.6.36}$$

(cf. (11.6.1)). It can be verified that T_1 is an $o(10,\mathbb{F})$-singular vector with weight ε_1 in End $\widehat{M}$. So it generates the 10-dimensional natural module. Set

$$T_2=\zeta_2(D-c-6)+\sum_{i=1}^{5}\zeta_i A_{2,i}+\sum_{r=1,3,4,5}\zeta_{5+r}B_{2,r}, \tag{11.6.37}$$

$$T_3=\zeta_3(D-c-6)+\sum_{i=1}^{5}\zeta_i A_{3,i}+\sum_{r=1,2,4,5}\zeta_{5+r}B_{3,r}, \tag{11.6.38}$$

$$T_4=\zeta_4(D-c-6)+\sum_{i=1}^{5}\zeta_i A_{4,i}+\sum_{r=1,2,3,5}\zeta_{5+r}B_{4,r}, \tag{11.6.39}$$

$$T_5 = \zeta_5(D - c - 6) + \sum_{i=1}^{5} \zeta_i A_{5,i} - \sum_{r=1}^{4} \zeta_{5+r} B_{r,5}, \tag{11.6.40}$$

$$T_6 = \zeta_6(D - c - 6) - \sum_{i=2}^{5} \zeta_i C_{i,1} - \sum_{r=1}^{5} \zeta_{5+r} A_{r,1}, \tag{11.6.41}$$

$$T_7 = \zeta_7(D - c - 6) - \sum_{i=1,3,4,5} \zeta_i C_{i,2} - \sum_{r=1}^{5} \zeta_{5+r} A_{r,2}, \tag{11.6.42}$$

$$T_8 = \zeta_8(D - c - 6) + \sum_{i=1,2,4,5} \zeta_i C_{3,i} - \sum_{r=1}^{5} \zeta_{5+r} A_{r,3}, \tag{11.6.43}$$

$$T_9 = \zeta_9(D - c - 6) + \sum_{i=1,2,3,5} \zeta_i C_{4,i} - \sum_{r=1}^{5} \zeta_{5+r} A_{r,4}, \tag{11.6.44}$$

$$T_{10} = \zeta_{10}(D - c - 6) + \sum_{i=1}^{4} \zeta_i C_{5,i} - \sum_{r=1}^{5} \zeta_{5+r} A_{r,5}. \tag{11.6.45}$$

Then $\mathscr{T} = \sum_{i=1}^{10} \mathbb{F}T_i$ forms the 10-dimensional natural module of $o(10, \mathbb{F})$ with the standard basis $\{T_1, \ldots, T_{10}\}$.

Denote

$$T_i' = T_i - \zeta_i(D - c - 6) \qquad \text{for } i \in \overline{1, 10}. \tag{11.6.46}$$

Easily see that $\mathscr{T}' = \sum_{i=1}^{10} \mathbb{F}T_i'$ forms the 10-dimensional natural module of $o(10, \mathbb{F})$ with the standard basis $\{T_1', \ldots, T_{10}'\}$. So we have the $o(10, \mathbb{F})$-module isomorphism from $U = \sum_{i=1}^{10} \mathbb{F}\zeta_i$ to $\mathscr{T}'$ determined by $\zeta_i \mapsto T_i'$ for $i \in \overline{1, 10}$. The weight set of U is

$$\Lambda(U) = \{\pm\varepsilon_1, \ldots, \pm\varepsilon_5\}. \tag{11.6.47}$$

Let $\lambda \in \Lambda^+$. Denote

$$\mathfrak{H}'(\lambda) = \{\lambda + \mu \mid \mu \in \Lambda(U), \ \ \lambda + \mu \in \Lambda^+\}. \tag{11.6.48}$$

Take $M = V(\lambda)$. It is known that

$$UV(\lambda) = U \otimes_{\mathbb{F}} V(\lambda) \cong \bigoplus_{\lambda' \in \mathfrak{H}'(\lambda)} V(\lambda'). \tag{11.6.49}$$

Given $\lambda' \in \mathfrak{H}'(\lambda)$, we pick a singular vector

$$u = \sum_{i=1}^{10} \zeta_i u_i \tag{11.6.50}$$

of weight λ' in $UV(\lambda)$, where $u_i \in V(\lambda)$. Moreover, any singular vector of weight λ' in $UV(\lambda)$ is a scalar multiple of u. Note that the vector

$$w = \sum_{i=1}^{10} T'_i(u_i) \tag{11.6.51}$$

is also a singular vector of weight λ' if it is not zero. Thus

$$w = \flat_{\lambda'} u, \qquad \flat_{\lambda'} \in \mathbb{F}. \tag{11.6.52}$$

Set

$$\flat(\lambda) = \min\{\flat_{\lambda'} \mid \lambda' \in \mathfrak{H}'(\lambda)\}. \tag{11.6.53}$$

Theorem 11.6.4 (Irreducibility of the Functor) *The $\mathcal{G}^{E_6}$-module $\widehat{V(\lambda)}$ is irreducible if*

$$c \in \mathbb{F} \setminus \{\flat(\lambda) - 6 + \mathbb{N}, 2\ell_\omega(\lambda) + 2\mathbb{N}\}. \tag{11.6.54}$$

Proof Recall that the $\mathcal{G}^{E_6}$-submodule $U(\mathcal{G}_-)(V(\lambda))$ is irreducible by Proposition 11.5.2. It is enough to prove $\widehat{V(\lambda)} = U(\mathcal{G}_-)(V(\lambda))$. It is obvious that

$$\widehat{V(\lambda)}_0 = V(\lambda) = (U(\mathcal{G}_-)(V(\lambda)))_0 \tag{11.6.55}$$

(cf. (11.6.3) and (11.6.4) with $M = V(\lambda)$). Moreover, Lemma 11.6.1 with $M = V(\lambda)$, (11.6.36) and (11.6.61) imply that $\varphi|_{\widehat{V(\lambda)}_1}$ is invertible, or equivalently,

$$\widehat{V(\lambda)}_1 = (U(\mathcal{G}_-)(V(\lambda)))_1. \tag{11.6.56}$$

Suppose that

$$\widehat{V(\lambda)}_i = (U(\mathcal{G}_-)(V(\lambda)))_i \tag{11.6.57}$$

for $i \in \overline{0, k}$ with $1 \leq k \in \mathbb{N}$.

For any $v \in V(\lambda)$ and $\alpha \in \mathbb{N}^{16}$ such that $|\alpha| = k - 1$, we have

$$T_r(x^\alpha v) = x^\alpha[(|\alpha| - c - 6)\zeta_r + T'_r](v) \in (U(\mathcal{G}_-)(V(\lambda)))_{k+1}, \qquad r \in \overline{1, 10} \tag{11.6.58}$$

by (11.6.57) with $i = k - 1, k$. But

$$V' = \text{Span}\{[(|\alpha| - c - 6)\zeta_r + T'_r](v) \mid r \in \overline{1, 10},\ v \in V(\lambda)\} \tag{11.6.59}$$

forms an $o(10, \mathbb{F})$-submodule of $UV(\lambda)$ with respect to the action in (11.6.1). Let u be a $o(10, \mathbb{F})$-singular vector in (11.6.50). Then

$$V' \ni \sum_{r=1}^{10}[(|\alpha| - c - 6)\zeta_r + T_r'](u_r) = (|\alpha| - c - 6)u + w = (|\alpha| - c - 6 + \flat_{\lambda'})u \tag{11.6.60}$$

by (11.6.51) and (11.6.52). Moreover, (11.6.53) and (11.6.54) yield $u \in V'$. Since $UV(\lambda)$ is an $o(10, \mathbb{F})$-module generated by all the singular vectors, we have $V' = UV(\lambda)$. So

$$x^{\alpha}UV(\lambda) \subset (U(\mathcal{G}_{-})(V(\lambda)))_{k+1}. \tag{11.6.61}$$

The arbitrariness of α implies

$$\zeta_r \widehat{V(\lambda)}_{k-1} \subset (U(\mathcal{G}_{-})(V(\lambda)))_{k+1} \qquad \text{for } r \in \overline{1, 10}. \tag{11.6.62}$$

Given any $f \in \mathcal{B}_k$ and $v \in V(\lambda)$, we have

$$\zeta_r \partial_{x_i}(f)v \in \zeta_r \widehat{V(\lambda)}_{k-1} \subset (U(\mathcal{G}_{-})(V(\lambda)))_{k+1} \qquad \text{for } r \in \overline{1, 10},\ i \in \overline{1, 16}. \tag{11.6.63}$$

Moreover,

$$\begin{aligned}\eta_s(fv) = \iota(\eta_s)(fv) &= P_s(fv) + f(\tilde{\omega} - c/2)(x_s v)\\ &\equiv f(k + \tilde{\omega} - c/2)(x_s v) \left(\text{mod} \sum_{r=1}^{10} \zeta_r \widehat{V(\lambda)}_{k-1}\right)\end{aligned} \tag{11.6.64}$$

for $s \in \overline{1, 16}$ by (11.4.16)–(11.4.32), (11.5.16)–(11.5.31) and Lemma 11.6.1. According to (11.6.28), (11.6.54), (11.6.62) and (11.6.64), we get

$$x_s f v \in (U(\mathcal{G}_{-})(V(\lambda)))_{k+1} \qquad \text{for } s \in \overline{1, 16}. \tag{11.6.65}$$

Thus (11.6.57) holds for $i = k + 1$. By induction on k, (11.6.57) holds for any $i \in \mathbb{N}$; that is, $\widehat{V(\lambda)} = U(\mathcal{G}_{-})(V(\lambda))$. □

When $\lambda = 0$, $V(0)$ is the one-dimensional trivial module and $\ell_{\omega}(0) = \flat(0) = 0$. So we have:

Corollary 11.6.5 *The $\mathcal{G}^{E_6}$-module $\widehat{V(0)}$ is irreducible if $c \in \mathbb{F} \setminus \{\mathbb{N} - 6\}$.*

Next we consider the case $\lambda = k\varepsilon_1 = k\lambda_1$ for some positive integer k, where λ_1 is the first fundamental weight. Note

$$\mathfrak{H}(k\varepsilon_1) = \{\lambda_4 + k\varepsilon_1, \lambda_4 + (k-1)\varepsilon_1 + \varepsilon_5\} \tag{11.6.66}$$

by (11.6.22) and (11.6.23). Thus (11.6.28) and (11.6.29) give

$$\ell_{\omega}(k\varepsilon_1) = -4 - k/2. \tag{11.6.67}$$

In order to calculate $\flat(k\varepsilon_1)$, we give a realization of $V(k\varepsilon_1)$. Observe that we have a representation of $o(10,\mathbb{F})$ on $\mathscr{B}=\mathbb{F}[y_1,\ldots,y_{10}]$ determined via

$$E_{i,j}|_{\mathscr{B}} = y_i\partial_{y_j} \qquad \text{for } i,j\in\overline{1,10}. \tag{11.6.68}$$

Denote by $\mathscr{B}_k$ the subspace of homogenous polynomials in $\mathscr{B}$ with degree k. Set

$$\mathscr{H}_k = \left\{h\in\mathscr{B}_k \mid \left(\sum_{i=1}^{5}\partial_{y_i}\partial_{y_{5+i}}\right)(h)=0\right\}. \tag{11.6.69}$$

Then $\mathscr{H}_k\cong V(k\varepsilon_1)$ and y_1^k is a highest-weight vector.

According to (11.6.47) and (11.6.48),

$$\mathfrak{H}'(k\varepsilon_1) = \{(k+1)\varepsilon_1,(k-1)\varepsilon_1,k\varepsilon_1+\varepsilon_2\}. \tag{11.6.70}$$

The vector $\zeta_1 y_1^k$ is a singular vector in $U\mathscr{H}_k$ with weight $(k+1)\varepsilon_1$, where we take $M=\mathscr{H}_k$ in the earlier settings. By (11.6.35) and (11.6.46),

$$T_1'(y_1^k) = k\zeta_1 y_1^k \Longrightarrow \flat_{(k+1)\varepsilon_1} = k. \tag{11.6.71}$$

Moreover, $\zeta_1 y_1^{k-1}y_2-\zeta_2 y_1^k$ is a singular vector in $U\mathscr{H}_k$ with weight $k\varepsilon_1+\varepsilon_2$. Furthermore, (11.6.35), (11.6.37) and (11.6.46) imply

$$\begin{aligned}T_1'(y_1^{k-1}y_2)-T_2'(y_1^k) &= (k-1)\zeta_1y_1^{k-1}y_2+\zeta_2y_1^k-k\zeta_1y_1^{k-1}y_2\\ &= \zeta_2y_1^k-\zeta_1y_1^{k-1}y_2 = -(\zeta_1y_1^{k-1}y_2-\zeta_2y_1^k).\end{aligned} \tag{11.6.72}$$

Thus $\flat_{k\varepsilon_1+\varepsilon_2}=-1$. Besides,

$$\varpi = (k+3)\sum_{i=1}^{5}[\zeta_iy_1^{k-1}y_{5+i}+\zeta_{5+i}y_1^{k-1}y_i]-(k-1)\zeta_1y_1^{k-2}\sum_{s=1}^{5}y_sy_{5+s} \tag{11.6.73}$$

is a singular vector in $U\mathscr{H}_k$ with weight $(k-1)\varepsilon_1$. Expressions (11.6.35) and (11.6.37)–(11.6.46) yield

$$\begin{aligned}&(k+3)\sum_{i=1}^{5}[T_i'(y_1^{k-1}y_{5+i})+T_{5+i}'(y_1^{k-1}y_i)]-(k-1)T_1'\left(x^\alpha y_1^{k-2}\sum_{s=1}^{5}y_sy_{5+s}\right)\\ &=(-8-k)x^\alpha\varpi \Longrightarrow \flat_{(k-1)\varepsilon_1}=-8-k.\end{aligned} \tag{11.6.74}$$

Therefore, $\flat(k\varepsilon_1)=-8-k$. By Theorem 11.6.4 and (11.6.67), we obtain:

Corollary 11.6.6 *The $\mathscr{G}^{E_6}$-module $\widehat{V(k\lambda_1)}$ is irreducible if $c\in\mathbb{F}\setminus\{\mathbb{N}-14-k\}$.*

By similar calculations, we obtain:

Corollary 11.6.7 *(a) The $\mathscr{G}^{E_6}$-module $\widehat{V(\lambda_2)}$ is irreducible if $c \in \mathbb{F} \setminus \{\mathbb{N} - 16\}$.*

(b) The $\mathscr{G}^{E_6}$-module $\widehat{V(\lambda_3)}$ is irreducible if $c \in \mathbb{F} \setminus \{\mathbb{N} - 15, -17, -19, -21\}$.

(c) The $\mathscr{G}^{E_6}$-module $\widehat{V(k\lambda_4)}$ is irreducible if $c \in \mathbb{F} \setminus \{\mathbb{N} - 10 - k/2, 2\mathbb{N} + k - 12\}$.

(d) The $\mathscr{G}^{E_6}$-module $\widehat{V(k\lambda_5)}$ is irreducible if $c \in \mathbb{F} \setminus \{\mathbb{N} - 10 - k/2, 2\mathbb{N} - k - 20\}$.

11.7 Representations on Exponential-Polynomial Functions

In this section, we want to study representations of $\mathscr{G}^{E_6}$ on exponential-polynomial functions.

Recall

$$D = \sum_{i=1}^{16} x_i \partial_{x_i}. \tag{11.7.1}$$

Fix $\mathfrak{c} \in \mathbb{F}$. Let $c = 2\mathfrak{c}$ and identify $\widehat{V(0)} = \mathscr{B} \otimes v_0$ with $\mathscr{B}$ by

$$f \otimes v_0 \leftrightarrow f \qquad \text{for } f \in \mathscr{B}, \tag{11.7.2}$$

where $V(0) = \mathbb{F}v_0$. Then we have the following one-parameter inhomogeneous first-order differential-operator representation $\pi_{\mathfrak{c}}$ of $\mathscr{G}^{E_6}$:

$$\pi_{\mathfrak{c}}(u) = \pi(\nu^{-1}(u)) \qquad \text{for } u \in \mathscr{G}^{D_5} \tag{11.7.3}$$

(cf. (11.3.30)–(11.3.34) and (11.3.36)–(11.3.57)),

$$\pi_{\mathfrak{c}}(\hat{\alpha}) = D + 4\mathfrak{c}, \qquad \pi_{\mathfrak{c}}(\xi_i) = \partial_{x_i} \qquad \text{for } i \in \overline{1, 16}, \tag{11.7.4}$$

$$\pi_{\mathfrak{c}}(\eta_1) = x_1(D - \mathfrak{c}) - \zeta_1\partial_{x_{11}} - \zeta_2\partial_{x_{13}} - \zeta_3\partial_{x_{14}} - \zeta_4\partial_{x_{15}} - \zeta_{10}\partial_{x_{16}}, \tag{11.7.5}$$

$$\pi_{\mathfrak{c}}(\eta_2) = x_2(D - \mathfrak{c}) - \zeta_1\partial_{x_9} - \zeta_2\partial_{x_{10}} - \zeta_3\partial_{x_{12}} + \zeta_5\partial_{x_{15}} - \zeta_9\partial_{x_{16}}, \tag{11.7.6}$$

$$\pi_{\mathfrak{c}}(\eta_3) = x_3(D - \mathfrak{c}) + \zeta_1\partial_{x_6} + \zeta_2\partial_{x_8} + \zeta_4\partial_{x_{12}} + \zeta_5\partial_{x_{14}} - \zeta_8\partial_{x_{16}}, \tag{11.7.7}$$

$$\pi_{\mathfrak{c}}(\eta_4) = x_4(D - \mathfrak{c}) - \zeta_1\partial_{x_5} + \zeta_3\partial_{x_8} - \zeta_4\partial_{x_{10}} - \zeta_5\partial_{x_{13}} + \zeta_7\partial_{x_{16}}, \tag{11.7.8}$$

$$\pi_{\mathfrak{c}}(\eta_5) = x_5(D - \mathfrak{c}) - \zeta_1\partial_{x_4} - \zeta_2\partial_{x_7} - \zeta_{10}\partial_{x_{12}} + \zeta_9\partial_{x_{14}} - \zeta_8\partial_{x_{15}}, \tag{11.7.9}$$

$$\pi_{\mathfrak{c}}(\eta_6) = x_6(D - \mathfrak{c}) + \zeta_1\partial_{x_3} - \zeta_3\partial_{x_7} + \zeta_{10}\partial_{x_{10}} - \zeta_9\partial_{x_{13}} + \zeta_7\partial_{x_{15}}, \tag{11.7.10}$$

$$\pi_{\mathfrak{c}}(\eta_7) = x_7(D - \mathfrak{c}) - \zeta_2\partial_{x_5} - \zeta_3\partial_{x_6} + \zeta_4\partial_{x_9} + \zeta_5\partial_{x_{11}} - \zeta_6\partial_{x_{16}}, \tag{11.7.11}$$

$$\pi_{\mathfrak{c}}(\eta_8) = x_8(D - \mathfrak{c}) + \zeta_2\partial_{x_3} + \zeta_3\partial_{x_4} - \zeta_{10}\partial_{x_9} + \zeta_9\partial_{x_{11}} - \zeta_6\partial_{x_{15}}, \tag{11.7.12}$$

$$\pi_{\mathfrak{c}}(\eta_9) = x_9(D - \mathfrak{c}) - \zeta_1\partial_{x_2} + \zeta_4\partial_{x_7} - \zeta_{10}\partial_{x_8} - \zeta_8\partial_{x_{13}} + \zeta_7\partial_{x_{14}}, \tag{11.7.13}$$

$$\pi_{\mathfrak{c}}(\eta_{10}) = x_{10}(D - \mathfrak{c}) - \zeta_2\partial_{x_2} - \zeta_4\partial_{x_4} + \zeta_{10}\partial_{x_6} + \zeta_8\partial_{x_{11}} - \zeta_6\partial_{x_{14}}, \tag{11.7.14}$$

$$\pi_{\mathfrak{c}}(\eta_{11}) = x_{11}(D - \mathfrak{c}) - \zeta_1\partial_{x_1} + \zeta_5\partial_{x_7} + \zeta_9\partial_{x_8} + \zeta_8\partial_{x_{10}} - \zeta_7\partial_{x_{12}}, \tag{11.7.15}$$

$$\pi_{\mathfrak{c}}(\eta_{12}) = x_{12}(D-\mathfrak{c}) - \zeta_3\partial_{x_2} + \zeta_4\partial_{x_3} - \zeta_{10}\partial_{x_5} - \zeta_7\partial_{x_{11}} + \zeta_6\partial_{x_{13}}, \quad (11.7.16)$$
$$\pi_{\mathfrak{c}}(\eta_{13}) = x_{13}(D-\mathfrak{c}) - \zeta_2\partial_{x_1} - \zeta_5\partial_{x_4} - \zeta_9\partial_{x_6} - \zeta_8\partial_{x_9} + \zeta_6\partial_{x_{12}}, \quad (11.7.17)$$
$$\pi_{\mathfrak{c}}(\eta_{14}) = x_{14}(D-\mathfrak{c}) - \zeta_3\partial_{x_1} + \zeta_5\partial_{x_3} + \zeta_9\partial_{x_5} + \zeta_7\partial_{x_9} - \zeta_6\partial_{x_{10}}, \quad (11.7.18)$$
$$\pi_{\mathfrak{c}}(\eta_{15}) = x_{15}(D-\mathfrak{c}) - \zeta_4\partial_{x_1} + \zeta_5\partial_{x_2} - \zeta_8\partial_{x_5} + \zeta_7\partial_{x_6} - \zeta_6\partial_{x_8}, \quad (11.7.19)$$
$$\pi_{\mathfrak{c}}(\eta_{16}) = x_{16}(D-\mathfrak{c}) - \zeta_{10}\partial_{x_1} - \zeta_9\partial_{x_2} - \zeta_8\partial_{x_3} + \zeta_7\partial_{x_4} - \zeta_6\partial_{x_7}. \quad (11.7.20)$$

Let $\vec{a} = (a_1, \ldots, a_{16}) \in \mathbb{F}^{16} \setminus \{\vec{0}\}$. Define

$$\vec{a}\cdot\vec{x} = \sum_{i=1}^{16} a_i x_i. \quad (11.7.21)$$

Recall $\mathscr{B} = \mathbb{F}[x_1, \ldots, x_{16}]$ and set

$$\mathscr{B}_{\vec{a}} = \{f e^{\vec{a}\cdot\vec{x}} \mid f \in \mathscr{B}\}. \quad (11.7.22)$$

Moreover, we write

$$\mathscr{F} = \{x_i\zeta_{r_1}\zeta_{r_2} \mid i \in \overline{1,16};\ (r_1, s_1), (r_2, s_2) \in \Upsilon_i,\ r_1 \neq r_2\}. \quad (11.7.23)$$

By Lemma 11.3.3, $|\mathscr{F}| = 160$. For any function $f(x_1, \ldots, x_{16})$, we define

$$f(\vec{b}) = f(b_1, \ldots, b_{16}) \qquad \text{for } \vec{b} = (b_1, \ldots, b_{16}) \in \mathbb{F}^{16}. \quad (11.7.24)$$

Now we define

$$\mathscr{V} = \{\vec{b} \in \mathbb{F}^{16} \setminus \{\vec{0}\} \mid f(\vec{b}) = 0 \text{ for } f \in \mathscr{F}\}, \quad (11.7.25)$$

which gives rise to a projective algebraic variety. The following is our main theorem in this section.

Theorem 11.7.1 *With respect to the representation $\pi_{\mathfrak{c}}$, $\mathscr{B}_{\vec{a}}$ forms an irreducible $\mathscr{G}^{E_6}$-module for any $\mathfrak{c} \in \mathbb{F}$ if $\vec{a} \notin \mathscr{V}$.*

Proof Let $\mathscr{B}_k$ be the subspace of homogeneous polynomials with degree k. Set

$$\mathscr{B}_{\vec{a},k} = \mathscr{B}_k e^{\vec{a}\cdot\vec{x}} \qquad \text{for } k \in \mathbb{N}. \quad (11.7.26)$$

Let M be a nonzero $\mathscr{G}^{E_6}$-submodule of $\mathscr{B}_{\vec{a}}$. Take any $0 \neq f e^{\vec{a}\cdot\vec{x}} \in M$ with $f \in \mathscr{B}$. By the second equation in (11.7.4),

$$(\xi_i - a_i)(f e^{\vec{a}\cdot\vec{x}}) = \partial_{x_i}(f) e^{\vec{a}\cdot\vec{x}} \in M \qquad \text{for } i \in \overline{1,16}. \quad (11.7.27)$$

Repeatedly applying (11.7.27) if necessarily, we obtain $e^{\vec{a}\cdot\vec{x}} \in M$; that is, $\mathscr{B}_{\vec{a},0} \subset M$.

Suppose $\mathscr{B}_{\vec{a},\ell} \subset M$ for some $\ell \in \mathbb{N}$. Let $g e^{\vec{a}\cdot\vec{x}}$ be any element in $\mathscr{B}_{\vec{a},\ell}$. Note that

$$\pi_{\mathfrak{c}}(\mathscr{G}^{E_6}) = \pi(o(10,\mathbb{F})) + \sum_{i=1}^{16}(\mathbb{F}\pi_{\mathfrak{c}}(\xi_i) + \mathbb{F}\pi_{\mathfrak{c}}(\eta_i)) + \mathbb{F}\pi_{\mathfrak{c}}(\hat{\alpha}) \tag{11.7.28}$$

by (11.3.20), (11.3.21), (11.4.2) and (11.7.3). Thus M is also an $o(10,\mathbb{F})$-module.

By Lemma 11.3.2, we can assume that $(x_1\zeta_{r_1}\zeta_{r_2})(\vec{a}) \neq 0$ for some $r_1, r_2 \in \{1,2,3,4,10\}$ with $r_1 \neq r_2$ by (11.3.111). Under the assumption, $a_1 \neq 0$.

Applying (11.3.116) and (11.3.117) to $ge^{\vec{a}\cdot\vec{x}}$, we get by (11.3.36)–(11.3.56) that

$$(-a_1x_2 + a_9x_{11} + a_{10}a_{13} + a_{12}x_{14})ge^{\vec{a}\cdot\vec{x}} \equiv 0 \pmod M, \tag{11.7.29}$$
$$(-a_1x_3 - a_6x_{11} - a_8x_{13} - a_{12}x_{15})ge^{\vec{a}\cdot\vec{x}} \equiv 0 \pmod M, \tag{11.7.30}$$
$$(a_1x_4 - a_5x_{11} + a_8x_{14} - a_{10}x_{15})ge^{\vec{a}\cdot\vec{x}} \equiv 0 \pmod M, \tag{11.7.31}$$
$$(-a_1x_5 + a_4x_{11} + a_7x_{13} + a_{12}x_{16})ge^{\vec{a}\cdot\vec{x}} \equiv 0 \pmod M, \tag{11.7.32}$$
$$(-a_1x_7 + a_5x_{13} + a_6x_{14} - a_9x_{15})ge^{\vec{a}\cdot\vec{x}} \equiv 0 \pmod M, \tag{11.7.33}$$
$$(a_1x_6 + a_3x_{11} - a_7x_{14} + a_{10}x_{16})ge^{\vec{a}\cdot\vec{x}} \equiv 0 \pmod M, \tag{11.7.34}$$
$$(-a_1x_8 - a_4x_{14} - a_3x_{13} + a_9x_{16})ge^{\vec{a}\cdot\vec{x}} \equiv 0 \pmod M, \tag{11.7.35}$$
$$(a_1x_9 - a_2x_{11} + a_7x_{15} - a_8x_{16})ge^{\vec{a}\cdot\vec{x}} \equiv 0 \pmod M, \tag{11.7.36}$$
$$(-a_1x_{10} + a_2x_{13} + a_4x_{15} - a_6x_{16})ge^{\vec{a}\cdot\vec{x}} \equiv 0 \pmod M, \tag{11.7.37}$$
$$(a_1x_{12} - a_2x_{14} + a_3x_{15} - a_5x_{16})ge^{\vec{a}\cdot\vec{x}} \equiv 0 \pmod M. \tag{11.7.38}$$

Multiplying a_1 to (11.3.119)–(11.3.123) and applying them to $ge^{\vec{a}\cdot\vec{x}}$ by (11.3.36)–(11.3.56), we obtain

$$a_1(a_7x_4 + a_8x_6 + a_{10}x_9 + a_{13}x_{11})ge^{\vec{a}\cdot\vec{x}} \equiv 0 \pmod M, \tag{11.7.39}$$
$$a_1(-a_7x_3 - a_8x_5 + a_{12}x_9 + a_{14}x_{11})ge^{\vec{a}\cdot\vec{x}} \equiv 0 \pmod M, \tag{11.7.40}$$
$$a_1(-a_7x_2 + a_{10}x_5 + a_{12}x_6 + a_{15}x_{11})ge^{\vec{a}\cdot\vec{x}} \equiv 0 \pmod M, \tag{11.7.41}$$
$$a_1(a_8x_2 + a_{10}x_3 + a_{12}x_4 + a_{16}x_{11})ge^{\vec{a}\cdot\vec{x}} \equiv 0 \pmod M, \tag{11.7.42}$$
$$a_1(a_4x_7 + a_6x_8 + a_9x_{10} + a_{11}x_{13})ge^{\vec{a}\cdot\vec{x}} \equiv 0 \pmod M, \tag{11.7.43}$$
$$a_1(a_4x_3 + a_6x_5 + a_{12}x_{10} + a_{14}x_{13})ge^{\vec{a}\cdot\vec{x}} \equiv 0 \pmod M, \tag{11.7.44}$$
$$a_1(a_4x_2 - a_9x_5 + a_{12}x_8 + a_{15}x_{13})ge^{\vec{a}\cdot\vec{x}} \equiv 0 \pmod M, \tag{11.7.45}$$
$$a_1(-a_6x_2 - a_9x_3 + a_{12}x_7 + a_{16}x_{13})ge^{\vec{a}\cdot\vec{x}} \equiv 0 \pmod M, \tag{11.7.46}$$
$$a_1(-a_3x_7 - a_5x_8 + a_9x_{12} + a_{11}x_{14})ge^{\vec{a}\cdot\vec{x}} \equiv 0 \pmod M, \tag{11.7.47}$$
$$a_1(a_3x_4 + a_5x_6 + a_{10}x_{12} + a_{13}x_{14})ge^{\vec{a}\cdot\vec{x}} \equiv 0 \pmod M, \tag{11.7.48}$$
$$a_1(-a_3x_2 - a_9x_6 - a_{10}x_8 + a_{15}x_{14})ge^{\vec{a}\cdot\vec{x}} \equiv 0 \pmod M, \tag{11.7.49}$$
$$a_1(a_5x_2 - a_9x_4 - a_{10}x_7 + a_{16}x_{14})ge^{\vec{a}\cdot\vec{x}} \equiv 0 \pmod M, \tag{11.7.50}$$
$$a_1(-a_2x_7 + a_5x_{10} + a_6x_{12} + a_{11}x_{15})ge^{\vec{a}\cdot\vec{x}} \equiv 0 \pmod M, \tag{11.7.51}$$
$$a_1(a_2x_4 - a_5x_9 + a_8x_{12} + a_{13}x_{15})ge^{\vec{a}\cdot\vec{x}} \equiv 0 \pmod M, \tag{11.7.52}$$
$$a_1(-a_2x_3 - a_6x_9 - a_8x_{10} + a_{14}x_{15})ge^{\vec{a}\cdot\vec{x}} \equiv 0 \pmod M, \tag{11.7.53}$$
$$a_1(-a_5x_3 - a_6x_4 - a_8x_7 + a_{16}x_{15})ge^{\vec{a}\cdot\vec{x}} \equiv 0 \pmod M, \tag{11.7.54}$$

$$a_1(a_2x_8+a_3x_{10}+a_4x_{12}+a_{11}x_{16})ge^{\vec{a}\cdot\vec{x}}\equiv 0 \pmod{M}, \tag{11.7.55}$$
$$a_1(-a_2x_6-a_3x_9+a_7x_{12}+a_{13}x_{16})ge^{\vec{a}\cdot\vec{x}}\equiv 0 \pmod{M}, \tag{11.7.56}$$
$$a_1(a_2x_5-a_4x_9-a_7x_{10}+a_{14}x_{16})ge^{\vec{a}\cdot\vec{x}}\equiv 0 \pmod{M}, \tag{11.7.57}$$
$$a_1(-a_3x_5-a_4x_6-a_7x_8+a_{15}x_{16})ge^{\vec{a}\cdot\vec{x}}\equiv 0 \pmod{M}. \tag{11.7.58}$$

According to (11.7.31), (11.7.34) and (11.7.36),

$$a_1x_4ge^{\vec{a}\cdot\vec{x}}\equiv (a_5x_{11}-a_8x_{14}+a_{10}x_{15})ge^{\vec{a}\cdot\vec{x}} \pmod{M}, \tag{11.7.59}$$
$$a_1x_6ge^{\vec{a}\cdot\vec{x}}\equiv (-a_3x_{11}+a_7x_{14}-a_{10}x_{16})ge^{\vec{a}\cdot\vec{x}} \pmod{M}, \tag{11.7.60}$$
$$a_1x_9ge^{\vec{a}\cdot\vec{x}}\equiv (a_2x_{11}-a_7x_{15}+a_8x_{16})ge^{\vec{a}\cdot\vec{x}} \pmod{M}. \tag{11.7.61}$$

Substituting them into (11.7.39), we get

$$\begin{aligned}
&[a_7(a_5x_{11}-a_8x_{14}+a_{10}x_{15})+a_8(-a_3x_{11}+a_7x_{14}-a_{10}x_{16})\\
&+a_{10}(a_2x_{11}-a_7x_{15}+a_8x_{16})+a_1a_{13}x_{11}]ge^{\vec{a}\cdot\vec{x}}\\
&=[a_1a_{13}+a_2a_{10}-a_3a_8+a_5a_7]x_{11}ge^{\vec{a}\cdot\vec{x}}\\
&=\zeta_2(\vec{a})x_{11}ge^{\vec{a}\cdot\vec{x}}\equiv 0 \pmod{\mathscr{M}}
\end{aligned} \tag{11.7.62}$$

by (11.3.65) and (11.7.24).

Similarly we substitute (11.7.29)–(11.7.38) into (11.7.40)–(11.7.58), and get by (11.3.59), (11.3.65)–(11.3.67) and (11.3.69) that

$$\zeta_i(\vec{a})x_{11}ge^{\vec{a}\cdot\vec{x}}\equiv 0 \pmod{\mathscr{M}} \quad \text{for } i=3,4,10; \tag{11.7.63}$$
$$\zeta_i(\vec{a})x_{13}ge^{\vec{a}\cdot\vec{x}}\equiv 0 \pmod{\mathscr{M}} \quad \text{for } i=1,3,4,10; \tag{11.7.64}$$
$$\zeta_i(\vec{a})x_{14}ge^{\vec{a}\cdot\vec{x}}\equiv 0 \pmod{\mathscr{M}} \quad \text{for } i=1,2,4,10; \tag{11.7.65}$$
$$\zeta_i(\vec{a})x_{15}ge^{\vec{a}\cdot\vec{x}}\equiv 0 \pmod{\mathscr{M}} \quad \text{for } i=1,2,3,10; \tag{11.7.66}$$
$$\zeta_i(\vec{a})x_{16}ge^{\vec{a}\cdot\vec{x}}\equiv 0 \pmod{\mathscr{M}} \quad \text{for } i=1,2,3,4. \tag{11.7.67}$$

Since $\zeta_{r_1}(\vec{a})\zeta_{r_2}(\vec{a})\neq 0$ for some $r_1,r_2\in\{1,3,4,5,10\}$ with $r_1\neq r_2$, (11.7.62)–(11.7.67) imply

$$x_ige^{\vec{a}\cdot\vec{x}}\in\mathscr{M} \quad \text{for } i=11,13,14,15,16. \tag{11.7.68}$$

Substituting (11.7.68) into (11.7.29)–(11.7.38), we obtain

$$x_ige^{\vec{a}\cdot\vec{x}}\in\mathscr{M} \quad \text{for } i\in\overline{2,16}. \tag{11.7.69}$$

By (11.3.57) and Table 11.3,

$$A_{1,1}(ge^{\vec{a}\cdot\vec{x}})\equiv a_1x_1ge^{\vec{a}\cdot\vec{x}}/2\equiv 0 \pmod{M}. \tag{11.7.70}$$

So $x_1ge^{\vec{a}\cdot\vec{x}}\in M$. Hence $\mathscr{B}_{\vec{a},\ell+1}\subset M$. By induction,

$$\mathscr{B}_{\vec{a},k} \subset M \qquad \text{for } k \in \mathbb{N}. \tag{11.7.71}$$

Therefore, $M = \sum_{k=0}^{\infty} \mathscr{B}_{\vec{a},k} = \mathscr{B}_{\vec{a}}$. So $\mathscr{B}_{\vec{a}}$ forms an irreducible $\mathscr{G}^{E_6}$-module. □

References

[D] L. Dickson, A class of groups in an arbitrary realm connected with the configuration of the 27 lines on a cubic surface. J. Math. **33**, 145–123 (1901)

[A] J. Adams, *Lectures Except. Lie Groups* (The University of Chicago Press Ltd., London, 1996)

[BK] R. Brylinski, B. Kostant, Minimal representations of E_6, E_7, and E_8 and the generalized Capelli identity. Proc. Nar. Acad. Sci. U.S.A. **91**(7), 2469–2472 (1994)

[Gd] D. Ginzburg, On standard L-functions for E_6 and E_7. J. Reine Angew. Math. **465**, 101–131 (1995)

[I] A. Iltyakov, On rational invariants of the group E_6. Proc. Math. Soc. **124**(12), 3637–3640 (1996)

[SW] K. Suzuki, M. Wakui, On the Turaev-Viro-Ocneanu invariant of 3-manifolds derived from the E_6-subfactor. Kyushu J. Math. **56**, 59–81 (2002)

[OM] Y. Okamoto, R. Marshak, A garnd unification preon model with E_6 metacolor. Nuclear Phys. B **268**(2), 397–405 (1986)

[HH] J.E.M. Homos, Y.M.M. Homos, Algebraic model for the evolution of the generic code. Phys. Rev. Lett. **71**, 4401–4404 (1991)

[Wa] X. Wang, Identification of Gepner's model with twisted LG model and E_6 singlets. Modern Phy. Lett. A **6**(23), 2155–2162 (1991)

[MPW] I. Morrison, P. Pieruschka, B. Wybourne, The interacting boson model with the exceptional groups G_2 and E_6. J. Math. Phys. **32**(2), 356–372 (1991)

[HaMt] N. Haba, T. Matsuoka, Large lepton flavor mixing and E_6-type unification models. Progr. Theoret. Phys. **99**(5), 831–842 (1998)

[GSA] A. Ghezelbash, A. Shafiekhani, M. Abolbasani, On the Picard-Fuchs equations of $N = 2$ supersymmetric E_6 Yang-Mills theory. Modern Phys. Lett. A **13**(7), 527–531 (1998)

[HK] R. Howl, S. King, Minimal E_6 supersymmetric standard model. J. High Ener. Phys. **01**, 030 (2008), 31 p

[DL] C. Das, L. Laperashvili, Preon model and family replicated E_6 unification. SIGMA **4**, 012 (2008), 15 p

[X17] X. Xu, A cubic E_6-generalization of the classical theorem on harmonic polynomials. J. Lie Theory **21**, 145–164 (2011)

[Sg] G. Shen, Graded modules of graded Lie algebras of Cartan type (I)-mixed product of modules. Sci. China A **29**, 570–581 (1986)

[X26] X. Xu, A new functor from D_5-**Mod** to E_6-**Mod**. Acta Math Sin. (Engl. Ser.) **32**(8), 867–892 (2016)

[Kb] B. Kostant, On the tensor product of a finite and an infinite dimensional representation. J. Fund. Anal. **20**, 257–285 (1975)

Chapter 12
Representations of E_7

The E_7 Lie algebra and group are important mathematical objects with broad applications. They are second most complicated finite-dimensional simple Lie algebra and group. The complexity indeed imply rich connotation. Ramond [R2] (1977) gave a group theoretical analysis of a symmetry breaking affected by Higgs fields for the vector-like unified theory based on E_7. Kato and Kitazawa [KaKy] (1989) studied the correlation functions in an E_7-type modular-invariant Wess–Zumino–Witten theory, which is related to the scheme of string compactification proposed by D'Auria, Ferrara and Trigiante [DFT] (2006). In [DFT], they considered M-theory compactified on seven-torus with fluxes when all the seven anti-symmetric tensor fields in four dimensions have been dualized into scalars and thus the $E_{7(7)}$ symmetry was recovered. Brink, Kim and Ramond [BKR] (2008) used the Cremmer–Julia $E_{7(7)}$ nonlinear symmetry of $N = 8$ supergravity to derive its order κ on-shell Hamiltonian in terms of the chiral light-cone superfield.

Brown [Br] (1968) constructed a central simple nonassociative algebraic structure on the Freudenthal triple system of E_7 and obtained a new interpretation of the E_7 Lie algebra. Faulkner [F] (1972) studied the geometry of the planes of the octave symplectic geometry of the ternary algebra related to the minimal representation of E_7. Ginzburg [Gd] (1995) showed that the twisted partial L-function on the 56-dimensional representation of $GE_7(\mathbb{C})$ is entire except the points 0 and 1. Cooperstein [C] (1995) proved that Cartan's quartic E_7-invariant is the unique fundamental invariant over the basic module. Plotkin [P] (1998) investigated the stability of the K_1-functor for the E_7 group. The E_7 root system was used by Sekiguchi [Se] (1999) to study the configurations of seven lines on the real projective plane. Garibaidi [Gr] (2001) gave explicit descriptions of the homogeneous projective varieties associated with the E_7 group with trivial Tits algebras. Weiss [W] (2006) solved a fundamental question about the structure of the automorphism group of the Moufang quadrangles of type E_7. Presented in this chapter are natural E_7 symmetries.

X. Xu, *Representations of Lie Algebras and Partial Differential Equations*,
DOI 10.1007/978-981-10-6391-6_12

By solving certain partial differential equations, we find the explicit decomposition of the polynomial algebra over the 56-dimensional basic irreducible module of the simple Lie algebra E_7 into a direct sum of irreducible submodules, which was due to our work [X22]. Weyl's Theorem of complete reducibility in Sect. 2.4 plays an important role. Moreover, we find a new representation of the simple Lie algebra of type E_7 on the polynomial algebra in 27 variables, which gives a fractional representation of the corresponding Lie group on 27-dimensional space. Using this representation and Shen's idea of mixed product (cf. [Sg]), we construct a new functor from the category of E_6-modules to the category of E_7-modules. A condition for the functor to map a finite-dimensional irreducible E_6-module to an infinite-dimensional irreducible E_7-module is obtained by Theorem 5.4.3 of decomposing the tensor module. Our general frame also gives a direct polynomial extension from irreducible E_6-modules to irreducible E_7-modules, which can be used to derive Gel'fand–Zetlin bases for E_7 from those for E_6 that can be obtained from those for D_5 in last chapter. Our results also yield explicit constructions of certain infinite-dimensional irreducible weight E_7-modules of finite-dimensional weight subspaces. In our approach, the idea of Kostant's characteristic identities and the well-known Dickson's E_6-invariant trilinear form play key roles. These results were due to our work [X23]. It is the fifth sequential work of constructing functors among the module categories of different simple Lie algebras after Sects. 6.7, 7.3, 8.3 and 11.3–11.6.

In the above work, we found a one-parameter ($\mathfrak{c}$) family of inhomogeneous first-order differential-operator representations of the simple Lie algebra of type E_7 in 27 variables. Letting these operators act on the space of exponential-polynomial functions that depend on a parametric vector $\vec{a} \in \mathbb{F}^{27} \setminus \{\vec{0}\}$, we prove that the space forms an irreducible E_7-module for any constant $\mathfrak{c}$ if $\vec{a}$ is not on an explicitly given projective algebraic variety. Certain equivalent combinatorial properties of the basic oscillator representation of E_6 play key roles in our proof. This part is taken from our work [X27].

In Sect. 12.1, we construct the basic oscillator representation of E_7 via the lattice construction of the simple Lie algebra E_8 included in (4.4.15)–(4.4.25). Section 12.2 is the constructions of certain singular vectors in the above representation. In Sect. 12.3, we decompose the oscillator representation of E_7 into a direct sum of irreducible ones. In Sect. 12.4, we give an inhomogeneous first-order differential-operator representation of E_7. A functor from the category of E_6-modules to the category of E_7-modules is constructed in Sect. 12.5. Moreover, a condition for the functor to map a finite-dimensional irreducible E_6-module to an infinite-dimensional irreducible E_7-module is obtained in Sect. 12.6. As a preparation for the later exponential-polynomial representation of E_7, we study in Sect. 12.7 the combinatorial properties of the oscillator representation of E_6. Finally in Sect. 12.8, we prove that the irreducibility of certain modules of E_7 is related to an algebraic variety. It is the fourth sequential work of connecting the irreducibility and algebraic varieties after Sects. 7.4, 8.4 and 10.7.

12.1 Basic Oscillator Representation of E_7

In this section, we construct the oscillator representation of E_7 lifted from its 27-dimensional basic irreducible representation.

First we go back to the construction of the simple Lie algebra $\mathscr{G}^X$ in (4.4.15)–(4.4.25) with $X = E_6,\ E_7$ and E_8. Recall the Dynkin diagram of E_8:

E_8:

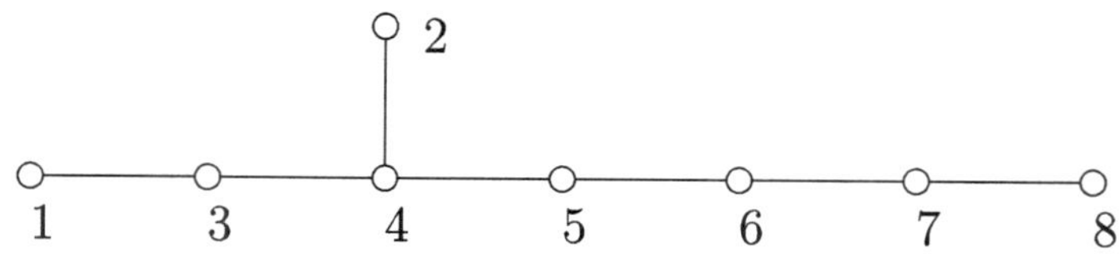

Let $\{\alpha_i \mid i \in \overline{1,8}\}$ be the simple positive roots corresponding to the vertices in the diagram, and let Φ_{E_8} be the root system of E_8. The simple Lie algebra of type E_8 is

$$\mathscr{G}^{E_8} = H \oplus \bigoplus_{\alpha\in\Phi_{E_8}} \mathbb{F}E_\alpha, \qquad H = H_{E_8} = \sum_{i=1}^{8} \mathbb{F}\alpha_i, \tag{12.1.1}$$

with the Lie bracket given in (4.4.24) and (4.4.25). Note that the Dynkin diagram of E_7 is a sub-diagram of that of E_8. Set

$$H_{E_7} = \sum_{i=1}^{7} \mathbb{F}\alpha_i, \qquad \Phi_{E_7} = \Phi_{E_8} \bigcap H_{E_7}. \tag{12.1.2}$$

We take the simple Lie algebra $\mathscr{G}^{E_7}$ of type E_7 as the Lie subalgebra

$$\mathscr{G}^{E_7} = H_{E_7} \oplus \bigoplus_{\alpha\in\Phi_{E_7}} \mathbb{F}E_\alpha. \tag{12.1.3}$$

Recall the notion in (4.4.43). Denote

$$\varrho_1 = E_{(2,3,4,6,5,4,3,1)}, \quad \varrho_2 = E_{(2,3,4,6,5,4,2,1)}, \quad \varrho_3 = E_{(2,3,4,6,5,3,2,1)}, \tag{12.1.4}$$

$$\varrho_4 = E_{(2,3,4,6,4,3,2,1)}, \quad \varrho_5 = E_{(2,3,4,5,4,3,2,1)}, \quad \varrho_6 = E_{(2,3,3,5,4,3,2,1)}, \tag{12.1.5}$$

$$\varrho_7 = E_{(2,2,4,5,4,3,2,1)}, \quad \varrho_8 = E_{(1,3,3,5,4,3,2,1)}, \quad \varrho_9 = E_{(2,2,3,5,4,3,2,1)}, \tag{12.1.6}$$

$$\varrho_{10} = E_{(2,2,3,4,4,3,2,1)}, \quad \varrho_{11} = E_{(1,2,3,5,4,3,2,1)}, \quad \varrho_{12} = E_{(2,2,3,4,3,3,2,1)}, \tag{12.1.7}$$

$$\varrho_{13} = E_{(1,2,3,4,4,3,2,1)}, \quad \varrho_{14} = E_{(2,2,3,4,3,2,2,1)}, \quad \varrho_{15} = E_{(1,2,2,4,4,3,2,1)}, \tag{12.1.8}$$

$$\varrho_{16} = E_{(1,2,3,4,3,3,2,1)}, \quad \varrho_{17} = E_{(2,2,3,4,3,2,1,1)}, \quad \varrho_{18} = E_{(1,2,2,4,3,3,2,1)}, \tag{12.1.9}$$

$$\varrho_{19} = E_{(1,2,3,4,3,2,2,1)}, \quad \varrho_{20} = E_{(1,2,2,3,3,3,2,1)}, \quad \varrho_{21} = E_{(1,2,2,4,3,2,2,1)}, \tag{12.1.10}$$

$$\varrho_{22} = E_{(1,2,3,4,3,2,1,1)}, \quad \varrho_{23} = E_{(1,1,2,3,3,3,2,1)}, \quad \varrho_{24} = E_{(1,2,2,3,3,2,2,1)}, \tag{12.1.11}$$

$$\varrho_{25} = E_{(1,2,2,4,3,2,1,1)}, \quad \varrho_{26} = E_{(1,1,2,3,3,2,2,1)}, \quad \varrho_{27} = E_{(1,2,2,3,2,2,2,1)}, \tag{12.1.12}$$

$$\varrho_{28} = E_{(1,2,2,3,3,2,1,1)}, \quad \varrho_{29} = E_{(1,1,2,3,2,2,2,1)}, \quad \varrho_{30} = E_{(1,1,2,3,3,2,1,1)}, \tag{12.1.13}$$

$$\varrho_{31} = E_{(1,2,2,3,2,2,1,1)}, \quad \varrho_{32} = E_{(1,1,2,2,2,2,2,1)}, \quad \varrho_{33} = E_{(1,1,2,3,2,2,1,1)}, \tag{12.1.14}$$

$$\varrho_{34} = E_{(1,2,2,3,2,1,1,1)}, \quad \varrho_{35} = E_{(1,1,1,2,2,2,2,1)}, \quad \varrho_{36} = E_{(1,1,2,2,2,2,1,1)}, \tag{12.1.15}$$

$$\varrho_{37} = E_{(1,1,2,3,2,1,1,1)}, \quad \varrho_{38} = E_{(1,1,1,2,2,2,1,1)}, \quad \varrho_{39} = E_{(1,1,2,2,2,1,1,1)}, \tag{12.1.16}$$

$$\varrho_{40} = E_{(0,1,1,2,2,2,2,1)}, \quad \varrho_{41} = E_{(1,1,1,2,2,1,1,1)}, \quad \varrho_{42} = E_{(1,1,2,2,1,1,1,1)}, \tag{12.1.17}$$

$$\varrho_{43} = E_{(0,1,1,2,2,2,1,1)}, \quad \varrho_{44} = E_{(1,1,1,2,1,1,1,1)}, \quad \varrho_{45} = E_{(0,1,1,2,2,1,1,1)}, \tag{12.1.18}$$

$$\varrho_{46} = E_{(1,1,1,1,1,1,1,1)}, \quad \varrho_{47} = E_{(0,1,1,2,1,1,1,1)}, \quad \varrho_{48} = E_{(0,1,1,1,1,1,1,1)}, \tag{12.1.19}$$

$$\varrho_{49} = E_{(1,0,1,1,1,1,1,1)}, \quad \varrho_{50} = E_{(0,1,0,1,1,1,1,1)}, \quad \varrho_{51} = E_{(0,0,1,1,1,1,1,1)}, \tag{12.1.20}$$

$$\varrho_{52} = E_{(0,0,0,1,1,1,1,1)}, \quad \varrho_{53} = E_{(0,0,0,0,1,1,1,1)}, \quad \varrho_{54} = E_{(0,0,0,0,0,1,1,1)}, \tag{12.1.21}$$

$$\varrho_{55} = E_{(0,0,0,0,0,0,1,1)}, \quad \varrho_{56} = E_{(0,0,0,0,0,0,0,1)}. \tag{12.1.22}$$

Then the subspace

$$V = \sum_{i=1}^{56} \mathbb{F}\varrho_i \tag{12.1.23}$$

forms an irreducible $\mathcal{G}^{E_7}$-module with respect to the adjoint representation of $\mathcal{G}^{E_8}$, ϱ_1 is a highest-weight vector of weight λ_7. Set

$$\theta = \alpha_{(2,3,4,6,5,4,3,2)} \tag{12.1.24}$$

(cf. (4.4.42)). Then

$$[E_\theta, \mathcal{G}^{E_7}] = \{0\} \tag{12.1.25}$$

and

$$\sum_{\alpha\in\Phi_{E_8}^+} \mathbb{C}E_\alpha = \sum_{\beta\in\Phi_{E_7}^+} \mathbb{C}E_\beta + V + \mathbb{C}E_\theta. \tag{12.1.26}$$

Write

$$[u, \varrho_i] = \sum_{j=1}^{56} \varphi_{i,j}(u)\varrho_j \qquad \text{for } u \in \mathcal{G}^{F_4}. \tag{12.1.27}$$

Set

$$\mathcal{B} = \mathbb{F}[x_1, \ldots, x_{56}] \tag{12.1.28}$$

and define the *basic oscillator representation of* $\mathcal{G}^{E_7}$ on $\mathcal{A}$ by

$$u(g) = \sum_{i,j=1}^{56} \varphi_{i,j}(u)x_j\partial_{x_i}(g) \qquad \text{for } u \in \mathcal{G}^{E_7},\ g \in \mathcal{B} \tag{12.1.29}$$

(cf. (2.2.17)–(2.2.20)). Then $\mathcal{B}$ forms a $\mathcal{G}^{E_7}$-module isomorphic to the symmetric tensor $S(V)$ over V. More explicitly, we have the following representation formulas for the positive root vectors in $\mathcal{G}^{E_7}$ by (4.4.54)–(4.4.58):

$$\begin{aligned} E_{\alpha_1}|_{\mathcal{B}} = {} & -x_6\partial_{x_8} - x_9\partial_{x_{11}} - x_{10}\partial_{x_{13}} - x_{12}\partial_{x_{16}} - x_{14}\partial_{x_{19}} - x_{17}\partial_{x_{22}} \\ & +x_{35}\partial_{x_{40}} + x_{38}\partial_{x_{43}} + x_{41}\partial_{x_{45}} + x_{44}\partial_{x_{47}} + x_{46}\partial_{x_{48}} + x_{49}\partial_{x_{51}}, \end{aligned} \tag{12.1.30}$$

$$\begin{aligned} E_{\alpha_2}|_{\mathcal{B}} = {} & x_5\partial_{x_7} + x_6\partial_{x_9} + x_8\partial_{x_{11}} - x_{20}\partial_{x_{23}} - x_{24}\partial_{x_{26}} - x_{27}\partial_{x_{29}} \\ & -x_{28}\partial_{x_{30}} - x_{31}\partial_{x_{33}} - x_{34}\partial_{x_{37}} + x_{46}\partial_{x_{49}} + x_{48}\partial_{x_{51}} + x_{50}\partial_{x_{52}}, \end{aligned} \tag{12.1.31}$$

$$\begin{aligned} E_{\alpha_3}|_{\mathcal{B}} = {} & -x_5\partial_{x_6} - x_7\partial_{x_9} - x_{13}\partial_{x_{15}} - x_{16}\partial_{x_{18}} - x_{19}\partial_{x_{21}} - x_{22}\partial_{x_{25}} \\ & +x_{32}\partial_{x_{35}} + x_{36}\partial_{x_{38}} + x_{39}\partial_{x_{41}} + x_{42}\partial_{x_{44}} + x_{48}\partial_{x_{50}} + x_{51}\partial_{x_{52}}, \end{aligned} \tag{12.1.32}$$

$$\begin{aligned} E_{\alpha_4}|_{\mathcal{B}} = {} & x_4\partial_{x_5} - x_9\partial_{x_{10}} - x_{11}\partial_{x_{13}} - x_{18}\partial_{x_{20}} - x_{21}\partial_{x_{24}} - x_{25}\partial_{x_{28}} \\ & -x_{29}\partial_{x_{32}} - x_{33}\partial_{x_{36}} - x_{37}\partial_{x_{39}} - x_{44}\partial_{x_{46}} - x_{47}\partial_{x_{48}} + x_{52}\partial_{x_{53}}, \end{aligned} \tag{12.1.33}$$

$$\begin{aligned} E_{\alpha_5}|_{\mathcal{B}} = {} & x_3\partial_{x_4} - x_{10}\partial_{x_{12}} - x_{13}\partial_{x_{16}} - x_{15}\partial_{x_{18}} - x_{24}\partial_{x_{27}} - x_{26}\partial_{x_{29}} \\ & -x_{28}\partial_{x_{31}} - x_{30}\partial_{x_{33}} - x_{39}\partial_{x_{42}} - x_{41}\partial_{x_{44}} - x_{45}\partial_{x_{47}} + x_{53}\partial_{x_{54}}, \end{aligned} \tag{12.1.34}$$

$$\begin{aligned} E_{\alpha_6}|_{\mathcal{B}} = {} & x_2\partial_{x_3} - x_{12}\partial_{x_{14}} - x_{16}\partial_{x_{19}} - x_{18}\partial_{x_{21}} - x_{20}\partial_{x_{24}} - x_{23}\partial_{x_{26}} \\ & -x_{31}\partial_{x_{34}} - x_{33}\partial_{x_{37}} - x_{36}\partial_{x_{39}} - x_{38}\partial_{x_{41}} - x_{43}\partial_{x_{45}} + x_{54}\partial_{x_{55}}, \end{aligned} \tag{12.1.35}$$

$$\begin{aligned} E_{\alpha_7}|_{\mathcal{B}} = {} & x_1\partial_{x_2} - x_{14}\partial_{x_{17}} - x_{19}\partial_{x_{22}} - x_{21}\partial_{x_{25}} - x_{24}\partial_{x_{28}} - x_{26}\partial_{x_{30}} \\ & -x_{27}\partial_{x_{31}} - x_{29}\partial_{x_{33}} - x_{32}\partial_{x_{36}} - x_{35}\partial_{x_{38}} - x_{40}\partial_{x_{43}} + x_{55}\partial_{x_{56}}, \end{aligned} \tag{12.1.36}$$

$$E_{(1,0,1)}|_{\mathscr{B}} = -x_5\partial_{x_8} - x_7\partial_{x_{11}} + x_{10}\partial_{x_{15}} + x_{12}\partial_{x_{18}} + x_{14}\partial_{x_{21}} + x_{17}\partial_{x_{25}} -x_{32}\partial_{x_{40}} - x_{36}\partial_{x_{43}} - x_{39}\partial_{x_{45}} - x_{42}\partial_{x_{47}} + x_{46}\partial_{x_{50}} + x_{49}\partial_{x_{52}}, \tag{12.1.37}$$

$$E_{(0,1,0,1)}|_{\mathscr{B}} = -x_4\partial_{x_7} - x_6\partial_{x_{10}} - x_8\partial_{x_{13}} - x_{18}\partial_{x_{23}} - x_{21}\partial_{x_{26}} - x_{25}\partial_{x_{30}} +x_{27}\partial_{x_{32}} + x_{31}\partial_{x_{36}} + x_{34}\partial_{x_{39}} + x_{44}\partial_{x_{49}} + x_{47}\partial_{x_{51}} + x_{50}\partial_{x_{53}}, \tag{12.1.38}$$

$$E_{(0,0,1,1)}|_{\mathscr{B}} = x_4\partial_{x_6} + x_7\partial_{x_{10}} - x_{11}\partial_{x_{15}} + x_{16}\partial_{x_{20}} + x_{19}\partial_{x_{24}} + x_{22}\partial_{x_{28}} +x_{29}\partial_{x_{35}} + x_{33}\partial_{x_{38}} + x_{37}\partial_{x_{41}} - x_{42}\partial_{x_{46}} + x_{47}\partial_{x_{50}} + x_{51}\partial_{x_{53}}, \tag{12.1.39}$$

$$E_{(0,0,0,1,1)}|_{\mathscr{B}} = -x_3\partial_{x_5} + x_9\partial_{x_{12}} + x_{11}\partial_{x_{16}} - x_{15}\partial_{x_{20}} + x_{21}\partial_{x_{27}} + x_{25}\partial_{x_{31}} -x_{26}\partial_{x_{32}} - x_{30}\partial_{x_{36}} + x_{37}\partial_{x_{42}} - x_{41}\partial_{x_{46}} - x_{45}\partial_{x_{48}} + x_{52}\partial_{x_{54}}, \tag{12.1.40}$$

$$E_{(0,0,0,0,1,1)}|_{\mathscr{B}} = -x_2\partial_{x_4} + x_{10}\partial_{x_{14}} + x_{13}\partial_{x_{19}} + x_{15}\partial_{x_{21}} - x_{20}\partial_{x_{27}} - x_{23}\partial_{x_{29}} +x_{28}\partial_{x_{34}} + x_{30}\partial_{x_{37}} - x_{36}\partial_{x_{42}} - x_{38}\partial_{x_{44}} - x_{43}\partial_{x_{47}} + x_{53}\partial_{x_{55}}, \tag{12.1.41}$$

$$E_{(0,0,0,0,0,1,1)}|_{\mathscr{B}} = -x_1\partial_{x_3} + x_{12}\partial_{x_{17}} + x_{16}\partial_{x_{22}} + x_{18}\partial_{x_{25}} + x_{20}\partial_{x_{28}} + x_{23}\partial_{x_{30}} -x_{27}\partial_{x_{34}} - x_{29}\partial_{x_{37}} - x_{32}\partial_{x_{39}} - x_{35}\partial_{x_{41}} - x_{40}\partial_{x_{45}} + x_{54}\partial_{x_{56}}, \tag{12.1.42}$$

$$E_{(1,0,1,1)}|_{\mathscr{B}} = x_4\partial_{x_8} + x_7\partial_{x_{13}} + x_9\partial_{x_{15}} - x_{12}\partial_{x_{20}} - x_{14}\partial_{x_{24}} - x_{17}\partial_{x_{28}} -x_{29}\partial_{x_{40}} - x_{33}\partial_{x_{43}} - x_{37}\partial_{x_{45}} + x_{42}\partial_{x_{48}} + x_{44}\partial_{x_{50}} + x_{49}\partial_{x_{53}}, \tag{12.1.43}$$

$$E_{(0,1,1,1)}|_{\mathscr{B}} = -x_4\partial_{x_9} + x_5\partial_{x_{10}} - x_8\partial_{x_{15}} + x_{16}\partial_{x_{23}} + x_{19}\partial_{x_{26}} + x_{22}\partial_{x_{30}} -x_{27}\partial_{x_{35}} - x_{31}\partial_{x_{38}} - x_{34}\partial_{x_{41}} + x_{42}\partial_{x_{49}} - x_{47}\partial_{x_{52}} + x_{48}\partial_{x_{53}}, \tag{12.1.44}$$

$$E_{(0,1,0,1,1)}|_{\mathscr{B}} = x_3\partial_{x_7} + x_6\partial_{x_{12}} + x_8\partial_{x_{16}} - x_{15}\partial_{x_{23}} + x_{21}\partial_{x_{29}} + x_{24}\partial_{x_{32}} +x_{25}\partial_{x_{33}} + x_{28}\partial_{x_{36}} - x_{34}\partial_{x_{42}} + x_{41}\partial_{x_{49}} + x_{45}\partial_{x_{51}} + x_{50}\partial_{x_{54}}, \tag{12.1.45}$$

$$E_{(0,0,1,1,1)}|_{\mathscr{B}} = -x_3\partial_{x_6} - x_7\partial_{x_{12}} + x_{11}\partial_{x_{18}} + x_{13}\partial_{x_{20}} - x_{19}\partial_{x_{27}} - x_{22}\partial_{x_{31}} +x_{26}\partial_{x_{35}} + x_{30}\partial_{x_{38}} - x_{37}\partial_{x_{44}} - x_{39}\partial_{x_{46}} + x_{45}\partial_{x_{50}} + x_{51}\partial_{x_{54}}, \tag{12.1.46}$$

$$E_{(0,0,0,1,1,1)}|_{\mathscr{B}} = x_2\partial_{x_5} - x_9\partial_{x_{14}} - x_{11}\partial_{x_{19}} + x_{15}\partial_{x_{24}} + x_{18}\partial_{x_{28}} - x_{23}\partial_{x_{32}} -x_{25}\partial_{x_{34}} + x_{30}\partial_{x_{39}} + x_{33}\partial_{x_{42}} - x_{38}\partial_{x_{46}} - x_{43}\partial_{x_{48}} + x_{52}\partial_{x_{55}}, \tag{12.1.47}$$

$$E_{(0,0,0,0,1,1,1)}|_{\mathscr{B}} = x_1\partial_{x_4} - x_{10}\partial_{x_{17}} - x_{13}\partial_{x_{22}} - x_{15}\partial_{x_{25}} + x_{20}\partial_{x_{31}} + x_{23}\partial_{x_{33}} +x_{24}\partial_{x_{34}} + x_{26}\partial_{x_{37}} - x_{32}\partial_{x_{42}} - x_{35}\partial_{x_{44}} - x_{40}\partial_{x_{47}} + x_{53}\partial_{x_{56}}, \tag{12.1.48}$$

$$E_{(1,1,1,1)}|_{\mathscr{B}} = -x_4\partial_{x_{11}} + x_5\partial_{x_{13}} + x_6\partial_{x_{15}} - x_{12}\partial_{x_{23}} - x_{14}\partial_{x_{26}} - x_{17}\partial_{x_{30}} + x_{27}\partial_{x_{40}} + x_{31}\partial_{x_{43}} + x_{34}\partial_{x_{45}} - x_{42}\partial_{x_{51}} - x_{44}\partial_{x_{52}} + x_{46}\partial_{x_{53}}, \tag{12.1.49}$$

$$E_{(1,0,1,1,1)}|_{\mathscr{B}} = -x_3\partial_{x_8} - x_7\partial_{x_{16}} - x_9\partial_{x_{18}} - x_{10}\partial_{x_{20}} + x_{14}\partial_{x_{27}} + x_{17}\partial_{x_{31}} - x_{26}\partial_{x_{40}} - x_{30}\partial_{x_{43}} + x_{37}\partial_{x_{47}} + x_{39}\partial_{x_{48}} + x_{41}\partial_{x_{50}} + x_{49}\partial_{x_{54}}, \tag{12.1.50}$$

$$E_{(0,1,1,1,1)}|_{\mathscr{B}} = x_3\partial_{x_9} - x_5\partial_{x_{12}} + x_8\partial_{x_{18}} + x_{13}\partial_{x_{23}} - x_{19}\partial_{x_{29}} - x_{22}\partial_{x_{33}} - x_{24}\partial_{x_{35}} - x_{28}\partial_{x_{38}} + x_{34}\partial_{x_{44}} + x_{39}\partial_{x_{49}} - x_{45}\partial_{x_{52}} + x_{48}\partial_{x_{54}}, \tag{12.1.51}$$

$$E_{(0,1,0,1,1,1)}|_{\mathscr{B}} = -x_2\partial_{x_7} - x_6\partial_{x_{14}} - x_8\partial_{x_{19}} + x_{15}\partial_{x_{26}} + x_{18}\partial_{x_{30}} + x_{20}\partial_{x_{32}} - x_{25}\partial_{x_{37}} - x_{28}\partial_{x_{39}} - x_{31}\partial_{x_{42}} + x_{38}\partial_{x_{49}} + x_{43}\partial_{x_{51}} + x_{50}\partial_{x_{55}}, \tag{12.1.52}$$

$$E_{(0,0,1,1,1,1)}|_{\mathscr{B}} = x_2\partial_{x_6} + x_7\partial_{x_{14}} - x_{11}\partial_{x_{21}} - x_{13}\partial_{x_{24}} - x_{16}\partial_{x_{28}} + x_{23}\partial_{x_{35}} + x_{22}\partial_{x_{34}} - x_{30}\partial_{x_{41}} - x_{33}\partial_{x_{44}} - x_{36}\partial_{x_{46}} + x_{43}\partial_{x_{50}} + x_{51}\partial_{x_{55}}, \tag{12.1.53}$$

$$E_{(0,0,0,1,1,1,1)}|_{\mathscr{B}} = -x_1\partial_{x_5} + x_9\partial_{x_{17}} + x_{11}\partial_{x_{22}} - x_{15}\partial_{x_{28}} - x_{18}\partial_{x_{31}} - x_{21}\partial_{x_{34}} + x_{23}\partial_{x_{36}} + x_{26}\partial_{x_{39}} + x_{29}\partial_{x_{42}} - x_{35}\partial_{x_{46}} - x_{40}\partial_{x_{48}} + x_{52}\partial_{x_{56}}, \tag{12.1.54}$$

$$E_{(1,1,1,1,1)}|_{\mathscr{B}} = x_3\partial_{x_{11}} - x_5\partial_{x_{16}} - x_6\partial_{x_{18}} - x_{10}\partial_{x_{23}} + x_{14}\partial_{x_{29}} + x_{17}\partial_{x_{33}} + x_{24}\partial_{x_{40}} + x_{28}\partial_{x_{43}} - x_{34}\partial_{x_{47}} - x_{39}\partial_{x_{51}} - x_{41}\partial_{x_{52}} + x_{46}\partial_{x_{54}}, \tag{12.1.55}$$

$$E_{(1,0,1,1,1,1)}|_{\mathscr{B}} = x_2\partial_{x_8} + x_7\partial_{x_{19}} + x_9\partial_{x_{21}} + x_{10}\partial_{x_{24}} + x_{12}\partial_{x_{28}} - x_{17}\partial_{x_{34}} - x_{23}\partial_{x_{40}} + x_{30}\partial_{x_{45}} + x_{33}\partial_{x_{47}} + x_{36}\partial_{x_{48}} + x_{38}\partial_{x_{50}} + x_{49}\partial_{x_{55}}, \tag{12.1.56}$$

$$E_{(0,1,1,2,1)}|_{\mathscr{B}} = -x_3\partial_{x_{10}} + x_4\partial_{x_{12}} - x_8\partial_{x_{20}} + x_{11}\partial_{x_{23}} + x_{19}\partial_{x_{32}} - x_{21}\partial_{x_{35}} + x_{22}\partial_{x_{36}} - x_{25}\partial_{x_{38}} - x_{34}\partial_{x_{46}} + x_{37}\partial_{x_{49}} - x_{45}\partial_{x_{53}} + x_{47}\partial_{x_{54}}, \tag{12.1.57}$$

$$E_{(0,1,1,1,1,1)}|_{\mathscr{B}} = -x_2\partial_{x_9} + x_5\partial_{x_{14}} - x_8\partial_{x_{21}} - x_{13}\partial_{x_{26}} - x_{16}\partial_{x_{30}} - x_{20}\partial_{x_{35}} + x_{22}\partial_{x_{37}} + x_{28}\partial_{x_{41}} + x_{31}\partial_{x_{44}} + x_{36}\partial_{x_{49}} - x_{43}\partial_{x_{52}} + x_{48}\partial_{x_{55}}, \tag{12.1.58}$$

$$E_{(0,1,0,1,1,1,1)}|_{\mathscr{B}} = x_1\partial_{x_7} + x_6\partial_{x_{17}} + x_8\partial_{x_{22}} - x_{15}\partial_{x_{30}} - x_{18}\partial_{x_{33}} - x_{21}\partial_{x_{37}} - x_{20}\partial_{x_{36}} - x_{24}\partial_{x_{39}} - x_{27}\partial_{x_{42}} + x_{35}\partial_{x_{49}} + x_{40}\partial_{x_{51}} + x_{50}\partial_{x_{56}}, \tag{12.1.59}$$

$$E_{(0,0,1,1,1,1,1)}|_{\mathscr{B}} = -x_1\partial_{x_6} - x_7\partial_{x_{17}} + x_{11}\partial_{x_{25}} + x_{13}\partial_{x_{28}} + x_{16}\partial_{x_{31}} + x_{19}\partial_{x_{34}} -x_{23}\partial_{x_{38}} - x_{26}\partial_{x_{41}} - x_{29}\partial_{x_{44}} - x_{32}\partial_{x_{46}} + x_{40}\partial_{x_{50}} + x_{51}\partial_{x_{56}}, \quad (12.1.60)$$

$$E_{(1,1,1,2,1)}|_{\mathscr{B}} = -x_3\partial_{x_{13}} + x_4\partial_{x_{16}} + x_6\partial_{x_{20}} - x_9\partial_{x_{23}} - x_{14}\partial_{x_{32}} + x_{21}\partial_{x_{40}} -x_{17}\partial_{x_{36}} + x_{25}\partial_{x_{43}} + x_{34}\partial_{x_{48}} - x_{37}\partial_{x_{51}} - x_{41}\partial_{x_{53}} + x_{44}\partial_{x_{54}}, \quad (12.1.61)$$

$$E_{(1,1,1,1,1,1)}|_{\mathscr{B}} = -x_2\partial_{x_{11}} + x_5\partial_{x_{19}} + x_6\partial_{x_{21}} + x_{10}\partial_{x_{26}} + x_{12}\partial_{x_{29}} - x_{17}\partial_{x_{37}} +x_{20}\partial_{x_{40}} - x_{28}\partial_{x_{45}} - x_{31}\partial_{x_{47}} - x_{36}\partial_{x_{51}} - x_{38}\partial_{x_{52}} + x_{46}\partial_{x_{55}}, \quad (12.1.62)$$

$$E_{(1,0,1,1,1,1,1)}|_{\mathscr{B}} = -x_1\partial_{x_8} - x_7\partial_{x_{22}} - x_9\partial_{x_{25}} - x_{10}\partial_{x_{28}} - x_{12}\partial_{x_{31}} - x_{14}\partial_{x_{34}} +x_{23}\partial_{x_{43}} + x_{26}\partial_{x_{45}} + x_{29}\partial_{x_{47}} + x_{32}\partial_{x_{48}} + x_{35}\partial_{x_{50}} + x_{49}\partial_{x_{56}}, \quad (12.1.63)$$

$$E_{(0,1,1,2,1,1)}|_{\mathscr{B}} = x_2\partial_{x_{10}} - x_4\partial_{x_{14}} + x_8\partial_{x_{24}} - x_{11}\partial_{x_{26}} + x_{16}\partial_{x_{32}} - x_{18}\partial_{x_{35}} -x_{22}\partial_{x_{39}} + x_{25}\partial_{x_{41}} - x_{31}\partial_{x_{46}} + x_{33}\partial_{x_{49}} - x_{43}\partial_{x_{53}} + x_{47}\partial_{x_{55}}, \quad (12.1.64)$$

$$E_{(0,1,1,1,1,1,1)}|_{\mathscr{B}} = x_1\partial_{x_9} - x_5\partial_{x_{17}} + x_8\partial_{x_{25}} + x_{13}\partial_{x_{30}} + x_{16}\partial_{x_{33}} + x_{19}\partial_{x_{37}} +x_{20}\partial_{x_{38}} + x_{24}\partial_{x_{41}} + x_{27}\partial_{x_{44}} + x_{32}\partial_{x_{49}} - x_{40}\partial_{x_{52}} + x_{48}\partial_{x_{56}}, \quad (12.1.65)$$

$$E_{(1,1,2,2,1)}|_{\mathscr{B}} = -x_3\partial_{x_{15}} + x_4\partial_{x_{18}} - x_5\partial_{x_{20}} + x_7\partial_{x_{23}} + x_{14}\partial_{x_{35}} - x_{19}\partial_{x_{40}} +x_{17}\partial_{x_{38}} - x_{22}\partial_{x_{43}} - x_{34}\partial_{x_{50}} + x_{37}\partial_{x_{52}} - x_{39}\partial_{x_{53}} + x_{42}\partial_{x_{54}}, \quad (12.1.66)$$

$$E_{(1,1,1,2,1,1)}|_{\mathscr{B}} = x_2\partial_{x_{13}} - x_4\partial_{x_{19}} - x_6\partial_{x_{24}} + x_9\partial_{x_{26}} - x_{12}\partial_{x_{32}} + x_{18}\partial_{x_{40}} +x_{17}\partial_{x_{39}} - x_{25}\partial_{x_{45}} + x_{31}\partial_{x_{48}} - x_{33}\partial_{x_{51}} - x_{38}\partial_{x_{53}} + x_{44}\partial_{x_{55}}, \quad (12.1.67)$$

$$E_{(1,1,1,1,1,1,1)}|_{\mathscr{B}} = x_1\partial_{x_{11}} - x_5\partial_{x_{22}} - x_6\partial_{x_{25}} - x_{10}\partial_{x_{30}} - x_{12}\partial_{x_{33}} - x_{14}\partial_{x_{37}} -x_{20}\partial_{x_{43}} - x_{24}\partial_{x_{45}} - x_{27}\partial_{x_{47}} - x_{32}\partial_{x_{51}} - x_{35}\partial_{x_{52}} + x_{46}\partial_{x_{56}}, \quad (12.1.68)$$

$$E_{(0,1,1,2,2,1)}|_{\mathscr{B}} = -x_2\partial_{x_{12}} + x_3\partial_{x_{14}} - x_8\partial_{x_{27}} + x_{11}\partial_{x_{29}} + x_{13}\partial_{x_{32}} - x_{15}\partial_{x_{35}} +x_{22}\partial_{x_{42}} - x_{25}\partial_{x_{44}} - x_{28}\partial_{x_{46}} + x_{30}\partial_{x_{49}} - x_{43}\partial_{x_{54}} + x_{45}\partial_{x_{55}}, \quad (12.1.69)$$

$$E_{(0,1,1,2,1,1,1)}|_{\mathscr{B}} = -x_1\partial_{x_{10}} + x_4\partial_{x_{17}} - x_8\partial_{x_{28}} + x_{11}\partial_{x_{30}} - x_{16}\partial_{x_{36}} + x_{18}\partial_{x_{38}} -x_{19}\partial_{x_{39}} + x_{21}\partial_{x_{41}} - x_{27}\partial_{x_{46}} + x_{29}\partial_{x_{49}} - x_{40}\partial_{x_{53}} + x_{47}\partial_{x_{56}}, \quad (12.1.70)$$

$$E_{(1,1,2,2,1,1)}|_{\mathscr{B}} = x_2\partial_{x_{15}} - x_4\partial_{x_{21}} + x_5\partial_{x_{24}} - x_7\partial_{x_{26}} + x_{12}\partial_{x_{35}} - x_{16}\partial_{x_{40}} -x_{17}\partial_{x_{41}} + x_{22}\partial_{x_{45}} - x_{31}\partial_{x_{50}} + x_{33}\partial_{x_{52}} - x_{36}\partial_{x_{53}} + x_{42}\partial_{x_{55}}, \tag{12.1.71}$$

$$E_{(1,1,1,2,2,1)}|_{\mathscr{B}} = -x_2\partial_{x_{16}} + x_3\partial_{x_{19}} + x_6\partial_{x_{27}} - x_9\partial_{x_{29}} - x_{10}\partial_{x_{32}} + x_{15}\partial_{x_{40}} -x_{17}\partial_{x_{42}} + x_{25}\partial_{x_{47}} + x_{28}\partial_{x_{48}} - x_{30}\partial_{x_{51}} - x_{38}\partial_{x_{54}} + x_{41}\partial_{x_{55}}, \tag{12.1.72}$$

$$E_{(1,1,1,2,1,1,1)}|_{\mathscr{B}} = -x_1\partial_{x_{13}} + x_4\partial_{x_{22}} + x_6\partial_{x_{28}} - x_9\partial_{x_{30}} + x_{12}\partial_{x_{36}} + x_{14}\partial_{x_{39}} -x_{18}\partial_{x_{43}} - x_{21}\partial_{x_{45}} + x_{27}\partial_{x_{48}} - x_{29}\partial_{x_{51}} - x_{35}\partial_{x_{53}} + x_{44}\partial_{x_{56}}, \tag{12.1.73}$$

$$E_{(0,1,1,2,2,1,1)}|_{\mathscr{B}} = x_1\partial_{x_{12}} - x_3\partial_{x_{17}} + x_8\partial_{x_{31}} - x_{11}\partial_{x_{33}} - x_{13}\partial_{x_{36}} + x_{15}\partial_{x_{38}} +x_{19}\partial_{x_{42}} - x_{21}\partial_{x_{44}} - x_{24}\partial_{x_{46}} + x_{26}\partial_{x_{49}} - x_{40}\partial_{x_{54}} + x_{45}\partial_{x_{56}}, \tag{12.1.74}$$

$$E_{(1,1,2,2,2,1)}|_{\mathscr{B}} = -x_2\partial_{x_{18}} + x_3\partial_{x_{21}} - x_5\partial_{x_{27}} + x_7\partial_{x_{29}} + x_{10}\partial_{x_{35}} - x_{13}\partial_{x_{40}} +x_{17}\partial_{x_{44}} - x_{22}\partial_{x_{47}} - x_{28}\partial_{x_{50}} + x_{30}\partial_{x_{52}} - x_{36}\partial_{x_{54}} + x_{39}\partial_{x_{55}}, \tag{12.1.75}$$

$$E_{(1,1,2,2,1,1,1)}|_{\mathscr{B}} = -x_1\partial_{x_{15}} + x_4\partial_{x_{25}} - x_5\partial_{x_{28}} + x_7\partial_{x_{30}} - x_{12}\partial_{x_{38}} + x_{16}\partial_{x_{43}} -x_{14}\partial_{x_{41}} + x_{19}\partial_{x_{45}} - x_{27}\partial_{x_{50}} + x_{29}\partial_{x_{52}} - x_{32}\partial_{x_{53}} + x_{42}\partial_{x_{56}}, \tag{12.1.76}$$

$$E_{(1,1,1,2,2,1,1)}|_{\mathscr{B}} = x_1\partial_{x_{16}} - x_3\partial_{x_{22}} - x_6\partial_{x_{31}} + x_9\partial_{x_{33}} + x_{10}\partial_{x_{36}} - x_{15}\partial_{x_{43}} -x_{14}\partial_{x_{42}} + x_{21}\partial_{x_{47}} + x_{24}\partial_{x_{48}} - x_{26}\partial_{x_{51}} - x_{35}\partial_{x_{54}} + x_{41}\partial_{x_{56}}, \tag{12.1.77}$$

$$E_{(0,1,1,2,2,2,1)}|_{\mathscr{B}} = -x_1\partial_{x_{14}} + x_2\partial_{x_{17}} - x_8\partial_{x_{34}} + x_{11}\partial_{x_{37}} + x_{13}\partial_{x_{39}} - x_{15}\partial_{x_{41}} +x_{16}\partial_{x_{42}} - x_{18}\partial_{x_{44}} - x_{20}\partial_{x_{46}} + x_{23}\partial_{x_{49}} - x_{40}\partial_{x_{55}} + x_{43}\partial_{x_{56}}, \tag{12.1.78}$$

$$E_{(1,1,2,3,2,1)}|_{\mathscr{B}} = x_2\partial_{x_{20}} - x_3\partial_{x_{24}} + x_4\partial_{x_{27}} - x_7\partial_{x_{32}} + x_9\partial_{x_{35}} - x_{11}\partial_{x_{40}} -x_{17}\partial_{x_{46}} + x_{22}\partial_{x_{48}} - x_{25}\partial_{x_{50}} + x_{30}\partial_{x_{53}} - x_{33}\partial_{x_{54}} + x_{37}\partial_{x_{55}}, \tag{12.1.79}$$

$$E_{(1,1,2,2,2,1,1)}|_{\mathscr{B}} = x_1\partial_{x_{18}} - x_3\partial_{x_{25}} + x_5\partial_{x_{31}} - x_7\partial_{x_{33}} - x_{10}\partial_{x_{38}} + x_{13}\partial_{x_{43}} +x_{14}\partial_{x_{44}} - x_{19}\partial_{x_{47}} - x_{24}\partial_{x_{50}} + x_{26}\partial_{x_{52}} - x_{32}\partial_{x_{54}} + x_{39}\partial_{x_{56}}, \tag{12.1.80}$$

$$E_{(1,1,1,2,2,2,1)}|_{\mathscr{B}} = -x_1\partial_{x_{19}} + x_2\partial_{x_{22}} + x_6\partial_{x_{34}} - x_9\partial_{x_{37}} - x_{10}\partial_{x_{39}} + x_{15}\partial_{x_{45}} -x_{12}\partial_{x_{42}} + x_{18}\partial_{x_{47}} + x_{20}\partial_{x_{48}} - x_{23}\partial_{x_{51}} - x_{35}\partial_{x_{55}} + x_{38}\partial_{x_{56}}, \tag{12.1.81}$$

$$E_{(1,2,2,3,2,1)}|_{\mathscr{B}} = -x_2\partial_{x_{23}} + x_3\partial_{x_{26}} - x_4\partial_{x_{29}} + x_5\partial_{x_{32}} - x_6\partial_{x_{35}} + x_8\partial_{x_{40}} -x_{17}\partial_{x_{49}} + x_{22}\partial_{x_{51}} - x_{25}\partial_{x_{52}} + x_{28}\partial_{x_{53}} - x_{31}\partial_{x_{54}} + x_{34}\partial_{x_{55}}, \quad (12.1.82)$$

$$E_{(1,1,2,3,2,1,1)}|_{\mathscr{B}} = -x_1\partial_{x_{20}} + x_3\partial_{x_{28}} - x_4\partial_{x_{31}} + x_7\partial_{x_{36}} - x_9\partial_{x_{38}} + x_{11}\partial_{x_{43}} -x_{14}\partial_{x_{46}} + x_{19}\partial_{x_{48}} - x_{21}\partial_{x_{50}} + x_{26}\partial_{x_{53}} - x_{29}\partial_{x_{54}} + x_{37}\partial_{x_{56}}, \quad (12.1.83)$$

$$E_{(1,1,2,2,2,2,1)}|_{\mathscr{B}} = -x_1\partial_{x_{21}} + x_2\partial_{x_{25}} - x_5\partial_{x_{34}} + x_7\partial_{x_{37}} + x_{10}\partial_{x_{41}} - x_{13}\partial_{x_{45}} +x_{12}\partial_{x_{44}} - x_{16}\partial_{x_{47}} - x_{20}\partial_{x_{50}} + x_{23}\partial_{x_{52}} - x_{32}\partial_{x_{55}} + x_{36}\partial_{x_{56}}, \quad (12.1.84)$$

$$E_{(1,2,2,3,2,1,1)}|_{\mathscr{B}} = x_1\partial_{x_{23}} - x_3\partial_{x_{30}} + x_4\partial_{x_{33}} - x_5\partial_{x_{36}} + x_6\partial_{x_{38}} - x_8\partial_{x_{43}} -x_{14}\partial_{x_{49}} + x_{19}\partial_{x_{51}} - x_{21}\partial_{x_{52}} + x_{24}\partial_{x_{53}} - x_{27}\partial_{x_{54}} + x_{34}\partial_{x_{56}}, \quad (12.1.85)$$

$$E_{(1,1,2,3,2,2,1)}|_{\mathscr{B}} = x_1\partial_{x_{24}} - x_2\partial_{x_{28}} + x_4\partial_{x_{34}} - x_7\partial_{x_{39}} + x_9\partial_{x_{41}} - x_{11}\partial_{x_{45}} -x_{12}\partial_{x_{46}} + x_{16}\partial_{x_{48}} - x_{18}\partial_{x_{50}} + x_{23}\partial_{x_{53}} - x_{29}\partial_{x_{55}} + x_{33}\partial_{x_{56}}, \quad (12.1.86)$$

$$E_{(1,2,2,3,2,2,1)}|_{\mathscr{B}} = -x_1\partial_{x_{26}} + x_2\partial_{x_{30}} - x_4\partial_{x_{37}} + x_5\partial_{x_{39}} - x_6\partial_{x_{41}} + x_8\partial_{x_{45}} -x_{12}\partial_{x_{49}} + x_{16}\partial_{x_{51}} - x_{18}\partial_{x_{52}} + x_{20}\partial_{x_{53}} - x_{27}\partial_{x_{55}} + x_{31}\partial_{x_{56}}, \quad (12.1.87)$$

$$E_{(1,1,2,3,3,2,1)}|_{\mathscr{B}} = -x_1\partial_{x_{27}} + x_2\partial_{x_{31}} - x_3\partial_{x_{34}} + x_7\partial_{x_{42}} - x_9\partial_{x_{44}} + x_{11}\partial_{x_{47}} -x_{10}\partial_{x_{46}} + x_{13}\partial_{x_{48}} - x_{15}\partial_{x_{50}} + x_{23}\partial_{x_{54}} - x_{26}\partial_{x_{55}} + x_{30}\partial_{x_{56}}, \quad (12.1.88)$$

$$E_{(1,2,2,3,3,2,1)}|_{\mathscr{B}} = x_1\partial_{x_{29}} - x_2\partial_{x_{33}} + x_3\partial_{x_{37}} - x_5\partial_{x_{42}} + x_6\partial_{x_{44}} - x_8\partial_{x_{47}} -x_{10}\partial_{x_{49}} + x_{13}\partial_{x_{51}} - x_{15}\partial_{x_{52}} + x_{20}\partial_{x_{54}} - x_{24}\partial_{x_{55}} + x_{28}\partial_{x_{56}}, \quad (12.1.89)$$

$$E_{(1,2,2,4,3,2,1)}|_{\mathscr{B}} = -x_1\partial_{x_{32}} + x_2\partial_{x_{36}} - x_3\partial_{x_{39}} + x_4\partial_{x_{42}} - x_6\partial_{x_{46}} + x_8\partial_{x_{48}} -x_9\partial_{x_{49}} + x_{11}\partial_{x_{51}} - x_{15}\partial_{x_{53}} + x_{18}\partial_{x_{54}} - x_{21}\partial_{x_{55}} + x_{25}\partial_{x_{56}}, \quad (12.1.90)$$

$$E_{(1,2,3,4,3,2,1)}|_{\mathscr{B}} = -x_1\partial_{x_{35}} + x_2\partial_{x_{38}} - x_3\partial_{x_{41}} + x_4\partial_{x_{44}} - x_5\partial_{x_{46}} + x_8\partial_{x_{50}} -x_7\partial_{x_{49}} + x_{11}\partial_{x_{52}} - x_{13}\partial_{x_{53}} + x_{16}\partial_{x_{54}} - x_{19}\partial_{x_{55}} + x_{22}\partial_{x_{56}}, \quad (12.1.91)$$

$$E_{(2,2,3,4,3,2,1)}|_{\mathscr{B}} = -x_1\partial_{x_{40}} + x_2\partial_{x_{43}} - x_3\partial_{x_{45}} + x_4\partial_{x_{47}} - x_5\partial_{x_{48}} + x_6\partial_{x_{50}} -x_7\partial_{x_{51}} + x_9\partial_{x_{52}} - x_{10}\partial_{x_{53}} + x_{12}\partial_{x_{54}} - x_{14}\partial_{x_{55}} + x_{17}\partial_{x_{56}}. \quad (12.1.92)$$

Table 12.1 Weights of the Basic Representation

i	$a_{i,1}$	$a_{i,2}$	$a_{i,3}$	$a_{i,4}$	$a_{i,5}$	$a_{i,6}$	$a_{i,7}$	i	$a_{i,1}$	$a_{i,2}$	$a_{i,3}$	$a_{i,4}$	$a_{i,5}$	$a_{i,6}$	$a_{i,7}$
1	0	0	0	0	0	0	1	2	0	0	0	0	0	1	−1
3	0	0	0	0	1	−1	0	4	0	0	0	1	−1	0	0
5	0	1	1	−1	0	0	0	6	1	1	−1	0	0	0	0
7	0	−1	1	0	0	0	0	8	−1	1	0	0	0	0	0
9	1	−1	−1	1	0	0	0	10	1	0	0	−1	1	0	0
11	−1	−1	0	1	0	0	0	12	1	0	0	0	−1	1	0
13	−1	0	1	−1	1	0	0	14	1	0	0	0	0	−1	1
15	0	0	−1	0	1	0	0	16	−1	0	1	0	−1	1	0
17	1	0	0	0	0	0	−1	18	0	0	−1	1	−1	1	0
19	−1	0	1	0	0	−1	1	20	0	1	0	−1	0	1	0
21	0	0	−1	1	0	−1	1	22	−1	0	1	0	0	0	−1
23	0	−1	0	0	0	1	0	24	0	1	0	−1	1	−1	1
25	0	0	−1	1	0	0	−1	26	0	−1	0	0	1	−1	1
27	0	1	0	0	−1	0	1	28	0	1	0	−1	1	0	−1

We define a symmetric linear operation τ on the space $\sum_{i,j=1}^{56} \mathbb{F} x_i \partial_{x_j}$ by

$$\tau(x_i \partial_{x_j}) = x_j \partial_{x_i}. \tag{12.1.93}$$

Denote

$$\bar{r} = 57 - r \qquad \text{for } r \in \overline{1,56}. \tag{12.1.94}$$

Then

$$E_{-\alpha}|_{\mathscr{B}} = -\tau(E_\alpha|_{\mathscr{B}}) \qquad \text{for } \alpha \in \Phi_{E_7}^+ \tag{12.1.95}$$

by (4.4.25). Moreover, (4.4.24) gives

$$\alpha_r|_V = \sum_{i=1}^{28} a_{i,r}(x_i \partial_{x_i} - x_{\bar{i}} \partial_{x_{\bar{i}}}) \qquad \text{for } r \in \overline{1,7}, \tag{12.1.96}$$

where $a_{i,r}$ are constants given by the Table 12.1.

12.2 Constructions of Singular Vectors

In this section, we construct special singular vectors in $\mathscr{B}$.

According to Table 12.1, we find the singular vector

$$\iota_1 = x_1 x_{17} + x_2 x_{14} + x_3 x_{12} + x_4 x_{10} + x_5 x_9 - x_6 x_7, \tag{12.2.1}$$

which generates an irreducible $\mathscr{G}^{E_7}$-module W isomorphic to the adjoint module. Set

$$\mathbb{B} = \mathscr{B}[\partial_{x_1}, \ldots, \partial_{x_{56}}]. \tag{12.2.2}$$

In terms of (12.1.94), we define an associative algebra isomorphism ν on $\mathbb{B}$ by

$$\nu(x_r) = x_{\bar{r}},\ \nu(\partial_{x_r}) = \partial_{x_{\bar{r}}} \qquad \text{for}\ r \in \overline{1, 56}. \tag{12.2.3}$$

Set

$$\mathfrak{l}_2 = x_1x_{22} + x_2x_{19} + x_3x_{16} + x_4x_{13} + x_5x_{11} - x_7x_8, \tag{12.2.4}$$

$$\mathfrak{l}_3 = x_1x_{25} + x_2x_{21} + x_3x_{18} + x_4x_{15} + x_6x_{11} - x_8x_9, \tag{12.2.5}$$

$$\mathfrak{l}_4 = -x_1x_{28} - x_2x_{24} - x_3x_{20} + x_5x_{15} - x_6x_{13} + x_8x_{10}, \tag{12.2.6}$$

$$\mathfrak{l}_5 = x_1x_{30} + x_2x_{26} + x_3x_{23} + x_7x_{15} - x_9x_{13} + x_{10}x_{11}, \tag{12.2.7}$$

$$\mathfrak{l}_6 = x_1x_{31} + x_2x_{27} - x_4x_{20} - x_5x_{18} + x_6x_{16} - x_8x_{12}, \tag{12.2.8}$$

$$\mathfrak{l}_7 = -x_1x_{33} - x_2x_{29} + x_4x_{23} - x_7x_{18} + x_9x_{16} - x_{11}x_{12}, \tag{12.2.9}$$

$$\mathfrak{l}_8 = -x_1x_{34} + x_3x_{27} + x_4x_{24} + x_5x_{21} - x_6x_{19} + x_8x_{14}, \tag{12.2.10}$$

$$\mathfrak{l}_9 = -x_2x_{34} - x_3x_{31} - x_4x_{28} - x_5x_{25} + x_6x_{22} - x_8x_{17}, \tag{12.2.11}$$

$$\mathfrak{l}_{10} = x_1x_{36} + x_2x_{32} + x_5x_{23} + x_7x_{20} - x_{10}x_{16} + x_{12}x_{13}, \tag{12.2.12}$$

$$\mathfrak{l}_{11} = x_1x_{37} - x_3x_{29} - x_4x_{26} + x_7x_{21} - x_9x_{19} + x_{11}x_{14}, \tag{12.2.13}$$

$$\mathfrak{l}_{12} = -x_1x_{38} - x_2x_{35} + x_6x_{23} + x_9x_{20} - x_{10}x_{18} + x_{12}x_{15}, \tag{12.2.14}$$

$$\mathfrak{l}_{13} = x_2x_{37} + x_3x_{33} + x_4x_{30} - x_7x_{25} + x_9x_{22} - x_{11}x_{17}, \tag{12.2.15}$$

$$\mathfrak{l}_{14} = -x_1x_{39} + x_3x_{32} - x_5x_{26} - x_7x_{24} + x_{10}x_{19} - x_{13}x_{14}, \tag{12.2.16}$$

$$\mathfrak{l}_{15} = -x_2x_{39} - x_3x_{36} + x_5x_{30} + x_7x_{28} - x_{10}x_{22} + x_{13}x_{17}, \tag{12.2.17}$$

$$\mathfrak{l}_{16} = x_1x_{41} - x_3x_{35} - x_6x_{26} - x_9x_{24} + x_{10}x_{21} - x_{14}x_{15}, \tag{12.2.18}$$

$$\mathfrak{l}_{17} = x_1x_{42} + x_4x_{32} + x_5x_{29} + x_7x_{27} - x_{12}x_{19} + x_{14}x_{16}, \tag{12.2.19}$$

$$\mathfrak{l}_{18} = x_2x_{41} + x_3x_{38} + x_6x_{30} + x_9x_{28} - x_{10}x_{25} + x_{15}x_{17}, \tag{12.2.20}$$

$$\mathfrak{l}_{19} = x_1x_{43} + x_2x_{40} + x_8x_{23} + x_{11}x_{20} - x_{13}x_{18} + x_{15}x_{16}, \tag{12.2.21}$$

$$\mathfrak{l}_{20} = x_2x_{42} - x_4x_{36} - x_5x_{33} - x_7x_{31} + x_{12}x_{22} - x_{16}x_{17}, \quad (12.2.22)$$

$$\mathfrak{l}_{21} = -x_1x_{44} - x_4x_{35} + x_6x_{29} + x_9x_{27} - x_{12}x_{21} + x_{14}x_{18}, \quad (12.2.23)$$

$$\mathfrak{l}_{22} = x_3x_{42} + x_4x_{39} + x_5x_{37} + x_7x_{34} - x_{14}x_{22} + x_{17}x_{19}, \quad (12.2.24)$$

$$\mathfrak{l}_{23} = -x_1x_{45} + x_3x_{40} - x_8x_{26} - x_{11}x_{24} + x_{13}x_{21} - x_{15}x_{19}, \quad (12.2.25)$$

$$\mathfrak{l}_{24} = -x_2x_{44} + x_4x_{38} - x_6x_{33} - x_9x_{31} + x_{12}x_{25} - x_{17}x_{18}, \quad (12.2.26)$$

$$\mathfrak{l}_{25} = x_1x_{46} - x_5x_{35} - x_6x_{32} - x_{10}x_{27} + x_{12}x_{24} - x_{14}x_{20}, \quad (12.2.27)$$

$$\mathfrak{l}_{26} = -x_2x_{45} - x_3x_{43} + x_8x_{30} + x_{11}x_{28} - x_{13}x_{25} + x_{15}x_{22}, \quad (12.2.28)$$

$$\mathfrak{l}_{27} = -x_3x_{44} - x_4x_{41} + x_6x_{37} + x_9x_{34} - x_{14}x_{25} + x_{17}x_{21}, \quad (12.2.29)$$

$$\mathfrak{l}_{28} = x_1x_{47} + x_4x_{40} + x_8x_{29} + x_{11}x_{27} - x_{16}x_{21} + x_{18}x_{19}, \quad (12.2.30)$$

$$\mathfrak{l}_{29} = x_2x_{46} + x_5x_{38} + x_6x_{36} + x_{10}x_{31} - x_{12}x_{28} + x_{17}x_{20}, \quad (12.2.31)$$

$$\mathfrak{l}_{30} = -x_1x_{48} + x_5x_{40} - x_8x_{32} - x_{13}x_{27} + x_{16}x_{24} - x_{19}x_{20}, \quad (12.2.32)$$

$$\mathfrak{l}_{31} = x_2x_{47} - x_4x_{43} - x_8x_{33} - x_{11}x_{31} + x_{16}x_{25} - x_{18}x_{22}, \quad (12.2.33)$$

$$\mathfrak{l}_{32} = x_3x_{46} - x_5x_{41} - x_6x_{39} - x_{10}x_{34} + x_{14}x_{28} - x_{17}x_{24}, \quad (12.2.34)$$

$$\mathfrak{l}_{33} = x_1x_{49} - x_7x_{35} - x_9x_{32} + x_{10}x_{29} - x_{12}x_{26} + x_{14}x_{23}, \quad (12.2.35)$$

$$\mathfrak{l}_{34} = -x_2x_{48} - x_5x_{43} + x_8x_{36} + x_{13}x_{31} - x_{16}x_{28} + x_{20}x_{22}, \quad (12.2.36)$$

$$\mathfrak{l}_{35} = x_3x_{47} + x_4x_{45} + x_8x_{37} + x_{11}x_{34} - x_{19}x_{25} + x_{21}x_{22}, \quad (12.2.37)$$

$$\mathfrak{l}_{36} = x_4x_{46} + x_5x_{44} + x_6x_{42} + x_{12}x_{34} - x_{14}x_{31} + x_{17}x_{27}, \quad (12.2.38)$$

$$\mathfrak{l}_{37} = x_1x_{50} + x_6x_{40} + x_8x_{35} - x_{15}x_{27} + x_{18}x_{24} - x_{20}x_{21}, \quad (12.2.39)$$

$$\mathfrak{l}_{38} = x_2x_{49} + x_7x_{38} + x_9x_{36} - x_{10}x_{33} + x_{12}x_{30} - x_{17}x_{23}, \quad (12.2.40)$$

$$\mathfrak{l}_{39} = -x_3x_{48} + x_5x_{45} - x_8x_{39} - x_{13}x_{34} + x_{19}x_{28} - x_{22}x_{24}, \quad (12.2.41)$$

$$\mathfrak{l}_{40} = -x_1x_{51} + x_7x_{40} - x_{11}x_{32} + x_{13}x_{29} - x_{16}x_{26} + x_{19}x_{23}, \quad (12.2.42)$$

$$\mathfrak{l}_{41} = x_2x_{50} - x_6x_{43} - x_8x_{38} + x_{15}x_{31} - x_{18}x_{28} + x_{20}x_{25}, \quad (12.2.43)$$

$$\iota_{42} = x_3x_{49} - x_7x_{41} - x_9x_{39} + x_{10}x_{37} - x_{14}x_{30} + x_{17}x_{26}, \tag{12.2.44}$$

$$\iota_{43} = -x_4x_{48} - x_5x_{47} + x_8x_{42} + x_{16}x_{34} - x_{19}x_{31} + x_{22}x_{27}, \tag{12.2.45}$$

$$\iota_{44} = x_1x_{52} + x_9x_{40} + x_{11}x_{35} + x_{15}x_{29} - x_{18}x_{26} + x_{21}x_{23}, \tag{12.2.46}$$

$$\iota_{45} = -x_2x_{51} - x_7x_{43} + x_{11}x_{36} - x_{13}x_{33} + x_{16}x_{30} - x_{22}x_{23}, \tag{12.2.47}$$

$$\iota_{46} = x_3x_{50} + x_6x_{45} + x_8x_{41} - x_{15}x_{34} + x_{21}x_{28} - x_{24}x_{25}, \tag{12.2.48}$$

$$\iota_{47} = x_4x_{49} + x_7x_{44} + x_9x_{42} - x_{12}x_{37} + x_{14}x_{33} - x_{17}x_{29}, \tag{12.2.49}$$

$$\iota_{48} = x_1x_{53} - x_{10}x_{40} - x_{13}x_{35} - x_{15}x_{32} + x_{20}x_{26} - x_{23}x_{24}, \tag{12.2.50}$$

$$\iota_{49} = x_2x_{52} - x_9x_{43} - x_{11}x_{38} - x_{15}x_{33} + x_{18}x_{30} - x_{23}x_{25}, \tag{12.2.51}$$

$$\iota_{50} = -x_3x_{51} + x_7x_{45} - x_{11}x_{39} + x_{13}x_{37} - x_{19}x_{30} + x_{22}x_{26}, \tag{12.2.52}$$

$$\iota_{51} = x_4x_{50} - x_6x_{47} - x_8x_{44} + x_{18}x_{34} - x_{21}x_{31} + x_{25}x_{27}, \tag{12.2.53}$$

$$\iota_{52} = x_5x_{49} - x_7x_{46} - x_{10}x_{42} + x_{12}x_{39} - x_{14}x_{36} + x_{17}x_{32}, \tag{12.2.54}$$

$$\iota_{53} = x_1x_{54} + x_{12}x_{40} + x_{16}x_{35} + x_{18}x_{32} - x_{20}x_{29} + x_{23}x_{27}, \tag{12.2.55}$$

$$\iota_{54} = x_2x_{53} + x_{10}x_{43} + x_{13}x_{38} + x_{15}x_{36} - x_{20}x_{30} + x_{23}x_{28}, \tag{12.2.56}$$

$$\iota_{55} = x_3x_{52} + x_9x_{45} + x_{11}x_{41} + x_{15}x_{37} - x_{21}x_{30} + x_{25}x_{26}, \tag{12.2.57}$$

$$\iota_{56} = -x_4x_{51} - x_7x_{47} + x_{11}x_{42} - x_{16}x_{37} + x_{19}x_{33} - x_{22}x_{29}, \tag{12.2.58}$$

$$\iota_{57} = x_5x_{50} + x_6x_{48} + x_8x_{46} - x_{20}x_{34} + x_{24}x_{31} - x_{27}x_{28}, \tag{12.2.59}$$

$$\iota_{58} = x_6x_{49} - x_9x_{46} + x_{10}x_{44} - x_{12}x_{41} + x_{14}x_{38} - x_{17}x_{35}, \tag{12.2.60}$$

$$\iota_{59} = x_1x_{55} - x_{14}x_{40} - x_{19}x_{35} - x_{21}x_{32} + x_{24}x_{29} - x_{26}x_{27}, \tag{12.2.61}$$

$$\iota_{60} = x_2x_{54} - x_{12}x_{43} - x_{16}x_{38} - x_{18}x_{36} + x_{20}x_{33} - x_{23}x_{31}, \tag{12.2.62}$$

$$\iota_{61} = x_3x_{53} - x_{10}x_{45} - x_{13}x_{41} - x_{15}x_{39} + x_{24}x_{30} - x_{26}x_{28}, \tag{12.2.63}$$

$$\iota_{62} = x_4x_{52} - x_9x_{47} - x_{11}x_{44} - x_{18}x_{37} + x_{21}x_{33} - x_{25}x_{29}, \tag{12.2.64}$$

$$\iota_{63} = -x_5x_{51} + x_7x_{48} - x_{13}x_{42} + x_{16}x_{39} - x_{19}x_{36} + x_{22}x_{32}, \tag{12.2.65}$$

$$\begin{aligned}\mathfrak{l}_{64} = &\ x_1x_{56} + x_2x_{55} + x_{14}x_{43} + x_{17}x_{40} + x_{19}x_{38} + x_{22}x_{35} \\ &+x_{21}x_{36} + x_{25}x_{32} - x_{24}x_{33} - x_{28}x_{29} + x_{26}x_{31} + x_{27}x_{30},\end{aligned} \tag{12.2.66}$$

$$\begin{aligned}\mathfrak{l}_{65} = &\ x_2x_{55} + x_3x_{54} + x_{12}x_{45} + x_{14}x_{43} + x_{16}x_{41} + x_{19}x_{38} \\ &+x_{18}x_{39} + x_{21}x_{36} - x_{20}x_{37} - x_{24}x_{33} + x_{23}x_{34} + x_{26}x_{31},\end{aligned} \tag{12.2.67}$$

$$\begin{aligned}\mathfrak{l}_{66} = &\ x_3x_{54} + x_4x_{53} + x_{10}x_{47} + x_{12}x_{45} + x_{13}x_{44} + x_{15}x_{42} \\ &+x_{16}x_{41} + x_{18}x_{39} - x_{24}x_{33} - x_{27}x_{30} + x_{26}x_{31} + x_{28}x_{29},\end{aligned} \tag{12.2.68}$$

$$\begin{aligned}\mathfrak{l}_{67} = &\ x_4x_{53} + x_5x_{52} + x_9x_{48} + x_{10}x_{47} + x_{11}x_{46} + x_{13}x_{44} \\ &+x_{18}x_{39} + x_{20}x_{37} - x_{21}x_{36} - x_{24}x_{33} + x_{25}x_{32} + x_{28}x_{29},\end{aligned} \tag{12.2.69}$$

$$\begin{aligned}\mathfrak{l}_{68} = &\ x_5x_{52} + x_6x_{51} + x_7x_{50} + x_8x_{49} + x_9x_{48} + x_{11}x_{46} \\ &+x_{20}x_{37} + x_{23}x_{34} - x_{24}x_{33} - x_{26}x_{31} + x_{27}x_{30} + x_{28}x_{29},\end{aligned} \tag{12.2.70}$$

$$\begin{aligned}\mathfrak{l}_{69} = &\ x_5x_{52} - x_6x_{51} - x_7x_{50} + x_9x_{48} + x_{13}x_{44} - x_{15}x_{42} \\ &-x_{16}x_{41} + x_{18}x_{39} + x_{19}x_{38} - x_{21}x_{36} - x_{22}x_{35} + x_{25}x_{32},\end{aligned} \tag{12.2.71}$$

$$\begin{aligned}\mathfrak{l}_{70} = &\ -x_6x_{51} + x_8x_{49} + x_9x_{48} - x_{10}x_{47} - x_{11}x_{46} + x_{13}x_{44} \\ &+x_{12}x_{45} - x_{14}x_{43} - x_{16}x_{41} + x_{17}x_{40} + x_{19}x_{38} - x_{22}x_{35},\end{aligned} \tag{12.2.72}$$

$$\mathfrak{l}_i = \nu(\mathfrak{l}_{134-i}) \qquad \text{for } i \in \overline{71, 133}. \tag{12.2.73}$$

Then

$$W = \sum_{i=1}^{133} \mathbb{F}\mathfrak{l}_i. \tag{12.2.74}$$

Moreover,

$$\alpha_r|_W = \sum_{i=1}^{63} b_{i,r}(\mathfrak{l}_i\partial_{\mathfrak{l}_i} - \mathfrak{l}_{134-i}\partial_{\mathfrak{l}_{134-i}}) \qquad \text{for } r \in \overline{1, 7}, \tag{12.2.75}$$

where $b_{i,r}$ are given by the Table 12.2.

Note for $\alpha, \beta \in \Phi_{E_7}$,

$$\alpha + \beta \in \Phi_{E_7} \Leftrightarrow (\alpha, \beta) = -1. \tag{12.2.76}$$

Table 12.2 Weights of the Adjoint Representation

i	$b_{i,1}$	$b_{i,2}$	$b_{i,3}$	$b_{i,4}$	$b_{i,5}$	$b_{i,6}$	$b_{i,7}$	i	$b_{i,1}$	$b_{i,2}$	$b_{i,3}$	$b_{i,4}$	$b_{i,5}$	$b_{i,6}$	$b_{i,7}$
1	1	0	0	0	0	0	0	2	−1	0	1	0	0	0	0
3	0	0	−1	1	0	0	0	4	0	1	0	−1	1	0	0
5	0	−1	0	0	1	0	0	6	0	1	0	0	−1	1	0
7	0	−1	0	1	−1	1	0	8	0	1	0	0	0	−1	1
9	0	1	0	0	0	0	−1	10	0	0	1	−1	0	1	0
11	0	−1	0	1	0	−1	1	12	1	0	−1	0	0	1	0
13	0	−1	0	1	0	0	−1	14	0	0	1	−1	1	−1	1
15	0	0	1	−1	1	0	−1	16	1	0	−1	0	1	−1	1
17	0	0	1	0	−1	0	1	18	1	0	−1	0	1	0	−1
19	−1	0	0	0	0	1	0	20	0	0	1	0	−1	1	−1
21	1	0	−1	1	−1	0	1	22	0	0	1	0	0	−1	0
23	−1	0	0	0	1	−1	1	24	1	0	−1	1	−1	1	−1
25	1	1	0	−1	0	0	1	26	−1	0	0	0	1	0	−1
27	1	0	−1	1	0	−1	0	28	−1	0	0	1	−1	0	1
29	1	1	0	−1	0	1	−1	30	−1	1	1	−1	0	0	1
31	−1	0	0	1	−1	1	−1	32	1	1	0	−1	1	−1	0
33	1	−1	0	0	0	0	1	34	−1	1	1	−1	0	1	−1
35	−1	0	0	1	0	−1	0	36	1	1	0	0	−1	0	0
37	0	1	−1	0	0	0	1	38	1	−1	0	0	0	1	−1
39	−1	1	1	−1	1	−1	0	40	−1	−1	1	0	0	0	1
41	0	1	−1	0	0	1	−1	42	1	−1	0	0	1	−1	0
43	−1	1	1	0	−1	0	0	44	0	−1	−1	1	0	0	1
45	−1	−1	1	0	0	1	−1	46	0	1	−1	0	1	−1	0
47	1	−1	0	1	−1	0	0	48	0	0	0	−1	1	0	1
49	0	−1	−1	1	0	1	−1	50	−1	−1	1	0	1	−1	0
51	0	1	−1	1	−1	0	0	52	1	0	1	−1	0	0	0
53	0	0	0	0	−1	1	1	54	0	0	0	−1	1	1	−1
55	0	−1	−1	1	1	−1	0	56	−1	−1	1	1	−1	0	0
57	0	2	0	−1	0	0	0	58	2	0	−1	0	0	0	0
59	0	0	0	0	0	−1	2	60	0	0	0	0	−1	2	−1
61	0	0	0	−1	2	−1	0	62	0	−1	−1	2	−1	0	0
63	−1	0	2	−1	0	0	0	64	0	0	0	0	0	0	0

Thus

$$E_{\alpha_r}(\mathfrak{l}_i) \neq 0 \Leftrightarrow b_{i,r} = -1 \qquad \text{for } r \in \overline{1,7},\ i \in \overline{1,63}. \tag{12.2.77}$$

Write

$$W_1 = \sum_{i=1}^{63} \mathbb{C}\mathfrak{l}_i. \tag{12.2.78}$$

Based on the above fact and Table 12.2, we find

$$\begin{aligned}E_{\alpha_1}|_{W_1} = & -\iota_1\partial_{\iota_2} - \iota_{12}\partial_{\iota_{19}} - \iota_{16}\partial_{\iota_{23}} - \iota_{18}\partial_{\iota_{26}} - \iota_{21}\partial_{\iota_{28}} - \iota_{25}\partial_{\iota_{30}} \\ & -\iota_{24}\partial_{\iota_{31}} - \iota_{29}\partial_{\iota_{34}} - \iota_{27}\partial_{\iota_{35}} - \iota_{32}\partial_{\iota_{39}} - \iota_{33}\partial_{\iota_{40}} \\ & -\iota_{36}\partial_{\iota_{43}} - \iota_{38}\partial_{\iota_{45}} - \iota_{42}\partial_{\iota_{50}} - \iota_{47}\partial_{\iota_{56}} - \iota_{52}\partial_{\iota_{63}},\end{aligned} \tag{12.2.79}$$

$$\begin{aligned}E_{\alpha_2}|_{W_1} = & \iota_4\partial_{\iota_5} + \iota_6\partial_{\iota_7} + \iota_8\partial_{\iota_{11}} + \iota_9\partial_{\iota_{13}} + \iota_{25}\partial_{\iota_{33}} + \iota_{29}\partial_{\iota_{38}} \\ & +\iota_{30}\partial_{\iota_{40}} + \iota_{32}\partial_{\iota_{42}} + \iota_{37}\partial_{\iota_{44}} + \iota_{34}\partial_{\iota_{45}} + \iota_{36}\partial_{\iota_{47}} \\ & +\iota_{41}\partial_{\iota_{49}} + \iota_{39}\partial_{\iota_{50}} + \iota_{46}\partial_{\iota_{55}} + \iota_{43}\partial_{\iota_{56}} + \iota_{51}\partial_{\iota_{62}},\end{aligned} \tag{12.2.80}$$

$$\begin{aligned}E_{\alpha_3}|_{W_1} = & -\iota_2\partial_{\iota_3} - \iota_{10}\partial_{\iota_{12}} - \iota_{14}\partial_{\iota_{16}} - \iota_{15}\partial_{\iota_{18}} - \iota_{17}\partial_{\iota_{21}} - \iota_{20}\partial_{\iota_{24}} \\ & -\iota_{22}\partial_{\iota_{27}} - \iota_{30}\partial_{\iota_{37}} - \iota_{34}\partial_{\iota_{41}} - \iota_{40}\partial_{\iota_{44}} - \iota_{39}\partial_{\iota_{46}} \\ & -\iota_{45}\partial_{\iota_{49}} - \iota_{43}\partial_{\iota_{51}} - \iota_{50}\partial_{\iota_{55}} - \iota_{52}\partial_{\iota_{58}} - \iota_{56}\partial_{\iota_{62}},\end{aligned} \tag{12.2.81}$$

$$\begin{aligned}E_{\alpha_4}|_{W_1} = & \iota_3\partial_{\iota_4} + \iota_7\partial_{\iota_{10}} + \iota_{11}\partial_{\iota_{14}} + \iota_{13}\partial_{\iota_{15}} + \iota_{21}\partial_{\iota_{25}} + \iota_{24}\partial_{\iota_{29}} \\ & +\iota_{28}\partial_{\iota_{30}} + \iota_{27}\partial_{\iota_{32}} + \iota_{31}\partial_{\iota_{34}} + \iota_{35}\partial_{\iota_{39}} + \iota_{44}\partial_{\iota_{48}} \\ & +\iota_{47}\partial_{\iota_{52}} + \iota_{49}\partial_{\iota_{54}} + \iota_{51}\partial_{\iota_{57}} + \iota_{55}\partial_{\iota_{61}} + \iota_{56}\partial_{\iota_{63}},\end{aligned} \tag{12.2.82}$$

$$\begin{aligned}E_{\alpha_5}|_{W_1} = & \iota_4\partial_{\iota_6} + \iota_5\partial_{\iota_7} + \iota_{14}\partial_{\iota_{17}} + \iota_{15}\partial_{\iota_{20}} + \iota_{16}\partial_{\iota_{21}} + \iota_{18}\partial_{\iota_{24}} \\ & +\iota_{23}\partial_{\iota_{28}} + \iota_{26}\partial_{\iota_{31}} + \iota_{32}\partial_{\iota_{36}} + \iota_{39}\partial_{\iota_{43}} + \iota_{42}\partial_{\iota_{47}} \\ & +\iota_{46}\partial_{\iota_{51}} + \iota_{48}\partial_{\iota_{53}} + \iota_{50}\partial_{\iota_{56}} + \iota_{54}\partial_{\iota_{60}} + \iota_{55}\partial_{\iota_{62}},\end{aligned} \tag{12.2.83}$$

$$\begin{aligned}E_{\alpha_6}|_{W_1} = & \iota_6\partial_{\iota_8} + \iota_7\partial_{\iota_{11}} + \iota_{10}\partial_{\iota_{14}} + \iota_{12}\partial_{\iota_{16}} + \iota_{20}\partial_{\iota_{22}} + \iota_{19}\partial_{\iota_{23}} \\ & +\iota_{24}\partial_{\iota_{27}} + \iota_{29}\partial_{\iota_{32}} + \iota_{31}\partial_{\iota_{35}} + \iota_{34}\partial_{\iota_{39}} + \iota_{38}\partial_{\iota_{42}} \\ & +\iota_{41}\partial_{\iota_{46}} + \iota_{45}\partial_{\iota_{50}} + \iota_{49}\partial_{\iota_{55}} + \iota_{53}\partial_{\iota_{59}} + \iota_{54}\partial_{\iota_{61}},\end{aligned} \tag{12.2.84}$$

$$\begin{aligned}E_{\alpha_7}|_{W_1} = & \iota_8\partial_{\iota_9} + \iota_{11}\partial_{\iota_{13}} + \iota_{14}\partial_{\iota_{15}} + \iota_{16}\partial_{\iota_{18}} + \iota_{17}\partial_{\iota_{20}} + \iota_{21}\partial_{\iota_{24}} \\ & +\iota_{23}\partial_{\iota_{26}} + \iota_{25}\partial_{\iota_{29}} + \iota_{28}\partial_{\iota_{31}} + \iota_{30}\partial_{\iota_{34}} + \iota_{33}\partial_{\iota_{38}} \\ & +\iota_{37}\partial_{\iota_{41}} + \iota_{40}\partial_{\iota_{45}} + \iota_{44}\partial_{\iota_{49}} + \iota_{48}\partial_{\iota_{54}} + \iota_{53}\partial_{\iota_{60}}.\end{aligned} \tag{12.2.85}$$

We define a linear transformation $\mathfrak{s}$ on the space

$$\sum_{i,j=1}^{133} \mathbb{C}\iota_i\partial_{\iota_j} \tag{12.2.86}$$

by

$$\mathfrak{s}(\iota_i \partial_{\iota_j}) = \iota_{134-j} \partial_{\iota_{134-i}}. \quad (12.2.87)$$

By symmetry, we have

$$E_{\alpha_1}|_W = (1 - \mathfrak{s})(E_{\alpha_1}|_{W_1}) + \iota_{70}\partial_{\iota_{76}} - \iota_{58}(2\partial_{\iota_{70}} + \partial_{\iota_{69}}), \quad (12.2.88)$$

$$E_{\alpha_2}|_W = (1 + \mathfrak{s})(E_{\alpha_2}|_{W_1}) + \iota_{68}\partial_{\iota_{77}} + \iota_{57}(2\partial_{\iota_{68}} + \partial_{\iota_{67}}), \quad (12.2.89)$$

$$E_{\alpha_3}|_W = (1 - \mathfrak{s})(E_{\alpha_3}|_{W_1}) + \iota_{69}\partial_{\iota_{71}} - \iota_{63}(\partial_{\iota_{70}} + 2\partial_{\iota_{69}} + \partial_{\iota_{67}}), \quad (12.2.90)$$

$$E_{\alpha_4}|_W = (1 + \mathfrak{s})(E_{\alpha_4}|_{W_1}) + \iota_{67}\partial_{\iota_{72}} + \iota_{62}(\partial_{\iota_{69}} + \partial_{\iota_{68}} + 2\partial_{\iota_{67}} + \partial_{\iota_{66}}), \quad (12.2.91)$$

$$E_{\alpha_5}|_W = (1 + \mathfrak{s})(E_{\alpha_5}|_{W_1}) + \iota_{66}\partial_{\iota_{73}} + \iota_{61}(\partial_{\iota_{67}} + 2\partial_{\iota_{66}} + \partial_{\iota_{65}}), \quad (12.2.92)$$

$$E_{\alpha_6}|_W = (1 + \mathfrak{s})(E_{\alpha_6}|_{W_1}) + \iota_{65}\partial_{\iota_{74}} + \iota_{60}(\partial_{\iota_{66}} + 2\partial_{\iota_{65}} + \partial_{\iota_{64}}), \quad (12.2.93)$$

$$E_{\alpha_7}|_W = (1 + \mathfrak{s})(E_{\alpha_7}|_{W_1}) + \iota_{64}\partial_{\iota_{75}} + \iota_{59}(\partial_{\iota_{65}} + 2\partial_{\iota_{64}}). \quad (12.2.94)$$

According to (12.1.30)–(12.1.36), (12.2.88)–(12.2.94) Tables 12.1 and 12.2, we find the following singular vectors

$$\begin{aligned}\vartheta = {} & \frac{x_1}{2}\iota_{64} - x_2\iota_{59} + x_3\iota_{53} - x_4\iota_{48} + x_5\iota_{44} - x_6\iota_{40} - x_7\iota_{37} \\ & + x_8\iota_{33} + x_9\iota_{30} + x_{10}\iota_{28} - x_{11}\iota_{25} + x_{12}\iota_{23} - x_{13}\iota_{21} + x_{14}\iota_{19} \\ & + x_{15}\iota_{17} - x_{16}\iota_{16} + x_{18}\iota_{14} - x_{19}\iota_{12} + x_{20}\iota_{11} + x_{21}\iota_{10} + x_{23}\iota_8 \\ & + x_{24}\iota_7 + x_{26}\iota_6 + x_{27}\iota_5 + x_{29}\iota_4 + x_{32}\iota_3 + x_{35}\iota_2 + x_{40}\iota_1 \end{aligned} \quad (12.2.95)$$

of weight λ_7, and

$$\varsigma = \iota_1\iota_{19} - \iota_2\iota_{12} + \iota_3\iota_{10} - \iota_4\iota_7 + \iota_5\iota_6 \quad (12.2.96)$$

of weight λ_6. Denote

$$I = \{\overline{1,6}, \overline{9,12}, 15, \overline{17,19}, 21, 22, \overline{26.29}, \overline{33,36}, \overline{40,44}, \overline{50,55}, 58, 63\}. \quad (12.2.97)$$

We can similarly obtain the Cartan's quartic invariant

$$\begin{aligned}\eta = {} & 4(\sum_{i\in I} \iota_i\iota_{134-i} - \sum_{r\in\overline{7,63}\setminus I} \iota_r\iota_{134-r}) + 4\iota_{70}(\iota_{70} - 2\iota_{68} - 3\iota_{69} + 4\iota_{67} - 3\iota_{66} \\ & + 2\iota_{65} - \iota_{64}) + \iota_{68}(7\iota_{68} + 16\iota_{69} - 24\iota_{67} + 18\iota_{66} - 12\iota_{65} + 6\iota_{64}) \\ & + 4\iota_{69}(3\iota_{69} - 8\iota_{67} + 6\iota_{66} - 4\iota_{65} + 2\iota_{64}) + 4\iota_{67}(6\iota_{67} - 9\iota_{66} + 6\iota_{65} \\ & - 3\iota_{64}) + \iota_{66}(15\iota_{66} - 20\iota_{65} + 10\iota_{64}) + 8\iota_{65}(\iota_{65} - \iota_{64}) + 3\iota_{64}^2. \end{aligned} \quad (12.2.98)$$

12.3 Decomposition of the Oscillator Representation

In this section, we want to prove the following theorem:

Theorem 12.3.1 *Any singular vector in $\mathscr{B}$ is a polynomial in x_1, $\mathfrak{l}_1$, ϑ, ς and η. Let $L(n_1, n_2, n_3, n_4, n_5)$ be the irreducible submodule generated by $\mathfrak{l}_1^{n_1}\varsigma^{n_2}x_1^{n_3}\vartheta^{n_4}\eta^{n_5}$ with highest weight $n_1\lambda_1 + n_2\lambda_6 + (n_3 + n_4)\lambda_7$. Then*

$$\mathscr{B} = \bigoplus_{n_1,n_2,n_3,n_4,n_5=0}^{\infty} L(n_1, n_2, n_3, n_4, n_5). \tag{12.3.1}$$

In particular,

$$(1-q)^{55} \sum_{n_1,n_2,n_3,n_4=0}^{\infty} (dim\ V(n_1\lambda_1 + n_2\lambda_6 + (n_3+n_4)\lambda_7))q^{2n_1+4n_2+n_3+3n_4}$$
$$= 1 + q + q^2 + q^3. \tag{12.3.2}$$

Let $\mathscr{D}$ be the invariant differential operator obtained from η by changing x_i to ∂_{x_i}. Then

$$\sum_{n_1,n_2,n_3=0}^{\infty} (L(n_1, n_2, n_3, 0, 0) + L(n_1, n_2, n_3, 1, 0)) \subset \{f \in \mathscr{B} \mid \mathscr{D}(f) = 0\}. \tag{12.3.3}$$

Proof Let f be a singular vector in $\mathscr{B}$. To eliminate the extra variables from f, we need certain change of variables. According to (12.2.1), (12.2.4)–(12.2.10), (12.2.12)–(12.2.14), (12.2.16), (12.2.18), (12.2.19), (12.2.23), (12.2.25), (12.2.27), (12.2.30), (12.2.32), (12.2.35), (12.2.39), (12.2.42), (12.2.46), (12.2.50), (12.2.55), (12.2.61), (12.2.66) and (12.2.96), we have

$$x_1x_{17} = \mathfrak{l}_1 - x_2x_{14} - x_3x_{12} - x_4x_{10} - x_5x_9 + x_6x_7. \tag{12.3.4}$$

$$x_1x_{22} = \mathfrak{l}_2 - x_2x_{19} - x_3x_{16} - x_4x_{13} - x_5x_{11} + x_7x_8, \tag{12.3.5}$$

$$x_1x_{25} = \mathfrak{l}_3 - x_2x_{21} - x_3x_{18} - x_4x_{15} - x_6x_{11} + x_8x_9, \tag{12.3.6}$$

$$x_1x_{28} = -\mathfrak{l}_4 - x_2x_{24} - x_3x_{20} + x_5x_{15} - x_6x_{13} + x_8x_{10}, \tag{12.3.7}$$

$$x_1x_{30} = \mathfrak{l}_5 - x_2x_{26} - x_3x_{23} - x_7x_{15} + x_9x_{13} - x_{10}x_{11}, \tag{12.3.8}$$

$$x_1x_{31} = \mathfrak{l}_6 - x_2x_{27} + x_4x_{20} + x_5x_{18} - x_6x_{16} + x_8x_{12}, \tag{12.3.9}$$

$$x_1x_{33} = -\mathfrak{l}_7 - x_2x_{29} + x_4x_{23} - x_7x_{18} + x_9x_{16} - x_{11}x_{12}, \tag{12.3.10}$$

$$x_1x_{34} = -\iota_8 + x_3x_{27} + x_4x_{24} + x_5x_{21} - x_6x_{19} + x_8x_{14}, \tag{12.3.11}$$

$$x_1x_{36} = \iota_{10} - x_2x_{32} - x_5x_{23} - x_7x_{20} + x_{10}x_{16} - x_{12}x_{13}, \tag{12.3.12}$$

$$x_1x_{37} = \iota_{11} + x_3x_{29} + x_4x_{26} - x_7x_{21} + x_9x_{19} - x_{11}x_{14}, \tag{12.3.13}$$

$$x_1x_{38} = -\iota_{12} - x_2x_{35} + x_6x_{23} + x_9x_{20} - x_{10}x_{18} + x_{12}x_{15}, \tag{12.3.14}$$

$$x_1x_{39} = \iota_{14} + x_3x_{32} - x_5x_{26} - x_7x_{24} + x_{10}x_{19} - x_{13}x_{14}, \tag{12.3.15}$$

$$x_1x_{41} = \iota_{16} + x_3x_{35} + x_6x_{26} + x_9x_{24} - x_{10}x_{21} + x_{14}x_{15}, \tag{12.3.16}$$

$$x_1x_{42} = \iota_{17} - x_4x_{32} - x_5x_{29} - x_7x_{27} + x_{12}x_{19} - x_{14}x_{16}, \tag{12.3.17}$$

$$\begin{aligned} x_1\iota_1x_{43} = {} & -\iota_1(x_2x_{40} + x_8x_{23} + x_{11}x_{20} - x_{13}x_{18} + x_{15}x_{16}) \\ & +\varsigma + \iota_2\iota_{12} - \iota_3\iota_{10} + \iota_4\iota_7 - \iota_5\iota_6, \end{aligned} \tag{12.3.18}$$

$$x_1x_{44} = -\iota_{21} - x_4x_{35} + x_6x_{29} + x_9x_{27} - x_{12}x_{21} + x_{14}x_{18}, \tag{12.3.19}$$

$$x_1x_{45} = -\iota_{23} + x_3x_{40} - x_8x_{26} - x_{11}x_{24} + x_{13}x_{21} - x_{15}x_{19}, \tag{12.3.20}$$

$$x_1x_{46} = \iota_{25} + x_5x_{35} + x_6x_{32} + x_{10}x_{27} - x_{12}x_{24} + x_{14}x_{20}, \tag{12.3.21}$$

$$x_1x_{47} = \iota_{28} - x_4x_{40} - x_8x_{29} - x_{11}x_{27} + x_{16}x_{21} - x_{18}x_{19}, \tag{12.3.22}$$

$$x_1x_{48} = -\iota_{30} + x_5x_{40} - x_8x_{32} - x_{13}x_{27} + x_{16}x_{24} - x_{19}x_{20}, \tag{12.3.23}$$

$$x_1x_{49} = -\iota_{33} + x_7x_{35} + x_9x_{32} - x_{10}x_{29} + x_{12}x_{26} - x_{14}x_{23}, \tag{12.3.24}$$

$$x_1x_{50} = \iota_{37} - x_6x_{40} - x_8x_{35} + x_{15}x_{27} - x_{18}x_{24} + x_{20}x_{21}, \tag{12.3.25}$$

$$x_1x_{51} = -\iota_{40} + x_7x_{40} - x_{11}x_{32} + x_{13}x_{29} - x_{16}x_{26} + x_{19}x_{23}, \tag{12.3.26}$$

$$x_1x_{52} = \iota_{44} - x_9x_{40} - x_{11}x_{35} - x_{15}x_{29} + x_{18}x_{26} - x_{21}x_{23}, \tag{12.3.27}$$

$$x_1x_{53} = \iota_{48} + x_{10}x_{40} + x_{13}x_{35} + x_{15}x_{32} - x_{20}x_{26} + x_{23}x_{24}, \tag{12.3.28}$$

$$x_1x_{54} = \iota_{53} - x_{12}x_{40} - x_{16}x_{35} - x_{18}x_{32} + x_{20}x_{29} - x_{23}x_{27}, \tag{12.3.29}$$

$$x_1x_{55} = \iota_{59} + x_{14}x_{40} + x_{19}x_{35} + x_{21}x_{32} - x_{24}x_{29} + x_{26}x_{27}. \tag{12.3.30}$$

Moreover, (12.2.1) and (12.2.4)–(12.2.73) imply that when $x_i = 0$ for $i \in \overline{3, 54}$,

$$\vartheta = \frac{x_1}{2}(x_1 x_{56} - x_2 x_{55}), \tag{12.3.31}$$

which is the homogeneous part of ϑ with degree 1 in x_{55} and x_{56} (cf. (12.2.95)), and

$$\eta = 3x_1^2 x_{56}^2 - 6x_1 x_2 x_{55} x_{56} - 5x_2^2 x_{55}^2, \tag{12.3.32}$$

which is the homogeneous part of η with degree 2 in x_{55} and x_{56} (cf. (12.2.98)). Under the assumption, we substitute $x_1 x_{56} = 2x_1^{-1}\vartheta + x_2 x_{55}$ into (12.3.32), we obtain

$$\eta = -8x_2^2 x_{55}^2 + 12x_1^{-2}\vartheta^2. \tag{12.3.33}$$

This shows that f can be written as a function in $\{x_i, \vartheta, \eta \mid i \in \overline{1, 54}$. Moreover, using (12.3.4)–(12.3.30), we obtain that f is a function f_1 in

$$\begin{aligned}&\{x_r, \iota_s, \varsigma, \vartheta, \eta \mid r \in \{\overline{1, 35}, 40\} \setminus \{17, 22, 25, 28, 30, 31, 33, 34\},\\ &s \in \{\overline{1, 17}, 21, 23, 25, 28, 30, 33, 37, 40, 44, 48, 53\} \setminus \{9, 13, 15\}\},\end{aligned} \tag{12.3.34}$$

which is rational in the above variable except ϑ and η.

Denote

$$\begin{aligned}U = &\sum_{i=1}^{8} \mathbb{F}\iota_i + \sum_{13,15 \neq r \in \overline{10,17}} \mathbb{F}\iota_r + \mathbb{F}\iota_{19} + \mathbb{F}\iota_{21} + \mathbb{F}\iota_{23} + \mathbb{F}\iota_{25} + \mathbb{F}\iota_{28}\\ &+\mathbb{F}\iota_{30} + \mathbb{F}\iota_{33} + \mathbb{F}\iota_{37} + \mathbb{F}\iota_{40} + \mathbb{F}\iota_{44} + \mathbb{F}\iota_{48} + \mathbb{F}\iota_{53}.\end{aligned} \tag{12.3.35}$$

By (12.2.79)–(12.2.85), we get

$$E_{\alpha_1}|_U = -\iota_1\partial_{\iota_2} - \iota_{12}\partial_{\iota_{19}} - \iota_{16}\partial_{\iota_{23}} - \iota_{21}\partial_{\iota_{28}} - \iota_{25}\partial_{\iota_{30}} - \iota_{33}\partial_{\iota_{40}}, \tag{12.3.36}$$

$$E_{\alpha_2}|_U = \iota_4\partial_{\iota_5} + \iota_6\partial_{\iota_7} + \iota_8\partial_{\iota_{11}} + \iota_{25}\partial_{\iota_{33}} + \iota_{30}\partial_{\iota_{40}} + \iota_{37}\partial_{\iota_{44}}, \tag{12.3.37}$$

$$E_{al_3}|_U = -\iota_2\partial_{\iota_3} - \iota_{10}\partial_{\iota_{12}} - \iota_{14}\partial_{\iota_{16}} - \iota_{17}\partial_{\iota_{21}} - \iota_{30}\partial_{\iota_{37}} - \iota_{40}\partial_{\iota_{44}}, \tag{12.3.38}$$

$$E_{\alpha_4}|_U = \iota_3\partial_{\iota_4} + \iota_7\partial_{\iota_{10}} + \iota_{11}\partial_{\iota_{14}} + \iota_{21}\partial_{\iota_{25}} + \iota_{28}\partial_{\iota_{30}} + \iota_{44}\partial_{\iota_{48}}, \tag{12.3.39}$$

$$E_{\alpha_5}|_U = \iota_4\partial_{\iota_6} + \iota_5\partial_{\iota_7} + \iota_{14}\partial_{\iota_{17}} + \iota_{16}\partial_{\iota_{21}} + \iota_{23}\partial_{\iota_{28}} + \iota_{48}\partial_{\iota_{53}}, \tag{12.3.40}$$

$$E_{\alpha_6}|_U = \iota_6\partial_{\iota_8} + \iota_7\partial_{\iota_{11}} + \iota_{10}\partial_{\iota_{14}} + \iota_{12}\partial_{\iota_{16}} + \iota_{19}\partial_{\iota_{23}}, \tag{12.3.41}$$

$$E_{\alpha_7}|_U = 0. \tag{12.3.42}$$

It turns out that

$$E_{\alpha_i}(U) \subset U \qquad \text{for } i \in \overline{1,7}. \tag{12.3.43}$$

By induction, we can prove that

$$[E_\alpha, E_\beta]|_U = [E_\alpha|_U, E_\beta|_U] \qquad \text{for } \alpha, \beta \in \Phi^+_{E_7}. \tag{12.3.44}$$

In particular,

$$E_\gamma|_U = 0 \qquad \text{for } \gamma \in \Phi^+_{E_7} \setminus \Phi^+_{E_6}. \tag{12.3.45}$$

According to (12.3.36)–(12.3.41) and (12.3.44), we have:

$$E_{(1,0,1)}|_U = \iota_1\partial_{\iota_3} - \iota_{10}\partial_{\iota_{19}} - \iota_{14}\partial_{\iota_{23}} - \iota_{17}\partial_{\iota_{28}} + \iota_{25}\partial_{\iota_{37}} + \iota_{33}\partial_{\iota_{44}}, \tag{12.3.46}$$

$$E_{(0,1,0,1)}|_U = -\iota_3\partial_{\iota_5} + \iota_6\partial_{\iota_{10}} + \iota_8\partial_{\iota_{14}} - \iota_{21}\partial_{\iota_{33}} - \iota_{28}\partial_{\iota_{40}} + \iota_{37}\partial_{\iota_{48}}, \tag{12.3.47}$$

$$E_{(0,0,1,1)}|_U = -\iota_2\partial_{\iota_4} + \iota_7\partial_{\iota_{12}} + \iota_{11}\partial_{\iota_{16}} - \iota_{17}\partial_{\iota_{25}} + \iota_{28}\partial_{\iota_{37}} - \iota_{40}\partial_{\iota_{48}}, \tag{12.3.48}$$

$$E_{(0,0,0,1,1)}|_U = \iota_3\partial_{\iota_6} - \iota_5\partial_{\iota_{10}} + \iota_{11}\partial_{\iota_{17}} - \iota_{16}\partial_{\iota_{25}} - \iota_{23}\partial_{\iota_{30}} + \iota_{44}\partial_{\iota_{53}}, \tag{12.3.49}$$

$$E_{(0,0,0,0,1,1)}|_U = \iota_4\partial_{\iota_8} + \iota_5\partial_{\iota_{11}} - \iota_{10}\partial_{\iota_{17}} - \iota_{12}\partial_{\iota_{21}} - \iota_{19}\partial_{\iota_{28}}, \tag{12.3.50}$$

$$E_{(1,0,1,1)}|_U = \iota_1\partial_{\iota_4} + \iota_7\partial_{\iota_{19}} + \iota_{11}\partial_{\iota_{23}} - \iota_{17}\partial_{\iota_{30}} - \iota_{21}\partial_{\iota_{37}} + \iota_{33}\partial_{\iota_{48}}, \tag{12.3.51}$$

$$E_{(0,1,1,1)}|_U = \iota_2\partial_{\iota_5} + \iota_6\partial_{\iota_{12}} + \iota_8\partial_{\iota_{16}} + \iota_{17}\partial_{\iota_{33}} - \iota_{28}\partial_{\iota_{44}} - \iota_{30}\partial_{\iota_{48}}, \tag{12.3.52}$$

$$E_{(0,1,0,1,1)}|_U = -\iota_3\partial_{\iota_7} - \iota_4\partial_{\iota_{10}} + \iota_8\partial_{\iota_{17}} + \iota_{16}\partial_{\iota_{33}} + \iota_{23}\partial_{\iota_{40}} + \iota_{37}\partial_{\iota_{53}}, \tag{12.3.53}$$

$$E_{(0,0,1,1,1)}|_U = -\iota_2\partial_{\iota_6} - \iota_5\partial_{\iota_{12}} + \iota_{11}\partial_{\iota_{21}} + \iota_{14}\partial_{\iota_{25}} - \iota_{23}\partial_{\iota_{37}} - \iota_{40}\partial_{\iota_{53}}, \tag{12.3.54}$$

$$E_{(0,0,0,1,1,1)}|_U = \iota_3\partial_{\iota_8} - \iota_5\partial_{\iota_{14}} - \iota_7\partial_{\iota_{17}} + \iota_{12}\partial_{\iota_{25}} + \iota_{19}\partial_{\iota_{30}}, \tag{12.3.55}$$

$$E_{(1,1,1,1)}|_U = -\iota_1\partial_{\iota_5} + \iota_6\partial_{\iota_{19}} + \iota_8\partial_{\iota_{23}} + \iota_{17}\partial_{\iota_{40}} + \iota_{21}\partial_{\iota_{44}} + \iota_{25}\partial_{\iota_{48}}, \tag{12.3.56}$$

$$E_{(1,0,1,1,1)}|_U = \iota_1\partial_{\iota_6} - \iota_5\partial_{\iota_{19}} + \iota_{11}\partial_{\iota_{28}} + \iota_{14}\partial_{\iota_{30}} + \iota_{16}\partial_{\iota_{37}} + \iota_{33}\partial_{\iota_{53}}, \tag{12.3.57}$$

$$E_{(0,1,1,1,1)}|_U = \iota_2\partial_{\iota_7} - \iota_4\partial_{\iota_{12}} + \iota_8\partial_{\iota_{21}} - \iota_{14}\partial_{\iota_{33}} + \iota_{23}\partial_{\iota_{44}} - \iota_{30}\partial_{\iota_{53}}, \tag{12.3.58}$$

$$E_{(0,1,0,1,1,1)}|_U = -\iota_3\partial_{\iota_{11}} - \iota_4\partial_{\iota_{14}} - \iota_6\partial_{\iota_{17}} - \iota_{12}\partial_{\iota_{33}} - \iota_{19}\partial_{\iota_{40}}, \tag{12.3.59}$$

$$E_{(0,0,1,1,1,1)}|_U = -\iota_2\partial_{\iota_8} - \iota_5\partial_{\iota_{16}} - \iota_7\partial_{\iota_{21}} - \iota_{10}\partial_{\iota_{25}} + \iota_{19}\partial_{\iota_{37}}, \tag{12.3.60}$$

$$E_{(1,1,1,1,1)}|_U = -\iota_1\partial_{\iota_7} - \iota_4\partial_{\iota_{19}} + \iota_8\partial_{\iota_{28}} - \iota_{14}\partial_{\iota_{40}} - \iota_{16}\partial_{\iota_{44}} + \iota_{25}\partial_{\iota_{53}}, \tag{12.3.61}$$

$$E_{(1,0,1,1,1,1)}|_U = \iota_1\partial_{\iota_8} - \iota_5\partial_{\iota_{23}} - \iota_7\partial_{\iota_{28}} - \iota_{10}\partial_{\iota_{30}} - \iota_{12}\partial_{\iota_{37}}, \tag{12.3.62}$$

$$E_{(0,1,1,1,1,1)}|_U = \iota_2\partial_{\iota_{11}} - \iota_4\partial_{\iota_{16}} - \iota_6\partial_{\iota_{21}} + \iota_{10}\partial_{\iota_{33}} - \iota_{19}\partial_{\iota_{44}}, \tag{12.3.63}$$

$$E_{(0,1,1,2,1)}|_U = \iota_2\partial_{\iota_{10}} + \iota_3\partial_{\iota_{12}} + \iota_8\partial_{\iota_{25}} + \iota_{11}\partial_{\iota_{33}} + \iota_{23}\partial_{\iota_{48}} + \iota_{28}\partial_{\iota_{53}}, \tag{12.3.64}$$

$$E_{(1,1,1,2,1)}|_U = -\iota_1\partial_{\iota_{10}} + \iota_3\partial_{\iota_{19}} + \iota_8\partial_{\iota_{30}} + \iota_{11}\partial_{\iota_{40}} - \iota_{16}\partial_{\iota_{48}} - \iota_{21}\partial_{\iota_{53}}, \tag{12.3.65}$$

$$E_{(1,1,1,1,1,1)}|_U = -\iota_1\partial_{\iota_{11}} - \iota_4\partial_{\iota_{23}} - \iota_6\partial_{\iota_{28}} + \iota_{10}\partial_{\iota_{40}} + \iota_{12}\partial_{\iota_{44}}, \tag{12.3.66}$$

$$E_{(0,1,1,2,1,1)}|_U = \iota_2\partial_{\iota_{14}} + \iota_3\partial_{\iota_{16}} - \iota_6\partial_{\iota_{25}} - \iota_7\partial_{\iota_{33}} - \iota_{19}\partial_{\iota_{48}}, \tag{12.3.67}$$

$$E_{(1,1,2,2,1)}|_U = -\iota_1\partial_{\iota_{12}} - \iota_2\partial_{\iota_{19}} + \iota_8\partial_{\iota_{37}} + \iota_{11}\partial_{\iota_{44}} + \iota_{14}\partial_{\iota_{48}} + \iota_{17}\partial_{\iota_{53}}, \tag{12.3.68}$$

$$E_{(1,1,1,2,1,1)}|_U = -\iota_1\partial_{\iota_{14}} + \iota_3\partial_{\iota_{23}} - \iota_6\partial_{\iota_{30}} - \iota_7\partial_{\iota_{40}} + \iota_{12}\partial_{\iota_{48}}, \tag{12.3.69}$$

$$E_{(0,1,1,2,2,1)}|_U = \iota_2\partial_{\iota_{17}} + \iota_3\partial_{\iota_{21}} + \iota_4\partial_{\iota_{25}} + \iota_5\partial_{\iota_{33}} - \iota_{19}\partial_{\iota_{53}}, \tag{12.3.70}$$

$$E_{(1,1,2,2,1,1)}|_U = -\iota_1\partial_{\iota_{16}} - \iota_2\partial_{\iota_{23}} - \iota_6\partial_{\iota_{37}} - \iota_7\partial_{\iota_{44}} - \iota_{10}\partial_{\iota_{48}}, \tag{12.3.71}$$

$$E_{(1,1,1,2,2,1)}|_U = -\iota_1\partial_{\iota_{17}} + \iota_3\partial_{\iota_{28}} + \iota_4\partial_{\iota_{30}} + \iota_5\partial_{\iota_{40}} + \iota_{12}\partial_{\iota_{53}}, \tag{12.3.72}$$

$$E_{(1,1,2,2,2,1)}|_U = -\iota_1\partial_{\iota_{21}} - \iota_2\partial_{\iota_{28}} + \iota_4\partial_{\iota_{37}} + \iota_5\partial_{\iota_{44}} - \iota_{10}\partial_{\iota_{53}}, \tag{12.3.73}$$

$$E_{(1,1,2,3,2,1)}|_U = -\iota_1\partial_{\iota_{25}} - \iota_2\partial_{\iota_{30}} - \iota_3\partial_{\iota_{37}} + \iota_5\partial_{\iota_{48}} + \iota_7\partial_{\iota_{53}}, \tag{12.3.74}$$

$$E_{(1,2,2,3,2,1)}|_U = -\iota_1\partial_{\iota_{33}} - \iota_2\partial_{\iota_{40}} - \iota_3\partial_{\iota_{44}} - \iota_4\partial_{\iota_{48}} - \iota_6\partial_{\iota_{53}}. \tag{12.3.75}$$

By (12.1.83)–(12.1.92) and (12.3.45), we have

$$0 = E_{(1,1,2,3,2,1,1)}(f_1) = -x_1\partial_{x_{20}}(f_1), \tag{12.3.76}$$

$$0 = E_{(1,1,2,2,2,2,1)}(f_1) = -x_1\partial_{x_{21}}(f_1), \tag{12.3.77}$$

$$0 = E_{(1,2,2,3,2,1,1)}(f_1) = x_1\partial_{x_{23}}(f_1), \tag{12.3.78}$$

$$0 = E_{(1,1,2,3,2,2,1)}(f_1) = x_1\partial_{x_{24}}(f_1), \tag{12.3.79}$$

$$0 = E_{(1,2,2,3,2,2,1)}(f_1) = -x_1\partial_{x_{26}}(f_1), \tag{12.3.80}$$

$$0 = E_{(1,1,2,3,3,2,1)}(f_1) = -x_1\partial_{x_{27}}(f_1), \tag{12.3.81}$$

$$0 = E_{(1,2,2,3,3,2,1)}(f_1) = x_1\partial_{x_{29}}(f_1), \tag{12.3.82}$$

$$0 = E_{(1,2,2,4,3,2,1)}(f_1) = -x_1\partial_{x_{32}}(f_1), \tag{12.3.83}$$

$$0 = E_{(1,2,3,4,3,2,1)}(f_1) = -x_1\partial_{x_{35}}(f_1), \tag{12.3.84}$$

$$0 = E_{(2,2,3,4,3,2,1)}(f_1) = -x_1\partial_{x_{40}}(f_1). \tag{12.3.85}$$

So f_1 is independent of $\{x_{20}, x_{21}, x_{23}, x_{24}, x_{26}, x_{27}, x_{29}, x_{32}, x_{35}, x_{40}\}$; that is, f_1 is a function in

$$\{x_r, \mathfrak{l}_s, \vartheta, \varsigma \mid 17 \neq r \in \overline{1,19};\ s \in \{\overline{1,17}, 21, 23, 25, 28, 30, 33, 37, 40, 44, 48, 53\} \setminus \{9, 13, 15\}\}. \tag{12.3.86}$$

Expressions (12.1.36), (12.1.42), (12.1.48), (12.1.54), (12.1.59), (12.1.60), (12.1.63), (12.1.64), (12.1.68), (12.1.70), (12.1.73), (12.1.74), (12.1.76)–(12.1.78), (12.1.80), (12.1.81) and (12.3.45) imply

$$0 = E_{\alpha_7}(f_1) = x_1\partial_{x_2}(f_1), \tag{12.3.87}$$

$$0 = E_{(0,0,0,0,0,1,1)}(f_1) = -x_1\partial_{x_3}(f_1), \tag{12.3.88}$$

$$0 = E_{(0,0,0,0,1,1,1)}(f_1) = x_1\partial_{x_4}(f_1), \tag{12.3.89}$$

$$0 = E_{(0,0,0,1,1,1,1)}(f_1) = -x_1\partial_{x_5}(f_1), \tag{12.3.90}$$

$$0 = E_{(0,1,0,1,1,1,1)}(f_1) = x_1\partial_{x_7}(f_1), \tag{12.3.91}$$

$$0 = E_{(0,0,1,1,1,1,1)}(f_1) = -x_1\partial_{x_6}(f_1), \tag{12.3.92}$$

$$0 = E_{(1,0,1,1,1,1,1)}(f_1) = -x_1\partial_{x_8}(f_1), \tag{12.3.93}$$

$$0 = E_{(0,1,1,1,1,1,1)}(f_1) = x_1\partial_{x_9}(f_1), \tag{12.3.94}$$

$$0 = E_{(1,1,1,1,1,1,1)}(f_1) = x_1\partial_{x_{11}}(f_1), \tag{12.3.95}$$

$$0 = E_{(0,1,1,2,1,1,1)}(f_1) = -x_1\partial_{x_{10}}(f_1), \tag{12.3.96}$$

$$0 = E_{(1,1,1,2,1,1,1)}(f_1) = -x_1\partial_{x_{13}}(f_1), \tag{12.3.97}$$

$$0 = E_{(0,1,1,2,2,1,1)}(f_1) = x_1\partial_{x_{12}}(f_1), \tag{12.3.98}$$

$$0 = E_{(1,1,2,2,1,1,1)}(f_1) = -x_1\partial_{x_{15}}(f_1), \tag{12.3.99}$$

$$0 = E_{(1,1,1,2,2,1,1)}(f_1) = x_1\partial_{x_{16}}(f_1), \tag{12.3.100}$$

$$0 = E_{(0,1,1,2,2,2,1)}(f_1) = -x_1\partial_{x_{14}}(f_1), \tag{12.3.101}$$

$$0 = E_{(1,1,2,2,2,1,1)}(f_1) = x_1\partial_{x_{18}}(f_1), \tag{12.3.102}$$

$$0=E_{(1,1,1,2,2,2,1)}(f_1)=-x_1\partial_{x_{19}}(f_1). \tag{12.3.103}$$

Hence f_1 is independent of $\{x_r,x_{18},x_{19}\mid \overline{2,16}\}$; that is, f_1 is a function in

$$\{x_1,\iota_s,\vartheta,\varsigma\mid s\in\{\overline{1,17},21,23,25,28,30,33,37,40,44,48,53\}\setminus\{9,13,15\}\}. \tag{12.3.104}$$

Now by (12.3.36)–(12.3.41), (12.3.53), (12.3.59), (12.3.64), (12.3.66), (12.3.68), (12.3.70), (12.3.71) and (12.3.73)–(12.3.75), we obtain

$$\iota_1\partial_{\iota_2}(f_1)+\iota_{16}\partial_{\iota_{23}}(f_1)+\iota_{21}\partial_{\iota_{28}}(f_1)+\iota_{25}\partial_{\iota_{30}}(f_1)+\iota_{33}\partial_{\iota_{40}}(f_1)=0, \tag{12.3.105}$$

$$\iota_4\partial_{\iota_5}(f_1)+\iota_6\partial_{\iota_7}(f_1)+\iota_8\partial_{\iota_{11}}(f_1)+\iota_{25}\partial_{\iota_{33}}(f_1)+\iota_{30}\partial_{\iota_{40}}(f_1)+\iota_{37}\partial_{\iota_{44}}(f_1)=0, \tag{12.3.106}$$

$$\iota_2\partial_{\iota_3}(f_1)+\iota_{10}\partial_{\iota_{12}}(f_1)+\iota_{14}\partial_{\iota_{16}}(f_1)+\iota_{17}\partial_{\iota_{21}}(f_1)+\iota_{30}\partial_{\iota_{37}}(f_1)+\iota_{40}\partial_{\iota_{44}}(f_1)=0, \tag{12.3.107}$$

$$\iota_3\partial_{\iota_4}(f_1)+\iota_7\partial_{\iota_{10}}(f_1)+\iota_{11}\partial_{\iota_{14}}(f_1)+\iota_{21}\partial_{\iota_{25}}(f_1)+\iota_{28}\partial_{\iota_{30}}(f_1)+\iota_{44}\partial_{\iota_{48}}(f_1)=0, \tag{12.3.108}$$

$$\iota_4\partial_{\iota_6}(f_1)+\iota_5\partial_{\iota_7}(f_1)+\iota_{14}\partial_{\iota_{17}}(f_1)+\iota_{16}\partial_{\iota_{21}}(f_1)+\iota_{23}\partial_{\iota_{28}}(f_1)+\iota_{48}\partial_{\iota_{53}}(f_1)=0, \tag{12.3.109}$$

$$\iota_6\partial_{\iota_8}(f_1)+\iota_7\partial_{\iota_{11}}(f_1)+\iota_{10}\partial_{\iota_{14}}(f_1)+\iota_{12}\partial_{\iota_{16}}(f_1)+\iota_{19}\partial_{\iota_{23}}(f_1)=0, \tag{12.3.110}$$

$$-\iota_3\partial_{\iota_7}(f_1)-\iota_4\partial_{\iota_{10}}(f_1)+\iota_8\partial_{\iota_{17}}(f_1)+\iota_{16}\partial_{\iota_{33}}(f_1)+\iota_{23}\partial_{\iota_{40}}(f_1)+\iota_{37}\partial_{\iota_{53}}(f_1)=0, \tag{12.3.111}$$

$$\iota_3\partial_{\iota_{11}}(f_1)+\iota_4\partial_{\iota_{14}}(f_1)+\iota_6\partial_{\iota_{17}}(f_1)+\iota_{12}\partial_{\iota_{33}}(f_1)+\iota_{19}\partial_{\iota_{40}}(f_1)=0, \tag{12.3.112}$$

$$\iota_2\partial_{\iota_{10}}(f_1)+\iota_3\partial_{\iota_{12}}(f_1)+\iota_8\partial_{\iota_{25}}(f_1)+\iota_{11}\partial_{\iota_{33}}(f_1)+\iota_{23}\partial_{\iota_{48}}(f_1)+\iota_{28}\partial_{\iota_{53}}(f_1)=0, \tag{12.3.113}$$

$$\iota_2\partial_{\iota_{14}}(f_1)+\iota_3\partial_{\iota_{16}}(f_1)-\iota_6\partial_{\iota_{25}}(f_1)-\iota_7\partial_{\iota_{33}}(f_1)-\iota_{19}\partial_{\iota_{48}}(f_1)=0, \tag{12.3.114}$$

$$-\iota_1\partial_{\iota_{12}}(f_1)+\iota_8\partial_{\iota_{37}}(f_1)+\iota_{11}\partial_{\iota_{44}}(f_1)+\iota_{14}\partial_{\iota_{48}}(f_1)+\iota_{17}\partial_{\iota_{53}}(f_1)=0, \tag{12.3.115}$$

$$\iota_2\partial_{\iota_{17}}(f_1)+\iota_3\partial_{\iota_{21}}(f_1)+\iota_4\partial_{\iota_{25}}(f_1)+\iota_5\partial_{\iota_{33}}(f_1)-\iota_{19}\partial_{\iota_{53}}(f_1)=0, \tag{12.3.116}$$

$$\iota_1\partial_{\iota_{16}}(f_1)+\iota_2\partial_{\iota_{23}}(f_1)+\iota_6\partial_{\iota_{37}}(f_1)+\iota_7\partial_{\iota_{44}}(f_1)+\iota_{10}\partial_{\iota_{48}}(f_1)=0, \tag{12.3.117}$$

$$-\,x_1\partial_{x_{21}}(f_1)-x_2\partial_{x_{28}}(f_1)+x_4\partial_{x_{37}}(f_1)+x_5\partial_{x_{44}}(f_1)-x_{10}\partial_{x_{53}}(f_1)=0, \qquad (12.3.118)$$

$$x_1\partial_{x_{25}}(f_1)+x_2\partial_{x_{30}}(f_1)+x_3\partial_{x_{37}}(f_1)-x_5\partial_{x_{48}}(f_1)-x_7\partial_{x_{53}}(f_1)=0, \qquad (12.3.119)$$

$$x_1\partial_{x_{33}}(f_1)+x_2\partial_{x_{40}}(f_1)+x_3\partial_{x_{44}}(f_1)+x_4\partial_{x_{48}}(f_1)+x_6\partial_{x_{53}}(f_1)=0. \qquad (12.3.120)$$

Thanks to (12.3.72), we have

$$-\,x_1\partial_{x_{17}}(f_1)+x_3\partial_{x_{28}}(f_1)+x_4\partial_{x_{30}}(f_1)+x_5\partial_{x_{40}}(f_1)+x_{12}\partial_{x_{53}}(f_1)=0. \qquad (12.3.121)$$

Moreover, $x_1\times(12.3.116)+x_2\times(12.3.121)$ gives

$$\begin{aligned}&x_3(x_1\partial_{x_{21}}(f_1)+x_2\partial_{x_{28}}(f_1))+x_4(x_1\partial_{x_{25}}(f_1)+x_2\partial_{x_{30}}(f_1))\\&+x_1x_5\partial_{x_{33}}(f_1)+x_2x_5\partial_{x_{40}}(f_1)+(x_2x_{12}-x_1x_{19})\partial_{x_{53}}(f_1)=0.\end{aligned} \qquad (12.3.122)$$

Furthermore, $(12.3.122)+x_3\times(12.3.118)-x_4\times(12.3.119)$ yields

$$\begin{aligned}&x_5(x_1\partial_{x_{33}}(f_1)+x_2\partial_{x_{40}}(f_1)+x_3\partial_{x_{44}}(f_1)+x_4\partial_{x_{48}}(f_1))\\&+(x_4x_7-x_3x_{10}+x_2x_{12}-x_1x_{19})\partial_{x_{53}}(f_1)=0.\end{aligned} \qquad (12.3.123)$$

Note that $x_5\times(12.3.120)-(12.3.123)$ shows

$$(x_5x_6-x_4x_7+x_3x_{10}-x_2x_{12}+x_1x_{19})\partial_{x_{53}}(f_1)=0\Longrightarrow\partial_{x_{53}}(f_1)=0. \qquad (12.3.124)$$

According to (12.3.69),

$$-\,x_1\partial_{x_{14}}(f_1)+x_3\partial_{x_{23}}(f_1)-x_6\partial_{x_{30}}(f_1)-x_7\partial_{x_{40}}(f_1)+x_{12}\partial_{x_{48}}(f_1)=0. \qquad (12.3.125)$$

Moreover, $x_1\times(12.3.114)+x_2\times(12.3.125)$ gives

$$\begin{aligned}&x_3(x_1\partial_{x_{16}}(f_1)+x_2\partial_{x_{23}}(f_1))-x_6(x_1\partial_{x_{25}}(f_1)+x_2\partial_{x_{30}}(f_1))\\&-x_7(x_1\partial_{x_{33}}(f_1)+x_2\partial_{x_{40}}(f_1))+(x_2x_{12}-x_1x_{19})\partial_{x_{48}}(f_1)=0.\end{aligned} \qquad (12.3.126)$$

Furthermore, $(12.3.126)-x_3\times(12.3.117)+x_6\times(12.3.119)+x_7\times(12.3.120)$ yields

$$(x_2x_{12}-x_3x_{10}+x_4x_7-x_5x_6-x_1x_{19})\partial_{x_{48}}(f_1)=0\Longrightarrow\partial_{x_{48}}(f_1)=0. \qquad (12.3.127)$$

Next (12.3.65) gives

$$-\,x_1\partial_{x_{11}}(f_1)-x_4\partial_{x_{23}}(f_1)-x_6\partial_{x_{28}}+x_{10}\partial_{x_{40}}(f_1)+x_{12}\partial_{x_{44}}(f_1)=0. \qquad (12.3.128)$$

Moreover, $\iota_1 \times$ (12.3.112) $+ \iota_3 \times$ (12.3.128) gives

$$\iota_4(\iota_1\partial_{\iota_{14}}(f_1) - \iota_3\partial_{\iota_{23}}(f_1)) + \iota_6(\iota_1\partial_{\iota_{17}}(f_1) - \iota_3\partial_{\iota_{28}}(f_1)) \\ +\iota_{12}(\iota_1\partial_{\iota_{33}}(f_1) + \iota_3\partial_{\iota_{44}}(f_1)) + (\iota_3\iota_{10} + \iota_1\iota_{19})\partial_{\iota_{40}}(f_1) = 0. \quad (12.3.129)$$

Furthermore, (12.3.129) $- \iota_{12} \times$ (12.3.120) $+ \iota_6 \times$ (12.3.121) $+ \iota_4 \times$ (12.3.125) yields

$$(\iota_3\iota_{10} - \iota_2\iota_{12} - \iota_4\iota_7 + \iota_5\iota_6 + \iota_1\iota_{19})\partial_{\iota_{40}}(f_1) = 0 \Longrightarrow \partial_{\iota_{40}}(f_1) = 0. \quad (12.3.130)$$

Using (12.3.63), we get

$$\iota_2\partial_{\iota_{11}}(f_1) - \iota_4\partial_{\iota_{16}}(f_1) - \iota_6\partial_{\iota_{21}}(f_1) + \iota_{10}\partial_{\iota_{33}}(f_1) - \iota_{19}\partial_{\iota_{44}}(f_1) = 0. \quad (12.3.131)$$

Moreover, $\iota_2 \times$ (12.3.112) $- \iota_3 \times$ (12.3.131) gives

$$\iota_4(\iota_2\partial_{\iota_{14}}(f_1) + \iota_3\partial_{\iota_{16}}(f_1)) + \iota_6(\iota_2\partial_{\iota_{17}}(f_1) + \iota_3\partial_{\iota_{21}}(f_1)) \\ +\iota_3\iota_{19}\partial_{\iota_{44}}(f_1) + (\iota_2\iota_{12} - \iota_3\iota_{10})\partial_{\iota_{33}}(f_1) = 0. \quad (12.3.132)$$

Furthermore, (12.3.132) $- \iota_4 \times$ (12.3.114) $- \iota_6 \times$ (12.3.116) $- \iota_{19} \times$ (12.3.120) yields

$$(\iota_2\iota_{12} - \iota_1\iota_{19} - \iota_3\iota_{10} + \iota_4\iota_7 - \iota_5\iota_6)\partial_{\iota_{33}}(f_1) = 0 \Longrightarrow \partial_{\iota_{33}}(f_1) = 0. \quad (12.3.133)$$

Substituting the second equations in (12.3.124), (12.3.127), (12.3.130) and (12.3.133) into (12.3.120), we get $\partial_{\iota_{44}}(f_1) = 0$.

Note that (12.3.62) gives

$$\iota_1\partial_{\iota_8}(f_1) - \iota_5\partial_{\iota_{23}}(f_1) - \iota_7\partial_{\iota_{28}}(f_1) - \iota_{10}\partial_{\iota_{30}}(f_1) - \iota_{12}\partial_{\iota_{37}}(f_1) = 0. \quad (12.3.134)$$

Moreover, $\iota_1 \times$ (12.3.110) $- \iota_6 \times$ (12.3.134) yields

$$\iota_1(\iota_7\partial_{\iota_{11}}(f_1) + \iota_{10}\partial_{\iota_{14}}(f_1) + \iota_{12}\partial_{\iota_{16}}(f_1)) + \iota_6\iota_7\partial_{\iota_{28}}(f_1) \\ +\iota_6\iota_{10}\partial_{\iota_{30}}(f_1) + (\iota_1\iota_{19} + \iota_5\iota_6)\partial_{\iota_{23}}(f_1) + \iota_6\iota_{12}\partial_{\iota_{37}}(f_1) = 0. \quad (12.3.135)$$

Furthermore, (12.3.135) $+ \iota_7 \times$ (12.3.128) says

$$\iota_1(\iota_{10}\partial_{\iota_{14}}(f_1) + \iota_{12}\partial_{\iota_{16}}(f_1)) + \iota_6\iota_{10}\partial_{\iota_{30}}(f_1) \\ +(\iota_1\iota_{19} - \iota_4\iota_7 + \iota_5\iota_6)\partial_{\iota_{23}}(f_1) + \iota_6\iota_{12}\partial_{\iota_{37}}(f_1) = 0 \quad (12.3.136)$$

by the fact $\partial_{\iota_{40}}(f_1) = \partial_{\iota_{44}}(f_1) = 0$. Now (12.3.136) $+ \iota_{10} \times$ (12.3.125) implies

$$\iota_1\iota_{12}\partial_{\iota_{16}}(f_1) + (\iota_1\iota_{19} + \iota_3\iota_{10} - \iota_4\iota_7 + \iota_5\iota_6)\partial_{\iota_{23}}(f_1) + \iota_6\iota_{12}\partial_{\iota_{37}}(f_1) = 0. \quad (12.3.137)$$

Observe that now (12.3.117) becomes

$$\mathfrak{l}_1\partial_{\mathfrak{l}_{16}}(f_1)+\mathfrak{l}_2\partial_{\mathfrak{l}_{23}}(f_1)+\mathfrak{l}_6\partial_{\mathfrak{l}_{37}}(f_1)=0. \tag{12.3.138}$$

Besides, (12.3.137) $-\ \mathfrak{l}_{12}\times$ (12.3.137) gives

$$(\mathfrak{l}_1\mathfrak{l}_{19}-\mathfrak{l}_2\mathfrak{l}_{12}+\mathfrak{l}_3\mathfrak{l}_{10}-\mathfrak{l}_4\mathfrak{l}_7+\mathfrak{l}_5\mathfrak{l}_6)\partial_{\mathfrak{l}_{23}}(f_1)=0\Longrightarrow\partial_{\mathfrak{l}_{23}}(f_1)=0. \tag{12.3.139}$$

According to (12.3.61),

$$-\mathfrak{l}_1\partial_{\mathfrak{l}_7}(f_1)+\mathfrak{l}_8\partial_{\mathfrak{l}_{28}}(f_1)=0. \tag{12.3.140}$$

Moreover, (12.3.111) now becomes

$$-\mathfrak{l}_3\partial_{\mathfrak{l}_7}(f_1)-\mathfrak{l}_4\partial_{\mathfrak{l}_{10}}(f_1)+\mathfrak{l}_8\partial_{\mathfrak{l}_{17}}(f_1)=0, \tag{12.3.141}$$

Subtracting $\mathfrak{l}_1\times$ (12.3.141) from $\mathfrak{l}_3\times$ (12.3.140), we get

$$\mathfrak{l}_1\mathfrak{l}_4\partial_{\mathfrak{l}_{10}}(f_1)+\mathfrak{l}_8(-\mathfrak{l}_1\partial_{\mathfrak{l}_{17}}(f_1)+\mathfrak{l}_3\partial_{\mathfrak{l}_{28}}(f_1))=0. \tag{12.3.142}$$

Moreover, (12.3.142) $-\ \mathfrak{l}_8\times$ (12.3.121) gives

$$\mathfrak{l}_1\partial_{\mathfrak{l}_{10}}(f_1)=\mathfrak{l}_8\partial_{\mathfrak{l}_{30}}(f_1). \tag{12.3.143}$$

Besides, now (12.3.115), (12.3.117), (12.3.125) and (12.3.128) become

$$\mathfrak{l}_1\partial_{\mathfrak{l}_{12}}(f_1)=\mathfrak{l}_8\partial_{\mathfrak{l}_{37}}(f_1),\qquad \mathfrak{l}_1\partial_{\mathfrak{l}_{16}}(f_1)=-\mathfrak{l}_6\partial_{\mathfrak{l}_{37}}(f_1), \tag{12.3.144}$$

$$\mathfrak{l}_1\partial_{\mathfrak{l}_{11}}(f_1)=-\mathfrak{l}_6\partial_{\mathfrak{l}_{28}},\qquad \mathfrak{l}_1\partial_{\mathfrak{l}_{14}}(f_1)=-\mathfrak{l}_6\partial_{\mathfrak{l}_{30}}(f_1). \tag{12.3.145}$$

Playing with (12.3.60), we find

$$-\mathfrak{l}_2\partial_{\mathfrak{l}_8}(f_1)-\mathfrak{l}_5\partial_{\mathfrak{l}_{16}}(f_1)-\mathfrak{l}_7\partial_{\mathfrak{l}_{21}}(f_1)-\mathfrak{l}_{10}\partial_{\mathfrak{l}_{25}}(f_1)+\mathfrak{l}_{19}\partial_{\mathfrak{l}_{37}}(f_1)=0. \tag{12.3.146}$$

On the other hand, substituting (12.3.144) and (12.3.145) into $\mathfrak{l}_1\times$ (12.3.110), we obtain

$$\mathfrak{l}_1\partial_{\mathfrak{l}_8}(f_1)-\mathfrak{l}_7\partial_{\mathfrak{l}_{28}}(f_1)-\mathfrak{l}_{10}\partial_{\mathfrak{l}_{30}}(f_1)-\mathfrak{l}_{12}\partial_{\mathfrak{l}_{37}}(f_1)=0. \tag{12.3.147}$$

Calculating $\mathfrak{l}_1\times$ (12.3.146) by (12.3.118), (12.3.119) and the second equation in (12.3.144), we find

$$\begin{aligned}&-\mathfrak{l}_1\mathfrak{l}_2\partial_{\mathfrak{l}_8}(f_1)+\mathfrak{l}_2\mathfrak{l}_{10}\partial_{\mathfrak{l}_{30}}(f_1)+\mathfrak{l}_2\mathfrak{l}_7\partial_{\mathfrak{l}_{28}}(f_1)\\&+(\mathfrak{l}_1\mathfrak{l}_{19}+\mathfrak{l}_3\mathfrak{l}_{10}-\mathfrak{l}_4\mathfrak{l}_7+\mathfrak{l}_5\mathfrak{l}_6)\partial_{\mathfrak{l}_{37}}(f_1)=0.\end{aligned} \tag{12.3.148}$$

Moreover, $\iota_2 \times$ (12.3.147) + (12.3.148) yields

$$(\iota_1\iota_{19} - \iota_2\iota_{12} + \iota_3\iota_{10} - \iota_4\iota_7 + \iota_5\iota_6)\partial_{\iota_{37}}(f_1) = 0 \Longrightarrow \partial_{\iota_{37}}(f_1) = 0. \qquad (12.3.149)$$

By (12.3.144),

$$\partial_{\iota_{12}}(f_1) = \partial_{\iota_{16}}(f_1) = 0. \qquad (12.3.150)$$

Furthermore, (12.3.118) and (12.3.119) become

$$\iota_1\partial_{\iota_{21}}(f_1) = -\iota_2\partial_{\iota_{28}}(f_1), \qquad \iota_1\partial_{\iota_{25}}(f_1) = -\iota_2\partial_{\iota_{30}}(f_1). \qquad (12.3.151)$$

Based on (12.3.57), we have

$$\iota_1\partial_{\iota_6}(f_1) + \iota_{11}\partial_{\iota_{28}}(f_1) + \iota_{14}\partial_{\iota_{30}}(f_1) = 0. \qquad (12.3.152)$$

Substituting (12.3.140) and (12.3.151) into $\iota_1 \times$ (12.3.109), we obtain

$$\iota_1\iota_4\partial_{\iota_6}(f_1) + \iota_1\iota_{14}\partial_{\iota_{17}}(f_1) + (\iota_1\iota_{23} - \iota_2\iota_{16} + \iota_5\iota_8)\partial_{\iota_{28}}(f_1) = 0. \qquad (12.3.153)$$

Moreover, we get

$$\iota_1\partial_{\iota_{17}}(f_1) = \iota_3\partial_{\iota_{28}}(f_1) + \iota_4\partial_{\iota_{30}}(f_1) \qquad (12.3.154)$$

by substituting (12.3.151) into (12.3.116). Again we substitute (12.3.154) into (12.3.153):

$$\iota_1\iota_4\partial_{\iota_6}(f_1) + \iota_4\iota_{14}\partial_{\iota_{30}}(f_1) + (\iota_1\iota_{23} - \iota_2\iota_{16} + \iota_3\iota_{14} + \iota_5\iota_8)\partial_{\iota_{28}}(f_1) = 0. \quad (12.3.155)$$

So (12.3.155) $- \iota_4 \times$ (12.3.152) gives

$$(\iota_1\iota_{23} - \iota_2\iota_{16} + \iota_3\iota_{14} - \iota_4\iota_{11} + \iota_5\iota_8)\partial_{\iota_{28}}(f_1) = 0 \Longrightarrow \partial_{\iota_{28}}(f_1) = 0. \qquad (12.3.156)$$

By (12.3.140), (12.3.145) and (12.3.151),

$$\partial_{\iota_7}(f_1) = \partial_{\iota_{11}}(f_1) = \partial_{\iota_{21}}(f_1) = 0. \qquad (12.3.157)$$

Now (12.3.56), (12.3.147) and (12.3.154) imply

$$\partial_{\iota_5}(f_1) = 0, \qquad \iota_1\partial_{\iota_8}(f_1) = \iota_{10}\partial_{\iota_{30}}(f_1), \qquad \iota_1\partial_{\iota_{17}}(f_1) = \iota_4\partial_{\iota_{30}}(f_1). \quad (12.3.158)$$

Applying (12.3.55) to f_1, we have

$$\iota_3\partial_{\iota_8}(f_1) - \iota_5\partial_{\iota_{14}}(f_1) - \iota_7\partial_{\iota_{17}}(f_1) + \iota_{12}\partial_{\iota_{25}}(f_1) + \iota_{19}\partial_{\iota_{30}}(f_1) = 0. \qquad (12.3.159)$$

Substituting (12.3.145), (12.3.151) and (12.3.158) into $\iota_1 \times$ (12.3.159), we obtain

$$(\iota_3\iota_{10}+\iota_5\iota_6-\iota_4\iota_7-\iota_2\iota_{12}+\iota_{19})\partial_{\iota_{30}}(f_1)=0 \Longrightarrow \partial_{\iota_{30}}(f_1)=0. \qquad (12.3.160)$$

By (12.3.143), (12.3.145), (12.3.151), (12.3.152) and (12.3.158), we get

$$\partial_{\iota_6}(f_1)=\partial_{\iota_8}(f_1)=\partial_{\iota_{10}}(f_1)=\partial_{\iota_{14}}(f_1)=\partial_{\iota_{17}}(f_1)=\partial_{\iota_{25}}(f_1)=0. \qquad (12.3.161)$$

Up to this stage, we have proved that f_1 is a function in $\{x_1,\iota_1,\iota_2,\iota_3,\iota_4,\vartheta,\varsigma,\eta\}$. Finally, (12.3.105), (12.3.107) and (12.3.108) give

$$\partial_{\iota_2}(f_1)=\partial_{\iota_3}(f_1)=\partial_{\iota_4}(f_1)=0. \qquad (12.3.162)$$

This proves that f_1 is a function in $\{x_1,\iota_1,\vartheta,\varsigma,\eta\}$, which is rational in x_1, ι_1 and ς.

Note that ϑ and η are functionally independent over the rational functions in $\{x_i \mid i\in\overline{3,54}\}$ by (12.3.31) and (12.3.32). Moreover, the right hand sides of (12.3.31) and (12.3.32) are coprime polynomials. Since $f_1=f$ is a polynomial in $\{x_i\mid i\in\overline{1,56}\}$, there exists a positive integer n such that $\partial_{x_i}^n(f)=0$ for any $i\in\overline{1,56}$. Applying $\partial_{x_{56}}^n$, $\partial_{x_{55}}^n$, $\partial_{x_{43}}^n$, $\partial_{x_{17}}^n$ and $\partial_{x_1}^n$ to f_1 in the above order, we find that $f=f_1$ must be a polynomial in $\{x_1,\iota_1,\vartheta,\varsigma,\eta\}$ by (12.2.1), (12.2.4)–(12.2.9), (12.2.12), (12.2.14), (12.3.17), (12.3.30) and (12.3.31). Since each subspace of homogeneous polynomials in $\mathscr{B}$ is a finite-dimensional $\mathscr{G}^{E_7}$-module, we get (12.3.1) by the Weyl's theorem of complete reducibility and the fact that all finite-dimensional irreducible $\mathscr{G}^{E_7}$-modules are of highest-weight type. Moreover, (12.3.1) gives

$$\left(\sum_{r=0}^{\infty}q^4\right)\sum_{n_1,n_2,n_3,n_4=0}^{\infty}(\dim L(n_1,n_2,n_3,n_4,0))q^{2n_1+4n_2+n_3+3n_4} = \frac{1}{(1-q)^{56}}; \qquad (12.3.163)$$

that is,

$$\frac{1}{1-q^4}\sum_{n_1,n_2,n_3,n_4=0}^{\infty}(\dim L(n_1,n_2,n_3,n_4,0))q^{2n_1+4n_2+n_3+3n_4} = \frac{1}{(1-q)^{56}}; \qquad (12.3.164)$$

which is equivalent to (12.3.2).

Observe that $\mathscr{D}(\iota_1^{m_1}\varsigma^{m_2}x_1^{m_3}\vartheta^{m_4})$ is also a singular vector of weight $m_1\lambda_1+m_2\lambda_6+(m_3+m_4)\lambda_7$ if it is nonzero. So is a linear combination of the elements $\iota_1^{m_1}\varsigma^{m_2}x_1^{n_3}\vartheta^{n_4}\eta^{n_5}$ such that

$$n_3+n_4=m_3+m_4, \qquad n_3+3n_4+4n_5=m_3+3m_4-4. \qquad (12.3.165)$$

Thus

$$m_4 = 2 + n_4 + 2n_5 \geq 2. \tag{12.3.166}$$

This shows

$$\mathscr{D}(\mathfrak{l}_1^{m_1}\varsigma^{m_2}x_1^{m_3}) = \mathscr{D}(\mathfrak{l}_1^{m_1}\varsigma^{m_2}x_1^{m_3}\vartheta) = 0 \tag{12.3.167}$$

for any nonnegative integers m_1, m_2 and m_3, or equivalently, (12.3.3) holds. The proof of our theorem is completed. □

12.4 Realization of E_7 in 27-Dimensional Space

In this section, we find a representation of the simple Lie algebra of type E_7 on the polynomial algebra in 27 variables, which gives a fractional representation of the corresponding Lie group on 27-dimensional space.

Let us go back to the construction of the basic oscillator representation $\mathfrak{r}$ of $\mathscr{G}^{E_6}$ via the lattice construction of $\mathscr{G}^{E_7}$ in Sect. 11.1.

Take the notations in (4.4.42)–(4.4.44) and (11.1.4)–(11.1.12). For any $i \in \overline{1,27}$, we write

$$\mathfrak{a}_i = E_\alpha \quad \text{if } \mathfrak{b}_i = E'_\alpha. \tag{12.4.1}$$

Set

$$\mathscr{G}_- = \sum_{i=1}^{27} \mathbb{F}\mathfrak{b}_i, \qquad \mathscr{G}_0 = \mathscr{G}^{E_6} + \mathbb{F}\alpha_7, \qquad \mathscr{G}_+ = \sum_{i=1}^{27} \mathbb{F}\mathfrak{a}_i. \tag{12.4.2}$$

Then $\mathscr{G}_\pm$ are abelian subalgebras of $\mathscr{G}^{E_7}$ and $\mathscr{G}_0$ is a maximal reductive Lie subalgebra of $\mathscr{G}^{E_7}$. Moreover,

$$[\mathscr{G}_+, \mathscr{G}_-] \subset \mathscr{G}_0, \qquad [\mathscr{G}_0, \mathscr{G}_\pm] \subset \mathscr{G}_\pm, \qquad \mathscr{G}^{E_7} = \mathscr{G}_- \oplus \mathscr{G}_0 \oplus \mathscr{G}_+. \tag{12.4.3}$$

Denote by λ_i the ith fundamental weight of $\mathscr{G}^{E_6}$. With respect to the adjoint representation of $\mathscr{G}^{E_7}$, $\mathscr{G}_+$ forms an irreducible $\mathscr{G}^{E_6}$-module with highest weight λ_1 and $\mathscr{G}_-$ forms an irreducible $\mathscr{G}^{E_6}$-module with highest weight λ_6.

Take the oscillator representation of $\mathscr{G}^{E_6}$ on

$$\mathscr{A} = \mathbb{F}[x_1, x_2, \ldots, x_{27}] \tag{12.4.4}$$

given in (11.1.18)–(11.1.56). Let

$$\widehat{\alpha} = 2\alpha_1 + 3\alpha_2 + 4\alpha_3 + 6\alpha_4 + 5\alpha_5 + 4\alpha_6 + 3\alpha_7. \tag{12.4.5}$$

Then

$$(\widehat{\alpha}, \alpha_r) = 0 \quad \text{for } r \in \overline{1,6} \tag{12.4.6}$$

by the Dynkin diagram of E_7, where $(\cdot,\cdot)$ is a symmetric $\mathbb{Z}$-bilinear form on Λ_r such that $(\alpha,\alpha)=2$ for $\alpha\in\Phi_{E_7}$. Thanks to (4.4.24),

$$[\widehat{\alpha},\mathcal{G}^{D_6}]=0. \tag{12.4.7}$$

By Schur's Lemma, $\widehat{\alpha}|_V=c\mathrm{Id}_V$ for some constant c, where $V=\mathcal{G}_-$ (cf. (11.1.13)). According to (11.1.15) and (11.1.17), $\mathfrak{r}(\widehat{\alpha})=\widehat{\alpha}|_{\mathcal{A}}=c\sum_{i=1}^{16}x_i\partial_{x_i}$. Expressions (4.4.24), (11.1.4)–(11.1.12), (11.1.15), (11.1.17) and the Dynkin diagram of E_7 yields

$$\mathfrak{r}(\alpha_7)=\alpha_7|_{\mathcal{A}}=-2x_1\partial_{x_1}-\sum_{i=2}^{13}x_i\partial_{x_i}-x_{15}\partial_{x_{15}}-x_{16}\partial_{x_{16}}-x_{18}\partial_{x_{18}}-x_{20}\partial_{x_{20}}. \tag{12.4.8}$$

According to (12.4.6), (12.4.8), the coefficients of $x_1\partial_{x_1}$ in (11.1.54) and Table 11.1, we have that

$$\mathfrak{r}(\widehat{\alpha})=-2D,\qquad \text{where } D=\sum_{i=1}^{27}x_i\partial_{x_i} \tag{12.4.9}$$

is the degree operator on $\mathcal{A}$.

Write

$$\mathcal{T}=\sum_{i=1}^{27}\mathbb{F}\partial_{x_i}. \tag{12.4.10}$$

Then $\mathcal{T}$ forms a $\mathcal{G}^{E_6}$-module with respect to the action

$$u(\partial)=[\mathfrak{r}(u),\partial]\qquad \text{for } u\in\mathcal{G}^{E_6},\ \partial\in\tilde{\Delta}. \tag{12.4.11}$$

On the other hand, $\mathcal{G}_\pm$ (cf. (11.1.4)–(11.1.12), (12.4.1) and (12.4.2)) form $\mathcal{G}^{E_6}$-modules with respect to the adjoint representation. According to (11.1.15) and (11.1.17), the linear map determined by $\mathfrak{b}_i\mapsto x_i$ for $i\in\overline{1,27}$ gives a $\mathcal{G}^{E_6}$-module monomorphism from $\mathcal{G}_-$ to $\mathcal{A}$. Define a bilinear form $(\cdot|\cdot)$ on $\mathcal{G}^{E_7}$ by

$$(h_1|h_2)=(h_1,h_2),\quad (h|E_\alpha)=0,\quad (E_\alpha|E_\beta)=-\delta_{\alpha+\beta,0} \tag{12.4.12}$$

for $h_1,h_2\in H$ and $\alpha,\beta\in\Phi_{E_7}$. It can be verified that $(\cdot|\cdot)$ is a $\mathcal{G}^{E_7}$-invariant form; that is,

$$([u,v]|w)=-(v|[u,w])\qquad \text{for } u,v\in\mathcal{G}^{E_7} \tag{12.4.13}$$

(cf. (4.4.24) and (4.4.25)). Hence the linear map determined by $\mathfrak{a}_i\mapsto\partial_{x_i}$ for $i\in\overline{1,27}$ gives a $\mathcal{G}^{E_6}$-module isomorphism from $\mathcal{G}_+$ to $\mathcal{T}$. Thus we define the action of $\mathcal{G}_+$ on $\mathcal{A}$ by

$$\mathfrak{a}_i|_{\mathcal{A}}=\partial_{x_i}\qquad \text{for } i\in\overline{1,27}. \tag{12.4.14}$$

Recall the Witt algebra

$$\mathbb{W}_{27} = \sum_{i=1}^{27} \mathscr{A}\partial_{x_i} \tag{12.4.15}$$

with the Lie bracket (6.7.9). Now we want to find the differential operators P_1, $P_2, \ldots, P_{27} \in \mathbb{W}_{27}$ such that the following action matches the structure of $\mathscr{G}^{E_7}$:

$$\mathfrak{b}_i|_{\mathscr{A}} = P_i \qquad \text{for } i \in \overline{1, 27}. \tag{12.4.16}$$

Comparing the weights in Tables 11.1 and 11.2, we use (11.2.1) and (11.2.3)–(11.2.28) to assume

$$\begin{aligned} P_1 = \; & x_1 D + c_1\zeta_1\partial_{x_{14}} + c_2\zeta_2\partial_{x_{17}} + c_3\zeta_3\partial_{x_{19}} + c_4\zeta_4\partial_{x_{21}} + c_5\zeta_5\partial_{x_{22}} \\ & + c_6\zeta_6\partial_{x_{23}} + c_7\zeta_7\partial_{x_{24}} + c_9\zeta_9\partial_{x_{25}} + c_{11}\zeta_{11}\partial_{x_{26}} + c_{14}\zeta_{14}\partial_{x_{27}}, \end{aligned} \tag{12.4.17}$$

where $c_i \in \mathbb{C}$. Imposing

$$\begin{aligned} [\partial_{x_1}, P_1] = \; & \mathfrak{r}([\mathfrak{a}_1, \mathfrak{b}_1]) = -\mathfrak{r}(\alpha_7) = 2x_1\partial_{x_1} + \sum_{i=2}^{13} x_i\partial_{x_i} \\ & + x_{15}\partial_{x_{15}} + x_{16}\partial_{x_{16}} + x_{18}\partial_{x_{18}} + x_{20}\partial_{x_{20}} \end{aligned} \tag{12.4.18}$$

by (4.4.25) and (12.4.8), we get

$$c_1 = c_2 = c_3 = -c_4 = c_5 = c_6 = -c_7 = c_9 = c_{11} = c_{14} = -1 \tag{12.4.19}$$

by (11.2.1), (11.2.3)–(11.2.8), (11.2.10), (11.2.12) and (11.2.15). Thus

$$\begin{aligned} P_1 = \; & x_1 D - \zeta_1\partial_{x_{14}} - \zeta_2\partial_{x_{17}} - \zeta_3\partial_{x_{19}} + \zeta_4\partial_{x_{21}} - \zeta_5\partial_{x_{22}} \\ & - \zeta_6\partial_{x_{23}} + \zeta_7\partial_{x_{24}} - \zeta_9\partial_{x_{25}} - \zeta_{11}\partial_{x_{26}} - \zeta_{14}\partial_{x_{27}}. \end{aligned} \tag{12.4.20}$$

According to (11.2.1), (11.2.3)–(11.2.8), (11.2.10), (11.2.12) and (11.2.15), we find

$$[\partial_{x_2}, P_1] = x_1\partial_{x_2} - x_{11}\partial_{x_{14}} - x_{13}\partial_{x_{17}} - x_{16}\partial_{x_{19}} - x_{18}\partial_{x_{21}} - x_{20}\partial_{x_{23}}, \tag{12.4.21}$$

$$[\partial_{x_3}, P_1] = x_1\partial_{x_3} - x_9\partial_{x_{14}} - x_{12}\partial_{x_{17}} - x_{15}\partial_{x_{19}} + x_{18}\partial_{x_{22}} + x_{20}\partial_{x_{24}}, \tag{12.4.22}$$

$$[\partial_{x_4}, P_1] = x_1\partial_{x_4} - x_7\partial_{x_{14}} - x_{10}\partial_{x_{17}} + x_{15}\partial_{x_{21}} + x_{16}\partial_{x_{22}} - x_{20}\partial_{x_{25}}, \tag{12.4.23}$$

$$[\partial_{x_5}, P_1] = x_1\partial_{x_5} + x_6\partial_{x_{14}} - x_{10}\partial_{x_{19}} - x_{12}\partial_{x_{21}} - x_{13}\partial_{x_{22}} + x_{20}\partial_{x_{25}}, \tag{12.4.24}$$

$$[\partial_{x_6}, P_1] = x_1\partial_{x_6} + x_5\partial_{x_{14}} + x_8\partial_{x_{17}} - x_{15}\partial_{x_{23}} - x_{16}\partial_{x_{24}} - x_{18}\partial_{x_{25}}, \tag{12.4.25}$$

$$[\partial_{x_7}, P_1] = x_1\partial_{x_7} - x_4\partial_{x_{14}} + x_8\partial_{x_{19}} + x_{12}\partial_{x_{23}} + x_{13}\partial_{x_{24}} + x_{18}\partial_{x_{26}}, \qquad (12.4.26)$$

$$[\partial_{x_8}, P_1] = x_1\partial_{x_8} + x_6\partial_{x_{17}} + x_7\partial_{x_{19}} + x_9\partial_{x_{21}} + x_{11}\partial_{x_{22}} + x_{20}\partial_{x_{27}}, \qquad (12.4.27)$$

$$[\partial_{x_9}, P_1] = x_1\partial_{x_9} - x_3\partial_{x_{14}} + x_8\partial_{x_{21}} - x_{10}\partial_{x_{23}} + x_{13}\partial_{x_{25}} - x_{16}\partial_{x_{26}}, \qquad (12.4.28)$$

$$[\partial_{x_{10}}, P_1] = x_1\partial_{x_{10}} - x_4\partial_{x_{17}} - x_5\partial_{x_{19}} - x_9\partial_{x_{23}} - x_{11}\partial_{x_{24}} + x_{18}\partial_{x_{27}}, \qquad (12.4.29)$$

$$[\partial_{x_{11}}, P_1] = x_1\partial_{x_{11}} - x_2\partial_{x_{14}} + x_8\partial_{x_{22}} - x_{10}\partial_{x_{24}} - x_{12}\partial_{x_{25}} + x_{15}\partial_{x_{26}}, \qquad (12.4.30)$$

$$[\partial_{x_{12}}, P_1] = x_1\partial_{x_{12}} - x_3\partial_{x_{17}} - x_5\partial_{x_{21}} + x_7\partial_{x_{23}} - x_{11}\partial_{x_{25}} - x_{16}\partial_{x_{27}}, \qquad (12.4.31)$$

$$[\partial_{x_{13}}, P_1] = x_1\partial_{x_{13}} - x_2\partial_{x_{17}} - x_5\partial_{x_{22}} + x_7\partial_{x_{24}} + x_9\partial_{x_{25}} + x_{15}\partial_{x_{27}}, \qquad (12.4.32)$$

$$[\partial_{x_{15}}, P_1] = x_1\partial_{x_{15}} - x_3\partial_{x_{19}} + x_4\partial_{x_{21}} - x_6\partial_{x_{23}} + x_{11}\partial_{x_{26}} + x_{13}\partial_{x_{27}}, \qquad (12.4.33)$$

$$[\partial_{x_{16}}, P_1] = x_1\partial_{x_{16}} - x_2\partial_{x_{19}} + x_4\partial_{x_{22}} - x_6\partial_{x_{24}} - x_9\partial_{x_{26}} - x_{12}\partial_{x_{27}}, \qquad (12.4.34)$$

$$[\partial_{x_{18}}, P_1] = x_1\partial_{x_{18}} - x_2\partial_{x_{21}} + x_3\partial_{x_{22}} - x_6\partial_{x_{25}} + x_7\partial_{x_{26}} + x_{10}\partial_{x_{27}}, \qquad (12.4.35)$$

$$[\partial_{x_{20}}, P_1] = x_1\partial_{x_{20}} - x_2\partial_{x_{23}} + x_3\partial_{x_{24}} - x_4\partial_{x_{25}} + x_5\partial_{x_{26}} + x_8\partial_{x_{27}}, \qquad (12.4.36)$$

$$[x_r, P_1] = 0 \qquad \text{for } r = 14, 17, 19, 21, 22, 23, 24, 25, 26, 27. \qquad (12.4.37)$$

On the other hand, if $[\mathfrak{a}_i, \mathfrak{b}_1] \neq 0$ for $i \geq 2$, then it is equal to the vector obtained by deleting the last coordinate 1 (cf. (11.1.4)–(11.1.12) and (12.4.1)). For instance, $\mathfrak{a}_{18} = E_{(1,1,2,3,2,1,1)}$ and $[\mathfrak{a}_{18}, \mathfrak{b}_1] = E_{(1,1,2,3,2,1)}$. By (12.4.21)–(12.4.36) and correspondingly (11.1.23), (11.1.28), (11.1.33), (11.1.38), (11.1.37), (11.1.42), (11.1.40), (11.1.45), (11.1.44), (11.1.48), (11.1.47), (11.1.50), (11.1.49), (11.1.51)–(11.1.53), we have:

$$[\partial_{x_i}, P_1] = [\mathfrak{a}_i, \mathfrak{b}_1]|_{\mathscr{A}} \qquad \text{for } i \in \overline{1, 27}. \qquad (12.4.38)$$

Expressions (6.7.9), (11.1.18)–(11.1.23) and (11.2.29)–(11.2.34) imply that P_1 is a $\mathscr{G}^{E_6}$-singular vector in $\mathbb{W}_{27}$ with weight λ_6. We set

$$\begin{aligned} P_2 = & -[E_{-\alpha_6}, P_1] = x_2 D - \zeta_1\partial_{x_{11}} - \zeta_2\partial_{x_{13}} - \zeta_3\partial_{x_{16}} + \zeta_4\partial_{x_{18}} \\ & -\zeta_6\partial_{x_{20}} - \zeta_8\partial_{x_{22}} + \zeta_{10}\partial_{x_{24}} - \zeta_{12}\partial_{x_{25}} - \zeta_{15}\partial_{x_{26}} + \zeta_{17}\partial_{x_{27}}, \end{aligned} \qquad (12.4.39)$$

$$\begin{aligned} P_3 = & -[E_{-\alpha_5}, P_2] = x_3 D - \zeta_1\partial_{x_9} - \zeta_2\partial_{x_{12}} - \zeta_3\partial_{x_{15}} + \zeta_5\partial_{x_{18}} \\ & -\zeta_7\partial_{x_{20}} - \zeta_8\partial_{x_{21}} + \zeta_{10}\partial_{x_{23}} - \zeta_{13}\partial_{x_{25}} - \zeta_{16}\partial_{x_{26}} + \zeta_{19}\partial_{x_{27}}, \end{aligned} \qquad (12.4.40)$$

$$P_4 = -[E_{-\alpha_4}, P_3] = x_4 D - \zeta_1 \partial_{x_7} - \zeta_2 \partial_{x_{10}} - \zeta_4 \partial_{x_{15}} + \zeta_5 \partial_{x_{16}} - \zeta_8 \partial_{x_{19}} - \zeta_9 \partial_{x_{20}} + \zeta_{12} \partial_{x_{23}} - \zeta_{13} \partial_{x_{24}} - \zeta_{18} \partial_{x_{26}} + \zeta_{21} \partial_{x_{27}}, \quad (12.4.41)$$

$$P_5 = -[E_{-\alpha_3}, P_4] = x_5 D + \zeta_1 \partial_{x_6} - \zeta_3 \partial_{x_{10}} + \zeta_4 \partial_{x_{12}} - \zeta_5 \partial_{x_{13}} + \zeta_8 \partial_{x_{17}} + \zeta_{11} \partial_{x_{20}} - \zeta_{15} \partial_{x_{23}} + \zeta_{16} \partial_{x_{24}} - \zeta_{18} \partial_{x_{25}} - \zeta_{22} \partial_{x_{27}}, \quad (12.4.42)$$

$$P_6 = -[E_{-\alpha_2}, P_4] = x_6 D + \zeta_1 \partial_{x_5} + \zeta_2 \partial_{x_8} - \zeta_6 \partial_{x_{15}} + \zeta_7 \partial_{x_{16}} - \zeta_9 \partial_{x_{18}} - \zeta_{10} \partial_{x_{19}} + \zeta_{12} \partial_{x_{21}} - \zeta_{13} \partial_{x_{22}} + \zeta_{20} \partial_{x_{26}} - \zeta_{23} \partial_{x_{27}}, \quad (12.4.43)$$

$$P_7 = -[E_{-\alpha_3}, P_6] = x_7 D - \zeta_1 \partial_{x_4} + \zeta_3 \partial_{x_8} + \zeta_6 \partial_{x_{12}} - \zeta_7 \partial_{x_{13}} + \zeta_{10} \partial_{x_{17}} + \zeta_{11} \partial_{x_{18}} - \zeta_{15} \partial_{x_{21}} + \zeta_{16} \partial_{x_{22}} + \zeta_{20} \partial_{x_{25}} + \zeta_{24} \partial_{x_{27}}, \quad (12.4.44)$$

$$P_8 = -[E_{-\alpha_2}, P_5] = x_8 D + \zeta_2 \partial_{x_6} + \zeta_3 \partial_{x_7} - \zeta_4 \partial_{x_9} + \zeta_5 \partial_{x_{11}} - \zeta_8 \partial_{x_{14}} + \zeta_{14} \partial_{x_{20}} + \zeta_{17} \partial_{x_{23}} - \zeta_{19} \partial_{x_{24}} + \zeta_{21} \partial_{x_{25}} + \zeta_{22} \partial_{x_{26}}, \quad (12.4.45)$$

$$P_9 = [E_{-\alpha_4}, P_7] = x_9 D - \zeta_1 \partial_{x_3} - \zeta_4 \partial_{x_8} - \zeta_6 \partial_{x_{10}} + \zeta_9 \partial_{x_{13}} - \zeta_{12} \partial_{x_{17}} - \zeta_{11} \partial_{x_{16}} + \zeta_{15} \partial_{x_{19}} - \zeta_{18} \partial_{x_{22}} - \zeta_{20} \partial_{x_{24}} + \zeta_{25} \partial_{x_{27}}, \quad (12.4.46)$$

$$P_{10} = -[E_{-\alpha_1}, P_7] = x_{10} D - \zeta_2 \partial_{x_4} - \zeta_3 \partial_{x_5} - \zeta_6 \partial_{x_9} + \zeta_7 \partial_{x_{11}} - \zeta_{10} \partial_{x_{14}} + \zeta_{14} \partial_{x_{18}} + \zeta_{17} \partial_{x_{21}} - \zeta_{19} \partial_{x_{22}} - \zeta_{23} \partial_{x_{25}} - \zeta_{24} \partial_{x_{26}}, \quad (12.4.47)$$

$$P_{11} = [E_{-\alpha_5}, P_9] = x_{11} D - \zeta_1 \partial_{x_2} + \zeta_5 \partial_{x_8} + \zeta_7 \partial_{x_{10}} - \zeta_9 \partial_{x_{12}} + \zeta_{11} \partial_{x_{15}} + \zeta_{13} \partial_{x_{17}} - \zeta_{16} \partial_{x_{19}} + \zeta_{18} \partial_{x_{21}} + \zeta_{20} \partial_{x_{23}} + \zeta_{26} \partial_{x_{27}}, \quad (12.4.48)$$

$$P_{12} = -[E_{-\alpha_1}, P_9] = x_{12} D - \zeta_2 \partial_{x_3} + \zeta_4 \partial_{x_5} + \zeta_6 \partial_{x_7} - \zeta_9 \partial_{x_{11}} + \zeta_{12} \partial_{x_{14}} - \zeta_{14} \partial_{x_{16}} - \zeta_{17} \partial_{x_{19}} + \zeta_{21} \partial_{x_{22}} + \zeta_{23} \partial_{x_{24}} - \zeta_{25} \partial_{x_{26}}, \quad (12.4.49)$$

$$P_{13} = [E_{-\alpha_5}, P_{12}] = x_{13} D - \zeta_2 \partial_{x_2} - \zeta_5 \partial_{x_5} - \zeta_7 \partial_{x_7} + \zeta_9 \partial_{x_9} - \zeta_{13} \partial_{x_{14}} + \zeta_{14} \partial_{x_{15}} + \zeta_{19} \partial_{x_{19}} - \zeta_{21} \partial_{x_{21}} - \zeta_{23} \partial_{x_{23}} - \zeta_{26} \partial_{x_{26}}, \quad (12.4.50)$$

$$P_{14} = [E_{-\alpha_6}, P_{11}] = x_{14} D - \zeta_1 \partial_{x_1} - \zeta_8 \partial_{x_8} - \zeta_{10} \partial_{x_{10}} + \zeta_{12} \partial_{x_{12}} - \zeta_{13} \partial_{x_{13}} - \zeta_{15} \partial_{x_{15}} + \zeta_{16} \partial_{x_{16}} - \zeta_{18} \partial_{x_{18}} - \zeta_{20} \partial_{x_{20}} + \zeta_{27} \partial_{x_{27}}, \quad (12.4.51)$$

$$P_{15} = -[E_{-\alpha_3}, P_{12}] = x_{15}D - \zeta_3\partial_{x_3} - \zeta_4\partial_{x_4} - \zeta_6\partial_{x_6} + \zeta_{11}\partial_{x_{11}} + \zeta_{14}\partial_{x_{13}} - \zeta_{15}\partial_{x_{14}} + \zeta_{17}\partial_{x_{17}} - \zeta_{22}\partial_{x_{22}} - \zeta_{24}\partial_{x_{24}} - \zeta_{25}\partial_{x_{25}}, \tag{12.4.52}$$

$$P_{16} = [E_{-\alpha_5}, P_{15}] = x_{16}D - \zeta_3\partial_{x_2} + \zeta_5\partial_{x_4} + \zeta_7\partial_{x_6} - \zeta_{11}\partial_{x_9} - \zeta_{14}\partial_{x_{12}} + \zeta_{16}\partial_{x_{14}} - \zeta_{19}\partial_{x_{17}} + \zeta_{22}\partial_{x_{21}} + \zeta_{24}\partial_{x_{23}} - \zeta_{26}\partial_{x_{25}}, \tag{12.4.53}$$

$$P_{17} = = -[E_{-\alpha_1}, P_{14}] = x_{17}D - \zeta_2\partial_{x_1} + \zeta_8\partial_{x_5} + \zeta_{10}\partial_{x_7} - \zeta_{12}\partial_{x_9} + \zeta_{13}\partial_{x_{11}} + \zeta_{17}\partial_{x_{15}} - \zeta_{19}\partial_{x_{16}} + \zeta_{21}\partial_{x_{18}} + \zeta_{23}\partial_{x_{20}} - \zeta_{27}\partial_{x_{26}}, \tag{12.4.54}$$

$$P_{18} = [E_{-\alpha_4}, P_{16}] = x_{18}D + \zeta_4\partial_{x_2} + \zeta_5\partial_{x_3} - \zeta_9\partial_{x_6} + \zeta_{11}\partial_{x_7} + \zeta_{14}\partial_{x_{10}} - \zeta_{18}\partial_{x_{14}} + \zeta_{21}\partial_{x_{17}} - \zeta_{22}\partial_{x_{19}} + \zeta_{25}\partial_{x_{23}} + \zeta_{26}\partial_{x_{24}}, \tag{12.4.55}$$

$$P_{19} = [E_{-\alpha_6}, P_{16}] = x_{19}D - \zeta_3\partial_{x_1} - \zeta_8\partial_{x_4} - \zeta_{10}\partial_{x_6} + \zeta_{15}\partial_{x_9} - \zeta_{17}\partial_{x_{12}} - \zeta_{16}\partial_{x_{11}} + \zeta_{19}\partial_{x_{13}} - \zeta_{22}\partial_{x_{18}} - \zeta_{24}\partial_{x_{20}} - \zeta_{27}\partial_{x_{25}}, \tag{12.4.56}$$

$$P_{20} = [E_{-\alpha_2}, P_{18}] = x_{20}D - \zeta_6\partial_{x_2} - \zeta_7\partial_{x_3} - \zeta_9\partial_{x_4} + \zeta_{11}\partial_{x_5} + \zeta_{14}\partial_{x_8} - \zeta_{20}\partial_{x_{14}} + \zeta_{23}\partial_{x_{17}} - \zeta_{24}\partial_{x_{19}} - \zeta_{25}\partial_{x_{21}} - \zeta_{26}\partial_{x_{22}}, \tag{12.4.57}$$

$$P_{21} = [E_{-\alpha_4}, P_{19}] = x_{21}D + \zeta_4\partial_{x_1} - \zeta_8\partial_{x_3} + \zeta_{12}\partial_{x_6} - \zeta_{15}\partial_{x_7} + \zeta_{17}\partial_{x_{10}} + \zeta_{18}\partial_{x_{11}} - \zeta_{21}\partial_{x_{13}} + \zeta_{22}\partial_{x_{16}} - \zeta_{25}\partial_{x_{20}} + \zeta_{27}\partial_{x_{24}}, \tag{12.4.58}$$

$$P_{22} = [E_{-\alpha_5}, P_{21}] = x_{22}D - \zeta_5\partial_{x_1} - \zeta_8\partial_{x_2} - \zeta_{13}\partial_{x_6} + \zeta_{16}\partial_{x_7} - \zeta_{19}\partial_{x_{10}} - \zeta_{18}\partial_{x_9} + \zeta_{21}\partial_{x_{12}} - \zeta_{22}\partial_{x_{15}} - \zeta_{26}\partial_{x_{20}} - \zeta_{27}\partial_{x_{23}}, \tag{12.4.59}$$

$$P_{23} = [E_{-\alpha_2}, P_{21}] = x_{23}D - \zeta_6\partial_{x_1} + \zeta_{10}\partial_{x_3} + \zeta_{12}\partial_{x_4} - \zeta_{15}\partial_{x_5} + \zeta_{17}\partial_{x_8} + \zeta_{20}\partial_{x_{11}} - \zeta_{23}\partial_{x_{13}} + \zeta_{24}\partial_{x_{16}} + \zeta_{25}\partial_{x_{18}} - \zeta_{27}\partial_{x_{22}}, \tag{12.4.60}$$

$$P_{24} = [E_{-\alpha_2}, P_{22}] = x_{24}D + \zeta_7\partial_{x_1} + \zeta_{10}\partial_{x_2} - \zeta_{13}\partial_{x_4} + \zeta_{16}\partial_{x_5} - \zeta_{19}\partial_{x_8} - \zeta_{20}\partial_{x_9} + \zeta_{23}\partial_{x_{12}} - \zeta_{24}\partial_{x_{15}} + \zeta_{26}\partial_{x_{18}} + \zeta_{27}\partial_{x_{21}}, \tag{12.4.61}$$

$$P_{25} = [E_{-\alpha_4}, P_{24}] = x_{25}D - \zeta_9\partial_{x_1} - \zeta_{12}\partial_{x_2} - \zeta_{13}\partial_{x_3} - \zeta_{18}\partial_{x_5} + \zeta_{21}\partial_{x_8} + \zeta_{20}\partial_{x_7} - \zeta_{23}\partial_{x_{10}} - \zeta_{25}\partial_{x_{15}} - \zeta_{26}\partial_{x_{16}} - \zeta_{27}\partial_{x_{19}}, \tag{12.4.62}$$

$$P_{26} = [E_{-\alpha_3}, P_{25}] = x_{26}D - \zeta_{11}\partial_{x_1} - \zeta_{15}\partial_{x_2} - \zeta_{16}\partial_{x_3} - \zeta_{18}\partial_{x_4} + \zeta_{22}\partial_{x_8} + \zeta_{20}\partial_{x_6} - \zeta_{24}\partial_{x_{10}} - \zeta_{25}\partial_{x_{12}} - \zeta_{26}\partial_{x_{13}} - \zeta_{27}\partial_{x_{17}}, \quad (12.4.63)$$

$$P_{27} = -[E_{-\alpha_1}, P_{26}] = x_{27}D - \zeta_{14}\partial_{x_1} + \zeta_{17}\partial_{x_2} + \zeta_{19}\partial_{x_3} + \zeta_{21}\partial_{x_4} - \zeta_{22}\partial_{x_5} - \zeta_{23}\partial_{x_6} + \zeta_{24}\partial_{x_7} + \zeta_{25}\partial_{x_9} + \zeta_{26}\partial_{x_{11}} + \zeta_{27}\partial_{x_{14}}. \quad (12.4.64)$$

Write

$$\mathscr{P} = \sum_{i=1}^{27} \mathbb{F}P_i, \qquad \mathfrak{C}_0 = \mathfrak{r}(\mathscr{G}^{E_6}) + \mathbb{F}D \quad (12.4.65)$$

(cf. (11.1.18)–(11.1.54) and (12.4.9)) and

$$\mathfrak{C} = \mathscr{P} + \mathfrak{C}_0 + \mathscr{T} \quad (12.4.66)$$

(cf. (12.4.10)). Then we have:

Theorem 12.4.1 *The space $\mathfrak{C}$ forms a Lie subalgebra of the Witt algebra $\mathbb{W}_{27}$ (cf.* (12.4.15)*). Moreover, the linear map ϑ determined by*

$$\vartheta(\mathfrak{a}_i) = \partial_{x_i}, \quad \vartheta(\mathfrak{b}_i) = P_i, \quad \vartheta(u) = u|_{\mathscr{A}} \qquad \text{for } i \in \overline{1,27},\ u \in \mathscr{G}_0 \quad (12.4.67)$$

(cf. (11.1.18)–(11.1.54) and (12.4.8)*) gives a Lie algebra isomorphism from $\mathscr{G}^{E_7}$ to $\mathfrak{C}$, which is a first-order inhomogeneous differential-operator representation of $\mathscr{G}^{E_7}$.*

Proof Since $\mathscr{T} \cong \mathscr{G}_+$ as $\mathscr{G}^{E_6}$-modules, we have

$$\mathscr{G}_0 + \mathscr{G}_+ \overset{\vartheta}{\cong} \mathfrak{C}_0 + \mathscr{T} \quad (12.4.68)$$

as Lie algebras. Recall that $U(\mathscr{G})$ stands for the universal enveloping algebra of a Lie algebra $\mathscr{G}$. Note that

$$\mathscr{B}_- = \mathscr{G}_0 + \mathscr{G}_-, \qquad \mathscr{B}_+ = \mathscr{G}_0 + \mathscr{G}_+ \quad (12.4.69)$$

are also Lie subalgebras of $\mathscr{G}^{E_7}$ and

$$\mathscr{G}^{E_7} = \mathscr{B}_- \oplus \mathscr{G}_+ = \mathscr{G}_- \oplus \mathscr{B}_+. \quad (12.4.70)$$

We define a one-dimensional $\mathscr{B}_-$-module $\mathbb{C}u_0$ by

$$w(u_0) = 0 \quad \text{for } w \in \mathscr{G}_- + \mathscr{G}^{E_6}, \ \ \widehat{\alpha}(u_0) = 54u_0 \quad (12.4.71)$$

(cf. (12.4.9)). Let

$$\Psi = U(\mathscr{G}^{E_7}) \otimes_{\mathscr{B}_-} \mathbb{C}u_0 \cong U(\mathscr{G}_+) \otimes_{\mathbb{F}} \mathbb{F}u_0 \tag{12.4.72}$$

be the induced $\mathscr{G}^{E_7}$-module.

Recall that $\mathbb{N}$ is the set of nonnegative integers. Let

$$\widehat{\mathscr{A}} = \mathbb{F}[\partial_{x_1}, \partial_{x_2}, \ldots, \partial_{x_{27}}]. \tag{12.4.73}$$

We define an action of the associative algebra $\mathbb{A}$ (cf. (11.2.35)) on $\widehat{\mathscr{A}}$ by

$$\partial_{x_i}\left(\prod_{j=1}^{27} \partial_{x_j}^{\beta_j}\right) = \partial_{x_i}^{\beta_i+1} \prod_{i\neq j\in\overline{1,27}} \prod_{j=1}^{27} \partial_{x_j}^{\beta_j} \tag{12.4.74}$$

and

$$x_i\left(\prod_{j=1}^{27} \partial_{x_j}^{\beta_j}\right) = -\beta_i \partial_{x_i}^{\beta_i-1} \prod_{i\neq j\in\overline{1,27}} \prod_{j=1}^{27} \partial_{x_j}^{\beta_j} \tag{12.4.75}$$

for $i \in \overline{1,27}$. Since

$$[-x_i, \partial_{x_j}] = [\partial_{x_i}, x_j] = \delta_{i,j} \qquad \text{for } i, j \in \overline{1,27}, \tag{12.4.76}$$

the above action gives an associative algebra representation of $\mathbb{A}$. Thus it also gives a Lie algebra representation of $\mathbb{A}$, whose Lie bracket is the commutator. It is straightforward to verify that

$$[d|_{\widehat{\mathscr{A}}}, \partial|_{\widehat{\mathscr{A}}}] = [d, \partial]|_{\widehat{\mathscr{A}}} \qquad \text{for } d \in \mathfrak{C}_0,\ \partial \in \mathscr{T}. \tag{12.4.77}$$

Define linear map $\varsigma : \Psi \to \mathscr{A}'$ by

$$\varsigma\left(\prod_{i=1}^{27} \xi_i^{\ell_i} \otimes u_0\right) = \prod_{i=1}^{27} \partial_{x_i}^{\ell_i} \qquad (\ell_1, \ldots, \ell_{27}) \in \mathbb{N}^{27}. \tag{12.4.78}$$

According to (11.1.18)–(11.1.54), (12.4.74) and (12.4.75),

$$D(1) = -27, \quad d(1) = 0 \qquad \text{for } d \in \mathfrak{r}(\mathscr{G}^{E_6}). \tag{12.4.79}$$

Moreover, (12.4.71), (12.4.72), (12.4.74), (12.4.75) and (12.4.79) imply

$$\varsigma(\xi(v)) = \vartheta(\xi)\varsigma(v) \qquad \text{for } \xi \in \mathscr{G}_0,\ v \in \Psi. \tag{12.4.80}$$

Now (12.4.72) and (12.4.74) imply

$$\varsigma(w(u)) = \vartheta(w)(\varsigma(u)) \qquad \text{for } w \in \mathscr{B}_+, \ \ u \in \Psi. \tag{12.4.81}$$

Thus we have

$$\varsigma w|_\Psi \varsigma^{-1} = \vartheta(w)|_{\widehat{\mathscr{A}}} \qquad \text{for } w \in \mathscr{B}_+. \tag{12.4.82}$$

On the other hand, the linear map

$$\psi(v) = \varsigma v|_\Psi \varsigma^{-1} \qquad \text{for } v \in \mathscr{G}^{E_7} \tag{12.4.83}$$

is a Lie algebra monomorphism from $\mathscr{G}^{E_7}$ to $\mathbb{A}|_{\widehat{\mathscr{A}}}$. According to (12.4.38) and (12.4.77),

$$\psi(\mathfrak{b}_1) = P_1|_{\widehat{\mathscr{A}}}. \tag{12.4.84}$$

By the constructions of $P_2, \ldots, P_{27}$ in (12.4.39)–(12.4.64), we have

$$\psi(\mathfrak{b}_i) = P_i|_{\widehat{\mathscr{A}}} \qquad \text{for } i \in \overline{2, 27}. \tag{12.4.85}$$

Therefore, we have

$$\psi(v) = \vartheta(v)|_{\widehat{\mathscr{A}}} \qquad \text{for } v \in \mathscr{G}^{E_7}. \tag{12.4.86}$$

In particular, $\mathfrak{C}|_{\widehat{\mathscr{A}}} = \vartheta(\mathscr{G}^{E_7})|_{\widehat{\mathscr{A}}} = \psi(\mathscr{G}^{E_7})$ forms a Lie algebra. Since the linear map $d \mapsto d|_{\widehat{\mathscr{A}}}$ for $d \in \mathfrak{C}$ is injective, we have that $\mathfrak{C}$ forms a Lie subalgebra of $\mathbb{A}$ and ϑ is a Lie algebra isomorphism. □

By the above theorem, a Lie group of type E_7 is generated by the linear transformations $\{e^{b\mathfrak{r}(E_\alpha)} \mid b \in \mathbb{R}, \ \alpha \in \Phi_{E_6}\}$ associated with (11.1.18)–(11.1.54), the real translations and dilations in $\mathbb{R}^{27}$ with x_r as the rth coordinate function, and the fractional transformations $\{e^{bP_s} : (x_1, x_2, \ldots, x_{27}) \mapsto (e^{bP_s}(x_1), e^{bP_s}(x_2), \ldots, e^{bP_s}(x_{27})) \mid b \in \mathbb{R}, \ s \in \overline{1, 27}\}$, where $e^{bP_s}(x_i)$ are of the forms as the following forms of the case $s = 1$:

$$e^{bP_1}(x_i) = \frac{x_i}{1 - bx_1}, \quad i \in \{\overline{1, 13}, 15, 16, 18, 20\}, \tag{12.4.87}$$

$$e^{bP_1}(x_{14}) = x_{14} - \frac{b(x_2x_{11} + x_3x_9 + x_4x_7 - x_5x_6)}{1 - bx_1}, \tag{12.4.88}$$

$$e^{bP_1}(x_{17}) = x_{17} - \frac{b(x_2x_{13} + x_3x_{12} + x_4x_{10} - x_6x_8)}{1 - bx_1}, \tag{12.4.89}$$

$$e^{bP_1}(x_{19}) = x_{19} - \frac{b(x_2x_{16} + x_3x_{15} + x_5x_{10} - x_7x_8)}{1 - bx_1}, \tag{12.4.90}$$

$$e^{bP_1}(x_{21}) = x_{21} - \frac{b(x_2x_{18} - x_4x_{15} + x_5x_{12} - x_8x_9)}{1 - bx_1}, \tag{12.4.91}$$

$$e^{bP_1}(x_{22}) = x_{22} + \frac{b(x_3x_{18} + x_4x_{16} - x_5x_{13} + x_8x_{11})}{1 - bx_1}, \tag{12.4.92}$$

$$e^{bP_1}(x_{23}) = x_{23} - \frac{b(x_2x_{20} + x_6x_{15} - x_7x_{12} + x_9x_{10})}{1 - bx_1}, \tag{12.4.93}$$

$$e^{bP_1}(x_{24}) = x_{24} + \frac{b(x_3x_{20} - x_6x_{16} + x_7x_{13} - x_{10}x_{11})}{1 - bx_1}, \tag{12.4.94}$$

$$e^{bP_1}(x_{25}) = x_{25} - \frac{b(x_4x_{20} + x_6x_{18} - x_9x_{13} + x_{11}x_{12})}{1 - bx_1}, \tag{12.4.95}$$

$$e^{bP_1}(x_{26}) = x_{26} + \frac{b(x_5x_{20} + x_7x_{18} - x_9x_{16} + x_{11}x_{15})}{1 - bx_1}, \tag{12.4.96}$$

$$e^{bP_1}(x_{27}) = x_{27} + \frac{b(x_8x_{20} + x_{10}x_{18} - x_{12}x_{16} + x_{13}x_{15})}{1 - bx_1} \tag{12.4.97}$$

by (11.2.1), (11.2.3)–(11.2.8), (11.2.10), (11.2.12), (11.2.15) and (12.4.20). This gives a fractional representation of the Lie group E_7.

12.5 Functor from E_6-Mod to E_7-Mod

In this section, we construct a new functor from E_6-**Mod** to E_7-**Mod**.

Note that

$$\mathcal{G}_{\mathcal{A}}^{E_6} = (\bigoplus_{i=1}^{6} \mathcal{A}\alpha_i) \oplus \bigoplus_{\alpha\in\Phi_{E_6}} \mathcal{A}E_\alpha \tag{12.5.1}$$

forms a Lie algebra with the Lie bracket:

$$[fu_1, gu_2] = fg[u_1, u_2] \quad \text{for } f, g \in \mathcal{A},\ u_1, u_2 \in \mathcal{G}^{E_6}. \tag{12.5.2}$$

Moreover, we define the Lie algebra

$$\mathcal{K} = \mathcal{G}_{\mathcal{A}}^{E_6} \oplus \mathcal{A}\kappa \tag{12.5.3}$$

with the Lie bracket:

$$[\xi_1 + f\kappa, \xi_2 + g\kappa] = [\xi_1, \xi_2] \quad \text{for } \xi_1, \xi_2 \in \mathcal{G}_{\mathcal{A}}^{E_6},\ f, g \in \mathcal{A}. \tag{12.5.4}$$

Recall the Witt algebra $\mathbb{W}_{27} = \sum_{i=1}^{27} \mathcal{A}\partial_{x_i}$, and Shen's monomorphism $\Im$ from the Lie algebra $\mathbb{W}_{27}$ to the Lie algebra of semi-product $\mathbb{W}_{27} + gl(27, \mathcal{A})$ (cf. (6.7.9)–(6.7.12)) defined by

$$\Im\left(\sum_{i=1}^{27} f_i\partial_{x_i}\right)=\sum_{i=1}^{27} f_i\partial_{x_i}+\Im_1\left(\sum_{i=1}^{27} f_i\partial_{x_i}\right),\quad \Im_1\left(\sum_{i=1}^{27} f_i\partial_{x_i}\right)=\sum_{i,j=1}^{27}\partial_{x_i}(f_j)E_{i,j}. \tag{12.5.5}$$

According to our construction of P_1–P_{27} in (12.4.20) and (12.4.39)–(12.4.64),

$$\Im_1(P_i)=\sum_{r=1}^{27} x_r\Im_1(\mathfrak{r}([\mathfrak{a}_r,\mathfrak{b}_i]))\qquad \text{for } i\in\overline{1,27}. \tag{12.5.6}$$

On the other hand,

$$\widehat{\mathscr{K}}=\mathbb{W}_{27}\oplus\mathscr{K} \tag{12.5.7}$$

becomes a Lie algebra with the Lie bracket

$$\begin{aligned}&[d_1+f_1u_1+f_2\kappa,d_2+g_1u_2+g_2\kappa]\\ =\;&[d_1,d_2]+f_1g_1[u_1,u_2]+d_1(g_2)u_2-d_2(g_1)u_1+(d_1(g_2)-d_2(g_1))\kappa\end{aligned} \tag{12.5.8}$$

for $f_1,f_2,g_1,g_2\in\mathscr{A}$, $u_1,u_2\in\mathscr{G}^{E_6}$ and $d_1,d_2\in\mathbb{W}_{27}$. Note

$$\mathscr{G}_0=\mathscr{G}^{E_6}\oplus\mathbb{F}\widehat{\alpha} \tag{12.5.9}$$

by (12.4.2) and (12.4.5). So there exists a Lie algebra monomorphism $\mathfrak{k}:\mathscr{G}_0\to\mathscr{K}$ determined by

$$\mathfrak{k}(\widehat{\alpha})=3\kappa,\quad \mathfrak{k}(u)=u\qquad \text{for } u\in\mathscr{G}^{E_6}. \tag{12.5.10}$$

Since $\Im$ is a Lie algebra monomorphism, our construction of P_1–P_{27} in (12.4.20) and (12.4.39)–(12.4.64) show that we have a Lie algebra monomorphism $\iota:\mathscr{G}^{E_7}\to\widehat{\mathscr{K}}$ given by

$$\iota(u)=\mathfrak{r}(u)+\mathfrak{k}(u)\qquad \text{for } u\in\mathscr{G}_0, \tag{12.5.11}$$

$$\iota(\mathfrak{a}_i)=\partial_{x_i},\quad \iota(\mathfrak{b}_i)=P_i+\sum_{r=1}^{27} x_r\mathfrak{k}([\mathfrak{a}_r,\mathfrak{b}_i])\qquad \text{for } i\in\overline{1,27}. \tag{12.5.12}$$

According to (4.4.18), (4.4.25), (11.1.4)–(11.1.12) and (12.4.1),

$$\begin{aligned}\iota(\mathfrak{b}_1)=\;&P_1+x_2E_{\alpha_6}+x_3E_{(0,0,0,0,1,1)}+x_4E_{(0,0,0,1,1,1)}+x_5E_{(0,0,1,1,1,1)}+x_6E_{(0,1,0,1,1,1)}\\ &-x_1\mathfrak{k}(\alpha_7)+x_7E_{(0,1,1,1,1,1)}+x_8E_{(1,0,1,1,1,1)}+x_9E_{(0,1,1,2,1,1)}+x_{10}E_{(1,1,1,1,1,1)}\\ &+x_{11}E_{(0,1,1,2,2,1)}+x_{12}E_{(1,1,1,2,1,1)}+x_{13}E_{(1,1,1,2,2,1)}+x_{15}E_{(1,1,2,2,1,1)}\\ &+x_{16}E_{(1,1,2,2,2,1)}+x_{18}E_{(1,1,2,3,2,1)}+x_{20}E_{(1,2,2,3,2,1)}.\end{aligned} \tag{12.5.13}$$

Moreover, (4.4.23), (11.1.23) and (11.1.56) yield

$$\begin{aligned}\iota(\mathfrak{b}_2) = & -\iota([E_{-\alpha_6}, \mathfrak{b}_1]) = -[\iota(E_{-\alpha_6}), \iota(\mathfrak{b}_1)] = -[\mathfrak{r}(E_{-\alpha_6}) + E_{-\alpha_6}, \iota(\mathfrak{b}_1)] \\ = & P_2 - x_1 E_{-\alpha_6} + x_3 E_{\alpha_5} + x_4 E_{(0,0,0,1,1)} + x_5 E_{(0,0,1,1,1)} + x_6 E_{(0,1,0,1,1)} \\ & -x_2 \mathfrak{k}(\alpha_{(0,0,0,0,0,1,1)}) + x_7 E_{(0,1,1,1,1)} + x_8 E_{(1,0,1,1,1)} + x_9 E_{(0,1,1,2,1)} + x_{10} E_{(1,1,1,1,1)} \\ & +x_{12} E_{(1,1,1,2,1)} - x_{14} E_{(0,1,1,2,2,1)} + x_{15} E_{(1,1,2,2,1)} - x_{17} E_{(1,1,1,2,2,1)} \\ & -x_{19} E_{(1,1,2,2,2,1)} - x_{21} E_{(1,1,2,3,2,1)} - x_{23} E_{(1,2,2,3,2,1)}. \end{aligned} \tag{12.5.14}$$

Similarly, we have:

$$\begin{aligned}\iota(\mathfrak{b}_3) = & P_3 - x_1 E'_{(0,0,0,0,1,1)} - x_2 E_{-\alpha_5} + x_4 E_{\alpha_4} + x_5 E_{(0,0,1,1)} + x_6 E_{(0,1,0,1)} \\ & -x_3 \mathfrak{k}(\alpha_{(0,0,0,0,1,1,1)}) + x_7 E_{(0,1,1,1)} + x_8 E_{(1,0,1,1)} + x_{10} E_{(1,1,1,1)} - x_{11} E_{(0,1,1,2,1)} \\ & -x_{13} E_{(1,1,1,2,1)} - x_{14} E_{(0,1,1,2,1,1)} - x_{16} E_{(1,1,2,2,1)} - x_{17} E_{(1,1,1,2,1,1)} \\ & -x_{19} E_{(1,1,2,2,1,1)} + x_{22} E_{(1,1,2,3,2,1)} + x_{24} E_{(1,2,2,3,2,1)}, \end{aligned} \tag{12.5.15}$$

$$\begin{aligned}\iota(\mathfrak{b}_4) = & P_4 - x_1 E'_{(0,0,0,1,1,1)} - x_2 E'_{(0,0,0,1,1)} - x_3 E_{-\alpha_4} + x_5 E_{\alpha_3} + x_6 E_{\alpha_2} \\ & -x_4 \mathfrak{k}(\alpha_{(0,0,0,1,1,1,1)}) - x_9 E_{(0,1,1,1)} + x_8 E_{(1,0,1)} - x_{12} E_{(1,1,1,1)} - x_{11} E_{(0,1,1,1,1)} \\ & -x_{13} E_{(1,1,1,1,1)} - x_{14} E_{(0,1,1,1,1,1)} - x_{17} E_{(1,1,1,1,1,1)} + x_{18} E_{(1,1,2,2,1)} \\ & +x_{21} E_{(1,1,2,2,1,1)} + x_{22} E_{(1,1,2,2,2,1)} - x_{25} E_{(1,2,2,3,2,1)}, \end{aligned} \tag{12.5.16}$$

$$\begin{aligned}\iota(\mathfrak{b}_5) = & P_5 - x_1 E'_{(0,0,1,1,1,1)} - x_2 E'_{(0,0,1,1,1)} - x_3 E'_{(0,0,1,1)} - x_4 E_{-\alpha_3} + x_7 E_{\alpha_2} \\ & -x_5 \mathfrak{k}(\alpha_{(0,0,1,1,1,1,1)}) + x_8 E_{\alpha_1} + x_9 E_{(0,1,0,1)} - x_{15} E_{(1,1,1,1)} + x_{11} E_{(0,1,0,1,1)} \\ & -x_{16} E_{(1,1,1,1,1)} + x_{14} E_{(0,1,0,1,1,1)} - x_{19} E_{(1,1,1,1,1,1)} - x_{18} E_{(1,1,1,2,1)} \\ & -x_{21} E_{(1,1,1,2,1,1)} - x_{22} E_{(1,1,1,2,2,1)} + x_{26} E_{(1,2,2,3,2,1)}, \end{aligned} \tag{12.5.17}$$

$$\begin{aligned}\iota(\mathfrak{b}_6) = & P_6 - x_1 E'_{(0,1,0,1,1,1)} - x_2 E'_{(0,1,0,1,1)} - x_3 E'_{(0,1,0,1)} - x_4 E_{-\alpha_2} + x_7 E_{\alpha_3} \\ & -x_6 \mathfrak{k}(\alpha_{(0,1,0,1,1,1,1)}) + x_9 E_{(0,0,1,1)} + x_{10} E_{(1,0,1)} + x_{12} E_{(1,0,1,1)} + x_{11} E_{(0,0,1,1,1)} \\ & +x_{13} E_{(1,0,1,1,1)} + x_{14} E_{(0,0,1,1,1,1)} + x_{17} E_{(1,0,1,1,1,1)} - x_{20} E_{(1,1,2,2,1)} \\ & -x_{23} E_{(1,1,2,2,1,1)} - x_{24} E_{(1,1,2,2,2,1)} - x_{25} E_{(1,1,2,3,2,1)}, \end{aligned} \tag{12.5.18}$$

$$\begin{aligned}\iota(\mathfrak{b}_7) = & P_7 - x_1 E'_{(0,1,1,1,1,1)} - x_2 E'_{(0,1,1,1,1)} - x_3 E'_{(0,1,1,1)} - x_5 E_{-\alpha_2} - x_6 E_{-\alpha_3} \\ & -x_7 \mathfrak{k}(\alpha_{(0,1,1,1,1,1,1)}) - x_9 E_{\alpha_4} + x_{10} E_{\alpha_1} + x_{15} E_{(1,0,1,1)} - x_{11} E_{(0,0,0,1,1)} \\ & +x_{16} E_{(1,0,1,1,1)} - x_{14} E_{(0,0,0,1,1,1)} + x_{19} E_{(1,0,1,1,1,1)} + x_{20} E_{(1,1,1,2,1)} \\ & +x_{23} E_{(1,1,1,2,1,1)} + x_{24} E_{(1,1,1,2,2,1)} + x_{26} E_{(1,1,2,3,2,1)}, \end{aligned} \tag{12.5.19}$$

$$\begin{aligned}\iota(\mathfrak{b}_8) = & P_8 - x_1 E'_{(1,0,1,1,1,1)} - x_2 E'_{(1,0,1,1,1)} - x_3 E'_{(1,0,1,1)} - x_4 E'_{(1,0,1)} - x_5 E_{-\alpha_1} \\ & -x_8 \mathfrak{k}(\alpha_{(1,0,1,1,1,1,1)}) + x_{10} E_{\alpha_2} + x_{12} E_{(0,1,0,1)} + x_{15} E_{(0,1,1,1)} + x_{13} E_{(0,1,0,1,1)} \\ & +x_{16} E_{(0,1,1,1,1)} + x_{17} E_{(0,1,0,1,1,1)} + x_{19} E_{(0,1,1,1,1,1)} + x_{18} E_{(0,1,1,2,1)} \end{aligned}$$

$$+x_{21}E_{(0,1,1,2,1,1)}+x_{22}E_{(0,1,1,2,2,1)}+x_{27}E_{(1,2,2,3,2,1)}, \tag{12.5.20}$$

$$\begin{aligned}\iota(\mathfrak{b}_9) = {} & P_9 - x_1E'_{(0,1,1,2,1,1)} - x_2E'_{(0,1,1,2,1)} + x_4E'_{(0,1,1,1)} - x_5E'_{(0,1,0,1)} - x_6E'_{(0,0,1,1)} \\ & -x_9\mathfrak{k}(\alpha_{(0,1,1,2,1,1,1)}) + x_7E_{-\alpha_4} + x_{12}E_{\alpha_1} - x_{15}E_{(1,0,1)} - x_{11}E_{\alpha_5} \\ & +x_{18}E_{(1,0,1,1,1)} - x_{14}E_{(0,0,0,0,1,1)} + x_{21}E_{(1,0,1,1,1,1)} - x_{20}E_{(1,1,1,1,1)} \\ & -x_{23}E_{(1,1,1,1,1,1)} + x_{25}E_{(1,1,1,2,2,1)} - x_{26}E_{(1,1,2,2,2,1)}, \end{aligned} \tag{12.5.21}$$

$$\begin{aligned}\iota(\mathfrak{b}_{10}) = {} & P_{10} - x_1E'_{(1,1,1,1,1,1)} - x_2E'_{(1,1,1,1,1)} - x_3E'_{(1,1,1,1)} - x_8E_{-\alpha_2} - x_6E'_{(1,0,1)} \\ & -x_{10}\mathfrak{k}(\alpha_{(1,1,1,1,1,1,1)}) - x_7E_{-\alpha_1} - x_{12}E_{\alpha_4} - x_{15}E_{(0,0,1,1)} - x_{13}E_{(0,0,0,1,1)} \\ & -x_{16}E_{(0,0,1,1,1)} - x_{17}E_{(0,0,0,1,1,1)} - x_{19}E_{(0,0,1,1,1,1)} - x_{20}E_{(0,1,1,2,1)} \\ & -x_{23}E_{(0,1,1,2,1,1)} - x_{24}E_{(0,1,1,2,2,1)} + x_{27}E_{(1,1,2,3,2,1)}, \end{aligned} \tag{12.5.22}$$

$$\begin{aligned}\iota(\mathfrak{b}_{11}) = {} & P_{11} - x_1E'_{(0,1,1,2,2,1)} + x_3E'_{(0,1,1,2,1)} + x_4E'_{(0,1,1,1,1)} - x_5E'_{(0,1,0,1,1)} \\ & -x_{11}\mathfrak{k}(\alpha_{(0,1,1,2,2,1,1)}) - x_6E'_{(0,0,1,1,1)} + x_7E'_{(0,0,0,1,1)} + x_9E_{-\alpha_5} \\ & +x_{13}E_{\alpha_1} - x_{16}E_{(1,0,1)} - x_{18}E_{(1,0,1,1)} - x_{14}E_{\alpha_6} + x_{20}E_{(1,1,1,1)} \\ & +x_{22}E_{(1,0,1,1,1,1)} - x_{24}E_{(1,1,1,1,1,1)} - x_{25}E_{(1,1,1,2,1,1)} + x_{26}E_{(1,1,2,2,1,1)}, \end{aligned} \tag{12.5.23}$$

$$\begin{aligned}\iota(\mathfrak{b}_{12}) = {} & P_{12} - x_1E'_{(1,1,1,2,1,1)} - x_2E'_{(1,1,1,2,1)} + x_4E'_{(1,1,1,1)} - x_6E'_{(1,0,1,1)} \\ & -x_{12}\mathfrak{k}(\alpha_{(1,1,1,2,1,1,1)}) - x_8E'_{(0,1,0,1)} - x_9E_{-\alpha_1} + x_{10}E_{-\alpha_4} \\ & -x_{13}E_{\alpha_5} + x_{15}E_{\alpha_3} - x_{17}E_{(0,0,0,0,1,1)} - x_{18}E_{(0,0,1,1,1)} + x_{20}E_{(0,1,1,1,1)} \\ & -x_{21}E_{(0,0,1,1,1,1)} + x_{23}E_{(0,1,1,1,1,1)} - x_{25}E_{(0,1,1,2,2,1)} - x_{27}E_{(1,1,2,2,2,1)}, \end{aligned} \tag{12.5.24}$$

$$\begin{aligned}\iota(\mathfrak{b}_{13}) = {} & P_{13} - x_1E'_{(1,1,1,2,2,1)} + x_3E'_{(1,1,1,2,1)} + x_4E'_{(1,1,1,1,1)} - x_8E'_{(0,1,0,1,1)} \\ & -x_{13}\mathfrak{k}(\alpha_{(1,1,1,2,2,1,1)}) - x_6E'_{(1,0,1,1,1)} + x_{10}E'_{(0,0,0,1,1)} - x_{11}E_{-\alpha_1} \\ & +x_{12}E_{-\alpha_5} + x_{16}E_{\alpha_3} - x_{17}E_{\alpha_6} + x_{18}E_{(0,0,1,1)} - x_{20}E_{(0,1,1,1)} \\ & -x_{22}E_{(0,0,1,1,1,1)} + x_{24}E_{(0,1,1,1,1,1)} + x_{25}E_{(0,1,1,2,1,1)} + x_{27}E_{(1,1,2,2,1,1)}, \end{aligned} \tag{12.5.25}$$

$$\begin{aligned}\iota(\mathfrak{b}_{14}) = {} & P_{14} + x_2E'_{(0,1,1,2,2,1)} + x_3E'_{(0,1,1,2,1,1)} + x_4E'_{(0,1,1,1,1,1)} - x_5E'_{(0,1,0,1,1,1)} \\ & -x_{14}\mathfrak{k}(\alpha_{(0,1,1,2,2,2,1)}) - x_6E'_{(0,0,1,1,1,1)} + x_7E'_{(0,0,0,1,1,1)} + x_9E'_{(0,0,0,0,1,1)} \\ & +x_{11}E_{-\alpha_6} + x_{17}E_{\alpha_1} - x_{19}E_{(1,0,1)} - x_{21}E_{(1,0,1,1)} - x_{22}E_{(1,0,1,1,1)} \\ & +x_{23}E_{(1,1,1,1)} + x_{24}E_{(1,1,1,1,1)} + x_{25}E_{(1,1,1,2,1)} - x_{26}E_{(1,1,2,2,1)}, \end{aligned} \tag{12.5.26}$$

$$\begin{aligned}\iota(\mathfrak{b}_{15}) = \;& P_{15} - x_1 E'_{(1,1,2,2,1,1)} - x_2 E'_{(1,1,2,2,1)} + x_5 E'_{(1,1,1,1)} - x_7 E'_{(1,0,1,1)} \\ & - x_{15}\mathfrak{k}(\alpha_{(1,1,2,2,1,1,1)}) - x_8 E'_{(0,1,1,1)} + x_9 E'_{(1,0,1)} + x_{10} E'_{(0,0,1,1)} \\ & - x_{12} E_{-\alpha_3} - x_{16} E_{\alpha_5} + x_{18} E_{(0,0,0,1,1)} - x_{19} E_{(0,0,0,0,1,1)} - x_{20} E_{(0,1,0,1,1)} \\ & + x_{21} E_{(0,0,0,1,1,1)} - x_{23} E_{(0,1,0,1,1,1)} + x_{26} E_{(0,1,1,2,2,1)} + x_{27} E_{(1,1,1,2,2,1)},\end{aligned} \tag{12.5.27}$$

$$\begin{aligned}\iota(\mathfrak{b}_{16}) = \;& P_{16} - x_1 E'_{(1,1,2,2,2,1)} + x_3 E'_{(1,1,2,2,1)} + x_5 E'_{(1,1,1,1,1)} - x_7 E'_{(1,0,1,1,1)} \\ & - x_{16}\mathfrak{k}(\alpha_{(1,1,2,2,2,1,1)}) - x_8 E'_{(0,1,1,1,1)} + x_{10} E'_{(0,0,1,1,1)} + x_{11} E'_{(1,0,1)} \\ & - x_{13} E_{-\alpha_3} + x_{15} E_{-\alpha_5} - x_{18} E_{\alpha_4} - x_{19} E_{\alpha_6} + x_{20} E_{(0,1,0,1)} + x_{22} E_{(0,0,0,1,1,1)} \\ & - x_{24} E_{(0,1,0,1,1,1)} - x_{26} E_{(0,1,1,2,1,1)} - x_{27} E_{(1,1,1,2,1,1)},\end{aligned} \tag{12.5.28}$$

$$\begin{aligned}\iota(\mathfrak{b}_{17}) = \;& P_{17} + x_2 E'_{(1,1,1,2,2,1)} + x_3 E'_{(1,1,1,2,1,1)} + x_4 E'_{(1,1,1,1,1,1)} - x_6 E'_{(1,0,1,1,1,1)} \\ & - x_{17}\mathfrak{k}(\alpha_{(1,1,1,2,2,2,1)}) - x_8 E'_{(0,1,0,1,1,1)} + x_{10} E'_{(0,0,0,1,1,1)} - x_{14} E_{-\alpha_1} \\ & + x_{12} E'_{(0,0,0,0,1,1)} + x_{13} E_{-\alpha_6} + x_{19} E_{\alpha_3} + x_{21} E_{(0,0,1,1)} - x_{23} E_{(0,1,1,1)} \\ & + x_{22} E_{(0,0,1,1,1)} - x_{24} E_{(0,1,1,1,1)} - x_{25} E_{(0,1,1,2,1)} - x_{27} E_{(1,1,2,2,1)},\end{aligned} \tag{12.5.29}$$

$$\begin{aligned}\iota(\mathfrak{b}_{18}) = \;& P_{18} - x_1 E'_{(1,1,2,3,2,1)} - x_4 E'_{(1,1,2,2,1)} + x_5 E'_{(1,1,1,2,1)} - x_8 E'_{(0,1,1,2,1)} \\ & - x_{18}\mathfrak{k}(\alpha_{(1,1,2,3,2,1,1)}) - x_9 E'_{(1,0,1,1,1)} + x_{11} E'_{(1,0,1,1)} + x_{12} E'_{(0,0,1,1,1)} \\ & - x_{13} E'_{(0,0,1,1)} - x_{15} E'_{(0,0,0,1,1)} + x_{16} E_{-\alpha_4} - x_{21} E_{\alpha_6} + x_{22} E_{(0,0,0,0,1,1)} \\ & - x_{20} E_{\alpha_2} - x_{25} E_{(0,1,0,1,1,1)} + x_{26} E_{(0,1,1,1,1,1)} + x_{27} E_{(1,1,1,1,1,1)},\end{aligned} \tag{12.5.30}$$

$$\begin{aligned}\iota(\mathfrak{b}_{19}) = \;& P_{19} + x_2 E'_{(1,1,2,2,2,1)} + x_3 E'_{(1,1,2,2,1,1)} + x_5 E'_{(1,1,1,1,1,1)} - x_7 E'_{(1,0,1,1,1,1)} \\ & - x_{19}\mathfrak{k}(\alpha_{(1,1,2,2,2,2,1)}) - x_8 E'_{(0,1,1,1,1,1)} + x_{10} E'_{(0,0,1,1,1,1)} + x_{14} E'_{(1,0,1)} \\ & + x_{15} E'_{(0,0,0,0,1,1)} + x_{16} E_{-\alpha_6} - x_{17} E_{-\alpha_3} - x_{21} E_{\alpha_4} + x_{23} E_{(0,1,0,1)} \\ & - x_{22} E_{(0,0,0,1,1)} + x_{24} E_{(0,1,0,1,1)} + x_{26} E_{(0,1,1,2,1)} + x_{27} E_{(1,1,1,2,1)},\end{aligned} \tag{12.5.31}$$

$$\begin{aligned}\iota(\mathfrak{b}_{20}) = \;& P_{20} - x_1 E'_{(1,2,2,3,2,1)} + x_6 E'_{(1,1,2,2,1)} - x_7 E'_{(1,1,1,2,1)} + x_{10} E'_{(0,1,1,2,1)} \\ & - x_{20}\mathfrak{k}(\alpha_{(1,2,2,3,2,1,1)}) + x_9 E'_{(1,1,1,1,1)} - x_{11} E'_{(1,1,1,1)} - x_{12} E'_{(0,1,1,1,1)} \\ & + x_{13} E'_{(0,1,1,1)} + x_{15} E'_{(0,1,0,1,1)} - x_{16} E_{(0,1,0,1)} + x_{18} E_{-\alpha_2} - x_{23} E_{\alpha_6} \\ & + x_{24} E_{(0,0,0,0,1,1)} - x_{25} E_{(0,0,0,1,1,1)} + x_{26} E_{(0,0,1,1,1,1)} + x_{27} E_{(1,0,1,1,1,1)},\end{aligned} \tag{12.5.32}$$

$$\begin{aligned}\iota(\mathfrak{b}_{21}) = {} & P_{21} + x_2 E'_{(1,1,2,3,2,1)} - x_4 E'_{(1,1,2,2,1,1)} + x_5 E'_{(1,1,1,2,1,1)} - x_8 E'_{(0,1,1,2,1,1)} \\ & - x_{21} \mathfrak{k}(\alpha_{(1,1,2,3,2,2,1)}) - x_9 E'_{(1,0,1,1,1,1)} + x_{12} E'_{(0,0,1,1,1,1)} + x_{14} E'_{(1,0,1,1)} \\ & - x_{15} E'_{(0,0,0,1,1,1)} - x_{17} E'_{(0,0,1,1)} + x_{18} E_{-\alpha_6} + x_{19} E_{-\alpha_4} - x_{23} E_{\alpha_2} \\ & - x_{22} E_{\alpha_5} + x_{25} E_{(0,1,0,1,1)} - x_{26} E_{(0,1,1,1,1)} - x_{27} E_{(1,1,1,1,1)}, \end{aligned} \tag{12.5.33}$$

$$\begin{aligned}\iota(\mathfrak{b}_{22}) = {} & P_{22} - x_3 E'_{(1,1,2,3,2,1)} - x_4 E'_{(1,1,2,2,2,1)} - x_{22} \mathfrak{k}(\alpha_{(1,1,2,3,3,2,1)}) + x_5 E'_{(1,1,1,2,2,1)} \\ & - x_8 E'_{(0,1,1,2,2,1)} - x_{11} E'_{(1,0,1,1,1,1)} + x_{13} E'_{(0,0,1,1,1,1)} + x_{14} E'_{(1,0,1,1,1)} \\ & - x_{16} E'_{(0,0,0,1,1,1)} - x_{17} E'_{(0,0,1,1.1)} - x_{18} E'_{(0,0,0,0,1,1)} + x_{19} E'_{(0,0,0,1,1)} \\ & + x_{21} E_{-\alpha_5} - x_{24} E_{\alpha_2} - x_{25} E_{(0,1,0,1)} + x_{26} E_{(0,1,1,1)} + x_{27} E_{(1,1,1,1)}, \end{aligned} \tag{12.5.34}$$

$$\begin{aligned}\iota(\mathfrak{b}_{23}) = {} & P_{23} + x_2 E'_{(1,2,2,3,2,1)} + x_6 E'_{(1,1,2,2,1,1)} - x_7 E'_{(1,1,1,2,1,1)} + x_9 E'_{(1,1,1,1,1,1)} \\ & - x_{23} \mathfrak{k}(\alpha_{(1,2,2,3,2,2,1)} + x_{10} E'_{(0,1,1,2,1,1)} - x_{12} E'_{(0,1,1,1,1,1)} - x_{14} E'_{(1,1,1,1)} \\ & + x_{15} E'_{(0,1,0,1,1,1)} + x_{17} E'_{(0,1,1,1)} + x_{20} E_{-\alpha_6} - x_{19} E_{(0,1,0,1)} + x_{21} E_{-\alpha_2} \\ & - x_{24} E_{\alpha_5} + x_{25} E_{(0,0,0,1,1)} - x_{26} E_{(0,0,1,1,1)} - x_{27} E_{(1,0,1,1,1)}, \end{aligned} \tag{12.5.35}$$

$$\begin{aligned}\iota(\mathfrak{b}_{24}) = {} & P_{24} - x_3 E'_{(1,2,2,3,2,1)} + x_6 E'_{(1,1,2,2,2,1)} - x_{24} \mathfrak{k}(\alpha_{(1,2,2,3,3,2,1)}) - x_7 E'_{(1,1,1,2,2,1)} \\ & + x_{10} E'_{(0,1,1,2,2,1)} + x_{11} E'_{(1,1,1,1,1,1)} - x_{13} E'_{(0,1,1,1,1,1)} - x_{14} E'_{(1,1,1,1,1)} \\ & + x_{16} E'_{(0,1,0,1,1,1)} + x_{17} E'_{(0,1,1,1,1)} - x_{19} E'_{(0,1,0,1,1)} - x_{20} E'_{(0,0,0,0,1,1)} \\ & + x_{22} E_{-\alpha_2} + x_{23} E_{-\alpha_5} - x_{25} E_{\alpha_4} + x_{26} E_{(0,0,1,1)} + x_{27} E_{(1,0,1,1)}, \end{aligned} \tag{12.5.36}$$

$$\begin{aligned}\iota(\mathfrak{b}_{25}) = {} & P_{25} + x_4 E'_{(1,2,2,3,2,1)} + x_6 E'_{(1,1,2,3,2,1)} - x_{25} \mathfrak{k}(\alpha_{(1,2,3,4,3,2,1)} - x_9 E'_{(1,1,1,2,2,1)} \\ & + x_{12} E'_{(0,1,1,2,2,1)} + x_{11} E'_{(1,1,1,2,1,1)} - x_{13} E'_{(0,1,1,2,1,1)} - x_{14} E'_{(1,1,1,2,1)} \\ & + x_{17} E'_{(0,1,1,2,1)} + x_{18} E'_{(0,1,0,1,1,1)} + x_{20} E'_{(0,0,0,1,1,1)} - x_{21} E'_{(0,1,0,1,1)} \\ & + x_{22} E'_{(0,1,0,1)} - x_{23} E'_{(0,0,0,1,1)} + x_{24} E_{-\alpha_4} - x_{26} E_{\alpha_3} - x_{27} E_{(1,0,1)}, \end{aligned} \tag{12.5.37}$$

$$\begin{aligned}\iota(\mathfrak{b}_{26}) = {} & P_{26} - x_5 E'_{(1,2,2,3,2,1)} - x_7 E'_{(1,1,2,3,2,1)} - x_{26} \mathfrak{k}(\alpha_{(1,2,3,4,3,2,1)}) + x_9 E'_{(1,1,2,2,2,1)} \\ & - x_{15} E'_{(0,1,1,2,2,1)} - x_{11} E'_{(1,1,2,2,1,1)} + x_{16} E'_{(0,1,1,2,1,1)} + x_{14} E'_{(1,1,2,2,1)} \\ & - x_{19} E'_{(0,1,1,2,1)} - x_{18} E'_{(0,1,1,1,1,1)} - x_{20} E'_{(0,0,1,1,1,1)} + x_{21} E'_{(0,1,1,1,1)} \\ & - x_{22} E'_{(0,1,1,1)} + x_{23} E'_{(0,0,1,1,1)} - x_{24} E'_{(0,0,1,1)} + x_{25} E_{-\alpha_3} + x_{27} E_{\alpha_1}, \end{aligned} \tag{12.5.38}$$

$$\begin{aligned}\iota(\mathfrak{b}_{27}) = {} & P_{27} - x_8 E'_{(1,2,2,3,2,1)} - x_{10} E'_{(1,1,2,3,2,1)} - x_{27} \mathfrak{k}(\alpha_{(2,2,3,4,3,2,1)}) + x_{12} E'_{(1,1,2,2,2,1)} \\ & - x_{15} E'_{(1,1,1,2,2,1)} - x_{13} E'_{(1,1,2,2,1,1)} + x_{16} E'_{(1,1,1,2,1,1)} + x_{17} E'_{(1,1,2,2,1)} \\ & - x_{19} E'_{(1,1,1,2,1)} - x_{18} E'_{(1,1,1,1,1,1)} - x_{20} E'_{(1,0,1,1,1,1)} + x_{21} E'_{(1,1,1,1,1)} \\ & - x_{22} E'_{(1,1,1,1)} + x_{23} E'_{(1,0,1,1,1)} - x_{24} E'_{(1,0,1,1)} + x_{25} E'_{(1,0,1)} - x_{26} E_{-\alpha_1}. \end{aligned} \tag{12.5.39}$$

Note that the coefficient of x_i in the above $\iota(\mathfrak{b}_i)$ is exactly $\mathfrak{k}(\beta)$ if $\mathfrak{b}_i = E_\beta$.

Recall $\mathscr{A} = \mathbb{F}[x_1, x_2, \ldots, x_{27}]$. Let M be a $\mathscr{G}^{E_6}$-module and set

$$\widehat{M} = \mathscr{A} \otimes_{\mathbb{C}} M. \tag{12.5.40}$$

We identify

$$f \otimes v = fv \qquad \text{for } f \in \mathscr{A},\ v \in M. \tag{12.5.41}$$

Recall the Lie algebra $\widehat{\mathscr{K}}$ defined via (12.5.1)–(12.5.8). Fix $c \in \mathbb{F}$. Then $\widehat{M}$ becomes a $\widehat{\mathscr{K}}$-module with the action defined by

$$d(fw) = d(f)w, \quad \kappa(fw) = cfw, \quad (gu)(fw) = fgu(w) \tag{12.5.42}$$

for $d \in \mathbb{W}_{27},\ f, g \in \mathscr{A},\ w \in M$ and $u \in \mathscr{G}^{E_6}$.

Since the linear map $\iota : \mathscr{G}^{E_7} \to \widehat{\mathscr{K}}$ defined in (12.5.10)–(12.5.12) is a Lie algebra monomorphism, $\widehat{M}$ becomes a $\mathscr{G}^{E_7}$-module with the action defined by

$$\xi(w) = \iota(\xi)(w) \qquad \text{for } \xi \in \mathscr{G}^{E_7},\ w \in \widehat{M}. \tag{12.5.43}$$

In fact, we have:

Theorem 12.5.1 *The map $M \mapsto \widehat{M}$ gives a functor from $\mathscr{G}^{E_6}$-**Mod** to $\mathscr{G}^{E_7}$-**Mod**, where the morphisms are $\mathscr{G}^{E_6}$-module homomorphisms and $\mathscr{G}^{E_7}$-module homomorphisms, respectively.*

We remark that the module $\widehat{M}$ is not a generalized module in general because it may not be equal to $U(\mathscr{G})(M) = U(\mathscr{G}_-)(M)$.

Proposition 12.5.2 *If M is an irreducible $\mathscr{G}^{E_6}$-module, then $U(\mathscr{G}_-)(M)$ is an irreducible $\mathscr{G}^{E_7}$-module.*

Proof Note that for any $i \in \overline{1, 27}$, $f \in \mathscr{A}$ and $v \in M$, (12.5.12), (12.5.42) and (12.5.43) imply

$$\mathfrak{a}_i(fv) = \partial_{x_i}(f)v. \tag{12.5.44}$$

Let W be any nonzero $\mathscr{G}^{E_7}$-submodule. The above expression shows that $W \bigcap M \neq \{0\}$. According to (12.5.42), $W \bigcap M$ is a $\mathscr{G}^{E_6}$-submodule. By the irreducibility of M, $M \subset W$. Thus $U(\mathscr{G}_-)(M) \subset W$. So $U(\mathscr{G}_-)(M) = W$ is irreducible. □

By the above proposition, the map $M \mapsto U(\mathscr{G}_-)(M)$ is a polynomial extension from irreducible $\mathscr{G}^{E_6}$-modules to irreducible $\mathscr{G}^{E_7}$-modules. We can use it to derive Gel'fand–Zetlin bases for E_7 from those for E_6.

12.6 Irreducibility of the Functor

In this section, we want to determine the irreducibility of $\mathscr{G}^{E_6}$-modules $\widehat{M}$.

Note that $\widehat{M}$ can be viewed as a $\mathscr{G}^{E_6}$-module. Indeed, (12.5.11) and (12.5.43) show

$$u(fv) = u(f)v + fu(v) \qquad \text{for } u \in \mathscr{G}^{E_6},\ f \in \mathscr{A},\ v \in M \tag{12.6.1}$$

(cf. (11.1.18)–(11.1.56)). So $\widehat{M} = \mathscr{A} \otimes_{\mathbb{C}} M$ is a tensor module of $\mathscr{G}^{E_6}$. Write

$$\mathfrak{b}^{\alpha} = \prod_{i=1}^{27} \mathfrak{b}_i^{\alpha_i}, \quad |\alpha| = \sum_{i=1}^{27} \alpha_i \ \text{ for } \alpha = (\alpha_1, \alpha_2, \ldots, \alpha_{27}) \in \mathbb{N}^{27} \tag{12.6.2}$$

(cf. (11.1.4)–(11.1.12)). Recall the Lie subalgebras $\mathscr{G}_{\pm}$ and $\mathscr{G}_0$ of $\mathscr{G}^{E_6}$ defined in (12.4.2). For $k \in \mathbb{N}$, we set

$$\mathscr{A}_k = \mathrm{Span}_{\mathbb{F}}\{x^{\alpha} \mid \alpha \in \mathbb{N}^{27}; |\alpha| = k\}, \quad \widehat{M}_k = \mathscr{A}_k M \tag{12.6.3}$$

(cf. (12.5.41)) and

$$(U(\mathscr{G}_-)(M))_k = \mathrm{Span}_{\mathbb{F}}\{\mathfrak{b}^{\alpha}(M) \mid \alpha \in \mathbb{N}^{27},\ |\alpha| = k\}. \tag{12.6.4}$$

Moreover,

$$(U(\mathscr{G}_-)(M))_0 = \widehat{M}_0 = M. \tag{12.6.5}$$

Furthermore,

$$\widehat{M} = \bigoplus_{k=0}^{\infty} \widehat{M}_k, \qquad U(\mathscr{G}_-)(M) = \bigoplus_{k=0}^{\infty} (U(\mathscr{G}_-)(M))_k. \tag{12.6.6}$$

Next we define a linear transformation φ on $\widehat{M}$ determined by

$$\varphi(x^{\alpha} v) = \mathfrak{b}^{\alpha}(v) \qquad \text{for } \alpha \in \mathbb{N}^{27},\ v \in M. \tag{12.6.7}$$

Note that $\mathscr{A}_1 = \sum_{i=1}^{27} \mathbb{F}x_i$ forms the 27-dimensional $\mathscr{G}_0$-module of highest weight λ_6. According to (4.4.24) and (4.4.25), $\mathscr{G}_-$ forms a $\mathscr{G}_0$-module with respect to the adjoint representation, and the linear map from $\mathscr{A}_1$ to $\mathscr{G}_-$ determined by $x_i \mapsto \mathfrak{b}_i$ for $i \in \overline{1, 27}$ gives a $\mathscr{G}_0$-module isomorphism. Thus φ can also be viewed as a $\mathscr{G}_0$-module homomorphism from $\widehat{M}$ to $U(\mathscr{G}_0)(M)$. Moreover,

$$\varphi(\widehat{M}_k) = (U(\mathscr{G}_-)(M))_k \qquad \text{for } k \in \mathbb{N}. \tag{12.6.8}$$

Observe that the fundamental weights of $\mathcal{G}^{E_6}$ are:

$$\lambda_1 = \frac{1}{3}(4\alpha_1 + 3\alpha_2 + 5\alpha_3 + 6\alpha_4 + 4\alpha_5 + 2\alpha_6), \tag{12.6.9}$$

$$\lambda_2 = \alpha_1 + 2\alpha_2 + 2\alpha_3 + 3\alpha_4 + 2\alpha_5 + \alpha_6, \tag{12.6.10}$$

$$\lambda_3 = \frac{1}{3}(5\alpha_1 + 6\alpha_2 + 10\alpha_3 + 12\alpha_4 + 8\alpha_5 + 4\alpha_6), \tag{12.6.11}$$

$$\lambda_4 = 2\alpha_1 + 3\alpha_2 + 4\alpha_3 + 6\alpha_4 + 4\alpha_5 + 2\alpha_6, \tag{12.6.12}$$

$$\lambda_5 = \frac{1}{3}(4\alpha_1 + 6\alpha_2 + 8\alpha_3 + 12\alpha_4 + 10\alpha_5 + 5\alpha_6), \tag{12.6.13}$$

$$\lambda_6 = \frac{1}{3}(2\alpha_1 + 3\alpha_2 + 4\alpha_3 + 6\alpha_4 + 5\alpha_5 + 4\alpha_6). \tag{12.6.14}$$

Using the above expressions, we can view $\lambda_i \in H_{E_6}$ by (11.1.2). Then Casimir element of $\mathcal{G}^{E_6}$ is

$$\omega = \sum_{i=1}^{6} \lambda_i \alpha_i - \sum_{\beta \in \Phi_{E_6}} E_\beta E_{-\beta} \in U(\mathcal{G}^{E_6}) \tag{12.6.15}$$

due to (12.4.12). The algebra $U(\mathcal{G}^{E_6})$ is imbedded into the tensor algebra $U(\mathcal{G}^{E_6}) \otimes_{\mathbb{F}} U(\mathcal{G}^{E_6})$ by the associative algebra homomorphism $\mathfrak{d} : U(\mathcal{G}^{E_6}) \to U(\mathcal{G}^{E_6}) \otimes_{\mathbb{F}} U(\mathcal{G}^{E_6}))$ determined by

$$\mathfrak{d}(u) = u \otimes 1 + 1 \otimes u \qquad \text{for } u \in \mathcal{G}^{E_6}. \tag{12.6.16}$$

Set

$$\tilde{\omega} = \frac{1}{2}(\mathfrak{d}(\omega) - \omega \otimes 1 - 1 \otimes \omega) \in U(\mathcal{G}^{E_6}) \otimes_{\mathbb{F}} U(\mathcal{G}^{E_6}). \tag{12.6.17}$$

Since $\sum_{i=1}^{6} \lambda_i \otimes \alpha_i$ is symmetric with respect to $\{\alpha_1, \dots, \alpha_6\}$ by (12.6.9)–(12.6.14),

$$\tilde{\omega} = \sum_{i=1}^{6} \lambda_i \otimes \alpha_i - \sum_{\beta \in \Phi_{E_6}} E_\beta \otimes E_{-\beta}. \tag{12.6.18}$$

Furthermore, $\tilde{\omega}$ acts on $\widehat{M}$ as a $\mathcal{G}^{E_6}$-module homomorphism via

$$(u_1 \otimes u_2)(f w) = u_1(f) u_2(w) \qquad \text{for } u_1, u_2 \in \mathcal{G}^{E_6},\ f \in \mathscr{A},\ w \in M. \tag{12.6.19}$$

Lemma 12.6.1 *We have* $\varphi|_{\widehat{M}_1} = (\tilde{\omega} - c)|_{\widehat{M}_1}$.

Proof By (11.1.54) with Table 11.1 and (12.6.9)–(12.6.14)

$$\lambda_r|_{\mathscr{A}} = \sum_{i=1}^{27} c_{i,r} x_i \partial_{x_i} \qquad \text{for } r \in \overline{1,6} \tag{12.6.20}$$

with $c_{i,r} \in \mathbb{F}$, for instance,

$$c_{1,1} = \frac{2}{3},\ c_{1,2} = 1,\ c_{1,3} = \frac{4}{3},\ c_{1,4} = 2,\ c_{1,5} = \frac{5}{3},\ c_{1,6} = \frac{4}{3}. \tag{12.6.21}$$

On the other hand,

$$\alpha_7 = \frac{1}{3}[\widehat{\alpha} - (2\alpha_1 + 3\alpha_2 + 4\alpha_3 + 6\alpha_4 + 5\alpha_5 + 4\alpha_6)] \tag{12.6.22}$$

by (12.4.5). According to (12.5.10), (12.5.42), (12.6.19) and (12.6.21),

$$\left(\sum_{i=1}^{6} \lambda_i \otimes \alpha_i\right)(x_1 w) = c x_1 w - x_1 \alpha_7(w) \qquad \text{for } w \in M. \tag{12.6.23}$$

Thus (11.1.18)–(11.1.56) yield

$$\begin{aligned}\tilde{\omega}(x_1 w) = \ & [c x_1 + x_2 E_{\alpha_6} + x_3 E_{(0,0,0,0,1,1)} + x_4 E_{(0,0,0,1,1,1)} + x_5 E_{(0,0,1,1,1,1)} + x_6 E_{(0,1,0,1,1,1)} \\ & - x_1 \alpha_7(w) + x_7 E_{(0,1,1,1,1,1)} + x_8 E_{(1,0,1,1,1,1)} + x_9 E_{(0,1,1,2,1,1)} + x_{10} E_{(1,1,1,1,1,1)} \\ & + x_{11} E_{(0,1,1,2,2,1)} + x_{12} E_{(1,1,1,2,1,1)} + x_{13} E_{(1,1,1,2,2,1)} + x_{15} E_{(1,1,2,2,1,1)} \\ & + x_{16} E_{(1,1,2,2,2,1)} + x_{18} E_{(1,1,2,3,2,1)} + x_{20} E_{(1,2,2,3,2,1)}](w).\end{aligned} \tag{12.6.24}$$

Comparing (12.5.13) and (12.6.24), we get $\varphi(x_1 w) = (\tilde{\omega} - c)(x_1 w)$. According to (11.1.23) and (11.1.56), $-E_{-\alpha_6}(x_1) = x_2$. So for $v \in M$,

$$\begin{aligned} & \varphi(x_2 v) - \varphi(x_1 E_{-\alpha_6}(v)) = -(E_{-\alpha_6})(\varphi(x_1 v)) \\ = \ & -(E_{-\alpha_6})[(\tilde{\omega} - c)(x_1 v)] = (\tilde{\omega} - c)[-E_{-\alpha_6}(x_1 v)] \\ = \ & (\tilde{\omega} - c)[x_2 v - x_1 E_{-\alpha_6}(v)] \\ = \ & (\tilde{\omega} - c)(x_2 v) - (\tilde{\omega} - c)(x_1 E_{-\alpha_6})(v)) \\ = \ & (\tilde{\omega} - c)(x_2 v) - \varphi(x_1 E_{-\alpha_6}(v)), \end{aligned} \tag{12.6.25}$$

or equivalently, $\varphi(x_2 v) = (\tilde{\omega} - c)(x_2 v)$. By (11.1.18)–(11.1.23) and (11.1.56), similar arguments as (12.6.25) and induction on i, we can prove

$$\varphi(x_i v) = (\tilde{\omega} - c)(x_i v) \qquad \text{for } i \in \overline{1,27},\ v \in M; \tag{12.6.26}$$

that is, the lemma holds. □

Recall that $\mathbb{N}$ is the set of nonnegative integers and the set of dominate integral weights is

$$\Lambda^+ = \sum_{r=1}^{6} \mathbb{N}\lambda_r. \tag{12.6.27}$$

Denote

$$\rho = \sum_{r=1}^{6} \lambda_r \tag{12.6.28}$$

(cf. (12.6.9)–(12.6.14). For any $\mu \in \Lambda^+$, we denote by $V(\mu)$ the finite-dimensional irreducible $\mathscr{G}^{E_6}$-module with the highest weight μ and have

$$\omega|_{V(\mu)} = (\mu + 2\rho, \mu)\mathrm{Id}_{V(\mu)} \tag{12.6.29}$$

by (12.6.15). According to (11.1.54) and Table 11.1, the weight set of the $\mathscr{G}^{E_6}$-module $\mathscr{A}_1$ is

$$\Pi(\mathscr{A}_1) = \left\{ \sum_{r=1}^{6} a_{i,r}\lambda_r \mid i \in \overline{1,27} \right\}. \tag{12.6.30}$$

Fixing $\lambda \in \Lambda^+$, we define

$$\Upsilon(\lambda) = \{\lambda + \mu \mid \mu \in \Pi(\mathscr{A}_1),\ \lambda + \mu \in \Lambda^+\}. \tag{12.6.31}$$

Lemma 12.6.2 *We have:*

$$\mathscr{A}_1 \otimes V(\lambda) \cong \bigoplus_{\lambda' \in \Upsilon(\lambda)} V(\lambda'). \tag{12.6.32}$$

Proof Note that all the weight subspaces of $\mathscr{A}_1$ are one-dimensional. Thus all the irreducible components of $\mathscr{A}_1 \otimes V(\lambda)$ are of multiplicity one. Since

$$\rho + \lambda + \mu \in \Lambda^+ \qquad \text{for } \mu \in \Pi(\mathscr{A}_1), \tag{12.6.33}$$

Theorem 5.4.3 says that $V(\lambda')$ is a component of $\mathscr{A}_1 \otimes V(\lambda)$ if and only if $\lambda' \in \Upsilon(\lambda)$. $\square$

Recall

$$\text{the highest weight of } \mathscr{A}_1 = \lambda_6 \tag{12.6.34}$$

by Table 11.1. Thus the eigenvalues of $\tilde{\omega}|_{\widehat{V(\lambda)_1}}$ are

$$\{[(\lambda' + 2\rho, \lambda') - (\lambda + 2\rho, \lambda) - (\lambda_6 + 2\rho, \lambda_6)]/2 \mid \lambda' \in \Upsilon(\lambda)\} \tag{12.6.35}$$

by (12.6.17) and (12.6.19). We remark that the above fact is equivalent to a special explicit version of Kostant's characteristic identity (cf. [Kb]). Define

$$\ell_\omega(\lambda)=\min\{[(\lambda'+2\rho,\lambda')-(\lambda+2\rho,\lambda)-(\lambda_6+2\rho,\lambda_6)]/2\mid\lambda'\in\Upsilon(\lambda)\},\tag{12.6.36}$$

which will be used to determine the irreducibility of $\widehat{V(\lambda)}$. If $\lambda'=\lambda+\lambda_6-\alpha\in\Upsilon(\lambda)$ with $\alpha\in\Phi^+_{D_6}$, then

$$(\lambda'+2\rho,\lambda')-(\lambda+2\rho,\lambda)-(\lambda_6+2\rho,\lambda_6)=2[(\lambda,\lambda_6)+1-(\rho+\lambda+\lambda_6,\alpha)].\tag{12.6.37}$$

Recall the differential operators $P_1,P_2,\ldots,P_{27}$ given in (12.4.20) and (12.4.39)–(12.4.64). We also view the elements of $\mathscr{A}$ as the multiplication operators on $\mathscr{A}$. Recall ζ_1 in (11.2.1). It turns out that we need the following lemma in order to determine the irreducibility of $\widehat{V(\lambda)}$. Recall the Dickson's invariant η given in (11.2.42).

Lemma 12.6.3 *As operators on* $\mathscr{A}$:

$$\begin{aligned}&P_{14}x_1+P_1x_{14}+P_{11}x_2+P_2x_{11}+P_9x_3+P_3x_9+P_7x_4+P_4x_7-P_6x_5-P_5x_6\\=&\ \zeta_1(D-8)+\eta\partial_{x_{27}}.\end{aligned}\tag{12.6.38}$$

Proof According to (12.4.20), (12.4.39)–(12.4.44), (12.4.46), (12.4.48) and (12.4.51), we find that

$$\begin{aligned}&P_{14}x_1+P_1x_{14}+P_{11}x_2+P_2x_{11}+P_9x_3+P_3x_9+P_7x_4+P_4x_7-P_6x_5-P_5x_6\\=&-8\zeta_1+x_1P_{14}+x_{14}P_1+x_2P_{11}+x_{11}P_2+x_3P_9\\&+x_9P_3+x_4P_7+x_7P_4-x_5P_6-x_6P_5\end{aligned}\tag{12.6.39}$$

and

$$\begin{aligned}&x_1P_{14}+x_{14}P_1+x_2P_{11}+x_{11}P_2+x_3P_9+x_9P_3+x_4P_7+x_7P_4-x_5P_6-x_6P_5\\&=x_1(x_{14}D-\zeta_1\partial_{x_1}-\zeta_8\partial_{x_8}-\zeta_{10}\partial_{x_{10}}+\zeta_{12}\partial_{x_{12}}-\zeta_{13}\partial_{x_{13}}-\zeta_{15}\partial_{x_{15}}+\zeta_{16}\partial_{x_{16}}-\zeta_{18}\partial_{x_{18}}\\&\quad-\zeta_{20}\partial_{x_{20}}+\zeta_{27}\partial_{x_{27}})+x_{14}(x_1D-\zeta_1\partial_{x_{14}}-\zeta_2\partial_{x_{17}}-\zeta_3\partial_{x_{19}}+\zeta_4\partial_{x_{21}}-\zeta_5\partial_{x_{22}}-\zeta_6\partial_{x_{23}}\\&\quad+\zeta_7\partial_{x_{24}}-\zeta_9\partial_{x_{25}}-\zeta_{11}\partial_{x_{26}}-\zeta_{14}\partial_{x_{27}})+x_2(x_{11}D-\zeta_1\partial_{x_2}+\zeta_5\partial_{x_8}+\zeta_7\partial_{x_{10}}-\zeta_9\partial_{x_{12}}\\&\quad+\zeta_{11}\partial_{x_{15}}+\zeta_{13}\partial_{x_{17}}-\zeta_{16}\partial_{x_{19}}+\zeta_{18}\partial_{x_{21}}+\zeta_{20}\partial_{x_{23}}+\zeta_{26}\partial_{x_{27}})+x_{11}(x_2D-\zeta_1\partial_{x_{11}}\\&\quad-\zeta_2\partial_{x_{13}}-\zeta_3\partial_{x_{16}}+\zeta_4\partial_{x_{18}}-\zeta_6\partial_{x_{20}}-\zeta_8\partial_{x_{22}}+\zeta_{10}\partial_{x_{24}}-\zeta_{12}\partial_{x_{25}}-\zeta_{15}\partial_{x_{26}}+\zeta_{17}\partial_{x_{27}})\\&\quad+x_3(x_9D-\zeta_1\partial_{x_3}-\zeta_4\partial_{x_8}-\zeta_6\partial_{x_{10}}+\zeta_9\partial_{x_{13}}-\zeta_{12}\partial_{x_{17}}-\zeta_{11}\partial_{x_{16}}+\zeta_{15}\partial_{x_{19}}-\zeta_{18}\partial_{x_{22}}\\&\quad-\zeta_{20}\partial_{x_{24}}+\zeta_{25}\partial_{x_{27}})+x_9(x_3D-\zeta_1\partial_{x_9}-\zeta_2\partial_{x_{12}}-\zeta_3\partial_{x_{15}}+\zeta_5\partial_{x_{18}}-\zeta_7\partial_{x_{20}}-\zeta_8\partial_{x_{21}}\\&\quad+\zeta_{10}\partial_{x_{23}}-\zeta_{13}\partial_{x_{25}}-\zeta_{16}\partial_{x_{26}}+\zeta_{19}\partial_{x_{27}})+x_4(x_7D-\zeta_1\partial_{x_4}+\zeta_3\partial_{x_8}+\zeta_6\partial_{x_{12}}-\zeta_7\partial_{x_{13}}\\&\quad+\zeta_{10}\partial_{x_{17}}+\zeta_{11}\partial_{x_{18}}-\zeta_{15}\partial_{x_{21}}+\zeta_{16}\partial_{x_{22}}+\zeta_{20}\partial_{x_{25}}+\zeta_{24}\partial_{x_{27}})+x_7(x_4D-\zeta_1\partial_{x_7}\\&\quad-\zeta_2\partial_{x_{10}}-\zeta_4\partial_{x_{15}}+\zeta_5\partial_{x_{16}}-\zeta_8\partial_{x_{19}}-\zeta_9\partial_{x_{20}}+\zeta_{12}\partial_{x_{23}}-\zeta_{13}\partial_{x_{24}}-\zeta_{18}\partial_{x_{26}}+\zeta_{21}\partial_{x_{27}})\end{aligned}$$

$$\begin{aligned}&-x_5(x_6D+\zeta_1\partial_{x_5}+\zeta_2\partial_{x_8}-\zeta_6\partial_{x_{15}}+\zeta_7\partial_{x_{16}}-\zeta_9\partial_{x_{18}}-\zeta_{10}\partial_{x_{19}}+\zeta_{12}\partial_{x_{21}}\\&-\zeta_{13}\partial_{x_{22}}+\zeta_{20}\partial_{x_{26}}-\zeta_{23}\partial_{x_{27}})-x_6(x_5D+\zeta_1\partial_{x_6}-\zeta_3\partial_{x_{10}}+\zeta_4\partial_{x_{12}}\\&-\zeta_5\partial_{x_{13}}+\zeta_8\partial_{x_{17}}+\zeta_{11}\partial_{x_{20}}-\zeta_{15}\partial_{x_{23}}+\zeta_{16}\partial_{x_{24}}-\zeta_{18}\partial_{x_{25}}-\zeta_{22}\partial_{x_{27}})\\&=\zeta_1D+\eta\partial_{x_{27}}.\qquad\Box\end{aligned}\tag{12.6.40}$$

Lemma 12.6.4 *As operators on* $\mathscr{A}$,

$$\sum_{14\neq i\in\overline{1,26}}P_i\zeta_{28-i}-P_{14}\zeta_{14}-P_{27}\zeta_1=\eta(24-5D).\tag{12.6.41}$$

Proof We calculate it by (11.2.1), (11.2.3)–(11.2.28), (12.4.20) and (12.4.39)–(12.4.64). In particular,

$$2(\zeta_1\zeta_{14}-\zeta_2\zeta_{11}-\zeta_3\zeta_9+\zeta_4\zeta_7-\zeta_5\zeta_6)=-2\eta x_1,\tag{12.6.42}$$

which is the coefficient of ∂_{x_1} in addition to the term containing D. According to (11.2.41), the operator on the left hand side of (12.6.41) is a $\mathscr{G}^{E_6}$-invariant differential operator. By symmetry,

$$\text{the coefficient of }\partial_{x_i}=-2\eta x_i\ \ \text{for}\ \ i\in\overline{1,27}.\qquad\Box\tag{12.6.43}$$

We define the multiplication

$$f(gv)=(fg)v\qquad\text{for }f,g\in\mathscr{A},\ v\in M.\tag{12.6.44}$$

Furthermore, we have:

Lemma 12.6.5 *As operators on* $\widehat{M}$,

$$\left[\sum_{14\neq i\in\overline{1,26}}\iota(\mathfrak{b}_i)\zeta_{28-i}-\iota(\mathfrak{b}_{14})\zeta_{14}-\iota(\mathfrak{b}_{27})\zeta_1\right]|_{\widehat{M}}=\eta(24-5D+3c).\tag{12.6.45}$$

Proof By (12.5.13)–(12.5.39), the coefficient of E_{α_6} on the left hand side of (12.6.45) is

$$x_2\zeta_{27}-x_{14}\zeta_{17}-x_{17}\zeta_{15}-x_{19}\zeta_{12}-x_{21}\zeta_{10}-x_{23}\zeta_8.\tag{12.6.46}$$

Moreover, we use (11.2.9), (11.2.11), (11.2.13), (11.2.16), (11.2.18) and (11.2.28) to find that (12.6.46) is equal to zero. Since the left hand side of (12.6.45) is a $\mathscr{G}^{E_6}$-invariant differential operator, it is invariant under the action of the E_6 Weyl group. The transitivity of the Weyl group on Φ_{E_6} yields that

$$\text{the coefficient of }E_\beta=0\ \ \text{for any }\beta\in\Phi_{E_6}.\tag{12.6.47}$$

By (12.5.10), (12.5.42), (12.6.22) and Lemma 12.6.4,

$$\sum_{14\neq i\in\overline{1,26}} \iota(\mathfrak{b}_i)\zeta_{28-i} - \iota(\mathfrak{b}_{14})\zeta_{14} - \iota(\mathfrak{b}_{27})\zeta_1 = \eta(24 - 5D + 3c) + \sum_{r=1}^{6} f_r\alpha_r \tag{12.6.48}$$

as the elements of $\widehat{\mathscr{K}}$ acting on $\widehat{M}$ (cf. (12.5.7)). The $\mathscr{G}^{E_6}$-invariancy implies

$$\left[E_\beta|_{\widehat{M}}, \sum_{r=1}^{6} f_r\alpha_r\right] = 0 \;\text{ for any } \beta \in \Phi_{E_6}. \tag{12.6.49}$$

Thus $\sum_{r=1}^{6} f_r\alpha_r = 0$; that is, (12.6.45) holds. □

Next we calculate

$$\begin{aligned} T_1 = {} & \iota(\mathfrak{b}_{14})x_1 + \iota(\mathfrak{b}_1)x_{14} + \iota(\mathfrak{b}_{11})x_2 + \iota(\mathfrak{b}_2)x_{11} + \iota(\mathfrak{b}_9)x_3 \\ & + \iota(\mathfrak{b}_3)x_9 + \iota(\mathfrak{b}_7)x_4 + \iota(\mathfrak{b}_4)x_7 - \iota(\mathfrak{b}_6)x_5 - \iota(\mathfrak{b}_5)x_6 \\ = {} & \zeta_1(D-8) + \eta\partial_{x_{27}} + \zeta_2 E_{\alpha_1} - \zeta_3 E_{(1,0,1)} + \zeta_4 E_{(1,0,1,1)} - \zeta_5 E_{(1,0,1,1,1)} \\ & + \zeta_6 E_{(1,1,1,1)} - \zeta_7 E_{(1,1,1,1,1)} + \zeta_8 E_{(1,0,1,1,1,1)} + \zeta_9 E_{(1,1,1,2,1)} \\ & + \zeta_{10} E_{(1,1,1,1,1,1)} - \zeta_{11} E_{(1,1,2,2,1)} - \zeta_{12} E_{(1,1,1,2,1,1)} + \zeta_{13} E_{(1,1,1,2,2,1)} \\ & + \zeta_{15} E_{(1,1,2,2,1,1)} - \zeta_{16} E_{(1,1,2,2,2,1)} + \zeta_{18} E_{(1,1,2,3,2,1)} + \zeta_{20} E_{(1,2,2,3,2,1)} \\ & - 2\zeta_1 c + \frac{\zeta_1}{3}(4\alpha_1 + 3\alpha_2 + 5\alpha_3 + 6\alpha_4 + 4\alpha_5 + 2\alpha_6) \end{aligned} \tag{12.6.50}$$

by (11.2.1), (11.2.3)–(11.2.14), (11.2.16), (11.2.17), (11.2.19), (11.2.21), (12.5.13)–(12.5.19), (12.5.21), (12.5.23), (12.5.26), (12.6.20) and Lemma 12.6.3.

We define a $\mathscr{G}^{E_6}$-module structure on the space End $\widehat{M}$ of linear transformations on $\widehat{M}$ by

$$\iota(u)(T) = [\iota(u), T] = \iota(u)T - T\iota(u) \qquad \text{for } u \in \mathscr{G}^{E_6},\; T \in \text{End}\,\widehat{M} \tag{12.6.51}$$

(cf. (12.6.1)). It can be verified that T_1 is a $\mathscr{G}^{E_6}$-singular vector with weight λ_1 in End $\widehat{M}$. So it generates the 27-dimensional module of highest weight λ_1. We set

$$\begin{aligned} T_2 = {} & -[\iota(E_{-\alpha_1}), T_1] = \zeta_2(D-8) - \eta\partial_{x_{26}} - \zeta_1 E_{-\alpha_1} + \zeta_3 E_{\alpha_3} - \zeta_4 E_{(0,0,1,1)} \\ & + \zeta_5 E_{(0,0,1,1,1)} - \zeta_6 E_{(0,1,1,1)} + \zeta_7 E_{(0,1,1,1,1)} - \zeta_8 E_{(0,0,1,1,1,1)} - \zeta_9 E_{(0,1,1,2,1)} \\ & - \zeta_{10} E_{(0,1,1,1,1,1)} - \zeta_{14} E_{(1,1,2,2,1)} + \zeta_{12} E_{(0,1,1,2,1,1)} - \zeta_{13} E_{(0,1,1,2,2,1)} \\ & - \zeta_{17} E_{(1,1,2,2,1,1)} + \zeta_{19} E_{(1,1,2,2,2,1)} - \zeta_{21} E_{(1,1,2,3,2,1)} - \zeta_{23} E_{(1,2,2,3,2,1)} \\ & - 2\zeta_2 c + \frac{\zeta_2}{3}(\alpha_1 + 3\alpha_2 + 5\alpha_3 + 6\alpha_4 + 4\alpha_5 + 2\alpha_6), \end{aligned} \tag{12.6.52}$$

$$\begin{aligned}T_3 = & -[\iota(E_{-\alpha_3}), T_2] = \zeta_3(D-8) - \eta\partial_{x_{25}} + \zeta_1 E'_{(1,0,1)} - \zeta_2 E_{-\alpha_3} + \zeta_4 E_{\alpha_4} \\ & -\zeta_5 E_{(0,0,0,1,1)} + \zeta_6 E_{(0,1,0,1)} - \zeta_7 E_{(0,1,0,1,1)} + \zeta_8 E_{(0,0,0,1,1,1)} + \zeta_{11} E_{(0,1,1,2,1)} \\ & +\zeta_{10} E_{(0,1,0,1,1,1)} + \zeta_{14} E_{(1,1,1,2,1)} - \zeta_{15} E_{(0,1,1,2,1,1)} + \zeta_{16} E_{(0,1,1,2,2,1)} \\ & +\zeta_{17} E_{(1,1,1,2,1,1)} - \zeta_{19} E_{(1,1,1,2,2,1)} + \zeta_{22} E_{(1,1,2,3,2,1)} + \zeta_{24} E_{(1,2,2,3,2,1)} \\ & -2\zeta_3 c + \frac{\zeta_3}{3}(\alpha_1 + 3\alpha_2 + 2\alpha_3 + 6\alpha_4 + 4\alpha_5 + 2\alpha_6),\end{aligned} \tag{12.6.53}$$

$$\begin{aligned}T_4 = & -[\iota(E_{-\alpha_4}), T_3] = \zeta_4(D-8) - \eta\partial_{x_{24}} - \zeta_1 E'_{(1,0,1,1)} + \zeta_2 E'_{(0,0,1,1)} - \zeta_3 E_{-\alpha_4} \\ & +\zeta_5 E_{\alpha_5} + \zeta_6 E_{\alpha_2} - \zeta_9 E_{(0,1,0,1,1)} - \zeta_8 E_{(0,0,0,0,1,1)} + \zeta_{11} E_{(0,1,1,1,1)} \\ & +\zeta_{12} E_{(0,1,0,1,1,1)} + \zeta_{14} E_{(1,1,1,1,1)} - \zeta_{15} E_{(0,1,1,1,1,1)} + \zeta_{18} E_{(0,1,1,2,2,1)} \\ & +\zeta_{17} E_{(1,1,1,1,1,1)} - \zeta_{21} E_{(1,1,1,2,2,1)} + \zeta_{22} E_{(1,1,2,2,2,1)} - \zeta_{25} E_{(1,2,2,3,2,1)} \\ & -2\zeta_4 c + \frac{\zeta_4}{3}(\alpha_1 + 3\alpha_2 + 2\alpha_3 + 3\alpha_4 + 4\alpha_5 + 2\alpha_6),\end{aligned} \tag{12.6.54}$$

$$\begin{aligned}T_5 = & -[\iota(E_{-\alpha_5}), T_4] = \zeta_5(D-8) - \eta\partial_{x_{23}} + \zeta_1 E'_{(1,0,1,1,1)} - \zeta_2 E'_{(0,0,1,1,1)} \\ & +\zeta_3 E'_{(0,0,0,1,1)} - \zeta_4 E_{-\alpha_5} + \zeta_7 E_{\alpha_2} - \zeta_9 E_{(0,1,0,1)} + \zeta_8 E_{\alpha_6} + \zeta_{11} E_{(0,1,1,1)} \\ & +\zeta_{13} E_{(0,1,0,1,1,1)} + \zeta_{14} E_{(1,1,1,1)} - \zeta_{16} E_{(0,1,1,1,1,1)} + \zeta_{18} E_{(0,1,1,2,1,1)} \\ & +\zeta_{19} E_{(1,1,1,1,1,1)} - \zeta_{21} E_{(1,1,1,2,1,1)} + \zeta_{22} E_{(1,1,2,2,1,1)} + \zeta_{26} E_{(1,2,2,3,2,1)} \\ & -2\zeta_5 c + \frac{\zeta_5}{3}(\alpha_1 + 3\alpha_2 + 2\alpha_3 + 3\alpha_4 + \alpha_5 + 2\alpha_6),\end{aligned} \tag{12.6.55}$$

$$\begin{aligned}T_6 = & -[\iota(E_{-\alpha_2}), T_4] = \zeta_6(D-8) - \eta\partial_{x_{22}} - \zeta_1 E'_{(1,1,1,1)} + \zeta_2 E'_{(0,1,1,1)} - \zeta_3 E'_{(0,1,0,1)} \\ & +\zeta_7 E_{\alpha_5} - \zeta_4 E_{-\alpha_2} + \zeta_9 E_{(0,0,0,1,1)} - \zeta_{10} E_{(0,0,0,0,1,1)} - \zeta_{11} E_{(0,0,1,1,1)} \\ & -\zeta_{12} E_{(0,0,0,1,1,1)} - \zeta_{14} E_{(1,0,1,1,1)} + \zeta_{15} E_{(0,0,1,1,1,1)} - \zeta_{20} E_{(0,1,1,2,2,1)} \\ & -\zeta_{17} E_{(1,0,1,1,1,1)} + \zeta_{23} E_{(1,1,1,2,2,1)} - \zeta_{24} E_{(1,1,2,2,2,1)} - \zeta_{25} E_{(1,1,2,3,2,1)} \\ & -2\zeta_6 c + \frac{\zeta_6}{3}(\alpha_1 + 2\alpha_3 + 3\alpha_4 + 4\alpha_5 + 2\alpha_6),\end{aligned} \tag{12.6.56}$$

$$\begin{aligned}T_7 = & -[\iota(E_{-\alpha_5}), T_6] = \zeta_7(D-8) - \eta\partial_{x_{21}} + \zeta_1 E'_{(1,1,1,1,1)} - \zeta_2 E'_{(0,1,1,1,1)} \\ & +\zeta_3 E'_{(0,1,0,1,1)} - \zeta_6 E_{-\alpha_5} - \zeta_5 E_{-\alpha_2} + \zeta_9 E_{\alpha_4} + \zeta_{10} E_{\alpha_6} - \zeta_{11} E_{(0,0,1,1)} \\ & -\zeta_{13} E_{(0,0,0,1,1,1)} - \zeta_{14} E_{(1,0,1,1)} + \zeta_{16} E_{(0,0,1,1,1,1)} - \zeta_{20} E_{(0,1,1,2,1,1)} \\ & -\zeta_{19} E_{(1,0,1,1,1,1)} + \zeta_{23} E_{(1,1,1,2,1,1)} - \zeta_{24} E_{(1,1,2,2,1,1)} + \zeta_{26} E_{(1,1,2,3,2,1)} \\ & -2\zeta_7 c + \frac{\zeta_7}{3}(\alpha_1 + 2\alpha_3 + 3\alpha_4 + \alpha_5 + 2\alpha_6),\end{aligned} \tag{12.6.57}$$

$$\begin{aligned}T_8 = & -[\iota(E_{-\alpha_6}), T_5] = \zeta_8(D-8) - \eta\partial_{x_{20}} - \zeta_1 E'_{(1,0,1,1,1,1)} + \zeta_2 E'_{(0,0,1,1,1,1)} \\ & -\zeta_3 E'_{(0,0,0,1,1,1)} + \zeta_4 E'_{(0,0,0,0,1,1)} - \zeta_5 E_{-\alpha_6} + \zeta_{10} E_{\alpha_2} - \zeta_{12} E_{(0,1,0,1)} \\ & +\zeta_{13} E_{(0,1,0,1,1)} + \zeta_{15} E_{(0,1,1,1)} - \zeta_{17} E_{(1,1,1,1)} - \zeta_{16} E_{(0,1,1,1,1)} + \zeta_{18} E_{(0,1,1,2,1)}\end{aligned}$$

$$+\zeta_{19}E_{(1,1,1,1,1)} - \zeta_{21}E_{(1,1,1,2,1)} + \zeta_{22}E_{(1,1,2,2,1)} - \zeta_{27}E_{(1,2,2,3,2,1)}$$
$$-2\zeta_8 c + \frac{\zeta_8}{3}(\alpha_1 + 3\alpha_2 + 2\alpha_3 + 3\alpha_4 + \alpha_5 - \alpha_6), \qquad (12.6.58)$$

$$T_9 = -[\iota(E_{-\alpha_4}), T_7] = \zeta_9(D-8) - \eta\partial_{x_{19}} - \zeta_1 E'_{(1,1,1,2,1)} + \zeta_2 E'_{(0,1,1,2,1)}$$
$$+\zeta_4 E'_{(0,1,0,1,1)} - \zeta_6 E'_{(0,0,0,1,1)} + \zeta_5 E'_{(0,1,0,1)} - \zeta_7 E_{-\alpha_4} - \zeta_{11}E_{\alpha_3} + \zeta_{12}E_{\alpha_6}$$
$$+\zeta_{13}E_{(0,0,0,0,1,1)} - \zeta_{14}E_{(1,0,1)} + \zeta_{18}E_{(0,0,1,1,1,1)} - \zeta_{20}E_{(0,1,1,1,1,1)}$$
$$-\zeta_{21}E_{(1,0,1,1,1,1)} + \zeta_{23}E_{(1,1,1,1,1,1)} + \zeta_{25}E_{(1,1,2,2,1,1)} + \zeta_{26}E_{(1,1,2,2,2,1)}$$
$$-2\zeta_9 c + \frac{\zeta_9}{3}(\alpha_1 + 2\alpha_3 + \alpha_5 + 2\alpha_6), \qquad (12.6.59)$$

$$T_{10} = -[\iota(E_{-\alpha_2}), T_8] = \zeta_{10}(D-8) - \eta\partial_{x_{18}} - \zeta_1 E'_{(1,1,1,1,1,1)} + \zeta_2 E'_{(0,1,1,1,1,1)}$$
$$-\zeta_3 E'_{(0,1,0,1,1,1)} + \zeta_6 E'_{(0,0,0,0,1,1)} - \zeta_7 E_{-\alpha_6} - \zeta_8 E_{-\alpha_2} + \zeta_{12}E_{\alpha_4}$$
$$-\zeta_{13}E_{(0,0,0,1,1)} - \zeta_{15}E_{(0,0,1,1)} + \zeta_{17}E_{(1,0,1,1)} + \zeta_{16}E_{(0,0,1,1,1)} - \zeta_{20}E_{(0,1,1,2,1)}$$
$$-\zeta_{19}E_{(1,0,1,1,1)} + \zeta_{23}E_{(1,1,1,2,1)} - \zeta_{24}E_{(1,1,2,2,1)} - \zeta_{27}E_{(1,1,2,3,2,1)}$$
$$-2\zeta_{10} c + \frac{\zeta_{10}}{3}(\alpha_1 + 2\alpha_3 + 3\alpha_4 + \alpha_5 - \alpha_6), \qquad (12.6.60)$$

$$T_{11} = [\iota(E_{-\alpha_3}), T_9] = \zeta_{11}(D-8) - \eta\partial_{x_{17}} + \zeta_1 E'_{(1,1,2,2,1)} - \zeta_3 E'_{(0,1,1,2,1)}$$
$$-\zeta_4 E'_{(0,1,1,1,1)} + \zeta_6 E'_{(0,0,1,1,1)} - \zeta_5 E'_{(0,1,1,1)} + \zeta_7 E'_{(0,0,1,1)} + \zeta_9 E_{-\alpha_3}$$
$$+\zeta_{14}E_{\alpha_1} + \zeta_{15}E_{\alpha_6} + \zeta_{16}E_{(0,0,0,0,1,1)} + \zeta_{18}E_{(0,0,0,1,1,1)} - \zeta_{20}E_{(0,1,0,1,1,1)}$$
$$-\zeta_{22}E_{(1,0,1,1,1,1)} + \zeta_{24}E_{(1,1,1,1,1,1)} + \zeta_{25}E_{(1,1,1,2,1,1)} + \zeta_{26}E_{(1,1,1,2,2,1)}$$
$$-2\zeta_{11} c + \frac{\zeta_{11}}{3}(\alpha_1 - \alpha_3 + \alpha_5 + 2\alpha_6), \qquad (12.6.61)$$

$$T_{12} = -[\iota(E_{-\alpha_4}), T_{10}] = \zeta_{12}(D-8) - \eta\partial_{x_{16}} + \zeta_1 E'_{(1,1,1,2,1,1)} - \zeta_2 E'_{(0,1,1,2,1,1)}$$
$$-\zeta_4 E'_{(0,1,0,1,1,1)} + \zeta_6 E'_{(0,0,0,1,1,1)} + \zeta_8 E'_{(0,1,0,1)} - \zeta_9 E_{-\alpha_6} - \zeta_{10}E_{-\alpha_4}$$
$$+\zeta_{13}E_{\alpha_5} - \zeta_{15}E_{\alpha_3} + \zeta_{17}E_{(1,0,1)} + \zeta_{18}E_{(0,0,1,1,1)} - \zeta_{20}E_{(0,1,1,1,1)}$$
$$-\zeta_{21}E_{(1,0,1,1,1)} + \zeta_{23}E_{(1,1,1,1,1)} + \zeta_{25}E_{(1,1,2,2,1)} - \zeta_{27}E_{(1,1,2,2,2,1)}$$
$$-2\zeta_{12} c + \frac{\zeta_{12}}{3}(\alpha_1 + 2\alpha_3 + \alpha_5 - \alpha_6), \qquad (12.6.62)$$

$$T_{13} = -[\iota(E_{-\alpha_5}), T_{12}] = \zeta_{13}(D-8) - \eta\partial_{x_{15}} - \zeta_1 E'_{(1,1,1,2,2,1)} - \zeta_2 E'_{(0,1,1,2,2,1)}$$
$$-\zeta_5 E'_{(0,1,0,1,1,1)} + \zeta_7 E'_{(0,0,0,1,1,1)} - \zeta_8 E'_{(0,1,0,1,1)} - \zeta_9 E'_{(0,0,0,0,1,1)} + \zeta_{10}E'_{(0,0,0,1,1)}$$
$$-\zeta_{12}E_{-\alpha_5} - \zeta_{16}E_{\alpha_3} + \zeta_{19}E_{(1,0,1)} + \zeta_{18}E_{(0,0,1,1)} - \zeta_{20}E_{(0,1,1,1)}$$
$$-\zeta_{21}E_{(1,0,1,1)} + \zeta_{23}E_{(1,1,1,1)} - \zeta_{26}E_{(1,1,2,2,1)} - \zeta_{27}E_{(1,1,2,2,1,1)}$$
$$-2\zeta_{13} c + \frac{\zeta_{13}}{3}(\alpha_1 + 2\alpha_3 - 2\alpha_5 - \alpha_6), \qquad (12.6.63)$$

$$\begin{aligned}T_{14} = & -[\iota(E_{-\alpha_1}), T_{11}] = \zeta_{14}(D-8) + \eta\partial_{x_{14}} + \zeta_2 E'_{(1,1,2,2,1)} - \zeta_3 E'_{(1,1,1,2,1)} \\ & -\zeta_4 E'_{(1,1,1,1,1)} + \zeta_6 E'_{(1,0,1,1,1)} - \zeta_5 E'_{(1,1,1,1,1)} + \zeta_7 E'_{(1,0,1,1)} + \zeta_9 E'_{(1,0,1)} \\ & -\zeta_{11} E_{-\alpha_1} - \zeta_{17} E_{\alpha_6} - \zeta_{19} E_{(0,0,0,0,1,1)} - \zeta_{21} E_{(0,0,0,1,1,1)} + \zeta_{23} E_{(0,1,0,1,1,1)} \\ & +\zeta_{22} E_{(0,0,1,1,1,1)} - \zeta_{24} E_{(0,1,1,1,1,1)} - \zeta_{25} E_{(0,1,1,2,1,1)} - \zeta_{26} E_{(0,1,1,2,2,1)} \\ & -2\zeta_{14} c + \frac{\zeta_{14}}{3}(-2\alpha_1 - \alpha_3 + \alpha_5 + 2\alpha_6), \end{aligned} \tag{12.6.64}$$

$$\begin{aligned}T_{15} = & [\iota(E_{-\alpha_3}), T_{12}] = \zeta_{15}(D-8) - \eta\partial_{x_{13}} - \zeta_1 E'_{(1,1,2,2,1,1)} + \zeta_3 E'_{(0,1,1,2,1,1)} \\ & +\zeta_4 E'_{(0,1,1,1,1,1)} - \zeta_6 E'_{(0,0,1,1,1,1)} - \zeta_8 E'_{(0,1,1,1)} + \zeta_{10} E'_{(0,0,1,1)} - \zeta_{11} E_{-\alpha_6} \\ & +\zeta_{12} E_{-\alpha_3} + \zeta_{16} E_{\alpha_5} - \zeta_{17} E_{\alpha_1} + \zeta_{18} E_{(0,0,0,1,1)} - \zeta_{20} E_{(0,1,0,1,1)} \\ & -\zeta_{22} E_{(1,0,1,1,1)} + \zeta_{24} E_{(1,1,1,1,1)} + \zeta_{25} E_{(1,1,1,2,1)} - \zeta_{27} E_{(1,1,1,2,2,1)} \\ & -2\zeta_{15} c + \frac{\zeta_{15}}{3}(\alpha_1 - \alpha_3 + \alpha_5 - \alpha_6), \end{aligned} \tag{12.6.65}$$

$$\begin{aligned}T_{16} = & [\iota(E_{-\alpha_3}), T_{13}] = \zeta_{16}(D-8) - \eta\partial_{x_{12}} + \zeta_1 E'_{(1,1,2,2,2,1)} + \zeta_3 E'_{(0,1,1,2,2,1)} \\ & +\zeta_5 E'_{(0,1,1,1,1,1)} - \zeta_7 E'_{(0,0,1,1,1,1)} + \zeta_8 E'_{(0,1,1,1,1)} - \zeta_{10} E'_{(0,0,1,1,1)} - \zeta_{11} E'_{(0,0,0,0,1,1)} \\ & +\zeta_{13} E_{-\alpha_3} - \zeta_{15} E_{-\alpha_5} - \zeta_{19} E_{\alpha_1} + \zeta_{18} E_{\alpha_4} - \zeta_{20} E_{(0,1,0,1)} \\ & -\zeta_{22} E_{(1,0,1,1)} + \zeta_{24} E_{(1,1,1,1)} - \zeta_{26} E_{(1,1,1,2,1)} - \zeta_{27} E_{(1,1,1,2,1,1)} \\ & -2\zeta_{16} c + \frac{\zeta_{16}}{3}(\alpha_1 - \alpha_3 - 2\alpha_5 - \alpha_6), \end{aligned} \tag{12.6.66}$$

$$\begin{aligned}T_{17} = & [\iota(E_{-\alpha_1}), T_{15}] = \zeta_{17}(D-8) - \eta\partial_{x_{11}} + \zeta_2 E'_{(1,1,2,2,1,1)} - \zeta_3 E'_{(1,1,1,2,1,1)} \\ & -\zeta_4 E'_{(1,1,1,1,1,1)} + \zeta_6 E'_{(1,0,1,1,1,1)} + \zeta_8 E'_{(1,1,1,1)} - \zeta_{10} E'_{(1,0,1,1)} - \zeta_{12} E'_{(1,0,1)} \\ & +\zeta_{14} E_{-\alpha_6} + \zeta_{15} E_{-\alpha_1} + \zeta_{19} E_{\alpha_5} + \zeta_{21} E_{(0,0,0,1,1)} - \zeta_{23} E_{(0,1,0,1,1)} \\ & -\zeta_{22} E_{(0,0,1,1,1)} + \zeta_{24} E_{(0,1,1,1,1)} + \zeta_{25} E_{(0,1,1,2,1)} - \zeta_{27} E_{(0,1,1,2,2,1)} \\ & -2\zeta_{17} c - \frac{\zeta_{17}}{3}(2\alpha_1 + \alpha_3 - \alpha_5 + \alpha_6), \end{aligned} \tag{12.6.67}$$

$$\begin{aligned}T_{18} = & -[\iota(E_{-\alpha_4}), T_{16}] = \zeta_{18}(D-8) - \eta\partial_{x_{10}} - \zeta_1 E'_{(1,1,2,3,2,1)} + \zeta_4 E'_{(0,1,1,2,2,1)} \\ & -\zeta_5 E'_{(0,1,1,2,1,1)} - \zeta_8 E'_{(0,1,1,2,1)} - \zeta_9 E'_{(0,0,1,1,1,1)} - \zeta_{11} E'_{(0,0,0,1,1,1)} - \zeta_{12} E'_{(0,0,1,1,1)} \\ & -\zeta_{13} E'_{(0,0,1,1)} - \zeta_{15} E'_{(0,0,0,1,1)} - \zeta_{16} E_{-\alpha_4} - \zeta_{20} E_{\alpha_2} - \zeta_{21} E_{\alpha_1} \\ & -\zeta_{22} E_{(1,0,1)} - \zeta_{25} E_{(1,1,1,1)} - \zeta_{26} E_{(1,1,1,1,1)} - \zeta_{27} E_{(1,1,1,1,1,1)} \\ & -2\zeta_{16} c + \frac{\zeta_{18}}{3}(\alpha_1 - \alpha_3 - 3\alpha_4 - 2\alpha_5 - \alpha_6), \end{aligned} \tag{12.6.68}$$

$$\begin{aligned}T_{19} = & [\iota(E_{-\alpha_1}), T_{16}] = \zeta_{19}(D-8) - \eta\partial_{x_9} - \zeta_2 E'_{(1,1,2,2,2,1)} - \zeta_3 E'_{(1,1,1,2,2,1)} \\ & -\zeta_5 E'_{(1,1,1,1,1,1)} + \zeta_7 E'_{(1,0,1,1,1,1)} - \zeta_8 E'_{(1,1,1,1,1)} + \zeta_{10} E'_{(1,0,1,1,1)} - \zeta_{13} E'_{(1,0,1)} \\ & +\zeta_{14} E'_{(0,0,0,0,1,1)} + \zeta_{16} E_{-\alpha_1} - \zeta_{17} E_{-\alpha_5} + \zeta_{21} E_{\alpha_4} - \zeta_{23} E_{(0,1,0,1)} \\ & -\zeta_{22} E_{(0,0,1,1)} + \zeta_{24} E_{(0,1,1,1)} - \zeta_{26} E_{(0,1,1,2,1)} - \zeta_{27} E_{(0,1,1,2,1,1)} \\ & -2\zeta_{19} c - \frac{\zeta_{19}}{3}(2\alpha_1 + \alpha_3 + 2\alpha_5 + \alpha_6), \end{aligned} \tag{12.6.69}$$

$$\begin{aligned}T_{20} = &\ [\iota(E_{-\alpha_2}), T_{18}] = \zeta_{20}(D-8) - \eta\partial_{x_8} - \zeta_1 E'_{(1,2,2,3,2,1)} - \zeta_6 E'_{(0,1,1,2,2,1)}\\ &+\zeta_7 E'_{(0,1,1,2,1,1)} + \zeta_{10} E'_{(0,1,1,2,1)} + \zeta_9 E'_{(0,1,1,1,1,1)} + \zeta_{11} E'_{(0,1,0,1,1,1)} + \zeta_{12} E'_{(0,1,1,1,1)}\\ &+\zeta_{13} E'_{(0,1,1,1)} + \zeta_{15} E'_{(0,1,0,1,1)} + \zeta_{16} E'_{(0,1,0,1)} + \zeta_{18} E_{-\alpha_2} - \zeta_{23} E_{\alpha_1}\\ &-\zeta_{24} E_{(1,0,1)} - \zeta_{25} E_{(1,0,1,1)} - \zeta_{26} E_{(1,0,1,1,1)} - \zeta_{27} E_{(1,0,1,1,1,1)}\\ &-2\zeta_{20} c + \frac{\zeta_{20}}{3}(\alpha_1 - 3\alpha_2 - \alpha_3 - 3\alpha_4 - 2\alpha_5 - \alpha_6),\end{aligned} \tag{12.6.70}$$

$$\begin{aligned}T_{21} = &\ -[\iota(E_{-\alpha_4}), T_{19}] = \zeta_{21}(D-8) - \eta\partial_{x_7} + \zeta_2 E'_{(1,1,2,3,2,1)} - \zeta_4 E'_{(1,1,1,2,2,1)}\\ &+\zeta_5 E'_{(1,1,1,2,1,1)} + \zeta_8 E'_{(1,1,1,2,1)} + \zeta_9 E'_{(1,0,1,1,1,1)} + \zeta_{12} E'_{(1,0,1,1,1)} + \zeta_{13} E'_{(1,0,1,1)}\\ &+\zeta_{14} E'_{(0,0,0,1,1,1)} - \zeta_{17} E'_{(0,0,0,1,1)} + \zeta_{18} E_{-\alpha_1} - \zeta_{19} E_{-\alpha_4} - \zeta_{22} E_{\alpha_3}\\ &-\zeta_{23} E_{\alpha_2} - \zeta_{25} E_{(0,1,1,1)} - \zeta_{26} E_{(0,1,1,1,1)} - \zeta_{27} E_{(0,1,1,1,1,1)}\\ &-2\zeta_{21} c - \frac{\zeta_{21}}{3}(2\alpha_1 + \alpha_3 + 3\alpha_4 + 2\alpha_5 + \alpha_6),\end{aligned} \tag{12.6.71}$$

$$\begin{aligned}T_{22} = &\ [\iota(E_{-\alpha_3}), T_{21}] = \zeta_{22}(D-8) - \eta\partial_{x_6} - \zeta_3 E'_{(1,1,2,3,2,1)} + \zeta_4 E'_{(1,1,2,2,2,1)}\\ &-\zeta_5 E'_{(1,1,2,2,1,1)} - \zeta_8 E'_{(1,1,2,2,1)} + \zeta_{11} E'_{(1,0,1,1,1,1)} + \zeta_{15} E'_{(1,0,1,1,1)} + \zeta_{16} E'_{(1,0,1,1)}\\ &-\zeta_{14} E'_{(0,0,1,1,1,1)} + \zeta_{17} E'_{(0,0,1,1,1)} + \zeta_{18} E'_{(1,0,1)} + \zeta_{19} E'_{(0,0,1,1)} + \zeta_{21} E_{-\alpha_3}\\ &-\zeta_{24} E_{\alpha_2} - \zeta_{25} E_{(0,1,0,1)} - \zeta_{26} E_{(0,1,0,1,1)} - \zeta_{27} E_{(0,1,0,1,1,1)}\\ &-2\zeta_{22} c - \frac{\zeta_{22}}{3}(2\alpha_1 + 4\alpha_3 + 3\alpha_4 + 2\alpha_5 + \alpha_6),\end{aligned} \tag{12.6.72}$$

$$\begin{aligned}T_{23} = &\ [\iota(E_{-\alpha_2}), T_{21}] = \zeta_{23}(D-8) - \eta\partial_{x_5} + \zeta_2 E'_{(1,2,2,3,2,1)} + \zeta_6 E'_{(1,1,1,2,2,1)}\\ &-\zeta_7 E'_{(1,1,1,2,1,1)} - \zeta_{10} E'_{(1,1,1,2,1)} - \zeta_9 E'_{(1,1,1,1,1,1)} - \zeta_{12} E'_{(1,1,1,1,1)} - \zeta_{13} E'_{(1,1,1,1)}\\ &-\zeta_{14} E'_{(0,1,0,1,1,1)} + \zeta_{17} E'_{(0,1,0,1,1)} + \zeta_{19} E'_{(0,1,0,1)} + \zeta_{20} E_{-\alpha_1} + \zeta_{21} E_{-\alpha_2}\\ &-\zeta_{24} E_{\alpha_3} - \zeta_{25} E_{(0,0,1,1)} - \zeta_{26} E_{(0,0,1,1,1)} - \zeta_{27} E_{(0,0,1,1,1,1)}\\ &-2\zeta_{23} c - \frac{\zeta_{23}}{3}(2\alpha_1 + 3\alpha_2 + \alpha_3 + 3\alpha_4 + 2\alpha_5 + \alpha_6),\end{aligned} \tag{12.6.73}$$

$$\begin{aligned}T_{24} = &\ [\iota(E_{-\alpha_3}), T_{23}] = \zeta_{24}(D-8) - \eta\partial_{x_4} - \zeta_3 E'_{(1,2,2,3,2,1)} - \zeta_6 E'_{(1,1,2,2,2,1)}\\ &+\zeta_7 E'_{(1,1,2,2,1,1)} + \zeta_{10} E'_{(1,1,2,2,1)} - \zeta_{11} E'_{(1,1,1,1,1,1)} + \zeta_{14} E'_{(0,1,1,1,1,1)} - \zeta_{15} E'_{(1,1,1,1,1)}\\ &-\zeta_{16} E'_{(1,1,1,1)} - \zeta_{17} E'_{(0,1,1,1,1)} - \zeta_{19} E'_{(0,1,1,1)} + \zeta_{20} E'_{(1,0,1)} + \zeta_{22} E_{-\alpha_2}\\ &+\zeta_{23} E_{-\alpha_3} - \zeta_{25} E_{\alpha_4} - \zeta_{26} E_{(0,0,0,1,1)} - \zeta_{27} E_{(0,0,0,1,1,1)}\\ &-2\zeta_{24} c - \frac{\zeta_{24}}{3}(2\alpha_1 + 3\alpha_2 + 4\alpha_3 + 3\alpha_4 + 2\alpha_5 + \alpha_6),\end{aligned} \tag{12.6.74}$$

$$\begin{aligned}T_{25} = &\ [\iota(E_{-\alpha_4}), T_{24}] = \zeta_{25}(D-8) - \eta\partial_{x_3} + \zeta_4 E'_{(1,2,2,3,2,1)} - \zeta_6 E'_{(1,1,2,3,2,1)}\\ &-\zeta_9 E'_{(1,1,2,2,1,1)} - \zeta_{11} E'_{(1,1,1,2,1,1)} - \zeta_{12} E'_{(1,1,2,2,1)} + \zeta_{14} E'_{(0,1,1,2,1,1)} - \zeta_{15} E'_{(1,1,1,2,1)}\\ &+\zeta_{18} E'_{(1,1,1,1)} - \zeta_{17} E'_{(0,1,1,2,1)} + \zeta_{20} E'_{(1,0,1,1)} + \zeta_{21} E'_{(0,1,1,1)} + \zeta_{22} E'_{(0,1,0,1)}\\ &+\zeta_{23} E'_{(0,0,1,1)} + \zeta_{24} E_{-\alpha_4} - \zeta_{26} E_{\alpha_5} - \zeta_{27} E_{(0,0,0,0,1,1)}\end{aligned}$$

$$-2\zeta_{25}c-\frac{\zeta_{25}}{3}(2\alpha_1+3\alpha_2+4\alpha_3+6\alpha_4+2\alpha_5+\alpha_6), \tag{12.6.75}$$

$$\begin{aligned}T_{26}=\ &[\iota(E_{-\alpha_5}),T_{25}]=\zeta_{26}(D-8)-\eta\partial_{x_2}-\zeta_5E'_{(1,2,2,3,2,1)}+\zeta_7E'_{(1,1,2,3,2,1)}\\&-\zeta_9E'_{(1,1,2,2,2,1)}-\zeta_{11}E'_{(1,1,1,2,2,1)}+\zeta_{13}E'_{(1,1,2,2,1)}+\zeta_{14}E'_{(0,1,1,2,2,1)}\\&+\zeta_{16}E'_{(1,1,1,2,1)}+\zeta_{18}E'_{(1,1,1,1,1)}+\zeta_{19}E'_{(0,1,1,2,1)}+\zeta_{20}E'_{(1,0,1,1,1)}\\&+\zeta_{21}E'_{(0,1,1,1,1)}+\zeta_{22}E'_{(0,1,0,1,1)}+\zeta_{23}E'_{(0,0,1,1,1)}+\zeta_{24}E'_{(0,0,0,1,1)}+\zeta_{25}E_{-\alpha_5}\\&-\zeta_{27}E_{\alpha_6}-2\zeta_{26}c-\frac{\zeta_{26}}{3}(2\alpha_1+3\alpha_2+4\alpha_3+6\alpha_4+5\alpha_5+\alpha_6),\end{aligned} \tag{12.6.76}$$

$$\begin{aligned}T_{27}=\ &[\iota(E_{-\alpha_6}),T_{26}]=\zeta_{27}(D-8)-\eta\partial_{x_1}+\zeta_8E'_{(1,2,2,3,2,1)}-\zeta_{10}E'_{(1,1,2,3,2,1)}\\&+\zeta_{12}E'_{(1,1,2,2,2,1)}+\zeta_{13}E'_{(1,1,2,2,1,1)}+\zeta_{15}E'_{(1,1,1,2,2,1)}+\zeta_{17}E'_{(0,1,1,2,2,1)}\\&+\zeta_{16}E'_{(1,1,1,2,1,1)}+\zeta_{18}E'_{(1,1,1,1,1,1)}+\zeta_{19}E'_{(0,1,1,2,1,1)}+\zeta_{20}E'_{(1,0,1,1,1,1)}\\&+\zeta_{21}E'_{(0,1,1,1,1,1)}+\zeta_{22}E'_{(0,1,0,1,1,1)}+\zeta_{23}E'_{(0,0,1,1,1,1)}+\zeta_{24}E'_{(0,0,0,1,1,1)}+\zeta_{26}E_{-\alpha_6}\\&+\zeta_{25}E'_{(0,0,0,0,1,1)}-2\zeta_{27}c-\frac{\zeta_{27}}{3}(2\alpha_1+3\alpha_2+4\alpha_3+6\alpha_4+5\alpha_5+4\alpha_6).\end{aligned} \tag{12.6.77}$$

Then $\mathfrak{T}=\sum_{r=1}^{27}\mathbb{F}T_r$ forms the 27-dimensional $\mathscr{G}^{E_6}$ of highest weight λ_1. Indeed, the map $\zeta_r\mapsto T_r$ for $r\in\overline{1,27}$ determines a $\mathscr{G}^{E_6}$-module isomorphism from U to $\mathfrak{T}$ (cf. (11.2.2)).

Denote

$$T'_r=T_r-\zeta_r(D-2c-8)+\eta\partial_{x_{28-r}}\qquad\text{for } r\in\overline{1,27}. \tag{12.6.78}$$

Easily see that $\mathfrak{T}'=\sum_{r=1}^{27}\mathbb{F}T'_r$ forms the 27-dimensional $\mathscr{G}^{E_6}$-module of highest weight λ_1. So we have the $\mathscr{G}^{E_6}$-module isomorphism from $U=\sum_{r=1}^{27}\mathbb{F}\zeta_r$ to $\mathfrak{T}'$ determined by $\zeta_r\mapsto T'_r$ for $r\in\overline{1,27}$. The weight set of U is

$$\Pi(U)=\left\{\sum_{s=1}^{6}b_{r,s}\lambda_s\mid r\in\overline{1,27}\right\} \tag{12.6.79}$$

(cf. (12.6.9)–(12.6.14) and Table 11.2). Let $\lambda\in\Lambda^+$. Denote

$$\Upsilon'(\lambda)=\{\lambda+\mu\mid\mu\in\Pi(U),\ \ \lambda+\mu\in\Lambda^+\} \tag{12.6.80}$$

(cf. (12.6.27)). Take $M=V(\lambda)$, the irreducible $\mathscr{G}^{E_6}$-module of highest weight λ. Theorem 5.4.3 implies

$$UV(\lambda)=U\otimes_{\mathbb{F}}V(\lambda)\cong\bigoplus_{\lambda'\in\Upsilon'(\lambda)}V(\lambda'). \tag{12.6.81}$$

Given $\lambda'\in\Upsilon'(\lambda)$, we pick a singular vector

$$u = \sum_{r=1}^{27} \zeta_r u_r \tag{12.6.82}$$

of weight λ' in $UV(\lambda)$, where $u_r \in V(\lambda)$. Moreover, any singular vector of weight λ' in $UV(\lambda)$ is a scalar multiple of u. Note that the vector

$$w = \sum_{r=1}^{27} T_r'(u_r) \tag{12.6.83}$$

is also a singular vector of weight λ' if it is not zero. Thus

$$w = \flat_{\lambda'} u, \qquad \flat_{\lambda'} \in \mathbb{F}. \tag{12.6.84}$$

Set

$$\flat(\lambda) = \min\{\flat_{\lambda'} \mid \lambda' \in \Upsilon'(\lambda)\}. \tag{12.6.85}$$

Recall the notion $\ell_\omega(\lambda)$ in (12.6.36).

Theorem 12.6.6 (Irreducibility of the Functor) *The* $\mathcal{G}^{E_7}$*-module* $\widehat{V(\lambda)}$ *is irreducible if*

$$c \in \mathbb{F} \setminus \{-8 + 5\mathbb{N}/3, (1/2)(\flat(\lambda) + \mathbb{N}) - 4, \ell_\omega(\lambda) + \mathbb{N}\}. \tag{12.6.86}$$

Proof Recall that the $\mathcal{G}^{E_7}$-submodule $U(\mathcal{G}_-)(V(\lambda))$ is irreducible by Proposition 12.5.2. It is enough to prove $\widehat{V(\lambda)} = U(\mathcal{G}_-)(V(\lambda))$. It is obvious that

$$\widehat{V(\lambda)}_0 = V(\lambda) = (U(\mathcal{G}_-)(V(\lambda)))_0 \tag{12.6.87}$$

(cf. (12.6.3) and (12.6.4) with $M = V(\lambda)$). Moreover, Lemma 12.6.1 with $M = V(\lambda)$, (12.6.36) and (12.6.86) imply that $\varphi|_{\widehat{V(\lambda)}_1}$ is invertible, or equivalently,

$$\widehat{V(\lambda)}_1 = (U(\mathcal{G}_-)(V(\lambda)))_1. \tag{12.6.88}$$

Suppose that

$$\widehat{V(\lambda)}_i = (U(\mathcal{G}_-)(V(\lambda)))_i \tag{12.6.89}$$

for $i \in \overline{0,k}$ with $1 \leq k \in \mathbb{N}$.

For any $v \in V(\lambda)$ and $\alpha \in \mathbb{N}^{16}$ such that $|\alpha| = k - 1$, we have

$$T_r(x^\alpha v) = x^\alpha[(|\alpha| - 2c - 8)\zeta_r + T_r'](v) + \eta\partial_{x_{28-r}}(x^\alpha)v \in (U(\mathcal{G}_-)(V(\lambda)))_{k+1} \tag{12.6.90}$$

for $r \in \overline{1,27}$ by (12.6.78), (12.6.89) with $i = k-1, k$. If $k = 1$, then $\alpha = 0$. So $\eta\partial_{x_{28-r}}(x^\alpha)v = 0$. When $k > 1$,

$$\partial_{x_{28-r}}(x^{\alpha})v \in \widehat{V(\lambda)}_{k-2} = (U(\mathscr{G}_-)(V(\lambda)))_{k-2} \tag{12.6.91}$$

and so

$$\left[\sum_{14\neq p\in\overline{1,26}} \iota(\mathfrak{b}_p)\zeta_{28-p} - \iota(\mathfrak{b}_{14})\zeta_{14} - \iota(\mathfrak{b}_{27})\zeta_1\right](\partial_{x_{28-r}}(x^{\alpha})v)$$
$$= \eta(24 - 5(|\alpha| - 1) + 3c)(\partial_{x_{28-r}}(x^{\alpha})v) \in (U(\mathscr{G}_-)(V(\lambda)))_{k+1} \tag{12.6.92}$$

by Lemma 12.6.5. Thus (12.6.86) gives

$$\eta\partial_{x_{28-r}}(x^{\alpha})v \in (U(\mathscr{G}_-)(V(\lambda)))_{k+1} \quad \text{for } r \in \overline{1,27}. \tag{12.6.93}$$

Hence in any case,

$$T_r(x^{\alpha}v) = x^{\alpha}[(|\alpha| - 2c - 8)\zeta_r + T_r'](v) \in (U(\mathscr{G}_-)(V(\lambda)))_{k+1} \quad \text{for } r \in \overline{1,27}. \tag{12.6.94}$$

On the other hand,

$$V' = \text{Span}\{[(|\alpha| - 2c - 8)\zeta_r + T_r'](v) \mid r \in \overline{1,27},\ v \in V(\lambda)\} \tag{12.6.95}$$

forms a $\mathscr{G}^{E_6}$-submodule of $UV(\lambda)$ with respect to the action in (12.6.1). Let u be a $\mathscr{G}^{E_6}$-singular vector in (12.6.82). Then

$$V' \ni \sum_{r=1}^{27}[(|\alpha| - 2c - 8)\zeta_r + T_r'](u_r) = (|\alpha| - 2c - 8)u + w = (|\alpha| - 2c - 8 + \flat_{\lambda'})u \tag{12.6.96}$$

by (12.6.83) and (12.6.84). Moreover, (12.6.85) and (12.6.86) yield $u \in V'$. Since $UV(\lambda)$ is a $\mathscr{G}^{E_6}$-module generated by all the singular vectors, we have $V' = UV(\lambda)$. So

$$x^{\alpha}UV(\lambda) \subset (U(\mathscr{G}_-)(V(\lambda)))_{k+1}. \tag{12.6.97}$$

The arbitrariness of α implies

$$\zeta_r\widehat{V(\lambda)}_{k-1} \subset (U(\mathscr{G}_-)(V(\lambda)))_{k+1} \quad \text{for } r \in \overline{1,27}. \tag{12.6.98}$$

Given any $f \in \mathscr{A}_k$ and $v \in V(\lambda)$, we have

$$\zeta_r\partial_{x_s}(f)v \in \zeta_r\widehat{V(\lambda)}_{k-1} \subset (U(\mathscr{G}_-)(V(\lambda)))_{k+1} \quad \text{for } r, s \in \overline{1,27}. \tag{12.6.99}$$

Moreover,

$$\mathfrak{d}_s(fv) = \iota(\mathfrak{d}_s)(fv) = P_s(fv) + f(\tilde{\omega} - c)(x_s v)$$
$$\equiv f(k + \tilde{\omega} - c)(x_s v) \left(\text{mod} \sum_{r=1}^{27} \zeta_r \widehat{V(\lambda)}_{k-1}\right) \quad (12.6.100)$$

for $s \in \overline{1,27}$ by (12.4.20), (12.4.39)–(12.4.64), (12.5.13)–(12.5.39) and Lemma 12.6.1. According to (12.6.36), (12.6.86), (12.6.96) and (12.6.98), we get

$$x_s f v \in (U(\mathscr{G}_-)(V(\lambda)))_{k+1} \quad \text{for } s \in \overline{1,27}. \quad (12.6.101)$$

Thus (12.6.89) holds for $i = k + 1$. By induction on k, (12.6.89) holds for any $i \in \mathbb{N}$; that is, $\widehat{V(\lambda)} = U(\mathscr{G}_-)(V(\lambda))$. □

When $\lambda = 0$, $V(0)$ is the one-dimensional trivial module and $\ell_\omega(0) = \flat(0) = 0$. So we have:

Corollary 12.6.7 *The $\mathscr{G}^{E_7}$-module $\widehat{V(0)}$ is irreducible if $c \in \mathbb{F} \setminus \{5\mathbb{N}/3 - 8, \mathbb{N}/2 - 4\}$.*

Next we consider $\lambda = k\lambda_1$. In this case,

$$\Upsilon(\lambda) = \{k\lambda_1 + \lambda_6, (k-1)\lambda_1 + \lambda_2, (k-1)\lambda_1\} \quad (12.6.102)$$

by (12.6.31) and Tables 11.1 and 11.2. Thus we have

$$\ell_\omega(\lambda) = -16 - \frac{4k}{3} \quad (12.6.103)$$

by (12.6.36). Moreover,

$$\Upsilon'(\lambda) = \{(k+1)\lambda_1, (k-1)\lambda_1 + \lambda_3, (k-1)\lambda_1 + \lambda_6\} \quad (12.6.104)$$

by Table 11.2 and (12.6.80).

We define a representation of $\mathscr{G}^{E_6}$ on $\mathscr{X} = \mathbb{F}[z_1, z_2, \ldots, z_{27}]$ determined via (11.2.36)–(11.2.40) with U replaced by $\mathscr{X}$ and ζ_i replaced by z_i. Then the $\mathscr{G}^{E_6}$-submodule $\mathscr{N}_k$ generated by z_1^k is isomorphic to $V(k\lambda_1)$.

We calculate:

$$z_1^{k-1} z_2 = -\frac{1}{k} E_{-\alpha_1}(z_1^k) \in \mathscr{N}_k, \quad z_1^{k-1} z_3 = -E_{-\alpha_3}(z_1^{k-1} z_2) \in \mathscr{N}_k, \quad (12.6.105)$$

$$z_1^{k-1} z_4 = -E_{-\alpha_4}(z_1^{k-1} z_3) \in \mathscr{N}_k, \quad z_1^{k-1} z_5 = -E_{-\alpha_5}(z_1^{k-1} z_4) \in \mathscr{N}_k, \quad (12.6.106)$$

$$z_1^{k-1} z_6 = -E_{-\alpha_2}(z_1^{k-1} z_4) \in \mathscr{N}_k, \quad z_1^{k-1} z_7 = -E_{-\alpha_2}(z_1^{k-1} z_5) \in \mathscr{N}_k, \quad (12.6.107)$$

$$z_1^{k-1}z_9 = -E_{-\alpha_4}(z_1^{k-1}z_7) \in \mathcal{N}_k, \quad z_1^{k-1}z_{11} = E_{-\alpha_3}(z_1^{k-1}z_9) \in \mathcal{N}_k, \tag{12.6.108}$$

$$-E_{-\alpha_1}(z_1^{k-1}z_{11}) = z_1^{k-1}z_{14} + (k-1)z_1^{k-2}z_2z_{11} \in \mathcal{N}_k, \tag{12.6.109}$$

$$-E_{-\alpha_3}(z_1^{k-1}z_{14} + (k-1)z_1^{k-2}z_2z_{11}) = (k-1)z_3z_{11} \in \mathcal{N}_k, \tag{12.6.110}$$

$$E_{\alpha_3}((k-1)z_3z_{11}) = (k-1)(z_2z_{11} - z_3z_9) \in \mathcal{N}_k, \tag{12.6.111}$$

$$E_{\alpha_4}E_{-\alpha_4}[(k-1)(z_2z_{11} - z_3z_9)] = (k-1)(z_3z_9 + z_4z_7) \in \mathcal{N}_k, \tag{12.6.112}$$

$$-E_{\alpha_5}E_{-\alpha_5}[(k-1)(z_3z_9 + z_4z_7)] = (k-1)(z_4z_7 + z_5z_6) \in \mathcal{N}_k. \tag{12.6.113}$$

Now we take $M = \mathcal{N}_k$ in our earlier settings. First $\zeta_1 z_1^k$ is a singular vector of weight $(k+1)\lambda_1$ in $U\mathcal{N}_k$. By (12.6.50), (11.2.40) and Table 11.2,

$$T_1'(z_1^k) = \frac{\zeta_1}{3}(4\alpha_1 + 3\alpha_2 + 5\alpha_3 + 6\alpha_4 + 4\alpha_5 + 2\alpha_6)(z_1^k) = \frac{4k}{3}\zeta_1 z_1^k. \tag{12.6.114}$$

So $\flat_{(k+1)\lambda_1} = 4k/3$. Next $\zeta_1 z_1^{k-1}z_2 - \zeta_2 z_1^k$ is a singular vector of weight $(k-1)\lambda_1 + \lambda_3$ in $U\mathcal{N}_k$. According to (12.6.50), (12.6.52), (11.2.40) and Table 11.2:

$$\begin{aligned} & T_1'(z_1^{k-1}z_2) - T_2'(z_1^k) \\ = \; & \zeta_2 z_1^k + \frac{4k-3}{3}\zeta_1 z_1^{k-1}z_2 - k\zeta_1 z_1^{k-1}z_2 - \frac{k}{3}\zeta_2 z_1^k = \frac{k-3}{3}(\zeta_1 z_1^{k-1}z_2 - \zeta_2 z_1^k), \end{aligned} \tag{12.6.115}$$

which gives $\flat_{(k-1)\lambda_1+\lambda_3} = k/3 - 1$.

Expressions (11.2.29)–(11.2.34) and (12.6.105)–(12.6.113) show that

$$\begin{aligned} u = \; & \zeta_1[4z_1^{k-1}z_{14} + (k-1)z_1^{k-2}(z_2z_{11} + z_3z_9 - z_4z_7 + z_5z_6)] - (k+3)z_1^{k-1} \\ & \times[\zeta_2z_{11} + \zeta_3z_9 - \zeta_4z_7 + \zeta_5z_6 + \zeta_6z_5 - \zeta_7z_4 + \zeta_9z_3 + \zeta_{11}z_2 - \zeta_{14}z_1] \end{aligned} \tag{12.6.116}$$

is a singular vector of weight $(k-1)\lambda_1 + \lambda_6$ in $U\mathcal{N}_k$. We find

$$\begin{aligned} & T_1'[4z_1^{k-1}z_{14} + (k-1)z_1^{k-2}(z_2z_{11} + z_3z_9 - z_4z_7 + z_5z_6)] \\ & -(k+3)[T_2'(z_1^{k-1}z_{11}) + T_3'(z_1^{k-1}z_9) - T_4'(z_1^{k-1}z_7) + T_5'(z_1^{k-1}z_6) \\ & +T_6'(z_1^{k-1}z_5) - T_7'(z_1^{k-1}z_4) + T_9'(z_1^{k-1}z_3) + T_{11}'(z_1^{k-1}z_2) - T_{14}'(z_1^k)] \\ = \; & (k+3)z_1^{k-1}(\zeta_2z_{11} + \zeta_3z_9 - \zeta_4z_7 + \zeta_5z_6 + \zeta_6z_5 - \zeta_7z_4 + \zeta_9z_3 + \zeta_{11}z_2) \\ & +(4k/3-2)\zeta_1[4z_1^{k-1}z_{14} + (k-1)z_1^{k-2}(z_2z_{11} + z_3z_9 - z_4z_7 + z_5z_6)] - (k+3)z_1^{k-2} \\ & \times[2(4\zeta_1z_1z_{14} + (k-1)\zeta_1(z_2z_{11} + z_3z_9 - z_4z_7 + z_5z_6)) + (2k/3+8)\zeta_{14}z_1^2 \\ & -(2k/3+7)z_1(\zeta_2z_{11} + \zeta_3z_9 - \zeta_4z_7 + \zeta_5z_6 + \zeta_6z_5 - \zeta_7z_4 + \zeta_9z_3 + \zeta_{11}z_2)] \\ = \; & -(2k/3+8)u \end{aligned} \tag{12.6.117}$$

by (11.2.29)–(11.2.34), (11.2.40) with Table 11.2, (12.6.50), (12.6.52)–(12.6.64), (12.6.78). Thus $\flat_{(k-1)\lambda_1+\lambda_6} = -(2k/3+8)$. Therefore,

$$\flat(k\lambda_1) = -(2k/3+8). \tag{12.6.118}$$

Corollary 12.6.8 *The* $\mathscr{G}^{E_7}$*-module* $\widehat{V(k\lambda_1)}$ *is irreducible if*

$$c \in \mathbb{F} \setminus \{-8+5\mathbb{N}/3, \mathbb{N}/2-8-k/3, \mathbb{N}-16-4k/3\}. \tag{12.6.119}$$

By (12.4.5), (12.5.10) and (12.5.42), the above irreducible $\mathscr{G}^{E_7}$-module $\widehat{V(k\lambda_1)}$ is a highest-weight module with highest weight $k\lambda_1 + (c-2k/3)\lambda_7$.

12.7 Combinatorics of the Representation of E_6

In this section, we want to study the symmetry of $\overline{1,27}$ with respect to the representation $\mathfrak{r}$ in (11.1.18)–(11.1.56).

Set

$$\mathfrak{W} = \{\mathfrak{r}(E_\alpha) \mid \alpha \in \Phi_{E_6}\} \tag{12.7.1}$$

(cf. (4.4.42)–(4.4.44), (11.1.18)–(11.1.53) and (11.1.56)). Write

$$\mathfrak{Z} = \{\zeta_i \mid i \in \overline{1,27}\} \tag{12.7.2}$$

(cf. (11.2.1) and (11.2.3)–(11.2.28)). First we have:

Lemma 12.7.1 *The set* $\overline{1,27}$ *is symmetric with respect to the sets* $\mathfrak{W}$ *and* $\mathfrak{Z}$.

Proof Let $\mathscr{G}_1$ be the Lie subalgebra of $\mathscr{G}^{E_6}$ generated by $\{E_{\pm\alpha_1}, E_{\pm\alpha_3}, E_{\pm\alpha_4}, E_{\pm\alpha_5}, E_{\pm\alpha_6}\}$. According to the Dynkin diagram, $\mathscr{G}_1 \cong sl(6,\mathbb{F})$. Table 11.1 tells us that x_1 and x_{20} are $\mathscr{G}_1$-singular vectors of weight λ_5 and x_6 is a $\mathscr{G}_1$-singular vector of weight λ_2. By (11.1.18), (11.1.20)–(11.1.23) and (11.1.56), the $\mathscr{G}_1$-submodule generated by x_1 is $\sum_{i=1}^{5}\mathbb{F}x_i + \mathbb{F}x_8$, the $\mathscr{G}_1$-submodule generated by x_{20} is $\mathbb{F}x_{20} + \sum_{r=23}^{27}\mathbb{F}x_r$ and the $\mathscr{G}_1$-submodule generated by x_6 is $\sum_{r=19}^{29}\mathbb{F}x_r + \sum_{s=6,7,21,22}\mathbb{F}x_s$. Thus the following subsets of indices:

$$\overline{1,5}\bigcup\{8\},\quad \overline{9,19}\bigcup\{6,7,21,22\},\quad \{20\}\bigcup\overline{23,27}. \tag{12.7.3}$$

are symmetric with respect to the restricted representation of $\mathscr{G}_1$

Let $\mathscr{G}_2$ be the Lie subalgebra of $\mathscr{G}^{E_6}$ generated by $\{E_{\pm\alpha_1}, E_{\pm\alpha_2}, E_{\pm\alpha_3}, E_{\pm\alpha_4}, E_{\pm\alpha_5}\}$. According to the Dynkin diagram, $\mathscr{G}_2 \cong o(10,\mathbb{F})$. Table 11.1 says that x_1 is a $\mathscr{G}_2$-singular vector of weight 0, x_2 is a $\mathscr{G}_2$-singular vector of weight λ_5 and x_{14} is a

$\mathscr{G}_2$-singular vector of weight λ_1. So $\mathbb{F}x_1$ is a trivial $\mathscr{G}_2$-module. The $\mathscr{G}_2$-submodule generated by x_{14} is $\mathbb{F}x_{14}+\mathbb{F}x_{17}+\mathbb{F}x_{19}+\sum_{i=21}^{27}\mathbb{F}x_i$ by (11.1.18)–(11.1.22) and (11.1.56). Now $\sum_{r=2}^{13}\mathbb{F}x_r+\sum_{s=15,16,18,20}\mathbb{F}x_s$ forms a spin module and x_{20} is its lowest-weight vector. Thus the subsets

$$\{1\},\ \{2\},\ \overline{3,10}\bigcup\{12,15\},\ \{11,13,16,18,20\},\ \{14,17,19\}\bigcup\overline{21,27} \tag{12.7.4}$$

are symmetric with respect to the restricted representation of $\mathscr{G}_2$. Since $\mathscr{G}^{E_6}$ is generated by $\mathscr{G}_1$ and $\mathscr{G}_2$, we have that the following subsets of indices:

$$\{1\},\ \{2\},\ \{3,4,5,8\},\ \{6,7,9,10,12,15\}, \tag{12.7.5}$$

$$\{11,13,16,18\},\ \{14,17,19,21,22\},\ \{20\},\ \overline{23,27} \tag{12.7.6}$$

are symmetric with respect to the representation $\mathfrak{r}$ of $\mathscr{G}^{E_6}$.

Denote

$$\epsilon_i=(0,\ldots,0,\overset{i}{1},0,\ldots,0)\in\mathbb{Z}^{27}. \tag{12.7.7}$$

If $d=\sum_{r=1}^{6}a_r x_{i_r}\partial_{x_{j_r}}\in\mathfrak{W}$ with $a_r\in\mathbb{F}$, we write

$$d^{\clubsuit}=\sum_{r=1}^{6}(\epsilon_{i_r}-\epsilon_{j_r})\in\mathbb{Z}^{27}. \tag{12.7.8}$$

It turns out that the map $d\mapsto d^{\clubsuit}$ is an injective map from $\mathfrak{W}$ to $\mathbb{Z}^{27}$. Moreover, if $[d_1,d_2]=bd_3$ for some $d_1,d_2,d_3\in\mathfrak{W}$ and $0\neq b\in\mathbb{F}$, then

$$d_3^{\clubsuit}=d_1^{\clubsuit}+d_2^{\clubsuit}. \tag{12.7.9}$$

According to (11.1.56),

$$(\mathfrak{r}(E_{-\alpha}))^{\clubsuit}=-(\mathfrak{r}(E_\alpha))^{\clubsuit}\qquad\text{for }\ \alpha\in\Phi_{E_6}^{+}. \tag{12.7.10}$$

Let σ be a permutation on $\overline{1,27}$ such that

$$\sigma(i)=i,\ \sigma(r)=28-r\qquad\text{for }\ i\in\overline{13,15},\ i\in\overline{1,12}\bigcup\overline{16,27}. \tag{12.7.11}$$

We define a linear transformation $\widehat{\sigma}$ on $\mathbb{Z}^{27}$ by

$$\widehat{\sigma}(\epsilon_i)=\epsilon_{\sigma(i)}\qquad\text{for }\ i\in\overline{1,27}. \tag{12.7.12}$$

We find that

$$\widehat{\sigma}[(\mathfrak{r}(E_{\alpha_1}))^{\clubsuit}] = (\mathfrak{r}(E_{-\alpha_6}))^{\clubsuit}, \quad \widehat{\sigma}[(\mathfrak{r}(E_{\alpha_2}))^{\clubsuit}] = (\mathfrak{r}(E_{-\alpha_2}))^{\clubsuit}, \tag{12.7.13}$$

$$\widehat{\sigma}[(\mathfrak{r}(E_{\alpha_3}))^{\clubsuit}] = (\mathfrak{r}(E_{-\alpha_5}))^{\clubsuit}, \quad \widehat{\sigma}[(\mathfrak{r}(E_{\alpha_4}))^{\clubsuit}] = (\mathfrak{r}(E_{-\alpha_4}))^{\clubsuit} \tag{12.7.14}$$

$$\widehat{\sigma}[(\mathfrak{r}(E_{\alpha_5}))^{\clubsuit}] = (\mathfrak{r}(E_{-\alpha_3}))^{\clubsuit}, \quad \widehat{\sigma}[(\mathfrak{r}(E_{\alpha_6}))^{\clubsuit}] = (\mathfrak{r}(E_{-\alpha_1}))^{\clubsuit}. \tag{12.7.15}$$

According to (12.7.9), (12.7.10) and (12.7.13)–(12.7.15), we find

$$\widehat{\sigma}(\mathfrak{W}^{\clubsuit}) = \mathfrak{W}^{\clubsuit}. \tag{12.7.16}$$

This shows that $\mathfrak{W}$ is symmetric under σ.

For $\zeta = \sum_{s=1}^{5} c_s x_{\ell_s} x_{k_s} \in \mathfrak{Z}$ and $0 \neq b \in \mathbb{F}$, we define

$$(b\zeta)^{\sharp} = \sum_{s=1}^{5} (\epsilon_{\ell_s} + \epsilon_{k_s}) \in \mathbb{Z}^{27}. \tag{12.7.17}$$

We find that the map $\zeta \mapsto \zeta^{\sharp}$ is an injective map from $\mathfrak{Z}$ to $\mathbb{Z}^{27}$. Moreover, if $d(\zeta) \neq 0$ for some $d \in \mathfrak{W}$ and $\zeta \in \mathfrak{Z}$, then we have

$$(d(\zeta))^{\sharp} = d^{\clubsuit} + \zeta^{\sharp}. \tag{12.7.18}$$

Furthermore, we define

$$\widehat{\sigma}(\zeta) = \sum_{s=1}^{5} c_s x_{\sigma(\ell_s)} x_{\sigma(k_s)} \quad \text{for } \zeta = \sum_{s=1}^{5} c_s x_{\ell_s} x_{k_s} \in \mathfrak{Z}. \tag{12.7.19}$$

We calculate

$$(\widehat{\sigma}(\zeta_i))^{\sharp} = \zeta_i^{\sharp}, \quad (\widehat{\sigma}(\zeta_r))^{\sharp} = \zeta_{28-r}^{\sharp} \quad \text{for } i \in \overline{13,15}, \quad r \in \overline{1,12} \bigcup \overline{16,27}. \tag{12.7.20}$$

Therefore, $\mathfrak{W}$ and $\mathfrak{Z}$ are symmetric under σ. Thus the subsets

$$\{1,27\}, \ \{2,26\}, \ \{3,25\}, \ \{6,22\}, \ \{11,17\}, \ \{12,16\} \tag{12.7.21}$$

are symmetric with respect to $\mathfrak{W}$ and $\mathfrak{Z}$. Expressions (12.7.5), (12.7.6) and (12.7.21) imply that $\overline{1,27}$ is symmetric with respect to $\mathfrak{W}$ and $\mathfrak{Z}$. □

For any $i \in \overline{1,27}$, we define

$$\Upsilon_i = \{(r,s) \mid \zeta_r \text{ contains } x_i x_s\}. \tag{12.7.22}$$

We have the following equivalent combinatorial property:

Lemma 12.7.2 *Each polynomial ζ_r contains exactly ten x_i's. Moreover,*

$$|\Upsilon_i| = 10 \quad \text{for } i \in \overline{27}. \tag{12.7.23}$$

Proof According to (11.2.1) and (11.2.3)–(11.2.28), we calculate

$$\Upsilon_1 = \{(1,14),(2,17),(3,19),(4,21),(5,22),\\ (6,23),(7,24),(9,25),(11,26),(14,27)\}. \tag{12.7.24}$$

Then (12.7.23) follows from Lemma 12.7.1. The first statement is obtained by checking (11.2.1) and (11.2.3)–(11.2.28) one-by-one. □

If we represent $\{x_i \mid i \in \overline{1,27}\}$ by 27 vertices and represent $\{\zeta_r \mid r \in \overline{1,27}\}$ by 27 lines, then we obtain a graph of 27 vertices and 27 lines such that each line contains 10 vertices and each vertex is exactly on 10 lines.

For $i \in \overline{1,27}$, we denote

$$I_i = \{r \in \overline{1,27} \mid \text{some } d \in \mathfrak{W} \text{ contains } x_r\partial_{x_i}\} \tag{12.7.25}$$

and

$$\mathfrak{W}_i = \{d \in \mathfrak{W} \mid d \text{ does not contain } x_i \text{ and } \partial_{x_i}\}. \tag{12.7.26}$$

Write

$$J_i = \{s \in \overline{1,27} \mid (r,s) \in \Upsilon_i \text{ for some } r \in \overline{1,27}\}. \tag{12.7.27}$$

The following equivalent combinatorial properties are crucial to our main result.

Lemma 12.7.3 *For any $i \in \overline{1,27}$, we have*

$$|I_i| = 16, \quad |\mathfrak{W}_i| = 40, \quad |J_i| = 10. \tag{12.7.28}$$

In fact,

$$I_i \bigcup J_i = \overline{1,27} \setminus \{i\}. \tag{12.7.29}$$

Moreover, every element in $\mathfrak{W}_i$ contains exactly two x_s with $s \in J_i$, and for any $r \in J_i$, x_r is contained in exactly eight elements in $\mathfrak{W}_i$.

Proof Note that the elements in $\mathfrak{W}$ containing ∂_{x_1} are

$$\mathfrak{r}(E_{-\alpha_6}),\ \mathfrak{r}(E'_{(0,0,0,0,1,1)}),\ \mathfrak{r}(E'_{(0,0,0,1,1,1)}),\ \mathfrak{r}(E'_{(0,1,0,1,1,1)}), \tag{12.7.30}$$

$$\mathfrak{r}(E'_{(0,0,1,1,1,1)}),\ \mathfrak{r}(E'_{(1,0,1,1,1,1)}),\ \mathfrak{r}(E'_{(0,1,1,1,1,1)}),\ \mathfrak{r}(E'_{(1,1,1,1,1,1)}), \tag{12.7.31}$$

$$\mathfrak{r}(E'_{(0,1,1,2,1,1)}),\quad \mathfrak{r}(E'_{(1,1,1,2,1,1)}),\quad \mathfrak{r}(E'_{(0,1,1,2,2,1)}),\quad \mathfrak{r}(E'_{(1,1,2,2,1,1)}), \tag{12.7.32}$$

$$\mathfrak{r}(E'_{(1,1,1,2,2,1)}),\quad \mathfrak{r}(E'_{(1,1,2,2,2,1)}),\quad \mathfrak{r}(E'_{(1,1,2,3,2,1)}),\quad \mathfrak{r}(E'_{(1,2,2,3,2,1)}). \tag{12.7.33}$$

Thus

$$I_1 = \overline{2,13}\bigcup\{15, 16, 18, 20\}. \tag{12.7.34}$$

According to (12.7.24),

$$J_1 = \{14, 17, 19\}\bigcup\overline{21,27}. \tag{12.7.35}$$

Since $\tau(\mathfrak{W}) = -\mathfrak{W}$ (cf. (11.1.55)), $|\mathfrak{W}_1| = 40$. So (12.7.27) and (12.7.28) hold for $i = 1$. The elements in $\mathfrak{W}_1$ containing x_{14} are

$$\mathfrak{r}(E_{\alpha_1}),\quad \mathfrak{r}(E_{(1,0,1)}),\quad \mathfrak{r}(E_{(1,0,1,1)}),\quad \mathfrak{r}(E_{(1,1,1,1)}), \tag{12.7.36}$$

$$\mathfrak{r}(E_{(1,0,1,1,1)}),\quad \mathfrak{r}(E_{(1,1,1,1,1}),\quad \mathfrak{r}(E_{(1,1,1,2,1)}),\quad \mathfrak{r}(E_{(1,1,2,2,1)}). \tag{12.7.37}$$

Lemma 12.7.1 implies that x_r is contained in exactly eight elements in $\mathfrak{W}_1$ for any $r \in J_1$. The fact that every element in $\mathfrak{W}_1$ contains exactly two x_s with $s \in J_1$ is checked case by case. By Lemma 12.7.1, the lemma holds for any $i \in \overline{1,27}$. □

If we view J_i as the set of vertices and $\mathfrak{W}_i$ as the set of edges, the last statement in Lemma 12.7.3 gives a graph consisting of 10 vertices and 40 edges such that each edge contain exactly two vertices and each vertex is contained exactly 8 edges. Finally, we have the following duality:

Lemma 12.7.4 *Let $i \in \overline{1,27}$. For any $r \in J_i$, there exists a unique $\Im_i(r) \in J_i$ such that an element in $\mathfrak{W}_i$ contains x_r if and only if it contains $\partial_{x_{\Im_i(r)}}$. Moreover, $\Im_i^2 = 1$.*

Proof According to Lemma 12.7.3, x_r is contained in exactly eight elements in $\mathfrak{W}_i$ for any $r \in J_i$. Since $\tau(\mathfrak{W}_i) = -\mathfrak{W}_i$, we have that ∂_{x_r} is contained in exactly eight elements in $\mathfrak{W}_i$ for any $r \in J_i$.

By Lemma 12.7.1, we only need to consider $i = 1$. Every differential operator in (12.7.36) and (12.7.37) contains $\partial_{x_{27}}$. So $\Im_1(14) = 27$ and $\Im_1(27) = 14$. By symmetry, there exists a unique $\Im_1(r) \in J_1$ for any $r \in J_1$ such that an element in $\mathfrak{W}_1$ contains x_r if and only if it contains $\partial_{x_{\Im_1(r)}}$. Moreover, $\Im_1^2 = 1$. So the lemma holds for $i = 1$. □

12.8 Representations on Exponential-Polynomial Functions

In this section, we study a family of representations of the simple Lie algebra of type E_7 on a space of exponential-polynomial functions and prove that their irreducibility is related to an explicit given algebraic variety.

Recall

$$D = \sum_{i=1}^{27} x_i \partial_{x_i}. \tag{12.8.1}$$

We identify $\widehat{V(0)} = \mathscr{A} \otimes v_0$ with $\mathscr{A}$ by

$$f \otimes v_0 \leftrightarrow f \qquad \text{for } f \in \mathscr{B}, \tag{12.8.2}$$

where $V(0) = \mathbb{F}v_0$. Then we have the following one-parameter inhomogeneous first-order differential-operator representation $\mathfrak{r}_c$ of $\mathscr{G}^{E_7}$:

$$\mathfrak{r}_c(u) = \mathfrak{r}(u) \qquad \text{for } u \in \mathscr{G}^{E_6} \tag{12.8.3}$$

(cf. (11.1.18)–(11.1.56)),

$$\mathfrak{r}_c(\hat{\alpha}) = -2D + 3c, \qquad \mathfrak{r}_c(\mathfrak{a}_i) = \partial_{x_i} \qquad \text{for } i \in \overline{1,27}, \tag{12.8.4}$$

$$\begin{aligned}\mathfrak{r}_c(\mathfrak{b}_1) = {} & x_1(D-c) - \zeta_1\partial_{x_{14}} - \zeta_2\partial_{x_{17}} - \zeta_3\partial_{x_{19}} + \zeta_4\partial_{x_{21}} - \zeta_5\partial_{x_{22}} \\ & -\zeta_6\partial_{x_{23}} + \zeta_7\partial_{x_{24}} - \zeta_9\partial_{x_{25}} - \zeta_{11}\partial_{x_{26}} - \zeta_{14}\partial_{x_{27}}.\end{aligned} \tag{12.8.5}$$

$$\begin{aligned}\mathfrak{r}_c(\mathfrak{b}_2) = {} & x_2(D-c) - \zeta_1\partial_{x_{11}} - \zeta_2\partial_{x_{13}} - \zeta_3\partial_{x_{16}} + \zeta_4\partial_{x_{18}} - \zeta_6\partial_{x_{20}} \\ & -\zeta_8\partial_{x_{22}} + \zeta_{10}\partial_{x_{24}} - \zeta_{12}\partial_{x_{25}} - \zeta_{15}\partial_{x_{26}} + \zeta_{17}\partial_{x_{27}},\end{aligned} \tag{12.8.6}$$

$$\begin{aligned}\mathfrak{r}_c(\mathfrak{b}_3) = {} & x_3(D-c) - \zeta_1\partial_{x_9} - \zeta_2\partial_{x_{12}} - \zeta_3\partial_{x_{15}} + \zeta_5\partial_{x_{18}} - \zeta_7\partial_{x_{20}} \\ & -\zeta_8\partial_{x_{21}} + \zeta_{10}\partial_{x_{23}} - \zeta_{13}\partial_{x_{25}} - \zeta_{16}\partial_{x_{26}} + \zeta_{19}\partial_{x_{27}},\end{aligned} \tag{12.8.7}$$

$$\begin{aligned}\mathfrak{r}_c(\mathfrak{b}_4) = {} & x_4(D-c) - \zeta_1\partial_{x_7} - \zeta_2\partial_{x_{10}} - \zeta_4\partial_{x_{15}} + \zeta_5\partial_{x_{16}} - \zeta_8\partial_{x_{19}} \\ & -\zeta_9\partial_{x_{20}} + \zeta_{12}\partial_{x_{23}} - \zeta_{13}\partial_{x_{24}} - \zeta_{18}\partial_{x_{26}} + \zeta_{21}\partial_{x_{27}},\end{aligned} \tag{12.8.8}$$

$$\begin{aligned}\mathfrak{r}_c(\mathfrak{b}_5) = {} & x_5(D-c) + \zeta_1\partial_{x_6} - \zeta_3\partial_{x_{10}} + \zeta_4\partial_{x_{12}} - \zeta_5\partial_{x_{13}} + \zeta_8\partial_{x_{17}} \\ & +\zeta_{11}\partial_{x_{20}} - \zeta_{15}\partial_{x_{23}} + \zeta_{16}\partial_{x_{24}} - \zeta_{18}\partial_{x_{25}} - \zeta_{22}\partial_{x_{27}},\end{aligned} \tag{12.8.9}$$

$$\begin{aligned}\mathfrak{r}_c(\mathfrak{b}_6) = {} & x_6(D-c) + \zeta_1\partial_{x_5} + \zeta_2\partial_{x_8} - \zeta_6\partial_{x_{15}} + \zeta_7\partial_{x_{16}} - \zeta_9\partial_{x_{18}} \\ & -\zeta_{10}\partial_{x_{19}} + \zeta_{12}\partial_{x_{21}} - \zeta_{13}\partial_{x_{22}} + \zeta_{20}\partial_{x_{26}} - \zeta_{23}\partial_{x_{27}},\end{aligned} \tag{12.8.10}$$

$$\begin{aligned}\mathfrak{r}_c(\mathfrak{b}_7) = {} & x_7(D-c) - \zeta_1\partial_{x_4} + \zeta_3\partial_{x_8} + \zeta_6\partial_{x_{12}} - \zeta_7\partial_{x_{13}} + \zeta_{10}\partial_{x_{17}} \\ & +\zeta_{11}\partial_{x_{18}} - \zeta_{15}\partial_{x_{21}} + \zeta_{16}\partial_{x_{22}} + \zeta_{20}\partial_{x_{25}} + \zeta_{24}\partial_{x_{27}},\end{aligned} \tag{12.8.11}$$

$$\begin{aligned}\mathfrak{r}_c(\mathfrak{b}_8) = {} & x_8(D-c) + \zeta_2\partial_{x_6} + \zeta_3\partial_{x_7} - \zeta_4\partial_{x_9} + \zeta_5\partial_{x_{11}} - \zeta_8\partial_{x_{14}} \\ & +\zeta_{14}\partial_{x_{20}} + \zeta_{17}\partial_{x_{23}} - \zeta_{19}\partial_{x_{24}} + \zeta_{21}\partial_{x_{25}} + \zeta_{22}\partial_{x_{26}},\end{aligned} \tag{12.8.12}$$

$$\begin{aligned}\mathfrak{r}_c(\mathfrak{b}_9) = {} & x_9(D-c) - \zeta_1\partial_{x_3} - \zeta_4\partial_{x_8} - \zeta_6\partial_{x_{10}} + \zeta_9\partial_{x_{13}} - \zeta_{12}\partial_{x_{17}} \\ & -\zeta_{11}\partial_{x_{16}} + \zeta_{15}\partial_{x_{19}} - \zeta_{18}\partial_{x_{22}} - \zeta_{20}\partial_{x_{24}} + \zeta_{25}\partial_{x_{27}},\end{aligned} \tag{12.8.13}$$

$$\begin{aligned}\mathfrak{r}_c(\mathfrak{b}_{10}) = {} & x_{10}(D-c) - \zeta_2\partial_{x_4} - \zeta_3\partial_{x_5} - \zeta_6\partial_{x_9} + \zeta_7\partial_{x_{11}} - \zeta_{10}\partial_{x_{14}} \\ & +\zeta_{14}\partial_{x_{18}} + \zeta_{17}\partial_{x_{21}} - \zeta_{19}\partial_{x_{22}} - \zeta_{23}\partial_{x_{25}} - \zeta_{24}\partial_{x_{26}},\end{aligned} \tag{12.8.14}$$

$$\begin{aligned}\mathfrak{r}_c(\mathfrak{b}_{11}) = {} & x_{11}(D-c) - \zeta_1\partial_{x_2} + \zeta_5\partial_{x_8} + \zeta_7\partial_{x_{10}} - \zeta_9\partial_{x_{12}} + \zeta_{11}\partial_{x_{15}} \\ & +\zeta_{13}\partial_{x_{17}} - \zeta_{16}\partial_{x_{19}} + \zeta_{18}\partial_{x_{21}} + \zeta_{20}\partial_{x_{23}} + \zeta_{26}\partial_{x_{27}},\end{aligned} \tag{12.8.15}$$

$$\begin{aligned}\mathfrak{r}_c(\mathfrak{b}_{12}) = {} & x_{12}(D-c) - \zeta_2\partial_{x_3} + \zeta_4\partial_{x_5} + \zeta_6\partial_{x_7} - \zeta_9\partial_{x_{11}} + \zeta_{12}\partial_{x_{14}} \\ & -\zeta_{14}\partial_{x_{16}} - \zeta_{17}\partial_{x_{19}} + \zeta_{21}\partial_{x_{22}} + \zeta_{23}\partial_{x_{24}} - \zeta_{25}\partial_{x_{26}},\end{aligned} \tag{12.8.16}$$

$$\begin{aligned}\mathfrak{r}_c(\mathfrak{b}_{13}) = {} & x_{13}(D-c) - \zeta_2\partial_{x_2} - \zeta_5\partial_{x_5} - \zeta_7\partial_{x_7} + \zeta_9\partial_{x_9} - \zeta_{13}\partial_{x_{14}} \\ & +\zeta_{14}\partial_{x_{15}} + \zeta_{19}\partial_{x_{19}} - \zeta_{21}\partial_{x_{21}} - \zeta_{23}\partial_{x_{23}} - \zeta_{26}\partial_{x_{26}},\end{aligned} \tag{12.8.17}$$

$$\begin{aligned}\mathfrak{r}_c(\mathfrak{b}_{14}) = {} & x_{14}(D-c) - \zeta_1\partial_{x_1} - \zeta_8\partial_{x_8} - \zeta_{10}\partial_{x_{10}} + \zeta_{12}\partial_{x_{12}} - \zeta_{13}\partial_{x_{13}} \\ & -\zeta_{15}\partial_{x_{15}} + \zeta_{16}\partial_{x_{16}} - \zeta_{18}\partial_{x_{18}} - \zeta_{20}\partial_{x_{20}} + \zeta_{27}\partial_{x_{27}},\end{aligned} \tag{12.8.18}$$

$$\begin{aligned}\mathfrak{r}_c(\mathfrak{b}_{15}) = {} & x_{15}(D-c) - \zeta_3\partial_{x_3} - \zeta_4\partial_{x_4} - \zeta_6\partial_{x_6} + \zeta_{11}\partial_{x_{11}} + \zeta_{14}\partial_{x_{13}} \\ & -\zeta_{15}\partial_{x_{14}} + \zeta_{17}\partial_{x_{17}} - \zeta_{22}\partial_{x_{22}} - \zeta_{24}\partial_{x_{24}} - \zeta_{25}\partial_{x_{25}},\end{aligned} \tag{12.8.19}$$

$$\begin{aligned}\mathfrak{r}_c(\mathfrak{b}_{16}) = {} & x_{16}(D-c) - \zeta_3\partial_{x_2} + \zeta_5\partial_{x_4} + \zeta_7\partial_{x_6} - \zeta_{11}\partial_{x_9} - \zeta_{14}\partial_{x_{12}} \\ & +\zeta_{16}\partial_{x_{14}} - \zeta_{19}\partial_{x_{17}} + \zeta_{22}\partial_{x_{21}} + \zeta_{24}\partial_{x_{23}} - \zeta_{26}\partial_{x_{25}},\end{aligned} \tag{12.8.20}$$

$$\begin{aligned}\mathfrak{r}_c(\mathfrak{b}_{17}) = {} & x_{17}(D-c) - \zeta_2\partial_{x_1} + \zeta_8\partial_{x_5} + \zeta_{10}\partial_{x_7} - \zeta_{12}\partial_{x_9} + \zeta_{13}\partial_{x_{11}} \\ & +\zeta_{17}\partial_{x_{15}} - \zeta_{19}\partial_{x_{16}} + \zeta_{21}\partial_{x_{18}} + \zeta_{23}\partial_{x_{20}} - \zeta_{27}\partial_{x_{26}},\end{aligned} \tag{12.8.21}$$

$$\begin{aligned}\mathfrak{r}_c(\mathfrak{b}_{18}) = {} & x_{18}(D-c) + \zeta_4\partial_{x_2} + \zeta_5\partial_{x_3} - \zeta_9\partial_{x_6} + \zeta_{11}\partial_{x_7} + \zeta_{14}\partial_{x_{10}} \\ & -\zeta_{18}\partial_{x_{14}} + \zeta_{21}\partial_{x_{17}} - \zeta_{22}\partial_{x_{19}} + \zeta_{25}\partial_{x_{23}} + \zeta_{26}\partial_{x_{24}},\end{aligned} \tag{12.8.22}$$

$$\begin{aligned}\mathfrak{r}_c(\mathfrak{b}_{19}) = & \ x_{19}(D-c) - \zeta_3\partial_{x_1} - \zeta_8\partial_{x_4} - \zeta_{10}\partial_{x_6} + \zeta_{15}\partial_{x_9} - \zeta_{17}\partial_{x_{12}} \\ & -\zeta_{16}\partial_{x_{11}} + \zeta_{19}\partial_{x_{13}} - \zeta_{22}\partial_{x_{18}} - \zeta_{24}\partial_{x_{20}} - \zeta_{27}\partial_{x_{25}},\end{aligned} \tag{12.8.23}$$

$$\begin{aligned}\mathfrak{r}_c(\mathfrak{b}_{20}) = & \ x_{20}(D-c) - \zeta_6\partial_{x_2} - \zeta_7\partial_{x_3} - \zeta_9\partial_{x_4} + \zeta_{11}\partial_{x_5} + \zeta_{14}\partial_{x_8} \\ & -\zeta_{20}\partial_{x_{14}} + \zeta_{23}\partial_{x_{17}} - \zeta_{24}\partial_{x_{19}} - \zeta_{25}\partial_{x_{21}} - \zeta_{26}\partial_{x_{22}},\end{aligned} \tag{12.8.24}$$

$$\begin{aligned}\mathfrak{r}_c(\mathfrak{b}_{21}) = & \ x_{21}(D-c) + \zeta_4\partial_{x_1} - \zeta_8\partial_{x_3} + \zeta_{12}\partial_{x_6} - \zeta_{15}\partial_{x_7} + \zeta_{17}\partial_{x_{10}} \\ & +\zeta_{18}\partial_{x_{11}} - \zeta_{21}\partial_{x_{13}} + \zeta_{22}\partial_{x_{16}} - \zeta_{25}\partial_{x_{20}} + \zeta_{27}\partial_{x_{24}},\end{aligned} \tag{12.8.25}$$

$$\begin{aligned}\mathfrak{r}_c(\mathfrak{b}_{22}) = & \ x_{22}(D-c) - \zeta_5\partial_{x_1} - \zeta_8\partial_{x_2} - \zeta_{13}\partial_{x_6} + \zeta_{16}\partial_{x_7} - \zeta_{19}\partial_{x_{10}} \\ & -\zeta_{18}\partial_{x_9} + \zeta_{21}\partial_{x_{12}} - \zeta_{22}\partial_{x_{15}} - \zeta_{26}\partial_{x_{20}} - \zeta_{27}\partial_{x_{23}},\end{aligned} \tag{12.8.26}$$

$$\begin{aligned}\mathfrak{r}_c(\mathfrak{b}_{23}) = & \ x_{23}(D-c) - \zeta_6\partial_{x_1} + \zeta_{10}\partial_{x_3} + \zeta_{12}\partial_{x_4} - \zeta_{15}\partial_{x_5} + \zeta_{17}\partial_{x_8} \\ & +\zeta_{20}\partial_{x_{11}} - \zeta_{23}\partial_{x_{13}} + \zeta_{24}\partial_{x_{16}} + \zeta_{25}\partial_{x_{18}} - \zeta_{27}\partial_{x_{22}},\end{aligned} \tag{12.8.27}$$

$$\begin{aligned}\mathfrak{r}_c(\mathfrak{b}_{24}) = & \ x_{24}(D-c) + \zeta_7\partial_{x_1} + \zeta_{10}\partial_{x_2} - \zeta_{13}\partial_{x_4} + \zeta_{16}\partial_{x_5} - \zeta_{19}\partial_{x_8} \\ & -\zeta_{20}\partial_{x_9} + \zeta_{23}\partial_{x_{12}} - \zeta_{24}\partial_{x_{15}} + \zeta_{26}\partial_{x_{18}} + \zeta_{27}\partial_{x_{21}},\end{aligned} \tag{12.8.28}$$

$$\begin{aligned}\mathfrak{r}_c(\mathfrak{b}_{25}) = & \ x_{25}(D-c) - \zeta_9\partial_{x_1} - \zeta_{12}\partial_{x_2} - \zeta_{13}\partial_{x_3} - \zeta_{18}\partial_{x_5} + \zeta_{21}\partial_{x_8} \\ & +\zeta_{20}\partial_{x_7} - \zeta_{23}\partial_{x_{10}} - \zeta_{25}\partial_{x_{15}} - \zeta_{26}\partial_{x_{16}} - \zeta_{27}\partial_{x_{19}},\end{aligned} \tag{12.8.29}$$

$$\begin{aligned}\mathfrak{r}_c(\mathfrak{b}_{26}) = & \ x_{26}(D-c) - \zeta_{11}\partial_{x_1} - \zeta_{15}\partial_{x_2} - \zeta_{16}\partial_{x_3} - \zeta_{18}\partial_{x_4} + \zeta_{22}\partial_{x_8} \\ & +\zeta_{20}\partial_{x_6} - \zeta_{24}\partial_{x_{10}} - \zeta_{25}\partial_{x_{12}} - \zeta_{26}\partial_{x_{13}} - \zeta_{27}\partial_{x_{17}},\end{aligned} \tag{12.8.30}$$

$$\begin{aligned}\mathfrak{r}_c(\mathfrak{b}_{27}) = & \ x_{27}(D-c) - \zeta_{14}\partial_{x_1} + \zeta_{17}\partial_{x_2} + \zeta_{19}\partial_{x_3} + \zeta_{21}\partial_{x_4} - \zeta_{22}\partial_{x_5} \\ & -\zeta_{23}\partial_{x_6} + \zeta_{24}\partial_{x_7} + \zeta_{25}\partial_{x_9} + \zeta_{26}\partial_{x_{11}} + \zeta_{27}\partial_{x_{14}}.\end{aligned} \tag{12.8.31}$$

Let $\vec{a} = (a_1, \ldots, a_{27}) \in \mathbb{F}^{27} \setminus \{\vec{0}\}$. Define

$$\vec{a} \cdot \vec{x} = \sum_{i=1}^{27} a_i x_i. \tag{12.8.32}$$

Recall $\mathscr{A}$ in (11.1.16) and set

$$\mathscr{A}_{\vec{a}} = \{f e^{\vec{a}\cdot\vec{x}} \mid f \in \mathscr{A}\}. \tag{12.8.33}$$

For $i \in \overline{1,27}$, we write

$$\Upsilon_i = \{(\iota(r), r) \mid r \in J_i\}. \tag{12.8.34}$$

Recall η in (11.2.42) and the map $\Im_i$ in Lemma 12.7.4. Denote

$$\mathscr{F} = \{x_i \zeta_{\iota(r)} \zeta_{\iota(s)} \eta \mid i \in \overline{1,27};\ r, s \in J_i,\ r \neq s, \Im_i(s)\}. \tag{12.8.35}$$

For any function $f(x_1, \dots, x_{27})$, we define

$$f(\vec{b}) = f(b_1, \dots, b_{27}) \qquad \text{for } \vec{b} = (b_1, \dots, b_{27}) \in \mathbb{F}^{27}. \tag{12.8.36}$$

Now we define

$$\mathscr{V} = \{\vec{b} \in \mathbb{F}^{27} \setminus \{\vec{0}\} \mid f(\vec{b}) = 0 \text{ for } f \in \mathscr{F}\}, \tag{12.8.37}$$

which gives rise to a projective algebraic variety. The following is our main theorem in this section.

Theorem 12.8.1 *With respect to the representation π_c, $\mathscr{A}_{\vec{a}}$ forms an irreducible $\mathscr{G}^{E_7}$-module if $\vec{a} \notin \mathscr{V}$.*

Proof Let $\mathscr{A}_k$ be the subspace of homogeneous polynomials with degree k. Set

$$\mathscr{A}_{\vec{a},k} = \mathscr{A}_k e^{\vec{a}\cdot\vec{x}} \qquad \text{for } k \in \mathbb{N}. \tag{12.8.38}$$

Let M be a nonzero $\mathscr{G}^{E_7}$-submodule of $\mathscr{A}_{\vec{a}}$. Take any $0 \neq f e^{\vec{a}\cdot\vec{x}} \in M$ with $f \in \mathscr{A}$. By the second equation in (12.8.4),

$$(\xi_i - a_i)(f e^{\vec{a}\cdot\vec{x}}) = \partial_{x_i}(f) e^{\vec{a}\cdot\vec{x}} \in M \qquad \text{for } i \in \overline{1,27}. \tag{12.8.39}$$

Repeatedly applying (12.8.39) if necessarily, we obtain $e^{\vec{a}\cdot\vec{x}} \in M$; that is, $\mathscr{A}_{\vec{a},0} \subset M$.

Suppose $\mathscr{A}_{\vec{a},\ell} \subset M$ for some $\ell \in \mathbb{N}$. Let $g e^{\vec{a}\cdot\vec{x}}$ be any element in $\mathscr{A}_{\vec{a},\ell}$. First we assume $a_1 \neq 0$. Applying (12.7.30)–(12.7.33) to $g e^{\vec{a}\cdot\vec{x}}$, we get by (11.1.18)–(11.1.53) and (11.1.56) that

$$(a_1 x_2 - a_{11} x_{14} - a_{13} x_{17} - a_{16} x_{19} - a_{18} x_{21} - a_{20} x_{23}) g e^{\vec{a}\cdot\vec{x}} \equiv 0 \pmod{M}, \tag{12.8.40}$$

$$(a_1 x_3 - a_9 x_{14} - a_{12} x_{17} - a_{15} x_{19} + a_{18} x_{22} + a_{20} x_{24}) g e^{\vec{a}\cdot\vec{x}} \equiv 0 \pmod{M}, \tag{12.8.41}$$

$$(a_1 x_4 - a_7 x_{14} - a_{10} x_{17} + a_{15} x_{21} + a_{16} x_{22} - a_{20} x_{25}) g e^{\vec{a}\cdot\vec{x}} \equiv 0 \pmod{M}, \tag{12.8.42}$$

$$(a_1 x_6 + a_5 x_{14} + a_8 x_{17} - a_{15} x_{23} - a_{16} x_{24} - a_{18} x_{25}) g e^{\vec{a}\cdot\vec{x}} \equiv 0 \pmod{M}, \tag{12.8.43}$$

$$(a_1x_5 + a_6x_{14} - a_{10}x_{19} - a_{12}x_{21} - a_{13}x_{22} + a_{20}x_{26})ge^{\vec{a}\cdot\vec{x}} \equiv 0 \pmod{M}, \tag{12.8.44}$$

$$(a_1x_8 + a_6x_{17} + a_7x_{19} + a_9x_{21} + a_{11}x_{22} + a_{20}x_{27})ge^{\vec{a}\cdot\vec{x}} \equiv 0 \pmod{M}, \tag{12.8.45}$$

$$(a_1x_7 - a_4x_{14} + a_8x_{19} + a_{12}x_{23} + a_{13}x_{24} + a_{18}x_{26})ge^{\vec{a}\cdot\vec{x}} \equiv 0 \pmod{M}, \tag{12.8.46}$$

$$(a_1x_{10} - a_4x_{17} - a_5x_{19} - a_9x_{23} - a_{11}x_{24} + a_{18}x_{27})ge^{\vec{a}\cdot\vec{x}} \equiv 0 \pmod{M}, \tag{12.8.47}$$

$$(a_1x_9 - a_3x_{14} + a_8x_{21} - a_{10}x_{23} + a_{13}x_{25} - a_{16}x_{26})ge^{\vec{a}\cdot\vec{x}} \equiv 0 \pmod{M}, \tag{12.8.48}$$

$$(a_1x_{12} - a_3x_{17} - a_5x_{21} + a_7x_{23} - a_{11}x_{25} - a_{16}x_{27})ge^{\vec{a}\cdot\vec{x}} \equiv 0 \pmod{M}, \tag{12.8.49}$$

$$(a_1x_{11} - a_2x_{14} + a_8x_{22} - a_{10}x_{24} - a_{12}x_{25} + a_{15}x_{26})ge^{\vec{a}\cdot\vec{x}} \equiv 0 \pmod{M}, \tag{12.8.50}$$

$$(a_1x_{15} - a_3x_{19} + a_4x_{21} - a_6x_{23} + a_{11}x_{26} + a_{13}x_{27})ge^{\vec{a}\cdot\vec{x}} \equiv 0 \pmod{M}, \tag{12.8.51}$$

$$(a_1x_{13} - a_2x_{17} - a_5x_{22} + a_7x_{24} + a_9x_{25} + a_{15}x_{27})ge^{\vec{a}\cdot\vec{x}} \equiv 0 \pmod{M}, \tag{12.8.52}$$

$$(a_1x_{16} - a_2x_{19} + a_4x_{22} - a_6x_{24} - a_9x_{26} - a_{12}x_{27})ge^{\vec{a}\cdot\vec{x}} \equiv 0 \pmod{M}, \tag{12.8.53}$$

$$(a_1x_{18} - a_2x_{21} + a_3x_{22} - a_6x_{25} + a_7x_{26} + a_{10}x_{27})ge^{\vec{a}\cdot\vec{x}} \equiv 0 \pmod{M}, \tag{12.8.54}$$

$$(a_1x_{20} - a_2x_{23} + a_3x_{24} - a_4x_{25} + a_5x_{26} + a_8x_{27})ge^{\vec{a}\cdot\vec{x}} \equiv 0 \pmod{M}. \tag{12.8.55}$$

Next we multiply a_1 to (12.7.36) and (12.7.37), and apply them to $ge^{\vec{a}\cdot\vec{x}}$:

$$a_1(a_8x_5 + a_{10}x_7 + a_{12}x_9 + a_{13}x_{11} + a_{17}x_{14} + a_{27}x_{26})ge^{\vec{a}\cdot\vec{x}} \equiv 0 \pmod{M}, \tag{12.8.56}$$

$$a_1(a_8x_4 + a_{10}x_6 - a_{15}x_9 - a_{16}x_{11} - a_{19}x_{14} - a_{27}x_{25})ge^{\vec{a}\cdot\vec{x}} \equiv 0 \pmod{M}, \tag{12.8.57}$$

$$a_1(a_8x_3 + a_{12}x_6 + a_{15}x_7 - a_{18}x_{11} - a_{21}x_{14} + a_{27}x_{24})ge^{\vec{a}\cdot\vec{x}} \equiv 0 \pmod{M}, \tag{12.8.58}$$

$$a_1(a_{10}x_3 - a_{12}x_4 - a_{15}x_5 + a_{20}x_{11} + a_{23}x_{14} + a_{27}x_{22})ge^{\vec{a}\cdot\vec{x}} \equiv 0 \pmod{M}, \tag{12.8.59}$$

$$a_1(a_8x_2 + a_{13}x_6 + a_{16}x_7 + a_{18}x_9 - a_{22}x_{14} - a_{27}x_{23})ge^{\vec{a}\cdot\vec{x}} \equiv 0 \pmod{M}, \tag{12.8.60}$$

$$a_1(a_{10}x_2 - a_{13}x_4 - a_{16}x_5 - a_{20}x_9 + a_{24}x_{14} - a_{27}x_{21})ge^{\vec{a}\cdot\vec{x}} \equiv 0 \pmod{M}, \tag{12.8.61}$$

$$a_1(a_{12}x_2 - a_{13}x_3 - a_{18}x_5 + a_{20}x_7 + a_{25}x_{14} + a_{27}x_{19})ge^{\vec{a}\cdot\vec{x}} \equiv 0 \pmod{M}, \tag{12.8.62}$$

$$a_1(a_{15}x_2 - a_{16}x_3 + a_{18}x_4 - a_{20}x_6 - a_{26}x_{14} - a_{27}x_{17})ge^{\vec{a}\cdot\vec{x}} \equiv 0 \pmod{M}. \tag{12.8.63}$$

Note that (12.8.44), (12.8.46), (12.8.48) and (12.8.50) yield

$$a_1x_5ge^{\vec{a}\cdot\vec{x}} \equiv -(a_6x_{14} - a_{10}x_{19} - a_{12}x_{21} - a_{13}x_{22} + a_{20}x_{26})ge^{\vec{a}\cdot\vec{x}} \pmod{M}, \tag{12.8.64}$$

$$a_1x_7ge^{\vec{a}\cdot\vec{x}} \equiv (a_4x_{14} - a_8x_{19} - a_{12}x_{23} - a_{13}x_{24} - a_{18}x_{26})ge^{\vec{a}\cdot\vec{x}} \pmod{M}, \tag{12.8.65}$$

$$a_1x_9ge^{\vec{a}\cdot\vec{x}} \equiv (a_3x_{14} - a_8x_{21} + a_{10}x_{23} - a_{13}x_{25} + a_{16}x_{26})ge^{\vec{a}\cdot\vec{x}} \pmod{M}, \tag{12.8.66}$$

$$a_1x_{11}ge^{\vec{a}\cdot\vec{x}} \equiv (a_2x_{14} - a_8x_{22} + a_{10}x_{24} + a_{12}x_{25} - a_{15}x_{26})ge^{\vec{a}\cdot\vec{x}} \pmod{M}. \tag{12.8.67}$$

Substituting (12.8.33)–(12.8.37) into (12.8.26), we get

$$\begin{aligned}
& [a_1a_{17}x_{14} - a_8(a_6x_{14} - a_{10}x_{19} - a_{12}x_{21} - a_{13}x_{22} + a_{20}x_{26}) \\
& +a_1a_{27}x_{26} + a_{10}(a_4x_{14} - a_8x_{19} - a_{12}x_{23} - a_{13}x_{24} - a_{18}x_{26}) \\
& +a_{12}(a_3x_{14} - a_8x_{21} + a_{10}x_{23} - a_{13}x_{25} + a_{16}x_{26}) \\
& +a_{13}(a_2x_{14} - a_8x_{22} + a_{10}x_{24} + a_{12}x_{25} - a_{15}x_{26})]ge^{\vec{a}\cdot\vec{x}} \\
= {} & [(a_1a_{17} + a_2a_{13} + a_3a_{12} + a_4a_{10} - a_6a_8)x_{14} \\
& +(a_1a_{27} - a_8a_{20} - a_{10}a_{18} + a_{12}a_{16} - a_{13}a_{15})x_{26}]ge^{\vec{a}\cdot\vec{x}} \\
= {} & (\zeta_2(\vec{a})x_{14} + \zeta_{14}(\vec{a})x_{26})ge^{\vec{a}\cdot\vec{x}} \equiv 0 \pmod{M}
\end{aligned} \tag{12.8.68}$$

by(11.2.3) and (11.2.15).

Substituting (12.8.40)–(12.8.55) into (12.8.57)–(12.8.63), we can similarly obtain

$$(\zeta_3(\vec{a})x_{14}+\zeta_{14}(\vec{a})x_{25})ge^{\vec{a}\cdot\vec{x}}\equiv 0 \pmod M, \tag{12.8.69}$$

$$(\zeta_4(\vec{a})x_{14}+\zeta_{14}(\vec{a})x_{24})ge^{\vec{a}\cdot\vec{x}}\equiv 0 \pmod M, \tag{12.8.70}$$

$$(\zeta_6(\vec{a})x_{14}+\zeta_{14}(\vec{a})x_{22})ge^{\vec{a}\cdot\vec{x}}\equiv 0 \pmod M, \tag{12.8.71}$$

$$(\zeta_5(\vec{a})x_{14}+\zeta_{14}(\vec{a})x_{23})ge^{\vec{a}\cdot\vec{x}}\equiv 0 \pmod M, \tag{12.8.72}$$

$$(\zeta_7(\vec{a})x_{14}+\zeta_{14}(\vec{a})x_{21})ge^{\vec{a}\cdot\vec{x}}\equiv 0 \pmod M, \tag{12.8.73}$$

$$(\zeta_9(\vec{a})x_{14}+\zeta_{14}(\vec{a})x_{19})ge^{\vec{a}\cdot\vec{x}}\equiv 0 \pmod M, \tag{12.8.74}$$

$$(\zeta_{11}(\vec{a})x_{14}+\zeta_{14}(\vec{a})x_{17})ge^{\vec{a}\cdot\vec{x}}\equiv 0 \pmod M. \tag{12.8.75}$$

By Lemma 12.7.4, we dually have

$$(\zeta_1(\vec{a})x_{17}+\zeta_{11}(\vec{a})x_{27})ge^{\vec{a}\cdot\vec{x}}\equiv 0 \pmod M, \tag{12.8.76}$$

$$(\zeta_1(\vec{a})x_{19}+\zeta_9(\vec{a})x_{27})ge^{\vec{a}\cdot\vec{x}}\equiv 0 \pmod M, \tag{12.8.77}$$

$$(\zeta_1(\vec{a})x_{21}+\zeta_7(\vec{a})x_{27})ge^{\vec{a}\cdot\vec{x}}\equiv 0 \pmod M, \tag{12.8.78}$$

$$(\zeta_1(\vec{a})x_{23}+\zeta_5(\vec{a})x_{27})ge^{\vec{a}\cdot\vec{x}}\equiv 0 \pmod M, \tag{12.8.79}$$

$$(\zeta_1(\vec{a})x_{22}+\zeta_6(\vec{a})x_{27})ge^{\vec{a}\cdot\vec{x}}\equiv 0 \pmod M, \tag{12.8.80}$$

$$(\zeta_1(\vec{a})x_{24}+\zeta_4(\vec{a})x_{27})ge^{\vec{a}\cdot\vec{x}}\equiv 0 \pmod M, \tag{12.8.81}$$

$$(\zeta_1(\vec{a})x_{25}+\zeta_3(\vec{a})x_{27})ge^{\vec{a}\cdot\vec{x}}\equiv 0 \pmod M, \tag{12.8.82}$$

$$(\zeta_1(\vec{a})x_{26}+\zeta_2(\vec{a})x_{27})ge^{\vec{a}\cdot\vec{x}}\equiv 0 \pmod M. \tag{12.8.83}$$

Now if $\zeta_2(\vec{a})\neq 0$, then (12.8.68) and (12.8.83) gives

$$\zeta_1(\vec{a})x_{14}\equiv\zeta_{14}(\vec{a})x_{27} \pmod M. \tag{12.8.84}$$

Moreover, (12.8.84) can be also derived from the following triples:

$$\{(12.8.69),(12.8.82),\zeta_3(\vec{a})\neq 0\},\quad \{(12.8.70),(12.8.81),\zeta_4(\vec{a})\neq 0\}, \tag{12.8.85}$$

$$\{(12.8.71),(12.8.80),\zeta_6(\vec{a})\neq 0\},\quad \{(12.8.42),(12.8.49),\zeta_5(\vec{a})\neq 0\}, \tag{12.8.86}$$

$$\{(12.8.43),(12.8.48),\zeta_7(\vec{a})\neq 0\},\quad \{(12.8.74),(12.8.77),\zeta_9(\vec{a})\neq 0\}, \tag{12.8.87}$$

$$\{(12.8.75), (12.8.76), \zeta_{11}(\vec{a}) \neq 0\}. \tag{12.8.88}$$

So

$$(12.8.83) \text{ holds if } \zeta_r(\vec{a}) \neq 0 \text{ for some } r \in \{2,3,4,5,6,7,9,11\}. \tag{12.8.89}$$

Note $\Im_1(14) = 27$. Moreover,

$$\iota(J_1) = \{1,2,3,4,5,6,7,9,11,14\} \tag{12.8.90}$$

by (12.7.24), where $\iota_1(14) = 1$ and $\iota_1(27) = 14$. Suppose

$$\{(x_1\zeta_r\zeta_{14}\vartheta)(\vec{a}) \mid r \in \{2,3,4,5,6,7,9,11\}\} \neq \{0\}. \tag{12.8.91}$$

By the above paragraph, (12.8.84) holds. Substituting (12.8.68)–(12.8.75) into $\zeta_{14}(\vec{a})\times$(12.8.40), we get

$$\begin{aligned}
a_1\zeta_{14}(\vec{a})x_2ge^{\vec{a}\cdot\vec{x}} &\equiv [a_{11}\zeta_{14}(\vec{a}) - a_{13}\zeta_{11}(\vec{a}) - a_{16}\zeta_9(\vec{a}) \\
&\quad -a_{18}\zeta_7(\vec{a}) - a_{20}\zeta_5(\vec{a})]x_{14}ge^{\vec{a}\cdot\vec{x}} \\
&\equiv [a_{11}(a_1a_{27} - a_8a_{20} - a_{10}a_{18} + a_{12}a_{16} - a_{13}a_{15}) \\
&\quad -a_{13}(a_1a_{26} - a_5a_{20} - a_7a_{18} + a_9a_{16} - a_{11}a_{15}) \\
&\quad -a_{16}(a_1a_{25} + a_4a_{20} + a_6a_{18} - a_9a_{13} + a_{11}a_{12}) \\
&\quad -a_{18}(-a_1a_{24} + a_3a_{20} - a_6a_{16} + a_7a_{13} - a_{10}a_{11}) \\
&\quad -a_{20}(a_1a_{22} - a_3a_{18} - a_4a_{16} + a_5a_{13} - a_8a_{11})]x_{14}ge^{\vec{a}\cdot\vec{x}} \\
&\equiv a_1(a_{11}a_{27} - a_{13}a_{26} - a_{16}a_{25} + a_{18}a_{24} - a_{20}a_{22})x_{14}ge^{\vec{a}\cdot\vec{x}} \\
&\equiv -a_1\zeta_{26}(\vec{a})x_{14}ge^{\vec{a}\cdot\vec{x}} \pmod{M}
\end{aligned} \tag{12.8.92}$$

by (11.2.6), (11.2.8), (11.2.10), (11.2.12), (11.2.15) and (11.2.27). Since $a_1 \neq 0$, we have

$$\zeta_{14}(\vec{a})x_2ge^{\vec{a}\cdot\vec{x}} \equiv -\zeta_{26}(\vec{a})x_{14}ge^{\vec{a}\cdot\vec{x}} \pmod{M}. \tag{12.8.93}$$

Moreover, we substitute (12.8.68)–(12.8.75) and (12.8.84) into $\zeta_{14}(\vec{a})\times$[(12.8.41)–(12.8.55)] and can similarly derive

$$\zeta_{14}(\vec{a})x_3ge^{\vec{a}\cdot\vec{x}} \equiv -\zeta_{25}(\vec{a})x_{14}ge^{\vec{a}\cdot\vec{x}} \pmod{M}, \tag{12.8.94}$$

$$\zeta_{14}(\vec{a})x_4ge^{\vec{a}\cdot\vec{x}} \equiv -\zeta_{24}(\vec{a})x_{14}ge^{\vec{a}\cdot\vec{x}} \pmod{M}, \tag{12.8.95}$$

$$\zeta_{14}(\vec{a})x_6ge^{\vec{a}\cdot\vec{x}} \equiv -\zeta_{22}(\vec{a})x_{14}ge^{\vec{a}\cdot\vec{x}} \pmod{M}, \tag{12.8.96}$$

$$\zeta_{14}(\vec{a})x_5ge^{\vec{a}\cdot\vec{x}} \equiv -\zeta_{23}(\vec{a})x_{14}ge^{\vec{a}\cdot\vec{x}} \pmod{M}, \tag{12.8.97}$$

$$\zeta_{14}(\vec{a})x_8ge^{\vec{a}\cdot\vec{x}} \equiv -\zeta_{20}(\vec{a})x_{14}ge^{\vec{a}\cdot\vec{x}} \pmod{M}, \tag{12.8.98}$$

$$\zeta_{14}(\vec{a})x_7 g e^{\vec{a}\cdot\vec{x}} \equiv -\zeta_{21}(\vec{a})x_{14} g e^{\vec{a}\cdot\vec{x}} \pmod{M}, \tag{12.8.99}$$

$$\zeta_{14}(\vec{a})x_{10} g e^{\vec{a}\cdot\vec{x}} \equiv -\zeta_{18}(\vec{a})x_{14} g e^{\vec{a}\cdot\vec{x}} \pmod{M}, \tag{12.8.100}$$

$$\zeta_{14}(\vec{a})x_9 g e^{\vec{a}\cdot\vec{x}} \equiv -\zeta_{19}(\vec{a})x_{14} g e^{\vec{a}\cdot\vec{x}} \pmod{M}, \tag{12.8.101}$$

$$\zeta_{14}(\vec{a})x_{12} g e^{\vec{a}\cdot\vec{x}} \equiv -\zeta_{16}(\vec{a})x_{14} g e^{\vec{a}\cdot\vec{x}} \pmod{M}, \tag{12.8.102}$$

$$\zeta_{14}(\vec{a})x_{11} g e^{\vec{a}\cdot\vec{x}} \equiv -\zeta_{17}(\vec{a})x_{14} g e^{\vec{a}\cdot\vec{x}} \pmod{M}, \tag{12.8.103}$$

$$\zeta_{14}(\vec{a})x_{15} g e^{\vec{a}\cdot\vec{x}} \equiv -\zeta_{13}(\vec{a})x_{14} g e^{\vec{a}\cdot\vec{x}} \pmod{M}, \tag{12.8.104}$$

$$\zeta_{14}(\vec{a})x_{13} g e^{\vec{a}\cdot\vec{x}} \equiv -\zeta_{15}(\vec{a})x_{14} g e^{\vec{a}\cdot\vec{x}} \pmod{M}, \tag{12.8.105}$$

$$\zeta_{14}(\vec{a})x_{16} g e^{\vec{a}\cdot\vec{x}} \equiv -\zeta_{12}(\vec{a})x_{14} g e^{\vec{a}\cdot\vec{x}} \pmod{M}, \tag{12.8.106}$$

$$\zeta_{14}(\vec{a})x_{18} g e^{\vec{a}\cdot\vec{x}} \equiv -\zeta_{10}(\vec{a})x_{14} g e^{\vec{a}\cdot\vec{x}} \pmod{M}, \tag{12.8.107}$$

$$\zeta_{14}(\vec{a})x_{20} g e^{\vec{a}\cdot\vec{x}} \equiv -\zeta_{8}(\vec{a})x_{14} g e^{\vec{a}\cdot\vec{x}} \pmod{M}. \tag{12.8.108}$$

Applying $\zeta_{14}(\vec{a})\alpha_6$ to $ge^{\vec{a}\cdot\vec{x}}$, we have

$$\begin{aligned}&\zeta_{14}(\vec{a})(a_1x_1 - a_2x_2 + a_{11}x_{11} + a_{13}x_{13} - a_{14}x_{14} + a_{16}x_{16} - a_{17}x_{17}\\ &+a_{18}x_{18} - a_{19}x_{19} + a_{20}x_{20} - a_{21}x_{21} - a_{23}x_{23})ge^{\vec{a}\cdot\vec{x}} \equiv 0 \pmod{M}\end{aligned} \tag{12.8.109}$$

by (11.1.54) and Table 11.1. According to (12.8.93), (12.8.103), (12.8.105), (12.8.106), (12.8.75), (12.8.107), (12.8.74), (12.8.108), (12.8.73) and (12.8.72), the above equation is equivalent to

$$\begin{aligned}&a_1\zeta_{14}(\vec{a})x_1 g e^{\vec{a}\cdot\vec{x}} + [a_2\zeta_{26}(\vec{a}) - a_{11}\zeta_{17}(\vec{a}) - a_{13}\zeta_{15}(\vec{a}) - a_{14}\zeta_{14}(\vec{a})\\ &-a_{16}\zeta_{12}(\vec{a}) + a_{17}\zeta_{11}(\vec{a}) - a_{18}\zeta_{10}(\vec{a}) + a_{19}\zeta_9(\vec{a}) - a_{20}\zeta_8(\vec{a})\\ &+a_{21}\zeta_7(\vec{a}) + a_{23}\zeta_5(\vec{a})]x_{14} g e^{\vec{a}\cdot\vec{x}} \equiv 0 \pmod{M}\end{aligned} \tag{12.8.110}$$

Moreover, (11.2.1) and (11.2.3)–(11.2.28) imply

$$\begin{aligned}&-a_{14}\zeta_{14}(\vec{a}) + a_{17}\zeta_{11}(\vec{a}) + a_{19}\zeta_9(\vec{a}) + a_{21}\zeta_7(\vec{a}) + a_{23}\zeta_5(\vec{a})\\ =\;&-a_{14}(a_1a_{27} - a_8a_{20} - a_{10}a_{18} + a_{12}a_{16} - a_{13}a_{15})\\ &+a_{17}(a_1a_{26} - a_5a_{20} - a_7a_{18} + a_9a_{16} - a_{11}a_{15})\\ &+a_{19}(a_1a_{25} + a_4a_{20} + a_6a_{18} - a_9a_{13} + a_{11}a_{12})\\ &+a_{21}(-a_1a_{24} + a_3a_{20} - a_6a_{16} + a_7a_{13} - a_{10}a_{11})\\ &+a_{23}(a_1a_{22} - a_3a_{18} - a_4a_{16} + a_5a_{13} - a_8a_{11})\end{aligned}$$

$$\begin{aligned}
&= a_1\zeta_{27}(\vec{a}) - a_{14}(-a_8a_{20} - a_{10}a_{18} + a_{12}a_{16} - a_{13}a_{15}) \\
&\quad + a_{17}(-a_5a_{20} - a_7a_{18} + a_9a_{16} - a_{11}a_{15}) \\
&\quad + a_{19}(a_4a_{20} + a_6a_{18} - a_9a_{13} + a_{11}a_{12}) \\
&\quad + a_{21}(a_3a_{20} - a_6a_{16} + a_7a_{13} - a_{10}a_{11}) \\
&\quad + a_{23}(-a_3a_{18} - a_4a_{16} + a_5a_{13} - a_8a_{11}),
\end{aligned} \tag{12.8.111}$$

$$\begin{aligned}
&a_2\zeta_{26}(\vec{a}) - a_{11}\zeta_{17}(\vec{a}) - a_{13}\zeta_{15}(\vec{a}) - a_{16}\zeta_{12}(\vec{a}) - a_{18}\zeta_{10}(\vec{a}) - a_{20}\zeta_8(\vec{a}) \\
&= a_2(-a_{11}a_{27} + a_{13}a_{26} + a_{16}a_{25} - a_{18}a_{24} + a_{20}a_{22}) \\
&\quad - a_{11}(-a_2a_{27} - a_8a_{23} - a_{10}a_{21} + a_{12}a_{19} - a_{15}a_{17}) \\
&\quad - a_{13}(a_2a_{26} + a_5a_{23} + a_7a_{21} - a_9a_{19} + a_{14}a_{15}) \\
&\quad - a_{16}(a_2a_{25} - a_4a_{23} - a_6a_{21} + a_9a_{17} - a_{12}a_{14}) \\
&\quad - a_{18}(-a_2a_{24} - a_3a_{23} + a_6a_{19} - a_7a_{17} + a_{10}a_{14}) \\
&\quad - a_{20}(a_2a_{22} + a_3a_{21} + a_4a_{19} - a_5a_{17} + a_8a_{14}) \\
&= -a_{11}(-a_8a_{23} - a_{10}a_{21} + a_{12}a_{19} - a_{15}a_{17}) \\
&\quad - a_{13}(+a_5a_{23} + a_7a_{21} - a_9a_{19} + a_{14}a_{15}) \\
&\quad - a_{16}(-a_4a_{23} - a_6a_{21} + a_9a_{17} - a_{12}a_{14}) \\
&\quad - a_{18}(-a_3a_{23} + a_6a_{19} - a_7a_{17} + a_{10}a_{14}) \\
&\quad - a_{20}(+a_3a_{21} + a_4a_{19} - a_5a_{17} + a_8a_{14}).
\end{aligned} \tag{12.8.112}$$

Thus we have

$$\zeta_{14}(\vec{a})x_1ge^{\vec{a}\cdot\vec{x}} \equiv -\zeta_{27}(\vec{a})x_{14}ge^{\vec{a}\cdot\vec{x}} \pmod{M}. \tag{12.8.113}$$

Applying $\zeta_{14}(\vec{a})\times$ [the first equation in (12.8.4)] to $ge^{\vec{a}\cdot\vec{x}}$, we get

$$\sum_{i=1}^{27} a_i\zeta_{14}(\vec{a})x_ige^{\vec{a}\cdot\vec{x}} \equiv 0 \pmod{M}. \tag{12.8.114}$$

By (12.8.68)–(12.8.75), (12.8.84), (12.8.93)–(12.8.108) and (12.8.113), the above equation is equivalent to

$$\left[a_{14}\zeta_{14}(\vec{a}) + a_{27}\zeta_1(\vec{a}) - \sum_{14\neq i\in\overline{1,26}} a_i\zeta_{28-i}(\vec{a})x_{14}\right]ge^{\vec{a}\cdot\vec{x}} \equiv 0 \pmod{M}. \tag{12.8.115}$$

Equivalently,

$$\eta(\vec{a})x_{14}ge^{\vec{a}\cdot\vec{x}} \equiv 0 \pmod{M} \tag{12.8.116}$$

by (11.2.41). Assumption (12.8.91) implies $\vartheta(\vec{a}) \neq 0$ and $\zeta_{14}(\vec{a}) \neq 0$. Thus $x_{14}ge^{\vec{a}\cdot\vec{x}} \in M$. Substituting it to (12.8.68)–(12.8.75), (12.8.84), (12.8.93)–(12.8.108) and (12.8.113), we get $x_i ge^{\vec{a}\cdot\vec{x}} \in M$ for any $i \in \overline{1,27}$. Hence $\mathscr{A}_{\vec{a},\ell+1} \subset M$. By induction,

$$\mathscr{A}_{\vec{a},k} \subset M \qquad \text{for } k \in \mathbb{N}. \tag{12.8.117}$$

Therefore, $M = \sum_{k=0}^{\infty} \mathscr{A}_{\vec{a},k} = \mathscr{A}_{\vec{a}}$. So $\mathscr{A}_{\vec{a}}$ forms an irreducible $\mathscr{G}^{E_7}$-module.

Now the theorem follows from Lemmas 12.7.1–12.7.4. □

References

[Br] R. Block, On torsion-free abelian groups and Lie algebras. Proc. Am. Soc. **9**, 613–620 (1958)

[BKR] L. Brink, S. Kim, P. Ramond, $E_{7(7)}$ on the light cone. J. High Ener. Phys. (6), 034, 18pp (2008)

[C] B. Cooperstein, The fifty-six-dimensional module for E_7.I. A four form for E_7. J. Algebra **173**, 361–389 (1995)

[DFT] R. D'Auria, S. Ferrara, M. Trigiante, $E_{7(7)}$ symmetry and dual gauge algebras of M-theory on a twisted even-torus. Nucl. Phys. B **732**, 389–400 (2006)

[F] J. Faulkner, A geometry for E_7. Trans. Am. Math. Soc. **167**, 49–58 (1972)

[Gr] R. Garibaldi, Structurable algebras and groups of type E_6 and E_7. J. Algebra **236**(2), 651–691 (2001)

[Gd] D. Ginzburg, On standard L-functions for E_6 and E_7. J. Reine Angew. Math. **465**, 101–131 (1995)

[KaKy] A. Kato, Y. Kitazawa, E_7-type modular invariant Wess-Zumino theory and Gepner's string compactification. Nucl. Phys. B **319**(2), 474–490 (1989)

[Kb] B. Kostant, On the tensor product of a finite and an infinite dimensional representation. J. Fund. Anal. **20**, 257–285 (1975)

[R2] E. Plotkin, On the stability of the K_1-functor for Chevalley groups of type E_7. J. Algebra **210**, 67–85 (1998)

[P] P. Ramond, Is there an exceptional group in your future? E_7 and the travails of the symmetry breaking. Nucl. Phys. B **126**(3), 509–524 (1977)

[Se] J. Sekiguchi, Configurations of seven lines on the real projective plane and the root system of type E_7. J. Math. Soc. Japan **51**(4), 987–1013 (1999)

[Sg] G. Shen, Graded modules of graded Lie algebras of Cartan type (I)–mixed product of modules. Sci. China A **29**, 570–581 (1986)

[W] R. Weiss, Moufang quadangles of type E_6 and E_7. J. Reine Angew. Math. **590**, 189–226 (2006)

[X22] X. Xu, Partial Differential equation approach to E_7. Acta Math Sin. (Engl. Ser.) **31**, 177–200 (2015)

[X23] X. Xu, A new functor from E_6-Mod to E_7-Mod. Commun. Algebra **43**(9), 3589–3636 (2015)

[X27] X. Xu, Representations of E_7, combinatorics and algebraic Varieties. J. Algebra Appl. (to appear)

Part III
Related Topics

Chapter 13
Oscillator Representations of $gl(n|m)$ and $osp(n|2m)$

A *supersymmetry* relating mesons and baryons was first proposed by Miyazawa [Mh] in 1966. It was largely ignored because it did not involve spacetime. In early 1970's, physicists rediscovered it in the context of quantum field theory, a radically new type of symmetry of spacetime and fundamental fields, which establishes a relationship among elementary particles of quantum nature, bosons and fermions, and unifies spacetime and internal symmetries of microscopic phenomena (e.g., cf. [GS]). Supersymmetry with a consistent Lie-algebraic graded structure on which the Gervais–Sakita rediscovery [GS] based directly first arose during 1971 in the context of a early version of string theory (cf. [R1]). It was later widely applied to nuclear physics, critical phenomena, quantum mechanics and statistical physics.

The aim of this chapter is to present some natural supersymmetries along the idea of generalizing classical theorem of harmonic polynomials. First we establish two-parameter $\mathbb{Z}^2$-graded supersymmetric oscillator generalizations of the above theorem for the Lie superalgebra $gl(n|m)$, where Theorems 6.2.5 and 6.4.3 play fundamental roles. Then we extend the result to two-parameter $\mathbb{Z}$-graded supersymmetric oscillator generalizations of the classical theorem of harmonic polynomials for the Lie superalgebras $osp(2n|2m)$ and $osp(2n+1|2m)$, where Lemma 6.1.1 plays an important role. The results can be interpreted as supersymmetric Howe dualities (cf. [Hr1, Hr2, Hr3, Hr4]). The chapter is a reformulation of Luo and the author's work [LX4].

In Sect. 13.1, we study the canonical supersymmetric orthogonal oscillator representation of general linear Lie superalgebras. Section 13.2 is a $\mathbb{Z}^2$-graded oscillator generalizations of the above representation obtained by partially swapping differential operators and multiplication operators. Using the results in the above two sections, we determine in Sect. 13.3 the structure of the corresponding representations for even ortho-symplectic Lie superalgebras. Finally in Sect. 13.4, we investigate the oscillator representations of odd ortho-symplectic Lie superalgebras based on the above three sections.

X. Xu, *Representations of Lie Algebras and Partial Differential Equations*,
DOI 10.1007/978-981-10-6391-6_13

13.1 Canonical Oscillator Representation of $gl(n|m)$

In this section, we study the canonical supersymmetric orthogonal oscillator representations of general linear Lie superalgebras.

Fix two positive integers m and n. Set

$$gl(n|m)_0 = \sum_{i,j=1}^{n} \mathbb{F}E_{i,j} + \sum_{r,s=1}^{m} \mathbb{F}E_{n+r,n+s} \tag{13.1.1}$$

and

$$gl(n|m)_1 = \sum_{i=1}^{n}\sum_{r=1}^{m}(\mathbb{F}E_{i,n+r} + \mathbb{F}E_{n+r,i}). \tag{13.1.2}$$

The general linear Lie superalgebra $gl(n|m) = gl(n|m)_0 + gl(n|m)_1$ with the algebraic operation $[\cdot,\cdot]$ defined by

$$[A, B] = AB - (-1)^{i_1 i_2} BA \qquad \text{for } A \in gl(n|m)_{i_1},\ B \in gl(n|m)_{i_2}. \tag{13.1.3}$$

Let $\mathscr{A}$ be the polynomial algebra in bosonic variables $\{x_i \mid i \in \overline{1,2n}\}$ and fermionic variables $\{\theta_j \mid j \in \overline{1,2m}\}$; i.e.,

$$x_r x_s = x_s x_r,\quad \theta_p\theta_q = -\theta_q\theta_p,\quad x_r\theta_p = \theta_p x_r \tag{13.1.4}$$

for $r, s \in \overline{1,2n}$ and $p, q \in \overline{1,2m}$. Set

$$\Theta = \sum_{p=1}^{2m} \mathbb{F}\theta_p. \tag{13.1.5}$$

Write

$$\mathscr{A}_{(0)} = \sum_{q=0}^{m} \mathbb{F}[x_1,\ldots,x_{2n}]\Theta^{2q},\qquad \mathscr{A}_{(1)} = \sum_{q=1}^{m-1} \mathbb{F}[x_1,\ldots,x_{2n}]\Theta^{2q+1}. \tag{13.1.6}$$

Then $\mathscr{A} = \mathscr{A}_{(0)} \oplus \mathscr{A}_{(1)}$ is a $\mathbb{Z}_2$-graded algebra.

For $r \in \overline{1,2n}$, the usual differential operator ∂_{x_r} acts on $\mathscr{A}$ as a derivation such that

$$\partial_{x_r}(x_s) = \delta_{r,s},\quad \partial_{x_r}(\theta_p) = 0 \qquad \text{for } s \in \overline{1,2n},\ p \in \overline{1,2m}. \tag{13.1.7}$$

Moreover, for $p \in \overline{1,2m}$, we define ∂_{θ_p} as a linear operator on $\mathscr{A}$ with

$$\partial_{\theta_p}(x_r) = 0,\quad \partial_{\theta_p}(\theta_q) = \delta_{p,q} \qquad \text{for } r \in \overline{1,n},\ q \in \overline{1,2m} \tag{13.1.8}$$

such that

$$\partial_{\theta_p}(fg) = \partial_{\theta_p}(f)g + (-1)^{\iota} f\partial_{\theta_p}(g) \qquad \text{for } f \in \mathscr{A}_{(\iota)},\ g \in \mathscr{A}. \tag{13.1.9}$$

For later notational convenience, we redenote

$$y_i = x_{n+i}, \qquad \vartheta_j = \theta_{m+j} \qquad \text{for } i \in \overline{1,n},\ j \in \overline{1,m}. \tag{13.1.10}$$

Define the *orthogonal oscillator representation* of $gl(n|m)$ on $\mathscr{A}$ determined by

$$E_{i,j}|_{\mathscr{A}} = x_i\partial_{x_j} - y_j\partial_{y_i}, \qquad E_{i,n+r}|_{\mathscr{A}} = x_i\partial_{\theta_r} - \vartheta_r\partial_{y_i}, \tag{13.1.11}$$

$$E_{n+r,i}|_{\mathscr{A}} = \theta_r\partial_{x_i} + y_i\partial_{\vartheta_r}, \qquad E_{n+r,n+s}|_{\mathscr{A}} = \theta_r\partial_{\theta_s} - \vartheta_s\partial_{\vartheta_r} \tag{13.1.12}$$

for $i, j \in \overline{1,n}$ and $r, s \in \overline{1,m}$.

13.1.1 Representations of the Even Subalgebra

Set

$$\bar{\mathscr{A}} = \mathbb{F}[x_1, \ldots, x_n, y_1, \ldots, y_n], \qquad \check{\mathscr{A}} = \sum_{i=0}^{2m} \Theta^i. \tag{13.1.13}$$

Then $\bar{\mathscr{A}}$ and $\check{\mathscr{A}}$ are subalgebras of $\mathscr{A}$, and

$$\mathscr{A} = \bar{\mathscr{A}}\check{\mathscr{A}}. \tag{13.1.14}$$

Write

$$\bar{\mathscr{G}} = \sum_{i,j=1}^{n} \mathbb{F}E_{i,j}, \qquad \check{\mathscr{G}} = \sum_{r,s=1}^{m} \mathbb{F}E_{n+r,n+s}. \tag{13.1.15}$$

Then they are Lie subalgebras of $gl(n|m)$. In fact, the even subalgebra $gl(n|m)_0 = \bar{\mathscr{G}} + \check{\mathscr{G}}$. Let

$$\bar{H} = \sum_{i=1}^{n} \mathbb{F}E_{i,i}, \qquad \check{H} = \sum_{r=1}^{m} \mathbb{F}E_{n+r,n+r}, \tag{13.1.16}$$

$$\bar{\mathscr{G}}_+ = \sum_{1\leq i<j\leq n} \mathbb{F}E_{i,j}, \qquad \check{\mathscr{G}}_+ = \sum_{1\leq r<s\leq m} \mathbb{F}E_{n+r,n+s}. \tag{13.1.17}$$

We take $\bar{H}$ as a Cartan subalgebra of $\bar{\mathscr{G}}$ and $\bar{\mathscr{G}}_+$ as the subalgebra spanned by positive root vectors in $\bar{\mathscr{G}}$. Similarly, we take $\check{H}$ as a Cartan subalgebra of $\check{\mathscr{G}}$ and $\check{\mathscr{G}}_+$ as the subalgebra spanned by positive root vectors in $\check{\mathscr{G}}$.

Let

$$\bar{\Delta} = \sum_{i=1}^{n} \partial_{x_i}\partial_{y_i}, \qquad \bar{\eta} = \sum_{i=1}^{n} x_i y_i. \tag{13.1.18}$$

Recall

$$x^\alpha = x_1^{\alpha_1}\cdots x_n^{\alpha_n}, \quad y^\beta = y_1^{\beta_1}\cdots y_n^{\beta_n} \text{ for } \alpha = (\alpha_1, , \alpha_n), \beta = (\beta_1, \ldots, \beta_n) \in \mathbb{N}^n. \tag{13.1.19}$$

For $\ell_1, \ell_2 \in \mathbb{N}$, we denote

$$\bar{\mathscr{A}}_{\ell_1,\ell_2} = \text{Span}\left\{x^\alpha y^\beta \mid \alpha, \beta \in \mathbb{N}^n; \sum_{i=1}^{n} \alpha_i = \ell_1; \sum_{i=1}^{n} \beta_i = \ell_2\right\}. \tag{13.1.20}$$

Define

$$\bar{\mathscr{H}}_{\ell_1,\ell_2} = \{f \in \bar{\mathscr{A}}_{\ell_1,\ell_2} \mid \bar{\Delta}(f) = 0\}. \tag{13.1.21}$$

For $i \in \overline{1,n}$, we define $\bar{\varepsilon}_i \in \bar{H}^*$ by

$$\bar{\varepsilon}_i(E_{j,j}) = \delta_{i,j} \qquad \text{for } j \in \overline{1,n}. \tag{13.1.22}$$

According to Theorem 6.2.4, we have:

Lemma 13.1.1 *Suppose $n > 1$. For any ℓ_1, ℓ_2, $\bar{\mathscr{H}}_{\ell_1,\ell_2}$ is a finite-dimensional irreducible $\bar{\mathscr{G}}$-module with highest-weight vector $x_1^{\ell_1} y_n^{\ell_2}$ of weight $\ell_1\varepsilon_1 + \ell_2\varepsilon_n$. Moreover,*

$$\bar{\mathscr{A}} = \bigoplus_{\ell_1,\ell_2,\ell_3=0}^{\infty} \bar{\eta}^{\ell_1} \bar{\mathscr{H}}_{\ell_2,\ell_3} \tag{13.1.23}$$

is a decomposition of irreducible $\bar{\mathscr{G}}$-submodules.

When $n = 1$, we have

$$\bar{\mathscr{H}}_{\ell,0} = \mathbb{F}x_1^\ell, \qquad \bar{\mathscr{H}}_{0,\ell} = \mathbb{F}y_1^\ell \qquad \text{for } \ell \in \mathbb{N}. \tag{13.1.24}$$

Moreover,

$$\bar{\mathscr{A}} = \bigoplus_{\ell_1,\ell_2=0}^{\infty} (\bar{\eta}^{\ell_1} \bar{\mathscr{H}}_{\ell_2,0} \oplus \bar{\eta}^{\ell_1} \bar{\mathscr{H}}_{0,\ell_2+1}). \tag{13.1.25}$$

Denote

$$\Theta_1 = \sum_{i=1}^{m} \mathbb{F}\theta_i, \qquad \Theta_2 = \sum_{i=1}^{m} \mathbb{F}\vartheta_i. \tag{13.1.26}$$

For $\ell_1, \ell_2 \in \overline{1,m}$, we define

$$\check{\mathscr{A}}_{\ell_1,\ell_2} = \Theta_1^{\ell_1}\Theta_2^{\ell_2}. \tag{13.1.27}$$

Then $\check{\mathscr{A}}_{\ell_1,\ell_2}$ is a finite-dimensional $\check{\mathscr{G}}$-module and

$$\check{\mathscr{A}} = \bigoplus_{\ell_1,\ell_2=0}^{m} \check{\mathscr{A}}_{\ell_1,\ell_2}. \tag{13.1.28}$$

Write

$$\check{\Delta} = \sum_{r=1}^{m} \partial_{\theta_r}\partial_{\vartheta_r}, \qquad \check{\eta} = \sum_{r=1}^{m} \theta_r\vartheta_r. \tag{13.1.29}$$

Recall the notions

$$\vec{\theta}_r = \theta_1\cdots\theta_r, \qquad \vec{\vartheta}_r = \vartheta_m\cdots\vartheta_r \qquad \text{for } r \in \overline{1,m} \tag{13.1.30}$$

and

$$\vec{\theta}_0 = 1 = \vec{\vartheta}_{m+1}. \tag{13.1.31}$$

For $\ell_1, \ell_2 \in \overline{0,m}$, we define

$$\check{\mathscr{H}}_{\ell_1,\ell_2} = \{f \in \check{\mathscr{A}}_{\ell_1,\ell_2} \mid \check{\Delta}(f) = 0\}. \tag{13.1.32}$$

For $r \in \overline{1,m}$, we define $\varepsilon_r' \in \check{H}^*$ by

$$\varepsilon_r'(E_{n+s,n+s}) = \delta_{r,s} \qquad \text{for } s \in \overline{1,m}. \tag{13.1.33}$$

Moreover, we treat $\varepsilon_0' = \varepsilon_{m+1}' = 0$. By Theorem 6.2.5, we have:

Lemma 13.1.2 *For $0 \leq r < s \leq m+1$, $\check{\mathscr{H}}_{r,m+1-s}$ is a finite-dimensional irreducible $\check{\mathscr{G}}$-module with the highest-weight vector $\vec{\theta}_r\vec{\vartheta}_s$ of weight $\sum_{p=0}^{r}\varepsilon_p' - \sum_{q=s}^{m+1}\varepsilon_q'$. Moreover,*

$$\check{\mathscr{A}} = \bigoplus_{0\leq r<s\leq m+1}\ \bigoplus_{\ell=0}^{s-r-1} \check{\eta}^{\ell}\check{\mathscr{H}}_{r,m+1-s}. \tag{13.1.34}$$

According to (6.2.38) and (6.2.57), we have:

$$\bar{\Delta}\bar{\eta} = \bar{\eta}\bar{\Delta} + n + \sum_{i=1}^{n}(x_i\partial_{x_i} + y_i\partial_{y_i}), \quad \check{\Delta}\check{\eta} = \check{\eta}\check{\Delta} - m + \sum_{r=1}^{m}(\theta_r\partial_{\theta_r} + \vartheta_r\partial_{\vartheta_r}). \tag{13.1.35}$$

Set

$$\Delta = \bar{\Delta} + \check{\Delta} = \sum_{i=1}^{n}\partial_{x_i}\partial_{y_i} + \sum_{r=1}^{m}\partial_{\theta_r}\partial_{\vartheta_r}, \quad \eta = \bar{\eta} + \check{\eta} = \sum_{i=1}^{n}x_iy_i + \sum_{r=1}^{m}\theta_r\vartheta_r. \tag{13.1.36}$$

For $\ell_1,\ell_2,\ell_3\in\mathbb{N}$, $0\le r<s\le m+1$ and $\ell\in\overline{0,s-r-1}$, $f\in\bar{\mathscr{H}}_{\ell_1,\ell_2}$ and $g\in\check{\mathscr{H}}_{r,m+1-s}$, we have

$$\Delta(\bar{\eta}^{\ell_3}\check{\eta}^{\ell}fg)=\ell_3(n+\ell_1+\ell_2+\ell_3-1)\bar{\eta}^{\ell_3-1}\check{\eta}^{\ell}fg+\ell(\ell+r-s)\bar{\eta}^{\ell_3}\check{\eta}^{\ell-1}fg. \quad (13.1.37)$$

Suppose $r+1<s$ and $\ell\in\overline{1,s-r-1}$. If

$$\Delta\left(\sum_{p=0}^{\ell}a_p\bar{\eta}^p\check{\eta}^{\ell-p}fg\right)=0, \quad (13.1.38)$$

then

$$(\ell-p)(p+s-r-\ell)a_p=(p+1)(n+\ell_1+\ell_2+p)a_{p+1}\qquad\text{for }p\in\overline{0,\ell-1}. \quad (13.1.39)$$

Thus we can take $a_0=(\ell+1)![\prod_{\iota_2=1}^{\ell+1}(\iota_2+n+\ell_1+\ell_2-1)]$ and

$$a_{p+1}=\left[\prod_{\iota_1=0}^{p}(\ell-\iota_1)(\iota_1+s-r-\ell)\right]\left[\prod_{\iota_2=p+2}^{\ell+1}\iota_2(\iota_2+n+\ell_1+\ell_2-1)\right] \quad (13.1.40)$$

for $p\in\overline{0,\ell-1}$. Denote

$$\begin{aligned}&\Im(\ell_1,\ell_2;r,s,\ell)\\&=(\ell+1)!\left[\prod_{\iota_2=1}^{\ell+1}(\iota_2+n+\ell_1+\ell_2-1)\right]\check{\eta}^{\ell}+\sum_{p=0}^{\ell-1}\left[\prod_{\iota_1=0}^{p}(\ell-\iota_1)(\iota_1+s-r-\ell)\right]\\&\quad\times\left[\prod_{\iota_2=p+2}^{\ell+1}\iota_2(\iota_2+n+\ell_1+\ell_2-1)\right]\bar{\eta}^{p+1}\check{\eta}^{\ell-p-1}.\end{aligned} \quad (13.1.41)$$

For convenience, we treat

$$\Im(\ell_1,\ell_2;r,s,0)=n+\ell_1+\ell_2. \quad (13.1.42)$$

Write

$$\mathscr{G}=\bar{\mathscr{G}}+\check{\mathscr{G}}\cong gl(n,\mathbb{F})\oplus gl(m,\mathbb{F}). \quad (13.1.43)$$

Then

$$\mathscr{A}=\bigoplus_{\ell_1,\ell_2,\ell_3=0}^{\infty}\ \bigoplus_{0\le r<s\le m+1}\ \bigoplus_{\ell=0}^{s-r-1}(\bar{\eta}^{\ell_1}\bar{\mathscr{H}}_{\ell_2,\ell_3})(\check{\eta}^{\ell}\check{\mathscr{H}}_{r,m+1-s}) \quad (13.1.44)$$

is a direct sum of irreducible $\mathscr{G}$-submodules. Moreover,

$$\mathscr{H} = \{f \in \mathscr{A} \mid \Delta(f) = 0\} \tag{13.1.45}$$

it is a $\mathscr{G}$-submodule. According to (13.1.37)–(13.1.42),

$$\mathscr{H} = \bigoplus_{\ell_2,\ell_3=0}^{\infty} \bigoplus_{0\leq r<s\leq m+1} \bigoplus_{\ell=0}^{s-r-1} \Im(\ell_2, \ell_3; r, s, \ell)(\bar{\mathscr{H}}_{\ell_2,\ell_3}\check{\mathscr{H}}_{r,m+1-s}) \tag{13.1.46}$$

is a direct sum of irreducible $\mathscr{G}$-submodules.

For $\ell, \ell' \in \mathbb{N}$, we let

$$\mathscr{A}_{\ell,\ell'} = \sum_{\ell_1,\ell_2\in\mathbb{N},\ \ell_3,\ell_4\in\overline{0,m};\ \ell_1+\ell_3=\ell,\ \ell_2+\ell_4=\ell'} \bar{\mathscr{A}}_{\ell_1,\ell_2}\check{\mathscr{A}}_{\ell_3,\ell_4} \tag{13.1.47}$$

and

$$\mathscr{H}_{\ell,\ell'} = \mathscr{A}_{\ell,\ell'} \bigcap \mathscr{H}. \tag{13.1.48}$$

It is straightforward to verify

$$E_{i,j}\Delta = \Delta E_{i,j}, \qquad E_{i,j}\eta = \eta E_{i,j} \qquad \text{for } i, j \in \overline{1, m+n} \tag{13.1.49}$$

by (13.1.11), (13.1.12) and (13.1.36). Thus $\mathscr{A}_{\ell,\ell'}$ and $\mathscr{H}_{\ell,\ell'}$ are $gl(n|m)$-submodules. Moreover,

$$\mathscr{H}_{\ell,\ell'} = \bigoplus_{\ell_1+r+\ell_3=\ell,\ \ell_2+\ell_3+m+1-s=\ell'} \Im(\ell_1, \ell_2; r, s, \ell_3)\bar{\mathscr{H}}_{\ell_1,\ell_2}\check{\mathscr{H}}_{r,m+1-s} \tag{13.1.50}$$

is a direct sum of irreducible $\mathscr{G}$-submodules. Take the Cartan subalgebra $H = \bar{H} + \check{H}$ of $\mathscr{G}$ (cf. (13.1.16)) and the subspace $\mathscr{G}_+ = \bar{\mathscr{G}}_+ + \check{\mathscr{G}}_+$ (cf. (13.1.17)) spanned by positive root vectors in $\mathscr{G}$. A *$\mathscr{G}$-singular vector* v is a nonzero weight vector of $\mathscr{G}$ such that $\mathscr{G}_+(v) = 0$. We count singular vector up to a nonzero scalar multiple. Hence we have:

Lemma 13.1.3 *The set*

$$\{\Im(\ell_1, \ell_2; r, s, \ell_3)(x_1^{\ell_1} y_n^{\ell_2}\vec{\theta}_r\vec{\vartheta}_s) \mid \ell_1, \ell_2 \in \mathbb{N}; 0 \leq r < s \leq m+1;$$
$$\ell_3 \in \overline{0, s-r-1}; \ell_1 + r + \ell_3 = \ell,\ \ell_2 + \ell_3 + m + 1 - s = \ell'\} \tag{13.1.51}$$

is the set of all the $\mathscr{G}$-singular vectors in $\mathscr{H}_{\ell,\ell'}$, where $\ell_1\ell_2 = \ell_1'\ell_2' = 0$ if $n = 1$.

13.1.2 Singular Vectors of $gl(n|m)$

Take $H = \bar{H} + \check{H}$ as a Cartan subalgebra of the Lie superalgebra $gl(n|m)$ and

$$gl(n|m)_+ = \mathscr{G}_+ + \sum_{r=1}^{n}\sum_{s=1}^{m}\mathbb{F}E_{r,n+s} \tag{13.1.52}$$

as the subalgebra generated by positive root vectors. A $gl(n|m)$-*singular vector* v is a nonzero weight vector of $gl(n|m)$ such that $gl(n|m)_+(v) = 0$. We count singular vector up to a nonzero scalar multiple. Assume that $x_1^{\ell_1'} y_n^{\ell_2'}\vec{\theta}_{r'}\vec{\vartheta}_{s'}$ is a $gl(n|m)$-singular vector, where $\ell_1'\ell_2' = 0$ when $n = 1$. If $r' \neq 0$, then

$$\begin{aligned} & E_{1,n+r'}(x_1^{\ell_1'} y_n^{\ell_2'}\vec{\theta}_{r'}\vec{\vartheta}_{s'}) \\ &= (x_1\partial_{\theta_{r'}} - \vartheta_{r'}\partial_{y_1})(x_1^{\ell_1'} y_n^{\ell_2'}\vec{\theta}_{r'}\vec{\vartheta}_{s'}) \\ &= (-1)^{r'-1}x_1^{\ell_1'+1} y_n^{\ell_2'}\vec{\theta}_{r'-1}\vec{\vartheta}_{s'} - \delta_{1,n}\ell_2' x_1^{\ell_1'} y_n^{\ell_2'-1}\vartheta_{r'}\vec{\theta}_{r'}\vec{\vartheta}_{s'} \neq 0 \end{aligned} \tag{13.1.53}$$

by the second equation in (13.1.11), which contradicts the definition of singular vector. So $r' = 0$. Suppose $\ell_2' > 0$ and $s' > 1$. Again the second equation in (13.1.11) gives

$$\begin{aligned} E_{n,n+s'-1}(x_1^{\ell_1'} y_n^{\ell_2'}\vec{\vartheta}_{s'}) &= (x_n\partial_{\theta_{s'-1}} - \vartheta_{s'-1}\partial_{y_n})(x_1^{\ell_1'} y_n^{\ell_2'}\vec{\vartheta}_{s'}) \\ &= -\ell_2' x_1^{\ell_1'} y_n^{\ell_2'-1}\vec{\theta}_{r'-1}\vec{\vartheta}_{s'-1} \neq 0, \end{aligned} \tag{13.1.54}$$

which is absurd. Thus $x_1^{\ell_1'} y_n^{\ell_2'}\vec{\theta}_{r'}\vec{\vartheta}_{s'}$ is a $gl(n|m)$-singular vector if and only if $r' = 0$ and $\ell_2'(s'-1) = 0$.

For $\ell_1, \ell_2 \in \mathbb{N}$, $1 \leq r < s-1 \leq m$ and $\ell \in \overline{1, s-r-1}$,

$$\begin{aligned} & E_{1,n+s-1}[\Im(\ell_1, \ell_2; r, s, \ell)] \\ &= (x_1\partial_{\theta_{s-1}} - \vartheta_{s-1}\partial_{y_1})[\Im(\ell_1, \ell_2; r, s, \ell)] \\ &= x_1\vartheta_{s-1}\{(\ell+1)!\left[\prod_{\iota_2=1}^{\ell+1}(\iota_2+n+\ell_1+\ell_2-1)\right]\ell\check{\eta}^{\ell-1} \\ &+ \sum_{p=0}^{\ell-2}\left[\prod_{\iota_1=0}^{p}(\ell-\iota_1)(\iota_1+s-r-\ell)\right] \\ &\times(\ell-p-1)\left[\prod_{\iota_2=p+2}^{\ell+1}\iota_2(\iota_2+n+\ell_1+\ell_2-1)\right]\bar{\eta}^{p+1}\check{\eta}^{\ell-p-2} \\ &- \sum_{p=0}^{\ell-1}(p+1)\left[\prod_{\iota_1=0}^{p}(\ell-\iota_1)(\iota_1+s-r-\ell)\right] \\ &\times\left[\prod_{\iota_2=p+2}^{\ell+1}\iota_2(\iota_2+n+\ell_1+\ell_2-1)\right]\bar{\eta}^{p}\check{\eta}^{\ell-p-1}\} \\ &= \ell(\ell+1)(n+\ell+\ell_1+\ell_2+r-s)\Im(\ell_1+1, \ell_2; r, s-1, \ell-1)x_1\vartheta_{s-1}. \end{aligned} \tag{13.1.55}$$

Moreover, (13.1.41) yields

$$\Im(\ell_1,\ell_2;r,s,\ell)=(\ell+1)!\left[\prod_{\iota_1=0}^{\ell}(\iota_1+s-r-\ell)\right]\eta^\ell \text{ if } n+\ell+\ell_1+\ell_2+r-s=0. \tag{13.1.56}$$

Suppose that $\Im(\ell_1,\ell_2;r,s,\ell_3)(x_1^{\ell_1}y_n^{\ell_2}\vec{\theta}_r\vec{\vartheta}_s)$ is a $gl(n|m)$-singular vector, then

$$n+\ell_3+\ell_1+\ell_2+r-s=0 \tag{13.1.57}$$

and

$$\Im(\ell_1,\ell_2;r,s,\ell_3)(x_1^{\ell_1}y_n^{\ell_2}\vec{\theta}_r\vec{\vartheta}_s)=c\eta^{\ell_3}x_1^{\ell_1}y_n^{\ell_2}\vec{\theta}_r\vec{\vartheta}_s, \tag{13.1.58}$$

where $c=(\ell_3+1)![\prod_{\iota_1=0}^{\ell}(\iota_1+s-r-\ell_3)]$ by (13.1.54). Since

$$E_{i,n+r}(c\eta^{\ell_3}x_1^{\ell_1}y_n^{\ell_2}\vec{\theta}_r\vec{\vartheta}_s)=c\eta^{\ell_3}E_{i,n+r}(x_1^{\ell_1}y_n^{\ell_2}\vec{\theta}_r\vec{\vartheta}_s) \tag{13.1.59}$$

by (13.1.11), the arguments in (13.1.53) and (13.1.54) show $r=\ell_2(s-1)=0$. On the other hand, $1\leq 1+r<s$. So $\ell_2=0$. According to (13.1.57),

$$\ell_3=s-\ell_1-n. \tag{13.1.60}$$

Thus we only get the singular vector

$$\eta^{s-\ell_1-n}x_1^{\ell_1}\vec{\vartheta}_s\in\mathscr{H}_{s-n,m+1-n-\ell_1} \text{ with } s>\ell_1+n. \tag{13.1.61}$$

Note $n\leq m$ by $\ell_3>0$. Moreover, $s\leq m+1$ implies

$$s-n,m+1-n-\ell_1\leq m+1-n. \tag{13.1.62}$$

Furthermore,

$$(s-n)+(m+1-n-\ell_1)=m+1-n+\ell_3>m+1-n. \tag{13.1.63}$$

This shows that

$$\begin{aligned}&\mathscr{H}_{\ell,\ell'} \text{ has a unique } gl(n|m)\text{-singular vector if}\\ &\ell>m+1-n \text{ or } \ell'>m+1-n \text{ or } \ell+\ell'\leq m+1-n.\end{aligned} \tag{13.1.64}$$

Suppose $n\leq m$ and $\ell,\ell'\in\overline{0,m+1-n}$ such that $\ell+\ell'>m+1-n$. We take

$$s=n+\ell,\qquad \ell_1=m+1-n-\ell',\qquad \ell_3=\ell+\ell'+n-m-1. \tag{13.1.65}$$

Then

$$\eta^{\ell+\ell'+n-m-1}(x_1^{m+1-n-\ell'}\vec{\vartheta}_{n+\ell})\in\mathscr{H}_{\ell,\ell'}. \tag{13.1.66}$$

Hence

$$\mathscr{H}_{\ell,\ell'} \text{ has exactly two } gl(n|m)\text{-singular vectors if } \ell+\ell' > m+1-n. \quad (13.1.67)$$

In summary, we have:

Lemma 13.1.4 *Let $\ell, \ell' \in \mathbb{N}$. If $\ell > m+1-n$ or $\ell' > m+1-n$ or $\ell+\ell' \leq m+1-n$, the $gl(n|m)$-module $\mathscr{H}_{\ell,\ell'}$ has a unique $gl(n|m)$-singular vector. When $\ell, \ell' \leq m+1-n$ and $\ell+\ell' > m+1-n$, the $gl(n|m)$-module $\mathscr{H}_{\ell,\ell'}$ has exactly two $gl(n|m)$-singular vectors.*

13.1.3 Structure of the Representation for $gl(n|m)$

Fix $\ell, \ell' \in \mathbb{N}$. Let

$$v_{\ell,\ell'} = x_1^{\ell} y_n^{\ell_1} \vec{\vartheta}_s, \qquad \ell_1+m+1-s = \ell', \ \ \ell_1(s-1) = 0, \ \ \delta_{n,1}\ell\ell_1 = 0. \quad (13.1.68)$$

Lemma 13.1.5 *The $gl(n|m)$-module $\mathscr{H}_{\ell,\ell'}$ is generated by $v_{\ell,\ell'}$.*

Proof Let M be the $gl(n|m)$-submodule of $\mathscr{H}_{\ell,\ell'}$ generated by $v_{\ell,\ell'}$. First consider $n > 1$ or $\ell = 0$. For any $s' \in \overline{s+1, m+1}$, we have

$$E_{n+s'-1,n} E_{n+s'-2,n} \cdots E_{n+s,n}(v_{\ell,\ell'}) = x_1^{\ell} y_n^{\ell_1+s'-s} \vec{\vartheta}_{s'} \in M \quad (13.1.69)$$

by (13.1.12). In other words,

$$\begin{aligned} &x_1^{\ell} y_n^{\ell_2} \vec{\vartheta}_{s'} \in M \ \text{ for any } \ell_2 \in \mathbb{N} \text{ and } s' \in \overline{1, m+1} \\ &\text{such that } \ \ell_2+m+1-s' = \ell' \text{ and } \delta_{n,1}\ell\ell_2 = 0. \end{aligned} \quad (13.1.70)$$

If $\ell > 0$, then for any $1 \leq r \leq \min\{\ell, s'-1\}$, we have

$$E_{n+1,1} E_{n+2,1} \cdots E_{n+r,1}(x_1^{\ell} y_n^{\ell_2} \vec{\vartheta}_{s'}) = \left[\prod_{p=0}^{r-1}(\ell-p)\right] x_1^{\ell-r} y_n^{\ell_2} \vec{\theta}_r \vec{\vartheta}_{s'} \in M \quad (13.1.71)$$

by (13.1.12) again. Thus we have showed that

$$x_1^{\ell_3} y_n^{\ell_2} \vec{\theta}_r \vec{\vartheta}_{s'} \in M \ \text{ whenever } \ r+\ell_3 = \ell, \ \ell_2+m+1-s' = \ell', \ \ \delta_{n,1}\ell_2\ell_3 = 0 \quad (13.1.72)$$

for $\ell_2, \ell_3 \in \mathbb{N}$ and $0 \leq r < s' \leq m+1$. Recall the Lie algebra $\mathscr{G}$ defined in (13.1.42). As $\mathscr{G}$-modules,

$$\sum_{r+\ell_3=\ell,\ \ell_2+m+1-s'=\ell'} \check{\mathscr{H}}_{\ell_3,\ell_2} \bar{\mathscr{H}}_{r,m+1-s'} \subset M. \quad (13.1.73)$$

For $i \in \mathbb{N}+1$, we define

$$\mathscr{H}_{\ell,\ell'}^{(i)} = \{f \in \mathscr{H}_{\ell,\ell'} \mid \bar{\Delta}^i(f) = 0.\} \tag{13.1.74}$$

First

$$\mathscr{H}_{\ell,\ell'}^{(1)} = \sum_{r+\ell_3=\ell,\ \ell_2+m+1-s'=\ell'} \check{\mathscr{H}}_{\ell_3,\ell_2}\bar{\mathscr{H}}_{r,m+1-s'} \subset M. \tag{13.1.75}$$

Denote

$$k_{\ell,\ell'} = \min\{\ell, \ell'\}. \tag{13.1.76}$$

We have

$$\mathscr{H}_{\ell,\ell'}^{(k_{\ell,\ell'}+1)} = \mathscr{H}_{\ell,\ell'} \tag{13.1.77}$$

by (13.1.48). Let $\ell_1, \ell_2, \in \mathbb{N}$, $0 \leq r'+1 < s' \leq m+1$ and $\ell_3 \in \overline{1, s'-r'-1}$ such that $\ell_1+\ell_3+r' = \ell$ and $\ell_2+\ell_3+m+1-s' = \ell'$. Then

$$\Im(\ell_1, \ell_2; r', s', \ell_3)(x_1^{\ell_1} y_n^{\ell_2} \vec{\theta}_{r'} \vec{\vartheta}_{s'}) \in \mathscr{H}_{\ell,\ell'}^{(\ell_3+1)} \tag{13.1.78}$$

and

$$\bar{\Delta}^{\ell_3}\Im(\ell_1, \ell_2; r', s', \ell_3)(x_1^{\ell_1} y_n^{\ell_2} \vec{\theta}_{r'} \vec{\vartheta}_{s'}) = c(\ell_1, \ell_2; r', s', \ell_3) x_1^{\ell_1} y_n^{\ell_2} \vec{\theta}_{r'} \vec{\vartheta}_{s'} \tag{13.1.79}$$

with

$$c(\ell_1, \ell_2; r', s', \ell_3) = \ell_3(n+\ell_1+\ell_3+\ell_3-1)(n+\ell_1+\ell_2+\ell_3)(\ell_3+1)! \left[\prod_{p=1}^{\ell_3}(s'-r'-p)\right] \tag{13.1.80}$$

by (13.1.35) and (13.1.41). On the other hand,

$$x_1^{\ell_1+\ell_3} y_n^{\ell_2} \vec{\theta}_{r'} \vec{\vartheta}_{s'-\ell_3} \in M \ \text{ with } \ \delta_{n,1}\ell_2 = 0 \tag{13.1.81}$$

by (13.1.73) and

$$\begin{aligned} M \ni f &= E_{n+s'-1,1}E_{n+s'-2,1}\cdots E_{n+s'-\ell_3,1}(x_1^{\ell_1+\ell_3} y_n^{\ell_2} \vec{\theta}_{r'} \vec{\vartheta}_{s'-\ell_3}) \\ &= (-1)^{r'\ell_3} x_1^{\ell_1+\ell_3} y_1^{\ell_3} y_n^{\ell_2} \vec{\theta}_{r'} \vec{\vartheta}_{s'} + y_n^{\ell_2} \sum_{i=0}^{\ell_3-1} \zeta_i y_1^i \end{aligned} \tag{13.1.82}$$

with $\zeta_0, \ldots, \zeta_{\ell_3-1} \in \mathbb{F}[x_1]\check{\mathscr{A}}$ (cf. (13.1.15)). Moreover,

$$\bar{\Delta}^{\ell_3}[(\ell_3!)^2\binom{\ell_1+\ell_3}{\ell_3}\Im(\ell_1,\ell_2;r',s',\ell_3)(x_1^{\ell_1}y_n^{\ell_2}\vec{\theta}_{r'}\vec{\vartheta}_{s'})$$
$$-(-1)^{r'\ell_3}c(\ell_1,\ell_2;r',s',\ell_3)f]=0. \tag{13.1.83}$$

Hence

$$\Im(\ell_1,\ell_2;r',s',\ell_3)(\bar{\mathscr{H}}_{\ell_1,\ell_2}\check{\mathscr{H}}_{r',m+1-s'})\subset\mathscr{H}_{\ell,\ell'}^{(\ell_3)}+M. \tag{13.1.84}$$

By (13.1.50) and induction on i, we have

$$\mathscr{H}_{\ell,\ell'}^{(i)}\subset M\qquad\text{for any }i\in\mathbb{N}+1. \tag{13.1.85}$$

According to (13.1.77), $\mathscr{H}_{\ell,\ell'}=M$. So $\mathscr{H}_{\ell,\ell'}$ is generated by $v_{\ell,\ell'}$. □

Now we have the following super-symmetric analogue of the classical theorem on harmonic polynomials:

Theorem 13.1.6 *Let $\ell,\ell'\in\mathbb{N}$. The space $\mathscr{H}_{\ell,\ell'}$ is an irreducible $gl(n|m)$-module if and only if $\ell>m+1-n$ or $\ell'>m+1-n$ or $\ell+\ell'\leq m+1-n$. When $|\ell-\ell'|>m+1-n$ or $\ell+\ell'\leq m+1-n$, $\mathscr{A}_{\ell,\ell'}=\bigoplus_{i=0}^{k_{\ell,\ell'}}\eta^i\mathscr{H}_{\ell-i,\ell'-i}$ is a decomposition of irreducible $gl(n|m)$-submodules.*

Proof According to (13.1.67), a necessary condition for $\mathscr{H}_{\ell,\ell'}$ to be an irreducible $gl(n|m)$-module is $\ell>m+1-n$ or $\ell'>m+1-n$ or $\ell+\ell'\leq m+1-n$. To prove the sufficiency, we suppose that $\ell>m+1-n$ or $\ell'>m+1-n$ or $\ell+\ell'\leq m+1-n$. Let V be a nonzero $gl(n|m)$-submodule of $\mathscr{H}_{\ell,\ell'}$. According to Lemma 13.1.4, $\mathscr{H}_{\ell,\ell'}$ has a unique singular vector $v_{\ell,\ell'}$ (cf. (13.1.68)). Since V is finite-dimensional, it contains a singular vector. So $v_{\ell,\ell'}\in V$. Lemma 13.1.5 says $V=\mathscr{H}_{\ell,\ell'}$. Hence $\mathscr{H}_{\ell,\ell'}$ is an irreducible $gl(n|m)$-module.

Let $\ell_1,\ell_2\in\mathbb{N}$ and $0\leq r<s\leq m+1$ such that $n+\ell_1+\ell_2+r-s\geq 0$ and $\delta_{n,1}\ell_1\ell_2=0$. For any $\ell_4\in\mathbb{N}+1$ and $\ell_3\in\overline{0,s-r-1}$,

$$\Delta^{\ell_4+1}(\eta^{\ell_4}\Im(\ell_1,\ell_2;r,s,\ell_3)x_1^{\ell_1}y_n^{\ell_2}\vec{\theta}_r\vec{\vartheta}_s)=0 \tag{13.1.86}$$

and

$$\begin{aligned}&\Delta^{\ell_4}(\eta^{\ell_4}\Im(\ell_1,\ell_2;r,s,\ell_3)x_1^{\ell_1}y_n^{\ell_2}\vec{\theta}_r\vec{\vartheta}_s)\\&=\ell_4!\left[\prod_{i=1}^{\ell_4}(n+i+\ell_1+2\ell_3+r-s)\right]\Im(\ell_1,\ell_2;r,s,\ell_3)(x_1^{\ell_1}y_n^{\ell_2}\vec{\theta}_r\vec{\vartheta}_s)\neq 0\end{aligned} \tag{13.1.87}$$

by (13.1.35). Thus the set

$$\{\eta^{\ell_4}\Im(\ell_1,\ell_2;r,s,\ell_3)x_1^{\ell_1}y_n^{\ell_2}\vec{\theta}_r\vec{\vartheta}_s\mid\ell_4\in\mathbb{N},\ \ell_3\in\overline{0,s-r-1}\} \tag{13.1.88}$$

is linearly independent.

Note that

$$\check{\eta}^{s-r}x_1^{\ell_1}y_n^{\ell_2}\vec{\theta}_r\vec{\vartheta}_s = 0 \tag{13.1.89}$$

by (13.1.29) and (13.1.30). So for any $k \in \mathbb{N}$,

$$\begin{aligned}&\text{Span}\{\eta^{\ell_4}\Im(\ell_1,\ell_2;r,s,\ell_3)x_1^{\ell_1}y_n^{\ell_2}\vec{\theta}_r\vec{\vartheta}_s \mid \ell_4 \in \mathbb{N},\ \ell_3 \in \overline{0,s-r-1};\ \ell_3+\ell_4 = k\}\\ &\subset \text{Span}\{\bar{\eta}^{\ell_5}\check{\eta}^{\ell_6}x_1^{\ell_1}y_n^{\ell_2}\vec{\theta}_r\vec{\vartheta}_s \mid \ell_5 \in \mathbb{N};\ \ell_6 \in \overline{0,s-r-1};\ \ell_5+\ell_6 = k\}.\end{aligned} \tag{13.1.90}$$

But the linear independency of (13.1.88) implies that the above subspaces have the same dimension. Thus

$$\begin{aligned}&\text{Span}\{\eta^{\ell_4}\Im(\ell_1,\ell_2;r,s,\ell_3)x_1^{\ell_1}y_n^{\ell_2}\vec{\theta}_r\vec{\vartheta}_s \mid \ell_4 \in \mathbb{N},\ \ell_3 \in \overline{0,s-r-1}\}\\ &= \text{Span}\{\bar{\eta}^{\ell_5}\check{\eta}^{\ell_6}x_1^{\ell_1}y_n^{\ell_2}\vec{\theta}_r\vec{\vartheta}_s \mid \ell_5 \in \mathbb{N};\ \ell_6 \in \overline{0,s-r-1}\}.\end{aligned} \tag{13.1.91}$$

Therefore, as $\mathscr{G}$-modules,

$$\bigoplus_{\ell_3=0}^{s-r-1}\sum_{\ell_4=0}^{\infty}\eta^{\ell_4}\Im(\ell_1,\ell_2;r,s,\ell_3)\bar{\mathscr{H}}_{\ell_1,\ell_2}\check{\mathscr{H}}_{r,m+1-s} = \bigoplus_{\ell_5=0}^{s-r-1}\sum_{\ell_6=0}^{\infty}\bar{\eta}^{\ell_5}\check{\eta}^{\ell_6}\bar{\mathscr{H}}_{\ell_1,\ell_2}\check{\mathscr{H}}_{r,m+1-s}. \tag{13.1.92}$$

Assume that $|\ell-\ell'| > m+1-n$ or $\ell+\ell' \leq m+1-n$. According to (13.1.44) and (13.1.91), the $\mathscr{G}$-module

$$\begin{aligned}\mathscr{A}_{\ell,\ell'} &= \text{Span}\{\bar{\eta}^{\ell_5}\check{\eta}^{\ell_6}\bar{\mathscr{H}}_{\ell_1,\ell_2}\check{\mathscr{H}}_{r,m+1-s} \mid \ell_1,\ell_2,\ell_6 \in \mathbb{N};\\ &\quad 0 \leq r < s \leq m+1;\ \ell_5 \in \overline{0,s-r-1};\ \delta_{n,1}\ell_1\ell_2 = 0;\\ &\quad \ell_1+\ell_5+\ell_6+r = \ell,\ \ell_2+\ell_5+\ell_6+m+1-s = \ell'\}\\ &= \text{Span}\{\eta^{\ell_4}\Im(\ell_1,\ell_2;r,s,\ell_3)\bar{\mathscr{H}}_{\ell_1,\ell_2}\check{\mathscr{H}}_{r,m+1-s} \mid \ell_1,\ell_2,\ell_6 \in \mathbb{N};\\ &\quad \delta_{n,1}\ell_1\ell_2 = 0;\ 0 \leq r < s \leq m+1;\ \ell_5 \in \overline{0,s-r-1};\\ &\quad \ell_1+\ell_3+\ell_4+r = \ell,\ \ell_2+\ell_3+\ell_4+m+1-s = \ell'\}.\end{aligned} \tag{13.1.93}$$

According to (13.1.50), (13.1.92) and (13.1.93),

$$\mathscr{A}_{\ell,\ell'} = \bigoplus_{i=0}^{k_{\ell,\ell'}}\eta^i\mathscr{H}_{\ell-i,\ell'-i} \tag{13.1.94}$$

(cf. (13.1.76)). Let $i \in \overline{0,k_{\ell,\ell'}}$. If $|\ell-\ell'| > m+1-n$, then

$$\ell-i \geq |\ell-\ell'| > m+1-n \quad\text{or}\quad \ell'-i \geq |\ell-\ell'| > m+1-n. \tag{13.1.95}$$

When $\ell+\ell' \leq m+1-n$,

$$(\ell-i)+(\ell'-i) = \ell+\ell'-2i \leq m+1-n. \tag{13.1.96}$$

Thus all $\mathscr{H}_{\ell-i,\ell'-i}$ are irreducible $gl(n|m)$-submodules. Hence (13.1.93) is a direct sum of irreducible $gl(n|m)$-submodules. □

Denote $\Gamma_0 = \emptyset$ and

$$\Gamma_\ell = \{\vec{j} = (j_1, j_2, .., j_\ell) \mid 1 \leq j_1 < j_2 < \cdots < j_\ell \leq m\} \quad \text{for } \ell \in \overline{1, m}. \tag{13.1.97}$$

Moreover, we set

$$\theta_\emptyset = \vartheta_\emptyset = 1, \quad \theta_{\vec{j}} = \theta_{j_1}\theta_{j_2}\cdots\theta_{j_\ell}, \quad \vartheta_{\vec{j}} = \vartheta_{j_1}\vartheta_{j_2}\cdots\vartheta_{j_\ell}. \tag{13.1.98}$$

Then the set

$$\Big\{\sum_{i=0}^{\infty} \frac{(-1)^i x_1^i y_1^i}{\prod_{r=1}^{i}(\alpha_1 + i)(\beta_1 + i)} \Big(\sum_{s=2}^{n} \partial_{x_2}\partial_{y_2} + \sum_{r=1}^{m} \partial_{\theta_r}\partial_{\vartheta_r}\Big)^i (x^\alpha y^\beta \theta_{\vec{j}} \vartheta_{\vec{k}}) \mid \alpha, \beta \in \mathbb{N}^n;$$
$$\vec{j} \in \Gamma_{\ell_1}; \vec{k} \in \Gamma_{\ell_2}; \alpha_1\beta_1 = 0; \ell_1, \ell_2 \in \overline{0, m}; |\alpha| + \ell_1 = \ell; |\beta| + \ell_2 = \ell'\Big\} \tag{13.1.99}$$

forms a basis of $\mathscr{H}_{\ell,\ell'}$ by Lemma 6.1.1 with $T_1 = \partial_{x_1}\partial_{y_1}$, $T_2 = \Delta - T_1$ and $T_1^- = \int_{(x_1)}\int_{(y_1)}$.

Remark 13.1.7 . If $\ell, \ell' \leq m + 1 - n$ and $\ell + \ell' > m + 1 - n$, then $\mathscr{H}_{\ell,\ell'}$ is an indecomposable $gl(n|m)$-module. In fact, $\mathscr{H}_{\ell,\ell'} \bigcap \eta\mathscr{A}_{\ell-1,\ell'-1} \neq \{0\}$. This also shows that $\mathscr{A}_{\ell,\ell'}$ is not completely reducible when $|\ell-\ell'| \leq m+1-n$ and $\ell+\ell' > m+1-n$. It can be verified that the space $\mathbb{F}\Delta + \mathbb{F}[\Delta, \eta] + \mathbb{F}\eta$ also forms an operator Lie algebra isomorphic to $sl(2, \mathbb{F})$. The above theorem establishes a supersymmetric $(sl(2, \mathbb{F}), gl(n|m))$ Howe duality on the homogeneous subspaces $\mathscr{A}_{\ell,\ell'}$ with $\ell, \ell' \in \mathbb{N}$ such that $|\ell - \ell'| > m + 1 - n$ or $\ell + \ell' \leq m + 1 - n$.

13.2 Noncanonical Oscillator Representations of $gl(n|m)$

In this section, we study $\mathbb{Z}^2$-graded oscillator generalizations of the canonical oscillator representations of $gl(n|m)$) presented in the last section.

Fix the integers $1 < n_1 + 1 < n_2 \leq n$. Recall the symmetry

$$[\partial_{x_r}, x_r] = 1 = [-x_r, \partial_{x_r}], \quad [\partial_{y_s}, y_s] = 1 = [-y_s, \partial_{y_s}]. \tag{13.2.1}$$

Changing operators $\partial_{x_r} \mapsto -x_r$, $x_r \mapsto \partial_{x_r}$ for $r \in \overline{1, n_1}$ and $\partial_{y_s} \mapsto -y_s$, $y_s \mapsto \partial_{y_s}$ for $s \in \overline{n_2 + 1, n}$ in (13.1.11) and (13.1.12), we get the following *noncanonical orthogonal oscillator representation* of $gl(n|m)$ on $\mathscr{A}$ determined by

$$E_{i,j}|_{\mathscr{A}} = E^x_{i,j} - E^y_{j,i}, \qquad E_{n+r,n+s}|_{\mathscr{A}} = \theta_r\partial_{\theta_s} - \vartheta_s\partial_{\vartheta_r} \tag{13.2.2}$$

with

$$E^x_{i,j}|_{\mathscr{A}} = \begin{cases} -x_j\partial_{x_i} - \delta_{i,j} & \text{if } i,\ j \in \overline{1, n_1}; \\ \partial_{x_i}\partial_{x_j} & \text{if } i \in \overline{1, n_1},\ \ j \in \overline{n_1+1, n}; \\ -x_i x_j & \text{if } i \in \overline{n_1+1, n},\ \ j \in \overline{1, n_1}; \\ x_i\partial_{x_j} & \text{if } i,\ j \in \overline{n_1+1, n} \end{cases} \tag{13.2.3}$$

and

$$E^y_{i,j}|_{\mathscr{A}} = \begin{cases} y_i\partial_{y_j} & \text{if } i,\ j \in \overline{1, n_2}; \\ -y_i y_j & \text{if } i \in \overline{1, n_2},\ \ j \in \overline{n_2+1, n}; \\ \partial_{y_i}\partial_{y_j} & \text{if } i \in \overline{n_2+1, n},\ \ j \in \overline{1, n_2}; \\ -y_j\partial_{y_i} - \delta_{i,j} & \text{if } i,\ j \in \overline{n_2+1, n}; \end{cases} \tag{13.2.4}$$

$$E_{i,n+r}|_{\mathscr{A}} = \begin{cases} \partial_{x_i}\partial_{\theta_r} - \vartheta_r\partial_{y_i} & \text{if } i \in \overline{1, n_1}; \\ x_i\partial_{\theta_r} - \vartheta_r\partial_{y_i} & \text{if } i \in \overline{n_1+1, n_2}; \\ x_i\partial_{\theta_r} + y_i\vartheta_r & \text{if } i \in \overline{n_2+1, n}; \end{cases} \tag{13.2.5}$$

$$E_{n+r,i}|_{\mathscr{A}} = \begin{cases} -x_i\theta_r + y_i\partial_{\vartheta_r} & \text{if } i \in \overline{1, n_1}; \\ \theta_r\partial_{x_i} + y_i\partial_{\vartheta_r} & \text{if } i \in \overline{n_1+1, n_2}; \\ \theta_r\partial_{x_i} + \partial_{y_i}\partial_{\vartheta_r} & \text{if } i \in \overline{n_2+1, n} \end{cases} \tag{13.2.6}$$

for $i,\ j \in \overline{1, n}$ and $r, s \in \overline{1, m}$.

The Laplace operator in (13.1.18) changes to

$$\bar{\Delta}_{n_1,n_2} = -\sum_{i=1}^{n_1} x_i\partial_{y_i} + \sum_{r=n_1+1}^{n_2} \partial_{x_r}\partial_{y_r} - \sum_{s=n_2+1}^{n} y_s\partial_{x_s} \tag{13.2.7}$$

and its dual changes to

$$\bar{\eta} = \sum_{i=1}^{n_1} y_i\partial_{x_i} + \sum_{r=n_1+1}^{n_2} x_r y_r + \sum_{s=n_2+1}^{n} x_s\partial_{y_s}. \tag{13.2.8}$$

We take (13.1.29) and then supersymmetric Laplace operator

$$\Delta_{n_1,n_2} = \bar{\Delta}_{n_1,n_2} + \check{\Delta} = -\sum_{i=1}^{n_1} x_i\partial_{y_i} + \sum_{r=n_1+1}^{n_2} \partial_{x_r}\partial_{y_r} - \sum_{s=n_2+1}^{n} y_s\partial_{x_s} + \sum_{r=1}^{m} \partial_{\theta_r}\partial_{\vartheta_r} \tag{13.2.9}$$

and its dual

$$\eta = \bar{\eta} + \check{\eta} = \sum_{i=1}^{n_1} y_i\partial_{x_i} + \sum_{r=n_1+1}^{n_2} x_r y_r + \sum_{s=n_2+1}^{n} x_s\partial_{y_s} + \sum_{r=1}^{m} \theta_r\vartheta_r. \tag{13.2.10}$$

Then with respect to the representation in (13.2.2)–(13.2.6), we have

$$E_{i,j}\Delta_{n_1,n_2}=\Delta_{n_1,n_2}E_{i,j},\qquad E_{i,j}\eta=\eta E_{i,j}\qquad \text{for } i,j\in\overline{1,m+n}. \tag{13.2.11}$$

Moreover, we take the settings in (13.1.13), (13.1.15)–(13.1.17) and (13.1.19).

Denote

$$\bar{\mathscr{A}}_{\langle \ell_1,\ell_2\rangle}=\text{Span}\{x^\alpha y^\beta \mid \alpha,\beta\in\mathbb{N}^n;\ \sum_{r=n_1+1}^{n}\alpha_r-\sum_{i=1}^{n_1}\alpha_i=\ell_1;\ \sum_{i=1}^{n_2}\beta_i-\sum_{r=n_2+1}^{n}\beta_r=\ell_2\} \tag{13.2.12}$$

for $\ell_1,\ell_2\in\mathbb{Z}$. Then $\bar{\mathscr{A}}_{\langle \ell_1,\ell_2\rangle}$ forms a $\bar{\mathscr{G}}$-submodule. Moreover, for $\ell,\ell'\in\mathbb{Z}$, we let

$$\mathscr{A}_{\langle \ell,\ell'\rangle}=\sum_{\ell_1,\ell_2\in\mathbb{Z},\ \ell_3,\ell_4\in\overline{0,m};\ \ell_1+\ell_3=\ell,\ \ell_2+\ell_4=\ell'}\bar{\mathscr{A}}_{\langle \ell_1,\ell_2\rangle}\check{\mathscr{A}}_{\ell_3,\ell_4}. \tag{13.2.13}$$

It can be verified that $\mathscr{A}_{\langle \ell,\ell'\rangle}$ forms a $gl(n|m)$-submodule. Define

$$\mathscr{H}=\{f\in\mathscr{A}\mid \Delta_{n_1,n_2}(f)=0\},\quad \mathscr{H}_{\langle \ell_1,\ell_2\rangle}=\mathscr{H}\bigcap\mathscr{A}_{\langle \ell_1,\ell_2\rangle}. \tag{13.2.14}$$

By (13.2.11), $\mathscr{H}_{\langle \ell,\ell'\rangle}$ forms a $gl(n|m)$-submodule of $\mathscr{A}_{\langle \ell,\ell'\rangle}$. According to (6.6.16) and (6.6.17), we have:

Lemma 13.2.1 *The nonzero vectors in*

$$\{\mathbb{F}[\bar{\eta}](x_i^{m_1}y_j^{m_2})\mid m_1,m_2\in\mathbb{N};\ i=n_1,n_1+1;\ j=n_2,n_2+1-\delta_{n_2,n}\} \tag{13.2.15}$$

are all the singular vectors of $\bar{\mathscr{G}}$ *(cf. (13.1.15)–(13.1.17)) in* $\bar{\mathscr{A}}$ *(cf. (13.1.13)).*

Recall $\mathscr{G}=\bar{\mathscr{G}}+\check{\mathscr{G}}$ (cf. (13.1.15)). Take the Cartan subalgebra $H=\bar{H}+\check{H}$ of $\mathscr{G}$ (cf. (13.1.16)) and the subspace $\mathscr{G}_+=\bar{\mathscr{G}}_++\check{\mathscr{G}}_+$ (cf. (13.1.17)) spanned by positive root vectors in $\mathscr{G}$. Then the nonzero vectors in

$$\{\mathbb{F}[\bar{\eta},\check{\eta}](x_i^{m_1}y_j^{m_2}\vec{\theta}_r\vec{\vartheta}_s)\mid m_1,m_2\in\mathbb{N};\ i=n_1,n_1+1;\ j=n_2,n_2+1-\delta_{n_2,n};\ 0\le r<s\le m+1\} \tag{13.2.16}$$

are all the $\mathscr{G}$-singular vectors in $\mathscr{A}$. Choose H as a Cartan subalgebra of the Lie superalgebra $gl(n|m)$ and $gl(n|m)_+=\mathscr{G}_++\sum_{r=1}^n\sum_{s=1}^m\mathbb{F}E_{r,n+s}$ as the subalgebra generated by positive root vectors.

Fix $\ell,\ell'\in\mathbb{Z}$. Then a $gl(n|m)$-singular vector in $\mathscr{A}_{\ell,\ell'}$ must be of the form

$$f=\sum_{p=0}^{\min\{s-r-1,\ell_1\}}b_p\bar{\eta}^{\ell_1-p}\check{\eta}^p(x_i^{m_1}y_j^{m_2}\vec{\theta}_r\vec{\vartheta}_s), \tag{13.2.17}$$

where $\ell_1,m_1,m_2\in\mathbb{N},\ \ 0\le r<s\le m+1,\ b_p\in\mathbb{F}$ and

$$(i,j)\in\{(n_1,n_2),(n_1,n_2+1-\delta_{n_2,n}),(n_1+1,n_2),(n_1+1,n_2+1-\delta_{n_2,n})\}. \quad (13.2.18)$$

Suppose that $\ell_1=0$ or $s-r-1=0$. Then we can assume $f=x_i^{m_1}y_j^{m_2}\vec{\theta}_r\vec{\vartheta}_s$. If $r\neq 0$, then (13.2.5) implies

$$E_{n_1+1,n+r}(f)=(-1)^{r-1}x_{n_1+1}x_i^{m_1}y_j^{m_2}\vec{\theta}_{r-1}\vec{\vartheta}_s\neq 0, \quad (13.2.19)$$

which is absurd. So $r=0$. Assume that $m_2>0,\ s>1$ and $j=n_2$. According to (13.2.5),

$$E_{n_2,n+1}(f)=-m_2x_i^{m_1}y_j^{m_2-1}\vartheta_1\vec{\vartheta}_s\neq 0, \quad (13.2.20)$$

which leads a contradiction. Hence $m_2(s-1)=0$ if $j=n_2$. If $n_2<n$, then we have

$$E_{n_2+1,n+1}(f)=x_i^{m_1}y_j^{m_2}y_{n_2+1}\vartheta_1\vec{\vartheta}_s\neq 0 \quad (13.2.21)$$

by (13.2.5), which is absurd. Thus $s=1$. In summary,

$$f=x_i^{m_1}y_j^{m_2}\vec{\vartheta}_s \qquad \text{with } s=1 \text{ or } n_2=n \text{ and } m_2=0. \quad (13.2.22)$$

Consider the case $\ell_1>0$ and $s-r-1>0$. By (13.2.5) and the fact $\check{\eta}^{s-r-1}\vartheta_{r+1}\vec{\theta}_r\vec{\vartheta}_s=0$, we have

$$\begin{aligned}0&=E_{n_1+1,n+r+1}(f)=(x_{n_1+1}\partial_{\theta_{r+1}}-\vartheta_{r+1}\partial_{y_{n_1+1}})(f)\\&=\left[\sum_{p=1}^{\min\{s-r-1,\ell_1\}}pb_p\bar{\eta}^{\ell_1-p}\check{\eta}^{p-1}-\sum_{p=0}^{\min\{s-r-1,\ell_1\}-1}(\ell_1-p)b_p\bar{\eta}^{\ell_1-p-1}\check{\eta}^p(x_i^{m_1}y_j^{m_2}\vec{\theta}_r\vec{\vartheta}_s)\right]\\&\quad\times x_{n_1+1}\vartheta_{r+1}x_i^{m_1}y_j^{m_2}\vec{\theta}_r\vec{\vartheta}_s,\end{aligned} \quad (13.2.23)$$

which implies

$$f=b_0(\bar{\eta}+\check{\eta})^{\ell_1}(x_i^{m_1}y_j^{m_2}\vec{\theta}_r\vec{\vartheta}_s)=b_0\eta^{\ell_1}(x_i^{m_1}y_j^{m_2}\vec{\theta}_r\vec{\vartheta}_s). \quad (13.2.24)$$

The arguments in the previous paragraph and (13.2.11) give:

Lemma 13.2.2 *Any $gl(n|m)$-singular vector in $\mathscr{A}_{\langle\ell,\ell'\rangle}$ must be of the form:*

$$\eta^{\ell_1}(x_i^{m_1}y_j^{m_2}\vec{\vartheta}_s)\ \ \textit{with}\ \ell_1,m_1,m_2\in\mathbb{N},\ s\in\overline{1,m+1}\ \textit{and}\ (13.2.18) \quad (13.2.25)$$

such that $s=1$ or $n_2=n$ and $m_2=0$.

We define

$$\tilde{D}=\sum_{r=n_1+1}^{n}x_r\partial_{x_r}-\sum_{i=1}^{n_1}x_i\partial_{x_i},\quad \tilde{D}'=\sum_{i=1}^{n_2}y_i\partial_{y_i}-\sum_{r=n_2+1}^{n}y_r\partial_{y_r}. \quad (13.2.26)$$

Then

$$\bar{\mathscr{A}}_{\langle \ell,\ell' \rangle} = \{f \in \bar{\mathscr{A}} \mid \tilde{D}(f) = \ell f;\ \tilde{D}'(f) = \ell' f\}. \tag{13.2.27}$$

We calculate

$$\bar{\Delta}_{n_1,n_2}\bar{\eta} = \bar{\eta}\bar{\Delta}_{n_1,n_2} + n_2 - n_1 + \tilde{D} + \tilde{D}'. \tag{13.2.28}$$

For $\ell_1 \in \mathbb{N}+1$ and $f \in \mathscr{H}_{\langle \ell,\ell' \rangle}$ (cf. (13.2.19)), (13.1.35) and (13.2.28) imply

$$\Delta_{n_1,n_2}\eta^{\ell_1}(f) = \ell_1(n_2 - n_1 - m + \ell + \ell' + \ell_1 - 1)\eta^{\ell_1 - 1}(f). \tag{13.2.29}$$

Thus

$$\Delta_{n_1,n_2}\eta^{\ell_1}(f) = 0 \Longleftrightarrow \ell + \ell' \leq n_1 + m - n_2 \ \text{ and } \ \ell_1 = n_1 + m - n_2 - \ell - \ell' + 1. \tag{13.2.30}$$

If the condition holds, then

$$\eta^{\ell_1}(f) \in \mathscr{H}_{\langle n_1+m-n_2-\ell'+1,\, n_1+m-n_2-\ell+1 \rangle}. \tag{13.2.31}$$

Moreover,

$$(n_1 + m - n_2 - \ell' + 1) + (n_1 + m - n_2 - \ell + 1) \geq n_1 + m - n_2 + 2. \tag{13.2.32}$$

Observe that

$$x_{n_1}^{m_1} y_{n_2}^{m_2} \vec{\vartheta}_s \in \mathscr{H}_{\langle -m_1, m+1+m_2-s \rangle}, \qquad x_{n_1}^{m_1} y_{n_2+1}^{m_2} \vec{\vartheta}_1 \in \mathscr{H}_{\langle -m_1, m-m_2 \rangle}, \tag{13.2.33}$$

$$x_{n_1+1}^{m_1} y_{n_2}^{m_2} \vec{\vartheta}_s \in \mathscr{H}_{\langle m_1, m+1+m_2-s \rangle}, \qquad x_{n_1+1}^{m_1} y_{n_2+1}^{m_2} \vec{\vartheta}_1 \in \mathscr{H}_{\langle m_1, m-m_2 \rangle}. \tag{13.2.34}$$

By Lemma 13.2.2,

$$\text{any nonzero } \mathscr{H}_{\langle \ell,\ell' \rangle} \text{ contains a singular vector of the form } x_i^{m_1} y_j^{m_2} \vec{\vartheta}_s, \tag{13.2.35}$$

where $s = 1$ or $n_2 = n$ and $m_2 = 0$.

Now we consider $f = x_i^{m_1} y_j^{m_2} \vec{\vartheta}_s$ with $m_1, m_2 \in \mathbb{N}$, $s \in \overline{1, m+1}$ and (13.2.18) such that $s = 1$ or $n_2 = n$ and $m_2 = 0$. Assume $\Delta_{n_1,n_2}\eta^{\ell_1}(f) = 0$ for some $\ell_1 \in \mathbb{N}+1$.

Case 1. $(i, j) = (n_1, n_2)$.

In this subcase, $\ell = -m_1$ and $\ell' = m_2 + m + 1 - s$ by (13.2.33). Thus $m_2 - m_1 + 1 - s \leq n_1 - n_2$ and $\ell_1 = n_1 + m_1 + s - n_2 - m_2$ by (13.2.30). So

$$\eta^{\ell_1}(f) \in \mathscr{H}_{\langle n_1+s-n_2-m_2,\, n_1+m_1-n_2+m+1 \rangle}. \tag{13.2.36}$$

Case 2. $(i, j) = (n_1, n_2 + 1)$.

In this subcase $s = 1$, $\ell = -m_1$ and $\ell' = m - m_2$ by (13.2.33). Thus $m_1 + m_2 \geq n_2 - n_1$ and $\ell_1 = n_1 + m_1 + m_2 - n_2 + 1$ by (13.2.30). Hence

$$\eta^{\ell_1}(f) \in \mathscr{H}_{\langle n_1+m_2-n_2+1, n_1+m_1-n_2+m+1\rangle}. \tag{13.2.37}$$

Case 3. $(i, j) = (n_1 + 1, n_2)$.

In this subcase, $\ell = m_1$ and $\ell' = m_2 + m + 1 - s$ by (13.2.34). Thus $m_2 + m_1 + 1 - s \leq n_1 - n_2$ and $\ell_1 = n_1 - m_1 + s - n_2 - m_2$ by (13.2.30). So

$$\eta^{\ell_1}(f) \in \mathscr{H}_{\langle n_1+s-n_2-m_2, n_1-m_1-n_2+m+1\rangle}. \tag{13.2.38}$$

Case 4. $(i, j) = (n_1 + 1, n_2 + 1)$.

In this subcase $s = 1$, $\ell = m_1$ and $\ell' = m - m_2$ by (13.2.34). Thus $m_1 - m_2 \leq n_1 - n_2$ and $\ell_1 = n_1 - m_1 + m_2 - n_2 + 1$ by (13.2.30). Hence

$$\eta^{\ell_1}(f) \in \mathscr{H}_{\langle n_1+m_2-n_2+1, n_1-m_1-n_2+m+1\rangle}. \tag{13.2.39}$$

Thus we obtain:

Lemma 13.2.3 *A nonzero $gl(n|m)$-module $\mathscr{H}_{\langle \ell,\ell'\rangle}$ has a unique singular vector if and only if $\ell + \ell' \leq n_1 + m + 1 - n_2$ or $\ell \notin \overline{n_1 + 1 - n, n_1 + m + 1 - n}$ and $n_2 = n$. If the condition holds, the unique singular vector is of the form $x_i^{m_1} y_j^{m_2} \vec{\vartheta}_s$ with (13.2.11), where $s = 1$ or $n_2 = n$ and $m_2 = 0$.*

Fix $\mathscr{H}_{\langle \ell,\ell'\rangle} \neq \{0\}$. Assume

$$v_{\ell,\ell'} = x_i^{m_1} y_j^{m_2} \vec{\vartheta}_s \in \mathscr{H}_{\langle \ell,\ell'\rangle} \tag{13.2.40}$$

for some (i, j) in (13.2.18), $m_1, m_2 \in \mathbb{N}$ and $s \in \overline{1, m+1}$ such that $s = 1$ or $n_2 = n$ and $m_2 = 0$.

Lemma 13.2.4 *As a $gl(n|m)$-module, $\mathscr{H}_{\langle \ell,\ell'\rangle}$ is generated by $v_{\ell,\ell'}$.*

Proof Denote

$$\tilde{\Gamma} = \{\tilde{\alpha} = (\alpha_{n_1+1}, \ldots, \alpha_{n_2}) \in \mathbb{N}^{n_2-n_1}\}, \quad |\tilde{\alpha}| = \sum_{i=1}^{n_2-n_1} \alpha_{n_1+i}. \tag{13.2.41}$$

Set

$$\tilde{\mathscr{A}} = \check{\mathscr{A}}[x_{n_1+1}, \ldots, x_{n_2}, y_{n_1+1}, \ldots, y_{n_2}], \tag{13.2.42}$$

$$\tilde{\mathscr{H}}_{\langle \ell_1,\ell_2\rangle} = \tilde{\mathscr{A}} \bigcap \mathscr{H}_{\langle \ell_1,\ell_2\rangle}, \tag{13.2.43}$$

$$\tilde{\Delta}_{n_1,n_2} = -\sum_{i=1}^{n_1} x_i \partial_{y_i} + \sum_{r=n_1+2}^{n_2} \partial_{x_r} \partial_{y_r} - \sum_{s=n_2+1}^{n} y_s \partial_{x_s} + \check{\Delta}. \tag{13.2.44}$$

Then $\Delta_{n_1,n_2} = \partial_{x_{n_1+1}}\partial_{y_{n_1+1}} + \tilde{\Delta}_{n_1,n_2}$. For any $k_1, k_2 \in \mathbb{N}$, we define the operator

$$T_{k_1,k_2} = \sum_{i=0}^{\infty} \frac{(-1)^i x_{n_1+1}^{k_1+i} y_{n_1+1}^{k_2+i}}{\prod_{r=1}^{i}(k_1+r)(k_2+r)} \tilde{\Delta}_{n_1,n_2}^i. \tag{13.2.45}$$

Moreover, Lemma 6.1.1 yields

$$\begin{aligned}\mathscr{H}_{\langle \ell,\ell'\rangle} = \operatorname{Span}\{ & T_{\alpha_{n_1+1},\beta_{n_1+1}}\Big(\Big[\prod_{i\neq n_1+1} x_i^{\alpha_i} y_i^{\beta_i}\Big]\Big)\theta_{\vec{j}}\vartheta_{\vec{k}}) \mid \alpha,\beta\in\mathbb{N}^n;\\ & \vec{j}\in\Gamma_{\ell_1};\vec{k}\in\Gamma_{\ell_2};\ \ell_1,\ell_2\in\overline{0,m};\ \alpha_{n_1+1}\beta_{n_1+1}=0;\\ & \sum_{s=n_1+1}^{n}\alpha_s - \sum_{r=1}^{n_1}\alpha_r + \ell_1 = \ell;\ \sum_{r=1}^{n_2}\beta_r - \sum_{s=n_2+1}^{n}\beta_s + \ell_2 = \ell'\},\end{aligned} \tag{13.2.46}$$

$$\begin{aligned}\tilde{\mathscr{H}}_{\langle \ell,\ell'\rangle} = \operatorname{Span}\{ & T_{\alpha_{n_1+1},\beta_{n_1+1}}\Big(\Big[\prod_{i=n_1+2}^{n_2} x_i^{\alpha_i} y_i^{\beta_i}\Big]\Big)\theta_{\vec{j}}\vartheta_{\vec{k}}) \mid \tilde{\alpha},\tilde{\beta}\in\tilde{\Gamma};\ \vec{j}\in\Gamma_{\ell_1};\ \vec{k}\in\Gamma_{\ell_2};\\ & \alpha_{n_1+1}\beta_{n_1+1}=0;\ \ell_1,\ell_2\in\overline{0,m};\ |\tilde{\alpha}|+\ell_1=\ell;\ |\tilde{\beta}|+\ell_2=\ell'\}.\end{aligned} \tag{13.2.47}$$

Write

$$\mathscr{G}_1 = \sum_{i,j=1}^{n_1}\mathbb{F}E_{i,j},\qquad \mathscr{G}_2 = \sum_{r,s=n_2+1}^{n}\mathbb{F}E_{r,s}, \tag{13.2.48}$$

$$\mathscr{G}_3 = \sum_{n_1+1\neq i,j\in\overline{1,n_2}}\mathbb{F}E_{i,j},\qquad \mathscr{G}_4 = \sum_{r,s=n_1+2}^{n}\mathbb{F}E_{r,s}. \tag{13.2.49}$$

Then

$$\xi\tilde{\Delta}_{n_1,n_2} = \tilde{\Delta}_{n_1,n_2}\xi \qquad \text{for } \xi\in\mathscr{G}_i,\ i\in\overline{1,4}. \tag{13.2.50}$$

Denote by V the $gl(n|m)$-submodule of $\mathscr{H}_{\langle \ell,\ell'\rangle}$ generated by $v_{\ell,\ell'}$. By (13.2.2)–(13.2.4)

$$-E_{n_1+1,n_1}|_{\mathscr{A}} = x_{n_1}x_{n_1+1} + y_{n_1}\partial_{y_{n_1+1}},\quad E_{n_2+1,n_2}|_{\mathscr{A}} = x_{n_2+1}\partial_{x_{n_2}} + y_{n_2}y_{n_2+1}. \tag{13.2.51}$$

According to (13.2.5) and (13.2.6),

$$E_{n_2,n+r} = -x_{n_2}\partial_{\theta_r} + \vartheta_r\partial_{y_{n_2}},\quad E_{n+r,n_2} = \theta_r\partial_{x_{n_2}} + y_{n_2}\partial_{\vartheta_r}\qquad \text{for } r\in\overline{1,m}. \tag{13.2.52}$$

Repeatedly applying the operators in (13.2.51) and (13.2.52) to $v_{\ell,\ell'}$, we obtain

$$x_{n_1}^{p_1}x_{n_1+1}^{p_2}y_{n_2}^{p_3}y_{n_2+1}^{p_4}\vec{\vartheta}_{s'}\in V \tag{13.2.53}$$

for $p_i\in\mathbb{N}$ and $s'\in\overline{1,m+1}$ such that

$$p_2-p_1=\ell,\;\; p_3-p_4+m+1-s'=\ell',\;\; p_3(s'-1)=0. \tag{13.2.54}$$

Lemma 13.1.5, (13.1.69), and (13.2.43) tell us that

$$x_{n_1}^{p_1}y_{n_2+1}^{p_4}\tilde{\mathscr{H}}_{p_2,p_3+m+1-s'}\subset V. \tag{13.2.55}$$

Recall that $U(\mathscr{G}_i)$ stands for the universal enveloping of the Lie algebra $\mathscr{G}_i$. Applying $U(\mathscr{G}_1)$ and $U(\mathscr{G}_2)$ to (13.2.55), we get

$$\left[\prod_{\iota_1=1}^{n_1}x_{\iota_1}^{\alpha_{\iota_1}}\right]\left[\prod_{\iota_2=n_2+1}^{n}y_{\iota_2}^{\beta_{\iota_2}}\right]\tilde{\mathscr{H}}_{\langle p_2,p_3+m+1-s'\rangle}\subset V \tag{13.2.56}$$

for any $(\alpha_1,\ldots,\alpha_{n_1})\in\mathbb{N}^{n_1}$ and $(\beta_{n_2+1},\ldots,\beta_n)\in\mathbb{N}^{n-n_2}$ such that $\sum_{\iota_1=1}^{n_1}\alpha_{\iota_1}=p_1$ and $\sum_{\iota_2=n_2+1}^{n}\beta_{\iota_2}=p_4$. Applying $U(\mathscr{G}_3)$ to (13.2.56), we obtain

$$T_{\alpha_{n_1+1},\beta_{n_1+1}}\left(\left[\prod_{i\neq n_1+1}x_i^{\alpha_i}y_i^{\beta_i}\right]\right)\theta_{\vec{j}}\vartheta_{\vec{k}})\in V \tag{13.2.57}$$

by (13.2.47) and (13.2.50), where $\alpha,\beta,\vec{j},\vec{k}$ are as those in (13.2.46) and $\alpha_{n_2+1}=\cdots=\alpha_n=0$. Finally, we get (13.2.57) with any $\alpha,\beta,\vec{j},\vec{k}$ in (13.2.46) by (13.2.50) and applying $U(\mathscr{G}_4)$. According to (13.2.46), $V=\mathscr{H}_{\langle\ell,\ell'\rangle}$. □

Now we have the following noncanonical super-symmetric analogue of the classical theorem on harmonic polynomials:

Theorem 13.2.5 *Let $\ell,\ell'\in\mathbb{Z}$ such that $\ell'\geq 0$ if $n_2=n$. The $gl(n|m)$-module $\mathscr{H}_{\langle\ell,\ell'\rangle}$ is irreducible if and only if $\ell+\ell'\leq n_1+m+1-n_2$ or $\ell\notin\overline{n_1+1-n,n_1+m+1-n}$ and $n_2=n$. When $\ell+\ell'\leq n_1+m+1-n_2$, $\mathscr{A}_{\langle\ell,\ell'\rangle}=\bigoplus_{i=0}^{\infty}\eta^i(\mathscr{H}_{\langle\ell-i,\ell'-i\rangle})$ is a decomposition of irreducible $gl(n|m)$-submodules if $n_2<n$, and $\mathscr{A}_{\langle\ell,\ell'\rangle}=\bigoplus_{i=0}^{\ell'}\eta^i(\mathscr{H}_{\langle\ell-i,\ell'-i\rangle})$ is a decomposition of irreducible $gl(n|m)$-submodules if $n_2=n$.*

Proof Suppose $\ell+\ell'\leq n_1+m+1-n_2$ or $\ell\notin\overline{n_1+1-n,n_1+m+1-n}$ and $n_2=n$. Let V be a nonzero submodule of $\mathscr{H}_{\langle\ell,\ell'\rangle}$. According to Lemma 13.2.3, the vector $v_{\ell,\ell'}$ in (13.2.40) is the unique singular vector of $\mathscr{H}_{\langle\ell,\ell'\rangle}$. Since $gl(n|m)_+$ in (13.1.52) is locally nilpotent by (13.2.2)–(13.2.6), V contains a singular vector. So $v_{\ell,\ell'}\in V$. By Lemma 13.2.4, $V=\mathscr{H}_{\langle\ell,\ell'\rangle}$; that is, $\mathscr{H}_{\langle\ell,\ell'\rangle}$ is irreducible. The necessity also follows from Lemma 12.2.3.

Assume $\ell + \ell' \leq n_1 + m + 1 - n_2$. Since Δ_{n_1,n_2} is locally nilpotent by (13.2.9) and (13.2.10), for any $0 \neq u \in \mathscr{A}_{\langle \ell,\ell' \rangle}$, there exists an element $\kappa(u) \in \mathbb{N}$ such that

$$\Delta_{n_1,n_2}^{\kappa(u)}(u) \neq 0 \text{ and } \Delta_{n_1,n_2}^{\kappa(u)+1}(u) = 0. \tag{13.2.58}$$

Set

$$\Psi = \begin{cases} \sum_{i=0}^{\infty} \eta^i(\mathscr{H}_{\langle \ell-i,\ell'-i \rangle}) & \text{if } n_2 < n, \\ \sum_{i=0}^{\ell'} \eta^i(\mathscr{H}_{\ell-i,\ell'-i}) & \text{if } n_2 = n. \end{cases} \tag{13.2.59}$$

Given $0 \neq u \in \mathscr{A}_{\langle \ell,\ell' \rangle}$, $\kappa(u) = 1$ implies $u \in \mathscr{H}_{\langle \ell,\ell' \rangle} \subset \Psi$. Suppose that $u \in \Psi$ whenever $\kappa(u) < r$ for some positive integer r. Assume $\kappa(u) = r$. First

$$v = \Delta_{n_1,n_2}^r(u) \in \mathscr{H}_{\langle \ell-r,\ell'-r \rangle} \subset \Psi. \tag{13.2.60}$$

Note

$$\Delta_{n_1,n_2}^r[\eta^r(v)] = r!\left[\prod_{i=1}^r (n_2 - n_1 - m + \ell + \ell' - r - i)\right] v \tag{13.2.61}$$

by (13.2.29). Thus we have either

$$u = \frac{1}{r![\prod_{i=1}^r (n_2 - n_1 - m + \ell + \ell' - r - i)]}\eta^r(v) \in \Psi \tag{13.2.62}$$

or

$$\kappa\left(u - \frac{1}{r![\prod_{i=1}^r (n_2 - n_1 - m + \ell + \ell' - r - i)]}\eta^r(v)\right) < r. \tag{13.2.63}$$

By induction,

$$u - \frac{1}{r![\prod_{i=1}^r (n_2 - n_1 - m + \ell + \ell' - r - i)]}\eta^r(v) \in \Psi, \tag{13.2.64}$$

which implies $u \in \Psi$. Therefore, we have $\Psi = \mathscr{A}_{\langle \ell,\ell' \rangle}$. Since all $\eta^i(\mathscr{H}_{\langle \ell-i,\ell'-i \rangle})$ have distinct highest weights, the sums in (13.2.59) are direct sums. $\square$

Remark 13.2.6 If $\ell + \ell' > n_1 + m + 1 - n_2$ and $\ell \in \overline{n_1 + 1 - n, n_1 + m + 1 - n}$ when $n_2 = n$, the $gl(n|m)$-module $\mathscr{H}_{\langle \ell,\ell' \rangle}$ is indecomposable. When $n_2 < n$ and $\ell + \ell' > n_1 + m + 1 - n_2$, $\mathscr{A}_{\langle \ell,\ell' \rangle}$ is not completely reducible. With the settings (13.2.9) and (13.2.10), the space $\mathbb{F}\Delta_{n_1,n_2} + \mathbb{F}[\Delta_{n_1,n_2}, \eta] + \mathbb{F}\eta$ again forms an operator Lie algebra isomorphic to $sl(2,\mathbb{F})$. The above theorem establishes a supersymmetric $(sl(2,\mathbb{F}), gl(n|m))$ Howe duality on the homogeneous subspaces $\mathscr{A}_{\langle \ell,\ell' \rangle}$ with $\ell, \ell' \in \mathbb{Z}$ such that $\ell + \ell' \leq n_1 + m + 1 - n_2$ and $\ell' \geq 0$ if $n_2 = n$.

Recall the notations in (13.1.97) and (13.1.98). The set

$$\Big\{\sum_{i=0}^{\infty}\frac{(-1)^i x_{n_1+1}^i y_{n_1+1}^i}{\prod_{r=1}^{i}(\alpha_{n_1+1}+r)(\beta_{n_1+1}+r)}(\Delta_{n_1,n_2}-\partial_{x_{n_1+1}}\partial_{y_{n_1+1}})^i(x^\alpha y^\beta \theta_{\vec{j}}\vartheta_{\vec{k}})$$
$$\mid \alpha,\beta\in\mathbb{N}^n;\ \vec{j}\in\Gamma_{\ell_1};\ \vec{k}\in\Gamma_{\ell_2};\ \alpha_{n_1+1}\beta_{n_1+1}=0;\ \ell_1,\ell_2\in\overline{0,m};$$
$$\sum_{r=n_1+1}^{n}\alpha_r-\sum_{i=1}^{n_1}\alpha_i+\ell_1=\ell;\ \sum_{i=1}^{n_2}\beta_i-\sum_{r=n_2+1}^{n}\beta_r+\ell_2=\ell'\Big\} \tag{13.2.65}$$

forms a basis of $\mathscr{H}_{\langle\ell,\ell'\rangle}$ by Lemma 6.1.1.

13.3 Oscillator Representations of $osp(2n|2m)$

In this section, we use the results in last two sections to study the corresponding the even ortho-symplectic Lie superalgebras.

Set

$$\mathscr{K}_0=\sum_{i,j=1}^{n}\mathbb{F}(E_{i,j}-E_{n+j,n+i})+\sum_{r,s=1}^{m}\mathbb{F}(E_{2n+r,2n+s}-E_{2n+m+s,2n+m+r}), \tag{13.3.1}$$

$$\mathscr{K}_1=\sum_{i=1}^{n}\sum_{r=1}^{m}[\mathbb{F}(E_{i,2n+r}-E_{2n+m+r,n+i})+\mathbb{F}(E_{2n+r,i}+E_{n+i,2n+m+r})]. \tag{13.3.2}$$

Then $\mathscr{K}=\mathscr{K}_0+\mathscr{K}_1$ forms a Lie sub-superalgebra of $gl(2n|2m)$ isomorphic to $gl(n|m)$. Let

$$\begin{aligned}osp(2n|2m)_0=\mathscr{K}_0&+\sum_{1\le i<j\le n}[\mathbb{F}(E_{i,n+j}-E_{j,n+i})+\mathbb{F}(E_{n+i,j}-E_{n+j,i})]\\&+\sum_{1\le r\le s\le m}[\mathbb{F}(E_{2n+r,2n+m+s}+E_{2n+s,2n+m+r})\\&+\mathbb{F}(E_{2n+m+r,2n+s}+E_{2n+m+s,2n+r})],\end{aligned} \tag{13.3.3}$$

$$osp(2n|2m)_1=\mathscr{K}_1+\sum_{i=1}^{n}\sum_{r=1}^{m}[\mathbb{F}(E_{i,2n+m+r}+E_{2n+r,n+i})+\mathbb{F}(E_{n+i,2n+r}-E_{2n+m+r,i})]. \tag{13.3.4}$$

The space $osp(2n|2m)=osp(2n|2m)_0+osp(2n|2m)_1$ forms a simple Lie sub-superalgebra of $gl(2n|2m)$, which is called an *even ortho-symplectic Lie superalgebra*. Moreover, its Lie subalgebra

$$osp(2n|2m)_0\cong o(2n,\mathbb{F})\oplus sp(2m,\mathbb{F}). \tag{13.3.5}$$

Take settings in (13.1.4)–(13.1.10). Define the *basic oscillator representation* of $osp(2n|2m)$ on $\mathscr{A}$ determined by

$$(E_{i,j} - E_{n+j,n+i})|_{\mathscr{A}} = x_i\partial_{x_j} - y_j\partial_{y_i}, \quad (E_{i,2n+r} - E_{2n+m+r,n+i})|_{\mathscr{A}} = x_i\partial_{\theta_r} - \vartheta_r\partial_{y_i}, \tag{13.3.6}$$

$$(E_{i,n+j} - E_{j,n+i})|_{\mathscr{A}} = x_i\partial_{y_j} - x_j\partial_{y_i}, \quad (E_{n+i,j} - E_{n+j,i})|_{\mathscr{A}} = y_i\partial_{x_j} - y_j\partial_{x_i}, \tag{13.3.7}$$

$$(E_{2n+r,i} + E_{n+i,2n+m+r})|_{\mathscr{A}} = \theta_r\partial_{x_i} + y_i\partial_{\vartheta_r}, \tag{13.3.8}$$

$$(E_{i,2n+m+r} + E_{2n+r,n+i})|_{\mathscr{A}} = x_i\partial_{\vartheta_r} + \theta_r\partial_{y_i}, \tag{13.3.9}$$

$$(E_{2n+r,2n+s} - E_{2n+m+s,2n+m+r})|_{\mathscr{A}} = \theta_r\partial_{\theta_s} - \vartheta_s\partial_{\vartheta_r}, \tag{13.3.10}$$

$$(E_{2n+m+r,2n+s} + E_{2n+m+s,2n+r})]|_{\mathscr{A}} = \vartheta_r\partial_{\theta_s} + \vartheta_s\partial_{\theta_r}, \tag{13.3.11}$$

$$(E_{2n+r,2n+m+s} + E_{2n+s,2n+m+r})|_{\mathscr{A}} = \theta_r\partial_{\vartheta_s} + \theta_s\partial_{\vartheta_r}, \tag{13.3.12}$$

$$(E_{n+i,2n+r} - E_{2n+m+r,i})|_{\mathscr{A}} = y_i\partial_{\theta_r} - \vartheta_r\partial_{x_i} \tag{13.3.13}$$

for $i, j \in \overline{1,n}$ and $r, s \in \overline{1,m}$.

Recall that we write $\Theta_1 = \sum_{r=1}^m \mathbb{F}\theta_r$ and $\Theta_2 = \sum_{s=1}^m \mathbb{F}\vartheta_s$. For $k \in \mathbb{N}$, we denote

$$\mathscr{A}_k = \text{Span}\{x^\alpha y^\alpha \Theta_1^{\ell_1'}\Theta_2^{\ell_2'} \mid \alpha, \beta \in \mathbb{N}^n; \ell_1', \ell_2' \in \mathbb{N}; |\alpha| + \ell_1' + |\beta| + \ell_2' = k\}. \tag{13.3.14}$$

Again we take

$$\Delta = \sum_{i=1}^n \partial_{x_i}\partial_{y_i} + \sum_{r=1}^m \partial_{\theta_r}\partial_{\vartheta_r}, \qquad \eta = \sum_{i=1}^n x_i y_i + \sum_{r=1}^m \theta_r\vartheta_r. \tag{13.3.15}$$

Set

$$\mathbb{W}_0 = \left[\sum_{i=1}^n(\mathbb{F}x_i + \mathbb{F}y_i) + \sum_{r=1}^m(\mathbb{F}\theta_r + \mathbb{F}\vartheta_r)\right]\left[\sum_{j=1}^n(\mathbb{F}\partial_{x_j} + \mathbb{F}\partial_{y_j}) + \sum_{s=1}^m(\mathbb{F}\partial_{\theta_s} + \mathbb{F}\partial_{\vartheta_s})\right]. \tag{13.3.16}$$

Then

$$osp(2n|2m)|_{\mathscr{A}} = \{T \in \mathbb{W}_0 \mid T(\eta) = 0\}. \tag{13.3.17}$$

Moreover,

$$\xi\Delta = \Delta\xi, \qquad \xi\eta = \eta\xi \qquad \text{for } \xi \in osp(2n|2m) \tag{13.3.18}$$

as operators on $\mathscr{A}$. For $k \in \mathbb{N}$, the subspace

$$\mathscr{H}_k = \{f \in \mathscr{A}_k \mid \Delta(f) = 0\} \tag{13.3.19}$$

forms an $osp(2n|2m)$-submodule. We have the following analogue of the classical theorem on harmonic polynomials:

Theorem 13.3.1 *Suppose $n > 1$. For $k \in \mathbb{N}$, $\mathscr{H}_k$ is an irreducible $osp(2n|2m)$-module if and only if $k \leq m+1-n$ or $k > 2(m+1-n)$. When $k \leq m+1-n$, $\mathscr{A}_k = \bigoplus_{i=0}^{[\![k/2]\!]} \eta^i \mathscr{H}_{k-2i}$ is a decomposition of irreducible $osp(2n|2m)$-submodules.*

Proof We take the subspace of diagonal matrices in $osp(2n|2m)$ as a Carten subalgebras and the subspace

$$\begin{aligned} osp(2n|2m)_+ = & \sum_{1\leq i<j\leq n} [\mathbb{F}(E_{i,j} - E_{n+j,n+i}) + \mathbb{F}(E_{i,n+j} - E_{j,n+i})] \\ & + \sum_{i=1}^{n}\sum_{r=1}^{m} [\mathbb{F}(E_{i,2n+r} - E_{2n+m+r,n+i}) + \mathbb{F}(E_{i,2n+m+r} + E_{2n+r,n+i})] \\ & + \sum_{1\leq r<s\leq m} \mathbb{F}(E_{2n+r,2n+s} - E_{2n+m+s,2n+m+r}) \\ & + \sum_{1\leq r\leq s\leq m} \mathbb{F}(E_{2n+r,2n+m+s} + E_{2n+s,2n+m+r}) \end{aligned} \tag{13.3.20}$$

as the space spanned by positive root vectors. An $osp(2n|2m)$-*singular vector* v is a nonzero weight vector of $osp(2n|2m)$ such that $osp(2n|2m)_+(v) = 0$. We count singular vector up to a nonzero scalar multiple.

Observe $\mathscr{K}|_{\mathscr{A}} = gl(n|m)|_{\mathscr{A}}$. According to the arguments in (13.1.52)–(13.1.61), the homogeneous singular vectors of $\mathscr{K}$ are

$$\{\eta^{\ell_3} x_1^{\ell_1} y_n^{\ell_2} \vec{\vartheta}_s \mid \ell_i \in \mathbb{N};\ s \in \overline{1, m+1};\ \ell_2(s-1) = 0\}. \tag{13.3.21}$$

By (13.3.9),

$$(E_{n,2n+m+s} + E_{2n+s,2n})|_{\mathscr{A}} = x_n \partial_{\vartheta_s} + \theta_s \partial_{y_n}. \tag{13.3.22}$$

Thus the homogeneous singular vectors of $osp(2n|2m)$ are

$$\{\eta^{\ell_2} x_1^{\ell_1} \mid \ell_1, \ell_2 \in \mathbb{N}\}. \tag{13.3.23}$$

Moreover, (13.1.35) imply

$$\Delta(\eta^{\ell_2} x_1^{\ell_1}) = \ell_2(n-m+\ell_1+\ell_2-1) = 0 \Longrightarrow \ell_1+n \leq m \text{ and } \ell_2 = m+1-n-\ell_1. \tag{13.3.24}$$

In this case,

$$\eta^{m+1-n-\ell_1} x_1^{\ell_1} \in \mathscr{H}_{2(m+1-n)-\ell_1}. \tag{13.3.25}$$

Thus

$$\begin{aligned}&\mathscr{H}_k \text{ has a unique singular vector if and only if}\\ &k \leq m+1-n \text{ or } k > 2(m+1-n),\end{aligned} \tag{13.3.26}$$

and

$$\mathscr{H}_k \text{ has two singular vectors when } m+1-n < k \leq 2(m+1-n). \tag{13.3.27}$$

Note $x_1^k \in \mathscr{H}_k$. Let U be the $osp(2n|2m)$-submodule generated by x_1^k. Repeatedly applying

$$(E_{n+1,2n+r} - E_{2n+m+r,1})|_{\mathscr{A}} = y_1\partial_{\theta_r} - \vartheta_r\partial_{x_1} \tag{13.3.28}$$

by (13.3.13), we get

$$x_1^\ell \vec{\vartheta}_s \in U \qquad \text{for } \ell + m + 1 - s = k. \tag{13.3.29}$$

According to (13.3.7),

$$(E_{n+1,n} - E_{2n,1})|_{\mathscr{A}} = y_1\partial_{x_n} - y_n\partial_{x_1}. \tag{13.3.30}$$

Thus

$$x_1^{\ell_1} y_n^{\ell_2} \vec{\vartheta}_1 \in U \qquad \text{for } \ell_1 + \ell_2 + m = k \tag{13.3.31}$$

when $k \geq m$. Since

$$\mathscr{H}_k = \sum_{i=0}^{k} \mathscr{H}_{i,k-i}, \tag{13.3.32}$$

Lemma 13.1.5, (13.3.29) and (13.3.31) imply $U = \mathscr{H}_k$. So $\mathscr{H}_k$ is an $osp(2n|2m)$-module generated by x_1^k.

Suppose $k \leq m+1-n$ or $k > 2(m+1-n)$. Let M be a nonzero $osp(2n|2m)$-submodule of $\mathscr{H}_k$. By (13.3.26), $\mathscr{H}_k$ contains a unique singular vector x_1^k. Thus $x_1^k \in M$. By the above paragraph, $\mathscr{H}_k \subset M$. Hence $\mathscr{H}_k$ is irreducible.

If $\mathscr{H}_k$ is irreducible, (13.3.27) implies $k \leq m+1-n$ or $k > 2(m+1-n)$.

Assume $k \leq m+1-n$. Note

$$\mathscr{A}_j = \sum_{\ell=0}^{j} \mathscr{A}_{\ell,j-\ell} \qquad \text{for } j \in \mathbb{N}. \tag{13.3.33}$$

By Theorem 13.1.6,

$$\mathscr{A}_k = \bigoplus_{\ell=0}^{k} \mathscr{A}_{\ell,k-\ell} = \bigoplus_{\ell=0}^{k}\bigoplus_{i=0}^{[\![k/2]\!]} \eta^i \mathscr{H}_{\ell-i,k-\ell-i} = \bigoplus_{i=0}^{[\![k/2]\!]} \eta^i \mathscr{H}_{k-2i}. \qquad \square \tag{13.3.34}$$

The above theorem establishes a supersymmetric $(sl(2,\mathbb{F}), osp(2n|2m))$ Howe duality on the homogeneous subspaces $\mathscr{A}_k$ with $k\in\mathbb{N}$ such that $k\leq m+1-n$.

We remark that if $m+1-n<k\leq 2(m+1-n)$, then $\mathscr{H}_k$ is an indecomposable $gl(n|m)$-module. In fact, $\mathscr{H}_k\bigcap\eta\mathscr{A}_{k-2}\neq\{0\}$. This also shows that $\mathscr{A}_k$ is not completely reducible when $k>m+1-n$. The conclusion with $n>m+1$ was first obtained by Zhang [Z].

Fix $n_1,n_2\in\overline{1,n}$ with $n_1+1<n_2$. Changing operators $\partial_{x_r}\mapsto -x_r,\ x_r\mapsto\partial_{x_r}$ for $r\in\overline{1,n_1}$ and $\partial_{y_s}\mapsto -y_s,\ y_s\mapsto\partial_{y_s}$ for $s\in\overline{n_2+1,n}$ in (13.3.6)–(13.3.13), we get the following *noncanonical oscillator representation* of $osp(2n|2m)$ on $\mathscr{A}$ determined by

$$E_{2n+r,2n+s}|_{\mathscr{A}}=\theta_r\partial_{\theta_s},\qquad E_{2n+m+r,2n+m+s}|_{\mathscr{A}}=\vartheta_r\partial_{\vartheta_s},\tag{13.3.35}$$

$$E_{2n+r,2n+m+s}|_{\mathscr{A}}=\theta_r\partial_{\vartheta_s},\qquad E_{2n+m+r,2n+s}|_{\mathscr{A}}=\vartheta_r\partial_{\theta_s},\tag{13.3.36}$$

$$E_{i,j}|_{\mathscr{A}}=\begin{cases}-x_j\partial_{x_i}-\delta_{i,j} & \text{if } i,j\in\overline{1,n_1};\\ \partial_{x_i}\partial_{x_j} & \text{if } i\in\overline{1,n_1},\ j\in\overline{n_1+1,n};\\ -x_ix_j & \text{if } i\in\overline{n_1+1,n},\ j\in\overline{1,n_1};\\ x_i\partial_{x_j} & \text{if } i,j\in\overline{n_1+1,n};\end{cases}\tag{13.3.37}$$

$$E_{n+i,n+j}|_{\mathscr{A}}=\begin{cases}y_i\partial_{y_j} & \text{if } i,j\in\overline{1,n_2};\\ -y_iy_j & \text{if } i\in\overline{1,n_2},\ j\in\overline{n_2+1,n};\\ \partial_{y_i}\partial_{y_j} & \text{if } i\in\overline{n_2+1,n},\ j\in\overline{1,n_2};\\ -y_j\partial_{y_i}-\delta_{i,j} & \text{if } i,j\in\overline{n_2+1,n};\end{cases}\tag{13.3.38}$$

$$E_{i,n+j}|_{\mathscr{A}}=\begin{cases}\partial_{x_i}\partial_{y_j} & \text{if } i\in\overline{1,n_1},\ j\in\overline{1,n_2};\\ -y_j\partial_{x_i} & \text{if } i\in\overline{1,n_1},\ j\in\overline{n_2+1,n};\\ x_i\partial_{y_j} & \text{if } i\in\overline{n_1+1,n},\ j\in\overline{1,n_2};\\ -x_iy_j & \text{if } i\in\overline{n_1+1,n},\ j\in\overline{n_2+1,n};\end{cases}\tag{13.3.39}$$

$$E_{n+i,j}|_{\mathscr{A}}=\begin{cases}-x_jy_i & \text{if } j\in\overline{1,n_1},\ i\in\overline{1,n_2};\\ -x_j\partial_{y_i} & \text{if } j\in\overline{1,n_1},\ i\in\overline{n_2+1,n};\\ y_i\partial_{x_j} & \text{if } j\in\overline{n_1+1,n},\ i\in\overline{1,n_2};\\ \partial_{x_j}\partial_{y_i} & \text{if } j\in\overline{n_1+1,n},\ i\in\overline{n_2+1,n};\end{cases}\tag{13.3.40}$$

$$E_{i,2n+r}|_{\mathscr{A}}=\begin{cases}\partial_{x_i}\partial_{\theta_r} & \text{if } i\in\overline{1,n_1};\\ x_i\partial_{\theta_r} & \text{if } i\in\overline{n_1+1,n};\end{cases}\tag{13.3.41}$$

$$E_{i,2n+m+r}|_{\mathscr{A}}=\begin{cases}\partial_{x_i}\partial_{\vartheta_r} & \text{if } i\in\overline{1,n_1};\\ x_i\partial_{\vartheta_r} & \text{if } i\in\overline{n_1+1,n};\end{cases}\tag{13.3.42}$$

$$E_{2n+r,i}|_{\mathscr{A}}=\begin{cases}-x_i\theta_r & \text{if } i\in\overline{1,n_1};\\ \theta_r\partial_{x_i} & \text{if } i\in\overline{n_1+1,n};\end{cases}\tag{13.3.43}$$

$$E_{2n+m+r,i}|_{\mathscr{A}} = \begin{cases} -x_i\vartheta_r & \text{if } i \in \overline{1,n_1}; \\ \vartheta_r\partial_{x_i} & \text{if } i \in \overline{n_1+1,n}; \end{cases} \tag{13.3.44}$$

$$E_{n+i,2n+r}|_{\mathscr{A}} = \begin{cases} y_i\partial_{\theta_r} & \text{if } i \in \overline{1,n_2}; \\ \partial_{y_i}\partial_{\theta_r} & \text{if } i \in \overline{n_2+1,n}; \end{cases} \tag{13.3.45}$$

$$E_{n+i,2n+m+r}|_{\mathscr{A}} = \begin{cases} y_i\partial_{\vartheta_r} & \text{if } i \in \overline{1,n_2}; \\ \partial_{y_i}\partial_{\vartheta_r} & \text{if } i \in \overline{n_2+1,n}; \end{cases} \tag{13.3.46}$$

$$E_{2n+r,n+i}|_{\mathscr{A}} = \begin{cases} \theta_r\partial_{y_i} & \text{if } i \in \overline{1,n_2}; \\ -y_i\theta_r & \text{if } i \in \overline{n_2+1,n}; \end{cases} \tag{13.3.47}$$

$$E_{2n+m+r,n+i}|_{\mathscr{A}} = \begin{cases} \vartheta_r\partial_{y_i} & \text{if } i \in \overline{1,n_2}; \\ -y_i\vartheta_r & \text{if } i \in \overline{n_2+1,n}; \end{cases} \tag{13.3.48}$$

for $i, j \in \overline{1,n}$ and $r, s \in \overline{1,m}$.

The related Laplace operator becomes

$$\Delta_{n_1,n_2} = -\sum_{i=1}^{n_1} x_i\partial_{y_i} + \sum_{r=n_1+1}^{n_2} \partial_{x_r}\partial_{y_r} - \sum_{s=n_2+1}^{n} y_s\partial_{x_s} + \sum_{r=1}^{m} \partial_{\theta_r}\partial_{\vartheta_r} \tag{13.3.49}$$

and its dual

$$\eta = \sum_{i=1}^{n_1} y_i\partial_{x_i} + \sum_{r=n_1+1}^{n_2} x_r y_r + \sum_{s=n_2+1}^{n} x_s\partial_{y_s} + \sum_{r=1}^{m} \theta_r\vartheta_r. \tag{13.3.50}$$

It can be verified that (13.3.18) holds again. Denote

$$\mathscr{A}_{\langle k\rangle} = \text{Span}\{x^\alpha y^\beta \Theta_1^{\ell_1'}\Theta_2^{\ell_2'} \mid \alpha, \beta \in \mathbb{N}^n; \ell_1', \ell_2' \in \mathbb{N}; \sum_{r=n_1+1}^{n} \alpha_r - \sum_{i=1}^{n_1} \alpha_i + \sum_{i=1}^{n_2} \beta_i - \sum_{r=n_2+1}^{n} \beta_r + \ell_1' + \ell_2' = \ell_2\} \tag{13.3.51}$$

for $k \in \mathbb{Z}$.

Again we set

$$\mathscr{H}_{\langle k\rangle} = \{f \in \mathscr{A}_{\langle k\rangle} \mid \Delta_{n_1,n_2}(f) = 0\}. \tag{13.3.52}$$

Now we have the following noncanonical super-symmetric analogue of the classical theorem on harmonic polynomials:

Theorem 13.3.2 *Let $k \in \mathbb{Z}$. The $osp(2n|2m)$-module $\mathscr{H}_{\langle k\rangle}$ is irreducible if and only if $k \leq n_1 + m + 1 - n_2$. When $k \leq n_1 + m + 1 - n_2$, $\mathscr{A}_{\langle k\rangle} = \bigoplus_{i=0}^{\infty} \eta^i(\mathscr{H}_{\langle k-2i\rangle})$ is the decomposition of irreducible $osp(2n|2m)$-submodules.*

Proof Observe $\mathscr{K}|_{\mathscr{A}} = gl(n|m)|_{\mathscr{A}}$ in terms of the representation of $gl(n|m)$ given in (13.2.2)–(13.2.6). Lemma 13.2.2 says that the homogeneous singular vectors of $\mathscr{K}$ are of the form:

$$\eta^{\ell_1}(x_i^{m_1} y_j^{m_2} \vec{\vartheta}_s) \text{ with } \ell_1, m_1, m_2 \in \mathbb{N},\; s \in \overline{1, m+1} \tag{13.3.53}$$

and

$$(i, j) \in \{(n_1, n_2), (n_1, n_2+1-\delta_{n_2,n}), (n_1+1, n_2), (n_1+1, n_2+1-\delta_{n_2,n})\}. \tag{13.3.54}$$

Claim 1. For $k \in \mathbb{N}$, $\mathscr{H}_{\langle k\rangle}$ is an $osp(2n|2m)$-module generated by $x_{n_1+1}^k$ and $\mathscr{H}_{\langle -k\rangle}$ is an $osp(2n|2m)$-module generated by $x_{n_1}^k$.

Let V be the $osp(2n|2m)$-module generated by $x_{n_1+1}^k \in \mathscr{H}_{\langle k\rangle}$. By (13.3.38),

$$(E_{n_2+1,n+n_1+1} - E_{n_1+1,n+n_2+1})|_{\mathscr{A}} = x_{n_2+1}\partial_{y_{n_1+1}} + x_{n_1+1}y_{n_2+1}. \tag{13.3.55}$$

Thus

$$(E_{n_2+1,n+n_1+1} - E_{n_1+1,n+n_2+1})^{k_1}(x_{n_1+1}^k) = x_{n_1+1}^{k+k_1} y_{n_2+1}^{k_1} \in V. \tag{13.3.56}$$

According to (13.3.40),

$$(E_{n+n_2,n_1+1} - E_{n+n_1+1,n_2})|_{\mathscr{A}} = y_{n_2}\partial_{x_{n_1+1}} - y_{n_1+1}\partial_{x_{n_2}}. \tag{13.3.57}$$

Repeatedly applying (13.3.57) to $x_{n_1+1}^k$, we obtain

$$x_{n_1+1}^{k_1} y_{n_2}^{k_2} \in V \qquad \text{for } k_1, k_2 \in \mathbb{N} \text{ such that } k_1 + k_2 = k. \tag{13.3.58}$$

Note that (13.3.44) and (13.3.45) imply

$$(E_{n+n_1+1,2n+r} - E_{2n+m+r,n_1+1})|_{\mathscr{A}} = y_{n_1+1}\partial_{\theta_r} - \vartheta_r\partial_{x_{n_1+1}} \tag{13.3.59}$$

Applying (13.3.59) with various r to (13.3.56) and (13.3.58), we obtain

$$x_{n_1+1}^{\ell_1} y_j^{\ell_2}\vec{\vartheta}_s \in \mathscr{H}_{\langle k\rangle} \text{ with } j \in \{n_2, n_2+1\} \Longrightarrow x_{n_1+1}^{\ell_1} y_j^{\ell_2}\vec{\vartheta}_s \in V. \tag{13.3.60}$$

In terms of (13.2.40),

$$v_{\ell,\ell'} \in V, \qquad \ell + \ell' = k. \tag{13.3.61}$$

By Lemma 13.2.4,

$$\mathscr{H}_{\langle \ell,\ell'\rangle} \subset V, \qquad \ell + \ell' = k. \tag{13.3.62}$$

Therefore

$$\mathscr{H}_{\langle k\rangle} = \bigoplus_{\ell,\ell'\in\mathbb{Z};\ \ell+\ell'=k} \mathscr{H}_{\langle \ell,\ell'\rangle} \subset V. \tag{13.3.63}$$

Suppose $k > 0$. Let U be the $osp(2n|2m)$-module generated by $x_{n_1}^k \in \mathscr{H}_{\langle -k\rangle}$. Observe

$$(E_{n_2+1,n+n_1} - E_{n_1,n+n_2+1})|_{\mathscr{A}} = x_{n_2+1}\partial_{y_{n_1}} + y_{n_2+1}\partial_{x_{n_1}} \tag{13.3.64}$$

by (13.3.39), and

$$(E_{n+n_1,n_2} - E_{n+n_2,n_1}|_{\mathscr{A}} = y_{n_1}\partial_{x_{n_1}} + x_{n_1}y_{n_1} \tag{13.3.65}$$

by (13.3.40). Repeatedly applying the above two equations to $x_{n_1}^k$, we have

$$x_{n_1}^{k_1}y_{n_2+1}^{k_2}, x_{n_1}^{k+k_3}y_{n_2}^{k_3} \in U \qquad \text{for } k_1, k_2, k_3 \in \mathbb{N} \text{ such that } k_1 + k_2 = k. \tag{13.3.66}$$

According to (13.3.44) and (13.3.45),

$$(E_{n+n_1,2n+r} - E_{2n+m+r,n_1})|_{\mathscr{A}} = y_{n_1}\partial_{\theta_r} + x_{n_1}\vartheta_r. \tag{13.3.67}$$

Applying (13.3.67) with various r to (13.3.66), we find

$$x_{n_1}^{\ell_1}y_j^{\ell_2}\vec{\vartheta}_s \in \mathscr{H}_{\langle -k\rangle} \text{ with } j \in \{n_2, n_2+1\} \Longrightarrow x_{n_1}^{\ell_1}y_j^{\ell_2}\vec{\vartheta}_s \in V. \tag{13.3.68}$$

In terms of (13.2.40),

$$v_{\ell,\ell'} \in V, \qquad \ell + \ell' = -k. \tag{13.3.69}$$

Lemma 13.2.4 gives

$$\mathscr{H}_{\langle \ell,\ell'\rangle} \subset V, \qquad \ell + \ell' = -k. \tag{13.3.70}$$

Thus

$$\mathscr{H}_{\langle -k\rangle} = \bigoplus_{\ell,\ell'\in\mathbb{Z};\ \ell+\ell'=-k} \mathscr{H}_{\langle \ell,\ell'\rangle} \subset V. \tag{13.3.71}$$

This prove Claim 1.

Claim 2. For $n_1 + m + 1 - n_2 \geq k \in \mathbb{Z}$, any nonzero $osp(2n|2m)$-submodule of $\mathscr{H}_{\langle k\rangle}$ contains $x_{n_1+1}^k$ if $k \geq 0$ or $x_{n_1}^{-k}$ when $k < 0$.

Note

$$x_{n_1}^{m_1}y_{n_2}^{m_2}\vec{\vartheta}_s \in \mathscr{H}_{\langle m+m_2+1-s-m_1\rangle}, \qquad x_{n_1}^{m_1}y_{n_2+1}^{m_2}\vec{\vartheta}_s \in \mathscr{H}_{\langle m+1-s-m_1-m_2\rangle}, \tag{13.3.72}$$

$$x_{n_1+1}^{m_1}y_{n_2}^{m_2}\vec{\vartheta}_s \in \mathscr{H}_{\langle m+m_1+m_2+1-s\rangle}, \qquad x_{n_1+1}^{m_1}y_{n_2+1}^{m_2}\vec{\vartheta}_s \in \mathscr{H}_{\langle m+m_1+1-s-m_2\rangle}. \tag{13.3.73}$$

For $\ell_1 \in \mathbb{N}+1$ and $f \in \mathscr{H}_{\langle k' \rangle}$ with $k' \in \mathbb{Z}$, (13.2.28) and the second equation in (13.1.35) imply

$$\Delta_{n_1,n_2}\eta^{\ell_1}(f) = \ell_1(n_2 - n_1 - m + k' + \ell_1 - 1)\eta^{\ell_1 - 1}(f). \tag{13.3.74}$$

Thus

$$\Delta_{n_1,n_2}\eta^{\ell_1}(f) = 0 \Longleftrightarrow k' \leq n_1 + m - n_2 \;\text{ and }\; \ell_1 = n_1 + m - n_2 - k' + 1. \tag{13.3.75}$$

If the condition holds, then

$$\eta^{\ell_1}(f) \in \mathscr{H}_{\langle 2(n_1+m+1-n_2)-k' \rangle}. \tag{13.3.76}$$

Moreover,

$$2(n_1+m+1-n_2) - k' = n_1+m+2-n_2+(n_1+m-n_2-k') \geq n_1+m+2-n_2. \tag{13.3.77}$$

This shows that

$$\mathscr{H}_{\langle k_1 \rangle} \bigcap \left(\bigcup_{i=1}^{\infty} \eta^i(\mathscr{H}) \right) \neq \{0\} \Longleftrightarrow k_1 \geq n_1 + m + 2 - n_2. \tag{13.3.78}$$

Suppose $k \leq n_1 + m + 1 - n_2$. Then the singular vectors of $\mathscr{K}$ in $\mathscr{H}_{\langle k \rangle}$ are of the form

$$x_i^{m_1} y_j^{m_2} \vec{\vartheta}_s \;\text{ with } m_1, m_2 \in \mathbb{N}, \; s \in \overline{1, m+1} \tag{13.3.79}$$

with (13.3.54). Observe that

$$(E_{n_1+1,n+n_2} - E_{n_2,n+n_1+1})|_{\mathscr{A}} = x_{n_1+1}\partial_{y_{n_2}} - x_{n_2}\partial_{y_{n_1+1}} \tag{13.3.80}$$

by (13.3.39), and

$$(E_{n+n_2+1,n_1} - E_{n+n_1,n_2+1})|_{\mathscr{A}} = -x_{n_1}\partial_{y_{n_2+1}} - y_{n_1}\partial_{x_{n_2+1}} \tag{13.3.81}$$

by (13.3.40). According to (13.3.42) and (13.3.47),

$$(E_{n_1+1,2n+m+r} + E_{2n+r,n+n_1+1})|_{\mathscr{A}} = x_{n_1+1}\partial_{\vartheta_r} + \theta_r\partial_{y_{n_1+1}}. \tag{13.3.82}$$

Let M be any nonzero $osp(2n|2m)$-submodule of $\mathscr{H}_{\langle k \rangle}$. Then M contains at least one of the $\mathscr{K}$-singular vectors in (13.3.79). Applying (13.3.70)–(13.3.82), we get

$$x_{n_1}^{k_1} x_{n_2}^{k_2} \in M \;\text{ for some } k_1, k_2 \in \mathbb{N} \text{ such that } k_2 - k_1 = k. \tag{13.3.83}$$

By (13.3.37) and (13.3.38),

$$(E_{n_1,n_1+1} - E_{n+n_1+1,n+n_1})|_{\mathscr{A}} = \partial_{x_{n_1}}\partial_{x_{n_1+1}} - y_{n_1+1}\partial_{y_{n_1}}. \tag{13.3.84}$$

Repeatedly applying (13.3.84) to (13.3.83), we obtain $x_{n_1+1}^k \in M$ if $k \geq 0$ or $x_{n_1}^{-k} \in M$ when $k < 0$.

The above claims show that $\mathscr{H}_{\langle k\rangle}$ is an irreducible $osp(2n|2m)$-module if $k \leq n_1 + m + 1 - n_2$.

Note

$$\mathscr{A}_{\langle j\rangle} = \sum_{\ell\in\mathbb{Z}} \mathscr{A}_{\langle \ell, j-\ell\rangle}, \qquad \mathscr{H}_{\langle j\rangle} = \sum_{\ell\in\mathbb{Z}} \mathscr{H}_{\langle \ell, j-\ell\rangle} \quad \text{for } j \in \mathbb{Z}. \tag{13.3.85}$$

By (13.3.78), $k \leq n_1 + m + 1 - n_2$ if $\mathscr{H}_{\langle k\rangle}$ is irreducible. When $k \leq n_1 + m + 1 - n_2$, Theorem 13.2.5 implies

$$\mathscr{A}_{\langle k\rangle} = \bigoplus_{\ell\in\mathbb{Z}} \mathscr{A}_{\langle \ell,k-\ell\rangle} = \bigoplus_{\ell\in\mathbb{Z}} \bigoplus_{i=0}^{\infty} \eta^i \mathscr{H}_{\langle \ell-i,k-\ell-i\rangle} = \bigoplus_{i=0}^{\infty} \eta^i \mathscr{H}_{\langle k-2i\rangle}. \qquad \square \tag{13.3.86}$$

The above theorem establishes a supersymmetric $(sl(2,\mathbb{F}), osp(2n|2m))$ Howe duality on the homogeneous subspaces $\mathscr{A}_{\langle k\rangle}$ with $k \in \mathbb{Z}$ such that $k \leq n_1 + m + 1 - n_2$.

We remark that $\mathscr{H}_{\langle k\rangle}$ is an indecomposable $osp(2n|2m)$-module if $k \geq n_1 + m + 2 - n_2$ by Claim 1 and (13.3.78). This also implies that $\mathscr{A}_{\langle k\rangle}$ is not completely reducible $osp(2n|2m)$-module when $k \geq n_1 + m + 2 - n_2$.

13.4 Oscillator Representations of $osp(2n+1|2m)$

In this section, we use the results in last three sections to study the corresponding representations of the odd ortho-symplectic Lie superalgebras.

Write

$$osp(2n+1|2m)_0 = osp(2n|2m)_0 + \sum_{i=1}^{n} [\mathbb{F}(E_{0,i} - E_{n+i,0}) + \mathbb{F}(E_{0,n+i} - E_{i,0})], \tag{13.4.1}$$

$$\begin{aligned} osp(2n+1|2m)_1 = osp(2n|2m)_1 + \sum_{r=1}^{m} [&\mathbb{F}(E_{0,2n+r} - E_{2n+m+r,0}) \\ &+ \mathbb{F}(E_{0,2n+m+r} + E_{2n+r,0})]. \end{aligned} \tag{13.4.2}$$

The *odd ortho-symplectic Lie superalgebra* $osp(2n+1|2m) = osp(2n+1|2m)_0 + osp(2n+1|2m)_1$ is a Lie sub-superalgebra of $gl(2n+1|2m)$. Let x_0 be a bosonic (commuting) variable. Set

$$\mathscr{B} = \mathscr{A}[x_0] = \bigoplus_{k=0}^{\infty} \mathscr{B}_k, \qquad \mathscr{B}_k = \sum_{i=0}^{k} \mathscr{A}_{k-i} x_0^i. \tag{13.4.3}$$

Moreover, the corresponding supersymmetric Laplace operator and invariant become

$$\Delta' = \partial_{x_0}^2 + 2\sum_{i=1}^{n} \partial_{x_i}\partial_{y_i} + 2\sum_{r=1}^{m} \partial_{\theta_r}\partial_{\vartheta_r}, \qquad \eta' = x_0^2 + 2\sum_{i=1}^{n} x_i y_i + 2\sum_{r=1}^{m} \theta_r \vartheta_r. \tag{13.4.4}$$

Now $osp(2n|2m)$ acts on $\mathscr{B}$ by the differential operators in (13.3.6)–(13.3.13); namely, we change the subindex $|_{\mathscr{A}}$ to $|_{\mathscr{B}}$. Extend the representation of $osp(2n|2m)$ on $\mathscr{B}$ to the *basic oscillator representation* of $osp(2n+1|2m)$ on $\mathscr{B}$ by:

$$(E_{0,i} - E_{n+i,0})|_{\mathscr{B}} = x_0\partial_{x_i} - y_i\partial_{x_0}, \quad (E_{0,2n+m+r} + E_{2n+r,0})|_{\mathscr{B}} = x_0\partial_{\vartheta_r} + \theta_r\partial_{x_0} \tag{13.4.5}$$

$$(E_{0,n+i} - E_{i,0})|_{\mathscr{B}} = x_0\partial_{y_i} - x_i\partial_{x_0},, \quad (E_{0,2n+r} - E_{2n+m+r,0})|_{\mathscr{B}} = x_0\partial_{\theta_r} - \vartheta_r\partial_{x_0}, \tag{13.4.6}$$

for $i \in \overline{1,n}$ and $r \in \overline{1,m}$.

Set

$$\mathbb{W}_0' = \left[\sum_{i=0}^{n}(\mathbb{F}x_i + \mathbb{F}y_i) + \sum_{r=1}^{m}(\mathbb{F}\theta_r + \mathbb{F}\vartheta_r)\right]\left[\sum_{j=0}^{n}(\mathbb{F}\partial_{x_j} + \mathbb{F}\partial_{y_j}) + \sum_{s=1}^{m}(\mathbb{F}\partial_{\theta_s} + \mathbb{F}\partial_{\vartheta_s})\right]. \tag{13.4.7}$$

Then

$$osp(2n+1|2m)|_{\mathscr{B}} = \{T \in \mathbb{W}_0' \mid T(\eta') = 0\}. \tag{13.4.8}$$

Define

$$\mathscr{H}' = \{f \in \mathscr{B} \mid \Delta'(f) = 0\}, \qquad \mathscr{H}_k' = \mathscr{H}' \bigcap \mathscr{B}_k. \tag{13.4.9}$$

Again we take the subspace of diagonal matrices in $osp(2n+1|2m)$ as a Cartan subalgebra and take the space spanned by positive roots:

$$osp(2n+1|2m)_+ = \mathbb{F}(E_{0,n+i} - E_{i,0}) + \mathbb{F}(E_{0,2n++m+r} + E_{2n+r,0}) + osp(2n|2m)_+. \tag{13.4.10}$$

An $osp(2n+1|2m)$-*singular vector* v is a nonzero weight vector of $osp(2n+1|2m)$ such that $osp(2n+1|2m)_+(v) = 0$. We count singular vector up to a nonzero scalar multiple. According to the proof of Theorem 13.3.1, any singular vector of $osp(2n+1|2m)$ must be in $\mathbb{F}[x_0, x_1, \eta]$, where

$$\eta = \sum_{i=1}^{n} x_i y_i + \sum_{r=1}^{m} \theta_r \vartheta_r. \tag{13.4.11}$$

Note that $\eta' = x_0^2 + 2\eta$. By (13.4.10), x_1^k is a singular vector of $osp(2n+1|2m)$ for any $k \in \mathbb{N}$. Thus a homogeneous singular vector of $osp(2n+1|2m)$ must be of the form

$$f = \sum_{i=0}^{\ell} b_i x_0^{2i+\iota} \eta^{\ell-i} x_1^k, \tag{13.4.12}$$

where $b_i \in \mathbb{F}$, $\ell, k \in \mathbb{N}$ and $\iota = 0, 1$. Note

$$(E_{0,n+i} - E_{i,0})(f) = (x_0 \partial_{y_i} - x_i \partial_{x_0})(f) = 0 \Longleftrightarrow f = b_0 \eta'^{\ell} x_1^k. \tag{13.4.13}$$

Thus $\{\eta'^{\ell} x_1^k \mid \ell, k \in \mathbb{N}\}$ are all the homogeneous singular vectors of $osp(2n+1|2m)$ in $\mathscr{B}$.

Observe that

$$[\Delta', \eta'] = 2+4(n-m)+4[x_0\partial_{x_0} + \sum_{i=1}^{n}(x_i\partial_{x_i} + y_i\partial_{y_i}) + \sum_{r=1}^{m}(\theta_r\partial_{\theta_r} + \vartheta_r\partial_{\vartheta_r})] \tag{13.4.14}$$

by (13.1.35). So

$$\Delta'(\eta'^{\ell} g) = 2\ell(1 + 2(n - m + k + \ell - 1))\eta'^{\ell-1} g \quad \text{for } g \in \mathscr{H}_k' \tag{13.4.15}$$

Thus

$$\mathscr{H}_k' \text{ has a unique singular vector } x_1^k \text{ for any } k \in \mathbb{N}. \tag{13.4.16}$$

Indeed we have the first main theorem in this section:

Theorem 13.4.1 *For any $k \in \mathbb{N}$, $\mathscr{H}_k'$ is an irreducible $osp(2n+1|2m)$-module. Moreover,*

$$\mathscr{B} = \bigoplus_{\ell,k=0}^{\infty} \eta'^{\ell} \mathscr{H}_k' \tag{13.4.17}$$

is a direct sum of irreducible $osp(2n+1|2m)$-submodules.

Proof By the arguments in (13.2.58)–(13.2.64), we only need to prove that $\mathscr{H}_k'$ is an $osp(2n+1|2m)$-module generated by x_1^k. For $\iota = 0, 1$, we define

$$T_\iota = \sum_{i=0}^{\infty} \frac{(-2)^i x_0^{2i+\iota}}{(2i+\iota)!} \Delta^i, \qquad \Delta = \sum_{i=1}^{n} \partial_{x_i}\partial_{y_i} + \sum_{r=1}^{m} \partial_{\theta_r}\partial_{\vartheta_r}. \tag{13.4.18}$$

We take the notations in (13.1.98) and (13.1.99). By Lemma 6.1.1,

$$\begin{aligned} \mathscr{H}_k' = \text{Span}\{T_\iota(x^\alpha y^\beta \theta_{\vec{j}} \vartheta_{\vec{j}'}) \mid \alpha, \beta \in \mathbb{N}^n;\ \vec{j} \in \Gamma_{k_1},\ \vec{j}' \in \Gamma_{k_2}; \\ \iota \in \{0, 1\};\ |\alpha| + |\beta| + k_1 + k_2 + \iota = k\}. \end{aligned} \tag{13.4.19}$$

Let U be the $osp(2n+1|2m)$-module generated by $x_1^k \in \mathscr{H}_k$. Since $o(2n+1,\mathbb{F})$ is a subalgebra of $osp(2n+1|2m)$, Theorem 8.1.1 yields

$$T_\iota(x^\alpha y^\beta) \in U \qquad \text{for } \alpha,\beta \in \mathbb{N}^n;\ |\alpha|+|\beta| = k. \tag{13.4.20}$$

Repeatedly applying

$$(E_{2n+r,i} + E_{n+i,2n+m+r})|_{\mathscr{A}} = \theta_r \partial_{x_i} + y_i \partial_{\vartheta_r} \tag{13.4.21}$$

to (13.4.20) with $i \in \overline{1,n}$ and $r \in \overline{1,m}$ (cf. (13.3.8)), we obtain

$$T_\iota(x^\alpha y^\beta \theta_{\vec{j}}) \in U \qquad \text{for } \alpha,\beta \in \mathbb{N}^n,\ \vec{j} \in \Gamma_{k_1};\ |\alpha|+|\beta|+k_1 = k. \tag{13.4.22}$$

Finally, we get $U = \mathscr{H}_k'$ by repeatedly applying $(E_{i,2n+r} - E_{2n+m+r,n+i})|_{\mathscr{B}} = x_i\partial_{\theta_r} - \vartheta_r\partial_{y_i}$ to (13.4.22) with $i \in \overline{1,n}$ and $r \in \overline{1,m}$ (cf. (13.3.6)). $\square$

It can be verified by (13.4.14) that the space $\mathbb{F}\Delta' + \mathbb{F}[\Delta',\eta'] + \mathbb{F}\eta'$ also forms an operator Lie algebra isomorphic to $sl(2,\mathbb{F})$. The above theorem establishes a supersymmetric $(sl(2,\mathbb{F}), osp(2n+1|2m))$ Howe duality. We remark that the conclusion in the above theorem with $n > m$ was first obtained by Zhang [Z].

Next $osp(2n|2m)$ acts on $\mathscr{B}$ via the differential operators in (13.3.35)–(13.3.48); namely, we change the subindex $|_{\mathscr{A}}$ to $|_{\mathscr{B}}$. Moreover, we extend the representation of $osp(2n|2m)$ on $\mathscr{B}$ to the *noncanonical oscillator representation* of $osp(2n+1|2m)$ on $\mathscr{B}$ determined by

$$(E_{0,i} - E_{n+i,0})|_{\mathscr{B}} = \begin{cases} -x_0x_i - y_i\partial_{x_0} & \text{if } i \in \overline{1,n_1}, \\ x_0\partial_{x_i} - y_i\partial_{x_0} & \text{if } i \in \overline{n_1+1,n_2}, \\ x_0\partial_{x_i} - \partial_{x_0}\partial_{y_i} & \text{if } i \in \overline{n_2+1,n}; \end{cases} \tag{13.4.23}$$

$$(E_{0,n+i} - E_{i,0})|_{\mathscr{B}} = \begin{cases} x_0\partial_{y_i} - \partial_{x_i}\partial_{x_0} & \text{if } i \in \overline{1,n_1}, \\ x_0\partial_{y_i} - x_i\partial_{x_0} & \text{if } i \in \overline{n_1+1,n_2}, \\ -x_0y_i - x_i\partial_{x_0} & \text{if } i \in \overline{n_2+1,n}; \end{cases} \tag{13.4.24}$$

$$(E_{0,2n+r} - E_{2n+m+r,0})|_{\mathscr{B}} = x_0\partial_{\theta_r} - \vartheta_r\partial_{x_0}, \tag{13.4.25}$$

$$(E_{0,2n++m+r} + E_{2n+r,0})|_{\mathscr{B}} = x_0\partial_{\vartheta_r} + \theta_r\partial_{x_0} \tag{13.4.26}$$

for $i \in \overline{1,n}$ and $r \in \overline{1,m}$.

Now the corresponding Laplace operator becomes

$$\Delta'_{n_1,n_2} = \partial_{x_0}^2 + 2\Delta_{n_1,n_2}, \quad \eta' = x_0^2 + 2\eta \tag{13.4.27}$$

with

$$\Delta_{n_1,n_2} = -\sum_{i=1}^{n_1} x_i \partial_{y_i} + \sum_{r=n_1+1}^{n_2} \partial_{x_r}\partial_{y_r} - \sum_{s=n_2+1}^{n} y_s \partial_{x_s} + \sum_{r=1}^{m} \partial_{\theta_r}\partial_{\vartheta_r} \tag{13.4.28}$$

and

$$\eta = \sum_{i=1}^{n_1} y_i \partial_{x_i} + \sum_{r=n_1+1}^{n_2} x_r y_r + \sum_{s=n_2+1}^{n} x_s \partial_{y_s} + \sum_{r=1}^{m} \theta_r \vartheta_r. \tag{13.4.29}$$

We take the notation in (13.3.51) and set

$$\mathscr{B}_{\langle k\rangle} = \sum_{i=0}^{\infty} \mathscr{A}_{\langle k-i\rangle} x_0^i, \qquad \mathscr{H}'_{\langle k\rangle} = \{f \in \mathscr{B}_k \mid \Delta'_{n_1,n_2}(f) = 0\}. \tag{13.4.30}$$

Then we have the second main theorem in this section:

Theorem 13.4.2 *For any $k \in \mathbb{Z}$, $\mathscr{H}'_{\langle k\rangle}$ is an irreducible $osp(2n+1|2m)$-module. Moreover,*

$$\mathscr{B} = \bigoplus_{\ell,k=0}^{\infty} \eta'^{\ell}(\mathscr{H}'_{\langle k\rangle}) \tag{13.4.31}$$

is a direct sum of irreducible $osp(2n+1|2m)$-submodules.

Proof We define T_ι as in (13.4.18) with Δ replaced by Δ_{n_1,n_2} in (13.4.28). By Lemma 6.1.1,

$$\mathscr{H}'_{\langle k\rangle} = T_0(\mathscr{A}_{\langle k\rangle}) + T_1(\mathscr{A}_{\langle k-1\rangle}) \qquad \text{for } k \in \mathbb{Z}. \tag{13.4.32}$$

Since $\Delta_{n_1,n_2}\xi = \xi\Delta_{n_1,n_2}$ for $\xi \in osp(2n|2m)$, we have

$$\xi(T_\iota(f)) = T_\iota(\xi(f)) \qquad \text{for } \xi \in osp(2n|2m),\ f \in \mathscr{A}. \tag{13.4.33}$$

First we consider $\mathscr{H}'_{\langle k\rangle}$ with $k \in \mathbb{N}$. Let V be any nonzero $osp(2n+1|2m)$-submodule of $\mathscr{H}'_{\langle k\rangle}$. According to the arguments in paragraph of (13.3.79)–(13.3.84), V contains some $T_\iota(\eta^\ell(x_{n_1+1}^{k-\iota-2\ell}))$. By (13.4.24),

$$(E_{n_1+1,0} - E_{0,n+n_1+1})|_{\mathscr{B}} = x_{n_1+1}\partial_{x_0} - x_0\partial_{y_{n_1+1}}. \tag{13.4.34}$$

Moreover, as operators on $\mathscr{B}$,

$$\begin{aligned}
&[E_{n_1+1,0} - E_{0,n+n_1+1}, T_0] \\
&= \left[x_{n_1+1}\partial_{x_0} - x_0\partial_{y_{n_1+1}}, \sum_{i=0}^{\infty} \frac{(-2)^i x_0^{2i}}{(2i)!} \Delta^i_{n_1,n_2}\right] \\
&= \left[x_{n_1+1}, \sum_{i=1}^{\infty} \frac{(-2)^i x_0^{2i}}{(2i)!} \Delta^i_{n_1,n_2}\right]\partial_{x_0} + x_{n_1+1}\sum_{i=1}^{\infty} \frac{(-2)^i x_0^{2i-1}}{(2i-1)!} \Delta^i_{n_1,n_2}
\end{aligned}$$

$$= -\left[\sum_{i=1}^{\infty}\frac{i(-2)^i x_0^{2i}}{(2i)!}\Delta_{n_1,n_2}^{i-1}\partial_{x_0} + \sum_{i=1}^{\infty}\frac{i(-2)^i x_0^{2i-1}}{(2i-1)!}\Delta_{n_1,n_2}^{i-1}\right]\partial_{y_{n_1+1}}$$
$$-2(T_1\Delta_{n_1,n_2})x_{n_1+1}, \tag{13.4.35}$$

$$[E_{n_1+1,0} - E_{0,n+n_1+1}, T_1]$$
$$= \left[x_{n_1+1}\partial_{x_0} - x_0\partial_{y_{n_1+1}}, \sum_{i=0}^{\infty}\frac{(-2)^i x_0^{2i+1}}{(2i+1)!}\Delta_{n_1,n_2}^{i}\right]$$
$$= \left[x_{n_1+1}, \sum_{i=1}^{\infty}\frac{(-2)^i x_0^{2i+1}}{(2i+1)!}\Delta_{n_1,n_2}^{i}\right]\partial_{x_0} + x_{n_1+1}\sum_{i=0}^{\infty}\frac{(-2)^i x_0^{2i}}{(2i)!}\Delta_{n_1,n_2}^{i}$$
$$= -\left[\sum_{i=1}^{\infty}\frac{i(-2)^i x_0^{2i+1}}{(2i+1)!}\Delta_{n_1,n_2}^{i-1}\partial_{x_0} + \sum_{i=1}^{\infty}\frac{i(-2)^i x_0^{2i}}{(2i)!}\Delta_{n_1,n_2}^{i-1}\right]\partial_{y_{n_1+1}} + T_0 x_{n_1+1}. \tag{13.4.36}$$

If $T_0(\eta^\ell(x_{n_1+1}^{k-2\ell})) \in V$ for some $\ell \in \mathbb{N}+1$, we have

$$(E_{n_1+1,0} - E_{0,n+n_1+1})T_0(\eta^\ell(x_{n_1+1}^{k-2\ell}))$$
$$= ([E_{n_1+1,0} - E_{0,n+n_1+1}, T_0] + T_0(E_{n_1+1,0} - E_{0,n+n_1+1}))(\eta^\ell(x_{n_1+1}^{k-2\ell}))$$
$$= -\left[\sum_{i=1}^{\infty}\frac{i(-2)^i x_0^{2i-1}}{(2i-1)!}\Delta_{n_1,n_2}^{i-1}\partial_{y_{n_1+1}} + x_0 T_0\partial_{y_{n_1+1}} + 2(T_1\Delta_{n_1,n_2})x_{n_1+1}\right](\eta^\ell(x_{n_1+1}^{k-2\ell}))$$
$$= [T_1\partial_{y_{n_1+1}} - 2(T_1\Delta_{n_1,n_2})x_{n_1+1}](\eta^\ell(x_{n_1+1}^{k-2\ell}))$$
$$= \ell[1 - 2(m + n_1 - n_2 + \ell - 1)](\eta^{\ell-1}(x_{n_1+1}^{k-2(\ell-1)-1}))$$
$$= \ell[3 - 2(m + n_1 - n_2 + \ell)]T_1(\eta^{\ell-1}(x_{n_1+1}^{k-2(\ell-1)-1})) \in V \tag{13.4.37}$$

by (13.3.74), (13.4.34) and (13.4.35). So $T_1(\eta^{\ell-1}(x_{n_1+1}^{k-2(\ell-1)-1})) \in V$. When $T_1(\eta^\ell(x_{n_1+1}^{k-2\ell-1})) \in V$ for some $\ell \in \mathbb{N}$, (13.4.34) and (13.4.36) yield

$$(E_{n_1+1,0} - E_{0,n+n_1+1})T_1(\eta^\ell(x_{n_1+1}^{k-2\ell-1})) = T_0(\eta^\ell(x_{n_1+1}^{k-2\ell})) \in V. \tag{13.4.38}$$

By induction on ℓ, we have $x_{n_1+1}^k = T_0(x_{n_1+1}^k) \in V$.

Note

$$(E_{n+i,n+n_1+1} - E_{n_1+1,i})|_{\mathscr{B}} = y_i\partial_{y_{n_1+1}} + x_i x_{n_1+1} \quad \text{for } i \in \overline{1,n_1} \tag{13.4.39}$$

and

$$(E_{n_2+r,n_2} - E_{n+n_2,n+m_2+r})|_{\mathscr{A}} = x_{n_2+r}\partial_{x_{n_2}} + y_{n_2}y_{n_2+r} \quad \text{for } r \in \overline{1,n-n_2} \tag{13.4.40}$$

if $n_2 < n$ by (13.3.37) and (13.3.38). Repeatedly applying (13.4.39) and (13.4.40) to (13.4.38) with various $i \in \overline{1, n_1}$ and $r \in \overline{1, n - n_2}$ if $n_2 < n$, we have

$$\left[\prod_{i=1}^{n_1+1} x_i^{\alpha_i}\right]\left[\prod_{j=n_2}^{n} y_j^{\beta_j}\right] \in V \quad \text{for } \alpha_i, \beta_j \in \mathbb{N};\ \alpha_{n_1+1}+\beta_{n_2}-\sum_{i=1}^{n_1}\alpha_i - \sum_{r=n_2+1}^{n}\beta_r = k. \tag{13.4.41}$$

Denote

$$I = \{0, \overline{n_1+1, n_2}, \overline{n+n_1+1, n+n_2}, \overline{2n+1, 2n+2m}\}. \tag{13.4.42}$$

Then the Lie subalgebra

$$\mathscr{G} = osp(2n+1|2m) \bigcap \left(\sum_{i,j\in I} \mathbb{F}E_{i,j}\right) \cong osp(2(n_2-n_1)+1|2m). \tag{13.4.43}$$

Applying Theorem 13.4.1 to $\mathscr{G}$ and $\mathbb{F}[x_0, x_{n_1+1}, \ldots, x_{n_2}, y_{n_1+1}, \ldots, y_{n_2}, \theta_1, \ldots, \theta_m, \vartheta_1, \ldots, \vartheta_m]$, we get

$$T_\iota(x^\alpha y^\beta \theta_{\vec{j}} \vartheta_{\vec{j}'}) \in V \tag{13.4.44}$$

for $\alpha, \beta \in \mathbb{N}^n$, $\vec{j} \in \Gamma_{k_1}$ and $\vec{j}' \in \Gamma_{k_2}$ such that $\beta_i = 0$ if $i \leq n_1$ and $\alpha_j = 0$ if $j > n_2$, and

$$\iota + k_1 + k_2 + \sum_{r=n_1+1}^{n_2}(\alpha_r + \beta_r) - \sum_{i=1}^{n_1}\alpha_i - \sum_{j=n_2+1}^{n}\beta_j = k. \tag{13.4.45}$$

Repeatedly applying (13.4.39) to (13.4.45) under above conditions with various $i \in \overline{1, n_1}$, we obtain (13.4.45) for $\alpha, \beta \in \mathbb{N}^n$, $\vec{j} \in \Gamma_{k_1}$ and $\vec{j}' \in \Gamma_{k_2}$ such that $\alpha_i = 0$ if $i > n_2$, and

$$\iota + k_1 + k_2 + \sum_{r=n_1+1}^{n_2}\alpha_r + \sum_{s=1}^{n_2}\beta_s - \sum_{i=1}^{n_1}\alpha_i - \sum_{j=n_2+1}^{n}\beta_j = k. \tag{13.4.46}$$

Observe

$$(E_{n_2+r,n_1+s} - E_{n+n_1+s,n+n_2+r})|_{\mathscr{B}} = y_{n_1+s}y_{n_2+r} + x_{n_2+r}\partial_{x_{n_1+s}} \tag{13.4.47}$$

for $r \in \overline{1, n-n_2}$ and $s \in \overline{1, n_2-n_1}$ by (13.3.37) and (13.3.38). Repeatedly applying (13.4.47) to (13.4.45) with $\alpha_i = 0$ if $i > n_2$, we obtain $\mathscr{H}'_{\langle k\rangle} \subset V$ by (13.4.32). So $\mathscr{H}'_{\langle k\rangle}$ is an irreducible $osp(2n+1|2m)$-module.

Next we consider $\mathscr{H}'_{\langle -k\rangle}$ with $k \in \mathbb{N}+1$. Let U be any nonzero $osp(2n+1|2m)$-submodule of $\mathscr{H}'_{\langle -k\rangle}$. According to the arguments in paragraph of (13.3.79)–(13.3.84), U contains some $T_\iota(\eta^\ell(x_{n_1}^{k+\iota+2\ell}))$. Observe Note

$$(E_{n_1,0} - E_{0,n+n_1}) = \partial_{x_0}\partial_{x_{n_1}} - x_0\partial_{y_{n_1}} \tag{13.4.48}$$

by (13.3.37) and (13.3.38). As operators on $\mathscr{B}$,

$$\begin{aligned}
&[E_{n_1,0} - E_{0,n+n_1}, T_0] \\
&= \left[\partial_{x_0}\partial_{x_{n_1}} - x_0\partial_{y_{n_1}}, \sum_{i=0}^{\infty}\frac{(-2)^i x_0^{2i}}{(2i)!}\Delta_{n_1,n_2}^i\right] \\
&= \sum_{i=1}^{\infty}\frac{(-2)^i x_0^{2i-1}}{(2i-1)!}\Delta_{n_1,n_2}^i\partial_{x_{n_1}} - \partial_{x_0}\sum_{i=1}^{\infty}\frac{i(-2)^i x_0^{2i}}{(2i)!}\Delta_{n_1,n_2}^{i-1}\partial_{y_{n_1}} \\
&= -2T_1\Delta_{n_1,n_2}\partial_{x_{n_1}} - \sum_{i=1}^{\infty}\frac{i(-2)^i x_0^{2i-1}}{(2i-1)!}\Delta_{n_1,n_2}^{i-1}\partial_{y_{n_1}} \\
&\quad - \sum_{i=1}^{\infty}\frac{i(-2)^i x_0^{2i}}{(2i)!}\Delta_{n_1,n_2}^{i-1}\partial_{y_{n_1}}\partial_{x_0},
\end{aligned} \tag{13.4.49}$$

$$\begin{aligned}
&[E_{n_1,0} - E_{0,n+n_1}, T_1] \\
&= \left[\partial_{x_0}\partial_{x_{n_1}} - x_0\partial_{y_{n_1}}, \sum_{i=0}^{\infty}\frac{(-2)^i x_0^{2i+1}}{(2i+1)!}\Delta_{n_1,n_2}^i\right] \\
&= \sum_{i=0}^{\infty}\frac{(-2)^i x_0^{2i}}{(2i)!}\Delta_{n_1,n_2}^i\partial_{x_{n_1}} - \partial_{x_0}\sum_{i=1}^{\infty}\frac{i(-2)^i x_0^{2i+1}}{(2i+1)!}\Delta_{n_1,n_2}^{i-1}\partial_{y_{n_1}} \\
&= T_0\partial_{x_{n_1}} - \sum_{i=1}^{\infty}\frac{i(-2)^i x_0^{2i}}{(2i)!}\Delta_{n_1,n_2}^{i-1}\partial_{y_{n_1}} - \sum_{i=1}^{\infty}\frac{i(-2)^i x_0^{2i+1}}{(2i+1)!}\Delta_{n_1,n_2}^{i-1}\partial_{y_{n_1}}\partial_{x_0}.
\end{aligned} \tag{13.4.50}$$

If $T_0(\eta^\ell(x_{n_1}^{k+2\ell})) \in U$ for some $\ell \in \mathbb{N}+1$, we have

$$\begin{aligned}
&(E_{n_1,0} - E_{0,n+n_1})T_0(\eta^\ell(x_{n_1+1}^{k-2\ell})) \\
&= [T_1\partial_{y_{n_1}} - 2T_1\Delta_{n_1,n_2}\partial_{x_{n_1}}](\eta^\ell(x_{n_1+1}^{k+2\ell})) \\
&= (k+2\ell)\ell[1+2(n_1+k+\ell-n_2)]T_1(\eta^{\ell-1}(x_{n_1+1}^{k+2\ell-1})) \in V
\end{aligned} \tag{13.4.51}$$

by (13.3.74), (13.4.48) and (13.4.49). So $T_1(\eta^{\ell-1}(x_{n_1}^{k+2(\ell-1)+1})) \in U$. When $T_1(\eta^\ell (x_{n_1}^{k+2\ell+1})) \in V$ for some $\ell \in \mathbb{N}$, (13.4.48) and (13.4.50) yield

$$(E_{n_1,0} - E_{0,n+n_1})T_1(\eta^\ell(x_{n_1}^{k+2\ell+1})) = (k+2\ell+1)T_0(\eta^\ell(x_{n_1}^{k+2\ell})) \in U. \tag{13.4.52}$$

By induction on ℓ, we have $x_{n_1}^k = T_0(x_{n_1}^k) \in U$.

According to (13.4.39) with $i = n_1$,

$$x_{n_1}^{k+k'} x_{n_1+1}^{k'} \in U \qquad \text{for } k' \in \mathbb{N}. \tag{13.4.53}$$

Moreover,

$$(E_{i,n_1} - E_{n+n_1,n+i})|_{\mathscr{B}} = x_i \partial_{x_{n_1}} - y_{n_1} \partial_{y_i} \qquad \text{for} i \in \overline{1, n_1 - 1} \tag{13.4.54}$$

by (13.3.37) and (13.3.38). Repeatedly applying (13.4.54) to (13.4.53) with various $i \in \overline{1, n_1 - 1}$, we have

$$\prod_{i=1}^{n_1+1} x_i^{\alpha_i} \in U \qquad \text{for } \alpha_i \in \mathbb{N};\ \alpha_{n_1+1} - \sum_{i=1}^{n_1} \alpha_i = -k. \tag{13.4.55}$$

Observe

$$(E_{n_2+r,n+1} - E_{1,n+n_2+r})|_{\mathscr{A}} = x_{n_2+r}\partial_{y_1} + y_{n_2+r}\partial_{x_1} \ \text{ for } r \in\in \overline{1, n - n_2} \text{ if } n_2 < n \tag{13.4.56}$$

by (13.3.39). Repeatedly applying (13.4.56) to (13.4.55) with various $r \in \overline{1, n - n_2}$ if $n_2 < n$, we find

$$\left[\prod_{i=1}^{n_1+1} x_i^{\alpha_i}\right]\left[\prod_{j=n_2+1}^{n} y_j^{\beta_j}\right] \in U \qquad \text{for } \alpha_i, \beta_j \in \mathbb{N};\ \alpha_{n_1+1} - \sum_{i=1}^{n_1} \alpha_i - \sum_{j=n_2+1}^{n} \ = -k. \tag{13.4.57}$$

By the same arguments from (13.4.43) to the end of the paragraph below (13.4.47) with k replaced by $-k$, we prove that $\mathscr{H}'_{\langle -k\rangle}$ is an irreducible $osp(2n+1|2m)$-module.

We calculate

$$\left[\Delta'_{n_1,n_2}, \eta'\right] = 2 + 4(n_2 - n_1 - m) + 4\left[x_0\partial_{x_0} + \sum_{i=1}^{n}(x_i\partial_{x_i} + y_i\partial_{y_i}) + \sum_{r=1}^{m}(\theta_r\partial_{\theta_r} + \vartheta_r\partial_{\vartheta_r})\right]. \tag{13.4.58}$$

By the arguments in (13.2.58)–(13.2.64), $\mathscr{B} = \bigoplus_{\ell,k=0}^{\infty} \eta'^{\ell}(\mathscr{H}'_{\langle k\rangle})$ is a direct sum of irreducible $osp(2n+1|2m)$-submodules for any $k \in \mathbb{Z}$. $\square$

It can be verified by (13.4.58) that the space $\mathbb{F}\Delta'_{n_1,n_2} + \mathbb{F}[\Delta'_{n_1,n_2}, \eta'] + \mathbb{F}\eta'$ in terms of (13.4.28) and (13.4.29) also forms an operator Lie algebra isomorphic to $sl(2, \mathbb{F})$. The above theorem establishes a supersymmetric $(sl(2, \mathbb{F}), osp(2n + 1|2m))$ Howe duality.

References

[GS] J. Gervais, B. Sakita, Field theory interpretation of supergauges in dual models. Nucl. Phys. B **34**(2), 632–639

[Hr1] R. Howe, Dual pairs in physics: harmonic oscillators, photons, electrons, and singletons, *Applications of Group Theory in Physics and Mathematical Physics (Chicago, 1982)*. Lectures in Applied Mathematics, vol. 21 (American Mathematical Society, Providence, 1985), pp. 179–207

[Hr2] R. Howe, Remarks on classical invariant theory. Trans. Am. Math. Soc. **313**, 539–570 (1989)

[Hr3] R. Howe, Transcending classical invariant theory. J. Am. Math. Soc. **2**, 535–552 (1989)

[Hr4] R. Howe, Perspectives on invariant theory: Schur duality, multiplicity-free actions and beyond, *The Schur lectures (1992) (Tel Aviv)*. Israel Mathematical Conference Proceedings, vol. 8 (Bar-Ilan University, Ramat Gan, 1995), pp. 1–182

[Mh] C. Luo, X. Xu, Supersymmetric analogues of the classical theorem on harmonic polynomials. J. Algebra Appl. **13**(6), 1450011 (2014). 42pp

[R1] H. Miyazawa, Baryon number changing currents. Prog. Theor. Phys. **36**(6), 1266–1276 (1966)

[LX4] P. Ramond, Dual theory of free fermions. Phys. Rev. D **3**(10), 2415–2418 (1971)

[Z] R.B. Zhang, Orthosymplectic Lie superalgebras in superspace analogues of quantum Kepler problems. Commun. Math. Phys. **280**, 545–562 (2008)

Chapter 14
Representation Theoretic Codes

Coding theory began with Shannon's famous paper "A mathematical theory of communication" (cf. [Sc]). Given a communication channel which may corrupt information sent over it, Shannon identified a number called the capacity of the channel and proved that arbitrary reliable communication is possible at any rate below the channel capacity. Examples of communication channels include telephone, telegraph, and magnetic storage devices or any kind of electronic communication device. The common feature of communication channels is that information is emanating from a source and is sent over the channel to a receiver at the other end. The channel is "noisy" in the sense that what is received is not always the same as what was sent. The fundamental problem is to determine what message was sent based on the approximation (due to the noise) that was received. The basic idea is to embellish the message by adding some redundancy to it so that the received message is a good approximation. Thereby error-correcting codes naturally arose in information theory.

Linear codes with large minimal distances are important error correcting codes in information theory. Orthogonal codes have more applications in the other fields of mathematics. The representation theory of finite-dimensional simple Lie algebras has been developing over one hundred years. Naturally, people had speculated the existence of some kind of relations between the representation theory and coding theory. However, it had not been found until our work [X20] published in 2012. We initially observed the coding phenomenon when we were studying the explicit structures of the basic oscillator representations of the exceptional simple Lie algebras of types F_4, E_6 and E_7. The results on these representations are presented in Sects. 10.3, 10.4, 11.1, 11.2 and 12.1–12.3. It turned out that error correcting codes are well dwelling in the representation theory.

The minimum distance of a code measures the error-correcting ability of the code. We found that certain weight matrices of $sl(n, \mathbb{C})$-modules in Chap. 6 and $o(2n, \mathbb{C})$-modules in Chap. 7 generate doubly-even binary orthogonal codes and ternary orthogonal codes with large minimal distances. Some of these codes are even optimal codes (e.g., cf. PHB]). Moreover, we proved that the weight matrices

X. Xu, *Representations of Lie Algebras and Partial Differential Equations*,
DOI 10.1007/978-981-10-6391-6_14

of F_4, E_6, E_7 and E_8 on their minimal irreducible modules and adjoint modules all generate ternary orthogonal codes with large minimal distances. In determining the minimal distances, we have used the Weyl groups (presented in Sect. 3.2) and branch rules of the irreducible representations of the related simple Lie algebras.

This chapter presents our theory of representation theoretic codes. In Sect. 14.1, we give the basic concepts on linear codes and the codes arisen from the representation theory. Section 14.2 is to study the binary and ternary codes related to representations of special linear Lie algebras. Section 14.3 is devoted to studying ternary codes related to certain representations of even orthogonal Lie algebras. Finally in Sect. 14.4, we determine the structure of the ternary weight codes of F_4, E_6, E_7, E_8 on its minimal irreducible module and adjoint module.

14.1 Basics Definitions

Let m be a positive integer and denote $\mathbb{Z}_m = \mathbb{Z}/m\mathbb{Z}$, which is a *ring* with the algebraic operations induced from $\mathbb{Z}$. A *code* $\mathscr{C}$ *of length n* is a subset of $(\mathbb{Z}_m)^n$ for some m, where the ring structure of $\mathbb{Z}_m$ may not be used. The elements of $\mathscr{C}$ are called *codewords*. The *(Hamming) distance* between two codewords is the number of different coordinates. The *minimal distance* of a code is the minimal number among the distances of all its pairs of codewords in the code. A code with minimal distance d can be used to correct $[\![(d-1)/2]\!]$ errors in signal transmissions.

A *linear code* $\mathscr{C}$ over the ring $\mathbb{Z}_m$ is a $\mathbb{Z}_m$-submodule of $(\mathbb{Z}_m)^n$. The *(Hamming) weight* of a codeword in a linear code $\mathscr{C}$ is the number of its nonzero coordinates. In this case, the minimal distance of $\mathscr{C}$ is exactly the minimal weight of the nonzero codewords in $\mathscr{C}$. The inner product in $(\mathbb{Z}_m)^n$ is defined by

$$(a_1, \ldots, a_n) \cdot (b_1, \ldots, b_n) = \sum_{i=1}^{n} a_i b_i. \tag{14.1.1}$$

Moreover, $\mathscr{C}$ is called *orthogonal* if

$$\mathscr{C} \subseteq \{\vec{a} \in (\mathbb{Z}_m)^n \mid \vec{\alpha} \cdot \vec{b} = 0 \text{ for } \vec{b} \in \mathscr{C}\}. \tag{14.1.2}$$

When the equality holds, we call $\mathscr{C}$ a *self-dual* code. Orthogonal linear codes (especially, self-dual codes) have important applications to the other mathematical fields such as sphere packing, integral linear lattices, finite group theory, etc. A code is called *binary* if $m = 2$ and *ternary* when $m = 3$. A binary linear code is called *even (doubly-even)* if the weights of all its codewords are divisible by 2 (by 4).

Let $\mathscr{G}$ be a finite-dimensional simple Lie algebras over $\mathbb{C}$, the field of complex numbers. Take a Cartan subalgebra H and simple positive roots $\{\alpha_1, \alpha_2, \ldots, \alpha_n\}$. Moreover, we denote by $\{h_1, h_2, \ldots, h_n\}$ the elements of H such that the matrix

$$(\alpha_i(h_j))_{n\times n} \text{ is the Cartan matrix of } \mathscr{G} \tag{14.1.3}$$

((2.5.13), (2.5.18), (3.1.8) and (3.3.2)). For a finite-dimensional $\mathscr{G}$-module V, V has a weight-subspace decomposition:

$$V = \bigoplus_{\mu\in H^*} V_\mu, \qquad V_\mu = \{v \in V \mid h(v) = \mu(h)v \text{ for } h \in H\}. \tag{14.1.4}$$

Take a maximal linearly independent set $\{u_1, u_2, \ldots, u_k\}$ of weight vectors with nonzero weights in V such that the order is compatible with the partial order $\lhd$ of weights in (3.4.49). Write

$$h_i(u_j) = c_{i,j}u_j, \qquad C(V) = (c_{i,j})_{n\times k}. \tag{14.1.5}$$

By the representation theory of simple Lie algebras, all $c_{i,j}$ are integers. We call $C(V)$ the *weight matrix of* $\mathscr{G}$ *on* V. Identify integers with their images in $\mathbb{Z}_m$ when the context is clear. Denote by $\mathscr{C}_m(V)$ the linear code over $\mathbb{Z}_m$ generated by $C(V)$.

Suppose that the weight of u_i is μ_i. Set

$$\bar{H}_m = \sum_{i=1}^n \mathbb{Z}_m h_i. \tag{14.1.6}$$

We define a map $\jmath : \bar{H}_m \to (\mathbb{Z}_m)^k$ by

$$\jmath(\sum_{i=1}^n l_i h_i) = (\sum_{i=1}^n l_i\mu_1(h_i), \sum_{i=1}^n l_i\mu_2(h_i), \ldots, \sum_{i=1}^n l_i\mu_k(h_i)). \tag{14.1.7}$$

Then

$$\mathscr{C}_m(V) = \jmath(\bar{H}_m). \tag{14.1.8}$$

Let $\mathscr{W}$ be the Weyl group of the simple Lie algebra $\mathscr{G}$. For any $\sigma \in \mathscr{W}$, there exists a linear automorphism $\hat{\sigma}$ of V such that

$$\hat{\sigma}(V_\mu) = V_{\sigma(\mu)}, \qquad \sigma(\mu)(\sigma(h)) = \mu(h) \qquad \text{for } h \in H \tag{14.1.9}$$

(cf. (5.2.41)), where $\sigma(t_\alpha) = t_{\sigma(\alpha)}$ in terms of (2.5.13). Moreover, we define an action of $\mathscr{W}$ on $\bar{H}_m$ by

$$\sigma(\sum_{i=1}^n l_i h_i) = \sum_{i=1}^n l_i\sigma(h_i) \qquad \text{for } \sigma \in \mathscr{W}. \tag{14.1.10}$$

According to (14.1.9),

$$\mathrm{wt}\, \jmath(\sigma(h)) = \mathrm{wt}\, \jmath(h) \qquad \text{for } \sigma \in \mathscr{W}(\mathscr{G}),\ h \in \bar{H}_m. \tag{14.1.11}$$

So the number of the distinct weights of codewords in $\mathscr{C}_m(V)$ is less than or equal to the number of $\mathscr{W}$-orbits in $\bar{H}_m$. Expression (14.1.11) will be used later in determining minimal distances.

Let $\Lambda(V)$ be the set of nonzero weights of V. The module V is called *self-dual* if $\Lambda(V) = -\Lambda(V)$. In this chapter, we are only interested in the binary and ternary codes. We call $\mathscr{C}_2(V)$ the *binary weight code* of $\mathscr{G}$ *on* V. If V is self-dual, then the weight matrix $C(V) = (-A, A)$ and $\mathscr{C}_3(V)$ is orthogonal if and only if A generates a ternary orthogonal code (e.g., cf. [P4]). For this reason, we call the ternary code generated by A the *ternary weight code* of $\mathscr{G}$ *on* V if V is self-dual. When V is not self-dual, then $\mathscr{C}_3(V)$ is the *ternary weight code* of $\mathscr{G}$ *on* V.

Denote by $V_X(\lambda)$ the finite-dimensional irreducible module of a simple Lie algebra of type X with the highest weight λ. Let p be a prime number. Then $\mathbb{Z}_p$ is a finite field, which is traditionally denoted as $\mathbb{F}_p$. A linear code $\mathscr{C}$ of length n over $\mathbb{F}_p$ is a vector subspace of $\mathbb{F}_p^n$ over $\mathbb{F}_p$. If $\dim \mathscr{C} = k$, we say that $\mathscr{C}$ is of *type* $[n, k]$. When d is the minimal distance of $\mathscr{C}$, we call $\mathscr{C}$ an $[n, k, d]$-*code*.

14.2 Codes Related to Representations of $sl(n, \mathbb{C})$

In this section, we study the binary and ternary codes related to representations of $sl(n, \mathbb{C})$, where $n > 1$ is an integer. Again we denote

$$\varepsilon_i = (0, \ldots, \overset{i}{1}, 0, \ldots, 0) \in \mathbb{R}^n. \tag{14.2.1}$$

So

$$\mathbb{R}^n = \sum_{i=1}^{n} \mathbb{R}\varepsilon_i. \tag{14.2.2}$$

Then inner product "$(\cdot, \cdot)$" is Euclidian; that is,

$$\left(\sum_{i=1}^{n} k_i\varepsilon_i, \sum_{j=1}^{n} l_j\varepsilon_j\right) = \sum_{i=1}^{n} k_i l_i. \tag{14.2.3}$$

Recall the special linear Lie algebra $sl(n, \mathbb{C})$ and related settings in (6.2.1)–(6.2.4). Its root system

$$\Phi_{A_{n-1}} = \{\varepsilon_i - \varepsilon_j \mid i, j \in \overline{1, n},\ i \neq j\}. \tag{14.2.4}$$

Take the simple positive roots

$$\alpha_i = \varepsilon_i - \varepsilon_{i+1} \qquad \text{for } i \in \overline{1, n-1}. \tag{14.2.5}$$

The Weyl group $\mathscr{W}_{A_{n-1}}$ of $sl(n, \mathbb{C})$ is exactly the full permutation group S_n on $\overline{1, n}$, which acts on $H_{A_{n-1}}$ and $\mathbb{R}^n$ by permuting sub-indices of $E_{i,i}$ and ε_i, respectively.

Let Ψ be the exterior algebra generated by $\{\theta_1, \theta_2, \dots, \theta_n\}$ (cf. (6.2.15)) and take the related settings (6.2.16)–(6.2.19). Proposition 6.2.2 says that Ψ_r forms an irreducible $sl(n, \mathbb{C})$-submodule of highest weight λ_r for $r \in \overline{1, n-1}$. The Weyl group $\mathscr{W}_{A_{n-1}}$ acts on $\mathscr{A}$ by permuting sub-indices of θ_i.

Two $k_1 \times k_2$ matrices A_1 and A_2 with entries in $\mathbb{Z}_m$ are called *equivalent* in the sense of coding theory if there exist an invertible $k_1 \times k_1$ matrix K_1 and an invertible $k_2 \times k_2$ monomial matrix K_2 such that $A_1 = K_1 A_2 K_2$. Equivalent matrices generate isomorphic codes. Take any order of the basis

$$\{x_{r,1}, x_{r,2}, \dots, x_{r,\binom{n}{r}}\} = \{\theta_{i_1}\theta_{i_2}\cdots\theta_{i_r} \mid 1 \le i_1 < i_2 < \cdots < i_r \le n\} \tag{14.2.6}$$

of Ψ_r. Then we have

$$h_i(x_{r,j}) = a_{i,j}(r) x_{r,j}, \qquad a_{i,j}(r) \in \mathbb{Z}. \tag{14.2.7}$$

Modulo equivalence, the weight matrix

$$C(\Psi_r) = [a_{i,j}(r)]_{(n-1)\times\binom{n}{r}}. \tag{14.2.8}$$

Theorem 14.2.1 *When $n = 2m \ge 4$ is even, $\mathscr{C}_2(\Psi_2)$ is a doubly-even binary orthogonal $[m(2m-1), 2(m-1), 4(m-1)]$-code.*

Proof Denote by ξ_i the ith row $C_2(\Psi_2)$. Then

$$\text{wt}\, \xi_i = 2(n-2) \qquad \text{for } i \in \overline{1, n-1}. \tag{14.2.9}$$

Moreover,

$$\sum_{i=0}^{m-1} \xi_{2i+1} = 0 \qquad \text{in } \mathscr{C}_2(\Psi_2). \tag{14.2.10}$$

Furthermore,

$$\xi_i \cdot \xi_j = 4 \equiv 0 \qquad \text{if } i + 1 < j \tag{14.2.11}$$

and

$$\xi_i \cdot \xi_{i+1} = 2(m-1) \equiv 0. \tag{14.2.12}$$

Write

$$E_{i,i}(x_{r,j}) = b_{i,j}(r)x_{r,j}, \qquad B_r = [b_{i,j}(r)]_{n\times\binom{n}{r}}. \tag{14.2.13}$$

Denote by ζ_i the ith row of B_2. By symmetry (cf. (14.1.9)–(14.1.11)), any nonzero codeword in $\mathscr{C}_2(\Psi_2)$ has the same weight as the codeword

$$u = \sum_{s=1}^{2t} \zeta_s \in \mathbb{F}_2^{n(n-1)/2} \quad \text{for some } t \in \overline{1, m-1}. \tag{14.2.14}$$

We calculate

$$\text{wt}\, u = 4t(m-t) = -4t^2 + 4mt. \tag{14.2.15}$$

Since the function $-4t^2 + t(4m-1)$ attains maximal at $t = m/2$, wt u is minimal at $t = 1$ or $m-1$. Note

$$\text{wt}\, u = 4(m-1) \quad \text{if } t = 1 \text{ or } m-1. \tag{14.2.16}$$

Thus the code $\mathscr{C}_2(\Psi_2)$ has the minimal distance $4(m-1)$. □

When $m = 2$, $\mathscr{C}_2(\Psi_2)$ is a doubly-even binary orthogonal [6, 2, 4]-code. If $m = 3$, $\mathscr{C}_2(\Psi_2)$ becomes a doubly-even binary orthogonal [15, 4, 8]-code. These two code are optimal linear codes (e.g., cf. [PHB]). In the case of $m = 4$, $\mathscr{C}_2(\Psi_2)$ is a doubly-even binary orthogonal [28, 6, 12]-code.

Theorem 2.2 *The code* $\mathscr{C}_2(\Psi_3)$ *is a doubly-even binary orthogonal* $[\binom{n}{3}, n-1, (n-2)(n-3)]$ *-code if* $n > 9$ *and* $n \equiv 2, 3$ *(mod* 4).

Proof Denote by ξ_i the ith row the weight matrix $C(\Psi_3)$. Then

$$\text{wt}\, \xi_i = (n-2)(n-3) \quad \text{for } i \in \overline{1, n-1}. \tag{14.2.17}$$

Moreover,

$$\xi_i \cdot \xi_j = 4(n-4) \quad \text{if } i+1 < j \tag{14.2.18}$$

and

$$\xi_i \cdot \xi_{i+1} = n - 3 + \binom{n-3}{2} = \frac{(n-2)(n-3)}{2}. \tag{14.2.19}$$

So $\mathscr{C}_2(\Psi_3)$ is a doubly-even binary orthogonal code under the assumption.

Denote by ζ_i the ith row of B_3 (cf. (14.2.13)). By symmetry (cf. (14.1.9)–(14.1.11)), any nonzero codeword in $\mathscr{C}_2(\Psi_3)$ has the same weight as the codeword

$$u(t) = \sum_{s=1}^{2t} \zeta_s \in \mathbb{F}_2^n \qquad \text{for some } t \in \overline{1, [\![n/2]\!]}. \tag{14.2.20}$$

We calculate

$$f(t) = 3\text{wt}\, u(t) = 3\binom{2t}{3} + 6t\binom{n-2t}{2} = t[16t^2 - 12nt + 3n(n-1) + 2]. \tag{14.2.21}$$

Moreover,

$$f'(t) = 48t^2 - 24nt + 3n(n-1) + 2 = 48\left(t - \frac{n}{4}\right)^2 - 3n + 2. \tag{14.2.22}$$

Thus

$$f'(t_0) = 0 \Longrightarrow t_0 = \frac{n}{4} \pm \frac{1}{4}\sqrt{n - \frac{2}{3}}. \tag{14.2.23}$$

Since $f'(0) = 3n(n-1) + 2 > 0$, $f(t)$ attains local maximum at

$$t = \frac{n}{4} - \frac{1}{4}\sqrt{n - \frac{2}{3}} \tag{14.2.24}$$

and local minimum at

$$t = \frac{n}{4} + \frac{1}{4}\sqrt{n - \frac{2}{3}}. \tag{14.2.25}$$

According to (14.2.17) and (14.2.21), $f(1) = 3(n-2)(n-3)$. Furthermore,

$$\begin{aligned} & f\left(\frac{n}{4} + \frac{1}{4}\sqrt{n - \frac{2}{3}}\right) \\ &= \left(\frac{n}{4} + \frac{1}{4}\sqrt{n - \frac{2}{3}}\right)\left[16\left(\frac{n}{4} + \frac{1}{4}\sqrt{n - \frac{2}{3}}\right)^2\right. \\ &\quad \left. -12n\left(\frac{n}{4} + \frac{1}{4}\sqrt{n - \frac{2}{3}}\right) + 3n(n-1) + 2\right] \\ &= \frac{1}{4}\left(n + \sqrt{n - \frac{2}{3}}\right)\left[\left(n + \sqrt{n - \frac{2}{3}}\right)^2\right. \\ &\quad \left. -3n\left(n + \sqrt{n - \frac{2}{3}}\right) + 3n(n-1) + 2\right] \end{aligned}$$

$$= \frac{1}{4}\left(n + \sqrt{n - \frac{2}{3}}\right)\left[n\left(n - \sqrt{n - \frac{2}{3}}\right) - 2n + \frac{4}{3}\right]$$
$$= \frac{1}{4}\left[n^3 - 3n^2 + 2n + \left(\frac{4}{3} - 2n\right)\sqrt{n - \frac{2}{3}}\right]$$
$$> \frac{1}{4}(n^3 - 5n^2 + 2n). \tag{14.2.26}$$

Thus

$$f\left(\frac{n}{4} + \frac{1}{4}\sqrt{n - \frac{2}{3}}\right) - f(1)$$
$$> \frac{1}{4}(n^3 - 5n^2 + 2n) - 3(n-2)(n-3) = \frac{1}{4}(n^3 - 17n^2 + 62n - 72)$$
$$> \frac{n^2(n-17)}{4}. \tag{14.2.27}$$

If $n \geq 17$, we have

$$f\left(\frac{n}{4} + \frac{1}{4}\sqrt{n - \frac{2}{3}}\right) > f(1) \tag{14.2.28}$$

and

$$f(n/2) - f(1)$$
$$= \frac{n}{2}[4n^2 - 6n^2 + 3n(n-1) + 2] - 3(n-2)(n-3)$$
$$= \frac{n(n-1)(n-2)}{2} - 3(n-2)(n-3)$$
$$= \frac{(n-2)(n^2 - 7n + 9)}{2} > 0 \text{ if } n \geq 6. \tag{14.2.29}$$

Thus the minimal weight is $f(1)/3 = (n-2)(n-3)$ when $n \geq 17$.

When $n = 10$, we calculate (Table 14.1)

Table 14.1 Weights of $\mathscr{C}_2(\Psi_3)$ with $n = 10$

t	1	2	3	4	5
wt u(t)	56	64	56	64	120

If $n = 11$, we find (Table 14.2)

Table 14.2 Weights of $\mathscr{C}_2(\Psi_3)$ with $n = 11$

t	1	2	3	4	5
wt u(t)	72	88	80	80	120

When $n = 14$, we obtain (Table 14.3)

Table 14.3 Weights of $\mathscr{C}_2(\Psi_3)$ with $n = 14$

t	1	2	3	4	5	6	7
wt u(t)	132	184	188	176	180	232	364

If $n = 15$, we get (Table 14.4)

Table 14.4 Weights of $\mathscr{C}_2(\Psi_3)$ with $n = 15$

t	1	2	3	4	5	6	7
wt u(t)	156	224	216	224	220	256	364

This prove the conclusion in the theorem. □

Note that when $n = 6$, we find (Table 14.5)

Table 14.5 Weights of $\mathscr{C}_2(\Psi_3)$ with $n = 6$

t	1	2	3
wt u(t)	12	8	20

So $\mathscr{C}_2(\Psi_3)$ is a doubly-even binary orthogonal $[20, 5, 8]$-code. Moreover, if $n = 7$, we find (Table 14.6)

Table 14.6 Weights of $\mathscr{C}_2(\Psi_3)$ with $n = 7$

t	1	2	3
wt u(t)	20	16	20

Hence $\mathscr{C}_2(\Psi_3)$ a doubly-even binary orthogonal $[35, 6, 16]$-code. In both cases, the above theorem fails and both codes are the best even codes among the binary codes with the same length and dimension (e.g., cf. [PHB]).

According to the above theorem, $\mathscr{C}_2(\Psi_3)$ is a doubly-even binary orthogonal $[120, 9, 56]$-code when $n = 10$, $[165, 10, 72]$-code if $n = 11$, $[364, 13, 132]$-code when $n = 14$ and $[455, 14, 156]$-code if $n = 15$.

Next let us consider the ternary codes. Again by symmetry, any nonzero codeword in $\mathscr{C}_3(\Psi_r)$ has the same weight as the codeword

$$u(s,t) = \sum_{r=1}^{s} \zeta_r - \sum_{i=1}^{t} \zeta_{s+i} \in \mathbb{F}_3^{\binom{n}{r}} \tag{14.2.30}$$

for some nonnegative integers s, t, where ζ_ι is the ιth row of the matrix B_r in (14.2.13). Moreover,

$$\mathrm{wt}\, u(s,t) = \mathrm{wt}\, u(t,s). \tag{14.2.31}$$

Furthermore, we have

$$\mathrm{wt}\, u(s,t) = (s+t)(n-s-t) + \binom{s}{2} + \binom{t}{2} \quad \text{in } \mathscr{C}_3(\Psi_2) \tag{14.2.32}$$

and

$$\mathrm{wt}\, u(s,t) = (s+t)\binom{n-s-t}{2} + (n-s)\binom{s}{2} + (n-t)\binom{t}{2} \quad \text{in } \mathscr{C}_3(\Psi_3). \tag{14.2.33}$$

For convenience, we denote

$$\begin{aligned} f(s,t) &= 2\mathrm{wt}\, u(s,t) = 2(s+t)(n-s-t) + s(s-1) + t(t-1) \\ &= (2n-1)(s+t) - s^2 - t^2 - 4st \end{aligned} \tag{14.2.34}$$

in $\mathscr{C}_3(\Psi_2)$ and

$$\begin{aligned} g(s,t) &= 2\mathrm{wt}\, u(s,t) \\ &= (s+t)(n-s-t)(n-s-t-1) \\ &\quad +(n-s)s(s-1) + (n-t)t(t-1) \\ &= (s+t)^3 - (2n-1)(s+t)^2 + n(n-1)(s+t) \\ &\quad -s^3 - t^3 + (n+1)(s^2+t^2) - n(s+t) \\ &= 3st^2 + 3s^2t + (2-n)(s^2+t^2) \\ &\quad -2(2n-1)st + n(n-2)(s+t) \end{aligned} \tag{14.2.35}$$

in $\mathscr{C}_3(\Psi_3)$.

Note

$$f(3,0) = 3(2n-1) - 9 = 6(n-2), \quad f(n,0) = n(2n-1) - n^2 = n(n-1), \tag{14.2.36}$$

$$f(1,1) = 2(2n-1) - 6 = 4(n-2), \quad f(1,n-1) = (n-1)(n-2). \tag{14.2.37}$$

Since geometrically $f(s,t)$ has only local minimum, it attains the absolute minimum at boundary points. Thus

$$\min\{f(s,t) \mid s \equiv t \pmod 3\} = 4(n-2) \quad \text{if } n \geq 5. \tag{14.2.38}$$

Now

$$g_s(s,t) = 3t^2 + 6st + 2(2-n)s - 2(2n-1)t + n(n-2), \tag{14.2.39}$$

$$g_t(s,t) = 3s^2 + 6st + 2(2-n)t - 2(2n-1)s + n(n-2). \tag{14.2.40}$$

Suppose that $g_s(s_0, t_0) = g_t(s_0, t_0) = 0$ for $s_0, t_0 \geq 0$; that is,

$$3t_0^2 + 6s_0t_0 + 2(2-n)s_0 - 2(2n-1)t_0 + n(n-2) = 0, \tag{14.2.41}$$

$$3s_0^2 + 6s_0t_0 + 2(2-n)t_0 - 2(2n-1)s_0 + n(n-2) = 0. \tag{14.2.42}$$

By (14.2.41)–(14.2.42), we get

$$(t_0 - s_0)(3t_0 + 3s_0 - 2(n+1)) = 0 \Longrightarrow t_0 = s_0 \text{ or } 3t_0 + 3s_0 = 2(n+1). \tag{14.2.43}$$

If $s_0 = t_0$, then we find

$$9s_0^2 - 2(n-1)s_0 + n(n-2) = 0 \sim 8s_0^2 + (s_0 - n + 1)^2 - 1 = 0, \tag{14.2.44}$$

which leads to a contradiction because $n > 1$. Thus $3t_0 + 3s_0 = 2(n-1)$. Denote $s_1 = 3t_0$ and $t_1 = 3t_0$. Then $s_1 + t_1 = 2(n+1)$ and (14.2.41) becomes

$$\begin{aligned} &t_1^2 + 2(2(n+1) - t_1)t_1 + 2(2-n)(2(n+1) - t_1) \\ &-2(2n-1)t_1 + 3n(n-2) = 0, \end{aligned} \tag{14.2.45}$$

or equivalently,

$$\begin{aligned} &t_1^2 - 2(n+1)t_1 + (n-2)(n+4) = 0 \\ &\sim (t_1 - n - 1)^2 - 9 = 0 \Longrightarrow t_1 = n+4,\ n-2. \end{aligned} \tag{14.2.46}$$

Therefore,

$$s_0 = \frac{n+4}{3}, \quad t_0 = \frac{n-2}{3} \quad \text{or} \quad t_0 = \frac{n+4}{3}, \quad s_0 = \frac{n-2}{3}. \tag{14.2.47}$$

We calculate

$$g(s_0, t_0) = \frac{2(n-2)(n^2 - n - 3)}{9}, \tag{14.2.48}$$

$$g(1,0) = g(n-1,0) = (n-1)(n-2), \tag{14.2.49}$$

$$g(3,0) = 3(n-2)(n-3), \quad g(n,0) = 0, \tag{14.2.50}$$

$$g(1,1) = g(n-2,1) = 2(n-2)(n-3), \quad g(n-2,0) = 2(n-2)^2. \tag{14.2.51}$$

Moreover,

$$g(s_0,t_0) \geq g(1,0), \; g(1,1) \qquad \text{if } n \geq 6. \tag{14.2.52}$$

When $n = 5$, we calculate

$$g(1,0) = g(1,1) = g(2,1) = g(2,2) = g(3,1) = g(4,0) = g(4,1) = 12, \tag{14.2.53}$$

$$g(2,0) = g(3,0) = g(3,2) = 18. \tag{14.2.54}$$

In summary, we have:

Theorem 14.2.3 *Let $n \geq 5$. The matrix B_3 (cf. (14.2.13)) generates a ternary $\left[\binom{n}{3}, n-1, \binom{n-1}{2}\right]$-code, which is equal to $\mathscr{C}_3(\Psi_3)$ if $n \not\equiv 0$ (mod 3). If $n = 3m+2$ for some positive integer m, $\mathscr{C}_3(\Psi_2)$ is a ternary orthogonal $[\binom{3m+2}{2}, 3m+1, 6m]$-code and $\mathscr{C}_3(\Psi_3)$ is a ternary orthogonal $[\binom{3m+2}{3}, 3m+1, 3m(3m+1)/2]$-code. The code $\mathscr{C}_3(\Psi_3)$ is a ternary orthogonal $\left[\binom{n}{3}, n-2, (n-2)(n-3)\right]$-code when $n \equiv 0$ (mod 3).*

Proof The part of minimal distances has been proved by the above arguments. We only need to prove orthogonality.

Suppose $n = 3m+2$. In $\mathscr{C}_3(\Psi_2)$, ξ_r stands for the rth row of the weight matrix $C(\Psi_2)$ and

$$\xi_i \cdot \xi_j = 2 - 2 = 0 \qquad \text{for } 1 \leq i < j-1 \leq n-2, \tag{14.2.55}$$

$$\xi_r \cdot \xi_{r+1} = -(n-2) = -3m, \quad \xi_s \cdot \xi_s = 2(n-2) = 6m \tag{14.2.56}$$

for $r \in \overline{1, n-2}$ and $s \in \overline{1, n-1}$. So $\mathscr{C}_3(\Psi_2)$ is orthogonal. Now ζ_r stands for the rth row of B_3 (cf. (14.2.18)). Observe

$$\sum_{i=1}^{n} \zeta_i = 0 \in \mathbb{C}_3^{\binom{n}{3}} \tag{14.2.57}$$

by (6.2.16) and (6.2.19). Moreover,

$$\zeta_i \cdot \zeta_j = n-2 = 3m, \quad \zeta_i \cdot \zeta_i = \binom{n-1}{2} = \frac{3m(3m+1)}{2}, \qquad i \neq j. \tag{14.2.58}$$

Thus B_3 generate a ternary orthogonal code.

Assume that $n = 3m$ for some nonnegative integer m. In $\mathscr{C}_3(\Psi_3)$, we also use ξ_r for the rth row of the weight code $C(\Psi_3)$ and

$$\xi_i \cdot \xi_j = 2(n-4) - 2(n-4) = 0 \qquad \text{for } 1 \le i < j-1 \le n-2, \tag{14.2.59}$$

$$\xi_s \cdot \xi_s = 2\xi_r \cdot \xi_{r+1} = (n-2)(n-3) = 3(3m-2)(m-1) \equiv 0 \tag{14.2.60}$$

for $r \in \overline{1, n-2}$ and $s \in \overline{1, n-1}$. So $\mathscr{C}_3(\Psi_3)$ is orthogonal. $\square$

According to the above theorem, $\mathscr{C}_3(\Psi_2)$ is a ternary orthogonal $[10, 4, 6]$-code when $n = 5$ (which is optimal (e.g., cf. [PHB])), $[28, 7, 12]$-code when $n = 8$, and $[55, 10, 18]$-code when $n = 11$. Moreover, $\mathscr{C}_3(\Psi_3)$ is a ternary orthogonal $[10, 4, 6]$-code when $n = 5$, $[15, 4, 12]$-code if $n = 6$, $[56, 7, 21]$-code when $n = 8$, $[84, 7, 42]$-code if $n = 9$, $[165, 10, 45]$-code when $n = 11$ and $[220, 10, 90]$-code when $n = 12$.

Finally, we consider the adjoint representation of $sl(n, \mathbb{C})$. Note that $\{E_{i,j} \mid 1 \le i < j \le n\}$ are positive root vectors. Given an order

$$\{y_1, \dots, y_{\binom{n}{2}}\} = \{E_{i,j} \mid 1 \le i < j \le n\}, \tag{14.2.61}$$

we write

$$[h_i, y_j] = k_{i,j} y_j, \qquad [E_{r,r}, y_j] = l_{r,j} y_j. \tag{14.2.62}$$

Denote

$$K = (k_{i,j})_{(n-1)\times\binom{n}{r}}, \qquad L = (l_{i,j})_{n\times\binom{n}{r}}. \tag{14.2.63}$$

Let $\mathscr{C}_K$ be the ternary code generated by K and let $\mathscr{C}_L$ be the ternary code generated by L. Moreover, $\vec{k}_i$ stands for the ith row of K and $\vec{l}_r$ stands for the rth row of L. Set

$$u(s,t) = \sum_{i=1}^{s} \vec{l}_i - \sum_{j=1}^{t} \vec{l}_{s+j}. \tag{14.2.64}$$

For any nonzero codeword $v \in \mathscr{C}_L$, using negative root vectors, we can prove

$$\text{wt}\,(v, -v) = \text{wt}\,(u(s,t), -u(s,t)) \tag{14.2.65}$$

for some s and t by symmetry (cf. (14.1.9)–(14.1.11)). Thus

$$\text{wt}\, v = \text{wt}\, u(s,t) = (s+t)(n-s-t) + st = \phi(s,t). \tag{14.2.66}$$

Note

$$\phi(s,t) = n^2 - \frac{1}{2}[(s-n)^2 + (t-n)^2 + (s-t)^2]. \tag{14.2.67}$$

So $\phi(s,t)$ has only local maximum. Thus it attains the absolute minimum at the boundary points. We calculate

$$\phi(1,0)=\phi(n-1,0)=n-1,\quad \phi(n-3,0)=3(n-3), \tag{14.2.68}$$

$$\phi(1,1)=2n-3,\qquad \phi(n-2,1)=2(n-1). \tag{14.2.69}$$

Since

$$\sum_{i=1}^{n}\vec{l}_i=0, \tag{14.2.70}$$

$$\mathscr{C}_K=\mathscr{C}_L\qquad \text{if}\ \ n\not\equiv 0\ (\text{mod }3). \tag{14.2.71}$$

$$\vec{k}_i\cdot\vec{k}_j=2-2=0\qquad 1\le i<j-1\le n, \tag{14.2.72}$$

$$\vec{k}_r\cdot\vec{k}_{r+1}=6-n,\quad \vec{k}_s\cdot\vec{k}_s=2n-3. \tag{14.2.73}$$

In summary, we have:

Theorem 14.2.4 *The code $\mathscr{C}_L$ is a ternary $[\binom{n}{2},n-1,n-1]$-code if $n\ge 4$, which is also the ternary weight code on the adjoint module $sl(n,\mathbb{C})$ when $n\not\equiv 0$ (mod 3). If $n=3m$ for some integer $m>1$, then the ternary weight code $\mathscr{C}_K$ on $sl(3m,\mathbb{C})$ is an orthogonal $[\binom{3m}{2},3m-2,3(m-1)]$-code.*

14.3 Codes Related to Representations of $o(2m,\mathbb{C})$

In this section, we only study ternary codes related to certain representations of $o(2m,\mathbb{C})$, some of which will be used to investigate the codes related to exceptional simple Lie algebras.

Let $n=2m$ be a positive even integer. Take the settings in (14.2.1)–(14.2.3) (with $n\to m$). The orthogonal Lie algebra

$$o(2m,\mathbb{C})=\sum_{1\le i<j\le m}[\mathbb{C}(E_{i,j}-E_{m+j,m+i})+\mathbb{C}(E_{j,i}-E_{m+i,m+j})+\mathbb{C}(E_{i,m+j}-E_{j,m+i})$$
$$+\mathbb{C}(E_{m+i,j}-E_{m+j,i})]+\sum_{r=1}^{m}\mathbb{C}h_r, \tag{14.3.1}$$

where

$$h_s=E_{s,s}-E_{s+1,s+1}-E_{m+s,m+s}+E_{m+s+1,m+s+1}\qquad \text{for}\ \ s\in\overline{1,m-1} \tag{14.3.2}$$

and

$$h_m = E_{m-1,m-1} + E_{m,m} - E_{2m-1,2m-1} - E_{2m,2m}. \tag{14.3.3}$$

Indeed, we take the Cartan subalgebra

$$H_{D_m} = \sum_{i=1}^{m} \mathbb{C}h_i \tag{14.3.4}$$

of $o(2m, \mathbb{C})$. The root system

$$\Phi_{D_m} = \{\pm\varepsilon_i \pm \varepsilon_j \mid i, j \in \overline{1, m},\ i \neq j\} \tag{14.3.5}$$

and simple positive roots are:

$$\alpha_i = \varepsilon_i - \varepsilon_{i+1},\quad \alpha_m = \varepsilon_{m-1} + \varepsilon_m, \qquad i \in \overline{1, m-1}. \tag{14.3.6}$$

The Weyl group is $S_m \ltimes \mathbb{Z}_2^{m-1}$, which acts H_{D_m} and $\mathbb{R}^m$ by permuting sub-indices of ε_i and $E_{i,i} - E_{m+i,m+i}$, and changing sign on even number of their coefficients.

Take the settings in (6.2.15)–(6.2.18) and (14.2.6)–(14.2.8). Moreover, the representation of $o(2m, \mathbb{C})$ on Ψ determined by (6.2.19). For any $\vec{\iota} = (\iota_1, \ldots, \iota_m)$ with $\iota_i \in \{0, 1\}$ and $\tau \in S_m$, we have an associative algebra automorphism $\sigma_{\tau,\vec{\iota}}$ of Ψ determined by

$$\sigma_{\tau,\vec{\iota}}(\theta_i) = \theta_{m\delta_{\iota_i,1}+\tau(i)},\quad \sigma_{\tau,\vec{\iota}}(\theta_{m+i}) = \theta_{m\delta_{\iota_i,0}+\tau(i)} \qquad \text{for } i \in \overline{1, m}. \tag{14.3.7}$$

Moreover, we define a linear map $\sigma_{\tau,\vec{\iota}}$ on $\mathscr{H}$ by

$$\sigma_{\tau,\vec{\iota}}(E_{i,i} - E_{m+i,m+i}) = (-1)^{\iota_i}(E_{\tau(i),\tau(i)} - E_{m+\tau(i),m+\tau(i)}) \qquad \text{for } i \in \overline{1, m}. \tag{14.3.8}$$

Then

$$\sigma_{\tau,\vec{\iota}}(h(w)) = \sigma_{\tau,\vec{\iota}}(h)[\sigma_{\tau,\vec{\iota}}(w)] \qquad \text{for } h \in \mathscr{H},\ w \in \Psi. \tag{14.3.9}$$

Note that all $\Psi_r \cong V_{D_m}(\lambda_r)$ are self-dual $o(2m, \mathbb{C})$-submodules for $r \in \overline{1, m-2}$ by Theorem 7.1.2. In particular, the ternary weight code $\mathscr{C}_2$ of $o(2m, \mathbb{C})$ on Ψ_2 is given by the weight matrix on its subspace

$$\Psi_{2,1} = \sum_{1 \leq i < j \leq m} (\mathbb{C}\theta_i\theta_j + \mathbb{C}\theta_i\theta_{m+j}). \tag{14.3.10}$$

We take any order

$$\{x_1, x_2, \cdots, x_{m(m-1)}\} = \{\theta_i\theta_j, \theta_i\theta_{m+j} \mid 1 \leq i < j \leq m\} \tag{14.3.11}$$

and write

$$(E_{i,i} - E_{m+i,m+i})(x_j) = c_{i,j}(2)x_j, \qquad C_2 = (c_{i,j}(2))_{m\times m(m-1)}. \tag{14.3.12}$$

Moreover,

$$\text{the weight matrix on } \Psi_2 \text{ is equivalent to } (C_2, -C_2). \tag{14.3.13}$$

Since

$$\sum_{i=1}^{m} \mathbb{C}_3 h_i = \sum_{i=1}^{m} \mathbb{C}_3(E_{i,i} - E_{m+i,m+i}), \tag{14.3.14}$$

C_2 is a generator matrix of the ternary code $\mathscr{C}_2$. Denote by ζ_i the ith row of C_2. By (14.3.9) and (14.3.13), any nonzero codeword in $\mathscr{C}_2$ has the same weight as the codeword

$$u(t) = \sum_{i=1}^{t} \zeta_t \qquad \text{for some } \ t \in \overline{1,m}. \tag{14.3.15}$$

Moreover,

$$f(t) = \text{wt}\, u(t) = \binom{t}{2} + 2t(m-t) = \frac{(4m-1)t - 3t^2}{2} \tag{14.3.16}$$

So $f(t)$ has only local maximum and it attains the absolute minimum at the boundary points. Note that

$$f(1) = 2(m-1), \qquad f(m) = \frac{m(m-1)}{2}. \tag{14.3.17}$$

Hence

$$\text{the minimal distance of } \mathscr{C}_2 \text{ is } 2(m-1) \text{ if } m \geq 4. \tag{14.3.18}$$

Theorem 14.3.1 *When $m = 3m_1 + 1$ for some positive integer m_1, the ternary weight code $\mathscr{C}_2$ of $o(2m, \mathbb{C})$ on Ψ_2 is an orthogonal $[m(m-1), m, 2(m-1)]$-code.*

Proof Note that for $i, j \in \overline{1,m}$ with $i \neq j$,

$$\zeta_i \cdot \zeta_i = f(1) = 6m_1, \tag{14.3.19}$$

$$(\zeta_i + \zeta_j) \cdot (\zeta_i + \zeta_j) = f(2) = 1 + 4(m-2) = 4m - 7 = 12(m_1 - 1). \tag{14.3.20}$$

Thus

$$\zeta_i \cdot \zeta_j = \frac{f(2) - 2f(1)}{2} = -6. \qquad (14.3.21)$$

Hence $\mathscr{C}_2$ is an orthogonal ternary code. $\square$

In particular, $\mathscr{C}_2$ is an orthogonal ternary [12, 4, 6]-code when $m_1 = 1$, [42, 7, 12]-code when $m_1 = 2$ and [90, 10, 18]-code when $m_1 = 3$. It can be proved that $\mathscr{C}_2$ is also the weight code on the adjoint module of $o(2m, \mathbb{C})$.

The ternary weight code $\mathscr{C}_3$ of $o(2m, \mathbb{C})$ on Ψ_3 is given by the weight matrix on its subspace

$$\Psi_{3,1} = \sum_{1\leq i<j<l\leq m} \mathbb{C}\theta_i\theta_j\theta_l + \sum_{1\leq i<j\leq m} \sum_{l=1}^{m} \mathbb{C}\theta_i\theta_j\theta_{m+l}. \qquad (14.3.22)$$

We take any order

$$\begin{aligned}&\{y_1, y_2, \cdots, y_{\binom{m}{3}+m\binom{m}{2}}\}\\ &= \{\theta_i\theta_j\theta_l, \theta_r\theta_s\theta_{m+q} \mid 1 \leq i < j < l \leq m;\ 1 \leq r < s \leq m;\ q \in \overline{1, m}\}\end{aligned} \qquad (14.3.23)$$

and write

$$(E_{i,i} - E_{m+i,m+i})(y_j) = c_{i,j}(3)y_j, \qquad C_3 = (c_{i,j}(3))_{m\times(\binom{m}{3}+m\binom{m}{2})}. \qquad (14.3.24)$$

Moreover,

$$\text{the weight matrix on } \Psi_3 \text{ is equivalent to } (C_3, -C_3). \qquad (14.3.25)$$

Denote by $\mathfrak{c}_i$ the ith row of C_3. By (14.3.9) and (14.3.25), any nonzero codeword in $\mathscr{C}_3$ has the same weight as the codeword

$$u(t) = \sum_{i=1}^{t} \mathfrak{c}_i \quad \text{for some } t \in \overline{1, m}. \qquad (14.3.26)$$

Moreover,

$$\begin{aligned} g(t) = \text{wt}\, u(t) &= (2m - t)\binom{t}{2} + 2t\binom{m-t}{2} + t(m-t)^2 \\ &= \frac{t(t-1)(2m-t) + 2t(m-t)(2m-2t-1)}{2} \\ &= \frac{t}{2}[3t^2 + 3(1-2m)t + 4(m^2 - m)]. \end{aligned} \qquad (14.3.27)$$

Observe that

$$g'(t) = \frac{1}{2}[9t^2 + 6(1-2m)t + 4(m^2-m)] = \frac{1}{2}[(3t+1-2m)^2 - 1]. \quad (14.3.28)$$

Thus

$$g'(t_0) = 0 \Longrightarrow t_0 = \frac{2(m-1)}{3}, \ \frac{2m}{3}. \quad (14.3.29)$$

Since $g'(0) = (m^2-m)/2 \geq 0$, $t = 2(m-1)/3$ is a point of local maximum and $t = 2m/3$ is a point of local minimum. We calculate

$$g(1) = (m-1)(2m-3), \qquad g(m) = \frac{m^2(m-1)}{2}, \qquad g(2m/3) = \frac{2}{9}m^2(2m-3). \quad (14.3.30)$$

Note that $g(m) \geq g(1)$ and $g(2m/3) \geq g(1)$ if $m \geq 3$.

Theorem 14.3.2 *Let $m \geq 3$. The ternary weight code $\mathscr{C}_3$ of $o(2m,\mathbb{C})$ on Ψ_3 is of type $[m(m-1)(2m-1)/3, m, (m-1)(2m-3)]$. Moreover, it is orthogonal if $m \not\equiv -1$ (mod 3).*

Proof Note

$$\mathfrak{c}_i \cdot \mathfrak{c}_i = g(1) = (m-1)(2m-3) \quad (14.3.31)$$

and

$$(\mathfrak{c}_i + \mathfrak{c}_j) \cdot (\mathfrak{c}_i + \mathfrak{c}_j) = g(2) = 2(2(m-2)^2 + 1) \quad (14.3.32)$$

for $i, j \in \overline{1,m}$ such that $i \neq j$. Thus

$$\mathfrak{c}_i \cdot \mathfrak{c}_j = \frac{g(2) - 2g(1)}{2} = 3(2-m). \quad (14.3.33)$$

So $\mathscr{C}_3$ is orthogonal if $m \not\equiv -1$ (mod 3). □

Remark that $\mathscr{C}_3$ is an orthogonal [10, 3, 6]-code when $m = 3$, [28, 4, 15]-code when $m = 4$, [110, 6, 45]-code when $m = 6$ and [182, 7, 66]-code when $m = 7$.

Let $\mathscr{B}$ be the subalgebra of Ψ generated by $\{1_{\mathscr{A}}, \theta_i \mid i \in \overline{1,m}\}$ and

$$\mathscr{B}_r = \Psi_r \bigcap \mathscr{B} \quad \text{for } r \in \overline{0,m}. \quad (14.3.34)$$

The spin representation of $o(2m,\mathbb{C})$ is given by

$$E_{i,j} - E_{m+j,m+i} = \theta_i \partial_{\theta_j} - \frac{\delta_{i,j}}{2} \quad \text{for } i, j \in \overline{1,m}, \quad (14.3.35)$$

$$E_{m+s,r} - E_{m+r,s} = \partial_{\theta_s}\partial_{\theta_r}, \qquad E_{r,m+s} - E_{s,m+r} = \theta_r\theta_s \tag{14.3.36}$$

for $1 \leq r < s \leq m$. Then the subspace

$$\mathcal{V} = \sum_{i=1}^{[\![m/2]\!]} \mathscr{B}_{m-i} \tag{14.3.37}$$

is the irreducible module with highest weight λ_m; that is, $\mathcal{V} \cong V_{D_m}(\lambda_m)$.

If $m = 2m_1 + 1$ is odd, then

$$\{\theta_{i_1}\cdots\theta_{i_{m-2r}} \mid r \in \overline{0, m_1};\ 1 \leq i_1 < \cdots < i_{m-2r} \leq m\} \tag{14.3.38}$$

forms a weight-vector basis of $\mathcal{V}$. When $m = 2m_1$ is even,

$$\{1, \theta_{i_1}\cdots\theta_{i_{m-2r}} \mid r \in \overline{0, m_1 - 1};\ 1 \leq i_1 < \cdots < i_{m-2r} \leq m\} \tag{14.3.39}$$

is a weight-vector basis of $\mathscr{V}$. Take any order $\{z_1, z_2, \ldots, z_{2^{m-1}}\}$ of the above base vectors. Denote

$$(E_{r,r} - E_{m+r,m+r})(z_i) = q_{r,i}z_i, \qquad C(\mathcal{V}) = (q_{r,i})_{m\times 2^{m-1}}. \tag{14.3.40}$$

Note that

$$\frac{1}{2} \equiv -1 \quad \text{in } \mathbb{F}_3. \tag{14.3.41}$$

Denote by ξ_r the rth row of the weight matrix $C(\mathcal{V})$. Set

$$\bar{u} = \sum_{r=1}^{m-1} \xi_r - \xi_m, \quad u(t) = \sum_{i=1}^{t} \xi_i \qquad \text{for } t \in \overline{1, m}. \tag{14.3.42}$$

Then any nonzero codeword in $\mathscr{C}_3(\mathcal{V})$ is conjugated to some $u(t)$ or $\bar{u}$ under the action of the Weyl group of $o(2m, \mathbb{C})$ (cf. (14.1.10) and (14.1.11)). It has the same weight as $u(t)$ or $\bar{u}$. We calculate

$$\text{wt}\, u(1) = 2^{m-1}, \qquad \text{wt}\, u(2) = 2^{m-2}. \tag{14.3.43}$$

Moreover, we have the following more general estimates. For any positive integer $k > 2$, we always have

$$\binom{k}{l-1} + \binom{k}{l+1} > \binom{k}{l} \qquad \text{for } l \in \overline{0, k}, \tag{14.3.44}$$

where we treat $\binom{k}{-1} = \binom{k}{k+1} = 0$. If $t = 3t_1$ for some positive integer t_1, we have

$$\begin{aligned}\operatorname{wt} u(t) &= 2^{m-3t_1-1} \sum_{i=0}^{t_1} \left[\binom{3t_1}{6i+1} + \binom{3t_1}{6i+2} + \binom{3t_1}{6i+4} + \binom{3t_1}{6i+5} \right] \\ &> 2^{m-3t_1-1} \sum_{i=0}^{t_1} \left[\binom{3t_1}{6i+1} + \binom{3t_1}{6i+3} + \binom{3t_1}{6i+5} \right] = 2^{m-2}.\end{aligned} \tag{14.3.45}$$

When $t = 3t_1 + 1$ for some positive integer t_1, we obtain

$$\begin{aligned}\operatorname{wt} u(t) &= 2^{m-3t_1-2} \sum_{i=0}^{t_1} \left[\binom{3t_1+1}{6i} + \binom{3t_1+1}{6i+1} + \binom{3t_1+1}{6i+3} + \binom{3t_1+1}{6i+4} \right] \\ &> 2^{m-3t_1-2} \sum_{i=0}^{t_1} \left[\binom{3t_1+1}{6i} + \binom{3t_1+1}{6i+2} + \binom{3t_1+1}{6i+4} \right] = 2^{m-2}.\end{aligned} \tag{14.3.46}$$

If $t = 3t_1 + 2$ for some positive integer t_1, we get

$$\begin{aligned}\operatorname{wt} u(t) &= 2^{m-3t_1-3} \sum_{i=0}^{t_1} \left[\binom{3t_1+2}{6i} + \binom{3t_1+2}{6i+2} + \binom{3t_1+2}{6i+3} + \binom{3t_1+2}{6i+5} \right] \\ &> 2^{m-3t_1-3} \sum_{i=0}^{t_1} \left[\binom{3t_1+2}{6i} + \binom{3t_1+2}{6i+2} + \binom{3t_1+2}{6i+4} \right] = 2^{m-2}.\end{aligned} \tag{14.3.47}$$

Let k be a positive integer. We have

$$\binom{2k}{i} + \binom{2k}{i+4} > \binom{2k}{i+1} \tag{14.3.48}$$

if $i \le k-3$ or $i \ge k$. Moreover,

$$\binom{2k}{k-2} + \binom{2k}{k+2} - \binom{2k}{k-1} = \frac{k-4}{k-1} \binom{2k}{k-2}, \tag{14.3.49}$$

$$\binom{2k}{k-1} + \binom{2k}{k+3} - \binom{2k}{k} = \frac{k^3 - 4k^2 - 3k - 6}{k(k-1)(k-2)} \binom{2k}{k-3}. \tag{14.3.50}$$

Thus (14.3.48) always holds if $k \geq 5$. Furthermore,

$$\binom{2k+1}{i} + \binom{2k+1}{i+4} > \binom{2k+1}{i+1} \tag{14.3.51}$$

if $i \neq k-1$. Observe that

$$\binom{2k+1}{k-1} + \binom{2k+1}{i+3} - \binom{2k+1}{k} = \frac{k^2-3k-6}{k(k-1)}\binom{2k+1}{k-2}. \tag{14.3.52}$$

So (14.3.51) holds whenever $k \geq 5$. Therefore,

$$\binom{k}{i} + \binom{k}{i+4} > \binom{k}{i+1} \quad \text{if } k \geq 10. \tag{14.3.53}$$

If $m = 3m_1$ for some positive integer m_1,

$$\begin{aligned} \text{wt}\,\bar{u} &= \sum_{i=0}^{m}\left[\binom{m}{6i} + \binom{m-1}{6i+1} + \binom{m-1}{6i+4}\right] \\ &= \sum_{i=0}^{m}\left[\binom{m-1}{6i} + \binom{m-1}{6i+5} + \binom{m-1}{6i+1} + \binom{m-1}{6i+4}\right], \end{aligned} \tag{14.3.54}$$

which is $> 2^{m-2}$ if $m_1 \geq 4$ by (14.3.53). When $m = 3m_1 + 1$ for some positive integer m_1,

$$\begin{aligned} \text{wt}\,\bar{u} &= \sum_{i=0}^{m}\left[\binom{m-1}{6i} + \binom{m}{6i+2} + \binom{m-1}{6i+3}\right] \\ &= \sum_{i=0}^{m}\left[\binom{m-1}{6i} + \binom{m-1}{6i+1} + \binom{m-1}{6i+2} + \binom{m-1}{6i+3}\right] \\ &= 1 + \sum_{i=0}^{m}\left[\binom{m-1}{6i+1} + \binom{m-1}{6i+3} + \binom{m-1}{6i+2} + \binom{m-1}{6i+6}\right], \end{aligned} \tag{14.3.55}$$

which is again $> 2^{m-2}$ if $m_1 \geq 4$ by (14.3.53). Assuming $m = 3m_1 + 2$ for some positive integer m_1, we have

$$\begin{aligned} \text{wt}\,\bar{u} &= \sum_{i=0}^{m}\left[\binom{m-1}{6i+2} + \binom{m}{6i+4} + \binom{m-1}{6i+5}\right] \\ &= \binom{m-1}{3} + \sum_{i=0}^{m}\left[\binom{m-1}{6i+2} + \binom{m-1}{6i+4} + \binom{m-1}{6i+5} + \binom{m-1}{6i+9}\right], \end{aligned} \tag{14.3.56}$$

which is $> 2^{m-2}$ if $m_1 \geq 3$ by (14.3.53). Moreover, we have the following table:

Table 14.7 Weights of $\bar{u}$

m	4	5	6	7	8	9	10
wt $\bar{u}$	8	11	12	43	112	171	260

In summary, we have:

Theorem 14.3.3 *Let $m > 3$ be an integer. The ternary code $\mathscr{C}_3(\mathcal{V})$ is of type $[2^{m-1}, m, 2^{m-2}]$ if $m \neq 6$ and of type $[32, 6, 12]$ when $n = 6$.*

We remark that the spin module $\mathcal{V}$ is self-dual if and only if m is even.

Corollary 14.3.4 *When $m = 6m_1 + 2$ for some positive integer m_1, the ternary weight code of $o(2m, \mathbb{C})$ on $o(2m, \mathbb{C}) + \mathcal{V}$ is an orthogonal ternary $[m(m-1) + 2^{m-2}, m, 4m - 7 + 2^{m-3}]$-code. If $m = 6m_1 + 3$ for some positive integer m_1, the ternary weight code of $o(2m, \mathbb{C})$ on $o(2m, \mathbb{C}) + \mathcal{V}$ is an orthogonal ternary $[2m(m-1) + 2^{m-1}, m, 8m - 14 + 2^{m-2}]$-code. In the case $m = 6m_1 + 5$ and $m = 6m_1 + 12$ for some nonnegative integer m_1, the code $\mathscr{C}_2 \oplus \mathscr{C}_3(\mathcal{V})$ is an orthogonal ternary $[m(m-1) + 2^{m-1}, m, 4m - 7 + 2^{m-2}]$-code. When $m = 6$, the code $\mathscr{C}_2 \oplus \mathscr{C}_3(\mathcal{V})$ is an orthogonal ternary $[62, 6, 27]$-code.*

Proof Suppose $m = 6m_1 + 2$ for some positive integer m_1. Then the weight matrix of $o(2m, \mathbb{C})$ on $o(2m, \mathbb{C}) + \mathcal{V}$ is equivalent to $(A, -A)$, where A generates the weight code $\mathscr{C}$ of $o(2m, \mathbb{C}) + \mathcal{V}$. Moreover, $\mathscr{C}$ is orthogonal if and only if the matrix $(A, -A)$ generates an orthogonal code. But

$$(A, -A) \sim (C_2, C_2, C(\mathcal{V})). \tag{14.3.57}$$

Note that

$$\mathrm{wt}\,(\zeta_i, \zeta_i, \xi_i) = 2f(1) + 2^{m-1} = 4(m-1) + 2^{m-1} \equiv 1 + (-1)^{6m_1+1} \equiv 0 \pmod 3, \tag{14.3.58}$$

$$\begin{aligned} &\mathrm{wt}\,(\zeta_i + \zeta_j, \zeta_i + \zeta_j, \xi_i + \xi_j) \\ &= 2f(2) + 2^{m-2} = 8m - 14 + 2^{m-2} \equiv 2 + (-1)^{6m_1} \equiv 0 \pmod 3 \end{aligned} \tag{14.3.59}$$

for $i, j \in \overline{1, m}$ with $i \neq j$ by (14.3.17) and (14.3.43). Thus

$$(\zeta_i, \zeta_i, \xi_i) \cdot (\zeta_i, \zeta_i, \xi_i) \equiv \mathrm{wt}\,(\zeta_i, \zeta_i, \xi_i) \equiv 0, \tag{14.3.60}$$

$$\begin{aligned}&(\zeta_i, \zeta_i, \xi_i) \cdot (\zeta_j, \zeta_j, \xi_j)\\ \equiv & -[\mathrm{wt}\,(\zeta_i + \zeta_j, \zeta_i + \zeta_j, \xi_i + \xi_j) - \mathrm{wt}\,(\zeta_i, \zeta_i, \xi_i) - \mathrm{wt}\,(\zeta_j, \zeta_j, \xi_j)] \equiv 0\end{aligned} \tag{14.3.61}$$

by (14.3.41). Thus $\mathscr{C}$ is orthogonal. Note

$$f(2) = 4m - 7 \leq \frac{m(m-1)}{2} = f(m) \quad \text{if } m \geq 7. \tag{14.3.62}$$

Thus

$$f(2) \leq f(t) \quad \text{for } t \in \overline{2, m}. \tag{14.3.63}$$

By (14.3.9),

$$\mathrm{wt}\,(\sum_{i=1}^{m-1} \zeta_i - \zeta_m) = f(m) \geq f(2). \tag{14.3.64}$$

Thus the minimum distance of $\mathscr{C}$ is

$$\min\{f(1) + 2^{m-2}, f(2) + 2^{m-3}\} = 4m - 7 + 2^{m-3} \quad \text{if } m \geq 6. \tag{14.3.65}$$

This proves the first conclusion. The other conclusions for $m \geq 7$ can be proved similarly.

In the case $m = 5$, we have

Table 14.8 Weights of the Code on the Adjoint Module ($m = 5$)

t	1	2	3	4	5
f(t)	8	13	15	14	10

and on the $\mathcal{V}$,

Table 14.9 Weights of the Code on the Spin Module ($m = 5$)

t	1	2	3	4	5
wt u(t)	16	8	12	10	11

By Tables 14.7, 14.8 and 14.9 and the fact $\mathrm{wt}\,(\sum_{i=1}^{4} \zeta_i - \zeta_5) = f(5)$ in $\mathscr{C}_3(\Psi_2)$, the third conclusion holds for $m = 5$.

If $m = 6$,

Table 14.10 Weights of the Code on the Adjoint Module ($m = 6$)

t	1	2	3	4	5	6
f(t)	10	17	21	22	20	15

and on the $\mathcal{V}$,

Table 14.11 Weights of the Code on the Spin Module ($m = 6$)

t	1	2	3	4	5	6
wt u(t)	32	16	24	20	22	21

By Tables 14.7, 14.10, and 14.11, and the fact wt $(\sum_{i=1}^{5} \zeta_i - \zeta_6) = f(6)$ in $\mathscr{C}_3(\Psi_2)$, the last conclusion holds. □

When $m = 8$, the ternary weight code of $o(16, \mathbb{C})$ on $o(16, \mathbb{C}) + \mathcal{V}$ is a ternary orthogonal $[120, 8, 57]$-code, which will later be proved also to be the ternary weight code of E_8 on its adjoint module. If $m = 9$, the ternary weight code of $o(18, \mathbb{C})$ on $o(18, \mathbb{C}) + \mathcal{V}$ is a ternary orthogonal $[400, 8, 186]$-code. When $m = 5$, the code $\mathscr{C}_2 \oplus \mathscr{C}_3(\mathcal{V})$ is a ternary orthogonal $[36, 5, 21]$-code, which will later be proved also to be the ternary weight code of E_6 on its adjoint module. In the case $m = 11$, the code $\mathscr{C}_2 \oplus \mathscr{C}_3(\mathcal{V})$ is a ternary orthogonal $[1134, 8, 549]$-code.

14.4 Exceptional Lie Algebras and Ternary Codes

In this section, we study the ternary weight codes of F_4, E_6, E_7, E_8 on its minimal irreducible module and adjoint module.

We go back to the settings in (14.2.1)–(14.2.3) with $n = 4$. The root system of F_4 is

$$\Phi_{F_4} = \left\{\pm\varepsilon_i, \pm\varepsilon_i \pm \varepsilon_j, \frac{1}{2}(\pm\varepsilon_1 \pm \varepsilon_2 \pm \varepsilon_3 \pm \varepsilon_4) \mid i \neq j\right\} \tag{14.4.1}$$

and the positive simple roots are

$$\bar{\alpha}_1 = \varepsilon_2 - \varepsilon_3, \bar{\alpha}_2 = \varepsilon_3 - \varepsilon_4, \bar{\alpha}_3 = \varepsilon_4, \bar{\alpha}_4 = \frac{1}{2}(\varepsilon_1 - \varepsilon_2 - \varepsilon_3 - \varepsilon_4). \tag{14.4.2}$$

The Weyl group $\mathscr{W}_{F_4}$ of F_4 contains the permutation group S_4 on the sub-indices of ε_i and all reflections with respect to the coordinate hyperplanes. Moreover, there is an identification:

$$h_1 \leftrightarrow \alpha_1,\ h_2 \leftrightarrow \alpha_2,\ h_3 \leftrightarrow 2\alpha_3,\ h_4 \leftrightarrow 2\alpha_4. \tag{14.4.3}$$

Thus

$$\mathscr{H}_3 = \sum_{i=1}^{4} \mathbb{F}_3 h_i = \sum_{i=1}^{4} \mathbb{F}_3 \varepsilon_i. \tag{14.4.4}$$

Moreover,

$$\mathscr{H}_3 = \{\mathscr{W}_{F_4}(h_1), \mathscr{W}_{F_4}(h_1 + h_3), \mathscr{W}_{F_4}(h_3), \mathscr{W}_{F_4}(h_4)\}. \tag{14.4.5}$$

Recall that the basic (minimal) irreducible module V_{F_4} of the 52-dimensional Lie algebra $\mathscr{G}^{F_4}$ has a basis $\{x_i \mid \overline{1,26}\}$ and with the representation determined by (10.3.44)–(10.3.74). The module V_{F_4} is self-dual. The weight matrix of V_{F_4} is $(A_{F_4}, -A_{F_4})$ with

$$A_{F_4} = \begin{bmatrix} 0 & 0 & 0 & 1 & 1 & -1 & 1 & -1 & -1 & 0 & 0 & 0 \\ 0 & 0 & 1 & -1 & 0 & 0 & 0 & 1 & 1 & -1 & -1 & 0 \\ 0 & 1 & -1 & 1 & -1 & 1 & 0 & -1 & 0 & 1 & 2 & -1 \\ 1 & -1 & 0 & 0 & 1 & 0 & -1 & 1 & -1 & 1 & -1 & 2 \end{bmatrix} \tag{14.4.6}$$

by (10.3.68)–(10.3.71).

Theorem 14.4.1 *The ternary weight code $\mathscr{C}_{F_4,1}$ (generated by A_{F_4}) of F_4 on V_{F_4} is an orthogonal [12,4,6]-code.*

Proof Denote by ξ_i the ith row of the matrix A_{F_4}. Then

$$\text{wt}\,\xi_1 = 6, \qquad \text{wt}\,(\xi_1 + \xi_3) = \text{wt}\,\xi_3 = \text{wt}\,\xi_4 = 9. \tag{14.4.7}$$

According to (14.4.5), any nonzero codeword in $\mathscr{C}_{F_4,1}$ has weight 6 or 9. By an argument as (14.3.31)–(14.3.33), $\mathscr{C}_{F_4,1}$ is orthogonal. □

Next we consider the adjoint representation of F_4. Its weight code $\mathscr{C}_{F_4,2}$ is determined by the set $\Phi_{F_4}^+$ of positive roots, the corresponding root vectors are constructed in (10.3.29)–(10.3.42). The weight matrix B_{F_4} on $\sum_{\alpha\in\Phi_{F_4}^+} \mathbb{C}E_\alpha$ is given by

$$\begin{bmatrix} 2 & -1 & 0 & 0 & 1 & -1 & 0 & 1 & -1 & -1 & 1 & -1 & 1 & 0 & 1 & -1 & 0 & 1 & 0 & 0 & 0 & 0 & -1 & 1 \\ -1 & 2 & -1 & 0 & 1 & 1 & -1 & 0 & 1 & 0 & -1 & 0 & 0 & 1 & -1 & 0 & 1 & -1 & 1 & 0 & 0 & -1 & 1 & 0 \\ 0 & -2 & 2 & -1 & -2 & 0 & 1 & 0 & -1 & 2 & 2 & 1 & -1 & 0 & 1 & 0 & -1 & 0 & -2 & 1 & 0 & 2 & 0 & 0 \\ 0 & 0 & -1 & 2 & 0 & -1 & 1 & -1 & 1 & -2 & -2 & 0 & 1 & -2 & 0 & 2 & 0 & 2 & 2 & -1 & 1 & 0 & 0 & 0 \end{bmatrix}. \tag{14.4.8}$$

Theorem 14.4.2 *The ternary weight code $\mathscr{C}_{F_4,2}$ (generated by B_{F_4}) of F_4 on its adjoint module is an orthogonal* [24, 4, 15]*-code.*

Proof Denote by η_i the ith row of the above matrix. Then

$$\mathrm{wt}\,\eta_i = 15, \qquad \mathrm{wt}\,(\eta_1 + \eta_3) = 18. \tag{14.4.9}$$

According to (14.4.5), any nonzero codeword in $\mathscr{C}_{F_4,2}$ has weight 15 or 18. By an argument as (14.3.31)–(14.3.33), $\mathscr{C}_{F_4,2}$ is orthogonal. □

Now we take the settings in (14.2.1)–(14.2.3) with $n = 7$ and have the following root system of E_6:

$$\Phi_{E_6} = \left\{\varepsilon_i - \varepsilon_j, \frac{1}{2}\left(\sum_{s=1}^{6} \iota_s\varepsilon_s \pm \sqrt{2}\varepsilon_7\right), \pm\sqrt{2}\varepsilon_7 \right. \left. \mid i, j \in \overline{1,6}, i \neq j; \iota_s = \pm 1; \sum_{i=1}^{6} \iota_i = 0\right\} \tag{14.4.10}$$

and the simple positive roots are

$$\alpha_1 = \varepsilon_1 - \varepsilon_2, \quad \alpha_2 = \frac{1}{2}\left(\sum_{j=1}^{3}(\varepsilon_{3+j} - \varepsilon_j) + \sqrt{2}\varepsilon_7\right), \quad \alpha_i = \varepsilon_{i-1} - \varepsilon_i, \qquad i \in \overline{3,6}. \tag{14.4.11}$$

Note that

$$\mathscr{H}_{E_6,3} = \sum_{i=1}^{6} \mathbb{F}_3 h_i = \left\{\sum_{i=1}^{6} \iota_i\varepsilon_i + \iota_7\sqrt{2}\varepsilon_7 \mid \iota_r \in \mathbb{F}_3, \sum_{i=1}^{6} \iota_i = 0\right\}. \tag{14.4.12}$$

Moreover, the Weyl group $\mathscr{W}_{E_6}$ contains the permutation group S_6 on the first six sub-indices of ε_i and the reflection

$$\sum_{i=1}^{6} \iota_i\varepsilon_i + \iota_7\sqrt{2}\varepsilon_7 \mapsto \sum_{i=1}^{6} \iota_i\varepsilon_i - \iota_7\sqrt{2}\varepsilon_7. \tag{14.4.13}$$

So

$$\mathscr{H}_{E_6,3} = \mathscr{W}_{E_6}\left(\left\{\sum_{i=1}^{s} \varepsilon_i - \sum_{j=1}^{t} \varepsilon_{s+j} + \iota\sqrt{2}\varepsilon_7, \sqrt{2}\varepsilon_7 \mid \iota = 0, 1; s - t \equiv 0 \ (\mathrm{mod}\ 3)\right\}\right). \tag{14.4.14}$$

Recall that the 27-dimensional basic irreducible module V_{E_6} of weight λ_1 for E_6 has a basis $\{x_i \mid i \in \overline{1,27}\}$ with the representation formulas determined by (11.1.31)–(11.1.70). Moreover,

$$E_{\alpha_r}(x_i) \neq 0 \Leftrightarrow a_{r,i} < 0, \quad E_{-\alpha_r}(x_i) \neq 0 \Leftrightarrow a_{r,i} > 0. \tag{14.4.15}$$

Theorem 14.4.3 *The ternary weight code $\mathscr{C}_{E_6,1}$ of E_6 on V_{E_6} is an orthogonal* $[27, 6, 12]$*-code.*

Proof According to Table 11.1 and (11.1.72), we write

$$A_{E_6} = (a_{r,i})_{6\times 27}. \tag{14.4.16}$$

Denote by ξ_r the rth row of the matrix A_{E_6}. Then

$$\text{wt}\,\xi_r = 12 \qquad \text{for } r \in \overline{1,6}. \tag{14.4.17}$$

Moreover,

$$\text{wt}\,(\xi_1 + \xi_3) = \text{wt}\,(\xi_2 + \xi_4) = 12, \quad \text{wt}\,(\xi_1 + \xi_4) = 18, \tag{14.4.18}$$

$$\text{wt}\,(\xi_1 + \xi_2) = \text{wt}\,(\xi_2 + \xi_3) = \text{wt}\,(\xi_2 + \xi_5) = \text{wt}\,(\xi_2 + \xi_6) = 18. \tag{14.4.19}$$

By an argument as (14.3.31)–(14.3.33) and symmetry, we have

$$\xi_i \cdot \xi_j \equiv 0 \pmod 3 \qquad \text{for } i, j \in \overline{1,6}; \tag{14.4.20}$$

that is $\mathscr{C}_{E_6,1}$ is orthogonal.

Note that the Lie subalgebra $\mathscr{G}^{E_6}_{A,1}$ generated by $\{E_{\pm\alpha_i} \mid 2 \neq i \in \overline{1,6}\}$ is isomorphic to $sl(6,\mathbb{C})$. Recall that a singular vector in a module of simple Lie algebra is a weight vector annihilated by its positive root vectors. By Table 11.1 and (14.4.15), the $\mathscr{G}^{E_6}_{A,1}$-singular vectors are x_1 of weight λ_1, x_6 of weight λ_4 and x_{20} of weight λ_1. So the $(\mathscr{G}^{E_6}, \mathscr{G}^{E_6}_{A,1})$-branch rule on V_{E_6} is

$$V_{E_6} \cong V_{A_5}(\lambda_1) \oplus V_{A_5}(\lambda_4) \oplus V_{A_5}(\lambda_1). \tag{14.4.21}$$

Denote by $\mathscr{G}^{E_6}_{A,2}$ the Lie subalgebra of $\mathscr{G}^{E_6}$ generated by $\{E_{\pm\alpha_r},\ E_{\pm(\alpha_2+\alpha_4)} \mid 2, 4 \neq r \in \overline{1,6}\}$. The algebra $\mathscr{G}^{E_6}_{A,2}$ is also isomorphic to $sl(6,\mathbb{C})$. According to Table 11.1 and (14.4.15), the $\mathscr{G}^{E_6}_{A,2}$-singular vectors are x_1 of weight λ_1, x_4 of weight λ_4 and x_{18} of weight λ_1. Hence (14.4.21) is also the $(\mathscr{G}^{E_6}, \mathscr{G}^{E_6}_{A,2})$-branch rule. Since the module $V_{A_5}(\lambda_2)$ is contragredient to $V_{A_5}(\lambda_4)$, they have the same ternary weight code. By (14.2.34) and (14.2.38) with $n = 6$, the minimal distances of the subcodes $\sum_{2\neq i\in\overline{1,6}} \mathbb{F}_3\xi_i$ and $\mathbb{F}_3(\xi_2 + \xi_4) + \sum_{2,4\neq i\in\overline{1,6}} \mathbb{F}_3\xi_i$ are $\text{wt}\,\xi_1 = 12$.

Recall $\frac{1}{2} = -1$ in $\mathbb{F}_3$. Moreover,

$$-(\alpha_2 + \alpha_4) = -\varepsilon_1 - \varepsilon_2 + \varepsilon_3 - \varepsilon_4 + \varepsilon_5 + \varepsilon_6 + \sqrt{2}\varepsilon_7 \text{ in } \mathscr{H}_{E_6,3}. \tag{14.4.22}$$

Thus in $\mathscr{H}_{E_6,3}$,

$$\alpha_1 - (\alpha_2 + \alpha_4) = \varepsilon_2 + \varepsilon_3 - \varepsilon_4 + \varepsilon_5 + \varepsilon_6 + \sqrt{2}\varepsilon_7, \tag{14.4.23}$$

$$\alpha_1 - \alpha_2 - (\alpha_2 + \alpha_4) = -\varepsilon_3 - \varepsilon_4 + \varepsilon_5 + \varepsilon_6 + \sqrt{2}\varepsilon_7, \tag{14.4.24}$$

$$\alpha_1 - \alpha_2 - (\alpha_2 + \alpha_4) + \alpha_6 = -\varepsilon_3 - \varepsilon_4 - \varepsilon_5 + \sqrt{2}\varepsilon_7, \tag{14.4.25}$$

$$\alpha_1 - \alpha_2 - (\alpha_2 + \alpha_4) - \alpha_5 + \alpha_6 = -\varepsilon_3 + \varepsilon_4 + \sqrt{2}\varepsilon_7. \tag{14.4.26}$$

Note that

$$\text{wt}\,(\xi_1 - (\xi_2 + \xi_4)),\ \text{wt}\,(\xi_1 - \xi_2 - (\xi_2 + \xi_4)) \geq 12, \tag{14.4.27}$$

$$\text{wt}\,(\xi_1 - \xi_2 - (\xi_2 + \xi_4) + \xi_6),\quad \text{wt}\,(\xi_1 - \xi_2 - (\xi_2 + \xi_4) - \xi_5 + \xi_6) \geq 12 \tag{14.4.28}$$

because the minimal distance of $\mathbb{F}_3(\xi_2 + \xi_4) + \sum_{2,4\neq i\in\overline{1,6}} \mathbb{F}_3\xi_i$ is 12. Furthermore,

$$-\sum_{i=1}^{6} \varepsilon_i + \sqrt{2}\varepsilon_7 = \alpha_1 - \alpha_2 - \alpha_3 \quad \text{in } \mathscr{H}_{E_6,3}. \tag{14.4.29}$$

We calculate

$$\text{wt}\,(\xi_1 - \xi_2 - \xi_3) = 21. \tag{14.4.30}$$

By (14.4.13), the minimal distance of the ternary code $\mathscr{C}_{E_6,1}$ is 12. □

Next we consider the ternary weight code $\mathscr{C}_{E_6,2}$ of E_6 on its adjoint module. Take any order

$$\{y_1, \ldots, y_{36}\} = \{E_\alpha \mid \alpha \in \Phi_{E_6}^+\}. \tag{14.4.31}$$

Write

$$[\alpha_i, y_j] = b_{i,j} y_j, \qquad B_{E_6} = (b_{i,j})_{6\times 36}. \tag{14.4.32}$$

Theorem 14.4.4 *The ternary weight code $\mathscr{C}_{E_6,2}$ (generated B_{E_6}) of E_6 on its adjoint module is an orthogonal* [36, 5, 21]*-code.*

Proof Denote by ζ_i the ith row of B_{E_6}. Note

$$(\alpha_1 - \alpha_3 + \alpha_5 - \alpha_6, \alpha_i) \equiv 0 \pmod 3 \qquad \text{for } i \in \overline{1,6}. \tag{14.4.33}$$

Thus

$$\zeta_1 - \zeta_3 + \zeta_5 - \zeta_6 \equiv 0 \quad \text{in } \mathbb{F}_3^{36}. \tag{14.4.34}$$

Hence

$$\mathscr{C}_{E_6,2} = \sum_{i=2}^{6} \mathbb{F}_3\zeta_i. \tag{14.4.35}$$

Denote by $\mathscr{G}_D^{E_6}$ the Lie subalgebra of $\mathscr{G}^{E_6}$ generated by $\{E_{\pm\alpha_r} \mid r \in \overline{2,6}\}$. According to the Dynkin diagram of E_6,

$$\mathscr{G}_D^{E_6} \cong o(10, \mathbb{C}). \tag{14.4.36}$$

Let $\mathscr{G}_+^{E_6} = \sum_{i=1}^{36} \mathbb{C}y_i$ and denote by $\mathscr{G}_{D,+}^{E_6}$ the subspace spanned by the root vectors $E_\alpha \in \mathscr{G}_D^{E_6}$ with $\alpha \in \Phi_{E_6}^+$. Then $[\mathscr{G}_{D,+}^{E_6}, \mathscr{G}_+^{E_6}] \subset \mathscr{G}_+^{E_6}$. Moreover, the space $\mathscr{G}_+^{E_6}$ contains $\mathscr{G}_D^{E_6}$-singular vectors $E_{\alpha_4+\alpha_5+\sum_{i=2}^{6}\alpha_i}$ of weight λ_2 (the highest root) and $E_{\alpha_2+\alpha_4+\sum_{r=3}^{5}\alpha_r+\sum_{i=1}^{6}\alpha_i}$ of weight λ_5. Hence we have the partial $(\mathscr{G}_{E_6}, \mathscr{G}_D^{E_6})$-branch rule on $\mathscr{G}_{E_6}$:

$$\mathscr{G}_{E_6}^+ \cong \mathscr{G}_{D+}^{E_6} \oplus V_{D_5}(\lambda_5). \tag{14.4.37}$$

Thus the ternary weight code $\mathscr{C}_{E_6,2}$ of E_6 on its adjoint module is exactly the code $\mathscr{C}_2 \oplus \mathscr{C}_3(\mathcal{V})$ with $n = 5$ in Corollary 14.3.4, which is a ternary orthogonal [36, 5, 21]-code. □

Take the settings in (14.2.1)–(14.2.3) with $n = 8$. We have the root system of E_7:

$$\Phi_{E_7} = \left\{\varepsilon_i - \varepsilon_j, \frac{1}{2}\sum_{s=1}^{8} \iota_s\varepsilon_s \mid i, j \in \overline{1,8},\ i \neq j;\ \iota_s = \pm 1,\ \sum_{s=1}^{8} \iota_s = 0\right\} \tag{14.4.38}$$

and the simple positive roots are:

$$\alpha_1 = \varepsilon_2 - \varepsilon_3,\ \alpha_2 = \frac{1}{2}\sum_{j=1}^{4}(\varepsilon_{4+j} - \varepsilon_j),\ \ \alpha_i = \varepsilon_i - \varepsilon_{i+1}, \qquad i \in \overline{3,7}. \tag{14.4.39}$$

The minimal module V_{E_7} of E_7 is of 56-dimensional and has a basis $\{x_i \mid i \in \overline{1,56}\}$ with the representation given by (12.1.30)–(12.1.96). Again we have

$$E_{\alpha_r}(x_i) \neq 0 \Leftrightarrow a_{r,i} < 0, \quad E_{-\alpha_r}(x_i) \neq 0 \Leftrightarrow a_{r,i} > 0. \tag{14.4.40}$$

According to (12.1.96) and Table 12.1.1, we define

$$A_{E_7} = (a_{r,i})_{7\times 28}. \tag{14.4.41}$$

Theorem 14.4.5 *The ternary weight code $\mathscr{C}_{E_7,1}$ of E_7 on V_{E_7} is an orthogonal [28, 7, 12]-code.*

Proof Note that the root system of A_7:

$$\Phi_{A_7} = \{\varepsilon_i - \varepsilon_j \mid i, j \in \overline{1,8},\ i \neq j\} \subset \Phi_{E_7}. \tag{14.4.42}$$

Thus we have the Lie subalgebra of $\mathscr{G}^{E_7}$ (cf. (12.1.1)–(12.1.3)):

$$\mathscr{G}_A^{E_7} = \sum_{i=1}^{7} \mathbb{C}\alpha_i + \sum_{\alpha\in\Phi_{A_7}} \mathbb{C}E_\alpha \cong sl(8,\mathbb{C}). \tag{14.4.43}$$

Moreover,

$$\alpha_1' = \varepsilon_1 - \varepsilon_2 = -2\alpha_2 - 2\alpha_1 - 3\alpha_3 - 4\alpha_4 - 3\alpha_5 - 2\alpha_6 - \alpha_7. \tag{14.4.44}$$

Note that x_{23} is a $\mathscr{G}_A^{E_7}$-singular vector of weight λ_6 and x_{49} is a $\mathscr{G}_A^{E_7}$-singular vector of weight λ_2 by (12.1.96), (14.4.40) and Table 12.1.1. Thus the $(\mathscr{G}^{E_7}, \mathscr{G}_A^{E_7})$-branch rule on V_{E_7} is

$$V_{E_7} \cong V_{A_7}(\lambda_2) \oplus V_{A_7}(\lambda_6). \tag{14.4.45}$$

Since $V_{A_7}(\lambda_6)$ is contragredient to $V_{A_7}(\lambda_2)$, they have the same ternary weight code of $\mathscr{G}_A^{E_7}$, which is the $\mathscr{C}_3(\Psi_2)$ with $m = 2$ in Theorem 14.2.3. Hence the weight matrix of $\mathscr{G}_A^{E_7}$ on V_{E_7} generates a ternary orthogonal [56, 7, 24]-code.

On the other hand,

$$\sum_{i=1}^{7} \mathbb{F}_3\alpha_i = \mathbb{F}_3\alpha_1' + \sum_{2\neq i\in\overline{1,7}} \mathbb{F}_3\alpha_i \tag{14.4.46}$$

by (14.4.44) and the fact $1/2 \equiv -1$ in $\mathbb{F}_3$. Thus the weight matrix $(A_{E_7}, -A_{E_7})$ of E_7 on V_{E_7} generates the same ternary code as the weight matrix of $\mathscr{G}_A^{E_7}$ on V_{E_7}. So $(A_{E_7}, -A_{E_7})$ generates a ternary orthogonal [56, 7, 24]-code. Hence the ternary code $\mathscr{C}_{E_7,1}$ generated by A_{E_7} is an orthogonal [28, 7, 12]-code. □

Next we consider the ternary weight code $\mathscr{C}_{E_7,2}$ of E_7 on its adjoint module. By (12.2.88)–(12.2.90) and Table 12.2, it can be proved that the $(\mathscr{G}^{E_7}, \mathscr{G}_A^{E_7})$-branch rule on $\mathscr{G}^{E_7}$ is

$$\mathscr{G}^{E_7} \cong \mathscr{G}_A^{E_7} \oplus V_{A_7}(\lambda_4). \tag{14.4.47}$$

The module $V_{A_7}(\lambda_4)$ of $sl(8,\mathbb{C})$ ($\cong \mathscr{G}_A^{E_7}$) is exactly Ψ_4 in (6.2.16) with $n = 8$, which is self-dual. For convenience, we study the ternary code generated by the weight matrix of $sl(8,\mathbb{C})$ on Ψ_4. Taking any order of its basis

$$\{z_1, \ldots, z_{70}\} = \{\theta_{i_1}\theta_{i_2}\theta_{i_3}\theta_{i_4} \mid 1 \le i_1 < i_2 < i_3 < i_4 \le 8\}, \tag{14.4.48}$$

we write

$$[E_{r,r}, z_i] = b_{r,i} z_i, \qquad B_{E_7} = (b_{r,i})_{7\times 70}. \tag{14.4.49}$$

Denote by η_r the rth row of B_{E_7} and by $\mathscr{C}'$ the ternary code generated by B_{E_7}. Set

$$v(s,t) = \sum_{i=1}^{s} \eta_i - \sum_{j=1}^{t} \eta_{s+j} \qquad \in \mathscr{C}'. \tag{14.4.50}$$

Moreover, we only calculate the related weights (Table 14.12):

Table 14.12 Weights of the Code on $V_{A_7}(\lambda_4)$

(s,t)	(1,1)	(2,2)	(3,3)	(4,4)	(3,0)	(6,0)	(4,1)	(5,2)
wt v(s,t)	40	44	48	34	60	30	46	50

Recall (14.2.65)–(14.2.70). We have (Table 14.13)

Table 14.13 Weights of the Code on $\mathscr{G}_A^{E_7}$

(s,t)	(1,1)	(2,2)	(3,3)	(4,4)	(3,0)	(6,0)	(4,1)	(5,2)
2wt u(s,t)	26	40	42	32	30	24	38	34

According to (14.4.1), the Weyl group $\mathscr{W}_{E_7}$ contains the permutation group S_8 on the sub-indices of ε_i. By (14.1.9), (14.1.11) and the values of wt $v(s,t)$ + 2wt $u(s,t)$ from the above tables, 54, 66, 84 and 90 are the only weights of the nonzero codewords in $\mathscr{C}_3(\mathscr{G}^{E_7})$, the ternary code generated by the weight matrix of $\mathscr{G}_A^{E_7}$ on $\mathscr{G}^{E_7}$. By (14.4.46) and an argument as (14.3.31)–(14.3.33), we have:

Theorem 14.4.6 *The ternary weight code $\mathscr{C}_{E_7,2}$ of E_7 on its adjoint module is an orthogonal* $[63, 7, 27]$*-code.*

The minimal representation of E_8 is its adjoint module given (12.1.1)–(12.1.3). Recall the settings in (14.2.1)–(14.2.3). We have the E_8 root system

$$\Phi_{E_8} = \left\{ \pm\varepsilon_i \pm \varepsilon_j, \frac{1}{2}\sum_{i=1}^{8} \iota_i\varepsilon_i \mid i, j \in \overline{1,8},\ i \neq j;\ \iota_i = \pm 1,\ \sum_{i=1}^{8} \iota_i \in 2\mathbb{Z} \right\} \tag{14.4.51}$$

and positive simple roots:

$$\alpha_1 = \frac{1}{2}\left(\sum_{j=2}^{7} \varepsilon_j - \varepsilon_1 - \varepsilon_8\right),\ \alpha_2 = -\varepsilon_1 - \varepsilon_2,\ \alpha_r = \varepsilon_{r-2} - \varepsilon_{r-1}, \qquad r \in \overline{3,8}. \tag{14.4.52}$$

Observe that the root system of $o(16, \mathbb{C})$:

$$\Phi_{D_8} = \{\pm\varepsilon_i \pm \varepsilon_j \mid i, j \in \overline{1,8},\ i \neq j\} \subset \Phi_{E_8}. \tag{14.4.53}$$

So the Lie subalgebra

$$\mathscr{G}_D^{E_8} = H_{E_8} + \sum_{\alpha \in \Phi_{D_8}} \mathbb{C}E_\alpha \tag{14.4.54}$$

of $\mathscr{G}^{E_8}$ is exactly isomorphic to $o(16, \mathbb{C})$. Moreover, it can be proved that the $(\mathscr{G}^{E_8}, \mathscr{G}_D^{E_8})$-branch rule on $\mathscr{G}^{E_8}$ is

$$\mathscr{G}^{E_8} \cong \mathscr{G}_D^{E_8} \oplus V_{D_8}(\lambda_8). \tag{14.4.55}$$

In fact, $V_{D_8}(\lambda_8)$ is exactly the spin module $\mathcal{V}$ in (14.3.37). Since

$$\sum_{i=1}^{8} \mathbb{F}_3\alpha_i = \sum_{\alpha \in \Phi_{D_8}} \mathbb{F}_3\alpha, \tag{14.4.56}$$

the ternary weight code $\mathscr{C}_{E_8}$ of E_8 on $\mathscr{G}^{E_8}$ is the same as that of $\mathscr{G}_D^{E_8}$ on $\mathscr{G}^{E_8}$. By Corollary 14.3.4 with $m = 8$, we have:

Theorem 14.4.7 *The ternary weight code $\mathscr{C}_{E_8}$ of E_8 on its adjoint module is an orthogonal* [120, 8, 57]*-code.*

References

[Sc] C. Shannon, A mathematical theory of communication. Bell System Tech. J. **27**, 379–423 and 623–656 (1948)

[X20] X. Xu, Representations of Lie algebras and coding theory. J. Lie Theory **22**(3), 647–682 (2012)

[P4] V. Pless, *Introduction to the Theory of Error-Correcting Codes*, 3rd edn. (Wiley, Inc., New Jersey, 1998

[PHB] V. Pless, W.C. Huffman, R.A. Brualdi, An introduction to algebraic codes, in *Handbook of Coding Theory* ed. by V. Pless, W. Huffman, Chap. 1 (Elsevier Science B.V., 1998)

Chapter 15
Path Hypergeometric Functions

The term of "hypergeometric" was first used by Wallis in Oxford as early as 1655 in his work *Arithmetrica Infinitorm* when referring to any series which could be regarded as a generalization of the ordinary geometric series $\sum_{n=0}^{\infty} z^n$. The well-known hypergeometric series of one-variable ${}_2F_1(a, b; c; z)$ was introduced and studied by Gauss in his thesis presented at Göttingen in 1812. Apell (1880) first systematically studied double hypergeometric functions and defined his famous four functions of types F_1–F_4. Horn more completely investigated all the double hypergeometric functions of second order, in 1889, 1931 and 1938, by studying corresponding second order hypergeometric partial differential systems with two independent variables. He found additional ten double hypergeometric functions of types G_1–G_3 and H_1–H_7, known as the Horn functions. Lauricella (1893) generalized Apell's four double hypergeometric functions to four families of multiple hypergeometric functions of types $F_A^{(n)}$, $F_B^{(n)}$, $F_C^{(n)}$ and $F_D^{(n)}$. Moreover, he conjectured the existence of fourteen complete triple hypergeometric functions of second order. Saran (1954) found the other ten triple hypergeometric functions of types F_E–F_T. However, Srivastava (1964) discovered three more complete triple hypergeometric functions of second order of types H_A, H_B and H_C. Saran (1955) gave integrals of Pochhammer type of the functions of types F_E–F_T. Pandey (1965) transformed the integral representations of F_F and F_G such that two interesting new triple hypergeometric functions of the generalized Horn type were obtained. Exton (1972, 1973) defined and studied twenty-six quadruple hypergeometric functions of types K_1–K_{21} and D_1–D_5, which are generalizations of the Horn functions. Furthermore, he introduced two families of multiple hypergeometric functions in the line of his quadruple functions. We refer [Eh] for the details of the hypergeometric functions of classical type and the references therein.

Heckman and Opdam [Hg1, Hg2, Hg3, HO, O1, O2, O3, O4, O5, O6, BO] introduced hypergeometric equations related to root systems and proved the existence of solutions (hypergeometric functions) of their equations. Gel'fand and Graev studied

X. Xu, *Representations of Lie Algebras and Partial Differential Equations*,
DOI 10.1007/978-981-10-6391-6_15

analogues of classical hypergeometric functions (so called GG-functions) by generalizing the differential property of the classical hypergeometric functions (e.g. cf. [GG]).

The Calogero–Sutherland model is an exactly solvable quantum many-body system in one-dimension (cf. [C, Sb]), which is related to a root system of A. Olshanestsky and Perelomov [OP] generalized it to the integrable systems related the root systems of types B, C and D. Etingof [Ep] found that the trace functions of certain intertwining operators among $gl(n, \mathbb{R})$-modules giving rise to certain solutions of the Calogero–Sutherland model. Moreover, Etingof and Styrkas [ES] proved a similar result for $sp(2n, \mathbb{R})$ that gives certain solutions of the Olshanestsky and Perelomov model of type C.

In [X14], we found that the Etingof trace functions are composite functions of certain rational functions with interesting master functions. In the case of $gl(2, \mathbb{R})$, the master function is exactly the classical Gauss hypergeometric function. When $n > 2$, the master functions are a new family of multiple hypergeometric functions, whose differential property and integral representation are dominated by certain polynomials of integral paths connecting pairs of positive integers. Similar results were also found for the Etingof–Styrkas trace functions. Based on the combinatorial feature of the path hypergeometric functions for $gl(n, \mathbb{R})$ and $sp(2n, \mathbb{R})$, we define and study analogous hypergeometric functions for orthogonal Lie algebras. The hypergeometric partial differential equations corresponding to the root systems of classical Lie algebras were obtained explicitly. Furthermore, we proved that certain variations of the Weyl functions of classical simple Lie algebras naturally give solutions of the Olshanestsky–Perelomov model with different parameters.

This chapter is a reformulation of our work [X14]. In Sect. 15.1, we prove that certain variations of Weyl denominator functions in (5.3.13) of classical types are solutions of the corresponding Olshanestsky–Perelomov models. In Sect. 15.2, we show that the Etingof's trace functions of types A and C give rise to the solutions of the Calogero–Sutherland model and the corresponding Olshanestsky–Perelomov model, respectively. In Sect. 15.3, we present a connection between the Calogero–Sutherland model of two particles and the classical Gauss hypergeometric function. A new family of multi-variable hypergeometric functions are found from the Calogero–Sutherland model of more particles via the Etingof's trace functions of types A. In Sect. 15.4, we determine another family of multi-variable hypergeometric functions from the Olshanestsky–Perelomov model via the Etingof's trace functions of types C. In Sect. 15.5, we find the differential properties, Euler integral representations and partial differential equations for the hypergeometric functions of type A, and the differential properties and partial differential equations for the hypergeometric functions of type C. Moreover, we define our multi-variable hypergeometric functions of type B and D analogously as those of type C and find the corresponding differential properties and partial differential equations.

15.1 Integrable Systems and Weyl Functions

In this section, we want to prove that certain variations of Weyl denominator functions of classical types are solutions of the corresponding Olshanestsky–Perelomov models.

The *Calogero–Sutherland model* is an exactly solvable quantum many-body system in one-dimension (cf. [C, Sb]), whose Hamiltonian is given by

$$H_{CS} = \sum_{i=1}^{n} \partial_{x_i}^2 + K \sum_{1\leq i<j\leq n} \frac{1}{\sinh^2(x_i - x_j)}, \tag{15.1.1}$$

where K is a constant. The model was used to study long-range interactions of n particles. It has become an important mathematical object partly because its potential is related to the root system of the special linear algebra $sl(n, \mathbb{C})$. Olshanestsky and Perelomov [OP] generalized the Calogero–Sutherland model to the system whose Hamiltonian is given by

$$\begin{aligned} H_{OP} = & \sum_{i=1}^{n} \partial_{x_i}^2 + \sum_{1\leq i<j\leq n} \left(\frac{K_1}{\sinh^2(x_i - x_j)} + \frac{K_2}{\sinh^2(x_i + x_j)} \right) \\ & + \sum_{i=1}^{n} \left(\frac{K_3}{\sinh^2 x_i} + \frac{K_4}{\sinh^2 2x_i} \right), \end{aligned} \tag{15.1.2}$$

where K_1, K_2, K_3, K_4 are constants. The potential of the Olshanestsky–Perelomov model is related to the root systems of all four families of finite-dimensional classical simple Lie algebras. Solving a system is to find eigenfunctions of its Hamiltonians.

Changing variables

$$z_i = e^{2x_i} \qquad \text{for } i \in \overline{1, n}, \tag{15.1.3}$$

we have the equation for the Calogero–Sutherland model (15.1.1):

$$\sum_{i=1}^{n} (z_i \partial_{z_i})^2(\Psi) + K \left(\sum_{1\leq i<j\leq n} \frac{z_i z_j}{(z_i - z_j)^2} \right) \Psi = \nu \Psi \tag{15.1.4}$$

and the equation for the Olshanestsky–Perelomov model (15.1.2):

$$\begin{aligned} & \sum_{i=1}^{n} \partial_{z_i}^2(\Psi) + \Bigg[\sum_{1\leq i<j\leq n} \left(K_1 \frac{z_i z_j}{(z_i - z_j)^2} + K_2 \frac{z_i z_j}{(z_i z_j - 1)^2} \right) \\ & + \sum_{i=1}^{n} \left(K_3 \frac{z_i}{(z_i - 1)^2} + K_4 \frac{z_i^2}{(z_i^2 - 1)^2} \right) \Bigg] \Psi = \nu \Psi. \end{aligned} \tag{15.1.5}$$

Let $f_{i,j}(z) \mid i, j \in \overline{1,n}\}$ be a set of one-variable differentiable functions and let d_i be a one-variable differential operator in z_i for $i \in \overline{1,n}$. It is easy to verified the following lemma:

Lemma 15.1.1 *We have the following equation on differentiation of determinants:*

$$\left(\sum_{i=1}^{n} d_i\right)\left(\begin{vmatrix} f_{1,1}(z_1) & f_{1,2}(z_2) & \cdots & f_{1,n}(z_n) \\ f_{2,1}(z_1) & f_{2,2}(z_2) & \cdots & f_{2,n}(z_n) \\ \vdots & \vdots & \vdots & \vdots \\ f_{n,1}(z_1) & f_{n,2}(z_2) & \cdots & f_{n,n}(z_n) \end{vmatrix}\right)$$
$$= \sum_{i=1}^{n} \begin{vmatrix} f_{1,1}(z_1) & f_{1,2}(z_2) & \cdots & f_{1,n}(z_n) \\ \vdots & \vdots & \vdots & \vdots \\ f_{i-1,1}(z_1) & f_{i-1,2}(z_2) & \cdots & f_{i-1,n}(z_n) \\ d_1(f_{i,1}(z_1)) & d_2(f_{i,2}(z_2)) & \cdots & d_n(f_{i,n}(z_n)) \\ f_{i+1,1}(z_1) & f_{i+1,2}(z_2) & \cdots & f_{i+1,n}(z_n) \\ \vdots & \vdots & \vdots & \vdots \\ f_{n,1}(z_1) & f_{n,2}(z_2) & \cdots & f_{n,n}(z_n) \end{vmatrix}. \tag{15.1.6}$$

Suppose that $R_X^+ \subset \mathbb{R}^n$ is the set of positive roots of the finite-dimensional simple Lie algebra of type X. We denote

$$z^\alpha = \prod_{i=1}^{n} z_i^{\alpha_i} \quad \text{for } \alpha \in \mathbb{R}^n. \tag{15.1.7}$$

The Weyl function

$$\mathcal{W}_X = \prod_{\alpha \in R_X^+} (z^{\alpha/2} - z^{-\alpha/2}). \tag{15.1.8}$$

First the *Weyl function of* $sl(n, \mathbb{C})$ is

$$\mathcal{W}_{A_{n-1}} = \prod_{1 \le i < j \le n} (z_i^{1/2} z_j^{-1/2} - z_i^{-1/2} z_j^{1/2}) = (z_1 z_2 \ldots z_n)^{(1-n)/2} \prod_{1 \le i < j \le n} (z_i - z_j), \tag{15.1.9}$$

Denote the Vandermonde determinant

$$W(z_1, z_2, \ldots, z_n) = \begin{vmatrix} 1 & 1 & \cdots & 1 \\ z_1 & z_2 & \cdots & z_n \\ z_1^2 & z_2^2 & \cdots & z_n^2 \\ \vdots & \vdots & \vdots & \vdots \\ z_1^{n-1} & z_2^{n-1} & \cdots & z_n^{n-1} \end{vmatrix} = \prod_{1 \le i < j \le n} (z_i - z_j). \tag{15.1.10}$$

Then $W(z_1, z_2, \ldots, z_n)$ is the fundamental part of $\mathcal{W}_{A_{n-1}}$. By Lemma 15.1.1, we have

$$\left(\sum_{i=1}^{n}(z_i\partial_{z_i})^2\right)(W(z_1,z_2,\ldots,z_n)) = \left(\sum_{i=1}^{n-1} i^2\right) W(z_1,z_2,\ldots,z_n)$$
$$= \frac{(n-1)n(2n-1)}{6} W(z_1,z_2,\ldots,z_n). \quad (15.1.11)$$

On the other hand,

$$\left(\sum_{i=1}^{n}(z_i\partial_{z_i})^2\right)(W(z_1,z_2,\ldots,z_n))$$
$$= \left(\sum_{r=1}^{n}(z_r\partial_{z_r})^2\right)\left(\prod_{1\le i<j\le n}(z_i-z_j)\right)$$
$$= \sum_{r=1}^{n} z_r\left[-\sum_{s=1}^{r-1}\left(\prod_{1\le i<j\le n;\ (i,j)\neq(s,r)}(z_i-z_j)\right) + \sum_{s=r+1}^{n}\left(\prod_{1\le i<j\le n;\ (i,j)\neq(r,s)}(z_i-z_j)\right)\right]$$
$$+2\sum_{r=1}^{n} z_r^2\left(\sum_{1\le s_1<s_2<r}\left[\prod_{1\le i<j\le n;\ (i,j)\neq(s_1,r),(s_2,r)}(z_i-z_j)\right]\right.$$
$$+\sum_{r<s_1<s_2\le n}\left[\prod_{1\le i<j\le n;\ (i,j)\neq(r,s_1),(r,s_2)}(z_i-z_j)\right]$$
$$\left.-\sum_{1\le s_1<r<s_2\le n}\left[\prod_{1\le i<j\le n;\ (i,j)\neq(s_1,r),(r,s_2)}(z_i-z_j)\right]\right)$$
$$= \sum_{1\le s<r\le n}(z_s-z_r)\prod_{1\le i<j\le n;\ (i,j)\neq(s,r)}(z_i-z_j)$$
$$+2W(z_1,z_2,\ldots,z_n)\sum_{r=1}^{n}\sum_{1\le s_1<s_2\le n;\ s_1,s_2\neq r}\frac{z_r^2}{(z_{s_1}-z_r)(z_{s_2}-z_r)}$$
$$= \left(\frac{n(n-1)}{2}+2\sum_{r=1}^{n}\sum_{1\le s_1<s_2\le n;\ s_1,s_2\neq r}\frac{z_r^2}{(z_{s_1}-z_r)(z_{s_2}-z_r)}\right)$$
$$\times W(z_1,z_2,\ldots,z_n). \quad (15.1.12)$$

Thus (15.1.11) and (15.1.12) yield

$$\sum_{r=1}^{n}\sum_{1\le s_1<s_2\le n;\ s_1,s_2\neq r}\frac{z_r^2}{(z_{s_1}-z_r)(z_{s_2}-z_r)}$$
$$= \frac{1}{2}\left[\frac{(n-1)n(2n-1)}{6}-\frac{n(n-1)}{2}\right]$$

$$= \frac{(n-1)n(n-2)}{6} = \binom{n}{3}. \tag{15.1.13}$$

Let

$$\phi^A_{\mu_1,\mu_2} = (z_1 z_2 \ldots z_n)^{\mu_1} W^{\mu_2}(z_1, z_2, \ldots, z_n) \qquad \text{for } \mu_1, \mu_2 \in \mathbb{C}, \tag{15.1.14}$$

which is a *variation of the Weyl function of* $sl(n, \mathbb{C})$. Then

$$z_r \partial_{z_r}(\phi^A_{\mu_1,\mu_2}) = \left(\mu_1 - \mu_2 \sum_{s=1}^{r-1} \frac{z_r}{z_s - z_r} + \mu_2 \sum_{s=r+1}^{n} \frac{z_r}{z_r - z_s}\right) \phi^A_{\mu_1,\mu_2} \tag{15.1.15}$$

for $r \in \overline{1,n}$. Thus

$$\begin{aligned}
&\sum_{r=1}^{n} (z_r \partial_{z_r})^2 (\phi^A_{\mu_1,\mu_2}) \\
&= \sum_{r=1}^{n} \Bigg[\mu_1^2 - 2\mu_1\mu_2 \sum_{s=1}^{r-1} \frac{z_r}{z_s - z_r} + 2\mu_1\mu_2 \sum_{s=r+1}^{n} \frac{z_r}{z_r - z_s} - \mu_2 \sum_{r \neq s \in \overline{1,n}} \frac{z_s z_r}{(z_s - z_r)^2} \\
&\quad + \mu_2^2 \sum_{r \neq s \in \overline{1,n}} \frac{z_r^2}{(z_s - z_r)^2} + 2\mu_2^2 \sum_{1 \leq s_1 < s_2 \leq n;\ s_1, s_2 \neq r} \frac{z_r^2}{(z_{s_1} - z_r)(z_{s_2} - z_r)} \Bigg] \phi^A_{\mu_1,\mu_2} \\
&= \Bigg[n\mu_1^2 + n(n-1)\mu_1\mu_2 + \mu_2^2 \binom{n}{3} - 2\mu_2 \sum_{1 \leq r < s \leq n} \frac{z_s z_r}{(z_s - z_r)^2} \\
&\quad + 2\mu_2^2 \sum_{1 \leq r < s \leq n} \frac{z_r^2 + z_s^2}{(z_s - z_r)^2} \Bigg] \phi^A_{\mu_1,\mu_2} \\
&= \Bigg[n\mu_1^2 + n(n-1)\mu_1\mu_2 + 2\mu_2^2 \binom{n}{3} - 2\mu_2 \sum_{1 \leq r < s \leq n} \frac{z_s z_r}{(z_s - z_r)^2} \\
&\quad + \mu_2^2 \sum_{1 \leq r < s \leq n} \frac{z_r^2 + z_s^2 - 2z_r z_s + 2z_r z_s}{(z_s - z_r)^2} \Bigg] \phi^A_{\mu_1,\mu_2} \\
&= \Bigg[n\mu_1^2 + n(n-1)(\mu_1 + \mu_2/2)\mu_2 + 2\mu_2^2 \binom{n}{3} \\
&\quad + 2\mu_2(\mu_2 - 1) \sum_{1 \leq r < s \leq n} \frac{z_s z_r}{(z_s - z_r)^2} \Bigg] \phi^A_{\mu_1,\mu_2}
\end{aligned} \tag{15.1.16}$$

by (15.1.13) and (15.1.15). Therefore, we have:

Theorem 15.1.2 *The function $\phi^A_{\mu_1,\mu_2}$ satisfies:*

$$\sum_{r=1}^n (z_r\partial_{z_r})^2(\phi^A_{\mu_1,\mu_2}) + 2\mu_2(1-\mu_2)\left(\sum_{1\le i<j\le n}\frac{z_iz_j}{(z_i-z_j)^2}\right)\phi^A_{\mu_1,\mu_2}$$
$$= \left[n\mu_1^2 + n(n-1)(\mu_1+\mu_2/2)\mu_2 + 2\binom{n}{3}\mu_2^2\right]\phi^A_{\mu_1,\mu_2}, \qquad (15.1.17)$$

which is a Calogero–Sutherland equation.

We remark that above result was known when $\mu_1 = \mu_2$ or $\mu_1 = 0$ before our work [X14]. Moreover, $\mathcal{W}_{A_{n-1}} = \phi^A_{(1-n)/2,1}$.

According to simplicity, we secondly consider the Weyl function of $o(2n,\mathbb{C})$. By the root structure, the *Weyl function of* $o(2n,\mathbb{C})$

$$\begin{aligned}\mathcal{W}_{D_n} &= \prod_{1\le i<j\le n}(z_i^{1/2}z_j^{-1/2} - z_i^{-1/2}z_j^{1/2})(z_i^{1/2}z_j^{1/2} - z_i^{-1/2}z_j^{-1/2})\\ &= \prod_{1\le i<j\le n}(z_i+z_i^{-1}-(z_j+z_j^{-1}))\\ &= W(z_1+z_1^{-1}, z_2+z_2^{-1},\ldots,z_n+z_n^{-1}) \qquad (15.1.18)\end{aligned}$$

(cf. (15.1.10)). Note that

$$\begin{aligned}(z_i\partial_{z_i})^2[(z_i+z_i^{-1})^k] &= k(z_i+z_i^{-1})^k + k(k-1)(z_i-z_i^{-1})^2(z_i+z_i^{-1})^{k-2}\\ &= k(z_i+z_i^{-1})^k + k(k-1)(z_i^2-2+z_i^{-2})(z_i+z_i^{-1})^{k-2}\\ &= k^2(z_i+z_i^{-1})^k - 4k(k-1)(z_i+z_i^{-1})^{k-2}. \qquad (15.1.19)\end{aligned}$$

By Lemma 15.1.1 and a similar calculation as (15.1.10),

$$\left(\sum_{i=1}^n (z_i\partial_{z_i})^2\right)(\mathcal{W}_{D_n}) = \frac{(n-1)n(2n-1)}{6}\mathcal{W}_{D_n}. \qquad (15.1.20)$$

On the other hand, by (15.1.12) and (15.1.19), we have

$$\sum_{r=1}^n \sum_{1\le s_1<s_2\le n;\ s_1,s_2\ne r}\frac{(z_r-z_r^{-1})^2}{(z_{s_1}+z_{s_1}^{-1}-z_r-z_r^{-1})(z_{s_2}+z_{s_2}^{-1}-z_r-z_r^{-1})}$$
$$= \frac{(n-1)n(n-2)}{6} = \binom{n}{3}. \qquad (15.1.21)$$

Moreover,

$$\frac{(z_r - z_r^{-1})^2 + (z_s - z_s^{-1})^2}{(z_s + z_s^{-1} - z_r - z_r^{-1})^2}$$
$$= \frac{z_r^2 + z_r^{-2} + z_s^2 + z_s^{-2} - 4}{(z_r z_s)^{-2}(z_s - z_r)^2(z_r z_s - 1)^2}$$
$$= \frac{z_r^4 z_s^2 + z_s^2 + z_r^2 z_s^4 + z_s^2 - 4z_r^2 z_s^2}{(z_s - z_r)^2(z_r z_s - 1)^2}$$
$$= \frac{1}{(z_s - z_r)^2(z_r z_s - 1)^2}[z_r^4 z_s^2 + z_s^2 + z_r^2 z_s^4 + z_s^2 - 2z_r z_s(z_r^2 + z_s^2 + (z_r z_s)^2$$
$$-2z_r z_s + 1) + 2z_r z_s(z_r^2 + z_s^2 + (z_r z_s)^2 - 4z_r z_s + 1)]$$
$$= \frac{(z_s - z_r)^2(z_r z_s - 1)^2 + 2z_r z_s[(z_r - z_s)^2 + (z_r z_s - 1)^2]}{(z_s - z_r)^2(z_r z_s - 1)^2}$$
$$= 1 + 2\frac{z_r z_s}{(z_r - z_s)^2} + 2\frac{z_r z_s}{(z_r z_s - 1)^2} \tag{15.1.22}$$

and

$$z_r \partial_{z_r}\left(\frac{z_r - z_r^{-1}}{z_r + z_r^{-1} - z_s - z_s^{-1}}\right)$$
$$= \frac{(z_r + z_r^{-1})(z_r + z_r^{-1} - z_s - z_s^{-1}) - (z_r - z_r^{-1})^2}{(z_r + z_r^{-1} - z_s - z_s^{-1})^2}$$
$$= \frac{(z_r + z_r^{-1})^2 - (z_r + z_r^{-1})(z_s + z_s^{-1}) - (z_r - z_r^{-1})^2}{(z_r + z_r^{-1} - z_s - z_s^{-1})^2}$$
$$= \frac{4 - z_r z_s - z_r z_s^{-1} - z_r^{-1} z_s - z_r^{-1} z_s^{-1}}{(z_r z_s)^{-2}(z_s - z_r)^2(z_r z_s - 1)^2}$$
$$= -z_r z_s \frac{(z_r z_s)^2 + z_r^2 + z_s^2 + 1 - 4z_r z_s}{(z_s - z_r)^2(z_r z_s - 1)^2}$$
$$= -\frac{z_r z_s}{(z_s - z_r)^2} - \frac{z_r z_s}{(z_r z_s - 1)^2} \tag{15.1.23}$$

for $r, s \in \overline{1, n}$ such that $r \neq s$.

Set

$$\phi_\mu^D = (\mathcal{W}_{D_n})^\mu = \prod_{1 \le i < j \le n} (z_i + z_i^{-1} - z_j - z_j^{-1})^\mu \quad \text{for } \mu \in \mathbb{C}, \tag{15.1.24}$$

which is a *variation of the Weyl function of* $o(2n, \mathbb{C})$. Then

$$\left(\sum_{r=1}^{n} (z_r \partial_{z_r})^2\right)(\phi_\mu^D)$$

$$= \sum_{r=1}^{n} \left[\sum_{r \neq s \in \overline{1,n}} \left[\mu z_r \partial_{z_r} \left(\frac{z_r - z_r^{-1}}{z_r + z_r^{-1} - z_s - z_s^{-1}} \right) + \mu^2 \frac{(z_r - z_r^{-1})^2}{(z_s + z_s^{-1} - z_r - z_r^{-1})^2} \right] \right.$$

$$\left. + 2\mu^2 \sum_{1 \le s_1 < s_2 \le n;\ s_1, s_2 \neq r} \frac{(z_r - z_r^{-1})^2}{(z_{s_1} + z_{s_1}^{-1} - z_r - z_r^{-1})(z_{s_2} + z_{s_2}^{-1} - z_r - z_r^{-1})} \right] \phi_\mu^D$$

$$= \left[\sum_{1 \le r < s \le n} \left[-2\mu \left(\frac{z_r z_s}{(z_s - z_r)^2} + \frac{z_r z_s}{(z_r z_s - 1)^2} \right) + \mu^2 \frac{(z_r - z_r^{-1})^2 + (z_s - z_s^{-1})^2}{(z_s + z_s^{-1} - z_r - z_r^{-1})^2} \right] \phi_\mu^D \right.$$

$$\left. + 2\mu^2 \sum_{r=1}^{n} \sum_{1 \le s_1 < s_2 \le n;\ s_1, s_2 \neq r} \frac{(z_r - z_r^{-1})^2}{(z_{s_1} + z_{s_1}^{-1} - z_r - z_r^{-1})(z_{s_2} + z_{s_2}^{-1} - z_r - z_r^{-1})} \right]$$

$$= \left[2\mu(\mu - 1) \sum_{1 \le r < s \le n} \left(\frac{z_r z_s}{(z_s - z_r)^2} + \frac{z_r z_s}{(z_r z_s - 1)^2} \right) + \mu^2 \left(\frac{n(n-1)}{2} + 2 \left(!! \begin{matrix} n \\ 3 \end{matrix} !! \right) \right) \right] \phi_\mu^D$$

$$= \left[2\mu(\mu - 1) \sum_{1 \le r < s \le n} \left(\frac{z_r z_s}{(z_s - z_r)^2} + \frac{z_r z_s}{(z_r z_s - 1)^2} \right) \right.$$

$$\left. + \frac{n(n-1)(2n-1)}{6} \mu^2 \right] \phi_\mu^D. \tag{15.1.25}$$

Thus we obtain:

Theorem 15.1.3 *The function ϕ_μ^D the equation:*

$$= \sum_{r=1}^{n} (z_r \partial_{z_r})^2 (\phi_\mu^D) + 2\mu(1 - \mu) \left(\sum_{1 \le i < j \le n} \frac{z_i z_j}{(z_i - z_j)^2} + \sum_{1 \le i < j \le n} \frac{z_i z_j}{(z_i z_j - 1)^2} \right) \phi_\mu^D$$

$$= \frac{n(n-1)(2n-1)\mu^2}{6} \phi_\mu^D, \tag{15.1.26}$$

which is an Olshanestsky–Perelomov equation.

We remark that the above result with $n = 4$ was given in [NFP].

The *Weyl function of* $sp(2n, \mathbb{C})$ is

$$\mathcal{W}_{C_n} = \left[\prod_{r=1}^{n} (z_r - z_r^{-1}) \right] \prod_{1 \le i < j \le n} (z_i + z_i^{-1} - (z_j + z_j^{-1})), \tag{15.1.27}$$

In terms of determinant,

$$\mathcal{W}_{C_n} = \begin{vmatrix} z_1 - z_1^{-1} & \cdots & z_n - z_n^{-1} \\ (z_1 - z_1^{-1})(z_1 + z_1^{-1}) & \cdots & (z_n - z_n^{-1})(z_n + z_n^{-1}) \\ (z_1 - z_1^{-1})(z_1 + z_1^{-1})^2 & \cdots & (z_n - z_n^{-1})(z_n + z_n^{-1})^2 \\ \vdots & \vdots & \vdots \\ (z_1 - z_1^{-1})(z_1 + z_1^{-1})^{n-1} & \cdots & (z_n - z_n^{-1})(z_n + z_n^{-1})^{n-1} \end{vmatrix}. \tag{15.1.28}$$

Moreover,

$$(z_r\partial_{z_r})^2[(z_r - z_r^{-1})(z_r + z_r^{-1})^k] = (k+1)^2(z_r - z_r^{-1})(z_r + z_r^{-1})^k \\ + 4k(k-1)(z_r - z_r^{-1})(z_r + z_r^{-1})^{k-2} \quad (15.1.29)$$

by (15.1.19). Thus

$$\left(\sum_{i=1}^{n}(z_i\partial_{z_i})^2\right)(\mathcal{W}_{C_n}) = \frac{n(n+1)(2n+1)}{6}\mathcal{W}_{C_n} \quad (15.1.30)$$

by Lemma 15.1.1 and (15.1.29).

Set

$$\psi_\mu^C = \prod_{r=1}^{n}(z_r - z_r^{-1})^\mu \qquad \text{for } \mu \in \mathbb{C}. \quad (15.1.31)$$

Then

$$\begin{aligned}\sum_{r=1}^{n}(z_r\partial_{z_r})^2(\psi_\mu^C) &= \mu\psi_\mu^C\sum_{r=1}^{n}\left[z_r\partial_{z_r}\left(\frac{z_r + z_r^{-1}}{z_r - z_r^{-1}}\right) + \mu\left(\frac{z_r + z_r^{-1}}{z_r - z_r^{-1}}\right)^2\right] \\ &= \mu\psi_\mu^C\sum_{r=1}^{n}\left[\frac{(z_r - z_r^{-1})^2 - (z_r + z_r^{-1})^2}{(z_r - z_r^{-1})^2} + \mu\frac{(z_r + z_r^{-1})^2}{(z_r - z_r^{-1})^2}\right] \\ &= \mu\sum_{r=1}^{n}\left[\mu + 4(\mu - 1)\frac{1}{(z_r - z_r^{-1})^2}\right]\psi_\mu^C \\ &= \left[n\mu^2 + 4\mu(\mu - 1)\sum_{r=1}^{n}\frac{z_r^2}{(z_r^2 - 1)^2}\right]\psi_\mu^C, \end{aligned} \quad (15.1.32)$$

which is a *variation of the Weyl function of* $sp(2n, \mathbb{C})$. Moreover,

$$\begin{aligned}&\sum_{r=1}^{n}(z_r\partial_{z_r})(\psi_{\mu_1}^C)(z_r\partial_{z_r})(\phi_{\mu_2}^D) \\ &= \mu_1\mu_2\sum_{r=1}^{n}\frac{z_r + z_r^{-1}}{z_r - z_r^{-1}}\psi_{\mu_1}^C\sum_{r\neq s\in\overline{1,n}}\frac{z_r - z_r^{-1}}{z_r + z_r^{-1} - z_s - z_s^{-1}}\phi_{\mu_2}^D \\ &= \mu_1\mu_2\psi_{\mu_1}^C\phi_{\mu_2}^D\sum_{r=1}^{n}\sum_{r\neq s\in\overline{1,n}}\frac{z_r + z_r^{-1}}{z_r + z_r^{-1} - z_s - z_s^{-1}} \\ &= \frac{n(n-1)}{2}\mu_1\mu_2\psi_{\mu_1}^C\phi_{\mu_2}^D. \end{aligned} \quad (15.1.33)$$

Set

$$\phi^C_{\mu_1,\mu_2} = \psi^C_{\mu_1}\phi^D_{\mu_2} = \left[\prod_{r=1}^{n}(z_r - z_r^{-1})^{\mu_1}\right] \prod_{1\leq i<j\leq n}(z_i + z_i^{-1} - z_j - z_j^{-1})^{\mu_2} \quad (15.1.34)$$

for $\mu_1, \mu_2 \in \mathbb{C}$. By Theorem 15.1.3 and (15.1.32)–(15.1.34), we have:

Theorem 15.1.4 *The function $\phi^C_{\mu_1,\mu_2}$ satisfies:*

$$\begin{aligned}
&\sum_{r=1}^{n}(z_r\partial_{z_r})^2(\phi^C_{\mu_1,\mu_2}) + \Bigg[4\mu_1(1-\mu_1)\sum_{r=1}^{n}\frac{z_r^2}{(z_r^2-1)^2} \\
&+ 2\mu_2(1-\mu_2)\left(\sum_{1\leq i<j\leq n}\frac{z_iz_j}{(z_i-z_j)^2} + \sum_{1\leq i<j\leq n}\frac{z_iz_j}{(z_iz_j-1)^2}\right)\Bigg]\phi^C_{\mu_1,\mu_2} \\
&= \left(n\mu_1^2 + n(n-1)\mu_1\mu_2 + \frac{n(n-1)(2n-1)\mu_2^2}{6}\right)\phi^C_{\mu_1,\mu_2},
\end{aligned} \quad (15.1.35)$$

which is an Olshanestsky–Perelomov equation.

Observe that $\mathcal{W}_{C_n} = \phi^C_{1,1}$. The *Weyl function of* $o(2n+1,\mathbb{C})$ is

$$\mathcal{W}_{B_n} = \left[\prod_{r=1}^{n}(z_r^{1/2} - z_r^{-1/2})\right] \prod_{1\leq i<j\leq n}(z_i + z_i^{-1} - (z_j + z_j^{-1})), \quad (15.1.36)$$

In terms of determinant,

$$\mathcal{W}_{B_n} = \begin{vmatrix} z_1^{1/2} - z_1^{-1/2} & \cdots & z_n^{1/2} - z_n^{-1/2} \\ (z_1^{1/2} - z_1^{-1/2})(z_1^{1/2} + z_1^{-1/2})^2 & \cdots & (z_n^{1/2} - z_n^{-1/2})(z_n^{1/2} + z_n^{-1/2})^2 \\ (z_1^{1/2} - z_1^{-1/2})(z_1^{1/2} + z_1^{-1/2})^4 & \cdots & (z_n^{1/2} - z_n^{-1/2})(z_n^{1/2} + z_n^{-1/2})^4 \\ \vdots & \vdots & \vdots \\ (z_1^{1/2} - z_1^{-1/2})(z_1^{1/2} + z_1^{-1/2})^{2(n-1)} & \cdots & (z_n^{1/2} - z_n^{-1/2})(z_n^{1/2} + z_n^{-1/2})^{2(n-1)} \end{vmatrix}. \quad (15.1.37)$$

Moreover,

$$\begin{aligned}
&(z_r\partial_{z_r})^2[(z_r^{1/2} - z_r^{-1/2})(z_r^{1/2} + z_r^{-1/2})^{2k}] \\
&= \left(k^2 + k + \frac{1}{4}\right)(z_r^{1/2} - z_r^{-1/2})(z_r^{1/2} + z_r^{-1/2})^{2k} \\
&\quad + 2k(2k-1)(z_r^{1/2} - z_r^{-1/2})(z_r^{1/2} + z_r^{-1/2})^{2(k-1)}.
\end{aligned} \quad (15.1.38)$$

Thus

$$\left(\sum_{i=1}^{n}(z_i\partial_{z_i})^2\right)(\mathcal{W}_{B_n}) = \frac{n(4n^2-1)}{12}\mathcal{W}_{B_n} \tag{15.1.39}$$

by Lemma 15.1.1 and (15.1.38).

Set

$$\psi_\mu^B = \prod_{r=1}^{n}(z_r^{1/2} - z_r^{-1/2})^\mu \qquad \text{for } \mu \in \mathbb{C}. \tag{15.1.40}$$

Then

$$\begin{aligned}
&\sum_{r=1}^{n}(z_r\partial_{z_r})^2(\psi_\mu^B) \\
&= \frac{\mu}{2}\psi_\mu^B \sum_{r=1}^{n}\left[z_r\partial_{z_r}\left(\frac{z_r^{1/2}+z_r^{-1/2}}{z_r^{1/2}-z_r^{-1/2}}\right) + \frac{\mu}{2}\left(\frac{z_r^{1/2}+z_r^{-1/2}}{z_r^{1/2}-z_r^{-1/2}}\right)^2\right] \\
&= \frac{\mu}{4}\psi_\mu^B \sum_{r=1}^{n}\left[\frac{(z_r^{1/2}-z_r^{-1/2})^2-(z_r^{1/2}+z_r^{-1/2})^2}{(z_r^{1/2}-z_r^{-1/2})^2} + \mu\frac{(z_r^{1/2}+z_r^{-1/2})^2}{(z_r^{1/2}-z_r^{-1/2})^2}\right] \\
&= \left[\frac{n\mu^2}{4} + \mu(\mu-1)\sum_{r=1}^{n}\frac{z_r}{(z_r-1)^2}\right]\psi_\mu^B.
\end{aligned} \tag{15.1.41}$$

Moreover,

$$\begin{aligned}
&\sum_{r=1}^{n}(z_r\partial_{z_r})(\psi_{\mu_1}^B)(z_r\partial_{z_r})(\phi_{\mu_2}^D) \\
&= \frac{\mu_1\mu_2}{2}\sum_{r=1}^{n}\frac{z_r^{1/2}+z_r^{-1/2}}{z_r^{1/2}-z_r^{-1/2}}\psi_{\mu_1}^B \sum_{r\neq s\in\overline{1,n}}\frac{z_r-z_r^{-1}}{z_r+z_r^{-1}-z_s-z_s^{-1}}\phi_{\mu_2}^D \\
&= \frac{\mu_1\mu_2}{2}\psi_{\mu_1}^B\phi_{\mu_2}^D\sum_{r=1}^{n}\sum_{r\neq s\in\overline{1,n}}\frac{(z_r^{1/2}+z_r^{-1/2})^2}{z_r+z_r^{-1}-z_s-z_s^{-1}} \\
&= \frac{n(n-1)}{4}\mu_1\mu_2\psi_{\mu_1}^B\phi_{\mu_2}^D.
\end{aligned} \tag{15.1.42}$$

Set

$$\phi_{\mu_1,\mu_2}^B = \psi_{\mu_1}^B\phi_{\mu_2}^D = \left[\prod_{r=1}^{n}(z_r^{1/2}-z_r^{-1/2})^{\mu_1}\right]\prod_{1\leq i<j\leq n}(z_i+z_i^{-1}-z_j-z_j^{-1})^{\mu_2} \tag{15.1.43}$$

for $\mu_1, \mu_2 \in \mathbb{C}$, which is a *variation of the Weyl function of* $o(2n+1, \mathbb{C})$. By Theorem 15.1.3 and (15.1.41)–(15.1.43), we obtain:

Theorem 15.1.5 *The function* $\phi^B_{\mu_1,\mu_2}$ *satisfies:*

$$\sum_{r=1}^{n}(z_r\partial_{z_r})^2(\phi^B_{\mu_1,\mu_2}) + \Bigg[\mu_1(1-\mu_1)\sum_{r=1}^{n}\frac{z_r}{(z_r-1)^2} + 2\mu_2(1-\mu_2)\left(\sum_{1\le i<j\le n}\frac{z_iz_j}{(z_i-z_j)^2} + \sum_{1\le i<j\le n}\frac{z_iz_j}{(z_iz_j-1)^2}\right)\Bigg]\phi^C_{\mu_1,\mu_2}$$
$$= \left(\frac{n\mu_1^2}{4} + \frac{n(n-1)}{2}\mu_1\mu_2 + \frac{n(n-1)(2n-1)\mu_2^2}{6}\right)\phi^B_{\mu_1,\mu_2}, \quad (15.1.44)$$

which is an Olshanestsky–Perelomov equation.

15.2 Etingof Traces

In this section, we present the trace functions of the intertwining operators among certain modules introduced by Etingof [Ep].

Recall that the general Lie algebra

$$gl(n,\mathbb{R}) = \sum_{i,j=1}^{n}\mathbb{R}E_{i,j}. \quad (15.2.1)$$

The subspace

$$H_A = \sum_{i=1}^{n}\mathbb{R}E_{i,i} \quad (15.2.2)$$

is a toral Cartan subalgebra of the Lie algebra $gl(n,\mathbb{R})$. For a weight $\lambda \in H_A^*$, we denote

$$\lambda(E_{i,i}) = \lambda_i \qquad \text{for } i \in \overline{1,n}. \quad (15.2.3)$$

Note that

$$\omega_A = \sum_{i,j=1}^{n}E_{i,j}E_{j,i} \quad (15.2.4)$$

is the Casimir element in $U(gl(n,\mathbb{R}))$. Let

$$M = \bigoplus_{\lambda\in H_A^*}M_\lambda \quad (15.2.5)$$

be any weight $gl(n,\mathbb{R})$-module with the representation $\pi_{_M}$ such that

$$\pi_{_M}(\omega_{_A}) = b\,\mathrm{Id}_M \qquad \text{for some } \ b \in \mathbb{C}, \tag{15.2.6}$$

where

$$M^{\lambda} = \{w \in M \mid h(w) = \lambda(h)w \text{ for } h \in H\}. \tag{15.2.7}$$

In particular, we can take M to be any highest weight module; say the Verma module $M(\lambda)$ given in (5.2.8)–(5.2.10). The module M is not necessarily irreducible.

Let $\{x_1, x_2, \ldots, x_n\}$ be n indeterminates. Set

$$x^{\vec{i}} = x_1^{i_1} x_2^{i_2} \ldots x_n^{i_n} \qquad \vec{i} = (i_1, i_2, \ldots, i_n) \in \mathbb{Z}^n. \tag{15.2.8}$$

Fix $\kappa \in \mathbb{R}$ and denote

$$\bar{x}^{\kappa} = (x_1 x_2 \ldots x_n)^{\kappa}. \tag{15.2.9}$$

Define the space of formal Laurent series in $\{x_1, x_2, \ldots, x_n\}$ with the coefficients in M:

$$\tilde{M} = \left\{ \sum_{\vec{i}\in\mathbb{Z}^n} w_{\vec{i}} x^{\vec{i}} \bar{x}^{\kappa} \mid w_{\vec{i}} \in M \right\}. \tag{15.2.10}$$

We extend the representation $\pi_{_M}$ to $\tilde{M}$ by

$$\pi_{_M}(\xi)\left(\sum_{\vec{i}\in\mathbb{Z}^n} w_{\vec{i}} x^{\vec{i}} \bar{x}^{\kappa}\right) = \sum_{\vec{i}\in\mathbb{Z}^n} \pi_{_M}(\xi)(w_{\vec{i}}) x^{\vec{i}} \bar{x}^{\kappa} \ \text{ for } \xi \in gl(n,\mathbb{R}). \tag{15.2.11}$$

Moreover, we define another representation π' of $gl(n,\mathbb{R})$ on $\tilde{M}$ by

$$\pi'(E_{r,s})\left(\sum_{\vec{i}\in\mathbb{Z}^n} w_{\vec{i}} x^{\vec{i}} \bar{x}^{\kappa}\right) = \sum_{\vec{i}\in\mathbb{Z}^n} [w_{\vec{i}} x_r \partial_{x_s}(x^{\vec{i}} \bar{x}^{\kappa}) - \delta_{r,s}\kappa w_{\vec{i}} x^{\vec{i}} \bar{x}^{\kappa}] \tag{15.2.12}$$

for $r, s \in \overline{1,n}$. Define the representation of $gl(n,\mathbb{R})$ on $\tilde{M}$ by

$$\pi = \pi_{_M} + \pi'. \tag{15.2.13}$$

Suppose that $\Phi : M \to \tilde{M}$ is a linear map such that

$$\Phi(\pi_{_M}(\xi)(w)) = \Delta(\xi)\Phi(w) \tag{15.2.14}$$

for $\xi \in gl(n,\mathbb{R})$ and $w \in M$. View Φ as a function in $\{x_1, x_2, \ldots, x_n\}$ taking value in the spaces of linear transformations on M. The *Etingof trace function of type A*

$$E_A(z_1, z_2, \dots, z_n) = \mathrm{tr}_{_M}\ \Phi z_1^{E_{1,1}} z_2^{E_{2,2}} \cdots z_n^{E_{n,n}} \tag{15.2.15}$$

(cf. [Ep]). By considering weights, we can prove

$$(\bar{x}^{\kappa})^{-1} E_A(z_1, z_2, \dots, z_n) \text{ is independent of } x_1, x_2, \dots, x_n. \tag{15.2.16}$$

Etingof [Ep] proved:

Theorem 15.2.1 *The function*

$$\Psi_A(z_1, z_2, \dots, z_n) = (\bar{x}^{\kappa})^{-1} \mathcal{W}_{A_{n-1}} E_A(z_1, z_2, \dots, z_n) \tag{15.2.17}$$

is a solution of (15.1.4) for suitable constants K and ν.

Proof Recall $\mathcal{W}_{A_{n-1}} = \phi^A_{(1-n)/2,1}$. By (15.1.15),

$$z_r \partial_{z_r}(\mathcal{W}_{A_{n-1}}) = \left(\frac{1-n}{2} + \sum_{s \neq r} \frac{z_r}{z_r - z_s} \right) \mathcal{W}_{A_{n-1}} = \left(\frac{n-1}{2} + \sum_{s \neq r} \frac{z_s}{z_r - z_s} \right) \mathcal{W}_{A_{n-1}} \tag{15.2.18}$$

for $r \in \overline{1,n}$. Set

$$E_A^{i_1,i_2} = \mathrm{tr}_M \Phi \pi_{_M}(E_{i_1,i_2} E_{i_2,i_1}) z_1^{\pi_M(E_{1,1})} z_2^{\pi_M(E_{2,2})} \cdots z_n^{\pi_M(E_{n,n})} \tag{15.2.19}$$

for $i_1, i_2 \in \overline{1,n}$. Observe that

$$\begin{aligned} &\mathrm{tr}_M \Phi \pi_{_M}(E_{i,i}) z_1^{\pi_M(E_{1,1})} z_2^{\pi_M(E_{2,2})} \dots z_n^{\pi_M(E_{n,n})} \\ &= \mathrm{tr}_M \Phi z_1^{\pi_M(E_{1,1})} z_2^{\pi_M(E_{2,2})} \dots z_n^{\pi_M(E_{n,n})} \pi_{_M}(E_{i,i})) \\ &= z_i \partial_{z_i} [\mathrm{tr}_M \Phi z_1^{\pi_M(E_{1,1})} z_2^{\pi_M(E_{2,2})} \dots z_n^{\pi_M(E_{n,n})}] \end{aligned} \tag{15.2.20}$$

for $i \in \overline{1,n}$. In particular,

$$E_A^{i,i} = (z_i \partial_{z_i})^2 (E_A) \tag{15.2.21}$$

and for $i_1 \neq i_2$,

$$\begin{aligned} E_A^{i_1,i_2} &= \mathrm{tr}_M \pi(E_{i_1,i_2}) \Phi \pi_{_M}(E_{i_2,i_1}) z_1^{\pi_M(E_{1,1})} z_2^{\pi_M(E_{2,2})} \dots z_n^{\pi_M(E_{n,n})} \\ &= \mathrm{tr}_M \pi'(E_{i_1,i_2}) \Phi \pi_{_M}(E_{i_2,i_1}) z_1^{\pi_M(E_{1,1})} z_2^{\pi_M(E_{2,2})} \dots z_n^{\pi_M(E_{n,n})} \\ &\quad + \mathrm{tr}_M \pi_{_M}(E_{i_1,i_2}) \Phi \pi_{_M}(E_{i_2,i_1}) z_1^{\pi_M(E_{1,1})} z_2^{\pi_M(E_{2,2})} \dots z_n^{\pi_M(E_{n,n})} \\ &= \mathrm{tr}_M \pi'(E_{i_1,i_2}) \Phi \pi_{_M}(E_{i_2,i_1}) z_1^{\pi_M(E_{1,1})} z_2^{\pi_M(E_{2,2})} \dots z_n^{\pi_M(E_{n,n})} \\ &\quad + \mathrm{tr}_M \Phi \pi_{_M}(E_{i_2,i_1}) z_1^{\pi_M(E_{1,1})} z_2^{\pi_M(E_{2,2})} \dots z_n^{\pi_M(E_{n,n})} \pi_{_M}(E_{i_1,i_2}) \end{aligned}$$

$$
\begin{aligned}
&= \mathrm{tr}_M \pi'(E_{i_1,i_2}) \Phi \pi_M(E_{i_2,i_1}) z_1^{\pi_M(E_{1,1})} z_2^{\pi_M(E_{2,2})} \dots z_n^{\pi_M(E_{n,n})} \\
&\quad + \frac{z_{i_1}}{z_{i_2}} \mathrm{tr}_M \Phi \pi_M(E_{i_2,i_1} E_{i_1,i_2}) z_1^{\pi_M(E_{1,1})} z_2^{\pi_M(E_{2,2})} \dots z_n^{\pi_M(E_{n,n})} \\
&= \mathrm{tr}_M \pi'(E_{i_1,i_2}) \Phi \pi_M(E_{i_2,i_1}) z_1^{\pi_M(E_{1,1})} z_2^{\pi_M(E_{2,2})} \dots z_n^{\pi_M(E_{n,n})} \\
&\quad + \frac{z_{i_1}}{z_{i_2}} \mathrm{tr}_M \Phi \pi_M(E_{i_1,i_2} E_{i_2,i_1} + [E_{i_2,i_1}, E_{i_1,i_2}]) z_1^{\pi_M(E_{1,1})} z_2^{\pi_M(E_{2,2})} \dots z_n^{\pi_M(E_{n,n})} \\
&= \mathrm{tr}_M \pi'(E_{i_1,i_2}) \Phi \pi_M(E_{i_2,i_1}) z_1^{\pi_M(E_{1,1})} z_2^{\pi_M(E_{2,2})} \dots z_n^{\pi_M(E_{n,n})} \\
&\quad + \frac{z_{i_1}}{z_{i_2}} \mathrm{tr}_M \Phi \pi_M(E_{i_1,i_2} E_{i_2,i_1} + E_{i_2,i_2} - E_{i_1,i_1}) z_1^{\pi_M(E_{1,1})} z_2^{\pi_M(E_{2,2})} \dots z_n^{\pi_M(E_{n,n})} \\
&= \mathrm{tr}_M \pi'(E_{i_1,i_2}) \Phi \pi_M(E_{i_2,i_1}) z_1^{\pi_M(E_{1,1})} z_2^{\pi_M(E_{2,2})} \dots z_n^{\pi_M(E_{n,n})} + \frac{z_{i_1}}{z_{i_2}} E_A^{i_1,i_2} \\
&\quad + \frac{z_{i_1}}{z_{i_2}} (z_{i_2} \partial_{z_{i_2}} - z_{i_1} \partial_{z_{i_1}})(E_A). \qquad (15.2.22)
\end{aligned}
$$

Thus

$$
\begin{aligned}
E_A^{i_1,i_2} &= \frac{z_{i_2}}{z_{i_2} - z_{i_1}} \mathrm{tr}_M \pi'(E_{i_1,i_2}) \Phi \pi_M(E_{i_2,i_1}) z_1^{\pi_M(E_{1,1})} z_2^{\pi_M(E_{2,2})} \dots z_n^{\pi_M(E_{n,n})} \\
&\quad + \frac{z_{i_1}}{z_{i_2} - z_{i_1}} (z_{i_2} \partial_{z_{i_2}} - z_{i_1} \partial_{z_{i_1}})(E_A). \qquad (15.2.23)
\end{aligned}
$$

Furthermore,

$$
\begin{aligned}
&\mathrm{tr}_M \pi'(E_{i_1,i_2}) \Phi \pi_M(E_{i_2,i_1}) z_1^{\pi_M(E_{1,1})} z_2^{\pi_M(E_{2,2})} \dots z_n^{\pi_M(E_{n,n})} \\
&= \mathrm{tr}_M \pi'(E_{i_1,i_2}) \pi(E_{i_2,i_1}) \Phi z_1^{\pi_M(E_{1,1})} z_2^{\pi_M(E_{2,2})} \dots z_n^{\pi_M(E_{n,n})} \\
&= \mathrm{tr}_M \pi'(E_{i_1,i_2}) \pi'(E_{i_2,i_1}) \Phi z_1^{\pi_M(E_{1,1})} z_2^{\pi_M(E_{2,2})} \dots z_n^{\pi_M(E_{n,n})} \\
&\quad + \mathrm{tr}_M \pi'(E_{i_1,i_2}) \pi_M(E_{i_2,i_1}) \Phi z_1^{\pi_M(E_{1,1})} z_2^{\pi_M(E_{2,2})} \dots z_n^{\pi_M(E_{n,n})} \\
&= x_{i_1} \partial_{x_{i_2}} x_{i_2} \partial_{x_{i_1}} (E_A) + \mathrm{tr}_M \pi'(E_{i_1,i_2}) \Phi z_1^{\pi_M(E_{1,1})} z_2^{\pi_M(E_{2,2})} \dots z_n^{\pi_M(E_{n,n})} \pi_M(E_{i_2,i_1}) \\
&= \frac{z_{i_2}}{z_{i_1}} \mathrm{tr}_M \pi'(E_{i_1,i_2}) \Phi \pi_M(E_{i_2,i_1}) z_1^{\pi_M(E_{1,1})} z_2^{\pi_M(E_{2,2})} \dots z_n^{\pi_M(E_{n,n})} \\
&\quad + \kappa(\kappa+1) E_A \qquad (15.2.24)
\end{aligned}
$$

by (15.2.18). Hence

$$
\mathrm{tr}_M \pi'(E_{i_1,i_2}) \Phi \pi_M(E_{i_2,i_1}) z_1^{\pi_M(E_{1,1})} z_2^{\pi_M(E_{2,2})} \dots z_n^{\pi_M(E_{n,n})} = \frac{\kappa(\kappa+1) z_{i_1}}{z_{i_1} - z_{i_2}} E_A. \qquad (15.2.25)
$$

Substituting it to (15.2.23), we obtain

$$
E_A^{i_1,i_2} = \frac{z_{i_1}}{z_{i_2} - z_{i_1}} (z_{i_2} \partial_{z_{i_2}} - z_{i_1} \partial_{z_{i_1}})(E_A) - \frac{\kappa(\kappa+1) z_{i_1} z_{i_2}}{(z_{i_1} - z_{i_2})^2} E_A \qquad (15.2.26)
$$

for $i_1, i_2 \in \overline{1,n}$ such that $i_1 \neq i_2$.

According to (15.1.17) with $\mu_1 = (1-n)/2$ and $\mu_2 = 1$, we have

$$\sum_{r=1}^{n}(z_r\partial_{z_r})^2(\mathcal{W}_{A_{n-1}}) = \left[\frac{1}{2} + 2\binom{n}{3}\right]\mathcal{W}_{A_{n-1}} \tag{15.2.27}$$

Now (15.2.6) yields

$$\begin{aligned}
b\Psi_A &= \mathcal{W}_{A_{n-1}}(x_1x_2\ldots x_n)^{-\kappa}\mathrm{tr}_{_M}\ \Phi\pi_{_M}(\omega_{_A})z_1^{E_{1,1}}z_2^{E_{2,2}}\ldots z_n^{E_{n,n}}\\
&= \sum_{i_1,i_2=1}^{n}\mathcal{W}_{A_{n-1}}(x_1x_2\ldots x_n)^{-\kappa}E_A^{i_1,i_2}\\
&= \sum_{i=1}^{n}\mathcal{W}_{A_{n-1}}(x_1x_2\ldots x_n)^{-\kappa}(z_i\partial_{z_i})^2(E_A) - \sum_{i_1\neq i_2}\frac{\kappa(\kappa+1)z_{i_1}z_{i_2}}{(z_{i_1}-z_{i_2})^2}\Psi_A\\
&\quad + \sum_{i_1\neq i_2}\mathcal{W}_{A_{n-1}}(x_1x_2\ldots x_n)^{-\kappa}\frac{z_{i_1}}{z_{i_2}-z_{i_1}}(z_{i_2}\partial_{z_{i_2}} - z_{i_1}\partial_{z_{i_1}})(E_A)\\
&= \sum_{i=1}^{n}\mathcal{W}_{A_{n-1}}(x_1x_2\ldots x_n)^{-\kappa}(z_i\partial_{z_i})^2(E_A) - \sum_{i_1\neq i_2}\frac{\kappa(\kappa+1)z_{i_1}z_{i_2}}{(z_{i_1}-z_{i_2})^2}\Psi_A\\
&\quad + \sum_{i_2=1}^{n}((1-n)/2 + z_{i_2}\partial_{z_{i_2}})(\mathcal{W}_{A_{n-1}})(x_1x_2\ldots x_n)^{-\kappa}z_{i_2}\partial_{z_{i_2}}(E_A)\\
&\quad + \sum_{i_1=1}^{n}((n-1)/2 + z_{i_1}\partial_{z_{i_1}})(\mathcal{W}_{A_{n-1}})(x_1x_2\ldots x_n)^{-\kappa}z_{i_1}\partial_{z_{i_1}}(E_A)\\
&= \sum_{i=1}^{n}\mathcal{W}_{A_{n-1}}(x_1x_2\ldots x_n)^{-\kappa}(z_i\partial_{z_i})^2(E_A) - \sum_{i_1\neq i_2}\frac{\kappa(\kappa+1)z_{i_1}z_{i_2}}{(z_{i_1}-z_{i_2})^2}\Psi_A\\
&\quad + \sum_{i=1}^{n}2z_i\partial_{z_i}(\mathcal{W}_{A_{n-1}})(x_1x_2\ldots x_n)^{-\kappa}z_i\partial_{z_i}(E_A)\\
&= \sum_{i=1}^{n}(z_i\partial_{z_i})^2(\Psi_A) - \left[\frac{1}{2}+2\binom{n}{3}\right]\Psi_A - \sum_{i_1\neq i_2}\frac{\kappa(\kappa+1)z_{i_1}z_{i_2}}{(z_{i_1}-z_{i_2})^2}\Psi_A
\end{aligned} \tag{15.2.28}$$

by (15.2.4), (15.2.6), (15.2.18), (15.2.19), (15.2.21), (15.2.26) and (15.2.27). Thus

$$\sum_{i=1}^{n}(z_i\partial_{z_i})^2(\Psi_A) - 2\kappa(\kappa+1)\left(\sum_{1\leq i<j\leq n}\frac{z_iz_j}{(z_i-z_j)^2}\right)\Psi_A = \left[b+\frac{1}{2}+2\binom{n}{3}\right]\Psi_A, \tag{15.2.29}$$

which shows that Ψ_A is a solution of (15.1.4) for

$$K = -2\kappa(\kappa+1), \quad \nu = b + \frac{1}{2} + 2\binom{n}{3}. \quad \square \tag{15.2.30}$$

For convenience, we denote

$$C_{i,j} = E_{i,j} - E_{n+j,n+i} \qquad \text{for } i, j \in \overline{1.n}, \tag{15.2.31}$$

$$C_{j,n+k} = E_{j,n+k} + E_{k,n+j}, \qquad C_{n+j,k} = E_{n+j,k} + E_{n+k,j} \tag{15.2.32}$$

for $j, k \in \overline{1,n}$ such that $j \neq k$, and

$$C_{r,n+r} = E_{r,n+r}, \qquad C_{n+r,r} = E_{n+r,r} \qquad \text{for } r \in \overline{1,n}. \tag{15.2.33}$$

Then the symplectic Lie algebra

$$sp(2n,\mathbb{R}) = \sum_{i,j=1}^{n} \mathbb{R}C_{i,j} + \sum_{1\leq r\leq s\leq n} (\mathbb{R}C_{r,n+s} + \mathbb{R}C_{n+r,s}). \tag{15.2.34}$$

Moreover,

$$\begin{aligned}\omega_{C} = &\sum_{i,j=1}^{n} C_{i,j}C_{j,i} + \sum_{1\leq s<r\leq n} (C_{n+r,s}C_{s,n+r} + C_{s,n+r}C_{n+r,s}) \\ &+ 2\sum_{p=1}^{n}(C_{n+p,p}C_{p,n+p} + C_{p,n+p}C_{n+p,p})\end{aligned} \tag{15.2.35}$$

is a Casimir element in $U(sp(2n,\mathbb{R}))$. Now

$$H_C = \sum_{i=1}^{n} \mathbb{R}C_{i,i} \tag{15.2.36}$$

is a toral Cartan subalgebra of $sp(2n,\mathbb{R})$. Let

$$M = \bigoplus_{\lambda\in H_C^*} M_\lambda \tag{15.2.37}$$

be a weight $sp(2n,\mathbb{R})$-module with the representation π_M such that

$$\pi_M(\omega_C) = b\,\mathrm{Id}_M \qquad \text{for some } b \in \mathbb{R}, \tag{15.2.38}$$

In particular, we can take M to be any highest weight module; say the Verma module $M(\lambda)$ given in (5.2.8)–(5.2.10). The module M is not necessarily irreducible.

Denote

$$x^* = x_1^{-1/2} x_2^{-1/2} \dots x_n^{-1/2}. \tag{15.2.39}$$

Denote the space of formal Laurent series in $\{x_1, x_2, \dots, x_n\}$ with the coefficients in M by

$$\tilde{M} = \left\{ \sum_{\vec{i} \in \mathbb{Z}^n} w_{\vec{i}} x^{\vec{i}} x^* \mid w_n \in M \right\}. \tag{15.2.40}$$

We extend the $\pi_{_M}$ to $\tilde{M}$ by

$$\pi_{_M}(u) \left(\sum_{\vec{i} \in \mathbb{Z}^n} w_{\vec{i}} x^{\vec{i}} x^* \right) = \sum_{\vec{i} \in \mathbb{Z}^n} \pi_{_M}(\xi)(w_{\vec{i}}) x^{\vec{i}} x^* \quad \text{for } \xi \in sp(2n, \mathbb{R}). \tag{15.2.41}$$

Moreover, we define another representation π' of $sp(2n, \mathbb{R})$ on $\tilde{M}$ by

$$\pi'(C_{p,q}) \left(\sum_{\vec{i} \in \mathbb{Z}^n} w_{\vec{i}} x^{\vec{i}} x^* \right) = \sum_{\vec{i} \in \mathbb{Z}^n} \left(C_{p,q}(w_{\vec{i}} x_p \partial_{x_q}(x^{\vec{i}} x^*) + \frac{\delta_{p,q}}{2} w_{\vec{i}} x^{\vec{i}} x^* \right), \tag{15.2.42}$$

$$\pi'(C_{p,n+q}) \left(\sum_{\vec{i} \in \mathbb{Z}^n} w_{\vec{i}} x^{\vec{i}} x^* \right) = -\frac{1}{1+\delta_{p,q}} \sum_{\vec{i} \in \mathbb{Z}^n} w_{\vec{i}} x_p x_q x^{\vec{i}} x^*, \tag{15.2.43}$$

$$\pi'(C_{n+p,q}) \left(\sum_{\vec{i} \in \mathbb{Z}^n} w_{\vec{i}} x^{\vec{i}} x^* \right) = \frac{1}{1+\delta_{p,q}} \sum_{\vec{i} \in \mathbb{Z}^n} w_{\vec{i}} \partial_{x_p} \partial_{x_q}(x^{\vec{i}} x^*) \tag{15.2.44}$$

for $p, q \in \overline{1n}$. Set

$$\pi = \pi_{_M} + \pi'. \tag{15.2.45}$$

Suppose that $\Phi : M \to \tilde{M}$ is a linear map such that

$$\Phi(\pi_{_M}(u)(w)) = \pi(u)\Phi(w) \tag{15.2.46}$$

for $u \in sp(2n, \mathbb{R})$ and $w \in M$. View Φ as a function in $\{x_1, x_2, \dots, x_n\}$ taking value in the spaces of linear transformations on M_λ. Define the *Etingof trace function of type C*:

$$E_C(z_1, z_2, \dots, z_n) = \text{tr}_M \Phi z_1^{\pi_M(C_{1,1})} z_2^{\pi_M(C_{2,2})} \cdots z_n^{\pi_M(C_{n,n})}. \tag{15.2.47}$$

By considering weights, we can prove

$$(x^*)^{-1}E_C(z_1, z_2, \dots, z_n) \text{ is independent of } x_1, x_2, \dots, x_n. \tag{15.2.48}$$

The following theorem was essentially proved by Etingof and Styrkas [ES].

Theorem 15.2.2 *The function*

$$\Psi_C = (x^*)^{-1}\mathcal{W}_{C_n}E_C(z_1, z_2, \dots, z_n) \tag{15.2.49}$$

satisfies the Olshanestsky–Perelomov equation (15.1.5) with $K_3 = 0$ *and proper constants* K_1, K_2, K_4.

Proof Note that (15.1.27) implies

$$\begin{aligned} z_i\partial_{z_i}(\mathcal{W}_{C_n}) &= \sum_{i=1}^n \left[\frac{z_i + z_i^{-1}}{z_i - z_i^{-1}} + \sum_{i\neq r\in\overline{1,n}} \frac{z_i - z_i^{-1}}{z_i + z_i^{-1} - z_r - z_r^{-1}}\right]\mathcal{W}_{C_n} \\ &= \sum_{i=1}^n \left[\frac{z_i^2+1}{z_i^2-1} + \sum_{i\neq r\in\overline{1,n}} \left(\frac{z_i}{z_i - z_r} + \frac{1}{z_iz_r - 1}\right)\right]\mathcal{W}_{C_n}. \end{aligned} \tag{15.2.50}$$

Set

$$E_C^{i_1,i_2} = (x^*)^{-1}\mathrm{tr}_M\Phi\pi_M(C_{i_1,i_2}C_{i_2,i_1})z_1^{\pi_M(C_{1,1})}z_2^{\pi_M(C_{2,2})}\cdots z_n^{\pi_M(C_{n,n})} \tag{15.2.51}$$

for $i_1, i_2 \in \overline{1,n}$. Then

$$E_C^{i,i} = (z_i\partial_{z_i})^2(E_C) \tag{15.2.52}$$

by (15.2.22), and

$$E_C^{i_1,i_2} = \frac{z_{i_1}}{z_{i_2} - z_{i_1}}(z_{i_2}\partial_{z_{i_2}} - z_{i_1}\partial_{z_{i_1}})(E_C) + \frac{z_{i_1}z_{i_2}}{4(z_{i_1} - z_{i_2})^2}E_C \tag{15.2.53}$$

for $i_1, i_2 \in \overline{1,n}$ such that $i_1 \neq i_2$ by (15.2.24)–(15.2.28).

Let

$$E_C^{s,n+r} = (x^*)^{-1}\mathrm{tr}_M\Phi\pi_M(C_{s,n+r}C_{n+r,s})z_1^{\pi_M(C_{1,1})}z_2^{\pi_M(C_{2,2})}\cdots z_n^{\pi_M(C_{n,n})} \tag{15.2.54}$$

and

$$E_C^{n+r,s} = (x^*)^{-1}\mathrm{tr}_M\Phi\pi_M(C_{n+r,s}C_{s,n+r})z_1^{\pi_M(C_{1,1})}z_2^{\pi_M(C_{2,2})}\cdots z_n^{\pi_M(C_{n,n})} \tag{15.2.55}$$

for $1 \leq s \leq r \leq n$. By similar calculations as (15.2.22)–(15.2.26), we have

$$E_C^{s,n+r} = \frac{z_r z_s}{z_r z_s - 1}(z_r \partial_{z_r} + z_s \partial_{z_s})E_C + \frac{z_r z_s}{4(z_r z_s - 1)^2}E_C. \tag{15.2.56}$$

$$E_C^{n+r,s} = \frac{1}{z_r z_s - 1}(z_r \partial_{z_r} + z_s \partial_{z_s})E_C + \frac{z_r z_s}{4(z_r z_s - 1)^2}E_C \tag{15.2.57}$$

for $1 \leq s < r \leq n$ and

$$E_C^{p,n+p} = \frac{z_p^2}{z_p^2 - 1}z_p \partial_{z_p} E_C + \frac{3z_p^2}{16(z_p^2 - 1)^2}E_C. \tag{15.2.58}$$

$$E_C^{n+p,p} = \frac{1}{z_p^2 - 1}z_p \partial_{z_p} E_C + \frac{3z_p^2}{16(z_p^2 - 1)^2}E_C \tag{15.2.59}$$

for $p \in \overline{1,n}$. Thus

$$\left(\sum_{i=1}^{n}(z_i \partial_{z_i}\right)^2 (\mathcal{W}_{C_n}) = \frac{n(n+1)(2n+1)}{6}\mathcal{W}_{C_n} \tag{15.2.60}$$

by (15.1.30).

Now (15.2.38) gives

$$\begin{aligned}
b\Psi &= \mathcal{W}_{C_n}(x^*)^{-1}\mathrm{tr}_M \Phi \pi_{_M}(\omega_{_C}) z_1^{\pi_M(C_{1,1})} z_2^{\pi_M(C_{2,2})} \ldots z_n^{\pi_M C_{n,n}} \\
&= \mathcal{W}_{C_n} x^*)^{-1}\left[\sum_{i_1,i_2=1}^{n} E_C^{i_1,i_2} + \sum_{1\leq s\leq r\leq n}(\delta_{r,s}+1)(E_C^{n+r,s} + E_C^{s,n+r})\right] \\
&= \mathcal{W}_{C_n}\left[\sum_{i_1,i_2\in\overline{1,n}\, i_1\neq i_2}\left(\frac{z_{i_1}}{z_{i_2} - z_{i_1}}(z_{i_2}\partial_{z_{i_2}} - z_{i_1}\partial_{z_{i_1}})(E_C) + \frac{z_{i_1}z_{i_2}}{4(z_{i_1} - z_{i_2})^2}E_C\right)\right. \\
&\quad + \sum_{1\leq s<r\leq n}\left(\frac{z_r z_s + 1}{z_r z_s - 1}(z_r\partial_{z_r} + z_s\partial_{z_s})(E_C) + \frac{z_r z_s}{2(z_r z_s - 1)^2}E_C\right) \\
&\quad \left. + 2\sum_{p=1}^{n}\left(\frac{z_p^2 + 1}{z_p^2 - 1}z_p\partial_{z_p}\mathcal{E} + \frac{3z_p^2}{8(z_p^2 - 1)^2}E_C\right) + \sum_{i=1}^{n}(z_i\partial_{z_i})^2(E_C)\right] \\
&= \left[\sum_{1\leq i_1<i_2\leq n}\frac{1}{2}\left(\frac{z_{i_1}z_{i_2}}{(z_{i_1} - z_{i_2})^2} + \frac{z_{i_1}z_{i_2}}{(z_{i_1}z_{i_2} - 1)^2}\right) + \frac{3}{4}\sum_{p=1}^{n}\frac{z_p^2}{(z_p^2 - 1)^2}\right]\Psi_C \\
&\quad + 2\mathcal{W}_{C_n}(x^*)^{-1}\sum_{r,s\in\overline{1,n};\; r\neq s}\left(\frac{z_r}{z_r - z_s} + \frac{1}{z_r z_s - 1}\right)z_r\partial_{z_r}(E_C)
\end{aligned}$$

$$+2\mathcal{W}_{C_n}(x^*)^{-1}\sum_{p=1}^{n}\frac{z_p^2+1}{z_p^2-1}z_p\partial_{z_p}(E_C)+\mathcal{W}_{C_n}(x^*)^{-1}\sum_{i=1}^{n}(z_i\partial_{z_i})^2(E_C)$$

$$=\left[\sum_{1<i_1<i_2\leq n}\frac{1}{2}\left(\frac{z_{i_1}z_{i_2}}{(z_{i_1}-z_{i_2})^2}+\frac{z_{i_1}z_{i_2}}{(z_{i_1}z_{i_2}-1)^2}\right)+\frac{3}{4}\sum_{p=1}^{n}\frac{z_p^2}{(z_p^2-1)^2}\right]\Psi_C$$

$$+\sum_{i=1}^{n}[\mathcal{W}_{C_n}(x^*)^{-1}(z_i\partial_{z_i})^2(E_C)+2z_i\partial_{z_i}(\mathcal{W}_{C_n})(x^*)^{-1}(z_i\partial_{z_i})(E_C)]$$

$$=\left[\sum_{1<i_1<i_2\leq n}\frac{1}{2}\left(\frac{z_{i_1}z_{i_2}}{(z_{i_1}-z_{i_2})^2}+\frac{z_{i_1}z_{i_2}}{(z_{i_1}z_{i_2}-1)^2}\right)+\frac{3}{4}\sum_{p=1}^{n}\frac{z_p^2}{(z_p^2-1)^2}\right]\Psi_C$$

$$+\sum_{i=1}^{n}\mathcal{W}_{C_n}(x^*)^{-1}(z_i\partial_{z_i})^2(\Psi_C)-\sum_{i=1}^{n}(z_i\partial_{z_i})^2(\mathcal{W}_{C_n})(x^*)^{-1}E_C$$

$$=\left[\sum_{1<i_1<i_2\leq n}\frac{1}{2}\left(\frac{z_{i_1}z_{i_2}}{(z_{i_1}-z_{i_2})^2}+\frac{z_{i_1}z_{i_2}}{(z_{i_1}z_{i_2}-1)^2}\right)+\frac{3}{4}\sum_{p=1}^{n}\frac{z_p^2}{(z_p^2-1)^2}\right]\Psi_C$$

$$+\sum_{i=1}^{n}(z_i\partial_{z_i})^2(\Psi_C)-\frac{n(n+1)(2n+1)}{6}\Psi_C \qquad (15.2.61)$$

by (15.2.50), (15.2.52), (15.2.53) and (15.2.56)–(15.2.60). Therefore, we have:

$$\left[\sum_{1<i_1<i_2\leq n}\frac{1}{2}\left(\frac{z_{i_1}z_{i_2}}{(z_{i_1}-z_{i_2})^2}+\frac{z_{i_1}z_{i_2}}{(z_{i_1}z_{i_2}-1)^2}\right)+\frac{3}{4}\sum_{p=1}^{n}\frac{z_p^2}{(z_p^2-1)^2}\right]\Psi_C$$

$$=\left(b+\frac{n(n+1)(2n+1)}{6}\right)\Psi_C-\sum_{i=1}^{n}(z_i\partial_{z_i})^2(\Psi_C). \qquad (15.2.62)$$

So Ψ_C satisfies the Olshanestsky–Perelomov equation (15.1.5) with $K_3=0$ and

$$K_1=K_2=\frac{1}{2},\quad K_4=\frac{3}{4},\quad \nu=b+\frac{n(n+1)(2n+1)}{6}. \quad \square \qquad (15.2.63)$$

15.3 Path Hypergeometric Functions of Type A

In this section, we present our path hypergeometric functions of type A found in [X14] via the Etingof trace functions of type A.

First we calculate the Etingof trace functions for $gl(2,\mathbb{R})$ in terms of classical Gauss hypergeometric functions. Let λ be a weight of $gl(2,\mathbb{R})$ such that

$$\sigma = \lambda_1 - \lambda_2 \notin \mathbb{N} \tag{15.3.1}$$

(cf. (15.2.3)). Choose $E_{1,2}$ as the positive root vector. The Verma $gl(2,\mathbb{R})$-module with the highest-weight vector v_λ of weight λ is given by

$$M(\lambda) = \text{Span}\{E_{2,1}^i v_\lambda \mid i \in \mathbb{N}\} \tag{15.3.2}$$

with the action

$$E_{1,,2}(E_{2,1}^i v_\lambda) = i(\sigma + 1 - i)E_{2,1}^{i-1} v_\lambda, \qquad E_{2,1}(E_{2,1}^i v_\lambda) = E_{2,1}^{i+1} v_\lambda, \tag{15.3.3}$$

$$E_{1,1}(E_{2,1}^i v_\lambda) = (\lambda_1 - i)E_{2,1}^i v_\lambda, \qquad E_{2,2}(E_{2,1}^i v_\lambda) = (\lambda_2 + i)E_{2,1}^i v_\lambda. \tag{15.3.4}$$

Under the assumption (15.4.1), $M(\lambda)$ is an irreducible $gl(2,\mathbb{R})$-module.

Let $\kappa \in \mathbb{R}$ be a fixed constant and recall

$$\tilde{M} = \left\{ \sum_{i_1,i_2\in\mathbb{Z}} v_{i_1,i_2} x_1^{i_1+\kappa} x_2^{i_2+\kappa} \mid v_{i_1,i_2} \in M(\lambda) \right\}, \tag{15.3.5}$$

a space of Laurent series with coefficients in $M(\lambda)$. Moreover, the representation π of $gl(2,\mathbb{R})$ on $\tilde{M}$ is given by

$$\begin{aligned} E_{j_1,j_2}\left(\sum_{i_1,i_2\in\mathbb{Z}} v_{i_1,i_2} x_1^{i_1+\kappa} x_2^{i_2+\kappa} \right) = &\sum_{i_1,i_2\in\mathbb{Z}} [E_{j_1,j_2}(v_{i_1,i_2}) x_1^{i_1+\kappa} x_2^{i_2+\kappa} \\ &+ v_{i_1,i_2}(x_{j_1}\partial_{x_{j_2}} - \kappa\delta_{j_1,j_2})(x_1^{i_1+\kappa} x_2^{i_2+\kappa})] \end{aligned} \tag{15.3.6}$$

for $j_1, j_2 \in \{1, 2\}$ by (15.2.11)–(15.2.13). Recall that a singular vector in a Verma module is a nonzero weight vector annihilated by positive root vectors. Any singular vector of weight λ in $\tilde{M}$ is of the form:

$$u_\lambda = (x_1 x_2)^\kappa \sum_{i=0}^{\infty} a_i E_{2,1}^i v_\lambda x_1^i x_2^{-i} \tag{15.3.7}$$

Observe that

$$\begin{aligned} 0 &= E_{1,2}(u_\lambda) \\ &= (x_1x_2)^\kappa \sum_{i=0}^{\infty} a_i[i(\sigma + 1 - i)E_{2,1}^{i-1} v_\lambda x_1^i x_2^{-i} + (\kappa - i)E_{2,1}^i v_\lambda x_1^{i+1} x_2^{-i-1}] \\ &= (x_1x_2)^\kappa \sum_{i=0}^{\infty} ((\kappa - i)a_i + (i+1)(\sigma - i)a_{i+1}) v_\lambda x_1^{i+1} x_2^{-i-1}. \end{aligned} \tag{15.3.8}$$

Thus

$$(\kappa - i)a_i + (i+1)(\sigma - i)a_{i+1} = 0 \qquad \text{for } i \in \mathbb{N}, \tag{15.3.9}$$

or equivalently,

$$a_{i+1} = -\frac{(\kappa - i)a_i}{(i+1)(\sigma - i)} \qquad \text{for } i \in \mathbb{N}, \tag{15.3.10}$$

For any constant c and nonnegative integer i, we denote

$$\langle c \rangle_i = c(c-1)\cdots(c-(i-1)). \tag{15.3.11}$$

We normalize u_λ by taking

$$a_0 = 1. \tag{15.3.12}$$

By induction,

$$a_i = \frac{(-1)^i \langle \kappa \rangle_i}{i!\langle \sigma \rangle_i}. \tag{15.3.13}$$

Thus

$$u_\lambda = \sum_{i=0}^{\infty} \frac{(-1)^i \langle \kappa \rangle_i}{i!\langle \sigma \rangle_i} E_{2,1}^i v_\lambda x_1^{i+\kappa} x_2^{-i+\kappa}. \tag{15.3.14}$$

Let $\Phi : M_\lambda \to \tilde{M}$ be the Lie algebra module homomorphism such that $\Phi(v_\lambda) = u_\lambda$. Note that

$$E_{2,1}^j (E_{2,1}^i v_\lambda x_1^{i+\kappa} x_2^{-i+\kappa}) = \sum_{k=0}^{j} \binom{j}{k} \langle \kappa + i \rangle_k E_{2,1}^{i+j-k} v_\lambda x_1^{i-k+\kappa} x_2^{-i+k+\kappa}. \tag{15.3.15}$$

Moreover, for any $\iota \in \mathbb{N}$, we have

$$\begin{aligned}\sum_{k=\iota}^{\infty} \binom{k}{\iota} z^k &= z^\iota \sum_{k=\iota}^{\infty} \binom{k}{\iota} z^{k-\iota} = \frac{z^\iota}{\iota!} \frac{d^\iota}{dz^\iota} \left(\sum_{k=0}^{\infty} z^k \right) \\ &= \frac{z^\iota}{\iota!} \frac{d^\iota}{dz^\iota} \left(\frac{1}{1-z} \right) = \frac{z^\iota}{(1-z)^{\iota+1}}.\end{aligned} \tag{15.3.16}$$

View Φ as a "function taking values in End M_λ." By (15.3.15) and (15.3.16), the Etingof trace function

$$\begin{aligned} E_A(z_1, z_2) &= \mathrm{tr}_{M_\lambda} \Phi z_1^{E_{1,1}} z_2^{E_{2,2}} \\ &= (x_1 x_2)^\kappa z_1^{\lambda_1} z_2^{\lambda_2} \sum_{j=0}^{\infty} \sum_{i=0}^{\infty} \binom{j}{i} \langle \kappa + i \rangle_i \frac{(-1)^i \langle \kappa \rangle_i}{i!\langle \sigma \rangle_i} \left(\frac{z_2}{z_1} \right)^j \end{aligned}$$

$$= (x_1x_2)^\kappa z_1^{\lambda_1} z_2^{\lambda_2} \sum_{i=0}^{\infty} \langle \kappa + i \rangle_i \frac{(-1)^i \langle \kappa \rangle_i}{i! \langle \sigma \rangle_i} \sum_{j=0}^{\infty} \binom{j}{i} \left(\frac{z_2}{z_1}\right)^j$$

$$= (x_1x_2)^\kappa z_1^{\lambda_1} z_2^{\lambda_2} \sum_{i=0}^{\infty} \langle \kappa + i \rangle_i \frac{(-1)^i \langle \kappa \rangle_i}{i! \langle \sigma \rangle_i} \left(\frac{z_2}{z_1}\right)^i \left(\frac{1}{1 - z_2/z_1}\right)^{i+1}$$

$$= (x_1x_2)^\kappa \frac{z_1^{\lambda_1+1} z_2^{\lambda_2}}{z_1 - z_2} \sum_{i=0}^{\infty} \frac{\langle \kappa + i \rangle_i \langle \kappa \rangle_i}{i! \langle \sigma \rangle_i} \left(\frac{z_2}{z_2 - z_1}\right)^i. \quad (15.3.17)$$

Denote

$$(c)_i = c(c+1)(c+2)\dots(c+i-1) \quad \text{for } c \in \mathbb{C},\ i \in \mathbb{N}. \quad (15.3.18)$$

The classical Gauss hypergeometric function

$${}_2F_1(a, b; c; z) = \sum_{m=0}^{\infty} \frac{(a)_m (b)_m}{m!(c)_m} z^m, \quad \text{where } a, b, c \in \mathbb{C},\ -c \notin \mathbb{N}. \quad (15.3.19)$$

Note

$$\langle c + n \rangle_n = (c+1)_n, \quad \langle c \rangle_n = (-1)^n (-c)_n \quad \text{for } c \in \mathbb{C},\ n \in \mathbb{N}. \quad (15.3.20)$$

By (15.3.1) and (15.3.17)–(15.4.20), we have:

Theorem 15.3.1 *The Etingof trace function for $gl(2, \mathbb{R})$ is*

$$E_A(z_1, z_2) = (x_1x_2)^\kappa \frac{z_1^{\lambda_1+1} z_2^{\lambda_2}}{z_1 - z_2} {}_2F_1\left(\kappa + 1, -\kappa; \lambda_2 - \lambda_1 : \frac{z_2}{z_2 - z_1}\right). \quad (15.3.21)$$

The function in (15.2.17)

$$\Psi_A(z_1, z_2) = z_1^{\lambda_1+1/2} z_2^{\lambda_2-1/2} {}_2F_1\left(\kappa + 1, -\kappa; \lambda_2 - \lambda_1 : \frac{z_2}{z_2 - z_1}\right) \quad (15.3.22)$$

Next we want to find the Etingof trace functions for the general case of $gl(n, \mathbb{R})$ with $n > 2$ under a certain condition and their master hypergeometric functions.

As the case of $sl(n, \mathbb{R})$, we choose

$$\{E_{i,j} \mid 1 \le i < j \le n\} \text{ as positive root vectors.} \quad (15.3.23)$$

In particular, we have

$$\{E_{i,i+1} \mid i = 1, 2, \ldots, n-1\} \text{ as positive simple root vectors.} \tag{15.3.24}$$

Accordingly,

$$\{E_{i,j} \mid 1 \leq j < i \leq n\} \text{ are negative root vectors} \tag{15.3.25}$$

and

$$\{E_{i+1,i} \mid i = 1, 2, \ldots, n-1\} \text{ are negative simple root vectors.} \tag{15.3.26}$$

Let

$$\Gamma_A = \sum_{1 \leq j < i \leq n} \mathbb{N}\varepsilon_{i,j} \tag{15.3.27}$$

be the torsion-free additive semigroup of rank $n(n-1)/2$ with $\varepsilon_{i,j}$ as base elements. For

$$\alpha = \sum_{1 \leq j < i \leq n} \alpha_{ij}\varepsilon_{i,j} \in \Gamma_A, \tag{15.3.28}$$

we denote

$$E^{\alpha} = E_{2,1}^{\alpha_{2,1}} E_{3,1}^{\alpha_{3,1}} E_{3,2}^{\alpha_{3,2}} E_{4,1}^{\alpha_{4,1}} \ldots E_{n,1}^{\alpha_{n,1}} \ldots E_{n,n-1}^{\alpha_{n,n-1}} \in U(gl(n)). \tag{15.3.29}$$

Denote by $\mathscr{G}_-$ the Lie subalgebra spanned by (15.3.25). Then

$$\{E^{\alpha} \mid \alpha \in \Gamma_A\} \text{ forms a PBW basis of } U(\mathscr{G}_-). \tag{15.3.30}$$

Let λ be a weight of $gl(n, \mathbb{R})$ such that

$$\lambda_1 - \lambda_2 = \cdots = \lambda_{n-2} - \lambda_{n-1} = \kappa \text{ and } \lambda_{n-1} - \lambda_n \notin \mathbb{N}, \tag{15.3.31}$$

for some constant κ (cf. (15.2.3)). Set

$$\sigma_i = \lambda_i - \lambda_{i+1} \quad \text{for } i \in \overline{1, n-1}. \tag{15.3.32}$$

The Verma $gl(n, \mathbb{R})$-module with the highest-weight vector v_λ of weight λ is given by

$$M(\lambda) = \text{Span}\{E^{\alpha}v_\lambda \mid \alpha \in \Gamma_A\}, \tag{15.3.33}$$

with the action determined by

$$E_{i,i+1}(E^\alpha v_\lambda) = \left(\sum_{p=1}^{i-1} \alpha_{i+1,p} E^{\alpha+\varepsilon_{i,p}-\varepsilon_{i+1,p}} - \sum_{p=i+2}^{n} \alpha_{p,i} E^{\alpha+\varepsilon_{p,i+1}-\varepsilon_{p,i}} \right.$$
$$\left. + \alpha_{i+1,i}\left(\sigma_i + 1 - \sum_{p=i+1}^{n} \alpha_{p,i} + \sum_{p=i+2}^{n} \alpha_{p,i+1}\right) E^{\alpha-\varepsilon_{i+1,i}}\right) v_\lambda, \tag{15.3.34}$$

$$E_{j,i}(E^\alpha v_\lambda) = \left(E^{\alpha+\varepsilon_{j,i}} + \sum_{p=1}^{i-1} \alpha_{i,p} E^{\alpha+\varepsilon_{j,p}-\varepsilon_{i,p}}\right) v_\lambda, \tag{15.3.35}$$

$$E_{k,k}(E^\alpha v_\lambda) = \left(\lambda_k + \sum_{p=1}^{k-1} \alpha_{k,p} - \sum_{q=k+1}^{n} \alpha_{q,k}\right) E^\alpha v_\lambda \tag{15.3.36}$$

for $1 \leq i < j \leq n$ and $k \in \overline{1,n}$.

Recall the notions in (15.2.7)–(15.2.9) and for $M = M(\lambda)$,

$$\tilde{M} = \left\{\sum_{\vec{i}\in\mathbb{Z}^n} v_{\vec{i}} x^{\vec{i}} \bar{x}^\kappa \mid v_{\vec{i}} \in M(\lambda)\right\}. \tag{15.3.37}$$

Moreover, the representation π of $gl(n,\mathbb{R})$ on $\tilde{M}$ is given by

$$E_{j_1,j_2}\left(\sum_{\vec{i}\in\mathbb{Z}^n} (v_{\vec{i}}) x^{\vec{i}} \bar{x}^\kappa\right) = \sum_{\vec{i}\in\mathbb{Z}^n} [E_{j_1,j_2}(v_{\vec{i}}) x^{\vec{i}} \bar{x}^\kappa + v_{\vec{i}}(\partial_{x_{j_2}} - \kappa\delta_{j_1,j_2})(x^{\vec{i}} \bar{x}^\kappa)] \tag{15.3.38}$$

for $j_1, j_2 \in \overline{1,n}$.

For

$$\vec{i} = (i_1, i_2, \ldots, i_{n-1}) \in \mathbb{N}^{n-1}, \tag{15.3.39}$$

we denote

$$x^{(\vec{i})} = x_1^{i_1} x_2^{i_2-i_1} \cdots x_{n-1}^{i_{n-1}-i_{n-2}} x_n^{-i_{n-1}}, \qquad \epsilon(\vec{i}) = \sum_{p=1}^{n-1} i_p \varepsilon_{p+1,p}. \tag{15.3.40}$$

Then we have:

Lemma 15.3.2 *The vector*

$$u_\lambda = \bar{x}^\kappa \sum_{i_1,\dots,i_{n-1}=0}^{\infty} \frac{(-1)^{i_1+i_2+\cdots+i_{n-1}}\langle\kappa\rangle_{i_{n-1}}}{i_1!i_2!\cdots i_{n-1}!\langle\sigma\rangle_{i_{n-1}}} E^{\epsilon(\vec{i})} v_\lambda x^{(\vec{i})} \tag{15.3.41}$$

is a singular vector of weight λ, *where* $\sigma = \sigma_{n-1} = \lambda_{n-1} - \lambda_n$.

Proof For $p \in \{1, 2, \dots, n-2\}$, we have

$$\begin{aligned}
&E_{p,p+1}(u_\lambda)\\
&= \bar{x}^\kappa \sum_{i_1,\dots,i_{n-1}=0}^{\infty} \frac{(-1)^{i_1+i_2+\cdots+i_{n-1}}\langle\kappa\rangle_{i_{n-1}}}{i_1!i_2!\cdots i_{n-1}!\langle\sigma\rangle_{i_{n-1}}} [(\kappa + i_{p+1} - i_p) E^{\epsilon(\vec{i})} v_\lambda x_p x_{p+1}^{-1} x^{(\vec{i})}\\
&\quad + (\kappa + 1 - i_p + i_{p+1}) i_p E^{\epsilon(\vec{i})-\varepsilon_p} v_\lambda x^{(\vec{i})}]\\
&= \bar{x}^\kappa \Bigg[\sum_{i_1,\dots,i_{n-1}=0}^{\infty} \frac{(-1)^{i_1+i_2+\cdots+i_{n-1}}\langle\kappa\rangle_{i_{n-1}}}{i_1!i_2!\cdots i_{n-1}!\langle\sigma\rangle_{i_{n-1}}} (\kappa + i_{p+1} - i_p) E^{\epsilon(\vec{i})} v_\lambda x_p x_{p+1}^{-1} x^{(\vec{i})}\\
&\quad + \sum_{i_1,\dots,i_{n-1}=0}^{\infty} \frac{(-1)^{i_1+i_2+\cdots+i_{n-1}}\langle\kappa\rangle_{i_{n-1}}}{i_1!i_2!\cdots i_{n-1}!\langle\sigma\rangle_{i_{n-1}}} (\kappa + 1 - i_p + i_{p+1}) i_p E^{\epsilon(\vec{i})-\varepsilon_p} v_\lambda x^{(\vec{i})} \Bigg]\\
&= \bar{x}^\kappa \Bigg[\sum_{i_1,\dots,i_{n-1}=0}^{\infty} \frac{(-1)^{i_1+i_2+\cdots+i_{n-1}}\langle\kappa\rangle_{i_{n-1}}(\kappa + i_{p+1} - i_p)}{i_1!i_2!\cdots i_{n-1}!\langle\sigma\rangle_{i_{n-1}}} E^{\epsilon(\vec{i})} v_\lambda x_p x_{p+1}^{-1} x^{(\vec{i})}\\
&\quad - \sum_{i_1,\dots,i_{n-1}=0}^{\infty} \frac{(-1)^{i_1+i_2+\cdots+i_{n-1}}\langle\kappa\rangle_{i_{n-1}}(\kappa + i_{p+1} - i_p)}{i_1!i_2!\cdots i_{n-1}!\langle\sigma\rangle_{i_{n-1}}} E^{\epsilon(\vec{i})} v_\lambda x_p x_{p+1}^{-1} x^{(\vec{i})} \Bigg] = 0.
\end{aligned} \tag{15.3.42}$$

Moreover,

$$\begin{aligned}
&E_{n-1,n}(u_\lambda)\\
&= \bar{x}^\kappa \sum_{i_1,\dots,i_{n-1}=0}^{\infty} \frac{(-1)^{i_1+i_2+\cdots+i_{n-1}}\langle\kappa\rangle_{i_{n-1}}}{i_1!i_2!\cdots i_{n-1}!\langle\sigma\rangle_{i_{n-1}}} [(\kappa - i_{n-1}) E^{\epsilon(\vec{i})} v_\lambda x_{n-1} x_n^{-1} x^{(\vec{i})}\\
&\quad + (\sigma + 1 - i_{n-1}) i_{n-1} E^{\epsilon(\vec{i})-\varepsilon_{n-1}} v_\lambda x^{(\vec{i})}]\\
&= \bar{x}^\kappa \Bigg[\sum_{i_1,\dots,i_{n-1}=0}^{\infty} \frac{(-1)^{i_1+i_2+\cdots+i_{n-1}}\langle\kappa\rangle_{i_{n-1}}}{i_1!i_2!\cdots i_{n-1}!\langle\sigma\rangle_{i_{n-1}}} (\kappa - i_{n-1}) E^{\epsilon(\vec{i})} v_\lambda x_{n-1} x_n^{-1} x^{(\vec{i})}\\
&\quad + \sum_{i_1,\dots,i_{n-1}=0}^{\infty} \frac{(-1)^{i_1+i_2+\cdots+i_{n-1}}\langle\kappa\rangle_{i_{n-1}}}{i_1!i_2!\cdots i_{n-1}!\langle\sigma\rangle_{i_{n-1}}} (\sigma + 1 - i_{n-1}) i_{n-1} E^{\epsilon(\vec{i})-\varepsilon_{n-1}} v_\lambda x^{(\vec{i})} \Bigg]
\end{aligned}$$

$$= \bar{x}^{\kappa}\left[\sum_{i_1,\dots,i_{n-1}=0}^{\infty}\frac{(-1)^{i_1+i_2+\cdots+i_{n-1}}\langle\kappa\rangle_{i_{n-1}}}{i_1!i_2!\cdots i_{n-1}!\langle\sigma\rangle_{i_{n-1}}}(\kappa-i_{n-1})E^{\epsilon(\vec{i})}v_\lambda x_{n-1}x_n^{-1}x^{(\vec{i})}\right.$$

$$\left.-\sum_{i_1,\dots,i_{n-1}=0}^{\infty}\frac{(-1)^{i_1+i_2+\cdots+i_{n-1}}\langle\kappa\rangle_{i_{n-1}+1}}{i_1!i_2!\cdots i_{n-1}!\langle\sigma\rangle_{i_{n-1}}}E^{\epsilon(\vec{i})}v_\lambda x_p x_{p+1}^{-1}x^{(\vec{i})}\right]=0. \quad (15.3.43)$$

Hence u_λ is a singular vector. □

Let $\Phi : M(\lambda) \to \tilde{M}$ be the Lie algebra module homomorphism such that $\Phi(v_\lambda) = u_\lambda$. Let $\vec{i}$ as in (15.2.8) and fix $\beta \in \Gamma_A$. By (15.3.35), we have

$$E_{n,n-1}^{\beta_{n,n-1}}(E^{\epsilon(\vec{i})}v_\lambda)=\sum_{p_{n,n-1}=0}^{\beta_{n,n-1}}\binom{\beta_{n,n-1}}{p_{n,n-1}}\langle i_{n-2}\rangle_{p_{n,n-1}}$$
$$E^{\epsilon(\vec{i})-p_{n,n-1}\varepsilon_{n-1,n-2}+p_{n,n-1}\varepsilon_{n,n-2}+(\beta_{n,n-1}-p_{n,n-1})\varepsilon_{n,n-1}}v_\lambda, \quad (15.3.44)$$

$$E_{n,n-2}^{\beta_{n,n-2}}E_{n,n-1}^{\beta_{n,n-1}}(E^{\epsilon(\vec{i})}v_\lambda)=\sum_{p_{n,n-2},p_{n,n-1}=0}^{\infty}\binom{\beta_{n,n-2}}{p_{n,n-2}}\binom{\beta_{n,n-1}}{p_{n,n-1}}\langle i_{n-3}\rangle_{p_{n,n-2}}\langle i_{n-2}\rangle_{p_{n,n-1}}$$
$$E^{\epsilon(\vec{i})+\sum_{r=n-2,n-1}[-p_{n,r}\varepsilon_{r,r-1}+p_{n,r}\varepsilon_{n,r-1}+(\beta_{n,r}-p_{n,r}\varepsilon_{n,r}]}v_\lambda, \quad (15.3.45)$$

$$\left[\prod_{r=1}^{n-1}E_{n,r}^{\beta_{n,r}}\right](E^{\epsilon(\vec{i})}v_\lambda)=\sum_{p_{n,2},\dots,p_{n,n-1}=0}^{\infty}\left[\prod_{r=2}^{n-1}\binom{\beta_{n,r}}{p_{n,r}}\langle i_{r-1}\rangle_{p_{n,r}}\right]$$
$$E^{\epsilon(\vec{i})+\beta_{n,1}\varepsilon_{n,1}+\sum_{s=2}^{n-1}(-p_{n,s}\varepsilon_{s,s-1}+p_{n,s}\varepsilon_{n,s-1}+(\beta_{n,s}-p_{n,s})\varepsilon_{n,s})}v_\lambda, \quad (15.3.46)$$

$$\left[\prod_{r=1}^{n-2}E_{n-1,r}^{\beta_{n-1,r}}\right]\left[\prod_{r=1}^{n-1}E_{n,r}^{\beta_{n,r}}\right](E^{\epsilon(\vec{i})}v_\lambda)=\sum_{i=n-1,n}\sum_{j=1}^{i-1}\sum_{p_{i,j}=0}^{\infty}$$
$$\left[\prod_{r=1}^{n-2}\binom{\beta_{n-1,r}}{p_{n-1,r}}\right]\left[\prod_{r=1}^{n-1}\binom{\beta_{n,r}}{p_{n,r}}\right]\left[\prod_{r=1}^{n-3}\langle i_r\rangle_{p_{n,r+1}+p_{n-1,r+1}}\right]\langle i_{n-2}\rangle_{p_{n,n-1}}$$
$$E^{\epsilon(\vec{i})+\sum_{r=n-1,n}\beta_{r,1}\varepsilon_{r,1}+\sum_{r=n-1,n}\sum_{s=2}^{r-1}(-p_{r,s}\varepsilon_{s,s-1}+p_{r,s}\varepsilon_{r,s-1}+(\beta_{r,s}-p_{r,s})\varepsilon_{r,s})}v_\lambda, \quad (15.3.47)$$

$$E^{\beta}(E^{\epsilon(\vec{i})}v_\lambda)=\sum_{2\le k<j\le n}\sum_{p_{j,k}=0}^{\infty}\left[\prod_{2\le m<l\le n}\binom{\beta_{l,m}}{p_{l,m}}\right]\left[\prod_{s=1}^{n-2}\langle i_s\rangle_{\sum_{r=s+2}^{n}p_{r,s+1}}\right]$$
$$E^{\epsilon(\vec{i})+\sum_{1\le m<l\le n}(\beta_{l,m}+(1-\delta_{l,m+1})p_{l,m+1}-p_{l,m}-\delta_{l,m+1}\sum_{s=l+1}^{n}p_{s,l})\varepsilon_{l,m}}v_\lambda. \quad (15.3.48)$$

where we treat $p_{2,1} = p_{3,1} = \cdots = p_{n,1} = 0$. Hence by (15.3.37) and (15.3.47), we obtain

$$\begin{aligned}
&E^{\alpha}(E^{\epsilon(\vec{i})}v_\lambda x^{(\vec{i})}\bar{x}^{\kappa})\\
&= \sum_{p^0_{2,1},\dots,p^0_{n,1}=0}^{\infty} \sum_{2\le k<j\le n} \sum_{p^0_{j,k},p_{j,k}=0}^{\infty} \left[\prod_{m=2}^{n}\binom{\alpha_{m,1}}{p^0_{m,1}}\right]\left[\prod_{2\le m<l\le n}\binom{\alpha_{l,m}}{p^0_{l,m},\, p_{l,m}}\right]\\
&\left[\prod_{s=1}^{n-2}\langle i_s\rangle_{\sum_{r=s+2}^{n} p_{r,s+1}}\right]\left[\prod_{s=1}^{n-1}\langle \kappa + i_s - i_{s-1}\rangle_{\sum_{r=s+1}^{n} p^0_{r,s}}\right]\\
&E^{\epsilon(\vec{i})+\sum_{1\le m<l\le n}(\alpha_{l,m}+p_{l,m+1}-p^0_{l,m}-p_{l,m}-\delta_{l,m+1}\sum_{s=l+1}^{n}p_{s,l})\epsilon_{l,m}}v_\lambda\\
&\bar{x}^{\kappa}\prod_{r=1}^{n} x_r^{i_r-i_{r-1}+\sum_{s=1}^{r-1}p^0_{r,s}-\sum_{q=r+1}^{n}p^0_{q,r}}
\end{aligned} \tag{15.3.49}$$

for $\alpha \in \Gamma_A$, where we treat

$$i_{-1} = i_n = p_{2,1} = p_{3,1} = \cdots = p_{n,1} = p_{2,2} = p_{3,3} = \cdots p_{n,n} = 0. \tag{15.3.50}$$

View Φ as a function in $\{x_1, x_2, \dots, x_n\}$ taking values in End $M(\lambda)$, the space of linear transformations on $M(\lambda)$. In order to calculate the Etingof trace function $E_A(z_1, z_2, \dots, z_n)$ in (15.2.15), we need to take

$$p_{l,m+1} = p^0_{l,m} + p_{l,m}, \quad i_k = p^0_{k+1,k} + p_{k+1,k} + \sum_{s=k+2}^{n} p_{s,k+1} \tag{15.3.51}$$

in (15.3.49) for $1 \le m < l-1 \le n-1$ and $k \in \overline{1, n-1}$. By (15.3.50), (15.3.51) and induction,

$$p_{l,m} = \sum_{r=1}^{m-1} p^0_{l,r} \quad \text{for } 1 \le m < l \le n. \tag{15.3.52}$$

Moreover,

$$i_k = \sum_{r=k+1}^{n}(p^0_{r,k} + p_{r,k}) = \sum_{r=k+1}^{n}\sum_{s=1}^{k} p^0_{r,s} \quad \text{for } k \in \overline{1, n-1}. \tag{15.3.53}$$

$$i_k - i_{k-1} = \sum_{r=k+1}^{n}\sum_{s=1}^{k} p^0_{r,s} - \sum_{r=k}^{n}\sum_{s=1}^{k-1} p^0_{r,s} = \sum_{r=k+1}^{n} p^0_{r,k} - \sum_{s=1}^{k-1} p^0_{k,s} \tag{15.3.54}$$

for $k \in \overline{1, k}$. Furthermore,

$$\binom{\alpha_{j,k}}{p^0_{j,k}, p_{j,k}} = \binom{p^0_{j,k} + p_{j,k}}{p_{j,k}} \binom{\alpha_{j,k}}{p^0_{j,k} + p_{j,k}} \tag{15.3.55}$$

and

$$\binom{p^0_{j,k} + p_{j,k}}{p_{j,k}} = \frac{(p^0_{j,k} + p_{j,k})!}{p^0_{j,k}! p_{j,k}!} = \frac{(p^0_{j,k} + p_{j,k})!}{p^0_{j,k}!(p^0_{j,k-1} + p_{j,k-1})!} \tag{15.3.56}$$

by (15.3.51).

For convenience, we denote

$$\alpha_{1*} = \alpha^*_n = 0, \quad \alpha_{i*} = \sum_{r=1}^{i-1} \alpha_{i,r}, \quad \alpha^*_j = \sum_{s=j+1}^{n} \alpha_{s,j} \tag{15.3.57}$$

for $\alpha \in \Gamma_A$, $i \in \overline{2, n}$ and $j \in \overline{1, n-1}$. By (15.3.40), (15.3.41) and (15.3.49)–(15.3.57), the Etingof's trace function

$$\begin{aligned}
&E_A(z_1, z_2, , z_n) \\
&= \sum_{\alpha \in \Gamma_A} \sum_{p^0_{r_1,r_2} \in \mathbb{N},\ 1 \leq r_2 < r_1 \leq n} (-1)^{\sum_{s=1}^{n-1} i_s} \left[\prod_{2 \leq k < j \leq n} \binom{p^0_{j,k} + p_{j,k}}{p_{j,k}} \right] \\
&\quad \times \frac{\prod_{s=1}^{n-1} \langle \kappa + i_s - i_{s-1} \rangle_{\sum_{r=s+1}^{n} p^0_{r,s}}}{\langle \sigma \rangle_{i_{n-1}} \prod_{s=1}^{n-1} (p^0_{s+1,s} + p_{s+1,s})!} \left[\prod_{1 \leq k < j \leq n} \binom{\alpha_{j,k}}{p^0_{j,k} + p_{j,k}} \right] \langle \kappa \rangle_{i_{n-1}} \\
&\quad \times \bar{x}^{\kappa} \left(\prod_{r=1}^{n} z_r^{\lambda_r} \right) \prod_{1 \leq k < j \leq n} \left(\frac{z_j}{z_k} \right)^{\alpha_{j,k}} \\
&= \sum_{p^0_{r_1,r_2} \in \mathbb{N},\ 1 \leq r_2 < r_1 \leq n} (-1)^{\sum_{s=1}^{n-1} i_s} \frac{\prod_{s=1}^{n-1} \langle \kappa + i_s - i_{s-1} \rangle_{\sum_{r=s+1}^{n} p^0_{r,s}} \langle \kappa \rangle_{i_{n-1}}}{\langle \sigma \rangle_{i_{n-1}} \prod_{1 \leq s_1 < s_2 \leq n} p^0_{s_2,s_1}!} \\
&\quad \times \left(\prod_{r=1}^{n} z_r^{\lambda_r} \right) \left[\prod_{1 \leq r_1 < r_2 \leq n} \left(\frac{z_{r_2}}{z_{r_1}} \right)^{p^0_{r_2,r_1} + p_{r_2,r_1}} \right] \\
&\quad \times \left[\prod_{1 \leq s_1 < s_2 \leq n} \left(\frac{1}{1 - z_{s_2}/z_{s_1}} \right)^{p^0_{s_2,s_1} + p_{s_2,s_1} + 1} \right] \bar{x}^{\kappa} \\
&= \frac{\bar{x}^{\kappa} \prod_{r=1}^{n} z_r^{\lambda_r + n - r}}{\prod_{1 \leq k < j \leq n} (z_k - z_j)} \sum_{\beta \in \Gamma_A} \frac{\left[\prod_{s=1}^{n-1} \langle \kappa + \beta^*_s - \beta_{s*} \rangle_{\beta^*_s} \right] \langle \kappa \rangle_{\beta_{n*}}}{\langle \sigma \rangle_{\beta_{n*}} \prod_{1 \leq s_1 < s_2 \leq n} \beta_{s_2,s_1}!}
\end{aligned}$$

$$\times\left[\prod_{1\leq k<j\leq n}\left(\frac{z_j}{z_j-z_k}\right)^{\sum_{r=1}^{k}\beta_{j,r}}\right]. \tag{15.3.58}$$

Set

$$\xi^A_{r_2,r_1}=\prod_{s=r_1}^{r_2-1}\frac{z_{r_2}}{z_{r_2}-z_s}\qquad\text{for } 1\leq r_1<r_2\leq n. \tag{15.3.59}$$

Motivated by (15.3.17)–(15.3.19) and (15.3.58), we define our $(n(n-1)/2)$-variable *path hypergeometric function of type A* by

$$\mathscr{X}_A(\tau_1,\ldots,\tau_n;\vartheta)\{z_{j,k}\}=\sum_{\beta\in\Gamma_A}\frac{\left[\prod_{s=1}^{n-1}(\tau_s-\beta_{s*})_{\beta_s^*}\right](\tau_n)_{\beta_{n*}}}{\beta!(\vartheta)_{\beta_{n*}}}z^\beta \tag{15.3.60}$$

(cf. (15.3.57)), where

$$\beta!=\prod_{1\leq k<j\leq n}\beta_{j,k}!,\qquad z^\beta=\prod_{1\leq k<j\leq n}z_{j,k}^{\beta_{j,k}}. \tag{15.3.61}$$

According to (15.3.20) and (15.3.58)–(15.3.60), we have:

Theorem 15.3.3 *Under the condition (15.3.31), the Etingof trace function is*

$$E_A(z_1,z_2,\ldots,z_n)=\left[\frac{\bar{x}^\kappa\prod_{r=1}^n z_r^{\lambda_r+n-r}}{\prod_{1\leq k<j\leq n}(z_k-z_j)}\right]\mathscr{X}_A(\kappa+1,\ldots,\kappa+1,-\kappa;-\sigma)\{\xi^A_{r_2,r_1}\}. \tag{15.3.62}$$

Moreover, (15.1.9) implies that the function in (15.2.17):

$$\Psi_A(z_1,z_2,\ldots,z_n)=\prod_{r=1}^n z_r^{\lambda_r+(n+1)/2-r}\mathscr{X}_A(\kappa+1,\ldots,\kappa+1,-\kappa;-\sigma)\{\xi^A_{r_2,r_1}\}. \tag{15.3.63}$$

15.4 Path Hypergeometric Functions of Type C

In this section, we want to introduce and calculate the Etingof trace function of $sp(2n,\mathbb{R})$.

Take the settings in (15.2.31)–(15.2.36). Set

$$h_n=C_{n,n},\quad h_i=C_{i,i}-C_{i+1,i+1} \tag{15.4.1}$$

for $i = 1, 2, \ldots, n-1$. We choose positive root vectors

$$\{C_{i,j},\ C_{i,n+j},\ C_{k,n+k} \mid 1 \leq i < j \leq n,\ k\overline{1,n}\}. \tag{15.4.2}$$

In particular, we take

$$\{C_{i,i+1},\ C_{n,2n} \mid i = 1, 2, \ldots, n-1\} \text{ as positive simple root vectors.} \tag{15.4.3}$$

Accordingly,

$$\{C_{i,j},\ C_{i,n+j},\ C_{n+k,k} \mid 1 \leq j < i \leq n,\ k \in \overline{1,n}\} \tag{15.4.4}$$

are negative root vectors and

$$\{C_{i+1,i},\ C_{2n,n} \mid i \in \overline{1, n-1}\} \text{ are negative simple root vectors.} \tag{15.4.5}$$

Let

$$\Gamma_C = \sum_{1\leq j<i\leq n} \mathbb{N}\varepsilon_{i,j} + \sum_{1\leq j\leq i\leq n} \mathbb{N}\varepsilon_{n+i,j} \tag{15.4.6}$$

be the torsion-free additive semigroup of rank n^2 with $\varepsilon_{p,q}$ as base elements, and let $U(sp(2n, \mathbb{R}))$ be the universal enveloping algebra of $sp(2n, \mathbb{R})$. For any

$$\alpha = \sum_{1\leq j<i\leq n} \alpha_{i,j}\varepsilon_{i,j} + \sum_{1\leq j\leq i\leq n} \alpha_{n+i,j}\varepsilon_{n+i,j} \in \Gamma_C, \tag{15.4.7}$$

we denote

$$\begin{aligned} C^\alpha = {} & C_{2,1}^{\alpha_{2,1}} C_{3,1}^{\alpha_{3,1}} C_{3,2}^{\alpha_{3,2}} C_{4,1}^{\alpha_{4,1}} \cdots C_{n,1}^{\alpha_{n,1}} \cdots C_{n,n-1}^{\alpha_{n,n-1}} \\ & \times C_{n+1,1}^{\alpha_{n+1,1}} C_{n+2,1}^{\alpha_{n+2,1}} C_{n+2,2}^{\alpha_{n+2,2}} C_{n+3,1}^{\alpha_{n+3,1}} \cdots C_{2n,1}^{\alpha_{2n,1}} \cdots C_{2n,n}^{\alpha_{2n,n}} \end{aligned} \tag{15.4.8}$$

Denote by $\mathscr{G}_-$ the Lie subalgebra spanned by (15.4.4). Then

$$\{C^\alpha \mid \alpha \in \Gamma_C\} \text{ forms a PBW basis of } U(\mathscr{G}_-)). \tag{15.4.9}$$

For convenience, we treat

$$\varepsilon_{n+i,j} = \varepsilon_{n+j,i} \qquad \text{for } i, j = 1, 2, \ldots, n. \tag{15.4.10}$$

Note that

$$[C_{n,2n}, C_{2n,n}] = h_n, \quad [C_{i,i+1}, C_{i+1,i}] = h_i, \tag{15.4.11}$$

$$[h_i, C_{n+i+1,i}] = 0, \quad [h_i, C_{n+i,i}] = -2C_{n+i,i}, \quad [h_i, C_{n+i+1,i+1}] = 2C_{n+i+1,i+1}, \tag{15.4.12}$$

$$[h_i, C_{n+k,i}] = -C_{n+k,i}, \quad [h_i, C_{n+k,i+1}] = C_{n+k,i+1}, \tag{15.4.13}$$

$$[C_{i,i+1}, C_{n+p,i}] = -C_{n+p,i+1}, \quad [C_{i,i+1}, C_{n+i+1,i}] = -2C_{n+i+1,i+1}, \tag{15.4.14}$$

$$[C_{n,2n}, C_{2n,i}] = C_{n,i}, \quad [C_{n+j,r}, C_{r,q}] = (1+\delta_{j,q})C_{n+j,q}, \quad [C_{n+r,r}, C_{r,q}] = C_{n+r,q} \tag{15.4.15}$$

for $i \in \overline{1, n-1}$ and $j, k, p, q, r \in \overline{1, n}$ such that $k \neq i, i+1,\ \ p \neq i+1$ and $q < r$.

Let λ be a weight, which is a linear function on H_C, such that

$$\lambda(h_i) = -\frac{1}{2} \quad \text{for } i = 1, 2, \ldots, n-1; \qquad \lambda(h_n) = \lambda_n \in \mathbb{R}. \tag{15.4.16}$$

Recall that $sp(2n, \mathbb{R})$ is generated by

$$\{C_{i,i+1}, C_{i+1,i}, C_{n,2n}, C_{2n,n}, \mid i = 1, 2, \ldots, n-1\} \tag{15.4.17}$$

as a Lie algebra. The Verma $sp(2n, \mathbb{R})$-module with the highest-weight vector v_λ of weight λ is given by

$$M(\lambda) = \text{Span}\{E^\alpha v_\lambda \mid \alpha \in \Gamma_C\}, \tag{15.4.18}$$

with the action determined by

$$\begin{aligned} &C_{i,i+1}(C^\alpha v_\lambda) \\ &= \Bigg[\sum_{j=1}^{i-1} \alpha_{i+1,j} C^{\alpha+\varepsilon_{i,j}-\varepsilon_{i+1,j}} - \sum_{j=i+2}^{n} \alpha_{j,i} C^{\alpha+\varepsilon_{j,i+1}-\varepsilon_{j,i}} - \sum_{k\neq i+1} \alpha_{n+k,i} C^{\alpha-\varepsilon_{n+k,i}+\varepsilon_{n+k,i+1}} \\ &\quad - 2\alpha_{n+i+1,i} C^{\alpha-\varepsilon_{n+i+1,i}+\varepsilon_{n+i+1,i+1}} + \alpha_{i+1,i}(1/2 - \sum_{j=i+1}^{n} \alpha_{j,i} + \sum_{j=i+2}^{n} \alpha_{j,i+1} \\ &\quad + \sum_{k\neq i,i+1} (\alpha_{n+k,i+1} - \alpha_{n+k,i}) - 2\alpha_{n+i,i} + 2\alpha_{n+i+1,i+1}) C^{\alpha-\varepsilon_{i+1,i}}\Bigg] v_\lambda \end{aligned} \tag{15.4.19}$$

for $i \in \overline{1, n-1}$,

$$\begin{aligned} C_{n,2n}(C^\alpha v_\lambda) = &\Bigg[\sum_{i=1}^{n-1} \alpha_{2n,i}(C^{\alpha-\varepsilon_{2n,i}+\varepsilon_{n,i}} + (\alpha_{2n,i}-1)C^{\alpha-2\varepsilon_{2n,i}+\varepsilon_{n+i,i}} \\ &+ \sum_{j=1}^{i-1} \alpha_{2n,j} C^{\alpha-\varepsilon_{2n,i}-\varepsilon_{2n,j}+\varepsilon_{n+i,j}}) + \alpha_{2n,n}(\lambda_n + 1 - \alpha_{2n,n}) C^{\alpha-\varepsilon_{2n,n}}\Bigg] v_\lambda \end{aligned} \tag{15.4.20}$$

and

$$C_{j,i}(E^{\alpha}v_{\lambda}) = \left(C^{\alpha+\varepsilon_{j,i}} + \sum_{p=1}^{i-1} \alpha_{i,p} C^{\alpha+\varepsilon_{j,p}-\varepsilon_{i,p}}\right) v_{\lambda}, \tag{15.4.21}$$

$$C_{n+j,i}(E^{\alpha}v_{\lambda}) = \left(C^{\alpha+\varepsilon_{n+j,i}} + \sum_{q=1}^{j-1} \alpha_{n+j,q} C^{\alpha+\varepsilon_{n+q,i}-\varepsilon_{j,q}} \right.$$
$$\left. + \sum_{p=1}^{i-1} \alpha_{i,p} \left(C^{\alpha+\varepsilon_{n+j,p}-\varepsilon_{i,p}} + \sum_{q=1}^{j-1} \alpha_{n+j,q} C^{\alpha+\varepsilon_{n+q,p}-\varepsilon_{i,p}-\varepsilon_{j,q}}\right)\right) \tag{15.4.22}$$

for $1 \leq i < j \leq n$,

$$C_{n+k,k}(C^{\alpha}v_{\lambda}) = \left[C^{\alpha+\varepsilon_{n+k,k}} + \sum_{r=1}^{k-1} \alpha_{k,r} (C^{\alpha-\varepsilon_{k,r}+\varepsilon_{n+k,r}} + (\alpha_{k,r}-1) C^{\alpha-2\varepsilon_{k,r}+\varepsilon_{n+r,r}} \right.$$
$$\left. + \sum_{s=r+1}^{k-1} \alpha_{k,s} C^{\alpha-\varepsilon_{k,r}-\varepsilon_{k,s}+\varepsilon_{n+s,r}})\right] v_{\lambda} \tag{15.4.23}$$

by (15.4.11)–(15.4.15).

Recall the notions in (15.2.8) and (15.2.39). Moreover, we take $M = M(\lambda)$ and

$$\tilde{M} = \left\{\sum_{\vec{i}\in\mathbb{Z}^n} w_{\vec{i}} x^{\vec{i}} x^* \mid w_{\vec{i}} \in M(\lambda)\right\}. \tag{15.4.24}$$

The representation π in (15.2.41)–(15.2.45) of $sp(2n,\mathbb{R})$ on $\tilde{M}$ is given by

$$C_{p,q}\left(\sum_{\vec{i}\in\mathbb{Z}^n} w_{\vec{i}} x^{\vec{i}} x^*\right) = \sum_{\vec{i}\in\mathbb{Z}^n} \left(C_{p,q}(w_{\vec{i}}) x^{\vec{i}} x^* + w_{\vec{i}} x_p \partial_{x_q}(x^{\vec{i}} x^*) + \frac{\delta_{p,q}}{2} w_{\vec{i}} x^{\vec{i}} x^*\right), \tag{15.4.25}$$

$$C_{p,n+q}\left(\sum_{\vec{i}\in\mathbb{Z}^n} w_{\vec{i}} x^{\vec{i}} x^*\right) = \sum_{\vec{i}\in\mathbb{Z}^n} \left(C_{p,n+q}(w_{\vec{i}}) x^{\vec{i}} x^* - \frac{1}{1+\delta_{p,q}} w_{\vec{i}} x_p x_q x^{\vec{i}} x^*\right), \tag{15.4.26}$$

$$C_{n+p,q}\left(\sum_{\vec{i}\in\mathbb{Z}^n} w_{\vec{i}} x^{\vec{i}} x^*\right) = \sum_{\vec{i}\in\mathbb{Z}^n} (C_{n+p,q}(w_{\vec{i}}) x^{\vec{i}} x^* + \frac{1}{1+\delta_{p,q}} w_{\vec{i}} \partial_{x_p} \partial_{x_q}(x^{\vec{i}} x^*)) \tag{15.4.27}$$

for $1 \leq p, q \leq n$.

For $\vec{i} \in \mathbb{N}^n$, we set

$$x^{(\vec{i})} = x_1^{i_1} x_2^{i_2-i_1} x_3^{i_3-i_2} \ldots x_{n-1}^{i_{n-1}-i_{n-2}} x_n^{2i_n-i_{n-1}}, \tag{15.4.28}$$

$$\epsilon(\vec{i}) = \sum_{p=1}^{n-1} i_p \varepsilon_{p+1,p} + i_n \varepsilon_{2n,n}. \tag{15.4.29}$$

Let

$$u_\lambda = \sum_{\vec{i}\in\mathbb{N}^n} \frac{(-1)^{i_1+i_2+\cdots+i_{n-1}}}{i_1!i_2!\cdots i_n!2^{i_n}\langle\lambda_n\rangle_{i_n}} C^{\epsilon(\vec{i})} v_\lambda x^{(\vec{i})} x^*. \tag{15.4.30}$$

It is verified that u_λ is a singular vector of weight λ in $\tilde{M}$; that is,

$$C_{i,j}(u_\lambda) = C_{p,n+q}(u_\lambda) = 0 \quad \text{for } 1 \leq i < j \leq n;\ 1 \leq p \leq q \leq n. \tag{15.4.31}$$

Hence there exists $sp(2n,\mathbb{R})$-module homomorphism $\Phi : M(\lambda) \to \tilde{M}$ such that

$$\Phi(v_\lambda) = u_\lambda. \tag{15.4.32}$$

View Φ as a function in $\{x_1, x_2, \ldots, x_n\}$ taking value in the spaces of linear transformations on $M(\lambda)$. In the rest of this section, we want to calculate $E_C(z_1, z_2, \ldots, z_n)$ in (15.2.47).

For $\vec{i} \in \mathbb{N}^n$ and $\beta \in \Gamma_C$ (cf. (15.4.6)), we have

$$\begin{aligned}
&C^\beta(C^{\epsilon(\vec{i})} v_\lambda) \\
&= \sum_{2\leq j_1<j_2\leq n} \sum_{k_1=1}^n \sum_{k_2=2}^n \sum_{k_3=k_2}^n \Bigg(\sum_{p_{j_2,j_1}, q_{k_1,k_2}, q^1_{k_2,k_2}=0}^\infty \\
&\left[\prod_{r=2}^n \binom{\beta_{n+r,r}}{q_{r,r},\, q^1_{r,r}} \binom{\beta_{m+r,1}}{q_{1,r}}\right] \langle i_{n-1}\rangle_{\sum_{r=1}^n q_{r,n} + \sum_{r=2}^n q^1_{n,r} + q^1_{n,n}} \\
&\times \left[\prod_{2\leq l_1<l_2\leq n} \binom{\beta_{l_2,l_1}}{p_{l_2,l_1}} \binom{\beta_{n+l_2,l_1}}{q_{l_1,l_2},\, q_{l_2,l_1},\, q^1_{l_2,l_1}}\right] \\
&\times \left[\prod_{s=1}^{n-2} \langle i_s\rangle_{\sum_{r=s+2}^n p_{r,s+1} + \sum_{r=1}^n q_{r,s+1} + \sum_{r=2}^{s+1} q^1_{s+1,r} + \sum_{r=s+1}^n q^1_{r,s+1}}\right] \\
&\times E^{-\delta_{l_2,l_1+1}(\sum_{r=l_2+1}^n p_{r,l_2} + \sum_{s=1}^n q_{s,l_2} + \sum_{s=2}^{l_2} q^1_{l_2,s} + \sum_{s=l_2}^n q^1_{s,l_2})\varepsilon_{l_2,l_1}} \\
&\times E^{\sum_{1\leq s_1<s_2\leq n}(q_{s_2,s_1+1} + q_{s_1,s_2+1} + q^1_{s_2+1,s_1+1} - q_{s_1,s_2} - q_{s_2,s_1} - q^1_{s_2,s_1})\varepsilon_{n+s_2,s_1}} \\
&\times E^{\epsilon(\vec{i}) + \beta + \sum_{r=1}^n (q_{r,r+1} + q^1_{r+1,r+1} - q_{r,r} - q^1_{r,r})\varepsilon_{n+r,r}} v_\lambda \Bigg) \\
&\times E^{\sum_{1\leq l_1<l_2\leq n}(p_{l_2,l_1+1} - p_{l_2,l_1})\varepsilon_{l_2,l_1}}
\end{aligned} \tag{15.4.33}$$

by the calculations in (15.3.44)–(15.3.49) and (15.4.19)–(15.4.23), where we treat

$$p_{r,r} = q_{1,1} = q^1_{n+1,n+1} = q^1_{r,1} = q^1_{n+1,r} = q_{s,n+1} = q_{n+1,s} = 0 \tag{15.4.34}$$

for $2 \le r \le n$ and $1 \le s \le n$. Hence by (15.4.25), (15.4.27) and (15.4.33), we obtain

$$
\begin{aligned}
&C^{\alpha}(C^{\epsilon(\vec{i})} v_{\lambda} x^{(\vec{i})} x^*) \\
&= \sum_{2\le j_1<j_2\le n} \sum_{k_1=1}^{n} \sum_{k_2=2}^{n} \sum_{k_3=k_1}^{n} \sum_{k_4=k_2}^{n} \Bigg\{ \sum_{p^0_{k_2,1}, p^0_{j_2,j_1}, p_{j_2,j_1}, q^0_{k_3,k_1}, q^1_{k_3,k_1}, q_{k_1,k_2}=0}^{\infty} \\
&\times \frac{\langle i_{n-1}\rangle_{\sum_{r=1}^n q_{r,n}+\sum_{r=2}^n q^1_{n,r}+q^1_{n,n}}}{2^{\sum_{r=1}^n q^0_{r,r}}} \langle 2i_n - i_{n-1} - 1/2\rangle_{2q^0_{n,n}+\sum_{r=1}^{n-1} q^0_{n,r}} \\
&\times \left[\prod_{2\le l_1<l_2\le n} \binom{\alpha_{l_2,l_1}}{p^0_{l_2,l_1}, p_{l_2,l_1}} \binom{\alpha_{n+l_2,l_1}}{q^0_{l_2,l_1}, q^1_{l_2,l_1}, q_{l_1,l_2}, q_{l_2,l_1}}\right] \\
&\times \left[\prod_{m=2}^{n} \binom{\alpha_{m,1}}{p^0_{m,1}}\right] \binom{\alpha_{n+1,1}}{q^0_{1,1}} \left[\prod_{r=2}^{n} \binom{\alpha_{n+r,r}}{q^0_{r,r}, q^1_{r,r}, q_{r,r}} \binom{\alpha_{n+r,1}}{q^0_{r,1}, q_{1,r}}\right] \\
&\times \left[\prod_{s=1}^{n-2} \langle i_s\rangle_{\sum_{r=s+2}^n p_{r,s+1}+\sum_{r=1}^n q_{r,s+1}+\sum_{r=2}^{s+1} q^1_{s+1,r}+\sum_{r=s+1}^n q^1_{r,s+1}}\right] \\
&\times \left[\prod_{s=1}^{n-1} \langle i_s - i_{s-1} - 1/2\rangle_{\sum_{r=s+1}^n p^0_{r,s}+\sum_{r=1}^s q^0_{s,r}+\sum_{r=s}^n q^0_{r,s}}\right] \\
&\times E^{-\sum_{1\le l_1<l_2\le n} \delta_{l_2,l_1+1}(\sum_{r=l_2+1}^n p_{r,l_2}+\sum_{s=1}^n q_{s,l_2}+\sum_{s=2}^{l_2} q^1_{l_2,s}+\sum_{s=l_2}^n q^1_{s,l_2})\varepsilon_{l_2,l_1}} \\
&\times E^{\sum_{1\le s_1<s_2\le n}(q_{s_2,s_1+1}+q_{s_1,s_2+1}+q^1_{s_2+1,s_1+1}-q_{s_1,s_2}-q_{s_2,s_1}-q^0_{s_2,s_1}-q^1_{s_2,s_1})\varepsilon_{n+s_2,s_1}} \\
&\times E^{\alpha+\epsilon(\vec{i})+\sum_{r=1}^n (q_{r,r+1}+q^1_{r+1,r+1}-q_{r,r}-q^0_{r,r}-q^1_{r,r})\varepsilon_{n+r,r}} \\
&\times E^{\sum_{1\le l_1<l_2\le n}(p_{l_2,l_1+1}-p_{l_2,l_1}-p^0_{l_2,l_1})\varepsilon_{l_2,l_1}} v_{\lambda} \\
&\left[\prod_{s=1}^{n-1} x_s^{i_s-i_{s-1}-1/2+\sum_{r=1}^{s-1} p^0_{s,r}-\sum_{r=s+1}^n p^0_{r,s}-\sum_{r=1}^s q^0_{s,r}-\sum_{r=s}^n q^0_{r,s}}\right] \\
&\times x_n^{2i_n-i_{n-1}-1/2+\sum_{r=1}^{n-1} p^0_{n,r}-2q^0_{n,n}-\sum_{r=1}^{n-1} q^0_{n,r}} \Bigg\},
\end{aligned}
\tag{15.4.35}
$$

where we treat $i_{-1} = 0$.

In order to calculate the Etingof trace (15.2.47), we have to let in (15.4.35):

$$p_{r_2,r_1} = p^0_{r_2,r_1-1} + p_{r_2,r_1-1} \quad \text{for } 2 \le r_1 < r_2 \le n, \tag{15.4.36}$$

$$q_{r,r+1} + q^1_{r+1,r+1} = q^0_{r,r} + q_{r,r} + +q^1_{r,r} \quad \text{for } r \in \overline{1, n-1}, \tag{15.4.37}$$

$$q_{s_2,s_1+1} + q_{s_1,s_2+1} + q^1_{s_2+1,s_1+1} = q_{s_1,s_2} + q_{s_2,s_1} + q^0_{s_2,s_1} + q^1_{s_2,s_1} \tag{15.4.38}$$

for $1 \leq s_1 < s_2 \leq n$ and

$$i_r = p^0_{r+1,r} + p_{r+1,r} + \sum_{s=r+2}^{n} p_{s,r+1} + \sum_{s=1}^{n} q_{s,r+1} + \sum_{s=2}^{r+1} q^1_{r+1,s} + \sum_{s=r+1}^{n} q^1_{s,r+1} \tag{15.4.39}$$

for $r \in \overline{1, n-1}$,

$$i_n = q^0_{n,n} + q_{n,n} + q^1_{n,n}. \tag{15.4.40}$$

In particular, (15.4.36) implies

$$p_{r_2,r_1} = \sum_{s=1}^{r_1-1} p^0_{r_2,s} \quad \text{for } 2 \leq r_1 < r_2 \leq n. \tag{15.4.41}$$

Set

$$\Im_r = \sum_{s=1}^{n} q_{s,r+1} + \sum_{s=2}^{r+1} q^1_{r+1,s} + \sum_{s=r+1}^{n} q^1_{s,r+1} \tag{15.4.42}$$

for $r \in \overline{1, n-1}$. By (15.4.34) and (15.4.37), we have

$$q_{1,2} + q^1_{2,2} = q^0_{1,1}. \tag{15.4.43}$$

Moreover, (15.4.38) implies

$$q_{i,2} + q_{1,i+1} + q^1_{i+1,2} = q_{1.i} + q^0_{i,1} \quad \text{for } i \in \overline{2, n}. \tag{15.4.44}$$

$$2(q_{1,2} + q^1_{2,2}) + \sum_{i=2}^{n} q_{i,2} + \sum_{r=3}^{n} (q_{1,r} + q^1_{r,2}) = 2q^0_{1,1} + \sum_{i=2}^{n} (q_{1,i} + q^0_{i,1}), \tag{15.4.45}$$

or equivalently,

$$\sum_{i=1}^{n} q_{i,2} + q^1_{2,2} + \sum_{r=2}^{n} q^1_{r,2} = q^0_{1,1} + \sum_{i=1}^{n} q^0_{i,1}. \tag{15.4.46}$$

Thus

$$\Im_1 = q^0_{1,1} + \sum_{i=1}^{n} q^0_{i,1}. \tag{15.4.47}$$

Let $r \in \overline{2, n-1}$. By (15.4.38), we have

$$q_{i,r+1} + q_{r,i+1} + q^1_{i+1,r+1} = q_{r,i} + q_{i,r} + q^0_{i,r} + q^1_{i,r} \tag{15.4.48}$$

for $i \in \overline{r+1,n}$ and

$$q_{r,j+1} + q_{j,r+1} + q^1_{r+1,j+1} = q_{j,r} + q_{r,j} + q^0_{r,j} + q^1_{r,j} \tag{15.4.49}$$

for $j \in \overline{1,r-1}$. Expressions (15.4.37), (15.4.48) and (15.4.49) imply

$$\begin{aligned}
&2(q_{r,r+1} + q^1_{r+1,r+1}) + \sum_{i=r+1}^{n} (q_{i,r+1} + q_{r,i+1} + q^1_{i+1,r+1}) \\
&\quad + \sum_{j=1}^{r-1} (q_{r,j+1} + q_{j,r+1} + q^1_{r+1,j+1}) \\
&= 2(q^0_{r,r} + q_{r,r} + +q^1_{r,r}) + q^1_{i,r}) + \sum_{j=1}^{r-1} (q_{j,r} + q_{r,j} + q^0_{r,j} + q^1_{r,j}) \\
&\quad + \sum_{s=r+1}^{n} (q_{r,s} + q_{s,r} + q^0_{s,r} + q^1_{s,r}),
\end{aligned} \tag{15.4.50}$$

or equivalently,

$$\begin{aligned}
&\sum_{i=1}^{n} q_{i,r+1} + \sum_{j=2}^{r+1} q^1_{r+1,j} + \sum_{i=r+1}^{n} q_{i,r+1} \\
&= \sum_{j=1}^{r} q^0_{r,j} + \sum_{s=r}^{n} q^0_{s,r} + \sum_{s=1}^{n} q_{s,r} + \sum_{j=2}^{r} q^1_{r,j} + \sum_{s=r}^{n} q^1_{s,r}
\end{aligned} \tag{15.4.51}$$

by (15.4.34). So

$$\Im_r - \Im_{r-1} = \sum_{j=1}^{r} q^0_{r,j} + \sum_{i=r}^{n} q^0_{i,r}. \tag{15.4.52}$$

Furthermore, by (15.4.39), (15.4.40), (15.4.42), (15.4.52) and induction, we have

$$i_r = \sum_{s=r+1}^{n} \sum_{j=1}^{r} p^0_{s,j} + \sum_{s=1}^{r} \left(\sum_{j=1}^{s} q^0_{s,j} + \sum_{j=s}^{n} q^0_{j,s} \right) \tag{15.4.53}$$

for $r \in \overline{1,n-1}$. In particular,

$$i_r - i_{r-1} = \sum_{s=r+1}^{n} p^0_{s,r} - \sum_{s=1}^{r-1} p^0_{r,s} + \sum_{j=1}^{r} q^0_{r,j} + \sum_{s=r}^{n} q^0_{s,r} \tag{15.4.54}$$

for $r \in \overline{1,n-1}$.

Next we want to calculate i_n. First we have

$$q_{n,n} + q^1_{n,n} = q_{n,n-1} + q_{n-1,n} + q^1_{n,n} + q^1_{n,n-1} + q^0_{n,n-1} \tag{15.4.55}$$

by (15.4.38). Moreover, for $r \in \overline{1, n-2}$, we have

$$\begin{aligned}
&\sum_{i=0}^{2r} q_{n-i,n-2r+i} + \sum_{i=0}^{r-1} q_{n-i,n-2r+1+i} \\
&= q_{n,n-2r} + \sum_{i=1}^{r}(q_{n-i,n-2r+i} + q_{n-2r-1+i,n-i+1} + q^1_{n-i+1,n-2r+i}) \\
&= \sum_{i=1}^{r}(q_{n-i,n-2r-1+i} + q_{n-2r-1+i,n-i} + q^0_{n-i,n-2r-1+i} + q^1_{n-i,n-2r-1+i}) \\
&\quad + q_{n,n-2r-1} + q_{n-2r-1,n} + q^0_{n,n-2r-1} + q^1_{n,n-2r-1} \\
&= \sum_{i=0}^{r}(q_{n-i,n-2r-1+i} + q_{n-2r-1+i,n-i}) + \sum_{i=0}^{r} q^1_{n-i,n-2r-1+i} + \sum_{i=0}^{r} q^0_{n-i,n-2r-1+i} \\
&= \sum_{i=0}^{2r+1} q_{n-i,n-2r-1+i} + \sum_{i=0}^{r} q^0_{n-i,n-2r-1+i} + \sum_{i=0}^{r} q^1_{n-i,n-2r-1+i}.
\end{aligned} \tag{15.4.56}$$

by (15.4.38). Furthermore, for $r \in \overline{0, n-2}$, we get

$$\begin{aligned}
&\sum_{i=0}^{2r+1} q_{n-i,n-2r-1+i} + \sum_{i=0}^{r} q_{n-i,n-2r+i} \\
&= q_{n,n-2r-1} + q_{n-r-1,n-r} + q^1_{n-r,n-r} + \sum_{i=1}^{r}(q_{n-i,n-2r-1+i} \\
&\quad + q_{n-2r-2+i,n+1-i} + q^1_{n-i+1,n-2r-1+i}) \\
&= q_{n,n-2r-2} + q_{n-2r-2,n} + q^0_{n,n-2r-2} + q^1_{n,n-2r-2} + q_{n-r-1,n-r-1} + q^0_{n-r-1,n-r-1} \\
&\quad + q^1_{n-r-1,n-r-1} + \sum_{i=1}^{r}(q_{n-i,n-2r-2+i} + q_{n-2r-2+i,n-i} + q^0_{n-i,n-2r-2+i} + q^1_{n-i,n-2r-2+i}) \\
&= q_{n-r-1,n-r-1} + \sum_{i=0}^{r}(q_{n-i,n-2r-2+i} + q_{n-2r-2+i,n-i}) \\
&\quad + \sum_{i=0}^{r+1} q^0_{n-i,n-2r-2+i} + \sum_{i=0}^{r+1} q^1_{n-i,n-2r-2+i} \\
&= \sum_{i=0}^{2r+2} q_{n-i,n-2r-2+i} + \sum_{i=0}^{r+1} q^0_{n-i,n-2r-2+i} + \sum_{i=0}^{r+1} q^1_{n-i,n-2r-2+i}
\end{aligned} \tag{15.4.57}$$

based on (15.4.37) and (15.4.38).

By (15.4.55)–(15.4.57) and induction, we obtain

$$\begin{aligned} q_{n,n} + q^1_{n,n} &= \sum_{r=0}^{n-1}\sum_{s=0}^{r} q^0_{n-s,n-2r-1+s} + \sum_{r=0}^{n-2}\sum_{s=0}^{r} q^0_{n-s,n-2r-2+s} \\ &= \sum_{1\le r_1\le r_2\le n} q^0_{r_2,r_1} - q^0_{n,n}. \end{aligned} \tag{15.4.58}$$

Therefore,

$$i_n = \sum_{1\le r_1\le r_2\le n} q^0_{s_2,s_1}. \tag{15.4.59}$$

Now

$$\begin{aligned} 2i_n - i_{n-1} &= \sum_{1\le r_1\le r_2\le n} 2q^0_{s_2,s_1} - \sum_{r=1}^{n-1}\left(p^0_{n,r} + \sum_{p=1}^{r} q^0_{r,i} + \sum_{s=r}^{n} q^0_{s,r}\right) \\ &= q^0_{n,n} + \sum_{r=1}^{n} q^0_{n,r} - \sum_{s=1}^{n-1} p^0_{n,s}. \end{aligned} \tag{15.4.60}$$

By (15.4.37) and induction on r, we get

$$q^1_{r,r} = \sum_{s=1}^{r-1}(q^0_{s,s} + q_{s,s} - q_{s,s+1}) \qquad \text{for } r \in \overline{2,n}. \tag{15.4.61}$$

Moreover, (15.4.38) and induction imply

$$q_{n,r} = \sum_{s=1}^{r-1}(q^0_{n,s} + q_{s,n} + q^1_{n,s}) \qquad \text{for } r \in \overline{2,n} \tag{15.4.62}$$

and

$$\begin{aligned} q^1_{r_2,r_1} &= \sum_{s=1}^{r_2-1}(q^0_{r_2-s,r_1-s} + q_{r_2-s,r_1-s} + q_{r_1-s,r_2-s}) \\ &\quad - \sum_{s=1}^{r_1-1}(q_{r_2-s,r_1-s+1} + q_{r_1-s,r_2-s+1}) \end{aligned} \tag{15.4.63}$$

for $2 \le r_1 < r_2 \le n$. Furthermore, (15.4.62) and (15.4.63) show

$$q_{n,r} = \sum_{s=1}^{r-1}(q^0_{n,i} + q_{i,n}) + \sum_{s=2}^{r-1}\sum_{j=1}^{s-1}(q^0_{n-j,s-j} + q_{n-j,s-j} + q_{s-j,n-j})$$

$$-\sum_{s=2}^{r-1}\sum_{j=1}^{s-1}(q_{n-j,s-j+1}+q_{s-j,n-j+1}) \tag{15.4.64}$$

for $r \in \overline{2,n}$. Set

$$\vec{p}^0! = \prod_{1\le r_1<r_2\le n} p^0_{r_2,r_1}!, \qquad \vec{q}^0! = \prod_{1\le r_1\le r_2\le n} q^0_{r_2,r_1}!. \tag{15.4.65}$$

By (15.2.47), (15.4.35), (15.4.53), (15.4.59), (15.4.61), (15.4.63) and (15.4.64), we have

$$\begin{aligned}
&E_C(z_1,z_2,\dots,z_n)\\
&= x^*\sum_{\alpha\in\Gamma_C}\ \sum_{1\le j_1<j_2\le n}\ \sum_{k_1=1}^{n}\sum_{k_2=2}^{n}\sum_{k_3=k_1}^{n}\sum_{k_4=1}^{n-1}\ \sum_{p^0_{j_2,j_1},q^0_{k_3,k_1},q_{k_4,k_2}=0}^{\infty}\\
&\frac{(-1)^{i_1+i_2+\cdots+i_{n-1}}}{i_1!i_2!\cdots i_n!2^{i_n+\sum_{r=1}^n q^0_{r,r}}\langle\lambda_n\rangle_{i_n}}\langle i_{n-1}\rangle_{\sum_{r=1}^n q_{r,n}+\sum_{r=2}^n q^1_{n,r}+q^1_{n,n}}\\
&\times\left[\prod_{2\le l_1\le l_2\le n}\binom{\alpha_{l_2,l_1}}{p^0_{l_2,l_1},\,p_{l_2,l_1}}\binom{\alpha_{n+l_2,l_1}}{q^0_{l_2,l_1},\,q^1_{l_2,l_1},\,q_{l_1,l_2},\,q_{l_2,l_1}}\right]\\
&\times\left[\prod_{m=2}^{n}\binom{\alpha_{m,1}}{p^0_{m,1}}\right]\binom{\alpha_{n+1,1}}{q^0_{1,1}}\left[\prod_{r=2}^{n}\binom{\alpha_{n+r,r}}{q^0_{r,r},\,q^1_{r,r},\,q_{r,r}}\binom{\alpha_{n+r,1}}{q^0_{r,1},\,q_{1,r}}\right]\\
&\times\left[\prod_{s=1}^{n-2}\langle i_s\rangle_{\sum_{r_1=s+2}^n p_{r_1,s+1}+\sum_{r=1}^n q_{r,s+1}+\sum_{r=2}^{s+1} q^1_{s+1,r}+\sum_{r=s+1}^n q^1_{r,s+1}}\right]\\
&\times\left[\prod_{s=1}^{n-1}\langle i_s-i_{s-1}-1/2\rangle_{\sum_{r=s+1}^n p^0_{r,s}+\sum_{r=1}^s q^0_{s,r}+\sum_{r=s}^n q^0_{r,s}}\right]\\
&\times\langle 2i_n-i_{n-1}-1/2\rangle_{2q^0_{n,n}+\sum_{r=1}^{n-1} q^0_{n,r}}(z_1z_2\cdots z_{n-1})^{-1/2}z_n^{\lambda_n}\\
&\times\left[\prod_{1\le r_1<r_2\le n}\left(\frac{z_{r_2}}{z_{r_1}}\right)^{\alpha_{r_2,r_1}}\frac{1}{(z_{r_1}z_{r_2})^{\alpha_{n+r_2,r_1}}}\right]\left[\prod_{r=1}^{n}\frac{1}{z_r^{\alpha_{n+r,r}}}\right]\\
&= x^*\sum_{1\le j_1<j_2\le n}\ \sum_{k_1=1}^{n}\sum_{k_2=2}^{n}\sum_{k_3=k_1}^{n}\sum_{k_4=1}^{n-1}\ \sum_{p^0_{j_2,j_1},q^0_{k_3,k_1},q_{k_4,k_2}=0}^{\infty}\frac{(-1)^{i_1+i_2+\cdots+i_{n-1}}}{\vec{p}^0!2^{i_n+\sum_{r=1}^n q^0_{r,r}}\langle\lambda_n\rangle_{i_n}}\\
&\times\frac{\langle 2i_n-i_{n-1}-1/2\rangle_{2q^0_{n,n}+\sum_{r=1}^{n-1} q^0_{n,r}}}{q^0_{n,n}!q_{n,n}!q^1_{n,n}!}\left[\prod_{r=2}^{n-1}\binom{q^0_{r,r}+q^1_{r,r}+q_{r,r}}{q^1_{r,r},\,q_{r,r}}\right]
\end{aligned}$$

$$\times \left[\prod_{1\leq l_1\leq l_2\leq n} \binom{q^0_{l_2,l_1} + q^1_{l_2,l_1} + q_{l_1,l_2} + q_{l_2,l_1}}{q^1_{l_2,l_1}, q_{l_1,l_2}, q_{l_2,l_1}} \right]$$
$$\times \left[\prod_{s=1}^{n-1} \langle i_s - i_{s-1} - 1/2 \rangle_{\sum_{r=s+1}^n p^0_{r,s} + \sum_{r=1}^s q^0_{s,r} + \sum_{r=s}^n q^0_{r,s}} \right]$$
$$\times \left[\prod_{1\leq r_1<r_2\leq n} \frac{z^2_{r_1} z_{r_2}^{1+\sum_{s=1}^{r_1} p^0_{r_2,s}}}{(z_{r_1} - z_{r_2})^{1+\sum_{s=1}^{r_1} p^0_{r_2,s}} (z_{r_1} z_{r_2} - 1)^{1+q^0_{r_2,r_1}+q^1_{r_2,r_1}+q_{r_1,r_2}+q_{r_2,r_1}}} \right]$$
$$\times \left[\prod_{r=1}^{n} \frac{z^2_r}{(z^2_r - 1)^{1+q^0_{r,r}+q^1_{r,r}+q_{r,r}}} \right] (z_1 z_2 \cdots z_{n-1})^{-1/2} z_n^{\lambda_n}. \tag{15.4.66}$$

Note that

$$\frac{\prod_{r=2}^{n-1} \binom{q^0_{r,r}+q^1_{r,r}+q_{r,r}}{q^1_{r,r}, q_{r,r}}}{q^0_{n,n}! q_{n,n}! q^1_{n,n}!} \left[\prod_{1\leq l_1\leq l_2\leq n} \binom{q^0_{l_2,l_1} + q^1_{l_2,l_1} + q_{l_1,l_2} + q_{l_2,l_1}}{q^1_{l_2,l_1}, q_{l_1,l_2}, q_{l_2,l_1}} \right]$$
$$= \frac{\prod_{r=2}^{n-1} \binom{q^1_{r+1,r+1}+q_{r,r+1}}{q^1_{r,r}, q_{r,r}}}{q^0_{n,n}! q_{n,n}! q^1_{n,n}!} \left[\prod_{1\leq l_1\leq l_2\leq n} \binom{q^1_{l_2+1,l_1+1} + q_{l_1,l_2+1} + q_{l_2,l_1+1}}{q^1_{l_2,l_1}, q_{l_1,l_2}, q_{l_2,l_1}} \right]$$
$$= \frac{\prod_{r=1}^{n-1} \binom{q^1_{r+1,r+1}+q_{r,r+1}}{q_{r,r+1}}}{\vec{q}^0!} \left[\prod_{1\leq l_1\leq l_2\leq n-1} \binom{q^1_{l_2+1,l_1+1} + q_{l_1,l_2+1} + q_{l_2,l_1+1}}{q_{l_1,l_2+1}, q_{l_2,l_1+1}} \right]$$
$$= \frac{\prod_{r=1}^{n-1} \binom{q^0_{r,r}+q^1_{r,r}+q_{r,r}}{q_{r,r+1}}}{\vec{q}^0!} \left[\prod_{1\leq l_1\leq l_2\leq n-1} \binom{q^0_{l_2,l_1} + q^1_{l_2,l_1} + q_{l_1,l_2} + q_{l_2,l_1}}{q_{l_1,l_2+1}, q_{l_2,l_1+1}} \right] \tag{15.4.67}$$

according to (15.4.37) and (15.4.38). Furthermore,

$$q^0_{r,r} + q^1_{r,r} + q_{r,r} = \sum_{s=1}^{r} (q^0_{s,s} + q_{s,s}) - \sum_{s=1}^{r-1} q_{s,s+1} \tag{15.4.68}$$

by (15.4.61) and

$$q^0_{n,r} + q^1_{n,r} + q_{r,n} + q_{n,r}$$
$$= \sum_{s=1}^{r} (q^0_{n,s} + q_{s,n}) + \sum_{s=0}^{r-2} \sum_{j=1}^{r-s-1} (q^0_{n-j,r-s-j} + q_{n-j,r-s-j} + q_{r-s-j,n-j})$$
$$- \sum_{s=0}^{r-2} \sum_{j=1}^{r-s-1} (q_{n-j,r-s-j+1} + q_{r-s-j,n-j+1})$$

$$= \sum_{s=0}^{r-1} \sum_{j=0}^{r-s-1} q^0_{n-j,r-s-j} + \sum_{j=0}^{r-1} q_{r-j,n-j} - \sum_{j=1}^{r-1} q_{r-j,n-j+1} \tag{15.4.69}$$

by (15.4.62) and (15.4.64) for $r \in \overline{1, n-1}$, and

$$q^0_{r_2,r_1} + q^1_{r_2,r_1} + q_{r_1,r_2} + q_{r_2,r_1} = \sum_{s=0}^{r_1-1} (q^0_{r_2-s,r_1-s} + q_{r_2-s,r_1-s} + q_{r_1-s,r_2-s}) - \sum_{s=1}^{r_1-1} (q_{r_2-s,r_1-s+1} + q_{r_1-s,r_2-s+1}) \tag{15.4.70}$$

by (15.4.63) for $1 \leq r_1 < r_2 \leq n-1$.

Let

$$y_{r_2,r_1} = \frac{1}{z_{r_1} z_{r_2} - 1} \quad \text{for } 1 \leq r_1 \leq r_2 \leq n. \tag{15.4.71}$$

Set

$$w_{r_2,r_1} = \frac{y_{r_2,r_1} y_{r_2+1,r_1+1} \cdots y_{n-1,n+r_1-r_2-1}}{y_{r_2+1,r_1} y_{r_2+2,r_1+1} \cdots y_{n,n+r_1-r_2-1}} \quad \text{for } 2 \leq r_1 \leq r_2 \leq n-1, \tag{15.4.72}$$

$$w_{k,n} = y_{n,k} \quad \text{for } k \in \overline{1, n-1} \tag{15.4.73}$$

and

$$w_{r_1,r_2} = \frac{y_{r_2,r_1} y_{r_2+1,r_1+1} \cdots y_{n,n+r_1-r_2}}{y_{r_2,r_1+1} y_{r_2+1,r_1+2} \cdots y_{n-1,n+r_1-r_2}} \quad \text{for } 1 \leq r_1 < r_2 \leq n-1. \tag{15.4.74}$$

Based on (15.4.68)–(15.4.70), we have

$$\prod_{1 \leq r_1 \leq r_2 \leq n} y_{r_2,r_1}^{q^0_{r_2,r_1} + q^1_{r_2,r_1} + q_{r_2,r_1} + q_{r_1,r_2}} = \left[\prod_{r_1=1}^{n-1} \prod_{r_2=2}^{n} w_{r_1,r_2}^{q_{r_1,r_2}} \right] \times \left[\prod_{r=1}^{n} (y_{r,r} y_{r+1,r+1} \cdots y_{n,n})^{q^0_{r,r}} \right] \left[\prod_{r=1}^{n-1} (y_{n,r} y_{n,r+1} \cdots y_{n,n})^{q^0_{n,r}} \right] \prod_{1 \leq r_1 < r_2 \leq n-1} (y_{r_2,r_1} y_{r_2+1,r_1+1} \cdots y_{n-1,n+r_1-r_2-1} y_{n,n+r_1-r_1} y_{n,n+r_1-r_1+1} \cdots y_{n,n})^{q^0_{r_2,r_1}}. \tag{15.4.75}$$

Next we have

$$\sum_{q_{n-1,n}=0}^{\infty} \left(!! \begin{matrix} q^0_{n-1,n-1} + q^1_{n-1,n-1} + q_{n-1,n-1} \\ q_{n-1,n} \end{matrix} !! \right) w_{n-1,n}^{q_{n-1,n}} = (1 + w_{n-1,n})^{q^0_{n-1,n-1} + q^1_{n-1,n-1} + q_{n-1,n-1}}$$

$$= (1 + w_{n-1,n})^{\sum_{s=1}^{n-1}(q_{s,s}^0 + q_{s,s}) - \sum_{s=1}^{n-2} q_{s,s+1}} \tag{15.4.76}$$

by (15.4.68). Set

$$w_{r,r+1}^{(2n-1)} = \frac{w_{r,r+1}}{1 + w_{n-1,n}}, \qquad w_{s,s}^{(2n-1)} = (1 + w_{n-1,n})w_{s,s} \tag{15.4.77}$$

for $r \in \overline{1, n-2}$, $s \in \overline{1, n-1}$ and

$$w_{r_1,r_2}^{(2n-1)} = w_{r_1,r_2} \qquad \text{for } r_1 \in \overline{1, n-1},\ r_1, r_1 + 1 \neq r_2 \in \overline{2, n}. \tag{15.4.78}$$

Suppose that we have defined $\{w_{r_1,r_2}^{(2k+1)} \mid r_1 + r_2 < 2k + 1\}$. Set

$$\ell = \max\{1, 2k - n\}. \tag{15.4.79}$$

Note

$$\begin{aligned}
&\sum_{q_{\ell,2k-\ell}, q_{\ell+1,2k-\ell-1}, \cdots, q_{2k-\ell-1,\ell+1}=0}^{\infty} \prod_{r=\ell}^{k-1} \binom{q_{2k-1-r,r}^0 + q_{2k-1-r,r}^1 + q_{2k-1-r,r} + q_{r,2k-1-r}}{q_{r,2k-r},\, q_{2k-r-1,r+1}} \\
&\times w_{r,2k-r}^{(2k+1)q_{r,2k-r}} w_{2k-r-1,r+1}^{(2k+1)q_{2k-r-1,r+1}} \\
&= \prod_{r=\ell}^{k-1} (1 + w_{r,2k-r}^{(2k+1)} + w_{2k-r-1,r+1}^{(2k+1)})^{q_{2k-1-r,r}^0 + q_{2k-1-r,r}^1 + q_{2k-1-r,r} + q_{r,2k-1-r}} \\
&= \prod_{r=\ell}^{k-1} (1 + w_{r,2k-r}^{(2k+1)} + w_{2k-r-1,r+1}^{(2k+1)})^{\sum_{s=1}^{r}(q_{2k-1-2r+s,s}^0 + q_{2k-1-2r+s,s} + q_{s,2k-1-2r+s})} \\
&\times (1 + w_{r,2k-r}^{(2k+1)} + w_{2k-r-1,r+1}^{(2k+1)})^{-\sum_{s=1}^{r}(q_{2k-1-2r+s,s+1} + q_{s,2k-2r+s})}
\end{aligned} \tag{15.4.80}$$

by (15.4.70). Set

$$w_{2k-1-2r+s,s}^{(2k)} = w_{2k-1-2r+s,s}^{(2k+1)} (1 + w_{r,2k-r}^{(2k+1)} + w_{2k-r-1,r+1}^{(2k+1)}), \tag{15.4.81}$$

$$w_{s,2k-1-2r+s}^{(2k)} = w_{s,2k-1-2r+s}^{(2k+1)} (1 + w_{r,2k-r}^{(2k+1)} + w_{2k-r-1,r+1}^{(2k+1)}) \tag{15.4.82}$$

for $1 \leq s \leq r \leq k - 1$;

$$w_{2k-1-2r+s,s+1}^{(2k)} = \frac{w_{2k-1-2r+s,s+1}^{(2k+1)}}{1 + w_{r,2k-r}^{(2k+1)} + w_{2k-r-1,r+1}^{(2k+1)}}, \tag{15.4.83}$$

$$w_{s,2k-2r+s}^{(2k)} = \frac{w_{s,2k-2r+s}^{(2k+1)}}{1 + w_{r,2k-r}^{(2k+1)} + w_{2k-r-1,r+1}^{(2k+1)}} \tag{15.4.84}$$

for $1 \le s \le r \le k-1$ and

$$w_{r_1,r_2}^{(2k)} = w_{r_1,r_2}^{(2k+1)} \text{ for the other pairs } (r_1,r_2) \text{ such that } r_1+r_2 \le 2k. \tag{15.4.85}$$

Let

$$\iota = \max\{1, 2k-1-n\}. \tag{15.4.86}$$

Moreover,

$$\begin{aligned}
&\sum_{q_{\iota,2k-\iota-1},q_{\iota+1,2k-\iota-2},\ldots,q_{2k-\iota-2,\iota+1}=0}^{\infty} \left[\prod_{r=\iota}^{k-2} \binom{q^0_{2k-2-r,r} + q^1_{2k-2-r,r} + q_{2k-2-r,r} + q_{r,2k-2-r}}{q_{r,2k-r-1},\, q_{2k-r-2,r+1}}\right.\\
&\left.\times\, w_{r,2k-r-1}^{(2k)q_{r,2k-r-1}} w_{2k-r-2,r+1}^{(2k)q_{2k-r-2,r+1}}\right] \binom{q^0_{k-1,k-1} + q^1_{k-1,k-1} + q_{k-1,k-1}}{q_{k-1,k}} w_{k-1,k}^{(2k)q_{k-1,k}}\\
&= \left[\prod_{r=\iota}^{k-2} (1 + w_{r,2k-r-1}^{(2k)} + w_{2k-r-2,r+1}^{(2k)})^{q^0_{2k-2-r,r} + q^1_{2k-2-r,r} + q_{2k-2-r,r} + q_{r,2k-2-r}}\right]\\
&\quad\times (1 + w_{k-1,k}^{(2k)})^{q^0_{k-1,k-1} + q^1_{k-1,k-1} + q_{k-1,k-1}}\\
&= \left[\prod_{r=\iota}^{k-2} (1 + w_{r,2k-r-1}^{(2k)} + w_{2k-r-2,r+1}^{(2k)})^{\sum_{s=1}^{r} (q^0_{2k-2-2r+s,s} + q_{2k-2-2r+s,s} + q_{s,2k-2-2r+s})}\right.\\
&\quad\left.\times (1 + w_{r,2k-r-1}^{(2k)} + w_{2k-r-2,r+1}^{(2k)})^{-\sum_{s=1}^{r-1} (q_{2k-2-2r+s,s+1} + q_{s,2k-1-2r+s})}\right]\\
&\quad\times (1 + w_{k-1,k}^{(2k)})^{\sum_{r=1}^{k-1} (q^0_{r,r} + q_{r,r}) - \sum_{r=1}^{k-1} q_{r,r+1}}
\end{aligned} \tag{15.4.87}$$

by (15.4.68) and (15.4.70). Set

$$w_{2k-2-2r+s,s}^{(2k-1)} = w_{2k-2-2r+s,s}^{(2k)} (1 + w_{r,2k-r-1}^{(2k)} + w_{2k-r-2,r+1}^{(2k)}), \tag{15.4.88}$$

$$w_{s,2k-2-2r+s}^{(2k-1)} = w_{s,2k-2-2r+s}^{(2k)} (1 + w_{r,2k-r-1}^{(2k)} + w_{2k-r-2,r+1}^{(2k)}) \tag{15.4.89}$$

for $1 \le s \le r \le k-2$;

$$w_{2k-2-2r+s,s+1}^{(2k-1)} = \frac{w_{2k-2-2r+s,s+1}^{(2k)}}{1 + w_{r,2k-r-1}^{(2k)} + w_{2k-r-2,r+1}^{(2k)}}, \tag{15.4.90}$$

$$w_{s,2k-1-2r+s}^{(2k-1)} = \frac{w_{s,2k-1-2r+s}^{(2k)}}{1 + w_{r,2k-r-1}^{(2k)} + w_{2k-r-2,r+1}^{(2k)}} \tag{15.4.91}$$

for $1 \le s \le r \le k-2$;

$$w_{r,r}^{(2k-1)} = w_{r,r}^{(2k)} (1 + w_{k-1,k}^{(2k)}), \qquad w_{s,s+1}^{(2k-1)} = \frac{w_{s,s+1}^{(2k)}}{1 + w_{k-1,k}^{(2k)}} \tag{15.4.92}$$

for $r \in \overline{1, k-1},\ s \in \overline{1, k-2}$; and

$$w_{r_1,r_2}^{(2k-1)} = w_{r_1,r_2}^{(2k)} \quad \text{for the other pairs } (r_1, r_2) \text{ such that } r_1 + r_2 \leq 2k-1. \tag{15.4.93}$$

By induction, we have defined

$$\{w_{r_1,r_2}^{(k)} \mid k \in \overline{1, 2n-1};\ r_1 \in \overline{1, n-1},\ r_2 \in \overline{1, n},\ r_1 + r_2 \leq k\}. \tag{15.4.94}$$

For convenience, we denote

$$w_{r_1,r_2}^{(2n)} = w_{r_1,r_2} \qquad \text{for } r_1 \in \overline{1, n-1},\ r_2 \in \overline{2, n}. \tag{15.4.95}$$

Based on (15.4.76), (15.4.80) and (15.4.87), we have

$$\begin{aligned}
&\sum_{r_1=1}^{n-1}\sum_{r_2=2}^{n}\sum_{q_{r_1,r_2}=0}^{\infty}\Bigg[\prod_{1\le l_1\le l_2\le n-1}\binom{!!\, q_{l_2,l_1}^0 + q_{l_2,l_1}^1 + q_{l_1,l_2} + q_{l_2,l_1}\, !!}{q_{l_1,l_2+1},\, q_{l_2,l_1+1}} \\
&\times w_{l_1,l_2+1}^{q_{l_1,l_2+1}} w_{l_2,l_1+1}^{q_{l_2,l_1+1}}\Bigg]\left[\prod_{r=1}^{n-1}\binom{!!\, q_{r,r}^0 + q_{r,r}^1 + q_{r,r}\, !!}{q_{r,r+1}} w_{r,r+1}^{q_{r,r+1}}\right] \\
=& \left[\prod_{k=2}^{n}(1 + w_{k-1,k}^{(2k)})^{\sum_{r=1}^{k-1} q_{r,r}^0}\right]\left[\prod_{k=2}^{n-1}\prod_{r=\ell}^{k-1}(1 + w_{r,2k-r}^{(2k+1)} + w_{2k-r-1,r+1}^{(2k+1)})^{\sum_{s=1}^{r} q_{2k-1-2r+s,s}^0}\right] \\
&\times\left[\prod_{k=3}^{n-1}\prod_{r=\iota}^{k-2}(1 + w_{r,2k-r-1}^{(2k)} + w_{2k-r-2,r+1}^{(2k)})^{\sum_{s=1}^{r} q_{2k-2-2r+s,s}^0}\right] \\
=& \Bigg[\prod_{1\le r_1<r_2\le n-1;\, r_2-r_1 \text{ is odd}}\Bigg(\prod_{k=(r_2-r_1+3)/2}^{n-1}(1 + w_{(2k+r_1-r_2-1)/2,(2k-r_1+r_2+1)/2}^{(2k+1)} \\
&+ w_{(2k-r_1+r_2-1)/2,(2k+r_1-r_2+1)/2}^{(2k+1)})^{q_{r_2,r_1}^0}\Bigg)\Bigg]\Bigg[\prod_{1\le r_1<r_2\le n-1;\, r_2-r_1 \text{ is even}} \\
&\Bigg(\prod_{k=(r_2-r_1+2)/2}^{n-1}(1 + w_{(2k+r_1-r_2)/2,(2k-r_1+r_2+2)/2}^{(2k+2)} + w_{(2k-r_1+r_2)/2,(2k+r_1-r_2+2)/2}^{(2k+2)})^{q_{r_2,r_1}^0}\Bigg)\Bigg] \\
&\times\left[\prod_{r=1}^{n-1}\left(\prod_{k=r}^{n-1}(1 + w_{k,k+1}^{(2k)})\right)^{q_{r,r}^0}\right].
\end{aligned} \tag{15.4.96}$$

According to (15.4.66), (15.4.75) and (15.4.96), we define

$$\xi_{n,i}^C = (-1)^{n+i+1}\frac{y_{n,i}y_{n,i+1}\cdots y_{n,n}}{2^{1+\delta_{n,i}}} \qquad \text{for } 1 \leq i \leq n, \tag{15.4.97}$$

$$\xi_{i,i}^C = -\frac{y_{i,i}\cdots y_{n,n}\prod_{k=i}^{n-1}(1 + w_{k,k+1}^{(2k)})}{4} \tag{15.4.98}$$

for $1 \leq i \leq n-1$,

$$\xi^C_{r_2,r_1} = \frac{(-1)^{r_1+r_2+1}}{2} y_{r_2,r_1} y_{r_2+1,r_1+1} \cdots y_{n-1,n+r_1-r_2-1} y_{n,n+r_1-r_2} \times y_{n,n+r_1-r_2+1} \cdots y_{n,n} \prod_{k=(r_2-r_1+3)/2}^{n-1} (1 + w^{(2k+1)}_{(2k+r_1-r_2-1)/2,(2k-r_1+r_2+1)/2} + w^{(2k+1)}_{(2k-r_1+r_2-1)/2,(2k+r_1-r_2+1)/2}) \quad (15.4.99)$$

for $1 \leq r_1 < r_2 \leq n-1$ such that $r_2 - r_1$ is odd, and

$$\xi^C_{r_2,r_1} = \frac{(-1)^{r_1+r_2+1}}{2} y_{r_2,r_1} y_{r_2+1,r_1+1} \cdots y_{n-1,n+r_1-r_2-1} y_{n,n+r_1-r_2} \times y_{n,n+r_1-r_2+1} \cdots y_{n,n} \prod_{k=(r_2-r_1+2)/2}^{n-1} (1 + w^{(2k+2)}_{(2k+r_1-r_2)/2,(2k-r_1+r_2+2)/2} + w^{(2k+2)}_{(2k-r_1+r_2)/2,(2k+r_1-r_2+2)/2}) \quad (15.4.100)$$

for $1 \leq r_1 < r_2 \leq n-1$ such that $r_2 - r_1$ is even.

For $\alpha \in \Gamma_C$, we set

$$\alpha^c_i = \sum_{r=1}^{i} \alpha_{n+r,i} + \sum_{s=i}^{n} \alpha_{n+s,i} \quad \text{for } i \in \overline{1,n} \quad (15.4.101)$$

and

$$\alpha^c = \sum_{1 \leq r_1 \leq r_2 \leq n} \alpha_{n+r_2,r_1}. \quad (15.4.102)$$

By (15.3.57), (15.3.62), (15.4.66), (15.4.67), (15.4.71)–(15.4.75) and (15.4.96)–(15.4.102), we have

$$E_C(z_1, z_2, \ldots, z_n) = \frac{x^* z_n^{n+\lambda_n+1} \prod_{i=1}^{n-1} z_i^{2n-i+1/2}}{[\prod_{1 \leq r_1 < r_2 \leq n} (z_{r_1} - z_{r_2})] [\prod_{1 \leq s_1 \leq s_2 \leq n} (z_{s_1} z_{s_2} - 1)]} \sum_{\gamma \in \Gamma_C} \frac{1}{\gamma! \langle \lambda_n \rangle_{\gamma^c}} \times \prod_{r=1}^{n} \langle \gamma^*_r - \gamma_{r*} + \gamma^c_r - 1/2 \rangle_{\gamma^*_r + \gamma^c_r} \xi^\gamma, \quad (15.4.103)$$

where

$$\gamma! = \left[\prod_{1 \leq r_1 < r_2 \leq n} \gamma_{r_2,r_1}! \right] \left[\prod_{1 \leq s_1 \leq s_2 \leq n} \gamma_{n+s_2,s_1}! \right] \quad (15.4.104)$$

and

$$\xi^{\gamma} = \left[\prod_{1\leq r_1<r_2\leq n} (\xi^A_{r_2,r_1})^{\gamma_{r_2,r_1}}\right]\left[\prod_{1\leq s_1\leq s_2\leq n} (\xi^C_{n+s_2,s_1})^{\gamma_{n+s_2,s_1}}\right]. \tag{15.4.105}$$

Recall the notations in (15.3.18) and (15.3.57). We define our *path hypergeometric function of type C* by

$$\mathscr{X}_C(\tau_1,\ldots,\tau_n;\vartheta)\{z_{r_2,r_1}, z_{n+s_2,s_1}\} = \sum_{\alpha\in\Gamma_C} \frac{\prod_{r=1}^n(\tau_r-\alpha_{r*})_{\alpha_r^*+\alpha_r^c}}{\alpha!(\vartheta)_{\alpha^c}} z^{\alpha}. \tag{15.4.106}$$

By (15.1.18), (15.1.27), (15.3.20), (15.4.102) and (15.4.106), we obtain the following main theorem in this section:

Theorem 15.4.1 *The Etingof trace function in (15.2.47):*

$$E_C(z_1,z_2,\ldots,z_n) = \frac{x^* z_n^{n+\lambda_n+1}\prod_{i=1}^{n-1} z_i^{2n-i+1/2}}{\left[\prod_{1\leq r_1<r_2\leq n}(z_{r_1}-z_{r_2})\right]\left[\prod_{1\leq s_1\leq s_2\leq n}(z_{s_1}z_{s_2}-1)\right]} \times \mathscr{X}_C(1/2,\ldots,1/2;-\lambda_n)\{\xi^A_{r_2,r_1},\xi^C_{n+s_2,s_1}\}. \tag{15.4.107}$$

Moreover, the function in (15.2.49):

$$\Psi_C = z_n^{\lambda_n+2-n}\left[\prod_{i=1}^{n-1} z_i^{-i-1/2}\right]\mathscr{X}_C(1/2,\ldots,1/2;-\lambda_n)\{\xi^A_{r_2,r_1},\xi^C_{n+s_2,s_1}\}. \tag{15.4.108}$$

15.5 Properties of Path Hypergeometric Functions

In this section, we find the differential properties, integral representations and partial differential equations for the hypergeometric functions of type A in (15.3.60), and the differential properties and partial differential equations for the hypergeometric functions of type C in (15.4.106). Moreover, we define our hypergeometric functions of type B and D analogously as those of type C.

Recall the differentiation property of the classical Gauss hypergeometric function:

$$\frac{d}{dz}\,{}_2F_1(\alpha,\beta;\gamma;z) = \frac{\alpha\beta}{\gamma}\,{}_2F_1(\alpha+1,\beta+1;\gamma+1;z) \tag{15.5.1}$$

(e.g., cf. [X21]). For two positive integers k_1 and k_2 such that $k_1 < k_2$, a *path* from k_1 to k_2 is a sequence $(m_0, m_1, \ldots, m_r)$ of positive integers such that

$$k_1 = m_0 < m_1 < m_2 < \cdots < m_{r-1} < m_r = k_2. \tag{15.5.2}$$

One can imagine a path from k_1 to k_2 is a way of a super man going from k_1th floor to k_2th floor through a stairway. Let

$$\mathscr{P}_{k_1}^{k_2} = \text{the set of all paths from } k_1 \text{ to } k_2. \tag{15.5.3}$$

The *path polynomial* from k_1 to k_2 is defined as

$$P_{[k_1,k_2]} = \sum_{(m_0,m_1,\ldots,m_r)\in\mathscr{P}_{k_1}^{k_2}} (-1)^r z_{m_1,m_0} z_{m_2,m_1} \cdots z_{m_{r-1},m_{r-2}} z_{m_r,m_{r-1}}. \tag{15.5.4}$$

Moreover, we set

$$P_{[k,k]} = 1 \quad \text{for } 0 < k \in \mathbb{N}. \tag{15.5.5}$$

For convenience, we simply denote

$$\mathscr{X}_A = \mathscr{X}_A(\tau_1, \ldots, \tau_n; \vartheta)\{z_{j,k}\}, \tag{15.5.6}$$

$$\mathscr{X}_A[i, j] = \mathscr{X}_A(\tau_1, \ldots, \tau_i + 1, \ldots, \tau_j - 1, \ldots, \tau_n; \vartheta)\{z_{r_2,r_1}\} \tag{15.5.7}$$

obtained from $\mathscr{X}_A$ by changing τ_i to $\tau_i + 1$ and τ_j to $\tau_j - 1$ for $1 \leq i < j \leq n - 1$ and

$$\mathscr{X}_A[k, n] = \mathscr{X}_A(\tau_1, \ldots, \tau_k + 1, \ldots, \tau_n + 1; \vartheta + 1)\{z_{r_2,r_1}\} \tag{15.5.8}$$

obtained from $\mathscr{X}_A$ by changing τ_i to $\tau_i + 1$, τ_n to $\tau_n + 1$ and ϑ to $\vartheta + 1$ for $k \in \overline{1, n-1}$. We have the following natural generalization of (15.5.1):

Theorem 15.5.1 *For* $1 \leq r_1 < r_2 \leq n - 1$ *and* $r \in \overline{1, n-1}$, *we have the differential property of* $\mathscr{X}_A$:

$$\partial_{z_{r_2,r_1}}(\mathscr{X}_A) = \sum_{s=1}^{r_1} \tau_s P_{[s,r_1]} \mathscr{X}_A[s, r_2], \tag{15.5.9}$$

$$\partial_{z_{n,r}}(\mathscr{X}_A) = \frac{\tau_n}{\vartheta} \sum_{s=1}^{r} \tau_s P_{[s,r]} \mathscr{X}_A[s, n]. \tag{15.5.10}$$

Proof Note that

$$\begin{aligned}
&\partial_{z_{r_2,r_1}}(\mathscr{X}_A) \\
&= \sum_{\beta\in\Gamma_A} \frac{\left[\prod_{s=1}^{n-1}(\tau_s - \beta_{s*})_{\beta_s^*}\right](\tau_n)_{\beta_{n*}}}{\beta!(\vartheta)_{\beta_{n*}}} \beta_{r_2,r_1} z^{\beta-\varepsilon_{r_2,r_1}} \\
&= \sum_{\beta\in\Gamma_A} \frac{(\tau_{r_1} - \beta_{r_1*})(\tau_{r_1} + 1 - \beta_{r_1*})_{\beta_{r_1}^*}(\tau_{r_2} - 1 - \beta_{r_2*})_{\beta_{r_2}^*}}{\beta!(\vartheta)_{\beta_{n*}}}
\end{aligned}$$

$$\times \left[\prod_{s \neq r_1, r_2} (\tau_s - \beta_{s*})_{\beta_s^*} \right] (\tau_n)_{\beta_{n*}} z^\beta$$

$$= \left(\tau_{r_1} - \sum_{s=1}^{r_1-1} z_{r_1,s} \partial_{z_{r_1,s}} \right) (\mathscr{X}_A[r_1, r_2]). \qquad (15.5.11)$$

In particular,

$$\partial_{z_{r,1}}(\mathscr{X}_A) = \tau_1 \mathscr{X}_A[1, r] \quad \text{for } r \in \overline{1, n-1}. \qquad (15.5.12)$$

By (15.5.7), (15.5.11), (15.5.12) and induction on r_1, we get (15.5.9). Moreover,

$$\partial_{z_{n,r}}(\mathscr{X}_A)$$

$$= \sum_{\beta \in \Gamma_A} \frac{\left[\prod_{s=1}^{n-1} (\tau_s - \beta_{s*})_{\beta_s^*} \right] (\tau_n)_{\beta_{n*}}}{\beta! (\vartheta)_{\beta_{n*}}} \beta_{n,r} z^{\beta - \varepsilon_{n,r}}$$

$$= \sum_{\beta \in \Gamma_A} \frac{\left[(\tau_r - \beta_{r*})(\tau_r + 1 - \beta_{r*})_{\beta_r^*} \prod_{s \neq r} (\tau_s - \beta_{s*})_{\beta_s^*} \right] \tau_n (\tau_n + 1)_{\beta_{n*}}}{\beta! \vartheta (\vartheta + 1)_{\beta_{n*}}} z^\beta$$

$$= \frac{\tau_n}{\vartheta} \left(\tau_r - \sum_{s=1}^{r} z_{r,s} \partial_{z_{r,s}} \right) (\mathscr{X}_A[r, n]), \qquad (15.5.13)$$

which leads to (15.5.10) by (15.5.7), (15.5.8) and (15.5.9). □

For any $z \in \mathbb{C} \setminus (-\mathbb{N})$, the gamma function

$$\Gamma(z) = \left[z e^{cz} \prod_{m=1}^{\infty} \left\{ \left(1 + \frac{z}{m} \right) e^{-z/m} \right\} \right]^{-1}, \qquad (15.5.14)$$

where c is Euler's constant given by

$$c = \lim_{m \to \infty} \left(\sum_{k=1}^{m} \frac{1}{k} - \ln m \right). \qquad (15.5.15)$$

When $\operatorname{Re} z > 0$, we have

$$\Gamma(z) = \int_0^\infty t^{z-1} e^{-t} dt. \qquad (15.5.16)$$

Recall the integral representation of the classical Gauss hypergeometric function:

$${}_2F_1(\alpha, \beta; \gamma; z) = \frac{\Gamma(\gamma)}{\Gamma(\beta)\Gamma(\gamma - \beta)} \int_0^1 t^{\beta-1} (1-t)^{\gamma-\beta-1} (1-zt)^{-\alpha} dt \qquad (15.5.17)$$

Now we have the following analogous integral representation for our path hypergeometric function of type A:

Theorem 15.5.2 *Suppose $Re\, \tau_n > 0$ and $Re\, (\vartheta - \tau_n) > 0$. We have the integral representation of $\mathscr{X}_A$:*

$$\mathscr{X}_A = \frac{\Gamma(\vartheta)}{\Gamma(\vartheta-\tau_n)\Gamma(\tau_n)}\int_0^1 \left[\prod_{r=1}^{n-1}\left(\sum_{s=r}^{n-1} P_{[r,s]} + tP_{[r,n]}\right)^{-\tau_r}\right] t^{\tau_n-1}(1-t)^{\vartheta-\tau_n-1}dt. \tag{15.5.18}$$

Proof For $\kappa_1, \kappa_2 \in \mathbb{C}$ with Re κ_1, Re $\kappa_1 > 0$, we have the following Euler integral

$$\int_0^1 t^{\kappa_1-1}(1-t)^{\kappa_2-1}dt = \frac{\Gamma(\kappa_1)\Gamma(\kappa_2)}{\Gamma(\kappa_1+\kappa_2)}. \tag{15.5.19}$$

Moreover,

$$(\kappa)_m = \frac{\Gamma(\kappa+m)}{\Gamma(\kappa)} \quad \text{for } m \in \mathbb{N},\ \kappa \in \mathbb{C}. \tag{15.5.20}$$

Recall

$$\mathscr{X}_A = \sum_{\beta\in\Gamma_A} \frac{\left[\prod_{s=1}^{n-1}(\tau_s - \beta_{s*})_{\beta_s^*}\right](\tau_n)_{\beta_{n*}}}{\beta!(\vartheta)_{\beta_{n*}}} z^\beta. \tag{15.5.21}$$

We have

$$\frac{(\tau_n)_{\beta_{n*}}}{(\vartheta)_{\beta_{n*}}} = \frac{\Gamma(\vartheta)}{\Gamma(\vartheta-\tau_n)\Gamma(\tau_n)} \frac{\Gamma(\tau_n+\beta_{n*})\Gamma(\vartheta-\tau_n)}{\Gamma(\vartheta+\beta_{n*})}$$
$$= \frac{\Gamma(\vartheta)}{\Gamma(\vartheta-\tau_n)\Gamma(\tau_n)} \int_0^1 t^{\tau_n+\beta_{n*}-1}(1-t)^{\vartheta-\tau_n-1}dt \tag{15.5.22}$$

by (15.5.19) and (15.5.20).

Denote

$$\Gamma_{(s)} = \sum_{1\le r_1<r_2\le s} \mathbb{N}\varepsilon_{r_2,r_1}, \quad \beta_r^s = \sum_{p=r+1}^{s} \beta_{p,r} \quad \text{for } \beta \in \Gamma_A,\ 1 \le r < s \le n-1. \tag{15.5.23}$$

Note that

$$\sum_{\beta\in\Gamma_A} \frac{\prod_{s=1}^{n-1}(\tau_s-\beta_{s*})_{\beta_s^*}}{\beta!} z^\beta$$
$$= \sum_{\beta\in\Gamma_{(n-1)}} \frac{\prod_{s=1}^{n-2}(\tau_s-\beta_{s*})_{\beta_s^{n-1}}}{\beta!} z^\beta \sum_{\beta_{n,1},\dots,\beta_{n,n-1}=0}^{n} \frac{(\tau_{n-1}-\beta_{(n-1)*})_{\beta_{n,n-1}} z_{n,n-1}^{\beta_{n,n-1}}}{\beta_{n,n-1}!}$$

$$\times \prod_{r=1}^{n-2} \frac{(\tau_r - \beta_{r*} + \beta_r^{n-1})_{\beta_{n,r}} z_{n,r}^{\beta_{n,r}}}{\beta_{n,r}!}$$

$$= \sum_{\beta \in \Gamma_{(n-1)}} \frac{\prod_{s=1}^{n-2} (\tau_s - \beta_{s*})_{\beta_s^{n-1}}}{\beta!} z^\beta (1 - z_{n,n-1})^{\beta_{(n-1)*} - \tau_{n-1}}$$
$$\times \prod_{r=1}^{n-2} (1 - z_{n,r})^{\beta_{r*} - \beta_r^{n-1} - \tau_r}$$
$$= \left[\prod_{s=1}^{n-1} (1 - z_{n,s})^{-\tau_s} \right] \sum_{\beta \in \Gamma_{(n-1)}} \frac{\prod_{s=1}^{n-2} (\tau_s - \beta_{s*})_{\beta_s^{n-1}}}{\beta!}$$
$$\times \prod_{1 \le r_1 < r_2 \le n-1} \left[\frac{(1 - z_{n,r_2}) z_{r_2,r_1}}{1 - z_{n,r_1}} \right]^{\beta_{r_2,r_1}}. \tag{15.5.24}$$

Observe that

$$1 - \frac{(1 - z_{n,n-1}) z_{n-1,r}}{1 - z_{n,r}} = \frac{1 - z_{n-1,r} - z_{n,r} + z_{n,n-1} z_{n-1,r}}{1 - z_{n,r}} \tag{15.5.25}$$

for $r \in \overline{1, n-1}$. By (15.5.24) and (15.5.25), we have

$$\sum_{\beta \in \Gamma_A} \frac{\prod_{s=1}^{n-1} (\tau_s - \beta_{s*})_{\beta_s^*}}{\beta!} z^\beta = (1 - z_{n,n-1})^{-\tau_{n-1}} (1 - z_{n-1,n-2} + P_{[n-2,n]})^{-\tau_{n-2}}$$
$$\times \left[\prod_{s=1}^{n-2} (1 - z_{n-1,s} - z_{n,s} + z_{n,n-1} z_{n-1,s})^{-\tau_s} \right] \sum_{\beta \in \Gamma_{(n-2)}} \frac{\prod_{s=1}^{n-3} (\tau_s - \beta_{s*})_{\beta_s^{n-2}}}{\beta!}$$
$$\times \prod_{1 \le r_1 < r_2 \le n-2} \left[\frac{(1 - z_{n-1,r_2} - z_{n,r_2} + z_{n,n-1} z_{n-1,r_2}) z_{r_2,r_1}}{1 - z_{n-1,r_1} - z_{n,r_1} + z_{n,n-1} z_{n-1,r_1}} \right]^{\beta_{r_2,r_1}}. \tag{15.5.26}$$

Now

$$1 - \frac{(1 - z_{n-1,n-2} - z_{n,r-2} + z_{n,n-1} z_{n-1,n-2}) z_{n-2,r}}{1 - z_{n-1,r} - z_{n,r} + z_{n,n-1} z_{n-1,r}}$$
$$= \frac{1}{1 - z_{n-1,r} - z_{n,r} + z_{n,n-1} z_{n-1,r}} [1 - z_{n-2,r} - z_{n-1,r} + z_{n-1,n-2} z_{n-2,r}$$
$$- z_{n,r} + z_{n,n-1} z_{n-1,r} + z_{n,r-2} z_{n-2,r} - z_{n,n-1} z_{n-1,n-2} z_{n-2,r}]. \tag{15.5.27}$$

By induction, we can prove that

$$\sum_{\beta\in\Gamma_A} \frac{\prod_{s=1}^{n-1}(\tau_s-\beta_{s*})_{\beta_s^*}}{\beta!} z^\beta = \prod_{r=1}^{n-1}\left(\sum_{s=r}^{n} P_{[r,s]}\right)^{-\tau_r}. \tag{15.5.28}$$

Thus

$$\sum_{\beta\in\Gamma_A} \frac{\prod_{s=1}^{n-1}(\tau_s-\beta_{s*})_{\beta_s^*}}{\beta!} z^\beta t^{\beta_{n*}} = \prod_{r=1}^{n-1}\left(\sum_{s=r}^{n-1} P_{[r,s]} + t P_{[r,n]}\right)^{-\tau_r}. \tag{15.5.29}$$

Hence we obtain (15.5.18) by (15.5.22) and (15.5.29). □

Remark 15.5.3 According to [M],

$$\begin{pmatrix} 1 & 0 & 0 & \cdots & 0 \\ z_{2,1} & 1 & 0 & \cdots & 0 \\ z_{3,1} & z_{3,2} & 1 & \ddots & \vdots \\ \vdots & \vdots & \ddots & \ddots & 0 \\ z_{n,1} & z_{n,2} & \cdots & z_{n,n-1} & 1 \end{pmatrix}^{-1} = \begin{pmatrix} 1 & 0 & 0 & \cdots & 0 \\ P_{[1,2]} & 1 & 0 & \cdots & 0 \\ P_{[1,3]} & P_{[2,3]} & 1 & \ddots & \vdots \\ \vdots & \vdots & \ddots & \ddots & 0 \\ P_{[1,n]} & P_{[2,n]} & \cdots & P_{[n-1,n]} & 1 \end{pmatrix}. \tag{15.5.30}$$

Recall the classical Gauss hypergeometric equation

$$z(1-z)y'' + [\gamma-(\alpha+\beta+1)z]y' - \alpha\beta y = 0. \tag{15.5.31}$$

Note

$$D = z\frac{d}{dz} \Longrightarrow D^2 = z^2\frac{d^2}{dz^2} + z\frac{d}{dz}. \tag{15.5.32}$$

Then (15.5.31) can be rewritten as

$$(\gamma + D)\frac{d}{dz}(y) = (\alpha + D)(\beta + D)(y). \tag{15.5.33}$$

Denote

$$\mathscr{D}_{i*}^A = \sum_{r=1}^{i-1} z_{i,r}\partial_{z_{i,r}}, \qquad \mathscr{D}_i^{A*} = \sum_{s=i+1}^{n} z_{s,i}\partial_{z_{s,i}} \qquad \text{for } i \in \overline{1,n}. \tag{15.5.34}$$

Then we have the following analogue of (15.5.33):

Theorem 15.5.4 *We have the hypergeometric equations of type A:*

$$(\tau_{r_2} - 1 - \mathscr{D}_{r_2*}^A + \mathscr{D}_{r_2}^{A*})\partial_{z_{r_2,r_1}}(\mathscr{X}_A) = (\tau_{r_2} - 1 - \mathscr{D}_{r_2*}^A)(\tau_{r_1} - \mathscr{D}_{r_1*}^A + \mathscr{D}_{r_1}^{A*})(\mathscr{X}_A) \tag{15.5.35}$$

for $1 \leq r_1 < r_2 \leq n-1$ *and*

$$(\vartheta + \mathscr{D}_{n*}^A)\partial_{z_{n,r}}(\mathscr{X}_A) = (\tau_n + \mathscr{D}_{n*}^A)(\tau_r - \mathscr{D}_{r*}^A + \mathscr{D}_r^{A*})(\mathscr{X}_A) \tag{15.5.36}$$

for $r \in \overline{1, n-1}$.

Proof Let

$$a_\beta = \frac{\left[\prod_{s=1}^{n-1}(\tau_s - \beta_{s*})_{\beta_s^*}\right](\tau_n)_{\beta_{n*}}}{\beta!(\vartheta)_{\beta_{n*}}} \quad \text{for } \beta \in \Gamma_A. \tag{15.5.37}$$

Then

$$\frac{a_{\beta+\varepsilon_{r_2,r_1}}}{a_\beta} = \frac{(\tau_{r_2} - \beta_{r_2*} - 1)(\tau_{r_1} - \beta_{r_1*} + \beta_{r_1}^*)}{(\tau_{r_2} - 1 - \beta_{r_2*} + \beta_{r_2}^*)(\beta_{r_2,r_1} + 1)}, \tag{15.5.38}$$

or equivalently,

$$\begin{aligned} &(\tau_{r_2} - 1 - \beta_{r_2*} + \beta_{r_2}^*)(\beta_{r_2,r_1} + 1)a_{\beta+\varepsilon_{r_2,r_1}} \\ &= (\tau_{r_2} - \beta_{r_2*} - 1)(\tau_{r_1} - \beta_{r_1*} + \beta_{r_1}^*)a_\beta. \end{aligned} \tag{15.5.39}$$

Thus

$$\begin{aligned} &\sum_{\beta\in\Gamma_A}(\tau_{r_2} - 1 - \beta_{r_2*} + \beta_{r_2}^*)(\beta_{r_2,r_1} + 1)a_{\beta+\varepsilon_{r_2,r_1}}z^\beta \\ &= \sum_{\beta\in\Gamma_A}(\tau_{r_2} - \beta_{r_2*} - 1)(\tau_{r_1} - \beta_{r_1*} + \beta_{r_1}^*)a_\beta z^\beta, \end{aligned} \tag{15.5.40}$$

which is equivalent to (15.5.35). Similarly, (15.5.36) follows from

$$\frac{a_{\beta+\varepsilon_{n,r}}}{a_\beta} = \frac{(\tau_n + \beta_{n*})(\tau_r - \beta_{r*} + \beta_r^*)}{(\vartheta + \beta_{n*})(\beta_{n,r} + 1)}. \quad \square \tag{15.5.41}$$

When $n = 2$, we do not have (15.5.35) and (15.5.36) is just (15.5.33). Next we study our hypergeometric functions of type C. For convenience, we simply denote

$$\mathscr{X}_C = \mathscr{X}_C(\tau_1, \ldots, \tau_n; \vartheta)\{z_{r_2,r_1}, z_{n+s_2,s_1}\} \tag{15.5.42}$$

(cf. (15.4.106)). First we have

$$\begin{aligned} &\partial_{z_{r_2,r_1}}(\mathscr{X}_C) \\ &= \sum_{\alpha\in\Gamma_C} \frac{\prod_{r=1}^n(\tau_r - \alpha_{r*})_{\alpha_r^*+\alpha_r^c}}{\alpha!(\vartheta)_{\alpha^c}}\alpha_{r_2,r_1}z^{\alpha-\varepsilon_{r_2,r_1}} \end{aligned}$$

$$= \sum_{\alpha\in\Gamma_C} \frac{\prod_{r\neq r_1,r_2}(\tau_r - \alpha_{r*})_{\alpha_r^*+\alpha_r^c}}{\alpha!(\vartheta)_{\alpha^c}}(\tau_{r_1} - \alpha_{r_1*})(\tau_{r_1} + 1 - \alpha_{r_1*})_{\alpha_{r_1}^*+\alpha_{r_1}^c}$$
$$\times (\tau_{r_2} - 1 - \alpha_{r_2*})_{\alpha_{r_2}^*+\alpha_{r_2}^c} z^\alpha. \tag{15.5.43}$$

or $1 \leq r_1 < r_2 \leq n$. Moreover,

$$\partial_{z_{n+r_2,r_1}}(\mathscr{X}_C)$$
$$= \sum_{\alpha\in\Gamma_C} \frac{\prod_{r=1}^{n-1}(\tau_r - \alpha_{r*})_{\alpha_r^*+\alpha_r^c}}{\alpha!(\vartheta)_{\alpha^c}}\alpha_{n+r_2,r_1} z^{\alpha-\epsilon_{n+r_2,r_1}}$$
$$= \frac{1}{\vartheta}\sum_{\alpha\in\Gamma_C} \frac{\prod_{r\neq r_1,r_2}(\tau_r - \alpha_{r*})_{\alpha_r^*+\alpha_r^c}}{\alpha!(\vartheta+1)_{\alpha^c}}(\tau_{r_1} - \alpha_{r_1*})(\tau_{r_2} - \alpha_{r_2*})$$
$$\times(\tau_{r_1} + 1 - \alpha_{r_1*})_{\alpha_{r_1}^*+\alpha_{r_1}^c}(\tau_{r_2} + 1 - \alpha_{r_2*})_{\alpha_{r_2}^*+\alpha_{r_2}^c} z^\alpha \tag{15.5.44}$$

for $1 \leq r_1 < r_2 \leq n$ and

$$\partial_{z_{n+s,s}}(\mathscr{X}_C) = \sum_{\alpha\in\Gamma_C} \frac{\prod_{r=1}^{n}(\tau_r - \alpha_{r*})_{\alpha_r^*+\alpha_r^c}}{\alpha!(\vartheta)_{\alpha^c}}\alpha_{n+s,s} z^{\alpha-\epsilon_{n+s,s}}$$
$$= \frac{1}{\vartheta}\sum_{\alpha\in\Gamma_C} \frac{\prod_{r\neq s}(\tau_r - \alpha_{r*})_{\alpha_r^*+\alpha_r^c}}{\alpha!(\vartheta+1)_{\alpha^c}}(\tau_s - \alpha_{s*})(\tau_s + 1 - \alpha_{s*})$$
$$\times(\tau_s + 2 - \alpha_{s*})_{\alpha_s^*+\alpha_s^c} z^\alpha \tag{15.5.45}$$

for $s \in \overline{1,n}$.

Expressions (15.5.43)–(15.5.45) motivate us to define

$$\mathscr{X}_C[i_1, i_2] = \mathscr{X}_C(\tau_1, \ldots, \tau_{i_1} + 1, \ldots, \tau_{i_2} - 1, \ldots, \tau_n; \vartheta)\{z_{r_2,r_1}, z_{n+s_2,s_1}\} \tag{15.5.46}$$

obtained from $\mathscr{X}_C$ by changing τ_{i_1} to $\tau_{i_1} + 1$ and τ_{i_2} to $\tau_{i_2} - 1$ for $1 \leq i_1 < i_2 \leq n$,

$$\mathscr{X}_C[j_1, j_2; 1] = \mathscr{X}_C(\tau_1, \ldots, \tau_{j_1} + 1, \ldots, \tau_{j_2} + 1, \ldots, \tau_n; \vartheta + 1)\{z_{r_2,r_1}, z_{n+s_2,s_1}\} \tag{15.5.47}$$

obtained from $\mathscr{X}_C$ by changing τ_{j_1} to $\tau_{j_1} + 1$, τ_{j_2} to $\tau_{j_2} + 1$ and ϑ to $\vartheta + 1$ for $1 \leq j_1 < j_2 \leq n$ and

$$\mathscr{X}_C[k(2)] = \mathscr{X}_C(\tau_1, \ldots, \tau_k + 2, \ldots, \tau_n; \vartheta + 1)\{z_{r_2,r_1}, z_{n+s_2,s_1}\} \tag{15.5.48}$$

obtained from $\mathscr{X}_C$ by changing τ_k to $\tau_k + 2$ and ϑ to $\vartheta + 1$ for $k \in \overline{1,n}$. Now (15.5.43) can be written as

$$\partial_{z_{r_2,r_1}}(\mathscr{X}_C) = \left(\tau_{r_1} - \sum_{s=1}^{r_1-1} z_{r_1,s}\partial_{z_{r_1,s}}\right)(\mathscr{X}_C[i_1, i_2]) \tag{15.5.49}$$

and (15.5.44) becomes

$$\partial_{z_{n+r_2,r_1}}(\mathscr{X}_C) = \frac{1}{\vartheta}\left(\tau_{r_1} - \sum_{s_1=1}^{r_1-1} z_{r_1,s_1}\partial_{z_{r_1,s_1}}\right)\left(\tau_{r_2} - \sum_{s_2=1}^{r_2-1} z_{r_2,s_2}\partial_{z_{r_2,s_2}}\right)(\mathscr{X}_C[i_1, i_2; 1]) \tag{15.5.50}$$

for $1 \leq r_1 < r_2 \leq n$. Moreover, (15.5.45) is equivalent to

$$\partial_{z_{n+s,s}}(\mathscr{X}_C) = \frac{1}{\vartheta}\left(\tau_s - \sum_{r=1}^{s-1} z_{s,r}\partial_{z_{s,r}}\right)\left(\tau_s + 1 - \sum_{r=1}^{s-1} z_{s,r}\partial_{z_{s,r}}\right)(\mathscr{X}_C[s(2)]) \tag{15.5.51}$$

for $s \in \overline{1,n}$. By (15.5.49)–(15.5.51) and induction, we obtain:

Theorem 15.5.5 *The following differential property of $\mathscr{X}_C$ holds:*

$$\partial_{z_{r_2,r_1}}(\mathscr{X}_C) = \sum_{s=1}^{r_1} \tau_s P_{[s,r_1]}\mathscr{X}_C[s, r_2], \tag{15.5.52}$$

$$\begin{aligned}\partial_{z_{n+r_2,r_1}}&(\mathscr{X}_C)\\ &= \frac{1}{\vartheta}\Bigg[\sum_{i=1}^{r_1} \tau_i^2 P_{[i,r_1]}P_{[i,r_2]}\mathscr{X}_C[i(2)] + \sum_{s_1=1}^{r_1}\sum_{s_2=r_1+1}^{r_2} \tau_{s_1}\tau_{s_2}P_{[s_1,r_1]}P_{[s_2,r_2]}\mathscr{X}_C[s_1, s_2; 1]\\ &\quad + \sum_{1\leq s_1<s_2\leq r_1} \tau_{s_1}\tau_{s_2}(P_{[s_1,r_1]}P_{[s_2,r_2]} + P_{[s_2,r_1]}P_{[s_1,r_2]})\mathscr{X}_C[s_1, s_2; 1]\Bigg]\end{aligned} \tag{15.5.53}$$

for $1 \leq r_1 < r_2 \leq n$ and

$$\begin{aligned}\partial_{z_{n+s,s}}(\mathscr{X}_C) = \frac{1}{\vartheta}\Bigg[&\sum_{i=1}^{s} \tau_i^2 P_{[i,s]}^2\mathscr{X}_C[i(2)] + \tau_s\mathscr{X}_C[s(2)] + \sum_{i=1}^{s-1}\tau_i P_{[i,s]}\mathscr{X}_C[i, s; 1]\\ &+2\sum_{1\leq r_1<r_2\leq s} \tau_{r_1}\tau_{r_2}P_{[r_1,s]}P_{[r_2,s]}\mathscr{X}_C[r_1, r_2; 1]\Bigg]\end{aligned} \tag{15.5.54}$$

for $s \in \overline{1,n}$.

Up to this stage, we have not found a nice integral representation for $\mathscr{X}_C$. In fact, there is no simple integral formula of Euler type with an elementary integrand for Lauricella third multiple hypergeometric function (e.g., cf. [Eh]). It might also be the case for our hypergeometric function $\mathscr{X}_C$.

By (15.4.106), we set

$$c_\alpha = \frac{\prod_{r=1}^n (\tau_r - \alpha_{r*})_{\alpha_r^* + \alpha_r^c}}{\alpha!(\vartheta)_{\alpha^c}} z^\alpha \quad \text{for } \alpha \in \Gamma_C. \tag{15.5.55}$$

Note that

$$\frac{c_{\alpha+\varepsilon_{r_2,r_1}}}{c_\alpha} = \frac{(\tau_{r_1} - \alpha_{r_1*} + \alpha_{r_1}^* + \alpha_{r_1}^c)(\tau_{r_2} - 1 - \alpha_{r_2*})}{(\alpha_{r_2,r_1} + 1)(\tau_{r_2} - 1 - \alpha_{r_2*} + \alpha_{r_2}^* + \alpha_{r_2}^c)}, \tag{15.5.56}$$

$$\frac{c_{\alpha+\varepsilon_{n+r_2,r_1}}}{c_\alpha} = \frac{(\tau_{r_1} - \alpha_{r_1*} + \alpha_{r_1}^* + \alpha_{r_1}^c)(\tau_{r_2} - \alpha_{r_2*} + \alpha_{r_2}^* + \alpha_{r_2}^c)}{(\alpha_{n+r_2,r_1} + 1)(\vartheta + \alpha^c)} \tag{15.5.57}$$

for $1 \leq r_1 < r_2 \leq n$ and

$$\frac{c_{\alpha+\varepsilon_{n+s,s}}}{c_\alpha} = \frac{(\tau_s - \alpha_{s*} + \alpha_s^* + \alpha_s^c)(\tau_s + 1 - \alpha_{s*} + \alpha_s^* + \alpha_s^c)}{(\alpha_{n+s,s} + 1)(\vartheta + \alpha^c)} \tag{15.5.58}$$

for $s \in \overline{1, n}$. Let

$$\mathscr{D}_r^C = \sum_{i=1}^r z_{n+r,i} \partial_{z_{n+r,i}} + \sum_{s=r}^n z_{n+s,r} \partial_{z_{n+s,r}} \quad \text{for } r \in \overline{1, n} \tag{15.5.59}$$

and

$$\mathscr{D}^C = \sum_{1 \leq r_1 \leq r_2 \leq n} z_{n+r_2,r_1} \partial_{z_{n+r_2,r_1}}. \tag{15.5.60}$$

By the proof of Theorem 15.5.4, we have:

Theorem 15.5.6 *The following hypergeometric equations of type C hold:*

$$\begin{aligned} &(\tau_{r_2} - 1 - \mathscr{D}_{r_2*}^A + \mathscr{D}_{r_2}^{A*} + \mathscr{D}_{r_2}^C)\partial_{z_{r_2,r_1}}(\mathscr{X}_C) \\ &= (\tau_{r_2} - 1 - \mathscr{D}_{r_2*}^A)(\tau_{r_1} - \mathscr{D}_{r_1*}^A + \mathscr{D}_{r_1}^{A*} + \mathscr{D}_{r_1}^C)(\mathscr{X}_C), \end{aligned} \tag{15.5.61}$$

$$\begin{aligned} &(\vartheta + \mathscr{D}^C)\partial_{z_{n+r_2,r_1}}(\mathscr{X}_C) \\ &= (\tau_{r_2} - \mathscr{D}_{r_2*}^A + \mathscr{D}_{r_2}^{A*} + \mathscr{D}_{r_2}^C)(\tau_{r_1} - \mathscr{D}_{r_1*}^A + \mathscr{D}_{r_1}^{A*} + \mathscr{D}_{r_1}^C)(\mathscr{X}_C) \end{aligned} \tag{15.5.62}$$

for $1 \leq r_1 < r_2 \leq n$ *and*

$$\begin{aligned} &(\vartheta + \mathscr{D}^C)\partial_{z_{n+s,s}}(\mathscr{X}_C) \\ &= (\tau_s - \mathscr{D}_{s*}^A + \mathscr{D}_s^{A*} + \mathscr{D}_s^C)(\tau_s + 1 - \mathscr{D}_{s*}^A + \mathscr{D}_s^{A*} + \mathscr{D}_s^C)(\mathscr{X}_C) \end{aligned} \tag{15.5.63}$$

for $s \in \overline{1, n}$.

Set

$$\Gamma_D = \sum_{1\le r_1<r_2\le n} (\mathbb{N}\varepsilon_{r_2,r_1} + \mathbb{N}\varepsilon_{n+r_2,r_1}) \subset \Gamma_C. \tag{15.5.64}$$

Moreover, we let

$$\beta_r^D = \sum_{i=1}^{r-1} \beta_{n+r,i} + \sum_{s=r+1}^{n} \beta_{n+s,r}, \quad \beta^D = \sum_{1\le r_1<r_2\le n} \beta_{n+r_2,r_1} \quad \text{for } \beta \in \Gamma_D \tag{15.5.65}$$

and

$$\beta_r^B = \sum_{i=1}^{r} \beta_{n+r,i} + \sum_{s=r+1}^{n} \beta_{n+s,r}, \quad \beta^B = \beta^c = \sum_{1\le r_1\le r_2\le n} \beta_{n+r_2,r_1} \quad \text{for } \beta \in \Gamma_C. \tag{15.5.66}$$

We define the following hypergeometric functions:

$$\mathscr{X}_D(\tau_1,\ldots,\tau_n;\vartheta)\{z_{r_2,r_1}, z_{n+r_2,r_1}\} = \sum_{\alpha\in\Gamma_D} \frac{\prod_{r=1}^n (\tau_r - \alpha_{r*})_{\alpha_r^*+\alpha_r^D}}{\alpha!(\vartheta)_{\alpha^D}} z^\alpha. \tag{15.5.67}$$

$$\mathscr{X}_B(\tau_1,\ldots,\tau_n;\vartheta)\{z_{r_2,r_1}, z_{n+s_2,s_1}\} = \sum_{\alpha\in\Gamma_C} \frac{\prod_{r=1}^n (\tau_r - \alpha_{r*})_{\alpha_r^*+\alpha_r^B}}{\alpha!(\vartheta)_{\alpha^B}} z^\alpha. \tag{15.5.68}$$

Furthermore, we define $\mathscr{X}_B$, $\mathscr{X}_B[i_1,i_2]$, $\mathscr{X}_B[j_1,j_2;1]$, $\mathscr{X}_B[k(2)]$ and $\mathscr{X}_D$, $\mathscr{X}_D[i_1,i_2]$, $\mathscr{X}_D[j_1,j_2;1]$, $\mathscr{X}_D[k(2)]$ as those of type C in (15.5.42) and (15.5.46)–(15.5.48). We let

$$\mathscr{X}_B[k] = \mathscr{X}_N(\tau_1,\ldots,\tau_k+1,\ldots,\tau_n;\vartheta+1)\{z_{r_2,r_1}, z_{n+s_2,s_1}\} \tag{15.5.69}$$

obtained from $\mathscr{X}_B$ by changing τ_k to τ_k+1 and ϑ to $\vartheta+1$ for $k \in \overline{1,n}$.

Theorem 15.5.7 *The differential property of $\mathscr{X}_B$ holds:*

$$\partial_{z_{r_2,r_1}}(\mathscr{X}_B) = \sum_{s=1}^{r_1} \tau_s P_{[s,r_1]} \mathscr{X}_B[s,r_2], \tag{15.5.70}$$

$$\begin{aligned}&\partial_{z_{n+r_2,r_1}}(\mathscr{X}_B)\\ &= \frac{1}{\vartheta}\Bigg[\sum_{i=1}^{r_1} \tau_i^2 P_{[i,r_1]} P_{[i,r_2]} \mathscr{X}_B[i(2)] + \sum_{s_1=1}^{r_1}\sum_{s_2=r_1+1}^{r_2} \tau_{s_1}\tau_{s_2} P_{[s_1,r_1]} P_{[s_2,r_2]} \mathscr{X}_B[s_1,s_2;1]\\ &\quad + \sum_{1\le s_1<s_2\le r_1} \tau_{s_1}\tau_{s_2} (P_{[s_1,r_1]} P_{[s_2,r_2]} + P_{[s_2,r_1]} P_{[s_1,r_2]}) \mathscr{X}_B[s_1,s_2;1]\Bigg]\end{aligned} \tag{15.5.71}$$

for $1 \le r_1 < r_2 \le n$ *and*

$$\partial_{z_{n+s,s}}(\mathscr{X}_B) = \sum_{r=1}^{s} \tau_r P_{[s,r]} \mathscr{X}_B[s] \tag{15.5.72}$$

for $s \in \overline{1,n}$. *Moreover, the following differential property of* $\mathscr{X}_D$ *holds:*

$$\partial_{z_{r_2,r_1}}(\mathscr{X}_D) = \sum_{s=1}^{r_1} \tau_s P_{[s,r_1]} \mathscr{X}_D[s, r_2], \tag{15.5.73}$$

$$\begin{aligned}
&\partial_{z_{n+r_2,r_1}}(\mathscr{X}_D)\\
&= \frac{1}{\vartheta}\Bigg[\sum_{i=1}^{r_1} \tau_i^2 P_{[i,r_1]} P_{[i,r_2]} \mathscr{X}_D[i(2)] + \sum_{s_1=1}^{r_1}\sum_{s_2=r_1+1}^{r_2} \tau_{s_1}\tau_{s_2} P_{[s_1,r_1]} P_{[s_2,r_2]} \mathscr{X}_D[s_1, s_2; 1]\\
&\quad + \sum_{1\le s_1<s_2\le r_1} \tau_{s_1}\tau_{s_2}(P_{[s_1,r_1]} P_{[s_2,r_2]} + P_{[s_2,r_1]} P_{[s_1,r_2]}) \mathscr{X}_D[s_1, s_2; 1]\Bigg],
\end{aligned} \tag{15.5.74}$$

for $1 \le r_1 < r_2 \le n$.

Let

$$\mathscr{D}_r^B = \sum_{i=1}^{r} z_{n+r,i}\partial_{z_{n+r,i}} + \sum_{s=r+1}^{n} z_{n+s,r}\partial_{z_{n+s,r}}, \tag{15.5.75}$$

$$\mathscr{D}_r^D = \sum_{i=1}^{r-1} z_{n+r,i}\partial_{z_{n+r,i}} + \sum_{s=r+1}^{n} z_{n+s,r}\partial_{z_{n+s,r}} \tag{15.5.76}$$

for $r \in \overline{1,n}$ and

$$\mathscr{D}^B = \mathscr{D}^C, \quad \mathscr{D}^D = \sum_{1\le r_1<r_2\le n} z_{n+r_2,r_1}\partial_{z_{n+r_2,r_1}}. \tag{15.5.77}$$

As Theorem 15.5.4, we have:

Theorem 15.5.8 *The following hypergeometric equations of type B hold:*

$$\begin{aligned}
&(\tau_{r_2} - 1 - \mathscr{D}_{r_2*}^A + \mathscr{D}_{r_2}^{A*} + \mathscr{D}_{r_2}^B)\partial_{z_{r_2,r_1}}(\mathscr{X}_B)\\
&= (\tau_{r_2} - 1 - \mathscr{D}_{r_2*}^A)(\tau_{r_1} - \mathscr{D}_{r_1*}^A + \mathscr{D}_{r_1}^{A*} + \mathscr{D}_{r_1}^B)(\mathscr{X}_B),
\end{aligned} \tag{15.5.78}$$

$$\begin{aligned}
&(\vartheta + \mathscr{D}^B)\partial_{z_{n+r_2,r_1}}(\mathscr{X}_B)\\
&= (\tau_{r_2} - \mathscr{D}_{r_2*}^A + \mathscr{D}_{r_2}^{A*} + \mathscr{D}_{r_2}^B)(\tau_{r_1} - \mathscr{D}_{r_1*}^A + \mathscr{D}_{r_1}^{A*} + \mathscr{D}_{r_1}^B)(\mathscr{X}_B)
\end{aligned} \tag{15.5.79}$$

for $1 \leq r_1 < r_2 \leq n$ *and*

$$(\vartheta + \mathscr{D}^B)\partial_{z_{n+s,s}}(\mathscr{X}_B) = (\tau_s - \mathscr{D}^A_{s*} + \mathscr{D}^{A*}_s + \mathscr{D}^B_s)(\mathscr{X}_B) \tag{15.5.80}$$

for $s \in \overline{1,n}$. *Moreover, the following hypergeometric equations of type D hold:*

$$\begin{aligned}&(\tau_{r_2} - 1 - \mathscr{D}^A_{r_2*} + \mathscr{D}^{A*}_{r_2} + \mathscr{D}^D_{r_2})\partial_{z_{r_2,r_1}}(\mathscr{X}_D)\\ &= (\tau_{r_2} - 1 - \mathscr{D}^A_{r_2*})(\tau_{r_1} - \mathscr{D}^A_{r_1*} + \mathscr{D}^{A*}_{r_1} + \mathscr{D}^D_{r_1})(\mathscr{X}_D),\end{aligned} \tag{15.5.81}$$

$$\begin{aligned}&(\vartheta + \mathscr{D}^D)\partial_{z_{n+r_2,r_1}}(\mathscr{X}_D)\\ &= (\tau_{r_2} - \mathscr{D}^A_{r_2*} + \mathscr{D}^{A*}_{r_2} + \mathscr{D}^D_{r_2})(\tau_{r_1} - \mathscr{D}^A_{r_1*} + \mathscr{D}^{A*}_{r_1} + \mathscr{D}^D_{r_1})(\mathscr{X}_D)\end{aligned} \tag{15.5.82}$$

for $1 \leq r_1 < r_2 \leq n$.

References

[BO] R. Beerends, E. Opdam, Certain hypergeometric series related to the root system BC. Trans. Am. Math. Soc. **119**, 581–609 (1993)

[C] B. Cooperstein, The fifty-six-dimensional module for E_7.I. A four form for E_7. J. Algebra **173**, 361–389 (1995)

[Ep] P. Etingof, Quantum integrable systems and representations of Lie algebras. J. Math. Phys. **36**(6), 2637–2651 (1995)

[ES] P. Etingof, K. Styrkas, Algebraic integrability of Schrödinger operators and representations of Lie algebras. Compos. Math. **98**(1), 91–112 (1995)

[Eh] H. Exton, *Multiple Hypergeometric Functions and Applications* (Halsted Press (Wiley), New York, 1976)

[GG] I.M. Gel'fand, M.I. Graev, GG-functions and their relation to general hypergeometric functions. Russ. Math. Surv. **52**(4), 639–684 (1997)

[Hg1] G. Heckman, Root systems and hypergeometric functions II. Compos. Math. **64**, 353–373 (1987)

[Hg2] G. Heckman, Heck algebras and hypergeometric functions. Invent. Math. **100**, 403–417 (1990)

[Hg3] G. Heckman, An elementary approach to the hypergeometric shift operators of Opdam. Invent. Math. **103**, 341–350 (1990)

[HO] G. Heckman, E. Opdam, Root systems and hypergeometric functions I. Compos. Math. **64**, 329–352 (1987)

[M] M. Marvan, Reducibility of zero curvature representations with application to recursion operators. Acta Appl. Math. **83**, 39–68 (2004)

[OP] M. Olslanetsky, A. Perelomov, Completely integrable Hamiltonian systems connected with semisimple Lie algebras. Invent. Math. **37**, 93–108 (1976)

[O1] E. Opdam, Root systems and hypergeometric functions III. Compos. Math. **67**, 21–49 (1988)

[O2] E. Opdam, Root systems and hypergeometric functions IVI. Compos. Math. **67**, 191–209 (1988)

[O3] E. Opdam, Some applications of hypergeometric shift operators. Invent. Math. **98**, 1–18 (1989)

[O4] E. Opdam, An analogue of the Gauss summation formula for hypergeometric functions related to root systems. Math. Z. **212**, 313–336 (1993)
[O5] E. Opdam, Harmonic analysis for certain representations of graded Hecke algebras. Acta Math. **175**, 75–121 (1995)
[O6] E. Opdam, Cuspital hypergeometric functions. Methods Appl. Anal. **6**, 67–80 (1999)
[X14] X. Xu, Path hypergeometric functions. J. Algebra Appl. **6**, 595–653 (2007)
[X21] X. Xu, *Algebraic Approaches to Partial Differential Equations* (Springer, Heidelberg, 2013)
[Sb] B. Sutherland, Exact results for a quantum many-body problem in one-dimension. Phys. Rev. A **5**, 1372–1376 (1972)

Index

X. Xu, *Representations of Lie Algebras and Partial Differential Equations*,
DOI 10.1007/978-981-10-6391-6

K

L

M

N

O

P

Q

R

S

MIX
Papier aus verantwortungsvollen Quellen
Paper from responsible sources
FSC® C105338

If you have any concerns about our products,
you can contact us on
ProductSafety@springernature.com

In case Publisher is established outside the EU,
the EU authorized representative is:
Springer Nature Customer Service Center GmbH
Europaplatz 3, 69115 Heidelberg, Germany

Printed by Libri Plureos GmbH
in Hamburg, Germany